WIRELESS VIDEO COMMUNICATIONS

IEEE Press Series on Digital & Mobile Communication

The IEEE Press Digital & Mobile Communication Series is written for research and development engineers and graduate students in communication engineering. The burgeoning wireless and personal communication fields receive special emphasis. Books are of two types -- graduate texts and the latest monographs about theory and practice.

John B. Anderson, *Series Editor*
Professor of Electrical Engineering
Rensselaer Polytechnic Institute

National Visiting Professor of Information Technology
Lund University, Sweden

Advisory Board

John B. Anderson
Rensselaer Polytechnic Institute
Troy, NY, U.S.A.
Dept. of Information Technology
Lund University, Sweden

Joachim Hagenauer
Dept. of Communications Engineering
Technical University
Munich, Germany

Rolf Johannesson
Dept. of Information Technology
Lund University, Sweden

Norman Beaulieu
Dept. of Electrical Engineering
Queen's University
Kingston, Ontario, Canada

Books in the IEEE Press Series on Digital & Mobile Communication

John B. Anderson, *Digital Transmission Engineering*

Rolf Johannesson and Kamil Sh. Zigangirov, *Fundamentals of Convolutional Coding*

Raj Pandya, *Mobile and Personal Communication Systems and Services*

Lajos Hanzo, Peter J. Cherriman, and Jürgen Streit, *Wireless Video Communications: Second to Third Generation Systems and Beyond*

Lajos Hanzo, F.Clare A. Somerville and Jason P. Woodard, *Voice Compression and Communications: Principles and Applications for Fixed and Wireless Channels*

WIRELESS VIDEO COMMUNICATIONS

Second to Third Generation Systems and Beyond

Lajos Hanzo
Department of Electronics and Computer Science
University of Southampton, U.K.

Peter J. Cherriman
Department of Electronics and Computer Science
University of Southampton, U.K.

Jürgen Streit
Department of Electronics and Computer Science
University of Southampton, U.K.

John B. Anderson, *Series Editor*

The Institute of Electrical and Electronics Engineers, Inc., New York

This book and other books may be purchased at a discount
from the publisher when ordered in bulk quantities. Contact:

IEEE Press Marketing
Attn: Special Sales
445 Hoes Lane
P.O. Box 1331
Piscataway, NJ 08855-1331
Fax: +1 732 981 9334

For more information about IEEE Press products, visit the
IEEE Online Catalog & Store: http://www.ieee.org/store.

© 2001 by the Institute of Electrical and Electronics Engineers, Inc.
3 Park Avenue, 17th Floor, New York, NY 10016-5997.

*All rights reserved. No part of this book may be reproduced in any form,
nor may it be stored in a retrieval system or transmitted in any form,
without written permission from the publisher.*

Printed in the United States of America.

10 9 8 7 6 5 4 3 2 1

ISBN 0-7803-6032-X
IEEE Order No. PC5880

Library of Congress Cataloging-in-Publication Data
Hanzo, Lajos.
  Wireless video communications : second to third generation systems and beyond /
Lajos Hanzo, P.J. Cherriman, J. Streit.
    p. cm.
  Includes bibliographical references and index.
  ISBN 0-7803-6032-X
  1. Wireless communication systems. 2. Digital video. 3. Multimedia systems. 4. Code division multiple access. I. Cherriman, P. J. (Peter J.), 1972- II. Streit, J. (Jürgen), 1968- III. Title.

TK5103.2 .H36 2000
621 .382--dc21

00-053853

We dedicate this book to the numerous contributors in the field, many of whom are listed in the Author Index of this book.

IEEE Press
445 Hoes Lane, P.O. Box 1331
Piscataway, NJ 08855-1331

IEEE Press Editorial Board
Stamatios V. Kartalopoulos, *Editor in Chief*

M. Akay	M. Eden	M. S. Newman
J. B. Anderson	M. E. El-Hawary	M. Padgett
R. J. Baker	R .J. Herrick	W. D. Reeve
J. E. Brewer	R. F. Hoyt	G. Zobrist
	D. Kirk	

Kenneth Moore, *Director of IEEE Press*
Catherine Faduska, *Senior Acquisitions Editor*
Linda Matarazzo, *Associate Acquisitions Editor*
Marilyn G. Catis, *Marketing Manager*
Surendra Bhimani, *Production Editor*

Cover design: William T. Donnelly, *WT Design*

Technical Reviewers

Raul Bruzzone, *Philips Consumer Communications, France*
Andrew Aftelak, *Motorola Labs, U.K.*
Sergio Lima Netto, *Universidade Federal do Rio de Janeiro, Brazil*
Petri Mahonen, *VTT Networking Research, Finland*
Eduardo Antonio Barros da Silva, *Universidade Federal do Rio de Janeiro, Brazil*

Books of Related Interest from the IEEE Press

VOICE COMPRESSION AND COMMUNICATIONS: Principles and Applications for Fixed and Wireless Channels
Lajos Hanzo, F. Clare A. Somerville, and Jason P. Woodard
2001   Hardcover   784 pp   IEEE Order No. PC5881   ISBN 0-7803-6033-8

SINGLE AND MULTICARRIER MODULATION: Principles and Applications
Lajos Hanzo, William Webb, and Thomas Keller
2000   Hardcover   712 pp   IEEE Order No. PC5864   ISBN 0-7803-6015-X

MOBILE RADIO COMMUNICATIONS: Second and Third Generation Cellular and WATM Systems, Second Edition
Edited by Raymond Steele and Lajos Hanzo
A John Wiley & Sons, Ltd. book published in cooperation with IEEE Press
1999   Hardcover   1,064 pp   IEEE Order No. PC5820   ISBN 0-471-97806-X

SMART ANTENNAS FOR WIRELESS CDMA
Joseph C. Liberti and Theodore Rappaport
A Prentice Hall book published in cooperation with IEEE Press
1999   Hardcover   400 pp   IEEE Order No. PC5783   ISBN 0-7803-4736-6

THE MOBILE COMMUNICATIONS HANDBOOK, Second Edition
Jerry D. Gibson, Editor in Chief
A CRC Press Handbook published in cooperation with IEEE Press
1999   Hardcover   726 pp   IEEE Order No. PC5772   ISBN 0-7803-4726-9

Contents

Preface xxiii
Acknowledgments xxix
Contributors xxxi

I Transmission Issues 1

1 Information Theory 3
1.1   Issues in Information Theory 3
1.2   Additive White Gaussian Noise Channel 7
    1.2.1   Background 7
    1.2.2   Practical Gaussian Channels 8
    1.2.3   Gaussian Noise 8
1.3   Information of a Source 11
1.4   Average Information of Discrete Memoryless Sources 12
    1.4.1   Maximum Entropy of a Binary Source 13
    1.4.2   Maximum Entropy of a q-ary Source 15
1.5   Source Coding for a Discrete Memoryless Source 15
    1.5.1   Shannon-Fano Coding 16
    1.5.2   Huffman Coding 18
1.6   Average Information of Discrete Sources Exhibiting Memory 22
    1.6.1   Two-State Markov Model for Discrete Sources Exhibiting Memory 22
    1.6.2 N-State Markov Model for Discrete Sources Exhibiting Memory 24
1.7   Examples 25
    1.7.1   Two-State Markov Model Example 25
    1.7.2   Four-State Markov Model for a 2-Bit Quantizer 27
1.8   Generating Model Sources 28
    1.8.1   Autoregressive Model 28
    1.8.2   AR Model Properties 29

		1.8.3 First-Order Markov Model	30
	1.9	Run-Length Coding for Discrete Sources Exhibiting Memory	31
		1.9.1 Run-Length Coding Principle	31
		1.9.2 Run-Length Coding Compression Ratio	32
	1.10	Information Transmission via Discrete Channels	34
		1.10.1 Binary Symmetric Channel Example	34
		1.10.2 Bayes' Rule	38
		1.10.3 Mutual Information	39
		1.10.4 Mutual Information Example	40
		1.10.5 Information Loss via Imperfect Channels	42
		1.10.6 Error Entropy via Imperfect Channels	43
	1.11	Capacity of Discrete Channels	49
	1.12	Shannon's Channel Coding Theorem	53
	1.13	Capacity of Continuous Channels	55
		1.13.1 Practical Evaluation of the Shannon-Hartley Law	58
		1.13.2 Shannon's Ideal Communications System for Gaussian Channels	62
	1.14	Shannon's Message and Its Implications for Wireless Channels	62
	1.15	Summary and Conclusions	65
2	**The Propagation Environment**		**67**
	2.1	The Cellular Concept	67
	2.2	Radio Wave Propagation	71
		2.2.1 Background	71
		2.2.2 Narrowband Fading Channels	73
		2.2.3 Propagation Path-loss Law	73
		2.2.4 Slow-Fading Statistics	76
		2.2.5 Fast-Fading Statistics	77
		2.2.6 Doppler Spectrum	83
		2.2.7 Simulation of Narrowband Fading Channels	85
		2.2.7.1 Frequency-Domain Fading Simulation	86
		2.2.7.2 Time-Domain Fading Simulation	86
		2.2.7.3 Box-Müller Algorithm of AWGN Generation	87
		2.2.8 Wideband Channels	87
		2.2.8.1 Modeling of Wideband Channels	87
	2.3	Summary and Conclusions	92
3	**Convolutional Channel Coding**		**93**
	3.1	Brief Channel Coding History	93
	3.2	Convolutional Encoding	94
	3.3	State and Trellis Transitions	96
	3.4	The Viterbi Algorithm	98
		3.4.1 Error-Free Hard-Decision Viterbi Decoding	98
		3.4.2 Erroneous Hard-Decision Viterbi Decoding	101
		3.4.3 Error-Free Soft-Decision Viterbi Decoding	104
	3.5	Summary and Conclusions	106

4 Block-Based Channel Coding — 107
- 4.1 Introduction — 107
- 4.2 Finite Fields — 108
  - 4.2.1 Definitions — 108
  - 4.2.2 Galois Field Construction — 111
  - 4.2.3 Galois Field Arithmetic — 113
- 4.3 Reed-Solomon and Bose-Chaudhuri-Hocquenghem Block Codes — 114
  - 4.3.1 Definitions — 114
  - 4.3.2 RS Encoding — 116
  - 4.3.3 RS Encoding Example — 118
  - 4.3.4 Linear Shift-Register Circuits for Cyclic Encoders — 122
    - 4.3.4.1 Polynomial Multiplication — 122
    - 4.3.4.2 Systematic Cyclic Shift-Register Encoding Example — 123
  - 4.3.5 RS Decoding — 126
    - 4.3.5.1 Formulation of the Key Equations [1–9] — 126
    - 4.3.5.2 Peterson-Gorenstein-Zierler Decoder — 130
    - 4.3.5.3 PGZ Decoding Example — 133
    - 4.3.5.4 Berlekamp-Massey Algorithm [1–9] — 138
    - 4.3.5.5 Berlekamp-Massey Decoding Example — 144
    - 4.3.5.6 Computation of the Error Magnitudes by the Forney Algorithm — 148
    - 4.3.5.7 Forney Algorithm Example — 151
    - 4.3.5.8 Error Evaluator Polynomial Computation — 153
- 4.4 RS and BCH Codec Performance — 156
- 4.5 Summary and Conclusions — 158

5 Modulation and Transmission Techniques — 161
- 5.1 Modulation Issues — 161
  - 5.1.1 Introduction — 161
  - 5.1.2 Quadrature Amplitude Modulation [10] — 164
    - 5.1.2.1 Background — 164
    - 5.1.2.2 Modem Schematic — 164
      - 5.1.2.2.1 Gray Mapping and Phasor Constellation — 167
      - 5.1.2.2.2 Nyquist Filtering — 168
      - 5.1.2.2.3 Modulation and Demodulation — 170
      - 5.1.2.2.4 Data Recovery — 171
    - 5.1.2.3 QAM Constellations — 172
    - 5.1.2.4 16QAM BER versus SNR Performance over AWGN Channels — 175
      - 5.1.2.4.1 Decision Theory — 175
      - 5.1.2.4.2 QAM Modulation and Transmission — 177
      - 5.1.2.4.3 16QAM Demodulation in AWGN — 178
    - 5.1.2.5 Reference-Assisted Coherent QAM for Fading Channels — 181
      - 5.1.2.5.1 PSAM System Description — 181
      - 5.1.2.5.2 Channel Gain Estimation in PSAM — 183
      - 5.1.2.5.3 PSAM Performance [11] — 185

		5.1.2.6	Differentially Detected QAM [10]	186
	5.1.3	Adaptive Modulation		190
		5.1.3.1	Background to Adaptive Modulation	190
		5.1.3.2	Bit Error Rate and Channel Capacity Optimization of Adaptive Modems	193
		5.1.3.3	Bit Error Rate and Channel Capacity of Adaptive Modulation	195
		5.1.3.4	Equalization Techniques	197

5.2 Orthogonal Frequency Division Multiplexing 197
5.3 Packet Reservation Multiple Access 201
5.4 Flexible Transceiver Architecture 202
5.5 Summary and Conclusions 204

6 Video Traffic Modeling and Multiple Access   205

6.1 Video Traffic Modeling 205
    6.1.1 Motivation and Background 205
    6.1.2 Markov Modeling of Video Sources 207
    6.1.3 Reduced-Length Poisson Cycles 210
    6.1.4 Video Model Matching 215
6.2 Multiple Access 223
    6.2.1 Background 223
    6.2.2 Classification of Multiple Access Techniques 225
    6.2.3 Multiframe Packet Reservation Multiple Access 227
          6.2.3.1 Performance of MF-PRMA 228
    6.2.4 Statistical Packet Assignment Multiple Access 237
          6.2.4.1 Statistical Packet Assignment Principles 237
          6.2.4.2 Performance of the SPAMA Protocol 242
6.3 Summary and Conclusions 243

7 Co-Channel Interference   247

7.1 Introduction 247
7.2 Factors Controlling Co-Channel Interference 248
    7.2.1 Effect of Fading on Co-Channel Interference 248
    7.2.2 Cell Shapes 249
    7.2.3 Position of Users and Interferers 250
7.3 Theoretical Signal-to-Interference Ratio 252
7.4 Simulation Parameters 255
7.5 Results for Multiple Interferers 258
    7.5.1 SIR Profile of a Cell 258
    7.5.2 Signal-to-Noise-Plus-Interference Ratio (SINR) 262
    7.5.3 Channel Capacity 263
7.6 Results for a Single Interferer 269
    7.6.1 Simple Model for SINR in a Single Interferer Situation 270
    7.6.2 Effect of SIR and SNR on Error Rates 272
    7.6.3 Time-Varying Effects of SIR and SINR 275
    7.6.4 Effect of Interference on the H.263 Videophone System 280

7.7		Summary and Conclusions	284
8	**Channel Allocation**		**287**
8.1		Introduction	287
8.2		Overview of Channel Allocation	288
	8.2.1	Fixed Channel Allocation	289
		8.2.1.1 Channel Borrowing	291
		8.2.1.2 Flexible Channel Allocation	292
	8.2.2	Dynamic Channel Allocation	292
		8.2.2.1 Centrally Controlled DCA Algorithms	294
		8.2.2.2 Distributed DCA Algorithms	295
		8.2.2.3 Locally distributed DCA algorithms	296
	8.2.3	Hybrid Channel Allocation	297
	8.2.4	The Effect of Handovers	298
	8.2.5	The Effect of Transmission Power Control	299
8.3		Simulation of the Channel Allocation Algorithms	299
	8.3.1	The Mobile Radio Network Simulator, "Netsim"	299
		8.3.1.1 Physical Layer Model	302
		8.3.1.2 Shadow Fading Model	302
	8.3.2	Channel Allocation Algorithms Investigated	304
		8.3.2.1 Fixed Channel Allocation Algorithm	304
		8.3.2.2 Distributed Dynamic Channel Allocation Algorithms	304
		8.3.2.3 Locally Distributed Dynamic Channel Allocation Algorithms	306
	8.3.3	Performance Metrics	306
	8.3.4	Nonuniform Traffic Model	309
8.4		Performance Comparisons	310
	8.4.1	System Parameters	310
	8.4.2	Carried Traffic with Quality Constraints	311
	8.4.3	Comparing the LOLIA with FCA	312
	8.4.4	Effect of the "Reuse Distance" Constraint on the LOLIA and LOMIA DCA Algorithms	314
	8.4.5	Comparison of the LOLIA and LOMIA with the LIA	317
	8.4.6	Interference Threshold-Based Distributed DCA Algorithms	318
	8.4.7	Performance Comparison of Fixed and Dynamic Channel Allocation Algorithms Using Nonuniform Traffic Distributions	321
	8.4.8	Effect of Shadow Fading on the FCA, LOLIA, and LOMIA	324
	8.4.9	Effect of Shadow Fading Frequency and Standard Deviation on the LOLIA	325
	8.4.10	Effect of Shadow Fading Standard Deviation on FCA and LOLIA	327
	8.4.11	SINR Profile across Cell Area	329
	8.4.12	Overview of Results	332
8.5		Summary and Conclusions	335

9 Second-Generation Mobile Systems — 339
- 9.1 The Wireless Communications Scene — 339
- 9.2 Global System for Mobile Communications — GSM — 342
  - 9.2.1 Introduction to GSM — 342
  - 9.2.2 Overview of GSM — 345
  - 9.2.3 Logical and Physical Channels in GSM — 346
  - 9.2.4 Speech and Data Transmission in GSM — 347
  - 9.2.5 Transmission of Control Signals in GSM — 351
  - 9.2.6 Synchronization Issues in GSM — 357
  - 9.2.7 Gaussian Minimum Shift Keying in GSM — 358
  - 9.2.8 Wideband Channel Models in GSM — 359
  - 9.2.9 Adaptive Link Control in GSM — 360
  - 9.2.10 Discontinuous Transmission in GSM — 363
  - 9.2.11 Summary and Conclusions — 363

10 CDMA Systems: Third-Generation and Beyond — 365
- 10.1 Introduction — 365
- 10.2 Basic CDMA System — 366
  - 10.2.1 Spread Spectrum Fundamentals — 366
    - 10.2.1.1 Frequency Hopping — 367
    - 10.2.1.2 Direct Sequence — 368
  - 10.2.2 The Effect of Multipath Channels — 371
  - 10.2.3 RAKE Receiver — 374
  - 10.2.4 Multiple Access — 378
    - 10.2.4.1 Down-Link Interference — 379
    - 10.2.4.2 Up-Link Interference — 380
    - 10.2.4.3 Gaussian Approximation — 383
  - 10.2.5 Spreading Codes — 385
    - 10.2.5.1 m-sequences — 385
    - 10.2.5.2 Gold Sequences — 386
    - 10.2.5.3 Extended m-sequences — 387
  - 10.2.6 Channel Estimation — 387
    - 10.2.6.1 Down-Link Pilot-Assisted Channel Estimation — 388
    - 10.2.6.2 Up-Link Pilot-Symbol Assisted Channel Estimation — 389
    - 10.2.6.3 Pilot-Symbol Assisted Decision-Directed Channel Estimation — 390
  - 10.2.7 Summary — 392
- 10.3 Third-Generation Wireless Mobile Communication Systems — 392
  - 10.3.1 Introduction — 392
  - 10.3.2 UMTS Terrestrial Radio Access (UTRA) — 395
    - 10.3.2.1 Characteristics of UTRA — 395
    - 10.3.2.2 Transport Channels — 399
    - 10.3.2.3 Physical Channels — 400
      - 10.3.2.3.1 Dedicated Physical Channels — 401
      - 10.3.2.3.2 Common Physical Channels — 404

10.3.2.3.2.1	Common Physical Channels of the FDD Mode	404
10.3.2.3.2.2	Common Physical Channels of the TDD Mode	408

10.3.2.4 Service Multiplexing and Channel Coding in UTRA . 410
    10.3.2.4.1 CRC Attachment 411
    10.3.2.4.2 Transport Block Concatenation 411
    10.3.2.4.3 Channel-Coding 411
    10.3.2.4.4 Radio Frame Padding 414
    10.3.2.4.5 First Interleaving 414
    10.3.2.4.6 Radio Frame Segmentation 414
    10.3.2.4.7 Rate Matching 414
    10.3.2.4.8 Discontinuous Transmission Indication . . . 415
    10.3.2.4.9 Transport Channel Multiplexing 415
    10.3.2.4.10 Physical Channel Segmentation 415
    10.3.2.4.11 Second Interleaving 415
    10.3.2.4.12 Physical Channel Mapping 415
    10.3.2.4.13 Mapping Several Multirate Services to the UL Dedicated Physical Channels in FDD Mode 416
    10.3.2.4.14 Mapping of a 4.1 Kbps Data Service to the DL DPDCH in FDD Mode 417
    10.3.2.4.15 Mapping Several Multirate Services to the UL Dedicated Physical Channels in TDD Mode 420
10.3.2.5 Variable-Rate and Multicode Transmission in UTRA 420
10.3.2.6 Spreading and Modulation 422
    10.3.2.6.1 Orthogonal Variable Spreading Factor Codes 423
    10.3.2.6.2 Up-Link Scrambling Codes 426
    10.3.2.6.3 Down-Link Scrambling Codes 426
    10.3.2.6.4 Up-Link Spreading and Modulation 426
    10.3.2.6.5 Down-Link Spreading and Modulation 428
10.3.2.7 Random Access 429
    10.3.2.7.1 Mobile-Initiated Physical Random Access Procedures 429
    10.3.2.7.2 Common Packet Channel Access Procedures 430
10.3.2.8 Power Control 430
    10.3.2.8.1 Closed-Loop Power Control in UTRA 430
    10.3.2.8.2 Open-Loop Power Control in TDD Mode . . 431
10.3.2.9 Cell Identification 432
    10.3.2.9.1 Cell Identification in the FDD Mode 432
    10.3.2.9.2 Cell Identification in the TDD Mode 434
10.3.2.10 Handover 436
    10.3.2.10.1 Intra-Frequency Handover or Soft Handover 436

     10.3.2.10.2 Inter-Frequency Handover or Hard Handover 436
    10.3.2.11 Intercell Time Synchronization in the UTRA TDD Mode 438
  10.3.3 The cdma2000 Terrestrial Radio Access 439
    10.3.3.1 Characteristics of cdma2000 439
    10.3.3.2 Physical Channels in cdma2000 441
    10.3.3.3 Service Multiplexing and Channel Coding 443
    10.3.3.4 Spreading and Modulation 445
      10.3.3.4.1 Down-Link Spreading and Modulation 446
      10.3.3.4.2 Up-link Spreading and Modulation 447
    10.3.3.5 Random Access 447
    10.3.3.6 Handover 450
  10.3.4 Performance-Enhancement Features 452
    10.3.4.1 Down-Link Transmit Diversity Techniques 452
      10.3.4.1.1 Space Time Block Coding-Based Transmit Diversity 452
      10.3.4.1.2 Time-Switched Transmit Diversity 452
      10.3.4.1.3 Closed-Loop Transmit Diversity 452
    10.3.4.2 Adaptive Antennas 453
    10.3.4.3 Multi-User Detection/Interference Cancellation 453
  10.3.5 Summary of 3G Systems 454
10.4 Summary and Conclusions 455

II Video Systems Based on Proprietary Video Codecs   457

11 Fractal Image Codecs   459
11.1 Fractal Principles 459
11.2 One-Dimensional Fractal Coding 462
  11.2.1 Fractal Codec Design 465
  11.2.2 Fractal Codec Performance 467
11.3 Error Sensitivity and Complexity 471
11.4 Summary and Conclusions 473

12 Very Low Bit-Rate DCT Codecs   475
12.1 Video Codec Outline 475
12.2 The Principle of Motion Compensation 477
  12.2.1 Distance Measures 481
  12.2.2 Motion Search Algorithms 482
    12.2.2.1 Full or Exhaustive Motion Search 483
    12.2.2.2 Gradient-Based Motion Estimation 484
    12.2.2.3 Hierarchical or Tree Search 485
    12.2.2.4 Subsampling Search 486
    12.2.2.5 Post-Processing of Motion Vectors 487
    12.2.2.6 Gain-Cost-Controlled Motion Compensation 487
  12.2.3 Other Motion Estimation Techniques 489
    12.2.3.1 Pel-Recursive Displacement Estimation 490

CONTENTS

    12.2.3.2 Grid Interpolation Techniques 490
    12.2.3.3 MC Using Higher Order Transformations 490
    12.2.3.4 MC in the Transform Domain 491
  12.2.4 Conclusion 491
12.3 Transform Coding 492
  12.3.1 One-Dimensional Transform Coding 492
  12.3.2 Two-Dimensional Transform Coding 493
  12.3.3 Quantizer Training for Single-Class DCT 496
  12.3.4 Quantizer Training for Multiclass DCT 497
12.4 The Codec Outline 499
12.5 Initial Intra-Frame Coding 502
12.6 Gain-Controlled Motion Compensation 502
12.7 The MCER Active/Passive Concept 503
12.8 Partial Forced Update of the Reconstructed Frame Buffers 504
12.9 The Gain/Cost-Controlled Inter-Frame Codec 506
  12.9.1 Complexity Considerations and Reduction Techniques 508
12.10 The Bit-Allocation Strategy 509
12.11 Results 510
12.12 DCT Codec Performance under Erroneous Conditions 512
  12.12.1 Bit Sensitivity 513
  12.12.2 Bit Sensitivity of Codec I and II 515
12.13 DCT-Based Low-Rate Video Transceivers 516
  12.13.1 Choice of Modem 516
  12.13.2 Source-Matched Transceiver 517
    12.13.2.1 System 1 517
      12.13.2.1.1 System Concept 517
      12.13.2.1.2 Sensitivity-Matched Modulation 518
      12.13.2.1.3 Source Sensitivity 518
      12.13.2.1.4 Forward Error Correction 518
      12.13.2.1.5 Transmission Format 519
    12.13.2.2 System 2 520
      12.13.2.2.1 Automatic Repeat Request 522
    12.13.2.3 Systems 3–5 523
12.14 System Performance 524
  12.14.1 Performance of System 1 524
  12.14.2 Performance of System 2 527
    12.14.2.1 FER Performance 527
    12.14.2.2 Slot Occupancy Performance 529
    12.14.2.3 PSNR Performance 530
  12.14.3 Performance of Systems 3–5 531
12.15 Summary and Conclusions 535

13 VQ Codecs and Multimode Video Transceivers — 537
- 13.1 Introduction — 537
- 13.2 The Codebook Design — 537
- 13.3 The Vector Quantizer Design — 541
  - 13.3.1 Mean and Shape Gain Vector Quantization — 544
  - 13.3.2 Adaptive Vector Quantization — 546
  - 13.3.3 Classified Vector Quantization — 548
  - 13.3.4 Algorithmic Complexity — 549
- 13.4 Performance under Erroneous Conditions — 550
  - 13.4.1 Bit-Allocation Strategy — 550
  - 13.4.2 Bit Sensitivity — 552
- 13.5 VQ-Based Low-Rate Video Transceivers — 554
  - 13.5.1 Choice of Modulation — 554
  - 13.5.2 Forward Error Correction — 554
  - 13.5.3 Architecture of System 1 — 555
  - 13.5.4 Architecture of System 2 — 557
  - 13.5.5 Architecture of Systems 3–6 — 558
- 13.6 System Performance — 558
  - 13.6.1 Simulation Environment — 558
  - 13.6.2 Performance of Systems 1 and 3 — 560
  - 13.6.3 Performance of Systems 4 and 5 — 561
  - 13.6.4 Performance of Systems 2 and 6 — 563
- 13.7 Summary and Conclusions — 564

14 Low Bit-Rate Parametric Quad-Tree-Based Codecs and Multimode Videophone Transceivers — 567
- 14.1 Introduction — 567
- 14.2 Quad-Tree Decomposition — 568
- 14.3 Quad-Tree Intensity Match — 571
  - 14.3.1 Zero-Order Intensity Match — 571
  - 14.3.2 First-Order Intensity Match — 573
  - 14.3.3 Decomposition Algorithmic Issues — 573
- 14.4 Model-Based Parametric Enhancement — 576
  - 14.4.1 Eye and Mouth Detection — 577
  - 14.4.2 Parametric Codebook Training — 580
  - 14.4.3 Parametric Encoding — 581
- 14.5 The Enhanced QT Codec — 582
- 14.6 Performance under Erroneous Conditions — 583
  - 14.6.1 Bit Allocation — 584
  - 14.6.2 Bit Sensitivity — 586
- 14.7 QT-Codec-Based Video Transceivers — 586
  - 14.7.1 Channel Coding and Modulation — 586
  - 14.7.2 QT-Based Transceiver Architectures — 588
- 14.8 QT-Based Video-Transceiver Performance — 591
- 14.9 Summary of QT-Based Video Transceivers — 595
- 14.10 Summary of Low-Rate Codecs/Transceivers — 595

III High-Resolution Image Coding   601

15 Low-Complexity Techniques   603
15.1 Introduction and Video Formats 603
15.2 Differential Pulse Code Modulation 608
    15.2.1 Basic Differential Pulse Code Modulation 608
    15.2.2 Intra/Inter-Frame Differential Pulse Code Modulation 610
    15.2.3 Adaptive Differential Pulse Code Modulation 611
15.3 Block Truncation Coding 613
    15.3.1 The Block Truncation Algorithm 613
    15.3.2 Block Truncation Codec Implementations 614
    15.3.3 Intra-Frame Block Truncation Coding 615
    15.3.4 Inter-Frame Block Truncation Coding 617
15.4 Subband Coding 618
    15.4.1 Perfect Reconstruction Quadrature Mirror Filtering 620
        15.4.1.1 Analysis Filtering 620
        15.4.1.2 Synthesis Filtering 623
        15.4.1.3 Practical QMF Design Constraints 624
    15.4.2 Practical Quadrature Mirror Filters 627
15.5 Run-Length-Based Intra-Frame Subband Coding 630
    15.5.1 Max-Lloyd-Based Subband Coding 633
15.6 Summary and Conclusions 637

16 High-Resolution DCT Coding   639
16.1 Introduction 639
16.2 Intra-Frame Quantizer Training 639
16.3 Motion Compensation for High-Quality Images 644
16.4 Inter-Frame DCT Coding 650
    16.4.1 Properties of the DCT transformed MCER 650
    16.4.2 Joint Motion Compensation and Residual Encoding 657
16.5 The Proposed Codec 658
    16.5.1 Motion Compensation 660
    16.5.2 The Inter/Intra-DCT Codec 661
    16.5.3 Frame Alignment 662
    16.5.4 Bit-Allocation 665
    16.5.5 The Codec Performance 666
    16.5.6 Error Sensitivity and Complexity 667
16.6 Summary and Conclusions 669

IV Video Systems Based on Standard Video Codecs   673

17 An ARQ-Assisted H.261-Based Reconfigurable Multilevel Videophone System   675
17.1 Introduction 675
17.2 The H.261 Video Coding Standard 675
    17.2.1 Overview 675

17.2.2	Source Encoder	676
17.2.3	Coding Control	679
17.2.4	Video Multiplex Coder	680
	17.2.4.1 Picture Layer	680
	17.2.4.2 Group of Blocks Layer	681
	17.2.4.3 Macroblock Layer	683
	17.2.4.4 Block Layer	684
17.2.5	Simulated Coding Statistics	686
	17.2.5.1 Fixed-Quantizer Coding	687
	17.2.5.2 Variable Quantizer Coding	689

17.3 Effect of Transmission Errors on the H.261 Codec 692
   17.3.1 Error Mechanisms 692
   17.3.2 Error Control Mechanisms 693
      17.3.2.1 Background 693
      17.3.2.2 Intra-Frame Coding 693
      17.3.2.3 Automatic Repeat Request 693
      17.3.2.4 Reconfigurable Modulations Schemes 694
      17.3.2.5 Combined Source/Channel Coding 694
   17.3.3 Error Recovery 695
   17.3.4 Effects of Errors 696
      17.3.4.1 Qualitative Effect of Errors on H.261 Parameters . . . 696
      17.3.4.2 Quantitative Effect of Errors on a H.261 Data Stream 699
         17.3.4.2.1 Errors in an Intra-Coded Frame 700
         17.3.4.2.2 Errors in an Inter-Coded Frame 702
         17.3.4.2.3 Errors in Quantizer Indices 705
         17.3.4.2.4 Errors in an Inter-Coded Frame with Motion Vectors 705
         17.3.4.2.5 Errors in an Inter-Coded Frame at Low Rate 708

17.4 A Wireless Reconfigurable Videophone System 710
   17.4.1 Introduction 710
   17.4.2 Objectives 710
   17.4.3 Bit-Rate Reduction of the H.261 Codec 711
   17.4.4 Investigation of Macroblock Size 711
   17.4.5 Error Correction Coding 715
   17.4.6 Packetization Algorithm 715
      17.4.6.1 Encoding History List 716
      17.4.6.2 Macroblock Compounding 717
      17.4.6.3 End of Frame Effect 719
      17.4.6.4 Packet Transmission Feedback 720
      17.4.6.5 Packet Truncation and Compounding Algorithms . . 720

17.5 H.261-Based Wireless Videophone System Performance 721
   17.5.1 System Architecture 721
   17.5.2 System Performance 725

17.6 Summary and Conclusions 731

18 Comparison of the H.261 and H.263 Codecs — 733
- 18.1 Introduction — 733
- 18.2 The H.263 Coding Algorithms — 735
  - 18.2.1 Source Encoder — 735
    - 18.2.1.1 Prediction — 735
    - 18.2.1.2 Motion Compensation and Transform Coding — 735
    - 18.2.1.3 Quantization — 736
  - 18.2.2 Video Multiplex Coder — 736
    - 18.2.2.1 Picture Layer — 738
    - 18.2.2.2 Group of Blocks Layer — 738
    - 18.2.2.3 H.261 Macroblock Layer — 739
    - 18.2.2.4 H.263 Macroblock Layer — 740
    - 18.2.2.5 Block Layer — 744
  - 18.2.3 Motion Compensation — 745
    - 18.2.3.1 H.263 Motion Vector Predictor — 745
    - 18.2.3.2 H.263 Subpixel Interpolation — 746
  - 18.2.4 H.263 Negotiable Options — 747
    - 18.2.4.1 Unrestricted Motion Vector Mode — 747
    - 18.2.4.2 Syntax-Based Arithmetic Coding Mode — 749
      - 18.2.4.2.1 Arithmetic coding [12] — 749
    - 18.2.4.3 Advanced Prediction Mode — 751
      - 18.2.4.3.1 Four Motion Vectors per Macroblock — 752
      - 18.2.4.3.2 Overlapped Motion Compensation for Luminance — 753
    - 18.2.4.4 P-B Frames Mode — 754
- 18.3 Performance Results — 757
  - 18.3.1 Introduction — 757
  - 18.3.2 H.261 Performance — 758
  - 18.3.3 H.261/H.263 Performance Comparison — 761
  - 18.3.4 H.263 Codec Performance — 764
    - 18.3.4.1 Gray-Scale versus Color Comparison — 765
    - 18.3.4.2 Comparison of QCIF Resolution Color Video — 767
    - 18.3.4.3 Coding Performance at Various Resolutions — 768
- 18.4 Summary and Conclusions — 776

19 A H.263 Videophone System for Use over Mobile Channels — 777
- 19.1 Introduction — 777
- 19.2 H.263 in a Mobile Environment — 777
  - 19.2.1 Problems of Using H.263 in a Mobile Environment — 777
  - 19.2.2 Possible Solutions for Using H.263 in a Mobile Environment — 778
    - 19.2.2.1 Coding Video Sequences Using Exclusively Intra-Coded Frames — 779
    - 19.2.2.2 Automatic Repeat Requests — 779
    - 19.2.2.3 Multimode Modulation Schemes — 779
    - 19.2.2.4 Combined Source/Channel Coding — 780
- 19.3 Design of an Error-Resilient Reconfigurable Videophone System — 781

19.3.1 Introduction 781
19.3.2 Controling the Bit Rate 782
19.3.3 Employing FEC Codes in the Videophone System 784
19.3.4 Transmission Packet Structure 785
19.3.5 Coding Parameter History List 786
19.3.6 The Packetization Algorithm 787
       19.3.6.1 Operational Scenarios of the Packetizing Algorithm . 788
19.4 H.263-Based Video System Performance 790
19.4.1 System Environment 790
19.4.2 Performance Results 792
       19.4.2.1 Error-Free Transmission Results 792
       19.4.2.2 Effect of Packet Dropping on Image Quality 793
       19.4.2.3 Image Quality versus Channel Quality without ARQ . 795
       19.4.2.4 Image Quality versus Channel Quality with ARQ . . 796
19.4.3 Comparison of H.263 and H.261-Based Systems 798
       19.4.3.1 Performance with Antenna Diversity 800
       19.4.3.2 Performance over DECT Channels 802
19.5 Transmission Feedback 806
19.5.1 ARQ Issues 811
19.5.2 Implementation of Transmission Feedback 811
       19.5.2.1 Majority Logic Coding 812
19.6 Summary and Conclusions 816

20 Error Rate Based Power Control   819
20.1 Background 819
20.2 Power Control Algorithm 819
20.3 Performance of the Power Control 824
20.3.1 Frame Error Rate Performance 824
20.3.2 Signal-to-Interference Ratio Performance 828
20.3.3 Signal-to-Interference-Plus-Noise Ratio Performance 828
20.4 Multimode Performance 832
20.5 Average Transmission Power 834
20.6 Optimization of Power Control Parameters 838
20.6.1 Joint Optimization of IPC and DPC Parameters 839
20.6.2 Joint Optimization of NEF and NFE 841
20.6.3 Joint Optimization of IPSS and DPSS 843
20.6.4 Conclusions from Optimizing the Power Control
       Algorithm Parameters 844
20.7 Power Control Performance at Various Speeds 845
20.7.1 Power Control Results for Pedestrians 845
20.7.2 Channel Fading 845
20.7.3 Tracking of Slow Fading 848
20.7.4 Power Control Error 851
20.8 Multiple Interferers 855
20.8.1 Frame Error Rate Performance 855
20.8.2 Further Effects of Power Control on System Performance . . . 857

20.9 Summary and Conclusions 859

21 Adaptive Single-Carrier, Multicarrier, and CDMA-based Video Systems   861
21.1 Turbo-equalised H.263-based videophony for GSM/GPRS 861
    21.1.1 Motivation and Background 861
    21.1.2 System Parameters 862
    21.1.3 Turbo Equalization 865
    21.1.4 Turbo-equalization Performance 868
        21.1.4.1 Video Performance 869
        21.1.4.2 Bit Error Statistics 873
    21.1.5 Summary and Conclusions 875
21.2 Adaptive QAM-based Wireless Videophony 875
    21.2.1 Introduction 875
    21.2.2 Adaptive Video Transceiver 877
    21.2.3 Burst-by-Burst Adaptive Videophone Performance 879
    21.2.4 Switching Thresholds 886
    21.2.5 Turbo-coded Video Performance 890
    21.2.6 Summary and Conclusions 892
21.3 UMTS-like Burst-by-burst Adaptive CDMA Videophony 894
    21.3.1 Motivation and Video Transceiver Overview 894
    21.3.2 Multimode Video System Performance 899
    21.3.3 Burst-by-Burst Adaptive Videophone System 902
    21.3.4 Summary and Conclusions 908
21.4 H.263/OFDM-Based Video Systems for Frequency-Selective Wireless Networks 908
    21.4.1 Background 908
    21.4.2 System Overview 909
        21.4.2.1 The WATM System 913
        21.4.2.2 The UMTS-type Framework 916
    21.4.3 The Channel Model 917
    21.4.4 Video-Related System Aspects 918
        21.4.4.1 Video Parameters of the WATM System 918
        21.4.4.2 Video parameters of the UMTS scheme 921
    21.4.5 System Performance 922
    21.4.6 Summary and Conclusions 926
21.5 Adaptive Turbo-coded OFDM-Based Videotelephony 927
    21.5.1 Motivation and Background 927
    21.5.2 AOFDM Modem Mode Adaptation and Signaling 929
    21.5.3 AOFDM Subband BER Estimation 929
    21.5.4 Video Compression and Transmission Aspects 930
    21.5.5 Comparison of Subband-Adaptive OFDM and Fixed Mode OFDM Transceivers 930
    21.5.6 Subband-Adaptive OFDM Transceivers Having Different Target Bit Rates 936
    21.5.7 Time-Variant Target Bit Rate OFDM Transceivers 941

21.5.8 Summary and Conclusions 950
21.6 Digital Terrestrial Video Broadcasting for Mobile Receivers 950
    21.6.1 Background and Motivation 950
    21.6.2 MPEG-2 Bit Error Sensitivity 951
    21.6.3 DVB Terrestrial Scheme 962
    21.6.4 Terrestrial Broadcast Channel Model 965
    21.6.5 Data Partitioning Scheme 966
    21.6.6 Performance of the Data Partitioning Scheme 972
    21.6.7 Nonhierarchical OFDM DVBP Performance 983
    21.6.8 Hierarchical OFDM DVB Performance 990
    21.6.9 Summary and Conclusions 993
21.7 Satellite-Based Video Broadcasting 996
    21.7.1 Background and Motivation 996
    21.7.2 DVB Satellite Scheme 997
    21.7.3 Satellite Channel Model 999
    21.7.4 The Blind Equalizers 999
    21.7.5 Performance of the DVB Satellite Scheme 1003
        21.7.5.1 Transmission over the Symbol-Spaced Two-Path Channel 1003
        21.7.5.2 Transmission over the Two-Symbol Delay Two-Path Channel 1010
        21.7.5.3 Performance Summary of the DVB-S System 1010
    21.7.6 Summary and Conclusions 1017
21.8 Summary and Conclusions 1018
21.9 Wireless Video System Design Principles 1020

Glossary 1023

Bibliography 1033

Subject Index 1065

Author Index 1081

About the Authors 1093

Preface

The Wireless Multimedia Communications Scene

Against the backdrop of the emerging third-generation wireless personal communications standards and broadband access network standard proposals, this book is dedicated to a range of topical wireless video communications aspects. The transmission of multimedia information over wireline based links can now be considered a mature area, where a range of interactive and distributive services are offered by various providers across the globe, such as Integrated Services Digital Network (ISDN) based on H.261/H.263 assisted videotelephony, video on demand services using the Motion Pictures Expert Group (MPEG) video compression standards, multimedia electronic mail, cable television and radio programs, and so on. A range of interactive mobile multimedia communications services are also realistic in technical terms at the time of writing, and their variety, quality, as well as market penetration are expected to exceed those of the wireline-oriented services during the next few years.

The wireless multimedia era is expected to witness a tremendous growth with the emergence of the third-generation (3G) personal communications networks (PCN) and wireless asynchronous transfer mode (WATM) systems, which constitute a wireless extension of the existing ATM networks. All three global 3G PCN standard proposals, which originated in the United States, Europe, and Japan, are based on Code Division Multiple Access (CDMA) and are capable of transmitting at bit-rates in excess of 2 Mbps. Furthermore, the European proposal was also designed to support multiple simultaneous calls and services. The WATM solutions often favor Orthogonal frequency Division Multiple Access (OFDM) as their modulation technique. Indeed, the Broadband Access Network (BRAN) standard also advocates OFDM. A range of WATM video aspects and mobile digital video broadcast (DVB) issues are also reviewed in Part IV of the book.

Research is also well under way toward defining a whole host of new modulation and signal processing techniques, and a further trend is likely to dominate this new era: namely, *the merger of wireless multimedia communications, multimedia consumer electronics, and multimedia computer technologies.* These trends are likely to hallmark the community's future research in the forthcoming years. This book is naturally limited in terms of its coverage of these aspects, simply because of space limitations. We endeavored, however, to provide the reader with a broad range of applications examples, which are pertinent to scenarios, such as transmitting low-latency interactive video as well as distributive or broadcast video signals over the existing second-generation (2G) wireless systems, 3G arrangements, and the forthcoming fourth-generation systems. We also characterized the video performance of

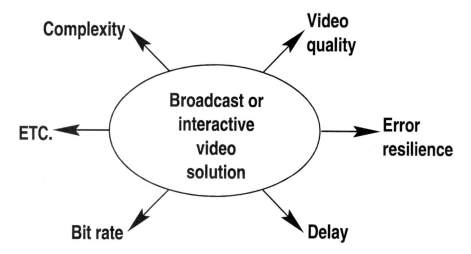

Figure 1: Contradictory system design requirements of various video communications systems.

a range of high bit-rate Local Area Network (LAN) type systems as well as various video broadcast systems, transmitting broadcast-quality video signals to mobile receivers, both within the home and farther afield, to demanding business customers on the move.

These enabling technologies facilitate a whole host of wireless services, such as videotelephony, electronic commerce, city guide, Internet access for games, electronic mail, and web browsing. Further attractive applications can be found in wireless in-home networks, DVB reception in busses, trains, cars, on board ships, and elsewhere — for example, using multimedia laptop PCs. Again, the book does not delve into the area of specific applications; rather it offers a range of technical solutions, which are applicable to various propagation and application environments.

We hope that the book offers you a range of interesting topics, sampling — and hopefully without gross aliasing errors, the current state of the art in the associated enabling technologies. In simple terms, finding a specific solution to a distributive or interactive video communications problem has to be based on a compromise in terms of the inherently contradictory constraints of video-quality, bit-rate, delay, robustness against channel errors, and the associated implementational complexity, as suggested by Figure 1. Analyzing these trade-offs and proposing a range of attractive solutions to various video communications problems are the basic aims of this book.

Video over Wireless Systems

Over the past decade second-generation (2G) wireless systems have been installed throughout the world and in some countries about half of the population possesses a mobile telephone. These systems typically exhibit a higher spectral efficiency than their analog counterparts and offer a significantly wider range of services, such as data,

Video Format	Luminance Dimensions	No. of Pels per Frame	Uncompressed Bit-Rate (Mbit/s)			
			10 frame/s		30 frame/s	
			Gray	Color	Gray	Color
SQCIF	128 x 96	12 288	0.983	1.47	2.95	4.42
QCIF	176 x 144	25 344	2.03	3.04	6.09	9.12
CIF	352 x 288	101 376	8.1	12.2	24.3	36.5
4CIF	704 x 576	405 504	32.4	48.7	97.3	146.0
16CIF	1408 x 1152	1 622 016	129.8	194.6	389.3	583.9
CCIR 601	720 x 480	345 600	27.65	41.472	82.944	124.416
HDTV 1440	1440 x 960	1 382 400	110.592	165.888	331.776	497.664
HDTV	1920 x 1080	2 073 600	165.9	248.832	497.664	746.496

SQCIF: Sub-Quarter Common Intermediate Format
QCIF: Quarter Common Intermediate Format
CIF: Common Intermediate Format
HDTV: High Definition Television

Table 1: Various video formats and their uncompressed bit-rate. Upon using compression 10-100 times lower average bit-rates are realistic.

fax, e-mail, short messages, and high-speed circuit switched data. However, because of their relatively low bit-rates, the provision of interactive wireless videotelephony has been hindered. Potentially there are two different options for transmitting video over the 2G systems, namely, over their data channel, or — provided that the standards can be amended accordingly — by allocating an additional speech channel for video transmissions. Considering the second option first, the low-rate speech channel of the 2G systems constrains the achievable bit-rate to such low values that the spatial video resolution supported is limited to the 174 × 144-pixel Quarter Common Intermediate Format (QCIF) or to the 128 × 96-pixel Sub-QCIF (SQCIF) at a video-frame scanning rate of 5 to 10 frames/s.

The range of standard video formats is summarized in Table 1, along with their uncompressed bit-rates at frame scanning rates of both at 10 and 30 frames/sec for both gray and color video signals. This table indicates the extremely wide range of potential bit-rate requirements. Clearly, the higher resolution formats can be realistically used only, for example, in the context of high-rate WATM systems.

The Cordless Telephone (CT) schemes of the second generation typically have a 32 kbit/s speech rate, which is more readily amenable to interactive videotelephony. For the sake of supporting a larger video-frame size, such as the 352 x 288-pixel Common Intermediate Format (QCIF), higher bit-rates must be supported, which is possible over the Digital European Cordless Telecommunications (DECT) system upon linking a number of slots at a rate in excess of 500 kbps.

By contrast, the data channel of the 2G systems can often offer a higher data rate than that of the speech channel, for example, by linking a number of time-slots, as was proposed in the DECT scheme or in the high-speed circuit switched data (HSCSD) mode of the Global System of Mobile communications known as GSM. CT schemes typically refrain from invoking channel coding, since they typically operate over benign channels. Hence, they do not employ channel interleavers, which is advantageous in video delay terms but disadvantageous in terms of error resilience. The data transmission mode of cellular systems, however, typically exhibits a high interleaving delay, which helps increase the system's robustness against channel errors. This is advantageous in terms of reducing the channel-induced video impairments but may result in "lip-synchronization" problems between the speech and video output signals.

Both the speech and data channels of the 2G systems tend to support a fixed constant bit-rate. However, the existing standard video codecs, such as the H.263 and MPEG2 codecs, generate a time-variant bit-rate. This is because they endeavor to reduce the bit-rate to near the lowest possible bit-rate constituted by the entropy of the source signal. Since this is achieved by invoking high-compression variable-length coding schemes, their time-variant bitstream becomes very sensitive to transmission errors. In fact, a single transmission error may render the video-quality of an entire video-frame unacceptable. Hence, the existing standard-based video codecs, such as the H.263 and MPEG2 schemes, require efficient system-level transport solutions in order to address the above-mentioned deficiencies. (This issue will be discussed in more depth in Part IV of the book). An alternative solution is invoking constant-rate proprietary video codecs, which — to a degree — sacrifice compression efficiency for the sake of an increased robustness against channel errors. This philosophy is pursued in Part II of the book, which relies on much of the compression and communications theory, as well as on the various error correction coding and transmission solutions presented in Part I.

At the time of writing, the standardization of 3G systems is approaching completion in Europe, the United States, and Japan. These systems, which are characterized in Part I of the book, along with their 2G counterparts, were designed to further enrich the range of services supported. They are more amenable to interactive wireless videotelephony, for example, than their 2G conterparts. This book aims to propose a range of video system solutions that bridge the evolutionary avenue between the second- and third-generation wireless systems.

Part I of the book provides an overview of the whole range of associated transmission aspects of the various video systems proposed and investigated. Specifically, Chapter 1 summarizes the necessary background on information, compression, and communications theory. This is followed by the characterization of wireless channels in Chapter 2. The impairments produced by these channels can be counteracted by the channel codecs of Chapters 3 and 4. Various modulation and transmission schemes are the topic of Chapter 5. We then provide a discussion of video traffic modeling and evaluate the proposed model's performance in the context of various statistical multiplexing and multiple access schemes in Chapter 6. The effects of co-channel interferences — which represent the most dominant performance-limiting factor of multiple access-based cellular systems — are described in Chapter 7. Dynamic channel allocation schemes — which rely on knowledge of co-channel interference and the multiple access scheme employed — are the topic of Chapter 8. The video transmission capabilities of 2G wireless systems are discussed in Chapter 9. These elaborations are followed by an in-depth discussion of various CDMA schemes in Chapter 10, including a variety of novel residual number system-based CDMA schemes and on the global 3G CDMA proposals, which concludes Part I of the book.

Part II is dedicated to a host of fixed but arbitrarily programmable rate video codecs based on fractal coding, on the discrete cosine transform (DCT), on vector quantized (VQ) codecs, and quad-tree-based codecs. These video codecs and their associated quadrature amplitude modulated (QAM) video systems are described in Chapters 11–14. Part III book focusses on high-resolution video coding, encompassing Chapters 15 and 16.

Part IV presents Chapters 17–21, which characterize the H.261 and H.263 video codecs, constituting one of the most important representatives of the family of state-of-the-art hybrid DCT codecs. Hence, the associated findings of these chapters can be readily applied in the context of other hybrid DCT codecs, such as the MPEG family, including the MPEG2 and MPEG4 codecs. Chapters 17–21 also portray the interactions of these hybrid DCT video codecs with reconfigurable multimode QAM transceivers. The book concludes in Chapter 21 by offering a range of system design studies related to wideband burst-by-burst adaptive TDMA/TDD, OFDM, and CDMA interactive as well as distributive mobile video systems and their performance characterization over highly dispersive transmission media.

Motivation

The rationale of this book has already been outlined from a technical perspective. Another important motivation of the book is to bring together two seemingly independent research communities, namely, the video compression and the wireless communications communities, by bridging the philosophical difference between them. These philosophical differences are based partially on the contradictory requirements portrayed and discussed in the context of Figure 1. Specifically, while a range of exciting developments have taken place in both the image compression and wireless communications communities, most of the video compression research has been cast in the context of wireline-based communications systems, such as ISDN and ATM links. These communications systems typically exhibit a low bit error rate (BER) and a low packet- or cell-loss rate. For example, ATM systems aim for a cell-loss rate of 10^{-9}. Thus, the error-resilience requirements of the video codecs were extremely relaxed.

In the increasingly pervasive wireless era, however, such extreme transmission integrity requirements are simply unrealistic, because they impose unreasonable constraints on the design of wireless systems, such as, for example, WATM systems. For example, the ATM cell-loss rate of 10^{-9} could only be maintained over wireless links at a high implementational cost, potentially invoking Automatic Repeat Requests (ARQs). ARQs, however, would increase the system delay, potentially precluding real-time interactive video communications, unless innovative design principles are invoked. Again, all these trade-offs are the subject of this book.

Part I of the book seeks to provide background for readers who require an overview of wireless communications, potentially, for example, video compression experts. Part II assumes a sound knowledge of the issues treated in Part I of the book, while offering an effortless introduction to the associated video compression aspects. Wireless experts may therefore skip Part I and begin reading Part II of the book. Part III focuses exclusively on video compression. Hopefully, readers from both the video compression and wireless communications communities will find Part IV informative and fun to read, since it integrates the knowledge base of both fields, aiming to design improved video systems.

It is our hope that the book underlines the range of contradictory system design trade-offs in an unbiased fashion and that you will be able to glean information from it in order to solve your own particular wireless video communications problem. Most of all however we hope that you will find it an enjoyable and relatively effortless reading, providing you with intellectual stimulation.

Lajos Hanzo, Peter J. Cherriman, and Jürgen Streit
Department of Electronics and Computer Science
University of Southampton

Acknowledgments

We are indebted to our many colleagues who have enhanced our understanding of the subject, in particular to Prof. Emeritus Raymond Steele. These colleagues and valued friends, too numerous to be mentioned, have influenced our views concerning various aspects of wireless multimedia communications. We thank them for the enlightenment gained from our collaborations on various projects, papers, and books. We are grateful to Jan Brecht, Jon Blogh, Marco Breiling, Marco del Buono, Clare Sommerville, Stanley Chia, Byoung Jo Choi, Joseph Cheung, Peter Fortune, Sheyam Lal Dhomeja, Lim Dongmin, Dirk Didascalou, Stephan Ernst, Eddie Green, David Greenwood, Hee Thong How, Thomas Keller, W. H. Lam, C. C. Lee, M. A. Nofal, Xiao Lin, Chee Siong Lee, Tong-Hooi Liew, Matthias Muenster, Vincent Roger-Marchart, Redwan Salami, David Stewart, Jeff Torrance, Spyros Vlahoyiannatos, William Webb, John Williams, Jason Woodard, Choong Hin Wong, Henry Wong, James Wong, Lie-Liang Yang, Bee-Leong Yeap, Mong-Suan Yee, Kai Yen, Andy Yuen, and many others with whom we enjoyed an association.

We also acknowledge our valuable associations with the Virtual Centre of Excellence in Mobile Communications, in particular with its chief executives, Dr. Tony Warwick and Dr. Walter Tuttlebee, and other members of its Executive Committee, namely Dr. Keith Baughan, Prof. Hamid Aghvami, Prof. Ed Candy, Prof. John Dunlop, Prof. Barry Evans, Dr. Mike Barnard, Prof. Joseph McGeehan, Prof. Peter Ramsdale and many other valued colleagues. Our sincere thanks are also due to the EPSRC, UK for supporting our research. We would also like to thank Dr. Joao Da Silva, Dr Jorge Pereira, Bartholome Arroyo, Bernard Barani, Demosthenes Ikonomou, and other valued colleagues from the Commission of the European Communities, Brussels, Belgium, as well as Andy Aftelak, Mike Philips, Andy Wilton, Luis Lopes, and Paul Crichton from Motorola ECID, Swindon, UK, for sponsoring some of our recent research. Further thanks are due to Tim Wilkinson at HP in Bristol for funding some of our research efforts.

We feel particularly indebted to Chee Siong Lee for his invaluable help with proofreading the manuscript and to Rita Hanzo as well as Denise Harvey for their skillful assistance in typesetting the manuscript in Latex. Similarly, our sincere thanks are due to Ken Moore, Cathy Faduska, Surendra Bhimani, Linda Matarazzo and a number of other staff members of IEEE Press for their kind assistance throughout the preparation of the camera-ready manuscript. The valuable support of Betty Pessagno and Suzanne Ingrao in copy editing and proof reading the manuscript is also gratefully acknowledged. We are also indebted to Professor John B. Anderson, the series editor, who incorporated the book in his prestigious IEEE communications series. Finally, our sincere gratitude is due to the numerous authors listed in the Author

Index — as well as to those, whose work was not cited due to space limitations — for their contributions to the state of the art, without whom this book would not have materialized.

Lajos Hanzo, Peter J. Cherriman, and Jürgen Streit
Department of Electronics and Computer Science
University of Southampton

Contributors

Chapter 6: Video Traffic Modelling and Multiple Access;
J. Brecht, M. del Buono, and L. Hanzo

Chapter 10: CDMA Systems: Third-Generation and Beyond;
K. Yen and L. Hanzo

Chapter 21: Adaptive Single-Carrier, Multicarrier, and CDMA-Based Video Systems;
P. Cherriman, L. Hanzo, B.L. Yeap, C.H. Wong, E.L. Kuan, T. Keller, C.S. Lee, S. Vlahoyiannatos

Chapter 21 is partially based on our research sponsored by the Virtual Center of Excellence in Mobile and Personal Communications (MVCE), UK.

Part I

Transmission Issues

Chapter 1

Information Theory

1.1 Issues in Information Theory

The ultimate aim of telecommunications is to communicate information between two geographically separated locations via a communications channel with adequate quality. The theoretical foundations of information theory accrue from Shannon's pioneering work [13–16], and hence most tutorial interpretations of his work over the past fifty years rely fundamentally on [13–16]. This chapter is no exception in this respect. Throughout this chapter we make frequent references to Shannon's seminal papers and to the work of various authors offering further insights into Shannonian information theory. Since this monograph aims to provide an all-encompassing coverage of video compression and communications, we begin by addressing the underlying theoretical principles using a light-hearted approach, often relying on worked examples.

Early forms of human telecommunications were based on smoke, drum or light signals, bonfires, semaphores, and the like. Practical *information sources* can be classified as analog and digital. The output of an analog source is a continuous function of time, such as, for example, the air pressure variation at the membrane of a microphone due to someone talking. The roots of Nyquist's sampling theorem are based on his observation of the maximum achievable telegraph transmission rate over bandlimited channels [17]. In order to be able to satisfy Nyquist's sampling theorem the analogue source signal has to be *bandlimited* before sampling. The analog source signal has to be transformed into a digital representation with the aid of time- and amplitude-discretization using *sampling and quantization*.

The output of a digital source is one of a finite set of ordered, discrete symbols often referred to as an alphabet. Digital sources are usually described by a range of characteristics, such as *the source alphabet, the symbol rate, the symbol probabilities, and the probabilistic interdependence of symbols*. For example, the probability of u following q in the English language is $p = 1$, as in the word "equation." Similarly, the joint probability of all pairs of consecutive symbols can be evaluated.

In recent years, electronic telecommunications have become prevalent, although most information sources provide information in other forms. For electronic telecommunications, the source information must be converted to electronic signals by a *transducer*. For example, a microphone converts the air pressure waveform $p(t)$ into voltage variation $v(t)$, where

$$v(t) = c \cdot p(t - \tau), \tag{1.1}$$

and the constant c represents a scaling factor, while τ is a delay parameter. Similarly, a video camera scans the natural three-dimensional scene using optics and converts it into electronic waveforms for transmission.

The electronic signal is then transmitted over the *communications channel* and converted back to the required form, which may be carried out, for example, by a loudspeaker. It is important to ensure that the channel conveys the transmitted signal with adequate quality to the receiver in order to enable information recovery. Communications channels can be classified according to their ability to support analog or digital transmission of the source signals in a *simplex, duplex, or half-duplex* fashion over *fixed or mobile* physical channels constituted by pairs of wires, Time Division Multiple Access (TDMA) time-slots, or a Frequency Division Multiple Access (FDMA) frequency slot.

The *channel impairments* may include superimposed, unwanted random signals, such as thermal noise, crosstalk via multiplex systems from other users, man-made interference from car ignition, fluorescent lighting, and other natural sources such as lightning. Just as the natural sound pressure wave between two conversing persons will be impaired by the acoustic background noise at a busy railway station, similarly the reception quality of electronic signals will be affected by the above unwanted electronic signals. In contrast, distortion manifests itself differently from additive noise sources, since no impairment is explicitly added. Distortion is more akin to the phenomenon of reverberating loudspeaker announcements in a large, vacant hall, where no noise sources are present.

Some of the channel impairments can be mitigated or counteracted; others cannot. For example, the effects of unpredictable additive random noise cannot be removed or "subtracted" at the receiver. Its effects can be mitigated by increasing the transmitted signal's power, but the transmitted power cannot be increased without penalties, since the system's nonlinear distortion rapidly becomes dominant at higher signal levels. This process is similar to the phenomenon of increasing the music volume in a car parked near a busy road to a level where the amplifier's distortion becomes annoyingly dominant.

In practical systems, the *Signal-to-Noise Ratio* (SNR) quantifying the wanted and unwanted signal powers at the channel's output is a prime channel parameter. Other important *channel parameters* are its *amplitude and phase response*, determining its usable bandwidth (B), over which the signal can be transmitted without excessive distortion. Among the most frequently used statistical noise properties are the *probability density function* (PDF), *cumulative density function* (CDF), and *power spectral density* (PSD).

The fundamental *communications system design considerations* are whether a high-fidelity (HI-FI) or just acceptable video or speech quality is required from a system, which predetermines, among other factors, its cost, bandwidth requirements, as

1.1. ISSUES IN INFORMATION THEORY

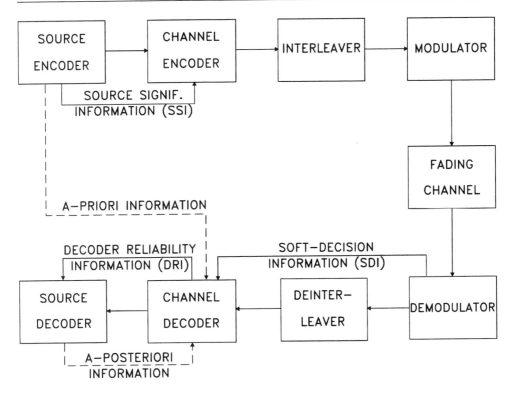

Figure 1.1: Basic transmission model of information theory.

well as the number of channels available, and has implementational complexity ramifications. Equally important are the issues of robustness against channel impairments, system delay, and so on. The required transmission range and worldwide roaming capabilities, the maximum available transmission speed in terms of symbols/sec, information confidentiality, reception reliability, convenient lightweight, solar-charged design, are similarly salient characteristics of a communications system.

Information theory deals with a variety of problems associated with the performance limits of the information transmission system, such as that depicted in Figure 1.1. The components of this system constitute the subject of this monograph; hence they will be treated in greater depth later in this volume. Suffice it to say at this stage that the transmitter seen in Figure 1.1 incorporates a source encoder, a channel encoder, an interleaver, and a modulator and their inverse functions at the receiver. The *ideal source encoder* endeavors to remove as much redundancy as possible from the information source signal without affecting its source representation fidelity (i.e., distortionlessly), and it remains oblivious of such practical constraints as a finite delay and limited signal processing complexity. In contrast, a practical source encoder will have to retain a limited signal processing complexity and delay while attempting to reduce the source representation bit rate to as low a value as possible. This operation seeks to achieve better transmission efficiency, which can be expressed in terms of bit-rate economy or bandwidth conservation.

The channel encoder re-inserts redundancy or parity information but in a controlled manner in order to allow error correction at the receiver. Since this component is designed to ensure the best exploitation of the re-inserted redundancy, it is expected to minimize the error probability over the most common channel, namely, the so-called *Additive White Gaussian Noise* (AWGN) channel, which is characterized by a memoryless, random distribution of channel errors. However, over wireless channels, which have recently become prevalent, the errors tend to occur in bursts due to the presence of deep received signal fades induced by the distructively superimposed multipath phenomena. This is why our schematic of Figure 1.1 contains an interleaver block, which is included in order to randomize the bursty channel errors. Finally, the modulator is designed to ensure the most bandwidth-efficient transmission of the source- and channel encoded, interleaved information stream, while maintaining the lowest possible bit error probability. The receiver simply carries out the corresponding inverse functions of the transmitter. Observe in the figure that besides the direct interconnection of the adjacent system components there are a number of additional links in the schematic, which will require further study before their role can be highlighted. Thus, at the end of this chapter we will return to this figure and guide the reader through its further intricate details.

Some fundamental problems transpiring from the schematic of Figure 1.1, which were addressed in depth by a range of references due to Shannon [13–16], Nyquist [17], Hartley [18], Abramson [19], Carlson [20], Raemer [21], and Ferenczy [22] and others are as follows:

- What is the true information generation rate of our information sources? If we know the answer, the efficiency of coding and transmission schemes can be evaluated by comparing the actual transmission rate used with the source's information emission rate. The actual transmission rate used in practice is typically much higher than the average information delivered by the source, and the closer these rates are, the better is the coding efficiency.

- Given a noisy communications channel, what is the maximum reliable information transmission rate? The thermal noise induced by the random motion of electrons is present in all electronic devices, and if its power is high, it can seriously affect the quality of signal transmission, allowing information transmission only at low-rates.

- Is the information emission rate the only important characteristic of a source, or are other message features, such as the probability of occurrence of a message and the joint probability of occurrence for various messages, also important?

- In a wider context, the topic of this whole monograph is related to the blocks of Figure 1.1 and to their interactions, but in this chapter we lay the theoretical foundations of source and channel coding as well as transmission issues and define the characteristics of an ideal Shannonian communications scheme.

Although numerous excellent treatises are available on these topics, which treat the same subjects with a different flavor [20, 22, 23], our approach is similar to that of the above classic sources; since the roots are in Shannon's work, references [13–16, 24, 25] are the most pertinent and authoritative sources.

In this chapter we consider mainly discrete sources, in which each source message is associated with a certain probability of occurrence, which might or might not be dependent on previous source messages. Let us now give a rudimentary introduction to the characteristics of the AWGN channel, which is the predominant channel model in information theory due to its simplicity. The analytically less tractable wireless channels of Chapter 2 will be modeled mainly by simulations in this monograph, although in Chapter 10 some analytical results are also provided in the context of Code Division Multiple Access (CDMA) systems.

1.2 Additive White Gaussian Noise Channel

1.2.1 Background

In this section, we consider the communications channel, which exists between the transmitter and the receiver, as shown in Figure 1.1. Accurate characterization of this channel is essential if we are to remove the impairments imposed by the channel using signal processing at the receiver. Here we initially consider only fixed communications links whereby both terminals are stationary, although mobile radio communications channels, which change significantly with time, are becoming more prevalent.

We define fixed communications channels to be those between a fixed transmitter and a fixed receiver. These channels are exemplified by twisted pairs, cables, wave guides, optical fiber and point-to-point microwave radio channels. Whatever the nature of the channel, its output signal differs from the input signal. The difference might be deterministic or random, but it is typically unknown to the receiver. Examples of channel impairments are dispersion, nonlinear distortions, delay, and random noise.

Fixed communications channels can often be modeled by a linear transfer function, which describes the channel dispersion. The ubiquitous additive Gaussian noise (AWGN) is a fundamental limiting factor in communications via linear time-invariant (LTI) channels. Although the channel characteristics might change due to factors such as aging, temperature changes, and channel switching, these variations will not be apparent over the course of a typical communication session. It is this inherent time invariance that characterizes fixed channels.

An ideal, distortion-free communications channel would have a flat frequency response and linear phase response over the frequency range of $-\infty \ldots +\infty$, although in practice it is sufficient to satisfy this condition over the bandwidth (B) of the signals to be transmitted, as seen in Figure 1.2. In this figure, $A(\omega)$ represents the magnitude of the channel response at frequency w, and $\phi(w) = wT$ represents the phase shift at frequency w due to the circuit delay T.

Practical channels always have some linear distortions due to their bandlimited, nonflat frequency response and nonlinear phase response. In addition, the group-delay response of the channel, which is the derivative of the phase response, is often given.

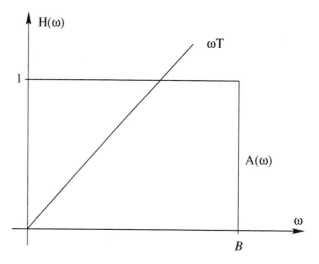

Figure 1.2: Ideal, distortion-free channel model having a linear phase and a flat magnitude response.

1.2.2 Practical Gaussian Channels

Conventional telephony uses twisted copper wire pairs to connect subscribers to the local exchange. The bandwidth is approximately 3.4 kHz, and the waveform distortions are relatively benign.

For applications requiring a higher bandwidth, coaxial cables can be used. Their attenuation increases approximately with the square root of the frequency. Hence, for wideband, long-distance operation, they require channel equalization. Typically, coaxial cables can provide a bandwidth of about 50 MHz, and the transmission rate they can support is limited by the so-called skin effect.

Point-to-point microwave radio channels typically utilize high-gain directional transmit and receive antennas in a line-of-sight scenario, where free-space propagation conditions may be applicable.

1.2.3 Gaussian Noise

Regardless of the communications channel used, random noise is always present. Noise can be broadly classified as natural or man-made. Examples of man-made noise are those due to electrical appliances, and fluorescent lighting, and the effects of these sources can usually be mitigated at the source. Natural noise sources affecting radio transmissions include galactic star radiations and atmospheric noise. There exists a low-noise frequency window in the range of 1–10 GHz, where the effects of these sources are minimized.

Natural thermal noise is ubiquitous. This is due to the random motion of electrons, and it can be reduced by reducing the temperature. Since thermal noise contains practically all frequency components up to some 10^{13} Hz with equal power, it is often referred to as white noise (WN) in an analogy to white light containing all colors with

1.2. ADDITIVE WHITE GAUSSIAN NOISE CHANNEL

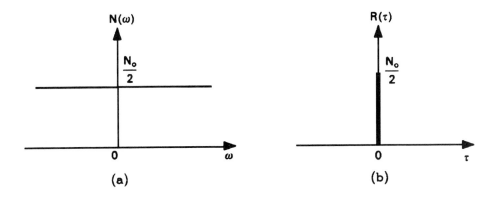

Figure 1.3: Power spectral density and autocorrelation of WN.

equal intensity. This WN process can be characterized by its uniform power spectral density (PSD) $N(\omega) = N_0/2$ shown together with its autocorrelation function (ACF) in Figure 1.3.

The power spectral density of any signal can be conveniently measured by the help of a selective narrowband power meter tuned across the bandwidth of the signal. The power measured at any frequency is then plotted against the measurement frequency. The autocorrelation function $R(\tau)$ of the signal $x(t)$ gives an average indication of how predictable the signal $x(t)$ is after a period of τ seconds from its present value. Accordingly, it is defined as follows:

$$R(\tau) = \lim_{T \to \infty} \frac{1}{T} \int_{-\infty}^{\infty} x(t)x(t+\tau)dt. \quad (1.2)$$

For a periodic signal $x(t)$, it is sufficient to evaluate the above equation for a single period T_0, yielding:

$$R(\tau) = \frac{1}{T_0} \int_{-T_0/2}^{T_0/2} x(t)x(t+\tau)dt. \quad (1.3)$$

The basic properties of the ACF are:

- The ACF is symmetric: $R(\tau) = R(-\tau)$.

- The ACF is monotonously decreasing: $R(\tau) \leq R(0)$.

- For $\tau = 0$ we have $R(0) = \overline{x^2(t)}$, which is the signal's power.

- The ACF and the PSD form a Fourier transform pair, which is formally stated as the Wiener-Khintchine theorem, as follows:

$$R(\tau) = \frac{1}{2\pi} \int_{-\infty}^{\infty} N(\omega)e^{j\omega\tau}d\omega$$

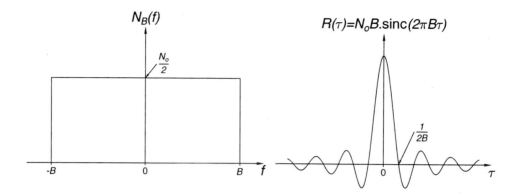

Figure 1.4: Power spectral density and autocorrelation of bandlimited WN.

$$\begin{aligned} &= \frac{1}{2\pi} \int_{-\infty}^{\infty} \frac{N_0 e^{j\omega\tau}}{2} d\omega \\ &= \frac{1}{2\pi} \frac{N_0}{2} \int_{-\infty}^{\infty} e^{j\omega\tau} d\omega = \frac{N_0}{2} \delta(\tau), \end{aligned} \quad (1.4)$$

where $\delta(\tau)$ is the Dirac delta function. Clearly, for any timed-domain shift $\tau > 0$, the noise is uncorrelated.

Bandlimited communications systems bandlimit not only the signal but the noise as well, and this filtering limits the rate of change of the time-domain noise signal, introducing some correlation over the interval of $\pm 1/2B$. The stylized PSD and ACF of bandlimited white noise are displayed in Figure 1.4.

After bandlimiting, the autocorrelation function becomes:

$$\begin{aligned} R(\tau) &= \frac{1}{2\pi} \int_{-B}^{B} \frac{N_0}{2} e^{j\omega\tau} d\omega = \frac{N_0}{2} \int_{-B}^{B} e^{j2\pi f\tau} df \\ &= \frac{N_0}{2} \left[\frac{e^{j2\pi f\tau}}{j2\pi\tau} \right]_{-B}^{B} \\ &= \frac{1}{j2\pi\tau} [\cos 2\pi B\tau + j\sin 2\pi B\tau - \cos 2\pi B\tau + j\sin 2\pi B\tau] \\ &= N_0 B \frac{\sin(2\pi B\tau)}{2\pi B\tau}, \end{aligned} \quad (1.5)$$

which is the well-known sinc-function seen in Figure 1.4.

In the time-domain, the amplitude distribution of the white thermal noise has a normal or Gaussian distribution, and since it is inevitably added to the received signal, it is usually referred to as additive white Gaussian noise (AWGN). Note that AWGN is therefore the noise generated in the receiver. The probability density function (PDF) is the well-known bell-shaped curve of the Gaussian distribution, given by

$$p(x) = \frac{1}{\sigma\sqrt{2\pi}} e^{-(x-m)/2\sigma^2}, \quad (1.6)$$

where m is the mean and σ^2 is the variance. The effects of AWGN can be mitigated by increasing the transmitted signal power and thereby reducing the relative effects of noise. The signal-to-noise ratio (SNR) at the receiver's input provides a good measure of the received signal quality. This SNR is often referred to as the channel SNR.

1.3 Information of a Source

Based on Shannon's work [13–16, 24, 25], let us introduce the basic terms and definitions of information theory by considering a few simple examples. Assume that a simple 8-bit analog-to-digital (ADC) converter emits a sequence of mutually independent source symbols that can take the values $i = 1, 2, \ldots 256$ with equal probability. One may wonder, how much information can be inferred upon receiving one of these samples? It is intuitively clear that this inferred information is definitely proportional to the "uncertainty" resolved by the reception of one such symbol, which in turn implies that the information conveyed is related to the number of levels in the ADC. More explicitly, the higher the number of legitimate quantization levels, the lower the relative frequency or probability of receiving any one of them and hence the more "surprising," when any one of them is received. Therefore, less probable quantized samples carry more information than their more frequent, more likely counterparts.

Not suprisingly, one could resolve this uncertainty by simply asking a maximum of 256 questions, such as "Is the level 1?" "Is it 2? ..." "Is it 256?" Following Hartley's approach [18], a more efficient strategy would be to ask eight questions, such as: "Is the level larger than 128?" No. "Is it larger than 64?" No. ... "Is it larger than 2?" No. "Is it larger than 1?" No. Clearly, the source symbol emitted was of magnitude one, provided that the zero level was not used. We could therefore infer that $\log_2 256 = 8$ "Yes/No" binary answers were needed to resolve any uncertainty as regards the source symbol's level.

In more general terms, the information carried by any one symbol of a q-level source, where all the levels are equiprobable with probabilities of $p_i = 1/q, i = 1 \ldots q$, is defined as

$$I = \log_2 q. \tag{1.7}$$

Rewriting Equation 1.7 using the message probabilities $p_i = \frac{1}{q}$ yields a more convenient form:

$$I = \log_2 \frac{1}{p_i} = -\log_2 p_i, \tag{1.8}$$

which now is also applicable in case of arbitrary, unequal message probabilities p_i, again, implying the plausible fact that the lower the probability of a certain source symbol, the higher the information conveyed by its occurrence. Observe, however, that for unquantized analog sources, where as regards to the number of possible source symbols we have $q \to \infty$ and hence the probability of any analog sample becomes infinitesimally low, these definitions become meaningless.

Let us now consider a sequence of N consecutive q-ary symbols. This sequence can take q^N number of different values, delivering q^N different messages. Therefore, the information carried by one such sequence is:

$$I_N = \log_2(q^N) = N \log_2 q, \tag{1.9}$$

which is in perfect harmony with our expectation, delivering N times the information of a single symbol, which was quantified by Equation 1.7. Doubling the sequence length to $2N$ carries twice the information, as suggested by:

$$I_{2N} = \log_2(q^{2N}) = 2N \cdot \log_2 q. \qquad (1.10)$$

Before we proceed, let us briefly summarize the basic *properties of information* following Shannon's work [13–16, 24, 25]:

- If for the probability of occurrences of the symbols j and k we have $p_j < p_k$, then as regards the information carried by them we have: $I(k) < I(j)$.

- If in the limit we have $p_k \to 1$, then for the information carried by the symbol k we have $I(k) \to 0$, implying that symbols, whose probability of occurrence tends to unity, carry no information.

- If the symbol probability is in the range of $0 \leq p_k \leq 1$, then as regards the information carried by it we have $I(k) \geq 0$.

- For independent messages k and j, their joint information is given by the sum of their information: $I(k,j) = I(k) + I(j)$. For example, the information carried by the statement "My son is 14 years old and my daughter is 12" is equivalent to that of the sum of these statements: "My son is 14 years old" and "My daughter is 12 years old."

- In harmony with our expectation, if we have two equiprobable messages 0 and 1 with probabilities, $p_1 = p_2 = \frac{1}{2}$, then from Equation 1.8 we have $I(0) = I(1) = 1$ bit.

1.4 Average Information of Discrete Memoryless Sources

Following Shannon's approach [13–16, 24, 25], let us now consider a source emitting one of q possible symbols from the alphabet $s = s_1, s_2, \ldots s_i \ldots s_q$ having symbol probabilities of p_i, $i = 1, 2, \ldots q$. Suppose that a long message of N symbols constituted by symbols from the alphabet $s = s_1, s_2, \ldots s_q$ having symbol probabilities of p_i is to be transmitted. Then the symbol s_i appears in every N-symbol message on the average $p_i \cdot N$ number of times, provided the message length is sufficiently long. The information carried by symbol s_i is $\log_2 1/p_i$ and its $p_i \cdot N$ occurrences yield an information contribution of

$$I(i) = p_i \cdot N \cdot \log_2 \frac{1}{p_i}. \qquad (1.11)$$

Upon summing the contributions of all the q symbols, we acquire the total information carried by the N-symbol sequence:

$$I = \sum_{i=1}^{q} p_i N \cdot \log_2 \frac{1}{p_i} \text{ [bits]}. \qquad (1.12)$$

1.4. AVERAGE INFORMATION OF DISCRETE MEMORYLESS SOURCES

Averaging this over the N symbols of the sequence yields the average information per symbol, which is referred to as the source's *entropy* [14]:

$$H = \frac{I}{N} = \sum_{i=1}^{q} p_i \cdot \log_2 \frac{1}{p_i} = -\sum_{i=1}^{q} p_i \log_2 p_i \quad \text{[bit/symbol]}. \tag{1.13}$$

Then the *average source information rate* can be defined as the product of the information carried by a source symbol, given by the entropy H and the source emission rate R_s:

$$R = R_s \cdot H \quad \text{[bits/sec]}. \tag{1.14}$$

Observe that Equation 1.13 is analogous to the discrete form of the first moment or in other words the mean of a random process with a probability density function (PDF) of $p(x)$, as in

$$\overline{x} = \int_{-\infty}^{\infty} x \cdot p(x) dx, \tag{1.15}$$

where the averaging corresponds to the integration, and the instantaneous value of the random variable x represents the information $\log_2 p_i$ carried by message i, which is weighted by its probability of occurrence p_i quantified for a continuous variable x by $p(x)$.

1.4.1 Maximum Entropy of a Binary Source

Let us assume that a binary source, for which $q = 2$, emits two symbols with probabilities $p_1 = p$ and $p_2 = (1-p)$, where the sum of the symbol probabilities must be unity. In order to quantify the maximum average information of a symbol from this source as a function of the symbol probabilities, we note from Equation 1.13 that the entropy is given by:

$$H(p) = -p \cdot \log_2 p - (1-p) \cdot \log_2(1-p). \tag{1.16}$$

As in any maximization problem, we set $\partial H(p)/\partial p = 0$, and upon using the differentiation chain rule of $(u \cdot v)' = u' \cdot v + u \cdot v'$ as well as exploiting that $(\log_a x)' = \frac{1}{x} \log_a e$ we arrive at:

$$\begin{aligned} \frac{\partial H(p)}{\partial p} &= -\log_2 p - \frac{p}{p} \cdot \log_2 e + \log_2(1-p) + \frac{(1-p)}{(1-p)} \log_2 e = 0 \\ \log_2 p &= \log_2(1-p) \\ p &= (1-p) \\ p &= 0.5. \end{aligned}$$

This result suggests that the entropy is maximum for equiprobable binary messages. Plotting Equation 1.16 for arbitrary p values yields Figure 1.5, in which Shannon suggested that the average information carried by a symbol of a binary source is low, if one of the symbols has a high probability, while the other a low probability.

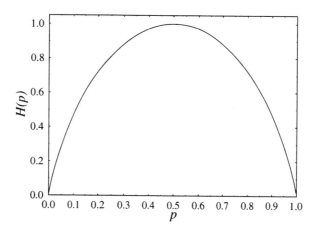

Figure 1.5: Entropy versus message probability p for a binary source. ©Shannon [14], BSTJ, 1948.

Example: *Let us compute the entropy of the binary source having message probabilities of $p_1 = \frac{1}{8}$, $p_2 = \frac{7}{8}$.*

The entropy is expressed as:

$$H = -\frac{1}{8}\log_2\frac{1}{8} - \frac{7}{8}\log_2\frac{7}{8}.$$

Exploiting the following equivalence:

$$\log_2(x) = \log_{10}(x) \cdot \log_2(10) \approx 3.322 \cdot \log_{10}(x) \qquad (1.17)$$

we have:

$$H \approx \frac{3}{8} - \frac{7}{8} \cdot 3.322 \cdot \log_{10}\frac{7}{8} \approx 0.54 \text{ [bit/symbol]},$$

again implying that if the symbol probabilities are rather different, the entropy becomes significantly lower than the achievable 1 bit/symbol. This is because the probability of encountering the more likely symbol is so close to unity that it carries hardly any information, which cannot be compensated by the more "informative" symbol's reception. For the even more unbalanced situation of $p_1 = 0.1$ and $p_2 = 0.9$ we have:

$$\begin{aligned} H &= -0.1\log_2 0.1 - 0.9 \cdot \log_2 0.9 \\ &\approx -(0.3322 \cdot \log_{10} 0.1 + 0.9 \cdot 3.322 \cdot \log_{10} 0.9) \\ &\approx 0.3322 + 0.1368 \\ &\approx 0.47 \text{ [bit/symbol]}. \end{aligned}$$

In the extreme case of $p_1 = 0$ or $p_2 = 1$ we have $H = 0$. As stated before, the *average source information rate* is defined as the product of the information carried by a source symbol, given by the entropy H and the source emission rate R_s, yielding $R = R_s \cdot H$ [bits/sec]. Transmitting the source symbols via a perfect noiseless channel yields the same received sequence without loss of information.

1.4.2 Maximum Entropy of a q-ary Source

For a q-ary source the entropy is given by:

$$H = -\sum_{i=1}^{q} p_i \log_2 p_i, \tag{1.18}$$

where, again, the constraint $\sum p_i = 1$ must be satisfied. When determining the extreme value of the above expression for the entropy H under the constraint of $\sum p_i = 1$, the following term has to be maximized:

$$D = \sum_{i=1}^{q} -p_i \log_2 p_i + \lambda \cdot \left[1 - \sum_{i=1}^{q} p_i\right], \tag{1.19}$$

where λ is the so-called Lagrange multiplier. Following the standard procedure of maximizing an expression, we set:

$$\frac{\partial D}{\partial p_i} = -\log_2 p_i - \frac{p_i}{p_i} \cdot \log_2 e - \lambda = 0$$

leading to

$$\log_2 p_i = -(\log_2 e + \lambda) = \text{Constant for } i = 1 \ldots q,$$

which implies that the maximum entropy of a q-ary source is maintained, if all message probabilities are identical, although at this stage the value of this constant probability is not explicit. Note, however, that the message probabilites must sum to unity, and hence:

$$\sum_{i=1}^{q} p_i = 1 = q \cdot a, \tag{1.20}$$

where a is a constant, leading to $a = 1/q = p_i$, implying that the entropy of any q-ary source is maximum for equiprobable messages. Furthermore, H is always bounded according to:

$$0 \leq H \leq \log_2 q. \tag{1.21}$$

1.5 Source Coding for a Discrete Memoryless Source

Interpreting Shannon's work [13–16, 24, 25] further, we see that source coding is the process by which the output of a q-ary information source is converted to a binary sequence for transmission via binary channels, as seen in Figure 1.1. When a discrete memoryless source generates q-ary equiprobable symbols with an average information rate of $R = R_s \log_2 q$, all symbols convey the same amount of information, and efficient signaling takes the form of binary transmissions at a rate of R bps. When the symbol probabilities are unequal, the minimum required source rate for distortionless transmission is reduced to

$$R = R_s \cdot H < R_s \log_2 q. \qquad (1.22)$$

Then the transmission of a highly probable symbol carries little information and hence assigning $\log_2 q$ number of bits to it does not use the channel efficiently. What can be done to improve transmission efficiency? *Shannon's source coding theorem* suggests that by using a *source encoder* before transmission the efficiency of the system with equiprobable source symbols can be arbitrarily approached.

Coding efficiency can be defined as the ratio of the source information rate and the average output bit rate of the source encoder. If this ratio approaches unity, implying that the source encoder's output rate is close to the source information rate, the source encoder is highly efficient. There are many source encoding algorithms, but the most powerful approach suggested was Shannon's method [13], which is best illustrated by means of the following example, portrayed in Table 1.1, Algorithm 1, and Figure 1.6.

1.5.1 Shannon-Fano Coding

The Shannon-Fano coding algorithm is based on the simple concept of encoding frequent messages using short codewords and infrequent ones by long codewords, while reducing the average message length. This algorithm is part of virtually all treatises dealing with information theory, such as, for example, Carlson's work [20]. The formal coding steps listed in Algorithm 1 and in the flowchart of Figure 1.6 can be readily followed in the context of a simple example in Table 1.1. The average codeword length is given by weighting the length of any codeword by its probability, yielding:

$$(0.27 + 0.2) \cdot 2 + (0.17 + 0.16) \cdot 3 + 2 \cdot 0.06 \cdot 4 + 2 \cdot 0.04 \cdot 4 \approx 2.73 \quad [\text{bit}].$$

The entropy of the source is:

$$\begin{aligned}
H &= -\sum_i p_i \log_2 p_i & (1.23) \\
&= -(\log_2 10) \sum_i p_i \log_{10} p_i \\
&\approx -3.322 \cdot [0.27 \cdot \log_{10} 0.27 + 0.2 \cdot \log_{10} 0.2 \\
&\quad + 0.17 \cdot \log_{10} 0.17 + 0.16 \cdot \log_{10} 0.16 \\
&\quad + 2 \cdot 0.06 \cdot \log_{10} 0.06 + 2 \cdot 0.04 \cdot \log_{10} 0.04] \\
&\approx 2.691 \quad [\text{bit/symbol}].
\end{aligned}$$

1.5. SOURCE CODING FOR A DISCRETE MEMORYLESS SOURCE

Algorithm 1 (Shannon-Fano Coding) *This algorithm summarizes the Shannon-Fano coding steps. (See also Figure 1.6 and Table 1.1.)*

1. The source symbols $S_0 \ldots S_7$ are first sorted in descending order of probability of occurrence.

2. Then the symbols are divided into two subgroups so that the subgroup probabilities are as close to each other as possible. This is symbolized by the horizontal divisions in Table 1.1.

3. When allocating codewords to represent the source symbols, we assign a logical zero to the top subgroup and logical one to the bottom subgroup in the appropriate column under ''coding steps.''

4. If there is more than one symbol in the subgroup, this method is continued until no further divisions are possible.

5. Finally, the variable-length codewords are output to the channel.

Symb.	Prob.	Coding Steps				Codeword
		1	2	3	4	
S_0	0.27	0	0			00
S_1	0.20	0	1			01
S_2	0.17	1	0	0		100
S_3	0.16	1	0	1		101
S_4	0.06	1	1	0	0	1100
S_5	0.06	1	1	0	1	1101
S_6	0.04	1	1	1	0	1110
S_7	0.04	1	1	1	1	1111

Table 1.1: Shannon-Fano Coding Example Based on Algorithm 1 and Figure 1.6

Since the average codeword length of 2.73 bit/symbol is very close to the entropy of 2.691 bit/symbol, a high coding efficiency is predicted, which can be computed as:

$$E \approx \frac{2.691}{2.73} \approx 98.6 \%.$$

The straightforward 3 bit/symbol binary coded decimal (BCD) assignment gives an efficiency of:

$$E \approx \frac{2.691}{3} \approx 89.69 \%.$$

In summary, Shannon-Fano coding allowed us to create a set of uniquely invertible mappings to a set of codewords, which facilitate a more efficient transmission of the source symbols, than straightforward BCD representations would. This was possible with no coding impairment (i.e., losslessly). Having highlighted the philosophy of the Shannon-Fano noiseless or distortionless coding technique, let us now concentrate on the closely related Huffman coding principle.

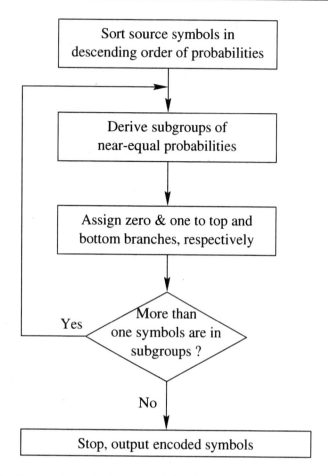

Figure 1.6: Shannon-Fano Coding Algorithm (see also Table 1.1 and Algorithm 1).

1.5.2 Huffman Coding

The Huffman Coding (HC) algorithm is best understood by referring to the flowchart of Figure 1.7 and to the formal coding description of Algorithm 2, while a simple practical example is portrayed in Table 1.2, which leads to the Huffman codes summarized in Table 1.3. Note that we used the same symbol probabilities as in our Shannon-Fano coding example, but the Huffman algorithm leads to a different codeword assignment. Nonetheless, the code's efficiency is identical to that of the Shannon-Fano algorithm.

The symbol-merging procedure can also be conveniently viewed using the example of Figure 1.8, where the Huffman codewords are derived by reading the associated 1 and 0 symbols from the end of the tree backward, that is, toward the source symbols $S_0 \ldots S_7$. Again, these codewords are summarized in Table 1.3.

In order to arrive at a fixed average channel bit rate, which is convenient in many communications systems, a long buffer might be needed, causing storage and delay problems. Observe from Table 1.3 that the Huffman coding algorithm gives code-

1.5. SOURCE CODING FOR A DISCRETE MEMORYLESS SOURCE 19

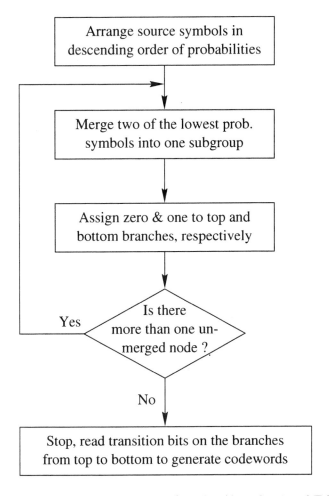

Figure 1.7: Huffman coding algorithm (see also Algorithm 2 and Table 1.2).

Algorithm 2 (Huffman Coding) *This algorithm summarizes the Huffman coding steps.*

1. Arrange the symbol probabilities p_i in decreasing order and consider them as ''leaf-nodes,'' as suggested by Table 1.2.

2. While there is more than one node, merge the two nodes having the lowest probability and assign 0/1 to the upper/lower branches, respectively.

3. Read the assigned ''transition bits'' on the branches from top to bottom in order to derive the codewords.

Symb.	Prob.	Step 1 & 2		Step 3 & 4		Group	Code
		Code	Prob.	Code	Prob.		
S_0	0.27					S_0	-
S_1	0.20					S_1	-
S_2	0.17			0	0.33	S_{23}	0
S_3	0.16			1			1
S_4	0.06	0	0.12	0			00
S_5	0.06	1		0	0.20	S_{4567}	01
S_6	0.04	0	0.08	1			10
S_7	0.04	1		1			11

Symb.	Prob.	Step 5 & 6		Step 7		Codeword
		Code	Prob.	Code	Prob.	
S_{23}	0.33	0	0.6	0		00
S_0	0.27	1			1.0	01
S_1	0.20	0	0.4	1		10
S_{4567}	0.20	1				11

Table 1.2: Huffman Coding Example Based on Algorithm 2 and Figure 1.7 (for final code assignment see Table 1.3)

Symbol	Probability	BCD	Huffman Code
S_0	0.27	000	01
S_1	0.20	001	10
S_2	0.17	010	000
S_3	0.16	011	001
S_4	0.06	100	1100
S_5	0.06	101	1101
S_6	0.04	110	1110
S_7	0.04	111	1111

Table 1.3: Huffman Coding Example Summary of Table 1.2

1.5. SOURCE CODING FOR A DISCRETE MEMORYLESS SOURCE

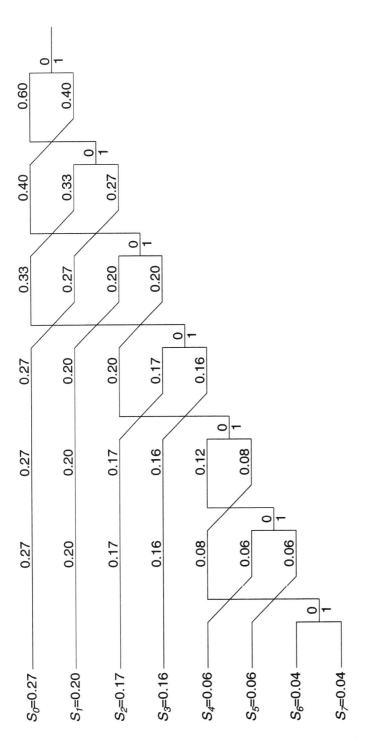

Figure 1.8: Tree-based Huffman coding example.

words that can be uniquely decoded, which is a crucial prerequisite for its practical employment. This is because no codeword can be a prefix of any longer one. For example, for the following sequence of codewords ..., 00, 10, 010, 110, 1111, ...the source sequence of ... S_0, S_1, S_2, S_3, S_8 ... can be uniquely inferred from Table 1.3.

In our discussions so far, we have assumed that the source symbols were completely independent of each other. Such a source is usually referred to as a memoryless source. By contrast, sources where the probability of a certain symbol also depends on what the previous symbol was are often termed *sources exhibiting memory*. These sources are typically bandlimited sample sequences, such as, for example, a set of correlated or "similar-magnitude" speech samples or adjacent video pixels. Let us now consider sources that exhibit memory.

1.6 Average Information of Discrete Sources Exhibiting Memory

Let us invoke Shannon's approach [13–16,24,25] in order to illustrate sources with and without memory. Let us therefore consider an uncorrelated random white Gaussian noise (WGN) process, which was passed through a low-pass filter. The corresponding autocorrelation functions (ACF) and power spectral density (PSD) functions were portrayed in Figures 1.3 and 1.4. Observe in the figures that through low-pass filtering a WGN process introduces correlation by limiting the rate at which amplitude changes are possible, smoothing the amplitude of abrupt noise peaks. This example suggests that all bandlimited signals are correlated over a finite interval. Most analog source signals, such as speech and video, are inherently correlated, owing to physical restrictions imposed on the analog source. Hence all practical analog sources possess some grade of memory, a property that is also retained after sampling and quantization. An important feature of sources with memory is that they are predictable to a certain extent, hence, they can usually be more efficiently encoded than unpredictable sources having no memory.

1.6.1 Two-State Markov Model for Discrete Sources Exhibiting Memory

Let us now introduce a simple analytically tractable model for treating sources that exhibit memory. Predictable sources that have memory can be conveniently modeled by *Markov processes*. A source having a memory of one symbol interval directly "remembers" only the previously emitted source symbol and depending on this previous symbol it emits one of its legitimate symbols with a certain probability, which depends explicitly on the state associated with this previous symbol. A one-symbol-memory model is often referred to as a *first-order model*. For example, if in a first-order model the previous symbol can take only two different values, we have two different states, and this simple two-state first-order Markov model is characterized by the state transition diagram of Figure 1.9. Previously, in the context of Shannon-Fano and Huffman coding of memoryless information sources, we used the notation of S_i, $i = 0, 1, \ldots$ for the various symbols to be encoded. In this section, we are dealing with sources

1.6. ENTROPY OF SOURCES EXHIBITING MEMORY

exhibiting memory and hence for the sake of distinction we use the symbol notation of $X_i, i = 1, 2, \ldots$. If, for the sake of illustration, the previous emitted symbol was X_1, the state machine of Figure 1.9 is in state X_1, and in the current signaling interval it can generate one of two symbols, namely, X_1 and X_2, whose probability depends explicitly on the previous state X_1. However, not all two-state Markov models are as simple as that of Figure 1.9, since the transitions from state X_1 to X_2 are not necessarily associated with emitting the same symbol as the transitions from state X_2 to X_1. Thus more elaborate example will be considered later in this chapter.

Observe in Figure 1.9 that the corresponding transition probabilities from state X_1 are given by the conditional probabilities $p_{12} = P(X_2/X_1)$ and $p_{11} = P(X_1/X_1) = 1 - P(X_2/X_1)$. Similar findings can be observed as regards state X_2. These dependencies can also be stated from a different point of view as follows. The probability of occurrence of a particular symbol depends not only on the symbol itself, but also on the previous symbol emitted. Thus, the symbol entropy for state X_1 and X_2 will now be characterized by means of the conditional probabilities associated with the transitions merging in these states. Explicitly, the symbol entropy for state $X_i, i = 1, 2$ is given by:

$$\begin{aligned} H_i &= \sum_{j=1}^{2} p_{ij} \cdot \log_2 \frac{1}{p_{ij}} \quad i = 1, 2 \\ &= p_{i1} \cdot \log_2 \frac{1}{p_{i1}} + p_{i2} \cdot \log_2 \frac{1}{p_{i2}}, \end{aligned} \quad (1.24)$$

yielding the symbol entropies, that is, the average information carried by the symbols emitted in states X_1 and X_2, respectively, as:

$$\begin{aligned} H_1 &= p_{11} \cdot \log_2 \frac{1}{p_{11}} + p_{12} \cdot \log_2 \frac{1}{p_{12}} \\ H_2 &= p_{21} \cdot \log_2 \frac{1}{p_{21}} + p_{22} \cdot \log_2 \frac{1}{p_{22}}. \end{aligned} \quad (1.25)$$

Both symbol entropies, H_1 and H_2, are characteristic of the average information conveyed by a symbol emitted in state X_1 and X_2, respectively. In order to compute the overall entropy H of this source, they must be weighted by the probability of occurrence, P_1 and P_2, of these states:

$$\begin{aligned} H &= \sum_{i=1}^{2} P_i H_i \\ &= \sum_{i=1}^{2} P_i \sum_{j=1}^{2} p_{ij} \log_2 \frac{1}{p_{ij}}. \end{aligned} \quad (1.26)$$

Assuming a highly predictable source having high adjacent sample correlation, it is plausible that once the source is in a given state, it is more likely to remain in that state than to traverse into the other state. For example, assuming that the state machine of Figure 1.9 is in state X_1 and the source is a highly correlated, predictable source, we are likely to observe long runs of X_1. Conversely, once in state X_2, long strings of X_2 symbols will typically follow.

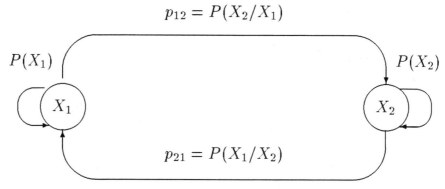

Figure 1.9: Two-state first-order Markov model.

1.6.2 N-State Markov Model for Discrete Sources Exhibiting Memory

In general, assuming N legitimate states, (i.e., N possible source symbols) and following similar arguments, Markov models are characterised by their state probabilities $P(X_i), i = 1 \ldots N$, where N is the number of states, as well as by the transition probabilities $p_{ij} = P(X_i/X_j)$, where p_{ij} explicitly indicates the probability of traversing from state X_j to state X_i. Their further basic feature is that they emit a source symbol at every state transition, as will be shown in the context of an example presented in Section 1.7. Similarly to the two-state model, we define the entropy of a source having memory as the weighted average of the entropy of the individual symbols emitted from each state, where weighting is carried out taking into account the probability of occurrence of the individual states, namely P_i. In analytical terms, the symbol entropy for state $X_i, i = 1 \ldots N$ is given by:

$$H_i = \sum_{j=1}^{N} p_{ij} \cdot \log_2 \frac{1}{p_{ij}} \quad i = 1 \ldots N. \tag{1.27}$$

The averaged, weighted symbol entropies give the source entropy:

$$\begin{aligned} H &= \sum_{i=1}^{N} P_i H_i \\ &= \sum_{i=1}^{N} P_i \sum_{j=1}^{N} p_{ij} \log_2 \frac{1}{p_{ij}}. \end{aligned} \tag{1.28}$$

Finally, assuming a source symbol rate of v_s, the average information emission rate R of the source is given by:

$$R = v_s \cdot H \quad [\text{bps}]. \tag{1.29}$$

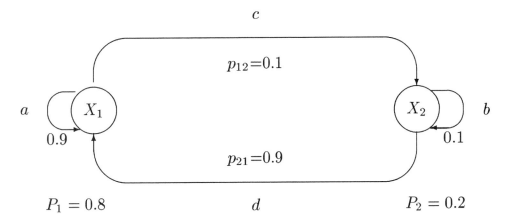

Figure 1.10: Two-state Markov model example.

1.7 Examples

1.7.1 Two-State Markov Model Example

As mentioned in the previous section, we now consider a slightly more sophisticated Markov model, where the symbols emitted upon traversing from state X_1 to X_2 are different from those when traversing from state X_2 to X_1. More explicitly:

- Consider a discrete source that was described by the two-state Markov model of Figure 1.9, where the transition probabilities are

$$p_{11} = P(X_1/X_1) = 0.9 \quad p_{22} = P(X_2/X_2) = 0.1$$
$$p_{12} = P(X_1/X_2) = 0.1 \quad p_{21} = P(X_2/X_1) = 0.9,$$

while the state probabilities are

$$P(X_1) = 0.8 \text{ and } P(X_2) = 0.2. \tag{1.30}$$

The source emits one of four symbols, namely, a, b, c, and d, upon every state transition, as seen in Figure 1.10. Let us find
 (a) the source entropy and
 (b) the average information content per symbol in messages of
  one, two, and three symbols.

- *Message Probabilities*

Let us consider two sample sequences acb and aab. As shown in Figure 1.10, the transitions leading to acb are $(1 \rightsquigarrow 1)$, $(1 \rightsquigarrow 2)$, and $(2 \rightsquigarrow 2)$. The probability of encountering this sequence is $0.8 \cdot 0.9 \cdot 0.1 \cdot 0.1 = 0.0072$. The sequence aab has a

Message Probabilities	Information conveyed (bit/message)
$P_a = 0.9 \times 0.8 = 0.72$	$I_a = 0.474$
$P_b = 0.1 \times 0.2 = 0.02$	$I_b = 5.644$
$P_c = 0.1 \times 0.8 = 0.08$	$I_c = 3.644$
$P_d = 0.9 \times 0.2 = 0.18$	$I_d = 2.474$
$P_{aa} = 0.72 \times 0.9 = 0.648$	$I_{aa} = 0.626$
$P_{ac} = 0.72 \times 0.1 = 0.072$	$I_{ac} = 3.796$
$P_{cb} = 0.08 \times 0.1 = 0.008$	$I_{cb} = 6.966$
$P_{cd} = 0.08 \times 0.9 = 0.072$	$I_{cd} = 3.796$
$P_{bb} = 0.02 \times 0.1 = 0.002$	$I_{bb} = 8.966$
$P_{bd} = 0.02 \times 0.9 = 0.018$	$I_{bd} = 5.796$
$P_{da} = 0.18 \times 0.9 = 0.162$	$I_{da} = 2.626$
$P_{dc} = 0.18 \times 0.1 = 0.018$	$I_{dc} = 5.796$

Table 1.4: Message Probabilities of Example

probability of zero because the transition from a to b is illegal. Further path (i.e., message) probabilities are tabulated in Table 1.4 along with the information of $I = -\log_2 P$ of all the legitimate messages.

- *Source Entropy*

  – According to Equation 1.27, the entropy of symbols X_1 and X_2 is computed as follows:

  $$\begin{align} H_1 &= -p_{12} \cdot \log_2 p_{12} - p_{11} \cdot \log_2 p_{11} \\ &= 0.1 \cdot \log_2 10 + 0.9 \cdot \log_2 \frac{1}{0.9} \\ &\approx 0.469 \text{ bit/symbol} \tag{1.31} \\ H_2 &= -p_{21} \cdot \log_2 p_{21} - p_{22} \cdot \log_2 p_{22} \\ &\approx 0.469 \text{ bit/symbol} \tag{1.32} \end{align}$$

  – Then their weighted average is calculated using the probability of occurrence of each state in order to derive the average information per message for this source:

  $$H \approx 0.8 \cdot 0.469 + 0.2 \cdot 0.469 \approx 0.469 \text{ bit/symbol}.$$

  – The average information per symbol I_2 in two-symbol messages is computed from the entropy h_2 of the two-symbol messages as follows:

  $$\begin{align} h_2 &= \sum_1^8 P_{symbol} \cdot I_{symbol} \\ &= P_{aa} \cdot I_{aa} + P_{ac} \cdot I_{ac} + \ldots + P_{dc} \cdot I_{dc} \\ &\approx 1.66 \text{ bits/2 symbols}, \tag{1.33} \end{align}$$

1.7. EXAMPLES

giving $I_2 = h_2/2 \approx 0.83$ bits/symbol information on average upon receiving a two-symbol message.

- There are eight two-symbol messages; hence, the maximum possible information conveyed is $\log_2 8 = 3$ bits/2 symbols, or 1.5 bits/symbol. However, since the symbol probabilities of $P_1 = 0.8$ and $P_2 = 0.2$ are fairly different, this scheme has a significantly lower conveyed information per symbol, namely, $I_2 \approx 0.83$ bits/symbol.

- Similarly, one can find the average information content per symbol for arbitrarily long messages of concatenated source symbols. For one-symbol messages we have:

$$\begin{aligned}
I_1 = h_1 &= \sum_1^4 P_{symbol} \cdot I_{symbol} \\
&= P_a \cdot I_a + \ldots + P_d \cdot I_d \\
&\approx 0.72 \times 0.474 + \ldots + 0.18 \times 2.474 \\
&\approx 0.341 + 0.113 + 0.292 + 0.445 \\
&\approx 1.191 \text{ bit/symbol}.
\end{aligned} \qquad (1.34)$$

We note that the maximum possible information carried by one-symbol messages is $h_{1max} = \log_2 4 = 2$ bit/symbol, since there are four one-symbol messages in Table 1.4.

- Observe the important tendency, in which, when sending longer messages of dependent sources, the average information content per symbol is reduced. This is due to the source's memory, since consecutive symbol emissions are dependent on previous ones and hence do not carry as much information as independent source symbols. This becomes explicit by comparing $I_1 \approx 1.191$ and $I_2 \approx 0.83$ bits/symbol.

- *Therefore, expanding the message length to be encoded yields more efficient coding schemes, requiring a lower number of bits, if the source has a memory. This is the essence of Shannon's source coding theorem.*

1.7.2 Four-State Markov Model for a 2-Bit Quantizer

Let us now augment the previously introduced two-state Markov-model concepts with the aid of a four-state example. Let us assume that we have a discrete source constituted by a 2-bit quantizer, which is characterized by Figure 1.11. Assume further that due to bandlimitation only transitions to adjacent quantization intervals are possible, since the bandlimitation restricts the input signal's rate of change. The probability of the signal samples residing in intervals 1–4 is given by:

$$P(1) = P(4) = 0.1, \quad P(2) = P(3) = 0.4.$$

The associated state transition probabilities are shown in Figure 1.11, along with the quantized samples a, b, c, and d, which are transmitted when a state transition

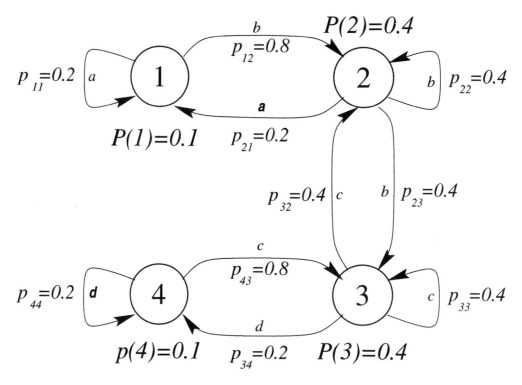

Figure 1.11: Four-state Markov model for a 2-bit quantizer.

takes place, that is, when taking a new sample from the analog source signal at the sampling-rate f_s.

Although we have stipulated a number of simplifying assumptions, this example attempts to illustrate the construction of Markov models in the context of a simple practical problem. Next we construct a simpler example for augmenting the underlying concepts and set aside the above four-state Markov-model example as a potential exercise for the reader.

1.8 Generating Model Sources

1.8.1 Autoregressive Model

In evaluating the performance of information processing systems, such as encoders and predictors, it is necessary to have "standardized" or easily described model sources. Although a set of semistandardized speech and images test sequences is widely used by researchers in codec performance testing, in contrast to analytical model sources, real speech or image sources cannot be used in analytical studies. A widely used analytical model source is the *Autoregressive (AR) model*. A zero mean random sequence $y(n)$

1.8. GENERATING MODEL SOURCES

is called *an AR process of order p*, if it is generated as follows:

$$y(n) = \sum_{k=1}^{p} a_k y(n-k) + \varepsilon(n), \forall n, \quad (1.35)$$

where $\varepsilon(n)$ is an uncorrelated zero-mean, random input sequence with variance σ^2; that is,

$$E\{\varepsilon(n)\} = 0$$
$$E\{\varepsilon^2(n)\} = \sigma^2$$
$$E\{\varepsilon(n) \cdot y(m)\} = 0. \quad (1.36)$$

From Equation 1.35 we surmise that an AR system recursively generates the present output from p number of previous output samples given by $y(n-k)$ and the present random input sample $\varepsilon(n)$.

1.8.2 AR Model Properties

AR models are very useful in studying information processing systems, such as speech and image codecs, predictors, and quantizers. They have the following basic properties:

1. The first term of Equation 1.35, which is repeated here for convenience,

$$\hat{y}(n) = \sum_{k=1}^{p} a_k y(n-k)$$

   defines a predictor, giving an estimate $\hat{y}(n)$ of $y(n)$, which is associated with the minimum mean squared error between the two quantities.

2. Although $\hat{y}(n)$ and $y(n)$ depend explicitly only on the past p number of samples of $y(n)$, through the recursive relationship of Equation 1.35 this entails the entire past of $y(n)$. This is because each of the previous p samples depends on their predecessors.

3. Then Equation 1.35 can be written in the form of:

$$y(n) = \hat{y}(n) + \varepsilon(n), \quad (1.37)$$

   where $\varepsilon(n)$ is the *prediction error* and $\hat{y}(n)$ is the minimum variance prediction estimate of $y(n)$.

4. Without proof, we state that for a random Gaussian distributed prediction error sequence $\varepsilon(n)$ these properties are characteristic of a pth order *Markov process* portrayed in Figure 1.12. When this model is simplified for the case of $p = 1$, we arrive at the schematic diagram shown in Figure 1.13.

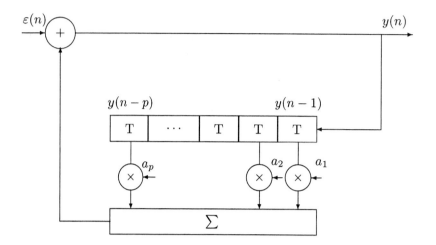

Figure 1.12: Markov model of order p.

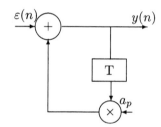

Figure 1.13: First-order Markov model.

5. The power spectral density (PSD) of the prediction error sequence $\varepsilon(n)$ is that of a random "white-noise" sequence, containing all possible frequency components with the same energy. Hence, its autocorrelation function (ACF) is the Kronecker delta function, given by the Wiener-Khintchine theorem:

$$E\{\varepsilon(n) \cdot \varepsilon(m)\} = \sigma^2 \delta(n - m). \qquad (1.38)$$

1.8.3 First-Order Markov Model

A variety of practical information sources are adequately modeled by the analytically tractable first-order Markov model depicted in Figure 1.13, where the prediction order is $p = 1$. With the aid of Equation 1.35 we have

$$y(n) = \varepsilon(n) + ay(n-1),$$

where a is the adjacent sample correlation of the process $y(n)$. Using the following recursion:

$$
\begin{aligned}
y(n-1) &= \varepsilon(n-1) + a_1 y(n-2) \\
&\vdots \quad \vdots \quad \vdots \\
y(n-k) &= \varepsilon(n-k) + a_1 y(n-k-1)
\end{aligned}
\tag{1.39}
$$

we arrive at:

$$
\begin{aligned}
y(n) &= \varepsilon(n) + a_1[\varepsilon(n-1) + a y(n-2)] \\
&= \varepsilon(n) + a_1 \varepsilon(n-1) + a^2 y(n-2),
\end{aligned}
$$

which can be generalized to:

$$
y(n) = \sum_{j=0}^{\infty} a^j \varepsilon(n-j).
\tag{1.40}
$$

Clearly, Equation 1.40 describes the first-order Markov process by the help of the adjacent sample correlation a_1 and the uncorrelated zero-mean random Gaussian process $\varepsilon(n)$.

1.9 Run-Length Coding for Discrete Sources Exhibiting Memory

1.9.1 Run-Length Coding Principle [20]

For discrete sources having memory, (i.e., possessing intersample correlation), the coding efficiency can be significantly improved by predictive coding, allowing the required transmission rate and hence the channel bandwidth to be reduced. Particularly amenable to run-length coding are binary sources with inherent memory, such as black and white documents, where the predominance of white pixels suggests that a Run-Length-Coding (RLC) scheme, which encodes the length of zero runs, rather than repeating long strings of zeros, provides high coding efficiency.

Following Carlson's interpretation [20], a predictive RLC scheme can be constructed according to Figure 1.14. The q-ary source messages are first converted to binary bit format. For example, if an 8-bit analog-digital converter (ADC) is used, the 8-bit digital samples are converted to binary format. This bit-stream, $x(i)$, is then compared with the output signal of the predictor, $\hat{x}(i)$, which is fed with the prediction error signal $e(i)$. The comparator is a simple mod-2 gate, outputting a logical 1, whenever the prediction fails; that is, the predictor's output is different from the incoming bit $x(i)$. If, however, $x(i) = \hat{x}(i)$, the comparator indicates this by outputting a logical 0. For highly correlated signals from sources with significant memory the predictions are usually correct, and hence long strings of 0 runs are emitted, interspersed with an occasional 1. Thus, the prediction error signal $e(i)$ can be efficiently run-length encoded by noting and transmitting the length of zero runs.

The corresponding binary run-length coding principle becomes explicit from Table 1.5 and from our forthcoming coding efficiency analysis.

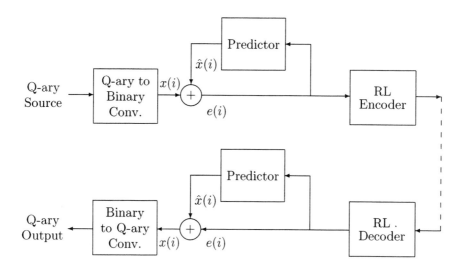

Figure 1.14: Predictive run-length codec scheme. ©Carlson [20].

Length of 0-run	Encoder Output	Decoder Output
	(n-bit codeword)	
0	00 ⋯ 000	1
1	00 ⋯ 001	01
2	00 ⋯ 010	001
3	00 ⋯ 011	0001
⋮	⋮	⋮
$N-1$	11 ⋯ 110	00 ⋯ 01
$\geq N = 2^n - 1$	11 ⋯ 111	00 ⋯ 00

Table 1.5: Run-length Coding Table © Carlson, 1975 [20]

1.9.2 Run-Length Coding Compression Ratio [26]

Following Jain's interpretation [26], let us now investigate the RLC efficiency by assuming that a run of r successive logical 0s is followed by a 1. Instead of directly transmitting these strings, we represent such a string as an n-bit word giving the length of the 0-run between successive logical ones. When a 0-run longer than $N = 2^n - 1$ bits occurs, this is signaled as the all 1 codeword, informing the decoder to wait for the next RLC codeword before releasing the decoded sequence. Again, the scheme's operation is characterized by Table 1.5. Clearly, data compression is achieved if the average number of 0 data bits per run d is higher than the number of bits, n, required to encode the 0-run length. Let us therefore compute the average number of bits per run without RLC. If a run of r logical zeros are followed by a 1, the run-length is $(r+1)$. The expected or mean value of $(r+1)$, namely, $d = \overline{(r+1)}$, is calculated by weighting each specific $(r+1)$ with its probability of occurrence that

1.9. RUN-LENGTH CODING

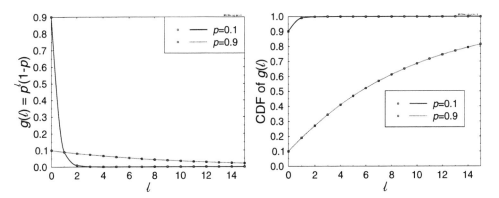

Figure 1.15: CDF and PDF of the geometric distribution of run-length l.

is, with its discrete PDF $c(r)$ and then averaging the weighted components, in:

$$d = \overline{(r+1)} = \sum_{r=0}^{N-1} (r+1) \cdot c(r) + Nc(N). \tag{1.41}$$

The PDF of a run of r zeros followed by a 1 is given by:

$$c(r) = \begin{cases} p^r(1-p) & 0 \leq r \leq N-1 \\ p^N & r = N, \end{cases} \tag{1.42}$$

since the probability of N consecutive zeros is p^N if $r = N$, while for shorter runs the joint probability of r zeros followed by a 1 is given by $p^r \cdot (1-p)$. The PDF and CDF of this distribution are shown in Figure 1.15 for $p = 0.9$ and $p = 0.1$, where p represents the probability of a logical zero bit. Substituting Equation 1.42 in Equation 1.41 gives:

$$\begin{aligned} d &= N \cdot p^N + \sum_{r=0}^{N-1} (r+1) \cdot p^r \cdot (1-p) \\ &= N \cdot p^N + 1 \cdot p^0 \cdot (1-p) + 2 \cdot p \cdot (1-p) + \ldots + N \cdot p^{N-1} \cdot (1-p) \\ &= N \cdot p^N + 1 + 2p + 3p^2 + \ldots + N \cdot p^{N-1} - p - 2p^2 \ldots - N \cdot p^N \\ &= 1 + p + p^2 + \cdots p^{N-1}. \end{aligned} \tag{1.43}$$

Equation 1.43 is a simple geometric progression, given in closed form as:

$$d = \frac{1 - p^N}{1 - p}. \tag{1.44}$$

RLC Example: *Using a run-length coding memory of $M = 31$ and a zero symbol probability of $p = 0.95$, characterize the RLC efficiency.*

Substituting N and p into Equation 1.44 for the average run-length we have:

$$d = \frac{1 - 0.95^{31}}{1 - 0.95} \approx \frac{1 - 0.204}{0.05} \approx 15.92. \tag{1.45}$$

The *compression ratio* C achieved by RLC is given by:

$$C = \frac{d}{n} = \frac{1-p^N}{n(1-p)} \approx \frac{15.92}{5} \approx 3.18. \qquad (1.46)$$

The achieved average bit rate is

$$B = \frac{n}{d} \approx 0.314 \text{ bit/pixel},$$

and the coding efficiency is computed as the ratio of the entropy (i.e., the lowest possible bit rate and the actual bit rate). The source entropy is given by:

$$\begin{aligned} H &\approx -0.95 \cdot 3.322 \cdot \log_{10} 0.95 - 0.05 \cdot 3.322 \cdot \log_{10} 0.05 \\ &\approx 0.286 \text{ bit/symbol}, \end{aligned} \qquad (1.47)$$

giving a coding efficiency of:

$$E = H/B \approx 0.286/0.314 \approx 91\%.$$

This concludes our RLC example.

1.10 Information Transmission via Discrete Channels

Let us now return to Shannon's classic references [13–16, 24, 25] and assume that both the channel and the source are discrete, and let us evaluate the amount of information transmitted via the channel. We define the channel capacity characterizing the channel and show that according to Shannon nearly error-free information transmission is possible at rates below the channel capacity via the binary symmetric channel (BSC). Let us begin our discourse with a simple introductory example.

1.10.1 Binary Symmetric Channel Example

Let us assume that a binary source is emitting a logical 1 with a probability of $P(1) = 0.7$ and a logical 0 with a probability of $P(0) = 0.3$. The channel's error probability is $p_e = 0.02$. This scenario is characterized by the binary symmetric channel (BSC) model of Figure 1.16. The probability of error-free reception is given by that of receiving 1, when a logical 1 is transmitted *plus* the probability of receiving a 0 when 0 is transmitted, which is also plausible from Figure 1.16. For example, the first of these two component probabilities can be computed with the aid of Figure 1.16 as the product of the *probability* $P(1)$ of a logical 1 being transmitted and the *conditional probability* $P(1/1)$ of receiving a 1, given the condition that a 1 was transmitted:

$$P(Y_1, X_1) = P(X_1) \cdot P(Y_1/X_1) \qquad (1.48)$$

$$P(1,1) = P(1) \cdot P(1/1) = 0.7 \cdot 0.98 = 0.686.$$

1.10. INFORMATION TRANSMISSION VIA DISCRETE CHANNELS

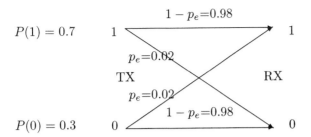

Figure 1.16: The binary symmetric channel. ©Shannon [15], BSTJ, 1948.

Similarly, the probability of the error-free reception of a logical 0 is given by:

$$P(Y_0, X_0) = P(X_0) \cdot P(Y_0/X_0)$$
$$P(0,0) = P(0) \cdot P(0/0) = 0.3 \cdot 0.98 = 0.294,$$

giving the total probability of error-free reception as:

$$P_{correct} = P(1,1) + P(0,0) = 0.98.$$

Following similar arguments, the probability of erroneous reception is also given by two components. For example, using Figure 1.16, the probability of receiving a 1 when a 0 was transmitted is computed by multiplying the probability $P(0)$ of a logical 0 being transmitted by the conditional probability $P(1/0)$ of receiving a logical 1, given the fact that a 0 is known to have been transmitted:

$$P(Y_1, X_0) = P(X_0) \cdot P(Y_1/X_0)$$
$$P(1,0) = P(0) \cdot P(1/0) = 0.3 \cdot 0.02 = 0.006.$$

Conversely,

$$P(Y_0, X_1) = P(X_1) \cdot P(Y_0/X_1)$$
$$P(0,1) = P(1) \cdot P(0/1) = 0.7 \cdot 0.02 = 0.014,$$

yielding a total error probability of:

$$P_{error} = P(1,0) + P(0,1) = 0.02,$$

which is constituted by the above two possible error events.

Viewing events from a different angle, we observe that *the total probability of receiving 1* is that of receiving a transmitted 1 correctly plus a transmitted 0 incorrectly:

$$P_1 = P(1) \cdot (1 - p_e) + P(0) \cdot p_e \qquad (1.49)$$
$$= 0.7 \cdot 0.98 + 0.3 \cdot 0.02 = 0.686 + 0.006 = 0.692.$$

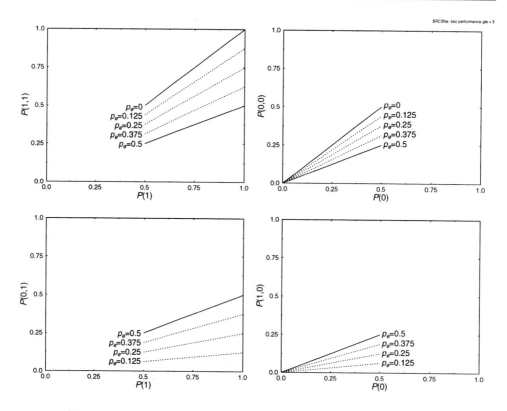

Figure 1.17: BSC performance for $p_e = 0$, 0.125, 0.25, 0.375, and 0.5.

On the same note, *the probability of receiving 0* is that of receiving a transmitted 0 correctly plus a transmitted 1 incorrectly:

$$\begin{align} P_0 &= P(0) \cdot (1 - p_e) + P(1) \cdot p_e \tag{1.50}\\ &= 0.3 \cdot 0.98 + 0.7 \cdot 0.02 = 0.294 + 0.014 = 0.308. \end{align}$$

In the next example, we further study the performance of the BSC for a range of different parameters in order to gain a deeper insight into its behavior.

> **Example:** *Repeat the above calculations for $P(1) = 1, 0.9, 0.5$, and $p_e = 0, 0.1, 0.2, 0.5$ using the BSC model of Figure 1.16. Compute and tabulate the probabilities $P(1,1)$, $P(0,0)$, $P(1,0)$, $P(0,1)$, $P_{correct}, P_{error}, P_1$, and P_0 for these parameter combinations, including also their values for the previous example, namely, for $P(1) = 0.7$, $P(0) = 0.3$ and $p_e = 0.02$.* Here we neglected the details of the calculations and summarized the results in Table 1.6. Some of the above quantities are plotted for further study in Figure 1.17, which reveals the interdependency of the various probabilities for the interested reader.

Having studied the performance of the BSC, the next question that arises is, how much information can be inferred upon reception of a 1 and a 0 over an imperfect

1.10. INFORMATION TRANSMISSION VIA DISCRETE CHANNELS

p_e	$P(1)$	$P(0)$	$(1-p_e)$ $=P(1/1)$ $=P(0/0)$	$P(1,1)$	$P(0,0)$	p_e $=P(1/0)$ $=P(0/1)$	$P(1,0)$	$P(0,1)$	P_1	P_0
0	1	0	1	1	0	0	0	0	1	0
	0.9	0.1	1	0.9	0.1	0	0	0	0.9	0.1
	0.7	0.3	1	0.7	0.3	0	0	0	0.7	0.3
	0.5	0.5	1	0.5	0.5	0	0	0	0.5	0.5
0.02	1	0	0.98	0.98	0	0.02	0	0.02	0.98	0.02
	0.9	0.1	0.98	0.882	0.098	0.02	0.002	0.018	0.884	0.116
	0.7	0.3	0.98	0.686	0.294	0.02	0.006	0.014	0.692	0.308
	0.5	0.5	0.98	0.49	0.49	0.02	0.01	0.01	0.491	0.509
0.1	1	0	0.9	0.9	0	0.1	0	0.1	0.9	0.1
	0.9	0.1	0.9	0.81	0.09	0.1	0.01	0.09	0.811	0.189
	0.7	0.3	0.9	0.63	0.27	0.1	0.03	0.07	0.723	0.277
	0.5	0.5	0.9	0.45	0.45	0.1	0.05	0.05	0.455	0.545
0.2	1	0	0.8	0.8	0	0.2	0	0.2	0.8	0.2
	0.9	0.1	0.8	0.72	0.08	0.2	0.02	0.18	0.722	0.278
	0.7	0.3	0.8	0.56	0.24	0.2	0.06	0.14	0.566	0.434
	0.5	0.5	0.8	0.40	0.40	0.2	0.1	0.1	0.5	0.5
0.5	1	0	0.5	0.5	0	0.5	0	0.5	0.5	0.5
	0.9	0.1	0.5	0.45	0.05	0.5	0.05	0.45	0.5	0.5
	0.7	0.3	0.5	0.35	0.15	0.5	0.15	0.35	0.5	0.5
	0.5	0.5	0.5	0.25	0.25	0.5	0.25	0.25	0.5	0.5

Table 1.6: BSC Performance Table

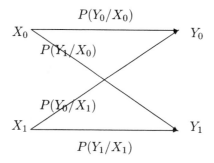

Figure 1.18: Forward transition probabilities of the nonideal binary symmetric channel.

(i.e., error-prone) channel. In order to answer this question, let us first generalize the above intuitive findings in the form of **Bayes' rule**.

1.10.2 Bayes' Rule

Let Y_j represent the received symbols and X_i the transmitted symbols having probabilities of $P(Y_j)$ and $P(X_i)$, respectively. Let us also characterize the forward transition probabilities of the binary symmetric channel as suggested by Figure 1.18.

Then in general, following from the previous introductory example, the joint probability $P(Y_j, X_i)$ of receiving Y_j, when the transmitted source symbol was X_i, is computed as the probability $P(X_i)$ of transmitting X_i, multiplied by the conditional probability $P(Y_j/X_i)$ of receiving Y_j, when X_i is known to have been transmitted:

$$P(Y_j, X_i) = P(X_i) \cdot P(Y_j/X_i), \tag{1.51}$$

a result that we have already intuitively exploited in the previous example. Since for the joint probabilities $P(Y_j, X_i) = P(X_i, Y_j)$ holds, we have:

$$\begin{aligned} P(X_i, Y_j) &= P(Y_j) \cdot P(X_i/Y_j) \\ &= P(X_i) \cdot P(Y_j/X_i). \end{aligned} \tag{1.52}$$

Equation 1.52 is often presented in the form:

$$\begin{aligned} P(X_i/Y_j) &= \frac{P(X_i, Y_j)}{P(Y_j)} \\ &= \frac{P(Y_j) \cdot P(X_i/Y_j)}{P(Y_j)}, \end{aligned} \tag{1.53}$$

which is referred to as *Bayes' rule*.

Logically, the probability of receiving a particular $Y_j = Y_{j_0}$ is the sum of all joint probabilities $P(X_i, Y_{j_0})$ over the range of X_i. This corresponds to the probability of receiving the transmitted X_i correctly, giving rise to the channel output Y_{j_0} plus the sum of the probabilities of all other possible transmitted symbols giving rise to Y_{j_0}:

$$P(Y_j) = \sum_X P(X_i, Y_j) = \sum_X P(X_i) P(Y_j/X_i). \qquad (1.54)$$

Similarly:

$$P(X_i) = \sum_Y P(X_i, Y_j) = \sum_Y P(Y_j) P(X_i/Y_j). \qquad (1.55)$$

1.10.3 Mutual Information

In this section, we elaborate further on the ramifications of Shannon's information theory [13–16, 24, 25]. Over nonideal channels impairments are introduced, and the received information might be different from the transmitted information. In this section, we quantify the amount of information that can be inferred from the received symbols over noisy channels. In the spirit of Shannon's fundamental work [13] and Carlson's classic reference [20], let us continue our discourse with the definition of mutual information. We have already used the notation $P(X_i)$ to denote the probability that the source symbol X_i was transmitted and $P(Y_i)$ to denote the probability that the symbol Y_j was received. The joint probability that X_i was transmitted and Y_j was received had been quantified by $P(X_i, Y_j)$, and $P(X_i/Y_j)$ indicated the conditional probability that X_i was transmitted, given that Y_j was received, while $P(Y_j/X_i)$ was used for the conditional probability that Y_j was received given that X_i was transmitted.

In case of $i = j$, the conditional probabilities $P(Y_j/X_j) j = 1 \cdots q$ represent the error-free transmission probabilities of the source symbols $j = 1 \cdots q$. For example, in Figure 1.18 the probabilities $P(Y_0/X_0)$ and $P(Y_1/X_1)$ are the probabilities of the error-free reception of a transmitted X_0 and X_1 source symbol, respectively. The probabilities $P(Y_j/X_i) j \neq i$, on the other hand, give the individual error probabilities, which are characteristic of error events that corrupted a transmitted symbol X_i to a received symbol of Y_j. The corresponding error probabilities in Figure 1.18 are $P(Y_0/X_1)$ and $P(Y_1/X_0)$.

Let us define the *mutual information* of X_i and Y_j as:

$$I(X_i, Y_j) = \log_2 \frac{P(X_i/Y_j)}{P(X_i)} = \log_2 \frac{P(X_i, Y_j)}{P(X_i) \cdot P(Y_j)} = \log_2 \frac{P(Y_j/X_i)}{P(Y_j)} \text{ bits}, \qquad (1.56)$$

which quantifies the amount of information conveyed, when X_i is transmitted and Y_j is received. Over a perfect, noiseless channel, each received symbol Y_j uniquely identifies a transmitted symbol X_i with a probability of $P(X_i/Y_j) = 1$. Substituting this probability in Equation 1.56 yields a mutual information of:

$$I(X_i, Y_j) = \log_2 \frac{1}{P(X_i)}, \qquad (1.57)$$

which is identical to the self-information of X_i and hence no information is lost over the channel. If the channel is very noisy and the error probability becomes 0.5, then the received symbol Y_j becomes unrelated to the transmitted symbol X_i, since for a binary system upon its reception there is a probability of 0.5 that X_0 was transmitted and the probability of X_1 is also 0.5. Then formally X_i and Y_j are independent and hence

$$P(X_i/Y_j) = \frac{P(X_i, Y_j)}{P(Y_j)} = \frac{P(X_i) \cdot P(Y_j)}{P(Y_j)} = P(X_i), \tag{1.58}$$

giving a mutual information of:

$$I(X_i, Y_j) = \log_2 \frac{P(X_i)}{P(X_i)} = \log_2 1 = 0, \tag{1.59}$$

implying that no information is conveyed via the channel. Practical communications channels perform between these extreme values and are usually characterized by the *average mutual information* defined as:

$$\begin{aligned} I(X, Y) &= \sum_{x,y} P(X_i, Y_j) \cdot I(X_i, Y_j) \\ &= \sum_{x,y} P(X_i, Y_j) \cdot \log_2 \frac{P(X_i/Y_j)}{P(X_i)} \quad [\text{bit/symbol}]. \end{aligned}$$

$$\tag{1.60}$$

Clearly, the average mutual information in Equation 1.60 is computed by weighting each component $I(X_i, Y_j)$ by its probability of occurrence $P(X_i, Y_j)$ and summing these contributions for all combinations of X_i and Y_j. The average mutual information $I(X, Y)$ defined above gives the average amount of source information acquired per received symbol, as distinguished from that per source symbol, which was given by the entropy $H(X)$. Let us now consolidate these definitions by working through the following numerical example.

1.10.4 Mutual Information Example

Using the same numeric values as in our introductory example as regards to the binary symmetric channel in Section 1.10.1, and exploiting that from Bayes' rule in Equation 1.53, we have:

$$P(X_i/Y_j) = \frac{P(X_i, Y_j)}{P(Y_j)}.$$

The following probabilities can be derived, which will be used at a later stage, in order to determine the mutual information:

$$P(X_1/Y_1) = P(1/1) = \frac{P(1,1)}{P_1} = \frac{0.686}{0.692} \approx 0.9913$$

and

$$P(X_0/Y_0) = P(0/0) = \frac{P(0,0)}{P_0} = \frac{0.294}{0.3080} \approx 0.9545,$$

where $P_1 = 0.692$ and $P_0 = 0.3080$ represent the total probability of receiving 1 and 0, respectively, which is the union of the respective events of error-free and erroneous receptions yielding the specific logical value concerned. *The mutual information* from Equation 1.56 is computed as:

$$\begin{aligned}
I(X_1, Y_1) &= \log_2 \frac{P(X_1/Y_1)}{P(X_1)} \\
&\approx \log_2 \frac{0.9913}{0.7} \approx 0.502 \text{ bit} \quad (1.61) \\
I(X_0, Y_0) &\approx \log_2 \frac{0.9545}{0.3} \approx 1.67 \text{ bit}. \quad (1.62)
\end{aligned}$$

These figures must be contrasted with the amount of source information conveyed by the source symbols X_0, X_1:

$$I(0) = \log_2 \frac{1}{0.3} \approx \log_2 3.33 \approx 1.737 \text{ bit/symbol} \quad (1.63)$$

and

$$I(1) = \log_2 \frac{1}{0.7} \approx \log_2 1.43 \approx 0.5146 \text{ bit/symbol}. \quad (1.64)$$

The amount of information "lost" in the noisy channel is given by the difference between the amount of information carried by the source symbols and the mutual information gained upon inferring a particular symbol at the noisy channel's output. Hence, the lost information can be computed from Equations 1.61, 1.62, 1.63, and 1.64, yielding $(1.737 - 1.67) \approx 0.067$ bit and $(0.5146 - 0.502) \approx 0.013$ bit, respectively. These values may not seem catastrophic, but in relative terms they are quite substantial and their values rapidly escalate, as the channel error probability is increased. For the sake of completeness and for future use, let us compute the remaining mutual information terms, namely, $I(X_0, Y_1)$ and $I(X_1, Y_0)$, which necessitate the computation of:

$$\begin{aligned}
P(X_0/Y_1) &= \frac{P(X_0, Y_1)}{P(Y_1)} \\
P(0/1) &= \frac{P(0,1)}{P_1} = \frac{0.3 \cdot 0.02}{0.692} \approx 0.00867 \\
P(X_1/Y_0) &= \frac{P(X_1, Y_0)}{P(Y_0)} \\
P(1/0) &= \frac{P(1,0)}{P_0} = \frac{0.7 \cdot 0.02}{0.308} \approx 0.04545 \\
I(X_0, Y_1) &= \log_2 \frac{P(X_0/Y_1)}{P(X_0)} \approx \log_2 \frac{0.00867}{0.3} \approx -5.11 \text{ bit} \quad (1.65) \\
I(X_1, Y_0) &= \log_2 \frac{P(X_1/Y_0)}{P(X_1)} \approx \log_2 \frac{0.04545}{0.7} \approx -3.945 \text{ bit}, \quad (1.66)
\end{aligned}$$

where the negative sign reflects the amount of "misinformation" as regards, for example, X_0 upon receiving Y_1. In this example we informally introduced the definition of mutual information. Let us now set out to formally exploit the benefits of our deeper insight into the effects of the noisy channel.

1.10.5 Information Loss via Imperfect Channels

Upon rewriting the definition of mutual information in Equation 1.56, we have:

$$\begin{aligned} I(X_i, Y_j) &= \log_2 \frac{P(X_i/Y_j)}{P(X_i)} \\ &= \log_2 \frac{1}{P(X_i)} - \log_2 \frac{1}{P(X_i/Y_j)} \\ &= I(X_i) - I(X_i/Y_j). \end{aligned} \qquad (1.67)$$

Following Shannon's [13–16, 24, 25] and Ferenczy's [22] approach and rearranging Equation 1.67 yields:

$$\underbrace{I(X_i)}_{\text{Source Inf.}} - \underbrace{I(X_i, Y_j)}_{\text{Inf. conveyed to rec.}} = \underbrace{I(X_i/Y_j)}_{\text{Inf. loss}}. \qquad (1.68)$$

Briefly returning to figure 1.18 assists the interpretation of $P(X_i/Y_j)$ as the probability or certainty/uncertainty that X_i was transmitted, given that Y_j was received, which justifies the above definition of the information loss. It is useful to observe from this figure that, as it was stated before, $P(Y_j/X_i)$ represents the probability of erroneous or error-free reception. Explicitly, if $j = i$, then $P(Y_j/X_i) = P(Y_j/X_j)$ is the probability of error-free reception, while if $j \neq i$, then $P(Y_j/X_i)$ is the probability of erroneous reception.

With the probability $P(Y_j/X_i)$ of erroneous reception in mind, we can actually associate an error information term with it:

$$I(Y_j/X_i) = \log_2 \frac{1}{P(Y_j/X_i)}. \qquad (1.69)$$

Let us now concentrate on the average mutual information's expression in Equation 1.60 and expand it as follows:

$$\begin{aligned} I(X, Y) &= \sum_{X,Y} P(X_i, Y_j) \cdot \log_2 \frac{1}{P(X_i)} \\ &\quad - \sum_{X,Y} P(X_i, Y_j) \log_2 \frac{1}{P(X_i/Y_j)}. \end{aligned} \qquad (1.70)$$

Considering the first term at the right-hand side (rhs) of the above equation and invoking Equation 1.55, we have:

$$\sum_X \left[\sum_Y P(X_i, Y_j) \right] \log_2 \frac{1}{P(X_i)} = \sum_X P(X_i) \log_2 \frac{1}{P(X_i)} = H(X). \qquad (1.71)$$

Then rearranging Equation 1.70 gives:

$$H(X) - I(X, Y) = \sum_{X,Y} P(X_i, Y_j) \log_2 \frac{1}{P(X_i/Y_j)}, \qquad (1.72)$$

1.10. INFORMATION TRANSMISSION VIA DISCRETE CHANNELS

where $H(X)$ is the average source information per symbol and $I(X,Y)$ is the average conveyed information per received symbol.

Consequently, the rhs term must be the average information per symbol lost in the noisy channel. As we have seen in Equation 1.67 and Equation 1.68, the information loss is given by:

$$I(X_i/Y_j) = \log_2 \frac{1}{P(X_i/Y_j)}. \quad (1.73)$$

The average information loss $H(X/Y)$ *equivocation*, which Shannon [15] terms is computed as the weighted sum of these components:

$$H(X/Y) = \sum_X \sum_Y P(X_i, Y_j) \cdot \log_2 \frac{1}{P(X_i/Y_j)}. \quad (1.74)$$

Following Shannon, this definition allowed us to express Equation 1.72 as:

$$\underbrace{H(X)}_{\text{(av. source inf/sym.)}} - \underbrace{I(X,Y)}_{\text{(av. conveyed inf/sym.)}} = \underbrace{H(X/Y)}_{\text{(av. lost inf/sym.)}} \quad (1.75)$$

1.10.6 Error Entropy via Imperfect Channels

Similarly to our previous approach and using the probability $P(Y_j/X_i)$ of erroneous reception associated with the information term of:

$$I(Y_j/X_i) = \log_2 \frac{1}{P(Y_j/X_i)} \quad (1.76)$$

we can define the average "error information" or error entropy. Hence, the above error information terms in Equation 1.76 are weighted using the probabilities $P(X_i, Y_j)$ and averaged for all X and Y values, defining the *error entropy*:

$$H(Y/X) = \sum_X \sum_Y P(X_i, Y_j) \log_2 \frac{1}{P(Y_j/X_i)}. \quad (1.77)$$

Using Bayes' rule from Equation 1.52, we have

$$\begin{aligned} P(X_i/Y_j) \cdot P(Y_j) &= P(Y_j/X_i) \cdot P(X_i) \\ \frac{P(X_i/Y_j)}{P(X_i)} &= \frac{P(Y_j/X_i)}{P(Y_j)}. \end{aligned} \quad (1.78)$$

Following from this, for the average mutual information in Equation 1.56 we have:

$$I(X,Y) = I(Y,X), \quad (1.79)$$

which, after interchanging X and Y in Equation 1.75, gives:

$$\underbrace{H(Y)}_{\text{destination entropy}} - \underbrace{I(Y,X)}_{\text{conveyed inf}} = \underbrace{H(Y/X)}_{\text{error entropy}} \quad (1.80)$$

Quantity	Definition
Source inf.	$I(X_i) = -\log_2 P(X_i)$
Received inf.	$I(Y_j) = -\log_2 P(Y_j)$
Joint inf.	$I_{X_i,Y_j} = -\log_2 P(X_i, Y_j)$
Mutual inf.	$I(X_i, Y_j) = \log_2 \frac{P(X_i/Y_j)}{P(X_i)}$
Av. Mut. inf.	$I(X,Y) = \sum_X \sum_Y P(X_i Y_j) \log_2 \frac{P(X_i/Y_j)}{P(X_i)}$
Source entropy	$H(X) = -\sum_X P(X_i) \cdot \log_2 P(X_i)$
Destination entr.	$H(Y) = -\sum_Y P(Y_j) \log_2 P(Y_j)$
Equivocation	$H(X/Y) = -\sum_X \sum_Y P(X_i, Y_j) \log_2 P(X_i/Y_j)$
Error entropy	$H(Y/X) = -\sum_X \sum_Y P(X_i Y_j) \log_2 P(Y_j/X_i)$

Table 1.7: Summary of Definitions ©Ferenczy [22]

Subtracting the conveyed information from the destination entropy gives the error entropy, which is nonzero, if the destination entropy and conveyed information are not equal due to channel errors. Let us now proceed following Ferenczy's approach [22] and summarize the most important definitions for future reference in Table 1.7 before we attempt to augment their physical interpretations using the forthcoming numerical example.

Example *Using the BSC model of Figure 1.16, as an extension of the worked examples of Subsections 1.10.1 and 1.10.4 and following Ferenczy's interpretation [22] of Shannon's elaborations [13–16, 24, 25], let us compute the following range of system characteristics:*

(a) *The* **joint information**, *as distinct from the mutual information introduced earlier, for all possible channel input/output combinations.*

(b) *The entropy, i.e., the average information of both the source and the sink.*

(c) *The average joint information $H(X,Y)$.*

(d) *The average mutual information per symbol conveyed.*

(e) *The average information loss and average error entropy.*

With reference to Figure 1.16 and to our introductory example from Section 1.10.1 we commence by computing further parameters of the BSC. Recall that the source information was:

$$I(X_0) = \log_2 \frac{1}{0.3} \approx 3.322 \log_{10} 3.333 \approx 1.737 \text{ bit}$$
$$I(X_1) = \log_2 \frac{1}{0.7} \approx 0.515 \text{ bit.}$$

The probability of receiving a logical 0 was 0.308 and that of logical 1 was 0.692, of whether 0 or 1 was transmitted. Hence, the information inferred upon the reception of 0 and 1, respectively, is given by:

$$I(Y_0) = \log_2 \frac{1}{0.308} \approx 3.322 \log_{10} 3.247 \approx 1.699 \text{ bit}$$

1.10. INFORMATION TRANSMISSION VIA DISCRETE CHANNELS 45

$$I(Y_1) = \log_2 \frac{1}{0.692} \approx 0.531 \text{ bit}.$$

Observe that because of the reduced probability of receiving a logical 1 from $0.7 \to 0.692$ as a consequence of channel-induced corruption, the probability of receiving a logical 0 is increased from $0.3 \to 0.308$. This is expected to increase the average destination entropy, since the entropy maximum of unity is achieved, when the symbols are equiprobable. We note, however, that this does not give more information about the source symbols, which must be maximized in an efficient communications system. In our example, the information conveyed increases for the reduced probability logical 1 from 0.515 bit $\to$ 0.531 bit and decreases for the increased probability 0 from 1.737 bit $\to$ 1.699 bit. Furthermore, the average information conveyed is reduced, since the reduction from 1.737 to 1.699 bit is more than the increment from 0.515 to 0.531. In the extreme case of an error probability of 0.5 we would have $P(0) = P(1) = 0.5$, and $I(1) = I(0) = 1$ bit, associated with receiving equiprobable random bits, which again would have a maximal destination entropy, but a minimal information concerning the source symbols transmitted. Following the above interesting introductory calculations, let us now turn our attention to the computation of the joint information.

a/ The *joint information*, as distinct from the mutual information introduced earlier in Equation 1.56, of all possible channel input/output combinations is computed from Figure 1.16 as follows:

$$\begin{align}
I_{X_i, Y_j} &= -\log_2 P(X_i, Y_j) \tag{1.81}\\
I_{00} &= -\log_2(0.3 \cdot 0.98) \approx -3.322 \cdot \log_{10} 0.294 \approx 1.766 \text{ bit}\\
I_{01} &= -\log_2(0.3 \cdot 0.02) \approx 7.381 \text{ bit}\\
I_{10} &= -\log_2(0.7 \cdot 0.02) \approx 6.159 \text{ bit}\\
I_{11} &= -\log_2(0.7 \cdot 0.98) \approx 0.544 \text{ bit}.
\end{align}$$

These information terms can be individually interpreted formally as the information carried by the simultaneous occurrence of the given symbol combinations. For example, as it accrues from their computation, I_{00} and I_{11} correspond to the favorable event of error-free reception of a transmitted 0 and 1, respectively, which hence were simply computed by formally evaluating the information terms. By the same token, in the computation of I_{01} and I_{10}, the corresponding source probabilities were weighted by the channel error probability rather than the error-free transmission probability, leading to the corresponding information terms. The latter terms, namely, I_{01} and I_{10}, represent low-probability, high-information events due to the low channel error probability of 0.02.

Lastly, a perfect channel with zero error probability would render the probability of the error-events zero, which in turn would assign infinite information contents to the corresponding terms of I_{01} and I_{10}, while I_{00} and I_{11} would be identical to the self-information of the 0 and 1 symbols. Then, if under zero error probability we evaluate the effect of the individual symbol probabilities on the remaining joint information terms, the less frequently a symbol is emitted by the source, the higher its associated joint information term becomes and vice versa, which is seen by comparing I_{00} and I_{11}. Their difference can be equalized by assuming an identical probability of 0.5 for both, which would yield $I_{00}=I_{11}=$ 1-bit. The unweighted average of I_{00} and I_{11} would then be lower than in case of the previously used probabilities of 0.3 and 0.7, respectively, since the maximum average would be associated with the case of 0 and 1, where the associated $\log_2$ terms would be 0 and $-\infty$, respectively. The appropriately weighted average joint information terms will be evaluted under paragraph **c/** during our later calculations. Let us now move on to evaluate the average information of the source and sink.

b/ *Calculating the entropy*, that is, the average information for both the source and the sink, is quite straightforward and ensues as follows:

$$\begin{aligned} H(X) &= \sum_{i=1}^{2} P(X_i) \cdot \log_2 \frac{1}{P(X_i)} \\ &\approx 0.3 \cdot \log_2 3.333 + 0.7 \cdot \log_2 1.429 \\ &\approx 0.5211 + 0.3605 \\ &\approx 0.8816 \text{ bit/symbol}. \end{aligned} \qquad (1.82)$$

For the computation of the sink's entropy, we invoke Equations 1.49 and 1.50, yielding:

$$\begin{aligned} H(Y) &= 0.308 \cdot \log_2 \frac{1}{0.308} + 0.692 \log_2 \frac{1}{0.692} \\ &\approx 0.5233 + 0.3676 \\ &\approx 0.8909 \text{ bit/symbol}. \end{aligned} \qquad (1.83)$$

Again, the destination entropy $H(Y)$ is higher than the source entropy $H(X)$ due to the more random reception caused by channel errors, approaching $H(Y) = 1$ bit/symbol for a channel bit error rate of 0.5. Note, however, that unfortunately this increased destination entropy does not convey more information about the source itself.

c/ Computing the *average joint information* $H(X,Y)$ gives:

$$\begin{aligned} H(X,Y) &= -\sum_{i=1}^{2}\sum_{j=1}^{2} P(X_i, Y_j) \log_2 P(X_i, Y_j) \\ &= -\sum_{i=1}^{2}\sum_{j=1}^{2} P(X_i, Y_j) I_{X_i, Y_j}. \end{aligned} \qquad (1.84)$$

Upon substituting the I_{X_i,Y_j} values calculated in Equation 1.81 into Equation 1.84, we have:

$$H(X,Y) \approx 0.3 \cdot 0.98 \cdot 1.766 + 0.3 \cdot 0.02 \cdot 7.381$$

1.10. INFORMATION TRANSMISSION VIA DISCRETE CHANNELS

$$+ 0.7 \cdot 0.02 \cdot 6.159 + 0.7 \cdot 0.98 \cdot 0.544$$
$$\approx 0.519 + 0.044 + 0.086 + 0.373$$
$$\approx 1.022 \text{ bit/symbol-combination.}$$

In order to interpret $H(X,Y)$, let us again scrutinize the definition given in Equation 1.84, which weights the joint information terms of Equation 1.81 by their probability of occurence. We have argued before that the joint information terms corresponding to erroneous events are high due to the low error probability of 0.02. Observe, therefore, that these high-information symbol combinations are weighted by their low-probability of occurrence, causing $H(X,Y)$ to become relatively low. It is also instructive to consider the above terms in Equation 1.84 for the extreme cases of zero and 0.5 error probabilities and for different source emission probabilities, which are left for the reader to explore. Here we proceed considering the average conveyed mutual information per symbol.

d/ The *average conveyed mutual information per symbol* was defined in Equation 1.60 in order to quantify the average source information acquired per received symbol, which is repeated here for convenience as follows:

$$I(X,Y) = \sum_X \sum_Y P(X_i, Y_j) \log_2 \frac{P(X_i/Y_j)}{P(X_i)}$$
$$= \sum_X \sum_Y P(X_i, Y_j) \cdot I(X_i, Y_j).$$

Using the individual mutual information terms from Equations 1.61–1.66 in Section 1.10.4, we get the average mutual information representing the average amount of source information acquired from the received symbols, as follows:

$$I(X,Y) \approx 0.3 \cdot 0.98 \cdot 1.67 + 0.3 \cdot 0.02 \cdot (-5.11)$$
$$+ 0.7 \cdot 0.02 \cdot (-3.945) + 0.7 \cdot 0.98 \cdot 0.502$$
$$\approx 0.491 - 0.03066 - 0.05523 + 0.3444$$
$$\approx 0.7495 \text{ bit/symbol.} \tag{1.85}$$

In order to interpret the concept of mutual information, in Section 1.10.4 we noted that the amount of information "lost" owing to channel errors was given by the difference between the amount of information carried by the source symbols and the mutual information gained upon inferring a particular symbol at the noisy channel's output. These were given in Equations 1.61–1.64, yielding $(1.737 - 1.67) \approx 0.067$ bit and $(0.5146 - 0.502) \approx 0.013$ bit, for the transmission of a 0 and 1, respectively. We also noted that the negative sign of the terms corresponding to the error-events reflected the amount of misinformation as regards, for example, X_0 upon receiving Y_1. Over a perfect channel, the cross-coupling transitions of Figure 1.16 are eliminated, since the associated error probabilities are 0, and hence there is no information loss over the channel. Consequently, the error-free mutual information terms become identical to the self-information of the source symbols, since exactly the same amount of information can be inferred upon reception of a symbol, as much is carried by its appearance at the output of the source.

It is also instructive to study the effect of different error probabilities and source symbol probabilities in the average mutual information definition of Equation 1.84 in order to acquire a better understanding of its physical interpretation and quantitative power as regards the system's performance. It is interesting to note, for example, that assuming an error probability of zero will therefore result in average mutual information, which is identical to the source and destination entropy computed above under paragraph **b/**. It is also plausible that $I(X,Y)$ will be higher than the previously computed 0.7495 bits/symbol, if the symbol probabilities are closer to 0.5, or in general in case of q-ary sources closer to $1/q$. As expected, for a binary symbol probability of 0.5 and error probability of 0, we have $I(X,Y)=1$ bit/symbol.

e/ Lastly, let us determine the *average information loss and average error entropy*, which were defined in Equations 1.74 and 1.80 and are repeated here for convenience. Again, we will be using some of the previously computed probabilities from Sections 1.10.1 and 1.10.4, beginning with computation of the average information loss of Equation 1.74:

$$\begin{aligned}
H(X/Y) &= -\sum_X \sum_Y P(X_i, Y_j) \log_2 P(X_i/Y_j) \\
&= -P(X_0,Y_0)\log_2 P(X_0/Y_0) - P(X_0,Y_1)\log_2 P(X_0/Y_1) \\
&\quad -P(X_1,Y_0)\log_2 P(X_1/Y_0) - P(X_1,Y_1)\log_2 P(X_1/Y_1) \\
&= P(0,0)\cdot\log_2 P(0/0) + P(0,1)\cdot\log_2 P(0/1) \\
&\quad P(1,0)\cdot\log_2 P(1/0) + P(1,1)\cdot\log_2 P(1/1) \\
&\approx -0.3\cdot 0.98\cdot\log_2 0.9545 - 0.3\cdot 0.02\cdot\log_2 0.00867 \\
&\quad -0.7\cdot 0.02\cdot\log_2 0.04545 - 0.7\cdot 0.98\cdot\log_2 0.9913 \\
&\approx 0.0198 + 0.0411 + 0.0624 + 0.0086 \\
&\approx 0.132 \ \text{bit/symbol}.
\end{aligned}$$

In order to augment the physical interpretation of the above-average information loss expression, let us examine the main contributing factors in it. It is expected to decrease as the error probability decreases. Although it is not straightforward to infer the clear effect of any individual parameter in the equation, experience shows that as the error probability increases, the two middle terms corresponding to the error events become more dominant. Again, the reader may find it instructive to alter some of the parameters on a one-by-one basis and study the way its influence manifests itself in terms of the overall information loss.

Moving on to the computation of the average error entropy, we find its definition equation is repeated below, and on inspecting Figure 1.16 we have:

$$\begin{aligned}
H(Y/X) &= -\sum_X \sum_Y P(X_i,Y_j)\cdot\log_2 P(Y_j/X_i) \\
&= -P(X_0,Y_0)\log_2 P(Y_0/X_0) - P(X_0,Y_1)\log_2 P(Y_1/X_0) \\
&\quad -P(X_1,Y_0)\log_2 P(Y_0/X_1) - P(X_1,Y_1)\log_2 P(Y_1/X_1) \\
P(Y_0/X_0) &= 0.98
\end{aligned}$$

$$
\begin{aligned}
P(Y_0/X_1) &= 0.02 \\
P(Y_1/X_0) &= 0.02 \\
P(Y_1/X_1) &= 0.98 \\
H(Y/X) &= P(0,0) \cdot \log_2 P(0/0) + P(0,1) \cdot \log_2 P(0/1) \\
&\quad P(1,0) \cdot \log_2 P(1/0) + P(1,1) \cdot \log_2 P(1/1) \\
&= -0.294 \cdot \log_2 0.98 - 0.014 \cdot \log_2 0.02 \\
&\quad -0.006 \cdot \log_2 0.02 - 0.686 \cdot \log_2 0.98 \\
&\approx 0.0086 + 0.079 + 0.034 + 0.02 \\
&\approx 0.141 \text{ bit/symbol.}
\end{aligned}
$$

The average error entropy in the above expression is expected to fall as the error probability is reduced and vice versa. Substituting different values into its definition equation further augments its practical interpretation. Using our previous results in this section, we see that the *average loss of information per symbol or equivocation* denoted by $H(X/Y)$ is given by the difference between the source entropy of Equation 1.82 and the average mutual information of Equation 1.85, yielding:

$$H(X/Y) = H(X) - I(X,Y) \approx 0.8816 - 0.7495 \approx 0.132 \text{ bit/symbol,}$$

which according to Equation 1.75, is identical to the value of $H(X/Y)$ computed earlier. In harmony with Equation 1.80, the error entropy can also be computed as the difference of the average entropy $H(Y)$ in Equation 1.83 of the received symbols and the mutual information $I(X,Y)$ of Equation 1.85, yielding:

$$H(Y) - I(X,Y) \approx 0.8909 - 0.7495 \approx 0.141 \text{ bit/symbol,}$$

as seen above for $H(Y/X)$.

Having defined the fundamental parameters summarized in Table 1.7 and used in the information-theoretical characterization of communications systems, let us now embark on the definition of channel capacity. Initially, we consider discrete noiseless channels, leading to a brief discussion of noisy discrete channels, and then we proceed to analog channels, before exploring the fundamental message of the Shannon-Hartley law.

1.11 Capacity of Discrete Channels [15, 22]

Shannon [15] defined the *channel capacity* C of a channel as the maximum achievable information transmission rate at which error-free transmission can be maintained over the channel.

Every practical channel is noisy, but transmitting at a sufficiently high power the channel error probability p_e can be kept arbitrarily low, providing us with a simple initial channel model for our further elaborations. Following Ferenczy's approach [22], assume that the transmission of symbol X_i requires a time interval of t_i, during which an average of

$$H(X) = \sum_{i=1}^{q} P(X_i) \log_2 \frac{1}{P(X_i)} \frac{\text{bit}}{\text{symbol}} \qquad (1.86)$$

information is transmitted, where q is the size of the source alphabet used. This approach assumes that a variable-length coding algorithm, such as the previously described Shannon-Fano or the Huffman coding algorithm may be used in order to reduce the transmission rate to as low as the source entropy. Then the average time required for the transmission of a source symbol is computed by weighting t_i with the probability of occurrence of symbol $X_i, i = 1\ldots q$:

$$t_{av} = \sum_{i=1}^{q} P(X_i) t_i \frac{\text{sec}}{\text{symbol}}. \qquad (1.87)$$

Now we can compute the average information transmission rate v by dividing the average information content of a symbol by the average time required for its transmission:

$$v = \frac{H(X)}{t_{av}} \frac{\text{bit}}{\text{sec}}. \qquad (1.88)$$

The maximum transmission rate v as a function of the symbol probability $P(X_i)$ must be found. This is not always an easy task, but a simple case occurs when the symbol duration is constant; that is, we have $t_i = t_0$ for all symbols. Then the maximum of v is a function of $P(X_i)$ only and we have shown earlier that the entropy $H(X)$ is maximized by equiprobable source symbols, where $P(X_i) = \frac{1}{q}$. Then from Equations 1.86 and 1.87 we have an expression for the channel's maximum capacity:

$$C = v_{max} = \frac{H(X)}{t_{av}} = \frac{\log_2 q}{t_0} \frac{\text{bit}}{\text{sec}}. \qquad (1.89)$$

Shannon [15] characterized the capacity of discrete noisy channels using the previously defined mutual information describing the amount of average conveyed information, given by:

$$I(X,Y) = H(Y) - H(Y/X), \qquad (1.90)$$

where $H(Y)$ is the average amount of information per symbol at the channel's output, while H(Y/X) is the error entropy. Here a unity symbol-rate was assumed for the sake of simplicity. Hence, useful information is transmitted only via the channel if $H(Y) > H(Y/X)$. Via a channel with $p_e = 0.5$, where communication breaks down, we have $H(Y) = H(Y/X)$, and the information conveyed becomes $I(X,Y) = 0$. The amount of information conveyed is maximum if the error entropy $H(Y/X) = 0$. Therefore, Shannon [15] defined the noisy channel's capacity as the maximum value of the conveyed information $I(X,Y)$:

$$C = I(X,Y)_{MAX} = [H(Y) - H(Y/X)]_{MAX}, \qquad (1.91)$$

where the maximization of Equation 1.91 is achieved by maximizing the first term and minimizing the second term.

1.11. CAPACITY OF DISCRETE CHANNELS

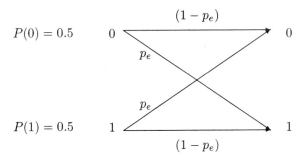

Figure 1.19: BSC model.

In general, the maximization of Equation 1.91 is an arduous task, but for the BSC seen in Figure 1.19 it becomes fairly simple. Let us consider this simple case and assume that the source probabilities of 1 and 0 are $P(0) = P(1) = 0.5$ and the error probability is p_e. The entropy at the destination is computed as:

$$H(Y) = -\frac{1}{2}\log_2 \frac{1}{2} - \frac{1}{2}\log_2 \frac{1}{2} = 1 \text{ bit/symbol,}$$

while the error entropy is given by:

$$H(Y/X) = -\sum_X \sum_Y P(X_i, Y_j) \cdot \log_2 P(Y_j/X_i). \tag{1.92}$$

In order to be able to compute the capacity of the BSC as a function of the channel's error probability, let us substitute the required joint probabilities of:

$$\begin{aligned} P(0,0) &= P(0)(1-p_e) \\ P(0,1) &= P(0)p_e \\ P(1,0) &= P(1)p_e \\ P(1,1) &= P(1)(1-p_e). \end{aligned} \tag{1.93}$$

and the conditional probabilities of:

$$\begin{aligned} P(0/0) &= (1-p_e) \\ P(0/1) &= p_e \\ P(1/0) &= p_e \\ P(1/1) &= (1-p_e). \end{aligned} \tag{1.94}$$

into Equation 1.92, yielding:

$$\begin{aligned} H(Y/X) = &-[P(0)(1-p_e) \cdot \log_2(1-p_e) + P(0) \cdot p_e \log_2 p_e \\ &+ P(1) \cdot p_e \log_2 p_e + P(1)(1-p_e)\log_2(1-p_e)] \end{aligned}$$

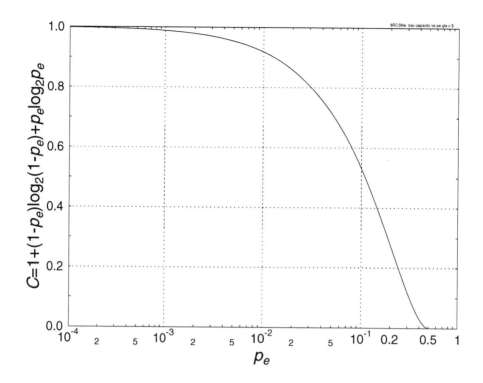

Figure 1.20: Channel capacity versus p_e for the BSC.

$$\begin{aligned}
&= -[P(0) + P(1)](1 - p_e)\log_2(1 - p_e) \\
&\quad + [P(0) + P(1)]p_e \log_2 p_e \\
&= -(1 - p_e) \cdot \log_2(1 - p_e) - p_e \cdot \log_2 p_e.
\end{aligned} \quad (1.95)$$

Finally, upon substituting $H(Y)$ and $H(Y/X)$ from above into Equation 1.91, the BSC's channel capacity becomes:

$$C = 1 + (1 - p_e)\log_2(1 - p_e) + p_e \log_2 p_e. \quad (1.96)$$

Following Ferenczy's [22] interpretation of Shannon's lessons [13–16,24,25], the graphic representation of the BSC's capacity is depicted in Figure 1.20 using various p_e error probabilities.

Observe, for example, that for $p_e = 10^{-2}$ the channel capacity is $C \approx 0.9$ bit/symbol, that is, close to its maximum of $C = 1$ bit/symbol, but for higher p_e values it rapidly decays, falling to $C = 0.5$ bit/symbol around $p_e = 10^{-1}$. If $p_e = 50\%$, we have $C = 0$ bit/symbol; since no useful information transmission takes place, the channel delivers random bits. Notice also that if $P(0) \neq P(1) \neq 0.5$, then $H(Y) < 1$ bit/symbol and hence $C < C_{max} = 1$ bit/symbol, even if $p_e = 0$.

1.12 Shannon's Channel Coding Theorem [19, 27]

In the previous section, we derived a simple expression for the capacity of the noisy BSC in Equation 1.96, which was depicted in Figure 1.20 as a function of the channel's error probability p_e. In this section, we focus on Shannon's *channel coding theorem*, which states that as long as the information transmission rate does not exceed the channel's capacity, the bit error rate can be kept arbitrarily low [24, 25]. In the context of the BSC channel capacity curve of Figure 1.20, this theorem implies that noise over the channel does not preclude the reliable transmission of information; it only limits the rate at which transmission can take place. Implicitly, this theorem prophesies the existence of an appropriate error correction code, which adds redundancy to the original information symbols. This reduces the system's useful information throughput but simultaneously allows error correction coding. Instead of providing a rigorous proof of this theorem, following the approach suggested by Abramson [19], which was also used by Hey and Allen [27] in their compilation of Feyman's lectures, we will make it plaussible.

The theorem is stated more formally as follows. Let us assume that a message of K useful information symbols is transmitted by assigning it to an N-symbol so-called block code, where the symbols are binary and the error probability is p_e. Then, according to Shannon, upon reducing the *coding rate* $R = \frac{K}{N}$ beyond every limit, the error probability obeys the following relationship:

$$R = \frac{K}{N} \leq C = 1 + (1 - p_e) \log_2(1 - p_e) + p_e \cdot \log_2 p_e. \quad (1.97)$$

As Figure 1.20 shows upon increasing the bit error rate p_e, the channel capacity reduces gradually toward zero, which forces the channel coding rate $R = \frac{K}{N}$ to zero in the limit. This inequality therefore implies that an arbitrarily low BER is possible only when the coding rate R tends to zero, which assumes an infinite-length block code and an infinite coding delay. By scrutinizing Figure 1.20, we can infer that, for example, for a BER of 10^{-1} an approximately $R = \frac{K}{N} \approx \frac{1}{2}$ so-called half-rate code is required in order to achieve asymptotically perfect communications, while for $BER = 10^{-2}$ an approximately $R \approx 0.9$ code is required.

Shannon's channel coding theorem does not specify how to create error correction codes, which can achieve this predicted performance; it merely states their existence. Hence, the error correction coding community has endeavored over the years to create such good codes but until 1993 had only limited success. Then in that year Berrou et al. [28] invented the family of iteratively decoded turbo-codes, which are capable of approaching the Shannonian predictions within a fraction of a dB.

Returning to the channel coding theorem, Hey and Feynman [27] offered a witty approach to deepening the physical interpretation of this theorem, which we briefly highlight below. Assuming that the block-coded sequences are long, in each block on the average there are $t = p_e \cdot N$ number of errors. In general, t number of errors can be allocated over the block of N positions in

$$C_N^t = \binom{N}{t} = \frac{N!}{t!(N-t)!}$$

different ways, which are associated with the same number of error patterns. The number of additional parity bits added during the coding process is $P = (N - K)$, which must be sufficiently high for identifying all the C_N^t number of error patterns, in order to allow inverting (i.e., correcting) the corrupted bits in the required positions. Hence, we have [27]:

$$\frac{N!}{t!(N-t)!} \leq 2^{(N-K)}. \tag{1.98}$$

Upon exploiting the Stirling formula of

$$N! \approx \sqrt{2\pi N} \cdot \left(\frac{N}{e}\right)^N = \sqrt{2\pi} \cdot \sqrt{N} \cdot N^N \cdot e^{-N}$$

and taking the logarithm of both sides, we have:

$$\log_e N! \approx \log_e \sqrt{2\pi} + \frac{1}{2} \log_e N + N \log_e N - N.$$

Furthermore, when N is large, the first and second terms are diminishingly small in comparison to the last two terms. Thus, we have:

$$\log_e N! \approx N \log_e N - N.$$

Then, after taking the logarithm, the factorial expression on the left-hand side (L) of Equation 1.98 can be written as:

$$L \approx [N \log_e N - N] - [t \log_e t - t] - [(N - t) \log_e (N - t) - (N - t)].$$

Now taking into account that $t \approx p_e \cdot N$, we have [27]:

$$\begin{aligned}
L &\approx [N \log_e N - N] - [p_e N \log_e (p_e N) - p_e N] \\
&\quad - [(N - p_e N) \log_e (N - p_e N) - (N - p_e N)] \\
&\approx [N \log_e N - N] - [p_e N \log_e p_e + p_e N \log_e N - p_e N] \\
&\quad - [N \log_e (N(1 - p_e)) - p_e N \log_e (N(1 - p_e)) - (N - p_e N)] \\
&\approx [N \log_e N - N] - [p_e N \log_e p_e + p_e N \log_e N - p_e N] \\
&\quad - [N \log_e N + N \log_e (1 - p_e) - p_e N \log_e N \\
&\quad - p_e N \log_e (1 - p_e) - (N - p_e N)] \\
&\approx N[\log_e N - 1 - p_e \log_e p_e - p_e \log_e N + p_e \\
&\quad - \log_e N - \log_e (1 - p_e) + p_e \log_e N \\
&\quad + p_e \log_e (1 - p_e) + 1 - p_e] \\
&\approx N[-p_e \log_e p_e - \log_e (1 - p_e) + p_e \log_e (1 - p_e)] \\
&\approx N[-p_e \log_e p_e - (1 - p_e) \log_e (1 - p_e)].
\end{aligned}$$

If we consider that $\log_e a = \log_2 a \cdot \log_e 2$, then we can convert the $\log_e$ terms to $\log_2$ as follows [27]:

$$L \approx N \log_e 2[-p_e \log_2 p_e - (1 - p_e) \log_2 (1 - p_e)].$$

Finally, upon equating this term with the logarithm of the right-hand side expression of Equation 1.98, we arrive at:

$$N \log_e 2[-p_e \log_2 p_e - (1-p_e) \log_2(1-p_e)] \leq (N-K) \log_e 2,$$

which can be simplified to:

$$-p_e \log_2 p_e - (1-p_e) \log_2(1-p_e) \leq 1 - \frac{K}{N}$$

or to a form, identical to Equation 1.97:

$$\frac{K}{N} \leq 1 + (1-p_e) \log_2(1-p_e) + p_e \log_2 p_e.$$

1.13 Capacity of Continuous Channels [16, 22]

During our previous discussions, it was assumed that the source emitted discrete messages with certain finite probabilities, which would be exemplified by an 8-bit analog-to-digital converter emitting one of 256 discrete values with a certain probability. However, after digital source encoding and channel encoding according to the basic schematic of Figure 1.1 the modulator typically converts the digital messages to a finite set of bandlimited analog waveforms, which are chosen for maximum "transmission convenience." In this context, transmission convenience can imply a range of issues, depending on the communications channel. Two typical constraints are predominantly power-limited or bandwidth-limited channels, although in many practical scenarios both of these constraints become important. Because of their limited solar power supply, satellite channels tend to be more severely power-limited than bandlimited, while typically the reverse situation is experienced in mobile radio systems.

The third part of Shannon's pioneering paper [16] considers many of these issues. Thus, in what follows we define the measure of information for continuous signals, introduce a concept for the continuous channel capacity, and reveal the relationships among channel bandwidth, channel capacity, and channel signal-to-noise ratio, as stated by the Shannon-Hartley theorem. Finally, the ideal communications system transpiring from Shannon's pioneering work is characterized, before concluding with a brief discussion of the ramifications of wireless channels as regards the applicability of Shannon's results.

Let us now assume that the channel's analog input signal $x(t)$ is bandlimited and hence that it is fully characterized by its Nyquist samples and by its probability density function (PDF) $p(x)$. The analogy of this continuous PDF and that of a discrete source are characterized by $P(X_i) \approx p(X_i)\Delta X$, which reflects the practical way of experimentally determining the histogram of a bandlimited analog signal by observing the relative frequency of events, when its amplitude resides in a ΔX wide amplitude bin-centered around X_i. As an analogy to the discrete average information or entropy expression of:

$$H(X) = -\sum_i P(X_i) \cdot \log_2 P(X_i), \tag{1.99}$$

Shannon [16] introduced the *entropy of analog sources*, as it was also noted and exploited, for example, by Ferenczy [22], as follows:

$$H(x) = -\int_{-\infty}^{\infty} p(x) \log_2 p(x) dx. \tag{1.100}$$

For our previously used discrete sources, we have shown that the source entropy is maximized for equiprobable messages. The question that arises is whether this is also true for continuous PDFs. Shannon [16] derived the maximum of the analog signal's entropy under the constraints of:

$$\int_{-\infty}^{\infty} p(x) dx = 1 \tag{1.101}$$

$$\sigma_x^2 = \int_{-\infty}^{\infty} x^2 \cdot p(x) dx = \text{Constant} \tag{1.102}$$

based on the calculus of variations. He showed that the entropy of a signal $x(t)$ having a constant variance of σ_x^2 is maximum, if $x(t)$ has a Gaussian distribution given by:

$$p(x) = \frac{1}{\sqrt{2\pi}\sigma} e^{-(x^2/2\sigma^2)}. \tag{1.103}$$

Then the maximum of the entropy can be derived upon substituting this PDF into the expression of the entropy. Let us first take the natural logarithm of both sides of the PDF and convert it to base two logarithm by taking into account that $\log_e a = \log_2 a \cdot \log_e 2$, in order to be able to use it in the entropy's $\log_2$ expression. Then the PDF of Equation 1.103 can be written as:

$$-\log_2 p(x) = +\log_2 \sqrt{2\pi}\sigma + (x^2/2\sigma^2) \cdot \frac{1}{\log_e 2}, \tag{1.104}$$

and upon exploiting that $\log_e 2 = 1/\log_2 e$, the entropy is expressed according to Shannon [16] and Ferenczy [22] as:

$$\begin{aligned}
H_{max}(x) &= -\int p(x) \cdot \log_2 p(x) dx \\
&= \int p(x) \cdot \log_2 \sqrt{2\pi}\sigma dx + \int p(x) \frac{x^2 \cdot \log_2 e}{2\sigma^2} dx \\
&= \log_2 \sqrt{2\pi}\sigma \int p(x) dx + \frac{\log_2 e}{2\sigma^2} \underbrace{\int x^2 p(x) dx}_{\sigma^2} \\
&= \log_2 \sqrt{2\pi}\sigma + \frac{\sigma^2}{2\sigma^2} \log_2 e \\
&= \log_2 \sqrt{2\pi}\sigma + \frac{\log_2 e}{2} \\
&= \log_2 \sqrt{2\pi}\sigma + \frac{1}{2} \log_2 e \\
&= \log_2 \sqrt{2\pi e}\sigma. \tag{1.105}
\end{aligned}$$

1.13. CAPACITY OF CONTINUOUS CHANNELS

Since the maximum of the entropy is proportional to the logarithm of the signal's average power $S_x = \sigma_x^2$, upon quadrupling the signal's power the entropy is increased by one bit because the range of uncertainty as regards where the signal samples can reside is expanded.

We are now ready to formulate the channel capacity versus channel bandwidth and versus channel SNR relationship of analog channels. Let us assume white, additive, signal-independent noise with a power of N via the channel. Then the received (signal+noise) power is given by:

$$\sigma_y^2 = S + N. \tag{1.106}$$

By the same argument, the channel's output entropy is maximum if its output signal $y(t)$ has a Gaussian PDF and its value is computed from Equation 1.105 as:

$$H_{max}(y) = \frac{1}{2}\log_2(2\pi e \sigma_y^2) = \frac{1}{2}\log_2 2\pi e(S+N). \tag{1.107}$$

We proceed by taking into account the channel impairments, reducing the amount of information conveyed by the amount of the error entropy $H(y/x)$ giving:

$$I(x,y) = H(y) - H(y/x), \tag{1.108}$$

where again the noise is assumed to be Gaussian and hence:

$$H(y/x) = \frac{1}{2}\log_2(2\pi e N). \tag{1.109}$$

Upon substituting Equation 1.107 and Equation 1.109 in Equation 1.108, we have:

$$\begin{aligned} I(x,y) &= \frac{1}{2}\log_2\left(\frac{2\pi e(S+N)}{2\pi e N}\right) \\ &= \frac{1}{2}\log_2\left(1 + \frac{S}{N}\right), \end{aligned} \tag{1.110}$$

where, again, both the channel's output signal and the noise are assumed to have Gaussian distribution.

The analog channel's capacity is then calculated upon multiplying the information conveyed per source sample by the Nyquist sampling rate of $f_s = 2 \cdot f_B$, yielding [24]:

$$C = f_B \cdot \log_2\left(1 + \frac{S}{N}\right) \frac{\text{bit}}{\text{sec}}. \tag{1.111}$$

Equation 1.111 is the well-known *Shannon-Hartley law*,[1] establishing the relationship among the channel capacity C, channel bandwidth f_B, and channel signal-to-noise ratio (SNR).

Before analyzing the consequences of the Shannon-Hartley law following Shannon's deliberations [24], we make it plausible from a simple practical point of view. As we

---
[1]Comment by the Authors: Although the loose definition of capacity is due to Hartley, the underlying relationship is entirely due to Shannon.

have seen, the root mean squared (RMS) value of the noise is $\sqrt{N}$, and that of the signal plus noise at the channel's output is $\sqrt{S+N}$. The receiver has to decide from the noisy channel's output signal what signal has been input to the channel, although this has been corrupted by an additive Gaussian noise sample. Over an ideal noiseless channel, the receiver would be able to identify what signal sample was input to the receiver. However, over noisy channels it is of no practical benefit to identify the corrupted received message exactly. It is more beneficial to quantify a discretized version of it using a set of decision threshold values, where the resolution is dependent on how corrupted the samples are. In order to quantify this SNR-dependent receiver dynamic range resolution, let us consider the following argument.

Having very densely spaced receiver detection levels would often yield noise-induced decision errors, while a decision-level spacing of $\sqrt{N}$ according to the RMS noise-amplitude intuitively seems a good compromise between high information resolution and low decision error rate. Then assuming a transmitted sample, which resides at the center of a $\sqrt{N}$ wide decision interval, noise samples larger than $\sqrt{N}/2$ will carry samples across the adjacent decision boundaries. According to this spacing, the number of receiver reconstruction levels is given by:

$$q = \frac{\sqrt{S+N}}{\sqrt{N}} = \left(1 + \frac{S}{N}\right)^{\frac{1}{2}}, \qquad (1.112)$$

which creates a scenario similar to the transmission of equiprobable q-ary discrete symbols via a discrete noisy channel, each conveying $\log_2 q$ amount of information at the Nyquist sampling rate of $f_s = 2 \cdot f_B$. Therefore, the channel capacity becomes [24]:

$$C = 2 \cdot f_B \cdot \log_2 q = f_B \cdot \log_2\left(1 + \frac{S}{N}\right), \qquad (1.113)$$

as seen earlier in Equation 1.111.

1.13.1 Practical Evaluation of the Shannon-Hartley Law

The Shannon-Hartley law of Equation 1.111 and Equation 1.113 reveals the fundamental relationship of the SNR, bandwidth, and channel capacity. This relationship can be further studied following Ferenczy's interpretation [22] upon referring to Figure 1.21.

Observe from the figure that a constant channel capacity can be maintained, even when the bandwidth is reduced, if a sufficiently high SNR can be guaranteed. For example, from Figure 1.21 we infer that at $f_B = 10\ KHz$ and SNR = 30 dB the channel capacity is as high as $C = 100$ kbps. Surprisingly, $C \approx 100$ kbps can be achieved even for $f_B = 5$ KHz, if SNR = 60 dB is guaranteed.

Figure 1.22 provides an alternative way of viewing the Shannon-Hartley law, where the SNR is plotted as a function of f_B, parameterized with the channel capacity C. It is important to notice how dramatically the SNR must be increased in order to maintain a constant channel capacity C, as the bandwidth f_B is reduced below $0.1 \cdot C$, where C and f_B are expressed in kbit/s and Hz, respectively. This is due to the $\log_2(1 + \text{SNR})$ function in Equation 1.111, where a logarithmically increasing SNR value is necessitated to compensate for the linear reduction in terms of f_B.

1.13. CAPACITY OF CONTINUOUS CHANNELS

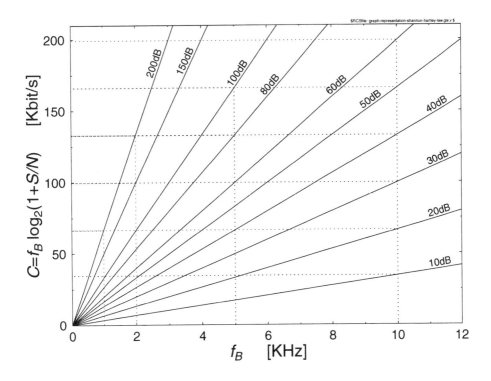

Figure 1.21: Graphical representation of the Shannon-Hartley law. ©Ferenczy [22].

From our previous discourse, the relationship between the relative channel capacity C/f_B expressed from Equation 1.113, and the channel SNR now becomes plausible. This relationship is quantified in Table 1.8 and Figure 1.23 for convenience. Notice that due to the logarithmic SNR scale expressed in dBs, the $C/f_B \left[\frac{bps}{Hz}\right]$ curve becomes near-linear, allowing a near-linearly proportional relative channel capacity improvement upon increasing the channel SNR. A very important consequence of this relationship is that if the channel SNR is sufficiently high to support communications using a high number of modulation levels, the channel is not exploited to its full capacity upon using C/f_B values lower than is afforded by the prevailing SNR. Proposing various techniques in order to exploit this philosophy was the motivation of reference [29].

The capacity C of a noiseless channel with $SNR = \infty$ is $C = \infty$, although noiseless channels do not exist. In contrast, the capacity of an ideal system with $f_B = \infty$ is finite [20, 23]. Assuming additive white Gaussian noise (AWGN) with a double-sided power spectral density (PSD) of $\eta/2$, we have $N = \frac{\eta}{2} \cdot 2 \cdot f_B = \eta \cdot f_B$, and applying the Shannon-Hartley law gives [20]:

$$C = f_B \cdot \log_2\left(1 + \frac{S}{\eta f_B}\right)$$

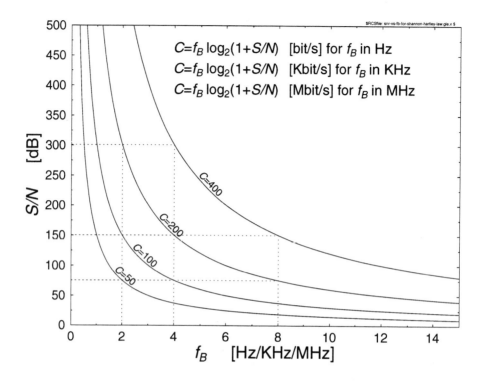

Figure 1.22: SNR versus f_B relations according to the Shannon-Hartley law. ©Ferenczy [22].

SNR		C/f_B
Ratio	dB	bit/sec/Hz
1	0	1
3	4.8	2
7	8.5	3
15	11.8	4
31	14.9	5
63	18.0	6
127	21.0	7

Table 1.8: Relative Channel Capacity versus SNR

1.13. CAPACITY OF CONTINUOUS CHANNELS

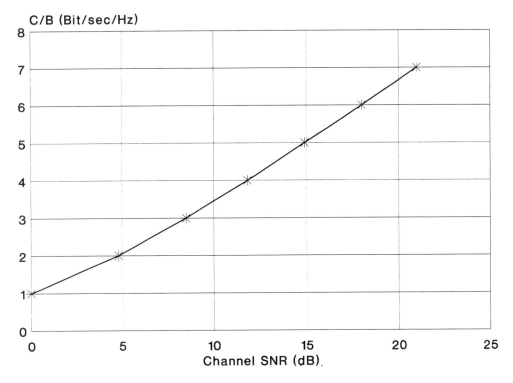

Figure 1.23: Relative channel capacity (C/f_B) versus SNR (dB).

$$\begin{aligned}
&= \left(\frac{S}{\eta}\right)\left(\frac{\eta f_B}{S}\right) \log_2\left(1 + \frac{S}{\eta f_B}\right) \\
&= \left(\frac{S}{\eta}\right) \log_2\left(1 + \frac{S}{\eta f_B}\right)^{\frac{\eta f_B}{S}}.
\end{aligned} \quad (1.114)$$

Our aim is now to determine $C_\infty = lim_{f_B \to \infty} C$. Upon exploiting that:

$$lim_{x \to 0}(1+x)^{\frac{1}{x}} = e \quad (1.115)$$

where $x = S/(\eta \cdot f_B)$, we have

$$C_\infty = lim_{f_B \to \infty} C = \frac{S}{\eta} \log_2 e = 1.45 \cdot \left(\frac{S}{\eta}\right), \quad (1.116)$$

which is the capacity of the channel with $f_B = \infty$. The practically achievable transmission rate R is typically less than the channel capacity C, although complex turbo-coded modems [28] can approach its value. For example, for a telephone channel with a signal-to-noise ratio of $S/N = 10^3 = 30\,dB$ and a bandwidth of $B = 3.4$ kHz from Equation 1.113, we have $C = 3.4 \cdot \log_2(1 + 10^3) \frac{kbit}{sec} \approx 3.4 \cdot 10 = 34\,kbit/s$, which is fairly close to the rate of the V.34 CCITT standard 28.8 kbit/s telephone-channel modem that was recently standardized.

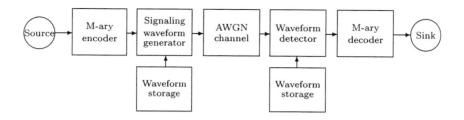

Figure 1.24: Shannon's ideal communications system for AWGN channels.

In this chapter, we have been concerned with various individual aspects of Shannon's information theory [13–16, 24, 25]. Drawing nearer to concluding our discourse on the foundations of information theory, let us now outline in broad terms the main ramifications of Shannon's work [13–16].

1.13.2 Shannon's Ideal Communications System for Gaussian Channels

The ideal Shannonian communications system shown in Figure 1.24 has the following characteristics. The system's information-carrying capacity is given by the information rate $C = f_B \log_2(1 + S/N)$, while as regards its error rate we have $p_e \to 0$. The transmitted and received signals are bandlimited Gaussian random variables, which facilitate communicating at the highest possible rate over the channel.

Information from the source is observed for T seconds, where T is the symbol duration and encoded as equiprobable M-ary symbols with a rate of $R = \frac{\log_2 M}{T}$. Accordingly, the signaling waveform generator of Figure 1.24 assigns a bandlimited AWGN representation having a maximum frequency of f_B from the set of $M = 2^{RT}$ possible waveforms to the source message, uniquely representing the signal $x(t)$ to be transmitted for a duration of T. The noisy received signal $y(t) = x(t) + n(t)$ is compared to all $M = 2^{RT}$ prestored waveforms at the receiver, and the most "similar" is chosen to identify the most likely transmitted source message. The observation intervals at both the encoder and decoder amount to T, yielding an overall coding delay of $2T$. Signaling at a rate equal to the channel capacity is only possible, if the source signal's observation interval is infinitely long, that is, $T \to \infty$.

Before concluding this chapter, we offer a brief discussion of the system-architectural ramifications of transmitting over wireless channels rather than over AWGN channels.

1.14 Shannon's Message and Its Implications for Wireless Channels

In wireless communications over power- and bandlimited channels it is always of prime concern to maintain an optimum compromise in terms of the contradictory requirements of low bit rate, high robustness against channel errors, low delay, and low complexity. The minimum bit rate at which distortionless communications is possible is determined by the entropy of the speech source message. Note, however, that in practical terms the source rate corresponding to the entropy is only asymptotically achievable as the encoding memory length or delay tends to infinity. Any further compression is associated with information loss or coding distortion. Note that the optimum source encoder generates a perfectly uncorrelated source-coded stream, where all the source redundancy has been removed; therefore, the encoded symbols are independent, and each one has the same significance. Having the same significance implies that the corruption of any of the source-encoded symbols results in identical source signal distortion over imperfect channels.

Under these conditions, according to Shannon's pioneering work [13], which was expanded, for example, by Hagenauer [30] and Viterbi [31], the best protection against transmission errors is achieved if source and channel coding are treated as separate entities. When using a block code of length N channel coded symbols in order to encode K source symbols with a coding rate of $R = K/N$, the symbol error rate can be rendered arbitrarily low, if N tends to infinity and the coding rate to zero. This condition also implies an infinite coding delay. Based on the above considerations and on the assumption of additive white Gaussian noise (AWGN) channels, source and channel coding have historically been separately optimized.

Mobile radio channels are subjected to multipath propagation and so constitute a more hostile transmission medium than AWGN channels, typically exhibiting path-loss, log-normal slow fading and Rayleigh fast-fading [32]. Furthermore, if the signaling rate used is higher than the channel's coherence bandwidth, over which no spectral-domain linear distortion is experienced, then additional impairments are inflicted by dispersion, which is associated with frequency-domain linear distortions. Under these circumstances the channel's error distribution versus time becomes bursty, and an infinite-memory symbol interleaver is required in Figure 1.1 in order to disperse the bursty errors and hence to render the error distribution random Gaussian-like, such as over AWGN channels. For mobile channels, many of the above mentioned, asymptotically valid ramifications of Shannon's theorems have limited applicability.

A range of practical limitations must be observed when designing mobile radio speech or video links. Although it is often possible to further reduce the prevailing typical bit rate of state-of-art speech or video codecs, in practical terms this is possible only after a concomitant increase of the implementational complexity and encoding delay. A good example of these limitations is the half-rate GSM speech codec, which was required to approximately halve the encoding rate of the 13 kbps full-rate codec, while maintaining less than quadrupled complexity, similar robustness against channel errors, and less than doubled encoding delay. Naturally, the increased algorithmic complexity is typically associated with higher power consumption, while the reduced number of bits used to represent a certain speech segment intuitively implies that each bit will have an increased relative significance. Accordingly, their corruption may inflict increasingly objectionable speech degradations, unless special attention is devoted to this problem.

In a somewhat simplistic approach, one could argue that because of the reduced source rate we could accommodate an increased number of parity symbols using a more powerful, implementationally more complex and lower rate channel codec, while maintaining the same transmission bandwidth. However, the complexity, quality, and robustness trade-off of such a scheme may not always be attractive.

A more intelligent approach is required to design better speech or video transceivers for mobile radio channels [30]. Such an intelligent transceiver is portrayed in Figure 1.1. Perfect source encoders operating close to the information-theoretical limits of Shannon's predictions can only be designed for stationary source signals, a condition not satisfied by most source signals. Further previously mentioned limitations are the encoding complexity and delay. As a consequence of these limitations the source-coded stream will inherently contain residual redundancy, and the correlated source symbols will exhibit unequal error sensitivity, requiring unequal error protection. Following Hagenauer [30], we will refer to the additional knowledge as regards the different importance or vulnerability of various speech-coded bits as source significance information (SSI). Furthermore, Hagenauer termed the confidence associated with the channel decoder's decisions as decoder reliability information (DRI). These additional links between the source and channel codecs are also indicated in Figure 1.1. A variety of such techniques have successfully been used in robust source-matched source and channel coding [33, 34].

The role of the interleaver and de-interleaver seen in Figure 1.1 is to rearrange the channel coded bits before transmission. The mobile radio channel typically inflicts bursts of errors during deep channel fades, which often overload the channel decoder's error correction capability in certain speech or video segments. In contrast other segments are not benefiting from the channel codec at all, because they may have been transmitted between fades and hence are error-free even without channel coding. This problem can be circumvented by dispersing the bursts of errors more randomly between fades so that the channel codec is always faced with an "average-quality" channel, rather than the bimodal faded/nonfaded condition. In other words, channel codecs are most efficient if the channel errors are near-uniformly dispersed over consecutive received segments.

In its simplest manifestation, an interleaver is a memory matrix filled with channel coded symbols on a row-by-row basis, which are then passed on to the modulator on a column-by-column basis. If the transmitted sequence is corrupted by a burst of errors, the de-interleaver maps the received symbols back to their original positions, thereby dispersing the bursty channel errors. An infinite memory channel interleaver is required in order to perfectly randomize the bursty errors and therefore to transform the Rayleigh-fading channel's error statistics to that of a AWGN channel, for which Shannon's information theoretical predictions apply. Since in interactive video or speech communications the tolerable delay is strictly limited, the interleaver's memory length and efficiency are also limited.

A specific deficiency of these rectangular interleavers is that in case of a constant vehicular speed the Rayleigh-fading mobile channel typically produces periodic fades and error bursts at traveled distances of $\lambda/2$, where λ is the carrier's wavelength, which may be mapped by the rectangular interleaver to another set of periodic bursts of errors. A range of more random rearrangement or interleaving algorithms exhibiting a higher performance than rectangular interleavers have been proposed for mobile channels in [35], where a variety of practical channel coding schemes have also been portrayed.

Returning to Figure 1.1, the soft-decision information (SDI) or channel state information (CSI) link provides a measure of confidence with regard to the likelihood that a specific symbol was transmitted. Then the channel decoder often uses this information in order to invoke maximum likelihood sequence estimation (MLSE) based on the Viterbi algorithm [35] and thereby improve the system's performance with respect to conventional hard-decision decoding. Following this rudimentary review of Shannon's information theory, let us now turn our attention to the characterization of wireless communications channels.

1.15 Summary and Conclusions

An overview of Shannonian information theory has been given, in order to establish a firm basis for our further discussions throughout the book. Initially we focussed our attention on the basic Shannonian information transmission scheme and highlighted the differences between Shannon's theory valid for ideal source and channel codecs as well as for Gaussian channels and its ramifications for Rayleigh channels. We also argued that practical finite-delay source codecs cannot operate at transmission rates as low as the entropy of the source. However, these codecs do not have to operate losslessly, since perceptually unobjectionable distortions can be tolerated. This allows us to reduce the associated bit rate.

Since wireless channels exhibit bursty error statistics, the error bursts can only be randomized with the aid of infinite-length channel interleavers, which are not amenable to real-time communications. Although with the advent of high-delay turbo channel codecs it is possible to operate near the Shannonian performance limits over Gaussian channels, over bursty and dispersive channels different information-theoretical channel capacity limits apply.

We considered the entropy of information sources both with and without memory and highlighted a number of algorithms, such as the Shannon-Fano, the Huffman and run-length coding algorithms, designed for the efficient encoding of sources exhibiting memory. This was followed by considering the transmission of information over noise-contaminated channels leading to Shannon's channel coding theorem. Our discussions continued by considering the capacity of communications channels in the context of the Shannon-Hartley law. The chapter was concluded by considering the ramifications of Shannon's messages for wireless channels.

Chapter 2

The Propagation Environment

2.1 The Cellular Concept

In mobile radio systems, communications take place between a stationary base station (BS) and a number of roaming mobile stations (MSs) or portable stations (PSs) [9, 36, 37]. The BS's and the MS's transmitter is expected to provide a sufficiently high received signal level for the far-end receivers in order to maintain the required communications integrity, which is usually ensured by power control. The geographical area in which this condition is satisfied is termed a traffic cell, which typically has an irregular shape, depending on the prevailing propagation environment determined by terrain and architectural features as well as the local paraphernalia. In theoretical studies, a simple hexagonal cell structure is often favored for its simplicity, where the BSs are located at the centers of the cells.

In an ideal situation, the total bandwidth available to a specific mobile radio system could be allocated within each cell, assuming no energy spills into the adjacent cell's coverage area. However, since wave propagation cannot be shielded at the cell boundary, PSs near the cell edge will experience approximately the same signal energy within their channel bandwidth from at least two BSs. This phenomenon is known as *co-channel interference*. A remedy for this problem is to divide the total bandwidth B_{total} in frequency slots of $B_{cell} = B_{total}/N$ and assign a mutually exclusive reduced bandwidth of $B_{cell} = B_{total}/N$ to each traffic cell within a *cluster* of N cells, as demonstrated in Figure 2.1 for $N = 7$. The seven-cell clusters are then tessellated in order to provide contiguous radio coverage. Observe from the figure that the phenomenon of *co-channel interference* between the black co-channel cells having an identical frequency set is not completely eliminated, but because of the increased co-channel BS distance or *frequency reuse distance*, the interference is significantly reduced. Note also that in analytical and simulation-based interference studies the "second tier" of interfering cells, which are hatched, is typically neglected.

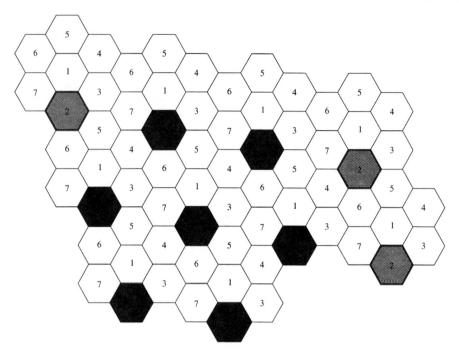

Figure 2.1: Hexagonal cells and seven-cell clusters.

A consequence of this cellular concept, however, is that the total number of MSs that can be supported simultaneously over a unit area is now reduced by a factor of N. This is because assuming a simple frequency division multiple access (FDMA) scheme, where each MS is assigned a radio frequency carrier and a user bandwidth of B_{user} now only $M = B_{cell}/B_{user}$ number of MS can be serviced rather than $N \cdot M = B_{total}/B_{user}$. This problem can be circumvented by making the clusters as small as the original cells, which is achieved by reducing the transmitted power. Further reduction of the cell size has the advantage of serving more and more users over a certain fixed-size geographical area, since the same frequency is reused within the same fixed-size area an increased number of times for a higher number of subscribers, while requiring a reduced transmitted signal power and hence lightweight batteries. A further favorable effect is that the smaller the traffic cell, the more benign the propagation environment due to the presence of a dominant line-of-sight (LOS) propagation path and due to the mitigated effects of the multipath propagation. These arguments lead to the concept of *micro- and picocells* [38–40], which are often confined to the size of a railway station or airport terminal and an office, respectively. Cluster sizes smaller than $N = 7$ are often used in practice in order to enhance the system's area spectral efficiency, but such schemes require modulation arrangements that are resilient against the increased co-channel interference, an issue to be addressed in more depth in Chapter 5. These different cells will have to coexist in practical systems, where, for example, an "over-sailing" macrocell can provide an emergency handover capability for the microcells, when the MS roams in a propagation blindspot but can-

2.1. THE CELLULAR CONCEPT

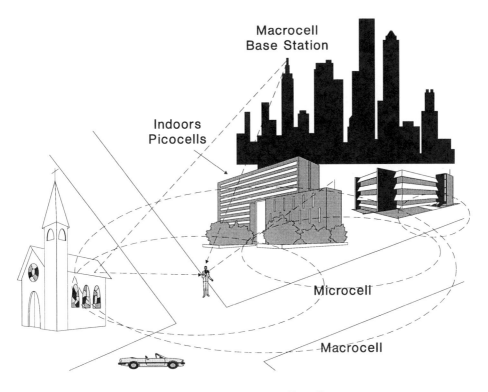

Figure 2.2: Various traffic cells.

not hand over to another microcell, since in the target microcell no traffic channels are available.

The various cell scenarios are exemplified by Figure 2.2, where the conventional macrocell BS is allocated to a high tower, illuminating a large area but providing a rather hostile propagation cell, since often there is no LOS path between the BS and PS. Hence, the communications are more prone to fading than in the stylized microcell illuminated by the antenna at the top of the lower buildings. The smaller microcells typically use lower transmit power, and they channel most of the energy in the street canyon, which mitigates the signal's variability and thus has more benign fading. Furthermore, microcells also reduce the signal's dispersion due to path-loss differences of the multipath components. Finally, indoor picocells provide typically even better channels and tend to mitigate co-channel interferences due to partitions and ceilings.

The previously mentioned handover process is crucial with regard to the perceived Grade of Service (GOS). A wide range of different complexity techniques have been proposed for the various existing and future systems, some of which are summarized in Figure 2.3. Explicitly, the BS and PS keep compiling the statistics of a range of communications quality parameters shown in the figure and weight them according to the prevalent optimization criterion, before a handover or operating mode recon-

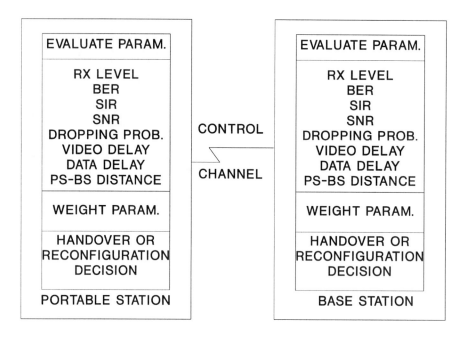

Figure 2.3: Handover control and transceiver reconfiguration parameters.

figuration command is issued, for example, in order to reconfigure the transceiver in a higher bit-rate but less error resilient modulation mode, when the channel conditions improve. Although the PS plays an active role in monitoring the various parameters, these are typically reported to the BS, which carries out the required decisions.

The first-generation Public Land Mobile Radio (PLMR) systems were designed for low traffic density, and the typical cell radius was often of the order of tens of miles. Even the second-generation GSM system [41] was contrived to be able to cope with the hostile large-cell environment of 35 km radius rural cells. These systems had to incorporate sophisticated and power-hungry signal processing in order to be able to combat a wide range of channel impairments. The less robust Pan-American Digital Advanced Phone System (DAMPS) [42] and the second-generation Japanese digital mobile radio (JDMR) systems [43] reflect the more recent trend of moving toward high-capacity small cells. The indoors-oriented cordless telephone or CT systems CT2 [44] and DECT [45] also adopted this tendency.

Following this rudimentary introduction, let us now concentrate on the propagation environment.

2.2 Radio Wave Propagation

2.2.1 Background

The mobile radio channel has been characterized in references [32,36,37,46,47]. Here we restrict our treatment to a rudimentary introduction so that we can arrive at the third-generation Personal Communications Network (PCN) concept.

The propagation laws obeyed by radio waves are not dissimilar to those known from our daily experience with light sources, whose beams propagate through direct LOS paths, through reflection, diffraction, and so on, although differences exist because of their rather different wavelenghts. Just as the light of a torch diminishes with increasing distance, so typically does the average received signal power of the MS, when moving away from the BS in a radial direction. This phenomenon is known as *path-loss*.

For the sake of illustration, let us assume that the MS orbits around the BS on a hypothetical orbit in free space. Then the average received signal power is constant, since no multipath propagation takes place owing to the lack of reflecting and scattering objects. In a typical real propagation scenario, when the BS transmits a sinusoidal radio frequency carrier, the MS receives a plethora of delayed, phase shifted, and attenuated replicas, which add vectorially at the MS's antenna. These so-called multipath components sometimes add constructively, sometimes destructively, which results in rapid received signal amplitude and phase fluctuations, known as *fast-fading*, often inflicting signal level variations up to 40 dB. This process is depicted vectorially for the simple case of a two-path channel in Figure 2.4, where both paths have an identical signal magnitude and one of them has a constant phase associated with a constant vector, while the other one rotates, as the MS moves along.

The third characteristic propagation phenomenon, referred to as *slow fading*, obeys medium-term path-loss variations due to the characteristic building patterns, obstructing large vehicles, and so on, which introduce shadowing effects.

When the length of the various propagation paths is approximately the same, because there are no significant far-field reflections due to mountains or high-rise buildings on the horizon, then the channel's impulse response is sharply decaying similarly to a Dirac delta function. This is referred to as a *nondispersive channel*, since the transmitted signaling pulses are not subjected to time dispersion along the time axis. As the signaling rate increases, the adjacent signaling pulses are more densely spaced and the effect of time dispersion on the reduced-duration signaling pulses becomes relatively more significant. When the signaling pulses start to overlap, *intersymbol interference* (ISI) is encountered, which rapidly degrades the communications integrity. Note that in contrast to the high-elevation antennas of the conventional PLMR systems, the previously introduced microcellular systems have low-elevation antennas, typically situated on lamp posts below the urban skyline. This property is advantageous in terms of channeling most of the transmitted energy into the street canyons and hence mitigating channel dispersion and supporting higher signaling rates than PLMR systems.

If the channel's impulse response is a sharply decaying Dirac delta, its frequency-domain transfer function given by the Fourier transform is constant across the entire

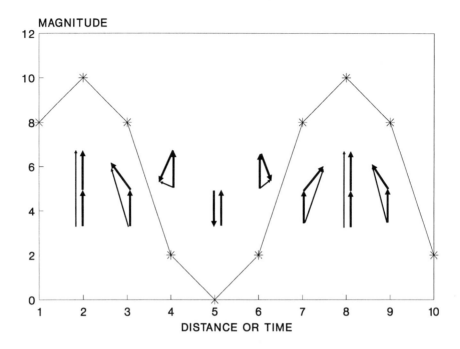

Figure 2.4: Two-path fading channel.

bandwidth. In practice, there are no such infinite bandwidth channels, but a low-spread impulse response is always associated with a frequency-domain transfer function, which is constant across a wide frequency range. If this near-constant passband domain is significantly wider than the transmitted signal's bandwidth, then the ISI due to dispersion is insignificant and the *narrowband propagation* conditions apply. Among these conditions every frequency component of the transmitted signal is subjected to the same fading, a condition that in wave propagation parlance is described as *frequency flat-fading* as distinct from dispersive wideband channel conditions or *frequency-selective fading*, where different frequencies fade differently.

If the channel's frequency-domain transfer function is not constant over the signal's bandwidth, the linear distortions produced must be equalized by a channel equalizer, before high-integrity demodulation can take place. The employment of channel equalizers in partial response modulation schemes [48], such as that of the large-cell GSM PLMR system [41], is fairly common, and, somewhat surprisingly, the wideband channel conditions often result in performance improvements because of their inherent diversity effect in terms of providing several staggered replicas of the transmitted signal. The channel equalizer can lock on to these staggered, differently fading multipath components, eliminate the dispersion introduced by them, and exploit the fact that the independently fading received paths are not likely to be in fade simultaneously. This scenario is briefly considered in the context of the Pan-European GSM system later in this chapter. In contrast, in order to avoid the use of complex and power-thirsty equalizers, third-generation PCNs are designed to have low-dispersion microcellular traffic cells. In order to underline the importance of this scenario for PCNs, narrowband propagation will be the subject of a more detailed discussion in the next section.

2.2.2 Narrowband Fading Channels

Since most existing public mobile radio systems use the 0.9-1.8 GHz band, we concentrate on channels in this range. These frequencies fall in the Ultra High Frequency (UHF) band. At these frequencies the signal level drops rapidly, once over the radio horizon, advantageously limiting co-channel interference, when the frequencies are reused in neighboring cell clusters. At these frequencies, even if there is no line-of-sight path between the transmitter and receiver, sufficient signal power may be received by means of wave scattering, reflection, and diffraction to allow communications.

The basic factors of path-loss, slow fading, and fast-fading influencing the wave propagation phenomena are illustrated in Figure 2.5.

The prediction of the expected mean or median received signal power plays a crucial role in planning the coverage area of a specific base station and in determining the closest acceptable reuse of the propagation frequency employed. For high antenna elevations and large rural cells, a more slowly decaying power exponent is expected than for low elevations and densely builtup urban areas. As mentioned, the received signal is also subjected to slow or shadow fading, which is governed mainly by terrain and topographical features in the vicinity of the mobile receiver, such as small hills and tall buildings. When designing the system's power budget and coverage area pattern, the slow-fading phenomenon is taken into account by adding a shadow fading margin to the required transmitted power.

Slow fading is compensated by increasing the transmitted power P_{tx} by the slow-fading margin L_{slow} illustrated in Figure 2.5, which is usually chosen to be the $1-2\%$ quantile of the slow-fading probability density function (PDF), in order to minimize the probability of unsatisfactorily low received signal power P_{rx}. In addition, the short-term fast signal fading due to multipath propagation is taken into account by employing the fast-fading margin L_{fast} of Figure 2.5, which is typically chosen to be a few percent quantile of the fast-fading distribution. There remains a certain probability that both of these fading margins are simultaneously exceeded by the superimposed slow and fast envelope fading. This situation is often referred to as fading margin overload and results in a very low received signal level, which may cause call dropping in mobile telephony, if this condition persists over a period of time. The probability of this worst-case event can be taken to be the sum of the individual margin overload probabilities.

2.2.3 Propagation Path-loss Law

When considering the propagation path-loss of Figure 2.5, the parameter of prime concern is normally the distance from the BS. Path-loss increases with distance due

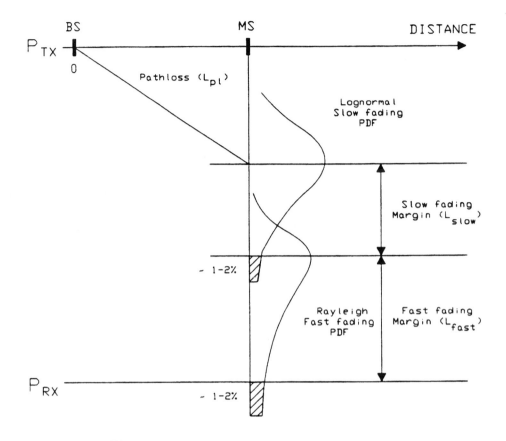

Figure 2.5: Power-budget design for mobile systems.

to the increasing illumination area of the spherical wavefront expanding from the BS, which reduces the power flux per unit area. An in-depth treatment of path-loss calculations can be found in references by Jakes [36], Lee [37], and Parsons [47]. Suffice to note here that path-loss calculations become significantly more complex when there is any form of obstruction between the transmitter and receiver. In large cells, these obstructions will often be hills or undulating terrain. In modelling tools, these obstacles are often considered to be "knife edges" at the point of the summit of the obstruction, and knife edge diffraction techniques can then be used [47] to predict the path-loss. The analysis becomes increasingly complex if there are a number of hills between the transmitter and receiver. An alternative approach is to determine which Fresnel zone the hill obstructs and to perform the prediction accordingly [49].

As mentioned earlier, in urban areas modeling becomes more complex, and in the case of large cells it is rarely possible to model each individual building, although this can be achieved in microcells. Typically, this situation is resolved by adding clutter-loss dependent on the density of the buildings in the urban area and also allowing for a shadowing loss behind buildings, when computing the link budget

2.2. RADIO WAVE PROPAGATION

according to Figure 2.5. This approximation leads to suboptimal network design, with overlapping cells and inefficient frequency reuse. Nevertheless, it seems unlikely that significantly more accurate tools will become available in the immediate future. In our probabilistic approach, it is difficult to give a worst-case path-loss exponent for any mobile channel. However, it is possible to specify the most optimistic scenario. This is given by propagation in free space. The free-space path-loss, L_{pl}, is given by [47]:

$$L_{pl} = 10 \log_{10} G_T + 10 \log_{10} G_R - 20 \log_{10} f^{MHz} - 20 \log_{10} d^{km} + 147.6 dB, \quad (2.1)$$

where G_T and G_R are the transmitter and receiver antenna gains, f is the propagation frequency in MHz, and d is the distance from the BS antenna in km. Observe that the free-space path-loss is increased by 6 dB every time, when the propagation frequency is doubled or the distance from the mobile is doubled. This corresponds to a 20 dB/decade decay when moving away from the BS, and at d=1 km, f=1 MHz, and $G_T = G_R = 1$ a path-loss of $L_F = 147.6$ dB is encountered. Not only technological difficulties, but also propagation losses discourage the deployment of higher frequencies. Nevertheless, spectrum is usually only available at increasingly higher frequency bands.

In practice, for the UHF mobile radio propagation channels of interest, the free-space conditions do not apply. However, a number of useful path-loss prediction models can be adopted to derive other prediction bounds. One such case is the "plane-earth" model. This is a two-path model constituted by a direct line-of-sight path and a ground-reflected one, which ignores the curvature of the earth's surface. Assuming transmitter base station (BS) and receiver mobile station (MS) antenna heights of $h_{BS}^m, h_{MS}^m \ll d$ expressed in meters, respectively, the plane-earth path-loss formula [47] can be derived:

$$\begin{aligned} L_{pl} = & 10 \log_{10} G_T + 10 \log_{10} G_R + 20 \log_{10} h_{BS}^m + \\ & 20 \log_{10} h_{MS}^m - 40 \log_{10} d^{km}, \end{aligned} \quad (2.2)$$

where the dependence on propagation frequency is now absent. Observe that a 6 dB path-loss reduction results when doubling the transmitter or receiver antenna elevations, and there is an inverse fourth power law decay with increasing the BS-MS distance d. In the close vicinity of the transmitter antenna, where the condition h_{BS}^m or $h_{MS}^m \ll d$ does not hold, Equation 2.2 is no longer valid.

Hata [50] developed three path-loss models for large cells, which are widely used, forming the basis for many modeling tools. These were developed from an extensive database derived by Okumura [51] from measurements in and around Tokyo. *The typical urban Hata model* is defined as:

$$\begin{aligned} L_{Hu} = & 69.55 + 26.16 \log_{10} f^{MHz} - 13.82 \log_{10} h_{BS}^m - a(h_{MS}^m) + \\ & (44.9 - 6.55 \log_{10} h_{BS}^m) \log_{10} d^{km} \; [dB], \end{aligned} \quad (2.3)$$

where f is the propagation frequency in MHz, h_{BS} and h_{MS} are the BS and MS antenna elevations in terms of meters, respectively, $a(h_{MS})$ is a terrain-dependent correction factor, while d is the BS-MS distance in km. The correction factor $a(h_{MS})$ for small- and medium-sized cities was found to be [50]:

$$a(h_{MS}) = (1.1\log_{10} f^{MHz} - 0.7)h_{MS}^m - (1.56\log_{10} f^{MHz} - 0.8), \quad (2.4)$$

while for large cities the connection factor is frequency-parametrized:

$$a(h_{MS}) = \begin{cases} 8.29[\log_{10}(1.54h_{MS}^m)]^2 - 1.1 & \text{if } f^{MHz} \leq 200MHz \\ 3.2[\log_{10}(11.75h_{MS}^m)]^2 - 4.97 & \text{if } f^{MHz} \geq 400MHz \end{cases}. \quad (2.5)$$

The typical suburban Hata model applies a correction factor to the urban model yielding:

$$L_{Hsuburban} = L_{Hu} - 2[\log_{10}(f^{MHz}/28)]^2 - 5.4 \ [dB]. \quad (2.6)$$

The rural Hata model modifies the urban formula differently:

$$L_{Hrural} = L_{Hu} - 4.78(\log_{10} f^{MHz})^2 + 18.33\log_{10} f^{MHz} - 40.94 \ [dB]. \quad (2.7)$$

The limitations of its parameters as listed by Hata are [50]:

$$\begin{array}{ll} f: & 150 - 1500 MHz \\ h_{BS}: & 30 - 200m \\ h_{MS}: & 1 - 10m \\ d: & 1 - 20km. \end{array}$$

For a 900 MHz PLMR system, these conditions can usually be satisfied, but for a 1.8 GHz typical PCN urban microcell all these limits have to be stretched.

For the sake of illustration in Figure 2.6 we have plotted a typical microcellular path-loss regression line fitting and the corresponding Hata path-loss characteristic for an antenna elevation of 6.4 m and propagation frequency of 1.9 GHz, as a function of the logarithmic distance from the BS. Observe that the measurement points show a denser clustering for higher distances than for logarithmic distances around 2, which corresponds to 100 m. This is a consequence of the limited number of streets in the BS's vicinity, and this fact gives a higher weighting to points allocated around 1000 m, corresponding to a logarithmic distance of 3. The path-loss exponent in this example appears somewhat lower than that of the corresponding urban Hata model.

2.2.4 Slow-Fading Statistics

Having highlighted a few propagation path-loss prediction models, we now briefly focus our attention on the characterization of the slow-fading phenomenon, which constitutes the second component of the overall power budget design of mobile radio links, as portrayed in Figure 2.5. In slow-fading analysis, the effects of fast-fading and path-loss have to be removed. The fast-fading fluctuations are removed by averaging the signal level over a distance of typically some 20 wavelengths. The slow-fading fluctuations are separated by subtracting the best-fit path-loss regression estimate from each individual 20-wavelength-spaced averaged received signal value. A slow-fading histogram derived this way is depicted in Figure 2.7. The figure suggests a log-normal distribution in terms of dBs caused by normally distributed random

2.2. RADIO WAVE PROPAGATION

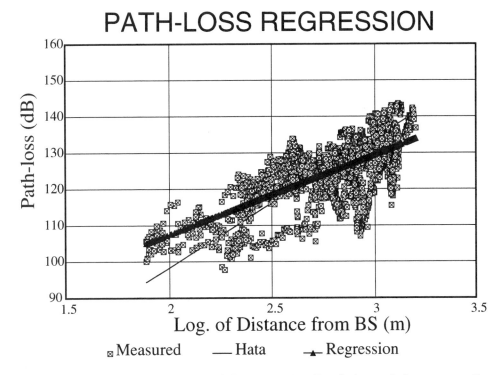

Figure 2.6: Typical microcellular path-loss regression line fitting and the corresponding Hata path-loss characteristic for an antenna elevation of 6.4 m and propagation frequency of 1.9 GHz [32]. ©Wiley 1999, Greenwood & Hanzo.

shadowing effects. Indeed, when subjected to rigorous distribution fitting using the log-normal hypothesis, the hypothesis is confirmed at a high confidence level. The associated standard deviation from the expected path-loss that is, from the long-term mean value in this particular case, is 6.5 dB.

2.2.5 Fast-Fading Statistics

Regardless of the distribution of the numerous individual constituent propagation paths of both the in-phase and quadrature component (a_i, a_q) of the received signal, their distribution will be normal or Gaussian due to the central limit theorem. Then the complex baseband equivalent signal's amplitude and phase characteristics are given by:

$$a(k) = \sqrt{a_i^2(k) + a_q^2(k)}, \tag{2.8}$$
$$\phi(k) = \arctan[a_q(k)/a_i(k)]. \tag{2.9}$$

Our aim is to determine the distribution of the amplitude $a(k)$, if $a_i(k)$ and $a_q(k)$ are known to have a normal or Gaussian distribution.

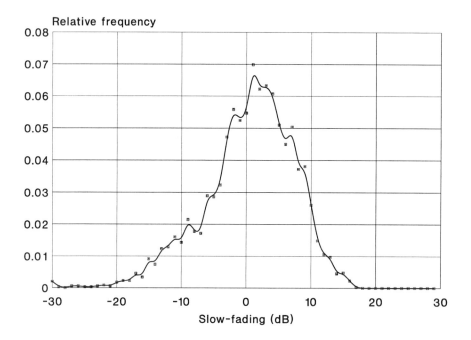

Figure 2.7: Typical microcellular slow-fading histogram [32]. ©Wiley 1999, Greenwood & Hanzo.

In general, for n normally distributed random constituent processes with means $\overline{a_i}$ and identical variances σ^2, the resultant process $y = \sum_{i=1}^{n} a_i^2$ has a χ^2 distribution with a PDF given below [52]:

$$p(y) = \frac{1}{2\sigma^2} \left(\frac{y}{s^2}\right)^{(n-2)/4} \cdot e^{-(s^2+y)/2\sigma^2} \cdot I_{(n/2)-1}\left(\sqrt{y}\frac{s}{\sigma^2}\right) \qquad (2.10)$$

where

$$y \geq 0 \qquad (2.11)$$

and

$$s^2 = \sum_{i=1}^{n} (\overline{a_i})^2 \qquad (2.12)$$

is the noncentrality parameter computed from the first moments of the component processes $a_1 \cdots a_n$. If the constituent processes have zero means, the χ^2 distribution is central; otherwise it is noncentral. Each of these processes has a variance of σ^2,

2.2. RADIO WAVE PROPAGATION

and $I_k(x)$ is the modified Kth order Bessel function of the first kind, given by

$$I_k(x) = \sum_{j=0}^{\infty} \frac{(x/2)^{k+2j}}{j!\Gamma(k+j+1)} \ , \ x \geq 0. \tag{2.13}$$

The Γ function is defined as

$$\begin{aligned}\Gamma(p) &= \int_0^{\infty} t^{p-1} e^{-t} dt \quad \text{if } p > 0 \\ \Gamma(p) &= (p-1)! \quad \text{if } p > 0 \text{ integer} \\ \Gamma(\frac{1}{2}) &= \sqrt{\pi} \ , \ \Gamma(\frac{3}{2}) = \frac{\sqrt{\pi}}{2}.\end{aligned} \tag{2.14}$$

In our case we have two quadrature components, that is, $n = 2$, $s^2 = (\overline{a_i})^2 + (\overline{a_q})^2$, the envelope is computed as $a = \sqrt{y} = \sqrt{a_i^2 + a_q^2}, a^2 = y, p(a)da = p(y)dy$, and hence $p(a) = p(y)dy/da = 2ap(y)$, yielding the Rician PDF [52]:

$$p_{\text{Rice}}(a) = \frac{a}{\sigma^2} e^{-(a^2+s^2)/2\sigma^2} I_o\left(\frac{as}{\sigma^2}\right) \quad a \geq 0. \tag{2.15}$$

Formally introducing the Rician K-factor as

$$K = s^2/2\sigma^2 \tag{2.16}$$

renders the Rician distribution's PDF dependent on one parameter only, yielding [52]:

$$p_{\text{Rice}}(a) = \frac{a}{\sigma^2} \cdot e^{-\frac{a^2}{2\sigma^2}} \cdot e^{-K} \cdot I_o\left(\frac{a}{\sigma} \cdot \sqrt{2K}\right), \tag{2.17}$$

where K physically represents the ratio of the power received in the direct line-of-sight path to the total power received via indirect scattered paths. Therefore, if there is no dominant propagation path, $K = 0$, $e^{-K} = 1$ and $I_0(0) = 1$, yielding the worst-case Rayleigh PDF [52]:

$$p_{\text{Rayleigh}}(a) = \frac{a}{\sigma^2} e^{-\frac{a^2}{2\sigma^2}}. \tag{2.18}$$

Conversely, in the clear direct line-of-sight situation with no scattered power, $K = \infty$, yielding a Dirac delta-shaped PDF, representing a step-function-like cumulative distribution function (CDF). The signal at the receiver then has a constant amplitude with a probability of one. Such a channel is referred to as a Gaussian channel. This is because although no fading is present, the receiver will still generate thermal additive white Gaussian noise (AWGN). If the K-factor is known, the fast-fading envelope's distribution is completely described. The set of Rician PDFs for K = 0, 1, 2, 4, 10, and 15 is plotted at the top of Figure 2.8, while the corresponding time-domain waveforms are portrayed in Figure 2.9.

Observe in the figures that for the worst-case Rician channel constituted by the Rayleigh fading channel we have $K = 0$ or $K = -\infty$ dB, although $K = -40$ dB practically corresponds to conditions similar to Rayleigh fading. In contrast, the best Rician channel is the Gaussian channel, where there is no envelope fading at all,

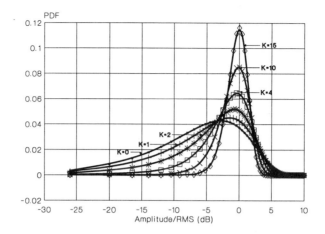

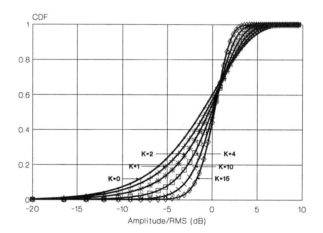

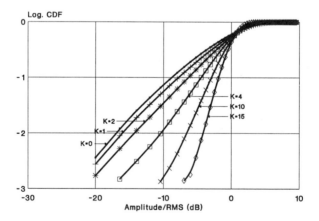

Figure 2.8: Rician PDFs. ©Webb & Hanzo [29].

2.2. RADIO WAVE PROPAGATION

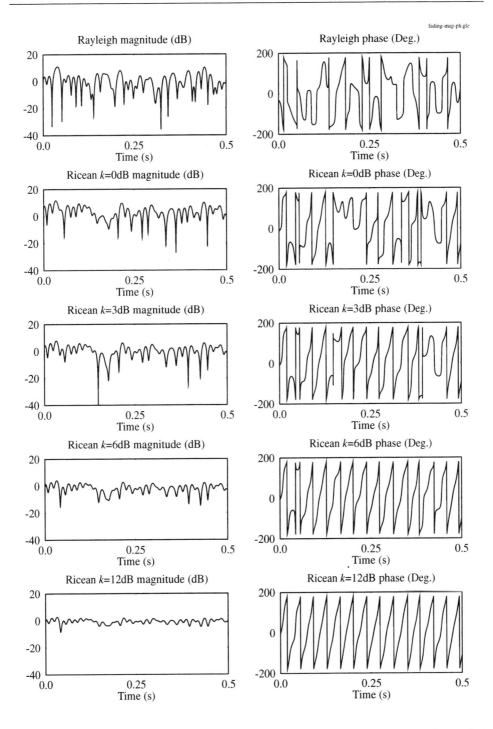

Figure 2.9: Typical Rayleigh and Ricean fast-fading and phase profiles for a MS speed of 30 mph and carrier frequency of 900 MHz.

since the received signal level is constant and the only impairment is the additive white Gaussian noise (AWGN). Accordingly, the fading PDF of an AWGN channel is a Dirac function at the average received signal level, while the receiver noise is represented by the usual Gaussian bell-shaped PDF. Expressed in terms of the CDF, the fading CDF of the AWGN channel is a step-function positioned at the mean received signal level, which also corresponds to the constant instantaneous received signal level. For the AWGN channel we have $K = \infty$, although Rician channels having K-factors in excess of about 15 can be considered near-Gaussian.

A range of typical Rician fading channels with K-factors between 0 and 15 are also shown in Figure 2.9, where for high K-factors the envelope fading is seen to be confined to a few dB, while the phase is virtually linear and restricted to $[\pm\pi]$. The probability density function (PDF), cumulative distribution function (CDF), and logarithmic CDF of a set of Rayleigh and Rician channels can be compared in Figure 2.8, where the K-factors are now expressed in actual ratios rather than in the logarithmic dB-domain. Observe at the bottom of Figure 2.8 that the Rician CDFs are plotted on a convenient logarithmic vertical probability scale, which reveals the enormous difference in terms of deep fades for the K values considered. When choosing the fading margin overload probability of Figure 2.5, defined as the probability of receiving a signal level below the designed received signal level of a system, the bottom curves of Figure 2.8 are particularly useful, since they expand the tails of the CDFs, where, for example, for a Rician CDF with $K = 1$ the 15 dB fading margin overload probability is seen to be approximately 10^{-2}.

Note that the relationship between the variances and means of the Rayleigh and the component Gaussian distributions is given by [52]:

$$\sigma_R^2 = (2 - \pi/2) \cdot \sigma^2$$

and

$$m_R = \sqrt{2} \cdot \sigma \cdot \sqrt{\pi}/2 = \sigma\sqrt{\pi/2},$$

respectively. The Rayleigh distribution's variance is not twice that of the composite Gaussian processes, since we are adding the squares of the Gaussian processes, not the processes themselves.

The Rician CDF takes the shape of [52]

$$\begin{aligned} C_{\text{Rice}}(a) &= 1 - e^{-\left(K + \frac{a^2}{2\sigma^2}\right)} \sum_{m=0}^{\infty} \left(\frac{s}{a}\right)^m \cdot I_m\left(\frac{a\,s}{\sigma^2}\right) \\ &= 1 - e^{-\left(K + \frac{a^2}{2\sigma^2}\right)} \sum_{m=0}^{\infty} \left(\frac{\sigma\sqrt{2K}}{a}\right)^m \cdot I_m\left(\frac{a\sqrt{2K}}{\sigma}\right). \end{aligned}$$

(2.19)

This formula is more difficult to evaluate than the PDF of Equation 2.17 because of the summation of an infinite number of terms, requiring double or quadruple computation precision, and it is avoided in numerical evaluations, if possible. However, in practical terms it is sufficient to increase m to a value, where the last term's contribution becomes less than 0.1%. Having characterized the path-loss as well as the slow-fading

2.2. RADIO WAVE PROPAGATION

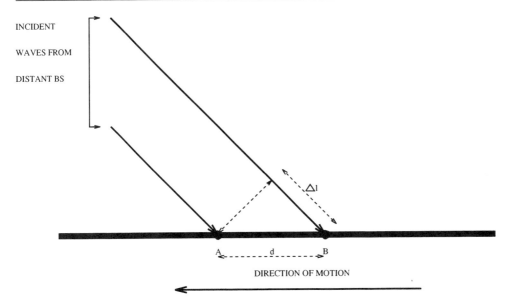

Figure 2.10: Relative path-length change due to the MS's movement. ©Webb & Hanzo [29].

and fast-fading statistics of narrowband mobile channels, let us now consider the rate or frequency of fast-fading, which are explicitly related to the velocity of motion. In the next section, we will introduce the Doppler frequency, which is a function of the vehicular speed, and highlight the influence of vehicular velocity on the correlation between consecutive Rayleigh-distributed fading samples.

2.2.6 Doppler Spectrum

Having described the fading statistics, let us now concentrate on the effects of the Doppler shift and consider again the transmission of an unmodulated carrier frequency f_c from a BS. Consider a MS traveling in a direction at an angle of α_i with respect to the signal received on the ith path, as seen in Figure 2.10, which advances a distance of $d = v \cdot \triangle t$ during $\triangle t$, when traveling at a velocity of v. This introduces a relative carrier phase change of $\triangle \Phi = 2\pi$, if the flight of the wave is shortened by an amount of $\triangle l = \lambda$, where the wavelength can be computed from $\lambda = c/f_c$, with c being the velocity of light.

For an arbitrary $\triangle l$ value we have $\triangle \Phi = -2\pi \cdot \triangle l/\lambda$, where the negative sign implies that the carrier wave's phase delay is reduced, if the MS is traveling toward the BS. From the simple geometry of Figure 2.10 we have $\triangle l = d \cdot \cos \alpha_i$; therefore the phase change becomes:

$$\triangle \Phi = -\frac{2\pi v \triangle t \cos \alpha_i}{\lambda}. \tag{2.20}$$

The Doppler frequency can be defined as the phase change due to the movement of the MS during the infinitesimal interval $\triangle t$:

$$f_D = -\frac{1}{2\pi}\frac{\triangle \Phi}{\triangle t}. \qquad (2.21)$$

When substituting Equation 2.20 into 2.21 we get:

$$f_D = \frac{v}{\lambda}\cos\alpha_i = f_m \cos\alpha_i, \qquad (2.22)$$

where $f_m = v/\lambda = v f_c/c$ is the maximum Doppler frequency deviation from the transmitted carrier frequency due to the MS's movement. Notice that a Doppler frequency can be positive or negative depending on α_i and that the maximum and minimum Doppler frequencies are $\pm f_m$. These extreme frequencies correspond to $\alpha_i = 0°$ and $180°$, when the ray is aligned with the street that the MS is traveling along, and correspond to the ray coming toward or from behind the MS, respectively. It is analogous to the change in the frequency of a whistle from a train perceived by a person standing on a railway line, when the train is bearing down or receding from the person, respectively.

According to Equation 2.22 and assuming that α_i is uniformly distributed, the Doppler frequency has a random cosine distribution. The received power in an angle of $d\alpha$ around α_i is given by $p(\alpha_i)d\alpha$, where $p(\alpha_i)$ is the probability density function (PDF) of the received power, which is assumed to be uniformly distributed over the range of $0 \leq \alpha_i \leq 2\pi$, giving $p(\alpha_i) = 1/(2\pi)$. The Doppler power spectral density $S(f_D)$ can be computed using Parseval's theorem by equating the incident received power $p(\alpha_i)d\alpha$ in an angle $d\alpha$ with the Doppler power $S(f_D)df_D$, yielding $S(f_D) = d\alpha_i/(2\pi \cdot df_D)$. Upon expressing α_i from $f_D = f_m \cos\alpha_i$ and exploiting that

$$\frac{d\cos^{-1} x}{dx} = -\frac{1}{\sqrt{1-x^2}}$$

we then have:

$$S(f_D) = \frac{d\alpha_i}{2\pi df_D} = \frac{d(\cos^{-1} f_D/f_m)}{2\pi df_D} = -\frac{1}{2\pi f_m \sqrt{1-(f_D/f_m)^2}} \qquad (2.23)$$

The incident received power at the MS depends on the power gain of the antenna and the polarization used. Thus, the transmission of an unmodulated carrier is received as a "smeared" signal whose spectrum is not a single carrier frequency f_c but contains frequencies up to $f_c \pm f_m$. In general, we can express the received RF spectrum $S(f_D)$ for a particular MS speed, propagation frequency, antenna design, and polarization as [53]:

$$S(f_D) = \frac{C}{\sqrt{1-(f_D/f_m)^2}}, \qquad (2.24)$$

where C is a constant that absorbs the $1/2\pi f_m$ multiplier in Equation 2.23. Notice that the Doppler spectrum of Equation 2.24 becomes $S(f_D = 0) = C$ at $f_D = 0$, while $S(f_D = f_m) = \infty$, when $f_D = f_m$. Between these extreme values $S(f)$ has a U-shaped characteristic, as it is portrayed in a stylized form in the simulation model of Figure 2.11, which will be developed in the next subsection.

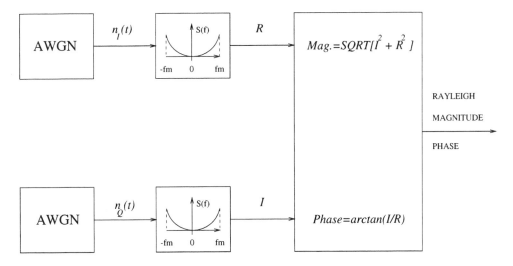

Figure 2.11: Baseband Rayleigh-fading simulation model. ©Webb & Hanzo [29].

2.2.7 Simulation of Narrowband Fading Channels

The Rayleigh-fading channel model used in the previous subsection can be represented by the quadrature arrangement shown in Figure 2.11, where the distribution of both received quadrature components (a_i, a_q) is normal due to the central limit theorem. These can be modeled as uncorrelated normally distributed AWGN sources. The outputs from the AWGN sources are applied to low-pass-type filters, having "U"-shaped frequency-domain transfer functions that represent the effects of Doppler frequency shifts, as will be demonstrated.

Observe that the maximum Doppler frequency $f_m = v/\lambda = v \cdot f_c/c$ depends on the product of the speed v of the MS and the propagation frequency f_c. Therefore, the higher the speed or the propagation frequency, the wider the frequency band over which the received carrier is "smeared." The effect of employing the Doppler filter in Figure 2.11 is that the originally uncorrelated quadrature components (n_I, n_Q) are now effectively low-pass filtered, which introduces some intersample correlation by restricting the maximum rate of change of (n_I, n_Q), while retaining their Gaussian distribution. This is true, because passing an AWGN process through a linear system, such as a low-pass filter, will reduce the variance of the signal but will retain the Gaussian nature of the distribution for both quadrature components. Having no low-pass Doppler filter at all would be equivalent to $f_m = \infty$, implying an infinitely high vehicular speed and hence allowing the reception of arbitrary uncorrelated frequencies, when transmitting the unmodulated carrier frequency of f_c.

The simulation of the Rayleigh-distributed fast-fading envelope is based on the concept portrayed in Figure 2.11, where the Rayleigh-distributed magnitude and uniformly distributed phase are then computed from the filtered real and imaginary AWGN components R and I as follows: Magnitude=SQRT$[I^2 + R^2]$, Phase=arctan (I/R). We note in closing that the above narrowband Rayleigh distributed fading channel model can be implemented in both time- and frequency-domain, as will be detailed in the next two paragraphs.

2.2.7.1 Frequency-Domain Fading Simulation

In the frequency-domain approach, one could exploit the fact that the Fourier transform of an AWGN process is another AWGN process. Hence the frequency-domain AWGN sequence generated by the Box-Müller algorithm to be described below can be multiplied by the frequency-domain Doppler transfer function of Equation 2.24 and then transformed back by IFFT to the time-domain in order to generate the real and imaginary components (R, I) of the filtered complex Gaussian process. The main difficulty associated with this frequency-domain approach is that the Doppler filter's bandwidth f_m is typically much lower than the sampling frequency, at which the faded samples are produced. The faded samples have to be generated exactly at the signaling rate at which modulation symbols are transmitted in order to supply a fading envelope and phase rotation value for their corruption.

A typical Doppler bandwidth f_m at a carrier frequency of $f_c = 1.8 \ GHz$ and vehicular speed of $v = 30 mph = 13.3 m/s$ is given by

$$f_m = \frac{v \cdot f_c}{c} = \frac{13.3 \frac{m}{s} \cdot 1.8 \cdot 10^9 \frac{1}{s}}{3 \cdot 10^8 \frac{m}{s}} \approx 80 \ Hz,$$

where $c = 3 \cdot 10^8 m/s$ is the speed of light. This f_m value becomes about 40 Hz at $f_c = 900 \ MHz$ and 8 Hz at 1.8 GHz and a pedestrian speed of 3 mph. Typical multiuser signaling rates in state-of-the-art Time Division Multiple Access (TDMA) systems are in excess of $f_s = 40 KBd$. Therefore, the pedestrian relative Doppler bandwidth D_r becomes $D_r = 8 \ Hz/40 \ KHz = 2 \cdot 10^{-4}$, suggesting that only a negligible proportion of the input samples will fall in the Doppler filter's passband. This fact may severely affect the statistical soundness of this approach, unless a sufficiently high number of frequency-domain AWGN samples is retained after removing those outside the Doppler filter's bandwidth.

2.2.7.2 Time-Domain Fading Simulation

When using the time-domain approach, the AWGN samples generated by the Box-Müller algorithm to be described in the next paragraph must be convolved with the impulse response $h_D(t)$ of the Doppler filter $S(f)$ of Equation 2.24 in both quadrature arms of Figure 2.11. Similarly to the frequency-domain method, we are faced with a difficult notch-filtering problem. Since the filter bandwidth is typically very narrow in comparision to the sampling frequency, the impulse response $h_D(t)$ will be very slowly decaying, requiring tens of thousands of tap values to be taken into account in the convolution in case of a low Doppler bandwidth. This problem is aggravated by the U-shaped nature of the Doppler's transfer function $S(f)$. The consequence of these difficulties is that the confidence measures delivered by rigorous goodness-of-fit techniques such as the χ^2-test or the Kolmogorov-Smirnov test described, for example, in Chapter 2 of reference [9] might become low, although the resulting fading envelope and phase trajectories are acceptable for simulation studies of mobile radio systems.

2.2.7.3 Box-Müller Algorithm of AWGN Generation

The previously mentioned Box-Müller algorithm is formulated as follows [54]:

1. Generate two random variables u_1, u_2 and let:
$$s = u_1^2 + u_2^2.$$

2. While $s \geq 1$, discard s and recompute u_1, u_2 and s.

3. If $s < 1$ is satisfied, compute the I and Q components of the noise as follows:

$$u_I = u_1 \sqrt{-\frac{2\sigma^2}{s} \cdot \log_e s}$$
$$u_Q = u_2 \sqrt{-\frac{2\sigma^2}{s} \cdot \log_e s},$$

where σ is the standard deviation of the AWGN. This algorithm can be used for generating both the Rayleigh-faded signal envelope and the thermal AWGN.

In conclusion, with regard to the narrowband propagation issues, the Doppler filter of Figure 2.11 acts as a low-pass filter, which simply limits the rate of change of both AWGN sources, while retaining a Gaussian distribution. Explicitly, for a Gaussian distributed source signal, the output of linear systems, such as the Doppler filter, also exhibits Gaussian distribution; only the variance of the process is altered by this linear system. We note, however, that this is only true for the Gaussian distribution in the context of linear systems. Furthermore, due to limiting the rate of change at their outputs, the Doppler filters introduce correlation into the originally uncorrelated AWGN sources. In other words, the complex Gaussion process of Figure 2.11 generates a Rayleigh distribution both with and without the Doppler filters. In the former case, however, there is no time-domain correlation between these Rayleigh-distributed samples, and hence the effect of the vehicular speed is not taken into account. By contrast, when incorporating the Doppler filters, the required velocity-dependent amount of adjacent-sample correlation is introduced in the Rayleigh-fading model.

As stated before, the impulse response of the nondispersive flat Rayleigh-fading mobile radio channel is constituted by a Dirac delta function whose weight has a Rayleigh PDF. This channel is encountered when there are no significant far-field reflectors on the horizon and hence no dispersion occurs, but a plethora of multipath components having similar delays is superimposed on the receiver antenna. As argued before, in this context the Gaussian channel may be represented by a unity-weight Dirac delta impulse response, where AWGN is superimposed. Before concluding our discourse on channel properties, let us briefly consider the properties of the wideband mobile channels in the next subsection.

2.2.8 Wideband Channels

2.2.8.1 Modeling of Wideband Channels

The impulse response of the flat Rayleigh-fading mobile radio channel consists of a single delta function, whose weight has a Rayleigh PDF. This occurs because all the multipath components arrive almost simultaneously and combine to have a Rayleigh PDF. If the signal's transmission bandwidth is narrower than the channel's coherence bandwidth $B_c = 1/(2 \cdot \pi \cdot \Delta)$, where Δ represents the time-dispersion interval over which significant multipath components are received, then all transmitted frequency components encounter nearly identical fading attenuations across the frequency-domain. Therefore, the so-called narrowband condition is met, and the signal is subjected to nonselective or flat fading. The channel's coherence bandwidth (B_c) is defined as the frequency separation, where the correlation of two different-frequency received signal components' attenuation becomes less than 0.5.

The effect of multipath propagation is to spread the received symbols in the time-domain. If the path-delay differences are longer than the symbol duration, several echoes of the same transmitted modulated symbol are received over a number of consecutive symbol periods. This is equivalent to saying that in wideband channels the symbol rate is sufficiently high that each symbol is spread into adjacent symbols, causing intersymbol interference (ISI). In order for the receiver to remove the ISI and regenerate the symbols correctly, it must determine the impulse response of the mobile radio channel by a process termed channel sounding. This response must be frequently measured, since the mobile channel may change rapidly in both time and space.

The magnitude of a typical impulse response is described by a continuous waveform, when plotting received amplitude against time-delay. If we partition the time-delay axis into equal-delay segments, usually called delay bins, then there will be, in general, a number of received signals in each bin corresponding to the different paths whose times of arrival are within the bin duration. These signals, when vectorially combined, can be represented by a delta function occurring in the center of the bin having a magnitude that is Rayleigh distributed. Impulses, which are sufficiently small in amplitude, such that they are of no significance to the receiver, can then be discarded.

As an example here, we describe a set of frequently used typical wideband channel impulses specified by the Group Speciale Mobile Committee during the definition of the Pan-European mobile radio system known as GSM [41]. These impulse responses describe typical urban, rural, and hilly terrain environments, as well as an artificially contrived equalizer test response.

The wideband propagation channel is the superposition of a number of dispersive fading paths, suffering from various attenuations and delays, aggravated by the phenomenon of Doppler shift caused by the MS's movement. The maximum Doppler shift (f_m) is given by $f_m = v/\lambda = v \cdot f_c/c$, where v is the vehicular speed, λ is the wavelength of the carrier frequency f_c, and c is the velocity of light. The momentary Doppler shift f_D depends on the angle of incidence α_i, which is uniformly distributed, that is, $f_D = f_m \cdot \cos \alpha_i$, which therefore has a random cosine distribution with a Doppler spectrum limited to $-f_m < f_D < f_m$. Because of time-frequency duality, this "frequency dispersive" phenomenon results in "time-selective" behavior and the wider the Doppler spread (i.e., the higher the vehicular speed), the faster is the time-domain impulse response fluctuation.

The set of 6-tap GSM impulse responses is depicted in Figure 2.12. In simple terms, the wideband channel's impulse response is measured by transmitting an im-

2.2. RADIO WAVE PROPAGATION

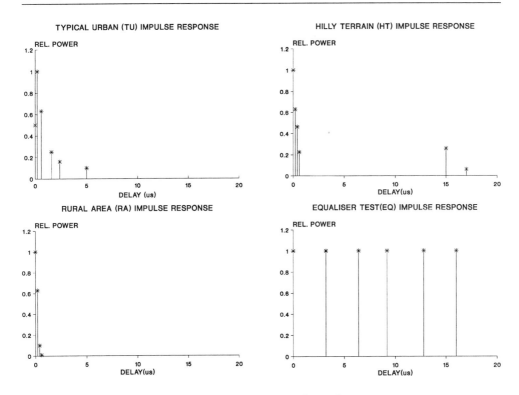

Figure 2.12: Typical GSM channel impulse responses.

pulse and detecting the received echoes at the channel's output in every D-spaced delay bin. In some bins, no delayed and attenuated multipath component is received, while in others significant energy is detected, depending on the typical reflecting objects and their distance from the receiver. The path-delay can be easily related to the distance of the reflecting objects, since radio waves are traveling at the speed of light. For example, at a speed of 300 000 km/s a reflecting object situated at a distance of 0.15 km yields a multipath component at a round-trip delay of 1 μs, corresponding to a path-length of 0.3 km.

The Typical Urban (TU) impulse response of Figure 2.12 spreads over a delay interval of 5 μs. Therefore, it definitely spills energy into adjacent signaling intervals for signaling rates in excess of 200 ksymbols/s, resulting in serious ISI. In practical terms, the transmissions can be considered nondispersive if the excess path-delay does not exceed 10% of the signaling interval, which in our example would correspond to 20 ksymbols/s or 20 kBaud. The Hilly Terrain (HT) model of Figure 2.12 has a sharply decaying short-delay section due to local reflections and a long-delay part around 15 μs due to distant reflections. Therefore, in practical terms, it can be considered a two- or three-path model having reflections associated with a path-length of $3 \cdot 10^8$ m/s $\cdot$ 15 μs = 4.5 km. The Rural Area (RA) response seems the least hostile among all standardized responses, decaying fast within 1 μs; hence up to signaling rates of 100 kBaud, this environment can be treated as a flat-fading narrowband channel.

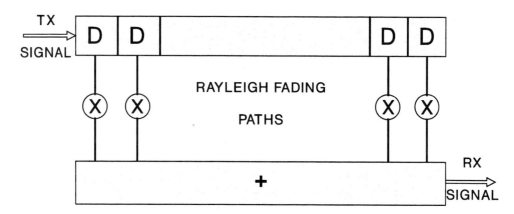

Figure 2.13: Generation of the baseband in-phase received signal via wideband channels.

The last one of the standardized GSM impulse responses in Figure 2.12 was artificially contrived in order to test the channel equalizer's performance, and it is constituted by six equidistant unit-amplitude impulses representing six equal-powered independent Rayleigh-fading paths with a delay-spread over 16 μs. Note that in practice such an impulse response would be unrealistic, since the delayed paths have a higher path-length, which is associated with a higher path-attenuation. With these impulse responses in mind, the required channel can be simulated by summing the appropriately delayed and weighted received signal components. In all but one case, the individual components are assumed to have Rayleigh amplitude distribution. In the RA model, the main tap at zero delay is supposed to have Rician distribution with the presence of a dominant path.

Once the impulse response of the channel is known, we can model a wideband impulse response using the simple finite-impulse response schematic of Figure 2.13, where the filter coefficients are determined by the impulse response and the shift-register delay D corresponds to the sampling frequency used in generating the impulse response. Accordingly, those impulse response coefficients, which were set to zero, since they were below the thresholding value will have to be set to zero in the schematic. We note again, however, that all the nonzero filter tap-weights correspond to a significant fading multipath component. When amalgamating the schematics of Figures 2.11 and 2.13, we arrive at the wideband channel simulator scheme of Figure 2.14.

The in-phase modulated signal $s_I(t)$ is applied to a series of delays equal to the width of a delay bin, namely D. At each delay $n \times D$, $S_I(t)$ is multiplied by the magnitude of the wideband channel impulse response at that delay, which is then faded obeying the Rayleigh distribution in the worst-case non-LOS scenario. An identical arrangement to that for $S_I(t)$ in Figure 2.14 is used for the quadrature component $s_Q(t)$, and the appropriate convolutional terms are combined to obtain the received signal's quadrature components $r_I(t)$ and $r_Q(t)$.

2.2. RADIO WAVE PROPAGATION

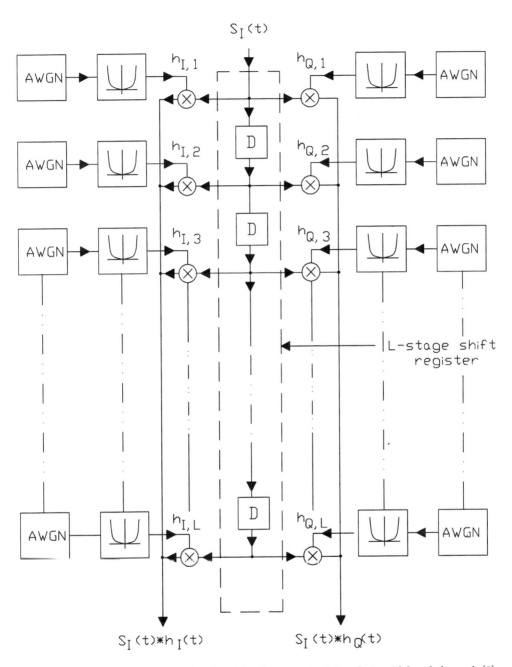

Figure 2.14: Generation of the baseband in-phase received signal via wideband channels [9]. ©1992, Steele.

2.3 Summary and Conclusions

In this chapter, we have introduced the cellular concept and demonstrated how radio coverage can be provided in propagation cells of various sizes. Specifically, we have argued that macrocells constitute a a hostile propagation environment, since the propagation properties vary across the geographic area of the cell. This results in less predictable propagation scenarios than those experienced in the typically smaller microcells often benefiting from LOS wave propagation. Finally, indoor coverage is often provided by picocells, where the LOS scenario is predominant. The required transmitted power is reduced in line with the cell size. This results in the reduction of far-field reflections, which ultimately reduces the channel-induced dispersion. Hence the effects of multipath propagation become less distructive, potentially supporting higher bit rates in picocells or allowing the system to dispense with employing high-complexity channel equalisers.

The range of parameters that may be used for controlling handovers and transceiver reconfiguration has also been reviewed. Path-loss, shadow fading, and fast-fading have been identified as the three most important factors influencing the power budget design of wireless systems. The Doppler phenomenon has been discussed briefly and a model has been introduced for the simulation of both narrowband and wideband Rayleigh and Rician fading channels.

Having characterized the mobile radio channel, we are now equipped to consider how the detrimental effects of this hostile transmission medium can be counteracted using channel coding techniques.

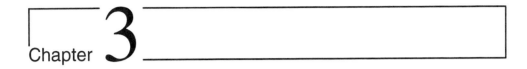

Chapter 3

Convolutional Channel Coding

3.1 Brief Channel Coding History

The history of channel coding or forward error correction (FEC) coding dates back to Shannon's pioneering work in which he predicted that arbitrarily reliable communications is achievable by redundant FEC coding, although he refrained from proposing explicit schemes for practical implementations. Historically, one of the first practical codes was the single error correcting Hamming code [55], which was a block code proposed in 1950. Convolutional FEC codes date back to 1955 [56], which were discovered by Elias, where as Wozencraft and Reiffen [57, 58], as well as Fano [59] and Massey [60], proposed various algorithms for their decoding. A major milestone in the history of convolutional error correction coding was the invention of a maximum likelihood sequence estimation algorithm by Viterbi [61] in 1967. A classic interpretation of the Viterbi algorithm (VA) can be found, for example, in Forney's often-quoted paper [62], and one of the first applications was proposed by Heller and Jacobs [63].

We note, however, that the VA does not result in minimum bit error rate (BER). The minimum BER decoding algorithm was proposed in 1974 by Bahl *et al.* [64], which was termed the Maximum *A posteriori* (MAP) algorithm. Although the MAP algorithm slightly outperforms the VA in BER terms, because of its significantly higher complexity it was rarely used, until turbo codes were contrived [28].

During the early 1970s, FEC codes were incorporated in various deep-space and satellite communications systems, and in the 1980s they also became common in virtually all cellular mobile radio systems. A further historic breakthrough was the invention of the turbo codes by Berrou, Glavieux, and Thitimajshima [28] in 1993, which facilitates the operation of communications systems near the Shannonian limits.

Focusing our attention on block codes, the single-error correcting Hamming block code was too weak, however, for practical applications. An important practical milestone was the discovery of the family of multiple error correcting Bose-Chaudhuri-

Hocquenghem (BCH) binary block codes [65] in 1959 and in 1960 [66, 67]. In 1960, Peterson [68] recognized that these codes exhibit a cyclic structure, implying that all cyclically shifted versions of a legitimate codeword are also legitimate codewords. Furthermore, in 1961 Gorenstein and Zierler [69] extended the binary coding theory to treat nonbinary codes as well, where code symbols were constituted by a number of bits, and this led to the birth of burst-error correcting codes. They also contrived a combination of algorithms, which are referred to as the Peterson-Gorenstein-Zierler (PGZ) algorithm. We will elaborate on this algorithm later in this chapter. In 1960 a prominent nonbinary subset of BCH codes were discovered by Reed and Solomon [70]; they were named after their inventors Reed-Solomon (RS) codes. These codes exhibit certain optimality properties, and they will also be treated in more depth in this chapter. We will show that the PGZ decoder can also be invoked for decoding nonbinary RS codes.

A range of powerful decoding algorithms for RS codes was found by Berlekamp [1, 71] and Massey [2, 72], which also constitutes the subject of this chapter. In recent years, these codes have found practical applications, for example, in Compact Disc (CD) players, in deep-space scenarios [73], and in the family of Digital Video Broadcasting (DVB) schemes, which were standardized by the European Telecommunications Standardization Institute (ETSI). We now consider the conceptually less complex class of convolutional codes, which will be followed by our discussions on block coding.

3.2 Convolutional Encoding

Both block codes and convolutional codes (CC) can be classified as systematic or nonsystematic codes, where the terminology suggests that in systematic codes the original information bits or symbols constitute part of the encoded codeword and hence they can be recognized explicitly at the output of the encoder. Their encoders can typically be implemented by the help of linear shift-register circuitries, an example of which can be seen in Figure 3.1. The figure will be explored in more depth after introducing some of the basic convolutional coding parameters.

Specifically, in general a k-bit information symbol is entered into the encoder, constituted by K shift-register stages. In our example of Figure 3.1, the corresponding two shift-register stages are s_1 and s_2. In general, the number of shift-register stages K is referred to as the constraint length of the code. An alternative terminology is to refer to this code as a memory three code, implying that the memory of the CC is given by $K + 1$. The current shift-register state s_1, s_2 plus the incoming bit b_i determine the next state of this state machine. The number of output bits is typically denoted by n, while the coding rate by $R = k/n$, implying that $R \leq 1$. In order to fully specify the code, we also have to stipulate the generator polynomial, which describes the topology of the modulo-2 gates generating the output bits of the convolutional encoder. For generating n bits, n generator polynomials are necessary. In general, a CC is denoted as a $CC(n, k, K)$ scheme, and given the n generator polynomials, the code is fully specified.

Once a specific bit enters the encoder's shift register in Figure 3.1, it has to tra-

3.2. CONVOLUTIONAL ENCODING

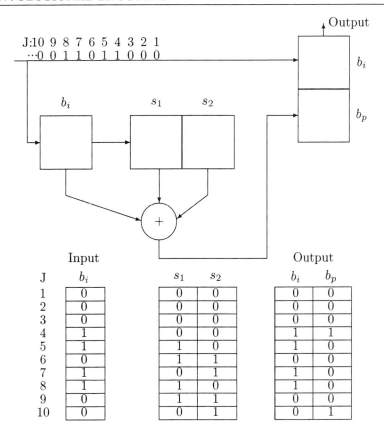

Figure 3.1: Systematic half-rate, constraint-length two convolutional encoder $CC(2,1,2)$.

verse through the register, and hence the register's sequence of state transitions is not arbitrary. Furthermore, the modulo-2 gates impose additional constraints concerning the output bit-stream. because of these constraints, the legitimate transmitted sequences are restricted to certain bit patterns, and if there are transmission errors, the decoder will conclude that such an encoded sequence could not have been generated by the encoder and that, it must be due to channel errors. In this case, the decoder will attempt to choose the most resemblent legitimately encoded sequence and output the corresponding bit-stream as the decoded string. These processes will be elaborated on in more detail later in the chapter.

The n generator polynomials $g_1, g_2, \ldots, g_n$, are described by the specific connections to the register stages. Upon clocking the shift register, a new information bit is inserted in the register, while the bits constituting the states of this state machine move to the next register stage and the last bit is shifted out of the register. The generator polynomials are constituted by a binary pattern, indicating the presence or absence of a specific link from a shift register stage by a binary one or zero, respectively. For example, in Figure 3.1 we observe that the generator polynomials are

constituted by:
$$g_1 = [1\ 0\ 0] \quad \text{and} \quad g_2 = [1\ 1\ 1]\ , \tag{3.1}$$
or, in an equivalent polynomial representation as:
$$g_1(z) = 1 + 0 \cdot z^1 + 0 \cdot z^2 \quad \text{and} \quad g_2(z) = 1 + z + z^2\ . \tag{3.2}$$

We note that in a nonsystematic CC, g_1 would also have more than one nonzero terms. It is intuitively expected that the more constraints are imposed by the encoder, the more powerful the code becomes, facilitating the correction of a higher number of bits, which renders nonsystematic convolution codes typically more powerful than their systematic counterparts.

Again, in a simple approach, we will demonstrate the encoding and decoding principles in the context of the systematic code specified as ($k = 1$), half-rate ($R = k/n = 1/2$), $CC(2, 1, 2)$ convolutional code, with a memory of three binary stages ($K = 2$). These concepts can then be extended to arbitrary codecs. At the commencement of the encoding, the shift register is typically cleared by setting it to the all-zero state, before the information bits are input to it. Figure 3.1 demonstrates the encoder's operation for the duration of the first ten clock cycles, tabulating the input bits, the shift-register states s_1, s_2, and the corresponding output bits. At this stage, the uninitiated reader is requested to follow the operations summarized in the figure before proceeding to the next stage of operations.

3.3 State and Trellis Transitions

An often used technique for characterizing the operations of a state machine, such as our convolutional encoder, is to refer to the state transition diagram of Figure 3.2. Given that there are two bits in the shift register at any moment, there are four possible states in the state machine and the state transitions are governed by the incoming bit b_i. A state transition due to a logical zero is indicated by a continuous line in the figure, while a transition activated by a logical one is represented by a dashed line. The inherent constraints imposed by the encoder manifest themselves here in that from any state there are only two legitimate state transitions, depending on the binary input bit. Similarly, in each state there are two merging paths. It is readily seen from the encoder circuit of Figure 3.1 that, for example, from state $(s_1, s_2)=(1,1)$ a logical one input results in a transition to (1,1), while an input zero leads to state (0,1). The remaining transitions can also be readily checked by the reader. A further feature of this figure is that the associated encoded output bits are also plotted in the boxes associated with each of the transitions. Hence, this diagram fully describes the operations of the encoder.

Another simple way of characterizing the encoder is to portray its trellis diagram, which is depicted in Figure 3.3. At the left of the figure, the four legitimate encoder states are portrayed. Commencing operations from the all-zero register state (0,0) allows us to mirror the encoder's actions seen in Figures 3.1 and 3.2 also in the trellis diagram, using the same input bit-stream. As before, the state transitions are governed by the incoming bits b_i and a state transition due to a logical zero is indicated

3.3. STATE AND TRELLIS TRANSITIONS

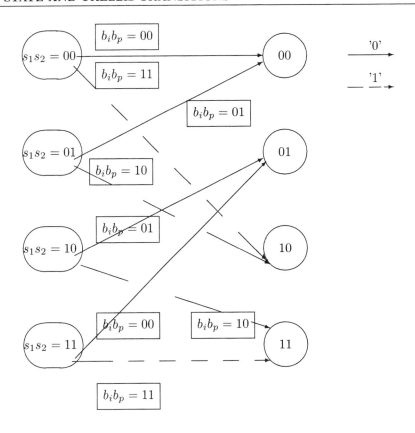

Figure 3.2: State-transition diagram of the $CC(2,1,2)$ systematic code, where broken lines indicate transitions due to an input one, while continuous lines correspond to input zeros.

by a continuous line, while a transition activated by a logical one is represented by a dashed line.

Again, the inherent constraints imposed by the encoder manifest themselves here in that from any state there are only two legitimate state transitions, depending on the binary input bit and in each state there are two merging paths. Given our specific input bit-stream, it is readily seen from the encoder circuit of Figure 3.1 and the state transition diagram of Figure 3.2 that, for example, from state $(s_1, s_2) = (0,0)$ a logical zero input bit results in a transition to (0,0), while an input one leads to state (1,0). The remaining transitions shown in the figure are associated with our specific input bit-stream, which can be readily explored by the reader. As before, the associated output bits are indicated in the boxes along each of the transitions. Hence, the trellis diagram gives a similarly unambiguous description of the encoder's operations to the state diagram of Figure 3.2. Armed with the above description of convolutional codes, we are now ready to give an informal description of the maximum likelihood Viterbi algorithm in the next section.

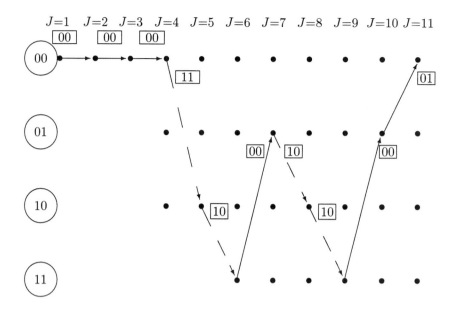

Figure 3.3: Trellis diagram of the $CC(2,1,2)$ systematic code, where broken lines indicate transitions due to a binary one, while continuous lines correspond to input zeros.

3.4 The Viterbi Algorithm

3.4.1 Error-Free Hard-Decision Viterbi Decoding

Given the received bit-stream, the decoder has to arrive at the best possible estimate of the original uncoded information sequence. Hence, the previously mentioned constraints imposed by the encoder on the legitimate bit sequences have to be exploited in order to eradicate illegitimate sequences and thereby remove the transmission errors. For the sake of computational simplicity, let us assume that the all-zero bit-stream has been transmitted and the received sequence of Figure 3.4 has been detected by the demodulator, which has been passed to the FEC decoder. We note here that if the demodulator carries out a binary decision concerning the received bit, this operation is referred to as *hard-decision demodulation*. By contrast, if the demodulator refrains from making a binary decision and instead it outputs a more finely graded multilevel confidence measure concerning the probability of a binary one and a binary zero, then it is said to invoke *soft-decision demodulation*.

For the time being we will consider only hard-decision demodulation. The decoder now has to compare the received bits with all the legitimate sequences of the trellis diagram of Figure 3.4 and quantify the probability of each of the associated paths, which ultimately assigns a probability-related quantity to specific decoded sequences, as we will show below.

Referring to Figure 3.4 and beginning the decoding operations from the all-zero state, we compute the Hamming distance of the received two-bit symbol with respect

3.4. THE VITERBI ALGORITHM

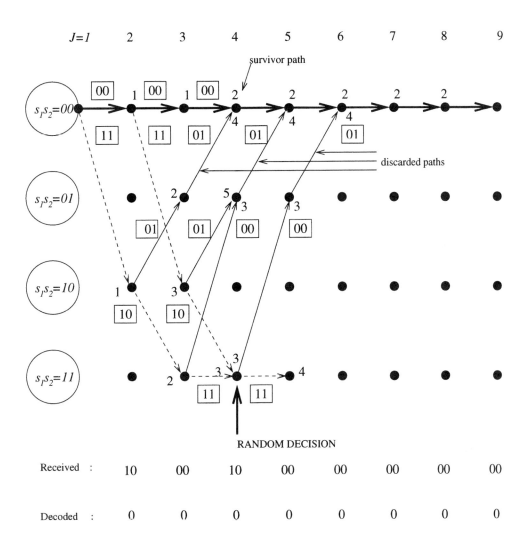

Figure 3.4: Trellis diagram based Viterbi decoding of the $CC(2,1,2)$ systematic code, where broken lines indicate transitions due to an input one, while continuous lines correspond to input zeros — **Error-free hard-decision decoding of two isolated bit errors.**

to both of the legitimate encoded sequences of the trellis diagram for the first trellis section (i.e., for the first trellis transition), noting that the Hamming distance is given by the number of different bit positions between two binary sequences. For example, for the first two-bit received symbol 10, the associated Hamming distances are 1 with respect to both the 00 and the 11 encoded sequences. Thus, at this stage the decoder is unable to express any preference as to whether 00 or 11 was the more likely transmitted symbol. We also note these Hamming distances in the trellis diagram of Figure 3.4, indicated at the top of the nodes corresponding to the new encoder states we arrived at, as a consequence of the state transitions due to a logical one and zero, respectively. These Hamming distances are termed in the context of Viterbi decoding as the *branch metric*. The power of the Viterbi decoding algorithm accrues from the fact that it carries out a maximum likelihood sequence estimation, as opposed to arriving at symbol-by-symbol decisions, and thereby exploits the constraints imposed by the encoder on the legitimate encoded sequences. Hence, the branch metrics will be accumulated over a number of consecutive trellis stages before a decision as to the most likely encoder path and information sequence can be released.

Proceeding to the next received two-bit symbol, namely, 00, the operations of the decoder are identical; namely, the Hamming distance between the encoded symbols of all four legitimate paths and the received symbol is computed. These distances yield the new branch metrics, associated with the second trellis stage. By now the encoded symbols of two original input bits were received, and this is why there are now four possible trellis states in which the decoder may reside. The branch metrics computed for these four legitimate transitions from top to bottom are 0, 2, 1, and 1, respectively. These are now added to the previous branch metrics of 1 in order to generate the *path metrics* of 1, 2, 3, and 2, respectively, quantifying the probability of each legitimate trellis path in terms of the accumulated Hamming distance. A low Hamming distance indicates a high similarity between the received sequence and the encoded sequence concerned, which is characteristic of the most likely encoded sequence, since the probability of a high number of errors is exponentially decreasing with the number of errors.

Returning to Figure 3.4 again, the corresponding accumulated Hamming distances or branch metrics from top to bottom are 1, 2, 3, and 2, respectively. At this stage, we can observe that the top branch has the lowest branch metric and hence it is the most likely encountered encoder path. The reader *knows* this, but the decoder can only *quantify the probability* of the corresponding paths and thus it cannot be sure of the validity of its decision. The other three encoder paths and their associated information bits also have a finite probability.

Continuing the legitimate paths of Figure 3.4 further, at trellis stage three the received sequence of 10 is compared to the four legitimate two-bit encoded symbols and the associated path metrics now become dependent on the actual path followed, since at this stage there are merging paths. For example, at the top node we witness the merger of the 00, 00, 00 path with the 11, 01, 01 path, where the associated original information bits were 0,0,0 and 1,0,0, respectively. On the basis of the associated path metrics, the decoder may "suspect" that the former one was the original information sequence, but it still refrains from carrying out a decision. Considering the two merging paths, future input bits would affect both of these in an identical fashion, resulting in an indistinguishable future set of transitions. Their path metrics will therefore also evolve indentically, suggesting that it is pointless to keep track of both of the merging paths, since the one with the lower metric will always remain the more likely encoder path. This is reflected in Figure 3.4 by referring to the path exhibiting the lower metric as the *survivor path*, while the higher metric merging path will be discarded.

We also note that at the bottom node of trellis stage three we ended up with two identical path metrics, namely 3, and in this case a random decision must be made as to which one becomes the survivor. This event is indicated in Figure 3.4 by the arrow. In this particular example, this decision does not affect the final outcome of the decoder's decision, since the top path appears to be the lowest metric path. Nonetheless, in some situations such random decisions will influence the decoded bit sequence and may indeed determine whether decoding errors are encountered. It is plausible that the confidence in the decoder's decision is increased, as the accumulation of the branch metrics continues. Indeed, one may argue that the "best" decision can be taken upon receiving the complete information sequence. However, deferring decisions for so long may not be acceptable in latency terms, in particular in delay-sensitive interactive speech or video communications. Nor is it necessary in practical terms, since experience shows that the decoder's bit error rate is virtually unaffected by curtailing the decision interval to about five times the encoder's memory, which was three in our example.

In our example, the received bit sequence does not contain any more transmission errors, and so it is plausible that the *winning path* remains the one at the top of Figure 3.4 and the associated branch metric of 2 reflects the actual number of transmission errors. We are now ready to release the error-free decoded sequence, namely, the all-zero sequence, as seen explicitly in terms of the corresponding binary bits at the bottom of Figure 3.4. The corresponding winning path was drawn in boldface in the figure.

3.4.2 Erroneous Hard-Decision Viterbi Decoding

Following the above double-error correction scenario, below we consider another instructive example where the number of transmission errors remains two and even their separation is increased. Yet the decoder may become unable to correct the errors, depending on the outcome of a random decision. This is demonstrated in Figure 3.5 at stage four of the top path. Observe furthermore that the corresponding received bits and path metrics of Figure 3.4 are also indicated in Figure 3.5, but they are crossed out and superseded by the appropriately updated values according to the current received pattern. Depending on the actual choice of the survivor path at stage four, the first decoded bit may become a logical one, as it was indicated at the bottom of Figure 3.5. The accumulated Hamming distance becomes 2, regardless of the random choice of the survivor path, which indicates that in the case of decoding errors the path-metric is not a reliable measure of the actual number of errors encountered. This will become even more evident in our next example.

Let us now consider a scenario in which there are more than two transmission errors in the received sequence, as seen in Figure 3.6. Furthermore, the bit errors are

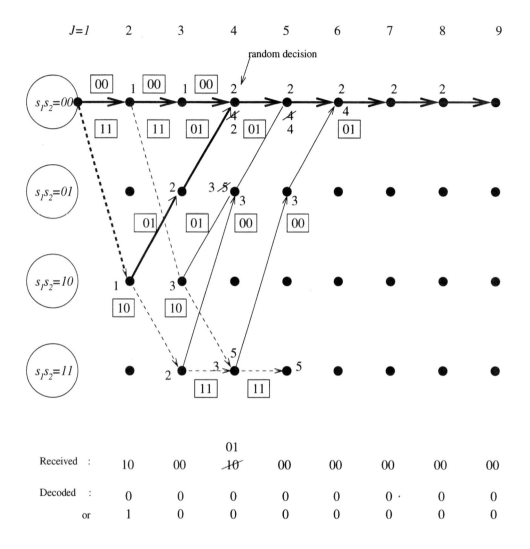

Figure 3.5: Trellis diagram based Viterbi decoding of the $CC(2,1,2)$ systematic code, where broken lines indicate transitions due to an input one, while continuous lines correspond to input zeros — **Erroneous hard-decision decoding of two isolated bit errors.**

3.4. THE VITERBI ALGORITHM

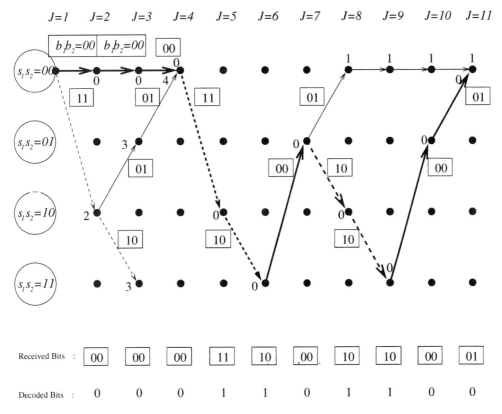

Figure 3.6: Trellis diagram based Viterbi decoding of the $CC(2,1,2)$ systematic code, where broken lines indicate transitions due to an input one, while continuous lines correspond to input zeros — **Erroneous hard-decision decoding of burst errors.**

more concentrated, forming a burst of errors, rather than remaining isolated error events. In this example, we show that the decoder becomes "overloaded" by the plethora of errors, and hence it will opt for an erroneous trellis path, associated with the wrong decoded bits.

Observe in Figure 3.4 that up to trellis stage three the lowest-metric path is the one at the top, which is associated with the error-free all-zero sequence. However, the double-error in the fourth symbol of Figure 3.6 results in a "dramatic" turn of events, since the survivor path deviates from the error-free all-zero sequence. Because of the specific received sequence encountered, the path metric remains 0 up to trellis stage $J = 11$, the last stage, accumulating a total of 0 Hamming distance, despite actually encountering a total of six transmission errors, resulting in four decoding errors at the bottom of Figure 3.6. Again, the winning path was drawn in bold in Figure 3.4.

From this experience we can infer two observations. First, the high-error rate scenario encountered is less likely than the previously considered double-error case, but it has a finite probability and hence it may be encountered in practice. Second, since the decoder carries out a maximum likelihood decision, in such cases it will opt for the wrong decoded sequence, in which case the accumulated path metric will not correctly reflect the number of transmission errors encountered. We therefore conclude that, in contrast to block codes, convolutional codes do not possess an ability to monitor the number of transmission errors encountered.

3.4.3 Error-Free Soft-Decision Viterbi Decoding

Having considered a number of hard-decision decoding scenarios, let us now demonstrate the added power of soft-decision decoding. Recall from our earlier discussions that if the demodulator refrains from making a binary decision and instead it outputs a finely graded *soft-decision confidence measure* related to the probability of a binary one and a binary zero, respectively, then it is said to invoke soft-decision demodulation. As an example, we may invoke an eight-level soft-decision demodulator output. This provides a more accurate indication, whether the demodulator's decision concerning a certain demodulated bit is a high- or low-reliability one. This clearly supplies the Viterbi decoder with substantially more information than the previous binary zero/one decisions. Hence, a higher error correction capability will be achieved, as we will demonstrate in Figure 3.7.

Specifically, let us assume that on our eight-level confidence scale +4 indicates the highest possible confidence concerning the demodulator's decision for a binary one and -4 the lowest possible confidence. In fact, if the demodulator outputs -4, the low confidence in a logical one implies a high confidence in a logical zero, and conversely, the demodulator output of +4 implies a very low probability of a binary zero. Bearing this eight-level confidence scale of [-4, -3 ... +3, +4] in mind, the previously erroneously decoded double-error scenario of Figure 3.5 can now be revisited in Figure 3.7, where we will demonstrate that the more powerful soft-decision decoding allows us to remove all transmission errors.

Let us first consider the soft-decision metrics provided by the demodulator, which now replaces the previously used hard-decision values at the bottom of Figure 3.5, appearing crossed out in Figure 3.7. For example, the first two values appear to be a high-confidence one and zero, respectively. The second two values are relatively high-confidence zeros, whereas the previously erroneously demodulated third symbol, namely 01, is now represented by the confidence values of -2,+1, indicating that these bits may well have been either one or zero. The rest of the soft-decision metrics appear to be of higher value, apart from the last-but-one.

The computation of the branch metrics and path metrics now has to be slightly modified. Previously, we were accumulating only "penalties" in terms of the Hamming distances encountered. By contrast, in soft-decision decoding we will have to accumulate both penalties and credits, since we now consider the possibility of all demodulated values being both a binary one and a zero and quantify their associated probabilities using the soft-decision (SD) metrics. Explicitly, in Figure 3.7 we replace the crossed-out hard-decision metrics by the corresponding soft-decision metrics.

Considering trellis stage one and the 00 encoded symbol, the first SD metric of +3 does not tally well with the bit zero; rather, it indicates a strong probability of a one; hence, we accumulate a "penalty" of -3. The second SD metric, however, indicates a strong probability of a zero, earning a credit of +3, which cancels the previous

3.4. THE VITERBI ALGORITHM

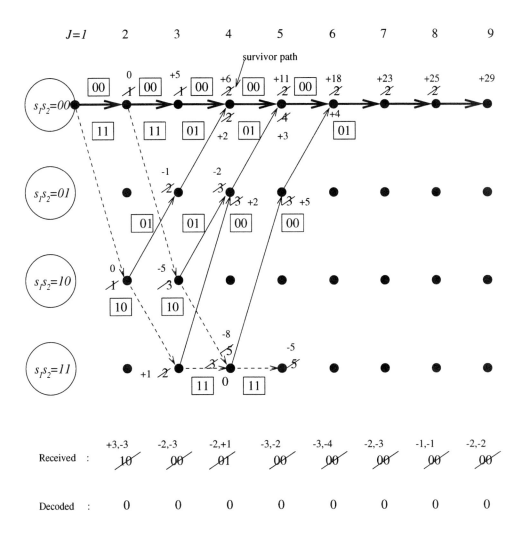

Figure 3.7: Trellis diagram based Viterbi decoding of the $CC(2,1,2)$ systematic code, where broken lines indicate transitions due to an input one, while continuous lines correspond to input zeros — **Error-free soft-decision decoding of two isolated bit errors.**

-3 penalty, yielding a SD branch metric of 0. Similar arguments are valid for the trellis branch from (0,0) to (1,0), which is associated with the encoded symbol 11, also yielding a SD branch metric of 0. During stage two, the received SD metrics of -2,-3 suggest a high probability of two zeros, earning an added credit of +5 along the branch (0,0) to (0,0). By contrast, these SD values do not tally well with the encoded symbol of 11 along the transition of (0,0) to (1,0), yielding a penalty of -5. During the third stage along the path from (0,0) to (0,0), we encounter a penalty of -1 and a credit of +2, bringing the total credits for the all-zero path to +6.

At this stage of the hard-decision decoder of Figure 3.7, we encountered a random decision, which now will not be necessary, since the merging path has a lower credit of +2. Clearly, at trellis stage nine we have a total credit of +29, allowing the decoder to release the correct original all-zero information sequence.

3.5 Summary and Conclusions

This brief chapter commenced by providing a historical perspective on convolution coding. The convolutional encoder has been characterized with the aid of its state transition diagram and trellis diagram. Then the classic Virebi algorithm has been introduced in the context of a simple decoding example, considering both hard- and soft-decision based scenarios. In the next chapter, we focus our attention on the family of block codes, which have found numerous applications in both standard and proprietary wireless communications systems. A variety of video schemes characterized in this monograph have also opted for employing Bose-Chaudhuri-Hocquenghem codes.

Chapter 4

Block-Based Channel Coding

4.1 Introduction

Having studied the properties of various communications channels, we now focus our attention on methods of combating transmission errors caused by channel impairments. After demodulating the received signal, transmission errors can be removed from the digital information if their number does not exceed the error correcting power of the error correction code used. The employment of forward error correction (FEC) coding techniques becomes important for transmissions over hostile mobile radio channels, where the violent channel fading precipitates bursts of error. This is particularly true when using vulnerable multilevel modulation schemes, such as Quadrature Amplitude Modulation (QAM). This chapter addresses issues of FEC coding in order to provide a self-contained reference for readers who want to delve into the theory of FEC coding. Those readers who are prepared to consider an FEC codec as a "black box" removing a given number of transmission errors at the cost of an increased transmission bit rate, bandwidth, transmitted power, and complexity can proceed to consecutive chapters.

The theory and practice of FEC coding has been well documented in classic references [1, 3–7, 74–76]; hence for an in-depth treatment, the reader is referred to these sources. Both convolutional and block FEC codes have been successfully utilized in various communications systems [8, 9]. In this brief overview, to a certain extent we will follow the philosophy of these references, which are also recommended for a more detailed discourse on the subject. Since the applications described in this book we often used the family of Reed-Solomon (RS) and Bose-Chaudhuri-Hocquenghem (BCH) block codes, in this chapter we concentrate on their characterization. We will draw the reader's attention to any differences between them, as they arise during our discussions. Both RS and BCH codes are defined over the mathematical structure of finite fields; therefore we briefly consider their construction. For a more detailed discourse on finite fields, the reader is referred to Lidl's work [77].

4.2 Finite Fields

4.2.1 Definitions

Loosely speaking, an *algebraic field* is any arithmetic structure in which addition, subtraction, multiplication, and division are defined and the associative, distributive, and commutative laws apply. In conventional algebraic fields, like the rational, real, or complex field, these operations are trivial, but in other fields they are somewhat different, such as *modulo* or *modulo polynomial* operations.

A more formal field definition can be formulated as follows [3]. An *algebraic field* is a set of elements that has addition and multiplication defined over it, such that the set is closed under both addition and multiplication; that is, the results of these operations are also elements in the same set. Furthermore, both addition and multiplication are associative and commutative. There is a zero element, such that $a + 0 = a$ and an additive inverse element, such that $a + (-a) = 0$. Subtraction is defined as $a - b = a + (-b)$. There is a one element, such that $1 \cdot a = a$ and a multiplicative inverse element for which we have $a \cdot (a^{-1}) = 1$. Division is defined as $a/b = a \cdot (b^{-1})$.

Digital signal processing (DSP) has been historically studied in the real algebraic field constituted by all finite and infinite decimals, or in the complex algebraic field with elements of the form $a + jb$, where a and b are real. In both the real and complex fields, the operations are fairly straightforward and well known, and the number of elements is infinite; consequently, they are *infinite algebraic fields*.

Finite algebraic fields or shortly finite fields are constituted by a finite number q of elements if there exists a field with q elements. They are also referred to as *Galois Fields GF(q)*. Every field must have a zero and a one element. Consequently, the smallest field is $GF(2)$ with mod2 addition and multiplication, because the field must be closed under the operations. That is, their result must be a field element, too. If the number of elements in a field is a prime p, the $GF(p)$ constructed from the elements $\{0, 1, 2, ...(p-1)\}$ is called a *prime field* with *modulo p* operations.

> **Example 1**: Galois prime-field modulo operation.
>
> As an illustrative example, let us briefly consider the basic operations, namely, addition and multiplication over the prime field $GF(7)$. The prime field $GF(7)$ is constituted by the elements $\{0,1,2,3,4,5,6\}$, and the corresponding *modulo7* addition and multiplication tables are summarized in Table 4.1.

Since every field must contain the unity element, we can define the *order n of a field element* α, such that $\alpha^n = 1$. In other words, to determine the order of a field element the exponent n yielding $\alpha^n = 1$ has to be found. Every $GF(p)$ has at least one element α of order $(p-1)$, which is called a *primitive element* and which always exists in a $GF(p)$ [1]. Hence, $\alpha^{(p-1)} = 1$ must apply. For practical FEC applications we must have sufficiently large finite fields, constituted by many field elements, since only this finite number of elements can be used in order to represent our codewords.

> **Example 2**: Extension field construction example.

4.2. FINITE FIELDS

+	0	1	2	3	4	5	6
0	0	1	2	3	4	5	6
1	1	2	3	4	5	6	0
2	2	3	4	5	6	0	1
3	3	4	5	6	0	1	2
4	4	5	6	0	1	2	3
5	5	6	0	1	2	3	4
6	6	0	1	2	3	4	5

×	0	1	2	3	4	5	6
0	0	0	0	0	0	0	0
1	0	1	2	3	4	5	6
2	0	2	4	6	1	3	5
3	0	3	6	2	5	1	4
4	0	4	1	5	2	6	3
5	0	5	3	1	6	4	2
6	0	6	5	4	3	2	1

Table 4.1: *Modulo* 7 Addition and Multiplication Tables for the Prime Field $GF(7)$

In order to augment the concept of *extension fields*, let us consider first the simple example of extending the real field to the complex field. Thus, we attempt to find a second-order polynomial over the real field that cannot be factorized over it, which is also referred to in finite field parlance as an *irreducible polynomial*. For example, it is well understood that the polynomial $x^2 + 1$ cannot be factorized over the real field because it has no real zeros. However, one can define a special zero j of the polynomial $x^2 + 1$, which we refer to as an imaginary number, for which $j^2 = -1$ and $x^2 + 1 = (x+j)(x-j)$. This way the real number system has been extended to the complex field, which has complex elements $\alpha_1 + j\alpha_2$, where α_1 and α_2 are real. The new number system derived from the real field fulfills all the criteria necessary for an extension field.

The elements $\alpha_1 + j\alpha_2$ of the complex field can also be interpreted as first-order polynomials $\alpha_1 x^0 + \alpha_2 x^1$ having real coefficients, if we replace j by x. In this context, multiplication of the field elements $(\alpha_1 + \alpha_2 x)$ and $(\alpha_3 + \alpha_4 x)$ is equivalent to modulo polynomial multiplication of the field elements:

$$\begin{align}
\alpha_5 + \alpha_6 x &= (\alpha_1 + \alpha_2 x)(\alpha_3 + \alpha_4 x) \pmod{x^2 + 1} \tag{4.1} \\
&= \alpha_1\alpha_3 + \alpha_2\alpha_3 x + \alpha_1\alpha_4 x + \alpha_2\alpha_4 x^2 \pmod{x^2 + 1} \\
&= \alpha_1\alpha_3 + (\alpha_2\alpha_3 + \alpha_1\alpha_4)x + \alpha_2\alpha_4 x^2 \pmod{x^2 + 1},
\end{align}$$

where $\alpha_1 \ldots \alpha_6$ are arbitrary elements of the original real field. But since $x^2 \pmod{x^2+1} = \text{Remainder}\{x^2 : (x^2+1)\} = -1$, this polynomial formulation delivers identical results to those given by the complex number formulation, where $j^2 = 1$. So we may write:

$$\begin{align}
\alpha_5 + \alpha_6 x &= (\alpha_1\alpha_3 - \alpha_2\alpha_4) + (\alpha_2\alpha_3 + \alpha_1\alpha_4)x \tag{4.2} \\
\alpha_5 &= \alpha_1\alpha_3 - \alpha_2\alpha_4 \tag{4.3} \\
\alpha_6 &= \alpha_2\alpha_3 + \alpha_1\alpha_4 . \tag{4.4}
\end{align}$$

Pursuing this polynomial representation, we find that the general framework for extending the finite prime field $GF(p)$ is to have m components rather than just two.

Explicitly, a convenient way of generating large algebraic fields is to extend the prime field $GF(p)$ to a so-called extension field $GF(p^m)$. In general, the elements of the extension field $GF(p^m)$ are all the possible m dimensional vectors, where all

m vector coordinates are elements of the original prime field $GF(p)$, and m is an integer. For example, an extension field $GF(p^m)$ of the original prime field $GF(p)$ with $p = 2$ elements $\{0, 1\}$ contains all the possible combinations of m number of $GF(p)$ elements, which simply means that there are p^m number of extension field elements. Consequently, an element of the extension field $GF(p^m)$ can be written as a polynomial of order $m - 1$, with coefficients from $GF(p)$:

$$GF(p^m) = \{(a_0 x^0 + a_1 x^1 + a_2 x^2 + \ldots + a_{m-1} x^{m-1})\}$$

with

$$\{a_0, a_1 \ldots a_{m-1}\} \in \{0, 1 \ldots (p-1)\}.$$

The operations in the extension field $GF(p^m)$ are *modulo polynomial* operations rather than conventional modulo operations. Hence, the addition is carried out as the addition of two polynomials:

$$\begin{align}
c(x) &= a(x) + b(x) \tag{4.5} \\
&= a_0 + a_1 x + a_2 x^2 + \ldots + a_{m-1} x^{m-1} + \tag{4.6} \\
&\quad + b_0 + b_1 x + b_2 x^2 + \ldots + b_{m-1} x^{m-1} \tag{4.7} \\
&= (a_0 + b_0) + (a_1 + b_1) x^1 + (a_2 + b_2) x^2 + \ldots \tag{4.8} \\
&\quad + (a_{m-1} + b_{m-1}) x^{m-1}. \tag{4.9}
\end{align}$$

Since

$$\{a_0, a_1, \ldots a_{m-1}\} \in \{0, 1, \ldots (p-1)\}$$

and

$$\{b_0, b_1, \ldots b_{m-1}\} \in \{0, 1, \ldots (p-1)\},$$

the component-wise addition of the polynomial coefficients must be mod p addition, so that the coefficients of the result polynomial are also elements of $GF(p)$.

The definition of the modulo polynomial multiplication is somewhat more complicated. First, one has to find an irreducible polynomial $p(x)$ of degree m, which cannot be factorized over $GF(p)$. In other words, $p(x)$ must be a polynomial with coefficients from $GF(p)$, and it must not have any zeros; that is, it cannot be factorized into polynomials of lower order with coefficients from $GF(p)$. Recall, for example, that the polynomial $x^2 + 1$ was irreducible over the real field. More formally, the polynomial $p(x)$ is *irreducible* if and only if it is divisible exclusively by $\alpha.p(x)$ or by α, where α is an arbitrary field element in $GF(p)$. Whenever the highest order coefficient of an irreducible polynomial is equal to one, the polynomial is called a *prime polynomial*.

Once we have found an appropriate prime polynomial $p(x)$, the multiplication of two extension field elements over the extension field is modulo polynomial multiplication of the polynomial representations of the field elements. Explicitly, the product of the two extension field elements given by their polynomial representations must be divided by $p(x)$, and the remainder must be retained as the final result. This is to ensure that the result is also an element of the extension field (i.e., a polynomial with an order less than p). Then the extension field is declared closed under the operations, yielding results that are elements of the original field. Accordingly, the nature of the extension field depends on the choice of the prime polynomial $p(x)$.

4.2. FINITE FIELDS

Sometimes there exist several prime polynomials, and some of them result in somewhat more advantageous extension field construction than others. Sometimes there is no prime polynomial of a given degree m over a given prime field $GF(p)$; consequently, $GF(p)$ cannot be extended to $GF(p^m)$. At this stage, we note that we have constructed a mathematical environment necessary for representing our information carrying signals, (i.e., codewords), and defined the operations over the extension field.

Since the source data to be encoded are usually binary, we represent the nonbinary information symbols of the extension field as a sequence of m bits. Logically, our prime field is the binary field $GF(2)$ with elements $\{0,1\}$ and $p = 2$. Then we extend $GF(2)$ to $GF(2^m)$ in order to generate a sufficiently large working field for signal processing, where an m-bit nonbinary information symbol constitutes a field element of $GF(2^m)$. If, for example, an information symbol is constituted by one byte of information, (i.e., $m = 8$), the extension field $GF(2^m) = GF(256)$ contains 256 different field elements. The appropriate prime polynomial can be chosen, for example, from Table 4.7 on p. 409 of reference [9], where primitive polynomials of degree less than 25 over $GF(2)$ are listed. These prime polynomials allow us to construct any arbitrary extension field from $GF(2^2)$ to $GF(2^{25})$.

It is useful for our further discourse to represent the field elements of $GF(q = 2^m)$ each as a unique power of an element α, which we refer to as the *primitive element*. The primitive element α was earlier defined to be the one that fulfills the condition $\alpha^{q-1} = \alpha^{2^m-1} = 1$. With this notation the elements of the extension field $GF(q = 2^m)$ can be written as $\{0, 1, \alpha, \alpha^2, \alpha^3 \ldots \alpha^{q-2}\}$, and their polynomial representation is given by the remainder of x^n upon division by the prime polynomial $p(x)$:

$$\alpha^n = Remainder\ \{x^n/p(x)\}. \qquad (4.10)$$

This relationship will become clearer in the practical example of the next subsection.

4.2.2 Galois Field Construction

Example 3: GF(16) construction example.

As an example, let us consider the construction of $GF(2^4) = GF(16)$ based on the prime polynomial $p(x) = x^4 + x + 1$ [9], p. 409, where $m = 4$. Each extension field element α can be represented as a polynomial of degree three with coefficients over $GF(2)$. Equivalently, each extension field element can be described by the help of the binary representation of the coefficients of the polynomial or by means of the decimal value of the binary representation.

Let us proceed with the derivation of all the representations mentioned above, which are summarized in Table 4.2. Since every field must contain the zero and one element, we have:

$$0 = 0$$
$$\alpha^0 = 1.$$

Further plausible assignments are:

$$\alpha^1 = x$$

Exponential Represent.	Polynomial Represent.	Binary Represent.	Decimal Represent.
0	0	0000	0
α^0	1	0001	1
α^1	x	0010	2
α^2	x^2	0100	4
α^3	x^3	1000	8
α^4	$x+1$	0011	3
α^5	x^2+x	0110	6
α^6	x^3+x^2	1100	12
α^7	x^3+x+1	1011	11
α^8	x^2+1	0101	5
α^9	x^3+x	1010	10
α^{10}	x^2+x+1	0111	7
α^{11}	x^3+x^2+x	1110	14
α^{12}	x^3+x^2+x+1	1111	15
α^{13}	x^3+x^2+1	1101	13
α^{14}	x^3+1	1001	9

Table 4.2: Different Representations of $GF(16)$ Elements ($\alpha^{15} \equiv \alpha^0 \equiv 1$) Generated Using the Prime Polynomial $p(x) = x^4 + x + 1$.

$$\alpha^2 = x^2$$
$$\alpha^3 = x^3,$$

because the remainders of x, x^2 and x^3 upon division by the primitive polynomial $p(x) = x^4 + x + 1$ are themselves. However, the polynomial representation of x^4 cannot be derived without polynomial division:

$$\alpha^4 = Rem\{x^4 : (x^4+x+1)\}_{p(x)=x^4+x+1}$$

The polynomial division:	$x^4 : (x^4+x+1) = 1$	(4.11)
	$-(x^4+x+1)$	(4.12)
The remainder:	$-(x+1)$	(4.13)

Hence we have:
$$\alpha^4 = x+1.$$

Following these steps the first two columns of Table 4.2 can easily be filled in. The binary notation simply comprises the coefficients of the polynomial representation, whereas the decimal notation is the decimal number computed from the binary representation.

4.2. FINITE FIELDS

4.2.3 Galois Field Arithmetic

Multiplication in the extension field is carried out by multiplying the polynomial representations of the elements, dividing the result by the prime polynomial $p(x) = x^4 + x + 1$, and finally taking the remainder. For example, for the field elements:

$$\alpha^4 \equiv x + 1$$
$$\alpha^7 \equiv x^3 + x + 1$$

we have:

$$\begin{aligned}
\alpha^4 \cdot \alpha^7 &= (x+1)(x^3+x+1) \bmod_{p(x)=x^4+x+1} \\
&= \text{Rem}\{(x+1)(x^3+x+1) : (x^4+x+1)\} \\
&= \text{Rem}\{(x^4+x^2+x+x^3+x+1) : (x^4+x+1)\} \\
&= \text{Rem}\{(x^4+x^3+x^2+1) : (x^4+x+1)\}.
\end{aligned}$$

Then the polynomial division is computed as follows:

$$(x^4 + x^3 + x^2 + 0 + 1) : (x^4 + x + 1) = 1$$
$$\underline{-(x^4 + x + 1)}$$

The remainder: $\quad x^3 + x^2 + x + 0 = x^3 + x^2 + x.$

Therefore, the required product is given by:

$$\alpha^4 \cdot \alpha^7 = x^3 + x^2 + x \equiv \alpha^{11}.$$

From Table 4.2 it is readily recognized that the exponential representation of the field elements allows us to simply add the exponents without referring back to their polynomial representations. Whenever the exponent happens to be larger than $q - 1 = 15$, it is simply collapsed back into the finite field by taking its value modulo $(q-1)$, that is, modulo 15. For example:

$$\alpha^{12} \cdot \alpha^6 = \alpha^{18} = \alpha^{15} \cdot \alpha^3 = 1 \cdot \alpha^3 = \alpha^3.$$

The addition of two field elements is also carried out easily by referring to the polynomial representation by the help of a component-wise addition, as follows:

$$\begin{aligned}
\alpha^{11} + \alpha^7 &\equiv (x^3 + x^2 + x + 0) \\
&+ \underline{(x^3 + 0 + x + 1)} \\
& 0 + x^2 + 0 + 1 = x^2 + 1 \equiv \alpha^8.
\end{aligned}$$

This is also equivalent to the modulo-2 addition of the binary representations:

$$\begin{aligned}
\alpha^{11} + \alpha^7 &\equiv 1110 \\
&+ \underline{1011} \\
& 0101 \equiv x^2 + 1 \equiv \alpha^8.
\end{aligned}$$

The fastest way to compute $GF(16)$ addition of exponentially represented field elements is to use the precomputed $GF(16)$ addition table, namely, Table 4.3. With our background in finite fields, we can now proceed to define and characterize the important so-called maximum-minimum distance family — a term to be clarified during our later discussion — of nonbinary block codes referred to as Reed-Solomon (RS) codes and their binary subclass, Bose-Chaudhuri-Hocquenghem (BCH) codes, which are often used in our prototype systems.

4.3 Reed-Solomon and Bose-Chaudhuri-Hocquenghem Block Codes

4.3.1 Definitions

As mentioned above, Reed-Solomon (RS) codes represent a nonbinary subclass of (BCH) multiple error correcting codes. Because of their nonbinary nature, RS codes pack the information represented by m consecutive bits into nonbinary symbols, which are elements in the extension field $GF(2^m)$ of $GF(2)$. In general, an RS code is defined as a block of n nonbinary symbols over $GF(2^m)$, constructed from k original information symbols by adding $n - k = 2t$ number of redundancy symbols from the same extension field, giving $n = k + 2t$. This code is often written as $RS(n, k, t)$ over $GF(2^m)$.

RS codes are cyclic codes, implying that any cyclically shifted version of a codeword is also a codeword of the same code. A further property of cyclic codes is that all of the codewords can be generated as a linear combination of any codeword and its cyclically shifted versions. The distance properties of a code are crucial in assessing its error correction capability. The *minimum distance* of a code is the minimum number of positions in which two arbitrary codewords differ. If we define the weight of a codeword as the number of nonzero symbol positions, then the minimum distance of a code is actually the weight of its minimum weight codeword, since the "all zero" word is always a codeword in a linear code. It is plausible that if in an (n, k) code less than half the minimum distance number of symbols are in error, it can be uniquely recognized, which k symbol's long information message has been sent. This is because the received message is still closer to the transmitted one than to any other one. On the other hand, if more than half the minimum distance number of symbols are in error, the decoder decides erroneously on the basis of the nearest legitimate codeword.

The *Singleton bound* imposes an upper limit on the maximum achievable minimum distance of a code on the following basis. If in a codeword one information symbol is changed, the highest possible distance for the newly computed codeword from the original one will be $d = (n - k) + 1$, provided that all the $n - k$ parity symbols also changed as a result, which is an extreme assumption. Consequently, for the code's minimum distance we have $d_{min} \leq (n-k)+1$. RS codes are one of the very few codes, which reach the maximum possible minimum distance of $d = (n-k)+1$. This is why they are referred to as a maximum-minimum distance code. In general, an $RS(n, k, t)$ code can correct up to $t = (n - k)/2$ symbol errors, or in other words a t error correcting code must contain $2t = (n - k)$ number of redundancy symbols. Therefore,

4.3. RS AND BCH CODES

Expon. Repr.	Binary Repr.	Hexad. Repr.	α^0 / 0001 / 1	α^1 / 0010 / 2	α^4 / 0011 / 3	α^2 / 0100 / 4	α^8 / 0101 / 5	α^5 / 0110 / 6	α^{10} / 0111 / 7	α^3 / 1000 / 8	α^{14} / 1001 / 9	α^9 / 1010 / A	α^7 / 1011 / B	α^6 / 1100 / C	α^{13} / 1101 / D	α^{11} / 1110 / E	α^{12} / 1111 / F
α^0	0001	1	0	α^4	α^1	α^8	α^2	α^{10}	α^5	α^{14}	α^3	α^7	α^9	α^{13}	α^6	α^{12}	α^{11}
α^1	0010	2	α^4	0	α^0	α^5	α^{10}	α^2	α^8	α^9	α^7	α^3	α^{14}	α^{11}	α^{12}	α^6	α^{13}
α^4	0011	3	α^1	α^0	0	α^{10}	α^5	α^8	α^2	α^7	α^9	α^{14}	α^3	α^{12}	α^{11}	α^{13}	α^6
α^2	0100	4	α^8	α^5	α^{10}	0	α^0	α^1	α^4	α^6	α^{13}	α^{11}	α^{12}	α^3	α^{14}	α^9	α^7
α^8	0101	5	α^2	α^{10}	α^5	α^0	0	α^4	α^1	α^{13}	α^6	α^{12}	α^{11}	α^{14}	α^3	α^7	α^9
α^5	0110	6	α^{10}	α^2	α^8	α^1	α^4	0	α^0	α^{11}	α^{12}	α^6	α^{13}	α^9	α^7	α^3	α^{14}
α^{10}	0111	7	α^5	α^8	α^2	α^4	α^1	α^0	0	α^{12}	α^{11}	α^{13}	α^6	α^7	α^9	α^{14}	α^3
α^3	1000	8	α^{14}	α^9	α^7	α^6	α^{13}	α^{11}	α^{12}	0	α^0	α^1	α^4	α^2	α^8	α^5	α^{10}
α^{14}	1001	9	α^3	α^7	α^9	α^{13}	α^6	α^{12}	α^{11}	α^0	0	α^4	α^1	α^8	α^2	α^{10}	α^5
α^9	1010	A	α^7	α^3	α^{14}	α^{11}	α^{12}	α^6	α^{13}	α^1	α^4	0	α^0	α^5	α^{10}	α^2	α^8
α^7	1011	B	α^9	α^{14}	α^3	α^{12}	α^{11}	α^{13}	α^6	α^4	α^1	α^0	0	α^{10}	α^5	α^8	α^2
α^6	1100	C	α^{13}	α^{11}	α^{12}	α^3	α^{14}	α^9	α^7	α^2	α^8	α^5	α^{10}	0	α^0	α^1	α^4
α^{13}	1101	D	α^6	α^{12}	α^{11}	α^{14}	α^3	α^7	α^9	α^8	α^2	α^{10}	α^5	α^0	0	α^4	α^1
α^{11}	1110	E	α^{12}	α^6	α^{13}	α^9	α^7	α^3	α^{14}	α^5	α^{10}	α^2	α^8	α^1	α^4	0	α^0
α^{12}	1111	F	α^{11}	α^{13}	α^6	α^7	α^9	α^{14}	α^3	α^{10}	α^5	α^8	α^2	α^4	α^1	α^0	0

Table 4.3: GF(16) Addition Table, ©Wong 1989 [8], 1999 Wong, Hanzo [9]

the minimum distance between codewords must be $d_{min} \geq 2t + 1$. Substituting $(n - k) = 2t$ into $d_{min} \geq 2t + 1$ and taking into account that previously we showed that $d_{min} \leq (n - k) + 1$, we find that the minimum distance of RS codes is exactly $d_{min} = 2t + 1$.

Before we proceed to define the RS encoding rules, the *generator polynomial* has to be introduced, which is defined to be the polynomial of order $2t$, which has its zeros at any $2t$ out of the 2^m possible field elements. For the sake of simplicity, but without any loss of generality, we always use the GF elements $\alpha^1, \alpha^2, \ldots \alpha^{2t}$ in order to determine the generator polynomial of an RS code:

$$g(x) = (x - \alpha)(x - \alpha^2) \ldots (x - \alpha^{2t}) = \prod_{j=1}^{2t}(x - \alpha^j) = \sum_{j=0}^{2t} g_j x^j. \tag{4.14}$$

In the next subsection, we highlight two algorithms using the above-mentioned generator polynomial for the encoding of the information symbols in order to generate RS-coded codewords.

4.3.2 RS Encoding

Since RS codes are cyclic, any cyclic encoding method can be used for their coding. When opting for the *nonsystematic encoding* rule, the information symbols are not explicitly recognizable in the encoded codeword — hence the terminology. Nonsystematic cyclic encoders generate the encoded word by multiplying the information polynomial $i(x)$ by the generator polynomial $g(x)$ using modulo plynomial algebra as follows:

$$c(x) = i(x) \cdot g(x) \tag{4.15}$$

$$i(x) = i_1 x + i_2 x^2 + \ldots + i_k x^k = \sum_{j=1}^{k} i_j x^j \tag{4.16}$$

$$c(x) = c_1 x + c_2 x^2 + \ldots + c_n x^n = \sum_{j=1}^{n} c_j x^j \tag{4.17}$$

where the information polynomial $i(x)$ is of order k and the encoded codeword polynomial $c(x)$ is of order $n = k + 2t$. The coefficients of the polynomials are the nonbinary information-carrying symbols, which are elements in $GF(2^m)$, and the powers of x can be thought of as place markers for the symbols in the codeword. Again, since the polynomial $c(x)$ does not contain explicitly the original k information symbols, this is called a nonsystematic encoder. The codeword is then modulated and sent over a nonideal transmission medium, where the code symbols or polynomial coefficients may become corrupted.

The channel effects can be modeled by the help of an additive error polynomial $e(x)$ of order n as follows:

$$\begin{aligned} r(x) = c(x) + e(x) &= c_1 x + c_2 x^2 + \ldots + c_n x^n \\ &+ e_1 x + e_2 x^2 + \ldots + e_n x^n, \end{aligned} \tag{4.18}$$

4.3. RS AND BCH CODES

where $r(x)$ is the corrupted received polynomial. The decoder has to determine up to t error positions and compute t error locations. In other words, the decoder has to compute $t+t = 2t$ unknowns from the $2t$ redundancy symbols and correct the errors to produce the error-free codeword $c(x)$. After error correction, the information symbols must be recovered by the help of the inverse operation of the encoding, using the following simple decoding rule:

$$i(x) = \frac{c(x)}{g(x)} \ . \tag{4.19}$$

However, if there are more than t transmission errors in the received polynomial $r(x)$, the decoder fails to produce the error-free codeword polynomial $c(x)$. This is plausible, since due to its excessive corruption the received codeword will become more similar to another legitimate codeword rather than the transmitted one. In algebraic terms, this is equivalent to saying that we cannot determine more than t unknown error positions and error values, when using $2 \times t$ redundancy symbols. We note that in case of binary BCH codes it is plausible that having determined the error positions, the corresponding bits are simply inverted in order to correct them, while for nonbinary RS codes a more complicated procedure will have to be employed. These statements will be made more explicit during our further discourse.

The *systematic RS encoding* rule is somewhat more sophisticated than the nonsystematic one, but the original information symbols are simply copied into the encoded word. This property is often attractive, because in some applications it is advantageous to know where the original information symbols reside in the encoded block. Explicitly, when a received codeword is deemed to be overwhelmed by transmission errors, a nonsystematic RS or BCH code has no other option than to attempt to compute the error locations and "magnitudes" for their correction, but this operation will be flawed by the plethora of transmission errors, and hence the decoding operation may actually corrupt more received symbols by carrying out a flawed decoding operation.

By contrast, in systematic RS or BCH codecs, instead of erroneously computing the error locations and "magnitudes" for their correction, this "code overload" condition can be detected. Hence, this flawed action can be avoided by simply extracting the k original information symbols from $c(x)$. Because of these differences, nonsystematic codes usually result in a bit error rate (BER) increase under hostile channel conditions, while powerful systematic codes can maintain a similar BER performance to the uncoded case under similar circumstances.

Again, in systematic RS codes the first k encoded symbols are chosen to be the original k information symbols. We simply multiply the information polynomial $i(x)$ by x^{n-k} in order to shift it into the highest order position of the codeword $c(x)$. Then we choose the parity symbols constituted by the polynomial $p(x)$ according to the systematic encoding rule in order to result in a legitimate codeword. Legitimate in this sense means that the remainder of the encoded word $c(x)$ upon division by the generator polynomial $g(x)$ is zero. Using the codeword $c(x)$ hosting the shifted information word $i(x)$ plus the parity segment $p(x)$, we have:

$$c(x) = x^{(n-k)} \cdot i(x) + p(x), \tag{4.20}$$

and according to the above definition of $c(x)$, we have:

$$Rem\left\{\frac{c(x)}{g(x)}\right\} = 0 \qquad (4.21)$$

$$Rem\left\{\frac{x^{(n-k)} \cdot i(x) + p(x)}{g(x)}\right\} = 0 \qquad (4.22)$$

$$Rem\left\{\frac{x^{(n-k)} \cdot i(x)}{g(x)}\right\} + Rem\left\{\frac{p(x)}{g(x)}\right\} = 0. \qquad (4.23)$$

Since the order of the parity polynomial $p(x)$ is less than $(n-k)$ and the order of $g(x)$ is $(n-k)$, we have:

$$Rem\left\{\frac{p(x)}{g(x)}\right\} = p(x). \qquad (4.24)$$

By substituting Equation 4.24 into Equation 4.23 and rearranging it, we get:

$$-Rem\left\{\frac{x^{(n-k)} \cdot i(x)}{g(x)}\right\} - p(x). \qquad (4.25)$$

Hence, if we substitute Equation 4.25 into Equation 4.20 and take into account that over GFs addition and subtraction are the same, the systematic encoding rule is as follows:

$$c(x) = x^{(n-k)} \cdot i(x) + Rem\left\{\frac{x^{(n-k)} \cdot i(x)}{g(x)}\right\}. \qquad (4.26)$$

The error correction ensues in a completely equivalent manner to that of the nonsystematic decoder, which will be the subject of our later discussion, but recovering the information symbols from the corrected received codeword is simpler. Namely, the first k information symbols of an n-symbol long codeword are retained as corrected decoded symbols. Let us now revise these definitions and basic operations with reference to the following example.

4.3.3 RS Encoding Example

Example 4:
Systematic and nonsystematic $RS(12,8,2)$ encoding example.

Consider a low-complexity double-error correcting RS code, namely, the RS(12,8,2) code over GF(16), and demonstrate the operation of both the systematic and nonsystematic encoder. We begin our example with the determination of the generator polynomial $g(x)$, which is a polynomial of order $2t = 4$, having zeros at the first four elements of GF(16), namely, at $\alpha^1, \alpha^2, \alpha^3,$ and α^4. Recall, however, that we could have opted for the last four GF(16) elements or any other four elements of it. Remember furthermore that multiplication is based on adding the exponents of the exponential representations, while addition is conveniently carried out using Table 4.3. Then the generator polynomial is arrived at as follows:

$$g(x) = (x - \alpha^1)(x - \alpha^2)(x - \alpha^3)(x - \alpha^4) \qquad (4.27)$$

4.3. RS AND BCH CODES 119

$$\begin{aligned}
&= (x^2 - \alpha^1 x - \alpha^2 x + \alpha^1 \alpha^2)(x^2 - \alpha^3 x - \alpha^4 x + \alpha^3 \alpha^4) \\
&= (x^2 - \underbrace{(\alpha^1 + \alpha^2)}_{\alpha^5} x + \alpha^3)(x^2 - \underbrace{(\alpha^3 + \alpha^4)}_{\alpha^7} x + \alpha^7) \\
&= (x^2 + \alpha^5 x + \alpha^3)(x^2 + \alpha^7 x + \alpha^7) \\
&= x^4 + \alpha^7 x^3 + \alpha^7 x^2 + \alpha^5 x^3 + \alpha^{12} x^2 + \alpha^{12} x + \alpha^3 x^2 + \alpha^{10} x + \alpha^{10} \\
&= x^4 + \underbrace{(\alpha^7 + \alpha^5)}_{\alpha^{13}} x^3 + \underbrace{(\alpha^7 + \alpha^{12} + \alpha^3)}_{\alpha^6} x^2 + \underbrace{(\alpha^{12} + \alpha^{10})}_{\alpha^3} x + \alpha^{10} \\
&= x^4 + \alpha^{13} . x^3 + \alpha^6 . x^2 + \alpha^3 . x + \alpha^{10}.
\end{aligned}$$

Let us now compute the codeword polynomial $c(x)$ for the $RS(12, 8, 2)$ double-error correcting code, when the "all one" information polynomial $i(x) = 11\ldots11$ is to be encoded. In hexadecimal format, this can be expressed as $i(x) = FFFF\,FFFF\#H$. Since $GF(2^m) = GF(16)$ for $m = 4$, $8.4 = 32$ bits; that is, 8 hexadecimal symbols must be encoded into 12 hexadecimal symbols. The exponential representation of the $1111 = F\#H$ hexadecimal symbol is α^{12}, as seen in Table 4.2. Hence, the seventh-order information polynomial is given by:

$$i(x) = \alpha^{12}\,x^7 + \alpha^{12}\,x^6 + \alpha^{12}\,x^5 + \alpha^{12}\,x^4 + \alpha^{12}\,x^3 + \alpha^{12}\,x^2 + \alpha^{12}\,x + \alpha^{12}.$$

Then the nonsystematic encoder simply computes the product of the information polynomial and the generator polynomial, as shown below:

$$\begin{aligned}
c(x) &= g(x)i(x) \quad &(4.28)\\
c(x) &= (x^4 + \alpha^{13} x^3 + \alpha^6 x^2 + \alpha^3 x + \alpha^{10}) \\
&\quad \cdot (\alpha^{12} x^7 + \alpha^{12} x^6 + \alpha^{12} x^5 + \alpha^{12} x^4 \\
&\qquad + \alpha^{12} x^3 + \alpha^{12} x^2 + \alpha^{12} x + \alpha^{12}) \\
&= \alpha^{12} x^{11} + \alpha^{12} x^{10} + \alpha^{12} x^9 + \alpha^{12} x^8 + \alpha^{12} x^7 + \alpha^{12} x^6 + \alpha^{12} x^5 \\
&+ \alpha^{12} x^4 + \alpha^{10} x^{10} + \alpha^{10} x^9 + \alpha^{10} x^8 + \alpha^{10} x^7 + \alpha^{10} x^6 + \alpha^{10} x^5 \\
&+ \alpha^{10} x^4 + \alpha^{10} x^3 + \alpha^3 x^9 + \alpha^3 x^8 + \alpha^3 x^7 + \alpha^3 x^6 + \alpha^3 x^5 \\
&+ \alpha^3 x^4 + \alpha^3 x^3 + \alpha^3 x^2 + x^8 + x^7 + x^6 + x^5 + x^4 + x^3 + x^2 + x \\
&+ \alpha^7 x^7 + \alpha^7 x^6 + \alpha^7 x^5 + \alpha^7 x^4 + \alpha^7 x^3 + \alpha^7 x^2 + \alpha^7 x + \alpha^7 \\
&= \alpha^{12} x^{11} + (\alpha^{12} + \alpha^{10})x^{10} + (\alpha^{12} + \alpha^{10} + \alpha^3)x^9 \\
&+ (\alpha^{12} + \alpha^{10} + \alpha^3 + 1)x^8 + (\alpha^{12} + \alpha^{10} + \alpha^3 + 1 + \alpha^7)x^7 \\
&+ (\alpha^{12} + \alpha^{10} + \alpha^3 + 1 + \alpha^7)x^6 + (\alpha^{12} + \alpha^{10} + \alpha^3 + 1 + \alpha^7)x^5 \\
&+ (\alpha^{12} + \alpha^{10} + \alpha^3 + 1 + \alpha^7)x^4 + (\alpha^{10} + \alpha^3 + 1 + \alpha^7)x^3 \quad &(4.29)\\
&+ (\alpha^3 + 1 + \alpha^7)x^2 + (1 + \alpha^7)x + \alpha^7 \\
&= \alpha^{12} x^{11} + \alpha^3 x^{10} + 0x^9 + \alpha^0 x^8 + \alpha^9 x^7 + \alpha^9 x^6 + \alpha^9 x^5 \\
&+ \alpha^9 x^4 + \alpha^8 x^3 + \alpha^1 x^2 + \alpha^9 x + \alpha^7.
\end{aligned}$$

Here it becomes clear that when using the nonsystematic encoding rule, the original information polynomial $i(x)$ cannot be directly recognized in $c(x)$.

We argued above that in case of systematic RS and BCH codes the bit error rate (BER) can be kept lower than that of the nonsystematic codes if a small additional signal processing complexity is tolerated. Thus, from now on we concentrate our attention on systematic codes. In order to compute the systematic codeword $c(x)$, Equations 4.20–4.24 must be used:

$$c(x) = x^4 \cdot i(x) + p(x),$$

where

$$p(x) = Rem \left\{ \frac{x^4 \, i(x)}{g(x)} \right\}.$$

The quotient $q(x)$ and remainder $p(x)$ of the above polynomial division are computed in Table 4.4, and the reader may find it beneficial at this stage to work through this example. Although the quotient polynomial is not necessary for our further elaborations, it is delivered by these operations, while the remainder appears in the bottom line of the table, which are given by:

$$q(x) = \alpha^{12} x^7 + \alpha^3 x^6 + \alpha^8 x^5 + \alpha^3 x^4 + \alpha^4 x^3 + \alpha^1 x^2 + \alpha^8 x + \alpha^{11}$$

$$p(x) = \alpha^{14} x^3 + \alpha^2 x^2 + \alpha^0 x + \alpha^6,$$

where

$$\frac{x^4 i(x)}{g(x)} = q(x)g(x) + p(x)$$

holds.

Since we now know the parity polynomial $p(x)$, the systematic codeword $c(x)$ is known from Equation 4.20 as well:

$$c(x) = x^4 i(x) + p(x)$$

$$\begin{aligned} c(x) &= \alpha^{12} x^{11} + \alpha^{12} x^{10} + \alpha^{12} x^9 + \alpha^{12} x^8 \\ &+ \alpha^{12} x^7 + \alpha^{12} x^6 + \alpha^{12} x^5 + \alpha^{12} x^4 \\ &+ \alpha^{14} x^3 + \alpha^2 x^2 + \alpha^0 x + \alpha^6. \end{aligned} \quad (4.30)$$

(4.31)

If there are no transmission errors in the received polynomial $r(x)$, then we have $r(x) = c(x)$, and decoding simply ensues by taking the first $k = 8$ information symbols of $r(x)$, concluding Example 4. Let us continue our discourse by considering an implementationally attractive linear shift register (LSR) encoder structure in the next subsection.

4.3. RS AND BCH CODES

$$\frac{x^4.i(x)}{g(x)} = \frac{\alpha^{12}.x^{11} + \alpha^{12}.x^{10} + \alpha^{12}.x^9 + \alpha^{12}.x^8 + \alpha^{12}.x^7 + \alpha^{12}.x^6 + \alpha^{12}.x^5 + \alpha^{12}.x^4}{x^4 + \alpha^{13}.x^3 + \alpha^6.x^2 + \alpha^3.x + \alpha^{10}}$$

$$= \underbrace{\alpha^{12}.x^7 + \alpha^3.x^6 + \alpha^8.x^5 + \alpha^3.x^4 + \alpha^4.x^3 + \alpha^1.x^2 + \alpha^8.x + \alpha^{11}}_{q(x)}$$

$(\alpha^{12}.x^{11}$	$+\alpha^{12}.x^{10}$	$+\alpha^{12}.x^9$	$+\alpha^{12}.x^8$	$+\alpha^{12}.x^7$	$+\alpha^{12}.x^6$	$+\alpha^{12}.x^5)$			
$-(\alpha^{12}.x^{11}$	$+\alpha^{10}.x^{10}$								
0	$+(\alpha^{12}+\alpha^{10}).x^{10}$	$+(\alpha^{12}+\alpha^3).x^9$	$+(\alpha^{12}+\alpha^0).x^8$	$+(\alpha^{12}+\alpha^7).x^7$	$+\alpha^{12}.x^6$	$+\alpha^{12}.x^5$	$+\alpha^{12}.x^4$		
	$\alpha^3.x^{10}$	$+\alpha^{10}.x^9$	$+\alpha^{11}.x^8$	$+\alpha^2.x^7$	$+\alpha^{12}.x^6$	$+\alpha^{12}.x^5$	$+\alpha^{12}.x^4$		
	$-(\alpha^3.x^{10}$	$+\alpha^1.x^9$	$+\alpha^9.x^8$	$+\alpha^6.x^7$	$+\alpha^{13}.x^6)$				
	0	$+(\alpha^1+\alpha^{10}).x^9$	$+(\alpha^9+\alpha^{11}).x^8$	$+(\alpha^2+\alpha^6).x^7$	$+(\alpha^{13}+\alpha^{12}).x^6$	$+\alpha^{12}.x^5$	$+\alpha^{12}.x^4$		
		$\alpha^8.x^9$	$+\alpha^2.x^8$	$+\alpha^3.x^7$	$+\alpha^1.x^6$	$+\alpha^{12}.x^5$	$+\alpha^{12}.x^4$		
		$-(\alpha^8.x^9$	$+\alpha^6.x^8$	$+\alpha^{14}.x^7$	$+\alpha^{11}.x^6)$				
		0	$+(\alpha^6+\alpha^2).x^8$	$+(\alpha^{14}+\alpha^3).x^7$	$+(\alpha^{11}+\alpha^1).x^6$	$+\alpha^{12}.x^5$	$+\alpha^{12}.x^4$		
			$\alpha^3.x^8$	$+\alpha^0.x^7$	$+\alpha^6.x^6$	$+\alpha^{10}.x^5$	$+\alpha^{12}.x^4$		
			$-(\alpha^3.x^8$	$+\alpha^1.x^7$	$+\alpha^9.x^6$	$+\alpha^6.x^5$	$+\alpha^{13}.x^4)$		
			0	$+(\alpha^1+\alpha^0).x^7$	$+(\alpha^9+\alpha^6).x^6$	$+(\alpha^6+\alpha^{10}).x^5$	$+(\alpha^{13}+\alpha^{12}).x^4$		
				$\alpha^4.x^7$	$+\alpha^7.x^5$	$+\alpha^5.x^5$	$+\alpha^1.x^4$	$+\alpha^{14}.x^3$	
				$-(\alpha^4.x^7$	$+\alpha^2.x^6$	$+\alpha^{14}.x^5$	$+\alpha^7.x^4$	$+\alpha^{14}.x^3$	
				0	$+(\alpha^2+\alpha^5).x^6$	$+(\alpha^{10}+\alpha^7).x^5$	$+(\alpha^7+\alpha^1).x^4$	$+\alpha^4.x^3$	
					$-(\alpha^1.x^6$	$+\alpha^{14}.x^5$	$+\alpha^7.x^4$	$+\alpha^4.x^3$	$+\alpha^{11}.x^2)$
					0	$+(\alpha^{14}+\alpha^6).x^5$	$+(\alpha^7+\alpha^{14}).x^4$	$+(\alpha^4+\alpha^{14}).x^3$	$+\alpha^{11}.x^2$
						$\alpha^8.x^5$	$+\alpha^6.x^4$	$+\alpha^{14}.x^3$	$+\alpha^{11}.x^2$
						$-(\alpha^8.x^5$	$+\alpha^6.x^4$	$+\alpha^{14}.x^3$	$+\alpha^{11}.x^2)$ + $\alpha^{3}.x$
						0	$+(\alpha^6+\alpha^1).x^4$	$+(\alpha^{14}+\alpha^9).x^3$	$+(\alpha^{11}+\alpha^{11}).x^2$ + $\alpha^{3}.x$
							$-(\alpha^{11}.x^4$	$+\alpha^9.x^3$	$+\alpha^2.x^2$ + $\alpha^{3}.x$
							0	$+(\alpha^9+\alpha^4).x^3$	$+(\alpha^2+0).x^2$ + $\alpha^{14}.x$
								$\alpha^{14}.x^3$	$+\alpha^2.x^2$ + $\alpha^{0}.x$ + α^{6}
									$+\alpha^6 = p(x)$

Table 4.4: Systematic RS(12,8,2) Encoding Example Using Polynomial Division

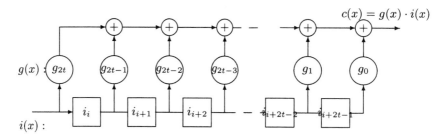

Figure 4.1: LSR circuit for multiplying polynomials in nonsystematic RS and BCH encoders.

4.3.4 Linear Shift-Register Circuits for Cyclic Encoders
4.3.4.1 Polynomial Multiplication

Since RS codes constitute a subclass of cyclic codes, any cyclic encoding circuitry can be used for RS and BCH encoding. The nonsystematic RS or BCH encoder can be implemented as a linear shift register (LSR) depicted in Figure 4.1, which multiplies the information polynomial $i(x)$ by the fixed generator polynomial $g(x)$ as follows:

$$c(x) = i(x) \cdot g(x),$$

where

$$i(x) = i_{k-1}x^{k-1} + i_{k-2}x^{k-2} + \ldots + i_1 x + i_0 \qquad (4.32)$$
$$g(x) = g_{2t}x^{2t} + g_{2t-1}x^{2t-1} + \ldots + g_1 x + g_0. \qquad (4.33)$$

The prerequisites for the circuit to carry out proper polynomial multiplications are as follows:

1. The LSR must be cleared before two polynomials are multiplied.
2. The k symbols of the information polynomial $i(x)$ must be followed by $2t$ zeros.
3. The highest order $i(x)$ coefficients must be entered into the LSR first.

When the first information symbol i_{k-1} appears at the input of the LSR of Figure 4.1, its output is $i_{k-1} \cdot g_{2t} \cdot x^{(k-1+2t)}$, since there is no contribution from its internal stages because its contents were cleared before entering a new codeword. After one clockpulse, the output is $(i_{k-2} \cdot g_{2t} + i_{k-1} \cdot g_{2t-1})x^{(k-2+2t)}$, and so on. After $(2t + k - 1)$ clockpulses, the LSR contains $0, 0 \ldots 0, i_0, i_1$, and the output is $(i_0 \cdot g_1 + i_1 \cdot g_0)x$, while the last nonzero output is $i_0 \cdot g_0$. Consequently, the product polynomial at the LSR's output is given by:

$$\begin{aligned} c(x) = i(x).g(x) &= i_{k-1}.g_{2t}.x^{k-1+2t} + \\ &+ (i_{k-2}.g_{2t} + i_{k-1}.g_{2t-1})x^{k-2+2t} + \\ &+ (i_{k-3}.g_{2t} + i_{k-2}.g_{2t-1} + i_{k-3}.g_{2t-2})x^{k-3+2t} + \end{aligned}$$

4.3. RS AND BCH CODES

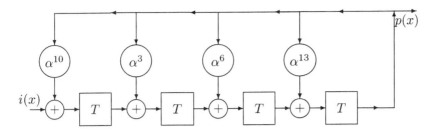

Figure 4.2: $RS(12, 8, 2)$ systematic encoder using polynomial division.

$$\vdots$$
$$+ (i_0.g_1 + i_1.g_0)x +$$
$$+ i_0.g_0. \qquad (4.34)$$

4.3.4.2 Systematic Cyclic Shift-Register Encoding Example

In this subsection we demonstrate how LSR circuits can be used in order to carry out the polynomial division necessary for computing the parity polynomial $p(x)$ of systematic RS or BCH codes. Let us attempt to highlight the operation of the LSR division circuit by referring to the example computed for the systematic $RS(12, 8)$ encoder previously.

Example 5
LSR polynomial division example for the $RS(12,8)$ GF(16)
code used for the generation of the parity polynomial
$p(x)$ in cyclic systematic encoders.

The corresponding division circuit is depicted in Figure 4.2, where the generator polynomial of Equation 4.27 was used. Note that the highest order information polynomial coefficients must be entered in the LSR first. Hence, $i(x)$ has to be arranged with the high-order coefficients at its right, ready for entering the LSR of Figure 4.2 from the left. By contrast, in the previously invoked polynomial division algorithm, $i(x)$ was arranged with the high-order coefficients at the left. In order to demonstrate the operation of the LSR encoder, we listed its contents during each clock-cycle in Table 4.5, which mimics the operation carried out during the polynomial division portrayed in Table 4.4. Explicitly, the feedback loop of the schematic in Figure 4.2 carries out the required multiplications by the generator polynomial after each division step, while the modulo-2 additions correspond to subtracting the terms of the same order during the polynomial division, where subtraction and addition are identical in our modulo algebra. Careful inspection of Tables 4.4 and 4.5 reveals the inherent analogy of the associated operations. The quotient polynomial exits the encoder at its output after twelve clock cycles, which is seen in the central column of Table 4.5, while the remainder appears at the bottom left section of the table. Observe in the

No. of shifts j	LSR content after j shifts				Output symbol after j shifts	Feedback				Input symbol
0	0	0	0	0	0	—	—	—	—	—
1	$\alpha^{12}+0=\alpha^{12}$	0	0	0	0	$0.\alpha^{10}=0$	$0.\alpha^{3}=0$	$0.\alpha^{6}=0$	$0.\alpha^{13}=0$	α^{12}
2	$\alpha^{12}+0=\alpha^{12}$	$\alpha^{12}+0=\alpha^{12}$	$0+0=0$	$0+0=0$	0	$0.\alpha^{10}=0$	$0.\alpha^{3}=0$	$0.\alpha^{6}=0$	$0.\alpha^{13}=0$	α^{12}
3	$\alpha^{12}+0=\alpha^{12}$	$\alpha^{12}+0=\alpha^{12}$	$\alpha^{12}+0=\alpha^{12}$	$0+0=0$	0	$0.\alpha^{10}=0$	$0.\alpha^{3}=0$	$0.\alpha^{6}=0$	$0.\alpha^{13}=0$	α^{12}
4	$\alpha^{12}+0=\alpha^{12}$	$\alpha^{12}+\alpha^{0}=\alpha^{11}$	$\alpha^{12}+\alpha^{3}=\alpha^{10}$	$\alpha^{12}+0=\alpha^{12}$	α^{12}	$\alpha^{12}.\alpha^{10}=\alpha^{7}$	$\alpha^{12}.\alpha^{3}=\alpha^{0}$	$\alpha^{12}.\alpha^{6}=\alpha^{3}$	$\alpha^{12}.\alpha^{13}=\alpha^{10}$	α^{12}
5	$\alpha^{12}+\alpha^{7}=\alpha^{2}$	$\alpha^{12}+\alpha^{0}=\alpha^{11}$	$\alpha^{12}+\alpha^{3}=\alpha^{10}$	α^{3}	α^{3}	$\alpha^{3}.\alpha^{10}=\alpha^{13}$	$\alpha^{3}.\alpha^{3}=\alpha^{6}$	$\alpha^{3}.\alpha^{6}=\alpha^{9}$	$\alpha^{3}.\alpha^{13}=\alpha^{1}$	α^{12}
6	$\alpha^{12}+\alpha^{13}=\alpha^{1}$	$\alpha^{2}+\alpha^{6}=\alpha^{3}$	$\alpha^{11}+\alpha^{9}=\alpha^{2}$	$\alpha^{10}+\alpha^{1}=\alpha^{8}$	α^{8}	$\alpha^{8}.\alpha^{10}=\alpha^{3}$	$\alpha^{8}.\alpha^{3}=\alpha^{11}$	$\alpha^{8}.\alpha^{6}=\alpha^{14}$	$\alpha^{8}.\alpha^{13}=\alpha^{6}$	α^{12}
7	$\alpha^{12}+\alpha^{3}=\alpha^{10}$	$\alpha^{1}+\alpha^{11}=\alpha^{6}$	$\alpha^{3}+\alpha^{14}+\alpha^{0}$	$\alpha^{2}+\alpha^{6}=\alpha^{3}$	α^{3}	$\alpha^{3}.\alpha^{10}=\alpha^{13}$	$\alpha^{3}.\alpha^{3}=\alpha^{6}$	$\alpha^{3}.\alpha^{6}=\alpha^{9}$	$\alpha^{3}.\alpha^{13}=\alpha^{1}$	α^{12}
8	$\alpha^{12}+\alpha^{13}=\alpha^{1}$	$\alpha^{10}+\alpha^{6}=\alpha^{7}$	$\alpha^{6}+\alpha^{9}=\alpha^{5}$	$\alpha^{0}+\alpha^{1}=\alpha^{4}$	α^{4}	$\alpha^{4}.\alpha^{10}=\alpha^{14}$	$\alpha^{4}.\alpha^{3}=\alpha^{7}$	$\alpha^{4}.\alpha^{6}=\alpha^{10}$	$\alpha^{4}.\alpha^{13}=\alpha^{2}$	α^{12}
9	$0+\alpha^{14}=\alpha^{14}$	$\alpha^{1}+\alpha^{7}=\alpha^{14}$	$\alpha^{7}+\alpha^{10}=\alpha^{6}$	$\alpha^{5}+\alpha^{2}=\alpha^{1}$	α^{1}	$\alpha^{1}.\alpha^{10}=\alpha^{11}$	$\alpha^{1}.\alpha^{3}=\alpha^{4}$	$\alpha^{1}.\alpha^{6}=\alpha^{7}$	$\alpha^{1}.\alpha^{13}=\alpha^{14}$	0
10	$0+\alpha^{11}=\alpha^{11}$	$\alpha^{14}+\alpha^{4}=\alpha^{9}$	$\alpha^{14}+\alpha^{7}=\alpha^{1}$	$\alpha^{6}+\alpha^{14}=\alpha^{8}$	α^{8}	$\alpha^{8}.\alpha^{10}=\alpha^{3}$	$\alpha^{8}.\alpha^{3}=\alpha^{11}$	$\alpha^{8}.\alpha^{6}=\alpha^{14}$	$\alpha^{8}.\alpha^{13}=\alpha^{6}$	0
11	$0+\alpha^{3}=\alpha^{3}$	$\alpha^{11}+\alpha^{1}=0$	$\alpha^{9}+\alpha^{14}=\alpha^{4}$	$\alpha^{1}+\alpha^{6}=\alpha^{11}$	α^{11}	$\alpha^{11}.\alpha^{10}=\alpha^{6}$	$\alpha^{11}.\alpha^{3}=\alpha^{14}$	$\alpha^{11}.\alpha^{6}=\alpha^{2}$	$\alpha^{11}.\alpha^{13}=\alpha^{9}$	0
12	$0+\alpha^{6}=\alpha^{6}$	$\alpha^{3}+\alpha^{14}=\alpha^{0}$	$0+\alpha^{2}=\alpha^{2}$	$\alpha^{4}+\alpha^{9}=\alpha^{14}$	α^{14}	$\alpha^{14}.\alpha^{10}=\alpha^{9}$	$\alpha^{14}.\alpha^{3}=\alpha^{2}$	$\alpha^{14}.\alpha^{6}=\alpha^{5}$	$\alpha^{14}.\alpha^{13}=\alpha^{12}$	0

Table 4.5: List of LSR Internal States for $RS(12, 8)$ Systematic Encoder

4.3. RS AND BCH CODES

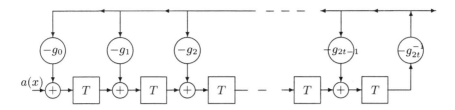

Figure 4.3: LSR circuit for dividing polynomials in nonsystematic RS and BCH encoders.

table that the first $2t = 4$ shifts of the LSR generate no output and consequently have no equivalent in the polynomial division algorithm. However, the following output symbols are identical to the quotient coefficients of $q(x)$ in the polynomial division algorithm of Table 4.4, and the LSR content after the twelfth clockpulse or shift is exactly the remainder polynomial $p(x)$ computed by polynomial division. The field additions at the inputs of the LSR cells are equivalent to subtracting the terms of identical order in the polynomial division, which actually represent the updating of the dividend polynomial in the course of the polynomial division after computing a new quotient coefficient.

Following similar generalized arguments, an arbitrary systematic encoder can be implemented by the help of the *division circuit* of Figure 4.3. As in case of the multiplier circuit, the LSR must again be cleared initially. Then the output of the LSR is zero for the first $2t$ clockpulses, and the first nonzero output is the first coefficient of the quotient, as we have seen it in Table 4.5 for the $RS(12, 8, 2)$ encoder. Again, as we have demonstrated in our systematic $RS(12, 8)$ encoder example, for each quotient coefficient q_j the product polynomial $q_j.g(x)$ must be subtracted from the dividend polynomial, which is contained in the LSR. This is carried out by arranging for the current quotient coefficient q_j, multiplied by $g(x)$, to manipulate cells of the LSR by the help of the modulo gates in order to update its contents to hold the current dividend polynomial. After n clockpulses, the entire quotient has appeared at the LSR's output, and the current dividend polynomial in the LSR has an order, which is lower than that of $g(x)$. Hence, it cannot be divided by $g(x)$; it is the remainder of the polynomial division operation. It therefore constitutes the required parity polynomial.

After this short overview of the analogies between GF field operations and their LSR implementations, we proceed with the outline of decoding and error correction algorithms for RS and BCH codes. We emphasize at this point that LSR implementations of GF arithmetics are usually computationally quite effective, though conceptually often a bit mysterious. However, one often accepts a conceptually more cumbersome solution for the sake of computational efficiency. Following the above portrayal of RS and BCH encoding techniques, let us now consider the corresponding decoding algorithms.

4.3.5 RS Decoding

4.3.5.1 Formulation of the Key Equations [1–9]

Since RS codes are nonbinary BCH codes and BCH codes are cyclic codes, any cyclic decoding technique can be deployed for their decoding. In this section we follow the philosophies of Peterson [6, 74], Clark and Cain [7], Blahut [3], Berlekamp [1], Lin and Costello [5], and a number of other prestigious authors, who have contributed substantially to the better understanding of algebraic decoding techniques and derive a set of nonlinear equations to compute the error locations and magnitudes, which can be solved by matrix inversion. This approach was originally suggested for binary BCH codes by Peterson [68], extended for nonbinary codes by Gorenstein and Zierler [78], and refined again by Peterson [6, 74].

Let us now attempt to explore what information is given by the parity symbols represented by the coefficients of $p(x)$ about the location and magnitude of error symbols represented by the error polynomial $e(x)$, which are corrupting the encoded codeword $c(x)$ into the received codeword $r(x)$. Since the encoded codeword is given by:

$$c(x) = i(x) \cdot x^{n-k} + p(x)$$

and the error polynomial $e(x)$ is added component-wise in the GF:

$$\begin{aligned} r(x) &= c(x) + e(x) \\ &= c_{n-1}.x^{n-1} + c_{n-2}.x^{n-2} + \ldots + c_1.x + c_0 \\ &+ e_{n-1}.x^{n-1} + e_{n-2}.x^{n-2} + \ldots + e_1 + e_0. \end{aligned} \quad (4.35)$$

May we remind the reader at this point that an $RS(n, k, t)$ code over $GF(2^m)$ contains in a codeword $c(x)$ n number of m-ary codeword symbols, constituted by k information and $(n - k) = 2t$ parity symbols of the polynomials $i(x)$ and $p(x)$, respectively. This code can correct up to t symbol errors, from which we suppose that $e(x)$ contains at most t nonzero coefficients. Otherwise, the error correction algorithm fails to compute the proper error positions and magnitudes. If we knew the error polynomial $e(x)$, we could remove its effect from the received codeword $r(x)$ resulting into $c(x)$, the error-free codeword, and we could recover the information $i(x)$ sent. Since effectively $e(x)$ carries merely $2t$ unknowns in the form of t unknown error positions and t unknown error magnitudes, it would be sufficient to know the value of $e(x)$ at $2t$ field elements to determine the $2t$ unknowns, required for error correction.

Recall that the generator polynomial has been defined as:

$$g(x) = \prod_{j=1}^{2t} (x - \alpha^j),$$

with zeros at the first $2t$ number of GF elements, and

$$c(x) = i(x) \cdot g(x).$$

Therefore, $c(x) = 0$ at all zeros of $g(x)$ even in case of systematic encoders. Hence, if we evaluate the received polynomial $r(x)$ at the zeros of $g(x)$, which are the first $2t$

4.3. RS AND BCH CODES

GF elements, we arrive at:

$$\begin{aligned} r(x)|_{\alpha^1\ldots\alpha^{2t}} &= [c(x)+e(x)]|_{\alpha^1\ldots\alpha^{2t}} \\ &= [i(x).g(x)+e(x)]|_{\alpha^1\ldots\alpha^{2t}} \\ &= [i(x).g(x)]|_{\alpha^1\ldots\alpha^{2t}} + e(x)|_{\alpha^1\ldots\alpha^{2t}} \\ &= 0 + e(x)|_{\alpha^1\ldots\alpha^{2t}}. \end{aligned} \quad (4.36)$$

Consequently, the received polynomial $r(x)$ evaluated at the zeros of $g(x)$, that is, at the first $2t$ GF elements, provides us with the $2t$ values of the error polynomial $e(x)$, which are necessary for computation of the t error positions and t error magnitudes. These $2t$ characteristic values are called the *syndromes* $S_1 \ldots S_{2t}$, which can be viewed as the algebraic equivalents of the symptoms of an illness.

If we now assume that actually $v \leq t$ errors have occurred and the error positions are $P_1 \ldots P_v$, while the error magnitudes are $M_1 \ldots M_v$, then an actual transmission error, that is, each nonzero component of the error polynomial $e(x)$, is characterized by the pair of GF elements (P_j, M_j). The $2t$ equations to be solved for error correction are as follows:

$$\begin{aligned} S_1 &= r(\alpha) = (e_{n-1}.x^{n-1} + e_{n-2}.x^{n-2} + \ldots + e_1.x + e_0)|_{x=\alpha} \\ S_2 &= r(\alpha^2) = (e_{n-1}.x^{n-1} + e_{n-2}.x^{n-2} + \ldots + e_1.x + e_0)|_{x=\alpha^2} \\ &\vdots \\ S_{2t} &= r(\alpha^{2t}) = (e_{n-1}.x^{n-1} + e_{n-2}.x^{n-2} + \ldots + e_1.x + e_0)|_{x=\alpha^{2t}}. \end{aligned} \quad (4.37)$$

There are only $v \leq t$ nonzero error magnitudes $M_1 \ldots M_v$ with the corresponding positions $P_1 \ldots P_v$, which can be thought of as place markers for the error magnitudes, and all the positions $P_1 \ldots P_v$ and magnitudes $M_1 \ldots M_v$ are $GF(2^m)$ elements, both of which can be expressed as powers of the primitive element α. Wherever on the right-hand side (r.h.s) of the first equation in the set of Equations 4.37 there is a nonzero error magnitude $M_j \neq 0$, $j = 1 \ldots v$, $x = \alpha$ is substituted. In the corresponding terms of the second equation, $x = \alpha^2$ is substituted, and so forth, while in the last equation $x = \alpha^{2t}$ is used. Regardless of the actual error positions, the nonzero terms in the syndrome equations $S_1 \ldots S_{2t}$ are always ordered in corresponding columns above each other with appropriately increasing powers of the original nonzero error positions. This is so whether they are expressed as powers of the primitive element α or that of the error positions $P_1 \ldots P_v$. When formulating this equation in terms of the error positions, we arrive at:

$$\begin{aligned} S_1 &= r(\alpha) = M_1.P_1 + M_2.P_2 + \ldots + M_v.P_v = \sum_{i=1}^{v} M_i.P_i \\ S_2 &= r(\alpha^2) = M_1.P_1^2 + M_2.P_2^2 + \ldots + M_v.P_v^2 = \sum_{i=1}^{v} M_i.P_i^2 \\ &\vdots \quad \vdots \end{aligned}$$

$$S_{2t} = r(\alpha^{2t}) = M_1.P_1^{2t} + M_2.P_2^{2t} + \ldots + M_v.P_v^{2t} = \sum_{i=1}^{v} M_i.P_i^{2t}. \tag{4.38}$$

Equation 4.38 can also be conveniently expressed in matrix form, as given below:

$$\begin{bmatrix} S_1 \\ S_2 \\ \vdots \\ S_{2t} \end{bmatrix} = \begin{bmatrix} P_1 & P_2 & P_3 & \ldots & P_v \\ P_1^2 & P_2^2 & P_3^2 & \ldots & P_v^2 \\ \vdots & & & & \\ P_1^{2t} & P_2^{2t} & P_3^{2t} & \ldots & P_v^{2t} \end{bmatrix} \begin{bmatrix} M_1 \\ M_2 \\ \vdots \\ M_v \end{bmatrix} \tag{4.39}$$

$$\underline{S} = \underline{\underline{P}} \cdot \underline{M}.$$

However, this set of equations is nonlinear, and hence a direct solution appears to be too complicated, since in general there are many solutions. All the solutions must be found, and the most likely error pattern is the one with the lowest number of errors, which in fact minimizes the probability of errors and the bit error rate.

Peterson suggested a simple method [68, 76] for binary BCH codes, which has been generalized by Gorenstein and Zierler [76, 78] for nonbinary RS codes. Hence, the corresponding algorithm is referred to as the Peterson-Gorenstein-Zierler (PGZ) decoder.

Their approach was based on the introduction of the *error locator polynomial* $L(x)$, which can be computed from the syndromes. The error locator polynomial can then be employed in order to linearize the set of equations, resulting in a more tractable solution for the error locations. Some authors define $L(x)$ as the polynomial with zeros at the error locations [4]. More frequently, however, [3, 5] it is supposed to have zeros at the multiplicative inverses of the error positions [3, 5], as suggested by the following equation:

$$L(x) = (1 - x.P_1)(1 - x.P_2)\ldots(1 - x.P_v) = \prod_{j=1}^{v} (1 - x.P_j) \tag{4.40}$$

$$L(x) = L_v.x^v + L_{v-1}.x^{v-1} + \ldots + L_1.x + 1 = \sum_{j=0}^{v} L_j.x^j. \tag{4.41}$$

Clearly, $L(x) = 0$ for $x = P_1^{-1}, P_2^{-1}, \ldots P_v^{-1}$. If $L(x)$, that is, its coefficients or zeros were known, we would know the error positions; therefore, we try to determine $L(x)$ from $S_1 \ldots S_{2t}$ [3], [6]. Upon multiplying Equation 4.41 by $M_i.P_i^{(j+v)}$, we get:

$$L(x).M_i.P_i^{j+v} = M_i.P_i^{j+v}(L_v.x^v + L_{v-1}.x^{v-1} + \ldots + L_1.x + 1) \tag{4.42}$$

and on substituting $x = P_i^{-1}$ into Equation 4.42 we arrive at:

$$0 = M_i.P_i^{j+v}(L_v.P_i^{-v} + L_{v-1}.P_i^{v-1} + \ldots + L_1.P_i^{-1} + 1), \tag{4.43}$$

$$0 = M_i(L_v.P_i^j + L_{v-1}.P_i^{j-1} + \ldots + L_1.P_i^{j+v-1} + P_i^{j+v}). \tag{4.44}$$

4.3. RS AND BCH CODES

There exists such an equation for all $i = 1 \ldots v$ and all j. If we sum the equations for $i = 1 \ldots v$, for each j $j = 1 \ldots 2t$ we get an equation of the form of Equation 4.45:

$$\sum_{i=1}^{v} M_i(L_v.P_i^j + L_{v-1}.P_i^{j-1} + \ldots + L_i.P_i^{j+v-1} + P_i^{j+v}) = 0. \quad (4.45)$$

Equivalently:

$$\sum_{i=1}^{v} M_i.L_v.P_i^j + \sum_{i=1}^{v} M_i.L_{v-1}.P_i^{j-1} + \ldots + \sum_{i=1}^{v} M_i.P_i^{j+v} = 0. \quad (4.46)$$

If we compare Equation 4.46 with Equation 4.38, we can recognize the syndromes in the sums. Therefore, we arrive at:

$$L_v.S_j + L_{v-1}.S_{j+1} + \ldots + L_1.S_{j+v-1} + S_{j+v} = 0. \quad (4.47)$$

The highest syndrome index is $j+v$, but since only the first $2 \cdot t$ syndromes $S_1 \ldots S_{2t}$ are specified, and since $v \leq t$, the condition $1 \leq j \leq t$ must be fulfilled. After rearranging Equation 4.47 we get a set of linear equations for the unknown coefficients $L_1 \ldots L_v$ as a function of the known syndromes $S_1 \ldots S_{2t}$, which is in fact the *key equation* for correcting RS or BCH codes. Any algorithm that delivers a solution to this set of equations can also be employed for correcting errors in RS codes. From Equation 4.47 we can also write:

$$L_v.S_j + L_{v-1}.S_{j+1} + \ldots + L_1.S_{j+v-1} = -S_{j+v} \quad \text{for } j = 1 \ldots v. \quad (4.48)$$

The key equation is more easily understood in a matrix form:

$$\begin{bmatrix} S_1 & S_2 & S_3 & \ldots & S_{v-1} & S_v \\ S_2 & S_3 & S_4 & \ldots & S_v & S_{v-1} \\ \vdots & & & & & \\ S_v & S_{v+1} & S_{v+2} & \ldots & S_{2v-2} & S_{2v-1} \end{bmatrix} \begin{bmatrix} L_v \\ L_{v-1} \\ \vdots \\ L_1 \end{bmatrix} = \begin{bmatrix} -S_{v+1} \\ -S_{v+2} \\ \vdots \\ -S_{2v} \end{bmatrix} \quad (4.49)$$

$$\underline{\underline{S}}.\underline{L} = \underline{S}. \quad (4.50)$$

Equation 4.49 can be solved for the unknown coefficients of the error locator polynomial $L(x)$ (i.e., for the error positions $P_1 \ldots P_v$) if the matrix of syndromes is nonsingular. Pless [75](p. 97) showed that the matrix $\underline{\underline{S}}$ has the form of the *Vandermonde matrix*, which plays a prominent role in the theory of error correction coding. $\underline{\underline{S}}$ can be shown to be nonsingular if it is of dimension $v \times v$, where v is the actual rather than the maximum number of errors that occurred, while it is singular if the dimension of $\underline{\underline{S}}$ is greater than v [6,78]. This theorem provides the basis for determining the actual number of errors v and determining the error positions $P_1 \ldots P_v$.

Before we proceed with our discourse on the various solutions to the key equation, we formulate it in different ways, from which the various solutions accrue. Notice that Equation 4.48 can also be interpreted in the form of a set of recursive formulas, which led to Massey's solution by the synthesis of an autoregressive filter [2]. Namely, if we assume that the coefficients $L_1 \ldots L_v$ are known for $j = 1 \ldots v$, Equation 4.48 generates recursively the next syndrome from the previous v number of syndromes as follows:

$$\begin{aligned} j=1 \quad -S_{v+1} &= L_v.S_1 + L_{v-1}.S_2 + L_{v-2}.S_3 + \ldots + L_1.S_v \\ j=2 \quad -S_{v+2} &= L_v.S_2 + L_{v-1}.S_3 + L_{v-2}.S_4 + \ldots + L_1.S_{v+1} \\ &\vdots \\ j=v \quad -S_{2v} &= L_v.S_v + L_{v-1}.S_{v+1} + L_{v-2}.S_{v+2} + \ldots + L_1.S_{2v-1} \end{aligned}$$
(4.51)

This set of equations can also be written as:

$$S_j = -\sum_{n=1}^{v} L_n.S_{j-n} \quad \text{for} \quad j = (v+1), (v+2)\ldots 2v, \tag{4.52}$$

which is reminiscent of a convolutional expression and hence can be implemented by the help of a linear feedback shift register or *autoregressive filter* having taps $L_1 \ldots L_v$, and with S_j fed back into the shift register. Massey's approach constitutes one of the most computationally effective alternatives to the solution of the key equations, in particular when the codeword length is high and the number of correctible errors is also high. Hence, we will demonstrate how it can be invoked after describing the conceptually simpler PGZ decoding algorithm.

A plethora of other solutions for the key equation in Equation 4.52 are also possible because in this form it is equivalent to the problem found, for example, in spectral estimation in speech coding, when solving a set of recursive equations of the same form for the *LPC* filter coefficients. The most powerful solution to the set of equations accrues, if one recognizes that the matrix of syndromes in Equation 4.49 can be shown to have not only *Vandermonde* structure, but also its values along the main diagonal are identical. In other words, it exhibits a symmetric Toeplitz structure. Efficient special algorithms for the solution of Toeplitz-type matrix equations have been proposed by Levinson, Robinson, Durbin, Berlekamp, Massey, Trench, Burg, and Schur in the literature for various prediction and equalization problems. All these techniques can be invoked for solving the key equations in Equation 4.52 when ensuring that all the operations are carried out over the previously described finite field $GF(2^m)$. Excellent discourses on the topic have been published in the literature by Makhoul [79], Blahut [80] (pp. 352–387), and Schur [81].

Having derived various representations of the key equation for error correction (see Equations 4.48–4.52), it is instructive to continue with the description of the conceptually most simple solution. This was originally proposed by Peterson [68] for binary BCH codes, and it was extended by Gorenstein and Zierler [78] for nonbinary RS codes on the basis of inverting the matrix of syndromes.

4.3.5.2 Peterson-Gorenstein-Zierler Decoder

As mentioned above, a number of solutions to the key equations have been suggested in [3, 4, 68, 78]. The following set of equations can be derived from Equation 4.50,

4.3. RS AND BCH CODES

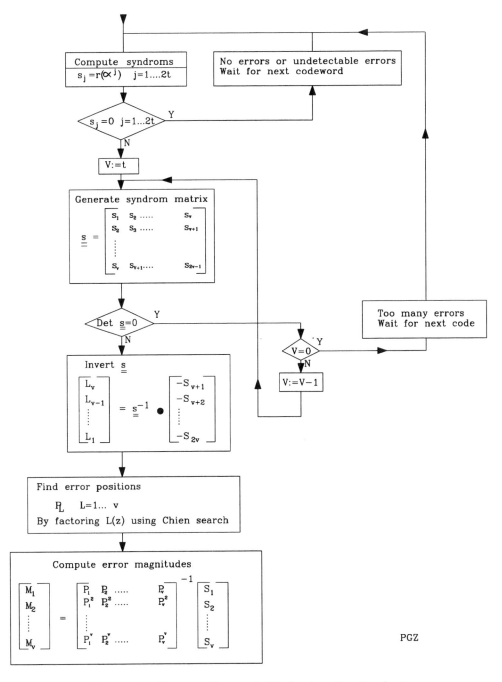

Figure 4.4: Peterson-Gorenstein-Zierler decoding flowchart.

whose solution is based on inverting the matrix of syndromes, as shown below:

$$\underline{L} = \underline{\underline{S}}^{-1}.\underline{S}. \qquad (4.53)$$

The solution is based on the following theorem. The *Vandermonde* matrix $\underline{\underline{S}}$ constituted by the syndromes is nonsingular and can be inverted if its dimension is $v \times v$, but it is singular and cannot be inverted if its dimension is greater than v, where v is the actual number of errors that occurred.

We have to determine v to be able to invert $\underline{\underline{S}}$ in order to solve Equation 4.53. Initially, we set $v = t$, since t is the maximum possible number of errors, and we compute the determinant of $\underline{\underline{S}}$. If $det(\underline{\underline{S}}) = 0$, then $v = t$, and we can proceed with the matrix inversion. Otherwise, we decrement v by one and tentatively try $v = t - 1$, and so forth, down to $v = 0$, until $det(\underline{\underline{S}}) \neq 0$ is found. If we have found the specific v value, for which $det(\underline{\underline{S}}) \neq 0$, we compute $\underline{\underline{S}}^{-1}$ by matrix inversion and derive $\underline{L} = \underline{\underline{S}}^{-1}.\underline{S}$.

Now the zeros of the error locator polynomial are determined by trial and error. This is carried out by substituting all nonzero field elements into $L(x)$, and finding those for which we have $L(x) = 0$. This method is called the *Chien search* [82]. The error positions $P_1 \ldots P_v$ can then be found by determining the multiplicative inverses of the zeros of $L(x)$. With this step, the solution is complete for binary BCH codes; the bits at the computed error positions must be inverted for error correction.

For nonbinary RS codes the error magnitudes $M_1 \ldots M_v$ must be determined in a further step. This is relatively straightforward now from the syndrome Equation 4.39; simply the known matrix of error positions $\underline{\underline{P}}$ has to be inverted to compute the vector $\underline{M}$ of error magnitudes:

$$\underline{M} = \underline{\underline{P}}^{-1}.\underline{S}. \qquad (4.54)$$

Similarly to the matrix of syndromes $\underline{\underline{S}}$, $\underline{\underline{P}}$ can also be shown to have $Vandermonde$ structure and consequently is nonsingular and can be inverted, if exactly v errors have occurred. Although Equation 4.54 represents a set of $2t$ equations, it is sufficient to solve v of them for the v unknown error magnitudes $M_1 \ldots M_v$. The Peterson--Gorenstein-Zierler method of RS decoding is summarized in the flowchart of Figure 4.4. Observe in the figure that the error correction problem was converted into a problem of inverting two matrices.

Before providing a worked numerical example for PGZ decoding, we briefly allude to the *error detection capability* of the RS and BCH codes. Explicitly, it is a very attractive property of RS and BCH codes that after error correction the syndromes can be recomputed in order to check whether the code was capable of removing the transmission errors. This is necessary, because without it the received codeword could have been more similar to some other legitimate codeword, and hence the decoder might have concluded that this other, more similar codeword was transmitted. If the error correction action of the decoder was successful, all recomputed syndromes must be zero. If, however, there are errors after the error correction, a systematic decoder can at least separate the original information part of the codeword. This measure allows the decoder to minimize the "damage," which would have been more catastrophic had the decoder attempted to correct the received symbols in the wrong positions, thereby actually corrupting potentially correct symbols.

4.3. RS AND BCH CODES

In order to appreciate the details of the PGZ decoding algorithm, we now offer a worked example using the previously invoked $RS(12, 8, 2)$ code.

4.3.5.3 PGZ Decoding Example

Example 6:
Consider the double-error correcting $RS(12, 8, 2)$ code over $GF(2^4)$ and carry out error correction by matrix inversion, using the Peterson-Gorenstein-Zierler decoder.

Let us assume that the "all one" information sequence has been sent in the form of the systematically encoded codeword of Section 4.3.3.

$$\begin{aligned} i(x) &= \alpha^{12}.x^{11} + \alpha^{12}.x^{10} + \alpha^{12}.x^9 + \alpha^{12}.x^8 \\ &+ \alpha^{12}.x^7 + \alpha^{12}.x^6 + \alpha^{12}.x^5 + \alpha^{12}.x^4 \\ p(x) &= \alpha^{14}.x^3 + \alpha^2.x^2 + \alpha^0.x + \alpha^6. \end{aligned}$$

Let us also assume that two errors have occurred in $c(x)$ in positions 11 and 3, resulting in the following received polynomial:

$$\begin{aligned} r(x) &= \underbrace{\alpha^4}_{error}.x^{11} + \alpha^{12}.x^{10} + \alpha^{12}.x^9 + \alpha^{12}.x^8 \\ &+ \alpha^{12}.x^7 + \alpha^{12}.x^6 + \alpha^{12}.x^5 + \alpha^{12}.x^4 \\ &+ \underbrace{\alpha^5}_{error}.x^3 + \alpha^2.x^2 + \alpha^0.x + \alpha^6. \end{aligned}$$

Since $r(x) = c(x) + e(x)$, based on Table 4.3, we infer that $e(x)$ contains two nonzero terms, in position 11, where the systematic data symbol was α^{12} and hence the additive error was:

$$e_{11} = r_{11} - c_{11} = \alpha^4 + \alpha^{12} = \alpha^6$$

and in position 3, where the parity symbol was α^{14}, requiring an error magnitude of:

$$e_3 = r_3 - c_3 = \alpha^5 + \alpha^{14} = \alpha^{12}.$$

Hence:

$$e(x) = \alpha^6.x^{11} + 0 + 0 + 0 + 0 + 0 + 0 + 0 + \alpha^{12}.x^3 + 0 + 0 + 0.$$

This exponential representation is easily translated into bit patterns by referring to Table 4.2, which contains the various representations of $GF(2^4)$ elements.

Syndrome calculation: Since the syndromes are the received polynomials evaluated at the $GF(2^4)$ elements $\alpha^1 \ldots \alpha^{2t} = \alpha^1 \ldots \alpha^4$, we have:

$$S_1 = r(\alpha^1)$$

$$
\begin{aligned}
&= \alpha^4\alpha^{11} + \alpha^{12}\alpha^{10} + \alpha^{12}\alpha^9 + \alpha^{12}\alpha^8 + \alpha^{12}\alpha^7 + \alpha^{12}\alpha^6 \\
&\quad + \alpha^{12}\alpha^5 + \alpha^{12}\alpha^4 + \alpha^5\alpha^3 + \alpha^2\alpha^2 + \alpha^0\alpha^1 + \alpha^6 \\
&= \underbrace{\alpha^0 + \alpha^7}_{\alpha^9} + \alpha^6 + \alpha^5 + \alpha^4 + \alpha^3 + \alpha^2 + \alpha^1 + \alpha^8 + \alpha^4 + \alpha^1 + \alpha^6 \\
&= \underbrace{\alpha^9 + \alpha^{11}}_{\alpha^2} + \alpha^0 \\
&= \alpha^2 + \alpha^0 \\
S_1 &= \alpha^8
\end{aligned}
$$

$$
\begin{aligned}
S_2 &= r(\alpha^2) \\
&= \alpha^4\alpha^{22} + \alpha^{12}\alpha^{20} + \alpha^{12}\alpha^{18} + \alpha^{12}\alpha^{16} + \alpha^{12}\alpha^{14} \\
&\quad + \alpha^{12}\alpha^{12} + \alpha^{12}\alpha^{10} + \alpha^{12}\alpha^8 + \alpha^5\alpha^6 + \alpha^2\alpha^4 + \alpha^0\alpha^2 + \alpha^6 \\
&= \alpha^{11} + \alpha^2 + \underbrace{\alpha^0 + \alpha^{13}}_{\alpha^6} + \alpha^{11} + \underbrace{\alpha^9 + \alpha^7}_{\alpha^0} + \underbrace{\alpha^5 + \alpha^{11}}_{\alpha^3} + \alpha^6 + \alpha^2 + \alpha^6 \\
&= \underbrace{\alpha^6 + \alpha^0}_{\alpha^{13}} + \alpha^3 \\
&= \alpha^{13} + \alpha^3 \\
S_2 &= \alpha^8
\end{aligned}
$$

$$
\begin{aligned}
S_3 &= r(\alpha^3) \\
&= \alpha^4\alpha^{33} + \alpha^{12}\alpha^{30} + \alpha^{12}\alpha^{27} + \alpha^{12}\alpha^{24} + \alpha^{12}\alpha^{21} \\
&\quad + \alpha^{12}\alpha^{18} + \alpha^{12}\alpha^{15} + \alpha^{12}\alpha^{12} + \alpha^5\alpha^9 + \alpha^2\alpha^6 + \alpha^0\alpha^3 + \alpha^6 \\
&= \alpha^7 + \alpha^{12} + \alpha^9 + \alpha^6 + \alpha^3 + \alpha^0 + \alpha^{12} + \alpha^9 + \underbrace{\alpha^{14} + \alpha^8}_{\alpha^6} + \alpha^3 + \alpha^6 \\
&= \alpha^9 + \alpha^6 \\
S_3 &= \alpha^5
\end{aligned}
$$

$$
\begin{aligned}
S_4 &= r(\alpha^4) \\
&= \alpha^4\alpha^{44} + \alpha^{12}\alpha^{40} + \alpha^{12}\alpha^{36} + \alpha^{12}\alpha^{32} + \alpha^{12}\alpha^{28} + \alpha^{12}\alpha^{24} \\
&\quad + \alpha^{12}\alpha^{20} + \alpha^{12}\alpha^{16} + \alpha^5\alpha^{12} + \alpha^2\alpha^8 + \alpha^0\alpha^4 + \alpha^6 \\
&= \alpha^3 + \underbrace{\alpha^7 + \alpha^3 + \alpha^{14}}_{\alpha^1} + \alpha^{10} + \alpha^6 + \alpha^2 + \alpha^{13} + \alpha^2 + \alpha^{10} + \alpha^4 + \alpha^6
\end{aligned}
$$

4.3. RS AND BCH CODES

$$S_4 \begin{array}{l} = \alpha^1 + \alpha^{11} \\ = \alpha^6. \end{array}$$

Since we have no information on the actual number of errors v in the receiver, we must determine v first. Let us suppose initially $v = t = 2$. The key equation to be solved according to Equation 4.49 is as follows:

$$\underbrace{\begin{bmatrix} S_1 & S_2 \\ S_2 & S_3 \end{bmatrix}}_{\underline{\underline{S}}} \underbrace{\begin{bmatrix} L_2 \\ L_1 \end{bmatrix}}_{\underline{L}} = -\underbrace{\begin{bmatrix} S_3 \\ S_4 \end{bmatrix}}_{\underline{S}}.$$

Let us now compute the determinant of $\underline{\underline{S}}$, as portrayed in:

$$det(\underline{\underline{S}}) = det \begin{vmatrix} S_1 & S_2 \\ S_2 & S_3 \end{vmatrix} = (S_1 S_3 - S_2^2).$$

By substituting the syndromes and using Table 4.3 we arrive at:

$$S_1 = \alpha^8,\ S_2 = \alpha^8,\ S_3 = \alpha^5,\ S_4 = \alpha^6$$

$$det(\underline{\underline{S}}) = (\alpha^8 . \alpha^5 - (\alpha^8)^2) = \alpha^{13} - \alpha^1 = \alpha^{12} \neq 0.$$

Since $det(\underline{\underline{S}}) \neq 0$, $v = 2$ errors occurred, and we can compute $\underline{\underline{S}}^{-1}$.

Let us use the *Gauss–Jordan* elimination [83] for the computation of $\underline{\underline{S}}^{-1}$. This can be achieved by transforming both the original matrix $\underline{\underline{S}}$, which has to be inverted, and the unit matrix in the same way, until the original $\underline{\underline{S}}$ is transformed to a unit matrix. The transformed unit matrix then becomes inverted matrix $\underline{\underline{S}}^{-1}$. The operations are naturally $GF(2^4)$ operations, and we commence the transformations with the following matrices:

$$\underline{\underline{S}} = \begin{bmatrix} \alpha^8 & \alpha^8 \\ \alpha^8 & \alpha^5 \end{bmatrix} \quad \underline{\underline{I}} = \begin{bmatrix} \alpha^0 & 0 \\ 0 & \alpha^0 \end{bmatrix}.$$

After adding row 1 of both matrices to their second rows, we get:

$$\begin{bmatrix} \alpha^8 & \alpha^8 \\ \alpha^8 + \alpha^8 = 0 & \alpha^5 + \alpha^8 = \alpha^4 \end{bmatrix} \quad \begin{bmatrix} \alpha^0 & 0 \\ \alpha^0 & \alpha^0 \end{bmatrix}.$$

In a second step we carry out the assignment row 1 = row 1 . α^{11} + row 2, yielding:

$$\begin{bmatrix} \alpha^4 & 0 \\ 0 & \alpha^4 \end{bmatrix} \quad \begin{bmatrix} \alpha^{11} + \alpha^0 = \alpha^{12} & \alpha^0 \\ \alpha^0 & \alpha^0 \end{bmatrix}.$$

Finally, we compute row 1 = row 1 . α^{11} and row 2 = row 2 . α^{11} in order to render $\underline{\underline{S}}$ a unit matrix, as seen in:

$$\begin{bmatrix} \alpha^0 & 0 \\ 0 & \alpha^0 \end{bmatrix} \quad \begin{bmatrix} \alpha^8 & \alpha^{11} \\ \alpha^{11} & \alpha^{11} \end{bmatrix}.$$

Therefore, the inverted matrix becomes:

$$\underline{\underline{S}}^{-1} = \begin{bmatrix} \alpha^8 & \alpha^{11} \\ \alpha^{11} & \alpha^{11} \end{bmatrix}$$

and hence

$$\underline{L} = \underline{\underline{S}}^{-1} \cdot \underline{S}$$

which gives:

$$\begin{bmatrix} L_2 \\ L_1 \end{bmatrix} = \begin{bmatrix} \alpha^8 & \alpha^{11} \\ \alpha^{11} & \alpha^{11} \end{bmatrix} \cdot \begin{bmatrix} \alpha^5 \\ \alpha^6 \end{bmatrix} = \begin{bmatrix} \alpha^{13} + \alpha^2 \\ \alpha^1 + \alpha^2 \end{bmatrix} = \begin{bmatrix} \alpha^{14} \\ \alpha^5 \end{bmatrix}.$$

Now we can compute the error positions from $L(x)$, where

$$L(x) = \alpha^{14}.x^2 + \alpha^5.x + 1$$

by trying all $GF(2^4)$ elements according to the Chien search:

$$\begin{aligned}
L(\alpha^0) &= \alpha^{14}\alpha^0 + \alpha^5\alpha^0 + 1 = \alpha^{14} + \alpha^5 + \alpha^0 = \alpha^{11} \\
L(\alpha^1) &= \alpha^{14}\alpha^2 + \alpha^5\alpha^1 + 1 = \alpha^1 + \alpha^6 + \alpha^0 = \alpha^{11} + \alpha^0 = \alpha^{12} \\
L(\alpha^2) &= \alpha^{14}\alpha^4 + \alpha^5\alpha^2 + 1 = \alpha^3 + \alpha^7 + \alpha^0 = \alpha^4 + \alpha^0 = \alpha^1
\end{aligned}$$

$$\begin{aligned}
L(\alpha^3) &= \alpha^{14}\alpha^6 + \alpha^5\alpha^3 + 1 = \alpha^5 + \alpha^8 + \alpha^0 = \alpha^4 + \alpha^0 = \alpha^1 \\
L(\alpha^4) &= \alpha^{14}\alpha^8 + \alpha^5\alpha^4 + 1 = \alpha^7 + \alpha^9 + \alpha^0 = \alpha^0 + \alpha^0 = 0 \longleftarrow \\
L(\alpha^5) &= \alpha^{14}\alpha^{10} + \alpha^5\alpha^5 + 1 = \alpha^9 + \alpha^{10} + \alpha^0 = \alpha^{13} + \alpha^0 = \alpha^6
\end{aligned}$$

$$\begin{aligned}
L(\alpha^6) &= \alpha^{14}\alpha^{12} + \alpha^5\alpha^6 + 1 = \alpha^{11} + \alpha^{11} + \alpha^0 = 0 + \alpha^0 = \alpha^0 \\
L(\alpha^7) &= \alpha^{14}\alpha^{14} + \alpha^5\alpha^7 + 1 = \alpha^{13} + \alpha^{12} + \alpha^0 = \alpha^1 + \alpha^0 = \alpha^4 \\
L(\alpha^8) &= \alpha^{14}\alpha^{16} + \alpha^5\alpha^8 + 1 = \alpha^0 + \alpha^{13} + \alpha^0 = 0 + \alpha^{13} = \alpha^{13}
\end{aligned}$$

$$\begin{aligned}
L(\alpha^9) &= \alpha^{14}\alpha^{18} + \alpha^5\alpha^9 + 1 = \alpha^2 + \alpha^{14} + \alpha^0 = \alpha^{13} + \alpha^0 = \alpha^6 \\
L(\alpha^{10}) &= \alpha^{14}\alpha^{20} + \alpha^5\alpha^{10} + 1 = \alpha^4 + \alpha^0 + \alpha^0 = \alpha^4 + 0 = \alpha^4 \\
L(\alpha^{11}) &= \alpha^{14}\alpha^{22} + \alpha^5\alpha^{11} + 1 = \alpha^6 + \alpha^1 + \alpha^0 = \alpha^{11} + \alpha^0 = \alpha^{12}
\end{aligned}$$

$$\begin{aligned}
L(\alpha^{12}) &= \alpha^{14}\alpha^{24} + \alpha^5\alpha^{12} + 1 = \alpha^8 + \alpha^2 + \alpha^0 = \alpha^0 + \alpha^0 = 0 \longleftarrow \\
L(\alpha^{13}) &= \alpha^{14}\alpha^{26} + \alpha^5\alpha^{13} + 1 = \alpha^{10} + \alpha^3 + \alpha^0 = \alpha^{12} + \alpha^0 = \alpha^{11} \\
L(\alpha^{14}) &= \alpha^{14}\alpha^{28} + \alpha^5\alpha^{14} + 1 = \alpha^{12} + \alpha^4 + \alpha^0 = \alpha^6 + \alpha^0 = \alpha^{13}.
\end{aligned}$$

4.3. RS AND BCH CODES

Since the error locator polynomial has its zeros at the inverse error positions, their multiplicative inverses have to be found in order to determine the actual error positions:

$$\begin{align}(\alpha^4)^{-1} &= \alpha^{11} = P_1 \\ (\alpha^{12})^{-1} &= \alpha^3 = P_2,\end{align}$$

which are indeed the positions, where the $c(x)$ symbols have been corrupted by $e(x)$.

If we had a binary BCH code, the error correction would simply be the inversion of the bits in positions $P_1 = \alpha^{11}$ and $P_2 = \alpha^3$. For the nonbinary RS(12, 8, 2) code over $GF(2^4)$, the error magnitudes still have to be determined. Now we know that there are $v = 2$ errors and so the matrix $\underline{P}$ of error positions in Equation 4.39 can be inverted and Equation 4.54 can be solved for the error positions $M_1 \ldots M_v$. But since $v = 2$, there are only two equations to be solved for M_1 and M_2, so we simply substitute P_1 and P_2 into Equation 4.38, which gives:

$$\begin{align}S_1 &= M_1 P_1 + M_2 P_2 \\ S_2 &= M_1 P_1^2 + M_2 P_2^2 \\ \alpha^8 &= M_1 \alpha^{11} + M_2 \alpha^3 \\ \alpha^8 &= M_1 \alpha^7 + M_2 \alpha^6.\end{align}$$

From the first equation we can express M_1 and substitute it into the second one as follows:

$$\begin{align}M_1 &= \frac{\alpha^8 + M_2 \alpha^3}{\alpha^{11}} = (\alpha^8 + M_2 \alpha^3)\alpha^4 \\ M_1 &= \alpha^{12} + M_2.\alpha^7\end{align}$$

$$\begin{align}\alpha^8 &= (\alpha^{12} + M_2.\alpha^7)\alpha^7 + M_2.\alpha^6 \\ \alpha^8 &= \alpha^4 + M_2.\alpha^{14} + M_2.\alpha^6 \\ \alpha^8 - \alpha^4 &= M_2(\alpha^{14} + \alpha^6)\end{align}$$

$$\begin{align}M_2 &= \frac{\alpha^8 + \alpha^4}{\alpha^{14} + \alpha^6} = \frac{\alpha^5}{\alpha^8} = \alpha^5.\alpha^7 = \alpha^{12} \\ M_1 &= \alpha^{12} + M_2.\alpha^7 = \alpha^{12} + \alpha^{12}.\alpha^7 = \alpha^{12} + \alpha^4 = \alpha^6.\end{align}$$

Now we are able to correct the errors by the help of the pairs:

$$\begin{align}(P_1, M_1) &= (\alpha^{11}, \alpha^6) \\ (P_2, M_2) &= (\alpha^3, \alpha^{12}),\end{align}$$

if we simply add the error polynomial $e(x) = \alpha^6.x^{11} + \alpha^{12}.x^3$ to the received polynomial $r(x)$ to recover the error-free codeword polynomial $c(x)$, carrying the original information $i(x)$.

As we have seen, the Peterson-Gorenstein-Zierler method involves the inversion of two matrices, one to compute the error positions $P_1 \ldots P_v$, and one to determine the error magnitudes $M_1 \ldots M_v$. For short codes over small GFs with low values of the n, k, t, and m parameters, the computational complexity is relatively low. This is because the number of multiplications is proportional to t^3. However, for higher t values the matrix inversions have to be circumvented. The first matrix inversion can, for example, be substituted by the algorithms suggested by Berlekamp [1] and Massey [2, 76], which pursue similar approaches. For the efficient computation of the error magnitudes, Forney has proposed a method [84] that can conveniently substitute the second matrix inversion. Let us now embark on highlighting the principles of these techniques, noting that at this stage we can proceed to the next chapter without jeopardizing the seamless flow of thought.

4.3.5.4 Berlekamp-Massey Algorithm [1–9]

In Section 4.3.5.1 we formulated the key equation first in matrix form and showed a solution based on matrix inversion. We also mentioned that by exploiting the Toeplitz structure of the syndrome matrix $\underline{\underline{S}}$, more efficient methods can be contrived. Instead of pursuing this matrix approach, we now concentrate on the form of the key equation described by Equation 4.52, which suggests the design of an autoregressive filter or *linear feedback shift register (LFSR)* in order to generate the required sequence of syndromes $S_1 \ldots S_{2t}$. Berlekamp suggested an efficient iterative heuristic solution to the problem [1] (p. 180), which can be better understood in Massey's original interpretation [2]. However, even Massey's approach is somewhat abstract, and several authors have attempted to give their own slants on its justification [3–7]. Some others find it instructive to introduce it after explaining a simpler algorithm, the Euclidean algorithm [85]. In this treatise, we will attempt to make the Berlekamp-Massey (BM) algorithm plausible by the help of a LFSR approach [3, 86], which shows that it is possible to design the taps of a LFSR, constituted in our case by the coefficients $L_0 \ldots L_v$ of the error locator polynomial $L(x)$ so that it produces a required sequence, which is represented by the syndromes $S_1 \ldots S_{2t}$ in our case. In a conventional signal processing context, this would be equivalent to designing the coefficients of a filter such that it would generate a required output sequence. However, in our current context all operations are over the finite GF.

According to this approach, Equation 4.52 can be modeled by the LFSR depicted in Figure 4.5. Clearly, our objective is to determine the length v and the feedback tap values L_n of the LFSR, such that it recursively generates the already known syndromes.

Naturally, there is only a limited number of LFSR designs, which produce the required sequence $S_1 \ldots S_{2t}$, and an error polynomial is associated with each possible LFSR design. Note that the order of the error locator polynomial predetermines the number of errors per codeword. The probability of receiving a transmitted codeword with one symbol error is always lower than the joint probability of having a double error within the codeword. In general terms, the most likely error pattern is the one with the lowest number of errors per codeword, which is equivalent to saying that we are looking for the minimum length LFSR design, that is, for the minimum order

4.3. RS AND BCH CODES

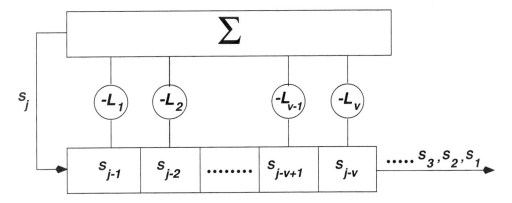

Figure 4.5: LFSR design to generate the syndromes $S_1 \ldots S_{2t}$ where $S_j = -\sum_{n=1}^{v} L_n \cdot S_{j-n}$ for $j = (v+1), \ldots 2v$.

error locator polynomial $L(x)$.

Since the BM algorithm is iterative, it is instructive to start from iteration one ($i = 1$), where the shortest possible LFSR length to produce the first syndrome S_1 is $l^{(i)} = l^{(1)} = 1$ and the corresponding LFSR connection polynomial, that is, the error locator polynomial at $i = 1$ is $L^i(x) = L^1(x) = 1$. Let us now assume tentatively that this polynomial also produces the second syndrome S_2, and let us attempt to compute the first estimated syndrome S_{2e} produced by this tentative feedback connection polynomial $L^1(x) = 1$ at $i = 1$ by referring to Equation 4.52 and Figure 4.5:

$$S_{i+1} = -\sum_{n=1}^{l^{(i)}} L_n^{(i)} S_{i+1-n}. \tag{4.55}$$

Since at the current stage of iteration we have $l^{(i)} = 1$, $j = 2$, $L^{(1)} = 1$, the estimated syndrome is given by:

$$S_{2e} = -1 \cdot S_1 = -S_1,$$

which means that the second estimated syndrome S_{2e} is actually approximated by the first one. This might be true in some special cases, but not in general, and a so-called *discrepancy* or error term $d^{(i)}$ is generated in order to check whether the present LFSR design adequately produces the next syndrome. This discrepancy is logically the difference of the required precomputed syndrome $S_{(i+1)}$ and its estimate $S_{(i+1)e}$, which is given by:

$$d^{(i)} = S_{i+1} - S_{(i+1)e} = S_{i+1} + \sum_{n=1}^{l^{(i)}} L_n^{(i)} \cdot S_{i+1-n} = \sum_{n=0}^{l^{(i)}} L_n^{(i)} S_{i+1-n}. \tag{4.56}$$

Upon exploiting that $L_0 \neq 1$, for the iteration index $i = 1$ from Equation 4.56, we have:

$$d^{(1)} = L_0^{(1)} \cdot S_2 + L_1^{(1)} \cdot S_1 = S_2 + S_1.$$

If the current discrepancy or syndrome estimation error is $d^{(1)} = 0$, then the present LFSR design correctly produces the next syndrome S_2. This means that the internal variables $l^{(i)}, L^{(i)}(x), i$ of the iterative process must remain tentatively unaltered for the duration of the next iteration. In other words, the present LFSR design will be initially employed in the next iteration as a trial design in order to estimate the next precomputed syndrome. Therefore, the following "update" has to take place:

$$\begin{aligned} l^{(2)} &:= l^{(1)} = 1 \\ L^{(2)}(x) &:= L^{(1)}(x) \\ i &:= i + 1 = 2. \end{aligned}$$

The above statements are also valid at other stages of the syndrome iteration and hence in general, if $d^{(i)} = 0$, then the following assignments become effective:

$$\begin{aligned} l^{(i+1)} &:= l^{(i)} \\ L^{(i+1)}(x) &:= L^{(i)}(x) \\ i &:= i + 1. \end{aligned}$$

However, if $d^{(i)} \neq 0$, the LFSR design must be modified until a solution is found, where $d^{(i)} = 0$. This means that the LFSR must be lengthened by the minimum possible number of stages and another connection polynomial has to be found, which produces the next desired syndrome S_{i+1} with an estimation error or discrepancy of $d_i = 0$. Furthermore, all the previous syndromes $S_1 \ldots S_i$ have to be also properly generated.

One possible way of doing this is to remember the last case at iteration m, when the LFSR has failed to produce the next syndrome S_m, and use the mth LFSR design with its associated discrepancy of d_m in order to modify the present design as follows. The linearity of the LFSR circuitry allows us to invoke superposition, where the nonzero discrepancy of the current LFSR design can be canceled by appropriately scaling the mth nonzero discrepancy and superimposing it on the current non-zero estimation error. This can be achieved by actually superimposing two LFSR designs exhibiting nonzero estimation errors. In terms of discrepancies we have:

$$d^{(i)} - d^{(i)} \cdot \frac{d^{(m)}}{d^{(m)}} = 0, \tag{4.57}$$

which suggests a solution for the choice of the proper connection polynomial in an equivalent form:

$$L^{(i+1)}(x) = L^{(i)}(x) - x^{i-m} \cdot \frac{d^{(i)}}{d^{(m)}} \cdot L^{(m)}(x). \tag{4.58}$$

In order to understand Equation 4.58, we emphasize again that two LFSRs can be superimposed to produce the required syndrome with zero discrepancy, since they are linear circuits, where superposition applies. Consequently, we can have a separate auxiliary LFSR of length $l^{(m)}$ with connection polynomial $L^{(m)}(x)$ and discrepancy d_m, the coefficients of which are scaled by the factor $\frac{d^{(i)}}{d^{(m)}}$ in order to compensate properly for the nonzero discrepancy. Note, however, that the auxiliary connection

4.3. RS AND BCH CODES

polynomial $L^{(m)}(x)$ has to be shifted to the required stage of the LFSR. This can be carried out using the multiplicative factor $x^{(i-m)}$. Having been shifted by $(i-m)$ positions, the output of the appropriately positioned auxiliary connection polynomial $x^{(i-m)} \cdot L^{(m)}(x)$ is added to that of the LFSR design $L^{(i)}(x)$, canceling the undesirable nonzero discrepancy at this iteration.

We note that because of its shift by $x^{(i-m)}$ the connection polynomial $L^{(m)}(x)$ actually contributes to $L^{(i+1)}(x)$ only in coefficients with indices in excess of $(i-m)$, which implies that the first m number of $L^{(i)}(x)$ coefficients are not altered by this operation, while those above the index m must be altered to result in $d^{(i)} = 0$. Since we have chosen the most recent mth iteration, at which $d^{(m)} \neq 0$ occurred and consequently LFSR lengthening was necessary, the LFSR length $l^{(m+1)}$ is higher than $l^{(m)}$, but only by the minimum required number of stages. Hence, we have a minimum-length LFSR design. Because of the linearity of the LFSR circuit, the auxiliary LFSR can be physically merged with the main LFSR. In order to assist in future modifications of the LFSR, when a nonzero discrepancy is produced, the most recent auxiliary LFSR with $d^{(m)} \neq 0$ must be stored as well.

Since we have made the inner workings of the BM algorithm plausible, we now attempt to summarize formally the set of iterative steps to be carried out to design the LFSR. The generation of the appropriate LFSR is synonymous with the determination of the minimum-length error locator polynomial $L(x)$, which generates the required precomputed sequence of syndromes. The algorithm had originally been stated in the form of a number of theorems, lemmas, and corollaries, for which the rigorous proofs can be found in Massey's original paper [2], but only its essence is presented here.

Theorem 1: An LFSR of length $l^{(i)}$, which generates the required syndromes $S_1, S_2 \ldots S_{i-1}$ and the required syndrome sequence $S_1, S_2 \ldots S_i$, does not have to be lengthened; hence, $l^{(i+1)} = l^{(i)}$ is satisfied. Conversely, the LFSR of length $l^{(i)}$ that generates $S_1, S_2 \ldots S_{i-1}$, but fails to generate the required syndrome sequence $S_1, S_2 \ldots S_i$, has to be lengthened. In general, the LFSR length has to obey $l^{(i+1)} := MAX\{l^{(i)}, (i+1-l^{(i)})\}$. Thus, the LFSR has to be lengthened if and only if:

$$i + 1 - l^{(i)} > l^{(i)} \tag{4.59}$$

$$\text{or} \quad i + 1 > 2l^{(i)}, \tag{4.60}$$

$$\text{ie} \quad i \geq 2l^{(i)}, \tag{4.61}$$

and the increased LFSR length is given by:

$$l^{(i+1)} = i + 1 - l^{(i)}. \tag{4.62}$$

Following the above introductory elaborations, the BM algorithm can be more formally summarized following Blahut's interpretation [3] as seen in Figure 4.6. The various stages of processing are numbered in the flowchart, and these steps are listed below.

1 Set the initial conditions:
    Iteration index: $\quad i = 0$
    LFSR length: $\quad l^{(1)} = 0$
    Connection polynomial: $\quad L^{(1)}(x) = 1$
    Auxiliary connection polynomial: $\quad A^{(1)}(x) = 1$

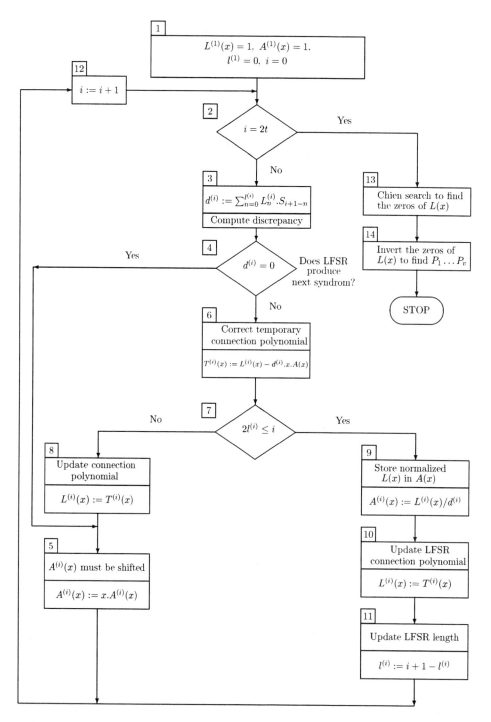

Figure 4.6: The BM algorithm: Computation of the error locator polynomial $L(x)$.

4.3. RS AND BCH CODES

For $i := 0 \ldots 2t - 1$ apply the set of recursive relations specified by *Steps* 2...11.

2 Check whether we reached the end of iteration, that is, whether $i = 2t$, and if so, branch to *Step* 13.

3 Compute the discrepancy or estimation error associated with the generation of the next syndrome from Equation 4.56:

$$d^{(i)} = \sum_{n=0}^{l^{(i)}} L_n^{(i)} . S_{i+1-n}$$

and go to *Step* 4.

4 Check whether the discrepancy is zero, and if $d^{(i)} = 0$, then the present LFSR design having a length of $l^{(i)}$ and connection polynomial $L^{(i)}(x)$ does produce the next syndrome S_{i+1}. Hence, go to *Step* 5; otherwise go to *Step* 6.

5 Simply shift the auxiliary LFSR by one position using the following operation:

$$A^{(i)}(x) := x . A^{(i)}(x) \tag{4.63}$$

and go to *Step* 12.

6 Since $d^{(i)} \neq 0$, we correct the temporary connection polynomial $T^{(i)}(x)$ by adding the properly shifted, normalized and scaled auxiliary connection polynomial $A^{(i)}(x)$ to the current connection polynomial $L^{(i)}(x)$, as seen below:

$$T^{(i)}(x) = L^{(i)}(x) - d^{(i)} . x . A^{(i)}(x). \tag{4.64}$$

7 Check whether the LFSR has to be lengthened, by using *Theorem 1* and Equation 4.61. If $2l^{(i)} \leq i$, go to *Step* 8; otherwise go to *Step* 9.

8 Update the connection polynomial according to the following formula:

$$L^{(i)}(x) := T^{(i)}(x) \tag{4.65}$$

and go to *Step* 5.

9 Normalize the most recent connection polynomial $L(x)$ by dividing it with $d^{(i)} \neq 0$ and store the results in the auxiliary LFSR $A^{(i)}(x)$, as follows:

$$A^{(i)}(x) := L^{(i)}(x)/d^{(i)} \tag{4.66}$$

and go to *Step* 10.

10 Now one can overwrite $L^{(i)}(x)$, since it has been normalized and stored in $A^{(i)}(x)$ during *Step* 9; hence, update the connection polynomial $L^{(i)}(x)$ according to:

$$L^{(i)}(x) := T^{(i)}(x) \tag{4.67}$$

and go to *Step* 11.

11 Since according to *Step* 7 the LFSR must be lengthened, we update the LFSR length by using *Theorem 1* and Equation 4.62, yielding:

$$l^{(i+1)} = i + 1 - l^{(i)} \tag{4.68}$$

and go to *Step* 12.

12 Increment the iteration index i and go to *Step 2*.

13 Invoke the Chien search in order to find the zeros of $L(x)$ by tentatively substituting all possible GF elements into $L(x)$ and finding those that render $L(x)$ zero.

14 Invert the zeros of $L(x)$ over the given GF, since the zeros of $L(x)$ are the inverse error locations and hence find the error positions $P_1 \ldots P_v$.

Having formalized the BM algorithm, let us now augment our exposition by working through our $RS(12,8,2)$ example assuming the same error pattern, as in our PGZ-decoder example.

4.3.5.5 Berlekamp-Massey Decoding Example

In order to familiarize ourselves with the somewhat heuristic nature of the BM algorithm, we return to our example in Section 4.3.5.3 and solve the same problem using the BM algorithm instead of the PGZ algortihm.

Example 7:
Computation of the error locator polynomial and error positions
in the $RS(12,8)$ $GF(2^m)$ code employing the BM algorithm.

The syndromes from Section 4.3.5.3 are as follows:

$$S_1 = \alpha^8, \quad S_2 = \alpha^8, \quad S_3 = \alpha^5, \quad S_4 = \alpha^6.$$

Let us now follow *Steps* $1 - 14$ of the formally stated BM algorithm:

1 Initialization:

$$i = 0, \quad l = 0, \quad A(x) = 1 \quad L(x) = L_0 = 1$$

Now we work through the flowchart of Figure 4.6 $2t = 4$ times:

2 End of iteration? No, because $0 \neq 4$; hence, go to *Step 3*.

3 Compute the discrepancy:

$$d = \sum_{n=0}^{l} L_n . S_{i+1-n} = L_0 . S_1 = 1.\alpha^8 = \alpha^8.$$

4 Check if $d = 0$? Since $d = \alpha^8 \neq 0$, a new LFSR design is required; hence, go to *Step 6*.

4.3. RS AND BCH CODES

6 The new temporary connection polynomial is given by:

$$T(x) = L(x) - d.x.A(x) = 1 - \alpha^8.x,$$

go to *Step 7*.

7 Check whether the LFSR has to be lengthened. Since $2l = 0$ and $i = 0$, $2l \leq i$ is true and the LFSR must be lenghtened; hence, go to *Step 9*.

9 Store the most recent connection polynomial $L(x)$ after normalizing it by its discrepancy d for later use as auxiliary LFSR $A(x)$:

$$A(x) = \frac{L(x)}{d} = \frac{1}{\alpha^8} = \alpha^7$$

and go to *Step 10*.

10 Update the connection polynomial $L(x)$ with the contents of the temporary connection polynomial as follows:

$$L(x) = T(x) = 1 - \alpha^8.x$$

and go to *Step 11*.

11 Update the LFSR length according to:

$$l = i + 1 - l = 0 + 1 - 0 = 1$$

and go to *Step 12*.

12 Increment the iteration index by assigning $i := i + 1 = 0 + 1 = 1$ and go to *Step 2*.

$$i = 1,\ l = 1,\ L(x) = 1 - \alpha^8.x,\ A(x) = \alpha^7$$

---

2 Continue iterating, and go to *Step 3*.

3

$$\begin{aligned} d &= \sum_{n=0}^{l} L_n.S_{i+1-n} = L_0.S_2 + L_1.S_1 \\ &= 1.\alpha^8 + (-\alpha^8).\alpha^8 = \alpha^8 + \alpha^1 = \alpha^{10} \end{aligned}$$

4 Since $d \neq 0$, go to *Step 6*.

6 Compute the temporary corrected LFSR connection polynomial:

$$\begin{aligned} T(x) &= L(x) - d.x.A(x) = 1 - \alpha^8.x - \alpha^{10}.x.\alpha^7 \\ &= 1 + (\alpha^8 + \alpha^2).x = 1 + \alpha^0.x \end{aligned}$$

and go to *Step 7*.

7 Since $2l = 2$ and $i = 1$, $2l \leq i$ is false and the LFSR does not have to be lengthened; therefore, go to *Step 8*.

8 Update the connection polynomial by the temporary connection polynomial:
$$L(x) = T(x) = 1 + \alpha^0.x,$$
and go to *Step 5*.

5 The contents of the auxiliary LFSR remains unchanged; it must simply be shifted:
$$A(x) = A(x).x = \alpha^7.x.$$
Go to *Step 12*.

12 $i := i + 1 = 2$ and go to *Step 2*.
$$i = 2,\ l = 1,\ L(x) = 1 + \alpha^0.x,\ A(x) = \alpha^7.x$$

2 Carry on iterating, and go to *Step 3*.

3
$$d = L_0.S_3 + L_1.S_2 = 1.\alpha^5 + \alpha^0.\alpha^8 = \alpha^5 + \alpha^8 = \alpha^4;$$
go to *Step 4*.

4 Since $d = \alpha^4 \neq 0$, go to *Step 6*.

6
$$T(x) = L(x) - d.x.A(x) = 1 + \alpha^0.x - \alpha^4.x.\alpha^7.x = 1 + \alpha^0.x + \alpha^{11}.x^2;$$
go to *Step 7*.

7 Since $2l \leq 2$ is true, because $2 \leq 2$, go to *Step 9*.

9 The LFSR must be lengthened; hence, we store the most recent normalized connection polynomial into the auxiliary LFSR:
$$A(x) = \frac{L(x)}{d} = \frac{1 + \alpha^0.x}{\alpha^4} = \alpha^{11} + \alpha^{11}.x;$$
go to *Step 10*.

10 Update the LFSR connection polynomial:
$$L(x) = T(x) = 1 + \alpha^0.x + \alpha^{11}.x^2;$$
go to *Step 11*.

11 Update the LFSR length:
$$l = i + 1 - l = 2 + 1 - 1 = 2;$$
go to *Step 12*.

4.3. RS AND BCH CODES

12 $i := i + 1 = 3$; go to *Step 2*.

$$i = 3, \quad l = 2, \quad L(x) = 1 + \alpha^0.x + \alpha^{11}.x^2, \quad A(x) = \alpha^{11} + \alpha^{11}.x$$

2 Continue iterating, go to *Step 3*.

3
$$d = L_0.S_4 + L_1.S_3 + L_2.S_1 = 1.\alpha^6 + \alpha^0.\alpha^5 + \alpha^{11}.\alpha^8 = \alpha^6 + \alpha^5 + \alpha^4$$
$$d = \alpha^9 + \alpha^4 = \alpha^{14};$$

go to *Step 4*.

4 Since $d = \alpha^{14} \neq 0$, go to *Step 6*.

6 Compute the new temporary connection polynomial:

$$\begin{aligned} T(x) &= L(x) - d.x.A(x) \\ &= 1 + \alpha^0.x + \alpha^{11}.x^2 - \alpha^{14}.x.(\alpha^{11} + \alpha^{11}.x) \\ &= 1 + \alpha^0.x + \alpha^{11}.x^2 - \alpha^{10}.x - \alpha^{10}.x^2 \\ &= 1 + (\alpha^0 + \alpha^{10}).x + (\alpha 11 + \alpha^{10}).x^2 \\ &= 1 + \alpha^5.x + \alpha^{14}.x^2; \end{aligned}$$

go to *Step 7*.

7 Because $2l \leq i$ is false, since $4 > 3$, the LFSR does not have to be lengthened; go to *Step 8*.

8 Simply the temporary connection polynomial is stored in $L(x)$:

$$L(x) = T(x) = 1 + \alpha^5.x + \alpha^{14}.x^2;$$

go to *Step 5*.

5
$$A(x) = x.A(x) = x.(\alpha^{11} + \alpha^{11}.x) = \alpha^{11}.x + \alpha^{11}.x^2;$$

go to *Step 12*.

12 $i := i + 1 = 3 + 1 = 4$; go to *Step 2*.

2 Since $i = 2t$ is true, we have computed the final error locator polynomial $L(x)$, which is identical to that computed in Section 4.3.5.3 using the Peterson-Gorenstein-Zierler decoder.

We can therefore use the error positions, determined previously by Chien search. The error positions computed from this second-order connection polynomial $L(x)$ are: $P_1 = \alpha^{11}$, $P_2 = \alpha^3$. Having determined the error positions, all we have to do for the sake of error correction now is to compute the error magnitudes. This was achieved by the PGZ decoder using matrix inversion, but for large matrices this becomes a computationally demanding operation. Fortunately, it can be circumvented, for example, by the Forney algorithm, which will be the subject of the next subsection.

4.3.5.6 Computation of the Error Magnitudes by the Forney Algorithm [1, 3–9, 84]

The Forney algorithm has been described in a number of classic references [1–9], and here we follow their philosophy. Once the error locator polynomial has been computed to give the error positions $P_1 \ldots P_v$, we concentrate on the determination of the error magnitudes $M_1 \ldots M_v$, which can be computed from the so-called error evaluator polynomial $E(x)$, defined as follows:

$$E(x) = S(x).L(x), \tag{4.69}$$

where $L(x)$ is the error locator polynomial from Equation 4.70:

$$L(x) = \prod_{l=1}^{v}(1 - x.P_l) \tag{4.70}$$

and $S(x)$ is the so-called syndrome polynomial defined here as:

$$S(x) = \sum_{j=1}^{2t} S_j.x^j. \tag{4.71}$$

By substituting the syndromes $S_1 \ldots S_{2t}$ from Equation 4.38 in the above equation, we arrive at:

$$S(x) = \sum_{j=1}^{2t}(\sum_{i=1}^{v} M_i.P_i^j).x^j . \tag{4.72}$$

The error evaluator polynomial $E(x)$ depends on both the error positions and the error magnitudes, as opposed to the error locator $L(x)$, which only depends on the error positions. This fact also reflects the parallelism to the PGZ decoder, where we had two matrix inversions to carry out. The first one has been circumvented by the BM algorithm to determine the error locator polynomial $L(x)$, that is, the error positions. The second matrix inversion is substituted by the Forney algorithm through the computation of the error evaluator polynomial by using the precomputed error positions $P_1 \ldots P_v$ as well.

Since the decoder is capable of computing only the first $2t$ syndromes $S_1 \ldots S_{2t}$ for a t error correcting code, the error evaluator polynomial has to be defined in $mod\ x^{2t}$. This is because, given the code construction, there are $2t$ parity symbols, allowing the determination of t error positions and t error magnitudes. Hence we have:

$$E(x) = S(x).L(x) \quad mod\ x^{2t}, \tag{4.73}$$

which is the key equation to be solved for the unknown error evaluator polynomial in order to determine the error magnitudes $M_1 \ldots M_v$. Let us substitute the polynomials $S(x)$ from Equation 4.72 and $L(x)$ from Equation 4.70 into Equation 4.73:

$$E(x) = (\sum_{j=1}^{2t}\sum_{i=1}^{v} M_i.P_i^j.x^j)(\prod_{l=1}^{v}(1 - x.P_l))\ (mod\ x^{2t}). \tag{4.74}$$

4.3. RS AND BCH CODES

By changing the order of summations and rearranging, we get:

$$\begin{align}
E(x) &= [\sum_{i=1}^{v} M_i \sum_{j=1}^{2t} P_i^j . x^j] \prod_{l=1}^{v} (1 - x.P_l) \ (mod \ x^{2t}) \tag{4.75}\\
&= [\sum_{i=1}^{v} M_i.P_i.x \sum_{j=1}^{2t} (P_i.x)^{j-1}](1 - x.P_i) \prod_{l \neq i} (1 - x.P_l) \ (mod \ x^{2t})\\
&= x. \sum_{i=1}^{v} M_i.P_i.[(1 - x.P_i) \sum_{j=1}^{2t} (P_i.x)^{j-1}] \prod_{l \neq i} (1 - x.P_l) \ (mod \ x^{2t}).
\end{align}$$

If we now expand the square bracketed term, we arrive at:

$$\begin{align}
(1 - xP_i) \sum_{j=1}^{2t} (P_i x)^{j-1} &= (1 - xP_i)[1 + P_i x + (P_i x)^2 + \ldots + (P_i x)^{2t-1}]\\
&= 1 + P_i x + (P_i x)^2 + (P_i x)^3 + \ldots + (P_i x)^{2t-1}\\
&\quad - P_i x - (P_i x)^2 - (P_i x)^3 \ldots - (P_i x)^{2t-1} - (P_i x)^{2t}\\
&= 1 - (P_i x)^{2t}. \tag{4.76}
\end{align}$$

If we substitute this simplification according to Equation 4.76 into Equation 4.75, we get:

$$E(x) = x. \sum_{i=1}^{v} M_i.P_i.[1 - (P_i x)^{2t}] \prod_{l \neq i} (1 - xP_l) \ (mod \ x^{2t}). \tag{4.77}$$

Since $(P_i x)^{2t}$ in the square bracket gives zero in $(mod \ x^{2t})$, Equation 4.77 yields:

$$E(x) = x. \sum_{i=1}^{v} M_i.P_i. \prod_{\substack{l=1 \\ l \neq i}}^{v} (1 - x.P_l), \tag{4.78}$$

which is a closed-form equation for the computation of the error evaluator polynomial $E(x)$ in terms of the error positions $P_1 \ldots P_v$ and error magnitudes $M_1 \ldots M_v$, enabling us to determine the error magnitudes by the help of the Forney algorithm [84].

Forney algorithm: If one evaluates the error evaluator polynomial at the inverse error positions $P_1^{-1} \ldots P_v^{-1}$, the error magnitudes are given by:

$$M_l = \frac{E(P_l^{-1})}{\prod_{j \neq l}(1 - P_j P_l^{-1})} = -\frac{E(P_l^{-1})}{P_l^{-1} L'(P_l^{-1})}, \tag{4.79}$$

where L' stands for the derivative of the error locator polynomial $L(x)$ with respect to x.

Proof of the Forney algorithm: Let us substitute $x = P_l^{-1}$ into Equation 4.78, which yields:

$$E(P_l^{-1}) = \sum_{i=1}^{v} M_i.P_i.P_l^{-1} \prod_{\substack{j=1 \\ j \neq i}}^{v} (1 - P_l^{-1}.P_j)$$

$$\begin{aligned}
&= M_1.P_1.P_l^{-1} \prod_{\substack{j=1 \\ j \neq 1}}^{v} (1 - P_l^{-1}.P_j) \\
&+ M_2.P_2.P_l^{-1} \prod_{\substack{j=1 \\ j \neq 2}}^{v} (1 - P_l^{-1}.P_j) \\
&\vdots \\
&+ M_v.P_v.P_l^{-1} \prod_{\substack{j=1 \\ j \neq v}}^{v} (1 - P_l^{-1}.P_j) \quad (4.80)
\end{aligned}$$

$$\begin{aligned}
E(P_l^{-1}) &= M_1.P_1.P_l^{-1}[(1 - P_l^{-1}.P_2)(1 - P_l^{-1}.P_3)\ldots(1 - P_l^{-1}.P_v)] \\
&+ M_2.P_2.P_l^{-1}[(1 - P_l^{-1}.P_1)(1 - P_l^{-1}.P_3)\ldots(1 - P_l^{-1}.P_v)] \\
&\vdots \\
&+ M_v.P_v.P_l^{-1}[(1 - P_l^{-1}.P_1)(1 - P_l^{-1}.P_2)\ldots(1 - P_l^{-1}.P_{v-1})].
\end{aligned} \quad (4.81)$$

Observe in the above equation that all the square bracketed terms contain all but one combination of the factors $(1 - P_l^{-1}.P_j)$, including $(1 - P_l^{-1}.P_l)$, which is zero. There is, however, one such square bracketed term, where the missing factor happens to be $(1 - P_l^{-1}.P_l) = 0$. Taking this into account, we arrive at:

$$E(P_l^{-1}) = M_l.P_l.P_l^{-1} \prod_{\substack{j=1 \\ j \neq l}}^{v} (1 - P_l^{-1}.P_j), \quad (4.82)$$

or after expressing the error magnitude M_l from the above equation we get:

$$M_l = \frac{E(P_l^{-1})}{\prod_{\substack{j=1 \\ j \neq l}}^{v} (1 - P_l^{-1}.P_j)}, \quad (4.83)$$

which is the first form of the Forney algorithm.

The second form of the Forney algorithm in Equation 4.80 can be proved by computing the derivative $L'(x)$ of the error locator polynomial $L(x)$, which ensues as follows:

$$L(x) = (1 - xP_1)(1 - xP_2)\ldots(1 - xP_v) = \prod_{j=1}^{v}(1 - xP_j)$$

4.3. RS AND BCH CODES

$$L'(x) = \frac{d}{dx}L(x) = -P_1(1-xP_2)\ldots(1-xP_v) \quad (4.84)$$
$$-(1-xP_1)(1-xP_3)\ldots(1-xP_v)$$
$$-(1-xP_1)(1-xP_2)\ldots(1-xP_v)$$
$$\vdots$$
$$-(1-xP_1)(1-xP_2)\ldots(1-xP_{v-1})$$

$$L'(x) = -\sum_{i=1}^{v} P_i \prod_{j=1}^{v}(1-xP_j). \quad (4.85)$$

By evaluating $L'(x)$ from Equation 4.85 at P_l^{-1} we get:

$$L'(P_l^{-1}) = -\sum_{i=1}^{v} P_i \prod_{\substack{j=1 \\ j \neq i}}^{v}(1-P_l^{-1}P_j), \quad (4.86)$$

where similarly to Equation 4.81 we have only one nonzero term left in the summation. Thus, we arrive at:

$$L'(P_l^{-1}) = -P_l \prod_{\substack{j=1 \\ j \neq l}}^{v}(1-P_l^{-1}P_j). \quad (4.87)$$

Hence, the denominator of Equation 4.80 can be formulated is as follows:

$$\prod_{\substack{j=1 \\ j \neq l}}^{v}(1-P_l^{-1}P_j) = -P_l^{-1}.L'(P_l^{-1}), \quad (4.88)$$

which proves the second form of the algorithm in Equation 4.80.

Let us now compute the error magnitudes in our standard example of the $RS(12,8)$ code.

4.3.5.7 Forney Algorithm Example

Example 8:
Computation of the error evaluator polynomial and error magnitudes in an $RS(12,8)$ $GF(2^4)$ code using the Forney algorithm.

The error locator polynomial from Section 4.3.5.5 is given by:

$$S(x) = \sum_{j=1}^{2t} S_j x^j = S_1 x + S_2 x^2 + S_3 x^3 + S_4 x^4$$
$$S(x) = \alpha^8 x + \alpha^8 x^2 + \alpha^5 x^3 + \alpha^6 x^4.$$

The error evaluator polynomial $E(x) = L(x)S(x)$ from the definition of Equation 4.73 is computed as follows:

$$E(x) = (1 + \alpha^5 x + \alpha^{14} x^2)(\alpha^8 x + \alpha^8 x^2 + \alpha^5 x^3 + \alpha^6 x^4) \ (mod \ x^4)$$

$$\begin{aligned}
= \alpha^8 x & \quad +\alpha^8 x^2 & +\alpha^5 x^3 & \quad +\alpha^6 x^4+ & & \\
& \quad +\alpha^{13} x^2 & +\alpha^{13} x^3 & \quad +\alpha^{10} x^4 & +\alpha^{11} x^5+ & \\
& & +\alpha^7 x^3 & \quad +\alpha^7 x^4 & +\alpha^4 x^5 & \quad +\alpha^5 x^6
\end{aligned}$$

$$\begin{aligned}
E(x) & = \alpha^8 x + (\alpha^8 + \alpha^{13}) x^2 + (\alpha^5 + \alpha^{13} + \alpha^7) x^3 \\
& + (\alpha^6 + \alpha^{10} + \alpha^7) x^4 + (\alpha^{11} + \alpha^4) x^5 + \alpha^5 x^6 \\
& = (\alpha^8 x + \alpha^3 x^2 + 0.x^3 + 0.x^4 + \alpha^{13} x^5 + \alpha^5 x^6) \ (mod \ x^4) \\
& = \alpha^3 x^2 + \alpha^8 x.
\end{aligned}$$

The error magnitudes can be computed from $E(x)$ by Forney's algorithm using Equation 4.79:

$$M_l = \frac{E(P_l^{-1})}{\prod_{j \neq l}(1 - P_j P_l^{-1})}. \tag{4.89}$$

The error positions are: $P_1 = \alpha^{11}$, $P_1^{-1} = \alpha^4$, $P_2 = \alpha^3$, $P_2^{-1} = \alpha^{12}$; hence, we have:

$$M_1 = \frac{E(P_1^{-1})}{(1 - P_2 P_1^{-1})} = \frac{\alpha^3(\alpha^4)^2 + \alpha^8 \alpha^4}{1 - \alpha^3 \alpha^4} = \frac{\alpha^{11} + \alpha^{12}}{1 - \alpha^7} = \frac{\alpha^0}{\alpha^9} = \alpha^6$$

$$M_2 = \frac{E(P_2^{-1})}{(1 - P_1 P_2^{-1})} = \frac{\alpha^3(\alpha^{12})^2 + \alpha^8 \alpha^{12}}{1 - \alpha^{11} \alpha^{12}} = \frac{\alpha^{12} + \alpha^5}{1 - \alpha^8} = \frac{\alpha^{14}}{\alpha^2} = \alpha^{12}.$$

The error magnitudes computed by the Forney algorithm are identical to those computed by the *PGZ* decoder. Thus, the quantities required for error correction are given by:

$$(P_1, M_1) = (\alpha^{11}, \alpha^6)$$
$$(P_2, M_2) = (\alpha^3, \alpha^{12}).$$

In order to compute the error-free information symbols, one simply has to add the error magnitudes to the received symbols at the error positions computed, as we have shown in our PGZ decoding example.

Clark and Cain [7] have noted that since $E(x) = L(x).S(x)$, $E(x)$ can be computed in the same iterative loop, as $L(x)$, which was portrayed using the BM algorithm in Figure 4.6. In order to achieve this, however, the original BM algorithm's flowchart in Figure 4.7 has to be properly initialized and slightly modified, as it will be elaborated on below. If we associate a separate normalized LFSR correction term $C(x)$ with $E(x)$, which is the counterpart of the auxiliary LFSR connection polynomial $A(x)$ in the BM algorithm, then the correction formula for $E(x)$ is identical to that of $A(x)$ in Equation 4.64, yielding:

$$E^{(i+1)}(x) = E^{(i)}(x) - d^{(i)}.x.C(x). \tag{4.90}$$

4.3. RS AND BCH CODES

Then, by using a second temporary storage polynomial $T_2(x)$, *Steps 6, 8, 5, 9, and 10* of the flowchart in Figure 4.6 are extended by the corresponding operations for $T_2(x)$, $E(x)$ and $C(x)$, as follows:

Step 6b:
$$T_2^{(i)}(x) := E^{(i)}(x) - d^{(i)}.x.C(x) \qquad (4.91)$$

Step 8b:
$$E^{(i)}(x) := T_2^{(i)}(x) \qquad (4.92)$$

Step 5b:
$$C^{(i)}(x) := x.C^{(i)}(x) \qquad (4.93)$$

Step 9b:
$$C^{(i)}(x) := E^{(i)}(x)/d^{(i)} \qquad (4.94)$$

Step 10b:
$$E^{(i)}(x) := T_2^{(i)}(x). \qquad (4.95)$$

The complete *Berlekamp–Massey* flowchart with these final refinements is depicted in Figure 4.7.

We now round off our discussion of the BM algorithm with the computation of the error evaluator polynomial $E(x)$ for our $RS(12,8)$ $GF(2^4)$ example using the flowchart of Figure 4.7.

4.3.5.8 Error Evaluator Polynomial Computation

Example 9:
```
Computation of the error evaluator polynomial E(x) in an
RS(12,8) GF(2^4 systematic code.
```

1. $E(x) = 0$, $C(x) = 1$, $l = 0$, $i = 0$; go to *Step 2*.
2. $0 \neq 4$; go to *Step 3*.
3. $d = L_0.S_1 = 1.\alpha^8 = \alpha^8$; go to *Step 4*.
4. $d \neq 0$; go to *Step 6*.
6b. $T_2(x) = E(x) - d.x.C(x) = 0 - \alpha^8.x.1 = \alpha^8.x$; go to *Step 7*.
7. $0 \leq 0$; go to *Step 9*.
9b. $C(x) = E(x)/d = 0/\alpha^8 = 0$; go to *Step 10*.
10b. $E(x) = T_2(x) = \alpha^8.x$; go to *Step 11*.
11. $l = i + 1 - l = 0 + 1 - 0 = 1$; go to *Step 12*.
12. $i = i + 1 = 0 + 1 = 1$; go to *Step 2*.

$$\boxed{i = 1,\ l = 1,\ E(x) = \alpha^8.x,\ C(x) = 0;}$$

2. $1 \neq 4$; go to *Step 3*.
3. $d = L_0.S_2 + L_1.S_1 = 1.\alpha^8 + \alpha^8\alpha^8 = \alpha^8 + \alpha^1 = \alpha^{10}$; go to *Step 4*.

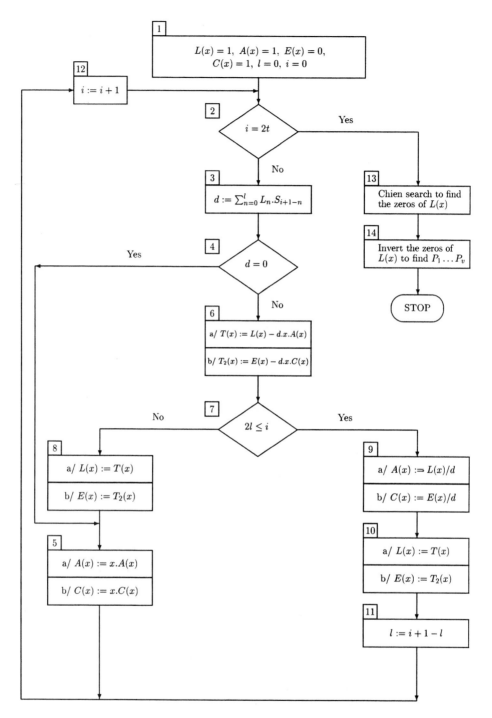

Figure 4.7: The BM algorithm: Computation of both the error locator polynomial $L(x)$ and the error evaluator polynomial $E(x)$.

4.3. RS AND BCH CODES

4 $d \neq 0$; go to *Step 6*.

6b $T_2(x) = E(x) - d.x.C(x) = \alpha^8.x - \alpha^{10}.x.0 = \alpha^8.x$; go to *Step 7*.

7 $2l \leq i$ is false; go to *Step 8*.

8b $E(x) = T_2(x) = \alpha^8.x$; go to *Step 5*.

5b $C(x) = x.C(x) = 0$; go to *Step 12*.

12 $i = i + 1 = 2$; go to *Step 2*.

$$\boxed{i = 2, \; l = 1, \; E(x) = \alpha^8.x, \; C(x) = 0;}$$

2 $2 \neq 4$; go to *Step 3*.

3 $d = L_0.S_3 + L_1.S_2 = 1.\alpha^5 + \alpha^0 \alpha^8 = \alpha^5 + \alpha^8 = \alpha^4$; go to *Step 4*.

4 $d \neq 0$; go to *Step 6*.

6b $T_2(x) = E(x) - d.x.C(x) = \alpha^8.x - \alpha^4.x.0 = \alpha^8.x$; go to *Step 7*.

7 $2l \leq i$ is true; go to *Step 9*.

9b $C(x) = E(x)/d = \alpha^8.x/\alpha^4 = \alpha^4.x$; go to *Step 10*.

10b $E(x) = T_2(x) = \alpha^8.x$; go to *Step 11*.

11 $l = i + 1 - l = 2 + 1 - 1 = 2$; go to *Step 12*.

12 $i = i + 1 = 2 + 1 = 3$; go to *Step 2*.

$$\boxed{i = 3, \; l = 2, \; E(x) = \alpha^8.x, \; C(x) = \alpha^4.x;}$$

2 $3 \neq 4$; go to *Step 3*.

3 $d = L_0.S_4 + L_1.S_3 + L_2.S_1 = 1.\alpha^6 + \alpha^0 \alpha^5 + \alpha^{11} \alpha^8 = \alpha^6 + \alpha^5 + \alpha^4$
$d = \alpha^9 + \alpha^4 = \alpha^{14}$; go to *Step 4*.

4 $d \neq 0$; go to *Step 6*.

6b $T_2(x) = E(x) - d.x.C(x) = \alpha^8.x - \alpha^{14}.x.\,\alpha^4.x = \alpha^8 + \alpha^3.x^2$; go to *Step 7*.

7 Since $2l \leq i$ is false, go to *Step 8*.

8b $E(x) = T_2(x) = \alpha^8.x + \alpha^3.x^2$; go to *Step 5*:

5b $C(x) = x.C(x) = x.\alpha^4.x = \alpha^4.x^2$; go to *Step 12*.

12 $i = i + 1 = 4$; go to *Step 2*.

2 Since $i = 2t$; go to *Step 13*.

13 *Chien* search; go to *Step 14*.

14 Find P_1, P_2 by inverting the zeros of $L(x)$. Stop.

We have found the error evaluator polynomial $E(x) = \alpha^8.x + \alpha^3.x^2$, which is the same as that in our previous example, and consequently results in the same error magnitudes. Having considered some of the algorithmic issues of RS and BCH codecs, in closing in the next section we provide some simulation results for a few practical codecs, in order to be able to gauge the BER reductions achieved. A more comprehensive set of results for a range of convolutional-, block- and concatenated codes can be found, for example, in [9].

4.4 RS and BCH Codec Performance

In our candidate systems throughout this chapter, we often used various nonbinary RS and binary BCH codecs in conjunction with various 1, 2, and 4 bits/symbol modems. In our illustrative example below, we transmit the FEC-coded information bits using differentially detected 16-level Quadrature Amplitude Modulation (16QAM). Our experiments were carried out over the best-case stationary Gaussian channel and the worst-case Rayleigh-fading channel, assuming a carrier frequency of 1.8 GHz. Both of these channels were characterized in our chapter dealing with the description of communication channels.

As an extreme example, we chose a very powerful and complex RS code in order to estimate the maximum practical performance potential achievable. Long RS codes inherently possess good error randomizing properties over bursty channels, since for their duration fading channels are more likely to exhibit a more or less average fading behavior. However, they are fairly complex in terms of their implementation. By contrast, when using short codes, the codewords are more likely to be transmitted either during a fade or during a high signal-to-noise ratio (SNR) interval. In the former case, the code's error correction capability is often overwhelmed by the plethora of transmission errors, while during the high-SNR spells typically there are hardly any transmission errors and hence the code's error correction potential is partially wasted. These problems associated with lower-complexity short codes can be mitigated using interleavers at the cost of increased latency, which can randomize the inherently bursty fading channel statistics. For a comparative description of various interleavers, the interested reader is referred to [9]. Suffice it to say here that in our experiments a simple block-diagonal interleaver was used, which is esentially a memory matrix, filled with FEC-coded symbols on a row-by-row basis that are then transmitted on a column-by-column basis. Hence, the bursts of errors over fading channels are dispersed in the inverse process referred to as de-interleaving, before FEC decoding takes place.

In our deliberations, we will represent the class of long codes using the powerful half-rate RS(380,190,95) code over the finite Galois field GF(512), which encodes 190 nine-bit nonbinary symbols or $190 \times 9 = 1710$ bits into 380 symbols or $380 \times 9 = 3420$ bits and can correct 95 symbol errors. The associated error correction capability in terms of bits is between 95 and $9 \times 95 = 855$ bits, depending on how many bit errors per symbol will occur. We note, therefore, that this code is not appropriate for speech communications, for example, since its associated latency is higher than that tolerated by speech signals, where a latency of typically a few hundred bits can be tolerated. We used an arbitrary input data rate of 86 kbits/s, resulting in an FEC-coded bit rate of about 171 kbits/s. Using our 4 bits/symbol noncoherent 16QAM modem, this yields a signaling rate of 42.75 kBd that requires a bandwidth of about 64 kHz, when employing a modulation excess bandwidth of 50%. The associated modulation issues are discussed in our modulation chapter.

A very good compromise in terms of implementational complexity and error correcting power is constituted by the family of binary BCH codes of 63 bits length. Often used members of this family are the BCH6 = BCH(63,30,6), BCH5 = BCH(63,36,5), BCH4 = BCH(63,39,4), BCH3 = BCH(63,45,3), BCH2 = BCH(63,51,2), and BCH1 =

4.4. RS AND BCH CODEC PERFORMANCE

Code	Coding Rate	Bits/Symb.	Interl. Depth
RS=RS(380,190,95)	0.5	9	1
BCH6=BCH(63,30,6)	0.48	1	57
BCH5=BCH(63,36,5)	0.57	1	48
BCH4=BCH(63,39,4)	0.62	1	44
BCH3=BCH(63,45,3)	0.71	1	38
BCH2=BCH(63,48,2)	0.76	1	36
BCH1=BCH(63,57,1)	0.9	1	30

Table 4.6: List of RS and BCH Codes Evaluated

BCH(63,57,1) codes, correcting 6, 5, 4, 3, 2 and 1 bits per frame, respectively, which are summarized in Table 4.6. The BCH(63,30,6) code has a coding rate of $R = 30/63 \approx 0.5$ and will be used in our experiments as a low-complexity alternative to the complex but similar rate RS(380,190,95) code. The associated bit rate for the BCH(63,30,6) code becomes 179.55 kbits/s, yielding a 16QAM signaling rate of 44.9 kBd and necessitating a bandwidth of about 67 kHz.

In conjunction with the RS(380,190,95), no interleaving was employed and a codeword was associated with $9 \cdot 190 = 1710$ bits. In case of the BCH(63,30,6) code, $1710/30 = 57$ codewords constitute an identical-length frame, allowing for an interleaving depth of 57 BCH codewords. When using the BCH(63,45,3) code, $1710/45 = 38$ codewords were interleaved. Despite interleaving, the BER curve of the modulation scheme often exhibits a residual BER at high SNR values due to the demodulator's inability to cope with the effects of fading, an issue considered in depth in our modulation chapter. This modem deficiency observed over fading channels can be mitigated by using so-called diversity techniques, which in their simplest manifestation are constituted by two antennas and a switch, selecting the arial delivering the higher received signal energy. This technique, which can improve the BER versus channel SNR performance typically by as much as 6 dB was used also in our experiments.

In Figure 4.8 we portray the BER performance of the proposed 16-StQAM modem for a signaling rate of 180 kBd with diversity (D) at a mobile speed of 30 mph, and propagation frequency of 1.8 GHz using the set of BCH2-BCH6 codes and the RS(380,190,95) code. Note the gradual BER improvements, when using stronger codes. It is also interesting to observe that the most complex RS codec is outperformed by the lower complexity BCH5 and BCH6 codecs, when they are interleaved over the duration of the RS code and the SNR is above about 22-24 dB. Similar tendencies are also true for the AWGN channel, depicted in Figure 4.9, which is characteristic of the best stationary scenario. This is explained by the fact that the interleaver depth is sufficiently high for the shorter codes to ensure sufficient fading randomization, while in case of code overload due to an excess number of transmission errors the shorter codes precipitate fewer errors than the long RS code. Hence, in practical systems these shorter codes are often preferred, unless the long code is used as an external layer in a so-called concatenated code, where the internal coding layer is typically a soft-decision assisted short convolutional code optimized for random error correction,

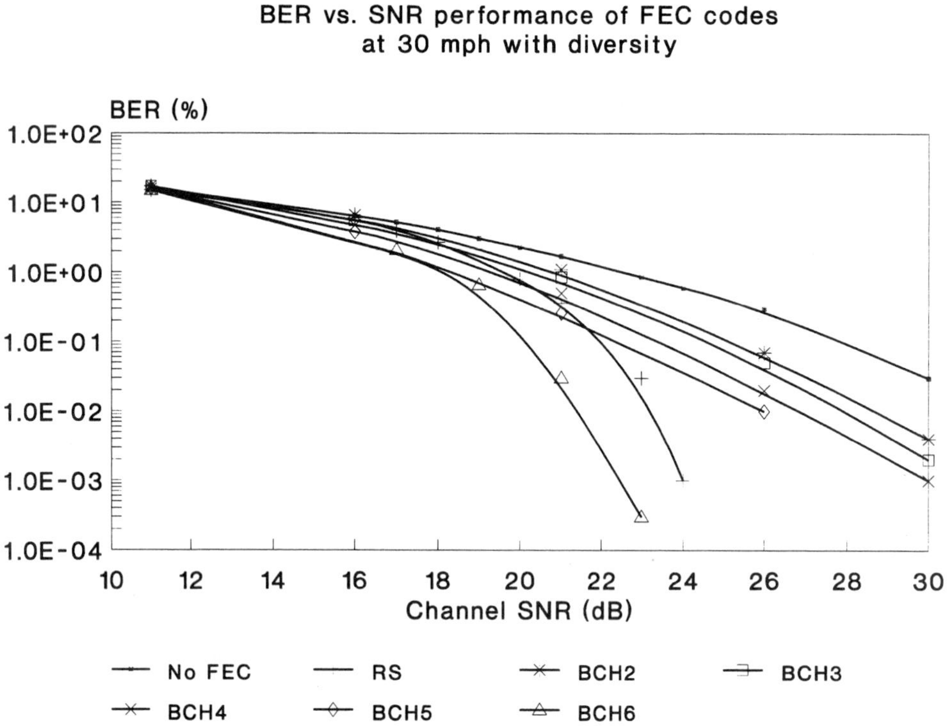

Figure 4.8: BER versus channel SNR performance of the proposed 16-StQAM/TDMA transceiver at 30 mph with diversity using various FEC codes.

an issue detailed in [9]. Let us now conclude this chapter and continue our discussion of modulation issues.

4.5 Summary and Conclusions

In this chapter, after a rudimentary introduction to finite fields, we have considered two decoding algorithms for RS and BCH codes. The *Peterson−Gorenstein−Zierler* decoder, which is based on matrix inversion techniques, and a combination of the *Berlekamp − Massey − Forney* algorithm were considered. The latter technique constitutes a computationally more effective iterative solution to the determination of the error positions and error magnitudes. The number of multiplications is proportional to t^3 in case of the matrix inversion method, as opposed to $6t^2$ in case of the BM algorithm, which means that whenever $t > 6$, the BM algorithm requires less computation than the matrix inversion-based PGZ decoder.

Unfortunately, because of lack of space, we could not delve into details of the Euclidean decoding algorithm [85] or into frequency decoding techniques [3]. Neither could we cover the exploitation of the Toeplitz structure of the syndrome matrix in the key equations [80] for decoding. The interested reader is referred to the references

4.5. SUMMARY AND CONCLUSIONS

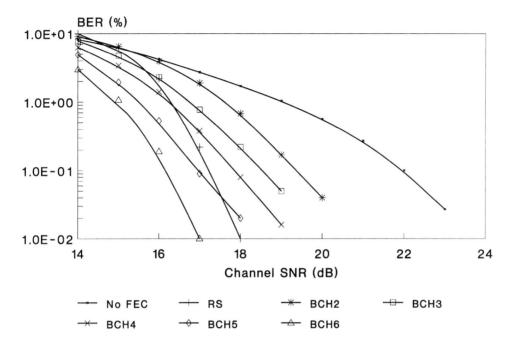

Figure 4.9: BER versus channel SNR performance of the proposed 16-StQAM/TDMA transceiver via AWGN channels using various FEC codes.

cited. This rudimentary introduction to RS and BCH block codes provides us with a practical basis for our further discussions in the forthcoming chapters on system optimization issues, where more performance results are presented.

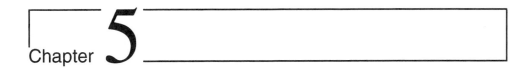

Modulation and Transmission Techniques

5.1 Modulation Issues

5.1.1 Introduction

The appropriate choice of the modem scheme [10] in a wireless system is based on the interplay of equipment complexity, power consumption, spectral efficiency, robustness against channel errors, co-channel and adjacent channel interference tolerance, as well as the propagation phenomena. Most of these factors are affected by the cell size [87]. Equally important are the associated issues of linear or nonlinear amplification and filtering, the applicability of noncoherent, differential detection, soft-decision detection, equalization, and associated issues [10], most of which are addressed in a certain depth in this chapter. The above mentioned, often conflicting factors led to the concept of intelligent multimode transceivers, which will be introduced in this chapter. This concept will then permeate our detailed discussions on a range of such video transceivers throughout this monograph.

Although the potential of adaptive modulation and transmission was recognized some 30 years ago by Cavers [88], the concept of intelligent multimode, multimedia transceivers (IMMT) has only emerged in the context of wireless systems in recent years [10, 89–93]. The range of various existing solutions that have found favor in already operational standard systems was summarized in the excellent overview by Nanda *et al.* [91]. *The aim of these adaptive transceivers is to provide mobile users with the best possible compromise amongst a number of contradicting design factors, such as the power consumption of the handheld portable station (PS), robustness against transmission errors, spectral efficiency, teletraffic capacity, audio/video-quality and so forth [90].*

The fundamental limitation of wireless systems is constituted by their time- and frequency-domain channel fading, as illustrated in Figure 5.1 in terms of the Signal-

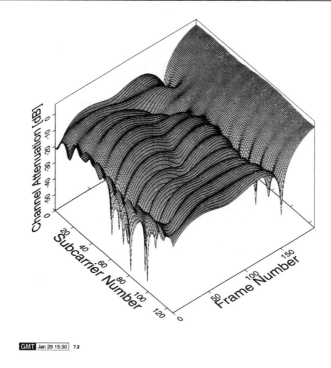

Figure 5.1: Frequency response in the bandwidth of $M \times f_0$ for the 128-channel OFDM system at 155 Mbps [10]. ©Hanzo, Webb, and Keller 2000.

to-Noise Ratio (SNR) fluctuations experienced by a modem over a dispersive channel. Suffice to say here that the violent SNR fluctuations observed both versus time and versus frequency suggest that over these channels no fixed-mode transceiver can be expected to provide an attractive performance, complexity and delay trade-off.

The channel transfer function seen in Figure 5.1 is characteristic of wireless local area networks (WLAN) communicating at a rate of 155 Mbps. This rate is used in Asynchronous Transfer Mode (ATM) systems. We assumed an indoors airport terminal or warehouse environment of dimensions 100m × 100m and a 7-path channel, corresponding to the four walls, ceiling, and floor plus the line-of-sight (LOS) path. The LOS path and the two reflections from floor and ceiling were combined into one single path in the impulse response. The worst-case impulse response associated with the highest path length and delay spread is experienced in the farthest corners of the hall, which was determined using inverse second power law attenuation and the speed of light for the computation of the path delays. The interpretation of this channel transfer function is further augmented in Section 5.2.

Motivated by the above mentioned performance limitations of fixed-mode transceivers, IMMTs have attracted considerable research interest in the past decade [10, 89–94]. Some of these research results are collated in this monograph, and the associated video performance is investigated.

The above mentioned calamities inflicted by the wireless channel can be mitigated by contriving a suite of near-instantaneously adaptive or Burst-by-Burst Adaptive

5.1. MODULATION ISSUES

(BbBA) wideband single-carrier [10, 93], multicarrier, or Orthogonal Frequency Division Multiplex [10, 92] (OFDM) as well as Code Division Multiple Access (CDMA) transceivers [95–99]. A plethora of such systems will be studied in Chapter 21. The aim of these IMMTs is to communicate over hostile mobile channels at a higher integrity or higher throughput, than conventional fixed-mode transceivers. A number of existing wireless systems already support some grade of adaptivity [91], and future research is likely to promote these principles further by embedding them into the already existing standards. For example, due to their high control channel rate and with the advent of the well-known Orthogonal Variable Spreading Factor (OVSF) codes the thrid-generation UTRA/IMT2000 systems are amenable to not only long-term spreading factor reconfiguration, but also to near-instantaneous reconfiguration on a 10ms transmission burst-duration basis.

With the advent of BbBA QAM, OFDM, or CDMA transmissions it becomes possible for mobile stations (MS) to invoke, for example, in indoor scenarios or in the central propagation cell region — where typically benign channel conditions prevail — a high-throughput modulation mode, such as 4 bit/symbol Quadrature Amplitude Modulation (16QAM). By contrast, a robust, but low-throughput modulation mode, such as 1 bit/symbol Binary Phase Shift Keying (BPSK) can be employed near the edge of the propagation cell, where hostile propagation conditions prevail. The BbBA QAM, OFDM, or CDMA mode switching regime is also capable of reconfiguring the transceiver at the rate of the channel's slow- or even fast-fading. This may prevent premature hand-overs and — more importantly — unnecessary powering up, which would inflict an increased interference upon co-channel users, resulting in further potential power increments. This detrimental process could result in all mobiles operating at unnecessarily high power levels.

A specific property of these transceivers is that their bit rate fluctuates, as a function of time. This is not an impediment in the context of data transmission. However, in interactive speech [94] or video communications appropriate source codecs have to be designed, which are capable of promptly reconfiguring themselves according to the near-instantaneous bit-rate budget provided by the transceiver. These issues will be studied at a later stage in Chapter 21.

The expected performance of our BbBA transceivers can be characterized with the aid of a whole plethora of performance indicators. In simple terms, adaptive modems out-perform their individual fixed-mode counterparts, since given an average number of transmitted bits per symbol (BPS), their average BER will be lower than that of the fixed-mode modems. From a different perspective, at a given BER their BPS throughput will be always higher. In general, the higher the tolerable BER, the closer the performance to that of the Gaussian channel capacity. Again, this fact underlines the importance of designing programmable-rate, error-resilient source codecs, - such as the Advanced Multi-Rate (AMR) speech codec to be employed in UMTS - which do not expect a low BER.

Similarly, when employing the above BbBA or AQAM principles in the frequency-domain in the context of OFDM [10] or in conjunction with OVSF spreading codes in CDMA systems, attractive system design trade-offs and a high over-all performance can be attained, as is detailed during our further discourse in Chapter 21. However, despite the extensive research in the field by the international community, there is a whole host of problems that remains to be solved and this monograph intends to contribute toward these efforts.

Although this introductory section inevitably relies on some of the parlance used by modulation experts, it does not assume any deep prior knowledge in the field. Our intention in this section is to simply raise the reader's interest in the topic and highlight the range of system design trade-offs encountered, when opting for a particular modulation and demodulation scheme. Modulation and transmission experts may wish to skip this chapter and proceed to the particular chapters of their interest, while those new to the field may wish to revisit this section after reading the rest of the chapter.

Following the above motivation notes, let us now turn our attention to the class of full-response modems, which can exploit the higher Shannonian channel capacity of high-SNR channels by transmitting several bits per information symbol and hence ensure high-bandwidth efficiency. These modems are also often referred to as multi-level or Quadrature Amplitude Modulation (QAM) schemes [10].

5.1.2 Quadrature Amplitude Modulation [10]

5.1.2.1 Background

Until quite recently, QAM developments focused on the benign AWGN telephone line and point-to-point radio applications [100], which led to the definition of the International Telecommunication Union's (ITU) telephone circuit modem standards V.29-V.34 based on various QAM constellations ranging from uncoded 16QAM to trellis coded (TC) 256QAM [10]. In recent years, QAM research for hostile fading mobile channels has been motivated by the ever-increasing bandwidth efficiency demand for mobile telephony [101, 102], although QAM schemes require power-inefficient class A or AB linear amplification [103–106]. However, the power consumption of the low-efficiency class-A amplifier [105, 106] is less critical than that of the digital speech, image, and channel codecs. Out-of-band emissions due to class AB amplifier non-linearities generating adjacent channel interferences can be reduced by some 15 dB using the adaptive predistorter proposed by Stapleton *et al.* [103–108]. The spectral efficiency of QAM in various macro- and microcellular frequency reuse structures was studied in comparison to a range of other modems in Chapter 16 of [10], while adaptive modem arrangements were proposed, for example, in [109–115]. Let us now highlight the basic concepts of quadrature amplitude modulation.

5.1.2.2 Modem Schematic

Multilevel full-response modulation schemes have been considered in depth in reference [10]. This chapter offers only a terse introduction, concentrating on the fundamental modem schematic of Figure 5.2. If an analog source signal must be transmitted, the signal is first low-pass filtered and analog-to-digital converted (ADC) using a sampling frequency satisfying the Nyquist criterion. The generated digital bit-stream is then mapped to complex modulation symbols by the MAP block, as seen in Figure 5.3 in case of mapping 4 bits/symbol to a 16QAM constellation.

5.1. MODULATION ISSUES

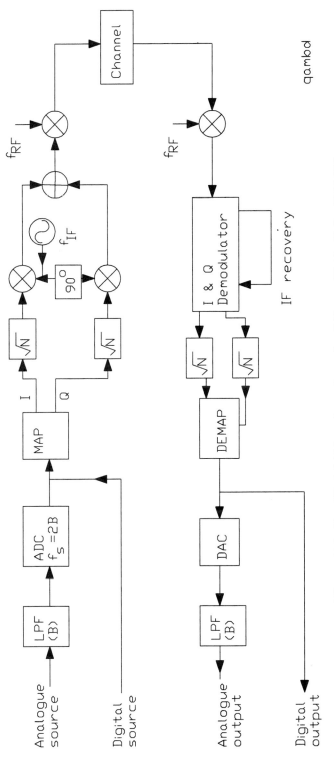

Figure 5.2: Simplified QAM modem schematic. ©Hanzo, Webb, Keller 2000 [10].

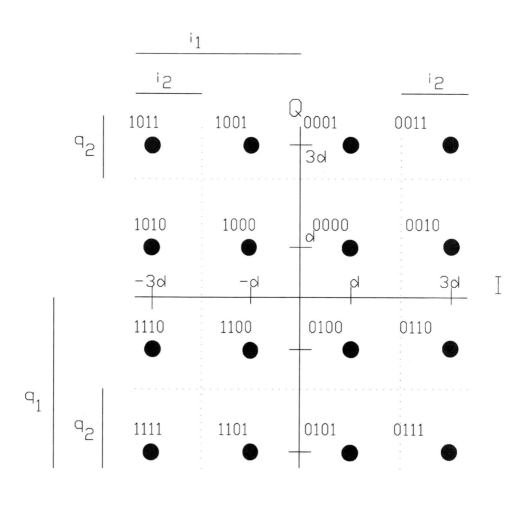

Figure 5.3: 16QAM square constellation. ©Hanzo, Webb, Keller 2000 [10].

5.1. MODULATION ISSUES

5.1.2.2.1 Gray Mapping and Phasor Constellation The process of mapping the information bits to the bit-streams modulating the in-phase (I) and quadrature (Q) carriers, which are orthogonal to each other and hence can be transmitted in the same bandwidth, plays a fundamental role in determining the properties of the modem, which will be elaborated on at a later stage in Section 5.1.2.3. The mapping can be represented by the constellation diagram of Figure 5.3. A phasor constellation is defined as the resulting two-dimensional plot, when the amplitudes of the I and Q levels of each of the points that could be transmitted (the constellation points) are drawn in a rectangular coordinate system. For a simple binary amplitude modulation scheme, the constellation diagram would be two points both on the positive x-axis. For a binary phase shift keying (BPSK) scheme, the constellation diagram would again consist of two points on the x-axis but both equidistant from the origin, one to the left and one to the right. The "negative amplitude" of the point to the left of the origin represents a phase shift of 180 degrees in the transmitted signal. If we allow phase shifts of angles other than 0 and 180 degrees, then the constellation points move off the x-axis. They can be considered to possess an amplitude and phase, the amplitude representing the magnitude of the transmitted carrier and the phase representing the phase shift of the carrier relative to the local oscillator in the transmitter. The constellation points may also be considered to have cartesian, or complex, coordinates, which are normally referred to as in-phase (I) and quadrature (Q) components corresponding to the x- and y-axes, respectively.

In the square-shaped 16QAM constellation of Figure 5.3, each phasor is represented by a 4-bit symbol, constituted by the in-phase bits i_1, i_2 and quadrature bits q_1, q_2, which are interleaved to yield the sequence i_1, q_1, i_2, q_2. The quaternary quadrature components I and Q are Gray encoded by assigning the bits *01, 00, 10,* and *11* to the levels *3d, d, -d,* and *-3d*, respectively. This constellation is widely used because it has equidistant constellation points arranged in a way that the average energy of the phasors is maximized. Using the geometry of Figure 5.3, we can compute the average energy as

$$E_0 = (2d^2 + 2 \times 10d^2 + 18d^2)/4 = 10 \times d^2. \qquad (5.1)$$

For any other phasor arrangement, the average energy will be less, and therefore, assuming a constant noise energy, the signal-to-noise ratio required to achieve the same bit error rate (BER) will be higher, a topic to be studied comparatively in the context of two different 16QAM constellations at a later stage in Section 5.1.2.3.

Notice from the mapping in Figure 5.3 that the Hamming distance among the constellation points, which are "closest neighbors" with a Euclidean distance of $2d$, is always one. The Hamming distance is defined as the number of different bits associated with the individual phasor constellation points. For example, the phasor points labeled 0101 and 0111 would have a Hamming distance of 1, and points labeled 0101 and 0011 would have a Hamming distance of 2. This is a fundamental feature of the Gray coding process and ensures that whenever a transmitted phasor is corrupted by noise sufficiently that it is incorrectly identified as a neighboring constellation point, the demodulator will choose a phasor with a single bit error. This minimizes the error probability.

It is plausible that the typical quaternary I or Q component sequence generated by the MAP block of Figure 5.2 would require an infinite transmission bandwidth due to the abrupt changes at the signaling interval boundaries. Hence, these signals must be bandlimited before transmission in order to contain the spectrum within a limited band and hence to minimize the adjacent channel interference inflicted on other users or systems in the neighboring spectral bands. This filtering is indicated in Figure 5.2 by the square root Nyquist-filter blocks denoted by $\sqrt{N}$, where the rationale behind the notation will become clear in the next section.

5.1.2.2.2 Nyquist Filtering A full theoretical treatment of Nyquist filtering was provided, for example, in [10]; hence, here we restrict our discussion to a rudimentary introduction. An ideal linear-phase low-pass filter (LPF) with a cutoff frequency of $f_N = f_s/2$, where $f_s = 1/T$ is the signaling frequency, T is the signaling interval duration, and $f_N = 1/(2T)$ is the Nyquist frequency, would be able to pass the main spectral lobe constituting most of the energy of the quadrature components I and Q within a compact frequency band. Because of linear phase response of the filter, all frequency components would exhibit the same group delay. Since such a filter has a $(\sin x)/x$ function-shaped impulse response with equidistant zero-crossings at the sampling instants $n \cdot T$, this ideal low-pass filter does not result in intersymbol interference (ISI) between consecutive signaling symbols, provided that they are spaced at intervals nT.

After its inventor Nyquist [17], this ideal low-pass transfer function and its derivatives about to be introduced in the next paragraph are referred to as the Nyquist characteristic. However, such an ideal low-pass filter is unrealizable, for all practical low-pass (LP) filters exhibit amplitude and phase distortions, particularly toward the transition between the pass- and stopband. Conventional Butterworth, Chebichev, or inverse-Chebichev LP filters have impulse responses with nonzero values at the equispaced sampling instants $n \cdot T$ and hence introduce ISI. They therefore degrade the bit error rate (BER) performance.

Nyquist's fundamental theoretical work [17] suggested that special pulse shaping filters must be employed, ensuring that the total transmission path, including the transmitter, receiver, and the channel, has an impulse response with a unity value at the current signaling instant and zero-crossings at all other consecutive sampling instants $n \cdot T$. He showed that any odd-symmetric frequency-domain extension characteristic fitted to the ideal LPF amplitude transfer function yields such an impulse response and is therefore free from ISI. Two examples of the corresponding filter characteristics are shown in Figure 5.4, which are described in [10]

A practical odd-symmetric extension characteristic is the raised-cosine (RC) characteristic fitted to the above-mentioned ideal low-pass filter characteristic [10]. The parameter controlling the bandwidth of the Nyquist filter is the roll-off factor α, which is unity, if the ideal LPF bandwidth is doubled by the extension characteristic. If $\alpha = 0.5$, a total bandwidth of $1.5 \times f_N = 1.5/(2T)$ results, and so on. The lower the value of the roll-off factor, the more compact the spectrum becomes, but the higher the complexity of the required filter and other receiver circuitry, such as clock and carrier recovery [10]. The stylized frequency response of these filters is shown in Figure 5.4 for $\alpha = 0.9$ and $\alpha = 0.1$. It follows from Fourier theory that the wider the transmission band, the more sharply decaying the impulse response. A sharply decaying impulse response has a favorable effect as regards to mitigating the

5.1. MODULATION ISSUES

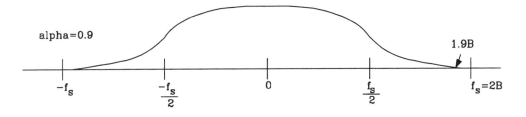

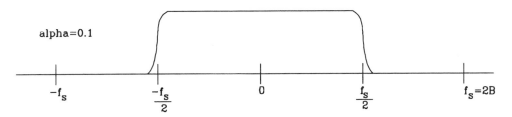

Figure 5.4: Stylized frequency response of two filters with $\alpha = 0.9$ and 0.1. ©Hanzo, Webb, Keller 2000 [10].

potential ISI in case of imperfect clock recovery, when there is a time-domain jitter superimposed on the optimum sampling instant. Hence, in terms of system performance an $\alpha = 1$ filtering scheme is more favorable than a more sharply filtered but more bandwidth-efficient scheme.

In case of Additive White Gaussian Noise (AWGN) with a uniform power spectral density (PSD), the noise power admitted to the receiver is proportional to its bandwidth. Therefore, it is also necessary to limit the received signal bandwidth at the receiver to a value close to the transmitter's bandwidth. Optimum detection theory [21] shows that the SNR is maximized, if matched filtering is used, where the Nyquist characteristic of Figure 5.4 is divided between two identical filters, a transmitter- and a receiver-filter, each characterized by the square root of the Nyquist shape, as suggested by the filters $\sqrt{N}$ in Figure 5.2.

In concluding our discussion on filtering issues, we note that Feher [116] proposed nonlinear filtering (NLF) as a low-complexity alternative to Nyquist filtering, which operates by simply fitting a time-domain quarter period of a sine wave between two symbols for both of the quadrature carriers. This technique can be simply implemented by using a lookup table, and there is no contribution from previous symbols at any sample point, which is advantageous when complex high-level QAM constellations are transmitted. The disadvantage of this form of filtering is that it is less spectrally efficient than optimal partial-response filtering schemes. Nevertheless, its implementational advantages often render this loss of spectral efficiency acceptable. The power spectrum of a NLF signal is given by [116]:

$$S(f) = T \left(\frac{\sin 2\pi fT}{2\pi fT} \frac{1}{1 - 4(fT)^2} \right)^2 \qquad (5.2)$$

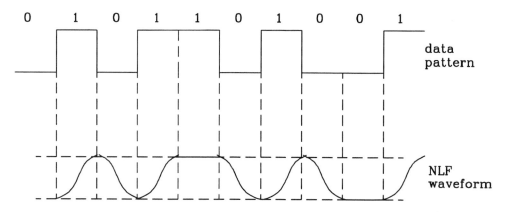

Figure 5.5: Stylized NLF waveforms. ©Hanzo, Webb, Keller 2000 [10].

and the corresponding original and NLF waveforms are given in Figure 5.5 for the I or Q component.

5.1.2.2.3 Modulation and Demodulation Once the analog I and Q signals have been generated and filtered, they are modulated by an I-Q modulator as shown in Figure 5.2. This modulator essentially consists of two mixers, one for the I channel and another for the Q channel. The I channel is mixed with an intermediate frequency (IF) signal that is in phase with respect to the carrier, and the Q channel is mixed with an IF that is 90 degrees out of phase. This process allows both signals to be transmitted over a single channel within the same bandwidth using quadrature carriers. In a similar fashion, the signal is demodulated at the receiver. Provided that the signal degradation is kept to a minimum, the orthogonality of the I and the Q channels will be retained, and their information sequences can be independently demodulated.

Following I-Q modulation, the signal is modulated by a radio frequency (RF) mixer, increasing its frequency to that used for transmission. Since the IF signal occurred at both positive and negative frequencies, it will occur at both the sum and difference frequencies when mixed up to the RF. Since there is no reason to transmit two identical sidebands, one is usually filtered out, as seen plotted in dashed lines in Figure 5.6. We also note that in theory one could dispense with the IF stage, mixing the baseband component directly to the transmission frequency, if it would be possible to design the required extremely narrowband so-called notch filters at the RF for removing the unwanted modulation products and out-of-band spectral spillage. Since this results in filter-design problems, in practical systems the signal is converted up to the RF usually in two or more mixing stages, which renders the design of the corresponding filters earlier. This is because the width of their passband selected to the carrier frequency becomes lower.

The transmission channel is often the most critical factor influencing the performance of any communications system. Here we consider only the addition of noise based on the signal-to-noise ratio (SNR). The noise is often the major contributing factor to signal degradation, and its effect is exhibited in terms of a noise floor, as

5.1. MODULATION ISSUES

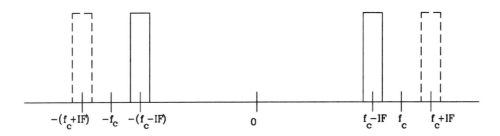

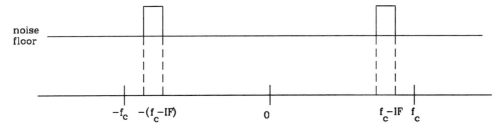

Figure 5.6: Stylized transmitted and received spectra. ©Hanzo, Webb, Keller 2000 [10].

portrayed in the received RF spectrum of Figure 5.6.

The RF demodulator mixes the received signal down to the IF for the I-Q demodulator. In order to accurately mix the signal back to the appropriate intermediate frequency, the RF mixer operates at the difference between the IF and RF frequencies. Since the I-Q demodulator includes IF recovery circuits, the accuracy of the RF oscillator frequency is not critical. However, it should be stable, exhibiting a low-phase noise, since any noise present in the down-conversion process will be passed on to the detected I and Q baseband signals, thereby adding to the possibility of bit errors. The recovered IF spectrum is similar to the transmitted one but with the additive noise floor seen in the RF spectrum of Figure 5.6.

Returning to Figure 5.2, we see that I-Q demodulation takes place in the reverse order to the modulation process. The signal is split into two paths, with each path being mixed down with IFs that are 90 degrees apart. Since the exact frequency of the original reference must be known to determine absolute phase, IF carrier recovery circuits are used to reconstruct the precise reference frequency at the receiver. The recovered I component should be nearly identical to that transmitted, with the only differences being caused by noise. The problem of carrier- and clock recovery is addressed, for example, in [10]

5.1.2.2.4 Data Recovery Once the analog I and Q components have been recovered, they must be digitized. This is carried out by the bit detector. The bit detector determines the most likely bit transmitted by sampling the I and Q signals

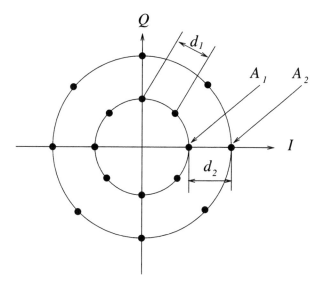

Figure 5.7: Star 16QAM constellation.

at the correct sampling instants and comparing them to the legitimate I and Q values of $-3d, -d, d, 3d$ in the case of a square 16QAM constellation. From each I and Q decision, 2 bits are derived, leading to a 4-bit 16QAM symbol. The four recovered bits are then passed on to the Digital-Analog Coverter (DAC). Although the process might sound simple, it is complicated by the fact that the "right time" to sample is a function of the clock frequency at the transmitter. The data clock must be regenerated upon recovery of the carrier. Any error in clock recovery will increase the BER. Again, these issues are treated in more depth in [10].

If there is no channel noise or the SNR is high, the reconstructed digital signal is identical to the original input signal. Provided the DAC operates at the same frequency and with the same number of bits as the input ADC, then the analog output signal after low-pass filtering with a cutoff frequency of B is also identical to the output signal of the LPF at the input of the transmitter. Hence, it is a close replica of the input signal. Following the above basic modem schematic description, let us now consider two often used 16QAM constellations.

5.1.2.3 QAM Constellations

A variety of different constellations have been proposed for QAM transmissions over Gaussian channels. However, in practice often the constellations shown in Figures 5.3 and 5.7 are preferred. The essential problem is to maintain a high minimum distance, d_{min}, between constellation points while keeping the average power required for the constellation to a minimum. Calculation of d_{min} and the average power is a straightforward geometric procedure and has been performed for a range of constellations by Proakis [52]. The results show that the square constellation of Figure 5.3 is optimal for Gaussian channels. We will show that the star constellation of Figure 5.7 requires a

5.1. MODULATION ISSUES

higher energy to achieve the same minimum distance d_{min} among constellation points than the square constellation of Figure 5.3, and hence the latter is often preferred for Gaussian channels. However, there may be implementational reasons for favoring circular constellations over the square ones.

When designing a constellation, consideration must be given to:

1. The minimum Euclidean distance among phasors, which is characteristic of the noise immunity of the scheme.

2. The minimum phase rotation among constellation points, determining the phase jitter immunity and hence the scheme's resilience against clock recovery imperfections and channel phase rotations.

3. The ratio of the peak-to-average phasor power, which is a measure of robustness against nonlinear distortions introduced by the power amplifier.

It is quite instructive to estimate the optimum ring ratio RR for the star constellation of Figure 5.7 in AWGN under the constraint of a constant average phasor energy E_0. Accordingly, a high ring ratio value implies that the Euclidean distance among phasors on the inner ring is reduced, while the distance among phasors on different rings is increased. In contrast, upon reducing the ring ratio the cross-ring distance is reduced, and the distances on the inner ring become larger.

Intuitively, one expects that there will be an optimum ring ratio, where the overall bit error rate (BER) constituted by detection errors on the same ring plus errors between rings is minimized. Suffice it to say here that the minimum Euclidean distance among phasors is maximized if $d_1 = d_2 = A_2 - A_1$ in the star constellation of Figure 5.7. Using the geometry of Figure 5.7, we can write that:

$$\cos 67.5° = \frac{d_1}{2} \cdot \frac{1}{A_1}$$
$$d_1 = 2 \cdot A_1 \cdot \cos 67.5°$$

and hence

$$A_2 - A_1 = d_1 = d_2 = 2 \cdot A_1 \cdot \cos 67.5°.$$

Upon dividing both sides by A_1 and introducing the ring ratio RR, we arrive at:

$$RR - 1 = 2 \cdot \cos 67.5°$$
$$RR \approx 1.77.$$

Simulation results using a variety of ring ratios in the interval of $1.5 < RR < 3.5$ both over Rayleigh and AWGN channels showed [29, 117] that the BER does not strongly depend on the ring ratio, exhibiting a flat BER minimum for RR values in the above range.

Under the constraint of having identical distances among constellation points, when $d_1 = d_2 = d$, the average energy E_0 of the star constellation can be computed as follows:

$$E_0 = \frac{8 \cdot A_1^2 + 8 \cdot A_2^2}{16} = \frac{1}{2}(A_1^2 + A_2^2)$$

where

$$A_1 = \frac{d}{2 \cdot \cos 67.5°} \approx \frac{d}{0.765} \approx 1.31d$$

Type	θ_{min}	d_{min}	r
Star	$45°$	$0.53\sqrt{E_0}$	1.5
Square	$< 45°$	$0.63 \cdot \sqrt{E_0}$	1.8

Table 5.1: Comparison of the Star and Square Constellations

and
$$A_2 \approx 1.77 \cdot A_1 \approx 2.3d$$
yielding
$$E_0 \approx 0.5 \cdot (5.3 + 1.72)d^2 \approx 3.5d^2.$$
The minimum distance of the constellation for an average energy of E_0 becomes:
$$d_{min} \approx \sqrt{E_0/3.5} \approx 0.53 \cdot \sqrt{E_0},$$
while the peak-to-average phasor energy ratio is:
$$r \approx \frac{(2.3d)^2}{3.5d^2} \approx 1.5.$$
The minimum phase rotation θ_{min}, the minimum Euclidean distance d_{min}, and the peak-to-average energy ratio r are summarized in Table 5.1 for both of the above constellations.

Let us now derive the above characteristic parameters for the square constellation. Observe from Figure 5.7 that $\theta_{min} < 45°$, while the distance between phasors is $2 \cdot d$. Hence, the average phasor energy becomes:
$$\begin{aligned} E_0 &= \frac{1}{16}\left[4 \cdot (d^2 + d^2) + 8(9d^2 + d^2) + 4 \cdot (9d^2 + 9d^2)\right] \\ &= \frac{1}{16}(8d^2 + 80 \cdot d^2 + 72d^2) \\ &= 10d^2. \end{aligned}$$
Assuming the same average phasor energy E_0 as for the star constellation, we now have a minimum distance of
$$d_{min} = 2d = 2 \cdot \sqrt{E_0/10} = \sqrt{E_0/2.5} \approx 0.63 \cdot \sqrt{E_0}.$$
Finally, the peak-to-average energy ratio r is given by:
$$r = \frac{18d^2}{10d^2} = 1.8.$$
The square constellation's characteristics are also summarized in Table 5.1. Observe that the star constellation has a higher jitter immunity and a slightly lower peak-to-average energy ratio than the square scheme. However, the square phasor constellation has an almost 20% higher minimum distance at the same average phasor energy, and hence it is very attractive for AWGN channels, where noise is the dominant channel impairment. Let us now consider the bit error rate (BER) versus channel signal-to-noise ratio (SNR) performance of the maximum-minimum distance square-constellation 16QAM over AWGN channels.

5.1. MODULATION ISSUES

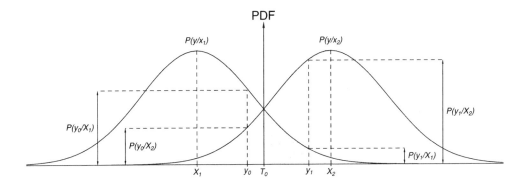

Figure 5.8: Transmitted samples and noisy received samples for BPSK.

5.1.2.4 16QAM BER versus SNR Performance over AWGN Channels

5.1.2.4.1 Decision Theory Before analyzing the effects of errors, let us briefly review the roots of decision theory in the spirit of Bayes' theorem formulated as follows:

$$P(X/Y) \cdot P(Y) = P(Y/X) \cdot P(X) = P(X,Y), \qquad (5.3)$$

where the random variables X and Y have probabilities of $P(X)$ and $P(Y)$, their joint probability is $P(X,Y)$, and their conditional probabilities are given by $P(X/Y)$ and $P(Y/X)$.

In decision theory, the above theorem is invoked in order to infer from the noisy analog received sample y, what the most likely transmitted symbol was, assuming that the so-called *a-priori* probability $P(x)$ of the transmitted symbols $x_n, n = 1\ldots M$ is known. Given that the received sample y is encountered at the receiver, the conditional probability $P(x_n/y)$ quantifies the chance that x_n has been transmitted:

$$P(x_n/y) = \frac{P(y/x_n) \cdot P(x_n)}{P(y)}, \qquad n = 1\ldots N \qquad (5.4)$$

where $P(y/x_n)$ is the conditional probability of the continuous-valued noise-contaminated sample y, given that $x_n, n = 1\ldots N$ was transmitted. The probability of encountering a specific y value will be the sum of all possible combinations of receiving y, given that $x_n, n = 1\ldots N$ was transmitted, which can be written as:

$$P(y) = \sum_{n=1}^{N} P(y/x_n) \cdot P(x_n) = \sum_{n=1}^{N} P(y, x_n). \qquad (5.5)$$

Let us now consider the case of binary phase shift keying (BPSK), where there are two legitimate transmitted values, x_1 and x_2 which are contaminated by noise, as portrayed in Figure 5.8. The conditional probability of receiving any particular noise-contaminated analog sample y, given that x_1 or x_2 was transmitted, is quantified by the Gaussian probability density functions (PDFs) seen in Figure 5.8, which are described by:

$$P(y/x) = \frac{1}{\sigma\sqrt{2\pi}} e^{\frac{-(y-x)^2}{2\sigma^2}}, \tag{5.6}$$

where $x = x_1$ or x_2 is the mean and σ^2 is the variance. Observe from the figure that the shaded area represents the probability of receiving values larger than the threshold T_0, when x_1 was transmitted, and this is equal to the probability of receiving a value below T_0, when x_2 was transmitted. As displayed in the figure, when receiving a specific $y = y_0$ sample, there is an ambiguity as to which symbol was transmitted. The corresponding conditional probabilities are given by $P(y_0/x_1)$, and $P(y_0/x_2)$, and their values are also marked on Figure 5.8. Given the knowledge that x_1 was transmitted, we are more likely to receive y_0 than on condition that x_2 was transmitted. Hence, upon observing $y = y_0$, statistically speaking it is advisable to decide that x_1 was transmitted. Following similar logic, when receiving y_1 as seen in Figure 5.8, it is logical to conclude that x_2 was transmitted.

Indeed, according to optimum decision theory [118], the optimum decision threshold above which x_2 is inferred is given by:

$$T_0 = \frac{x_1 + x_2}{2} \tag{5.7}$$

and below this threshold x_1 is assumed to have been transmitted. If $x_1 = -x_2$, then $T_0 = 0$ is the optimum decision threshold minimizing the bit error probability.

In order to compute the error probability in case of transmitting x_1, the PFD $P(y/x_1)$ of Equation 5.6 has to be integrated from x_1 to ∞, which gives the shaded area under the curve in Figure 5.8. In other words, the probability of a zero-mean noise sample exceeding the magnitude of x_1 is sought, which is often referred to as *noise protection distance*, given by the so-called Gaussian Q-function:

$$Q(x_1) = \frac{1}{\sigma\sqrt{2\pi}} \int_{x_1}^{\infty} e^{\frac{-y^2}{2\sigma^2}} dy, \tag{5.8}$$

where σ^2 is the noise variance. Notice that since $Q(x_1)$ is the probability of exceeding the value x_1, it is actually the complementary cumulative density function (CDF) of the Gaussian distribution.

Assuming that $x_1 = -x_2$, the probability that the noise can carry x_1 across $T_0 = 0$ is equal to that of x_2 being corrupted in the negative direction. Hence, assuming that $P(x_1) = P(x_2) = 0.5$, the overall error probability is given by:

$$\begin{aligned} P_e &= P(x_1) \cdot Q(x_1) + P(x_2) \cdot Q(x_2) \\ &= \frac{1}{2}Q(x_1) + \frac{1}{2}Q(x_1) = Q(x_1). \end{aligned} \tag{5.9}$$

The values of the Gaussian Q-function plotted in Figure 5.9 are tabulated in many textbooks [118], along with values of the Gaussian PDF in case of zero-mean, unit-variance processes. For abscissa values of $y > 4$, the following approximation can be used:

$$Q(y) \approx \frac{1}{y\sqrt{2\pi}} e^{\frac{-y^2}{2}} \quad \text{for } y > 4. \tag{5.10}$$

5.1. MODULATION ISSUES

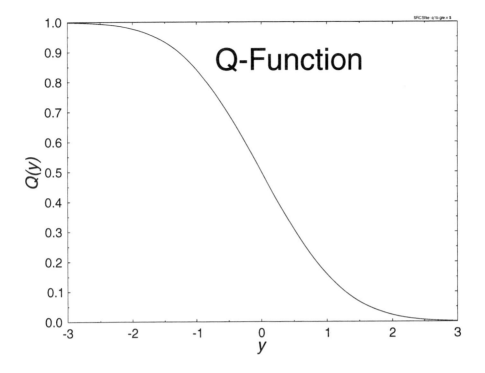

Figure 5.9: Gaussian Q-function.

Having provided a rudimentary introduction to decision theory, let us now focus our attention on the demodulation of 16QAM signals in AWGN.

5.1.2.4.2 QAM Modulation and Transmission In general, the modulated signal can be represented by

$$s(t) = a(t)\cos[2\pi f_c t + \Theta(t)] = Re[a(t)e^{j[w_c t + \Theta(t)]}] \qquad (5.11)$$

where the carrier $\cos(w_c t)$ is said to be amplitude modulated if its amplitude $a(t)$ is adjusted in accordance with the modulating signal, and it is said to be phase modulated if $\Theta(t)$ is varied in accordance with the modulating signal. In QAM the amplitude of the baseband modulating signal is determined by $a(t)$ and the phase by $\Theta(t)$. The in-phase component I is then given by

$$I = a(t)\cos\Theta(t) \qquad (5.12)$$

and the quadrature component Q by

$$Q = a(t)\sin\Theta(t). \qquad (5.13)$$

This signal is corrupted by the channel. Here we will only consider AWGN. The received signal is given by

$$r(t) = a(t)\cos[2\pi f_c t + \Theta(t)] + n(t) \quad (5.14)$$

where $n(t)$ represents the AWGN, which has both an in-phase and quadrature component. It is this received signal which we will attempt to demodulate.

5.1.2.4.3 16QAM Demodulation in AWGN

The demodulation of the received QAM signal is achieved by performing quadrature amplitude demodulations using the decision boundaries constituted by the coordinate axes and the dotted lines portrayed in Figure 5.3 for the I and Q components, as shown below for the bits i_1 and q_1:

$$\begin{array}{llll} \text{if} & I,Q \geq 0 & \text{then} & i_1, q_1 = 0 \\ \text{if} & I,Q < 0 & \text{then} & i_1, q_1 = 1 \end{array} \quad (5.15)$$

The decision boundaries for the third and fourth bits i_2 and q_2, respectively, are again shown in Figure 5.3, and thus:

$$\begin{array}{lllll} \text{if} & & I,Q \geq 2d & \text{then} & i_2, q_2 = 1 \\ \text{if} & -2d \leq & I,Q < 2d & \text{then} & i_2, q_2 = 0 \\ \text{if} & -2d > & I,Q & \text{then} & i_2, q_2 = 1. \end{array} \quad (5.16)$$

We will show that, in the process of demodulation, the positions of the bits in the QAM symbols associated with each point in the QAM constellation have an effect on the probability of them being in error. In the case of the two most significant bits (MSBs) of the 4-bit symbol i_1, q_1, i_2, q_2, that is, i_1 and q_1, the distance from a demodulation decision boundary of each received phasor in the absence of noise is $3d$ for 50% of the time and d for 50% of the time, if each phasor occurs with equal probability. The average protection distance for these bits is therefore $2d$, although the bit error probability for a protection distance of $2d$ would be dramatically different from that calculated. Indeed, the average protection distance is never encountered; we only use this term to aid our investigations. The two least significant bits (LSB), that is, i_2 and q_2, are always at a distance of d from the decision boundary, and consequently the average protection distance is d. We may consider our QAM system as a class one ($C1$) and as a class two ($C2$) subchannel, where bits transmitted via the $C1$ subchannel are received with a lower probability of error than those transmitted via the $C2$ subchannel.

Observe in the phasor diagram of Figure 5.3 that upon demodulation in the $C2$ subchannel, a bit error will occur if the noise exceeds d in one direction or $3d$ in the opposite direction, where the latter probability is insignificant. Hence, the $C2$ bit error probability becomes

$$P_{2G} = Q\{\frac{d}{\sqrt{N_0/2}}\} = \frac{1}{\sqrt{2\pi}} \int_{\frac{d}{\sqrt{N_0/2}}}^{\infty} \exp(-x^2/2)dx \quad (5.17)$$

where $N_0/2$ is the double-sided spectral density of the AWGN, $\sqrt{N_0/2}$ is the corresponding noise voltage, and the $Q\{\}$ function was given in Equation 5.8 and Figure 5.9.

5.1. MODULATION ISSUES

As the average symbol energy of the 16-level QAM constellation computed for the phasors in Figure 5.3 is

$$E_0 = 10d^2, \tag{5.18}$$

then we have that

$$P_{2G} = Q\{\sqrt{\frac{E_0}{5N_0}}\}. \tag{5.19}$$

For the $C1$ subchannel data, the bits i_1, q_1 are at a protection distance of d from the decision boundaries for half the time, and their protection distance is $3d$ for the remaining half of the time. Therefore, the probability of a bit error is

$$P_{1G} = \frac{1}{2}Q\{\frac{d}{\sqrt{N_0/2}}\} + \frac{1}{2}Q\{\frac{3d}{\sqrt{N_0/2}}\} = \frac{1}{2}[Q\{\sqrt{\frac{E_0}{5N_0}}\} + Q\{3\sqrt{\frac{E_0}{5N_0}}\}]. \tag{5.20}$$

The $C1$ and $C2$ error probabilities P_{1G} and P_{2G} as a function of E_b/N_0 are given by Equation 5.19 and 5.20 and displayed in Figure 5.10 as a function of the channel SNR in contrast to a range of other modulation schemes. Note that for 1 bit/symbol uncoded transmissions the E_b/N_0 and SNR values are identical, but, for example, for 2 bit/symbol transmissions for a given signal and noise energy (i.e., channel SNR), the E_b/N_0 value must be reduced by a factor of two or 3.01 dB. Viewing this observation from a different angle, 2 bit/symbol transmissions, require a 3 dB higher signal energy or SNR for maintaining a constant E_b/N_0 value. Similarly, for 4 bit/symbol transmissions a factor four E_b/N_0 reduction is necessary for a fixed SNR value, which corresponds to a 6.02 dB higher channel SNR. Returning to the figure, the BER versus channel SNR performance of binary phase shift keying (BPSK), quaternary phase shift keying (QPSK), 16QAM and 64QAM BER are portrayed. For 16QAM the two protection classes differ by a factor of two in terms of their BER. Similarly to our above deliberations, in reference [10] we also showed that 64QAM exhibits three subchannels, whose BERs are also shown in the figure. Observe in the figure that given a certain channel SNR (i.e., a constant signal power), in harmony with our expectations, 2 bit/symbol transmissions require about 3 dB higher channel SNR for a given BER than binary signaling. A further 6 dB is necessitated by 16QAM and an additional 6 dB by 64QAM transmissions. The average probability P_{AV} of bit error for the 16-level QAM system is then computed as:

$$P_{AV} = (P_{1G} + P_{2G})/2. \tag{5.21}$$

Simulation results gave practically identical curves to those in Figure 5.10, exhibiting a BER advantage in using the $C1$ subchannel over using the $C2$. The computation of the error rate over Rayleigh-fading channels is more involved; for the square 16QAM constellation, Cavers [120] provided symbol error rate formulas, while for the star constellation analytical error rate formulas were disseminated by Adachi [121]. With the above considerations in mind, let us now concentrate on multilevel communications over Rayleigh-fading channels, which were described in Chapter 2.

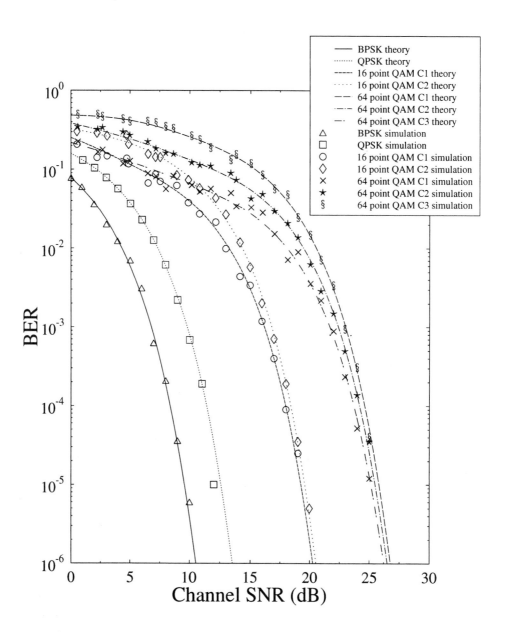

Figure 5.10: BPSK, QPSK, 16QAM, and 64QAM BER versus channel SNR performance over AWGN channels. ©Torrance, 1996 [119].

5.1.2.5 Reference-Assisted Coherent QAM for Fading Channels

Over fading channels, a number of additional measures have to be taken in order to be able to invoke bandwidth-efficient multilevel QAM schemes. The major difficulty is that over fading channels the magnitude of the transmitted phasors is attenuated, and their phase is rotated by the channel, as was shown in Figure 1.1. Two powerful methods have been proposed to ensure adequate QAM operation in fading environments. Both techniques deliver channel measurement information in terms of attenuation and phase shift due to fading. The first is transparent tone in band (TTIB) assisted modulation proposed by McGeehan, Bateman, et al. [122–129], where a pilot carrier is typically inserted in the center of the modulated spectrum. At the receiver, the signal is extracted and used to estimate the channel-induced attenuation and phase rotation, as it was also detailed in [10]. A disadvantage of TTIB schemes is their relatively high complexity and expanded spectral occupancy, since the pilot tone is inserted in one or several spectral gaps created by segmenting the signal spectrum and shifting the contiguous spectrum segments apart.

An alternative lower complexity technique is pilot symbol-assisted modulation (PSAM) [120], where known channel sounding phasors are periodically inserted into the transmitted time-domain signal sequence. Similarly to the frequency-domain pilot tone, these known symbols deliver channel measurement information. Let us here focus on the latter, implementationally less complex technique.

5.1.2.5.1 PSAM System Description Following Cavers' approach [120], the block diagram of a general PSAM scheme is depicted in Figure 5.11, where the pilot symbols p are cyclically inserted into the data sequence prior to pulse shaping, as demonstrated by Figure 5.12. A frame of data is constituted by M symbols, and the first symbol in every frame is assumed to be the pilot symbol $b(0)$, followed by $(M-1)$ useful data symbols $b(1), b(2) \ldots b(M-1)$.

Detection can be carried out by matched filtering, and the output of the matched filter is split into data and pilot paths, as seen in Figure 5.11. The set of pilot symbols can be extracted by decimating the matched filter's sampled output sequence using a decimation factor of M. The extracted sequence of pilot symbols must then be interpolated in order to derive a channel estimate $v(k)$ for every useful received information symbol $r(k)$. Decision is carried out against a decision-level reference grid, scaled and rotated according to the instantaneous channel estimate $v(k)$.

Observe in Figure 5.11 that the received data symbols must be delayed according to the interpolation and prediction delay incurred. This delay becomes longer if interpolation is carried out using a longer history of the received signal to yield better channel estimates. Consequently, there is a trade-off between processing delay and accuracy, an issue documented by Torrance [11] for a wide range of parameters. The interpolation coefficients can be kept constant over a whole pilot-period of length M, but better channel estimates can be obtained if the interpolator's coefficients are optimally updated for every received symbol.

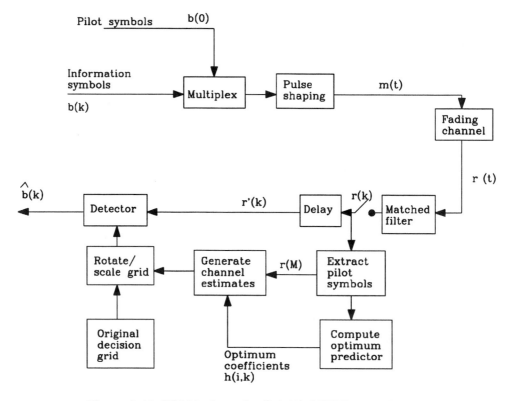

Figure 5.11: PSAM schematic. © [120] ©IEEE, 1991, Cavers.

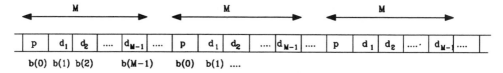

Figure 5.12: Insertion of pilot symbols in PSAM. ©Hanzo, Webb, Keller 2000 [10].

The complex envelope of the modulated signal can be formulated as:

$$m(t) = \sum_{k=-\infty}^{\infty} b(k)p(t-kT), \qquad (5.22)$$

where $b(k) = -3, -1, 1,$ or 3 represents the quaternary I or Q components of the 16QAM symbols to be transmitted, T is the symbol duration, and $p(t)$ is a bandlimited unit-energy signaling pulse, for which we have:

$$\int_{-\infty}^{\infty} |p(t)|^2 dt = 1. \qquad (5.23)$$

The value of the pilot symbols $b(kM)$ can be arbitrary, although sending a sequence of known pseudorandom symbols instead of always using the same phasor avoids

5.1. MODULATION ISSUES

the transmission of a periodic tone, which would increase the detrimental adjacent channel interference [130].

The narrowband Rayleigh channel is assumed to be "flat"-fading, which implies that all frequency components of the transmitted signal suffer the same attenuation and phase shift. This condition is met if the transmitted signal's bandwidth is much lower than the channel's coherence bandwidth. The received signal is then given by:

$$r(t) = c(t) \cdot m(t) + n(t), \tag{5.24}$$

where $n(t)$ is the AWGN and $c(t)$ is the channel's complex gain. Assuming a Rayleigh-fading envelope $\alpha(t)$, uniformly distributed phase $\phi(t)$, and a residual frequency offset of f_0, we have:

$$c(t) = \alpha(t) e^{j\phi(t)} \cdot e^{j\omega_0 t}. \tag{5.25}$$

The matched filter's output symbols at the sampling instants kT are then given as:

$$r(k) = b(k) \cdot c(k) + n(k). \tag{5.26}$$

Without imposing limitations on the analysis, Cavers [120] assumed that in every channel sounding block $b(0)$ was the pilot symbol and considered the detection of the useful information symbols in the range $\lfloor -M/2 \rfloor \leq k \leq \lfloor (M-1)/2 \rfloor$, where $\lfloor \bullet \rfloor$ is the integer of $\bullet$. Optimum detection is achieved if the corresponding channel gain $c(k)$ is estimated for every received symbol $r(k)$ in the above range. The channel gain estimate $v(k)$ can be derived as a weighted sum of the surrounding K received pilot symbols $r(iM)$, $\lfloor -K/2 \rfloor \leq i \leq \lfloor K/2 \rfloor$, as shown below:

$$v(k) = \sum_{i=\lfloor -K/2 \rfloor}^{\lfloor K/2 \rfloor} h(i,k) \cdot r(iM), \tag{5.27}$$

and the weighting coefficients $h(i,k)$ explicitly depend on the symbol position k within the frame of M symbols.

The estimation error $e(k)$ associated with the gain estimate $v(k)$ is computed as:

$$e(k) = c(k) - v(k). \tag{5.28}$$

Let us now consider the computation of the optimum channel gains.

5.1.2.5.2 Channel Gain Estimation in PSAM While previously proposed PSAM schemes used either a low-pass interpolation filter [130] or an approximately Gaussian filter [131], Cavers employed an optimum Wiener filter [132] to minimize the channel estimation error variance $\sigma^2_e(k) = E\{e^2(k)\}$, where $E\{\ \}$ represents the expectation. This well-known estimation error variance minimization problem can be formulated as follows:

$$\begin{aligned} \sigma_e^2(k) &= E\{e^2(k)\} = E\{[c(k) - v(k)]^2\} \\ &= E\left\{ \left[c(k) - \sum_{i=\lfloor -K/2 \rfloor}^{\lfloor K/2 \rfloor} h(i,k) \cdot r(iM) \right]^2 \right\}. \end{aligned} \tag{5.29}$$

In order to find the optimum interpolator coefficients $h(i,k)$, minimizing the estimation error variance $\sigma^2_e(k)$ we consider estimating the kth sample and set:

$$\frac{\partial \sigma_e^2(k)}{\partial h(i,k)} = 0 \quad \text{for} \quad \lfloor -K/2 \rfloor \leq i \leq \lfloor K/2 \rfloor. \tag{5.30}$$

Then using Equation 5.29 we have:

$$\frac{\partial \sigma^2_e(k)}{\partial h(i,k)} = E\left\{ 2\left[c(k) - \sum_{i=\lfloor -K/2 \rfloor}^{\lfloor K/2 \rfloor} h(i,k) \cdot r(iM) \right] \cdot r(jM) \right\} = 0. \tag{5.31}$$

After multiplying both square bracketed terms with $r(jM)$, and computing the expected value of both terms separately, we arrive at

$$E\{c(k) \cdot r(jM)\} = E\left\{ \sum_{i=\lfloor -K/2 \rfloor}^{\lfloor K/2 \rfloor} h(i,k) \cdot r(iM) \cdot r(jM) \right\}. \tag{5.32}$$

Observe that
$$\Phi(j) = E\{c(k) \cdot r(jM)\} \tag{5.33}$$

is the cross-correlation of the received pilot symbols and complex channel gain values, while

$$R(i,j) = E\{r(iM) \cdot r(jM)\} \tag{5.34}$$

represents the pilot symbol autocorrelations. Hence, Equation 5.32 yields:

$$\sum_{i=\lfloor -K/2 \rfloor}^{\lfloor K/2 \rfloor} h(i,k) \cdot R(i,j) = \Phi(j), \quad j = \lfloor -\frac{k}{2} \rfloor \ldots \lfloor \frac{k}{2} \rfloor. \tag{5.35}$$

If the fading statistics can be considered stationary, the autocorrelations $R(i,j)$ will only depend on the difference $|i-j|$, giving $R(i,j) = R(|i-j|)$. Therefore, Equation 5.35 can be written as:

$$\sum_{i=\lfloor -K/2 \rfloor}^{\lfloor K/2 \rfloor} h(i,k) \cdot R(|i-j|) = \Phi(j), \quad j = \lfloor -K/2 \rfloor \ldots \lfloor K/2 \rfloor, \tag{5.36}$$

which is a form of the well-known Wiener-Hopf equations [132], often used in estimation and prediction theory, as we have shown with reference to optimum linear prediction of speech signals in Chapter 3.

This set of K equations contains K unknown prediction coefficients $h(i,k)$, $i = \lfloor -K/2 \rfloor \ldots \lfloor K/2 \rfloor$, which must be determined in order to arrive at a minimum error variance estimate of $c(k)$ by $v(k)$. First, the correlation terms $\Phi(j)$ and $R(|i-j|)$ must be computed, and to do this the expectation value computations in Equations 5.33 and 5.34 need to be restricted to a finite duration window. This approach is referred to as the *autocorrelation method*, which was detailed in the context of speech coding in Chapter 3. The pilot autocorrelation, $R(i,j)$, may then be calculated from the fading

5.1. MODULATION ISSUES

estimates at the pilot positions within this window. Calculation of the received pilots' and the complex channel gains' cross-correlation is less straightforward because in order to calculate the cross-correlation the complex channel gains have to be known at the position of the data symbols as well as the pilot symbols. However, the channel gains are only known at the pilot positions, while for the data symbol positions they must be derived by interpolation. Hence, Torrance [11] fitted a polynomial to the known samples of $R(|i-j|)$ and then estimated the values of $\Phi(j)$ for the unknown positions in order to provide a wide range of PSAM modem BER versus channel SNR performance figures for the IS-54 system using 1, 2, and 4 bits/symbol signaling.

The set of Equations 5.36 can also be expressed in a convenient matrix form as:

$$\begin{bmatrix} R(0) & R(1) & R(2) & \ldots & R(K) \\ R(1) & R(0) & R(1) & \ldots & R(K-1) \\ R(2) & R(1) & R(0) & \ldots & R(K-2) \\ \vdots & \vdots & \vdots & \ldots & \vdots \\ R(K) & R(K-1) & R(K-2) & \ldots & R(0) \end{bmatrix} \quad (5.37)$$

$$\cdot \begin{bmatrix} h\left(\lfloor -\frac{K}{2}\rfloor, k\right) \\ h\left(\lfloor -\frac{K}{2}+1\rfloor, k\right) \\ h\left(\lfloor -\frac{K}{2}+2\rfloor, k\right) \\ \vdots \\ h\left(\lfloor \frac{K}{2}\rfloor, k\right) \end{bmatrix} = \begin{bmatrix} \Phi\left(\lfloor -\frac{K}{2}\rfloor\right) \\ \Phi\left(\lfloor -\frac{K}{2}+1\rfloor\right) \\ \Phi\left(\lfloor -\frac{K}{2}+2\rfloor\right) \\ \vdots \\ \Phi\left(\lfloor \frac{K}{2}\rfloor\right) \end{bmatrix},$$

which can be solved for the optimum predictor coefficients $h(i, k)$ by matrix inversion using Gauss-Jordan elimination or the recursive Levinson-Durbin algorithm highlighted, for example, in Chapter 3 of [87]. Once the optimum predictor coefficients $h(i, k)$ are known, the minimum error variance channel estimate $v(k)$ can be derived from the received pilot symbols using Equation 5.27, as also demonstrated by Figure 5.11.

5.1.2.5.3 PSAM Performance [11] Torrance et al. [11] also compared the performance of the above Cavers-interpolator with that of the conventional linear, low-pass, and a higher-order polynomial interpolator using 1, 2, and 4 bit/symbol modems. They concluded that in the fast-fading IS-54 environment investigated, the highest-complexity minimum mean-squared error Cavers-interpolator did not significantly outperform the above low-complexity lienar, low-pass, or polynomial interpolators in terms of reduced residual BER. In these experiments, the propagation frequency was increased from the 900 MHz IS-54 frequency to the 1.8 GHz propagation frequency of the current generation of systems, the vehicular speed was fixed at 50 km/h or approximately 30 mph, and the signaling rate was set to 20 kBd, which corresponded to a modulation excess bandwidth of 50%, when using the standard IS-54 bandwidth of 30 kHz. The corresponding Doppler frequency f_d was

$$f_d = (v \cdot f_p)/c = (13.88 m/s \cdot 1.8 \cdot 10^9 Hz)/(3 \cdot 10^8 m/s) \approx 83.3 Hz,$$

where v is the vehicular speed and f_p is the propagation frequency. The corresponding normalized Doppler frequency is

$$f_d \cdot T = 83.3 Hz \cdot 1/(20 \cdot 10^3 Baud) \approx 0.0042.$$

Because of its approximately 13 times higher signaling rate of 271 kbps the GSM-like DCS1800 system under identical propagation conditions results in a relative Doppler frequency of 0.0003, which is associated with a less dramatically fading signal envelope and hence better fade tracking properties. The 1 bit/symbol, 2 bit/symbol, and 4 bit/symbol modulation schemes were combined with all four interpolators, and their bit error rate (BER) performance was evaluated at channel SNRs of 20, 30, and 40 dB, which yielded $3 \cdot 4 \cdot 3 = 36$ sets of results. In each set of results, pilot Buffer lengths of 3, 5, 7, 9, 11 PSAM frames and pilot separation or Gap values of 10, 20, 40, 60, 80, 100, 116 were employed, leading to a plethora of performance curves, which allowed us to generate a corresponding set of three-dimensional (3D) graphs of BER versus Buffer and Gap. These results are presented in Figure 5.13 as a set of three-dimensional (3D) graphs of BER versus Buffer and Gap for pilot-assisted square-constellation 16QAM. The corresponding graphs for BPSK and QPSK were presented by Torrance in [119], while a variety of further results can be found in [11]. As an alternative to the above coherent PSAM scheme, let us now consider the advantages and disadvantages of noncoherent differential detection using the star constellation of Figure 5.7.

5.1.2.6 Differentially Detected QAM [10]

We have shown above that the maximum minimum distance square-shaped QAM constellation [10] is optimum for transmissions over Additive White Gaussian (AWGN) channels, since it has the highest average distance among its constellation points, yielding the highest noise protection distances for a given average power. We have also introduced the star QAM constellation in Figure 5.7 and compared some of its properties with those of the square constellation of Figure 5.3 in Table 5.1. In this subsection, we introduce a differentially encoded version of the star constellation shown in Figure 5.7, which can often be advantageously employed over fading channels.

When using the previously discussed square-shaped 16QAM constellation, it is essential to be able to separate the information modulated onto the in-phase (I) and quadrature-phase (Q) carriers by the help of coherent demodulation, invoking the Transparent-tone-in-band (TTIB) principle invented by McGeehan and Bateman [123, 129, 133] or employing the above PSAM schemes [120]. In order to achieve this, a perfectly phase-coherent replica of the transmitter's I and Q carrier has to be recovered by the carrier recovery circuitry. In contrast, in the differentially encoded schemes, it is not necessary to derive this phase-coherent reference carrier, an issue that is elaborated on below.

The pivotal point of differentially encoded noncoherent QAM demodulation is to find a rotationally symmetric QAM constellation, in which all constellation points are rotated by the same amount. Such a rotationally symmetric, differentially encoded "star constellation" was proposed by Webb et al. [102], and is similar to the star scheme shown in Figure 5.7 in terms of the location of its phasors, but differentially

5.1. MODULATION ISSUES

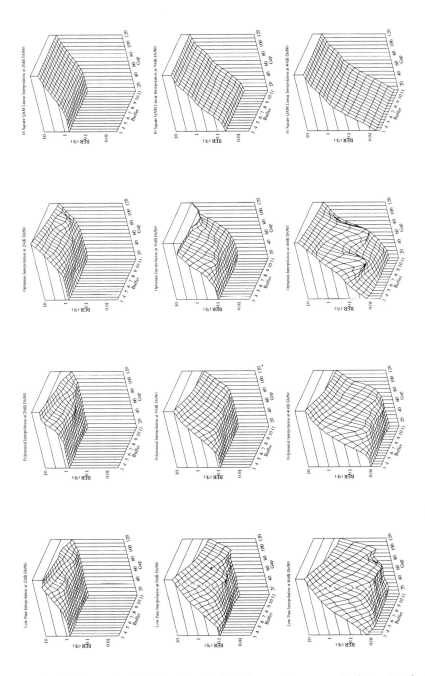

Figure 5.13: BER versus Buffer length and pilot Gap performance of pilot-assisted 16QAM over Rayleigh channels at 20 kBd, 50 km/h, 1.8 GHz. Top: SNR = 20 dB, Middle: SNR = 30 dB, Bottom: SNR = 40 dB; Left to right: low-pass, polynomial, optimum, and linear interpolator. ©IEE 1995 Torrance, Hanzo [11].

encoded, as is described below. We have seen in Table 5.1 that a disadvantage of the proposed star 16QAM (16StQAM) constellation is its lower average energy.

Our differential encoder obeys the following rules. The first bit $b1$ of a 4-bit symbol is differentially encoded onto the phasor magnitude in Figure 5.7, yielding a ring-swap for an input logical one and maintaining the current magnitude, that is, ring for $b1 = 0$. Bits $(b2, b3, b4)$ are then differentially Gray-coded onto the phasors of the particular ring pinpointed by $b1$ in Figure 5.7. Accordingly, $(b2, b3, b4) = (0, 0, 0)$ implies no phase change, $(0, 0, 1)$ a change of $45°$, $(0, 1, 1)$ a change of $90°$, and so on.

The corresponding noncoherent differential 16-StQAM demodulation is equally straightforward, having decision boundaries at a concentric ring of radius $B = (A_1 + A_2)/2$ and at phase rotations of $(22.5° + n.45°)$ $n = 0 \ldots 7$. Assuming received phasors of P_t and P_{t+1} at consecutive sampling instants of t and $t + 1$, respectively, bit $b1$ is inferred by evaluating the condition:

$$|\frac{P_{t+1}}{P_t}| \geq (A_1 + A_2)/2. \tag{5.38}$$

If this condition is met, $b1 = 1$ is assigned; otherwise, $b1 = 0$ is demodulated. Bits $(b2, b3, b4)$ are then recovered by computing the phase difference

$$\Delta\Theta = (\Theta_{t+1} - \Theta_t) \pmod{2\pi} \tag{5.39}$$

and comparing it against the decision boundaries $(22.5° + n.45°)$ $n = 0 \ldots 7$. Having decided which rotation interval the received phase difference $\Delta\Theta$ belongs to, Gray-decoding delivers the bits $(b2, b3, b4)$.

From our previous discourse it is plausible that the less dramatic the fading envelope and phase trajectory fluctuation between adjacent signaling instants, the better this differential scheme works. This implies that lower vehicular speeds are preferred by this arrangement, if the signaling rate is fixed. Therefore, the modem's performance improves for low pedestrian speeds, when compared to typical vehicular scenarios. Alternatively, for a fixed vehicular speed higher signaling rates are favorable, since the relative amplitude and phase changes introduced by the fading channel between adjacent information symbols are less drastic.

Similar differentially encoded and noncoherently detected constellations can be used in conjunction with any number bits/symbol. Torrance et al. [11] presented the BER of pilot-symbol assisted 1, 2, and 4 bits/symbol BPSK, QPSK, and 16QAM for channel SNRs of 20, 30, and 40 dB in contrast to their lower-complexity, differentially detected counterparts. These comparative results are reproduced in Figure 5.14 [11] as a set of BER versus number of bits per symbol curves for both the pilot-assisted and differential schemes using channel SNRs of 20, 30, and 40 dB. The bold symbols in the figure represent the PSAM schemes, while the hollow symbols correspond to the differentially detected schemes. Observe in the figure that as the modulation constellation becomes less complex (i.e., the number of bits per symbol is reduced), the benefits of coherent modulation are reduced, although this is also a function of the channel SNR. In contrast, for higher order constellations, such as QPSK and 16QAM, PSAM does reduce the residual BER of the slightly less complex, differentially detected schemes while having a somewhat higher delay.

5.1. MODULATION ISSUES

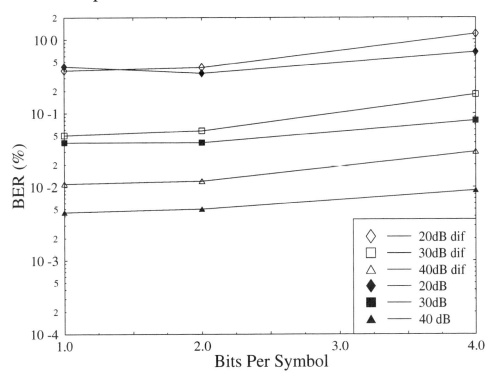

Figure 5.14: Residual BER for 1,2, and 4 bits per symbol PSAM modulation compared with equivalent differential schemes. ©IEE [11] Torrance, Hanzo, 1995.

For a full treatise on various aspects of QAM, the interested reader is referred to [10], where the modem performance was documented for various constellations and channel conditions. As a brief performance comparison, we remind the reader that in Figure 5.14 we portrayed the coherent and noncoherent modem's residual BER performances for 1, 2, and 4 bits/symbol signaling under identical conditions. Observe in the figure that the differentially detected scheme typically has a factor two higher BER because in case of an erroneous decision errors occur in both the current and the forthcoming signaling interval, where the current phasor is used as reference in deriving the next one. Some further BER degradation is expected owing to the reduced distance of the constellation points in the star constellation, although this does not appear to be a significant factor in fading channels. In conclusion, star- and square-constellation QAM have a high bandwidth efficiency in exchange for a typically higher channel SNR. The BER versus channel SNR performance of coherently detected pilot-assisted QAM is slightly higher than that of the lower complexity star-QAM, providing the system designer with a choice of implementational options.

In closing, we note that in cellular frequency-reuse structures, where the co-channel interference is a dominant impairment, the bandwidth efficiency of these schemes often erodes because of the increased frequency reuse factor required by the higher interference sensitivity of these schemes. In Chapter 17 of [10], we have shown that the true spectral efficiency of a modulation scheme, taking into account the effect of the required frequency reuse factor, is dependent on the bit error ratio (BER) targeted. The required BER in turn is dependent on the robustness of the source codecs used. However, for example, in indoor picocells the partitioning walls and floors mitigate the co-channel interference, and this facilitates the employment of more bandwidth-efficient, but less interference-resilient 16QAM. In this rudimentary introduction to QAM techniques, we assumed perfect clock recovery and dispensed with a range of important aspects of the transceiver design, such as clock and carrier recovery, wideband aspects and channel equalization, the effects of co- and adjacent-channel interference, trellis coding, and so on, which are treated in depth in the corresponding chapters of [10]. The analytical error rate performance of square and star QAM was characterized in [120] and [121] by Cavers and Adachi, respectively.

5.1.3 Adaptive Modulation
J. M. Torrance and L. Hanzo

5.1.3.1 Background to Adaptive Modulation

Steele and Webb [89, 134] proposed adaptive differentially encoded, noncoherently detected star-constellation Quadrature Amplitude Modulation [10] (Star-QAM) for exploiting the time-variant Shannonian channel capacity of fading channels. Their contribution has stimulated further pioneering work, in particular by Kamio, Sampei, Sasaoka, Morinaga, Morimoto, Harada, Okada, Komaki, and Otsuki *et al.* at Osaka University and the Ministry of Post in Japan [109, 135, 136], as well as by Goldsmith *et al.* [137–143] at Stanford University in the United States or by Pearce, Burr, and Tozer [144] in the United Kingdom.

Following Shannon's fundamental work and Lee's seminal contribution on the channel capacity of fading channels [145], in this chapter we propose a scheme that will allow us to approach the predicted channel capacity potential of fading wireless channels more closely than conventional nonadaptive modems. The rapid fluctuations in received power result from multipath propagation in a mobile radio environment. In the past, systems operated at increased average transmit powers in order to account for these fluctuations and hence to obtain the desired bit error rate (BER). However, in a mobile radio environment increasing the average transmit power has undesirable consequences in terms of co- and adjacent-channel interference and power consumption. A more attractive alternative to mitigate this fast-fading is to employ a more robust modulation scheme, when a low instantaneous SNR is expected at the receiver. Moreover, when the instantaneous SNR increases again, a less rugged scheme exhibiting a higher throughput should be employed. This research was initiated by Steele and Webb and presented in a keynote paper [134], which discussed adaptive differential modulation and considered the effect of SNR and co-channel interference. Morinaga [146] considered the employment of such adaptive modulation schemes in terms of its future within mobile multimedia apparatus. A range of fur-

5.1. MODULATION ISSUES

ther aspects of such adaptive systems were considered by a number of authors in [109–115, 135–143, 146], and this summary is based mainly on [111, 113–115].

Specifically, Chua and Goldsmith [137] proposed the employment of the concomitant variation of both the modulation scheme and the transmitted power, and in [138] they showed that there was a BER-dependent required power discrepancy between their adaptive regime and the channel capacity. This corroborates the experience that the higher the acceptable BER, the better the system's capacity approximates the Shannonian chanel capacity. They showed that by using five to six different constellations, it is possible to achieve a performance very close to that of the best possible system, imposing no restrictions on the number of modulation constellations.

In contrast to previous studies on modem spectral efficiency, where conventional fixed modems were used, in [10] the radio capacity of adaptive differentially detected noncoherent modems was studied by simulation using various radio cell constellations and a propagation prediction tool, while Alouini and Goldsmith in [142] derived theoretical expressions for the [bits/s/Hz/m^2] capacity of adaptive modems. Further capacity-related work by the CalTech team was disseminated in [137–143], while their contributions on adaptive coded modulation were crystallized in [143]. Specifically, the simulation and theoretical results by Goldsmith and Chua showed that a 3 dB coding gain was achievable at a BER of 10^{-6} for a 4-state trellis code and 4 dB by an 8-state trellis code in the context of the adaptive scheme over Rayleigh-fading channels, while a 128-state code performed within 5 dB of the Shannonian capacity limit. The benefits of Automatic Repeat Request (ARQ) in adaptive modems were quantified by the Japanese teams from Osaka and Tokyo [136], while some of the latency aspects were addressed in [115].

For a transmitting radio to select the appropriate modulation scheme, there must be some information available to it upon the instantaneous quality of the channel. This can be achieved by using Time Division Duplex (TDD) schemes, which provide a convenient framework for exploiting the correlation between the up- and down-link complex envelope of the channel, given that the normalized Doppler frequency is sufficiently low. Therefore, adaptive modulation within a TDD scenario relies on the correlation between received and transmitted frames to estimate s, the instantaneous power that will be received for a given set of symbols. Figure 5.15 shows this principle and demonstrates that it is reasonable to assume that the channel does not fluctuate rapidly over the duration of a received slot. The estimation of s is compared against a set of n switching levels, l_n, and the appropriate modulation scheme is selected accordingly. Following Kamio et al. [109] Binary Phase Shift Keying (BPSK), Quaternary Phase Shift Keying (QPSK), and multilevel Quadrature Amplitude Modulation (QAM) can be invoked on the basis of the following thresholding operations [113, 119]:

$$\text{MS} = \begin{cases} \text{No Transmission} & \text{if } l_1 < s \\ \text{BPSK} & \text{if } l_1 \leq s < l_2 \\ \text{QPSK} & \text{if } l_2 \leq s < l_3 \\ \text{Square 16 Point QAM} & \text{if } l_3 \leq s < l_4 \\ \text{Square 64 Point QAM} & \text{if } s \leq l_4 \end{cases} \quad (5.40)$$

where the threshold values of l_n will be discussed later.

The thresholds l_1, l_2, l_3, and l_4 predetermine the performance of an adaptive

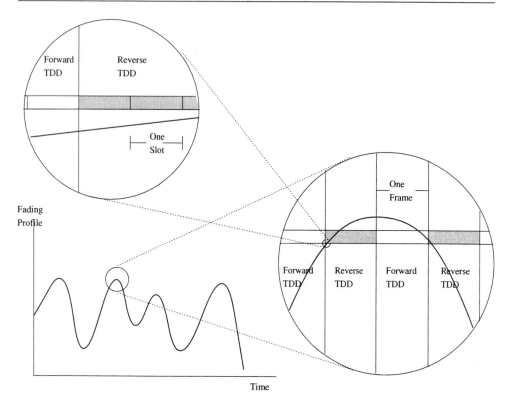

Figure 5.15: Received power fluctuation over the duration of one TDD slot, when compared with the deviation over the whole channel results in a single slot exhibiting near-Gaussian channel characteristics. The correlation between corresponding slots in adjacent TDD frames may be exploited. ©Torrance, Hanzo 1996 [113].

modulation scheme under given channel conditions. The lower the threshold values, the higher the throughput; that is, on average more bits will be transmitted using a higher number of bits per symbol (BPS). Conversely, the higher the values of l_1, l_2, l_3, and l_4, the lower the BER of the overall adaptive modulation scheme for given channel conditions. Clearly, BER can be traded with BPS performance and vice versa. The optimum trade-off will depend on the type of information being transmitted and on the source as well as channel coders used.

The trade-off is dependent on the type of information that is to be transmitted in order to satisfy the different network characteristics required by video, voice and computer data. Generally, interactive video and voice information cannot sustain as much latency across the link as computer data information. However, computer data information is less robust to channel errors. The BER performance of adaptive modulation can be optimized, depending on how robust the information transmitted over the link has to be. The latency introduced depends on whether the information is buffered at the transmitter, when the instantaneous channel SNR is low. As the

5.1. MODULATION ISSUES

instantaneous channel SNR increases, the buffer may be emptied but latency has been incurred. The latency is therefore dependent on the average BPS performance of the system.

The choice of source encoder and adaptive modulation scheme has to be harmonized. Appropriate selection of the source encoder [147] can support an improved robustness against transmission impairments in poor channel conditions by reducing the bit rate and consequently invoking a more robust modem mode, experiencing a reduced BER. By contrast, in good channel conditions, the overall source representation quality can be improved by increasing the bit rate, while maintaining the required target BER.

5.1.3.2 Bit Error Rate and Channel Capacity Optimization of Adaptive Modems [110]

The upper-bound performance of the underlying adaptive modem scheme was characterized in [110]. The BER performance of coherent modulation schemes with 1, 2, 4, and 6 bits per symbol (BPS) assuming perfect clock and carrier recovery in a Gaussian channel is known [10], and the corresponding expressions are as follows:

$$P_b(\gamma) = Q\left(\sqrt{2\gamma}\right), \tag{5.41}$$

$$P_q(\gamma) = Q\left(\sqrt{\gamma}\right), \tag{5.42}$$

$$P_{16}(\gamma) = \frac{1}{4}\left[Q\left(\sqrt{\frac{\gamma}{5}}\right) + \left(3Q\sqrt{\frac{\gamma}{5}}\right)\right] + \frac{1}{2}Q\left(\sqrt{\frac{\gamma}{5}}\right), \tag{5.43}$$

$$\begin{aligned}
P_{64}(\gamma) &= \frac{1}{12}\left[\begin{array}{c} Q\left(\sqrt{\frac{\gamma}{21}}\right) + Q\left(3\cdot\sqrt{\frac{\gamma}{21}}\right) \\ +Q\left(5\cdot\sqrt{\frac{\gamma}{21}}\right) + Q\left(7\cdot\sqrt{\frac{\gamma}{21}}\right) \end{array}\right] \\
&+ \frac{1}{6}Q\left(\sqrt{\frac{\gamma}{21}}\right) + \frac{1}{6}Q\left(3\cdot\sqrt{\frac{\gamma}{21}}\right) \\
&+ \frac{1}{12}Q\left(5\cdot\sqrt{\frac{\gamma}{21}}\right) + \frac{1}{12}Q\left(7\cdot\sqrt{\frac{\gamma}{21}}\right) \\
&+ \frac{1}{3}Q\left(\sqrt{\frac{\gamma}{21}}\right) + \frac{1}{4}Q\left(3\cdot\sqrt{\frac{\gamma}{21}}\right) \\
&- \frac{1}{4}Q\left(5\cdot\sqrt{\frac{\gamma}{21}}\right) - \frac{1}{6}Q\left(7\cdot\sqrt{\frac{\gamma}{21}}\right) \\
&+ \frac{1}{6}Q\left(9\cdot\sqrt{\frac{\gamma}{21}}\right) + \frac{1}{12}Q\left(11\cdot\sqrt{\frac{\gamma}{21}}\right) \\
&- \frac{1}{12}Q\left(13\cdot\sqrt{\frac{\gamma}{21}}\right),
\end{aligned} \tag{5.44}$$

where $Q(x) = \frac{1}{\sqrt{2\pi}}\int_x^\infty e^{-x^2/2}dx$, γ is the signal-to-noise ratio (SNR), while $P_b(\gamma)$, $P_q(\gamma)$, $P_{16}(\gamma)$, and $P_{64}(\gamma)$, are the mean BERs of BPSK, QPSK, square 16 point square QAM, and square 64 point QAM, respectively. The PDF of the fluctuations in instantaneous received power, s, over a Rayleigh channel are given by:

$$F(s, S) = \frac{2s}{S} \cdot e^{-s^2/S}; \quad s \geq 0, \tag{5.45}$$

where S is the average signal power. Assuming a sufficiently low normalized Doppler frequency in order to maintain a near-constant fading envelope and hence near-Gaussian conditions for the duration of a modulation symbol and employing Pilot Symbol Assisted Modulation (PSAM) [11], upper-bound BER performances can be obtained for the above four modulation schemes over a Rayleigh channel. For any of the modulation schemes, if $X_g(\gamma)$ is the Gaussian BER performance, as given in Equations 5.41, 5.42, 5.43, or 5.44, then $X_r(S/N)$ given below will be the upper bound for the BER performance in a Rayleigh channel:

$$X_r(S/N) = \int_0^\infty X_g(s/N) \cdot F(s, S)\, ds. \tag{5.46}$$

Therefore, the narrowband upper-bound BER performance of an adaptive modulation scheme similar to that described in [109] may be computed from:

$$P_a(S/N) = B^{-1} \begin{bmatrix} 1\int_{l_1}^{l_2} P_b(s/N) \cdot F(s, S) & ds \\ +2\int_{l_2}^{l_3} P_q(s/N) \cdot F(s, S) & ds \\ +4\int_{l_3}^{l_4} P_{16}(s/N) \cdot F(s, S) & ds \\ +6\int_{l_4}^{\infty} P_{64}(s/N) \cdot F(s, S) & ds \end{bmatrix}, \tag{5.47}$$

where l_1, l_2, l_3, l_4, and B are the reconfiguration thresholds between transmission of BPSK, QPSK, square 16 point, and square 64 point QAM, while B is the mean number of BPS, given by:

$$\begin{aligned} B &= 1 \cdot \int_{l_1}^{l_2} F(s, S)\, ds + 2 \cdot \int_{l_2}^{l_3} F(s, S)\, ds \\ &+ 4 \cdot \int_{l_3}^{l_4} F(s, S)\, ds + 6 \cdot \int_{l_4}^{\infty} F(s, S)\, ds. \end{aligned} \tag{5.48}$$

Adaptive modulation schemes may select the appropriate modulation level for the transmitted frame on the basis of either the number of errors encountered in the received frame or the received signal strength [10]. The former is only possible in a system that includes some error detection. Here, the received signal strength relative to the switching values of l_1, l_2, l_3, and l_4 is used to select the appropriate modulation level for each frame. The difference between this upper-bound performance and a certain desired performance can be used as a cost function for given modulation switching levels of l_1, l_2, l_3, and l_4. Minimization of this cost function may be achieved iteratively by varying the switching levels [111] l_1, l_2, l_3, and l_4 using Powell's optimization algorithm [148].

The BER and BPS performances were evaluated for average channel SNRs in the range of 0 dB to 50 dB in 1 dB intervals. The optimization cost function was defined as:

$$\text{Total Cost} = \sum_{i=0}^{50} \text{BER Cost}(i) + \text{BPS Cost}(i) \tag{5.49}$$

5.1. MODULATION ISSUES

where

$$\text{BER Cost}(i) = \begin{cases} 10 log_{10}\left(\dfrac{\text{BER}_m(i)}{\text{BER}_d(i)}\right) & \text{if } \text{BER}_m(i) > \text{BER}_d(i) \\ 0 & \text{otherwise} \end{cases} \quad (5.50)$$

$$\text{BPS Cost}(i) = \begin{cases} \text{BPS}_d(i) - \text{BPS}_m(i) & \text{if } \text{BPS}_d(i) > \text{BPS}_m(i) \\ 0 & \text{otherwise} \end{cases} \quad (5.51)$$

and $\text{BER}_m(i)$, $\text{BER}_d(i)$, $\text{BPS}_m(i)$, and $\text{BPS}_d(i)$ are, respectively, the measured and desired BER and BPS at an average channel SNR of i. It can be seen from Equations 5.49, 5.50, and 5.51 that the cost function can only be positive and increases when either the BER or the BPS performance becomes inferior to their desired performance at an average channel SNR of i. The cost function cannot be negative; therefore, at high average channel SNRs, where both the BER and the BPS outperform their respective desired performance targets, the combined performance will be suboptimum. The advantage of this approach is that it reduces the minimum average channel SNR above which both the desired BER and BPS performance criteria are achieved. However, the performance does improve at higher average SNRs because of the favorable prevailing channel conditions. Equation 5.50 utilizes the logarithm function to increase the significance of small BERs. A weighting factor of 10 is employed in order to bias the optimization toward achieving the desired BER performance in preference to the BPS performance.

5.1.3.3 Bit Error Rate and Channel Capacity of Adaptive Modulation

Two desired system performance profiles were considered. One of them was optimized for a speech codec with a target BER and BPS of 0.01 and 4.5, respectively. The other performance profile was intended for computer data transfer with target BER and BPS performances of 0.0001 and 3, respectively. The important speech interactivity aspects of buffer size, delay, and latency issues were studied in [115]. It is convenient to refer to these systems as the speech and data schemes, which here are synonymous with the higher and lower BER systems. Naturally, the portrayed techniques can be invoked in the context of arbitrary desired BERs. The optimization was performed for Rayleigh channel conditions, and the initial condition for both minimizations was $l_1 = 5$ dB, $l_2 = 8$ dB, $l_3 = 14$ dB, and $l_4 = 20$ dB. After optimization, the values of $l_1 = 3.31$, $l_2 = 6.48$, $l_3 = 11.61$, and $l_4 = 17.64$ dB were registered for the speech system. For the computer data system, the values of $l_1 = 7.98$, $l_2 = 10.42$, $l_3 = 16.76$, and $l_4 = 26.33$ dB were recorded, which are summarized in Table 5.2.

Considering Figure 5.16, the desired BER is achieved between 0 and 50 dBs for both the speech and computer data switching level schemes. The targeted BPS performance is achieved at about 18 dB and 19 dB average channel SNRs for the speech and computer data schemes, respectively. Observe in the figure that both the speech and data BER profiles outperform the BER requirements for average channel SNRs greater than these values. The system was capable of maintaining the target BER

BER	l_1	l_2	l_3	l_4
$1 \cdot 10^{-2}$	3.31 (dB)	6.48 (dB)	11.61 (dB)	17.64 (dB)
$1 \cdot 10^{-4}$	7.98 (dB)	10.42 (dB)	16.76 (dB)	26.33 (dB)

Table 5.2: Optimized Switching Levels for Speech and Computer Data Systems through a Rayleigh Channel, Shown in dB ©Torrance, Hanzo 1996 [113, 119]

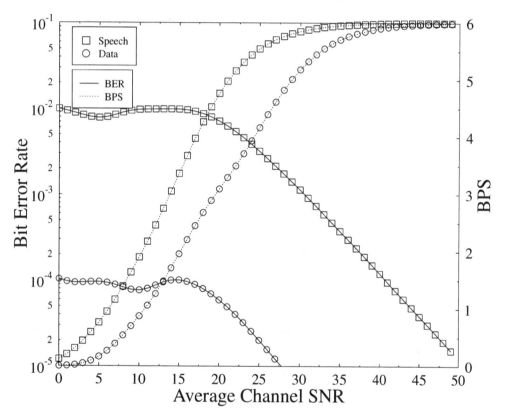

Figure 5.16: Upper-bound BER and BPS performance of adaptive QAM in Rayleigh channel optimized separately for speech and data transfer. ©Torrance, Hanzo 1996 [113, 119].

performances at extremely low average SNR values. This robust performance was achieved at the cost of reducing the BPS channel capacity below that of BPSK, which was possible due to disabling transmissions for low instantaneous SNR values. The associated latency, networking, as well as channel-coding aspects were addressed in [115, 149–152].

5.1.3.4 Equalization Techniques

In conclusion of the previous elaborations on various modulation techniques, we note that when the signaling rate is increased, the relative effects of signal dispersion due to the multipath channel's path-length difference become more detrimental. In this case, channel equalizers have to be invoked in order to mimic the inverse of the channel's transfer function and hence to remove its linear distortion inflicted.

The performance of wideband wireless channel equalizers was studied by a large cohort of researchers, such as Narayanan and Cimini [153], Wu and Aghvami [154] as well as Gu and Le-Ngoc [155]. In order to achieve fast equalizer coefficient convergence, these contributions typically invoked the Kalman algorithm [10] and its diverse incarnations, such as the Square Root Kalman scheme, although Clark and Harun [156] argued that there were only marginal performance differences between the Kalman algorithm and the Least Mean Squared [10] (LMS) algorithm in typical practical situations. Maximum likelihood sequence estimator (MLSE) type receivers typically outperform Decision Feedback Equalizers (DFE) at the cost of higher complexity. A range of hybrid compromise-schemes were proposed, for example, by Wu and Aghvami [154] as well as by Gu and Le-Ngoc [155]. For a full discussion of various channel equalization techniques, the interested reader is referred to [10].

Below we provide a brief discourse on a recently very popular alternative technique, namely, Orthogonal Frequency Division Multiplexing (OFDM). Instead of invoking a channel equalizer, OFDM modems split the transmitted signal's spectrum into a high number of low-rate, narrowband subchannels. The result of this is that each subchannel signal has a substantially extended symbol duration in comparison to the full-rate, full-band signal and hence the relative effects of signal, dispersion are mitigated by a factor corresponding to the number of subchannels. We consider this technique briefly in the next section.

5.2 Orthogonal Frequency Division Multiplexing [10, 92]

In this section, we briefly introduce Orthogonal Frequency Division Multiplexing (OFDM) as a means of mitigating the channel-induced linear distortion problems encountered, when transmitting over a dispersive radio channel, such as that characterised previously in Figure 5.1. The fundamental principle of orthogonal multiplexing originates from Chang [157], and over the years a number of researchers have investigated this technique [158–175]. Despite its conceptual elegance, until recently its employment has been mostly limited to military applications due to implementational difficulties. However, it has recently been adopted as the new European digital au-

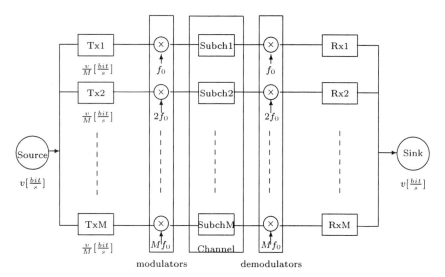

Figure 5.17: Simplified blockdiagram of the orthogonal parallel modem.

dio broadcasting (DAB) standard [176] for digital terrestrial television broadcast (DTTB), now known as DVB-T [177] and for a range of other high-rate applications, such as 155 Mbps wireless Asynchronous Transfer Mode (ATM) local area networks. It has also been proposed by Lindner *et al.* [178] as a specific form of Code Division Multiple Access (CDMA) using sinusoidal signature codes or by Kammeyer *et al.* [166] for the transmission of audio signals over mobile radio channels. These wide-ranging applications underline its significance as an alternative technique to conventional channel equalization in order to combat signal dispersion [29, 167–170, 179].

In the OFDM scheme of Figure 5.17 the serial data-stream of a traffic channel is passed through a serial-to-parallel converter, which splits the data into a number of parallel subchannels. The data in each subchannel are applied to a modulator, such that for M channels there are M modulators whose carrier frequencies are f_0, f_1, ...f_M. The difference between adjacent channels is Δf, and the overall bandwidth W of the N modulated carriers is $M\Delta f$.

These M modulated carriers are then combined to give an OFDM signal. We may view the serial-to-parallel converter, as applying every Mth symbol to a modulator. This has the effect of interleaving the symbols into each modulator; hence, symbols S_0, S_M, S_{2M}, ... are applied to the modulator whose carrier frequency is f_1. At the receiver the received OFDM signal is demultiplexed into M frequency bands, and the M modulated signals are demodulated. The baseband signals are then recombined using a parallel-to-serial converter.

The main advantage of the OFDM concept is that because the symbol period has been increased, the channel delay spread is a significantly shorter fraction of a symbol period than in the serial system, potentially rendering the system less sensitive to ISI than the conventional serial system. In other words, in the low-rate subchannels the signal is no longer subject to frequency-selective fading; thus, no channel equalization

5.2. ORTHOGONAL FREQUENCY DIVISION MULTIPLEXING 199

is necessary.

A disadvantage of the OFDM approach shown in Figure 5.17 is the increased complexity over the conventional system caused by employing M modulators and filters at the transmitter and M demodulators and filters at the receiver. It can be shown, however, that this complexity can be reduced through use of the discrete Fourier transform (DFT), typically implemented as a Fast Fourier Transform (FFT) [10]. The rather plausible idea behind the FFT-based OFDM modem implementation is that since in the schematic of Figure 5.17 M harmonically related sinusoidal carriers are used, the Fourier transform with its harmonically related set of kernels or basis functions is exactly the operation required. Clearly, a block of M consecutive binary or nonbinary information symbols are transformed by the FFT using its M subcarriers in order to arrive at the modulated signal. For a rigorous proof and a more detailed treatment of various OFDM techniques, the interested reader is referred to [10, 92]. The subchannel modems can use almost any modulation scheme, coherent or noncoherent, and 4- or 16-level QAM is an attractive choice in many situations.

The FFT-based QAM/FDM modem's more detailed schematic is shown in Figure 5.18. The bits provided by the source are serial/parallel converted in order to form the n-level Gray-coded symbols, M of which are collected in TX buffer 1, while the contents of TX buffer 2 are being transformed by the IFFT in order to form the time-domain modulated signal. The digital-to-analog (D-A) converted, low-pass filtered modulated signal is then transmitted via the channel, and its received samples are collected in RX buffer 1, while the contents of RX buffer 2 are being transformed to derive the demodulated signal. The twin buffers are alternately filled with data to allow for the finite FFT demodulation time. Before the data bits are Gray-coded and passed to the data sink, they may be equalized by a low-complexity frequency-domain pilot-based method, if there is some residual dispersion within the narrow subbands.

Again, the frequency response of a 155 Mbit/s WATM system was plotted in Figure 5.1 for a 128-channel OFDM system, as a function of both the OFDM symbol index and that of the frequency-domain subchannel index. Observe the hostile frequency selective nature of the fading in Figure 5.1. The bursty channel effects of this frequency-selective fading channel transfer function are efficiently combated by the OFDM modem, since in the narrow OFDM subchannels the transfer function of the propagation channel can be considered more or less flat-fading, associated with a near-constant frequency-domain transfer function and a near-linear phase response. The residual fading can be equalized using simple pilot-assisted fading compensation [10]. Again, for a deeper tutorial exposure, on OFDM the interested reader is referred to [10].

Following the above discussions on various modulation techniques, let us now briefly consider the principles of Packet Reservation Multiple Access (PRMA) in the next section.

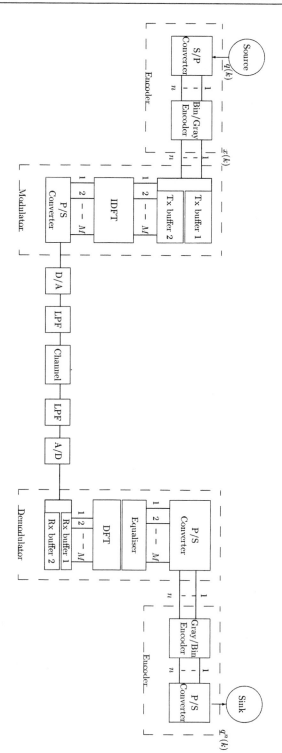

Figure 5.18: FFT-based OFDM modem schematic.

5.3 Packet Reservation Multiple Access

PRMA is a relative of slotted ALOHA contrived for conveying speech, video, and other signals on a flexible demand basis via time division multiple access (TDMA) systems. PRMA was documented in a series of excellent treatises by Goodman *et al.* [180], while a PRMA-assisted adaptive differential pulse code modulation (ADPCM) transceiver was proposed in reference [181]. The voice activity detector (VAD) [181] queues the active speech spurts to contend for an up-link TDMA time-slot for transmission to the BS. Alternatively, a VAD similar to that of the GSM system [41] can be employed. Inactive users' TDMA time-slots are offered by the BS to other users, who become active and who are allowed to contend for the unused time-slots with a less than unity permission probability. This measure prevents previously colliding users from consistently keep colliding in their further attempts to attain a time-slot reservation.

If several users contend for an available slot, neither of them will be granted it, while if only one user requires the time-slot, it can be reserved for his or her uncontended future communications. When many users are contending for a reservation, the collision probability is increased and hence a speech packet might have to contend for a number of consecutive slots, until its maximum contention delay of typically 32 ms expires. In this case the speech packet must be dropped, but the packet-dropping probability must be kept below 1%, a value inflicting minimal degradation in perceivable speech quality in contemporary speech codecs. As an example, the 8 kbps G.729 CCITT/ITU Algerbraic Code Excited Linear Predictive (ACELP) speech codec's target was to inflict less than 0.5 Mean Opinion Score (MOS) degradation in case of a speech frame error rate of 3% [182]. Here we curtail our PRMA-related discussions and note that a more detailed discourse on PRMA will be provided in a high-rate wireless local area network-based multimedia scenario in Chapter 6.

The performance of communications systems is often evaluated in terms of the teletraffic carried, while maintaining a set of communications quality measures. In conventional TDMA mobile systems the grade of service (GOS) degrades due to speech impairments caused by call blocking, handover failures and speech frame interference engendered by noise, as well as co- and adjacent-channel interference. In PRMA-assisted systems, calls are not blocked due to the lack of an idle time-slot, but the packet-dropping probability is increased gracefully. Handovers will be performed in the form of contention for an idle time-slot provided by the specific BS offering the highest signal quality among the potential handover target BSs.

The specific physical up-link to the BS offering the best signal quality during decoding the packet header is not likely to substantially degrade during the lifetime of an active speech spurt having a typical mean duration of 1s or some thirty consecutive 30 ms speech frames. If, however the link degrades, before the next active speech or video spurt is due for transmission, the subsequent contention phase is likely to establish a link with another BS. Hence this process will have a favorable effect on the channel's quality, effectively simulating a diversity system having independent fading channels and limiting the time spent by the MS in deep fades, thereby avoiding channels with high noise or interference.

This advantageous property can be exploited to train a self-adjusting adaptive system using the channel segregation scheme proposed for PRMA systems in [183].

Accordingly, each BS evaluates and ranks the quality of its idle physical channels constituted by the unused time-slots on a frame-by-frame basis and identifies a certain number of slots, N, with the highest quality, i.e., lowest noise and interference. These high-quality, low-interference channels are segregated for contention, while the lower quality idle slots contaminated by noise and interference are temporarily disabled. Hence, upon a new access request the BS is likely to receive a signal having low-interference, which maximizes the chances of successful packet decoding, unless a collision caused by a simultaneous MS attempt to attain a reservation has occurred. When a successful, uncontended reservation takes place, the BS promotes the highest quality disabled time-slot to the set of N segregated channels, unless its quality is unacceptably low. It appears plausible that if N is high, the packet-dropping probability becomes low, but the physical channels constituted by the time-slots might become heavily interfered, while if N is low, we have a packet-dropping-dominated scenario, which equally limits the GOS.

Clearly, the main cause of GOS degradation in PRMA systems is limited to speech or video packet corruption due to noise or interference and packet-dropping [184]. They both result in different subjective speech or video GOS degradation. With the system elements described in previous sections of this chapter, we now focus our attention on the amalgamated PCS transceiver proposed.

5.4 Flexible Transceiver Architecture

The schematic of a flexible, toolbox-based multimedia system is portrayed in Figure 5.19. The pivotal implementational point of such a multimedia PS is that of finding the best compromise among a number of contradicting design factors, such as power consumption, robustness against transmission errors, spectral efficiency, audio/video-quality, and so forth. The time-variant optimization criteria of a flexible multimedia system can only be met by an adaptive scheme, comprising the firmware of a suite of system components and loading that combination of speech codecs, video codecs, embedded channel codecs, voice activity detector (VAD), and modems, which fulfills the prevalent criterion [10]. A few examples are maximizing the teletraffic carried or the robustness against channel errors, while in other cases minimization of the bandwidth occupancy, the call blocking probability, or the power consumption is of prime concern.

Focusing our attention on the speech and video links displayed in Figure 5.19, the voice activity detector (VAD) is employed to control the packet reservation multiple access (PRMA) slot allocator [10], multiplexing speech and video. Control traffic and system information is carried by packet headers added to the composite signal by the bit mapper before K-class source sensitivity-matched forward error correction coding (FEC) takes place. Observe that the Video Encoder supplies its bits to an adaptive buffer (BUF) having a feedback loop. If the PRMA video packet delay becomes too high or the buffer fullness exceeds a certain threshold, the video encoder is instructed to reduce its bit rate, implying a concomitant dropping of the image quality.

The bit mapper assigns the most significant source coded bits (MSB) to the input of the strongest FEC codec, FEC K, while the least significant bits (LSB) are

5.4. FLEXIBLE TRANSCEIVER ARCHITECTURE

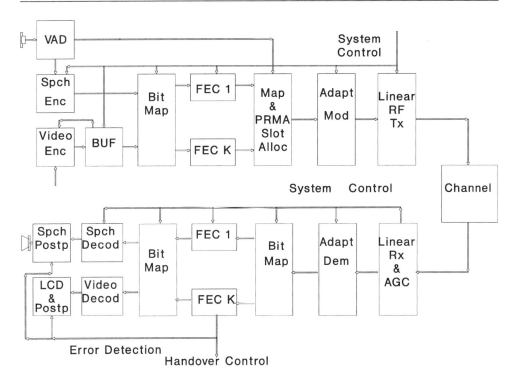

Figure 5.19: Flexible multimedia communicator schematic.

protected by the weakest one, FEC 1. K-class FEC coding is used after mapping the speech and video bits to their appropriate bit protection classes, which ensures source sensitivity-matched transmission. Adaptive Modulation originally proposed by Steele and Webb [134] is employed [10] with the number of modulation levels, the FEC coding power and the speech/video source coding algorithm adjusted by the System Control according to the dominant propagation conditions, bandwidth and power efficiency requirements, channel blocking probability, or PRMA packet-dropping probability. If the communications quality or the prevalent system optimization criterion cannot be improved by adaptive transceiver reconfiguration, the serving BS will hand the PS over to another BS providing a better grade of service.

One of the most important and reliable parameters used to control these algorithms is the Error Detection flag of the FEC decoder of the most significant bit (MSB) class of speech and video bits, namely FEC K. This flag can also be invoked to control the speech and video Postprocessing algorithms. The adaptive modulator transmits the user bursts from the PS to the BS using the specific PRMA slot allocated by the BS for the PS's speech, data, or video information via the linear radio frequency (RF) transmitter (Tx). Although the linear RF transmitter has a low power efficiency, its power consumption is less critical due to the low transmitted power requirement of the multimedia microcellular network than that of the digital signal processing (DSP) hardware.

The receiver structure essentially follows that of the transmitter. After linear class-

A amplification and automatic gain control (AGC) the System Control information characterizing the type of modulation and the number of modulation levels must be extracted from the received signal, before demodulation can take place. This information also controls the various internal bit mapping algorithms and invokes the appropriate speech and video decoding as well as FEC decoding procedures. After Adaptive Demodulation at the BS the source bits are mapped back to their original bit protection classes and FEC decoded. As mentioned, the error detection flag of the strongest FEC decoder, FEC K, is used to control handovers or speech and video postprocessing. The FEC decoded speech and video bits are finally source decoded and the recovered speech arrives at the earpiece while the video information is displayed on a flat liquid crystal display (LCD). Following the above brief overview of the system architecture of IMMTs, in the next section we conclude this chapter.

5.5 Summary and Conclusions

This chapter commenced by arguing that conventional fixed-mode transceivers are incapable of accommodating the time- and frequency-domain channel quality fluctuations of wireless channels. The concept of IMMTs was introduced as a new system design paradigm, where the transceivers are designed for maintaining a given target performance expressed, for example, in terms of the BER, while attaining the highest possible throughput. QAM was introduced as a powerful means of controlling the number of bits per modulation symbols in this context without affecting the system's bandwidth requirement. The chapter is concluded by a comparison of coherent versus noncoherent demodulation techniques, indicating that coherent detection requires typically 3 dB lower channel SINRs, than its lower-complexity noncoherent counterpart. We alluded to the fact that the concept of IMMTs is also applicable to OFDM [10, 92] as well as to CDMA [95–99]. These systems are characterized in Chapter 6.

The system control algorithms of the reconfigurable mobile multimedia communicator will dynamically evolve over the years. PSs of widely varying complexity will coexist, where newer ones provide backward compatibility with existing ones, while offering more intelligent new services and more convenient features. Again, a range of adaptively reconfigurable modem schemes have been proposed by Torrance *et al.* in [110–115, 149–152]. Let us consider in the next chapter, how channel-coding can be invoked, in order to mitigate the transmission errors encountered.

Chapter 6

Video Traffic Modeling and Multiple Access

J. Brecht, L. Hanzo, and M. Del Buono [1] [2]

6.1 Video Traffic Modeling

6.1.1 Motivation and Background

In recent years, there have been increased worldwide research activities in the area of modeling video codecs, mainly for studying the performance of multiple access (MAC) schemes [187, 188], where a simple, but sufficiently accurate, video source model is needed. This is motivated by the fact that in studying the performance of various MAC schemes, it is unnecessary and extremely time consuming to conduct bit-true simulations of the physical layer, including video compression, channel-coding, and modulation, as long as these physical layer operations can be adequately modeled by an appropriate source model. This is particularly so, for example, for high-rate, 155 Mbps Wireless Asynchronous Transfer Mode (WATM) and other local area networks (LANs), where typically excessive simulation times are encountered in modeling the complex, high-rate physical-layer simulations. With this motivation, [189] provides a review of different models commonly used to mimic the bit-generation process of voice, data, and video sources.

In the majority of cases, Markov models or their derivatives have been favored for their simplicity. Other common models are the autoregressive models. A comparison beetween these and a range of other models can be found, for example, in [190].

---

[1] This chapter is based on [185, 186].
[2] © 1998 Elsevier. Personal use of this material is permitted. However, permission to reprint/republish this material for advertizing or promotional purposes or for creating new collective works for resale or redistribution to servers or lists, or to refuse any copyrighted component of this work in other works, must be obtained from Elsevier.

Heymann and Lakshman in [191] employed discrete autoregressive (DAR) and Markov models, while reference [188] has studied the problems associated with the bit-rate fluctuation of a video source.

The motivation of this chapter is twofold. Firstly, a novel video model is proposed, and its ability to characterize various video sources is studied in diverse practical scenarios. The proposed source-modeling technique was investigated in the context of the H.263 video codec [192, 193] of Chapter 18, since this is to date the most successful standard video codec. Although this codec extensively employs error-sensitive run-length coding techniques for achieving maximum coding efficiency, since it was originally designed for low error-rate Gaussian channels, in recent years a variety of techniques have been proposed for facilitating its employment in wireless communications.

In particular, Khansari, Jalali, Dubois, and Mermelstein from INRS in Canada investigated the employment of the H.261 codec, which was the predecessor of the H.263 scheme [194], while Färber, Steinbach, and Girod [195] at Erlangen University in Germany as well as Cherriman and Hanzo in the United Kingdom [196–200] considered various H.263-based video systems. It was found advantageous to superimpose a simple acknowledgment flag on the video packets in both directions of a duplex video link. This can be used for informing the local and remote encoder of the success or failure of the transmitted packet. Then the contents of the local and remote reconstruction frame buffers can remain identical, since corrupted packets can be dropped at both ends of the link. This measure prevents the potential propagation of transmission errors, which would be encountered if the reconstructed frame buffers were allowed to become "misaligned," that is, different. The result was an extremely robust operation, tolerating packet-dropping rates as high as 5%.[3] The video-frame areas represented by the dropped packets can then be updated in future frames. The associated latency or delay of this technique is particularly low in short-frame duration LANs, since the acknowledgment flags become available almost instantaneously. We note furthermore that the proposed modeling technique is also applicable to modeling more complex sources, such as, for example, the Intra- (I), Bidirectional- (B) and Predicted- (P) frames of the Motion Pictures Expert Group (MPEG2) codec.

The second motivation of this chapter is to employ the video model advocated in two different multiple access schemes, namely, in multiframe packet reservation multiple access (MF-PRMA) and in another novel access scheme, which we refer to as Statistical Packet Assignment Multiple Access (SPAMA). Both of the proposed access schemes can be used to support variable-rate multimedia traffic in a wireless local area network (WLAN) scenario. Finally, a number of applications scenarios based on the proposed video models and MF-PRMA as well as SPAMA are characterized in terms of their throughput and delay performances.

In Section 6.1.2, the Markov modeling of video sequences is introduced, and the model limitations are highlighted. Section 6.1.3 proposes a range of practical improvements to the basic Markov model, while the performance of the algorithm is characterized in Section 6.1.4. Various multiple access schemes are introduced and classified

---

[3]Various H.263 video demonstrations over a range of channel conditions and packet-dropping rates can be viewed under http://www-mobile.ecs.soton.ac.uk by means of an MPEG player. Further contributions on the corresponding wireless H.263 systems are also downloadable from this address.

6.1. VIDEO TRAFFIC MODELING

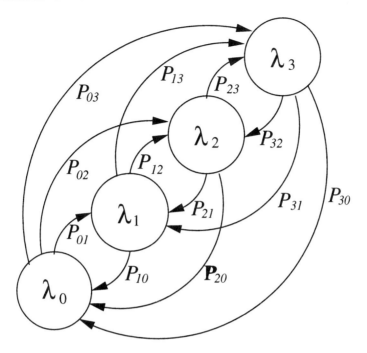

Figure 6.1: Four-state Markov-model example, where λ_i is the packet generation rate of state i, while P_{ij} denotes the state transition probabilities. © Elsevier [185].

in Section 6.2. Specifically, our MF-PRMA scheme is outlined in Section 6.2.3, and its performance is characterized using the proposed video model in Section 6.2.3.1. Section 6.2.4 outlines another efficient resource allocation scheme, which we refer to here as the statistical packet assignment multiple access (SPAMA) algorithm, while Section 6.2.4.2 characterizes its performance, before concluding in the last section.

6.1.2 Markov Modeling of Video Sources

In the spirit of our previous discussions, we opted for a so-called Markov modulated process, which can adequately model both the first and the second moment of the bit-rate fluctuation of various sources, that is both the average bit rate, and the variance of the bit rate. Furthermore, we found that it was possible to superimpose a number of Markov chains in order to account for particular features of the sequence, such as, for example, spikes in the bit-rate histogram often used in the characterization of variable-rate sources and for modeling other bit-rate irregularities, as will be discussed later.

In a Markov Modulated Poisson Process (MMPP) [189], the instantaneous transmitted packet "arrival rate" or in other words, the rate at which transmitted packets are generated is "modulated" — influenced — by the current state of the continuous-time discrete state Markov model of Figure 6.1, a process that will be augmented later in this chapter. This process is characterized by the transmitted packet arrival

0.000000	0.500000	0.250000	0.250000
0.333333	0.333333	0.333334	0.000000
0.000000	0.000000	0.666667	0.333333
0.100000	0.725000	0.075000	0.100000

Table 6.1: Markov transition probability matrix

or generation rate λ_i per each Markov state and the mean sojourn time $1/r_i$, spent in each Markov state. The sojourn time has a negative exponential distribution [201]. The packet arrival or generation rate λ_i is proportional to the mean bit rate in state i of Figure 6.1, while the state transition probabilities are denoted by P_{ij}.

A practical video model generates a certain number of video packets or video bits describing each video-frame. The frames follow each other at a fixed frame repetition rate, which is modeled by a fixed sojourn time in each Markov state, generating a certain number of bits. Hence, in our model we assumed that the sojourn time has a constant value, which correponds to the video-frame duration and at the end of each video-frame a Markov state-transition takes place. These state transitions are characterized by the associated transition probabilities. Furthermore, while residing in any of the legitimate Markov states, that is, within a frame duration, the number of bits generated obeys a Poissonian process [201] rather than a constant arrival rate model. Again, transitions are permitted between all the states, as shown in Figure 6.1, according to the so-called transition probability matrix, where the element P_{ij} represents the probability of having a transition between state i and state j. A typical transition probability matrix example is shown in Table 6.1. However, this specific matrix is used only as a representative example, and it is not related to any particular practical model used in our experiments, where we typically employed models having a higher number of states.

In video source modeling, the first problem is the choice of the number of states in the Markov chain. In order to match the bit-rate histogram of the original source by that of the model with sufficient accuracy, a high number of states is required. However, upon increasing the number of states; we found many more bit-rate histogram spikes in the simulated sequence than there were in the original (an issue to be augmented in more depth during our further discourse). This indicates that a high number of Markov states requires a sufficiently long training sequence for generating an accurate state transition matrix in order to arrive at a statistically meaningful number of state transitions among all possible states. When experimenting with limited-duration practical video sequences, it was impractical to choose a sufficiently high number of states; since then the statistical credibility of the investigations became questionable as a result of the associated low number of state transitions among certain low-probability states. We addressed this problem by constructing mosaic sequences, constituted, for example, by four different-length sequences combined to form a quadruple-sized sequence, which will be invoked in Section 6.1.4 for algorithmic performance testing. In order to find a good compromise between these two conflicting requirements, after a range of experiments we opted for use of twenty 20-state Markov model for the ITU's H.263 codec.

We will return to this issue in the context of various source rates, since for high-rate

6.1. VIDEO TRAFFIC MODELING

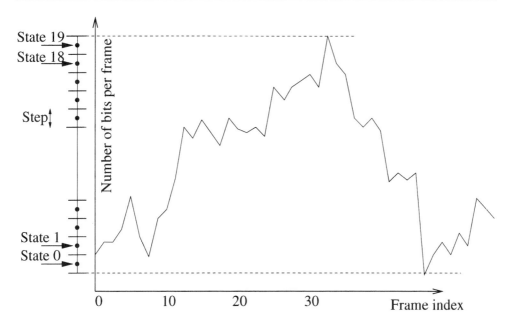

Figure 6.2: Quantization of the video source-rate fluctuation using a 20-state Markov model. © Elsevier [185].

systems typically more states are required for maintaining a low bit-rate granularity, an issue that becomes plausible in the context of Figures 6.2 and 6.8. By comparison, in [202] only eight states were used.

According to the above considerations, we investigated a variety of different video resolution sequences to be modeled and after identifying the maximum and minimum bit rate (i.e., the bit-rate range of the sequences), we divided this range in twenty uniform bit-rate ranges. At the center of each bit-rate interval we allocated a state of the Markov chain, as seen in Figure 6.2. The transition probability from state i to state j has been found by simulation upon observing the sequence after assigning the actual measured bit rate to one of the twenty states.

Our tentative bit-generation model has the following construction. In each bit-generation cycle — initially corresponding to a video-frame duration — a random generator is used to determine the next state of the Markov model, which can be any of the twenty states. These transitions are governed by the transition matrix, generated by evaluating the relative transition frequencies, approximating the probabilities of all possible Markov state transitions using simulations. Then in each state the actual number of bits generated obeys the Poisson distribution, and the corresponding probability density function (PDF) typically overlaps with those of the adjacent states. Having stipulated the basic video model, let us now scrutinize its behavior in the next section.

6.1.3 Reduced-Length Poisson Cycles

From a practical point of view, it is inconvenient to operate directly on the basis of the number of bits generated per video-frame, since observing the Poisson PDF of [201]:

$$P(k, \lambda_i) = \frac{(\lambda_i T)^k \exp(-\lambda_i T)}{k!} \qquad (6.1)$$

we found that the factorial function of the denominator results in an excessive computational demand, when determining the number of bits generated. Hence, a potential solution is to divide the video-frame duration into a number of bit-generation cycles with the advantage that in this way we can find the number of bits generated on a more convenient scale, using a granularity that is more compatible with the typical burst-length of conventional wireless networks. More explicitly, we have to generate the number of bits to be transmitted on a transmission packet-by-packet basis.

We therefore decided to divide the video-frame into a number of shorter segments in order to reduce the number of bits generated per Poisson cycle to around 300 or less, simply because in case of a higher average number of bits per generation cycle the probability that we must calculate a factorial higher than 1000! is not negligible. Therefore, we opted to invoke a *division factor* of $D = 5000$, which is used to divide the target bit number per video-frame in $D = 5000$ smaller *bit generation cycles*. We will show that this choice constitutes a good compromise for source rates from 10 Kbps to 10 Mbps. For sources at higher bit rates, we have to increase the value of D.

As a consequence, now the number of bits generated per video-frame is the accumulation of the number of bits generated per Poissonian cycle. This implies that the distribution is the convolution of 5000 Poisson distributions. We observed that in this case the number of bits per video-frame was not sufficiently spread around the average in order to provide a statistically sound model of the bit-rate fluctuation for the H.263 scheme or for other practical variable-rate codecs. This is demonstrated in Figure 6.3, in comparison to the actual number of bits generated by the H.263 codec for the high-resolution 704×576 pixel so-called 4-times Common Intermediate Format (4CIF) "Susie" video sequence seen in Figure 15.1, which was encoded with a target bit rate of 7 Mbps. Observe that apart from a single excursion to a state corresponding to about 220 000 bits per frame, which occurs at frame index 105, the process resides in a state obeying a distribution given by the convolution of the 5000 Poisson PDFs. The resultant bit-rate fluctuation exhibits a low variance around 207 000 bits per frame. Again, these rate fluctuations appear quite limited.

In order to avoid this near-constant bit-rate problem, we introduce an *oscillation factor* O, whose role and terminology become explicit below. In order to generate adequate bit-rate fluctuation statistics at a practical complexity, the number of convolved Poissonian variables has to be reduced, namely, with the aid of the above *oscillation factor*. The effective number of bit generation cycles is now computed as D/O, but at the end of each modeling cycle we multiply the number of bits generated, namely, k, by the value of O. This measure allows us to maintain an improved bit-rate granularity as will be demonstrated later. The operation of the model is illustrated in the flowchart of Figure 6.4, which will be described with reference to

6.1. VIDEO TRAFFIC MODELING

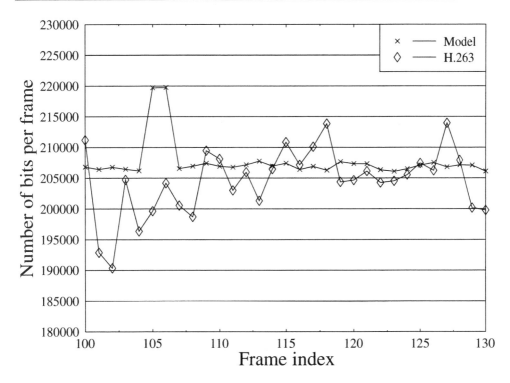

Figure 6.3: The number of bits per video-frame versus frame index when modeling the H.263-encoded 4CIF "Susie" sequence of Figure 15.1 at a target bit rate of 7 Mbps. For many frames the model resides in the same state resulting in a near-constant bit rate exhibiting a low variance. © Elsevier [185].

the relationship between the Poissonian probability density function (PDF) and the cumulative probability density function (CDF) in Figure 6.5.

For the sake of augmenting the inner working of the somewhat heuristic flowchart in Figure 6.4, let us consider the stylized Poissonian CDF and PDF of Figure 6.5 and initially let us assume an oscillation factor of $O = 1$. Thus, $D/O = 5000$ Poissonian generation cycles are required to generate a whole video-frame that is, the main loop at the top of the flowchart, in which the index C is incremented, and is passed through 5000 times. At the commencement of each bit-generation cycle, a uniformly distributed random number is generated and assigned to P, where $0 < P < 1$ as seen in the flowchart of Figure 6.4. This corresponds to choosing a uniformly distributed random value on the vertical axis of the CDF of Figure 6.5, which is 0.45 in our example. Then the Poisson PDF of Equation 6.1 is evaluated for the iteration index of $k = 0$, which will be incremented at the bottom of Figure 6.4, and the returned Poissonian PDF value $Pss(k, \lambda_i)$ of Equation 6.1 is assigned to the variable Ds. The number of bits generated in the current bit-generation cycle obeys the Poissonian CDF, when the values of the Poisson PDF, $Pss(k, \lambda_i)$, are accumulated in the variable Ds for incremented values of k, until the condition $Ds < P$, is satisfied in Figure 6.4, which is true below $k = M$ in our example of Figure 6.5. At this stage, the loop at

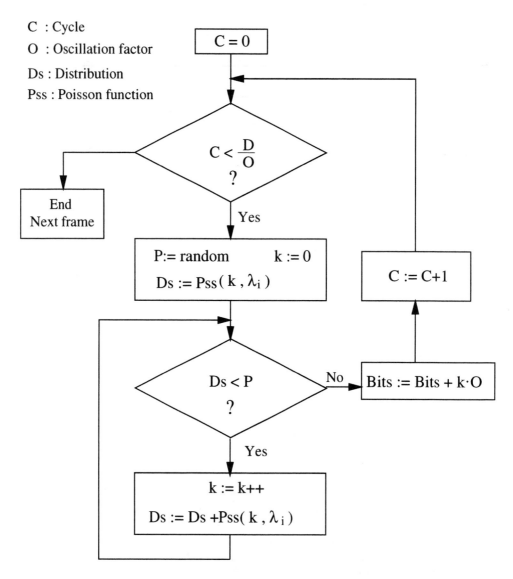

Figure 6.4: Flowchart of the bit-generation algorithm for a video-frame. © Elsevier [185].

6.1. VIDEO TRAFFIC MODELING

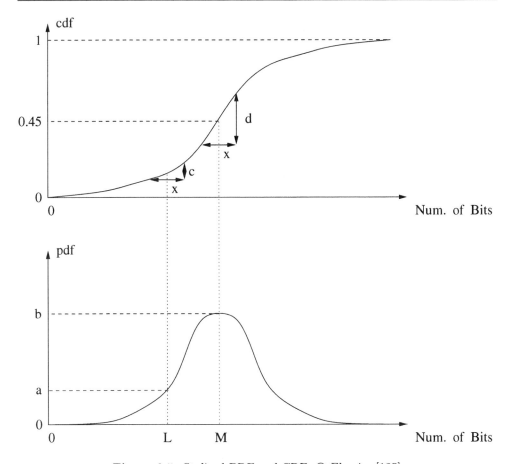

Figure 6.5: Stylized PDF and CDF. © Elsevier [185].

the bottom of Figure 6.4 is exited, which corresponds to accumulating the Poissonian PDF values $Pss(k, \lambda_i)$ in Ds, until the CDF value Ds reached the random value P, namely, the CDF value representing the probability of generating M bits in the current bit-generation cycle, as exemplified in Figure 6.5. The index $k = M$, multiplied by the oscillation factor O, represents the number of bits generated in the current bit-generation cycle. This number is then added to the contents of the counter Bits at the right of the flowchart, where Bits was initialized to zero at the commencement of each new bit-generation cycle. In order to augment our exposition further, here we note that the bit-rate granularity or resolution is one bit without using our proposed oscillation factor, but for nonunity oscillation factors it becomes identical to the factor itself. This is attractive, since in high-rate scenarios, where O is high, a higher absolute bit-rate granularity is maintained.

In order to study the effect of different oscillation factors O, the following experiment was carried out. Initially, $O = 1$ was stipulated, resulting in $D = 5000$ Poissonian cycles per video-frame, yielding 5000 numbers. Each were generated by a Poisson process using Equation 6.1, where λ_i is the mean bit rate of the Markov

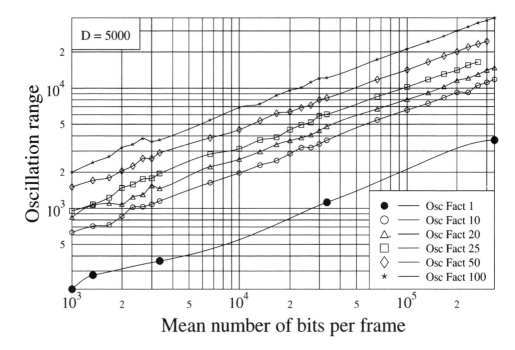

Figure 6.6: Oscillation range ΔR versus mean number of bits per frame R_a: different ranges can be selected using different O values in order to ensure a better fit of the bit-rate histogram of the model to that of the codec modeled. © Elsevier [185].

chain in state i and the generation cycle duration is $T = Frame - Duration/D$. These simulations were conducted for 3000 video-frames. The lowest and the highest number of bits generated were recorded, and we refer to their difference ΔR as the *bit-rate oscillation range*. This range was then recorded for various average bit rates or mean number of bits per frame. The results are plotted in Figure 6.6 for a range of O values between 1 and 100.

Explicitly, the curve plotted for $O = 1$ is the original curve, where we used $D = 5000$ Poisson cycles per video-frame. The figure shows, for example, that for an average source rate around 5 Mbps, corresponding to about 167 000 bits per video-frame at 30 fps, a fluctuation of 3000 bits around the mean value of 167 000 bits is almost negligible. This phenomenon was observed earlier in Figure 6.3.

Observe furthermore in Figure 6.6 that because of the introduction of an oscillation factor of $O = 100$ for a 5 Mbps, 30 frames/s, 167 000 bits per frame scenario, the fluctuation range now becomes significantly higher,– approximately 30 000 bits around the mean value. This is because for $D = 5000$ and $O = 1$ the number of bits per frame was the cumulative value of 5000 Poissonian variables, yielding a near-constant value. By contrast, for $O = 100$ the higher oscillation range is a consequence of accumulating only 50 such variables.

At this stage, a further step was required to complete the model design. Specifically, given an overall average bit rate of R_a, a minimum and maximum bit rate of

R_{min} and R_{max}, respectively, as well as a set of N Markov-model states, the resulting target bit rate R_i (or λ_i) of state i with $0 \leq i \leq N$ is given by:

$$R_i = R_{min} + i \frac{R_{max} - R_{min}}{N} = R_{min} + i\,\Delta R. \tag{6.2}$$

While residing in any of the Markov states, the model will ensure that the range of the instantaneous bit-rate fluctuations is limited to $\Delta R = [R_{min}; R_{max}]$, and the specific bit-rate values of each state associated with a certain mean bit rate obey the Poissonian distribution. Given a certain oscillation range ΔR, we can invoke Figure 6.6 in order to determine the required oscillation factor O.

In order to assist in this effort, we found an empirical relation between the quantities involved. From Figure 6.6, which is plotted on a log-log axis, we inferred that the relationship between the average value of bits per frame R_a and the oscillation range ΔR can be approximated by:

$$\Delta R = K\sqrt{R_a}. \tag{6.3}$$

Minimum mean-squared fitting of the experimental R_a and ΔR values for various K values revealed the following approximate dependence of K on the oscillation factor O:

$$K = a \cdot \sqrt{O}, \tag{6.4}$$

where we have $a = 6.48$, yielding

$$\Delta R = 6.48 \cdot \sqrt{O \cdot R_a}. \tag{6.5}$$

The goodness-of-fit of this matching process is characterized in informal terms by Figure 6.7.

The above-mentioned experimental relationship has been used in our simulations, and the corresponding bit-rate histograms are depicted in Figure 6.8 for two different O factors, namely, for $O = 1$ and 8, as well as for our experimental data generated by the H.263 codec for the 4CIF "Susie" sequence, while maintaining an average bit rate of 5 Mbps. Observe in the sub-fgure seen in the middle that in accordance with our previous experience for $O = 1$ there is only a very limited bit-rate fluctuation or spread within the Markov states around the target bit rates of the individual Markov states. However, for $O = 8$ a more appropriately spread Poissonian bit-rate distribution is observed in each state, where ΔR is a factor $\sqrt{8}$ higher. Observe for $O = 8$ at the bottom of the Figure that the PDFs are slightly more spread toward the top end of the bit rate range than in the lower-rate Markov states. This is because for the Poisson distribution the value of the variance is equal to the mean value, which is clearly higher for the states closer to the top end of the bit-rate scale.

6.1.4 Video Model Matching

From our simulation results we found that for a source bit rate around 1Mb/s or less an oscillation factor between $O = 1$ and 3 was appropriate. For source rates around 10 Mb/s, a value around $O = 50$ was required, depending on the target source rate.

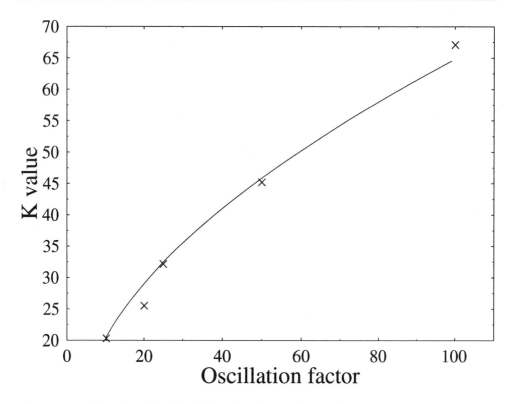

Figure 6.7: The value of K for different oscillation factors O, where the curve was found by minimum mean-squared fitting. © Elsevier [185].

Furthermore, we found that the model was quite flexible and allowed us to emulate a range of different video scenes adequately, always obeying Equations 6.3–6.5.

Figures 6.9–6.12 show a number of model characteristics for various video sequences. Specifically, at the bottom of each of these figures the typical bit-rate fluctuation of the original H.263 codec and that of the model can be seen as an easily interpreted illustrative example. The frame indices chosen represent arbitrary typical segments of four different video clips at different resolutions. In the center of each illustration the normalized correlation between the bit rates of consecutive frames was plotted, while at the top the bit-rate histogram of both the original experimental data and that of the model is displayed.

A representative range of low, medium, and high bit-rate scenarios at 64 kbps, 1 Mbps, and 3 Mbps, respectively, were studied using various video sequences encoded at various bit rates. Although the bit-rate histograms would not be acceptable at a high confidence level using rigorous goodness-of-fit distribution testing techniques, such as the Kolmogorov-Smirnov or the χ^2-test [148], for practical network modeling purposes they were deemed informally adequate. The bit-rate correlation functions also exhibited an adequate informally judged match. The observed deviations from experimental features were deemed to be a consequence of the limited-duration train-

6.1. VIDEO TRAFFIC MODELING

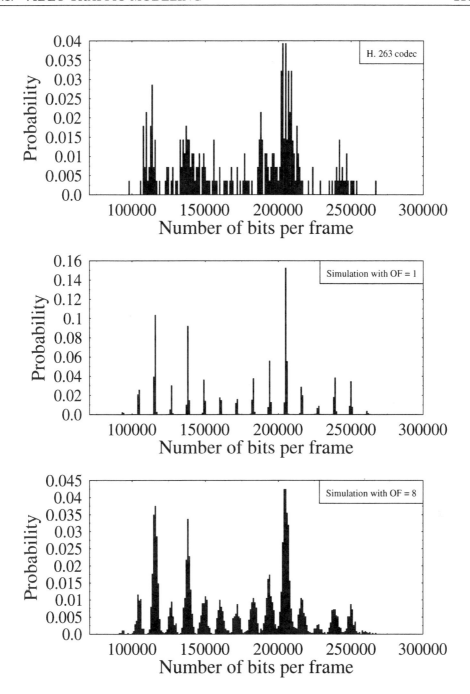

Figure 6.8: Bit-rate histograms for the 4CIF "Susie" sequence at a 5 Mbps target bit rate generated by the H.263 codec and by the proposed model with $O = 1$ and $O = 8$. Observe the better histogram fit due to a higher value of the O. © Elsevier [185].

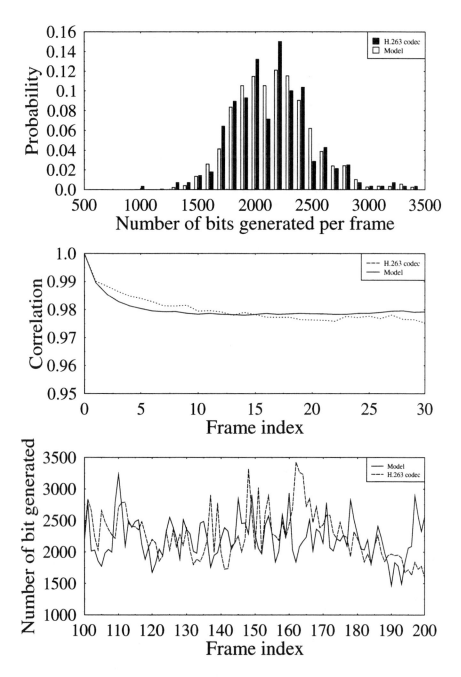

Figure 6.9: Bit-rate histogram, correlation, and typical bit-rate fluctuation of the QCIF "Carphone" sequence of Figure 15.2 comparing the H.263 codec at a 64 kbps average target bit rate and the Markov model. © Elsevier [185].

6.1. VIDEO TRAFFIC MODELING

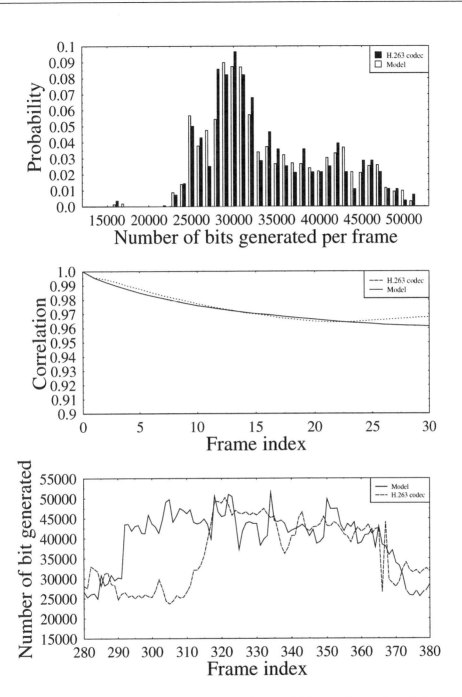

Figure 6.10: Bit-rate histogram, correlation, and typical bit-rate fluctuation of the 4CIF "Susie" sequence of Figure 15.1, comparing the H.263 codec at a 1 Mbps average target bit rate and the Markov model. © Elsevier [185].

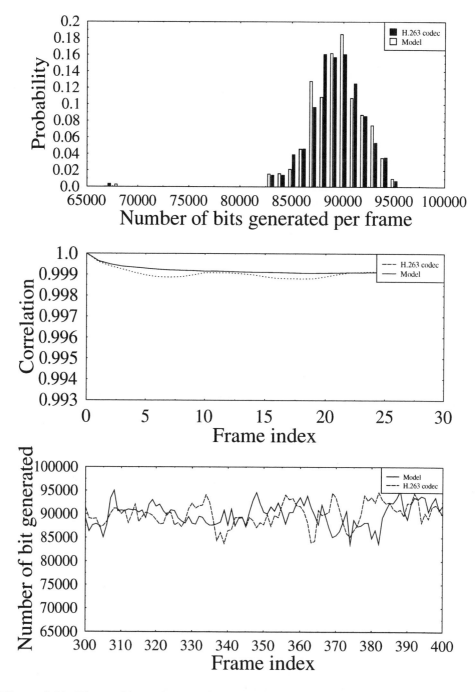

Figure 6.11: Bit-rate histogram, correlation, and typical bit-rate fluctuation of the CIF "Miss America" sequence, comparing the H.263 codec at a 3 Mbps average target bit rate and the Markov model. © Elsevier [185].

6.1. VIDEO TRAFFIC MODELING

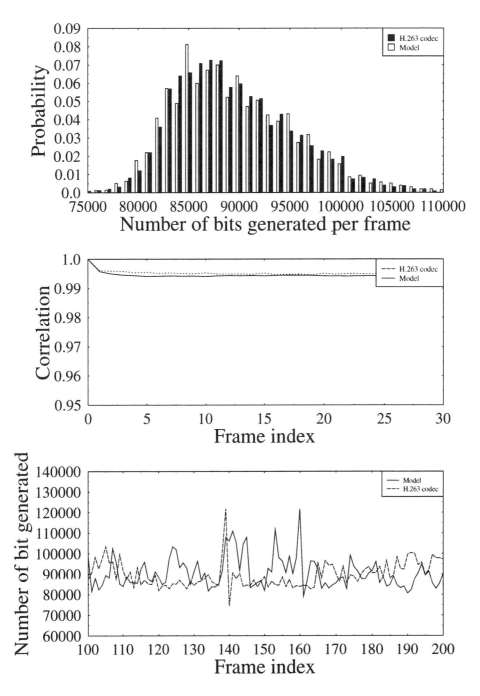

Figure 6.12: Bit-rate histogram, correlation, and typical bit-rate fluctuation of the CIF "mosaic" sequence of Figure 6.13, comparing the H.263 codec at a 2.6 Mbps average target bit rate and the 30-state Markov model. © Elsevier [185].

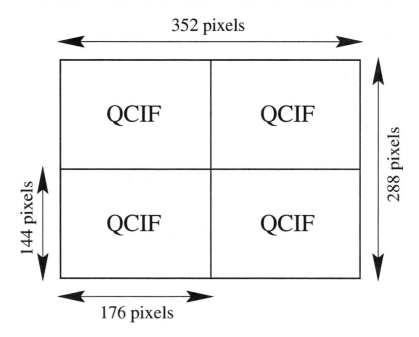

Figure 6.13: Creating mosaic video clips. © Elsevier [185].

ing data for the model, which adopted the transition matrix probabilities found from our experimental video data. Figure 6.11 shows a very high correlation due to the particular features of the "Miss America" sequence, which exhibits a rather limited motion activity. Hence, the number of bits generated per frame is almost constant without large excursions around the mean value.

In order to overcome the problem of having limited-duration training sequences, we attempted to create longer sequences. Four QCIF sequences have been combined to obtain a quadruple-sized CIF sequence, as illustrated in Figure 6.13. Since each QCIF component sequence has a different length in terms of the number of picture frames, it is possible to create an arbitrarily long CIF sequence by playing the constituent sequences back and forth. These experiments have been carried out for 3000 video-frames using the sequences "Sales-man, Carphone, Newscaster, and Foreman" in order to mix high-dynamic and low-dynamic sequences. The results are shown in Figure 6.12. A 30-state Markov model has been used, since the high number of frames available allowed us to find a sound set of transition matrix statistics.

In conclusion, we note that we tested the applicability of our model for mimicking the MPEG-2 codec using the superposition of three Markov chains, one for each different frame type, namely, for the so-called Intra-coded (I), Predicted (P), and Bidirectionally predicted (B) frames, which will be introduced in more depth in Section 18.2.4.4. Eight Markov states were assigned for each of the three Markov chains. The transition between the three corresponding Markov chains was calculated according to the frame sequence IBBPBBPBBPBBI. for an MPEG transmission, although different transition probabilities can be readily used. Our results are similar

6.2. MULTIPLE ACCESS

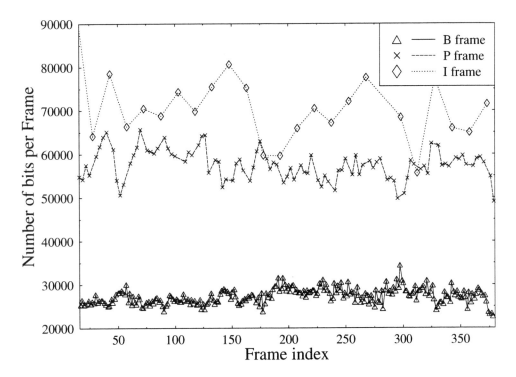

Figure 6.14: Bit-rate fluctuation profile of the MPEG-2 Carphone sequence for an IBBPBBPBBPBBPBBI sequence, in QCIF resolution at 30 frames/s for an average bit rate of 1 Mbps. © Elsevier [185].

to those in [203]. In Figure 6.14 we plotted the bit-rate evolution of the three different frame types for the original source, while Figure 6.15 portrays both the correlation and the bit-rate profile of the original MPEG-2 codec and that of the model. Having characterized the proposed video model, let us now focus on a range of multiple access schemes.

6.2 Multiple Access

6.2.1 Background

In the following sections, we will focus on a range of multiple access techniques, which are applicable to both wireless local area networks (WLAN) and personal communications systems. Media access control (MAC) attempts to efficiently and equitably allocate a shared communications channel to a number of users. The simplest well-known examples are Time Division Multiple Access (TDMA) and Frequency Division Multiple Access (FDMA). In TDMA the channel is typically divided in transmission frames, constituted by time-slots; once a time-slot is allocated to a specific user, this user has a uncontended access to this slot all the time. This effectively corresponds

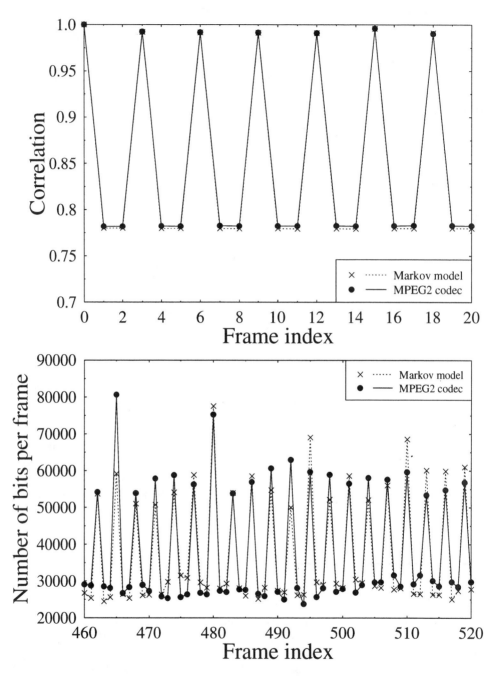

Figure 6.15: Correlation of the MPEG2 codec's bit-rate profile with that of the model using the Carphone sequence of Figure 15.2 (top). Bit-rate profile for the same sequence and the model (bottom). © Elsevier [185].

to transmitting each user's information discontinuously at an increased multi-user transmission rate during one time-slot rather than transmitting at a lower single-user rate continuously.

In case of TDMA-based voice-oriented duplex conversations, the momentarily silent users may disable transmitting the silent speech segments using a voice activity detector (VAD). Since during these passive speech segments no energy is transmitted, the total interference load of the system is reduced; hence, potentially more users can be supported, when using VADs. Another potential exploitation of the passive speech spurts can be found in *statistical multiplexing* schemes, where the momentarily passive speech users can surrender their time-slots under the control of the VAD in order to accommodate additional users, who are requesting transmission resources for their active speech spurts. This, however, results in an increased amount of co-channel interference in comparison to the scenario, where the passive time-slots remain unused.

In contrast to TDMA, in FDMA all users transmit all the time at their single-user rate but in a different frequency slot. Hence, in TDMA the users transmit in orthogonal time-slots, while in FDMA in orthogonal frequency slots. In Code Division Multiple Access (CDMA), which is the topic of Chapter 10, each user is assigned a user-specific orthogonal code, which is also often referred to as a user signature code. Hence, in CDMA all active users transmit all the time in the entire frequency band, and the orthogonality of their codes provides the multi-user access.

Following the above brief introduction, let us now classify the range of MAC protocols, noting that the rest of this chapter is dedicated to *statistical multiplexing* techniques, which attempt to allocate transmission resources using a multiple access channel to a multiplicity of users on an equitable basis, while taking into account their different delay, bit rate, integrity, and other transmission requirements.

6.2.2 Classification of Multiple Access Techniques

Multiple access protocols can be divided into two main classes: *contention-based access* and *controlled-access* protocols. These two classes, which can be further divided into subclasses, draw a distinction between those schemes that allow data packets to collide during contention and those that do not. Access schemes of the first class may have a collision detection mechanism, which was already pointed out in our brief introduction to packet reservation multiple access (PRMA) in Chapter 5, and hence the colliding data packets can be retransmitted.

In general, the medium access process can be partitioned in four phases: *request*, *access control*, *approval*, and *holding time* phases, as shown in Figure 6.16. A more detailed classification of access protocols can be derived by classifying each phase of the protocol, as was suggested by Panken [204]. The following brief description of the four access phases and their classification will provide a general context for the forthcoming discussions.

In order to gain access to a multi-user system, a terminal issues a resource allocation for declaring its transmission needs, conveyed by one or several data packets or signaling messages to the resource allocation controller. The fulfillment of requests may either be guaranteed (i.e., the transmission rights are unconditional) or by contrast, conditionally granted or uncertain, as indicated in Figure 6.16. In other words,

226 CHAPTER 6. VIDEO TRAFFIC MODELING AND MULTIPLE ACCESS

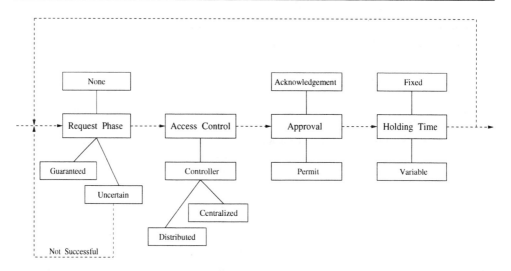

Figure 6.16: Classification of multiple access protocols. Panken [204]. © Elsevier [185].

in the latter case the transmission rights may be refused, or the request may collide with other data, leading to an unsuccessful access attempt. The latter class includes contention-based request protocols, such as, for example, the PRMA scheme already highlighted in Chapter 5, which tentatively transmits the first packet of the data in an attempt to acquire one or more of the slots, which were free, when the base station last broadcast the reserved/free status of the slots. This procedure is elaborated in more detail later in the chapter.

In the second protocol phase of Figure 6.16, an *access control* function is invoked in order to avoid or resolve conflicts between different terminals, which contend for transmission resource allocation. This can be based on a distributed agreement between all autonomous terminals sharing the medium, as in PRMA, or it can be carried out under the supervision of a controller. The first situation is referred to as distributed access control, while the second case is termed centralized access control.

The right to transmit has to be granted for a terminal by means of an *approval* process, which is also indicated in Figure 6.16. Either this can be based on a permission to transmit type of approval, or alternatively, the terminal may tentatively attempt transmission and await an acknowledgment, confirming that no access conflicts occurred during the previous transmission. As argued in Chapter 5, PRMA belongs to the latter type of protocols.

Finally, the time interval during which a terminal can transmit data is referred to as the *holding time*. As indicated in Figure 6.16, the holding time can be of fixed or variable duration. The fixed holding time is assigned by the access control function, while PRMA's holding time is variable, expiring when the terminal ceases transmission.

An excellent in-depth overview of various wireless multiple access schemes is provided by Li and Qiu [205]. In our later deliberations, we initially concentrate on Packet Reservation Multiple Access (PRMA), which Goodman *et al.* proposed [180, 206] for

6.2. MULTIPLE ACCESS

wireless networks, where the base station (BS) informs the portable stations (PS) as to the status of the slots in each PRMA frame in its broadcast message. The PSs contend for unused slots but since they are unaware of each other's contentions, collisions may occur in case of simultaneous transmission of more than one packet in the same slot. In this case, none of the contending PSs acquires the slot for future transmissions. Consistent future collisions are prevented by introducing a so-called permission probability that allows terminals to re-contend according to this system parameter, which is optimized as a function of the number of slots to contend for. This optimization process is demonstrated in the next section. By contrast, in case of uncontended transmission, the PS reserves the slot in each PRMA frame, as long as it has packets to transmit. In case of high traffic loads many collisions occur, leading to wasting a large fraction of the slots and hence degrading the system's throughput. The PRMA protocol was further developed by Dunlop *et al.* [207] under the name of PRMA++ by allowing contention only in the so-called contention slots (i.e., in a fraction of the slots) a measure, that protected the information slots from collisions.

The above PRMA schemes were contrived mainly for for constant-rate speech transmission. When supporting different multimedia services, including speech, video, and data, a range of problems arise in the context of PRMA, which will be addressed in the next section.

6.2.3 Multiframe Packet Reservation Multiple Access

A possible approach to supporting multirate users in Packet Reservation Multiple Access (PRMA) is the introduction of a hierarchical structure of PRMA multiframes, which comprise a certain number of frames. Here we refer to this protocol as Multiframe PRMA (MF-PRMA). In conventional Time Division Multiple Access (TDMA) systems, a terminal can only reserve one slot per frame. In MF-PRMA, however, if M is the number of frames per multiframe, as seen in Figure 6.17, a reservation of $n_R < M$ slots per multiframe is possible and corresponds to a *reservation rate* lower than that of one slot per frame. The term reservation rate describes the rate of a temporary virtual channel occupied by a terminal holding a reservation. The reservation rate can be reduced to $(1/M)th$ of the reservation rate of the original single-frame-based PRMA structure. Initially, we assume that a certain reservation rate is associated with each column of the multiframe structure of Figure 6.17. This will be referred to as the reservation rate of the column. Figure 6.17 displays an example for two columns, where a high-rate video user would choose the column with a high reservation rate, namely, one slot per frame, while a speech user or a low-rate video user only needs one slot every fourth frame, and hence it is transmitting in slots of the appropriate column. A terminal seeking to transmit packets for a certain service will thus contend in a column with the reservation rate that is most suitable for the required service. The introduction of a finite reservation expiry time, allows a delayed reservation cancellation during short inactive transmission periods, hence reducing the number of contentions and improving the performance of a MF-PRMA system. Let us now evaluate the performance of such a scheme.

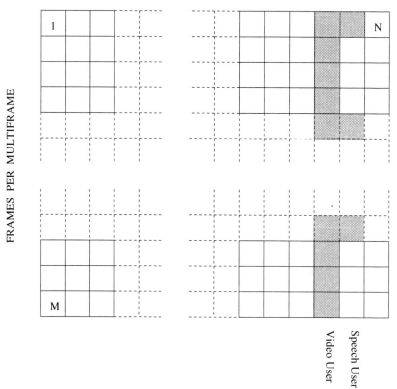

Figure 6.17: PRMA multiframe comprising M frames — slots occupied, for example, by a speech and a video user. © Elsevier [185].

6.2.3.1 Performance of MF-PRMA

In this section, video services of different rates are studied in the context of the proposed MF-PRMA protocol. We investigated medium access for mean FEC-coded video target rates of 64 kbps, 1 Mbps, and 3 Mbps. Video sequences were generated using the video source model described during our previous discussions. Assuming that the services considered are interactive, a delay constraint of 30 ms was imposed, a latency, that is tolerated by both speech and video signals. For calculating the average delay, only the successfully received packets were taken into account. Thus, the delay curves must always be considered in conjunction with the packet-dropping probability performance. In order to avoid the unrealistic situation that all terminals start contending at the same time, the packet generation process of a terminal starts randomly in one of the first 120 video-frames. All simulations correspond to 260 seconds of video transmission.

Table 6.2 summarizes the parameters of our 1 Mbps single-user rate, 30 ms-latency, 50 Mbps multi-user-rate WLAN video service. The up- and down-link channels are

6.2. MULTIPLE ACCESS

Definition	Notation	Unit	Value
Channel rate	R_C	Mbps	50.0
Source target rate	R_S	Mbps	1.0
Gross slot size	S_G	bit	1024
Net slot size	S_N	bit	424
Slots per frame	N		64
Slots per up-link partition	N_U		29
Maximum delay	D_{max}	sec.	0.03
Max. number of reserved slots/frame/terminal	n_R		10
Frames per multiframe	M		16
Reservation expiry time	T_{exp}	slots	0
Permission probability	p		0.1 .. 0.4
Number of video terminals	U		7
Frame duration	T_F	ms	1.31

Table 6.2: MF-PRMA Parameters for a 1 Mbps Video Service.

accommodated in a Time Division Duplex (TDD) frame. It is assumed that twenty nine slots out of the total of sixty four slots are available for the up-link partition, while the remaining $32 - 29 = 3$ slots are used for system control. Each packet carries 424 bits of payload. FEC coding, the packet header, and the quasiperiodic extension of the employed Orthogonal Frequency Division Multiplex (OFDM) modulation technique of Chapter 5 require the remaining bits in each 1024-bit packet. Therefore, the previously mentioned reservation rate of one slot per frame is equivalent to 424 bits/1.31 ms = 323.66 Kbps, and the available maximum up-link channel rate is 29 × 323.66 Kbps = 9.386 Mbps, where the frame duration T_F is 1.31 ms, as seen in Table 6.2. We will show in the context of Figure 6.18 that this allows seven video users to transmit with a packet-dropping probability smaller than 0.2%, if the corresponding set of parameters in Table 6.2 is used. Figure 19:7 in Chapter 19 portrays the expected video performance of an H.263-based video transceiver for various packet-loss or frame error rates (FER), demonstrating that FERs as high as 5% can be tolerated by the H.263 scheme, if appropriate transmission techniques are invoked. At the previously mentioned 0.2% FER, no perceived video degradation was noted. Hence, this value can be treated as a very conservative design target.[4] Considering the MF-PRMA protocol, the target rate of 1 Mbps is higher than the 323 kbps reservation rate of one slot per frame, requiring on average of three to four slots per frame. As the bit rate fluctuates, additional slots must be assigned and released sufficiently

---

[4] Various H.263 video demonstrations over a range of channel conditions and packet-dropping rates can be viewed under http://www-mobile.ecs.soton.ac.uk by means of an MPEG player. Further contributions on the corresponding wireless H.263 systems are also downloadable from this address.

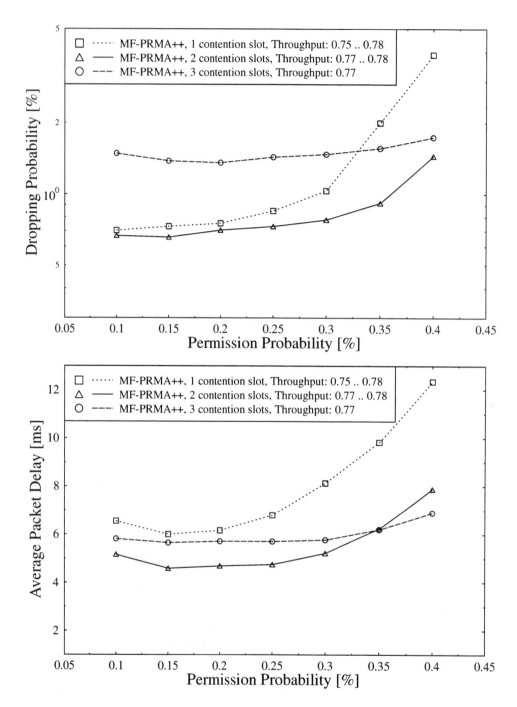

Figure 6.18: MF-PRMA++ performace for 1 Mbps video services, for different numbers of contention slots. © Elsevier [185].

6.2. MULTIPLE ACCESS

promptly, hence, no reservation expiry time is imposed.

In our experiments, three different types of PRMA protocols were examined: the proposed MF-PRMA scheme, MF-PRMA++, and MF-PRMA++ with adaptive permission probability. In MF-PRMA++, a certain number of slots per frame is used only for contentions, as in the PRMA++ protocol proposed by Dunlop [207]. The remaining slots of the up-link partition carry noncontending packets. Apart from this, the MF-PRMA++ structure is equivalent to the MF-PRMA scheme. In a modified version of the MF-PRMA++ protocol, the permission probability was decreased with an increasing number of reserved slots, as we will explain.

In order to characterize MF-PRMA++, simulations were carried out for reserving one, two, and three contention slots, considering seven video users and an overall up-link partition of twenty nine slots out of a total of sixty four. Figure 6.18 displays the associated packet-dropping probability and average packet delay versus contention permission probability performances.

The system with two contention slots, that is, $2/29 \times 100\% = 6.9\%$ contention resource allocation, gives the best performance in terms of both packet-dropping and average packet delay. Because of an increased number of collisions for high permission probabilities p and low contention resource allocation, the delay and dropping probability performances degrade significantly for $p > 0.3$ in the case of two or fewer contention slots. For values of p between 0.1 and 0.3, the protocol's performance is remarkably steady.

In the next series of investigations, the MF-PRMA++ performance will be compared to that of a MF-PRMA system. In addition, as mentioned earlier a novel variant of MF-PRMA++ is examined, which involves an adaptively decreasing permission probability for an increasing number of reserved slots. This approach, which does not require any additional signaling, is chosen owing to the following observations. Initial simulations showed that although only each terminal on average needs three to four slots for the 1 Mbps video service, the system performance increased significantly if the terminal was allowed to reserve more slots when its buffer queue was significantly longer than that of other terminals. For this reason, each terminal was allowed to reserve up to ten slots per frame. With acknowledgments for successful contentions being given only in the broadcast cell once a sixty four slot TDD/MF-PRMA frame, however, a terminal keeps contending until the next broadcast cell arrives, even if contentions earlier in the same frame were successful; hence, it does not have any more packets to transmit. In order to mitigate this effect without reducing the maximum number of reserved slots per frame, above we proposed a decreasing permission probability. Initial simulations showed that a slower than linear decline with an exponent of about 0.7 yielded a good performance. Results for the MF-PRMA++ approach with declining permission probability are presented in Figure 6.19, along with the corresponding results for the MF-PRMA system without contention slots.

Figure 6.19 shows that in terms of both packet delay and dropping probability performance, the MF-PRMA++ approach is inferior to the MF-PRMA scheme. This finding was not anticipated, but it is the consequence of occupying slots for the contentions, a disadvantage, which outweighs the advantage of ensuring no contentions for traffic slots. Whereas the average packet delay for this MF-PRMA protocol is nearly

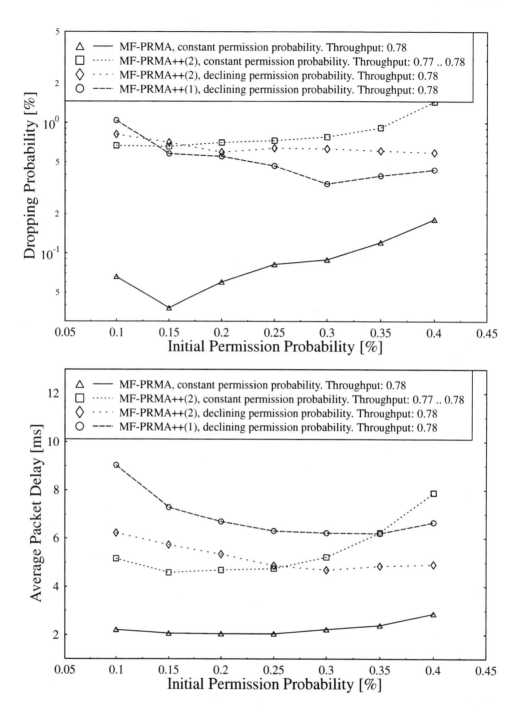

Figure 6.19: MF-PRMA and MF-PRMA++ performance for 1 Mbps video services. The figure in brackets indicates the number of contention slots. © Elsevier [185].

6.2. MULTIPLE ACCESS

Definition	Notation	Unit	Value
Channel rate	R_C	Mbps	50.0
Source target rate	R_S	kbps	64.0
Gross slot size	S_G	bit	1024
Net slot size	S_N	bit	424
Slots per frame	N		64
Maximum delay	D_{max}	sec.	0.03
Max. no. of reserved slots/frame/terminal	n_R		1
No. of considered upl. slots	N_U		2
Frames per multiframe	M		16
Reservation expiry time	T_{exp}	slots	0 or 1
Permission probability	p		0.1 .. 0.6
Number of video terminals	U		8

Table 6.3: Parameters for the 64 kbps Video Service

independent of the permission probability in the simulated range of $0.10 < p < 0.45$, the packet-dropping probability is clearly minimized for $p = 0.15$. In comparison to MF-PRMA++ and MF-PRMA++ with reducing permission probability, the latter scheme yields a slightly better performance. However, further work is needed in order to exploit the potential of this idea, which was also investigated in [208] using explicit signaling.

For the next set of experiments, namely, for a 64 kbps video service, the parameters are displayed in Table 6.3. Again, a 30 frames-per-second video service was considered and packets were discarded after 30 ms. Assuming a high traffic load scenario, eight video users shared two columns of the multiframe, all contending for acquiring a reservation rate associated with reserving one slot every fourth frame, which corresponds to a reservation rate of 80.87 kbps. In our simulations, we compared standard PRMA, MF-PRMA, and MF-PRMA with an adequate reservation expiry time. Without a reservation expiry time, the reservation is cancelled as soon as a slot is unused, which happens frequently because the source target rate of 64 kbps is lower than the bit rate of the virtual 80.87 kbps channel. Therefore, frequent contentions are necessary in order to "re-obtain" a reservation, leading to an increased number of collisions. Hence, a reservation expiry time of one slot significantly decreases the number of contentions. Figure 6.20 shows the superiority of our proposed MF-PRMA scheme having an adequate reservation expiry, which exhibits itself in terms of both packet delay and dropping probability, especially in comparison to the standard PRMA scheme. In this experiment, the deficiency of PRMA becomes explicit even in comparison to MF-PRMA without expiry time, which is explained as follows. At 64 kbps, where three time-slots/MF have to be assigned on the average

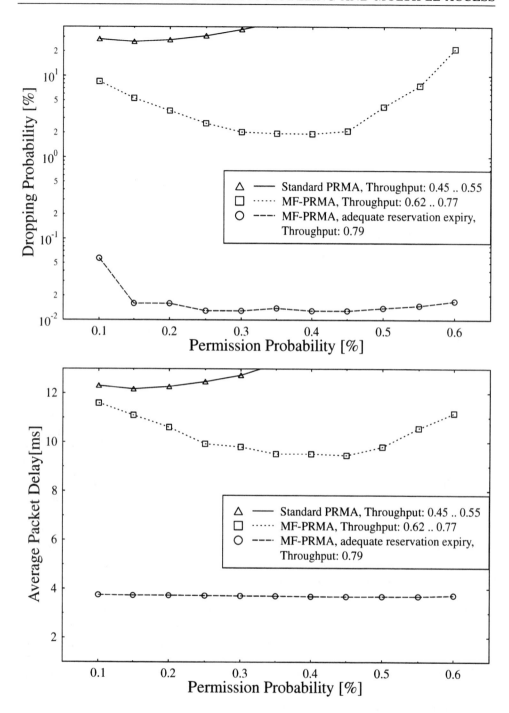

Figure 6.20: PRMA and MF-PRMA performance for a 64 kbps video service. A reservation expiry time of 1 slot is adequate in these scenarios. Table 6.3 shows the system parameters. © Elsevier [185].

6.2. MULTIPLE ACCESS

Definition	Notation	Unit	Value
Channel rate	R_C	Mbps	50.0
Source target rate	R_S	Mbps	3.0
Gross slot size	S_G	bit	1024
Net slot size	S_N	bit	424
Slots per frame	N		64
Maximum delay	D_{max}	sec.	0.03
Max. number of reserved slots per frame per terminal	n_R		15
No. of considered upl. slots	N_U		29 .. 31
Frames per multiframe	M		16
Reservation expiry time	T_{exp}	slots	0
Permission probability	p		0.2 .. 0.6
Number of video terminals	U		3

Table 6.4: Parameters for the 3 Mbps Video Service

for MF-PRMA, which corresponds to a reservation rate of 80.87 kbps, in PRMA a perpetual reserve/release cycle is maintained, resulting in frequent collisions. Note that according to the corresponding performance curves for the proposed MF-PRMA scheme, the dropping probability and the average delay are nearly independent of the permission probability in the interval of $0.15 < p < 0.60$.

Finally, we evaluated the system performance for a 3 Mbps video service scenario, for which the system parameters are displayed in Table 6.4. In order to quantify the average packet delay and dropping probability for different throughput values, the number of up-link slots was varied between twenty nine and thirty one, since varying the number of users would result in a coarse throughput variation, which would be too high for obtaining results near the maximum system throughput. Figure 6.21 shows the system performance in terms of packet-dropping and average packet delay. A remarkably high MF-PRMA system performance is feasible for this video service. Specifically, with an average delay of not more than 2.5 ms, a dropping probability of less than 0.15% is achieved for a throughput of nearly 82%. In addition, all results are highly independent of the permission probability, an evidence of system stability even for throughput values around 87%.

Having considered the Markov modeling of video codecs in multiple access studies in the context of the proposed MF-PRMA procedure, let us now examine another attractive procedure, namely, Statistical Packet Assignment Multiple Access (SPAMA).

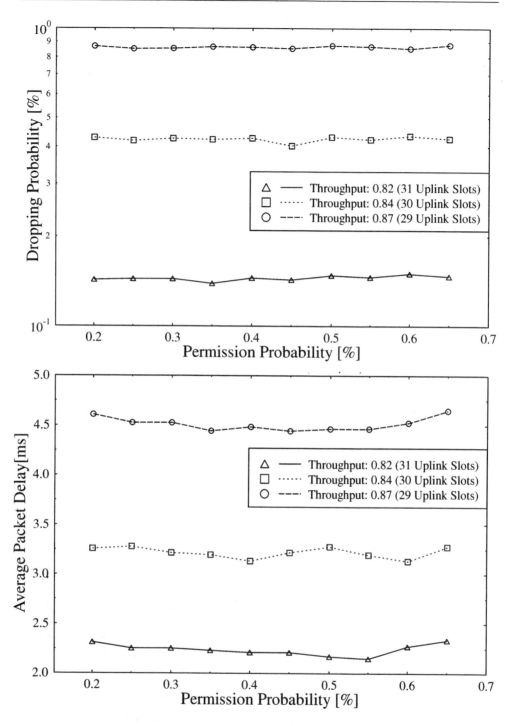

Figure 6.21: MF-PRMA performance for a 3 Mbps video service. The system parameters are displayed in Table 6.4. © Elsevier [185].

6.2.4 Statistical Packet Assignment Multiple Access

6.2.4.1 Statistical Packet Assignment Principles

In this section, a novel multiple access scheme is proposed, which we refer to as the Statistical Packet Assignment Multiple Access (SPAMA) protocol. The new scheme attempts to achieve an efficient and simple bandwidth allocation for multirate users in high-rate systems, such as Wireless Asynchronous Transfer Mode (ATM) schemes [209]. The basic principle of SPAMA accrues from a nondeterministic bandwidth assignment, performed by the base station on the basis of knowing only the average rate of a certain service. A time division duplex (TDD) system is assumed in which the base station broadcasts packetized information in the down-link channel, whereas geographically dispersed terminals transmit packets to the base station in the up-link slots, according to slot allocations controlled by the considered medium access (MAC) protocol [208].

At the end of each TDD frame, assuming that all up-link slots for the following frame are initialized as *available*, slot assignments are performed by the base station according to the SPAMA algorithm. We first describe the general principle of the algorithm, and we then formally describe its detailed implementation, as it is displayed in Figure 6.22.

The new approach of the SPAMA algorithm is its nondeterministic assignment of slots according to the average rate of a certain service. Assuming that an active spurt of the service has an average bit rate of R_S and that the assignment of one slot per frame corresponds to a virtual channel having a bit rate of R_{VC}, the required number of slots per TDD frame is calculated as $x_S = R_S/R_{VC}$. Note that an active spurt is a transmission phase, which is long in comparison to the frame duration, that is, usually one or more seconds, such as a talk spurt in speech transmission or a whole call in a video service without passive gaps. Given x_S, the interval $\tilde{I}_S$ expressed in terms of the number of frames, in which the considered service has to transmit one packet, equals $1/x_S$. In previous approaches, such as all Packet Reservation Multiple Access (PRMA)-based access schemes [185,206], these requirements had to be translated into integer expressions, allowing the users to occupy only an integer number of slots per frame on the average. These translations always implied complex algorithms and a suboptimal bandwidth allocation. In the SPAMA algorithm, however, the proposed statistical assignment allows us to meet noninteger slot allocation needs. Namely, since only an integer number of slots can be assigned, an assignment of x_S slots per frame can be interpreted as an assignment of one slot with a probability of $p_A = x_S$. For example, in a wireless ATM system, such as that represented by the parameters in Table 6.5, a one slot per frame virtual channel provides a bit rate of $R_{VC} = 963.264$ kbps. Considering a speech service in accordance with the GSM standard [210], that is, an FEC-coded source-rate of $R_S = 22.8$ kbps during a talkspurt, the required number of slots per frame is calculated as:

$$x_S = \frac{R_S}{R_{VC}} = \frac{22.8\,kbps}{962.264\,kbps} = 23.694 \cdot 10^{-3}. \tag{6.6}$$

This corresponds to an assignment of one slot per frame with a probability of $p_A = x_S = 0.02369$ or to an assignment of one slot in $\tilde{I}_S = 42.205$ frames. Note, however,

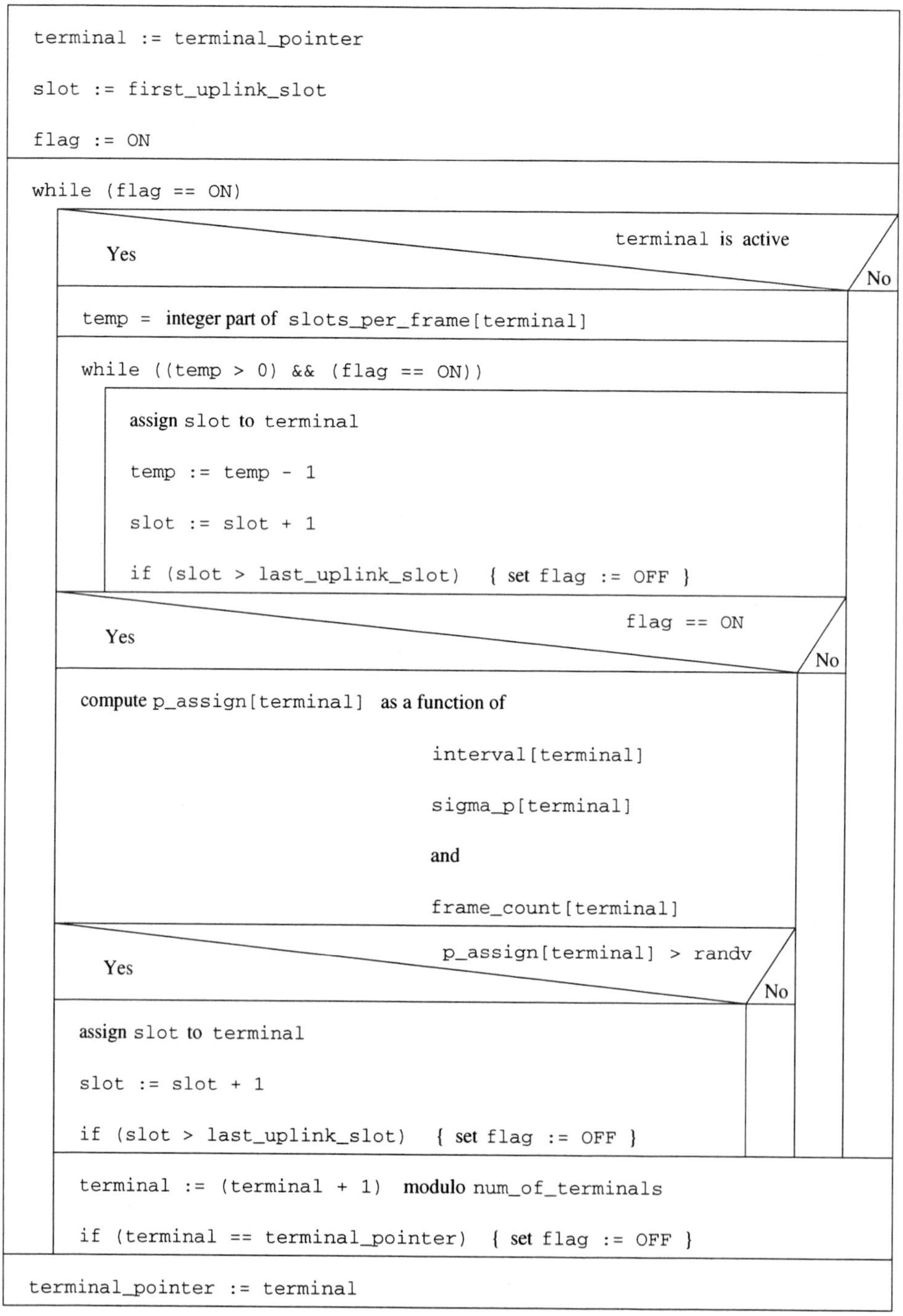

CHAPTER 6. VIDEO TRAFFIC MODELING AND MULTIPLE ACCESS

```
terminal := terminal_pointer

slot := first_uplink_slot

flag := ON

while (flag == ON)

    |  terminal is active  → Yes / No
    |
    |   temp = integer part of slots_per_frame[terminal]
    |
    |   while ((temp > 0) && (flag == ON))
    |
    |       assign slot to terminal
    |       temp := temp - 1
    |       slot := slot + 1
    |       if (slot > last_uplink_slot)   { set flag := OFF }
    |
    |   flag == ON  → Yes / No
    |
    |   compute p_assign[terminal] as a function of
    |                               interval[terminal]
    |                               sigma_p[terminal]
    |                           and
    |                               frame_count[terminal]
    |
    |   p_assign[terminal] > randv  → Yes / No
    |
    |       assign slot to terminal
    |       slot := slot + 1
    |       if (slot > last_uplink_slot)   { set flag := OFF }
    |
    |   terminal := (terminal + 1) modulo num_of_terminals
    |   if (terminal == terminal_pointer)  { set flag := OFF }

terminal_pointer := terminal
```

Figure 6.22: Formal description of the SPAMA algorithm, displaying one assignment cycle. © Elsevier [185].

6.2. MULTIPLE ACCESS

Definition	Notation	Unit	Value
Channel rate	R_C	Mbps	164.226
Frame duration	T_F	msec.	0.3991
Slot duration	T_S	msec.	0.0062
Speech source rate	R_{SS}	kbps	22.8
Video source rate (target)	R_{SV}	Mbps	2.0
Range of video rate		Mbps	1.62 - 2.40
Gross slot size	S_G	bit	1024
Net slot size	S_N	bit	384
Slots per frame	N		64
Maximum speech delay	$D_{max,S}$	sec.	0.03
Maximum video delay	$D_{max,V}$	sec.	0.03
Number of speech users	U_S		15
Number of video users	U_V		15
Ageing packet constraint	t_{age}	sec.	0.015

Table 6.5: SPAMA Parameters for a 34 Mbps WATM System

that slot assignments with a probability of p_A in each frame lead to a high variance in terms of the assignment interval duration, or in terms of the number of slots between two assignments, which implies an inefficient bandwidth allocation. This is due to potential premature assignment of slots that cannot be used by a terminal, since it does not have a ready-to-send packet in its buffer. In order to avoid this problem, we do base the slot allocation regime of the SPAMA algorithm not on the computed assignment probabilities for each frame but on the cumulated assignment probability for each interval $\tilde{I}_S$. In our example above a slot should be assigned with a cumulated probability of one in $\tilde{I}_S = 42.205$ frames, that is, with a cumulated probability of $\sigma_p = 42/42.205 = 0.995$ in an interval of $I_S = 42$ frames. In other words, it is necessary to guarantee an assignment of one slot in 42 frames with a cumulated probability of $\sigma_p = 0.995$. The distribution of the "per-frame" assignment probabilities in each frame of the interval I_S can theoretically be chosen without any restrictions, bearing in mind, however, that the choice will significantly influence the system's performance. In order to avoid the above-mentioned early assignments of slots, the per-frame assignment probabilities have to increase with the number of frames encountered since the last slot assignment. In a first approach, we assume a simple *assignment probability growth function*:

$$p_A(n) = q_k \, n^k, \qquad (6.7)$$

where n is the number of elapsed frames since the last slot assignment, and the parameter k, defining the constant q_k, remains to be chosen in order to obtain good system performance. For reasons of simplicity, we assume an integer value for k and then formulate the cumulated probability constraint, which was given in our example above by $\sigma_p = 42/42.205 = 0.995$. In general, we have:

$$\sigma_p = \sum_{n=1}^{I_S} p_A(n) = \sum_{n=1}^{I_S} q_k\, n^k = q_k \sum_{n=1}^{I_S} n^k, \qquad (6.8)$$

which yields the constant q_k given by:

$$q_k = \frac{\sigma_p}{\sum_{n=1}^{I_S} n^k}. \qquad (6.9)$$

Still keeping k as a parameter, we obtain the per-frame assignment probability as a function of the number of frames n encountered since the last slot assignment by inserting Equation 6.9 in Equation 6.7, yielding:

$$p_A(n) = \frac{\sigma_p}{\sum_{n=1}^{I_S} n^k}\, n^k. \qquad (6.10)$$

Equation 6.10 explicitly defines the assignment probability growth function. The assumption of an integer value for k allows us to express the denominator of Equation 6.10 in a simpler form, for example, for $k = 1, 3, 5$, and 7, we have:

$$\sum_{n=1}^{I_S} n = \frac{I_S\,(I_S + 1)}{2} \qquad (6.11)$$

$$\sum_{n=1}^{I_S} n^3 = \frac{I_S^{\,2}\,(I_S + 1)^2}{4} \qquad (6.12)$$

$$\sum_{n=1}^{I_S} n^5 = \frac{I_S^{\,2}\,(I_S + 1)^2\,(2\,I_S^{\,2} + 2\,I_S - 1)}{12} \qquad (6.13)$$

$$\sum_{n=1}^{I_S} n^7 = \frac{I_S^{\,2}\,(I_S + 1)^2\,(3\,I_S^{\,4} + 6\,I_S^{\,3} - I_S^{\,2} - 4\,I_S + 2)}{24}. \qquad (6.14)$$

Referring to the example introduced above, we use Equation 6.10 and obtain the assignment probabilities displayed in Figure 6.23.

Once the base station has calculated the assignment probabilities p_A for each of the different-rate services supported, a binary random decision based on p_A defines whether the service is granted a slot assignment in the current transmission frame. This computation is performed once a frame and is then communicated to all terminals in the broadcast cell at the beginning of the following frame, so that each terminal is informed as to the slots it can use for transmitting packets. Note that if a service requires more than one slot per frame, that is, $x_S > 1.0$, it is assigned a number of slots, corresponding to the integer part of x_S in each frame. In order to cater for the noninteger fraction of a slot, the algorithm is applied as described above.

In order to give a more detailed description of the proposed SPAMA algorithm, its Nassi-Shneiderman diagram [211] is shown in Figure 6.22. Assuming that each

6.2. MULTIPLE ACCESS

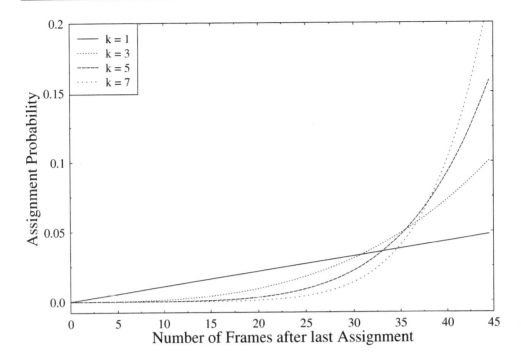

Figure 6.23: Assignment probability growth function versus transmission frame index for our example of $\sigma_p = 0.995$, $I_S = 42$, $k = 1, 3, 5,$ and 7. © Elsevier [185].

terminal is involved in only one service, the interval I_S and the cumulated probability σ_p are denoted by interval[terminal] and sigma_p[terminal], which can be calculated for each terminal as described above. All terminals are consecutively numbered, so that the terms *previous* and *next* terminal can be used. At the beginning of each assignment cycle, the variable terminal is then incremented by one, in order to consider the next terminal. Each assignment cycle commences in the first up-link slot, implying the assignment of slot:=first_up-link_slot and ends, if either all up-link slots have been assigned, corresponding to (slot > last_up-link_slot) in Figure 6.22, or all terminals have been served, satisfying the condition (terminal == terminal_pointer). For a specific terminal, in a first step, the integer part of the required number of slots_per_frame[terminal] is assigned, if necessary, as reflected by Figure 6.22. Then the assignment probability p_assign[terminal] is computed, as described above, and a slot assignment is performed according to the binary random decision if (p_assign[terminal] > randv), where randv is randomly chosen from a set of uniformly distributed numbers in the interval (0,1). Then the next terminal is considered. At the end of the assignment cycle of a certain transmission frame, terminal_pointer is set to the next terminal to be considered, which is represented by terminal_pointer := terminal in Figure 6.22.

For call setup, a terminal has to signal the average required bit rate of the service. In a first approach, each terminal contends with an initialization packet containing the necessary information. Assuming 64 slots per frame and 15 speech, plus 15 video users, as in our investigations below, it proved to be sufficient to guarantee one contention slot in every second frame. At the end of a call, another signaling packet is transmitted to the base station, this time in an assigned slot, indicating that assignments are no longer necessary.

Applying the assignment algorithm described above, not all up-link slots are necessarily used. In order to keep the average packet delay low, these unallocated slots can be assigned to users whose packets already exceeded a certain delay. In our efforts to minimize the amount of signaling, we assume that terminals indicate aging packets by using a one-bit flag in the packet-header they are sending. If this bit is set to one and not all slots have been used in the current assignment cycle, the terminal gets an assignment for a number of slots depending on its average bit rate, which is already known to the base station. Both the time after which a packet is considered as aging and the relation between the average bit rate and the number of assigned slots will be identified according to our simulation results. Performance figures for the SPAMA algorithm are given in the next section.

6.2.4.2 Performance of the SPAMA Protocol

Having presented the structure of the proposed SPAMA algorithm, we now characterize its performance. In a first set of simulations, we examine the influence of the assignment probability growth function, which was presented in Equation 6.10. The throughput performance determined by the access protocol's efficiency is then examined by varying the number of up-link slots for a given traffic load.

For our simulations, we assume a frame length of $N = 64$ slots, where each slot carries $S_N = 384$ bits of payload, as for the ATM standard [212]. Let U_S and U_V denote the number of speech and video users present in the system. Their source rates are referred to as R_{SS} and R_{SV}, where R_{SS} is fixed at the FEC-coded speech rate of 22.8 kbps, according to the GSM standard [210]. The video source model of Section 6.1 emulated the H.263 encoder of Chapter 18. The channel rate R_C was chosen in order to guarantee duplex communications at a 34 Mbps ATM rate. Specifically, ATM cells comprising $(384 + 40)$ bits were transmitted at the 34 Mbps bit rate in both directions. The remaining bits of the 1024-bit packets in each slot were required for FEC coding, for the packet header and for the quasi-periodic extension necessary in the OFDM modem highlighted in Chapter 5. For both speech and video services, a maximum delay of 30 ms was tolerated, before a packet was dropped. All investigations were carried out for fifteen speech and fifteen video users. We refrain from simulating a constant rate data service in these initial experiments in order to create a randomistic worst-case scenario for the SPAMA algorithm. Specifically, since the protocol's efficiency increases with a decreasing variance of the average rate, a constant rate data service implies good system performance. The span of time after which a packet is considered as aging, triggering additional slot assignments, as mentioned above, was set to 15 ms. If any packet in the buffer of a terminal was older than 15 ms, then the additional request bit was transmitted to the base stations in the header of the forthcoming packet that the terminal was transmitting. In our investigations, this span of aging time proved to yield good performances for different sets of parameters. The number of additionally assigned slots for aging packets was limited by the number of slots, which were still available in the frame after the standard assignment cycle. In order to prevent low-rate users having only a few aging packets in the buffer to get assignments for too many slots, the number of additionally allocated slots per terminal corresponded to 5% of the average bit rate of the service considered. All system parameters are summarized in Table 6.5.

We now examine the influence of the assignment probability growth function on the delay performance of our WATM system characterized by Table 6.5. Figure 6.23 displays the function given in Equation 6.10 for different values of k. For increasing exponents, the packet assignment probability growth function has a very high positive gradient from a certain number of frames onward, and then it is reset to zero after the assignment. In order to model a strongly impulse-like growth function, hereafter indicated by the index i, we fixed the assignment probability in frame $(I_S - 1)$ at 0.25 and from frame I_S on at 0.75, meeting the cumulated assignment probability constraint. This choice gives an approximation of Equation 6.10 for high exponents k. Figure 6.24 shows performance results for growth functions with $k = 1, 3, 5, 7$, and for the aforementioned i-function. As observed in Figure 6.24, the packet delay and dropping performances for the low-rate speech service increased for increasing values of k, $k \leq 7$, but both the speech packet-dropping and delay curves suggest significantly reduced performances for the previously introduced impulse-like i-function. The high-rate video service, however, remains nearly unaffected. Note that even for a throughput around 93%, both the speech and video services show dropping characteristics better than the required quality limits.

In the next set of simulations, the throughput performance of the SPAMA protocol was evaluated. In order to vary the traffic load in small equidistant steps without changing the one-to-one proportion of speech and video users, the number of up-link slots (i.e., the available channel bandwidth) was varied, while maintaining a constant number of fifteen video and fifteen speech users. Results for these simulations are displayed in Figure 6.25. For a throughput up to 90%, the video packet-dropping rate was virtually zero, and the speech dropping rate was at least a factor ten better than the required quality limit of 1%. Considering the associated packet delay performances, the video packet delay was very low at less than 4 ms, and with approximately 12 ms, the average speech packet delay was still less than half the tolerable delay of 30 ms. The associated throughput values can rise up to 93% without infringing the imposed quality limits. For a higher traffic load, although the packet-dropping probability is unacceptable for video users, the system still exhibited a stable behavior, with just a graduall increase in dropping probability.

6.3 Summary and Conclusions

In conclusion, oscillation-scaled Markov models have been proposed for modeling various video sources for networking studies. A twenty-state model was found to reproduce most video source features for various video resolutions and frame rates quite accurately. This video model then was invoked in order to evaluate the performance of MF-PRMA, supporting a variety of different-rate users in a WLAN environment.

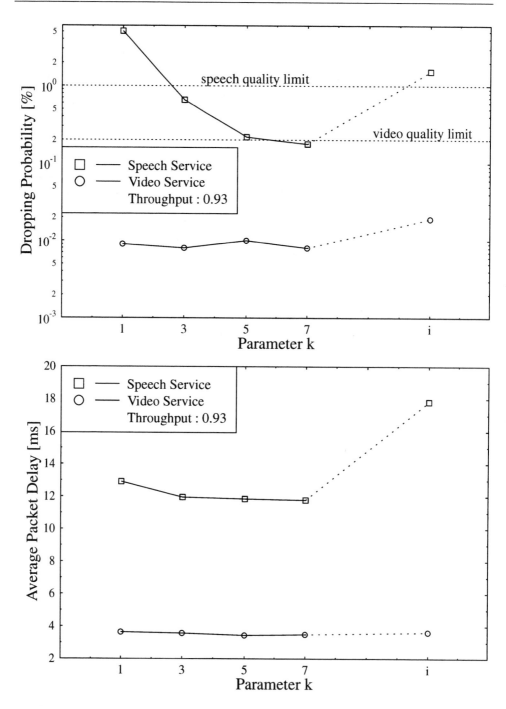

Figure 6.24: Delay-induced packet-dropping and average packet delay performance versus exponent k of the assignment probability growth function for the WATM system defined in Table 6.5. © Elsevier [185].

6.3. SUMMARY AND CONCLUSIONS

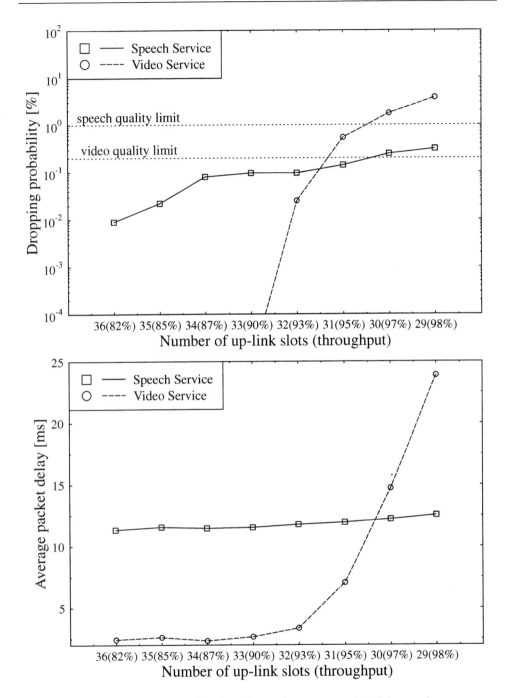

Figure 6.25: Delay-induced packet-dropping and average packet delay performance as a function of the number of up-link slots. The numbers in brackets display the measured throughput. The WATM system of Table 6.5 was used. © Elsevier [185].

MF-PRMA was shown in Figure 6.20 to outperform conventional PRMA in terms of both its delay and packet-dropping performance, reaching throughput values up to 79%.

In the second half of this chapter, Statistical Packet Assignment Multiple Access (SPAMA) was proposed, which is an algorithm of comparatively low-complexity, yielding remarkably good performance results. In addition, because of the central control of all slot assignments, SPAMA-based medium access shows stable behavior even in overloaded scenarios, when the system's throughput reached about 93% in the studied scenarios, significantly outperforming MF-PRMA.

Chapter 7

Co-Channel Interference

7.1 Introduction

As mobile radio systems develop toward microcellular and picocellular architectures, the effects of co-channel interference are becoming more important, owing mainly to the reduced cell sizes. This has the advantage that handsets can become simpler and their power consumption can be reduced. Furthermore, the reduced frequency reuse distance results in increasing the user capacity of the system. However, as the co-channel reuse distance is reduced, the co-channel interference increases. A significant body of research has been published on frequency reuse structures for cellular radio systems, as exemplified by the work of Safak [213] on optimal frequency reuse with interference. Also, Wang and Lea [214] investigated the effects of co-channel interference in microcellular radio systems under a variety of channel conditions.

The co-channel interference performance and capacity of various cellular systems were investigated, for example, by Lee and Steele [215]. Our co-channel interference studies have concentrated on the up-link of hexagonal cells with a reuse factor of 7, using an omnidirectional antenna at the center of each cell. This is a commonly investigated cellular cluster type, in which each base station (BS) has six first-tier co-channel interferers, as shown in Figure 7.1(a).

In this chapter, we provide performance results obtained from our simulations in order to quantify the co-channel interference and its effects. Our simulations model a TDMA system with embedded binary Bose-Chaudhuri-Hocquenghem (BCH) coding [9] combined with a reconfigurable Pilot Symbol Assisted (PSA) and Automatic Repeat Request (ARQ) aided Quadrature Amplitude Modulation (QAM) modem [29]. In this chapter, we do not use transmission power control; this is covered in Chapter 20. The simulation models the system at the transmission symbol level; hence, results can be obtained both for long-term characterization and on a frame-by-frame basis. This allows us to characterize the time-varying effects of co-channel interference, as well as its long-term effects.

We will initially describe the factors that control co-channel interference.

7.2 Factors Controlling Co-Channel Interference

Co-channel interference is caused by the simultaneous use of the same frequency channel in an adjacent reuse cluster. The interference caused is dependent on many parameters such as:

- The voice/video activity factor of the frequency channel in the adjacent cluster
- The geographical position of the co-channel user in the adjacent cluster
- The transmission power of the users and base stations
- The effects of slow and fast-fading in the various radio channels
- The path-loss attenuation
- The cell shape and frequency reuse distances

The voice/video activity factor of the adjacent clusters has a significant effect on the amount of interference. Our simulation models the effect of constant nonbursty transmissions as required by video users. Also, speech users are simulated using a two-state Markov speech model [216, 217]. The geographical position of the served users and interfering users also has an effect on the amount of interference. This is one of the factors we have investigated most, which is described in Section 7.2.3. The transmission power of users and base stations was initially fixed, but at a later stage a power control algorithm was implemented, which will be elaborated on in Chapter 20.

7.2.1 Effect of Fading on Co-Channel Interference

A radio signal in a cellular system is attenuated by three main factors. The first is known as path-loss, which describes how the attenuation varies with distance between the transmitter and receiver. For our simulations we have used a power path-loss exponent of 3.5. The path-loss attenuation in decibels between two points r meters apart is shown in Equation 7.1:

$$Attenuation(dB) = 35\log_{10}(r). \tag{7.1}$$

The two remaining major factors that can attenuate a radio signal are fast- and slow-fading. However, due to constructively or destructively superimposing the energy of several signal paths, fading can boost the radio signal as well as attenuate it.

Slow fading is caused by shadowing of the radio signal, owing mainly to building and terrain features. The shadow fading varies slowly and has been found by measurements to exhibit a log-normal PDF [32, 47].

Fast fading is caused by the multipath effects and scattering, where the superposition of multiple paths generates a rapidly fluctuating fading envelope. Fast-fading channels having a dominant path between transmitter and receiver are referred to as a Rician channel. When there is no dominant path, the channel is referred to as a Rayleigh channel [9].

7.2. FACTORS CONTROLLING CO-CHANNEL INTERFERENCE

When the co-channel interference is calculated, it should be remembered that both the user and interferer signals are faded by independent channels. Therefore, fast and slow fading typically have an effect on the co-channel interference. The fading of both the user and interferer signals results in a larger variance in the instantaneous signal to interference plus noise ratio (SINR). This can be particularly adverse, when the up-link signal is highly attenuated by a deep fade and coincidentally the interference is boosted by the fading, resulting in a low instantaneous SINR. The instantaneous SINR is a good measure of the likelihood that a particular symbol will be corrupted by the radio channel.

7.2.2 Cell Shapes

Our co-channel interference studies have concentrated on the up-link of hexagonal cells with a reuse factor of seven, using an omnidirectional antenna at the center of each cell. This is a commonly investigated cellular cluster type, in which each base station has six first-tier co-channel interferers. We have also simulated a cellular structure using 120 degree directional antennas and a reuse factor of three. However, this cell type has been investigated in less depth. The reuse patterns we have employed were derived from those suggested by Appleby and Ko [218] and those of Lee [215, 219]. We have also assumed perfect radiation patterns for the antennas in our simulations. An arbitrary radiation pattern model could easily be incorporated, but it would add another level of complexity.

The seven-cell reuse cluster shown in Figure 7.1(a) is a commonly employed structure. The shaded cells are co-channel interfering cells. Therefore, each base station has six first-tier co-channel interferers. Simulations were carried out using these six first-layer interferers.

The application of directional antennas allows the reuse cluster size to be reduced, thereby increasing the maximum system capacity. Directional antennas are used in the three-cell cluster shown in Figure 7.1(b). In this figure, the co-channel interfering cells have the same shading pattern. This cellular reuse structure has two so-called first-tier co-channel interfering cells and three second-tier interfering cells. These five interfering cells were used for the simulations.

A radio signal is attenuated with increased distance from the transmitter. Therefore, the further the interfering users can be from the receiver, the less interference there will be. The signal and interference paths for the frequency reuse clusters are shown in Figures 7.2 and 7.3 for seven and three-cell clusters.

The maximum interference is higher in the up-link than in the down-link, since interfering users can roam at the edge of their cell, transmitting at full power, in order to be able to communicate with their base stations. This causes maximum interference in a neighboring co-channel cell, due to the associated shorter path length and increased transmission power. In the down-link the position of the interferers (base stations) is fixed.

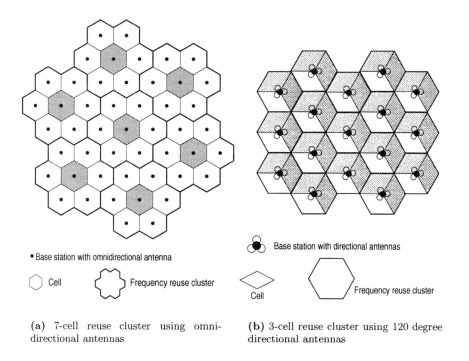

(a) 7-cell reuse cluster using omni-directional antennas

(b) 3-cell reuse cluster using 120 degree directional antennas

Figure 7.1: Frequency reuse cluster shapes.

7.2.3 Position of Users and Interferers

In the previous section, we have seen that the position of a user in a co-channel cell is important, since it has an effect on the amount of interference on the up-link. The user positions that would inflict the maximum and minimum interference during an up-link slot for a seven-cell reuse cluster without transmission power control are shown in Figure 7.4.

The maximum signal-to-interference ratio (SIR) occurs when the interfering users are as far as possible from the receiving base station, and the transmitting user is as close as possible to his base station, as shown in Figure 7.4(a). The minimum SIR is maintained when the interfering users are as close as possible to the served user's base station, and the transmitting user is as far as possible from his serving base station, hence having a low signal power. This is shown in Figure 7.4(b). When transmission power control is used, the positions inflicting maximum interference are the same. However, the positions causing minimum interference are those where the interfering users are close to their base stations, thereby transmitting at a low power.

Hence, in the up-link the SIR is dependent on the position of both the user and interferers. However, on the down-link the SIR is dependent only on the user position within the cell. Using a simple SIR model based only on a path-loss model, the SIR was estimated for both the up-link and down-link in a seven-cell cluster in the next section.

7.2. FACTORS CONTROLLING CO-CHANNEL INTERFERENCE

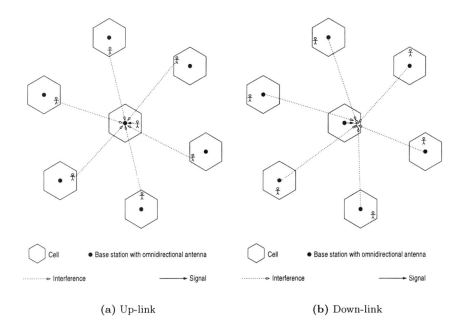

Figure 7.2: Signal and interference paths for the up- and down-links in a seven-cell reuse cluster.

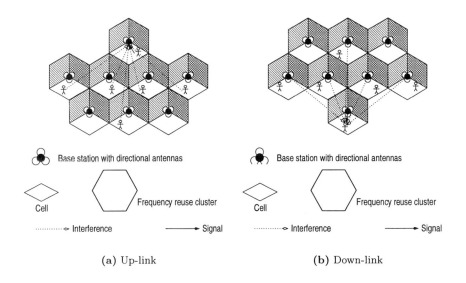

Figure 7.3: Signal and interference paths for the up- and down-links in a three-cell reuse cluster when employing 120 degree directional antennas.

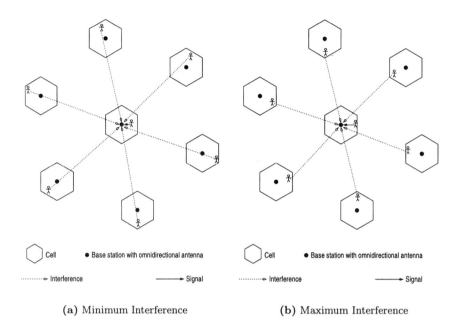

(a) Minimum Interference (b) Maximum Interference

Figure 7.4: Signal and Interference paths for up-link simulations, when interfering users are in the positions to inflict minimum or maximum interference, for a seven-cell reuse cluster.

7.3 Theoretical Signal-to-Interference Ratio

The theoretical SIR model we used is based solely on path-loss, and it takes no account of fading of any kind. In a simple case, where there is just one interferer and both the user and interferer are transmitting at the same power, the signal-to-interference ratio (SIR) is expressed as:

$$SIR(dB) = Pathloss\_in\_dB(d_s) - Pathloss\_in\_dB(d_i), \qquad (7.2)$$

where d_s is the distance between the user and its base station, which is interfered, while the distance between the interfering user in a co-channel cell and the interfered base station is denoted by d_i. Assuming a 35 dB/decade inverse power law path-loss approximation, the SIR for a single interferer can be simplified to:

$$SIR(dB) = 35 \log_{10}(d_i/d_s). \qquad (7.3)$$

When there is more than one interferer, the SIR is defined by Equation 7.4, where S is the signal power and I_k is the power of the kth interferer:

$$SIR = \frac{S}{I} = \frac{S}{\sum_k I_k}. \qquad (7.4)$$

In the best and worst interferer positions, all the interferers are at the same distance from the interfered base station. This simplifies Equation 7.4, so that the

7.3. THEORETICAL SIGNAL-TO-INTERFERENCE RATIO

summation becomes a multiplication. This is shown in Equation 7.5, where n is the number of interferers and I is the power of the interference from one of the interferers:

$$SIR = \frac{S}{I} = \frac{S}{n \times I}. \tag{7.5}$$

This approximation is valid only when the interferers are approximately at the same distance from the base station and when all the interferers are continuously transmitting. Therefore, this model is suitable only for video users but not speech users. The signal-to-interference ratio in decibels can be simplified, as shown in Equation 7.6, where n is the number of interferers and SIR_{SI} is the SIR for a single interferer:

$$SIR(dB) = SIR_{SI} - 10\log_{10}(n). \tag{7.6}$$

Therefore, the SIR decreases by 7.8 decibels, when the number of interferers increases from 1 to 6, for the same interferer distance. Using a power path-loss exponent of 3.5, the multiple interferer SIR expression of Equation 7.6 can be written as:

$$SIR(dB) = 35\log_{10}(d_i/d_s) - 10\log_{10}(n), \tag{7.7}$$

where d_s is the user distance and d_i is the distance of the interferer from the interfered base station. Having derived the equations for the theoretical SIR values, we will now provide the interferer distances to use in conjunction with them. According to Lee [220], the inter-base station or reuse distance (D) to cell radius (r) ratio (D/r) is defined as:

$$D/r = \sqrt{3k}, \tag{7.8}$$

for regular hexagonal structures, where k is the number of cells in the reuse pattern. This is shown for the seven-cell reuse cluster in Figure 7.5. It can be seen from the geometry that the D/r ratio is $\sqrt{4.5^2 + \left(\frac{\sqrt{3}}{2}\right)^2} = \sqrt{21}$, which verifies Equation 7.8 for a reuse cluster of size 7 ($k = 7$).

Lee [220] suggested that an estimate of the minimum SIR in a cellular system could be made by assuming that the distance of the user was the radius of the cell and all the interferers were at a distance D (reuse distance) from the interfered base station. The interferer distance can vary in the range $D \pm r$, and if $D \gg r$, then all the distances can be approximated by D, an assumption that Lee [220] used. However, for a seven-cell cluster, the interferer distance range $D \pm r$ is approximately $D \pm 20\%$ and therefore, assuming an interferer distance of D, is not a good approximation. Lee's [220] assumption is valid only if all the interferers are close to their own base stations or when the cluster size is large.

Using the theoretical SIR model that we derived earlier in Equation 7.7 and Lee's assumption for user and interferer distances, we can derive an estimate for the minimum SIR in the cell. This implies that the user's distance from the base station (d_s), is set to the cell radius r, and the interferer distance from the interfered base station d_i is set to $D = \sqrt{21} \times r$, the reuse distance, which gives an SIR of 15.4 dB for a reuse cluster of seven cells.

Figure 7.6 shows the positions, where an interferer can inflict the most and least amounts of interference. It can be seen from the geometry of this figure that the

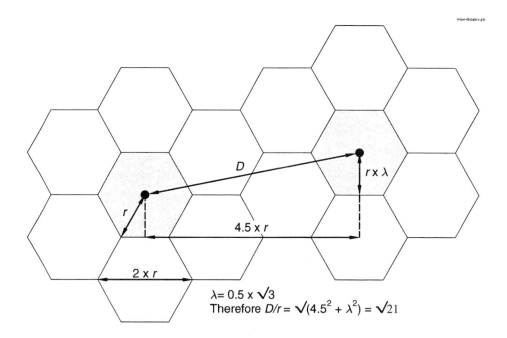

Figure 7.5: Reuse distance to cell radius ratio calculation.

shortest interferer distance is $r \times \sqrt{3.5^2 + \left(\frac{\sqrt{3}}{2}\right)^2} = r \times \sqrt{13}$. The longest distance between an interferer and the interfered base station is $r \times \sqrt{5.5^2 + \left(\frac{\sqrt{3}}{2}\right)^2} = r \times \sqrt{31}$.

Substituting the shortest interferer distance into Equation 7.7 gives the minimum SIR a user at the edge of the cell can expect, which amounts to an SIR of 11.7 dB for a reuse cluster of seven cells. The least amount of interference users at the edge of their cell can expect in a seven cell cluster is when all the interferers are as far away from the interfered base station as possible. When this longest interferer distance is substituted into Equation 7.7, the maximum SIR of 18.3 dB is maintained. Therefore, the SIR a user at the edge of the cell can expect can vary by up to 6.6 dB, depending on the position of the users within the co-channel cells.

Using Equation 7.7, we plotted the graph of SIR versus user distance shown in Figure 7.7. The curves show the SIR for both the up-link and down-link of a seven-cell reuse cluster, where there are six co-channel video interferers. Since the down-link SIR is dependent only on the user's distance from the BS, it can easily be plotted on the graph. However, the up-link SIR is dependent on both the user distance and the distance of the various interferers. Therefore, the up-link SIR is plotted using three curves to show the range of possible SIR values for a specific user distance. The best SIR occurs when all the interferers are as far as possible from the interfered base station, as shown in Figure 7.4(a). The worst-case SIR occurs when all the interferers are at the edge of their cell at the closest point to the interfered base station, as shown

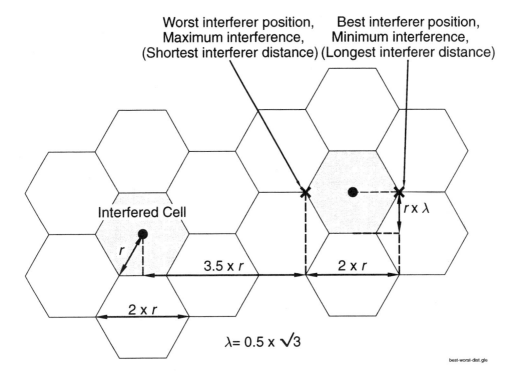

Figure 7.6: Calculation of interferer distance, for best- and worst-case interferer positions.

in Figure 7.4(b). The graph in Figure 7.7 shows that when all the interferers are at, or very close to, their base stations, the up-link SIR is similar to the down-link SIR.

7.4 Simulation Parameters

For our co-channel simulations, the system parameters we employed previously were changed. A new system was designed, based loosely on a DCS1800 system, accommodating the TDMA transmitted signal in a 200 KHz bandwidth used by both the GSM and the DCS1800 [41] systems.

The system proposed uses embedded binary Bose-Chaudhuri-Hocquenghem (BCH) coding [9] combined with a reconfigurable Pilot Symbol Assisted (PSA) Quadrature Amplitude Modulation (QAM) modem [29]. The system can operate, under network control, in one of four modes, each mode corresponding to a different modulation scheme. This allows the scheme to span a wide range of operating conditions in terms of video-quality, bit rate, robustness against channel errors, and implementational complexity, while exploiting the higher SNR and hence higher channel capacity of the central region of the cell. For example, the transceiver operates using highly bandwidth-efficient 64-level Pilot Symbol Assisted Quadrature Amplitude Modulation (64-PSAQAM) in a benign indoors cordless environment, where high SNRs and

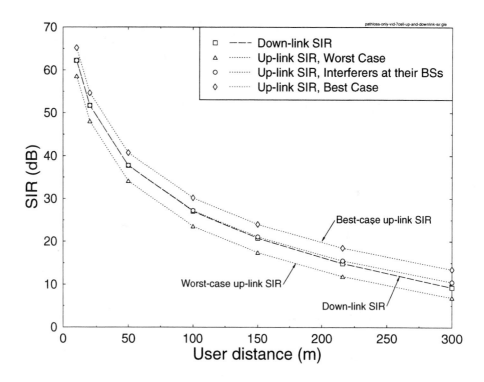

Figure 7.7: Path-loss-based SIR model devised from Equation 7.7 for the down-link and for various up-link scenarios, in a seven-cell reuse cluster with six co-channel video interferers. The co-channel base station separation was 1 Km. Therefore, the cell radius is 218 m.

SIRs prevail. The number of modulation levels drops from sixty four to sixteen, when the portable station (PS) is handed over to an outdoors street microcell, and it can be further reduced to four or even two in less friendly propagation scenarios. The system parameters are summarized in Table 7.1.

Observe in the table that the different QAM modes have different integrity subchannels, where the number of the subchannels ranges from 1 to 3 [29]. In general, these modulation subchannels can provide source-sensitivity-matched protection; however, in the case of the H.263 scheme, to be used in our transceiver, all bits are very sensitive to transmission errors and hence have to be equally well protected. Therefore, we adjusted the error correction power of the various subchannel codes in order to equalize the different subchannel integrities, as suggested by the different BCH codecs of Table 7.1. The propagation conditions are also listed in Table 7.1.

The system was capable of operating over a bit-rate range of up to six times the BPSK rate, corresponding to the 1–6 bit/symbol QAM modem modes. Hence the forward error correction coded signaling rate became 7.3 kBaud in all modes. When opting for a modulation excess bandwidth of 50% and a system bandwidth of 200 kHz, as in the Pan-European GSM system, the maximum signaling rate becomes 133.33

7.4. SIMULATION PARAMETERS

Modem-mode specific system parameters				
Features	Multirate System			
Modem	BPSK	4QAM	16QAM	64QAM
PSAM	yes	yes	yes	yes
Bits/Symbol	1	2	4	6
Number of sub-channels	1	1	2	3
C1 FEC	BCH (127,85,6)	BCH (255,171,11)	BCH (255,191,8)	BCH (255,199,7)
C2 FEC	N/A	N/A	BCH (255,147,14)	BCH (255,163,12)
C3 FEC	N/A	N/A	N/A	BCH (255,131,18)
Bits / TDMA frame	85	171	338	493
Source bitrate (kbit/s)	4.25	8.55	16.9	24.65
Min. AWGN SINR (dB)	4	10	15	20
Min. Rayleigh SINR (dB)	10	15	20	30

Modem-mode independent system parameters	
Features	General System
User Data Symbols / TDMA frame	128
User Pilot Symbols / TDMA frame	14
User Ramping Symbols / TDMA frame	2
User Padding Symbols / TDMA frame	2
User Symbols / TDMA frame	146
TDMA frame length (ms)	20
User Symbol Rate (kBd)	7.3
Slots/Frame	18
No. of Users	9
System Symbol Rate (kBd)	131.4
System Bandwidth (kHz)	200
Eff. User Bandwidth (kHz)	11.1
Vehicular Speed	13.4m/s or 30mph
Propagation Frequency (GHz)	1.8
Fast-Fading Normalized Doppler Frequency	6.2696×10^{-4}
Log-Normal Shadowing standard deviation (dB)	6
Path-loss Model	Power law 3.5
Base station Separation (km)	1

Table 7.1: Summary of System Features for the Reconfigurable Mobile Radio System. Minimum Required SINR Derived from Figures 7.33 and 7.34

kBaud. At this signaling rate INT(133.33/7.3)=18 time-slots can be created, where INT indicates integer division. Assuming an identical speech signaling rate of 7.3 kBd, nine videophone users can be supported by the proposed scheme in the GSM system's 200 kHz bandwidth [41]. Each video user has one time-slot for speech and another time-slot for video transmissions. A range of further system aspects can be inferred from Table 7.1.

7.5 Results for Multiple Interferers

In this section, we discuss the effects of co-channel interference in mobile cellular radio systems, elaborating on the SIR profile of a cell. We then tabulate the effects of slow fading and of the interferers position, the signal-to-interference ratio (SIR). These results are then compared with the theoretical results. Based on the same simulations, the normalized Shannonian channel capacity profile across the cell was plotted.

7.5.1 SIR Profile of a Cell

In our first experiments, we simulated the signal-to-interference ratio (SIR) profile across the cell area of a seven-cell reuse cluster. In order to find how different parameters affect the co-channel interference and SIR profile, we simulated different situations, as described below.

Specifically, in order to quantify the effect of the interferer position on the co-channel interference, we simulated the following up-link transmission scenarios:

1. The least detrimental, that is, "best" interferer position, where all the interferers are placed as far as possible from the interfered BS.

2. The worst-case interferer position, where all the interferers are positioned as close as possible to the interfered BS.

3. Random interferer positions, where interferers are allocated randomly within their cell, representing the most typical scenario.

These scenarios were simulated using constant bit-rate video users for all the interferers. The constant bit-rate transmissions of video/data users causes more interference than the bursty transmissions of speech users. Figure 7.8 shows the SIR profile for the cell with interference from six video users, and the interferers are in positions to cause the least amount of interference. When the interferers are randomly distributed within their cell, which is a typical situation, the SIR profile is as shown in Figure 7.9. Figure 7.10 is the SIR profile for the same parameters, except the interferers are in positions to inflict maximum interference. The SIR profiles are shown in two forms; a three-dimensional (3D) view, showing the general trend; and a more precisely readable contour form. The simulation parameters were defined in Table 7.1.

7.5. RESULTS FOR MULTIPLE INTERFERERS

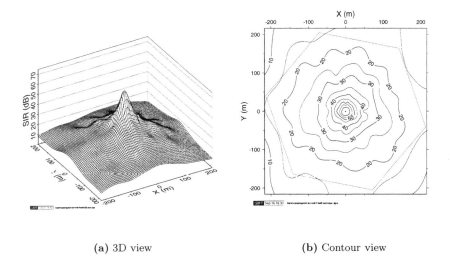

(a) 3D view (b) Contour view

Figure 7.8: Simulated average SIR (dB) profile of a hexagonal cell, employing a reuse factor of 7, path-loss exponent of 3.5, fast Rayleigh fading, log-normal slow fading with a frequency of 1 Hz as well as standard deviation of 6 dB, 4QAM modulation, interference from six video users in co-channel cells. *The interferers are in the best position* (as far as possible from the interfered base station).

These three SIR profiles show the typical SIR behavior over the cell area. Each point in the cell is characterized by the average SIR over 1000 TDMA frames, corresponding to 20 s of talk time. Since the log-normal slow-fading frequency was approximately 1Hz, each SIR value was only averaged over 20 slow-fading cycles. Ideally, the averaging should be over at least 100 slow-fading cycles. This was impractical, however, due to the associated computational complexity. Nonetheless, this averaging period is sufficiently long to smooth out the effects of the Rayleigh fast-fading.

Using the same deterministic shadow fading profile, when changing other simulation parameters, allows the effects of these parameter changes to be compared under these controlled conditions. Figure 7.11 shows a comparison of the SIR profiles, when random and fixed shadow fading profiles are used. The fixed fading (Figure 7.11(b)) uses the same shadow fading sequence for each measured (x,y) point in the cell. The fast-fading is not affected, since the evaluation period is sufficiently long to average out the effects of the Rayleigh fast-fading.

Upon using the fixed shadow fading profiles, we can evaluate and compare the effects of changing various parameters on the SIR. By comparing simulations with and without log-normal shadow fading the effect of shadowing was found to be in the range of 5 to 7 dB. The position of the interferers and its effect on the SIR were found by comparing simulations, where the interferers were in a position to cause maximum interference with simulations positioning the interferers to inflict the minimum amount of interference.

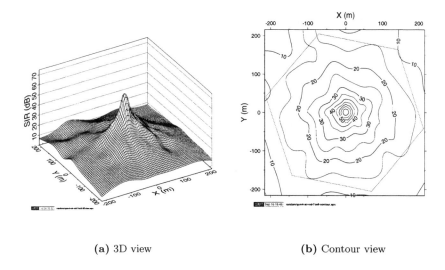

(a) 3D view (b) Contour view

Figure 7.9: Simulated average SIR (dB) profile of a hexagonal cell, employing a reuse factor of 7, path-loss exponent of 3.5, fast Rayleigh fading, log-normal slow fading with a frequency of 1 Hz as well as standard deviation of 6 dB, 4QAM modulation, interference from six video users in co-channel cells. *The interferers are randomly distributed* within their cells.

Figure 7.12 shows the SIR profiles with and without slow fading for the fixed fading profiles. A comparison between the SIR profile of video and speech users can be carried out using Figure 7.13. For this experiment we assumed that the video users were constantly transmitting. Speech users can reduce the amount of interference they inflict by turning off transmissions during the natural silent periods within speech. From experiments it has been found that the activity ratio of speech is only about 40% on the average [217]. When transmitting speech information, then, the transmitter can be typically turned off for about 60% of the time. Table 7.2 tabulates the SIR difference when using video or speech interfering users for different interferer positions.

We will now compare the simulated SIR with a simple theoretical model that we derived earlier in Section 7.3. Using the theoretical model, we found that the expected SIR with the interfering users in positions to induce maximum interference was 11.7 dB, and for the minimum interference case it was 18.3 dB. Therefore, the SIR difference between the interferer positions causing the maximum and minimum interference is 6.6 dB. Consequently, the theoretical model matches our simulated results in this case.

The increase of SIR, when using purely speech users instead of video users, is caused by the discontinuous transmissions of the speech users. We used a two-state speech model [216] with a mean talk period of 1 s and a mean silent period of 1.35 s,

7.5. RESULTS FOR MULTIPLE INTERFERERS

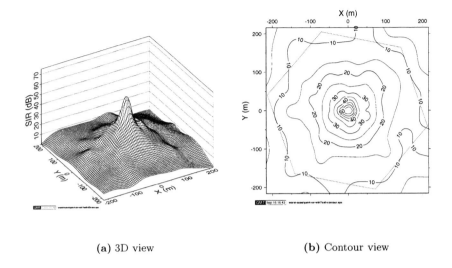

(a) 3D view   (b) Contour view

Figure 7.10: Simulated average SIR (dB) profile of a hexagonal cell, employing a reuse factor of 7, path-loss exponent of 3.5, fast Rayleigh fading, log-normal slow fading with a frequency of 1 Hz as well as standard deviation of 6 dB, 4QAM modulation, interference from six video users in co-channel cells. *The interferers are in the worst position* (as close as possible to the interfered base station).

Comparison	SIR Improvement (dB)
Improvement without 6dB, 1Hz log-normal shadow fading, with video interferers	3.9
Improvement without 6dB, 1Hz log-normal shadow fading, with speech interferers	4.0
Difference between interferer positions causing the maximum and minimum interference	6.6
Improvement with speech interferers compared to video interferers	3.6

Table 7.2: SIR Improvement for Various Parameters Affecting the Amount of Interference

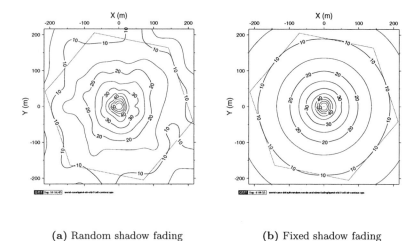

(a) Random shadow fading (b) Fixed shadow fading

Figure 7.11: Simulated average SIR (dB) profile of a hexagonal cell, employing a reuse factor of 7, path-loss exponent of 3.5, fast Rayleigh fading, log-normal slow fading with a frequency of 1 Hz as well as standard deviation of 6 dB, 4QAM modulation, interference from six video users in co-channel cells. *The interferers are in the worst-case position* (as close as possible to the interfered basestation), *using both random and fixed shadow fading profiles.*

corresponding to an average voice activity factor (VAF):

$$VAF = \frac{1.0s}{1.0s + 1.35s} = \frac{20}{47} = 43\%. \tag{7.9}$$

The average voice activity factor of speech is usually in the range of 40 to 50%, and as expected, the increase in SIR is approximately 3 dB. For the voice activity factor of 43% that we simulated, the SIR increase should be $10\log_{10}(1/\text{VAF}) = 3.7\text{dB}$. This compares well with the measured SIR reduction of 3.6 dB.

7.5.2 Signal-to-Noise-Plus-Interference Ratio (SINR)

While the Signal-to-Noise Ratio (SNR) and the Signal-to-Interference Ratio (SIR) are useful metrics to describe system performance, neither gives a reliable measure of system performance in both noise- and interference-limited scenarios. The Signal-to-Interference-plus-Noise Ratio (SINR) gives a more reliable measure of system performance in both noise- and interference-limited situations. The SINR is defined below in Equation 7.10, where S is the received signal power, I is the received interference power, and N is the AWGN power within the channel's bandwidth:

$$SINR = \kappa = \frac{S}{I + N}. \tag{7.10}$$

7.5. RESULTS FOR MULTIPLE INTERFERERS

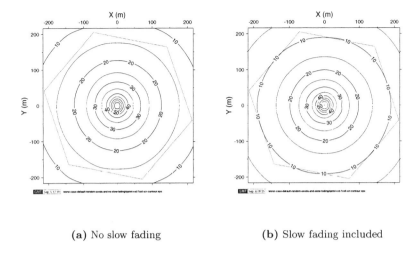

(a) No slow fading

(b) Slow fading included

Figure 7.12: Simulated average SIR (dB) profile of a hexagonal cell, employing a reuse factor of 7, path-loss exponent of 3.5, fast Rayleigh fading, 4QAM modulation, interference from six video users in co-channel cells. The interferers are in the worst-case position (as close as possible to the interfered base station), *comparing performance with and without a fixed shadow fading profile*, where the log-normal slow-fading frequency was 1 Hz with a standard deviation of 6 dB.

In interference-limited situations, $SINR \simeq SIR$, while for noise-limited scenarios $SINR \simeq SNR$. In our multiple interferer simulations, we have only investigated interference-limited scenarios. Therefore, only one SINR profile is shown for the multiple interferer scenarios in Figure 7.14, since they appear identical to the SIR profiles of Section 7.5.1. This profile was plotted by calculating the SINR (κ) using Equation 7.10 and the averaged signal, interference, and noise power from the simulations whose parameters were defined in Table 7.1.

In the next section, we introduce the normalized channel capacity, and then we plot its profile across the cell area, as we have done for the SIR and SINR profiles.

7.5.3 Channel Capacity

The Shannon-Hartley law states that the theoretical upper-bound channel capacity of a bandlimited Additive White Gaussian Noise (AWGN) channel can be expressed as:

$$C = B \log_2(1 + \gamma), \qquad (7.11)$$

where B is the channel bandwidth, γ is the signal-to-noise ratio (SNR), and C is the channel capacity in bits/sec. The SNR, γ is defined as $\gamma = \frac{S}{N}$, where S is the received signal power and N is the AWGN power within the channel bandwidth.

In most mobile radio systems, however, the channel exhibits Rayleigh fast-fading, aggravated by typically log-normally distributed shadowing or slow fading, resulting

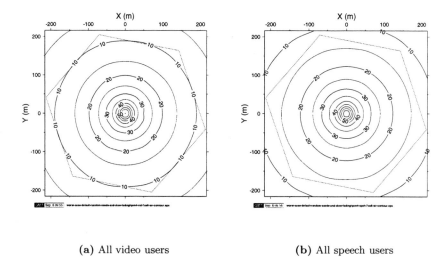

(a) All video users   (b) All speech users

Figure 7.13: Simulated average SIR (dB) profile of a hexagonal cell, employing a reuse factor of 7, path-loss exponent of 3.5, fast Rayleigh fading, log-normal slow-fading with a frequency of 1 Hz as well as standard deviation of 6 dB, 4QAM modulation, interference from 6 users in co-channel cells. *The interferers are in the worst-case position* (as close as possible to the interfered base station), using a fixed shadow fading profile. Comparing performance with all video users and all speech users.

in a time-variant channel capacity. Lee [145] derived an estimate of the channel capacity in Rayleigh-fading environments and argued that when using diversity in a Rayleigh-fading environment, the average channel capacity can approach that for a Gaussian channel. The normalized Shannonian channel capacity of bandlimited AWGN channels can be expressed as:

$$\eta = \frac{C}{B} = log_2(1+\gamma), \qquad (7.12)$$

an upper-bound approximation, which has to be replaced by Lee's upper-bound estimate [145] in case of Rayleigh channels:

$$\eta \approx \log_2 e \cdot e^{-1/\gamma}(-E + \ln\gamma + \frac{1}{\gamma}), \qquad (7.13)$$

where $E \approx 0.577$ is the Euler constant and γ is the average SNR. Evaluation of this upper-bound formula shows a 32% channel capacity reduction in comparison to the Gaussian channel at an SNR of 10 dB.

In a cellular reuse structure, the effect of co-channel interference must be included in the channel capacity estimate. Hence, the definition of γ in Equations 7.11–7.13 must be modified by replacing the SNR (γ) by the signal-to-noise-plus-interference ratio (SINR), κ. The SINR was defined before in Equation 7.10 in Section 7.5.2. This

7.5. RESULTS FOR MULTIPLE INTERFERERS

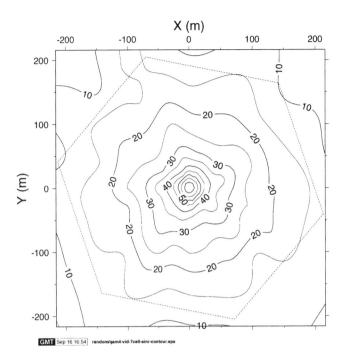

Figure 7.14: Simulated SINR contours of a hexagonal cell, employing a reuse factor of 7, path-loss exponent of 3.5, fast Rayleigh fading, log-normal slow fading with a frequency of 1 Hz as well as standard deviation of 6 dB and random 4QAM video user positions within cell boundaries.

implies the assumption that the interference obeys Gaussian statistics. This is true, due to the central limit theorem, if a high number of interferers are present. It is optimistic for a single interferer scenario. However, the corresponding results can be used as a loose upper-bound estimate of the cellular channel capacity.

Therefore, the normalized channel capacity for a bandlimited, interference-contaminated Gaussian channel is defined as:

$$\eta = \frac{C}{B} \approx log_2(1+\kappa) = log_2\left(\frac{S+I+N}{I+N}\right), \qquad (7.14)$$

where S is the received signal power, I is the received interference power, and N is the AWGN power within the channel's bandwidth. Furthermore, the estimated upper-bound normalized channel capacity for a bandlimited, interference contaminated Rayleigh-faded channels is defined as:

$$\eta \approx \log_2 e \cdot e^{-1/\kappa}(-E + \ln\kappa + \frac{1}{\kappa}), \qquad (7.15)$$

where E is the Euler constant and κ is the SINR defined in Equation 7.10. In a noise-limited radio system without power control, one would expect the SINR to reduce with distance from the transmitter, when using an omnidirectional aerial. However, in an interference-limited system the pattern of SINR is less regular. The normalized channel capacity for a typical hexagonal cell in a simulated system, with Rayleigh fast and log-normal shadow fading having a standard deviation of 6 dB and a frequency of 1 Hz, is shown in Figure 7.15(c), again using the parameters defined in Table 7.1. This typical normalized channel capacity profile is for the situation when the interfering users are randomly distributed within their own cells. When the interferers are causing the maximum amount of interference, the normalized channel capacity profile is as shown in Figure 7.15(b). Figure 7.15(a) represents the normalized channel capacity with the interferers in a position to induce the least interference.

These figures provide an estimate of the maximum possible normalized channel capacity within the cell area. For areas, where the normalized channel capacity is higher than 6, we may be able to use 64QAM. However, the mobile terminal would have to be sufficiently intelligent to change modulation scheme, when it notices it is in a high-capacity region of the cell, that is, near the base station.

The previous three normalized channel capacity (NCC) profiles were simulated using randomly assigned slow fading profiles. These random fading profiles provided realistic NCC contours; however, because of their random fluctuations they are not conducive to comparing different profiles, while quantifying the effect of changing a certain simulation parameter. This problem was circumvented using the same technique, as was used for the SIR profiles described earlier in Section 7.5.1, by employing a fixed slow-fading profile for every simulation point. Figure 7.16 shows a comparison of NCC profiles, when random and fixed fading profiles are used, for the simulation parameters of Table 7.1.

Using fixed fading profiles allows the effect of various simulation parameters to be measured exactly. The effect of slow fading can be assessed from Figure 7.17, which is a pair of NCC profiles with and without slow fading for the fixed fading profiles, using the simulation parameters of Table 7.1. A comparison between the NCC profile for video and speech users can be made in Figure 7.18.

Table 7.3 tabulates the normalized channel capacity (NCC) for a user 100 meters from the base station. The fixed fading profile simulation results were used for this comparison. We note, however, that these C/B estimates are significantly higher than those one could exploit in practical terms.

Recently, many papers have been presented on spectrum efficiency and channel capacity for cellular radio systems. Apart from the excellent papers by Lee [145,220], there have been many others of note. Gejji [221] rewrote the cellular radio capacity equation derived by Lee [220] in a more useful form. More recent work by Chiani *et al.* [222] and Haas *et al.* [223] has also been informative.

In the results we have presented thus far in this chapter, the co-channel interference has been generated by multiple interferers. In the next section, we limit the number of interferers to one. Due to the central limit theorem, the interference from multiple interferers is more noise-like. However, when inflicted by a single dominant interferer, the interference becomes more detrimental to the transceiver's performance.

7.5. RESULTS FOR MULTIPLE INTERFERERS

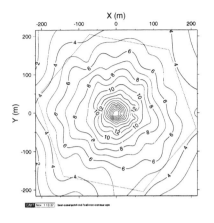

(a) The interferers are in the best position (as far as possible from the interfered base station)

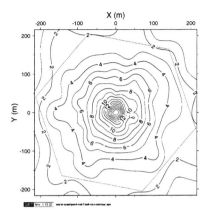

(b) The interferers are in the worst position (as close as possible to the interfered base station)

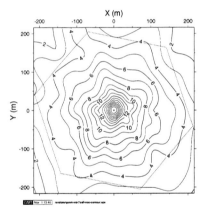

(c) Interfering video users randomly positioned within cell boundaries

Figure 7.15: Simulated average normalized channel capacity profile of a hexagonal cell, employing a reuse factor of 7, path-loss exponent of 3.5, fast Rayleigh fading, log-normal slow fading with a frequency of 1 Hz as well as standard deviation of 6 dB, 4QAM modulation, interference from six video users in co-channel cells.

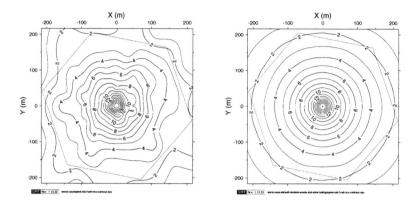

(a) Random Fading (b) Fixed Fading

Figure 7.16: Simulated average normalized channel capacity across a hexagonal cell, employing a reuse factor of 7, path-loss exponent of 3.5, fast Rayleigh fading, log-normal slow fading with a frequency of 1 Hz as well as standard deviation of 6 dB, 4QAM modulation, interference from six video users in co-channel cells. *The interferers are in the worst-case position* (as close as possible to the interfered base station), using both random and fixed fading profiles.

Comparison	Normalized Channel Capacity (bits/s/Hz)	
	Without Slow Fading	With Slow Fading
Video interfering users (maximum interference)	6.97	5.67
Speech interfering users (maximum interference)	8.19	6.83

Comparison	Normalized Channel Capacity (bits/s/Hz)	
	Maximum Interference	Minimum Interference
Interfering user positions (no slow fading and video interferers)	6.97	9.18
Interfering user positions (no slow fading and speech interferers)	8.19	10.40

Table 7.3: Normalized Channel Capacity (NCC) at a Position 100 Meters from the Base Station for Various Parameters Affecting the Amount of Interference

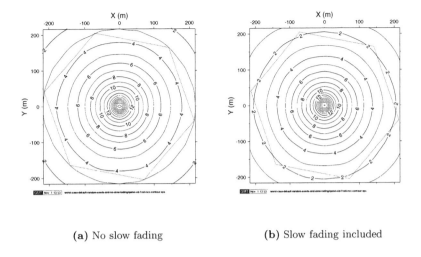

(a) No slow fading (b) Slow fading included

Figure 7.17: Simulated average normalized channel capacity profile of a hexagonal cell, employing a reuse factor of 7, path-loss exponent of 3.5, fast Rayleigh fading, 4QAM modulation, and interference from six video users in co-channel cells. The interferers are in the worst-case position (as close as possible to the interfered base station), quantifying the performance with and without a fixed log-normal shadow-fading profile, where the slow-fading frequency is 1 Hz with a standard deviation of 6 dB.

7.6 Results for a Single Interferer

In this section, we present our results for simulations, where the number of interferers was limited to one. By limiting the number of interferers, the results become more predictable, allowing more complex analysis to be used with greater reliability. Furthermore, because of the lower simulation complexity, a larger range of SNR and SIR can be simulated.

First, we will derive a simple model for the signal-to-noise-plus-interference ratio (SINR). We then show how closely this matches the simulated results. Next we present results of transmission frame error rate (FER) and channel bit error rates (CBER) for various SNR and SIR values. In addition to showing average results for various parameters, we portray the cumulative distribution functions (CDF) of both the SIR and SINR for various average SNR and SIR values. This will help interpret the effect of various parameters on the system performance. Finally, we present performance results when using the robust H.263 video system described in Chapter 19. These results show the effect of the co-channel interference due to a single interferer on the videophone performance.

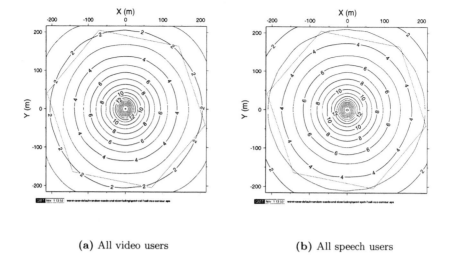

(a) All video users   (b) All speech users

Figure 7.18: Simulated average normalized channel capacity profile of a hexagonal cell, employing a reuse factor of 7, path-loss exponent of 3.5, fast Rayleigh fading, log-normal slow fading with a frequency of 1 Hz as well as standard deviation of 6 dB, 4QAM modulation, interference from six users in co-channel cells. *The interferers are in the worst-case position* (as close as possible to the interfered base station), using a fixed fading profile. The performance is quantified using exclusively video users and purely speech users.

7.6.1 Simple Model for SINR in a Single Interferer Situation

In this section, we derive a simple model for the signal-to-noise-plus-interference ratio (SINR). This model gives an approximation of the SINR in terms of the signal-to-interference ratio (SIR) and the signal-to-noise ratio (SNR). The definitions of the SIR, SNR, and SINR are:

$$SIR = \nu = \frac{S}{I}, \qquad (7.16)$$

$$SNR = \gamma = \frac{S}{N}, \qquad (7.17)$$

$$SINR = \kappa = \frac{S}{I+N}, \qquad (7.18)$$

where S is the signal power, I is the interference power, and N is the additive white Gaussian noise (AWGN) power in the channel bandwidth. The SINR can be represented in terms of the SIR and SNR, as:

$$SINR = \kappa = \frac{S}{I+N} = \frac{1}{\frac{I}{S} + \frac{N}{S}} = \left(\nu^{-1} + \gamma^{-1}\right)^{-1}, \qquad (7.19)$$

7.6. RESULTS FOR A SINGLE INTERFERER

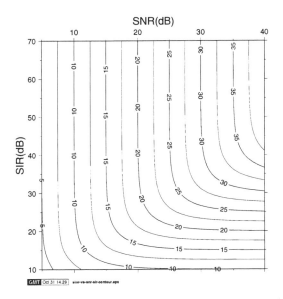

Figure 7.19: Simple SINR (dB) model in terms of SIR and SNR according to Equation 7.20.

which can be rewritten in dBs as:

$$SINR(dB) = -10\log_{10}\left(10^{-\left(\frac{SIR(dB)}{10}\right)} + 10^{-\left(\frac{SNR(dB)}{10}\right)}\right). \quad (7.20)$$

This model for the signal-to-interference-plus-noise ratio is plotted as a function of the SNR and SIR in the form of a contour graph in Figure 7.19. This graph clearly shows that the SINR tends to the SIR in interference-limited scenarios or to the SNR in a noise-limited situation. The SINR is particularly useful, since it gives a reliable measure of system performance in both noise- and interference-limited scenarios. The model is symmetrical with respect to SIR and SNR. However, in practice a certain level of interference is more detrimental to the transceiver's performance than the same level of noise. This model compares well with our simulated results shown in Figures 7.20 and 7.21. Figure 7.20 portrays the measured SINR versus measured SNR and SIR for simulations without slow fading. When slow fading was introduced into the simulations, the SINR did not change dramatically, as evidenced by Figure 7.21. Both of these simulated SINR figures were generated by calculating the SINR, SIR, and SNR using Equations 7.16–7.18 from the average signal, interference, and noise power of the simulations. The simulation parameters were defined in Table 7.1.

We will show that the SINR is a good indicator as to whether a symbol is received in error. In the next section, we show how the frame error rate and channel bit error rate vary with different ratios of SNR and SIR, given a fixed SINR value.

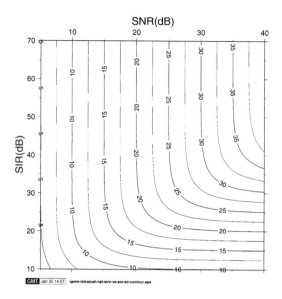

Figure 7.20: Measured SINR (dB) in terms of measured SIR and SNR from simulations using a single video interferer, 4QAM modulation, fast Rayleigh fading, and *no slow fading.*

7.6.2 Effect of SIR and SNR on Error Rates

In this section, we demonstrate the effect of SNR and SIR on the error rate of the mobile system. We show how both the frame error rate (FER) and channel bit error rate (CBER) or pre-FEC bit error rate are affected. For our simulations a single interferer is used, which is more detrimental at a given SIR than multiple interferers. With a large number of interferers, the interference becomes more like Gaussian noise due to the central limit theorem. By simulating the effect of a single interferer, the range of SNRs and SIRs, for which a minimum error rate constraint can be maintained, can be found. As with previous simulations using multiple interferers (Section 7.5.1), a fixed slow-fading profile is used. This means that the same slow-fading sequence is used for every combination of SNR and SIR, which ensures more consistent results. The remainder of the simulation parameters was defined in Table 7.1.

Figure 7.22 shows the contour plots of the frame error rate (FER) and the channel bit error rate (CBER) for a single video interferer, using 4QAM modulation, fast Rayleigh fading and no slow fading, for various SNR and SIR values. The frame is considered to be in error if the 4QAM-specific FEC decoder becomes overloaded and hence cannot correct all the bits in error. The 4QAM/BCH frame error rate gives a good measure of system performance, especially in conjunction with speech/video source codecs, which drop the received speech/video packet if the transmission errors cannot be corrected.

The effect of interference on the less robust, but higher capacity, 16QAM mode

7.6. RESULTS FOR A SINGLE INTERFERER

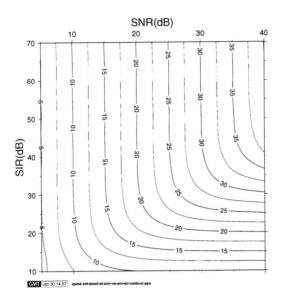

Figure 7.21: Measured SINR (dB) in terms of measured SIR and SNR from simulations using a single video interferer, 4QAM modulation, fast Rayleigh fading and log-normal slow fading with a frequency of 1 Hz as well as standard deviation of 6 dB.

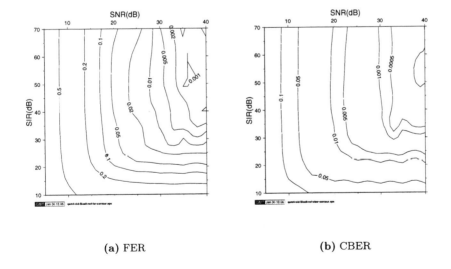

(a) FER

(b) CBER

Figure 7.22: Simulated FER and CBER versus SNR and SIR using a single video interferer, *4QAM*, BCH(255,171,11) coding, fast Rayleigh fading, and *no slow fading*, using the parameters of Table 7.1.

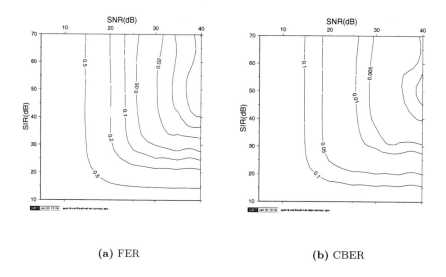

(a) FER (b) CBER

Figure 7.23: Simulated FER and CBER versus SNR and SIR using a single video interferer, *16QAM*, using the 16QAM/BCH codes of Table 7.1, fast Rayleigh fading, and *no slow fading*.

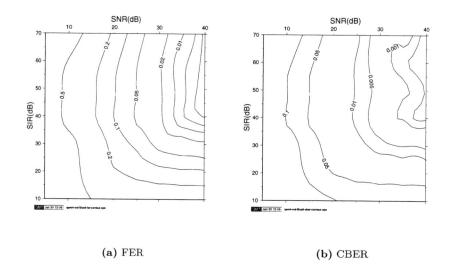

(a) FER (b) CBER

Figure 7.24: Simulated FER and CBER versus SNR and SIR using a single video interferer, *4QAM*, BCH(255,171,11) coding, fast Rayleigh fading and log-normal *slow fading* with a frequency of 1 Hz as well as standard deviation of 6 dB, using the parameters of Table 7.1.

7.6. RESULTS FOR A SINGLE INTERFERER

can be judged using Figure 7.23. It can be seen that 16QAM is much less robust than 4QAM, as we would expect. When shadowing was added to the simulations, the system performance decreased. The results of our simulations using 4QAM and shadowing with a slow-fading frequency of 1 Hz and standard deviation of 6 dB are shown in Figure 7.24.

Using the graphs described previously, we can estimate the minimum SIR and SNR constraints required for maintaining a specified error rate. The shape of the FER and CBER contours is similar to the SINR contours described in Section 7.6.1. This implies that the error rates are dependent on the signal-to-interference-plus-noise ratio (SINR). By plotting the error rates (FER, CBER) versus the measured SINR in Figure 7.25, it was found that the error rates were commensurate with the SINR. The advantage of this representation is that the error rate becomes a function of one parameter, namely, the SINR rather than two (SNR and SIR). The channel bit error rate (CBER) versus SINR graph is shown in Figure 7.25(a).

Observe in Figure 7.25 that for a given SINR value the CBER/FER curves can take different values. This is because a certain SINR can be given by different combinations of SNR and SIR values, leading to different CBERs and FERs. A higher CBER/FER results from a given SINR by a combination of low SIR and high SNR values than in the case of high SIR and low SNR values, thereby emphasizing the more detrimental nature of interference in comparison to noise. When there is no shadowing, these curves overlap very closely and allow a simple line to be drawn through all the curves. However, for low SINR simulations, which include slow fading, some of the curves diverge from the main group of curves. This implies that at a low SINR, better performance is achieved with a higher SIR and a lower SNR, which is a consequence of the more detrimental nature of the interference in comparison to noise. Since this effect only occurs when shadow fading is used, it is deemed to be caused by the effect of the slow fading on the co-channel interferers. Figure 7.25(b) is the graph of FER versus SINR. This graph is not so accurately approximated by a single line. This is probably due to the "quantizing"-type effect that the FEC decoding has on the channel bit error rate, manifesting itself in a bimodal, error-free, or badly corrupted frames.

In this section, we have produced graphs from simulations to show the effect of SIR, SNR, and SINR on the error rate in the system. In the next section, we investigate the distributions of how the SIR and SINR vary with time.

7.6.3 Time-Varying Effects of SIR and SINR

In this section, we investigate how the SIR and SINR vary with time for different amounts of noise and co-channel interference. All previous simulations have used measured SNR, SIR, and SINR values averaged over the length of the simulation. In this section, we use simulation-based measurements for these ratios, which have been averaged over just one TDMA time-slot. These simulations produced a sequence of SIR values in order to represent the SIR distribution. Ideally, a measurement of the SIR, SNR, and SINR should be made for each symbol, giving the true time-varying statistics, but the storage requirements are excessive.

Figure 7.26 shows the cumulative distribution function (CDF) of the SIR averaged

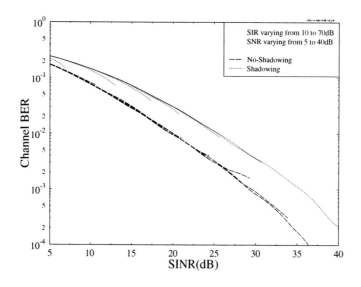

(a) CBER versus SINR

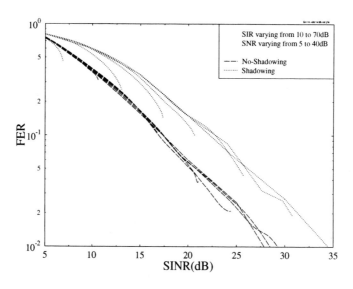

(b) FER versus SINR

Figure 7.25: Simulated channel bit error rate or frame error rate versus SINR for a single video interferer, 4QAM, fast Rayleigh fading and shadowing with a slow-fading frequency of 1 Hz as well as a standard deviation of 6 dB, using the parameters of Table 7.1.

7.6. RESULTS FOR A SINGLE INTERFERER

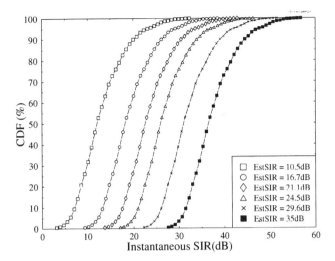

Figure 7.26: Cumulative distribution function of simulated "instantaneous" SIR averaged over one TDMA time-slot for a single video interferer, 4QAM, fast Rayleigh fading and no shadowing, using the parameters of Table 7.1.

over each time-slot for various average SIRs. The SIRs specified for each curve are the values stipulated for the simulations. We found that the averaged experimental values were within a few percent of the values stipulated. When shadowing is incorporated, the range of the SIR values encountered increased due to the signal and interference being faded by different uncorrelated fading channels. The CDFs of the SIR with and without shadowing are shown in Figures 7.26 and 7.27. The parameters for these simulations were summarized in Table 7.1.

Figure 7.28 shows the CDF of SIR with and without shadowing on the same axis. It should be noted that the CDF with and without shadowing cross at approximately 40% on the CDF axis. Furthermore, the SIR value at the crossover point is approximately the SIR average over all the simulated time-slots (long-term average). It has also been confirmed by other simulations that the long-term average SIR corresponds approximately to 40% on the vertical CDF axis. Recall that in the graphs of earlier sections, the SIR value plotted was the long-term average. However, for 40% of the time the instantaneous SIR is less than the plotted average value, which has detrimental effects on the overall system performance.

Even though the CDFs of the SIR are useful, the parameter that has the greatest effect on system performance is the SINR. Therefore, the CDFs of the SINR are of particular importance. However, the CDF of the SINR is dependent on two parameters — the long-term average SIR and SNR, as we have seen in Figure 7.25. Therefore, the CDFs of the SINR for different SNRs are shown on different graphs for clarity. Figures 7.29(a) and 7.29(b) show the CDF of the instantaneous SINR averaged over one TDMA time-slot for various average SIRs at two different SNRs. The limiting effect of the long-term average SNR can easily be seen. As with the SIR CDFs, the

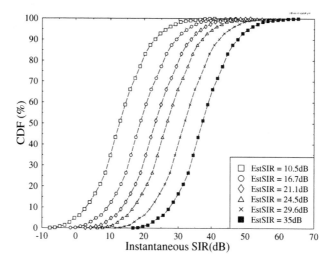

Figure 7.27: Cumulative distribution function of simulated "instantaneous" SIR averaged over one TDMA time-slot for a single video interferer, 4QAM, fast Rayleigh fading, and shadowing with a slow-fading frequency of 1 Hz, a standard deviation of 6 dB, using the parameters of Table 7.1.

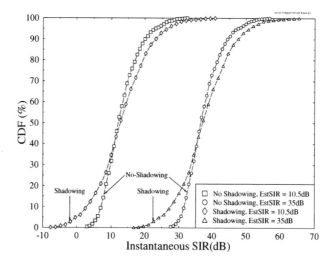

Figure 7.28: Cumulative distribution function of simulated "near-instantaneous" SIR averaged over one TDMA time-slot for a single video interferer, 4QAM, fast Rayleigh fading, and shadowing with a slow-fading frequency of 1 Hz, a standard deviation of 6 dB, as well as without shadowing, using the parameters of Table 7.1.

7.6. RESULTS FOR A SINGLE INTERFERER

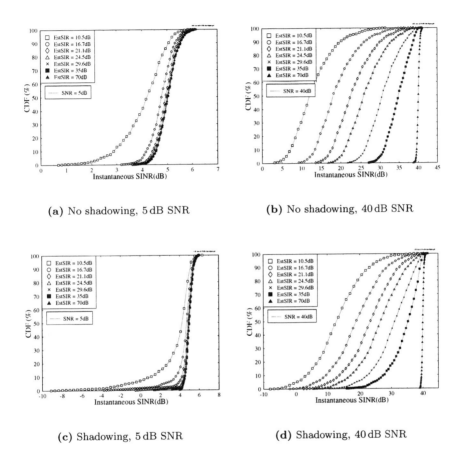

(a) No shadowing, 5 dB SNR

(b) No shadowing, 40 dB SNR

(c) Shadowing, 5 dB SNR

(d) Shadowing, 40 dB SNR

Figure 7.29: Cumulative distribution function of simulated "near-instantaneous" SINR averaged over one TDMA time-slot for a single video interferer, 4QAM, fast Rayleigh fading, and shadowing with a slow-fading frequency of 1 Hz and standard deviation of 6 dB, using the parameters of Table 7.1.

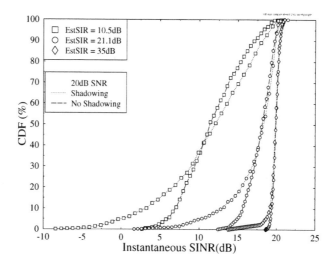

Figure 7.30: Cumulative distribution function of simulated "near-instantaneous" SINR averaged over one TDMA time-slot for a single video interferer, 4QAM, fast Rayleigh fading, and shadowing with a slow-fading frequency of 1 Hz, standard deviation of 6 dB, using the parameters of Table 7.1.

range of the instantaneous SINR increases, when shadowing is added, as shown in Figures 7.29(c) and 7.29(d).

The CDFs of the SINR with and without shadowing are plotted on the same axis in Figure 7.30, for the average SNR of 20dB. As in the above SIR investigations, the shadowing and non shadowing CDFs cross at approximately 40% in the interference-limited cases, and the corresponding SINR at the crossover point happens to be approximately the long-term average SINR measured on the horizontal axis.

There has been much research in similar areas by Zorti et al. [224], Senarath et al. [225], Abu-Dayya et al. [226], and Zhang [227]. In this section, we have characterized the statistics of the time-varying SIR and SINR parameters. In the next section we use the robust H.263 video system described in Chapter 19 for videophone transmissions over our simulated mobile radio system.

7.6.4 Effect of Interference on the H.263 Videophone System

In Chapter 19 we describe a robust, H.263-based videophone system suitable for mobile radio channels. The system was tested over both Gaussian and Rayleigh-fading channels. The resulting average peak signal-to-noise ratio (PSNR) of the decoded video was plotted against the radio channel's signal-to-noise ratio (SNR). In this section, we present the results of subjecting the robust videophone system to co-channel interference. The co-channel interference was induced by a single interferer that was constantly transmitting random symbols, which corresponded to a worst-case scenario.

The videophone system was subjected to a range of different combinations of SNR

7.6. RESULTS FOR A SINGLE INTERFERER

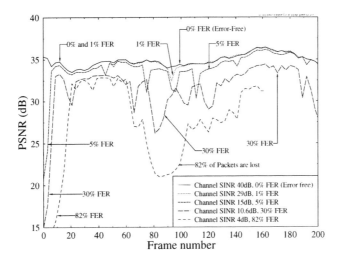

Figure 7.31: Decoded video PSNR versus video-frame index for transmission over Rayleigh-fading channels, with a single video co-channel interferer, for various FERs, using the parameters of Table 7.1.

and SIR values. The decoded average video PSNR could then be plotted against the SINR. Several combinations of SIR and SNR can produce similar SINR values. Hence, there were several possible average PSNRs for similar signal-to-noise-plus-interference ratios (SINR). To overcome this problem, the range of SINR was split into 5 dB bands, and the worst average PSNR was picked from each band and plotted.

For an error-free channel, the decoded video quality evaluated in terms of average PSNR would remain constant and would be the same as in a fixed network. As the QAM/BCH transmission frame error rate increases, more and more of the video packets have to be dropped. As more packets are dropped, error concealment is used to reduce the visual effect of the packet-losses. However, the average PSNR will decrease with increased packet-dropping. Figure 7.31 shows the PSNR of the decoded video versus the video-frame index (time) for various packet-dropping rates or frame error rates (FER). The effects of different amounts of packet-dropping can be studied. When the FER becomes too high, the video-quality visibly degrades, and a more robust mode of operation should be invoked; that is, a more robust modulation scheme is invoked in the multirate system. It was found that typically a FER of 5% was the maximum dropping rate, which the system could endure before the video degraded drastically. The decoded video sequences with packet-dropping can be viewed on the World Wide Web,[1] to help judge the subjective quality reduction caused by packet-dropping.

The effect of different types of mobile radio channels on the performance of the videophone can be judged using Figure 7.32. The graphs show the average PSNR versus SINR for Gaussian, Rayleigh-fading, and Rayleigh plus shadow fading channels.

---

[1] http://www-mobile.ecs.soton.ac.uk/peter/robust-h263/robust.html#DROPPING

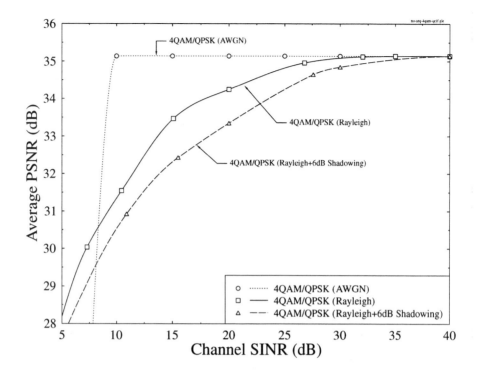

Figure 7.32: Average decoded video PSNR versus channel SINR for transmission over *Gaussian, Rayleigh fading, and Rayleigh plus shadow fading channels*, with a single video user co-channel interferer, using 4QAM, slow-fading frequency of 1 Hz, log-normal standard deviation of 6 dB, using the parameters of Table 7.1.

Figures 7.33–7.35 display the behavior of our multirate videophone system over a variety of mobile radio channels and interference conditions. Specifically, Figure 7.33 shows the average decoded video PSNR versus channel SINR for each of the four modes of operation in Table 7.1 over Gaussian radio channels. Each mode of operation is represented by a different modulation scheme. Figure 7.34 depicts the average decoded video PSNR versus channel SINR over Rayleigh-fading channels for the four modes of operation. The average decoded video PSNR versus channel SINR, for each of the four modes of operation in Table 7.1, over Rayleigh plus shadow faded channels is shown in Figure 7.35.

Having shown the effects of interference on our robust H.263-based videophone system, let us now summarize the message of this chapter.

7.6. RESULTS FOR A SINGLE INTERFERER

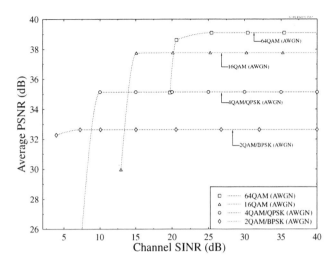

Figure 7.33: Average decoded video PSNR versus channel SINR for transmission over *Gaussian channels*, with a single video user co-channel interferer, using BPSK, 4QAM, 16QAM and 64QAM, using the parameters of Table 7.1.

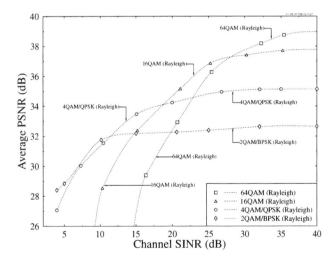

Figure 7.34: Average decoded video PSNR versus channel SINR for transmission over Rayleigh-fading channels, with a single video user co-channel interferer, using BPSK, 4QAM, 16QAM, and 64QAM, using the parameters of Table 7.1.

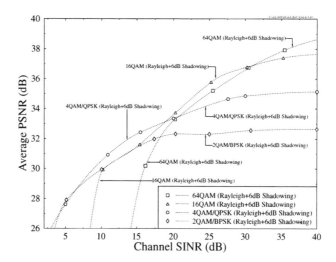

Figure 7.35: Average decoded video PSNR versus channel SINR for transmission over *Rayleigh and shadow-fading channels*, with a single video user co-channel interferer, using BPSK, 4QAM, 16QAM, and 64QAM, slow-fading frequency of 1Hz, log-normal standard deviation of 6dB, and the parameters of Table 7.1.

7.7 Summary and Conclusions

In this chapter we have discussed co-channel interference and its effects. Initially, we described the factors that control the amount of co-channel interference, such as cell shape, frequency reuse pattern, transmission power, user position, and channel fading. We then discussed the theoretical SIR measure that is commonly used to quantify the effects of a frequency reuse pattern. We showed that this metric was inaccurate in predicting the potential co-channel interference of a reuse pattern, except for large reuse cluster sizes.

In Section 7.5 we plotted the SIR profile across the cell area using multiple interferers. We showed how the position of mobile users could affect this SIR profile. In addition, we demonstrated that the use of video transmissions instead of speech has a detrimental effect on the SIR profile, since speech transmissions are discontinuous. Hence, the interference can be reduced by disabling transmissions during the silent periods using voice activity detection.

In Section 7.5.2 we defined the signal-to-interference-plus-noise ratio (SINR). This performance metric is a combination of the signal-to-noise ratio (SNR) and the signal-to-interference ratio (SIR), which gives a reliable measure of channel quality in both noise- and interference-limited environments. Using the SINR and Lee's approximation [145], an upper-bound estimate of the channel capacity in interference-limited environments was derived and plotted, which showed the potential high capacity available near a base station. We argued that this could be exploited by a reconfigurable multimode system.

In Section 7.6.1 we derived a simple model to estimate the signal-to-interference-

7.7. SUMMARY AND CONCLUSIONS

plus noise ratio (SINR) from the signal-to-noise ratio (SNR) and the signal-to-interference ratio (SIR). We showed that this model matched the measured results very well. In Section 7.6.2 we investigated the effects of the SNR and SIR on the error rate of our mobile radio system. We found that the results were similar to our SINR results, which suggested that the SINR was a good measure for estimating the channel error rate.

In Section 7.6.3 we investigated how the range of instantaneous SIR and SINR values varied for different average SIR and SNR values. We also studied the effects of shadow fading on the range of instantaneous SIR and SINR values. Finally, in Section 7.6.4 we investigated the performance of our error-resilient H.263-based videophone system in an interference-limited environment. We also evaluated the videophone system in an environment under shadow fading.

In the next chapter, we suggest a method of increasing the system capacity for our videophone system using a variety of channel allocation algorithms.

Chapter 8

Channel Allocation[1]

8.1 Introduction

With the growth in the number of subscribers to mobile telecommunications systems worldwide and the expected introduction of multimedia services in handheld wireless terminals, a tremendous demand for bandwidth has arisen. Since bandwidth is scarce and becoming increasingly expensive, it must be utilized in an efficient manner.

The main limiting factor in radio spectrum reuse is co-channel interference. In reduced cell-size micro/picocellular architectures, the frequency reuse distance is reduced, thereby increasing the capacity and area spectral efficiency of the system. However, as the channel reuse distance is reduced, the co-channel interference increases. In Chapter 7 the effects and nature of co-channel interference were investigated. Co-channel interference caused by frequency reuse is the most severe limiting factor of the overall system capacity of mobile radio systems. The most important technique for reducing co-channel interference is power control, which is the topic of Chapter 20. Interference cancellation techniques [230] or adaptive antennas [231–233] can also be used to reduce co-channel interference. However, a simpler and more effective technique used in current systems is employing sectorized antennas [234].

Although handovers are necessary in mobile radio systems, they often cause several problems, and they constitute the major cause of calls being forcibly terminated. As the cell size is decreased, the average sojourn time or cell-crossing time for a user is reduced. This results in an increased number of handovers, requiring more rapid handover completion. In practice a seamless handover is not always possible except when soft-handovers [235] are used in CDMA-based systems. Rapid and numerous handovers require a fast backbone network between the base stations and the mobile switching centers, or they necessitate an increased number of mobile switching centers. Clearly, the handover process is crucial with regard to the perceived Grade of Service (GOS), and a wide range of different complexity techniques have been proposed, for example, by Tekinay and Jabbari [236] and Pollini [237] for the forth-

---

[1]This chapter is based on [228, 229] and was prepared in collaboration with Francesca Romiti.

coming future systems. The related issue of time-slot reassignment was investigated by Bernhardt [238].

8.2 Overview of Channel Allocation

The purpose of channel allocation algorithms is to exploit the variability of the radio channel propagation characteristics in order to allow increased efficiency radio spectrum utilization, while maintaining required signal quality. The most commonly used signal quality measure is the signal-to-interference ratio (SIR), also known as the carrier-to-interference ratio (CIR). The signal quality measure that we have used previously was the signal-to-interference+noise ratio (SINR), which was defined in Equation 7.10. The SINR is approximately equal to the signal-to-noise ratio (SNR) in a noise-limited environment and approximately equal to the SIR in an interference-limited environment.

The radio spectrum is divided into sets of noninterfering physical radio channels, which can be achieved using orthogonal time or frequency slots, orthogonal user signature codes, and so on. The channel allocation algorithm attempts to assign these physical channels to mobiles requesting a channel, such that the required signal quality constraints are met. There are three main techniques for dividing the radio spectrum into radio channels. The first is frequency division (FD), in which the radio spectrum is divided into several nonoverlapping frequency bands. However, in practice the spectral spillage from one frequency band to another causes adjacent channel interference, which can be reduced by introducing frequency guard bands. However, these guard bands waste radio spectrum, and hence there is a compromise between adjacent channel interference and frequency band-packing efficiency. Tighter filtering can help reduce adjacent channel interference, allowing the guard bands to be reduced.

The second technique is time division (TD), in which the radio spectrum is divided into disjunct timeperiods, which are usually termed time-slots. However, straightforward rectangular windowing of the time-domain signal corresponds to convolving the signal spectrum with a frequency-domain sinc-function, resulting in Gibbs-oscillation. Hence, in practical systems a smooth time-domain ramp-up and ramp-down function associated with a time-domain guard period is employed. Therefore, there is a trade-off between complex synchronization, time-domain guard periods, and adjacent channel interference.

The third technique for dividing the radio spectrum into channels is code division (CD). Code division multiple access (CDMA) [239–242] has been used in military applications, in the IS-95 mobile radio system [243], and in the recently standardized Universal Mobile Telecommunications System (UMTS) [242, 244]. In code division, the physical channels are created by encoding different users with different user signature sequences.

In most systems a combination of these techniques is used. For example, the Pan-European GSM system [41] uses frequency division duplexing for up- and down-link transmissions, while accommodating eight TDMA users per carrier. In this chapter, the term "channel" typically implies a physical channel, constituted by a time-slot of

8.2. OVERVIEW OF CHANNEL ALLOCATION

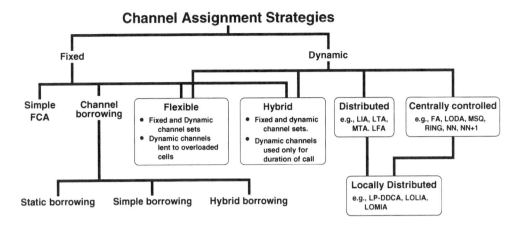

Figure 8.1: Family tree of channel allocation algorithms.

a given carrier frequency.

A wide variety of channel allocation algorithms have been suggested for mobile radio systems. The majority of these techniques can be classified into one of three main classes: fixed channel allocation (FCA), dynamic channel allocation (DCA), and hybrid channel allocation (HCA). Hybrid channel allocation is constituted by a combination of fixed and dynamic channel allocation, which is designed to amalgamate the best features of both, in order to achieve better performance or efficiency than DCA or FCA can provide. Several channel allocation schemes and the associated trade-offs in terms of performance and complexity are discussed in detail in the excellent overview papers of Katzela and Naghshineh [245] and those by Jabbari and Tekinay et al. [246, 247]. Figure 8.1 portrays the family tree for the main types of channel allocation algorithms, where the acronyms are introduced during our further discourse. Zander [248] investigated the requirements and limitations of radio resource management in general for future wireless networks. Everitt [249] compared various fixed and dynamic channel assignment techniques and investigated the effect of handovers in the context of CDMA-based systems.

8.2.1 Fixed Channel Allocation

In fixed channel allocation (FCA), the available radio spectrum is divided into sets of frequencies. One or more of these sets is then assigned to each base station on a semipermanent basis. The minimum distance between two base stations, they have been assigned the same set of frequencies is referred to as the frequency reuse distance.

This distance is chosen such that the co-channel interference is within acceptable limits, when interferers are at least the reuse distance away from each other. The assignment of frequency sets to base stations is based on a predefined reuse pattern. The group of cells that contain one of each of the frequency sets is referred to as the frequency reuse cluster. The grade of frequency reuse is usually characterized in terms of the number of cells in the frequency reuse cluster. The lower the number of cells in a reuse cluster, the more bandwidth-efficient the frequency reuse pattern and the higher the so-called area spectral efficiency, since this implies partitioning the available total bandwidth in a lower number of frequency subsets used in the different cells, thereby supporting more users across a given cell area. However, small reuse clusters exhibit increased co-channel interference, which has to be tolerated by the transceiver.

In FCA, the assignment of frequencies to cells is considered semipermanent. However, the assignment can be modified in order to accommodate teletraffic demand changes. Although FCA schemes are very simple, modifying them to adapt to changing traffic conditions or user distributions can be problematic. Hence, FCA schemes have to be designed carefully, in order to remain adaptable and scalable, as the number of mobile subscribers increases. In this context, adaptability implies the ability to rearrange the network to provide increased capacity in a particular area on a long- or short-term basis, where scalability refers to the ability of easily increasing capacity across the whole network via tighter frequency reuse. For example, Dahlin et al. [250] suggested a reuse pattern structure for the GSM system that can be scaled to meet increased capacity requirements, as the number of subscribers increases. This is discussed in more detail in the overview paper by Madfors et al. [251]. Each measure invoked, in order to further increase the network capacity, increases the system's complexity and hence becomes expensive. Furthermore, such systems cannot be easily modified to provide increased capacity in the specific area of a traffic hot-spot on a short-term basis.

A commonly invoked reuse cluster/pattern is the seven-cell reuse cluster, providing coverage over regular hexagonal shaped cells, which is shown in Figure 7.1(a) Each cell in the seven-cell reuse cluster has six first-tier co-channel interfering cells at a distance D, the reuse distance. It is shown in Figure 7.5 that for the seven-cell cluster the reuse distance, D, is 4.58 times the cell radius r. This reuse pattern supports the same number of channels at each cell site, and hence the same system capacity. Therefore, the teletraffic capacity is distributed uniformly across all the cells. Since traffic distributions usually are not uniform in practice, such a system can lead to inefficiencies. For example, under nonuniform traffic loading, some cells may have no spare capacity; hence, new calls in these cells are blocked. However, nearby cells may have spare capacity.

Several studies have suggested techniques to find the optimal reuse pattern for particular traffic and users distributions, as exemplified by the work of Safak [213], on optimal frequency reuse with interference. While such contributions are useful, a practical system would need to modify the whole network configuration every time the traffic or user distributions changed significantly. Therefore, suboptimal but adaptable and scalable solutions are more desirable for practical implementations. When the traffic distribution changes, an alternative technique to modifying the reuse pat-

8.2.1.1 Channel Borrowing

In channel-borrowing schemes, a cell that has a call setup request but no available channels (which is termed an acceptor cell), can borrow free channels from neighboring cells referred to as donor cells in order to accommodate new calls, which would otherwise have been blocked. A channel can be borrowed only if its use will not interfere with existing ongoing calls. When a channel is borrowed, several cells are then prohibited from using the borrowed channel because it would cause interference. The process of prohibiting the use of borrowed channels is referred to as channel locking [252]. The various channel-borrowing algorithms differ in the way the free channel is chosen from a donor cell to be borrowed by an acceptor cell.

There are three main types of channel-borrowing algorithms: static, simple, and hybrid borrowing; a good overview of these algorithms can be found in [245–247]. Static borrowing could be described as a fixed channel re-allocation strategy rather than channel borrowing. In static borrowing, channels are reassigned from lightly loaded cells to heavily loaded cells, which are at distances in excess of the reuse distance. This reassignment is semipermanent and can be done based on measured or predicted changes in traffic. The other two types of channel borrowing (simple and hybrid) are different from static borrowing in that borrowed channels are returned when the call using the channels ends or is handed off to another base station. Therefore, the simple and hybrid channel borrowing schemes use short-term borrowing in order to cope with traffic excesses.

Simple channel- borrowing schemes allow any of the channels in a donor cell to be lent to an acceptor cell. Hybrid channel borrowing schemes split the channels assigned to each cell into two subsets. One subset of channels cannot be lent to other cells; hence, these are referred to as standard or local channels. The other subset can be lent to other cells, and so they are termed nonstandard or borrowable channels.

Simple borrowing [246, 253, 254] can reduce new call blocking, but it can cause increased interference in other cells; it can also prevent handovers of future calls in these cells. Experiments have shown that simple channel-borrowing algorithms perform better than static fixed channel allocation under light- and moderate traffic loads. However, at high traffic loads the borrowing of channels leads to channel locking, which reduces the channel utilization and therefore results in an increase in new call blocking and in failed handovers. The various simple channel-borrowing algorithms differ in terms of flexibility, complexity and their reduction of channel locking. Some algorithms [253, 254] pick the channel to borrow, while taking into account the associated "cost" in terms of channel locking for each candidate channel. Other algorithms [254] invoke channel reassignment in order to reduce channel locking. The innovative technique used by Jiang and Rappaport [252] to reduce channel locking is to limit the transmission power of borrowed channels.

Hybrid channel borrowing [245, 246] is a hybrid of simple channel borrowing and static fixed channel allocation. By dividing the channels at each base station into two subsets, and only allowing channels of one of the subsets to be borrowed, the chance of channel locking or failed handovers can be mitigated under high traffic loads. A range of algorithms is discussed in the literature, each having different objectives in terms of improving performance in a particular area of operation. Some algorithms [255] have the ratio of channels in each subset assigned a priori, while others dynamically adapt the size of the subsets based on traffic measurements or predictions [256]. The algorithm may also check whether the candidate borrowed channel is free in the co-channel cells [257]. A common technique [254, 258] is to reassign calls using a borrowed channel to another borrowed channel in order to reduce channel locking. A better policy is to reassign a call currently using a borrowed channel to a local channel, thereby returning the borrowed channel to the donor cell. Another procedure [255, 257] to reduce channel locking is to estimate the direction of movement of the mobile in an attempt to reduce future channel locking and interference. A simple technique [259] is to subdivide cells into sectors and only allow borrowed channels to be used in particular sectors of the acceptor cell, thereby reducing channel locking.

8.2.1.2 Flexible Channel Allocation

Flexible channel allocation schemes [245, 246, 260] are similar to hybrid channel allocation schemes (which are described in Section 8.2.3) in that they divide the available channels into fixed and dynamic allocation subsets. However, flexible channel allocation is similar to a fixed channel allocation strategy, such as that used in static channel borrowing. In flexible channel allocation, the fixed channel set is assigned to cells in the same way as in fixed and hybrid channel allocation. The dynamic or flexible channels can be assigned to cells depending on traffic measurements or predictions. The difference between so-called hybrid and flexible channel allocation schemes is that in hybrid channel allocation the dynamic channels are assigned to cells only for the duration of the call. In flexible channel allocation the dynamic channels are assigned to cells, when the blocking probability in these cells becomes intolerable. Flexible channel allocation requires much more centralized control than hybrid channel allocation.

8.2.2 Dynamic Channel Allocation

Although fixed channel allocation schemes are common in most existing cellular radio systems, the cost of increasing their teletraffic capacity can become high. In theory, the use of dynamic channel allocation allows the employment of all carrier frequencies in every cell, thereby ensuring much higher capacity, provided the transceiver-specific interference constraints can be met. Therefore, it is feasible to design a mobile radio system, which configures itself to meet the required capacity demands as and when they arise. However, in practice there are many complications, which make this simplistic view hard to implement in practice. Dynamic channel allocation is used, for example, in the Digital European Cordless Telephone (DECT) standard [45, 261–264]. Law and Lopes [265] used the DECT system to compare the performance of two distributed DCA algorithms. However, DECT is a low-capacity system, where the time-slot utilization is expected to be comparatively low. For low slot utilization DCA is ideally suited. Dynamic channel allocation becomes more difficult to use in large-cell systems, which have higher channel utilization. Salgado-Galicia et al. [266]

8.2. OVERVIEW OF CHANNEL ALLOCATION

discussed the practical problems that may be encountered in designing a DCA-based mobile radio system.

Even though much research has been carried out into channel allocation algorithms, particularly dynamic channel allocation, many unknowns remain. For example, the trade-offs and range of achievable capacity gains are not clearly understood. Furthermore, it is not known how to combine even two simple algorithms in order to produce a hybrid that has the best features of both. One reason that the issues of dynamic channel allocation are not well understood is the computational complexity encountered in investigating such algorithms. In addition, the algorithms have to be compared to others in a variety of scenarios. Furthermore, changing one algorithmic parameter in order to improve the performance in one respect usually has some effect on another aspect of the algorithm's performance, due to the parameters highly interrelated nature. This is particularly true, since experience showed that some handover algorithms are better suited for employment in certain dynamic channel allocation algorithms [236]. Therefore the various channel allocation algorithms have to be compared in conjunction with a variety of handover algorithms in order to ensure that the performance is not degraded significantly by a partially incompatible handover algorithm. The large number of parameters and the associated high computational complexity of implementing channel allocation algorithms complicate study of the trade-offs of the various algorithms.

Again, in dynamic channel allocation, typically all channels can be used at any base station as long as they satisfy the associated quality requirements. Channels are then allocated from this pool as and when they are required. This solution provides maximum flexibility and adaptability at the cost of higher system complexity. The various dynamic channel allocation algorithms have to balance allocating new channels to users against the potential co-channel interference they could inflict upon users already in the system. Dynamic channel allocation is better suited to microcellular systems [267] because it can handle the more nonuniform traffic distributions, the increased handover requests, and the more variable co-channel interference better than fixed channel allocation due to its higher flexibility. The physical implementation of DCA is more complex than that of FCA. However, with DCA the complex and labor-intensive task of frequency planning is no longer required.

The majority of DCA algorithms choose the channel to be used based on received signal quality measurements. This information is then used to decide which channel to allocate or whether to allocate a channel at all. It is sometimes better not to allocate a channel if it is likely to inflict severe interference on another user, forcibly terminating existing calls or preventing the setup of other new calls. Ideally, the channel quality measurements should be made at both the mobile and base station. If measurements are made only at the mobile or only at the base station, the channel allocation is partially blind [268]. Channel allocation decisions that are based on blind channel measurements can in some circumstances cause severe interference, leading to the possible termination of the new call as well as curtailing another user's call, who is using the same channel. If measurements are made at both the mobile and the base station, then the measurements need to be compared, requiring additional signaling, which increases the call setup time. The call setup time is longer in DCA algorithms than in FCA due to the time required to make measurements and to compare them.

This can be a problem, when, for example, a handover is urgently required.

Probably the simplest dynamic channel allocation algorithm is to allocate the least interfered channel available to users requesting a channel. By measuring the received power within unused channels, effectively the noise plus interference on that channel can be measured. By allocating the least interfered channel, the new channel is not likely to encounter interference, and, due to semireciprocity, it is not likely to cause too much interference to channels already allocated. This works well for lightly loaded systems. However, this algorithm's performance is seriously impaired in high-load scenarios, where FCA would work better. However, the above is a very simple dynamic channel allocation algorithm. In Section 8.4 we will demonstrate that it is possible to achieve better performance and efficiency than that of FCA even at high loads, when using certain channel allocation algorithms. For these reasons, some channel allocation algorithms use a combination of FCA and DCA to achieve better performance than simple DCA, and better reuse efficiency than FCA. These algorithms are classified as hybrid channel allocation (HCA) algorithms.

The difference between the various dynamic channel allocation algorithms is, essentially, how the allocated channel is chosen. All the algorithms assign a so-called cost to allocating each of the possible candidate channels, and the one with the lowest cost is allocated. The difference between the algorithms is how the "cost" is calculated using the cost function. The cost function can be calculated on the basis of one or more of the following aspects: future call blocking probability; usage frequency of the channel; distance to where the channel is already being used, that is, the actual reuse distance; channel occupancy distribution; radio signal quality measurements; and so on. Some algorithms may give better performance than others, but only in certain conditions. Most DCA algorithms' objectives can be classified into two types, where most of them attempt to reduce interference, while others try to maximizee channel utilization in order to achieve spectral compactness.

There are three main types of DCA algorithms, namely:

- Centrally controlled algorithms
- Distributed algorithms
- Locally distributed algorithms (hybrid)

8.2.2.1 Centrally Controlled DCA Algorithms

Centrally controlled DCA algorithms are also often referred to as centrally located or centralized DCA algorithms. These algorithms use interference measurements that are made by the mobiles and base stations that are then passed to a central controller, which in most cases would be a mobile switching center. The algorithm that determines the channel allocation is located at the central controller, and it decides on the allocation of channels based on the interference measurements provided by all the base stations and mobiles under its control. These algorithms provide very good performance even at high traffic loads. However, they are complex to implement and require a fast backbone network between the base stations and the central controller. The central controller can become a "bottleneck" and increase the call setup time, which may be critical for "emergency" handovers.

Centralized algorithms [255, 257, 269–271] have been researched actively for over twenty years. One of the simplest is referred to as the First Available (FA) [269, 272] algorithm, which allocates the first channel found that is not reused within a given preset reuse distance. The Locally Optimized Dynamic Assignment (LODA) [255, 257] algorithm bases its allocation decisions on the future blocking probability in the vicinity of the cell. Some algorithms exploit the amount of channel usage to make allocation decisions. The RING algorithm [245, 271], for example, allocates the most often used channel within the cells, which are approximately at the reuse distance, and the terminology RING is justified by the fact that these cells effectively form a ring. There are also several algorithms, which attempt to optimize the reuse distance constraint. The Mean Square (MSQ) algorithm [272] attempts to minimize the mean square distance between cells using the same channel while maintaining the required signal quality. The Nearest Neighbor (NN) and Nearest Neighbor plus One (NN+1) algorithms [269, 272] pick a channel used by the nearest cell, which is at least at a protection distance amounting to the reuse distance (or reuse distance plus one cell radius for NN+1). Other algorithms [271] use channel reassignments to maintain the reuse distance constraint. Recall again that these algorithms were summarized in Figure 8.1.

8.2.2.2 Distributed DCA Algorithms

In contrast to centrally controlled algorithms, distributed algorithms are the least complex DCA techniques, in which the same algorithm is used by each mobile or base station in order to determine the best channel for setting up a call. Each mobile and/or base station makes channel allocation decisions independently using the same algorithm — hence the name distributed algorithms. The algorithmic decisions are usually based on the interference measurements made by the mobile or the base station. These algorithms are easy to implement, and they perform well for low-slot occupancy systems. However, in high-load systems their performance is degraded. Distributed algorithms require less signaling than centralized algorithms. However, the allocation is generally suboptimal owing to their locally based decisions. One real advantage of distributed algorithms is that base stations can easily be added, moved, or removed because the system automatically reorganizes and reconfigures itself. However, the cost of this flexibility is that the local decision making generally leads to a suboptimal channel allocation solution and to a higher probability of interference in neighboring cells. Furthermore, generally distributed algorithms are based on signal strength measurements and estimates of interference. However, these interference estimates can sometimes be poor, which can lead to bad channel allocation decisions. When a new allocation is made, the co-channel interference it inflicts may lead to an ongoing call to experience low-service quality, often termed a service interruption. If a service interruption leads to the ongoing call being terminated prematurely, this is referred to deadlock [245]. Successive service interruptions are termed as instability. A further problem with distributed algorithms is that the same channel can be allocated at the same time to two or more different users in adjacent cells. However, when the mobiles attempt to use the channel, they may find the quality unacceptably low. Therefore, distributed algorithms have to be able to check the quality of an allocation, before it is made permanent, which increases the call setup time further.

Chuang et al. [273] investigated the performance of several distributed DCA algorithms, arguing that under certain conditions these techniques can converge to a local minimum of the total interference averaged over the network. Grandhi et al. [274] and Chuang et al. [275] also evaluated the performance of combining dynamic channel allocation with transmission power control.

Examples of distributed algorithms are the Sequential Channel Search (SCS) and the least interference algorithm (LIA). The SCS algorithm [276] searches the available channels in a predetermined order, picking the first channel found, which meets the interference constraints. The LIA algorithm, alluded to earlier, picks the channel with the lowest measured interference that is available. One of the most complex distributed algorithms is the Channel segregation technique [277], which is a fully distributed, autonomous, self-organizing assignment scheme. Each cell maintains a measure of the relative frequency of channel usage for each channel. This probability-based measure is modified every time an attempt to access a specific channel is made. The channel assigned to the new call is the one with the highest probability of being or having been idle. The algorithm has been shown to reduce blocking and adapt to traffic changes. Although the channel allocation may rapidly converge to a near-optimal solution, it may take a long time to reach a globally optimal solution. As before, for the family tree of these techniques, please refer to Figure 8.1.

8.2.2.3 Locally distributed DCA algorithms

The third and final class of DCA algorithms are the locally distributed algorithms, which constitute a hybrid of distributed and centralized algorithms. These algorithms provide the greatest number of performance benefits of the centralized algorithms at a much lower complexity. Examples of locally distributed DCA algorithms are those proposed by Delli Priscoli et al. [278, 279] as an evolution of the Pan-European GSM system [41]. Locally distributed DCA algorithms use information from nearby base stations to augment their local channel quality information in order to make a more informed channel allocation decision. Most of the locally distributed algorithms maintain an Augmented Channel Occupancy (ACO) matrix [280]. This matrix contains the channel occupancy for the local and surrounding base stations from which information is received. After every channel allocation, the information to update the ACO matrices is sent to the nearby base stations. This signaling requires a fast backbone network, but it is far less complex than the signaling required for the centralized algorithms.

The Local Packing Dynamic Distributed Channel Assignment (LP-DDCA) algorithm, proposed in [280], maintains an ACO matrix for every base station for all surrounding cells within the co-channel interference distance or reuse distance from the base station. The LP-DDCA algorithm assigns the first channel available that is not used by the surrounding base stations, whose information is contained in the ACO matrix. There are several algorithms similar to this one, including those by Del Re et al. [281], and the Locally Optimized Least/Most Interference Algorithms (LOLIA/LOMIA) that we will use in Section 8.3.2.3 in the context of our performance

8.2. OVERVIEW OF CHANNEL ALLOCATION

Fixed Channel Allocation (FCA)	Dynamic Channel Allocation (DCA)
• Better under heavy traffic loads	• Better under light/moderate traffic loads
• Low call setup delay	• Moderate to high call setup delay
• Suited to large-cell environment	• Suited to microcellular environment
• Low flexibility in channel assignment	• Highly flexible channel assignment
• Sensitive to time and spatial changes in traffic load	• Insensitive to time and spatial changes in traffic load
• Low computational complexity	• High computational complexity
• Labor-intensive and complex frequency planning	• No frequency planning required
• Radio equipment only covers channels assigned to cell	• Radio equipment may have to cover all possible channels available
• Low signaling load	• High signaling load
• Centralized control	• Control dependent on the specific scheme from centralized to fully distributed
• Low implementational complexity	• Medium to high implementational complexity
• Increasing system capacity is expensive and time-consuming	• Simple and quick to increase system capacity

Table 8.1: FCA and DCA Features

comparisons.

An overview of the main differences between fixed and dynamic channel allocation is shown in Table 8.1; exploration of its detailed contents is left to the reader. However, this table does not show the increase in spectral efficiency and channel utilization that becomes possible with dynamic schemes, as will be demonstrated during our performance comparisons.

8.2.3 Hybrid Channel Allocation

Hybrid channel allocation schemes constitute a compromise between fixed and dynamic channel allocation schemes. They have been suggested in order to combine the benefits of DCA at low and medium traffic loads with the more stable performance of FCA at high traffic loads. Furthermore, hybrid schemes have been proposed as possible extensions to the fixed channel allocation used in second-generation mobile radio systems. In hybrid channel allocation schemes, the channels are divided into fixed and dynamic subsets. The fixed channels are assigned to the cells, as would be done for fixed channel allocation, and they are the preferred choice for channel allocation. When a cell exhausts all its fixed channels, it attempts to allocate a dynamically assigned channel from the central pool of channels. The algorithm used to pick the dynamically allocated channel depends on the hybrid scheme, but it can be any arbitrary DCA algorithm. The ratio of fixed and dynamic channels could be fixed [282] or varied dynamically, depending on the traffic load. At high loads, best performance is achieved, when the hybrid scheme behaves like FCA, by having none or a limited number of dynamically allocated channels [282, 283]. Some hybrid channel allocation algorithms reallocate fixed channels, which become free to calls using dynamic channels in order to free up the dynamic channels. This technique is known as channel reordering [271].

8.2.4 The Effect of Handovers

A handover or handoff event occurs when the quality of the channel being used degrades, and hence the call is switched to a newly allocated channel. If the new channel belongs to the same base station, then this is called an intra-cell handover. If the new channel belongs to a different base station, it is referred to as an inter-cell handover. Generally intra-cell handovers occur when the channel quality degrades due to interference or when the channel allocation algorithm decides that a channel reallocation will help increase the system's performance and capacity. Inter-cell handovers occur mainly because the mobile moves outside the cell area; hence, the signal strength degrades, requiring a handover to a nearer base station.

Handovers have a substantial effect on the performance of channel allocation algorithms. At high traffic loads, the majority of forced call terminations are due to the lack of channels available for handover rather than to interference. This can be a particular problem in microcellular systems, where the rate of handovers is significantly higher than that in normal cellular systems.

There are several known solutions to reduce the performance penalty caused by handovers. One of the simplest solutions is to reserve some channels exclusively for handovers, commonly referred to as cutoff priority [236, 284, 285] or guard channel [286] schemes. However, this solution reduces the maximum amount of carried traffic or system capacity and hence yields increased new call blocking. The guard or handover channels do not need to be permanently assigned to cells; they are invoked from an "emergency pool."

Algorithms that give higher priority to requests for handovers than to new calls are called Handover prioritization schemes. Guard channel schemes are therefore a type of handover prioritization arrangement. Another type of handover prioritization is constituted by handover queuing schemes [245, 246, 284, 285]. Normally, when an allocation request for handoff is rejected, the call is forcibly terminated. By allowing handover allocation requests to be queued temporarily, the forced termination probability can be reduced. The simplest handover queuing schemes use a First-In First-Out (FIFO) queuing regime [285]. Tekinay et al. [236] have suggested a nonpreemptive priority handover queuing scheme in which handover requests in the queue that are the most urgent ones are served first.

A further alternative to help reduce the probability of handover failure is to allow allocation requests for new calls to be queued [286]. New call allocation requests can be queued more readily than handovers because they are less sensitive to delay. Handover queuing reduces the forced termination probability owing to handover failures but increases the new call blocking probability. New call queuing reduces the new call blocking probability and also increases the carried teletraffic. This is because the new calls are not immediately blocked but queued, and in most cases they receive an allocation later.

8.2.5 The Effect of Transmission Power Control

Transmission power control is an effective way of reducing co-channel interference while also reducing the power consumption of the mobile handset. Jointly optimizing transmission power control with the channel allocation decisions is promising in terms of increasing spectral efficiency. However, little research has been done into this area, apart from a contribution by Chuang and Sollenberger [275] showing the potential benefits. Transmission power control, like channel allocation, can be implemented in a centralized [287, 288] or distributed [289] manner.

An alternative fixed channel allocation strategy, referred to as Reuse partitioning [245], relies on transmission power control. In reuse partitioning, a cell is divided into two or more concentric subcells or zones. If a channel is used in the inner zone with transmission power control, the interference is reduced due to the reduced transmission power. Therefore, the interference from channels used in the inner zones is less than that by those channels, used in the outer zones. Channels used in the inner zones can thus be reused at much shorter distances than those utilized in the outer zones.

By combining transmission power control with dynamic channel allocation, the additional performance gains of reuse partitioning can be achieved. Using reuse partitioning with DCA is far simpler to implement than using FCA, since the system is self-configuring and does not require network reuse pattern planning.

8.3 Simulation of the Channel Allocation Algorithms

In this section, we highlight how we simulated the various channel allocation algorithms we investigated. Section 8.3.1 describes the simulation program, "Netsim," which was developed to simulate the performance of the channel allocation algorithms. The channel allocation algorithms that we simulated are described in detail in Section 8.3.2. In Section 8.3.3, we describe the performance metrics we have used to compare the performance of the channel allocation algorithms. Finally, in Section 8.3.4, we describe the model used to generate the nonuniform traffic distributions we used in our simulations.

8.3.1 The Mobile Radio Network Simulator, "Netsim"

In order to characterize the performance of the various channel allocation algorithms, we simulated a mobile radio network. The simulator program we developed is referred to as Netsim. The simulated base stations can be placed in a regular pattern or at arbitrary positions within the simulation area. Mobiles are distributed randomly across the simulation area. Each mobile can have different characteristics, such as a particular mobility model or velocity.

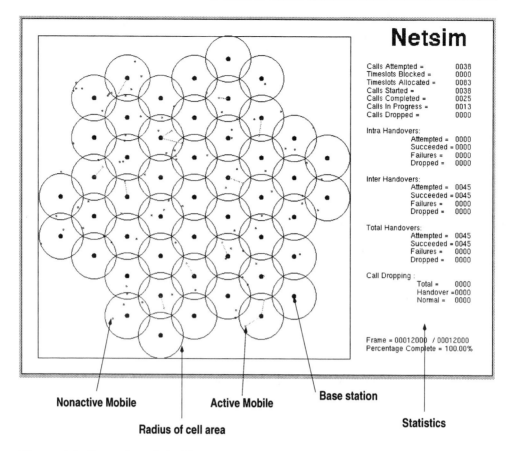

Figure 8.2: Screenshot of the Netsim program, showing 100 users in a 49-cell simulation. Each base station is located at the center of each cell, and the large circles represent the radius of the cell area. The connection between an active mobile and a base station is represented by a line.

A screenshot from the simulator is shown in Figure 8.2. The figure shows a forty-nine-base station simulation, where the cell areas are represented by circles. The mobiles are shown as small squares, and when they become active, they change color on the video screen. The connection between an active mobile and a base station is represented by a line linking the base station and the mobile. The simulator has the following features:

New Call Queuing Channel allocation requests for new calls are queued if they cannot immediately be served [286]. The new call request is blocked if its request cannot be served within a preset timeout period, referred to as the Maximum new-call queue time.

Handover Prioritization Channel allocation requests for handovers are given priority over new calls, supporting Handover Prioritization [245].

Handover Urgency Prioritization Channel allocation requests for handovers are processed by each base station, so that the more urgent handovers are served first [236].

Handover Hysteresis A call will not be handed over to another base station or channel unless the new channel has a signal quality better than the current channel by at least the preset handover hysteresis level. The only exception is when the current channel quality is below the signal quality level required to maintain the call and the new channel is above this quality level, but the difference between the quality of the new and current channel is less than the hysteresis threshold.

Channel Models The simulator models each propagation channel using one of several path-loss models and a shadow fading model. The shadow fading model can be turned off if necessary.

Call Generation Model Each mobile's activity is described by how much of the time the mobile is active (i.e., making a call). The activity of each mobile is controlled by two parameters, average call duration and average intercall time. The average call duration is the long-term mean of the length of all the calls made by the mobile. The duration of all the calls made by the mobile is Poisson-distributed [201, 275]. The average intercall time is the long-term mean duration of time between calls being made. Similarly to the call durations, the time between calls is also Poisson distributed [201, 275].

Edge effects The cells at the edge of the simulation area behave differently from cells near the center of the simulation area. This is because the cells near the edge have fewer neighboring cells and hence less interference. Therefore, in order to reduce the effect of these edge cells, the statistical results can be gathered only from the cells near the center of the grid (i.e., from the active cells). Furthermore, when a mobile reaches the edge of the simulation area, it is randomly repositioned somewhere else in the simulation area. In order that this does not cause handover problems, active mobiles reaching the edge of the simulation area finish their calls before they are repositioned.

Extensive Statistical data gathering The Netsim simulator stores a large range of statistics from each simulation. For example, the probability density function of the number of simultaneous calls at each base station is stored. Furthermore, the simulation area can be divided into a fine grid, the resolution of which depends on the required accuracy of the statistical evaluation aimed for. Statistics can be gathered separately for each grid square, allowing coverage maps of the simulation area to be generated.

Warmup period When the simulation is first begun, the number of active calls is far below the normal level. There is a latency, before the number of active calls is built up to the correct level, owing to the nature of the Poisson distributed call generation models [201, 275]. Therefore, in order not to bias the results, simulations are conducted for a sufficiently long period of time before the simulation statistics can be gathered. This period of time is referred to as a warmup period.

The Netsim simulator is a network layer-based framework employing a simple physical layer model in order to reduce the complexity of the simulations, which is described in the next section.

8.3.1.1 Physical Layer Model

The physical layer, that is, the modulator and demodulator, are modeled using two parameters, Outage SINR and Reallocation SINR. The Reallocation SINR threshold is always set above the Outage SINR threshold. When the signal quality measured as signal-to-interference+noise ratio (SINR) (defined in Equation 7.10) drops below the reallocation SINR level, the mobile requests a new channel to hand over to. This handover request can be asking for another channel from the same base station to which the mobile is currently connected and is called an intra-cell handover. Alternatively, the handover can be initiated to a channel from a different base station and is called an inter-cell handover.

If, while waiting for a reallocation handover, the signal quality drops further, below the so-called Outage SINR threshold, the signal is deemed to be lost for that time period. This is referred to as an outage. If a channel is in outage for several consecutive time periods, then the call is forcibly terminated. The parameter termed the Maximum Consecutive Outage reflects the number of consecutive outages that need to occur to cause a call to be forcibly terminated.

The Reallocation SINR threshold should be set at the average SINR required to maintain marginal signal quality. The Outage SINR threshold should be set as the SINR, below which the demodulated signal cannot be decoded error free. This twin-threshold physical layer model is similar to those described by Tekinay *et al.* [246] and by Katzela *et al.* [245]. The difference is that our model is based on SINR thresholds instead of received power thresholds used in these references. Since the computational complexity would be too high to simulate fast Rayleigh fading in a network-layer simulation, the SINR threshold of the physical layer model should include a margin to emulate the effects of fast fading, thereby increasing the required outage level.

The simulator calculates the probability of outage as the proportion of time in which a channel was below the Outage SINR threshold (i.e., in outage). The simulator can also calculate the low signal quality probability, as the proportion of time a channel is below the Reallocation SINR threshold.

The next section describes the model used to simulate shadow fading of the radio channels.

8.3.1.2 Shadow Fading Model

The channel model used by the Netsim simulator is fairly simple in order to reduce the computational complexity of the simulations. The channel can be modeled using a variety of path-loss models and an optional shadow fading model. This section is concerned with the shadow fading model. Network simulations are particularly

8.3. SIMULATION OF THE CHANNEL ALLOCATION ALGORITHMS

complex, since all the possible interfering channels may need to be modeled, that is, from each transmitter to every receiver tuned to the same carrier frequency at the same time.

Shadow fading can be modeled using a correlated signal, which is log-normally distributed [32]. In our previous chapters, shadow fading was modeled by using pre-calculated shadow fading signal envelopes. However, because of the high number of interfering channels, where the channels should be uncorrelated, a large number of precalculated shadow fading envelopes would be needed. This is impractical because of the associated high storage requirements, and the increased simulation time resulting from storage access delays.

We decided to invoke a method originally used to generate Rayleigh fading rather than shadow fading in order to produce the correlated log-normally distributed shadow fading envelope required. Jakes' method [36] was originally proposed to produce Rayleigh-distributed correlated signal envelope and phase. Jakes' technique is also often called the sum of sinusoids method, which uses the summation of several low-frequency sinusoids with regularly spaced phase differences in order to produce the desired signal. A signal, $r(t)$, exhibiting Rayleigh-distributed envelope or magnitude fluctuations can be produced from the complex summation of two independent Gaussian random variables, which is formulated as:

$$r(t) = X_1 + jX_2. \tag{8.1}$$

Jakes' method produces the required pair of correlated independent Gaussian distributed random variables, X_1, X_2, which are approximated by $x_1(t)$ and $x_2(t)$, given by:

$$x_1(t) = 2\left[\sum_{n=1}^{N_o} \cos(\beta_n) \cos(\omega_n t)\right] + \sqrt{2}\cos(a)\cos(\omega_m t) \tag{8.2}$$

$$x_2(t) = 2\left[\sum_{n=1}^{N_o} \sin(\beta_n) \cos(\omega_n t)\right] + \sqrt{2}\sin(a)\cos(\omega_m t) \tag{8.3}$$

$$\beta_n = \frac{n\pi}{(N_o + 1)} \tag{8.4}$$

$$N = 2(2N_o + 1) \tag{8.5}$$

$$\omega_n = \omega_m \cos\left(\frac{2\pi n}{N}\right) \tag{8.6}$$

$$\omega_m = 2\pi f_d, \tag{8.7}$$

where the functions $x_1(t)$ and $x_2(t)$ produce the in-phase and quadrature components of the Rayleigh-fading signal, $r(t)$. Both the in-phase and quadrature components are the sum of $(N_o + 1)$ oscillators, yielding the sum of sinusoids. The maximum Doppler frequency (f_d) sets the highest oscillator's frequency (ω_m), the phase of which is set by a. The remaining N_o oscillators have frequencies of less than ω_m set by ω_n, the phase of which is set by β_n. Therefore, $x_1(t)$ and $x_2(t)$ are functions of t, with parameters f_d and N_o.

Either one of the variables $x_1(t)$ or $x_2(t)$ can be used to produce the log-normally distributed shadow fading envelope $s(t)$, given by:

$$s(t) = 10^{[x_1(t)/10]} \quad \text{or} \quad s(t) = 10^{[x_2(t)/10]}. \tag{8.8}$$

In the next sections, we describe the investigated algorithms in detail.

8.3.2 Channel Allocation Algorithms Investigated

In this section, we describe the channel allocation algorithms that we have investigated in order to identify the most attractive performance trade-offs. Our simulations have concentrated on dynamic channel allocation (DCA) algorithms (Section 8.2.2). However, we have also performed experiments using a basic fixed channel allocation (FCA) algorithm (Section 8.2.1) as a benchmarker.

We investigated two classes of dynamic channel allocation (DCA) algorithms, namely, distributed and locally distributed algorithms, described previously in Sections 8.2.2.2 and 8.2.2.3. We studied four distributed DCA algorithms, which are characterized in Section 8.3.2.2, while Section 8.3.2.3 portrays the two locally distributed DCA algorithms that we investigated. In the next section, we introduce the fixed channel allocation algorithm employed.

8.3.2.1 Fixed Channel Allocation Algorithm

In order to benchmark our dynamic channel assignment (DCA) algorithms, a fixed channel allocation (FCA) scheme was required. We decided to employ a basic fixed channel assignment algorithm, which uses omnidirectional antennas and a reuse cluster size of seven cells. This structure is commonly used to provide coverage over a grid of regular hexagonally shaped cells. The frequency spectrum was divided into seven frequency sets, and one set was assigned to each cell.

Figure 8.3 shows such a reuse structure, where the shaded cells represent cells assigned the same set of carrier frequencies. The figure shows the center cell and its six first-tier interfering cells. This fixed channel allocation reuse structure provides uniform capacity across all cells, since each cell site has the same number of carrier frequencies. In the next section we describe the distributed DCA algorithms investigated.

8.3.2.2 Distributed Dynamic Channel Allocation Algorithms

In this section we highlight four well-known distributed DCA algorithms that we have studied comparatively. The most plausible technique is the Least Interference Algorithm (LIA) [273], which allocates the channel suffering from the least received instantaneous interference power; hence, it attempts to minimize the total interference within the system. More specifically, this algorithm minimizes the interference at low traffic loads but increases it at high loads. This is because at high loads the LIA algorithm will still attempt to allocate a channel to a new call, even when all the slots have a high level of interference. Again, this increases the total interference load of the system.

8.3. SIMULATION OF THE CHANNEL ALLOCATION ALGORITHMS

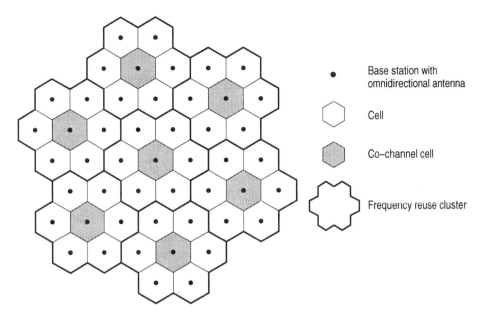

Figure 8.3: A commonly employed frequency reuse pattern for fixed channel assignment (FCA) algorithms. The frequency spectrum divided in seven frequency sets, one set assigned to each cell, yielding a seven-cell reuse cluster. Omnidirectional antennae were used, and the shaded cells represent cells assigned the same frequency set.

The second distributed DCA algorithm we studied is a refinement of the LIA algorithm, which is referred to as the Least interference below Threshold Algorithm (LTA) [273]. This algorithm attempts to reduce the interference caused by the LIA algorithm at high loads by blocking calls from using those channels, where the interference measured is deemed excessive for the transceiver to sustain adequate communications quality. The algorithm allocates the least interfered channel, whose interference is below a preset maximum tolerable interference threshold. Therefore, the LTA algorithm attempts to minimize the overall interference in the system, while maintaining the quality of each call above the minimum acceptable level.

The third algorithm we investigated attempts to utilize the frequency spectrum more efficiently while maintaining acceptable call quality. This algorithm works in a similar way to the LTA algorithm, and it is termed the Highest (or Most) interference below Threshold Algorithm (HTA or MTA) [273]. Since its goal is not to reduce the interference, but to maximize the spectral efficiency, it allocates the most interfered channel, whose interference is below the maximum tolerable interference threshold. The interference threshold is determined by the transceiver's interference resilience.

The final distributed DCA algorithm can be characterized as the Lowest Frequency below Threshold Algorithm (LFA) [273]. This algorithm is a derivative of the LTA algorithm, the difference being that the LFA algorithm attempts to reduce the number of carrier frequencies being used concurrently. This has the advantage that, statistically speaking, fewer transceivers may then be required at each base station. The algorithm allocates the least interfered channel below the maximum tolerable interference threshold, while also attempting to reduce the number of carrier frequencies used. Therefore, no new carrier frequency is invoked from the set of carriers, unless all the available time-slots on the currently used carrier frequencies are considered too interfered. In the next section, we describe the two locally distributed DCA algorithms, whose performance we have compared to the above algorithms using simulations.

8.3.2.3 Locally Distributed Dynamic Channel Allocation Algorithms

We have investigated the performance of two locally distributed dynamic channel allocation algorithms, both of which are quite similar. The Locally Optimized Least Interference Algorithm (LOLIA) attempts to reduce the overall interference in a system, like the LIA and LTA algorithms, while the Locally Optimized Most Interference Algorithm (LOMIA) attempts to increase the spectral efficiency in a similar way to the HTA algorithm.

Specifically, the locally distributed DCA algorithms constitute a hybrid of distributed and centralized channel allocation decisions. They exploit the information provided by neighboring base stations in order to improve the channel allocation decisions, which constitute the centrally controlled part of the distributed/centralized hybrid solution. Their complexity is therefore somewhere between that required for centralized and distributed algorithms.

The LOLIA algorithm carries out its channel allocation decisions in the same way as the distributed LIA algorithm. However, it will not allocate a channel, if it is used in the nearest "n," neighboring cells by another subscriber. Therefore, the nearby base stations exchange information concerning the channels that are currently being used. This requires a fast backbone network but does not rely on central control. The overall level of interference in the system can be reduced by increasing the number of cells, which are classed as neighboring cells. However, the larger "n." the more calls are blocked, since there will be fewer available channels, which are not being used in the nearest "n" base stations. Figure 8.4 shows the arrangement of neighboring cells for $n = 7$ and $n = 19$. The "n" parameter of the algorithm effectively imposes a minimum reuse distance constraint on the algorithm.

The second locally distributed DCA algorithm we consider is similar to LOLIA, but it is based on the HTA and not the LIA distributed algorithm. The LOMIA algorithm picks the most interfered channel, provided that this channel is not used in the nearest "n" neighboring cells. The LOLIA and LOMIA algorithms are similar to those proposed by De Re et al. [281] and ChihLin et al. [280].

Having described the algorithms that we have simulated in order to identify the performance trade-offs of the various channel allocation algorithms, in the next section we describe the metrics used to compare the performance of the various algorithms.

8.3.3 Performance Metrics

Several performance metrics can be used to quantify the performance or quality of service provided by a particular channel allocation algorithm. The five performance metrics defined below have been widely used in the literature [273], and we also opted

8.3. SIMULATION OF THE CHANNEL ALLOCATION ALGORITHMS

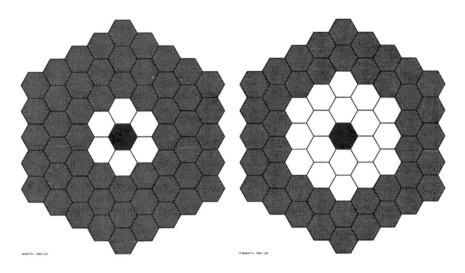

(a) 7 nearest base stations monitored

(b) 19 nearest base stations monitored

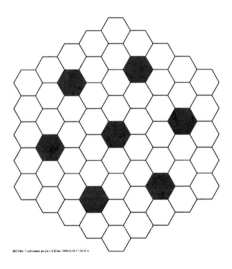

(c) 7-cell cluster FCA

Figure 8.4: The nearest neighbor constraint for $n = 7$ and $n = 19$ for the locally optimized algorithms, LOLIA and LOMIA, compared to a seven-cell reuse cluster for FCA.

for their employment:

- New call blocking probability, P_B
- Call dropping or forced termination probability, P_D or P_{FT}
- Probability of low-quality connection, P_{low}
- Probability of outage, P_{out}
- Grade of Service, GOS

The new call blocking probability, P_B, is defined as the probability that a new call is denied access to the network. This may be the case because there are no available channels or the channel allocation algorithm decided that to allow the new call to access any of the available channels would cause increased interference, which might lead to loss of the new call or calls in progress. Ideally, a low call blocking probability is desired. However, it is even more undesirable when calls in progress are lost, and this is where the second performance metric, namely, P_{FT} is useful.

The call dropping probability, P_D, also widely known as the forced termination probability, P_{FT}, is the probability that a call is forced to terminate prematurely. This can be caused by excessive interference. However, generally when a channel becomes excessively interfered with, the mobile or base station will request a new channel. If no channels are available and the quality of the call degrades significantly because of interference or low signal strength, then the call may be forcibly terminated. Calls can also be forcibly terminated when a mobile moves across a cell boundary into a heavily loaded cell. If there are no available channels in the new cell to hand over to, then the call may be lost prematurely. Since premature call termination is annoying to mobile subscribers, the channel allocation algorithm should attempt to keep the call dropping probability low.

The third performance metric we have used is the probability of a low-quality connection or access, P_{low}. This is the probability that either the up-link or down-link signal quality is below the level required by the specific transceiver to maintain a good-quality connection. A low-quality access could be due to low signal strength or high interference, which is defined as:

$$\begin{aligned} P_{low} &= P\{SINR_{up-link} < SINR_{req} \text{ or } SINR_{down-link} < SINR_{req}\} \\ &= P\{min(SINR_{up-link}, SINR_{down-link}) < SINR_{req}\}. \end{aligned} \quad (8.9)$$

This metric allows different channel allocation algorithms, which may have similar call dropping and blocking probability to be compared, in order to identify which is better, when calls are in progress. The quantity $SINR_{req}$ is the required reallocation SINR threshold described in Section 8.3.1.1. The probability of outage is similar to the probability of low communications quality metric (P_{low}), which was defined in Equation 8.9, except in this case the quantity $SINR_{req}$ is the required SINR value, below which the call is deemed to be in outage, as described in Section 8.3.1.1.

The final metric we have used to evaluate the performance of various channel allocation algorithms is the grade of service (GOS). The definition we have used is that proposed by Cheng and Chuang [273] which is stated as follows:

$$\begin{aligned} GOS &= P\{\text{unsuccessful or low-quality call accesses}\} \\ &= P\{\text{call is blocked}\} + P\{\text{call is admitted}\} \times \\ & P\{\text{low signal quality and call is admitted}\} \\ &= P_B + (1 - P_B)P_{low}. \end{aligned} \qquad (8.10)$$

The grade of service is the probability of unsuccessful network access (blocking, P_B) or low-quality access, when a call is admitted into the system (P_{low}). This performance metric is a hybrid of the new call blocking probability (P_B) and the low-quality access probability (P_{low}), when calls are not blocked and it is therefore an important performance metric. Now that we have described the algorithms and the metrics used to compare their performance, the next section describes the model used to generate nonuniform traffic distributions.

8.3.4 Nonuniform Traffic Model

Generally, investigations using fixed channel allocation assume a uniform traffic distribution and therefore a uniform carrier frequency allocation per base station. In practice some base stations have more channels, where demand is expected to be increased, for example, at airports and railway stations. However, fixed channel allocation cannot cope with unexpected traffic demand peaks [290], which are sometimes referred to as traffic "hot spots" [259]. Dynamic channel allocation algorithms are better equipped to cope with these unexpected traffic demands, since a DCA system is effectively self-adapting. Furthermore, DCA schemes typically have more potential channels available at each base station. This is an area in which DCA algorithms have a clear advantage over FCA.

Therefore we defined a model to generate a sudden unexpected traffic "hot spot" in order to measure the performance benefits that DCA algorithms provide over FCA. The model we developed is very simple and causes an increase in teletraffic in the cells affected. The model simply limits the maximum velocity of mobile terminals within a particular geographical area. Mobile users can still enter and leave a "hot spot" cell. However, since the users slow down as they enter the cell, the average cell crossing time is increased. This leads to a higher mobile terminal density in the cell, which in turn leads to increased generated teletraffic.

As an example, we refer to Figure 8.13, presented later in this chapter, in which the speed of mobiles in the gray cells is not limited by the model. For our simulations, however, the mobiles all travel at 30 mph. Upon roaming and entering the white cells, these mobiles reduced their speed to 20 mph. The white cells could represent the outskirts of a city. Upon entering the black cell, which could represent a city center, the speed of mobiles is again reduced to 9 mph. The next section presents the results obtained for the previously described channel allocation algorithms.

Parameter	Value
Noisefloor	-104 dB
Path-loss exponent	-3.5
Multiple access	TDMA
No. of time-slots	8
No. of carriers	7
Frame length	4.615 ms
BS TX power	10 dBm
MS TX power	10 dBm
Average call-length	1 min
Average inter-call-time	6 min
MS speed (uniform traffic)	13.4 m/s (30 mph)
MS speed (nonuniform traffic)	$\leq$13.4 m/s ($\leq$30 mph)
No. of base stations simulated	49
Max new-call queue-time	5 s
Handover hysteresis	2 dB
Cell radius (see Figure 8.2)	218 m

Table 8.2: GSM-like DCA System Parameters

8.4 Performance Comparisons

In Section 8.3.1, we described the mobile radio simulator program that we developed. In Section 8.3.2, we described the channel allocation algorithms we have investigated, the simulation results of which are presented in this section. In order to compare the channel allocation algorithms, the performance metrics discussed in Section 8.3.3 were used. However, let us first present the simulation parameters that we used in our comparisons.

8.4.1 System Parameters

The performance of the various channel allocation algorithms was investigated in a GSM-like microcellular system, the parameters of which are defined in Table 8.2. We used a power path-loss model with an exponent of -3.5. The number of carrier frequencies in the whole system was limited to seven, each with eight time-slots since using more carriers would further slow down our experiments. This meant that the DCA system could theoretically handle a maximum of $7 \times 8 = 56$ instantaneous calls at one base station, provided that their quality was adequate. The simulations did not use transmission power control, and both the mobile and the base stations were limited to a maximum transmission power of 10 dBm or 10 mW. Channel allocation requests for new calls were queued by the base stations, if they could not immediately be satisfied, but they were blocked if the request had not been satisfied within 5s.

All the mobiles could move in random directions at a fixed speed of 30 mph, apart from the scenario, when the nonuniform traffic model was used. The mobiles could roam freely within the simulation area (shown in Figure 8.2), which comprised regular

8.4. PERFORMANCE COMPARISONS

hexagonal forty-nine-cell grid, with a cell radius of 218 m.

We investigated the algorithms with uniform and nonuniform traffic distributions. When the nonuniform traffic model described in Section 8.3.4 was used, the speed of the mobiles in the central cells of Figure 8.13 was reduced. The call length and intercall time periods were Poisson distributed [201, 275], with the mean values shown in the table. The activity rate of the users was fairly high. On average, a user would make a one minute call every six minutes, a measure that allowed us to speed up our experiments.

For some of our experiments, the shadow fading model described in Section 8.3.1.2 was used in order to evaluate the performance over shadow fading radio channels. Every radio path between a transmitter and receiver was modeled using the path-loss model and the optional shadow fading model. The shadow fading model of different subscribers used random time offsets.

In Section 8.3.1.1, we described the physical layer model used by the Netsim mobile radio simulator. The physical layer model used two signal-to-interference-plus-noise (SINR) thresholds. These thresholds were found for BPSK, QPSK/4QAM, and 16QAM modems from performance curves, which were similar to those in Figure 19.19 that were related to the Pan-European DECT system. By contrast, here we used the thresholds required by our video system of Chapter 19 and the parameters, which were summarized in Table 7.1. The physical layer simulations used Rayleigh fading and channel coding, the parameters of which are described in that table.

The reallocation SINR threshold was determined by simulations as the average SINR required to maintain a 5% transmission FER. The loss of 5% of the video packets was found in Chapter 19 to be the maximum amount of loss, before the video quality degraded noticeably. Therefore, by setting the reallocation threshold at this level, the system asked for a new channel before the video quality degradation became perceivable.

The other SINR threshold in the physical layer model sets the level, below which the system declares that the radio channel has degraded to such a level as to cause a service outage. If the radio channel continues to be in outage, then the call is forcibly terminated. The call dropping or outage SINR threshold was determined by simulations to be the average SINR required to maintain a 10% FER. This would mean that if the radio channel degraded so that at least 10% of the video packets were lost for some period of time, then the call would be forcibly terminated. The SINR thresholds for the three modems investigated in our channel allocation simulations are shown in Table 8.3. Having described the simulation parameters, in the next section we present our simulation results quantifying the amount of traffic that can be carried by each channel allocation algorithm.

8.4.2 Carried Traffic with Quality Constraints

The first comparison of the various channel allocation algorithms was based on the carried traffic for a range of offered traffic values, which is shown in Table 8.4, quantifying how much extra capacity each DCA algorithm can provide in comparison to fixed channel allocation (FCA) within the preset quality constraints. We decided on using two different scenarios for comparing our results:

Modulation Scheme	Reallocation SINR Threshold (dB) for 5% FER	Outage SINR Threshold (dB) for 10% FER
BPSK	17	13
4QAM	21	17
16QAM	27	24

Table 8.3: Physical Layer Model (Section 8.3.1.1) Minimum Required SINR Thresholds, Found by Simulation for BPSK, QPSK/4QAM, and 16QAM modems. The Physical layer parameters are summarized in Table 7.1. The reallocation threshold is the SINR, below which a channel reallocation will be requested. The Outage threshold is the SINR, below which a service outage is declared. Successive service outages cause the call to be forcibly terminated.

- A *conservative scenario*, where the maximum acceptable value was 1% for the forced termination probability P_{FT} and 1% for P_{low}.

- A *lenient scenario*, in which the forced termination probability P_{FT} still must be less than 1%, but the maximum tolerable percentage for P_{low} was 2%.

These scenarios allow us to determine the maximum average carried traffic each algorithm can support, while maintaining the required system quality. The offered traffic is commensurate with the number of mobile users in the system. The carried traffic in the table is displayed as the normalized mean traffic, which is defined as the mean Erlang-traffic per square kilometer per megahertz ($Erlang/km^2/MHz$). The mean Erlang-traffic was found by simulation, from the mean number of simultaneous calls within the forty-nine-cell simulation area. Table 8.4 portrays the maximum carried traffic and the maximum number of mobile subscribers that can be supported, while maintaining the quality constraints described above.

The table shows that the DCA algorithms can carry more teletraffic than the FCA algorithm. Some of the DCA algorithms are limited by the forced termination (P_{FT}) constraint, and others by the low communications quality (P_{low}) constraint. It can be seen that the locally optimized least interference algorithms (LOLIA) achieve the highest carried traffic in both the conservative and lenient scenarios. It can be seen in the table that the large exclusion zone of nineteen base stations gives the $n = 19$ LOLIA a slight performance advantage over the $n = 7$ LOLIA. However, this is at the expense of a higher new call blocking probability.

8.4.3 Comparing the LOLIA with FCA

We began our investigations by comparing the fixed channel allocation (FCA) and the locally optimized least interference DCA algorithm under uniform traffic conditions. The fixed channel allocation algorithm employed a seven-cell reuse cluster, corresponding to one carrier frequency per base station. The LOLIA used seven nearest base stations, that is, $n = 7$, which statistically speaking has the promise of tighter channel reuse than the FCA scheme. Figure 8.5 shows the new call blocking and forced call termination probability for different uniform traffic loads, measured in terms of

8.4. PERFORMANCE COMPARISONS

Algorithm	Conservative $P_{FT}=1\%, P_{low}=1\%$			Lenient $P_{FT}=1\%, P_{low}=2\%$		
	Users	Traffic	Limiting Factor	Users	Traffic	Limiting Factor
FCA	1100	92	P_{FT}	1100	92	P_{FT}
LFA	1450	130	P_{low}	1800	155	P_{low}
LOMIA (n = 19)	1500	132	P_{low}	>2000	>200	P_{low}
MTA/HTA	1600	138	P_{low}	1780	151	P_{FT}
LTA	1800	155	P_{FT}	1800	155	P_{FT}
LIA	1810	160	P_{FT}	1810	160	P_{FT}
LOLIA (n = 7)	1860	165	P_{low}	>2000	>200	P_{low}
LOLIA (n = 19)	>2000	>200	P_{low}	>2000	>200	P_{low}

Table 8.4: Maximum Mean Carried Traffic and Maximum Number of Mobile Users That Can Be Supported by Each Algorithm, While Meeting the Preset Quality Constraints for a Uniform Traffic Distribution. Carried traffic was expressed in normalized terms ($Erlang/km^2/MHz$), considering the GSM-like system of Table 8.2. The acronyms are summarized in Figure 8.1.

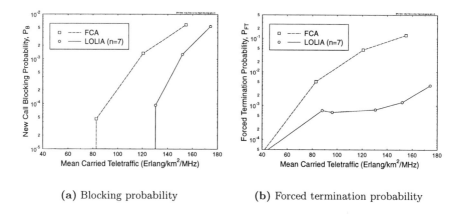

(a) Blocking probability (b) Forced termination probability

Figure 8.5: Blocking and forced termination performance versus mean carried traffic for comparison of the locally optimized least interference algorithm with seven "local" base stations and of Fixed Channel Allocation (FCA) using a seven-cell reuse cluster under uniform traffic. LOLIA has lower blocking and lower forced termination probabilities than FCA, due to its adaptability and due to having more channels available at each base station.

the mean normalized carried traffic expressed in terms of $Erlangs/Mhz/km^2$. The figure shows that LOLIA has lower blocking and lower forced termination probabilities than FCA at the same traffic load. This is due to its adaptability and to having more possible channels available at each base station.

Figure 8.6 shows the grade of service and probability of low-quality access versus a range of uniform traffic loads. Again, LOLIA outperforms the FCA algorithm in terms of a better (lower) grade of service and fewer low-quality accesses. This is again because of the adaptability of the DCA algorithm and because of the higher number of potential channels available at each base station.

Figure 8.7 portrays the mean carried teletraffic versus the number of mobiles in the simulated system. The figure shows that at low traffic loads, the FCA and LOLIA algorithm carry nearly identical amounts of traffic. However, as the mobile density and hence the traffic load increased, the LOLIA consistently carried more traffic than the FCA algorithm. This is due to the lower new call blocking and reduced forced termination probabilities experienced by the LOLIA at higher loads.

8.4.4 Effect of the "Reuse Distance" Constraint on the LOLIA and LOMIA DCA Algorithms

In this section, we quantify the effect of modifying the number of nearest base stations from which channel usage information is obtained when a channel allocation request is made in the LOLIA and LOMIA DCA algorithms. In the LOLIA and LOMIA, a channel cannot be allocated if it is being used in the local "neighborhood" of base stations. Therefore, by increasing the number of excluded nearest base stations in this

8.4. PERFORMANCE COMPARISONS

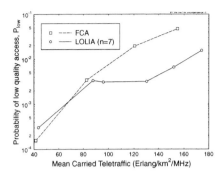
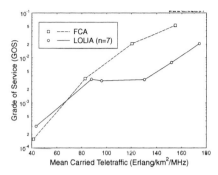

(a) Probability of low-quality access

(b) Grade of service (smaller is better)

Figure 8.6: Probability of low-quality signal (P_{low}) and Grade of Service (GOS) versus mean carried traffic for comparison of the locally optimized least interference algorithm with seven "local" base stations and fixed channel allocation (FCA) using a seven-cell reuse cluster under uniform traffic. LOLIA has fewer low-quality accesses and better (i.e., lower) grade of service than FCA, due to its adaptability and to its higher number of channels available at each base station.

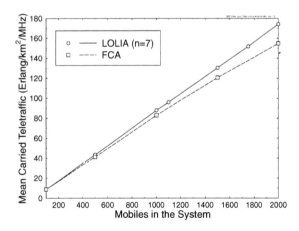

Figure 8.7: Mean carried teletraffic versus number of mobile users in the system for the locally optimized least interference algorithm with seven "local" base stations and fixed channel allocation (FCA) using a seven-cell reuse cluster under uniform traffic. LOLIA carries more traffic consistently as the offer traffic load increases.

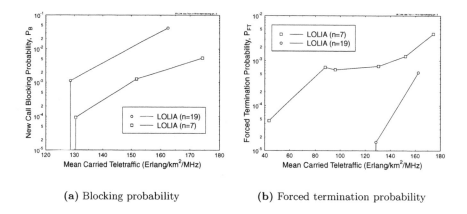

(a) Blocking probability   (b) Forced termination probability

Figure 8.8: Blocking and forced termination performance versus mean carried traffic for comparison of the locally optimized least interference algorithm with the number of "local" base stations equal to 7 and to 19 under uniform traffic. Increased P_B and reduced P_{FT} is observed for NBS = 19 rather than for NBS = 7 due to the higher reuse distance and reduced number of candidate channels.

neighborhood, the minimum reuse distance for a channel is effectively made higher.

Figure 8.8 displays the call blocking and forced termination probabilities for a variety of uniform traffic loads with a nearest base station constraint of 7 and 19, when using the LOLIA DCA algorithm. It can be seen that the new call blocking probability is higher for the larger exclusion zone scenarios. This is because when the exclusion zone is made larger, the number of available channels not used in the neighborhood is lower, causing a higher new call blocking. However, use of the larger exclusion zone does reduce the forced call termination probability significantly. This is because by increasing the size of the exclusion zone, the effective minimum reuse distance of the co-channel users is increased, thereby reducing the interference and yielding a reduced forced termination of calls due to interference. The reduced forced termination probability is also due partly to fewer calls being allowed to access the system, which is caused by the higher new call blocking probability.

Figure 8.9 shows the probability of low-quality access and the grade of service (GOS) for a range of uniform traffic loads, measured in terms of the carried teletraffic for the LOLIA DCA algorithm. The figure shows results for the nearest base station constraint of seven and nineteen cells. It can be seen that the number of low-quality accesses is higher, when the exclusion zone is small because the effective reuse distance is lower, inflicting more co-channel interference. The grade of service is better (i.e., lower) for larger exclusion zones when the traffic load is low, which is reversed for high traffic loads. This is attributable mainly to the higher call blocking probability of the larger exclusion zone of nineteen cells, particularly in the region of the highest traffic load, corresponding to 2000 users, as seen in Figure 8.8(a).

The LOMIA algorithm required a larger exclusion zone, that is, a higher effective reuse distance, in order to obtain similar call blocking and forced termination

8.4. PERFORMANCE COMPARISONS

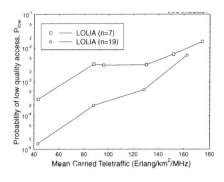
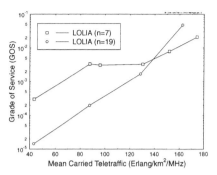

(a) Probability of low-quality access

(b) Grade of service (smaller is better)

Figure 8.9: Probability of low-quality signal (P_{low}) and Grade of Service (GOS) versus mean carried traffic for comparison of the locally optimized least interference algorithm with the number of "local" base stations equal to 7 and to 19 under uniform traffic.

probabilities. This is because the LOMIA algorithm picks the most interfered channel available within the effective reuse distance. Simulations were carried out with a nineteen-cell exclusion zone, within a forty-nine-cell structure, as for the LOLIA algorithm. When the exclusion zone was increased to thirty-seven cells for the LOMIA algorithm, the number of cells in the simulation area was increased from 49 to 133 in order to reduce edge effects. In Figure 8.10 the results are shown versus the normalized carried traffic, which is defined in terms of Erlangs/Km^2/MHz. Therefore, the results are independent of the size of the simulation area used for the LOMIA ($n = 37$) investigations.

Figure 8.10 shows the call blocking and forced termination probabilities versus carried teletraffic for exclusion zones of nineteen and thirty-seven cells, using the LOMIA algorithm. As expected, the LOMIA algorithm exhibits lower call blocking but more forced terminations, when the exclusion zone is smaller. The call blocking is higher when the exclusion zone is larger, since there are fewer channels available that are not used in the local neighborhood. The forced call termination probability is lower when the exclusion zone is larger because the effective reuse distance is larger, and hence the co-channel interference is reduced.

8.4.5 Comparison of the LOLIA and LOMIA with the LIA

In this section, we compare the locally distributed DCA algorithms, LOLIA and LOMIA, with the simplest distributed DCA algorithm, the least interference algorithm (LIA). The locally distributed algorithms are expected to perform better than the corresponding distributed algorithm, because they have additional knowledge in order to assist in their decisions. The locally distributed algorithms are aware of the channel allocations made by the nearby base stations. However, this additional knowledge

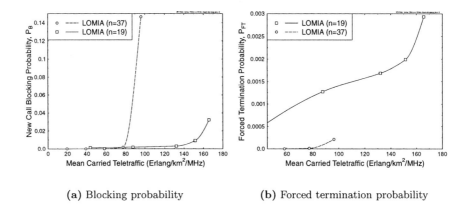

(a) Blocking probability   (b) Forced termination probability

Figure 8.10: Blocking and forced termination performance versus mean carried teletraffic for comparison of the locally optimized most interference algorithm with the number of "local" base stations equal to 19 and to 37 under uniform traffic. Lower P_B and higher P_{FT} is observed for NBS=37 than for NBS=19 due to the high reuse distance and reduced number of candidate channels.

comes at the expense of additional complexity and cost required in order to signal all the channel allocations between the neighborhood of base stations. Therefore, the LOLIA should perform better than the corresponding distributed algorithm, the LIA.

Figure 8.11 shows the call blocking and forced termination probabilities versus traffic load for the LOLIA, LOMIA, and LIA DCA algorithms. Although the LOLIA and LIA have similar call blocking performances, the LIA is slightly better, especially at higher traffic loads. This is because the LOLIA will have fewer available channels at higher traffic loads, leading to an increased call blocking probability. The LOMIA algorithm has higher call blocking (i.e., lower performance) than both the LOLIA and LIA algorithms. This is due to the high reuse distance required by the LOMIA for maintaining an adequate call quality. The higher reuse distance is implemented by having a larger exclusion zone from which the BS receives channel allocation information. Hence, fewer channels will be available, leading to the increased call blocking probability experienced.

The figure also shows that both LOMIA and LOLIA achieve lower forced call termination probabilities than the LIA algorithm. This is because the interference is more easily controlled in the LOLIA and LOMIA algorithms by the exclusion zone constraint. It should be noted that LOLIA achieves a better call termination performance than LOMIA.

8.4.6 Interference Threshold-Based Distributed DCA Algorithms

Recall that channel allocation algorithms were classified in Figure 8.1. Interference threshold-based DCA algorithms rely on an interference threshold in order to main-

8.4. PERFORMANCE COMPARISONS

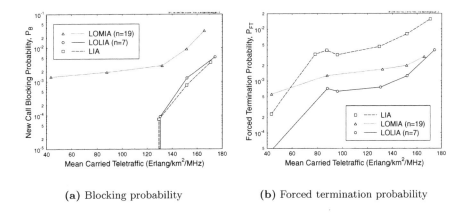

(a) Blocking probability (b) Forced termination probability

Figure 8.11: Comparison of DCA algorithms: Locally Optimized Most Interference Algorithm (LOMIA) , Locally Optimized Least Interference Algorithm (LOLIA), and Least Interference Algorithm (LIA) under uniform traffic.

tain the minimum required signal quality. An interference threshold-based channel allocation algorithm will not allocate a channel if the measured interference in that channel is higher than the maximum tolerable interference threshold. The interference threshold must be above the noise floor; otherwise no channel can be allocated. The closer the interference threshold is to the noise floor, the less the interference in the system. Increasing the interference threshold allows more interfered channels to be allocated to channel requests. The effective reuse distance of interference threshold-based DCA algorithms is dependent on the interference threshold. As the interference threshold is increased, the effective reuse distance is decreased, allowing more calls to be handled at the expense of higher interference.

We found that the optimal interference threshold for the LTA, MTA, and LFA algorithms was extremely sensitive to traffic loads and propagation conditions. This meant that the interference threshold would need to be adaptable to changing conditions. This would render the interference threshold-based distributed DCA algorithms more complex. The threshold-based algorithms require two interference thresholds – one for admitting new calls and one for handovers of calls in progress. Generally, the interference threshold is adjusted to give a higher priority to handovers than to new calls, which is another form of Handover prioritization. A range of interference thresholds was investigated for the three interference threshold-based algorithms, namely, for LTA, MTA, and LFA, using various traffic loads in order to find the optimal threshold values. To simplify the optimization process, the difference between the handover and new call interference threshold was fixed at 5dB.

Figure 8.12 shows the new call blocking and forced termination probabilities versus traffic load for the above three interference-based distributed DCA algorithms, namely, for LTA, LFA, and MTA and for the two locally distributed algorithms, namely, for LOLIA and LOMIA. The LTA and LOLIA have the best new call blocking performance. This is because the LTA and LOLIA attempt to reduce interference

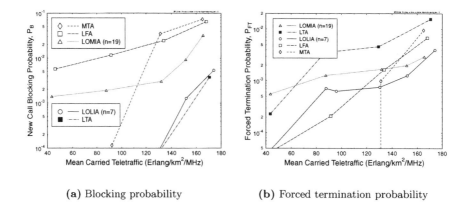

(a) Blocking probability (b) Forced termination probability

Figure 8.12: Comparison of interference threshold-based distributed DCA algorithms (LTA, MTA, LFA) with locally distributed DCA algorithms (LOLIA, LOMIA) using a uniform traffic distribution. Acronyms: Locally Optimized Most Interference Algorithm (LOMIA), Locally Optimized Least Interference Algorithm (LOLIA), Least interference below Threshold Algorithm (LTA), Most interference below Threshold Algorithm (MTA), and Lowest Frequency below threshold Algorithm (LFA).

in the system. Channel allocations are blocked if the interference is above the the maximum tolerable interference threshold. Therefore, by reducing the interference in the system, more channel allocations can be made. The LFA has a fewer new call blocking performance than the LOMIA, though statistically speaking the LFA is using a lower number of carrier frequencies. The MTA algorithm attempts to allocate the most interfered channel available, having an interference level below the acceptable threshold. When the traffic load is low, there are lots of unused channels, which exhibit an interference level below the interference threshold, and these are allocated to requests by the MTA. Therefore, at low traffic loads the MTA behaves like the LIA. This explains its low call blocking probability at low traffic loads, which degrades to the worst new call blocking performance at high loads.

Figure 8.12(b) shows the forced termination probability versus traffic load. It should be noted at high traffic loads that the locally distributed DCA algorithms, LOLIA and LOMIA, outperform the interference threshold-based distributed DCA algorithms. The spectrum compacting nature of the MTA and LFA algorithms provides a better forced termination performance at low loads, when compared to LOLIA and LOMIA. The LTA algorithm performs the worst in terms of forced termination probability at all but very low loads.

8.4. PERFORMANCE COMPARISONS

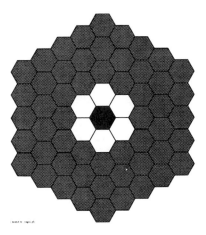

Figure 8.13: Nonuniform traffic conditions exhibiting a traffic "hot spot" in the central cell (black), and a "warm spot" (white) surrounding it. Mobiles in the gray cells move at the standard speed of 13.4 m/s (30 mph). Mobiles in the white ("warm-spot cells") can move at a speed of 9 m/s (20 mph). Mobiles in the black "hot-spot cell" are limited to a speed of 4 m/s (9 mph).

8.4.7 Performance Comparison of Fixed and Dynamic Channel Allocation Algorithms Using nonuniform Traffic Distributions

In order to show the performance benefits of DCA under nonuniform traffic conditions, we investigated the performance of FCA and some DCA algorithms using the nonuniform traffic model described earlier in Section 8.3.4. We decided to have a "hot-spot" cell, surrounded by some less heavily loaded cells for our nonuniform traffic distribution model. This is shown in Figure 8.13, where the black cell in the center is the most heavily loaded cell, since mobile terminals are limited to a maximum speed of 4 m/s (9 mph). The black hot-spot cell is surrounded by six white cells in which the maximum speed is limited to 9 m/s (20 mph). All the other cells (gray) serve mobile terminals that move with a constant velocity of 13.4 m/s (30 mph). The effect of these hot-spot cells is that mobile terminals stay longer in such cells, increasing the terminal density and hence the teletraffic.

Table 8.5 shows the maximum number of mobile subscribers that can be accommodated in the whole system while maintaining the quality constraints defined in Section 8.4.2. The table shows results for various channel allocation algorithms, for uniform (UT) and nonuniform traffic (NUT) distributions. The uniform traffic results were copied for convenience from Table 8.4. The nonuniform traffic results use the traffic generation model of Figure 8.13. The users in the system are distributed over the forty-nine-cell simulation area, which is 6 Km^2. The simulation parameters were described in Table 8.2.

Table 8.5 shows that the performance of both the fixed and dynamic channel allocation schemes is reduced under the assumption of nonuniform traffic. However,

Algorithm	Conservative $P_{FT} = 1\%$, $P_{low} = 1\%$		Lenient $P_{FT} = 1\%$, $P_{low} = 2\%$	
	UT	NUT	UT	NUT
FCA	1100	1020	1100	1020
LIA	1810	1375	1810	1715
LOMIA (n = 19)	1500	1430	> 2000	> 2000
LOLIA (n = 7)	1860	1510	> 2000	1750

Table 8.5: Maximum Number of Mobile Users That Can Be Supported Using the Algorithms Investigated While Meeting the Preset Quality Constraints for Uniform (UT) and Nonuniform (NUT) Traffic Distributions. The conservative and lenient quality constraints were defined in Section 8.4.2. The users in the system are distributed over the forty-nine-cell simulation area, which is $6Km^2$. The simulation parameters were described in Table 8.2.

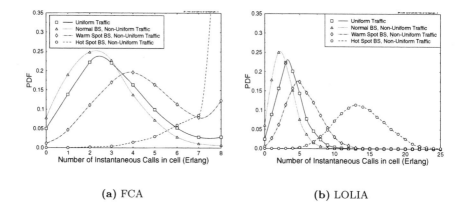

(a) FCA (b) LOLIA

Figure 8.14: PDF of the number of instantaneous calls at a base station in terms of Erlangs, for uniform traffic and for the three possible cell types in nonuniform traffic. Results are shown for the Fixed Channel Assignment algorithm (FCA) and for the Locally Optimized Least Interference algorithm (LOLIA).

all the dynamic allocation schemes achieve better performance than fixed channel allocation under uniform- and nonuniform traffic conditions.

Another way of showing the benefits of DCA algorithms over FCA under nonuniform traffic conditions is to study the PDF of instantaneous traffic at a range of base stations. Figure 8.14 shows the PDF of instantaneous traffic measured in Erlang at a range of base stations using uniform and nonuniform traffic distributions for both fixed and dynamic channel allocation algorithms. Under a nonuniform traffic distribution, the figure shows the histogram of the number of channels used at each of the three types of base stations. It can be seen that with FCA, the maximum carried traffic is limited to 8 Erlangs, as is clearly shown by the PDF for the hot- and warm-spot cells. The FCA algorithm is limited to a maximum of eight instantaneous calls (8 Erlangs), since each base station has only one carrier frequency and eight time-slots. It can be seen that the DCA algorithm can cope with traffic demands in excess of

8.4. PERFORMANCE COMPARISONS

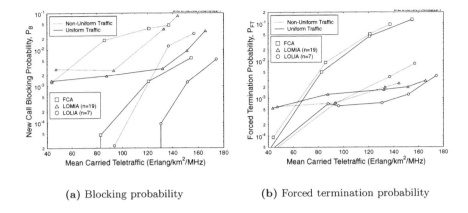

(a) Blocking probability

(b) Forced termination probability

Figure 8.15: Blocking and forced termination performance versus mean carried traffic for comparison of the Locally Optimized Least Interference algorithm (LOLIA), Locally Optimized Most Interference algorithm (LOMIA) and fixed channel allocation (FCA) under both uniform and nonuniform traffic distributions.

20 Erlangs in the hot-spot cell under nonuniform traffic. Clearly, these results have shown the benefits of dynamic channel assignment (DCA) in terms of being able to cope with unexpected peaks of traffic demand.

These results have shown that dynamic channel allocation (DCA) schemes can carry more teletraffic than fixed channel allocation (FCA) when the traffic distribution is nonuniform. However, we will now show how the performance of the various channel allocation algorithms changes in comparison to our previous results throughout Section 8.4, when subjected to nonuniform traffic distributions.

Figure 8.15 portrays the performance comparison of the FCA, LOLIA, and LOMIA under uniform and nonuniform traffic distributions. The new call blocking probability shown in Figure 8.15(a) suggests that the call blocking probability is always higher under a nonuniform traffic loading. Under a uniform traffic distribution, the LOLIA algorithm has the best blocking performance, followed by FCA and LOMIA with the worst performance. The new call blocking probability increases significantly for both the FCA and LOLIA algorithms under nonuniform traffic loading. The LOMIA algorithm is less affected by a nonuniform traffic distribution at low traffic loads, albeit it exhibits an inherently high P_B value at low traffic loads.

The comparison of the forced termination probability under uniform and nonuniform traffic loading is shown in Figure 8.15(b). The FCA algorithm experiences a small increase in forced terminations, when the traffic distribution changes from a uniform to a nonuniform distribution, although both probabilities are inherently high. The LOLIA and LOMIA algorithms have reduced forced terminations at low traffic loads for nonuniform traffic loadings. This is probably due to the algorithms being able to handle the increases in traffic, while maintaining sufficient channel reuse separation due to the low traffic load. At high traffic loads, the LOLIA and LOMIA have higher forced terminations under nonuniform traffic loads.

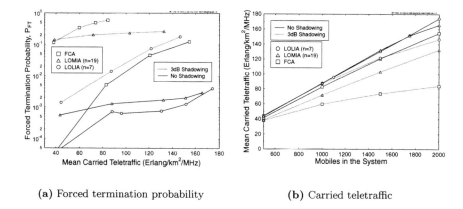

(a) Forced termination probability (b) Carried teletraffic

Figure 8.16: Forced termination performance versus mean carried traffic, and mean carried traffic versus number of mobiles for comparison of the Locally Optimized Least Interference Algorithm (LOLIA), Locally Optimized Most Interference Algorithm (LOMIA), and fixed channel allocation (FCA) with and without shadowing. The shadow fading was generated using the model defined in Section 8.3.1.2 with a frequency of 1 Hz and a standard deviation of 3 dB.

8.4.8 Effect of Shadow Fading on the FCA, LOLIA, and LOMIA

In this section, we quantify the effect of shadow fading on three channel allocation algorithms. For our shadow fading investigations, we used the log-normally distributed shadow fading model of Section 8.3.1.2, with a fading frequency of 1 Hz and a standard deviation of 3 dB. We performed simulations with and without shadow fading using the fixed channel allocation (FCA) algorithm and the two locally distributed DCA algorithms, LOLIA and LOMIA.

Figure 8.16(a) shows the forced termination probability versus mean carried teletraffic with and without shadow fading for the FCA, LOLIA, and LOMIA. As expected, for all traffic loads, the forced termination is higher when the radio channels are shadow-faded. The FCA algorithm has the worst forced termination probability with and without shadowing, except at very low loads. The best forced termination probability is obtained using the LOLIA, regardless of whether or not shadow fading is used. At low loads, despite its higher "exclusion zone" of $n = 19$, the LOMIA exhibits the worst forced termination probability. However, at high loads its performance is similar to that of the LOLIA. This figure shows that the shadow fading caused more calls to be forcibly terminated due to the more time-variant nature of the interference.

In Figure 8.16(b) the mean carried teletraffic is plotted against the number of users in the system. The figure shows that the LOLIA and LOMIA carry the most traffic. As the traffic load increases, the fixed channel allocation (FCA) algorithm cannot carry as much traffic as the LOLIA and LOMIA. When the radio channels are shadow faded, the LOLIA has the least loss of carried traffic for increasing traffic loads, which is a consequence of allocating the least interfered channels that are likely to withstand the irregular channel impairments of shadowed channels. For shadow faded radio channels of the FCA algorithm experiences the highest loss of carried teletraffic.

These results show that the performance of all three channel allocation algorithms is severely affected by shadow fading. However, the dynamic channel allocation (DCA) algorithms, particularly the LOLIA, copes best with shadow fading. In the next section, we study the effect of the shadow fading frequency and standard deviation on the LOLIA.

8.4.9 Effect of Shadow Fading Frequency and Standard Deviation on the LOLIA

In this section, we investigate the performance effects of the locally optimized least interference (LOLIA) channel allocation algorithm among different shadow fading conditions. We investigated the LOLIA algorithm with a seven-cell exclusion zone ($n = 7$) and 1100 users using BPSK modulation for a range of shadow fading frequencies and standard deviations. The BPSK users were using the BPSK SINR thresholds in Table 8.3. The shadow fading frequencies investigated were 0.25, 0.5, 1, and 2 Hz. The standard deviation of the shadow fading was varied in the range of 1–6 dB.

The forced termination probability is plotted versus the standard deviation of the shadow fading for the range of shadow fading frequencies used in Figure 8.17(a). The figure shows the expected increase of the forced termination probability, as the standard deviation of the shadow fading increases. Except at small standard deviations, where the forced termination probability is very small, the forced termination probability is higher if the shadow fading frequency is higher, due to the highly variable interference profile.

Figure 8.17(b) shows the mean carried teletraffic plotted versus the standard deviation of the shadow fading for the range of shadow fading frequencies. The figure indicates that the carried teletraffic is approximately constant, until the standard deviation of the shadow fading exceeds about 4.5 dB. For large standard deviations the carried teletraffic is reduced. This is because the shadow fading varies over a wider range, inducing much larger changes in the amount of interference inflicted upon other users and, hence causing calls to be forcibly terminated and eventually reducing the carried teletraffic. The carried teletraffic performance is worse at higher shadow fading frequencies, as observed for the forced termination performance earlier.

Figure 8.18 shows the average number of handovers per call and the grade of service (GOS) for a range of shadow fading frequencies and fading standard deviations in the context of the LOLIA. The average number of handovers per call is shown to increase with the standard deviation and frequency of the shadow fading. Every time a handover is required, either because of moving between cells or because of increased interference, the chance of the call being forcibly terminated increases. Therefore, it is best to restrict handovers to those calls that are most likely to succeed. The figure shows that the number of handovers increases with both the frequency and the standard deviation of the shadow fading. However, it is the frequency of the shadow

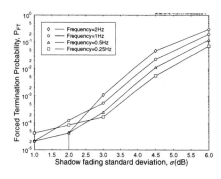 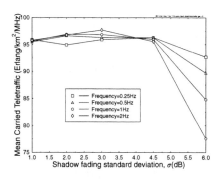

(a) Forced termination probability

(b) Carried teletraffic

Figure 8.17: Forced termination performance and mean carried traffic versus shadow fading standard deviation and frequency, using the Locally Optimized Least Interference algorithm (LOLIA) with an exclusion zone of seven cells ($n = 7$). A total of 1100 mobile users were simulated in a 49-cell area, all using the BPSK modulation thresholds of Table 8.3.

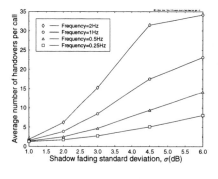 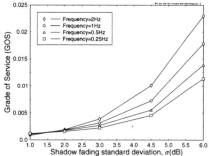

(a) Average handovers per call

(b) Grade of Service (GOS)

Figure 8.18: Average number of handovers per call and the grade of service versus shadow fading standard deviation and frequency. The Locally Optimized Least Interference algorithm (LOLIA) with an exclusion zone of seven cells ($n = 7$) was used. A total of 1100 mobile users were simulated in a 49-cell simulation area, all using BPSK modulation thresholds.

8.4. PERFORMANCE COMPARISONS

fading that has a more pronounced effect because it requires interference-induced handovers more often.

Figure 8.18(b) shows the grade of service for a range of shadow fading frequencies and standard deviations. The grade of service is degraded (i.e., increased) for higher frequencies and standard deviations of the shadow fading in the same way as the other performance metrics. In the next section, we investigate the effect of the standard deviation of the shadow fading on the FCA and LOLIA.

8.4.10 Effect of Shadow Fading Standard Deviation on FCA and LOLIA

In this section, we studied the effect of shadow fading on both fixed and dynamic channel allocation algorithms. We employed a fixed channel allocation scheme with the seven-cell reuse cluster described in Section 8.3.2.1 and the LOLIA of Section 8.3.2.3. As previously stated, the LOLIA uses information from nearby base stations in order to aid its decision making. For the LOLIA we used an exclusion zone of seven nearest base stations ($n = 7$) and 19 nearest base stations ($n = 19$). The LOLIA with $n = 19$ has approximately the same reuse distance as the FCA algorithm with the seven-cell reuse cluster, as was demonstrated earlier in Figure 8.4. However, in highly loaded situations the number of interferers that are at a distance of approximately the reuse distance could be far higher for the LOLIA than the limit of six in the case of FCA.

Figure 8.19 shows the forced termination and carried traffic performance versus the standard deviation of the shadow fading for FCA and LOLIA with $n = 7$ and 19. As the shadow fading standard deviation increases, so does the range of the interference fluctuations. In other words, as the standard deviation increases, the interference will become more aggressive and variable. Furthermore, the chance of a handover to base stations, which are not geographically the closest ones, becomes more likely. Handovers to a base station, when the mobile is not in that cell's designed coverage area, is not a problem in the context of DCA algorithms since they can adapt. However, such handovers can spuriously reduce the designed reuse distance in fixed channel allocation. A standard deviation of zero indicates the case where shadow fading was not used.

Figure 8.19(a) shows the forced termination probability for the three-channel allocation algorithms. As expected, fixed channel allocation (FCA) is outperformed by the dynamic channel assignment (DCA) algorithms. Furthermore, the LOLIA algorithm with the more stringent reuse distance constraint ($n = 19$) performs better than the LOLIA algorithm with an exclusion zone of $n = 7$. Although not shown in the figure, the FCA simulations exhibit a marked increase in the number of intra-cell handovers or handovers within the same cell. This is a sign of the interference causing call outages and requesting an emergency intra-cell handover.

The high forced termination probability of the FCA algorithm causes a drastic reduction in carried traffic, as shown in Figure 8.19(b). The figure shows that the carried traffic using the FCA algorithm is lower than that of the LOLIA, even when there is no shadowing ($\sigma = 0$ dB). However, as the shadowing standard deviation (σ) increases, the carried traffic reduces far more rapidly for the fixed channel allocation (FCA). The carried traffic remains nearly constant for the LOLIA for standard

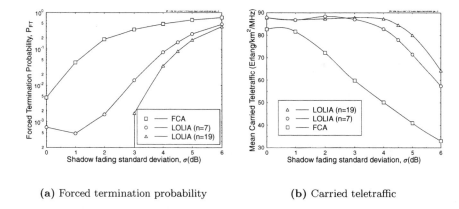

(a) Forced termination probability   (b) Carried teletraffic

Figure 8.19: Forced termination probability and mean carried traffic versus shadow fading standard deviation. Simulations use the Locally Optimized Least Interference (LOLIA) dynamic channel allocation algorithm with an exclusion zone of 7 and 19 cells ($n = 7, 19$), and fixed channel allocation. A total of 1000 mobile users were simulated in a 49-cell simulation area, all using 4QAM/QPSK modulation thresholds.

deviations less than 3 or 4 dB. The LOLIA with the more stringent minimum reuse distance ($n = 19$) carries a little more traffic than the $n = 7$ LOLIA at higher standard deviations.

Figure 8.20 displays some quality-of-service performance metrics. The grade of service is the probability that a new call is blocked, or if it is allowed access, the quality of the channel provided is poor, which is shown in Figure 8.20(a). The figure demonstrates that for small shadowing standard deviations, the LOLIAs have a better (lower) grade of service than the FCA algorithm. This is because the LOLIA algorithms are more flexible and can cope better with nonlocal inter-cell handovers.

At higher standard deviations, however, the fixed channel allocation (FCA) algorithm has a better performance. This is because the LOLIAs are more flexible and have more channels available at each cell site and therefore can perform handovers, where FCA would not be able to. In FCA, the lack of handovers causes increased interference, and thus more calls are forcibly terminated rather than degrading the GOS and the outage performance. In contrast, the LOLIA can perform such handovers, reducing the number of forcible terminations but increasing the interference, and so the number of outages.

The probability of outage, which was defined in Section 8.3.3, is shown in Figure 8.20(b), exhibiting a similar tendency to the grade of service for the same reasons. In terms of grade of service (GOS) and outage probability, the LOLIA with the increased minimum reuse distance ($n = 19$) always outperforms the smaller reuse distance scenario ($n = 7$). In the next section, we analyze the distribution of the signal-to-interference+noise ratio (SINR) profiles across the cells.

8.4. PERFORMANCE COMPARISONS

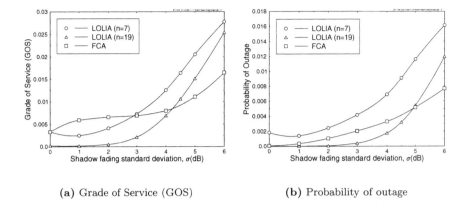

(a) Grade of Service (GOS)

(b) Probability of outage

Figure 8.20: Grade of Service (GOS) performance and probability of outage versus shadow fading standard deviation. Simulations use the Locally Optimized Least Interference algorithm (LOLIA) with an exclusion zone of 7 and 19 cells ($n = 7, 19$), and fixed channel allocation. A total of 1000 mobile users were simulated in a 49-cell simulation area, all using the 4QAM/QPSK modulation thresholds of Table 8.3.

8.4.11 SINR Profile across Cell Area

In addition to the performance metrics we derived in Section 8.3.3, the mobile radio simulator Netsim can provide a range of other statistical characteristics, such as the signal quality within the cells. The simulator divides the service area into a fine grid, and for every time interval when a mobile is in a specific grid square, the statistics for that grid square are updated. For each grid square, the minimum, maximum, and average values for the SIR, SINR, SNR, and interference signal strength are recorded. In this section, we compare the up-link SINR profiles of the FCA and LOLIA algorithms. For the LOLIA algorithm we used an exclusion zone of nineteen cells ($n = 19$). This LOLIA setup uses information from the nineteen nearest base stations and hence has approximately the same reuse distance as the FCA algorithm with a seven-cell reuse cluster that was used. However, again in highly loaded situations, the number of interferers that are at a distance of approximately the reuse distance could be far higher for the LOLIA than the limit of six in the case of FCA.

Figure 8.21 shows the average up-link SINR profile across several cells in the middle of a forty-nine-cell simulation. Results are presented for FCA and for the LOLIA ($n = 19$), when there are 1000 mobiles in the system area of forty-nine-cells. The mobiles do not use transmission power control. Therefore, the peaks in the SINR profiles represent the positions of base stations. The results are shown both in a 3D form to portray the macroscopic nature of the cell coverage and in a contour form to allow more exact measurements and comparisons to be made. In the contour plots, the contours are 5 dB apart and are labeled at every 20 dB. Figure 8.21(b) and 8.21(d) show that, for example, the 35 dB contours are farther from the base station for the LOLIA than for FCA. This implies that in the area between cells, where the coverage is normally bad, the LOLIA algorithm is typically able to maintain a higher SINR and thus a better signal quality.

A comparison between the algorithms can be seen more clearly in the shaded contour plots shown in Figure 8.22. The figure displays the same average up-link SINR profile as in Figure 8.21. However, the contour intervals have been shaded to allow a comparison to be more easily made. Each shaded contour interval represents a range of SINR values. For example, the base stations are within the 42–146 dB SINR contour interval. The contours were found using the histogram equalization[2] technique [291] in order to show the fine detail of the SINR profile between base stations. It can be seen in Figure 8.22(b) that the LOLIA has a tiny fraction of the area, where the average SINR was below 30 dB. However, the FCA algorithm shown in Figure 8.22(a) has a large area, where the SINR is below 30 dB. Furthermore, the FCA algorithm has some considerable area between base stations, where the SINR is even lower, between 23 and 26 dB. This figure shows the performance improvement in terms of average SINR that the LOLIA achieves over fixed channel allocation (FCA). So far we have shown the average up-link SINR experienced by each tiny grid element in the simulation area. Next, we present results for the minimum up-link SINR experienced in each tiny grid element.

Figure 8.23 shows the minimum up-link SINR profile for the same simulations, which produced Figure 8.21. The minimum SINR profile identifies the areas where the minimum SINR is very low, and these are the areas where outages are more likely to occur. The figure shows that the minimum SINR is less predictable for the LOLIA, since the contours are not as smooth as for FCA, and the 3D view is fuzzier, than with the FCA algorithm. However, the LOLIA seems to achieve a higher minimum SINR than the FCA algorithm, as evidenced by the contour plots in Figure 8.23(b) and 8.23(d). In these figures the contours are at 5 dB intervals, with every 20 dB interval being labeled. The 40 dB contours are so small that they are not labeled in the FCA results shown in Figure 8.23(b), whereas the 40 dB contour area is considerably larger for the LOLIA as characterized in Figure 8.23(d). Furthermore, the 20 dB contour is more erratic in the LOLIA simulations, but it is farther from the base stations, indicating a better signal quality between the base stations than for FCA.

The performance advantage of the LOLIA over the FCA algorithm is more clearly demonstrated in the shaded contour plots of Figure 8.24. The figure shows the same minimum up-link SINR profile as in Figure 8.23. In Figure 8.24(a), the minimum SINR between base stations is within the 13–19 dB contour for the FCA simulation. However, in the LOLIA, whose results are shown in Figure 8.24(b), the fraction of the simulation area covered by the 13–19 dB contour is much smaller. It should be further noted that the good signal quality regions near the base stations are larger for the LOLIA.

The results in this section have shown that the LOLIA achieves better SINR performance across the cell area than the FCA algorithm. This has been demonstrated in terms of both average and minimum SINR cell profiles. In the next section, we

---

[2]Histogram equalization is an image transformation for contrast enhancement in order to produce an image whose brightness levels are equally distributed over the brightness scale. In other words, the brightness levels in the image are changed such that each brightness level is equally probable. This generally leads to enhancement of details, which previously had similar brightness levels.

8.4. PERFORMANCE COMPARISONS

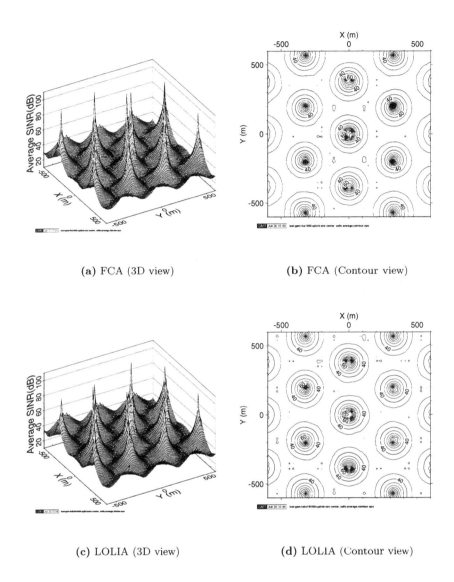

(a) FCA (3D view)　　　(b) FCA (Contour view)

(c) LOLIA (3D view)　　　(d) LOLIA (Contour view)

Figure 8.21: Average up-link SINR profile across the cell area, for several cells in the center of a forty-nine-cell simulation. The results correspond to 1000 4QAM mobiles under uniform traffic distribution and without shadow fading, using fixed channel allocation (FCA) and the locally optimized least interference algorithm (LOLIA) processing information for the nearest ninteen base stations ($n = 19$). The contour lines are 5 dB apart.

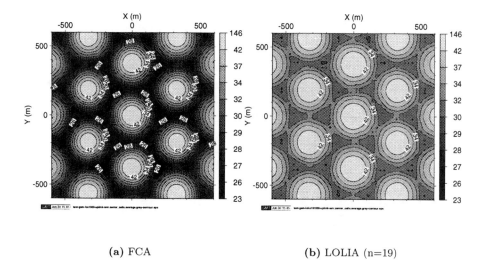

Figure 8.22: Average up-link SINR profile across the cell area, for several cells in the center of a forty-nine-cell simulation. The results correspond to 1000 4QAM mobiles under uniform traffic distribution and without shadow fading, using fixed channel allocation (FCA) and the locally optimized least interference algorithm (LOLIA) processing information for the nearest nineteen base stations ($n = 19$).

attempt to combine the different performance metrics in order to produce an overall comparison between the various channel allocation algorithms that we have investigated.

8.4.12 Overview of Results

In our previous investigations, we have simulated several channel allocation algorithms. However, no single algorithm performs best in terms of every performance metric. Therefore, in order to compare our results for the fixed and the various dynamic channel allocation algorithms, it is necessary to consider a combination of performance metrics. Sometimes an algorithm may provide excellent performance in terms of one metric but poor performance in another. Therefore, we decided to modify the conservative and lenient scenarios defined in Section 8.4.2 in order to produce a method for comparing all the results obtained. This was done by adding the grade of service (GOS) and new call blocking probability (P_B) constraints to the conservative and lenient scenarios. The full set of constraints for the two scenarios is as follows.

- A *conservative scenario*, where the maximum acceptable value for the new call blocking probability P_B is 3%, for the forced termination probability P_{FT} is 1%, for P_{low} is 1%, and for the GOS is 4%.

- A *lenient scenario*, in which the forced termination probability P_{FT} still must be

8.4. PERFORMANCE COMPARISONS

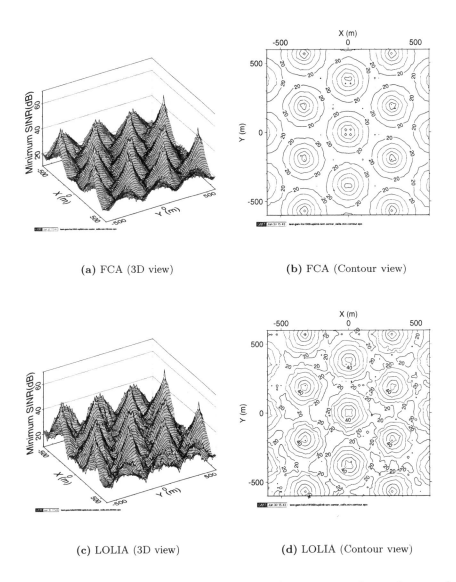

(a) FCA (3D view) (b) FCA (Contour view)

(c) LOLIA (3D view) (d) LOLIA (Contour view)

Figure 8.23: Minimum up-link SINR profile across the experimental area, for several cells in the center of a forty-nine-cell simulation. The results correspond to 1000 4QAM mobiles under uniform traffic distribution and without shadow fading, using fixed channel allocation (FCA) and the locally optimized least interference algorithm (LOLIA) processing information for the nearest nineteen base stations ($n = 19$). The contour lines are 5 dB apart.

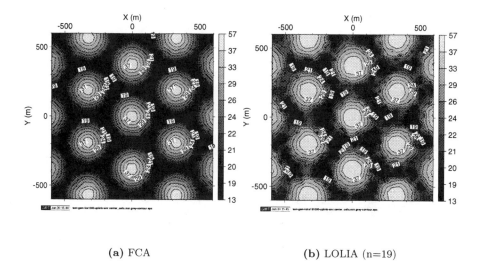

(a) FCA **(b)** LOLIA (n=19)

Figure 8.24: Minimum up-link SINR profile across the experimental area, for several cells in the center of a forty-nine-cell simulation. The results correspond to 1000 4QAM mobiles under uniform traffic distribution and without shadow fading, using fixed channel allocation (FCA) and the locally optimized least interference algorithm (LOLIA) processing information for the nearest nineteen base stations ($n = 19$).

less than 1%, but the maximum tolerable percentage for the blocking probability P_B is 5%, for P_{low} is 2%, and for the GOS is 6%.

The maximum traffic load, measured in terms of the number of users in the system that could be served, while maintaining the constraints imposed by the above two scenarios, is shown in Table 8.6, portraying results for uniform and nonuniform traffic distributions. The number of users in the table are distributed over the forty-nine-cell simulation area, which is $6km^2$ therefore, 1200 users represent a user density of 198 users/Km^2. The table shows that the DCA algorithm can cope with more users than fixed channel allocation (FCA) under both uniform and nonuniform traffic distributions.

The additional new call blocking probability (P_B) and grade of service (GOS) constraints were added to the conservative and lenient scenarios of Section 8.4.2. As seen in Table 8.6, this accounts for the reduced performance of the LFA and MTA algorithms in the lenient scenario, and that of the LOLIA ($n=19$) and MTA algorithms in the conservative scenario, when compared to Table 8.4. In addition, the two new constraints reduced the performance of the LOMIA ($n=19$) algorithm under nonuniform traffic loading, when compared to Table 8.5.

Although this table is useful for finding the amount of traffic that can be handled under certain constraints, it can give misleading impressions. For example, the two LOLIAs with different exclusion zones ($n = 7, 19$) seem to have a similar performance.

Algorithm	Conservative $P_{FT}=1\%$, $P_{low}=1\%$ GOS=4%, $P_B=3\%$		Lenient $P_{FT}=1\%$, $P_{low}=2\%$ GOS=6%, $P_B=5\%$	
	UT	NUT	UT	NUT
FCA	1100	1020	1100	1020
MTA/HTA	1425	—	1650	—
LFA	1450	—	1760	—
LOMIA (n = 19)	1500	1430	> 2000	1600
LTA	1800	—	1800	—
LIA	1810	1375	1810	1715
LOLIA (n = 7)	1860	1510	> 2000	1750
LOLIA (n = 19)	1900	—	> 2000	—

Table 8.6: Maximum Number of Mobile Users That Can Be Supported Using the Algorithms Investigated While Meeting the Desired System Constraints for Uniform (UT) and Nonuniform (NUT) Traffic Distributions. The users in the system are distributed over the forty-nine-cell simulation area, which is $6Km^2$. The simulation parameters were described in Table 8.2.

However, this is a somewhat simplistic view since the LOLIA with a large exclusion zone ($n = 19$) has a better quality of service than the other LOLIA ($n = 7$) at the expense of higher probability of blocking new calls. The two LOLIAs seem to have similar performance in the table because one of them is limited by the new call blocking constraint ($n = 19$), and the smaller exclusion zone scenario ($n = 7$) is limited by the probability of low-quality access constraint.

Table 8.6 also demonstrates that the LOLIA can maintain the required performance constraints for the highest number of users in both the conservative and lenient scenarios, and with both uniform and nonuniform traffic distributions. In addition, the table shows that the DCA algorithms that attempt to reduce the interference, like LIA, LTA, and LOLIA, typically achieve a better overall performance than the radio spectrum packing algorithms, such as LOMIA and MTA.

8.5 Summary and Conclusions

In this chapter, we have discussed various channel allocation techniques that have been proposed. We then compared the performance of a range of channel allocation algorithms and showed the potential benefits of dynamic channel allocation (DCA). Initially, in Section 8.2 we provided an overview of the various channel allocation techniques. Many channel allocation schemes are derivatives or hybrids of older schemes, and their position in this hierarchy can be seen in the family tree of Figure 8.1. In this overview, we also touched on the associated issues of call handover and power control. These issues are not trivial; hence, they are not discussed in detail in this chapter. However, power control is the subject of Chapter 20.

Having provided an overview of the various channel allocation techniques in Section 8.2, we compared the performance of some of the techniques in Section 8.3. In Section 8.3.1, we described the mobile radio network simulator, Netsim, that we developed, including the various modeling techniques used. A subset of the various channel allocation techniques was chosen for a detailed comparative study, as discussed in Section 8.3.2. A variety of performance metrics was used to compare the performance of the algorithms, and these were summarized in Section 8.3.3. The various channel allocation algorithms have different and sometimes opposite aims. Therefore, a variety of performance metrics are required to quantify their performance in particular areas. In order to show the flexibility of some of the channel allocation algorithms, a nonuniform traffic model was developed, as described in Section 8.3.4. Nonuniform traffic is handled best by the more flexible channel allocation techniques, such as LOLIA.

In Section 8.4 we presented the results obtained by comparing the various channel allocation techniques. The system parameters used to make these comparisons were discussed in Section 8.4.1. First, in Section 8.4.2 we quantified the maximum number of subscribers that could be supported by each channel allocation technique, while maintaining some preset quality constraints. This demonstrated the additional system capacity advantage that dynamic channel allocation (DCA) techniques can provide instead of fixed channel allocation (FCA). In Section 8.4.7, we compared the performance of fixed and dynamic channel allocation techniques in uniform and nonuniform traffic conditions, and we found that dynamic channel allocation techniques maintain a system capacity advantage over fixed allocation schemes. In addition, the more flexible dynamic schemes allow much higher traffic loading at an individual cell-site to be handled.

We compared all the channel allocation algorithms under shadow fading conditions. It was found that the more flexible, dynamic schemes could cope with fading conditions more readily than the fixed allocation schemes. In Section 8.4.11, we produced coverage maps of cell areas. These showed the areas where the SINR, and hence signal quality, was high. These maps highlighted how much better the average and minimum signal quality was in those systems, where dynamic rather than fixed channel allocation was used. Finally, in Section 8.4.12 we defined a set of conservative and lenient performance metrics and calculated the maximum number of mobile subscribers each channel allocation technique could sustain, while meeting these performance constraints. Again, these dynamic channel assignment schemes outperformed the fixed channel assignment scheme.

We have shown the potential benefits of dynamic channel allocation, which can cope with unexpected large increased teletraffic demands. The advantages of dynamic channel allocation are:

- Better frequency reuse.

- Ability to cope with nonuniform traffic distributions and unexpected traffic demand peaks.

- Carrier/time-slot classification can be used to aid multimode modem operation and reconfiguration.

- No frequency planning is required.

The disadvantages of dynamic channel allocation that have to be overcome in practical high-capacity systems are:

- Longer call setup and complex call control are required.

8.5. SUMMARY AND CONCLUSIONS

- Some algorithms require fast backbone networks for base station to base station signaling.

- Transceivers have to be more frequency agile, particularly at the base station, and may need more transceivers per base station.

Of all the dynamic allocation algorithms we characterized, the LOLIA gave the best performance. However, this is based on the very simple distributed LIA algorithm, and hence slightly more complex algorithms similar to LOLIA may give even better performance and increased capacity. In the next chapter, we suggest a method of increasing the system capacity using transmission power control.

A range of interesting further research problems arise, when these algorithms are combined with statistical multiplexing algorithms or when burst-by-burst adaptive modems and slot classification algorithms are invoked, where more robust modem modes can be employed in strongly interfered channels, while supporting more bandwidth-efficient modem modes in uninterfered slots — a challenging new research area.

Chapter 9

Second-Generation Mobile Systems

9.1 The Wireless Communications Scene

While the second-generation digital mobile radio systems are now widespread across the globe, researchers endeavor worldwide to define the *third-generation* personal communications network (PCN), which is referred to as a personal communications system (PCS) in North America. The European Community's Research in Advanced Communications Equipment (RACE) program [292,293] and the follow-up framework referred to as the Advanced Communications Technologies and Services (ACTS) program [294] spearheaded these initiatives. The RACE program comprised two dedicated projects; endeavoring to resolve the ongoing debate with regard to the most appropriate multiple access scheme, and studying Time Division Multiple Access (TDMA) [292,293,295] and Code Division Multiple Access (CDMA) [240,292,293, 296]. Parts II and IV of the book are focused on TDMA/TDD-based systems, while Chapter 10 is dedicated to CDMA. A range of statistical multiplexing-based multiple access techniques, such as Packet Reservation Multiple Access (PRMA), are the topics of Chapter 6.

In the pan-European ACTS program Orthogonal Frequency Division Multiplex (OFDM) [29,297], hybrid TDMA-CDMA and wideband CDMA were considered [298]. The ACTS workplan [294] included a number of projects dealing with multimedia source and channel coding, modulation, and multiple access techniques for both cellular and wireless local area networks (WLANs). A range of adaptive video systems based on TDMA/TDD, OFDM, and CDMA is portrayed in Chapter 21. These European studies designed the architecture and produced demonstration models of the various candidate systems for the universal mobile telecommunications system (UMTS) that the Europeans standardized in 1998 [244]. The corresponding Japanese initiative is known as Intelligent Mobile Telecommunications in the year 2000 (IMT2000) [299–302], while the third-generation Pan-American proposal is the cdma2000 system [303–

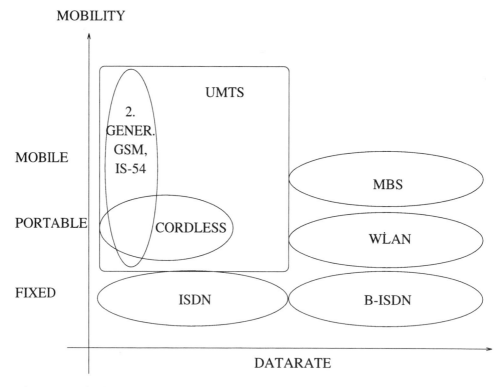

Figure 9.1: Stylized mobility versus bit-rate plane classification of existing and future wireless systems.

305]. These third-generation (3G) systems are characterized in depth in Chapter 10, while the video performance of a multimode CDMA-based video system is quantified in Section 21.3.

A common feature of the third-generation systems is the ability to support video services on a more flexible basis than the somewhat rigid second-generation (2G) standards of Table 9.1, allocating bit rates up to 2 Mbps on a demand basis. Here we refrain from detailing the various system parameters of the 2G systems, although many of the associated features are discussed in some depth throughout the book. Furthermore, the most widespread global second-generation system, namely, GSM is discussed in slightly more depth in Section 9.2. The third-generation systems are expected to be more amenable to wireless video and multimedia transmission than second-generation systems. A wide variety of further associated system aspects were treated by the following overviews [292, 293, 306–313], while various TDMA/TDD, OFDM, and TDMA-based video systems are portrayed throughout this book.

The range of existing and future systems can be characterized with the help of Figure 9.1 in terms of their expected grade of mobility and bit rate, which are the two most fundamental parameters in terms of determining the systems' potential and flexibility in wireless video multimedia applications.

Specifically, the fixed networks are evolving from the basic 2.048 Mbit/s Integrated

9.1. THE WIRELESS COMMUNICATIONS SCENE

Services Digital Network (ISDN) toward higher-rate broadband ISDN or B-ISDN. In comparison to these fixed networks, a higher grade of mobility, which we refer to here as portability, is a feature of cordless telephones (CTs), such as the Digital European Cordless Telephone (DECT) [45], the British (CT2) [44], the American PACS scheme, and the Japanese Personal Handyphone (PHP) [314] systems, although their transmission rate is more limited than that of the fixed ISDN network. The basic features of these systems are summarized on the right-hand side of Table 9.1, which will be referred to frequently during this discussion. All of the CT systems exhibit a speech coding rate of 32 kbit/s and employ a low-complexity Adaptive Differential Pulse Code Modulated (ADPCM) speech codec. At this bit rate most of the video systems proposed in Parts II and IV of the book are capable of delivering 176×144-pixel Quarter Common Intermediate Format (QCIF) or sub-QCIF interactive videotelephone signals, provided that an additional speech channel can be allocated for the video signal. The various standard video formats are summarized in Table 18.1 along with their uncompressed video bit rates.

The DECT system is the most flexible CT scheme among those in Table 9.1, allowing the multiplexing of twenty-three single-user channels in one direction, which provides bit rates up to 23×32 kbps = 736 kbps for advanced services, although only about 500 kbps can be dedicated to useful information transmission. Because of its specific importance, we have quantified the video performance of the basic 32 kbit/s DECT system in Section 19.4.3.2. However, because of their identical bit rate, the other CT systems of Table 9.1 are expected to deliver a similar video quality to DECT, provided there are no channel impairments. Naturally, the robustness of the other CT systems of Table 9.1 is different owing to their different transmission schemes. We note furthermore that the video codecs of Parts II and IV have to be programmed to operate at a fixed video coded rate of 32 kbit/s. If Forward Error Correction (FEC) coding is employed, either the video source rate must be dropped to 16 kbit/s or an additional 32 kbit/s DECT channel has to be reserved.

As suggested by Figure 9.1, wireless local area networks (WLANs) can support higher bit rates of up to 155 Mbits/s in order to extend existing Asynchronous Transfer Mode (ATM) fixed links to portable terminals, such as laptop computers. However, at the time of writing, WLANs usually do not support full mobility functions, such as location update or handover from one base station (BS) to another, particularly not across different global networks. We note here that Part III of the book is dedicated to various high-resolution, high-quality proprietary video compression schemes, while in Section 21.4 we will provide video performance results in the context of such high-rate WLAN systems, which are capable of supporting high-quality, high-resolution video signals with the aid of the H.263 standard video codec. Again, the various standard video formats are summarized in Table 18.1 along with their uncompressed video bit rates. Another ambitious European initiative is targeted at high-rate, high-mobility system studies hallmarked by the Mobile Broadband System (MBS), which is also featured in Figure 9.1. We note, however, that at the time of writing such ambitious systems are in their conceptual design phase.

By contrast, as seen in the figure, contemporary second-generation Public Land Mobile Radio (PLMR) systems, such as the Pan-European GSM [41] system of Table 9.1 and Section 9.2 or the American IS-54 [42] and the Japanese Digital Cellular

(JDC) [43] systems of Table 9.1 cannot support high bit-rate services, since they typically have to communicate over lower quality, dispersive mobile channels, but they exhibit the highest grade of mobility, including high-speed international roaming capabilities. As seen in Table 9.1, the speech coding rate of these 2G cellular systems is typically low — between 6.7 and 13 kbit/s — and definitely significantly lower than that of the 32 kbit/s rate of the CT schemes. Hence, their ability to carry video is more limited. Nonetheless, in Parts II and IV of the book we provide a wide range of video performance results at similarly low bit rates, under the assumption that an additional speech channel is provided for the transmission of the video signal. The severe bit-rate constraints imposed by these low-rate 2G cellular systems can be mitigated by transmitting sub-QCIF (SQCIF) video signals, which are constituted by half as many pixels, as QCIF video sequences, as seen in Table 18.1.

A further constraint of these 2G systems is that their speech channel operates at a constant bit rate, while the standard video codecs of Part IV typically generate a time-variant bit rate. This bit-rate incompatibility problem can be circumvented with the aid of the fixed but programmable-rate video codecs of Part II or upon invoking the adaptive packetization, buffering, and rate-control regime of Part IV. Another alternative for supporting video services over GSM is, for example, invoking the High Speed Circuit Switched Data (HSCSD) mode of GSM according to the recommendations of [315], provided that the associated delay does not affect "lip-synchronization" between speech and video. Since the speech delay of the GSM systems is about 57.25 ms, this is the ideal target video delay.

Let us now consider the most widespread second-generation system, namely, GSM in slightly more depth. The global third-generation system proposals are discussed in detail after an introduction to CDMA in Section 10.3.

9.2 Global System for Mobile Communications — GSM

9.2.1 Introduction to GSM

Here we elaborate in slightly more depth on the features of the Pan-European digital mobile cellular radio system known as GSM. Since GSM operating licenses have been allocated to well over 100 service providers in nearly 100 countries, it is justifiable that the GSM system is often referred to as the Global System of Mobile communications.

The GSM specifications were released as thirteen sets of Recommendations [210], which are summarized in Table 9.2, covering various aspects of the system [41].

After a brief system overview in Section 9.2.2 and the introduction of physical and logical channels in Section 9.2.3, we describe aspects of mapping logical channels onto physical resources for speech and control channels in Sections 9.2.4 and 9.2.5, respectively. These details can be found in Recommendations [R.05.02] and [R.05.03]. Synchronization issues are considered in Section 9.2.6. Modulation [R.05.04], transmission via the standardized wideband GSM channel models [R.05.05] as well as adaptive radio link control [R.05.06], [R.05.08], discontinuous transmission (DTX) [R.06.31] and voice activity detection (VAD) [R.06.32] are highlighted in Sections 9.2.7– 9.2.10,

9.2. GLOBAL SYSTEM FOR MOBILE COMMUNICATIONS — GSM

System	TACS	GSM [41]	DCS-1800	IS-95 CDMA [296]	IS-54 DAMPS [295]	JDC	CT2 [44]	DECT [45, 263]	PHS	PACS [316]
Origin	UK	Europe	Europe	USA	USA	Japan	UK	Europe	Japan	USA
Forward Band (MHz)	935-950	935-960	1805-1880	869-894	869-894	810-826 1477-1489 1501-1513 940-956 1429-1441 1453-1465	864-868	1880-1900	1895-1918	1930-1990
Reverse Band (MHz)	890-905	890-915	1710-1785	824-849	824-849		(TDD)	(TDD)	(TDD)	1850-1910
Multiple Access	FDMA	TDMA	TDMA	CDMA	TDMA	TDMA	FDMA	TDMA	TDMA	TDMA
Duplex	FDD	FDD	FDD	FDD	FDD	FDD	TDD	TDD	TDD	FDD
Carrier Spacing	25	200	200	1250	30	25	100	1728	300	300
Channels/carrier	1/pair	8/pair	8/pair	55-62	3	3	1	12	4	8/pair
Bandwidth/channel (kHz)	50	50	50	21	20	16.66	100	144	75	75
Modulation	FM	GMSK	GMSK	DL:QPSK;UL:64-ary	$\frac{\pi}{4}$-DQPSK	$\frac{\pi}{4}$-DQPSK	FSK	GMSK	$\frac{\pi}{4}$-DQPSK	$\frac{\pi}{4}$-QPSK
Modulation Rate (kBd)	N/A	271	271	1228	48.6	42	72	1152	192	192
Voice+FEC Rate(kbps)	N/A	22.8	22.8	8/Var.	11.2	13	32	32	32	32
Speech codec	N/A	RPE-LTP	RPE-LTP	CELP	VSELP	VSELP	ADPCM	ADPCM	ADPCM	ADPCM
Unprotected Voice Rate(kbps)		13	13	1.2-9.6	7.95	6.7	32	32	32	32
Control Chan. Name		SACCH	SACCH	SACCH	SACCH	SACCH	D	C	SACCH	
Control Chan. Rate (bps)		967	967	800	600		2000	6400		4000
Control Message Size (bits)		184	184	1	65		64	64		
Control Delay (ms)		480	480	1.25	240		32	10		10/2.5ms
Peak Power (Mobile) (W)	0.6-10	2-20	0.25-2	0.6-3	0.6-3	0.3-3	10mW	250mW	80mW	200mW
Mean Power (Mobile) (W)	0.6-10	0.25-2.5	0.03-0.25	0.2-1	0.6-3	0.1-1	5mW	10mW	10mW	
Power Control	Yes	Yes	Yes	Yes	Opt.	Yes	No	No	Opt.	
Voice Activity Detection	Yes	Yes	Yes	Yes	Yes	Opt.	No	No		
Handover	Yes	Yes	Yes	Yes	Yes	Yes	No	Yes	Yes	Yes
Dynamic Channel Allocation	No	No	No	N/A	No	Opt.	Yes	Yes	Yes	autonom.
Min. Cluster Size	7	3	3	1	7	4	N/A	N/A	N/A	N/A
Capacity(Dpx ch/cell/MHz)	2.8	6.7	6.7	16.5 †	7		N/A	N/A	N/A	N/A
Frame duration(ms)	N/A	4.615	4.615	20	40	40	2	10	5	2.5
Speech FEC		Conv. (2,1,5)	Conv. (2,1,5)	Fwd:(2,1,9) Rev: R = 1/3	Conv. (2,1,5)	Conv. R = 9/17	No	No	CRC	CRC
Channel Eq.	N/A	Yes	Yes	Yes	Opt.	Opt.	No	No	No	No
Half-rate Codec (kbps)	N/A	5.6	5.6	No	No	3.2	No	No	16	No
Half-rate+FEC (kbps)	N/A	11.4	11.4			5.6				
Enhanced Full-rate (kbps)	N/A	12.2	12.2	Yes	7.4	No	No	No	No	No

DCS-1800: GSM-like European system in the 1800 MHz band
IS-95: American CDMA system
PHS: Japanese Personal Handyphone System
TACS: Total Access Communications System
CDMA: Code Division Multiple Access
FDMA: Frequency Division Multiple Access
TDMA: Time Division Multiple Access
FM: Frequency Modulation
DQPSK: Differential Quadrature Phase Shift Keying
RPE-LTP: Regular Pulse Excited - Long Term Predicted
VSELP: Vector Sum Excited Linear Predictive
SACCH: Slow Associated Control Channel

IS-54: American Digital Advanced Mobile Phone System (DAMPS)
CT2: British Cordless Telephone System
PACS: Personal Access Communications System
GSM: Global System of Mobile Communications
DECT: Digital European Cordless Telephone
FDD: Frequency Division Duplex
TDD: Time Division Duplex
GMSK: Gaussian Minimum Shift Keying
GFSK: Gaussian Phase Shift Keying
CELP: Code Excited Linear Predicted
ADPCM: Adaptive Differential Pulse Code Modulation
JDC: Japanese Digital Cellular

N/A means not applicable.

†Assumes $\frac{2}{3}$ frequency re-use

Table 9.1: Summary of Second-Generation Mobile Systems

R.00 *Preamble* to the GSM Recommendations.

R.01 *General structure* of the Recommendations, description of a GSM network, associated recommendations, vocabulary, etc.

R.02 *Service aspects:* bearer-, tele- and supplementary services, use of services, types and features of mobile stations (MS), licensing and subscription, as well as transferred and international accounting, etc.

R.03 *Network aspects,* including network functions and architecture, call routing to the MS, technical performance, availability and reliability objectives, handover, and location registration procedures as well as discontinuous reception and cryptological algorithms, etc.

R.04 *Mobile/base station (BS) interface and protocols,* including specifications for layer 1 and three aspects of the open systems interconnection (OSI) seven-layer structure.

R.05 *Physical layer on the radio path,* incorporating issues of multiplexing and multiple access, channel coding and modulation, transmission and reception, power control, frequency allocation and synchronization aspects, etc.

R.06 *Speech coding specifications,* such as functional, computational, and verification procedures for the speech codec and its associated voice activity detector (VAD) and other optional features.

R.07 *Terminal adapters for MSs,* including circuit and packet mode as well as voiceband data services.

R.08 *Base station (BS) and mobile switching center (MSC) interface* and transcoder functions.

R.09 *Network interworking* with the public switched telephone network (PSTN), integrated services digital network (ISDN), and packet data networks.

R.10 *Service interworking, short message service.*

R.11 *Equipment specification and type approval specification* as regards MSs, BSs, MSCs, home (HLR) and visited location register (VLR) as well as system simulator.

R.12 *Operation and maintenance,* including subscriber, routing tariff, and traffic administration, as well as BS, MSC, HLR, and VLR maintenance issues.

Table 9.2: GSM Recommendations [R.01.01]

9.2. GLOBAL SYSTEM FOR MOBILE COMMUNICATIONS — GSM

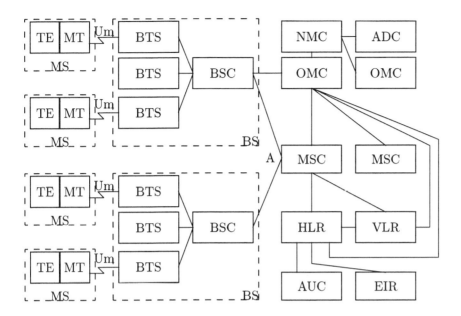

Figure 9.2: Simplified structure of GSM PLMN. ©ETT [Hanzo & Steele, 1994].

while a summary of the fundamental GSM features is offered in Section 9.2.11.

9.2.2 Overview of GSM

The system elements of a GSM public land mobile network (PLMN) are portrayed in Figure 9.2, where their interconnections via the standardized interfaces A and Um are indicated as well. The mobile station (MS) communicates with the serving and adjacent base stations (BS) via the radio interface Um, while the BSs are connected to the mobile switching centre (MSC) through the network interface A. As seen in Figure 9.2, the MS includes a Mobile Termination (MT) and a Terminal Equipment (TE). The TE may be constituted, for example, by a telephone set and fax machine. The MT performs functions needed to support the physical channel between the MS and the base station, such as radio transmissions, radio channel management, channel coding/decoding, and speech encoding/decoding.

The base station (BS) is divided functionally into a number of base transceiver stations (BTS) and a base station controller (BSC). The BS is responsible for channel allocation [R.05.09], link quality and power budget control [R.05.06], [R.05.08], signaling and broadcast traffic control, frequency hopping (FH) [R.05.02], handover (HO) initiation [R.03.09], [R.05.08], and so on. The MSC represents the gateway to other networks, such as the public switched telephone network (PSTN), integrated services digital network (ISDN), and packet data networks using the interworking functions standardized in [R.09]. The MSC's further functions include paging, MS location updating [R.03.12], and HO control [R.03.09]. The MS's mobility management is assisted by the home location register (HLR) [R.03.12], storing part of the

MS's location information and routing incoming calls to the visitor location register (VLR) [R.03.12] in charge of the area, where the paged MS roams. The MS asks for location update, whenever it detects from the received and decoded broadcast control channel (BCCH) messages that it entered a new location area. The HLR contains, among other parameters, the International Mobile Subscriber Identity (IMSI), which is used for the authentication [R.03.20] of the subscriber by his AUthentication Center (AUC). This enables the system to confirm that the subscriber is allowed to access it. Every subscriber belongs to a home network, and the specific services that the subscriber is allowed to use are entered into his HLR. The Equipment Identity Register (EIR) allows for stolen, fraudulent, or faulty mobile stations to be identified by the network operators. The VLR is the functional unit that attends to an MS operating outside the area of its HLR. The visiting MS is automatically registered at the nearest MSC, and the VLR is informed of the MS's arrival. A roaming number is then assigned to the MS and this enables calls to be routed to it. The Operations and Maintenance Center (OMC), Network Management Center (NMC) and ADministration Center (ADC) are the functional entities through which the system is monitored, controlled, maintained, and managed [R.12].

The MS initiates a call by searching for a BS with a sufficiently high received signal level on the BCCH carrier; it will await and recognize a frequency correction burst and synchronize to it [R.05.08]. Now the BS allocates a bidirectional signaling channel and also sets up a link with the MSC via the network. How the control frame structure assists in this process is highlighted in Section 9.2.5. The MSC uses the IMSI received from the MS to interrogate its HLR and sends the data obtained to the serving VLR. After authentication [R.03.20], the MS provides the destination number, the BS allocates a traffic channel, and the MSC routes the call to its destination. If the MS moves to another cell, it is reassigned to another BS and a handover occurs. If both BSs in the handover process are controlled by the same BSC, the handover takes place under the control of the BSC. Otherwise it is performed by the MSC. In case of incoming calls, the MS must be paged by the BSC. A paging signal is transmitted on a paging channel (PCH) monitored continuously by all MSs, and it covers the location area in which the MS roams. In response to the paging signal, the MS performs an access procedure identical to that employed when the MS initiates a call.

9.2.3 Logical and Physical Channels in GSM

The GSM logical traffic and control channels are standardized in Recommendation [R.05.02], while their mapping onto physical channels is the subject of [R.05.02] and [R.05.03]. The GSM system's prime objective is to transmit the logical traffic channel's (TCH) speech or data information. Their transmission via the network requires a variety of logical control channels. The set of logical traffic and control channels defined in the GSM system is summarized in Table 9.3. There are two general forms of speech and data traffic channels: the full-rate traffic channels (TCH/F), which carry information at a gross rate of 22.8 kbit/s, and the half-rate traffic channels (TCH/H), which communicate at a gross rate of 11.4 kbit/s. A physical channel carries either a full-rate traffic channel or two half-rate traffic channels. In the former, the traffic channel occupies one time slot, while in the latter the two half-rate traffic channels are mapped onto the same time slot, but in alternate frames.

For a summary of the logical control channels carrying signaling or synchronization data, see Table 9.3. There are four categories of logical control channels: the broadcast control channel (BCCH), the common control channel (CCCH), the stand-alone dedicated control channel (SDCCH), and the associated control channel (ACCH). The purpose and way of deployment of the logical traffic and control channels will be explained by highlighting how they are mapped onto physical channels in assisting high-integrity communications.

A physical channel in a Time Division Multiple Access (**TDMA**) system is defined as a time slot with a time slot number (TN) in a sequence of TDMA frames. However, the GSM system deploys TDMA combined with frequency hopping (FH), and hence the physical channel is partitioned in both time and frequency. Frequency hopping [R.05.02] combined with interleaving is known to be very efficient in combating channel fading, and it results in near-Gaussian performance even over hostile Rayleigh-fading channels. The principle of frequency hopping (FH) is that each TDMA burst is transmitted via a different RF CHannel (**RFCH**). If the present TDMA burst happens to be in a deep fade, then the next burst most probably will not be. Consequently, the physical channel is defined as a sequence of radio frequency channels and time slots. Each carrier frequency supports eight physical channels mapped onto eight time slots within a TDMA frame. A given physical channel always uses the same time slot number TN in every TDMA frame. Therefore, a time slot sequence is defined by a time slot number TN and a TDMA frame number (FN) sequence.

9.2.4 Speech and Data Transmission in GSM

The speech coding standard is [R.06.10], while issues of mapping the logical speech traffic channel's information onto the physical channel constituted by a time slot of a certain carrier are specified in [R.05.02]. Since the error correction coding represents part of this mapping process, [R.05.03] is also relevant to these discussions. The example of the full-rate speech traffic channel (TCH/FS) is used here to highlight how this logical channel is mapped onto the physical channel constituted by a Normal Burst (NB) of the TDMA frame structure. This mapping is explained by referring to Figure 9.3 and Figure 9.4. Then this example will be extended to other physical bursts such as the Frequency Correction- (**FCB**), Synchronization- (SB), Access- (AB), and Dummy-Burst (DB) carrying logical control channels, as well as to their TDMA frame structures, as seen in Figure 9.3 and Figure 9.7.

The Regular Pulse Excited (RPE) speech encoder is fully characterized in [41, 317, 318]. Because of its complexity, its description is beyond the scope of this treatise. As it can be seen in Figure 9.4, it delivers 260 bits/20 ms at a bit rate of 13 kbit/s, which are divided into three significance classes: Class 1a (50 bits), Class 1b (132 bits), and Class 2 (78 bits). The Class 1a bits are encoded by a systematic (53,50) cyclic error detection code by adding three parity bits. Then the bits are reordered, and four zero tailing bits are added to periodically reset the memory of the subsequent half-rate, constraint length five convolutional codec (CC) CC(2,1,5), as portrayed in Figure 9.4. Now the unprotected 78 Class 2 bits are concatenated to yield a block of 456 bits/20 ms, which implies an encoded bit rate of 22.8 kbit/s. This frame is

Logical Channels					
Duplex BS ↔ MS Traffic Channels: TCH		Control Channels: CCH			
FEC-coded Speech	FEC-coded Data	Broadcast CCH BCCH BS → MS	Common CCH CCCH	Stand-alone Dedicated CCH SDCCH BS ↔ MS	Associated CCH ACCH BS ↔ MS
TCH/F 22.8 kbit/s	TCH/F9.6 TCH/F4.8 TCH/F2.4 22.8 kbit/s	Freq.Corr.Ch: FCCH	Paging Ch: PCH BS → MS	SDCCH/4	Fast ACCH: FACCH/F FACCH/H
TCH/H 11.4 kbit/s	TCH/H4.8 TCH/H2.4 11.4 kbit/s	Synchron. Ch: SCH	Random Access Ch: RACH MS → BS	SDCCH/8	Slow ACCH: SACCH/TF SACCH/TH SACCH/C4 SACCH/C8
		General Inf.	Access Grant Ch: AGCH BS → MS		

Table 9.3: GSM Logical Channels. ©ETT Hanzo and Steele, 1994.

9.2. GLOBAL SYSTEM FOR MOBILE COMMUNICATIONS — GSM 349

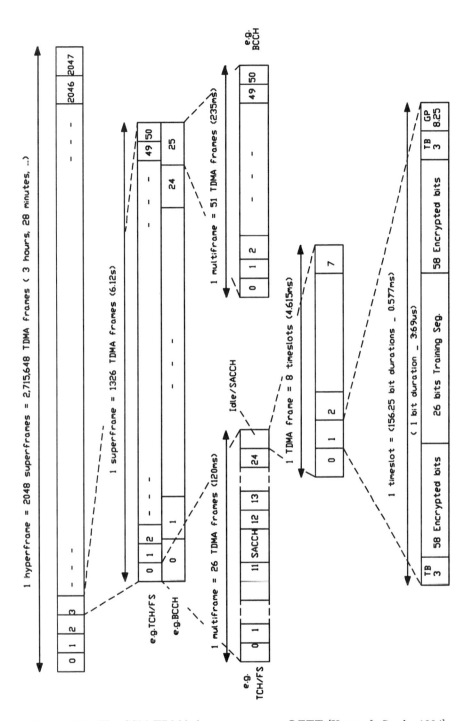

Figure 9.3: The GSM TDMA frame structure. ©ETT [Hanzo & Steele, 1994].

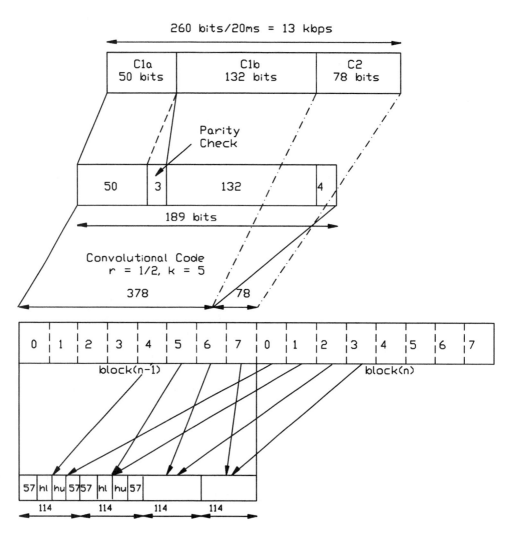

Figure 9.4: Mapping the TCH/FS logical channel onto a physical channel. ©ETT [Hanzo & Steele, 1994].

partitioned into eight 57-bit subblocks that are blockdiagonally interleaved before undergoing intraburst interleaving. At this stage, each 57-bit subblock is combined with a similar subblock of the previous 456-bit frame to construct a 116-bit burst, where the flag bits hl and hu are included to classify whether the current burst is really a TCH/FS burst or it was "stolen" by an urgent fast associated (FACCH) control channel message. Now the bits are encrypted and positioned in a normal burst (NB), as depicted at the bottom of Figure 9.3, where three tailing bits (TB) are added at both ends of the burst to reset the memory of the Viterbi channel equalizer (VE), which is responsible for removing both the channel-induced and the intentional controlled intersymbol interference [9].

The 8.25 bit-interval duration guard period (GP) at the bottom of Figure 9.3 is provided to prevent burst overlapping due to propagation delay fluctuations. Finally, a 26-bit equalizer training segment is included in the center of the normal traffic burst. This segment is constructed by a 16-bit Viterbi channel equalizer training pattern surrounded by five quasiperiodically repeated bits on both sides. Since the MS has to be informed about which BS it communicates with, for neighboring BSs, one of eight different training patterns is used, associated with the BS color codes that assist in identifying the BSs.

This 156.25-bit duration TCH/FS normal burst (NB) constitutes the basic time slot of the TDMA frame structure, which is input to the Gaussian minimum shift keying (GMSK) modulator to be highlighted in Section 9.2.7, at a bit rate of approximately 271 kbit/s. Since the bit interval is 1/(271 kbps) = 3.69 μs, the time slot duration is $156.25 \cdot 3.69 \approx 0.577$ ms. Eight such normal bursts of eight appropriately staggered TDMA users are multiplexed onto one (RF) carrier, giving a TDMA frame of $8 \cdot 0.577 \approx 4.615$ ms duration, as shown in Figure 9.3. The physical channel as characterized above provides a physical time slot with a throughput of 114 bits/4.615 ms = 24.7 kbit/s, which is sufficiently high to transmit the 22.8 kbit/s TCH/FS information. It even has a "reserved" capacity of 24.7-22.8 = 1.9 kbit/s, which can be exploited to transmit slow control information associated with this specific traffic channel — that is, to construct a Slow Associated Control Channel (SACCH), constituted by the SACCH TDMA frames, interspersed with traffic frames at the multiframe level of the hierarchy, as seen in Figure 9.3.

Mapping logical data traffic channels onto a physical channel is essentially carried out by the channel codecs [35], as specified in [R.05.03]. Convolutional and block-based channel codecs were the topic of Chapters 3 and 4. The full- and half-rate data traffic channels standardized in the GSM system are: **TCH/F9.6**, **TCH/F4.8**, **TCH/F2.4**, as well as **TCH/H4.8**, **TCH/H2.4**, as was portrayed earlier in Table 9.3. Note that the numbers in these acronyms represent the data transmission rate in kbps. Without considering the details of these mapping processes, we now focus our attention on control signal transmission issues.

9.2.5 Transmission of Control Signals in GSM

The exact derivation, FEC coding, and mapping of logical control channel information are beyond the scope of this chapter. The interested reader is referred to [R.05.02], [R.05.03], and [Hanzo, Stefanov, 1992] for a detailed discussion. As an example,

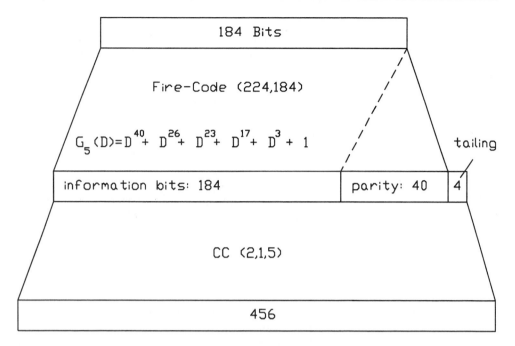

Figure 9.5: FEC in SACCH, FACCH, BCCH, SDCCH, PCH, and AGCH. ©ETT [Hanzo & Steele, 1994].

the mapping of the 184-bit slow associated control channel (SACCH), fast associated control channel (FACCH), broadcast control channel (**BCCH**), stand-alone dedicated control channel (**SDCCH**), paging channel (PCH), and access grant control channel (AGCH) messages onto a 456-bit block, that is, onto four 114-bit bursts, is demonstrated in Figure 9.5. A double-layer concatenated FIRE-code/convolutional code scheme generates 456 bits, using an overall coding rate of $R = 184/456$, which gives a stronger protection for control channels than the error protection of traffic channels.

Upon returning to Figure 9.3, we will show how the SACCH is accommodated by the TDMA frame structure. The TCH/FS TDMA frames of the eight users are multiplexed into multiframes of 24 TDMA frames, but the 13th frame will carry a SACCH message rather than the 13th TCH/FS frame, while the 26th frame will be an idle or dummy frame, as seen at the left-hand side of Figure 9.3 at the multiframe level of the traffic channel hierarchy. The general control channel frame structure shown at the right of Figure 9.3 is discussed later. In this way, 24 TCH/FS frames are sent in a 26-frame multiframe during $26 \cdot 4.615 = 120$ ms. This reduces the traffic throughput to $\frac{24}{26} \cdot 24.7 = 22.8$ kbit/s required by TCH/FS, allocates $\frac{1}{26} \cdot 24.7 = 950$ bps to the SACCH, and "wastes" 950 bps in the idle frame. Observe that the SACCH frame has eight time slots to transmit the eight 950 bps SACCHs of the eight users on the same carrier. The 950 bps idle capacity will be used in case of half-rate channels, where sixteen users will be multiplexed onto alternate frames of the TDMA structure to increase system capacity. Then sixteen 11.4-kbit/s encoded half-rate speech TCHs will be transmitted in a 120 ms multiframe, where sixteen SACCHs are also available.

9.2. GLOBAL SYSTEM FOR MOBILE COMMUNICATIONS — GSM

The Fast Associated Control Channel (FACCH) messages are transmitted via the physical channels provided by bits "stolen" from their own host traffic channels. The construction of the FACCH bursts from 184 control bits is identical to that of the SACCH, as also shown in Figure 9.5, but its 456-bit frame is mapped onto eight consecutive 114-bit TDMA traffic bursts, exactly as specified for TCH/FS. This is carried out by stealing the even bits of the first four and the odd bits of the last four bursts, which is signaled by setting $hu = 1$, $hl = 0$ and $hu = 0$, $hl = 1$ in the first and last bursts, respectively. The unprotected FACCH information rate is 184 bits/20 ms = 9.2 kbit/s, which is transmitted after concatenated error protection at a rate of 22.8 kbit/s. The repetition delay is 20 ms, and the interleaving delay is $8 \cdot 4.615 = 37$ ms, resulting in a total of 57 ms delay.

In Figure 9.3, at the next hierarchical level 51 TCH/FS multiframes are multiplexed into one superframe lasting $51 \cdot 120\ ms = 6.12$ s, which contains $26 \cdot 51 = 1326$ TDMA frames. However, in case of 1326 TDMA frames, the frame number would be limited to $0 \leq FN \leq 1326$, and the encryption rule relying on such a limited range of FN values would not be sufficiently secure. Hence 2048 superframes were amalgameted to form a hyperframe of $1326 \cdot 2048 = 2\ 715\ 648$ TDMA frames lasting $2048 \cdot 6.12$ s $\approx$ 3 h 28 min, allowing a sufficiently high FN value to be used in the encryption algorithm. The up-link and down-link traffic-frame structures are identical with a shift of three time slots between them, which relieves the MS from having to transmit and receive simultaneously, preventing high-level transmitted power leakage back to the sensitive receiver. The received power of adjacent BSs can be monitored during unallocated time slots.

In contrast to duplex traffic and associated control channels, the simplex BCCH and CCCH logical channels of all MSs roaming in a specific cell share the physical channel provided by time slot zero of the BCCH carriers available in the cell. Furthermore, as demonstrated by the right-hand side of Figure 9.3, 51 BCCH and CCCH TDMA frames are mapped onto a $51 \cdot 4.615 = 235$ ms duration multiframe rather than on a 26-frame, 120 ms duration multiframe. In order to compensate for the extended multiframe length of 235 ms, 26 multiframes constitute a 1326-frame superframe of 6.12 s duration. Note in Figure 9.6 that the allocation of the up-link and down-link frames is different, for these control channels exist in only one direction.

Specifically, the random access channel (RACH) is only used by the MSs in the up-link direction if they request, for example, a bidirectional stand-alone dedicated control channel (SDCCH) to be mapped onto an RF channel to register with the network and set up a call. The up-link RACH has a low capacity, carrying messages of 8 bits per 235 ms multiframe, which is equivalent to an unprotected control information rate of 34 bps. These messages are concatenated forward error correction (FEC) coded to a rate of 36 bits/235 ms = 153 bps. They are not transmitted by the normal bursts (NB) derived for TCH/FS, SACCH, or FACCH logical channels, but by the access bursts (AB) depicted in Figure 9.7 in comparison to an NB and other types of bursts to be described later. The FEC coded encrypted 36-bit AB messages of Figure 9.7, containing among other parameters the encoded 6-bit BS identifier code (BSIC) constituted by the 3-bit PLMN color code and 3-bit BS color code for unique BS identification. These 36 bits are positioned after the 41-bit synchronization sequence, which has a high wordlength in order to ensure reliable access burst

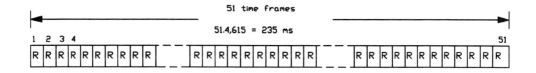

(a) Uplink Direction

(b) Downlink Direction

R: Random Access Channel
F: Frequency Correction Channel
S: Synchronisation Channel
B: Broadcast Control Channel
C: Access Grant/Paging Channel
I: Idle Frame

Figure 9.6: The control multiframe. ©ETT [Hanzo & Steele, 1994].

9.2. GLOBAL SYSTEM FOR MOBILE COMMUNICATIONS — GSM

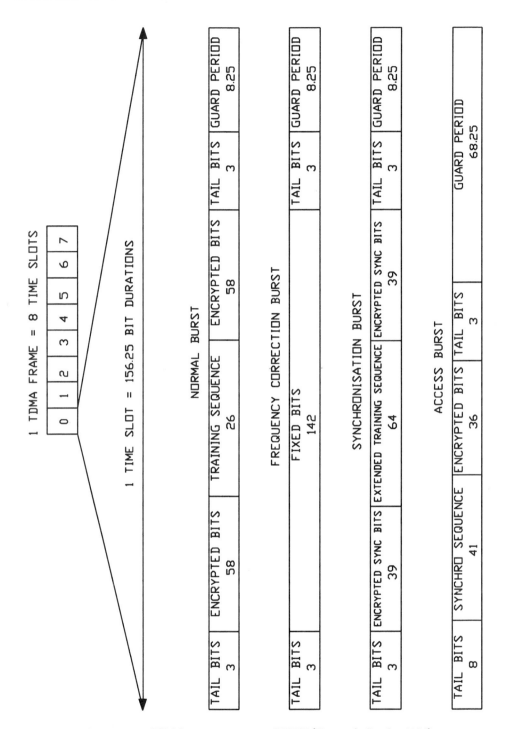

Figure 9.7: GSM burst structures. ©ETT [Hanzo & Steele, 1994].

recognition and a low probability of being emulated by interfering stray data. These messages have no interleaving delay, while they are transmitted with a repetition delay of one control multiframe length, that is, 235 ms.

Adaptive time frame alignment is a technique designed to equalize propagation delay differences between MSs at different distances. The GSM system is designed to allow for cell sizes up to 35 km radius. The time a radio signal takes to travel the 70 km from the base station to the mobile station and back again is 233.3 μs. As signals from all the mobiles in the cell must reach the base station without overlapping each other, a long guard period of 68.25 bits (252 μs) is provided in the access burst, which exceeds the maximum possible propagation delay of 233.3 μs. This long guard period in the access burst is needed when the mobile station attempts its first access to the base station or after a handover has occurred. When the base station detects a 41-bit random access synchronization sequence with a long guard period, it measures the received signal delay relative to the expected signal from a mobile station of zero range. This delay, called the timing advance, is signaled using a 6-bit number to the mobile station, which advances its time base over the range of 0 to 63 bits, that is, in units of 3.69 μs. By this process, the TDMA bursts arrive at the BS in their correct time slots and do not overlap with adjacent ones. This process allows the guard period in all other bursts to be reduced to $8.25 \cdot 3.69$ μs ≈ 30.46 μs (8.25 bits) only. During normal operation the BS continuously monitors the signal delay from the MS, and if necessary, it will instruct the MS to update its time advance parameter. In very large traffic cells there is an option to actively utilize every second time slot only to cope with higher propagation delays, which is spectrally inefficient, but in these large, low-traffic rural cells it is admissible.

As demonstrated by Figure 9.3, the down-link multiframe transmitted by the BS is shared among a number of BCCH and CCCH logical channels. In particular, the last frame is an idle frame (I), while the remaining fifty frames are divided in five blocks of ten frames, where each block starts with a frequency correction channel (FCCH) followed by a synchronization channel (SCH). In the first block of ten frames, the FCH and SCH frames are followed by four broadcast control channel (BCCH) frames and by either four access grant control channels (AGCH) or four paging channels (PCH). In the remaining four blocks of ten frames, the last eight frames are devoted to either PCHs or AGCHs, which are mutually exclusive for a specific MS being either paged or granted a control channel.

The frequency correction channel (FCCH), synchronization channel (SCH), and random access channel (RACH) require special transmission bursts, tailored to their missions, as depicted in Figure 9.7. The FCCH uses frequency correction bursts (FCB) hosting a specific 142-bit pattern. In partial response to GMSK, it is possible to design a modulating data sequence that results in a near-sinusoidal modulated signal imitating an unmodulated carrier exhibiting a fixed frequency offset from the RF carrier utilized. The synchronization channel transmits synchronization bursts (SB) hosting a $16 \cdot 4 = 64$ bit extended sequence exhibiting a high correlation peak in order to allow frame alignment with a quarter bit accuracy. Furthermore, the SB contains $2 \cdot 39 = 78$ encrypted FEC-coded synchronization bits, hosting the BS and PLMN color codes, each representing one of eight legitimate identifiers. Lastly, the access bursts (AB) contain an extended 41-bit synchronization sequence, and they are

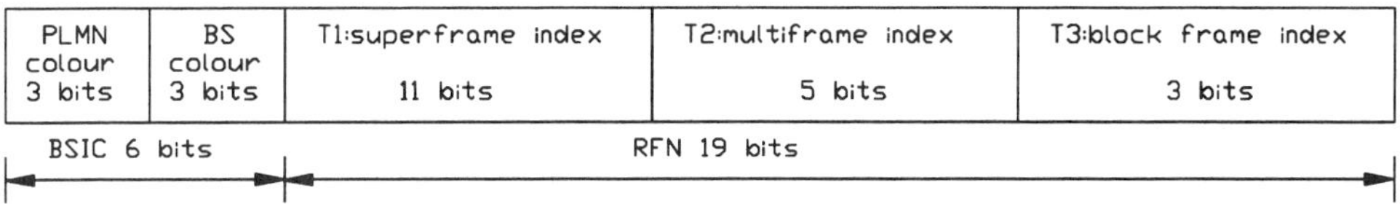

9.2. GLOBAL SYSTEM FOR MOBILE COMMUNICATIONS — GSM

PLMN colour 3 bits	BS colour 3 bits	T1:superframe index 11 bits	T2:multiframe index 5 bits	T3:block frame index 3 bits
BSIC 6 bits		RFN 19 bits		

Figure 9.8: Synchronization channel(SCH) message format. ©ETT [Hanzo & Steele, 1994].

invoked to facilitate initial access to the system. Their long guard space of 68.25 bit duration prevents frame overlap, before the MS's distance, that is, the propagation delay becomes known to the BS and could be compensated for by adjusting the MS's timing advance.

9.2.6 Synchronization Issues in GSM

Although some synchronization issues are standardized in [R.05.02] and [R.05.03], the GSM Recommendations do not specify the exact BS-MS synchronization algorithms to be used. These are left to the equipment manufacturers. However, a unique set of time base counters is defined in order to ensure perfect BS-MS synchronism. The BS sends frequency correction bursts (FCB) and synchronization bursts (SB) on specific time slots of the BCCH carrier to the MS to ensure that the MS's frequency standard is perfectly aligned with that of the BS, as well as to inform the MS about the required initial state of its internal counters. The MS transmits its uniquely numbered traffic and control bursts staggered by three time slots with respect to those of the BS to prevent simultaneous MS transmission and reception. It also takes into account the required timing advance (TA) to cater for different BS-MS-BS round-trip delays.

The time-base counters used to uniquely describe the internal timing states of BSs and MSs are the Quarter bit Number ($QN = 0 \ldots 624$) counting the quarter bit intervals in bursts, Bit Number ($BN = 0 \ldots 156$), Time-slot Number ($TN = 0 \ldots 7$), and TDMA Frame Number ($FN = 0 \ldots 26 \cdot 51 \cdot 2048$), given in the order of increasing interval duration. The MS sets up its time base counters after receiving an SB by determining QN from the 64-bit extended training sequence in the center of the SB, setting $TN = 0$ and decoding the 78 encrypted, protected bits carrying the 25 SCH control bits.

The SCH carries frame synchronization information as well as BS identification information to the MS, as seen in Figure 9.8, and it is provided solely to support the operation of the radio subsystem. The first six bits of the 25-bit segment consist of three PLMN color code bits and three BS color code bits supplying a unique BS Identifier Code (BSIC) to inform the MS which BS it is communicating with. The second 19-bit segment is the Reduced TDMA Frame Number (RFN), derived from the full TDMA Frame Number (FN), constrained to the range of $[0 \ldots (26 \cdot 51 \cdot 2048) - 1] = [0 \ldots 2,715,647]$ in terms of three subsegments T1, T2, and T3. These subsegments are computed as follows: T1(11 bits) = [FN div (26·51)], T2(5 bits) = [FN mod 26] and T3'(3 bits) = [(T3-1) div 10], where T3 = [FN mod 5], where div and mod represent the integer division and modulo operations, respectively. In Figure 9.8,

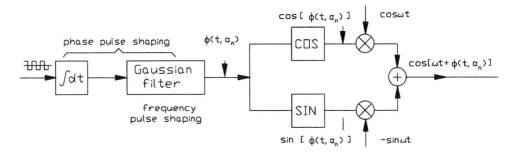

Figure 9.9: GMSK modulator schematic diagram. ©ETT [Hanzo & Steele, 1994].

T1 determines the superframe index in a hyperframe, T2 the multiframe index in a superframe, T3 the frame index in a multiframe, while T3' is the signaling block index [1 ... 5] of a frame in a specific 51-frame control multiframe, and their role is best understood by referring to Figure 9.3. Once the MS has received the Synchronization Burst (SB), it readily computes the FN required in various control algorithms, such as encryption and handover, as shown below:

$$FN = 51[(T3 - T2) \bmod 26] + T3 + 51 \cdot 26 \cdot T1, \quad \text{where} \quad T3 = 10 \cdot T3' + 1. \quad (9.1)$$

9.2.7 Gaussian Minimum Shift Keying in GSM

The GSM system uses constant envelope partial response GMSK modulation [87] specified in Recommendation [R.05.04]. Constant envelope, continuous-phase modulation schemes are robust against signal fading as well as interference and have good spectral efficiency. The slower and smoother are the phase changes, the better is the spectral efficiency, since the signal is allowed to change less abruptly, requiring lower frequency components. However, the effect of an input bit is spread over several bit periods, leading to a partial response system, which requires a channel equalizer in order to remove this controlled, intentional intersymbol interference (ISI) even in the absence of uncontrolled channel dispersion.

The widely employed partial response GMSK scheme is derived from the full response Minimum Shift Keying (MSK) scheme. In MSK the phase changes between adjacent bit periods are piecewise linear, which results in discontinuous phase derivative (i.e., instantaneous frequency at the signaling instants), and hence widens the spectrum. However, smoothing these phase changes by a filter having a Gaussian impulse response [87], which is known to have the lowest possible bandwidth, this problem is circumvented using the schematic of Figure 9.9, where the GMSK signal is generated by modulating and adding two quadrature carriers. The key parameter of GMSK in controlling both bandwidth and interference resistance is the 3 dB-down filter-bandwidth × bit interval product (B·T) referred to as normalized bandwidth. It was found that as the B·T product is increased from 0.2 to 0.5, the interference resistance is improved by approximately 2 dB at the cost of increased bandwidth occupancy, and best compromise was achieved for B·T = 0.3. This corresponds to

9.2. GLOBAL SYSTEM FOR MOBILE COMMUNICATIONS — GSM

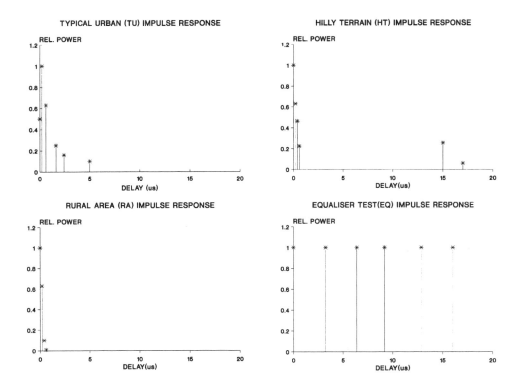

Figure 9.10: Typical GSM channel impulse responses. ©ETT [Hanzo & Steele, 1994].

spreading the effect of 1 bit over approximately 3-bit intervals. The spectral efficiency gain due to higher interference tolerance, and hence more dense frequency reuse, was found to be more significant than the spectral loss caused by wider GMSK spectral lobes.

The channel separation at the TDMA burst-rate of 271 kbit/s is 200 kHz, and the modulated spectrum must be 40 dB down at both adjacent carrier frequencies. When TDMA bursts are transmitted in an on-off keyed mode, further spectral spillage arises. This spillage is mitigated by a smooth power ramp-up and down envelope at the leading and trailing edges of the transmission bursts, attenuating the signal by 70 dB during a 28 μs and 18 μs interval, respectively.

9.2.8 Wideband Channel Models in GSM

The set of 6-tap GSM impulse responses [87] specified in Recommendation [R.05.05] is depicted in Figure 9.10, where the individual propagation paths are independent Rayleigh-fading paths, weighted by the appropriate coefficients h_i corresponding to their relative powers portrayed in the figure. In simple terms, the wideband channel's impulse response is measured by transmitting an impulse and detecting the received echoes at the channel's output in every D-spaced delay bin. In some bins no delayed and attenuated multipath component is received, while in others significant energy is detected, depending on the typical reflecting objects and their distance from the receiver. The path-delay can be easily related to the distance of the reflecting objects, since radio waves are traveling at the speed of light. For example, at a speed of 300 000 km/s, a reflecting object situated at a distance of 0.15 km yields a multipath component at a round-trip delay of 1 μs.

The Typical Urban (TU) impulse response spreads over a delay interval of 5 μs, which is almost two 3.69 μs bit intervals duration and therefore results in serious intersymbol interference (ISI). In simple terms, it can be treated as a two-path model, where the reflected path has a length of 0.75 km, corresponding to a reflector located at a distance of about 375 m. The Hilly Terrain (HT) model has a sharply decaying short-delay section due to local reflections and a long-delay path around 15 μs due to distant reflections. Therefore, in practical terms, it can be considered a two- or three-path model having reflections from a distance of about 2 km. The Rural Area (RA) response seems the least hostile among all standardized responses, decaying rapidly inside one bit interval and therefore is expected to be easily combated by the channel equalizer. Although the type of the equalizer is not standardized, partial response systems typically use Viterbi Equalizers (VE). Since the RA channel effectively behaves as a single-path nondispersive channel, it would not require an equalizer. The fourth standardized impulse response is artificially contrived in order to test the equalizer's performance, and it is constituted by six equidistant unit-amplitude impulses representing six equal-powered independent Rayleigh-fading paths with a delay-spread over 16 μs. With these impulse responses in mind the required channel is simulated by summing the appropriately delayed and weighted received signal components. In all but one case the individual components are assumed to have Rayleigh amplitude distribution, while in the RA model the main tap at zero delay is supposed to have Rician distribution with the presence of a dominant line-of-sight path.

9.2.9 Adaptive Link Control in GSM

The adaptive link control algorithm portrayed in Figure 9.11 and specified in [R.05.08] allows the MS to favor that specific traffic cell, which provides the highest probability of reliable communications associated with the lowest possible path-loss. It also decreases interference with other co-channel users and, through dense frequency reuse, improves spectral efficiency, while maintaining an adequate communications quality and facilitates a reduction in power consumption, which is particularly important in hand-held MSs. The handover process maintains a call in progress as the MS moves between cells, or when there is an unacceptable transmission quality degradation caused by interference, in which case an intracell handover to another carrier in the same cell is performed. A radio link failure occurs when a call with an unacceptable voice or data quality cannot be improved either by RF power control or by handover. The reasons for the link failure may be loss of radio coverage or very high interference levels. The link control procedures rely on measurements of the received RF signal strength (RXLEV), the received signal quality (RXQUAL), and the absolute distance between base and mobile stations (DISTANCE).

RXLEV is evaluated by measuring the received level of the broadcast control channel (BCCH) carrier, which is continuously transmitted by the BS on all time

9.2. GLOBAL SYSTEM FOR MOBILE COMMUNICATIONS — GSM

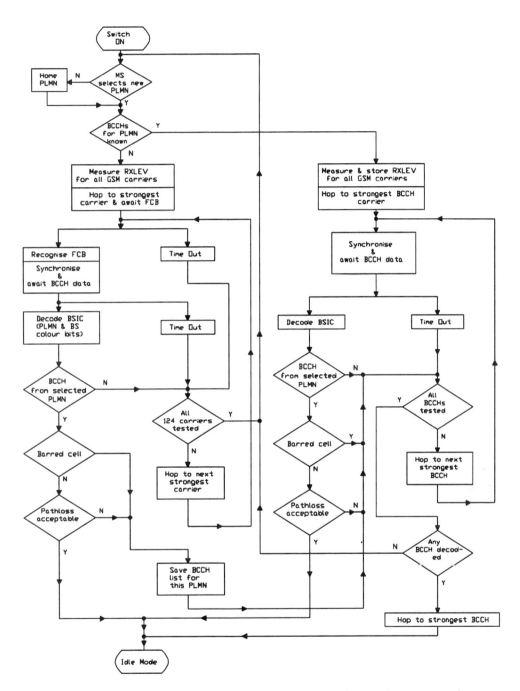

Figure 9.11: Initial cell selection by the MS. ©ETT [Hanzo & Steele, 1994].

slots of the B frames in Figure 9.6 and without variations of the RF level. An MS measures the received signal level from the serving cell and from the BSs in all adjacent cells by tuning and listening to their BCCH carriers. The root mean squared (RMS) level of the received signal is measured over a dynamic range of -103 to -41 dBm for intervals of one SACCH multiframe (480 ms). The received signal level is averaged over at least 32 SACCH frames ($\approx$15 s) and mapped to give RXLEV values between 0 and 63 to cover the range $-103\ldots-41$ dBm in steps of 1 dB. The RXLEV parameters are then coded into 6-bit words for transmission to the serving BS via the SACCH.

RXQUAL is estimated by measuring the BER before channel decoding, using the Viterbi channel equalizer's metrics [87] and/or those of the Viterbi convolutional decoder of Chapter 3 [35]. Eight values of RXQUAL span the logarithmically scaled BER range of $0.2\%\ldots12.8\%$ before channel decoding.

The absolute DISTANCE between base and mobile stations is measured using the "timing advance" parameter. The timing advance is coded as a 6 bit number corresponding to a propagation delay from 0 to $63\cdot 3.69$ μs $= 232.6$ μs, characteristic of a cell radius of 35 km.

While roaming, the MS needs to identify which potential target BS it is measuring. The BCCH carrier frequency may not be sufficient for this purpose, because in small cluster sizes the same BCCH frequency may be used in more than one surrounding cell. To avoid ambiguity, a 6-bit Base Station Identity Code (BSIC) is transmitted on each BCCH carrier in the synchronization burst (SB) of Figure 9.7. Two other parameters transmitted in the BCCH data provide additional information about the BS. The binary flag called PLMN_PERMITTED indicates whether the measured BCCH carrier belongs to a PLMN which the MS is permitted to access. The second Boolean flag, CELL_BAR_ACCESS, indicates whether the cell is barred for access by the MS, although it belongs to a permitted PLMN. An MS in idle mode, that is, after it has just been switched-on, or after it has lost contact with the network, searches all 125 RF channels and takes readings of RXLEV on each of them. Then it tunes to the carrier with the highest RXLEV and searches for frequency correction bursts (FCB) in order to determine whether or not the carrier is a BCCH carrier. If it is not, then the MS tunes to the next highest carrier, and so on, until it finds a BCCH carrier, synchronizes to it, and decodes the parameters BSIC, PLMN_PERMITTED, and CELL_BAR_ACCESS in order to decide whether to continue the search. The MS may store the BCCH carrier frequencies used in the network accessed, in which case the search time would be reduced. Again, the process described is summarized in the flowchart of Figure 9.11.

The adaptive power control is based on RXLEV measurements. In every SACCH multiframe the BS compares the RXLEV readings reported by the MS, or obtained by the base station, with a set of thresholds. The exact strategy for RF power control is determined by the network operator with the aim of providing an adequate quality of service for speech and data transmissions while keeping interferences low. Clearly, "adequate" quality must be achieved at the lowest possible transmitted power to keep co-channel interferences low, which implies contradictory requirements in terms of transmitted power. The criteria for reporting radio link failure are based on the measurements of RXLEV and RXQUAL performed by both the mobile and base stations, and the procedures for handling link failures result in the reestablishment

9.2. GLOBAL SYSTEM FOR MOBILE COMMUNICATIONS — GSM

or the release of the call, depending on the network operator's strategy.

The handover process involves the most complex set of procedures in the radio link control. Handover decisions are based on results of measurements performed by both the base and mobile stations. The base station measures RXLEV, RXQUAL, and DISTANCE, as well as the interference level in unallocated time slots, while the MS measures and reports to the BS the values of RXLEV and RXQUAL for the serving cell and RXLEV for the adjacent cells. When the MS moves away from the BS, the RXLEV and RXQUAL parameters for the serving station become lower, while RXLEV for one of the adjacent cells increases.

9.2.10 Discontinuous Transmission in GSM

Discontinuous transmission (DTX) issues are standardized in the [R.06.31] Recommendation, while the associated problems of voice activity detection (VAD) are specified by [R.06.32]. Assuming an average speech activity of 50% and a high number of interferers combined with frequency hopping to randomize the interference load, significant spectral efficiency gains can be achieved when deploying discontinuous transmissions due to decreasing interferences, while reducing power dissipation as well. Because of the reduction in power consumption, full DTX operation is mandatory for MSs, but in BSs only receiver DTX functions are compulsory.

The fundamental problem in voice activity detection is how to differentiate between speech and noise, while keeping false noise triggering and speech spurt clipping as low as possible. In vehicle-mounted MSs the severity of the speech/noise recognition problem is aggravated by the excessive vehicle background noise. This problem is resolved by deploying a combination of threshold comparisons and spectral-domain techniques [41, 210]. Another important associated problem is the introduction of noiseless inactive segments, which is mitigated by comfort noise insertion (CNI) in these segments at the receiver.

9.2.11 Summary and Conclusions

The key features of the GSM system were summarized in this brief review with reference to Table 9.1. Time Division Multiple Access (TDMA) with eight users per carrier is used at a multi-user rate of 271 kbit/s, demanding a channel equalizer to combat dispersion in large-cell environments. The error protected chip-rate of the full-rate traffic channels is 22.8 kbit/s, while in half-rate channels it is 11.4 kbit/s. Apart from the full- and half-rate speech traffic channels there are five different-rate data traffic channels and fourteen various control and signaling channels to support the system's operation. A moderately complex, 13 kbit/s Regular Pulse Excited speech codec with long-term predictor (LTP) is used, combined with an embedded three-class error correction codec and multilayer interleaving to provide sensitivity-matched unequal error protection for the speech bits. An overall speech delay of 57.5 ms is maintained. Slow frequency hopping at 217 hops/sec yields substantial performance gains for slowly moving pedestrians.

Constant envelope partial response GMSK with a channel spacing of 200 kHz is deployed to support 125 duplex channels in the 890−915 MHz up-link and 935−960 MHz down-link bands, respectively. At a transmission rate of 271 kbit/s, a spectral efficiency of 1.35 bit/s/Hz is achieved. The controlled GMSK-induced and uncontrolled channel-induced intersymbol interferences are removed by the channel equalizer. The set of standardized wideband GSM channels was introduced in order to provide benchmarkers for performance comparisons. Efficient power budgeting and minimum co-channel interferences are ensured by the combination of adaptive power and handover-control based on weighted averaging of up to eight up-link and down-link system parameters. Discontinuous transmissions assisted by reliable spectral-domain voice activity detection and comfort-noise insertion further reduce interferences and power consumption. Because of ciphering, no unprotected information is sent via the radio link. As a result, spectrally efficient, high-quality mobile communications with a variety of services and international roaming is possible in cells of up to 35 km radius for signal-to-noise- and interference ratios in excess of $10 - 12$ dBs. Again, the key system features are summarized in Table 9.1.

We noted earlier in the context of Table 9.1 that GSM and the other 2G systems have constant speech bit rates between 6.7 and 13 kbps, while the speech rate is 32 kbit/s for the CT systems. By contrast, the standard H.263 video codec of Part IV generates a time-variant bit rate. This bit-rate incompatibility problem can be circumvented with the aid of the fixed, but programmable-rate video codecs of Part II or upon invoking the adaptive packetization, buffering, and rate-control regime of Part IV. As noted before, another alternative for supporting video services over GSM is, for example, invoking the High Speed Circuit Switched Data (HSCSD) mode according to the recommendations of [315], provided that the associated delay does not affect the lip-synchronization between speech and video. Since the speech delay of the GSM systems is about 57.25 ms, this is the ideal target video delay.

In the next chapter, we turn our attention to CDMA systems and highlight the features of the forthcoming 3G wireless systems.

Chapter 10

CDMA Systems: Third-Generation and Beyond

K. Yen and L. Hanzo[1]

10.1 Introduction

Although the number of cellular subscribers continues to grow worldwide [319], the predominantly speech-, data- and e-mail-oriented services are expected to be enriched by a whole host of new services in the near future. Thus the performance of the recently standardized Code Division Multiple Access (CDMA) third-generation (3G) mobile systems is expected to become comparable to, if not better than, that of their wired counterparts.

These ambitious objectives are beyond the capabilities of the present second-generation (2G) mobile systems such as the Global System for Mobile Communications known as GSM [41], the Interim Standard-95 (IS-95) Pan-American system, or the Personal Digital Cellular (PDC) system [320] in Japan. The basic features of these systems are summarized in Table 9.1. Thus, in recent years, a range of new system concepts and objectives were defined, and these will be incorporated in the 3G mobile systems. Both the European Telecommunications Standards Institute (ETSI) and the International Telecommunication Union (ITU) are defining a framework for these systems under the auspices of the Universal Mobile Telecommunications System (UMTS) [319–324] and the International Mobile Telecommunications scheme in the year 2000 (IMT-2000)[2] [321, 322, 325].

Their objectives and the system concepts will be discussed in more detail in later sections. CDMA is the predominant multiple access technique proposed for the 3G

---

[1] University of Southampton and Multiple Access Communications Ltd., UK.
[2] Formerly known as Future Public Land Mobile Telecommunication Systems.

wireless communications systems worldwide. CDMA was already employed in some 2G systems, such as the IS-95 system of Table 9.1, and it has proved to be a success. Partly motivated by this successer, both the Pan-European UMTS and the IMT-2000 initiatives have opted for a CDMA-based system, although the European system also incorporates an element of TDMA. In this chapter, we provide a rudimentary introduction to a range of CDMA concepts. Then the European, American and Japanese CDMA-based 3G mobile system concepts are considered, follwed by a research-oriented outlook on potential future systems.

The chapter is organized as follows. Section 10.2 offers a rudimentary introduction to CDMA in order to make this chapter self-contained, whereas Section 10.3 focuses on the basic objectives and system concepts of the 3G mobile systems, highlighting the European, American and Japanese CDMA-based third-generation system concepts. Finally, our conclusions are presented in Section 10.4.

10.2 Basic CDMA System

CDMA is a spread spectrum communications technique that supports simultaneous digital transmission of several users' signals in a multiple access environment. Although the development of CDMA was motivated by user capacity considerations, the system capacity provided by CDMA is similar to that of its more traditional counterparts, frequency division multiple access (FDMA), and time division multiple access (TDMA) [326]. However, CDMA has the unique property of supporting a multiplicity of users in the same radio channel with a graceful degradation in performance due to multi-user interference. Hence, any reduction in interference can lead to an increase in capacity [327]. Furthermore, the frequency reuse factor in a CDMA cellular environment can be as high as unity, and being a so-called wideband system, it can coexist with other narrowband microwave systems, which may corrupt the CDMA signal's spectrum in a narrow frequency band without inflicting significant interference [328]. This eases the problem of frequency management as well as allowing a smooth evolution from narrowband systems to wideband systems. But perhaps the most glaring advantage of CDMA is its ability to combat or in fact to benefit from multipath fading, as it will become explicit during our further discourse.

In the forthcoming sections, we introduce our nomenclature, which will be used throughout the subsequent sections. Further in-depth information on CDMA can be found in a range of excellent research papers [326, 328, 329] and textbooks [239–241, 330].

10.2.1 Spread Spectrum Fundamentals

In spread spectrum transmission, the original information signal, which occupies a bandwidth of B Hz, is transmitted after spectral spreading to a bandwidth N times higher, where N is known as the processing gain. In practical terms the processing gain is typically in the range of $10 - 30$ dB [328]. This frequency-domain spreading concept is illustrated in Figure 10.1. The power of the transmitted spread spectrum signal is spread over N times the original bandwidth, while its spectral density is

10.2. BASIC CDMA SYSTEM

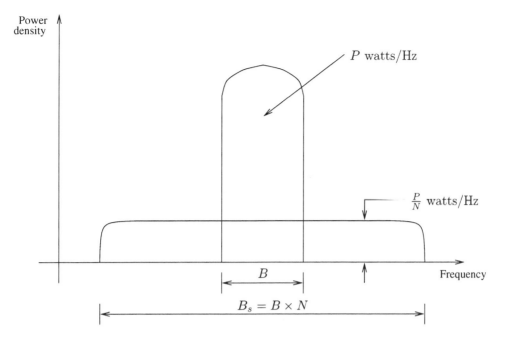

Figure 10.1: Power spectral density of signal before and after spreading.

correspondingly reduced by the same amount. Hence, the processing gain is given by :

$$N = \frac{B_s}{B}, \qquad (10.1)$$

where B_s is the bandwidth of the spread spectrum signal while B is the bandwidth of the original information signal. As we shall see during our further discourse, this unique technique of spreading the information spectrum is the key to improving its detection in a mobile radio environment, and it also allows narrowband signals exhibiting a significantly higher spectral density to share the same frequency band [328].

There are basically two main types of spread spectrum (SS) systems [326] :

- Direct Sequence (DS) SS systems and
- Frequency Hopping (FH) SS systems.

10.2.1.1 Frequency Hopping

In FH spreading, which was invoked in the 2G GSM system of Table 9.1, the narrowband signal is transmitted using different carrier frequencies at different times. Thus, the data signal is effectively transmitted over a wide spectrum. There are two classes of frequency hopping patterns. In fast frequency hopping, the carrier frequency changes several times per transmitted symbol, while in slow frequency hopping, the carrier frequency changes typically after a number of symbols or a transmission burst. In the GSM system of Table 9.1, each transmission burst of 114 channel-coded speech

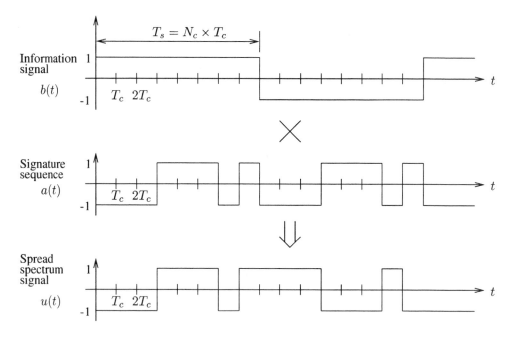

Figure 10.2: Time-domain waveforms involved in generating a direct sequence spread signal.

bits was transmitted on a different frequency and since the TDMA frame duration was 4.615 ms, the associated hopping frequency was its reciprocal, that is, 217 hops/sec. The exact sequence of frequency hopping will be made known only to the intended receiver so that the frequency hopped pattern may be dehopped in order to demodulate the signal [328]. Direct sequence (DS) spreading is more commonly used in CDMA. Hence, our forthcoming discussions will be in the context of direct sequence spreading.

10.2.1.2 Direct Sequence

In DS spreading, the information signal is multiplied by a high-frequency signature sequence, also known as a spreading code or spreading sequence. This user signature sequence facilitates the detection of different users' signals in order to achieve a multiple access capability in CDMA. Although in CDMA this user "separation" is achieved using orthogonal spreading codes, in FDMA and TDMA orthogonal frequency slots or time-slots are provided, respectively.

We can see from Figure 10.2 that each information symbol of duration T_s is broken into N_c equi-spaced subintervals of duration T_c, each of which is multiplied with a different chip of the spreading sequence. Hence, $N_c = \frac{T_s}{T_c}$. The resulting output is a high-frequency sequence.

For binary signaling $T_s = T_b$, where T_b is the data bit duration. Hence, N_c is equal to the processing gain N. However, for M-ary signaling, where $M > 2$, $T_s \neq T_b$ and hence $N_c \neq N$. An understanding of the distinction between N_c and N is important,

10.2. BASIC CDMA SYSTEM

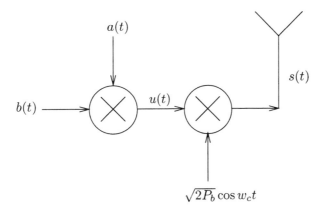

Figure 10.3: BPSK modulated DS-SS transmitter.

since the values of N_c and N have a direct effect on the bandwidth efficiency and performance of the CDMA system.

The block diagram of a typical binary phase shift keying (BPSK) modulated DS-SS transmitter is shown in Figure 10.3. We will now express the associated signals mathematically.

The binary data signal may be written as:

$$b(t) = \sum_{j=-\infty}^{\infty} b_j \Gamma_{T_b}(t - jT_b), \qquad (10.2)$$

where T_b is the bit duration, $b_j \in \{+1, -1\}$ denotes the jth data bit, and $\Gamma_{T_b}(t)$ is the pulse shape of the data bit. In practical applications, $\Gamma_\tau(t)$ has a bandlimited waveform, such as a raised cosine Nyquist pulse, which was introduced in Chapter 5. However, for analysis and simulation simplicity, we will assume that $\Gamma_\tau(t)$ is a rectangular pulse throughout this chapter, which is defined as:

$$\Gamma_\tau(t) = \begin{cases} 1, & 0 \leq t < \tau, \\ 0, & \text{otherwise.} \end{cases} \qquad (10.3)$$

Similarly, the spreading sequence may be written as

$$a(t) = \sum_{h=-\infty}^{\infty} a_h \Gamma_{T_c}(t - hT_c), \qquad (10.4)$$

where $a_h \in \{+1, -1\}$ denotes the hth chip and $\Gamma_{T_c}(t)$ is the chip-pulse with a chip duration of T_c. The energy of the spreading sequence over a bit duration of T_b is normalized according to:

$$\int_0^{T_b} |a(t)|^2 dt = T_b. \qquad (10.5)$$

As seen in Figure 10.3, the data signal and spreading sequence are multiplied, and the resultant spread signal is modulated on a carrier in order to produce the wideband

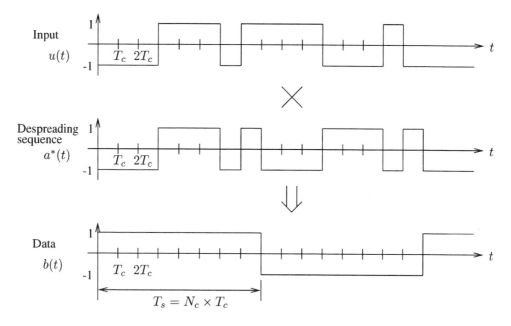

Figure 10.4: Time-domain waveforms involved in decoding a direct sequence signal.

signal $s(t)$ at the output:

$$s(t) = \sqrt{2P_b} b(t) a(t) \cos w_c t, \qquad (10.6)$$

where P_b is the average transmitted power. At the intended receiver, the signal is multiplied by the conjugate of the transmitter's spreading sequence, which is known as the despreading sequence, in order to retrieve the information. Ideally, in a single-user, nonfading, noiseless environment, the original information can be decoded without errors. This is seen in Figure 10.4.

In reality, however, the conditions are never so perfect. The received signal will be corrupted by noise, interfered by both multipath fading — resulting in intersymbol interference (ISI) — and by other users, generating multi-user interference. Furthermore, this signal is delayed by the time-dispersive medium. It is possible to reduce the interference due to multipath fading and other users by innovative signal processing methods, which will be discussed in more detail in later sections.

Figure 10.5 shows the block diagram of the receiver for a noise-corrupted channel using a correlator for detecting the transmitted signal, yielding:

$$\begin{aligned} \hat{b}_i &= \operatorname{sgn}\left\{ \frac{1}{\sqrt{T_b}} \int_{iT_b}^{(i+1)T_b} a^*(t)[s(t) + n(t)] \cos w_c t \, dt \right\} \\ &= \operatorname{sgn}\left\{ \sqrt{\frac{\xi_b}{2}} b_i + \frac{1}{\sqrt{T_b}} \int_{iT_b}^{(i+1)T_b} a^*(t) n(t) \cos w_c t \, dt \right\}, \qquad (10.7) \end{aligned}$$

where $\xi_b = T_b \times P_b$ is the bit energy and $\operatorname{sgn}(x)$ is the signum function of x, which returns a value of 1, if $x > 0$ and returns a value of -1, if $x < 0$. In a single-user

10.2. BASIC CDMA SYSTEM

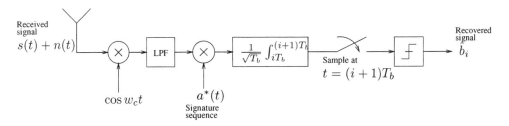

Figure 10.5: BPSK DS-SS receiver for AWGN channel.

Additive White Gaussian Noise (AWGN) channel, the receiver shown in Figure 10.5 is optimum. In fact, the performance of the DS-SS system discussed so far is the same as that of a conventional BPSK modem in an AWGN channel, whereby the probability of bit error $Pr_b(\epsilon)$ is given by:

$$Pr_b(\epsilon) = Q\left(\sqrt{\frac{2\xi_b}{N_0}}\right), \tag{10.8}$$

where

$$Q(x) = \frac{1}{\sqrt{2\pi}} \int_x^\infty e^{-y^2/2} dy \tag{10.9}$$

is the Gaussian Q-function. The advantages of spread spectrum communications and CDMA will only be appreciated in a multipath multiple access environment. The multipath aspects and how the so-called RAKE receiver [52, 331] can be used to overcome the multipath effects will be highlighted in the next section.

10.2.2 The Effect of Multipath Channels

In this section, we present an overview of the effects of the multipath wireless channels encountered in a digital mobile communication system, which was treated in depth in Chapter 2. Interested readers may also refer to the recent articles written by Sklar in [332, 333] for a brief overview on this subject.

Since the mobile station is usually close to the ground, the transmitted signal is reflected, refracted, and scattered from objects in its vicinity, such as buildings, trees, and mountains [326]. Therefore, the received signal is comprised of a succession of possibly overlapping, delayed replicas of the transmitted signal. Each replica is unique in its arrival time, power, and phase [334]. As the receiver or the reflecting objects are not stationary, such reflections will be imposed fading on the received signal, where the fading causes the signal strength to vary in an unpredictable manner. This phenomenon was referred to as multipath propagation in Chapter 2 [332].

There are typically two types of fading in the mobile radio channel [332]:

- Long-term fading

- Short-term fading

As argued in Chapter 2, long-term fading is caused by the terrain configuration between the base station and the mobile station, such as hills and clumps of buildings, which result in an average signal power attenuation as a function of distance. For our purposes the channel can be described in terms of its average path-loss, typically obeying an inverse fourth power law [326] and a log-normally distributed variation around the mean. Thus, long-term shadow fading was also referred to as log-normal fading in Chapter 2 [332].

On the other hand, short-term fading refers to the dramatic changes in signal amplitude and phase as a result of small changes in the spatial separation between the receiver and transmitter, as we noted in Chapter 2 [332]. Furthermore, the motion between the transmitter and receiver results in propagation path changes, such that the channel appears to be time-variant. The time-variant frequency-selective channel was modeled as a tapped delay line in Chapter 2, where the complex low-pass impulse response can be modeled as:

$$\tilde{h}(t) = \sum_{l=1}^{L} |\alpha_l(t)| e^{j\phi_l(t)} \delta(t - \tau_l), \qquad (10.10)$$

where $|\alpha_l(t)|, \phi_l(t)$ and τ_l are the amplitude, phase, and delay of the lth path, respectively, and L is the total number of multipath components. We argued in Chapter 2 that the rate of signal level fluctuation is determined by the Doppler frequency, f_D, which in turn is dependent on the carrier frequency, f_c, and the speed of the mobile station v according to (see also page 16 of [36]):

$$f_D = v \frac{f_c}{c}, \qquad (10.11)$$

where c is the speed of light.

Typically, the short-term fading phenomenon is modeled statistically by a Rayleigh, Rician, or Nakagami-m distribution [335]. The Rayleigh and Rician distributions were characterized in Chapter 2. There have been some contrasting views in the literature as to which of these distributions best describes the fast-fading channel statistically. Although empirical results have shown that the fading statistics are best described by a Nakagami distribution [336], in most cases a Rayleigh-distributed fading is used for analysis and simulation because of simplicity, and it serves as a useful illustrative example in demonstrating the effects of fading on transmission. Moreover, the Rayleigh distribution is a special case of the Nakagami distribution, when m, known as the fading parameter, is equal to unity (see page 48 in [52]). The Rician distribution is more applicable to satellite communication, due to the presence of a dominant signal component known as the specular component [332], than to large-cell terrestrial communication, where often there is no Line-of-Sight (LOS) path between the terrestrial base station and the mobile station. However, in small microcells often the opposite is true. In our investigations in this chapter, Rayleigh-distributed frequency selective fading is assumed.

The delay is proportional to the length of the corresponding signal path between the transmitter and receiver. The delay spread due to the path-length differences between the multipath components causes Intersymbol Interference (ISI) in data transmission, which becomes particularly dominant for high data rates.

10.2. BASIC CDMA SYSTEM

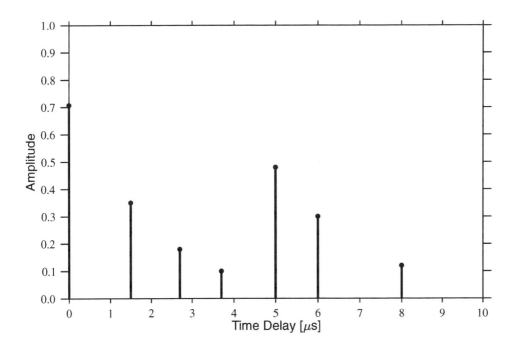

Figure 10.6: COST 207 BU impulse response.

A typical radio channel impulse response is shown in Figure 10.6. This channel impulse response is known as the COST 207 bad urban (BU) impulse response [337]. It can be clearly seen that the response consists of two main groups of delayed propagation paths: a main profile and a smaller echo profile following the main profile at a delay of 5 μs. The main profile is caused by reflections of the signal from structures in the vicinity of the receiver with shorter delay times. On the other hand, the echo profile could be caused by several reflections from a larger but more distant object, such as a hill [338]. In either case, we can see that both profiles approximately follow a negative exponentially decaying function with respect to the time-delay.

Figure 10.7 shows the impairments of the spread spectrum signal traveling over a multipath channel with L independent paths, yielding the equivalent baseband received signal of:

$$r(t) = \sum_{l=1}^{L} \alpha_l(t)\tilde{s}(t - \tau_l) + n(t), \tag{10.12}$$

where $\alpha_l(t)$ is the time-variant complex channel gain, which is given by $|\alpha_l(t)|e^{j\phi_l(t)}$ in Equation 10.10 with a Rayleigh-distributed amplitude, uniformly distributed phase over $[-\pi \ldots \pi]$ and $\tilde{s}(t - \tau_l)$ is the equivalent baseband transmitted spread spectrum signal from Equation 10.6 delayed by τ_l. The above equation shows that the lth path is attenuated by the channel coefficient $\alpha_l(t)$ and delayed by τ_l. Without intelligent diversity techniques [52], these paths are added together at the receiver and any phase or delay difference between the paths may result in a severely multipath interfered

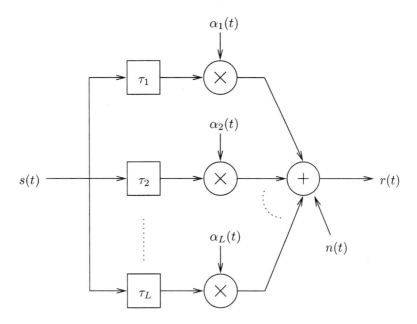

Figure 10.7: Multipath propagation model of the transmitted signal.

signal, corrupted by dispersion-induced intersymbol interference (ISI).

Figure 10.8 shows the effect of a nonfading channel and a fading channel on the bit error probability of BPSK-modulated CDMA. Without diversity, the bit error rate (BER) in a fading channel decreases approximately according to $Pr_b(\epsilon) \approx \frac{1}{4\bar{\gamma}_c}$, where $\bar{\gamma}_c$ is the average Signal-to-Noise Ratio (SNR), and hence plotted on a logarithmic scale according to $\log Pr_b(\epsilon) = -\log 4\bar{\gamma}_c$ we have a near-linear curve [52]. This is different from a nonfading, or AWGN, channel, whereby the BER decreases exponentially with increasing the SNR. Thus, in a fading channel, a high transmitted power is required to obtain a low probability of error. As we shall see in the next section, diversity techniques can be used to overcome this impediment.

10.2.3 RAKE Receiver

As mentioned previously, spread spectrum techniques can take advantage of the multipath nature of the mobile channel in order to improve reception. This is possible due to the signal's wideband nature, which has a significantly higher bandwidth than the multipath channel's coherence bandwidth [32]. In this case, the channel was termed a frequency selective fading channel, since different transmitted frequencies faded differently if their separation was higher than the previously mentioned coherence bandwidth. Suppose that the spread spectrum has a bandwidth of B_s and the channel's coherence bandwidth is B_c, such that $B_s \gg B_c$. Thus, the number of resolvable independent paths — that is, the paths that fade near-independently — L_R

10.2. BASIC CDMA SYSTEM

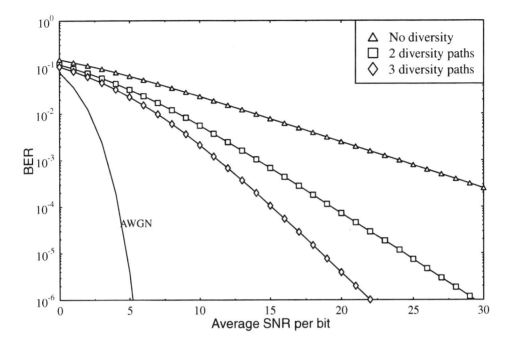

Figure 10.8: Performance of BPSK modulated CDMA over various Rayleigh-fading channels. The curves were obtained using perfect channel estimation, and there was no self-interference between diversity paths.

is equal to

$$L_R = \left\lfloor \frac{B_s}{B_c} \right\rfloor + 1, \qquad (10.13)$$

where $\lfloor x \rfloor$ is the largest integer that is less than or equal to x. The number of resolvable paths L_R varies according to the environment, and it is typically higher in urban than in suburban areas, since in urban areas the coherence bandwidth is typically lower due to the typically higher delay-spread of the channel [326]. More explicitly, this is a consequence of the more dipersive impulse response, since the coherence bandwidth is proportional to the reciprocal of the impulses responses delay spread, as it was argued in [32]. Similarly to frequency diversity or space diversity, these L_R resolvable paths can be employed in multipath diversity schemes, which exploit the fact that statistically speaking, the different paths cannot be in deep fades simultaneously; hence, there is always at least one propagation path, which provides an unattenuated channel. These multipath components are diversity paths.

Multipath diversity can only be exploited in conjunction with wideband signals. From Equation 10.13, for a narrowband signal, where no deliberate signal spreading takes place, the signal bandwidth B_s is significantly lower than B_c. In this case, the channel was termed frequency nonselective in [32]. Hence, no resolvable diversity paths can be observed, unlike in a wideband situation, and this renders TDMA and FDMA potentially less robust in a narrowband mobile radio channel than CDMA.

Multipath diversity is achieved, for example, by a receiver referred to as the RAKE receiver invented by Price and Green [331]. This is the optimum receiver for wideband fading multipath signals. It inherited its name from the analogy of a garden rake, whereby the fingers constitute the resolvable paths. The point where the handle and fingers meet is where diversity combining takes place. There are four basic methods of diversity combining, namely [339]:

- Selection Combining (SC).
- Maximal Ratio Combining (MRC).
- Equal Gain Combining (EGC).
- Combining of the n best signals (SCn).

The performance analysis of selection combining in CDMA can be found in [340, 341], while a general comparison of the various diversity combining techniques can be found in [339] for Rayleigh-fading channels. Maximal ratio combining gives the best performance, while selection combining is the simplest to implement. The number of resolvable paths that are combined at the receiver, represents the order of diversity of the receiver, which is denoted here as L_P. We note, however, that in practical receivers not all resolvable multipath components are combined due to complexity reasons, that is, $L_P \leq L_R$.

There are two basic demodulation techniques: coherent and noncoherent; both of them were exemplified in Chapter 5 in the context of quadrature amplitude modulation (QAM), [52]. We will highlight the basics of coherent demodulation in this section in the context of CDMA. However, before demodulation can take place, synchronization between the transmitter and the intended receiver has to be achieved.

Synchronization in DS-CDMA is performed by a process known as code acquisition and tracking. Acquisition is usually carried out by invoking correlation techniques between the receiver's own copy of the signature sequence and the received signature sequence and searching for the displacement between them — associated with a specific chip epoch — that results in the high correlation [328, 342, 343]. Once acquisition has been accomplished, usually a code tracking loop [344] is employed to achieve fine alignment of the two sequences and to maintain their alignment. The details of code acquisition and tracking are beyond the scope of this chapter. Interested readers may refer to [345–348] and the references therein for an in-depth treatise on this subject. Hence, in this chapter, we will assume that the transmitter and the intended receiver are perfectly synchronized.

For optimum performance of the RAKE receiver using coherent demodulation, the path attenuation and phase must be accurately estimated. This estimation is performed by another process known as channel estimation, which will be elaborated on in Section 10.2.6. In typical low-complexity applications, known pilot symbols can be inserted in the transmitted sequence in order to estimate the channel's attenuation and phase rotation, as we demonstrated it in the context of QAM in Chapter 5. However, for now, let us assume perfect channel estimation in order to assess the performance of the RAKE diversity combiner.

Figure 10.9 shows the block diagram of the BPSK RAKE receiver. The received

10.2. BASIC CDMA SYSTEM

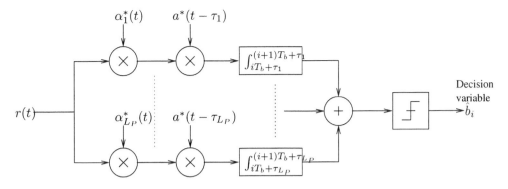

Figure 10.9: RAKE receiver.

signal is first multiplied by the estimated channel coefficients $\alpha_1(t), \ldots, \alpha_{L_P}(t)$ in each RAKE branch tuned to each resolvable path. For optimum performance of the RAKE receiver using maximal ratio combining, these channel coefficient estimates should be the conjugates of the actual coefficients of the appropriate paths in order to invert the channel effects.[3] Note that for equal gain combining only the phase is estimated, and the various path contributions are multiplied by a unity gain before summation. The resulting signals in each RAKE branch are then multiplied by the conjugate signature sequences as we have seen in Figure 10.3, delayed accordingly by the code acquisition process. After despreading by the conjugate signature sequences $a^*(t - \tau_1), \ldots, a^*(t - \tau_{L_P})$, the outputs of the correlators in Figure 10.9 are combined in order to obtain the decoded symbol of:[4]

$$\begin{aligned}
\hat{b}_i &= \mathrm{sgn}\left\{\sum_{l=1}^{L_P}\left[\frac{1}{\sqrt{T_b}}\int_{iT_b+\tau_l}^{(i+1)T_b+\tau_l} \alpha_l^*(t)r(t)a^*(t-\tau_l)\,dt\right]\right\} \\
&= \mathrm{sgn}\left\{\sum_{l=1}^{L_P}\left[\sqrt{\frac{P_b}{T_b}}\int_{iT_b+\tau_l}^{(i+1)T_b+\tau_l} |\alpha_l(t)|^2 b(t-\tau_l)a(t-\tau_l)a^*(t-\tau_l)\,dt\right.\right. \\
&\qquad\left.\left. +\frac{1}{\sqrt{T_b}}\int_{iT_b+\tau_l}^{(i+1)T_b+\tau_l}\alpha_l^*(t)n(t)a^*(t-\tau_l)\,dt\right]\right\} \\
&= \mathrm{sgn}\left\{\sum_{l=1}^{L_P}\left[|\alpha_l(t)|^2 \sqrt{\xi_b}b_i\right.\right. \\
&\qquad\left.\left. +\frac{1}{\sqrt{T_b}}\int_{iT_b+\tau_l}^{(i+1)T_b+\tau_l}\alpha_l^*(t)n(t)a^*(t-\tau_l)\,dt\right]\right\}. \quad (10.14)
\end{aligned}$$

Normally, the first term of Equation 10.14, which contains the useful information, is much larger than the despread, noise-related second term. This is because the first

---

[3] $\alpha_l e^{j\phi_l} \times \alpha_l e^{-j\phi_l} = \alpha_l^2$

[4] Here we assumed that there is no multipath interference. This interference can be considered as part of multi-user interference, which will be discussed in the next section.

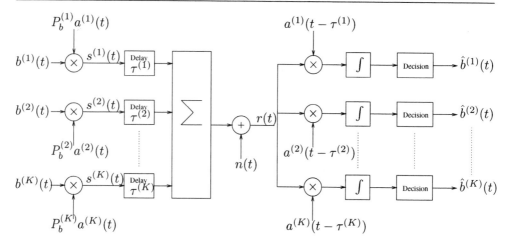

Figure 10.10: CDMA system model.

term is proportional to the sum of the absolute values of the channel coefficients, whereas the second term in Equation 10.14 is proportional to the vectorial sum of the complex-valued channel coefficients. Hence, the real part of the first term is typically larger than that of the second term. Thus, the RAKE receiver can enhance the detection of the data signal in a multipath environment.

Referring back to the BER curves of Figure 10.8, we can see that the performance of the system is improved when multipath diversity is used. Better performance is observed by increasing the number of diversity paths L_P. However, this also increases the complexity of the receiver, since the number of correlators has to be increased, which is shown in Figure 10.9.

10.2.4 Multiple Access

So far, only single-user transmission was considered. The system is simple and straightforward to implement. Let us now consider how multiple user transmission can affect the performance of the system.

Multiple access is achieved in DS-CDMA by allowing different users to share a common frequency band. Each transmitter and its intended receiver are assigned a distinct user signature sequence. Only the receivers having the explicit knowledge of this distinct sequence are capable of detecting the transmitted signal. Consider a CDMA scenario with K number of active users, transmitting simultaneously. The baseband equivalent system model is shown in Figure 10.10. For simplicity, it is assumed that there is no multipath propagation and perfect power control is maintained.

The mathematical representation of the kth user's data signal is similar to that shown in Equation 10.2, except for an additional superscript, denoting multi-user transmission. Hence, it is written as:

$$b^{(k)}(t) = \sum_{j=-\infty}^{\infty} b_j^{(k)} \Gamma_{T_b}(t - jT_b), \qquad (10.15)$$

10.2. BASIC CDMA SYSTEM

where $b_j^{(k)} \in \{+1, -1\}$. There is a distinct user signature sequence $a^{(k)}(t)$ associated with the kth user, which is similar to that of Equation 10.4, with the exception of a superscript, differentiating between users:

$$a^{(k)}(t) = \sum_{h=-\infty}^{\infty} a_h^{(k)} \Gamma_{T_c}(t - hT_c). \qquad (10.16)$$

The kth user's data signal $b^{(k)}(t)$ and signature sequence $a^{(k)}(t)$ are multiplied in order to produce an equivalent baseband wideband signal, namely,

$$s^{(k)}(t) = \sqrt{P_b^{(k)}} b^{(k)}(t) a^{(k)}(t), \qquad (10.17)$$

where $P_b^{(k)}$ is the average transmit power of the kth user's signal. The composite multi-user baseband received signal is:

$$r(t) = \sum_{k=1}^{K} \sqrt{P_b^{(k)}} b^{(k)}(t - \tau^{(k)}) a^{(k)}(t - \tau^{(k)}) + n(t), \qquad (10.18)$$

where $\tau^{(k)}$ is the propagation delay plus the relative transmission delay of the kth user with respect to other users, and $n(t)$ is the AWGN with a double-sided power spectral density of $\frac{N_0}{2}$ W/Hz.

10.2.4.1 Down-Link Interference

In the down-link (base station to mobile), the base station is capable of synchronizing the transmission of all the users' signals, such that the symbol durations are aligned with each other. Hence the composite signal is received at each mobile station with $\tau^{(k)} = 0$ for $k = 1, 2, \ldots, K$. This scenario is also known as symbol-synchronous transmission. Using the conventional so-called single-user detector, each symbol of the jth user is retrieved from the received signal $r(t)$ by correlating it with the jth user's spreading code in order to give:

$$\hat{b}_i^{(j)} = \mathrm{sgn}\left\{ \frac{1}{\sqrt{T_b}} \int_{iT_b}^{(i+1)T_b} r(t) a^{(j)*}(t) dt \right\}. \qquad (10.19)$$

Substituting Equation 10.18 into Equation 10.19 yields:

$$\begin{aligned}
\hat{b}_i^{(j)} &= \mathrm{sgn}\left\{ \frac{1}{\sqrt{T_b}} \int_{iT_b}^{(i+1)T_b} \left[\sum_{k=1}^{K} \sqrt{P_b^{(k)}} b^{(k)}(t) a^{(k)}(t) + n(t) \right] a^{(j)*}(t) \, dt \right\} \\
&= \mathrm{sgn}\left\{ \frac{1}{\sqrt{T_b}} \int_{iT_b}^{(i+1)T_b} \sqrt{P_b^{(j)}} b^{(j)}(t) a^{(j)}(t) a^{(j)*}(t) \, dt \right. \\
&\quad \left. + \frac{1}{\sqrt{T_b}} \int_{iT_b}^{(i+1)T_b} \sum_{\substack{k=1 \\ k \neq j}}^{K} \sqrt{P_b^{(k)}} b^{(k)}(t) a^{(k)}(t) a^{(j)*}(t) \, dt \right.
\end{aligned}$$

$$+ \frac{1}{\sqrt{T_b}} \int_{iT_b}^{(i+1)T_b} n(t)a^{(j)^*}(t)\, dt \Bigg\}$$

$$= \text{sgn} \Bigg\{ \underbrace{\sqrt{\xi_b^{(j)}} b_i^{(j)}}_{\text{wanted signal}} + \underbrace{\sum_{\substack{k=1 \\ k \neq j}}^{K} \sqrt{\xi_b^{(k)}} b_i^{(k)} R_{jk}}_{\text{multiple access interference}} + \underbrace{n^{(j)}}_{\text{white noise}} \Bigg\},$$

(10.20)

where R_{jk} is the cross-correlation of the spreading codes of the kth and jth user for $iT_b \leq t \leq (i+1)T_b$, which is given by:

$$R_{jk} = \frac{1}{T_b} \int_0^{T_b} a^{(j)}(t) a^{(k)}(t)\, dt. \qquad (10.21)$$

There will be no interference from the other users if the spreading codes are perfectly orthogonal to each other. That is, $R_{jk} = 0$ for all $k \neq j$. However, designing orthogonal codes for a large number of users is extremely complex. The so-called Walsh-Hadamard codes [349] used in the IS-95 system excel in terms of achieving orthogonality.

10.2.4.2 Up-Link Interference

In contrast to the previously considered down-link scenario, in practical systems perfect orthogonality cannot be achieved in the up-link (mobile to base station), since there is no coordination in the transmission of the users' signals. In CDMA, all users transmit in the same frequency band in an uncoordinated fashion. Hence, $\tau^{(k)} \neq 0$, and the corresponding scenario is referred to as an asynchronous transmission scenario. In this case, the time-delay $\tau^{(k)}$, $k = 1, ..., K$, has to be included in the calculation. Without loss of generality, it can be assumed that $\tau^{(1)} = 0$ and that $0 < \tau^{(2)} < \tau^{(3)} < ... < \tau^{(K)} < T_b$. In contrast to the synchronous down-link scenario of Equation 10.19, the demodulation of the ith symbol of the jth user is performed by correlating the received signal $r(t)$ with $a^{(j)^*}(t)$ delayed by $\hat{\tau}^{(j)}$, yielding:

$$\hat{b}_i^{(j)} = \text{sgn} \Bigg\{ \frac{1}{\sqrt{T_b}} \int_{iT_b + \hat{\tau}^{(j)}}^{(i+1)T_b + \hat{\tau}^{(j)}} r(t) a^{(j)^*}(t - \hat{\tau}^{(j)})\, dt \Bigg\}, \qquad (10.22)$$

where $\hat{\tau}^{(j)}$ is the estimated time-delay at the receiver.

Substituting Equation 10.18 into Equation 10.22 and assuming perfect code acquisition and tracking yield:[5]

$$\hat{b}_i^{(j)} = \text{sgn} \Bigg\{ \frac{1}{\sqrt{T_b}} \int_{iT_b + \tau^{(j)}}^{(i+1)T_b + \tau^{(j)}} \Bigg[\sum_{k=1}^{K} \sqrt{P_b^{(k)}} b^{(k)}(t - \tau^{(k)}) a^{(k)}(t - \tau^{(k)})$$

---
[5]For perfect code acquisition and tracking, $\hat{\tau}^{(j)} = \tau^{(j)}$.

10.2. BASIC CDMA SYSTEM

$$
\begin{aligned}
&\quad + n(t)] \cdot a^{(j)^*}(t - \tau^{(j)}) dt \bigg\} \\
&= \operatorname{sgn}\bigg\{ \frac{1}{\sqrt{T_b}} \bigg[\int_{iT_b+\tau^{(j)}}^{(i+1)T_b+\tau^{(j)}} \sqrt{P_b^{(j)}} b^{(j)}(t - \tau^{(j)}) a^{(j)}(t - \tau^{(j)}) \\
&\quad \times a^{(j)^*}(t - \tau^{(j)}) dt \\
&\quad + \sum_{k=1}^{j-1} \int_{(i+1)T_b+\tau^{(j)}}^{(i+1)T_b+\tau^{(k)}} \sqrt{P_b^{(k)}} b^{(k)}(t - \tau^{(k)}) a^{(k)}(t - \tau^{(k)}) \\
&\quad \times a^{(j)^*}(t - \tau^{(j)}) dt \\
&\quad + \sum_{k=1}^{j-1} \int_{iT_b+\tau^{(k)}}^{(i+1)T_b+\tau^{(j)}} \sqrt{P_b^{(k)}} b^{(k)}(t + T_b - \tau^{(k)}) a^{(k)}(t + T_b - \tau^{(k)}) \\
&\quad \times a^{(j)^*}(t - \tau^{(j)}) dt \\
&\quad + \sum_{k=j+1}^{K} \int_{iT_b+\tau^{(j)}}^{iT_b+\tau^{(k)}} \sqrt{P_b^{(k)}} b^{(k)}(t - T_b - \tau^{(k)}) a^{(k)}(t - T_b - \tau^{(k)}) \\
&\quad \times a^{(j)^*}(t - \tau^{(j)}) dt \\
&\quad + \sum_{k=j+1}^{K} \int_{iT_b+\tau^{(k)}}^{(i+1)T_b+\tau^{(j)}} \sqrt{P_b^{(k)}} b^{(k)}(t - \tau^{(k)}) a^{(k)}(t - \tau^{(k)}) \\
&\quad \times a^{(j)^*}(t - \tau^{(j)}) dt + \int_{iT_b+\tau^{(j)}}^{(i+1)T_b+\tau^{(j)}} n(t) a^{(j)^*}(t - \tau^{(j)}) dt \bigg] \bigg\} \\
&= \operatorname{sgn}\bigg\{ \underbrace{\sqrt{\xi_b^{(j)}} b_i^{(j)}}_{\text{wanted signal}} + \underbrace{\sum_{k=1}^{j-1} \sqrt{\xi_b^{(k)}} b_i^{(k)} R_{jk}(0)}_{\text{multiple access interference}} \\
&\quad + \underbrace{\sum_{k=1}^{j-1} \sqrt{\xi_b^{(k)}} b_{i+1}^{(k)} \hat{R}_{jk}(+1) + \sum_{k=j+1}^{K} \sqrt{\xi_b^{(k)}} b_{i-1}^{(k)} R_{jk}(-1)}_{\text{multiple access interference}} \\
&\quad + \underbrace{\sum_{k=j+1}^{K} \sqrt{\xi_b^{(k)}} b_i^{(k)} \hat{R}_{jk}(0)}_{\text{multiple access interference}} + \underbrace{n^{(j)}}_{\text{white noise}} \bigg\}, \quad (10.23)
\end{aligned}
$$

where $R_{jk}(i)$ and $\hat{R}_{jk}(i)$, $i \in \{+1, 0, -1\}$ represent the cross-correlation of the spreading codes due to asynchronous transmissions, which are given by [350]:

$$
R_{jk}(i) = \frac{1}{T_b} \int_{\tau^{(j)}}^{\tau^{(k)}} a^{(j)}(t - \tau^{(j)}) a^{(k)}(t + iT_b - \tau^{(k)}) dt \quad (10.24)
$$

and

$$\hat{R}_{jk}(i) = \frac{1}{T_b} \int_{\tau^{(k)}}^{T_b+\tau^{(j)}} a^{(j)}(t-\tau^{(j)})a^{(k)}(t+iT_b-\tau^{(k)})dt \qquad (10.25)$$

and is limited to $+1, 0, -1$, since the maximum path delay is assumed to be limited to one symbol duration, as mentioned in Section 10.2.2.

Equations 10.23 and 10.20 represent the estimated demodulated data symbol of the jth user at the base station and mobile station, respectively. Both contain the desired symbol of the jth user. However, this is corrupted by noise and interference from the other users. This interference is known as multiple access interference (MAI). It contains the undesired interfering signals from the other $(K-1)$ users. The MAI arises due to the nonzero cross-correlation of the spreading codes. Ideally, the spreading codes should satisfy the orthogonality property such that

$$R_{jk}(\tau) = \frac{1}{T_b} \int_0^{T_b} a^{(k)}(t)a^{(j)}(t-\tau)dt = \begin{cases} 1 & \text{for } k=j, \tau=0 \\ 0 & \text{for all } k \text{ and all } \tau. \end{cases} \qquad (10.26)$$

However, it is impossible to design codes that are orthogonal for all possible time offsets imposed by the asynchronous up-link transmissions. Thus there will always be MAI in the up-link. These observations are augmented by comparing the terms of Equations 10.20 and 10.23.

On the other hand, multipath interference is always present in both the forward and reverse link. Multipath interference is due to the different arrival times of the same signal via the different paths at the receiver. This is analogous to the signals transmitted from other users; hence, multipath interference is usually analyzed in the same way as MAI.

As the number of users increases, the MAI increases too. Thus, the capacity of CDMA is known to be interference limited. CDMA is capable of accommodating additional users at the expense of a gradual degradation in performance in a fixed bandwidth, whereas TDMA or FDMA would require additional bandwidth to accommodate additional users. Intensive research has been carried out to find ways of mitigating the effects of MAI. Some of the methods include voice activity control, spreading code design, power control schemes, and sectored/adaptive antennas [351]. These methods reduce the MAI to a certain extent.

The most promising up-link method so far has been in the area of *multi-user detection*, which was first proposed by Verdu [352]. Multi-user detection [99,353,354] invokes the knowledge of all users' signature sequences and all users' channel impulse response estimates in order to improve the detection of each individual user. The employment of this algorithm is more feasible for the up-link, because all mobiles transmit to the base station and the base station has to detect all the users' signals anyway. Multi-user detection is beyond the scope of this chapter, and hence interested readers are referred to [355] for a comprehensive discussion on the topic. Its concept and practical employment will be augmented in the context of a multi-user video system in Chapter 21. A general review of the various multi-user detection schemes and further references can be found, for example, in [351].

10.2. BASIC CDMA SYSTEM

Another shortcoming of CDMA systems is their susceptibility to the near-far problem to be highlighted below. If all users transmit at equal power, then signals from users near the base station are received at a higher power than those from users at a higher distance due to their different path-losses. The effects of fading highlighted in Section 10.2.2 also contribute to the power variation. Hence, according to Equation 10.23, if the jth user is transmitting from the cell border and all other users are transmitting near the base station, then the desired jth user's signal will be masked by the other users' stronger signals, which results in a high bit error rate. In order to mitigate this so-called near-far problem, power control is used to ensure that all signals from the users are received at near-equal power, regardless of their distance from the base station.

There are typically two basic types of power control [329]:

- open-loop power control
- closed-loop power control

Open-loop power control is usually used to overcome the variation in power caused by path-loss. On the other hand, closed-loop power control is used to overcome shadow fading caused by multipath. The details of the various power control techniques will not be elaborated on in this chapter. Readers may refer to [356] for more information.

10.2.4.3 Gaussian Approximation

In order to simplify any analysis involving multi-user transmission in CDMA, the MAI is usually assumed to be Gaussian distributed by virtue of the central limit theorem [357–359]. This assumption is fairly accurate even for $K < 10$ users, when the BER is 10^{-3} or higher. We will use the standard Gaussian approximation theory presented by Pursley [357] to represent the MAI. When the desired user sequence is chip- and phase-synchronous with all the interfering sequences, where the phase-synchronous relationship is defined as in the absence of noise, the worst-case probability of error $Pr_b(\epsilon)$ performance was given by Pursley [357] as:

$$Pr_b(\epsilon) = Q\left[\sqrt{\frac{N_c}{(K-1)}}\right], \qquad (10.27)$$

where $Q(\cdot)$ is the Gaussion Q-function of Equation 10.9, since the synchronous transitions do not generate pure random Gaussian-like impairments. This formula would be characteristic of the synchronous down-link scenario of Section 10.2.4.1. However, in practical up-link situations as augmented in Section 10.2.4.2, there is always some delay among the users, and each received signal will be phase-shifted independently. In this case, according to Pursley, the probability of error in the absence of noise will be [357]:

$$Pr_b(\epsilon) = Q\left[\sqrt{\frac{3N_c}{(K-1)}}\right]. \qquad (10.28)$$

Equation 10.28 represents the best performance corresponding to Gaussian-like impairments. In between these two extremes are situations whereby, in the first case,

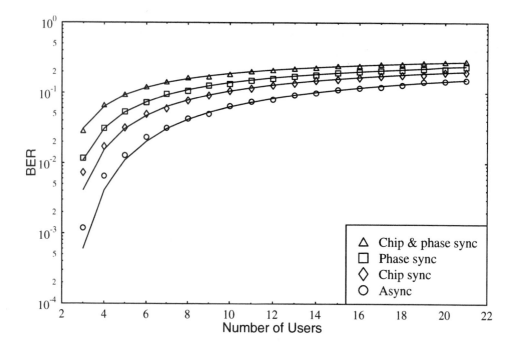

Figure 10.11: Probability of error against number of users using Equations 10.27, 10.28, 10.29, and 10.30. Markers: Simulation; solid line - Numerical computation. The processing gain is 7.

the desired sequence and the interfering sequence are chip synchronous but not phase synchronous. The probability of error in the absence of noise is given by [357]:

$$Pr_b(\epsilon) = Q\left[\sqrt{\frac{2N_c}{(K-1)}}\right]. \tag{10.29}$$

In the second case, the desired sequence and interfering sequence are phase synchronous but not chip synchronous. Hence, the probability of error in the absence of noise is given by [357]:

$$Pr_b(\epsilon) = Q\left[\sqrt{\frac{3N_c}{2(K-1)}}\right]. \tag{10.30}$$

Analyzing the above equations, it can be seen that by increasing the number of chips N_c per symbol, the performance of the system will be improved. However, there is a limitation to the rate of the spreading sequence based on Digital Signal Processing (DSP) technology. Figure 10.11 compares the simulated results with the numerical results given by Equations 10.27 to 10.30 for a binary system with a processing gain of 7. The figure shows that the assumption of Gaussian distributed MAI is valid, especially for a high number of users. It also demonstrates that *CDMA attains its best possible performance in an asynchronous multi-user transmission system.* This

10.2. BASIC CDMA SYSTEM

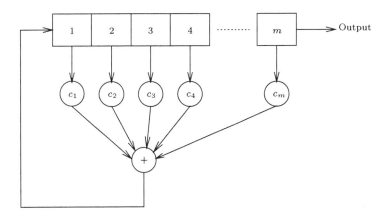

Figure 10.12: m-stage shift register with linear feedback.

is an advantage over TDMA and FDMA because TDMA and FDMA require some coordination among the transmitting users, which increases the complexity of the system.

10.2.5 Spreading Codes

As seen previously, the choice of spreading codes plays an important role in DS-CDMA. The main criteria for selecting a particular set of user signature sequences in CDMA applications are that the number of possible different sequences in the set for any sequence length must be high in order to accommodate a high number of users in a cell. The spreading sequences must also exhibit low cross-correlations for the sake of reducing the multi-user interference during demodulation. A high autocorrelation main-peak to secondary-peak ratio — as indicated by Equation 10.26 — is also essential, in order to minimize the probability of so-called false alarms during code acquisition. This also reduces the self-interference among the diversity paths. Below we provide a brief overview of a few different spreading sequences.

10.2.5.1 m-sequences

Perhaps the most popular set of codes known are the m-sequences [52]. An m-sequence with a periodicity of $n = 2^m - 1$ can be readily generated by an m-stage shift register with linear feedback, as shown in Figure 10.12.

The tap coefficients $c_1, c_2, \ldots, c_m$ can be either 1 (short circuit) or 0 (open circuit). Information on the shift register feedback polynomials, describing the connections between the register stages and the modulo-2 adders can be found, for example, in [52]. Note that in spread spectrum applications, the output binary sequences of 0,1 are mapped into a bipolar sequence of -1,1, respectively. Table 10.1 shows the total number of m-sequences and the associated chip-synchronous peak cross-correlation for m = 3, 4, 5, 6, 7, and 8. In this context, the peak cross-correlation quantifies the maximum number of identical chips in a pair of different spreading codes. It is desirable

m	Number of m-sequences	Peak Cross-Correlation	Number of Gold Sequences	Peak Cross-Correlation
3	2	5	$2^m + 1 = 9$	5
4	2	9	$2^m + 1 = 17$	9
5	6	11	$2^m + 1 = 33$	9
6	6	23	$2^m + 1 = 65$	17
7	18	41	$2^m + 1 = 129$	17
8	16	95	$2^m + 1 = 257$	33

Table 10.1: Properties of m- and Gold-Sequences. ©McGraw-Hill, 1995 [52]

to have as low a number of code pairs as possible, which exhibit this peak cross-correlation. Furthermore, the peak cross-correlation has to be substantially lower than the codes' autocorrelation, which is given by the length of the code. In general, the cross-correlations of m-sequences are too high to be useful in CDMA. Another set of spreading codes, which exhibit fairly low chip-synchronous cross-correlations are the Gold sequences [52], which will be elaborated on in the next section.

10.2.5.2 Gold Sequences

Gold sequences [52] with a period of $n = 2^m - 1$ are derived from a pair of m-sequences having the same period. Out of the total number of possible m-sequences having a periodicity or length of n, there exists a pair of m-sequences, whose chip-synchronous cross-correlation equals to either $-1, -t(m)$ or $[t(m) - 2]$, where

$$t(m) = \begin{cases} 2^{(m+1)/2} + 1 & \text{odd } m \\ 2^{(m+2)/2} + 1 & \text{even } m. \end{cases} \quad (10.31)$$

This unique pair of m-sequences is commonly known as the pair of preferred codes. A set of $n = 2^m - 1$ sequences can be constructed by cyclically shifting a preferred code one chip at a time and then taking the modulo-2 summation with the other code for every chip shift. The resulting set of $n = 2^m - 1$ sequences together with the two preferred codes constitute a set of Gold sequences. Table 10.1 compares the total number of Gold sequences for $m = 3, 4, 5, 6, 7$, and 8, and their corresponding peak cross-correlation with the same parameters of m-sequences.

Table 10.1, shows that the Gold sequences exhibit equal or lower peak cross-correlation between different sequences of the set, in comparison to m-sequences for all m. There are also more Gold sequences than m-sequences for all values of m. Thus, Gold sequences are always preferred to m-sequences in CDMA applications, despite having a poorer asynchronous autocorrelation peak, which is a disadvantage in terms of both code acquisition and detection by correlators. Pseudo Noise (PN) sequences, such as m-sequences and Gold sequences, have periods of $N = 2^l - 1$ where l is the sequence length, which is a rather awkward number to match to the system clock requirements. Extended m-sequences having periods of 2^l solved this problem, an issue augmented below.

10.2.5.3 Extended m-sequences [360]

Extended m-sequences are derived from an m-sequence, generated by a linear feedback shift register, by adding an element into each period of the m-sequence. We will follow the notation, whereby the binary sequences of 0 and 1 are mapped to the corresponding bipolar sequences of -1 and +1, respectively. In order to arrive at zero-balanced extended m-sequences, which have a zero DC-component, the element to be inserted must be chosen so that the number of -1s and +1s within a period is the same. There are $2^m - 1$ positions in a period, where the additional element can be inserted. In [360], the element is inserted into the longest run of -1s in a period. In an m-sequence of period $2^m - 1$, the longest run of -1s is $n - 1 = 2^m - 2$. It was shown in [360] that the off-peak autocorrelation of extended sequences was similar to that of Gold sequences. However, the cross-correlation of different extended m-sequences at even-indexed chip-positions — that is, time-domain displacements — is similar to that of the m-sequences, which is much higher than that of the Gold sequences in Table 10.1. Thus, the extended m-sequences are not suitable in a multi-user environment, where the cross-correlation between the codes of different users is required to be as low as possible. Since this has a high impact on the user-capacity of cellular mobile systems, the additional hardware needed to synchronize the $N = 2^l - 1$ chip-duration m-sequences or Gold sequences with the system clock has to be tolerated and hence, extended m-sequences are not recommended in CDMA. Section 10.3.2.6 highlights the various spreading codes proposed for employment in the forthcoming 3G systems. In the next section we provide a rudimentary introduction to channel estimation for CDMA systems.

10.2.6 Channel Estimation

As mentioned earlier, accurate estimation of the channel parameters is vital in optimizing the coherent demodulation. This channel parameter estimation process is an integral part of coherent demodulation, particularly in a multipath mobile radio environment. This is because the mobile radio channel changes randomly as a function of time, and thus the channel estimates have to be continuously estimated. This section describes various techniques used to estimate the channel path gains and phases, which will be referred to as channel coefficients. There are basically three practical channel coefficient estimation methods, each with their advantages and disadvantages, namely :

- Pilot-channel assisted, [122, 125, 361]

- Pilot-symbol assisted and [130]

- Pilot-symbol assisted decision-directed channel estimation [362],

which we briefly characterized in the following subsections. Recall at this stage that pilot-symbol assisted QAM was also the subject of Chapter 5.

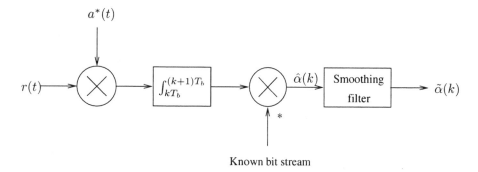

Figure 10.13: Structure of the channel estimator using known transmitted pilot symbols or bits.

10.2.6.1 Down-Link Pilot-Assisted Channel Estimation

Channel estimation using a pilot channel/tone was proposed, for example, in [122,125, 361], where a channel is dedicated solely for the purpose of estimating the multipath channel attenuations and delays. In order to prevent the pilot channel from interfering with the data channel, the pilot channel must either be allocated to a dedicated portion of the spectrum or share the spectrum with the data channel, but a spectral notch has to be created for accommodating the pilot. The former technique is known as the pilot tone-above-band (TAB) regime, while the latter is referred to as the transparent tone-in-band (TTIB) technique [122], both of which have been used in conventional single-carrier modems [29].

However, CDMA is more amenable to employing the TAB or TTIB techniques and their various derivatives, since the pilot signal can be transmitted in the same frequency band as the data signal by invoking orthogonal or quasi-orthogonal spreading codes. Hence, the pilot signal is treated as part of the MAI, and no notch filtering or additional pilot frequency band is required. In some 2G mobile systems, such as the IS-95 system of Table 9.1, this method is used on the down-link but not on the up-link. This is because it would be inefficient to have every mobile station transmitting their own pilot channel.

In 3G mobile systems, however, it was proposed [363] that a separate dedicated user control channel be transmitted simultaneously with the information channel, which could also be used as an alternative to the pilot channel - an issue to be elaborated on at a later stage. Suffice to say here that the main advantage of pilot-channel based channel estimation is that since the pilot channel is always present, the channel coefficients can be continuously estimated for every data symbol's demodulation. Hence, it is particularly useful for channels that are highly time-variant.

The block diagram of the channel estimator is shown in Figure 10.13, where $r(t)$ is the received signal and $a(t)$ is the spreading code. Assume that the known bit-stream is a continuous sequence of binary 1s, then

$$\hat{\alpha}(k) = \frac{1}{T_b} \int_{kT_b}^{(k+1)T_b} r(t) a^*(t) dt$$

10.2. BASIC CDMA SYSTEM

$$= \frac{1}{T_b} \int_{kT_b}^{(k+1)T_b} [\alpha(t)a(t) + n(t)]a^*(t)dt$$

$$= \alpha(k) + \frac{1}{T_b} \int_{kT_b}^{(k+1)T_b} n(t)a^*(t)dt, \qquad (10.32)$$

where $\alpha(k)$ is the complex channel coefficient in the bit interval $kT_b \leq t < (k+1)T_b$. The variable $\hat{\alpha}(k)$ is termed the noisy channel estimate derived from the received signal contaminated by the noise element in the second term of Equation 10.32, while $\tilde{\alpha}(k)$ are estimates obtained from the output of the smoothing filter in Figure 10.13, which assists in averaging out the random effects of channel noise. Assuming that $n(t)$ is the AWGN having a zero mean (any MAI can be fairly accurately modeled also as AWGN [364]), averaging a large number of these noisy estimates will suppress the noise's influence. Several proposals have been published in the literature regarding the smoothing algorithm used in channel estimation, such as moving average [365, 366], least squares line fitting [11], low-pass filtering [11, 130, 362], and adaptive linear smoothing [120]. A more in-depth discourse on the TTIB technique was also given in Section 10.3.1 of [29] in the context of QAM. A compromise in terms of complexity and accuracy has to be made in selecting a particular algorithm. So far, only the down-link channel estimation has been elaborated on. The associated up-link issues are discussed next.

10.2.6.2 Up-Link pilot-symbol assisted Channel Estimation

Pilot-symbol assisted channel estimation was first proposed by Moher anand Lodge [130], and the first detailed analysis of this technique was carried out by Cavers [120]. Since then, several papers have been published, which analyzed its effect on system performance [11, 366, 367]. This technique is the time-domain equivalent of the frequency-domain pilot channel-assisted TTIB method mentioned in Section 10.2.6.1, which was detailed in Section 10.3.2 of [29]. The advantage of this technique is that it dispensed with the use of a notch filter in the context of QAM modems, and so it did not result in an expanded bandwidth. However, for this technique, several parameters such as the number of pilot symbols or their periodicity has to be carefully chosen in order to trade-off the accuracy of estimation against the required pilot overhead. More explicitly, the pilot-spacing has to be sufficiently low to satisfy the Nyquist sampling theorem for the fading Doppler frequency encountered. This technique can be used for efficient coherent demodulation on the up-link, and Section 10.3.2.3 highlights how up-link channel estimation is carried out in the context of 3G systems.

The pilot symbols are multiplexed with the data stream periodically, as shown in Figure 10.14. This multiplexed stream is then transmitted to the base station from every communicating mobile station. The base station will extract the channel estimates from the known demodulated pilot symbols, and using, for example, ideal low-pass or simple linear interpolation [11], it will generate a channel magnitude and phase estimate for each up-link symbol. These channel estimates will then be used to "de-fade," "de-rotate," and demodulate the data symbols.

If the channel has a slow fading characteristic, such that it is more or less constant between consecutive pilot symbols, this method can be fairly accurate and of low

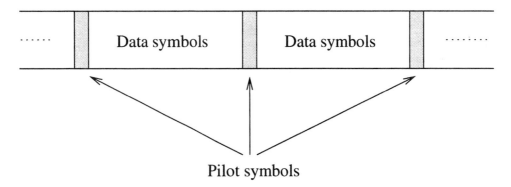

Figure 10.14: Data stream with embedded pilot symbols.

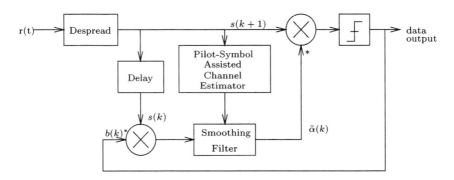

Figure 10.15: Receiver structure of PSA decision-directed channel estimation.

complexity. However, the bandwidth efficiency is slightly compromised, since again, a sufficiently high number of pilots has to be incorporated in order to satisfy the Nyquist sampling criterion corresponding to the normalized Doppler frequency of the fading channel. For more information on this subject we refer to Chapter 5 or to Section 10.3.2 of [29]. The above pilot-symbol assisted (PSA) concept is further developed in the next section.

10.2.6.3 Pilot-Symbol Assisted Decision-Directed Channel Estimation

Pilot-symbol assisted decision-directed channel estimation was first proposed by Irvine and McLane [362], and it was shown that it improves the accuracy of the estimation as compared with the original pilot symbol-assisted method of Section 10.2.6.2. It extends the concept of the pilot-symbol assisted channel estimation technique by using the detected data symbols in order to obtain the subsequent channel parameters, since in the absence of channel errors these demodulated data symbols can be considered to be known pilot symbols.

A decision-directed pilot-symbol assisted (PSA) scheme is illustrated in Figure 10.15, where $s(k)$ is the kth received symbol and $b(k)$ is the kth detected symbol. The signal is still transmitted in a transmission burst or frame format, similarly to that shown

10.2. BASIC CDMA SYSTEM

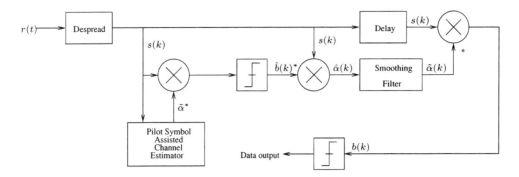

Figure 10.16: Receiver structure using decision-feedforward PSA channel estimation.

in Figure 10.14. At the beginning of the frame, the pilot symbols will be used to estimate the channel parameters in order to demodulate the data symbol immediately following the pilot symbol. This is performed by the pilot symbol-assisted channel estimator block of Figure 10.15. The detected data symbol $b(k)$ is then fed back and multiplied with its original but delayed received version $s(k)$, as seen in Figure 10.15. If this symbol is detected correctly, then it is analogous to a known pilot symbol and the channel coefficient corresponding to this received symbol can be estimated in the same way. This estimated channel coefficient is then passed through the smoothing filter of Figure 10.15 in order to obtain a smoothed estimate $\tilde{a}(k)$ to be used in its conjugate form for de-fading and de-rotating the next symbol, as portrayed in Figure 10.15.

If the decision is wrong, obviously the estimated channel coefficient would be inaccurate. The effect of erroneous decisions is mitigated by the smoothing filter, which will suppress the effects of an occasional glitch due to the incorrect channel estimates. In the event that the smoothing filter is unable to average out the channel coefficient errors and its output is a complex channel coefficient, which is far from the actual value, then this error may propagate through the data stream, since the correct decoding of each data symbol is dependent on the accuracy of the previous channel coefficient estimates. In order to prevent this from happening, the smoothing process is reset, when the next block of pilot symbols arrives. The averaging process will recommence with the pilot symbol-assisted channel estimates.

The schematic shown in Figure 10.15 is only one of the few possibilities of implementing a decision-directed PSA channel estimation arrangement. This structure is also known as a decision-feedback PSA channel estimator because the estimated channel coefficient is used for compensating the channel's effects for the next symbol. In another version of this algorithm, shown in Figure 10.16, a tentative decision, $\hat{b}(k)$, is carried out concerning the current symbol, $s(k)$, using the pilot symbol-assisted estimate, $\tilde{a}^*(k)$. Using this tentative decision concerning the received symbol $s(k)$, its corresponding channel coefficient estimate, $\hat{a}(k)$, is derived from the product of $\hat{b}^*(k)$ and $s(k)$ in Figure 10.15 and averaged or smoothed with the aid of the previous estimates. The output of the smoothing filter is then multiplied with the received signal $s(k)$ again, in order to compensate the channel attenuation and phase rotation and hence to obtain the final decision, $b(k)$. Such an estimator is known as a feedforward estimator. This implementation is slightly more complicated but has the advantage of using the current estimate on the current symbol rather than tolerating a latency in the channel estimation process.

10.2.7 Summary

In this section we have briefly studied the fundamentals of a CDMA system. We have seen that several processes are vital in optimizing the performance, such as spreading, channel estimation, code synchronization, and power control. In the subsequent sections, we will make certain assumptions that will ease our analysis and simulation. These assumptions are :

- Perfect code acqusition and tracking. Hence, the transmitter and the intended receiver will always be synchronized for every path.

- Perfect channel estimation. This assumption will be used unless our focus is on the effects of imperfect estimation.

- Gaussian approximation of multi-user and multipath interference. This assumption will be used only in analysis and numerical computation, and will be validated by simulations performed in actual multi-user and multipath transmission scenarios. This also implies that random sequences will be considered instead of the deterministic sequences introduced in Section 10.2.5.

- On the up-link, the number of paths encountered by each user's signal is equal.

- Perfect power control is used. This implies that all users' signals will be received at the base station with equal power.

Following the above rudimentary considerations on PSA channel estimation, let us now review the third-generation (3G) mobile system proposals in the next section.

10.3 Third-Generation Wireless Mobile Communication Systems

10.3.1 Introduction

The evolution of third-generation (3G) wireless systems began in the late 1980s when the International Telecommunication Union's Radiocommunication Sector (ITU-R) Task Group 8/1 defined the requirements for the 3G mobile radio systems. This initiative was then known as Future Public Land Mobile Telecommunication System (FPLMTS) [319, 325]. The frequency spectrum for FPLMTS was identified on a worldwide basis during the World Administrative Radio Conference (WARC) in 1992 [325], as the bands 1885–2025 MHz and 2110–2200 MHz.

The tongue-twisting acronym of FPLMTS was also aptly changed to IMT-2000, which refers to the International Mobile Telecommunications system in the year 2000. Besides possessing the ability to support services from rates of a few kbps to as high as

10.3. THIRD-GENERATION SYSTEMS

2 Mbps in a spectrally efficient way, IMT-2000 aimed to provide a seamless global radio coverage for global roaming. This implied the ambitious goal of aiming to connect virtually any two mobile terminals worldwide. The IMT-2000 system was designed to be sufficiently flexible in order to operate in any propagation environment, such as indoor, outdoor to indoor, and vehicular scenarios. It is also aiming to be sufficiently flexible to handle circuit as well as packet mode services and to handle services of variable data rates. In addition, these requirements must be fulfilled with a quality of service (QoS) comparable to that of the current wired network at an affordable cost.

Several regional standard organizations — led by the European Telecommunications Standards Institute (ETSI) in Europe, the Association of Radio Industries and Businesses (ARIB) in Japan, and the Telecommunications Industry Association (TIA) in the United States — have been dedicating their efforts to specifying the standards for IMT-2000. A total of 15 Radio Transmission Technology (RTT) IMT-2000 proposals were submitted to ITU-R in June 1998, five of which are satellite-based solutions, while the rest are terrestrial solutions. Table 10.2 shows a list of the terrestrial-based proposals submitted by the various organizations and their chosen radio access technology.

As shown in Table 10.2 most standardization bodies have based their terrestrial oriented solutions on *Wideband-CDMA* (W-CDMA), due to its advantageous properties, which satisfy most of the requirements specified for 3G mobile radio systems. W-CDMA is aiming to provide improved coverage in most propagation environments in addition to an increased user capacity. Furthermore, it has the ability to combat — or to benefit from — multipath fading through RAKE multipath diversity combining [239–241]. W-CDMA also simplifies frequency planning due to its unity frequency reuse. A rudimentary discourse on the RTT proposals submitted by ETSI, ARIB, and TIA can be found in [368].

Recently, several of the regional standard organizations have agreed to cooperate and jointly prepare the Technical Specifications (TS) for the 3G mobile systems in order to assist as well as to accelerate the ITU process for standardization of IMT-2000. This led to the formation of two Partnership Projects (PPs), which are known as 3GPP1 [369] and 3GPP2 [370]. 3GPP1 was officially launched in December 1998 with the aim of establishing the TS for IMT-2000 based on the evolved Global System for Mobile Telecommunications (GSM) [41] core networks and the UMTS[6] Terrestrial Radio Access (UTRA) RTT proposal. There are six organizational partners in 3GPP1: ETSI, ARIB, the China Wireless Telecommunication Standard (CWTS) group, the Standards Committee T1 Telecommunications (T1, USA), the Telecommunications Technology Association (TTA, Korea), and the Telecommunication Technology Committee (TTC, Japan). The first set of specifications for UTRA was released in December 1999, which contained detailed information on not just the physical layer aspects for UTRA, but also on the protocols and services provided by the higher layers. Here we will concentrate on the UTRA physical layer specifications, and a basic familiarity with CDMA principles is assumed.

In contrast to 3GPP1, the objective of 3GPP2 is to produce the TS for IMT-2000 based on the evolved ANSI-41 core networks, the cdma2000 RTT. 3GPP2 is

---

[6]UMTS, an abbreviation for Universal Mobile Telecommunications System, is a term introduced by ETSI for the 3G wireless mobile communication system in Europe.

Proposal	Description	Access Technology	Source
DECT	Digital Enhanced Cordless Telecommunications	Multicarrier TDMA (TDD)	ETSI Project (EP) DECT
UWC-136	Universal Wireless Communications	TDMA (FDD and TDD)	USA TIA TR45.3
WIMS W-CDMA	Wireless Multimedia and Messaging Services Wideband CDMA	Wideband CDMA (FDD)	USA TIA TR46.1
TD-CDMA	Time Division Synchronous CDMA	Hybrid with TDMA/CDMA/SDMA (TDD)	Chinese Academy of Telecommunication Technology (CATT)
W-CDMA	Wideband CDMA	Wideband DS-CDMA (FDD and TDD)	Japan ARIB
CDMA II	Asynchronous DS-CDMA	DS-CDMA (FDD)	South Korean TTA
UTRA	UMTS Terrestrial Radio Access	Wideband DS-CDMA (FDD and TDD)	ETSI SMG2
NA: W-CDMA	North America Wideband CDMA	Wideband DS-CDMA (FDD and TDD)	USA T1P1-ATIS
cdma2000	Wideband CDMA (IS-95)	DS-CDMA (FDD and TDD)	USA TIA TR45.5
CDMA I	Multiband synchronous DS-CDMA	Multiband DS-CDMA	South Korean TTA

Table 10.2: Proposals for the Radio Transmission Technology of Terrestrial IMT-2000 (obtained from ITU's web site: http://www.itu.int/imt)

spearheaded by TIA, and its members include ARIB, CWTS, TTA, and TTC. Despite evolving from completely diversified core networks, members from the two PPs have agreed to cooperate closely in order to produce a globally applicable TS for the 3G mobile systems.

This chapter serves as an overview of the UTRA specifications, which is based on the evolved GSM core network. However, information given here is by no means the final specifications for UTRA or indeed for IMT-2000. It is very likely that the parameters and technologies presented in this chapter will evolve further. Readers may also want to refer to a recent book by Ojanperä and Prasad [371], which addresses W-CDMA 3G mobile radio systems in more depth.

10.3.2 UMTS Terrestrial Radio Access (UTRA) [301, 302, 323, 369, 371–376]

Research activities for UMTS [298, 319, 320, 322, 324, 372, 373] within ETSI have been spearheaded by the European Union's (EU) sponsored programs, such as the Research in Advanced Communication Equipment (RACE) [363, 377] and the Advanced Communications Technologies and Services (ACTS) [298, 372, 377] initiative. The RACE program, which is comprised of two phases, commenced in 1988 and ended in 1995. The objective of this program was to investigate and develop testbeds for the air interface technology candidates. The ACTS program succeeded the RACE programe in 1995. Within the ACTS Future Radio Wideband Multiple Access System (FRAMES) project, two multiple access modes have been chosen for intensive study, as the candidates for UMTS terrestrial radio access (UTRA). They are based on Time Division Multiple Access (TDMA) with and without spreading, and on W-CDMA [321, 323, 378].

As early as January 1997, ARIB decided to adopt W-CDMA as the terrestrial radio access technology for its IMT-2000 proposal and proceeded to focus its activities on the detailed specifications of this technology [322]. Driven by a strong support behind W-CDMA worldwide and this early decision from ARIB, ETSI reached a consensus agreement in January 1998 to adopt W-CDMA as the terrestrial radio access technology for UMTS. In this section, we highlight the key features of the physical layer aspects of UTRA that have been developed since then. Most of the material in this section is based on an amalgam of [301, 302, 323, 369, 371–376].

10.3.2.1 Characteristics of UTRA

The proposed spectrum allocation for UTRA is shown in Figure 10.17. As can be seen, UTRA is unable to utilize the full frequency spectrum allocated for the 3G mobile radio systems during the WARC'92, since those frequency bands have also been partially allocated to the Digital Enhanced Cordless Telecommunications (DECT) systems. Also, the allocated frequency spectrum was originally based on the assumption that speech and low data rate transmission would be the dominant services offered by IMT-2000. However, this assumption has become invalid, as the trend has shifted toward services that require high-speed data transmission, such as Internet access and multimedia services. A study conducted by the UMTS Forum [379] forecasted that

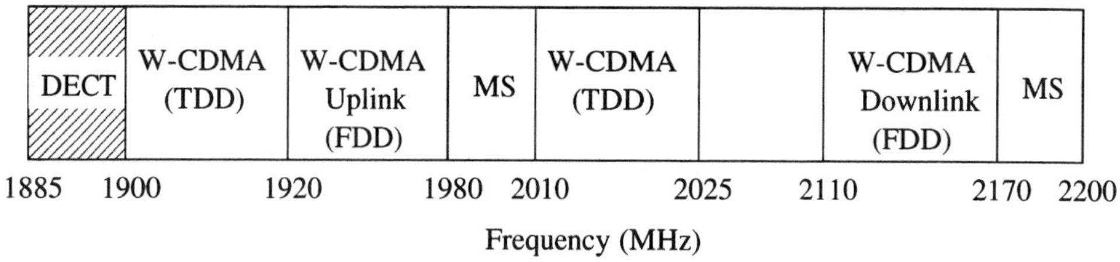

| DECT | W-CDMA (TDD) | W-CDMA Uplink (FDD) | MS | W-CDMA (TDD) | | W-CDMA Downlink (FDD) | MS |

1885   1900   1920   1980 2010   2025   2110   2170 2200

Frequency (MHz)

MS : Mobile satellite application
DECT : Digital Enhanced Cordless Telecommunications
FDD : Frequency Division Duplex
TDD : Time Division Duple
DECT frequency band : 1880 - 1900 MHz

Figure 10.17: The proposed spectrum allocation in UTRA.

the current frequency bands allocated for IMT-2000 are only sufficient for the initial deployment until the year 2005. According to the current demand estimates, it was foreseen that an additional frequency spectrum of 187 MHz is required for IMT-2000 in high-traffic demand areas by the year 2010. This extension band will be identified during the World Radio Conference (WRC)-2000. Among the many candidate extension bands, the band 2520-2670 MHz has been deemed by many people to be the most likely to be chosen. Unlike other bands, which have already been allocated for use in other applications, this band was allocated to mobile services in all regions. Furthermore, the 150 MHz bandwidth available is sufficiently wide to satisfy most of the forecasted spectrum requirements.

The radio access supports both *Frequency Division Duplex* (FDD) and *Time Division Duplex (TDD)* operations. The operating principles of these two schemes are augmented here in the context of Figure 10.18.

Specifically, the up-link (UL) and down-link (DL) signals are transmitted using different carrier frequencies f_1 and f_2, respectively, separated by a frequency guard band in FDD mode. On the other hand, the UL and DL messages in the TDD mode are transmitted using the same carrier frequency f_c, but in different time-slots, separated by a guard period. As seen from the spectrum allocation in Figure 10.17, the paired bands of 1920–1980 MHz and 2110–2170 MHz are allocated for FDD operation in the UL and DL, respectively, whereas the TDD mode is operated in the remaining unpaired bands [372]. The parameters designed for FDD and TDD operations are mutually compatible so as to ease the implementation of a dual-mode terminal capable of accessing the services offered by both FDD and TDD operators.

We note furthermore that recent research advocates the TDD mode quite strongly in the context of burst-by-burst adaptive CDMA modems [99,354], in order to adjust the modem parameters, such as the spreading factor or the number of bits per symbol on a burst-by-burst basis. This allows the system to more efficiently exploit the time-variant wireless channel capacity, hence maintaining a higher bits/s/Hz bandwidth efficiency. Furthermore, there have been proposals in the literature for allowing TDD operation in certain segments of the FDD spectrum as well, since FDD is incapable

10.3. THIRD-GENERATION SYSTEMS

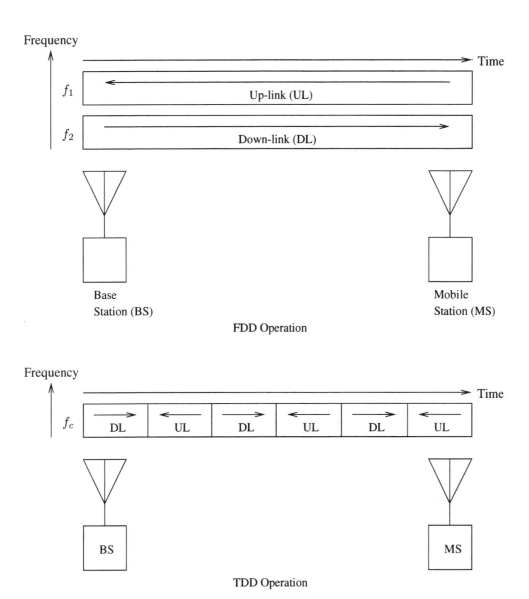

Figure 10.18: Principle of FDD and TDD operation.

Radio Access Technology	FDD : DS-CDMA
	TDD : TDMA/CDMA
Operating environments	Indoor/Outdoor to indoor/Vehicular
Chip rate (Mcps)	3.84
Channel bandwidth (MHz)	5
Nyquist rolloff factor	0.22
Duplex modes	FDD and TDD
Channel bit rates (kbps)	FDD (UL) : 15/30/60/120/240/480/960
	FDD (DL) : 15/30/60/120/240/480/960/1920
	TDD (UL)† : variable, from 366 to 6624
	TDD (DL)† : 366/414/5856/6624
Frame length	10 ms
Spreading factor	FDD (UL) : variable, 4 to 256
	FDD (DL) : variable, 4 to 512
	TDD (UL) : variable, 1 to 16
	TDD (DL) : 1, 16
Detection scheme	Coherent with time-multiplexed pilot symbols
	Coherent with common pilot channel
Intercell operation	FDD : Asynchronous
	TDD : Synchronous
Power control	Inner-loop
	Open loop (TDD UL)
Transmit power dynamic range	80 dB (UL), 30 dB (DL)
Handover	Soft handover
	Inter-frequency handover

† Channel bit rate per time-slot.

Table 10.3: UTRA Basic Parameters

of surrendering the UL or DL frequency band of the duplex link, when the traffic demand is basically simplex. In fact, segmenting the spectrum in FDD/TDD bands inevitably results in some inefficiency in bandwidth utilization terms, especially in case of asymmetric or simplex traffic, when only one of the FDD bands is required. Hence, the more flexible TDD link could potentially double the link's capacity by allocating all time-slots in one direction. The idea of eliminating the dedicated TDD band was investigated [380], where TDD was invoked within the FDD band by simply allowing TDD transmissions in either the UL or DL frequency band, depending on which one was less interfered. This flexibility is unique to CDMA, since as long as the amount of interference is not excessive, FDD and TDD can share the same bandwidth. This would be particularly feasible in the indoor scenario of [380], where the surrounding outdoor cell could be using FDD, while the indoor cell would reuse the same frequency band in TDD mode. The buildings' walls and partitions could mitigate the interference between the FDD/TDD schemes.

Table 10.3 shows the basic parameters of the UTRA. Some of these parameters are discussed during our further discourse, but significantly more information can be gleaned concerning the UTRA system by carefully studying the table.

The UTRA system is operated at a basic chip rate of 3.84 Mcps,[7] giving a nominal bandwidth of 5 MHz, when using root-raised cosine Nyquist pulse-shaping filters with

---

[7]In the UTRA RTT proposal submitted by ETSI to ITU, the chip rate was actually set at 4.096 Mcps.

Dedicated Transport Channel	Common Transport Channel
Dedicated CHannel (DCH) (UL/DL)	Broadcast CHannel (BCH) (DL)
	Forward Access CHannel (FACH) (DL)
	Paging CHannel (PCH) (DL)
	Random Access CHannel (RACH) (UL)
	Common Packet CHannel (CPCH) (UL)
	Down-link Shared CHannel (DSCH) (DL)

Table 10.4: UTRA Transport Channels

a rolloff factor of 0.22. UTRA fulfilled the requirements of 3G mobile radio systems by offering a range of user bit rates up to 2 Mbps. Various services having different bit rates and QoS can be readily supported using Orthogonal Variable Spreading Factor (OVSF) codes [299], which will be highlighted in Section 10.3.2.6.1, and service multiplexing, which will be discussed in Section 10.3.2.4. A key feature of the UTRA system, which was absent in the second-generation (2G) IS-95 system [349] was the use of a dedicated pilot sequence embedded in the users' data stream. These can be invoked in order to support the operation of adaptive antennas at the base station (BS), which was not facilitated by the common pilot channel of the IS-95 system. However, a common pilot channel was still retained in UTRA in order to provide the demodulator's phase reference for certain common physical channels, when embedding pilot symbols for each user is not feasible.

Regardless of whether a common pilot channel is used or dedicated pilots are embedded in the data, they facilitate the employment of *coherent detection*. Coherent detection is known to provide better performance than noncoherent detection [52]. Furthermore, the inclusion of short spreading codes enables the implementation of various performance enhancement techniques, such as interference cancelers and joint-detection algorithms, which results in excessive complexity in conjunction with long spreading codes. In order to support flexible system deployment in indoor and outdoor environments, *intercell-asynchronous operation* is used in the FDD mode. This implies that no external timing source, such as a reference signal or the Global Positioning System (GPS) is required. However, in the TDD mode intercell synchronization is required in order to be able to seamlessly access the time-slots offered by adjacent BSs during handovers. This is achieved by maintaining synchronization between the BSs.

10.3.2.2 Transport Channels

Transport channels are offered by the physical layer to the higher Open Systems Interconnection (OSI) layers, and they can be classified into two main groups, as shown in Table 10.4 [323, 372]. The Dedicated transport CHannel (DCH) is related to a specific Mobile Station (MS)-BS link, and it is used to carry user and control information between the network and an MS. Hence, the DCHs are bidirectional channels. There are six transport channels within the common transport channel group, as shown in Table 10.4. The Broadcast CHannel (BCH) is used to carry system- and cell-specific information on the DL to all MSs in the entire cell. This channel conveys information, such as the initial UL transmit power of the MS during a random access transmission and the cell-specific scrambling code, as we shall see in

Section 10.3.2.7. The Forward Access CHannel (FACH) of Table 10.4 is a DL common channel used for carrying control information and short user data packets to MSs, if the system knows the serving BS of the MS. On the other hand, the Paging CHannel (PCH) of Table 10.4 is used to carry control information to an MS, if the serving BS of the MS is unknown, in order to page the MS, when there is a call for the MS. The Random Access CHannel (RACH) of Table 10.4 is UL channel used by the MS to carry control information and short user data packets to the BS, in order to support the MS's access to the system, when it wishes to set up a call. The Common Packet CHannel (CPCH) is an UL channel used for transmitting bursty data traffic in a contention-based random access manner. Lastly, as its name implies, the Down-link Shared CHannel (DSCH) is a DL channel that is shared by several users.

The philosophy of these channels is fairly plausible, and it is informative as well as enlightening to explore the differences between the somewhat less flexible control regime of the 2G GSM [41] system and the more advanced 3G proposals, which we leave for the motivated reader due to lack of space. Unfortunately it is not feasible to design the control regime of a sophisticated mobile radio system by "direct synthesis" and so some of the solutions reviewed throughout this section in the context of the 3G proposals may appear somewhat heuristic and quite ingenious. These solutions constitute an amalgam of the wireless research community's experience in the design of the existing 2G systems and of the lessons learned from their operation. Further contributing factors in the design of the 3G systems were based on solving the signaling problems specific to the favored physical layer traffic channel solutions, namely, CDMA. In order to mention only one of them, the TDMA-based GSM system [41] was quite robust against power control inaccuracies, while the Pan-American IS-95 CDMA system [349] required an accurate power control. As we will see in Section 10.3.2.8, the power control problem was solved quite elegantly in the 3G proposals. We will also see that statistical multiplexing schemes — such as ALOHA, the original root of the recently more familiar Packet Reservation Multiple Access (PRMA) procedure — found their way into public mobile radio systems. A variety of further interesting solutions have also found applications in these 3G proposals, which are the results of the past decade of wireless system research. Let us now review the range of physical channels in the next section.

10.3.2.3 Physical Channels

The transport channels are transmitted using the physical channels. The physical channels are typically organized in terms of radio frames and time-slots, as shown in Figure 10.19. The philosophy of this hierarchical frame structure is also reminiscent to a certain degree of the GSM TDMA frame hierarchy of [41]. However, while in GSM each TDMA user had an exclusive slot allocation, in W-CDMA the number of simultaneous users supported is dependent on the users' required bit rate and their associated spreading factors. The MSs can transmit continuously in all slots or discontinuously, for example, when invoking a voice activity detector (VAD). Some of these issues will be addressed in Section 10.3.2.4.

As seen in Figure 10.19, there are 15 time-slots within each radio frame. The duration of each time-slot is $\frac{2}{3}$ ms, which gives a duration of 10 ms for the radio

10.3. THIRD-GENERATION SYSTEMS

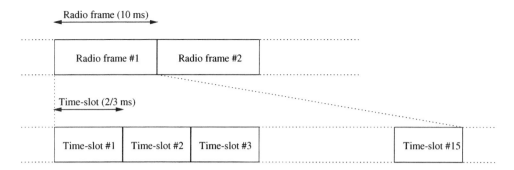

Figure 10.19: UTRA physical channel structure.

frame. As we shall see later in this section, the configuration of the information in the time-slots of the physical channels differs from one another in the UL and DL, as well as in the FDD and TDD modes. The 10 ms frame duration also conveniently coincides, for example, with the frame length of the ITU's G729 speech codec for speech communications, while it is a "submultiple" of the GSM system's various full- and half-rate speech codecs' frame durations [41]. We also note that a convenient mapping of the video stream of the H.263 videophone codec can be arranged on the 10 ms-duration radio frames for supporting interactive video services, while on the move. Furthermore, the spreading factor (SF) can be varied on a 10 ms burst-by-burst (BbB) basis, in order to adapt the transmission mode in harmony with channel quality fluctuations, while maintaining a given target bit error rate. Although it is not part of the standard proposal, we found that it was more beneficial to adapt the number of bits per symbol on a BbB basis than varying the SF [354].

In the FDD mode, a DL physical channel is defined by its spreading code and frequency. Furthermore, in the UL, the modem's orthogonal in-phase (I) and quadrature-phase (Q) branches are used to deliver the data and control information simultaneously in parallel (as will be augmented in Figure 10.37). Thus, knowledge of the relative carrier phase, namely whether the I or Q branch is involved, constitutes part of the physical channel's identifier. On the other hand, in the TDD mode, a physical channel is defined by its spreading code, frequency, and time-slot.

Similarly to the transport channels of Table 10.4, the physical channels in UTRA can also be classified as dedicated and common channels. Table 10.5 shows the type of physical channels and the corresponding mapping of transport channels on the physical channels in UTRA.

10.3.2.3.1 Dedicated Physical Channels The dedicated physical channels of UTRA shown in Table 10.5 consist of the Dedicated Physical Data CHannel (DPDCH) and Dedicated Physical Control CHannel (DPCCH), both of which are bidirectional. The time-slot structures of the UL and DL dedicated physical channels are shown in Figures 10.20 and 10.21, respectively. Notice that on the DL, as illustrated by Figure 10.21, the DPDCH and DPCCH are interspersed by time-multiplexing to form a single Dedicated Physical CHannel (DPCH), as will be augmented in the context of

Dedicated Physical Channels	Transport Channels
Dedicated Physical Data CHannel (DPDCH) (UL/DL) [†] ————————	DCH
Dedicated Physical Control CHannel (DPCCH) (UL/DL)	

Common Physical Channels	Transport Channels
Physical Random Access CHannel (PRACH) (UL) ————————	RACH
Physical Common Packet CHannel (PCPCH) (UL) ————————	CPCH
Common PIlot CHannel (CPICH) (DL)	
Primary Common Control Physical CHannel (P-CCPCH) (DL) ————	BCH
Secondary Common Control Physical CHannel (S-CCPCH) (DL) ⟵	FACH
Synchronization CHannel (SCH) (DL)	PCH
Physical Down-link Shared CHannel (PDSCH) (DL) ————————	DSCH
Acquisition Indication CHannel (AICH) (DL)	
Page Indication CHannel (PICH) (DL)	

[†] On the DL, the DPDCH and DPCCH are time-multiplexed in each time slot to form a single Dedicated Physical CHannel (DPCH).

Table 10.5: Mapping the Transport Channels of Table 10.4 to the UTRA Physical Channels

Figure 10.38. On the other hand, the DPDCH and DPCCH on the UL are transmitted in parallel on the I and Q branches of the modem, as will become more explicit in the context of Figure 10.37 [323]. The reason for the parallel transmission on the UL is to avoid Electromagnetic Compatibility (EMC) problems due to Discontinuous Transmission (DTX) of the DPDCH of Table 10.5 [322]. DTX occurs when temporarily there are no data to transmit, but the link is still maintained by the DPCCH. If the UL DPCCH is time-multiplexed with the DPDCH, as in the DL of Figure 10.21, this can create short, sharp energy spikes. Since the MS may be located near sensitive electrical equipment, these spikes may affect this equipment.

The DPDCH is used to transmit the DCH information between the BS and MS, while the DPCCH is used to transmit the Layer 1 information, which includes the pilot bits, Transmit Power Control (TPC) commands, and an optional Transport-Format Combination Indicator (TFCI), as seen in Figures 10.20 and 10.21. In addition, on the UL the Feedback Information (FBI) is also mapped to the DPDCH in Figure 10.20. The pilot bits are used to facilitate coherent detection on both the UL and DL as well as to enable the implementation of performance enhancement techniques, such as adaptive antennas and interference cancellation. Since the pilot sequences are known, they can also be used as frame synchronization words in order to maintain transmission frame synchronization between the BS and MS. The TPC commands support an agile and efficient power control scheme, which is essential in DS-CDMA using the techniques to be highlighted in Section 10.3.2.8. The TFCI carries information concerning the instantaneous parameters of each transport channel multiplexed on the physical channel in the associated radio frame. The FBI is used to provide the capability to support certain transmit diversity techniques. The FBI field is further

10.3. THIRD-GENERATION SYSTEMS

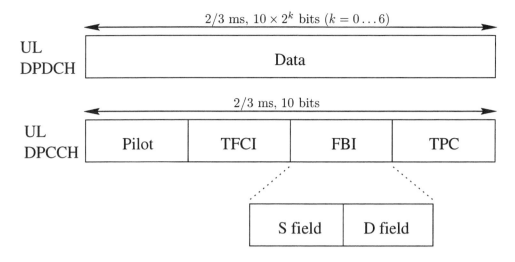

Figure 10.20: UTRA UL FDD dedicated physical channels time-slot configuration, which is mapped to the time-slots of Figure 10.19. The UL DPDCH and DPCCH messages are transmitted in parallel on the I and Q branches of the modem of Figure 10.37. By contrast, the DPDCH and DPCCH bursts are time-multiplexed on the DL as shown in Figure 10.21.

divided into two smaller fields as shown in Figure 10.20, which are referred to as the S field and D field. The S field is used to support the *Site Selection DiversiTy* (SSDT), which can reduce the amount of interference caused by multiple transmissions during a soft handover operation, while assisting in fast cell selection. On the other hand, the D field is used to provide attenuation and phase information in order to facilitate *closed-loop transmit diversity*, a technique highlighted in Section 10.3.4.1.3. Given that the TPC and TFCI segments render the transmission packets "self-descriptive," the system becomes very flexible, supporting burst-by-burst adaptivity, which substantially improves the system's performance [99, 354], although this side-information is vulnerable to transmission errors.

The parameter k in Figures 10.20 and 10.21 determines the number of bits in each time-slot, which in turn corresponds to the bit rate of the physical channel. Therefore, the channel bit rates available for the UL DPDCH are 15/30/60/120/240/480/960 kbps, due to the associated "payload" of 10×2^k bits per $\frac{2}{3}$ ms burst in Figure 10.20, where $k = 0 \ldots 6$. Note that the UL DPCCH has a constant channel bit rate of 15 kbps. Similarly, the channel bit rates available for the DL DPCH are 15/30/60/120/-

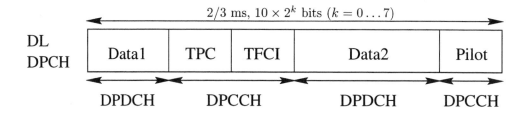

Figure 10.21: UTRA DL FDD dedicated physical channels time-slot configuration, which is mapped to the time-slots of Figure 10.19. The DPDCH and DPCCH messages are time-multiplexed on the DL, as it will be augmented in Figure 10.38. By contrast, the UL DPDCH and DPCCH bursts are transmitted in parallel on the I and Q branches of the modem as shown in Figure 10.20.

240/480/960/1920 kbps. However, since the user data are time-multiplexed with the Layer 1 control information, the actual user data rates on the DL will be slightly lower than those mentioned above. Even higher channel bit rates can be achieved using a technique known as multicode transmission [300], which will be highlighted in more detail in the context of Figure 10.35 in Section 10.3.2.5. Let us now consider the common physical channels summarized in Table 10.5.

10.3.2.3.2 Common Physical Channels

10.3.2.3.2.1 Common Physical Channels of the FDD Mode

The Physical Random Access CHannel (PRACH) of Table 10.5 is used to carry the RACH message on the UL. A random access transmission is activated whenever the MS has data to transmit and wishes to establish a connection with the local BS. Although the procedure of this transmission will be elaborated on in Section 10.3.2.7, here we will briefly highlight the structure of a random access transmission burst. Typically, a random access burst consists of one or several so-called preambles and a message. Each preamble contains a signature that is constructed of 256 repetitions of a 16-chip Hadamard code, which yields a $256 \times 16 = 4096$-chip long signature. Similarly to the UL dedicated physical channels of Figure 10.20, the message part of the random access transmission consists of data information and control information that are transmitted in parallel on the I/Q channels of the modulator, as shown in Figure 10.22. The channel bit rates available for the data part of the message are 15/30/60/120 kbps. By contrast, the control information, which contains an 8-bit pilot and a 2-bit TFCI, is transmitted at a fixed rate of 15 kbps. Obviously in this

10.3. THIRD-GENERATION SYSTEMS

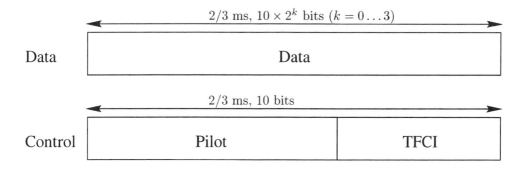

TFCI : Transport-Format Combination Indicator

Figure 10.22: The time-slot configuration of the message part during a random access transmission in UTRA, which are mapped to the frame structure of Figure 10.19. The data and control information are multiplexed on the I/Q channels of the modulator and the frame is transmitted at the beginning of an access slot, as it will be augmented in Section 10.3.2.7.1.

case, no FBI and TPC commands are required, since transmission is initiated by the MS.

The Physical Common Packet CHannel (PCPCH) of Table 10.5 is used to carry the CPCH message on the UL, based on a Digital Sense Multiple Access-Collision Detection (DSMA-CD) random access technique. A CPCH random access burst consists of one or several Access Preambles (A-P), one Collision Detection Preamble (CD-P), a DPCCH Power Control Preamble (PC-P), and a message. The length of both the A-P and CD-P spans a total of 4096 chips, while the duration of the PC-P can be equivalent to either 0 or 8 time-slots. Each time-slot of the PC-P contains the pilot, the FBI, and the TPC bits. The message part of the CPCH burst consists of a data part and a control part, which is identical to the UL dedicated physical channel shown in Figure 10.20 in terms of its structure and available channel bit rates. A 15 kbps DL DPCH is always associated with an UL PCPCH. Hence, both the FBI and TPC information are included in the message of a CPCH burst in order to facilitate a DL transmit diversity and power control, unlike a RACH burst. The procedure of a CPCH transmission will be further elaborated in Section 10.3.2.7.

The DL Primary Common Control Physical CHannel (P-CCPCH) of Table 10.5 is used by the BS in order to broadcast the BCH information at a fixed rate of 30 kbps to all MSs in the cell. The P-CCPCH is transmitted only after the first 256 chips of each slot, as shown in Figure 10.23. During the first 256 chips of each slot, the Synchronization CHannel (SCH) message is transmitted instead, as will be discussed in Section 10.3.2.9. The P-CCPCH is used as a timing reference directly for all the DL physical channels and indirectly for all the UL physical channels. Hence, as long as the MS is synchronized to the DL P-CCPCH of a specific cell, it is capable of detecting any DL messages transmitted from that BS by listening at the predefined times. For example, the DL DPCH will commence transmission at an offset, which is a

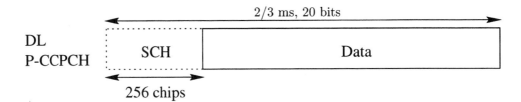

P-CCPCH: Primary Common Control Physical CHannel
SCH: Synchronization CHannel

Figure 10.23: UTRA DL FDD Primary Common Control Physical CHannel (P-CCPCH) time-slot configuration, which is mapped to the time-slots of Figure 10.19.

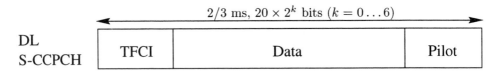

S-CCPCH: Secondary Common Control Physical CHannel
TFCI: Transport-Format Combination Indicator

Figure 10.24: UTRA DL FDD Secondary Common Control Physical CHannel (S-CCPCH) time-slot configuration, which is mapped to the time-slots of Figure 10.19.

multiple of 256 chips from the start of the P-CCPCH radio frame seen in Figure 10.23. Upon synchronization with the P-CCPCH, the MS will know precisely when to begin receiving the DL DPCH. The UL DPDCH/DPCCH is transmitted 1024 chips after the reception of the corresponding DL DPCH.

The Secondary Common Control Physical CHannel (S-CCPCH) of Table 10.5 carries the FACH and PCH information of Table 10.4 on the DL, and they are transmitted only when data are available for transmission. The S-CCPCH will be transmitted at an offset, which is a multiple of 256 chips from the start of the P-CCPCH message seen in Figure 10.23. This will allow the MS to know exactly when to detect the S-CCPCH, as long as the MS is synchronized to the P-CCPCH. The time-slot configuration of the S-CCPCH is shown in Figure 10.24. Notice that the S-CCPCH message can be transmitted at a variable bit rate, namely, at 30/60/120/240/480/960/1920 kbps.

At this stage it is worth mentioning that the available control channel rates are significantly higher in the 3G systems than in their 2G counterparts. For example, the maximum BCH signaling rate in GSM [41] is more than an order of magnitude lower than the above-mentioned 30 kbps UTRA BCH rate. In general, this increased control channel rate will support a significantly more flexible system control than the 2G systems.

The Physical Down-link Shared CHannel (PDSCH) of Table 10.5 is used to carry

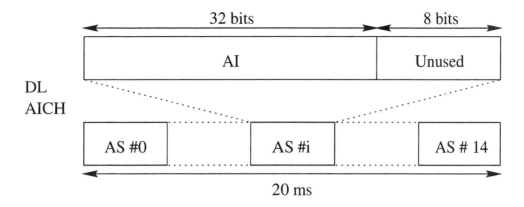

AICH: Acquisition Indicator CHannel
AI: Acquisition Indicator
AS: Access Slot

Figure 10.25: UTRA DL Acquisition Indicator CHannel (AICH) Access Slot (AS) configuration, which is mapped to the corresponding AS of the AICH. Due to its duration of 20 ms, it is mapped to every other 10 ms frame in Figure 10.19.

the DSCH message at rates of 30/60/120/240/480/960/1920 kbps. The PDSCH is shared among several users based on code multiplexing. The Layer 1 control information is transmitted using the associated DL DPCH.

The Acquisition Indicator CHannel (AICH) of Table 10.5 and the Page Indicator CHannel (PICH) are used to carry Acquisition Indicator (AI) and Page Indicator (PI) messages, respectively. More specifically, the AI is a response to a PRACH or PCPCH transmission, and it corresponds to the signature used by the associated PRACH preamble, a PCPCH A-P or a PCPCH CD-P, which were defined above. The AICH consists of a repeated sequence of 15 consecutive Access Slots (AS). Each AS consists of a 32-symbol AI part and an eight-symbol unused part, as shown in Figure 10.25. The AS#0 will commence at the start of every other 10 ms P-CCPCH radio frame seen in Figure 10.19, since its duration is 20 ms.

A PI message is used to signal to the MS on the associated S-CCPCH that there are data addressed to it, in order to facilitate a power-efficient sleep-mode operation. A PICH, illustrated in Figure 10.26, is a 10 ms frame consisting of 300 bits, out of which 288 bits are used to carry PIs, while the remaining 12 bits are unused. Each PICH frame can carry a total of N PIs, where $N = 18, 36, 72$, and 144. The PICH is also transmitted at an offset with respect to the start of the P-CCPCH, which is a multiple of 256 chips. The associated S-CCPCH will be transmitted 7680 chips later.

Finally, the Common PIlot CHannel (CPICH) of Table 10.5 is a 30 kbps DL physical channel that carries a predefined bit sequence. It provides a phase reference for the SCH, P-CCPCH, AICH, and PICH, since these channels do not carry pilot bits, as shown in Figures 10.23, 10.25, and 10.26, respectively. The PICH is transmitted

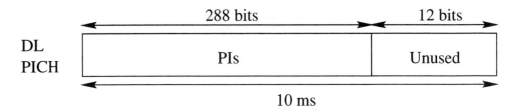

PICH: Page Indicator CHannel
PI: Page Indicator

Figure 10.26: UTRA DL Page Indicator CHannel (PICH) configuration. Each PICH frame can carry a total of N PIs, where $N = 18, 36, 72$, and 144.

synchronously with the P-CCPCH.

10.3.2.3.2.2 Common Physical Channels of the TDD Mode

In contrast to the previous FDD structures of Figures 10.20–10.26, in TDD operation the burst structure of Figure 10.27 is used for all the physical channels, where each time-slot's transmitted information can be arbitrarily allocated to the DL or UL, as shown in the three possible TDD allocations in Figure 10.28. Hence, this flexible allocation of the UL and DL burst in the TDD mode enables the use of an adaptive modem [93, 99, 354] whereby the modem parameters, such as the spreading factor or the number of bits per symbol can be adjusted on a burst-by-burst basis to optimize the link quality. A symmetric UL/DL allocation refers to a scenario in

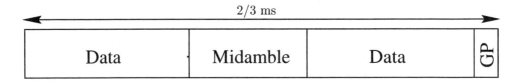

Burst Type 1 : Data = 976 chips, Midamble = 512 chips
Burst Type 2 : Data = 1104 chips, Midamble = 256 chips
GP : Guard Period = 96 chips

Figure 10.27: Burst configuration mapped on the TDD burst structure of Figure 10.28 in the UTRA TDD mode. Two different types of TDD bursts are defined in UTRA, namely, Burst Type 1 and Burst Type 2.

which an approximately equal number[8] of DL and UL bursts are allocated within a

---
[8] Since there are 15 time-slots per frame, there will always be one more additional DL or UL burst per frame in a symmetric allocation.

10.3. THIRD-GENERATION SYSTEMS

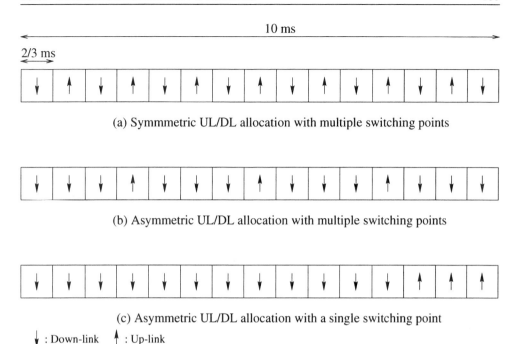

Figure 10.28: Up-link/down-link allocation examples for the 15 slots in UTRA TDD operation using the time-slot configurations of Figure 10.27.

TDD frame, while in asymmetric UL/DL allocation, there is an unequal number of UL and DL bursts, such as, for example, in "near-simplex" file download from the Internet or video-on-demand.

In UTRA, two different TDD burst structures, known as *Burst Type 1* and *Burst Type 2*, are defined, as shown in Figure 10.27. The Type 1 burst has a longer midamble (512 chips) than the Type 2 burst (256 chips). However, both types of bursts have an identical *Guard Period* (GP) of 96 chips. The midamble sequences that are allocated to the different TDD bursts in each time-slot belong to a so-called *midamble code set*. The codes in each midamble code set are derived from a unique *Basic Midamble Code*. Adjacent cells are allocated different midamble code sets, that is, different basic midamble code. This can be exploited to assist in cell identification.

Unlike in the FDD mode, there is only one type of Dedicated Physical CHannel (DPCH) in the TDD mode. Hence, the Layer 1 control information — such as the TPC command and the TFCI information — will be transmitted in the data field of Figure 10.27, if required. The TDD burst structures that incorporate the TFCI information as well as the TFCI+TPC information are shown in Figure 10.29. This should be contrasted with their corresponding FDD allocations in Figures 10.20 and 10.21. The TFCI field is divided into two parts, which reside immediately before and after the midamble (or after the TPC command, if power control is invoked) in the data field. The TPC command is always transmitted immediately after the midamble, as portrayed in Figure 10.29. As a result of these control information segments, the

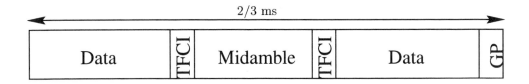

Burst structure with TFCI information only

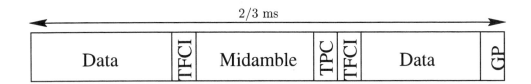

Burst structure with TFCI and TPC information

GP : Guard Period
TFCI : Transport Format Combination Indicator
TPC : Transmit Power Control

Figure 10.29: Burst configuration mapped on the TDD burst configuration of Figure 10.28 in the UTRA TDD mode incorporating TFCI and/or TPC information.

amount of user data is reduced in each time-slot. Note that the TPC command is only transmitted on the UL and only once per 10 ms frame for each MS.

In contrast to the FDD mode, the SCH in the TDD mode is not time-multiplexed onto the P-CCPCH of Table 10.5. Instead, the SCH messages are transmitted on one or two time-slots per frame.[9] The P-CCPCH will be code-multiplexed with the first SCH time-slot in each frame.

Having highlighted the basic features of the various UTRA channels, let us now consider how the various services are error protected, interleaved, and multiplexed on to the physical channels. This issue is discussed with reference to Figures 10.30 and 10.31 in the context of UTRA.

10.3.2.4 Service Multiplexing and Channel Coding in UTRA

Service multiplexing is employed when multiple services of identical or different bit rates requiring different QoS belonging to the same user's connection are transmitted.

---

[9]If two time-slots are allocated to the SCH per frame, they will be spaced seven slots apart.

10.3. THIRD-GENERATION SYSTEMS

An example would be the simultaneous transmission of a voice and video service for a multimedia application. Each service is represented by its corresponding transport channels, as described in Section 10.3.2.2. The coding and multiplexing of the transport channels are performed in sets of transport blocks that arrived from the higher layers at fixed intervals of 10, 20, 40 or 80 ms. These intervals are known as the *Transmission Time Interval* (TTI). Note that the number of bits on each transport channel can vary between different TTIs, as well as between different transport channels. A possible method of transmitting multiple services is by using code-multiplexing with the aid of orthogonal codes. Every service could have its own DPDCH and DPCCH, each assigned to a different orthogonal code. This method is not very efficient, however, since a number of orthogonal codes would be reserved by a single user, while on the UL it would also inflict self-interference when the multiple DPDCH and DPCCH codes' orthogonality is impaired by the fading channel. Alternatively, these services can be time-multiplexed into one or several DPDCHs, as shown in Figures 10.30 and 10.31 for the UL and DL, respectively. The function of the individual processing steps is detailed below.

10.3.2.4.1 CRC Attachment A Cyclic Redundancy Checksum (CRC) is first calculated for each incoming transport block within a TTI. The CRC consists of either 24, 16, 12, 8, or 0 parity bits, which is decided by the higher layers. The CRC is then attached to the end of the corresponding transport block in order to facilitate reliable error detection at the receiver. This facility is very important, for example, for generating the video packet acknowledgment flag in wireless video telephony using standard video codecs, such as H.263 [199].

10.3.2.4.2 Transport Block Concatenation Following the CRC attachment, the incoming transport blocks within a TTI are serially concatenated in order to form a code block. If the number of bits exceeds the maximum code block length, denoted as Z, then the code block is segmented into shorter ones and filler bits (zeros) are added to the last code block, if neccessary, in order to generate code blocks of the same length. The maximum code block length Z is dependent on the type of channel-coding invoked. For convolutional coding $Z = 504$, while for turbo coding $Z = 5114$, since turbo codes require a long coded block length [381]. If no channel-coding is invoked, then the code block can be of unlimited length.

10.3.2.4.3 Channel-Coding Each of the code blocks is then delivered to the channel-coding unit. Several Forward Error Correction (FEC) techniques are proposed for channel-coding. The FEC technique used is dependent on the QoS requirement of that specific transport channel. Table 10.6 shows the various types of channel-coding techniques invoked for different transport channels. Typically, *convolutional coding* is used for services with a bit error rate requirement on the order of 10^{-3}, for example, for voice services. For services requiring a lower BER, namely, on the order of 10^{-6}, *turbo coding* is applied. Turbo coding is known to guarantee a high performance [382] over AWGN channels at the cost of increased interleaving-induced latency or delay. The implementational complexity of the turbo codec (TC) does not necessarily have to be higher than that of the convolutional codes (CC), since a

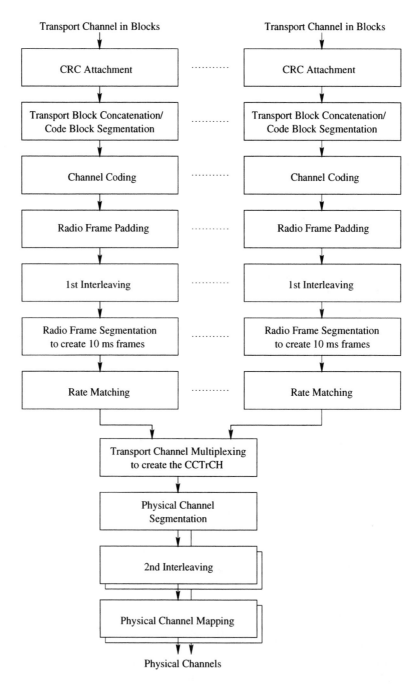

Figure 10.30: Transport channel-coding/multiplexing flowchart for the UL in UTRA.

10.3. THIRD-GENERATION SYSTEMS

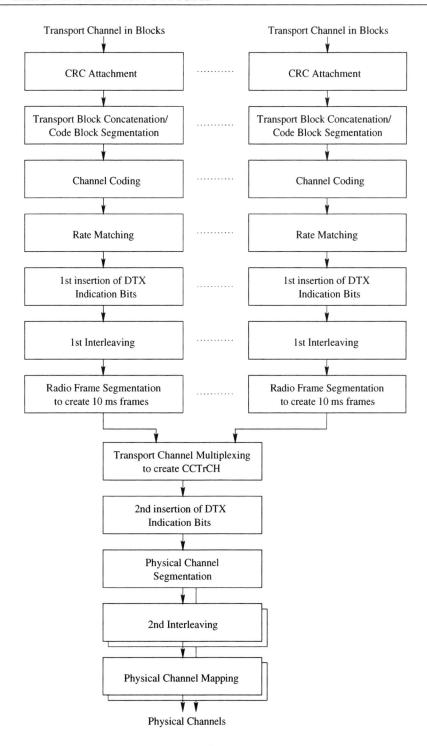

Figure 10.31: Transport channel-coding/multiplexing flowchart for the DL in UTRA.

Transport Channels	Channel-Coding Schemes	Coding Rate
BCH, PCH, RACH	Convolutional code	1/2
CPCH,DCH,DSCH,FACH	Convolutional code Turbo code No coding	1/3, 1/2 1/3

Table 10.6: UTRA Channel-Coding Parameters for the Channels of Table 10.4

constraint-length $K = 7$ or $K = 9$ CC is often invoked, while the constraint-length of the turbo codes employed may be as low as $K = 3$. In somewhat simplistic but plausible terms, one could argue that a $K = 3$ TC using two decoding steps per iteration and employing four iterations has a similar complexity to a $K = 6$ CC, since they are associated with the same number of trellis states. The encoded code blocks within a TTI are then serially concatenated after the channel-coding unit, as seen in Figures 10.30 and 10.31.

10.3.2.4.4 Radio Frame Padding Radio frame padding is only performed on the UL whereby the input bit sequence (the concatenated encoded code blocks from the channel-coding unit) is padded in order to ensure that the output can be segmented equally into (TTI/10ms) number of 10 ms radio frames. Note that radio frame padding is not required on the DL, since DTX is invoked, as seen in Figure 10.31. This process was termed Radio Frame Equalization in the standard. However, in order to avoid confusion with channel equalization, we used the terminology "padding."

10.3.2.4.5 First Interleaving The depth of this first interleaving seen in Figures 10.30 and 10.31 can range from one radio frame (10 ms) to as high as 80 ms, depending on the TTI.

10.3.2.4.6 Radio Frame Segmentation The input bit sequence after the first interleaving is then segmented into consecutive radio frames of 10 ms duration, as highlighted in Section 10.3.2.3. The number of radio frames required is equivalent to (TTI/10). Because of the Radio Frame Padding step performed prior to the segmentation on the UL in Figure 10.30 and also because of the Rate Matching step on the DL in Figure 10.31, the input bit sequence can be conveniently divided into the required number of radio frames.

10.3.2.4.7 Rate Matching The rate matching process of Figures 10.30 and 10.31 implies that bits on a transport channel are either repeated or punctured in order to ensure that the total bit rate after multiplexing all the associated transport channels will be identical to the channel bit rate of the corresponding physical channel, as highlighted in Section 10.3.2.3. Thus, rate matching must be coordinated among the different coded transport channels, so that the bit rate of each channel is adjusted to a level that fulfills its minimum QoS requirements [372]. On the DL, the bit rate is also adjusted so that the total instantanenous transport channel bit rate approximately matches the defined bit rate of the physical channel, as listed in Table 10.3.

10.3.2.4.8 Discontinuous Transmission Indication On the DL, the transmission is interrupted if the bit rate is less than the allocated channel bit rate. This is known as discontinuous transmission (DTX). DTX indication bits are inserted into the bit sequence in order to indicate when the transmission should be turned off. The first insertion of the DTX indication bits shown in Figure 10.31 is performed only if the position of the transport channel in the radio frame is fixed. In this case, a fixed number of bits is reserved for each transport channel in the radio frame. For the second insertion step shown in Figure 10.31, the DTX indication bits are inserted at the end of the radio frame.

10.3.2.4.9 Transport Channel Multiplexing One radio frame from each transport channel that can be mapped to the same type of physical channel is delivered to the transport channel multiplexing unit of Figures 10.30 and 10.31, where they are serially multiplexed to form a *Coded Composite Transport CHannel* (CCTrCH). At this point, it should be noted that the bit rate of the multiplexed radio frames may be different for the various transport channels. In order to successfully de-multiplex each transport channel at the receiver, the TFCI — which contained information about the bit rate of each multiplexed transport channel — can be transmitted together with the CCTrCH information (which will be mapped to a physical channel), as highlighted in Section 10.3.2.3. Alternatively, **blind transport format detection** can be performed at the receiver without the explicit knowledge of the TFCI, where the receiver acquires the transport format combination through some other means, such as, for example, the received power ratio of the DPDCH to the DPCCH.

10.3.2.4.10 Physical Channel Segmentation If more than one physical channel is required in order to accommodate the bits of a CCTrCH, then the bit sequence is segmented equally into different physical channels, as seen in Figures 10.30 and 10.31. A typical example of this scenario would be where the bit rate of the CCTrCH exceeds the maximum allocated bit rate for the particular physical channel. Thus, multiple physical channels are required for its transmission. Furthermore, restrictions are imposed on the number of transport channels that can be multiplexed onto a CCTrCH. Hence, several physical channels are required to carry any additional CCTrCHs.

10.3.2.4.11 Second Interleaving The depth of the second interleaving stage shown in Figures 10.30 and 10.31 is equivalent to one radio frame. Hence, this process does not increase the system's delay.

10.3.2.4.12 Physical Channel Mapping Finally, the bits are mapped to their respective physical channels summarized in Table 10.5, as portrayed in Figures 10.30 and 10.31.

Having highlighted the various channel-coding and multiplexing techniques as well as the structures of the physical channels illustrated by Figures 10.20–10.27, let us now consider how the services of different bit rates are mapped on the UL and DL dedicated physical data channels (DPDCH) of Figures 10.20 and 10.21, respectively. In order to augment the process, we will present three examples. Specifically, we consider the mapping of two multirate services on an UL DPDCH and an example of the mapping of a 4.1 kbps data service on a DL DPDCH in the FDD mode. We will then use the same parameters as employed in the first example and show how the multirate services can be mapped to the corresponding UL DPCH in TDD mode.

10.3.2.4.13 Mapping Several Multirate Services to the UL Physical Channels in FDD Mode [369] In this example, we assume that a 4.1 kbps speech service and a 64 kbps video service are to be transmitted simultaneously on the UL. The parameters used for this example are shown in Table 10.7. As illustrated in Fig-

	Service 1, DCH#1	Service 2, DCH#2
Transport Block Size	640 bits	164 bits
Transport Block Set Size	4 * 640 bits	1 * 164 bits
TTI	40 ms	40 ms
Bit Rate	64 kbps	4.1 kbps
CRC	16 bits	16 bits
Coding	Turbo Rate: 1/3	Convolutional Rate: 1/3

Table 10.7: Parameters for the Multimedia Communication Example of Section 10.3.2.4.13

ure 10.32, a 16-bit CRC checksum is first attached to each transport block of DCH#1, that is, #1a,...,#1d, as well as the transport block of DCH#2 for the purpose of error detection. As a result, the number of bits in the transport block of Service 1 and Service 2 is increased to $640 + 16 = 656$ bits and $164 + 16 = 180$ bits, respectively. The four transport blocks of Service 1 are then concatenated, as illustrated in Figure 10.32. Notice that no code block segmentation is invoked, since the total number of bits in the concatenated transport block is less than $Z = 5114$ for turbo coding, as highlighted in Section 10.3.2.4.2. Since the video service typically requires a low BER — unless specific measures are invoked for mitigating the video effects of transmission errors [93]–, turbo coding is invoked, using a coding rate of $\frac{1}{3}$. Hence, after turbo coding and the attachment of tailing bits, the resulting 40 ms segment would contain $(656 \times 4) \times 3 + 12 = 7884$ bits, as shown in Figure 10.32. By contrast, the speech service can tolerate a higher BER. Hence, convolutional coding is invoked. First, a block of $4 + 4 = 8$ tail bits is concatenated to the transport block in order to flush the assumed constraint-length $K = 5$ shift registers of the convolutional encoder. Thus, a total of $180 + 8 = 188$ bits are conveyed to the convolutional encoder of DCH#1, as shown in Figure 10.32. Again, no code block segmentation is invoked, since the total number of bits in the transport channel is less than $Z = 504$ for convolutional coding, as highlighted in Section 10.3.2.4.2. A coding rate of $\frac{1}{3}$ is used for the convolutional encoding of DCH#1, as exemplified in Table 10.7. The output of the convolutional encoder of DCH#1 will have a total of $188 \times 3 = 564$ bits per 40 ms segment. Since the TTI of these transport channels is 40 ms, four radio frames are required to transmit the associated data. At this stage, notice that there are a total of 7884 bits and 564 bits for DCH#1 and DCH#2, respectively. Since these numbers are divisible by four, they can be divided equally into four radio frames. Thus, no padding is required as illustrated in the Radio Frame Padding step of Figure 10.32. Interleaving is then performed across the 40 ms segment for each transport channel before being

10.3. THIRD-GENERATION SYSTEMS

segmented into four 10 ms radio frames.

At this point, we note that these two transport channels can be mapped to the same DPDCH, since they belong to the same MS. Hence, the 10 ms radio frames, marked "A" in Figure 10.32 will be multiplexed, in order to form a CCTrCH. Similarly, the frames marked "B," "C" (not shown in Figure 10.7 due to lack of space), and "D" will be multiplexed, in order to form another three CCTrCHs. The rate of these CCTrCHs must be matched to the allocated channel bit rate of the physical channel. Without rate matching, the bit rate of these CCTrCHs is $(1971 + 141)/10$ ms $= 211.2$ kbps, which does not fit any of the available channel bit rates of the UL DPDCH, as listed in Table 10.3. Hence, the Rate Matching step of Figures 10.30, 10.31, and 10.32 must be invoked in order to adapt the multiplexed bit rate to one of the available UL DPDCH bit rates of Table 10.3. Let us assume that the allocated channel bit rate is 240 kbps. Thus, a number of bits must be punctured or repeated for each service, in order to increase the total number of bits per 10 ms segment after multiplexing from 2171 to 2400. This would require coordination among the different services, as it was highlighted in Section 10.3.2.4.7. After multiplexing the transport channels, a second interleaving is performed across the 10 ms radio frame before finally mapping the bits to the UL DPDCH.

10.3.2.4.14 Mapping of a 4.1 Kbps Data Service to the DL DPDCH in FDD Mode
The parameters for this example are shown in Table 10.8. In this

	Service 1, DCH#1
Transport Block Size	164 bits
TTI	40 ms
Bit Rate	4.1 kbps
CRC	16 bits
Coding	Convolutional Rate: 1/3

Table 10.8: Parameters for the Example of Section 10.3.2.4.14

context, we assume that a single DCH consisting of one transport block within a TTI duration of 40 ms is to be transmitted on the DL. As illustrated in Figure 10.33, a 16-bit CRC sum segment is appended to the transport block. A $4 + 4 = 8$-bit tailing block is then attached to the end of the segment in order to form a 188-bit code block. Similarly to the previous example, the length of the code block is less than $Z = 504$, since CC is used. Hence, no segmentation is invoked. The 188-bit data block is convolutional coded at a rate of $\frac{1}{3}$, which results in a $3 \times 188 = 564$ bit segment. According to Figure 10.31, rate matching is invoked for the encoded block. Since the TTI duration is 40 ms, four radio frames are required to transmit the data. Without rate matching, the bit rate per radio frame is $564/40$ ms $= 14.1$ kbps, which does not fit any of the available bit rates listed in Table 10.4 for the DL. Note that for the case of the DL dedicated physical channels, the channel bit rate will include the additional bits required for the pilot and TPC, as shown explicitly in Figure 10.21. Since there is only one transport channel in this case, no TFCI bits are required. We assume that an 8-bit pilot and a 2-bit TPC per slot are assigned to this tranmission, which yields a total rate of 15 kbps for the DPCCH. Hence all the bits in the encoded block will be repeated in order to increase its bit rate of 15 kbps to 30 kbps for the DL

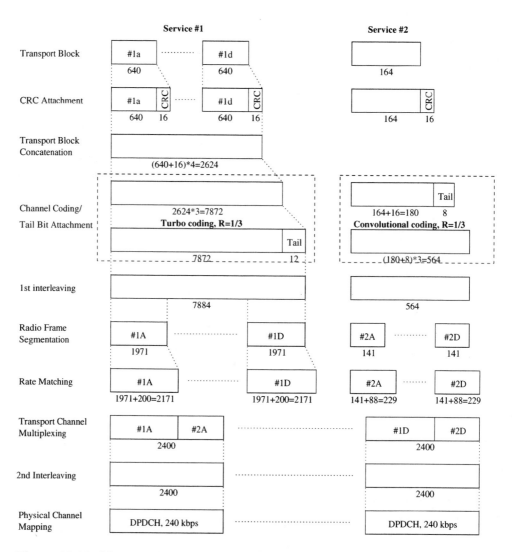

Figure 10.32: Mapping of several multimedia services to the UL dedicated physical data channel of Figure 10.20 in FDD mode. The corresponding schematic is seen in Figure 10.30.

10.3. THIRD-GENERATION SYSTEMS

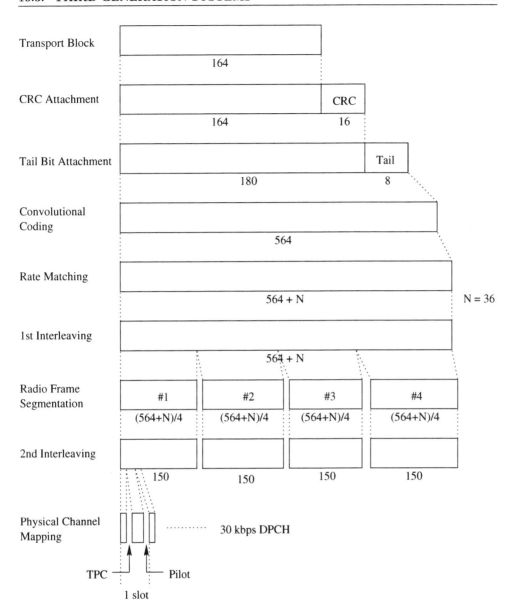

Figure 10.33: Mapping of a 4.1 kbps data service to the DL dedicated physical channel of Figure 10.21 in FDD mode. The corresponding schematic is seen in Figure 10.31.

DPCH. In this case the number of padding bits appended becomes $N = 36$. After the second interleaving stage of Figure 10.31, the segmented radio frames are mapped to the corresponding DPDCH, which are then multiplexed with the DPCCH, as shown in Figure 10.33.

10.3.2.4.15 Mapping Several Multirate Services to the UL Physical Channels in TDD Mode [369]

In this example, we will demonstrate how the multirate multimedia services, considered previously in the example of Section 10.3.2.4.13 in an FDD context, are mapped to the corresponding dedicated physical channels (DPCH) in the TDD mode. The channel-coding/multiplexing process is identical in the FDD and TDD mode, and so both are based on Figures 10.30 and 10.31. The only difference is in the mapping of the transport channels to the corresponding physical channels seen at the bottom of Figures 10.30 and 10.31, since the FDD and TDD modes have a different frame structure, as shown previously in Figures 10.20–10.26 and Figure 10.27, respectively. In this example, we are only interested in the process of service mapping to the physical channel, which follows the second interleaving stage of Figure 10.34. Here we assumed that for the TDD UL scenario of Table 10.7 the total number of bits per segment after DCH multiplexing is 2186 as a result of rate matching. In the FDD example of Section 10.3.2.4.13, this was 2400. Each segment is divided into two bursts, which can be transmitted either by orthogonal code multiplexing onto a single time-slot, or using two time-slots within a 10 ms radio frame. Note that only one burst in each segment is required to carry the TFCI and the TPC information.

Following these brief discussions on service multiplexing, channel coding, and interleaving, let us now concentrate on the aspects of variable-rate and multicode transmission in UTRA in the next section.

10.3.2.5 Variable-Rate and Multicode Transmission in UTRA

Three different techniques have been proposed in the literature for supporting variable-rate transmission, namely, multicode-, modulation-division multiplexing- (MDM), and multiple processing gain (MPG) based techniques [383]. UTRA employs a number of different processing gains, or variable spreading factors, in order to transmit at different channel bit rates, as highlighted previously in Section 10.3.2.3. The spreading factor (SF) has a direct effect on the performance and capacity of a DS-CDMA system. Since the chip rate is constant, the SF — which is defined as the ratio of the spread bandwidth to the original information bandwidth — becomes lower, as the bit rate increases. Hence, there is a limit to the value of the SF used, which is SF = 4 in FDD mode in the proposed UTRA standards. Multicode transmission [300, 383, 384] is used if the total bit rate to be transmitted exceeds the maximum bit rate supported by a single DPDCH, which was stipulated as 960 kbps for the UL and 1920 kbps for the DL. When this happens, the bit rate is split among a number of spreading codes and the information is transmitted using two or more codes. However, only one DPCCH is transmitted during this time. Thus, on the UL one DPCCH and several DPDCH are code-multiplexed and transmitted in parallel, as it will be augmented in the context of Figure 10.37. On the DL, the DPDCH and DPCCH are time-multiplexed on the first physical channel associated with the first spreading code as seen in Figure 10.35. If more physical channels are required, the DPCCH part in the slot will be left blank again, as shown in Figure 10.35. The transmit power of the DPDCH is also reduced.

10.3. THIRD-GENERATION SYSTEMS

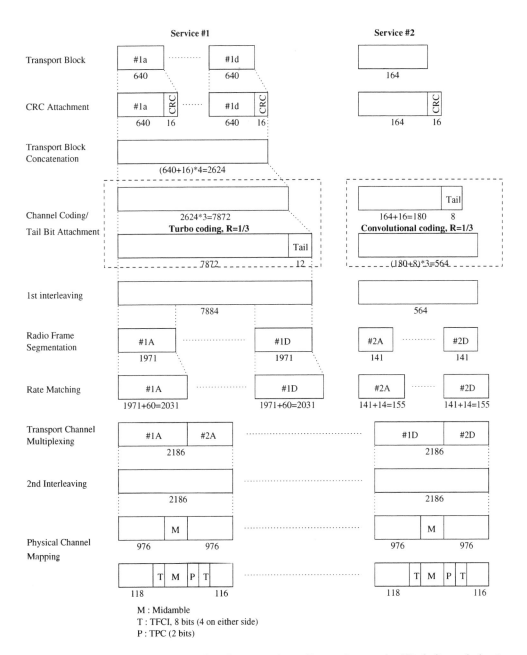

Figure 10.34: Mapping of several multirate multimedia services to the UL dedicated physical data channel of Figure 10.20 in TDD mode. The corresponding schematic is seen in Figure 10.30.

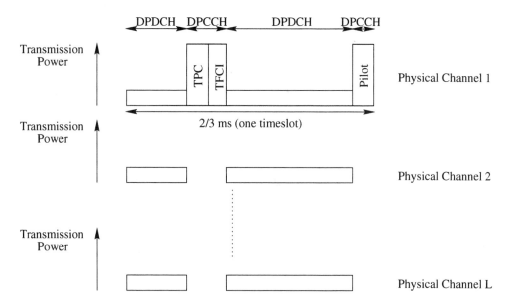

Figure 10.35: DL FDD slot format for multicode transmission in UTRA, based on Figure 10.21, but dispensing with transmitting DPCCH over all multicode physical channels.

10.3.2.6 Spreading and Modulation

It is well-known that the performance of DS-CDMA is interference limited [357]. The majority of the interference originates from the transmitted signals of other users within the same cell, as well as from neighboring cells. This interference is commonly known as *Multiple Access Interference* (MAI). Another source of interference, albeit less dramatic, is a result of the wideband nature of CDMA, yielding several delayed replicas of the transmitted signal, which reach the receiver at different time instants, thereby inflicting what is known as *interpath interference*. However, the advantages gained from wideband transmissions, such as multipath diversity and the noise-like properties of the interference, outweigh the drawbacks.

The choice of the spreading codes [385, 386] used in DS-CDMA will have serious implications for the amount of interference generated. Suffice to say that the traditional measures used in comparing different codes are their *cross-correlations* (CCL) and *autocorrelation* (ACL). If the CCL of the spreading codes of different users is nonzero, this will increase their interference, as perceived by the receiver. Thus a low CCL reduces the MAI. The so-called out-of-phase ACL of the codes, on the other hand, plays an important role during the initial synchronization between the BS and MS, which has to be sufficiently low to minimize the probability of synchronizing to the wrong ACL peak.

In order to reduce the MAI and thereby improve the system's performance and capacity, the UTRA physical channels are spread using two different codes, namely, the *channelization code* and a typically longer so-called *scrambling code*. In general, the channelization codes are used to maintain orthogonality between the different physical channels originating from the same source. On the other hand, the scrambling codes are used to distinguish between different cells, as well as between different MSs. All the scrambling codes in UTRA are in complex format. Complex-valued scrambling balances the power on the I and Q branches. This can be shown by letting c_s^I and c_s^Q be the I and Q branch scrambling codes, respectively. Let $d(t)$ be the complex-valued data of the transmitter, which can be written as:

$$d(t) = d_I + jd_Q, \tag{10.33}$$

where d_I and d_Q represent the data on the I and Q branches, respectively. Let us assume for the sake of argument that the power level in the I and Q branches is unbalanced due to, for instance, their different bit rates or different QoS requirements. If only real-valued scrambling is used, then the output becomes:

$$s(t) = c_s^I (d_I + jd_Q), \tag{10.34}$$

which is also associated with an unbalanced power level on the I and Q branches. By contrast, if complex-valued scrambling is used, then the output would become:

$$\begin{aligned} s(t) &= (d_I + jd_Q) \cdot (c_s^I + jc_s^Q) & (10.35) \\ &= c_s^I \cdot d_I - c_s^Q \cdot d_Q + j(c_s^Q \cdot d_I + c_s^I \cdot d_Q). & (10.36) \end{aligned}$$

As can be seen, the power on the I and Q branches after complex scrambling is the same, regardless of the power level of the unscrambled data on the I and Q branches. Hence, complex scrambling potentially improves the power amplifier's efficiency by reducing the peak-to-average power fluctuation. This also relaxes the linearity requirements of the UL power amplifier used.

Table 10.9 shows the parameters and techniques used for spreading and modulation in UTRA, which will be discussed in depth in the following sections.

	Channelization Codes	**Scrambling Codes**
Type of codes	OVSF (Section 10.3.2.6.1)	UL : Gold codes, S(2) codes (Section 10.3.2.6.2) DL : Gold codes (Section 10.3.2.6.3)
Code length	Variable	UL : 10 ms of $(2^{25} - 1)$-chip Gold code DL : 10 ms of $(2^{18} - 1)$-chip Gold code
Type of spreading	BPSK (UL/DL)	QPSK (UL/DL)

Table 10.9: UL/DL Spreading and Modulation Parameters in UTRA

10.3.2.6.1 Orthogonal Variable Spreading Factor Codes

The UTRA channelization codes are derived from a set of orthogonal codes known as *Orthogonal Variable Spreading Factor* (OVSF) codes [299]. OVSF codes are generated from a tree-structured set of orthogonal codes, such as the Walsh-Hadamard codes, using the procedure shown in Figure 10.36. Specifically, each channelization code is denoted by $c_{N,n}$, where $n = 1, 2, \ldots, N$ and $N = 2^x, x = 2, 3, \ldots 8$. Each code $c_{N,n}$ is

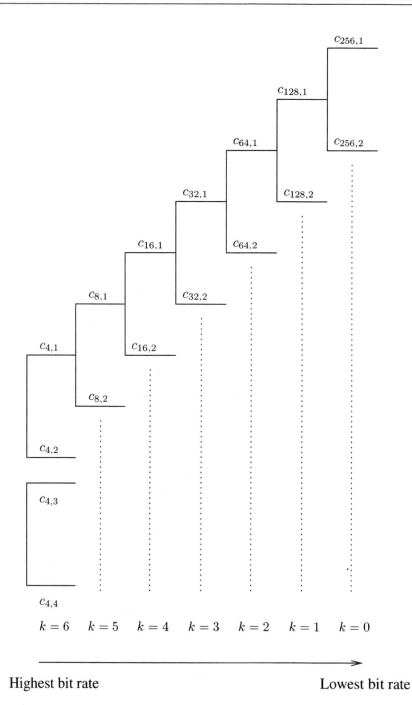

Figure 10.36: Orthogonal variable-spreading factor code tree in UTRA according to Equation 10.37. The parameter k in the figure is directly related to that found in Figures 10.20–10.24.

10.3. THIRD-GENERATION SYSTEMS

derived from the previous code $c_{(N/2),n}$ as follows [299]:

$$\begin{bmatrix} c_{N,1} \\ c_{N,2} \\ c_{N,3} \\ \vdots \\ c_{N,N} \end{bmatrix} = \begin{bmatrix} c_{(N/2),1} | c_{(N/2),1} \\ c_{(N/2),1} | \bar{c}_{(N/2),1} \\ c_{(N/2),2} | c_{(N/2),2} \\ \vdots \\ c_{(N/2),(N/2)} | \bar{c}_{(N/2),(N/2)} \end{bmatrix}, \qquad (10.37)$$

where [|] at the right-hand side of Equation 10.37 denotes an augmented matrix and $\bar{c}_{(N/2),n}$ is the binary complement of $c_{(N/2),n}$. For example, according to Equation 10.37 and Figure 10.36, $c_{N,1} = c_{8,1}$ is created by simply concatenating $c_{(N/2),1}$ and $c_{(N/2),1}$, which doubles the number of chips per bit. By contrast, $c_{N,2} = c_{8,2}$ is generated by attaching $\bar{c}_{(N/2),1}$ to $c_{(N/2),1}$. From Equation 10.37, we see that, for example, $c_{N,1}$ and $c_{N,2}$ at the left-hand side of Equation 10.37 are not orthogonal to $c_{(N/2),1}$, since the first half of both was derived from $c_{(N/2),1}$ in Figure 10.36, but they are orthogonal to $c_{(N/2),n}, n = 2, 3, \ldots, (N/2)$. The code $c_{(N/2),1}$ in Figure 10.36 is known as the mother code of the codes $c_{N,1}$ and $c_{N,2}$, since these two codes are derived from $c_{(N/2),1}$. The codes on the "highest"-order branches ($k = 6$) of the tree at the left of Figure 10.36 have a spreading factor of 4, and they are used for transmission at the highest possible bit rate for a single channel, which is 960 kbps. On the other hand, the codes on the "lowest"-order branches ($k = 0$) of the tree at the right of Figure 10.36 result in a spreading factor of 256, and these are used for transmission at the lowest bit rate, which is 15 kbps. It is worth noting here that an intelligent BbB adaptive scheme may vary its SF on a 10 ms frame basis in an attempt to adjust the SF on a near-instantaneous channel-quality motivated basis [93,354]. Orthogonality between parallel transmitted channels of the same bit rate is preserved by assigning each channel a different orthogonal code accordingly. For channels with different bit rates transmitting in parallel, orthogonal codes are assigned, ensuring that no code is the mother-code of the other. Thus, OVSF channelization codes provide total isolation between different users' physical channels on the DL that are transmitted synchronously and hence eliminate MAI among them. OVSF channelization codes also provide orthogonality between the different physical channels seen in Figure 10.35 during multicode transmission.

Since there is only a limited set of OVSF codes, which is likely to be insufficient to support a large user-population, while also allowing identification of the BSs by the MSs on the DL, *each cell will reuse the same set of OVSF codes*. Statistical multiplexing schemes such as packet reservation multiple access (PRMA) can be used to allocate and de-allocate the OVSF codes on a near-instantaneous basis, for example, depending on the users' voice activity in the case of DTX-based communications [387]. However, orthogonal codes, such as the orthogonal OVSF codes, in general exhibit poor out-of-phase ACL and CCL properties [388]. Therefore, the correlations of the OVSFs of adjacent asynchronous BSs will become unacceptably high, degrading the correlation receiver's performance at the MS. On the other hand, certain long codes such as Gold codes exhibit low CCL, which is advantageous in CDMA applications [239]. Hence in UTRA, cell-specific long codes are used in order to reduce the intercell interference on the DL. On the UL, MAI is reduced by assigning different scrambling codes to different users.

10.3.2.6.2 Up-Link Scrambling Codes The UL scrambling codes in UTRA can be classified into long scrambling codes and short scrambling codes. A total of 2^{24} UL scrambling codes can be generated for both the long and short codes.

Long scrambling codes are constructed from two m-sequences using the polynomials of $1+X^3+X^{25}$ and $1+X+X^2+X^3+X^{25}$, following the procedure highlighted by Proakis [52] in order to produce a set of **Gold codes** for the I branch. The Q-branch Gold code is a shifted version of the I-branch Gold code, where a shift of 16,777,232 chips was recommended. Gold codes are rendered different from each other by assigning a unique initial state to one of the shift registers of the m-sequence. The initial state of the other shift register is a sequence of logical 1. Although the Gold codes generated have a length of $2^{25}-1$ chips, only 38,400 chips (10 ms at 3.84 Mcps) are required in order to scramble a radio frame.

Short scrambling codes are defined from a family of periodically extended S(2) codes. This 256-chip S(2) code was introduced to ease the implementation of multi-user detection at the BS [322]. The multi-user detector has to invert the so-called system matrix [353], the dimension of which is proportional to the sum of the channel impulse response duration and the spreading code duration. Thus, using a relatively short scrambling code is an important practical consideration in reducing the size of the system-matrix to be inverted.

10.3.2.6.3 Down-Link Scrambling Codes Unlike the case for the UL, only Gold codes are used on the DL. The DL Gold codes on the I branch are constructed from two m-sequences using the polynomials of $1+X^7+X^{18}$ and $1+X^5+X^7+X^{10}+X^{18}$. These Gold codes are shifted by 131,072 chips in order to produce a set of Gold codes for the Q branch.

Although a total of $2^{18}-1=262,143$ Gold codes can be generated, only 8192 of them will be used as the DL scrambling code. These codes are divided into 512 groups, each of which contains a *primary scrambling code* and 15 *secondary scrambling codes*. Altogether there are 512 primary scrambling codes and $8192-512=7680$ secondary scrambling codes. Each cell is allocated one primary scrambling code, which is used on the CPICH and P-CCPCH channels of Table 10.5. This primary scrambling code will be used to identify the BS for the MS. All the other physical channels belonging to this cell can use either the primary scrambling code or any of the 15 secondary scrambling codes that belong to the same group, as the primary scrambling code. In order to facilitate fast cell or BS identification, the set of 512 primary scrambling codes is further divided into 64 subsets, each consisting of eight primary scrambling codes, as will be shown in Section 10.3.2.9.

10.3.2.6.4 Up-Link Spreading and Modulation A model of the UL transmitter for a single DPDCH is shown in Figure 10.37 [323]. We have seen in Figure 10.20 that the DPDCH and DPCCH are transmitted in parallel on the I and Q branches of the UL, respectively. Hence, to avoid I/Q channel interference in case of I/Q imbalance of the quadrature carriers, different orthogonal spreading codes are assigned to the DPDCH and DPCCH on the I and Q branch, respectively. These two channelization

10.3. THIRD-GENERATION SYSTEMS

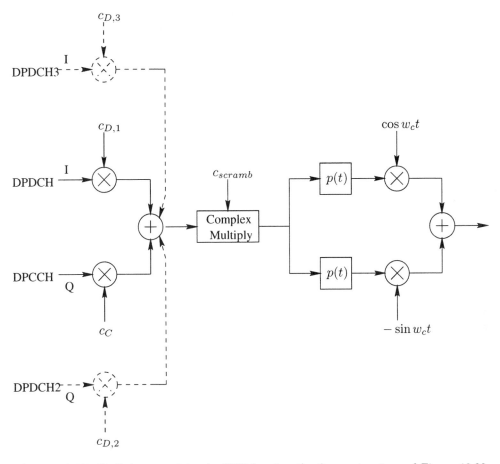

Figure 10.37: Up-link transmitter in UTRA using the frame structure of Figure 10.20. Multicode transmissions are indicated by the dashed lines.

codes for DPDCH and DPCCH, denoted by $c_{D,1}$ and c_C in Figure 10.37, respectively, are allocated in a predefined order. From Figure 10.20, we know that the SF of the DPCCH is 256. Hence, $c_C = c_{256,1}$ in the context of Figure 10.36. This indicates that the high SF of the DPCCH protects the vulnerable control channel message against channel impairments. On the other hand, we have $c_{D,1} = c_{SF,2}$, depending on the SF of the DPDCH. In the event of multicode transmission portrayed by the dashed lines in Figure 10.37, different additional orthogonal channelization codes, namely, $c_{D,2}$ and $c_{D,3}$, are assigned to each DPDCH for the sake of maintaining orthogonality, and they can be transmitted on either the I or Q branch. In this case, the BS and MS have to agree on the number of channelization codes to be used. After spreading, the BPSK modulated I and Q branch signals are summed in order to produce a complex Quadrature Phase Shift Keying (QPSK) signal. The signal is then scrambled by the complex scrambling code, c_{scramb}. The pulse-shaping filters, $p(t)$, are root-raised cosine Nyquist filters using a roll-off factor of 0.22.

The transmitter of the UL PRACH and PCPCH message part is also identical to that shown in Figure 10.37. As we have mentioned in Section 10.3.2.3.2 in the context of Figure 10.22, the PRACH and the CPCH message consist of a data part and a control part. In this case, the data part will be transmitted on the I branch, and the control part on the Q branch. The choice of the channelization codes for the data and control part depends on the signature of the preambles transmitted beforehand. As highlighted in Section 10.3.2.3.2, the preamble signature is a 256-chip sequence generated by the repetition of a 16-chip Hadamard code. This 16-chip code actually corresponds to one of the OVSF codes, namely, to $c_{16,n}$, where $n = 1, \ldots, 16$. The codes in the subtree of Figure 10.36 below this specific 16-chip code n will be used as the channelization codes for the data part and control part.

10.3.2.6.5 Down-Link Spreading and Modulation

The schematic of the DL transmitter is shown in Figure 10.38. All the DL physical channel bursts (except for the SCH) are first QPSK modulated in order to form the I and Q branches, before spreading to the chip rate. In contrast to the UL of Figure 10.37, the same OVSF channelization code c_{ch} is used on the I and Q branches. Different physical channels are assigned different channelization codes in order to maintain their orthogonality. For instance, the channelization codes for the CPICH and P-CCPCH of Table 10.5 are fixed to the codes $c_{256,1}$ and $c_{256,2}$ of Figure 10.36, respectively. The channelization codes for all the other physical channels are assigned by the network.

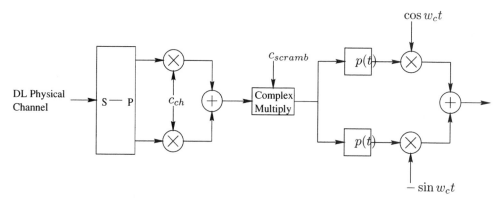

Figure 10.38: Down-link transmitter in UTRA using the frame structure of Figure 10.21.

The resulting signal in Figure 10.38 is then scrambled by a cell-specific scrambling code c_{scramb}. Similarly to the DL, the pulse-shaping filters are root-raised cosine Nyquist filters using a rolloff factor of 0.22.

In TDD mode, the transmitter structure for both the UL and DL are similar to that of a FDD DL transmitter of Figure 10.38. Since each time-slot can be used for transmitting several TDD bursts from the same source or from different sources, the OVSF codes are invoked in order to maintain orthogonality between the burst of different TDD/CDMA users/messages. An advantage of the TDD/CDMA mode is that the user population is separated in both the time and the code domain. In other words, only a small number of CDMA users/services will be supported within a TDD

10.3. THIRD-GENERATION SYSTEMS

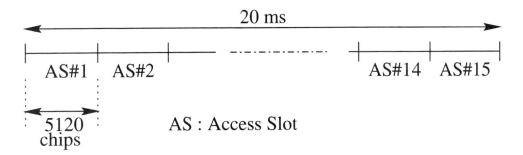

Figure 10.39: ALOHA-based physical UL random access slots in UTRA.

time-slot, which dramatically reduces the complexity of the multi-user detector that can be used in both the UL and DL for mitigating the MAI or multi-code interference.

10.3.2.7 Random Access

10.3.2.7.1 Mobile-Initiated Physical Random Access Procedures If data transmission is initiated by an MS, it is required to send a random access request to the BS. Since such requests can occur at any time, collisions may result when two or more MSs attempt to access the network simultaneously. Hence, in order to reduce the probability of a collision, the random access procedure in UTRA is based on the slotted ALOHA technique [372].

Random access requests are transmitted to the BS via the PRACH of Table 10.5. Each random access transmission request may consist of one or several preambles and a message part, whose time-slot configuration was shown in Figure 10.22. According to the regime of Figure 10.39, the preambles and the message part can only be transmitted at the beginning of one of those 15 so-called *access slots*, which span two radio frames (i.e., 20 ms). Thus, each access slot has a length equivalent to 5120 chips or $\frac{4}{3}$ ms.

Before any random access request can be transmitted, the MS has to obtain certain information via the DL BCH transmitted on the P-CCPCH of Table 10.5 according to the format of Figure 10.23. This DL BCH/PCCPCH information includes the identifier of the cell-specific scrambling code for the preamble and message part of Figure 10.22, the available preamble signatures, the available access slots of Figure 10.39, which can be contended for in ALOHA mode, the initial preamble transmit power, the preamble power ramping factor, and the maximum number of preamble retransmissions necessitated by their decoding failure due to collisions at the BS. All this information may become available once synchronization is achieved, as will be discussed in Section 10.3.2.9. After acquiring all the necessary information, the MS will randomly select a preamble signature from the available signatures and transmit a preamble at the specific power level specified by the BS on a randomly selected access slot chosen from the set of available access slots seen in Figure 10.39. Note that the preamble is formed by multiplying the selected signature with the preamble scrambling code.

After the preamble is transmitted, the MS will listen for the acknowledgment of reception transmitted from the BS on the AICH of Table 10.5. Note that the AICH is also transmitted at the beginning of an access slot and the phase reference for coherent detection is obtained from the DL CPICH of Table 10.5. The acknowledgment is represented by an AI in the AICH of Table 10.5 that corresponds to the selected preamble signature. If a negative acknowledgment is received, the random access transmission will recommence in a later access slot. If a positive acknowledgment is received, the MS will proceed to transmit the message part at the beginning a predefined access slot. However, if the MS fails to receive any acknowledgment after a predefined time-out, it will retransmit the preamble in another randomly selected access slot of Figure 10.39 with a newly selected signature, provided that the maximum number of preamble retransmissions was not exceeded. The transmit power of the preamble is also increased, as specified by the above-mentioned preamble power ramping factor. This procedure is repeated until either an acknowledgment is received from the BS or the maximum number of preamble retransmissions is reached.

10.3.2.7.2 Common Packet Channel Access Procedures The transmission of the CPCH of Table 10.5 is somewhat similar to that of the RACH transmission regime highlighted in Figure 10.39. Before commencing any CPCH transmission, the MS must acquire vital information from the BCH message transmitted on the P-CCPCH. This information includes the scrambling codes, the available signatures and the access slots for both the A-P and CD-P messages introduced in Section 10.3.2.3.2.1, the scrambling code of the message part, the DL AICH and the associated DL DPCCH channelization code, the initial transmit power of the preambles, the preamble power ramping factor, and the maximum allowable number of retransmissions.

The procedure of the A-P transmission is identical to that of the random access transmission highlighted in Section 10.3.2.7.1. We will accordingly omit the details here.

Once a positive acknowledgment is received from the BS on the DL AICH, the MS will transmit the CD-P on a randomly selected access slot of Figure 10.39 using a randomly selected signature. Upon receiving a positive acknowledgment from the BS on the AICH, the MS will begin transmitting the PC-P followed immediately by the message part shown in Figure 10.20 at a predefined access slot of Figure 10.39.

10.3.2.8 Power Control

Accurate power control is essential in CDMA in order to mitigate the so-called *near-far problem* [389, 390]. Furthermore, power control has a dramatic effect on the coverage and capacity of the system: we will therefore consider the UTRA power control issues in detail.

10.3.2.8.1 Closed-Loop Power Control in UTRA Closed-loop power control is employed on both the UL and DL of the FDD mode through the TPC commands that are conveyed in the UL and DL according to the format of Figures 10.20

10.3. THIRD-GENERATION SYSTEMS

and 10.21, respectively. Since the power control procedure is the same on both links, we will only elaborate further on the UL procedure.

UL closed-loop power control is invoked in order to adjust the MS's transmit power such that the received Signal-to-Interference Ratio (SIR) at the BS is maintained at a given target SIR. The value of the target SIR depends on the required quality of the connection. The BS measures the received power of the desired UL transmitted signal for both the DPDCH and the DPCCH messages shown in Figure 10.20 after RAKE combining, and it also estimates the total received interference power in order to obtain the estimated received SIR. This SIR estimation process is performed every $\frac{2}{3}$ ms, or a time-slot duration, in which the SIR estimate is compared to the target SIR. According to the values of the estimated and required SIRs, the BS will generate a TPC command, which is conveyed to the MS using the burst of Figure 10.21. If the estimated SIR is higher than the target SIR, the TPC command will instruct the MS to lower the transmit power of the DPDCH and DPCCH of Figure 10.20 by a step size of Δ_{TPC} dB. Otherwise, the TPC command will instruct the MS to increase the transmit power by the same step size. The step size Δ_{TPC} is typically 1 dB or 2 dB. Transmitting at an unnecessarily high power reduces the battery life, while degrading other users' reception quality, who — as a consequence — may request a power increment, ultimately resulting in an unstable overall system operation.

In some cases, BS-diversity combining may take place, whereby two or more BSs transmit the same information to the MS in order to enhance its reception quality. These BSs are known as the active BS set of the MS. The received SIR at each BS will be different and so the MS may receive different TPC commands from its active set of BSs. In this case, the MS will adjust its transmit power according to a simple algorithm, increasing the transmit power only if the TPC commands from all the BSs indicate an "increase power" instruction. Similarly, the MS will decrease its transmit power if all the BSs issue a "decrease power" TPC command. Otherwise, the transmit power remains the same. In this way, the multi-user interference will be kept to a minimum without significant deterioration of the performance, since at least one BS has a good reception. Again, the UL and DL procedures are identical, obeying the TPC transmission formats of Figures 10.20 and 10.21, respectively.

10.3.2.8.2 Open-Loop Power Control in TDD Mode As mentioned in Section 10.3.2.3, in contrast to the closed-loop power control regime of the FDD mode, no TPC commands are transmitted on the DL in TDD mode. Instead, open loop power control is used to adjust the transmit power of the MS. Prior to any data burst transmission, the MS would have acquired information about the interference level measured at the BS and also about the BS's P-CCPCH transmitted signal level, which are conveyed to the MS via the BCH according to the format of Figure 10.27. At the same time, the MS would also measure the power of the received P-CCPCH. Hence, with knowledge of the transmitted and received power of the P-CCPCH, the DL path-loss can be found. Since the interference level and the estimated path-loss are now known, the required transmitted power of the TDD burst can be readily calculated based on the required SIR level. Let us now consider how the MS identifies the different cells or BSs with which it is communicating.

10.3.2.9 Cell Identification

10.3.2.9.1 Cell Identification in the FDD Mode System- and cell-specific information is conveyed via the BCH transmitted by the P-CCPCH of Table 10.5 in the context of Figure 10.23 in UTRA. This information has to be obtained before the MS can access the network. The P-CCPCH information broadcast from each cell is spread by the system-specific OVSF channelization code $c_{256,2}$ of Figure 10.36. However, each P-CCPCH message is scrambled by a cell-specific primary scrambling code as highlighted in Section 10.3.2.6.3 in order to minimize the intercell interference as well as to assist in identifying the corresponding cell. Hence, the first step for the MS is to recognize this primary scrambling code and to synchronize with the corresponding BS.

As specified in Section 10.3.2.6.3, there are a total of 512 DL primary scrambling codes available in the network. Theoretically, it is possible to achieve scrambling code identification by cross-correlating the P-CCPCH broadcast signal with all the possible 512 primary scrambling codes. However, this would be an extremely tedious and slow process, unduly delaying the MS's access to the network. In order to achieve a fast cell identification by the MS, UTRA adopted a three-step approach [391], which invoked the SCH broadcast from all the BSs in the network. The SCH message is transmitted during the first 256 chips of the P-CCPCH, as illustrated in Figure 10.23. The concept behind this three-step approach is to divide the set of 512 possible primary scrambling codes into 64 subsets, each containing a smaller set of primary scrambling codes, namely, eight codes. Once knowledge of which subset the primary scrambling code of the selected BS belongs to is acquired, the MS can proceed to search for the correct primary scrambling code from a smaller subset of the possible codes.

The frame structure of the DL SCH message seen in Figure 10.23 is shown in more detail in Figure 10.40. It consists of two subchannels, the *Primary SCH* and *Secondary SCH*, transmitted in parallel using code multiplexing. As seen in Figure 10.40, in the Primary SCH a so-called *Primary Synchronization Code* (PSC), based on a generalized hierarchical Golay sequence [392] of length 256 chips, is transmitted periodically at the beginning of each slot, which is denoted by c_p in Figure 10.40. The same PSC is used by all the BSs in the network. This allows the MS to establish slot-synchronization and to proceed to the frame-synchronization phase with the aid of the secondary SCH. On the secondary SCH, a sequence of 15 *Secondary Synchronization Codes* (SSCs), each of length 256 chips, is transmitted with a period of one 10 ms radio frame duration, that is, 10 ms, as seen in Figure 10.40. An example of this 15-SSC sequence would be:

$$c_1^1 \; c_1^2 \; c_2^3 \; c_8^4 \; c_9^5 \; c_{10}^6 \; c_{15}^7 \; c_8^8 \; c_{10}^9 \; c_{16}^{10} \; c_2^{11} \; c_7^{12} \; c_{15}^{13} \; c_7^{14} \; c_{16}^{15}, \qquad (10.38)$$

where each of these 15 SSCs is selected from a set of 16 legitimate SSCs. The specific sequence of 15 SSCs denoted by $c_i^1, \ldots, c_i^{15}$ — where $i = 1, \ldots, 16$ in Figure 10.40 — is used as a code in order to identify and signal to the MS which of the 64 subsets the primary scrambling code used by the particular BS concerned belongs to. The parameter a in Figure 10.40 is a binary flag used to indicate the presence ($a = +1$) or absence ($a = -1$) of a Space Time Block Coding Transmit Diversity (STTD) encoding scheme [393] in the P-CCPCH, as will be discussed in Section 10.3.4.1.1. Specifically, when each of the 16 legitimate 256-chip SSCs can be picked for any of the 15 positions

10.3. THIRD-GENERATION SYSTEMS

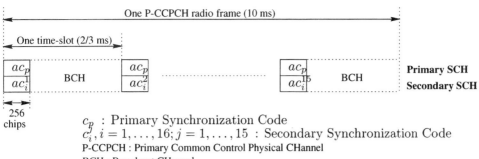

Figure 10.40: Frame structure of the UTRA DL synchronization channel (SCH), which is mapped to the first 256 chips of the P-CCPCH of Figure 10.23. The primary and secondary SCH are transmitted in parallel using code multiplexing. The parameter a is a gain factor used to indicate the presence ($a = +1$) or absence ($a = -1$) of STTD encoding in the P-CCPCH.

in Figure 10.40 and assuming no other further constraints, one could construct

$$\begin{aligned}
c_{i,j}^{\text{repeated}} &= \binom{i+j-1}{j} \\
&= \frac{(i+j-1)!}{j!(i-1)!} \\
&= \frac{30!}{15! \cdot 15!} \\
&= 155,117,520
\end{aligned} \quad (10.39)$$

different such sequences, where $i = 16$ and $j = 15$. However, the 15 different 256-chip SSCs of Figure 10.40 must be constructed so that their cyclic shifts are also unique, since these sequences have to be uniquely recognized before synchronization. In other words, none of the cyclic shifts of the 64 required $15 \times 256 = 3840$-chip sequences can be identical to any of the other sequences' cyclic shifts. Provided that these conditions are satisfied, the 15 specific 256-chip secondary SCH sequences can be recognized within one 10 ms-radio frame-duration of 15 slots. Thus, both slot and frame synchronization can be established within the particular 10 ms frame received. Using this technique, initial cell identification and synchronization can be carried out in the following three basic steps.

Step 1: The MS uses the 256-chip PSC of Figure 10.40 to perform cross-correlation with all the received Primary SCHs of the BSs in its vicinity. The BS with the highest correlator output is then chosen, which constitutes the best cell site associated with the lowest path-loss. Several periodic correlator output peaks have to be identified in order to achieve a high BS detection reliability, despite the presence of high-level interference. *Slot synchronization is also achieved* in this step by recognizing the 15 consecutive c_p sequences, providing 15 periodic correlation peaks.

Step 2: Once the best cell site is identified, the primary scrambling code subset of that cell site is found by cross-correlating the Secondary SCH with the 16 possible SSCs in each of the 15 time-slots of Figure 10.40. This can be easily implemented using 16 correlators, since the timing of the SSCs is known from Step 1. Hence, there are a total of $15 \times 16 = 240$ correlator outputs. From these outputs, a total of $64 \times 15 = 960$ decision variables corresponding to the 64 possible secondary SCH sequences and 15 cyclic shifts of each $15 \times 256 = 3840$-chip sequence are obtained. The highest decision variable determines the primary scrambling code subset. Consequently, *frame synchronization is also achieved.*

Step 3: With the primary scrambling code subset identified and frame synchronization achieved, the primary scrambling code itself is acquired in UTRA by cross-correlating the received CPICH signal — which is transmitted synchronously with the P-CCPCH — on a symbol-by-symbol basis with the eight possible primary scrambling codes belonging to the identified primary scrambling code subset. Note that the CPICH is used in this case, because it is scrambled by the same primary scrambling code as the P-CCPCH and also uses a predefined pilot sequence and so it can be detected more reliably. By contrast, the P-CCPCH carries the unknown BCH information. Once the exact primary scrambling code is identified, the BCH information of Table 10.5, which is conveyed by the P-CCPCH of Figure 10.23, can be detected.

10.3.2.9.2 Cell Identification in the TDD Mode The procedure of cell identification in the TDD mode is somewhat different from that in FDD mode. In the TDD mode, a combination of three 256-chip SSCs out of 16 unique SSCs are used to identify one of 32 SSC code groups allocated to that cell. Each code group contains four different scrambling codes and four corresponding long (for Type 1 burst) and short (for Type 2 burst) basic midamble codes, which were introduced in the context of Figure 10.27. Each code group is also associated with a specific time offset, t_{offset}. The three SSCs, c_i^1, c_i^2, and c_i^3, are transmitted in parallel with the PSC, c_p, at a time offset t_{offset} measured from the start of a time-slot, as shown in Figure 10.41. Similarly to the FDD mode, the PSC is based on a so-called generalized hierarchical Golay sequence [392], which is common to all the cells in the system. Initial cell identification and synchronization in the TDD mode can also be carried out in three basic steps.

Step 1: The MS uses the 256-chip PSC of Figure 10.41 to perform cross-correlation with all the received PSC of the BSs in its vicinity. The BS associated with the highest correlator output is then chosen, which constitutes the best cell site exhibiting the lowest path-loss. Slot synchronization is also achieved in this step. If only one time-slot per frame is used to transmit the SCH as outlined in the context of Figure 10.27, then frame synchronization is also achieved.

Step 2: Once the PSC of the best cell site is identified, the three SSCs transmitted in parallel with the PSC in Figure 10.41 can be identified by cross-correlating the received signal with the 16 possible prestored SSCs. The specific combination of the three SSCs will identify the code group used by the corresponding cell. The specific frame timing of that cell also becomes known from the time offset t_{offset} associated with that code group.

If two time-slots per frame are used to transmit the SCH as outlined in the context

10.3. THIRD-GENERATION SYSTEMS

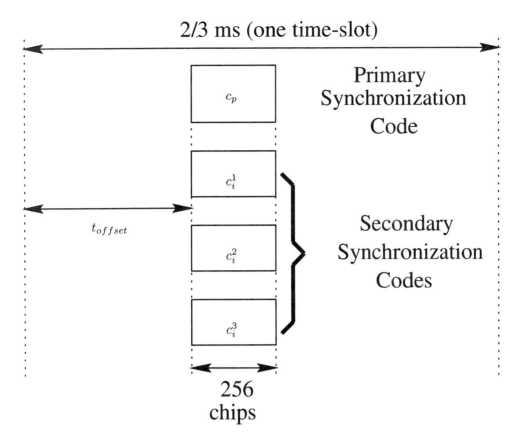

Figure 10.41: Time-slot structure of the UTRA TDD DL synchronization channel (SCH), which obeys the format of Figure 10.19. The primary and three secondary synchronization codes are transmitted in parallel at a time offset t_{offset} from the start of a time-slot.

of Figure 10.27, then the second PSC must be detected at an offset of seven or eight time-slots with respect to the first one in order to achieve frame synchronization.

Step 3: As mentioned in Section 10.3.2.3, each basic midamble code defined in the context of Figure 10.27 is associated with a midamble code set. The P-CCPCH of Table 10.5 is always associated with the first midamble of that set. Hence, with the code group identified and frame synchronization achieved, the cell-specific scrambling code and the associated basic midamble code are acquired in the TDD mode of UTRA by cross-correlating the four possible midamble codes with the P-CCPCH. Once the exact basic midamble code is identified, the associated scrambling code will be known, and the BCH information of Table 10.5, which is conveyed by the P-CCPCH of Figure 10.23, can be detected. Having highlighted the FDD and TDD UTRA cell-selection and synchronization solutions, let us now consider some of the associated handover issues.

10.3.2.10 Handover

In this section, we consider the handover issues in the context of the FDD mode, since the associated procedures become simpler in the TDD mode, where the operations can be carried out during the unused time-slots. Theoretically, DS-CDMA has a frequency reuse factor of one [394]. This implies that neighboring cells can use the same carrier frequency without interfering with each other, unlike in TDMA or FDMA. Hence, seamless uninterrupted handover can be achieved when mobile users move between cells, since no switching of carrier frequency and synthesizer retuning is required. However, in *hierarchical cell structures* (HCS)[10] catering, for example, for high-speed mobiles with the aid of a macrocell oversailing a number of microcells, using a different carrier frequency is necessary in order to reduce the intercell interference. In this case, inter-frequency handover is required. Furthermore, because the various operational GSM systems used different carrier frequencies, handover from UTRA systems to GSM systems will have to be supported during the transitory migration phase, while these systems will coexist. Thus, handovers in terrestrial UMTSs can be classified into inter-frequency and intra-frequency handovers.

10.3.2.10.1 Intra-Frequency Handover or Soft Handover Soft handover [395, 396] involves no frequency switching because the new and old cell use the same carrier frequency. The MS will continuously monitor the received signal levels from the neighboring cells and compares them against a set of thresholds. This information is fed back to the network. Based on this information, if a weak or strong cell is detected, the network will instruct the MS to drop or add the cell from/to its active BS set. In order to ensure a seamless handover, a new link will be established before relinquishing the old link, using the *make before break* approach.

10.3.2.10.2 Inter-Frequency Handover or Hard Handover In order to achieve handovers between different carrier frequencies without affecting the data flow, a technique known as *compressed mode* can be used [397]. With this technique, the UL data, which normally occupies the entire 10 ms frame of Figure 10.19 is time-compressed, so that it only occupies a portion of the frame, that is, slot#1-slot#M and slot#N-slot#15, while no data is transmitted during the remaining portion, that is, slot#(M+1)-slot#(N-1). The latter interval is known as the idle period, as shown in Figure 10.42. There are two types of frame structures for the DL compressed mode, as shown in Figure 10.43. In the Type A structure, shown at the top of Figure 10.43, no data is transmitted after the pilot field of slot#M until the start of the pilot field of slot#(N-1) in order to maximize the transmission gap length. By contrast, in the Type B structure shown at the bottom of Figure 10.43, a TPC command is transmitted in slot#(M+1) during the idle period in order to optimize the power control.

The idle period has a variable duration, but the maximum period allowable within a 15-slot, 10 ms radio frame is seven slots. The idle period can occur either at the center of a 10 ms frame or at the end and the beginning of two consecutive 10 ms frames, such that the idle period spans over two frames. However, in order to maintain

---

[10] Microcells overlaid by a macrocell.

10.3. THIRD-GENERATION SYSTEMS

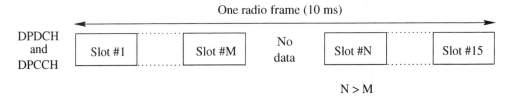

Figure 10.42: Up-link frame structure in compressed mode operation during UTRA handovers.

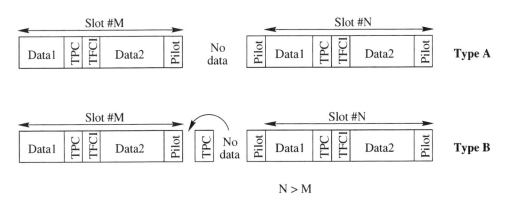

Figure 10.43: Down-link frame structure in compressed mode operation during UTRA handovers using the transmission formats of Figure 10.19.

the seamless operation of all MSs occupying the uncompressed 15-slot, 10 ms frame, the duration of all time-slots has to be shortened by "compressing" their data. The compression of data can be achieved by channel-code puncturing, a procedure that obliterates some of the coded parity bits, thereby slightly reducing the code's error correcting power, or by adjusting the spreading factor. In order to maintain the quality of the link, the instantaneous power is also increased during the compressed mode operation. After receiving the data, the MS can use this idle period in the 10 ms frame, to switch to other carrier frequencies of other cells and to perform the necessary link-quality measurements for handover.

Alternatively, a twin-receiver can be used in order to perform inter-frequency handovers. One receiver can be tuned to the desired carrier frequency for reception, while the other receiver can be used to perform handover link-quality measurements at other carrier frequencies. This method, however, results in a higher hardware complexity at the MS.

The 10 ms frame length of UTRA was chosen so that it is compatible with the multiframe length of 120 ms in GSM. Hence, the MS is capable of receiving the Frequency Correction Channel (FCCH) and Synchronization Channel (SCH) messages in the GSM [41] frame using compressed mode transmission and to perform the necessary handover link-quality measurements [371].

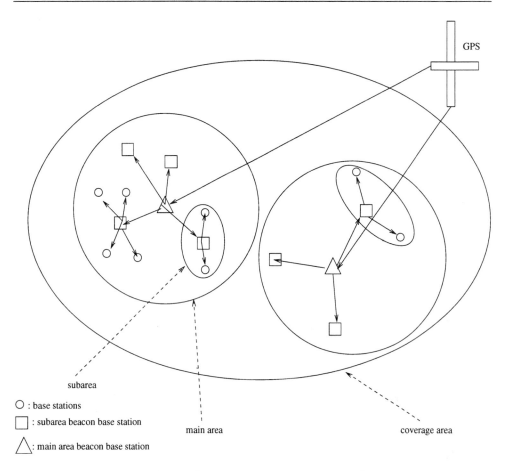

Figure 10.44: Intercell time synchronization in UTRA TDD mode.

10.3.2.11 Intercell Time Synchronization in the UTRA TDD Mode

Time synchronization between BSs is required when operating in the TDD mode in order to support seamless handovers. A simple method of maintaining intercell synchronization is by periodically broadcasting a reference signal from a source to all the BSs. The propagation delay can be easily calculated, and hence compensated, from the fixed distance between the source and the receiving BSs. There are three possible ways of transmitting this reference signal, namely, via the terrestrial radio link, via the physical wired network, or via the Global Positioning System (GPS).

Global time synchronization in 3G mobile radio systems is achieved by dividing the synchronous coverage region into three areas, namely, the so-called subarea, main area and coverage area, as shown in Figure 10.44. Intercell synchronization within a sub-area is provided by a subarea reference BS. Since the subarea of Figure 10.44 is smaller than the main area, transmitting the reference signal via the terrestrial radio link or the physical wired network is more feasible. All the subarea reference BSs in a main area are in turn synchronized by a main-area reference BS. Similarly, the

10.3. THIRD-GENERATION SYSTEMS

reference signal can be transmitted via the terrestrial radio link or the physical wired network. Finally, all the main-area reference BSs are synchronized using the GPS. The main advantage of dividing the coverage regions into smaller areas is that each lower hierarchical area can still operate on its own, even if the synchronization link with the higher hierarchical areas is lost.

10.3.3 The cdma2000 Terrestrial Radio Access [303–305]

The current 2G mobile radio systems standardized by TIA in the United States are IS-95-A and IS-95-B [303]. The radio access technology of both systems is based on narrowband DS-CDMA with a chip rate of 1.2288 Mcps, which gives a bandwidth of 1.25 MHz. The basic features of this system were summarized in Table 9.1. IS-95-A was commercially launched in 1995, supporting circuit and packet mode transmissions at a maximum bit rate of only 14.4 kbps [303]. An enhancement to the IS-95-A standards, known as IS-95-B, was developed and introduced in 1998 in order to provide higher data rates, on the order of 115.2 kbps [322]. This was feasible without changing the physical layer of IS-95-A. However, this still falls short of the 3G mobile radio system requirements. Hence, the technical committee TR45.5 within TIA has proposed cdma2000, a 3G mobile radio system that is capable of meeting all the requirements laid down by ITU. One of the problems faced by TIA is that the frequency bands allocated for the 3G mobile radio system, identified during WARC'92 to be 1885–2025 MHz and 2110–2200 MHz, have already been allocated for Personal Communications Services (PCS) in the United States from 1.8 GHz to 2.2 GHz. In particular, the CDMA PCS based on the IS-95 standards has been allocated the frequency bands of 1850–1910 MHz and 1930–1990 GHz. Hence, the 3G mobile radio systems have to fit into the allocated bandwidth without imposing significant interference on the existing applications. Thus, the framework for cdma2000 was designed so that it can be overlaid on IS-95 and it is backwards compatible with IS-95. Most of this section is based on [303–305].

10.3.3.1 Characteristics of cdma2000

The basic parameters of cdma2000 are shown in Table 10.10. The cdma2000 system has a basic chip rate of 3.6864 Mcps, which is accommodated in a bandwidth of 3.75 MHz. This chip rate is in fact three times the chip rate used in the IS-95 standards, which is 1.2288 Mcps. Accordingly, the bandwidth was also trebled. Hence, the existing IS-95 networks can also be used to support the operation of cdma2000. Higher chip rates on the order of $N \times 1.2288$ Mcps, $N = 6, 9, 12$ are also supported. These are used to enable higher bit rate transmission. The value of N is an important parameter in determining the channel-coding rate and the channel bit rate. In order to transmit the high chip-rate signals ($N > 1$), two modulation techniques are employed. In the *direct-spread modulation mode*, the symbols are spread according to the chip rate and transmitted using a single carrier, giving a bandwidth of $N \times 1.25$ MHz. This method is used on both the up-link and down-link. In *multicarrier (MC) modulation*, the symbols to be transmitted are de-multiplexed into separate signals, each of which is then spread at a chip rate of 1.2288 Mcps. N different carrier frequencies are used

Radio Access Technology	DS-CDMA, Multicarrier CDMA
Operating environments	Indoor/Outdoor to indoor/Vehicular
Chip rate (Mcps)	1.2288/3.6864/7.3728/11.0592/14.7456
Channel bandwidth (MHz)	1.25/3.75/7.5/11.25/15
Duplex modes	FDD and TDD
Frame length	5 and 20 ms
Spreading factor	variable, 4 to 256
Detection scheme	Coherent with common pilot channel
Intercell operation	FDD : Synchronous TDD : Synchronous
Power control	Open and closed loop
Handover	Soft-handover Inter-frequency handover

Table 10.10: The cdma2000 Basic Parameters

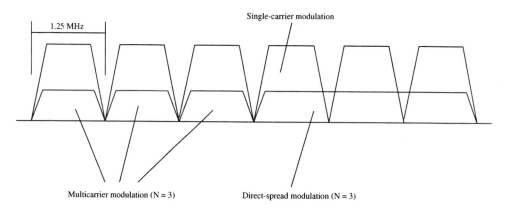

Figure 10.45: Example of an overlay deployment in cdma2000. The multicarrier mode is only used in the down-link.

to transmit these spread signals, each of which has a bandwidth of 1.25 MHz. This method is used for the down-link only, because in this case, transmit diversity can be achieved by transmitting the different carrier frequencies over spatially separated antennas.

By using multiple carriers, cdma2000 is capable of overlaying its signals on the existing IS-95 1.25 MHz channels and its own channels, while maintaining orthogonality. An example of an overlay scenario is shown in Figure 10.45. Higher chip rates are transmitted at a lower power than lower chip rates, thereby keeping the interferences to a minimum.

Similarly to UTRA and IMT-2000, cdma2000 also supports TDD operation in unpaired frequency bands. In order to ease the implementation of a dual-mode FDD/TDD terminal, most of the techniques used for FDD operation can also be applied in TDD operation. The difference between these two modes is in the frame structure, whereby an additional guard time has to be included for TDD operation.

In contrast to UTRA and IMT-2000, where the pilot symbols of Figure 10.21 are time-multiplexed with the dedicated data channel on the down-link, cdma2000

10.3. THIRD-GENERATION SYSTEMS

Dedicated Physical Channels (DPHCH)	Common Physical Channels (CPHCH)
Fundamental Channel (FCH) (UL/DL)	Pilot Channel (PICH) (DL)
Supplemental Channel (SCH) (UL/DL)	Common Auxiliary Pilot Channel (CAPICH) (DL)
Dedicated Control Channel (DCCH) (UL/DL)	Forward Paging Channel (PCH) (DL)
Dedicated Auxiliary Pilot Channel (DAPICH) (DL)	Sync Channel (SYNC) (DL)
Pilot Channel (PICH) (UL)	Access Channel (ACH) (UL)
	Common Control Channel (CCCH) (UL/DL)

Table 10.11: The cdma2000 Physical Channels

employs a common code multiplexed continuous pilot channel on the down-link, as in the IS-95 system of Table 9.1. The advantage of a common down-link pilot channel is that no additional overhead is incurred for each user. However, if adaptive antennas are used, then additional pilot channels have to be transmitted from each antenna.

Another difference with respect to UTRA and IMT-2000 is that the base stations are operated in synchronous mode in cdma2000. As a result, the same PN code but with different phase offsets can be used to distinguish the base stations. Using one common PN sequence can expedite cell acquisition as compared to a set of PN sequences, as we have seen in Section 10.3.2.9 for IMT-2000/UTRA. Let us now consider the cdma2000 physical channels.

10.3.3.2 Physical Channels in cdma2000

The physical channels (PHCH) in cdma2000 can be classified into two groups, namely Dedicated Physical Channels (DPHCH) and Common Physical Channels (CPHCH). DPHCHs carry information between the base station and a single mobile station, while CPHCHs carry information between the base station and several mobile stations. Table 10.11 shows the collection of physical channels in each group. These channels will be elaborated on during our further discourse. Typically, all physical channels are transmitted using a frame length of 20 ms. However, the control information on the so-called Fundamental Channel (FCH) and Dedicated Control Channel (DCCH) can also be transmitted in 5 ms frames.

Each base station transmits its own down-link Pilot Channel (PICH), which is shared by all the mobile stations within the coverage area of the base station. Mobile stations can use this common down-link PICH in order to perform channel estimation for coherent detection, soft handover, and fast acquisition of strong multipath rays for RAKE combining. The PICH is transmitted orthogonally along with all the other down-link physical channels from the base station by using a unique orthogonal code (Walsh code 0) as in the IS-95 system of Table 9.1. The optional Common Auxiliary Pilot Channels (CAPICH) and Dedicated Auxiliary Pilot Channels (DAPICH) are used to support the implementation of antenna arrays. CAPICHs provide spot coverage shared among a group of mobile stations, while a DAPICH is directed toward a particular mobile station. Every mobile station also transmits an orthogonal code-multiplexed up-link pilot channel (PICH), which enables the base station to perform coherent detection in the up-link as well as to detect strong multipaths and to invoke power control measurements. This differs from IS-95, which supports only noncoherent detection in the up-link due to the absence of a coherent up-link reference. In addition to the pilot symbols, the up-link PICH also contains time-multiplexed power control bits assisting in down-link power control. A power control bit is multiplexed

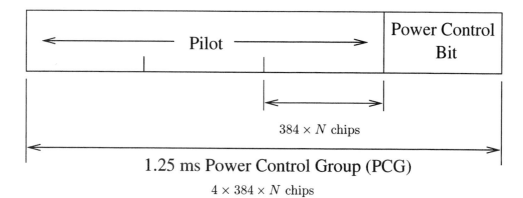

Figure 10.46: Up-link pilot channel structure in cdma2000 for a 1.25 ms duration PCG, where $N = 1, 3, 6, 9, 12$ is the rate-control parameter.

onto the 20 ms frame every 1.25 ms, giving a total of 16 power control bits per 20 ms frame or 800 power updates per second, implying a very agile, fast response power control regime. Each 1.25 ms duration is referred to as a Power Control Group, as shown in Figure 10.46.

The use of two dedicated data physical channels, namely, the so-called Fundamental (FCH) and Supplemental (SCH) channels, optimizes the system during multiple simultaneous service transmissions. Each channel carries a different type of service and is coded and interleaved independently. However, in any connection, there can be only one FCH, but several SCHs can be supported. For a FCH transmitted in a 20 ms frame, two sets of uncoded data rates, denoted as Rate Set 1 (RS1) and Rate Set 2 (RS2), are supported. The data rates in RS1 and RS2 are 9.6/4.8/2.7/1.5 kbps and 14.4/7.2/3.6/1.8 kbps, respectively. Regardless of the uncoded data rates, the coded data rate is 19.2 kbps and 38.4 kbps for RS1 and RS2, respectively, when the rate-control parameter is $N = 1$. The 5 ms frame only supports one data rate, which is 9.6 kbps. The SCH is capable of transmitting higher data rates than the FCH. The SCH supports variable data rates ranging from 1.5 kbps for $N=1$ to as high as 2073.6 kbps, when $N=12$. Blind rate detection [398] is used for SCHs not exceeding 14.4 kbps, while rate information is explicitly provided for higher data rates. The dedicated control physical channel has a fixed uncoded data rate of 9.6 kbps on both 5 ms and 20 ms frames. This control channel rate is more than an order of magnitude higher than that of the IS-95 system in Table 9.1, so it supports a substantially enhanced system control.

The Sync Channel (SYCH) — note the different acronym in comparison to the SCH abbreviation in UTRA/IMT-2000 — is used to aid the initial synchronization of a mobile station to the base station and to provide the mobile station with system-related information, including the Pseudo Noise (PN) sequence offset, which is used to identify the base stations and the long code mask, which will be defined explicitly in Section 10.3.3.4. The SYCH has an uncoded data rate of 1.2 kbps and a coded data rate of 4.8 kbps.

Paging functions and packet data transmission are handled by the down-link Pag-

10.3. THIRD-GENERATION SYSTEMS

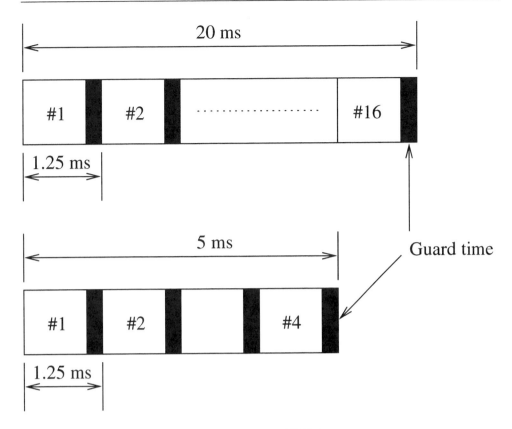

Figure 10.47: The cdma2000 TDD frame structure.

ing Channel (PCH) and the down-link Common Control Channel (CCCH). The uncoded data rate of the PCH can be either 4.8 kbps or 9.6 kbps. The CCCH is an improved version of the PCH, which can support additional higher data rates, such as 19.2 and 38.4 kbps. In this case, a 5 ms or 10 ms frame length will be used. The PCH is included in cdma2000 in order to provide IS-95-B functionality.

In TDD mode, the 20 ms and 5 ms frames are divided into 16 and 4 time-slots, respectively. This gives a duration of 1.25 ms per time-slot, as shown in Figure 10.47. A guard time of 52.08 μs and 67.44 μs is used for the down-link in multicarrier modulation and for direct-spread modulation, respectively. In the up-link, the guard time is 52.08 μs. Having described the cdma2000 physical channels of Table 10.11, let us now consider the service multiplexing and channel-coding aspects.

10.3.3.3 Service Multiplexing and Channel Coding

Services of different data rates and different QoS requirements are carried by different physical channels, namely, by the FCH and SCH of Table 10.11. This differs from UTRA and IMT-2000, whereby different services were time-multiplexed onto one or more physical channels, as highlighted in Section 10.3.2.4. These channels in

	Convolutional	Turbo
Rate	1/2 or 1/3 or 1/4	1/2 or 1/3 or 1/4
Constraint length	9	4

Table 10.12: The cdma2000 Channel-Coding Parameters

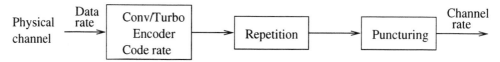

Physical Channel	Data Rate	Code Rate	Repetition	Puncturing	Channel Rate
SYCH	1.2 kbps	1/2	×2	0	4.8 ksps
PCH	4.8 kbps	1/2	×2	0	19.2 ksps
	9.6 kbps	1/2	×1	0	19.2 ksps
CCCH	9.6 kbps	1/2	×1	0	19.2 ksps
	19.2 kbps	1/2	×1	0	38.4 ksps
	38.4 kbps	1/2	×1	0	76.8 ksps
FCH	1.5 kbps	1/2	×8	1 of 5	19.2 ksps
	2.7 kbps	1/2	×4	1 of 9	19.2 ksps
	4.8 kbps	1/2	×2	0	19.2 ksps
	9.6 kbps	1/2	×1	0	19.2 ksps
	1.8 kbps	1/3	×8	1 of 9	38.4 ksps
	3.6 kbps	1/3	×4	1 of 9	38.4 ksps
	7.2 kbps	1/3	×2	1 of 9	38.4 ksps
	14.4 kbps	1/3	×1	1 of 9	38.4 ksps
SCH	9.6 kbps	1/2	×1	0	19.2 ksps
	19.2 kbps	1/2	×1	0	38.4 ksps
	38.4 kbps	1/2	×1	0	76.8 ksps
	76.8 kbps	1/2	×1	0	153.6 ksps
	153.6 kbps	1/2	×1	0	307.2 ksps
	307.2 kbps	1/2	×1	0	614.4 ksps
	14.4 kbps	1/3	×1	1 of 9	38.4 ksps
	28.8 kbps	1/3	×1	1 of 9	76.8 ksps
	57.6 kbps	1/3	×1	1 of 9	153.6 ksps
	115.2 kbps	1/3	×1	1 of 9	307.2 ksps
	230.4 kbps	1/3	×1	1 of 9	614.4 ksps
DCCH	9.6 kbps	1/2	×1	0	19.2 ksps

Table 10.13: The cdma2000 Down-Link Physical Channel (see Table 10.11) Coding Parameters for $N = 1$, Where Repetition × 2 Implies Transmitting a Total of Two Copies

cdma2000 are code-multiplexed using Walsh codes. Two types of coding schemes are used in cdma2000, as shown in Table 10.12. Basically, all channels use convolutional codes for forward error correction. However, for SCHs at rates higher than 14.4 kbps, turbo coding [382] is preferable. The rate of the input data stream is matched to the given channel rate by either adjusting the coding rate or using symbol repetition with and without symbol puncturing, or alternatively, by sequence repetition. Tables 10.13 and 10.14 show the coding rate and the associated rate matching procedures for the various down-link and up-link physical channels, respectively, when $N = 1$. Following the above brief notes on the cdma2000 channel coding and service multiplexing issues, let us now turn to the spreading and modulation processes.

10.3. THIRD-GENERATION SYSTEMS

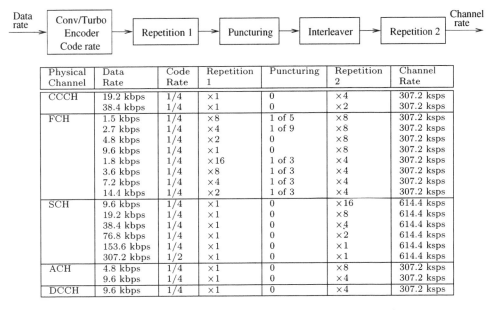

Physical Channel	Data Rate	Code Rate	Repetition 1	Puncturing	Repetition 2	Channel Rate
CCCH	19.2 kbps	1/4	×1	0	×4	307.2 ksps
	38.4 kbps	1/4	×1	0	×2	307.2 ksps
FCH	1.5 kbps	1/4	×8	1 of 5	×8	307.2 ksps
	2.7 kbps	1/4	×4	1 of 9	×8	307.2 ksps
	4.8 kbps	1/4	×2	0	×8	307.2 ksps
	9.6 kbps	1/4	×1	0	×8	307.2 ksps
	1.8 kbps	1/4	×16	1 of 3	×4	307.2 ksps
	3.6 kbps	1/4	×8	1 of 3	×4	307.2 ksps
	7.2 kbps	1/4	×4	1 of 3	×4	307.2 ksps
	14.4 kbps	1/4	×2	1 of 3	×4	307.2 ksps
SCH	9.6 kbps	1/4	×1	0	×16	614.4 ksps
	19.2 kbps	1/4	×1	0	×8	614.4 ksps
	38.4 kbps	1/4	×1	0	×4	614.4 ksps
	76.8 kbps	1/4	×1	0	×2	614.4 ksps
	153.6 kbps	1/4	×1	0	×1	614.4 ksps
	307.2 kbps	1/2	×1	0	×1	614.4 ksps
ACH	4.8 kbps	1/4	×1	0	×8	307.2 ksps
	9.6 kbps	1/4	×1	0	×4	307.2 ksps
DCCH	9.6 kbps	1/4	×1	0	×4	307.2 ksps

Table 10.14: The cdma2000 Up-Link Physical Channel (see Table 10.11) Coding Parameters for $N = 1$, Where Repetition × 2 Implies Transmitting a Total of Two Copies

	Channelization Codes (UL/DL)	User-specific Scrambling Codes (UL)	Cell-specific Scrambling Codes (DL)
Type of codes	Walsh codes	Different offsets of a real m-sequence	Different offsets of a complex m-sequence
Code length	Variable	$2^{42} - 1$ chips	2^{15} chips
Type of Spreading	BPSK	BPSK	QPSK
Data Modulation	DL : QPSK UL : BPSK		

Table 10.15: Spreading Parameters in cdma2000

10.3.3.4 Spreading and Modulation

There are generally three layers of spreading in cdma2000, as shown in Table 10.15. Each user's up-link signal is identified by different offsets of a long code, a procedure that is similar to that of the IS-95 system portrayed in [9]. As seen in Table 10.15, this long code is an m-sequence with a period of $2^{42} - 1$ chips. The construction of m-sequences was highlighted by Proakis [52]. Different user offsets are obtained using a long code mask. Orthogonality between the different physical channels of the same user belonging to the same connection in the up-link is maintained by spreading using Walsh codes.

In contrast to the IS-95 down-link of Figure 1.42 of [9], whereby Walsh code spreading is performed prior to QPSK modulation, the data in cdma2000 is first QPSK modulated before spreading the resultant I and Q branches with the same Walsh code. In this way, the number of Walsh codes available is increased twofold

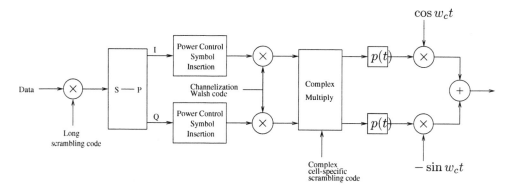

Figure 10.48: The cdma2000 down-link transmitter. The long scrambling code is used for the purpose of improving user privacy. Hence, only the paging channels and the traffic channels are scrambled with the long code. The common pilot channel and the SYNC channel are not scrambled by this long code (the terminology of Table 10.15 is used).

due to the orthogonality of the I and Q carriers. The length of the up-link/down-link (UL/DL) channelization Walsh codes of Table 10.15 varies according to the data rates. All the base stations in the system are distinguished by different offsets of the same complex down-link m-sequence, as indicated by Table 10.15. This down-link m-sequence code is the same as that used in IS-95, which has a period of $2^{15} = 32768$, and it is derived from m-sequences. The feedback polynomials of the shift registers for the I and Q sequences are $X^{15} + X^{13} + X^9 + X^8 + X^7 + X^5 + 1$ and $X^{15} + X^{12} + X^{11} + X^{10} + X^6 + X^5 + X^4 + X^3 + 1$, respectively. The offset of these codes must satisfy a minimum value, which is equal to $N \times 64 \times \text{Pilot\_Inc}$, where Pilot_Inc is a code reuse parameter, which depends on the topology of the system, analogously to the frequency reuse factor in FDMA. Let us now focus on down-link spreading issues more closely.

10.3.3.4.1 Down-Link Spreading and Modulation Figure 10.48 shows the structure of a down-link transmitter for a physical channel. In contrast to the IS-95 down-link transmitter shown in [9], the data in the cdma2000 down-link transmitter shown in Figure 10.48 are first QPSK modulated before spreading using Walsh codes. As a result, the number of Walsh codes available is increased twofold due to the orthogonality of the I and Q carriers, as mentioned previously. The user data are first scrambled by the long scrambling code by assigning a different offset to different users for the purpose of improving user privacy, which is then mapped to the I and Q channels. This long, scrambling code is identical to the up-link user-specific scrambling code given in Table 10.15. The down-link pilot channels of Table 10.11 (PICH, CAPICH, DAPICH) and the SYNC channel are not scrambled with a long code since there is no need for user-specificity. The up-link power control symbols are inserted into the FCH at a rate of 800 Hz, as shown in Figure 10.48. The I and Q channels are then spread using a Walsh code and complex multiplied with the cell-specific complex PN sequence of Table 10.15, as portrayed in Figure 10.48.

Each base station's down-link channel is assigned a different Walsh code in order to eliminate any intracell interference since all Walsh codes transmitted by the serving base station are received synchronously. The length of the down-link channelization Walsh code of Table 10.15 is determined by the type of physical channel and its data rate. Typically for $N = 1$, down-link FCHs with data rates belonging to RS1, that is, those transmitting at 9.6/4.8/2.7/1.5 kbps, use a 128-chip Walsh code, and those in RS2, transmitting at 14.4/7.2/3.6/1.8 kbps, use a 64-chip Walsh code. Walsh codes for down-link SCHs can range from 4-chip to 128-chip Walsh codes. The down-link PICH is an unmodulated sequence (all 0s) spread by Walsh code 0. Finally, the complex spread data in Figure 10.48 are baseband filtered using the Nyquist filter impulse responses $p(t)$ in Figure 10.48 and modulated on a carrier frequency.

For the case of multicarrier modulation, the data is split into N branches immediately after the long code scrambling of Figure 10.48 which was omitted in the figure for the sake of simplicity. Each of the N branches is then treated as a separate transmitter and modulated using different carrier frequencies.

10.3.3.4.2 Up-link Spreading and Modulation The up-link cdma2000 transmitter is shown in Figure 10.49. The up-link PICH and DCCH of Table 10.11 are mapped to the I data channel, while the up-link FCH and SCH of Table 10.11 are mapped to the Q channel in Figure 10.49. Each of these up-link physical channels belonging to the same user is assigned different Walsh channelization codes in order to maintain orthogonality, with higher rate channels using shorter Walsh codes. The I and Q data channels are then spread by complex multiplication with the user-specifically offset real m-sequence based scrambling code of Table 10.15 and a complex scrambling code, which is the same for all the mobile stations in the system, as seen at the top of Figure 10.49. However, this latter complex scrambling code is not explicitly shown in Table 10.15, since it is identical to the down-link cell-specific scrambling code. This complex scrambling code is only used for the purpose of quadrature spreading. Thus, in order to reduce the complexity of the base station receiver, this complex scrambling code is identical to the cell-specific scrambling code of Table 10.15 used on the down-link by all the base stations.

10.3.3.5 Random Access

The mobile station initiates an access request to the network by repeatedly transmitting a so-called access probe until a request acknowledgment is received. This entire process of sending a request is known as an access attempt. Within a single access attempt, the request may be sent to several base stations. An access attempt addressed to a specific base station is known as a subattempt. Within a subattempt, several access probes with increasing power can be sent. Figure 10.50 shows an example of an access attempt. The access probe transmission follows the slotted ALOHA algorithm, which is a relative of PRMA. An access probe can be divided into two parts, as shown in Figure 10.51. The access preamble carries a nondata-bearing pilot channel at an increased power level. The so-called access channel message capsule carries the data-bearing Access Channel (ACH) or up-link Common Control Channel (CCCH) messages of Table 10.11 and the associated nondata-bearing pilot channel.

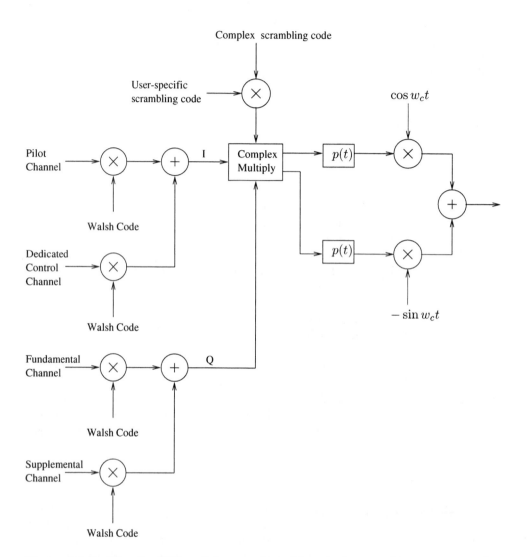

Figure 10.49: The cdma2000 up-link transmitter. The complex scrambling code is identical to the down-link cell-specific complex scrambling code of Table 10.15 used by all the base stations in the system (the terminology of Table 10.15 is used).

10.3. THIRD-GENERATION SYSTEMS

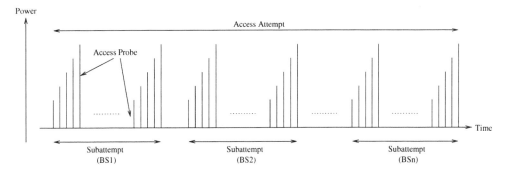

Figure 10.50: An access attempt by a mobile station in cdma2000 using the access probe of Figure 10.51.

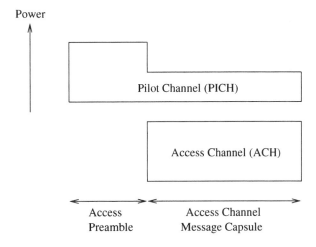

Figure 10.51: A cdma2000 access probe transmitted using the regime of Figure 10.50.

The structure of the pilot channel is similar to that of the up-link pilot channel (PICH) of Figure 10.46 except that in this case there are no time-multiplexed power control bits. The preamble length in Figure 10.51 is an integer multiple of the 1.25 ms slot intervals. The specific access preamble length is indicated by the base station, which depends on how fast the base station can search the PN code space in order to recognize an access attempt. The ACH is transmitted at a fixed rate of either 9.6 or 4.8 kbps, as seen in Table 10.14. This rate is constant for the duration of the access probe of Figure 10.50. The ACH or CCCH and their associated pilot channel are spread by the spreading codes of Table 10.15, as shown in Figure 10.52. Different ACHs or CCCHs and their associated pilot channels are spread by different long codes.

The access probes of Figures 10.50 and 10.51 are transmitted in predefined slots, where the slot length is indicated by the base station. Each slot is sufficiently long in order to accommodate the preamble and the longest message of Figure 10.51. The transmission must begin at the start of each 1.25 ms slot. If an acknowledgment of the most recently transmitted probe is not received by the mobile station after a time-out

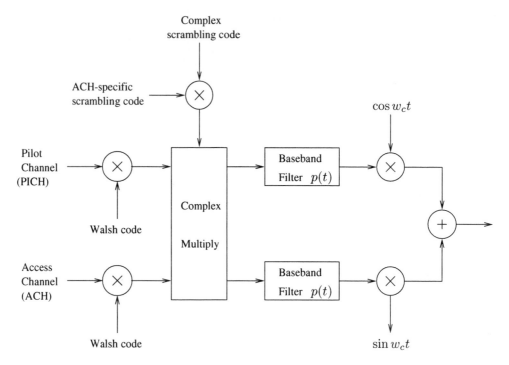

Figure 10.52: The cdma2000 access channel modulation and spreading. The complex scrambling code is identical to the down-link cell-specific complex scrambling code of Table 10.15 used by all the base stations in the system (the terminology of Table 10.15 is used).

period, another probe is transmitted in another randomly chosen slot, obeying the regime of Figure 10.50.

Within a subattempt of Figure 10.50, a sequence of access probes is transmitted until an acknowledgment is received from the base station. Each successive access probe is transmitted at a higher power compared to the previous access probe, as shown in Figure 10.53. The initial power (IP) of the first probe is determined by the open-loop power control plus a nominal offset power that corrects for the open-loop power control imbalance between up-link and down-link. Subsequent probes are transmitted at a power level higher than the previous probe. This increased level is indicated by the Power Increment (PI). Let us now highlight some of the cdma2000 handover issues.

10.3.3.6 Handover

Intra-frequency or soft-handover is initiated by the mobile station. While communicating, the mobile station may receive the same signal from several base stations. These base stations constitute the Active Set of the mobile station. The mobile station will continuously monitor the power level of the received pilot channels (PICH) transmitted from neighboring base stations, including those from the mobile station's

10.3. THIRD-GENERATION SYSTEMS 451

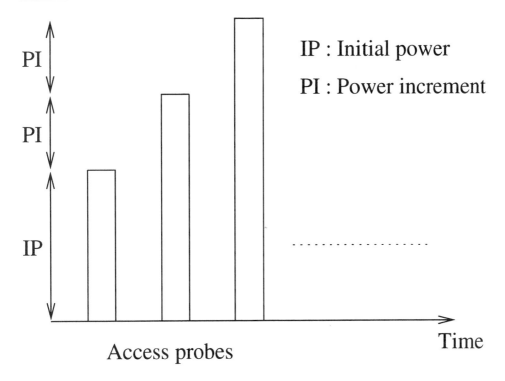

Figure 10.53: Access probes within a subattempt of Figure 10.50.

active set. The power levels of these base stations are then compared to a set of thresholds according to an algorithm, which will be highlighted later in this chapter. The set of thresholds consists of the static thresholds, which are maintained at a fixed level, and the dynamic thresholds, which are dynamically adjusted based on the total received power. Subsequently, the mobile station will inform the network when any of the monitored power levels exceed the thresholds.

Whenever the mobile station detects a PICH, whose power level exceeds a given static threshold, denoted as T_1, this PICH will be moved to a candidate set and will be searched and compared more frequently against a dynamically adjusted threshold denoted as T_2. This value of T_2 is a function of the received power levels of the PICHs of the base stations in the active set. This process will determine whether the candidate base station is worth adding to the active set. If the overall power level in the active set is weak, then adding a base station of higher power will improve the reception. By contrast, if the overall power level in the active set is relatively high, then adding another high-powered base station may not only be unnecessary, but may actually utilize more network resources.

For the base stations that are already in the active set, the power level of their corresponding PICH is compared to a dynamically adjusted threshold, denoted as T_3, which is also a function of the total power of the PICH in the active set, similar to T_2. This is to ensure that each base station in the active set is contributing sufficiently

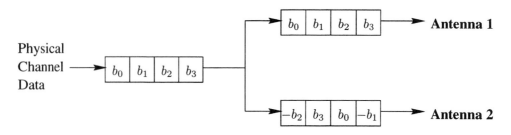

Figure 10.54: Transmission of a physical channel using Space Time block coding Transmit Diversity (STTD).

to the overall power level. If any of the PICH's power level dropped below T_3 after a specified period of time allowed in order to eliminate any uncertainties due to fading which may have caused fluctuations in the power level, the base station will again be moved to the candidate set where it will be compared with a static threshold T_4. At the same time, the mobile station will report to the network the identity of the low-powered base station in order to allow the corresponding base station to increase its transmit power. If the power level decreases further below a static threshold, denoted as T_4, then the mobile station will again report this to the network and the base station will subsequently be dropped from the candidate set.

Inter-frequency or hard-handovers can be supported between cells having different carrier frequencies. Here we conclude our discussions on the cdma2000 features and provide some rudimentary notes on a number of advanced techniques, which can be invoked in order to improve the performance of the 3G W-CDMA systems.

10.3.4 Performance-Enhancement Features

The treatment of adaptive antennas, multi-user detection, interference cancellation, or the portrayal of transmit diversity techniques is beyond the scope of this chapter. Here we simply provide a few pointers to the associated literature.

10.3.4.1 Down-Link Transmit Diversity Techniques

10.3.4.1.1 Space Time Block Coding-Based Transmit Diversity Space Time block coding-based Transmit Diversity (STTD) [393] can be applied to all the DL physical channels with the exception of the SCH. Typically the data of a physical channels are encoded and transmitted using two antennas, as shown in Figure 10.54.

10.3.4.1.2 Time-Switched Transmit Diversity Time-Switched Transmit Diversity (TSTD) [399] is only applicable to the SCH, and its operation becomes explicit in Figure 10.55.

10.3.4.1.3 Closed-Loop Transmit Diversity Closed-loop transmit diversity is only applicable to the DPCH and PDSCH messages of Table 10.5 on the DL, which

10.3. THIRD-GENERATION SYSTEMS

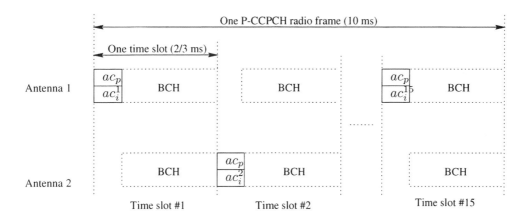

c_p : Primary Synchronization Code
$c_i^j, i = 1, \ldots, 16; j = 1, \ldots, 15$: Secondary Synchronization Code
BCH : Broadcast CHannel
P-CCPCH : Primary Common Control Physical CHannel

Figure 10.55: Frame structure of the UTRA DL synchronization channel (SCH), transmitted by a TSTD scheme. The primary and secondary SCH are transmitted alternatively from Antennas 1 and 2. The parameter a is a binary flag used to indicate the presence ($a = +1$) or absence ($a = -1$) of STTD encoding in the P-CCPCH.

is illustrated in Figure 10.56. The weights w_1 and w_2 are related to the DL channel's estimated phase and attenuation information, which are determined and transmitted by the MS to the BS using the FBI D field, as portrayed in Figure 10.20. The weights for each antenna are independently measured by the MS using the corresponding pilot channels CPICH1 and CPICH2.

10.3.4.2 Adaptive Antennas

The transmission of time-multiplexed user-specific pilot symbols on both the UL and DL as seen for UTRA in Figures 10.20–10.24 facilitates the employment of adaptive antennas. Adaptive antennas are known to enhance the capacity and coverage of the system [400, 401].

10.3.4.3 Multi-User Detection/Interference Cancellation

Following Verdu's seminal paper [352], extensive research has shown that Multi-user Detection (MUD) [351, 353, 402–407] and Interference Cancellation techniques [350, 408–418] can substantially improve the performance of the CDMA link in comparison to conventional RAKE receivers. However, using long scrambling codes increases the complexity of the MUD [322]. As a result, UTRA introduced an optional short scrambling code, namely, the S(2) code of Table 10.9, as mentioned in Section 10.3.2.6.4, in

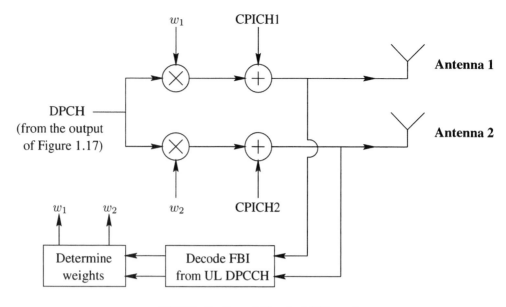

Figure 10.56: Transmission of the DL DPCH using a closed-loop transmit diversity technique.

order to reduce the complexity of MUD [372]. Another powerful technique is invoking burst-by-burst adaptive CDMA [93, 99, 354] in conjunction with MUD.

However, interference cancellation and MUD schemes require accurate channel estimation, in order to reproduce and deduct or cancel the interference. Several stages of cancellation are required in order to achieve a good performance, which in turn increases the canceller's complexity. It was shown that recursive channel estimation in a multistage interference canceller improved the accuracy of the channel estimation and hence gave improved BER performance [366].

Because of the complexity of the multi-user or interference canceller detectors, they were originally proposed for the UL. However, recently reduced-complexity DL MUD techniques have also been proposed [419].

10.3.5 Summary of 3G Systems

We have presented an overview of the terrestrial radio transmission technology of 3G mobile radio systems proposed by ETSI, ARIB, and TIA. All three proposed systems are based on Wideband-CDMA. Despite the call for a common global standard, there are some differences in the proposed technologies, notably, the chip rates and intercell operation. These differences are partly due to the existing 2G infrastructure already in use all over the world, and are specifically due to the heritage of the GSM and the IS-95 systems. Huge capital has been invested in these current 2G mobile radio systems. Therefore, the respective regional standard bodies have endeavored to ensure that the 3G systems are compatible with the 2G systems. Because of the diversified nature of these 2G mobile radio systems, it is not an easy task to reach a common 3G standard that can maintain perfect backwards compatibility. Non-coherent M-ary orthogonal CDMA is described in the next chapter.

10.4 Summary and Conclusions

Following the rudimentary introduction of Sections 10.1–10.2.6, Section 10.3 reviewed the 3G WB-CDMA standard proposals. The 3G systems are more amenable to the transmission of interactive video signals than their more rigid 2G counterparts. This is due partly to the higher supported bit rate and partly to the higher variety of available transmission integrities and bit rates. In Chapter 21, we will rely on this chapter and quantify the performance of various joint-detection-based CDMA video systems.

Part II

Video Systems Based on Proprietary Video Codecs

Chapter 11

Fractal Image Codecs

11.1 Fractal Principles

Fractal image codecs have attracted considerable interest in recent years through the valuable contributions, theses, of Barnsley [420], Beaumont [421], Jacquin [422], and Monroe [423, 424]. Fractals are geometrical objects with endless self-similar details, resulting in efficient parametric image representations [420]. An often-quoted example for self-similarity is a fern, the leaves of which can be viewed as small ferns, and the same is valid for the leaf of the leaf, and so on, as seen in Figure 11.1.

This self-similarity expressed in technical terms is intra-frame redundancy, which can be removed by expressing regions of the image to be encoded as transformed versions of other image segments using so-called *contractive affine* transforms. This transform can be expressed in two dimensions (2D) [420], as follows:

$$\begin{bmatrix} X \\ Y \end{bmatrix} = \begin{bmatrix} r\cos\phi & -s\sin\theta \\ r\sin\phi & s\cos\theta \end{bmatrix} \begin{bmatrix} x \\ y \end{bmatrix} + \begin{bmatrix} X_0 \\ Y_0 \end{bmatrix}. \tag{11.1}$$

$$= \begin{bmatrix} A & B \\ C & D \end{bmatrix} \begin{bmatrix} x \\ y \end{bmatrix} + \begin{bmatrix} X_0 \\ Y_0 \end{bmatrix}. \tag{11.2}$$

or, alternatively, as:

$$\begin{aligned} X &= (r\cos\phi)x - (s\sin\theta)y + X_0 \\ &= Ax + By + X_0 \\ Y &= (r\sin\phi)x + (s\cos\theta)y + Y_0 \\ &= Cx + Dy + Y_0 \end{aligned}$$

These transforms allow us to represent each pixel (X, Y) of a 2D two-tone object to be encoded as a transformed version of the pixels x, y of another object by scaling the x and y coordinates using the factors $(r, s) < 1$, rotating them by angles (ϕ, θ), as well as linearly translating the original object in both directions by (X_0, Y_0). This contractivity of the affine transformations, which is imposed by setting $(r, s) < 1$, is

Figure 11.1: Affine transformation example.

important, for these transformations are invoked repetitively a high number of times in order to be able to resolve arbitrarily fine details in images. The effect of this contractive affine transformation is shown in Figure 11.1 using the example of a fern. Note, however, that from an image-coding point of view, it is preferable to find a transformation that translates each pixel of an object into the corresponding pixels of another object in a single object-transformation step, rather than carrying out the transformation on a pixel-by-pixel basis, since this would incur a high computational complexity and require storing all the transformation parameters for the sake of facilitating the inverse mapping. These remarks will be augmented later in this chapter.

For nonbinary gray-scale images, a third dimension must be introduced in order to represent the luminance information. The simple principle of fractal coding requires us first to partition the original image into small regions, which are referred to as range blocks (RB)11.2 that perfectly tile the original image. This is particularly necessary for large image segments, for would be difficult to find other similar segments, which could be used for their encoding. The larger the image segment, the more difficult to find matching objects. Then a pool of domain blocks (DB) is defined in Figure 11.2, which can be thought of as a two-dimensional (2D) codebook. Each image segment constituted by the legitimate domain blocks is tentatively contractive affine transformed to the specific RB about to be encoded. The fidelity of the match expressed typically in terms of the Mean Squared Error (MSE) is remembered, and the coordinates of the best matching DB for each RB are stored and transmitted to the decoder for image reconstruction. This process is illustrated in Figure 11.2, and the specific techniques to be used will be detailed later.

11.1. FRACTAL PRINCIPLES

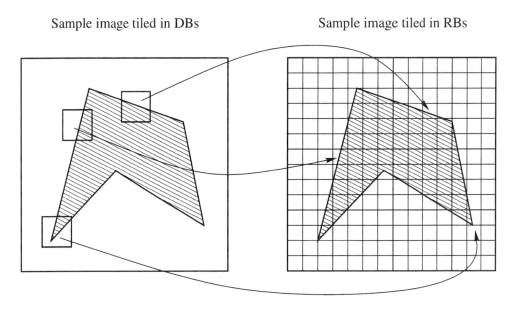

Figure 11.2: Example of mapping three RBs on DBs.

A	B	C	D	X_0	Y_0
0	0	0	0.16	0.0	0
0.2	-0.26	0.23	0.22	0.0	1.6
-0.15	0.28	0.26	0.24	0.0	0.44
0.85	0.04	-0.04	0.85	0.0	1.6

Table 11.1: The IFS Code of a Fern

The affine transform associated with the best MSE match is described by means of its Iterated Function System (IFS) code [420] given by $(r, s, \phi, \theta, X_0, Y_0)$, which can be used at the decoder to reconstruct the original image. The technique of image reconstruction will be highlighted after exemplifying the concept of IFS codes with reference to our previous "fern example," whose IFS code is given in Table 11.1. A more detailed discussion on one-dimensional fractal coding is given in Section 11.2.

Returning to the problem of image reconstruction from the IFS codes, the Random Collage Theorem [420] must be invoked, which states that from any arbitrary initial state the image can be reconstructed by the iterative application of the contractive affine transform described by the IFS code. As previously mentioned, contractivity is required in order to consistently reduce the size of the transformed objects, and thus resolve more and more fine detail to ensure algorithmic convergence. Algorithmic details of the above techniques will be augmented in the next section, first in the context of a one-dimensional fractal coding example, before delving into details of two-dimensional fractal video coding.

11.2 One-Dimensional Fractal Coding

One-dimensional signals can be acoustic signals or line-by-line scanned images. In order to allow reduced-complexity processing, the signal waveform is divided into short one-dimensional segments, constituting the range blocks. The Collage Theorem requires that the range blocks (RBs) perfectly tile the waveform segment to be encoded. Because of the required contractivity for each RB, a longer domain block (DB) has to be found, which has a similar waveform shape. Finally, each DB is mapped to the RB using a specific affine transform. A set of legitimate contractive affine transforms is as follows:

- Scale the amplitude of the block.
- Add an offset to the amplitude.
- Reflect the waveform of the block about the x-axis or y-axis.
- Rotation by 180 degrees.

The larger the set of legitimate transforms of the block, the better the match becomes. However, an increasing number of possible transformations results in a higher number of IFS codes and ultimately in a higher bit rate, as well as higher computational complexity.

The encoding technique is explained with reference to Figure 11.3, following the Beaumont approach [421]. At the top of the figure, a segment of an image scanline luminance is divided into eight RBs. In order to encode the first RB, a similar pattern in the signal has to be found. In general, it is convenient to use a DB size, that is an integer multiple of the RB size, and here we opted for a double-length DB. Observe in the figure that RB1 could be represented by DB4 reasonably well, if DB4 were appropriately rotated and scaled using contractive affine transformations as follows. After contracting the length of DB4 to the RB size, as portrayed in Figures 11.3(a) and (b), the waveform segment has to be amplitude-scaled and vertically shifted, as displayed in Figure 11.3(c), before it is finally rotated in Figure 11.3(d) in order to produce a waveform closely matching RB1. The amplitude-scaling, offset, and reflection parameters represent the IFS code of RB1. By calculating these values for all RBs, the IFS code of the scanline is fully described.

Let us now consider the more specific example of Figure 11.4, for which we can readily compute the associated IFS codes. The description of RB "a" by the help of the appropriately transformed RB "A" is illustrated step-by-step in Figure 11.5, resulting in the IFS codes given in the first line of Table 11.2. Specifically, DB "A" is first subjected to "contraction" by a factor of 2, as seen at the consecutive stages of Figure 11.5, and then its magnitude is scaled by 0.5. Lastly, an offset of -0.25 is applied, yielding exactly RB "a." Similar fractal transformation steps result in the IFS codes of Table 11.2 for the remaining RBs, namely, RB b, c, and d of Figure 11.4. The decoder has to rely on the IFS codes of each RB in order to reproduce them. However, the decoder is oblivious of the DB pool of the encoder, which was used to generate the IFS codes. Simple logic would suggest that a repeated application of the fractal transforms conveyed by the IFS codes and applied to the DBs is the only

11.2. ONE-DIMENSIONAL FRACTAL CODING 463

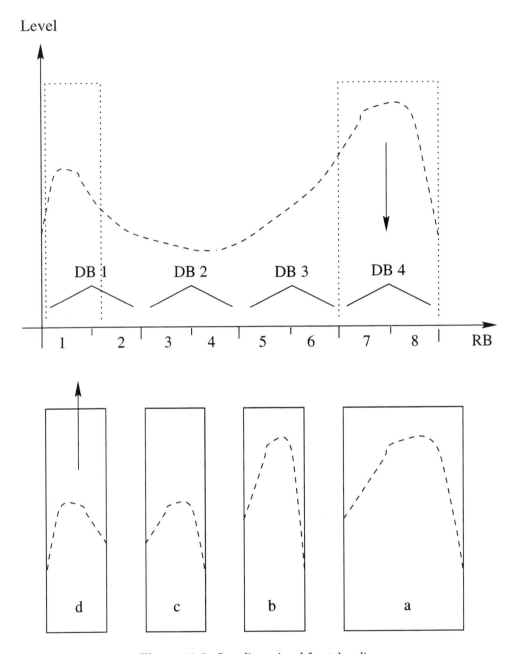

Figure 11.3: One-dimensional fractal coding.

RB	DB	Scaling	Offset	Rotation
a	A	0.5	-0.25	0
b	A	0.5	0.25	0
c	B	0.5	0.25	0
d	A	0.25	0.125	180

Table 11.2: IFS Codes for the Example of Figure 11.4

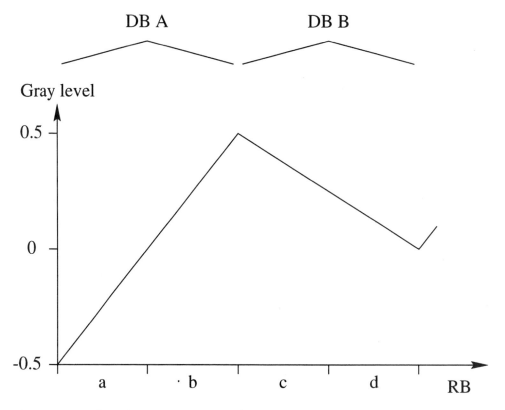

Figure 11.4: DBs and RBs for IFS code computation example for which the IFS codes are given in Table 11.2.

way of reproducing the RBs. *The surprising fact is that repeated application of the IFS-coded contractive affine transforms to an arbitrary starting pattern will converge on the encoded RB, which was stated before as the Random Collage Algorithm.*

The application of the Random Collage Algorithm is demonstrated in Figure 11.6. As seen in the figure, the decoding operation commences with a simple "flat" signal, since the original DBs are not known to the decoder. According to Table 11.2, the first RB can be obtained by scaling the decoder's DB A by 0.5 and shifting it by -0.25. During the first iteration, the scaling does not have any influence, the initialization with zero makes all DBs "flat," Thus, at this point, every RB contains a constant DC level, corresponding to the offset information. In the second step, the DBs are

11.2. ONE-DIMENSIONAL FRACTAL CODING

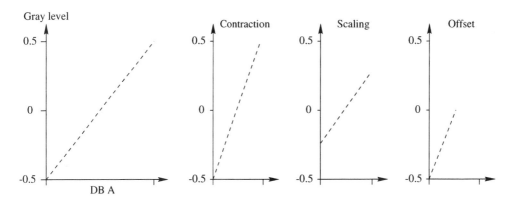

Figure 11.5: Contractive affine transform steps for the IFS code computation example related to RB "a" of Figure 11.4 for which the IFS codes are given in Table 11.2.

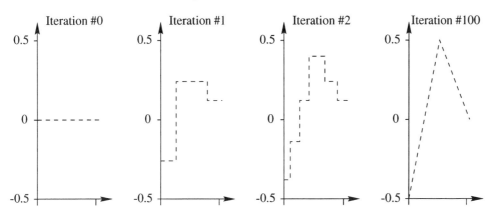

Figure 11.6: Application of the one-dimensional random collage algorithm to the example of Figure 11.4 for which the IFS codes are given in Table 11.2.

already different. DB A now contains a step function. It is scaled and shifted again, and after this second iteration the original waveform is already recognizable. With every iteration the number of steps approximating the ramp-signal increases, whereas the step size shrinks. In this example, the encoder was able to find a perfect match for every RB, and hence after an infinite number of iterations the decoded signal yields the source signal.

Having considered the basic fractal coding principles, let us now examine a range of fractal codec designs contrived for QCIF "head-and-shoulders" videophone sequences.

11.2.1 Fractal Codec Design

Previously proposed fractal codec designs were targeted at high-resolution images, having large intra-frame domain-block pools [422, 423], which will be introduced later in this chapter. Following the approaches proposed by pioneering researchers of the field, such as Barnsley [420], Jacquin [422], Monroe [423,424], Ramamurthi and Gersho [425], and Beaumont [421], in this study we explored the range of design tradeoffs available using a variety of different head-and-shoulders fractal videophone codecs (Codecs A-E) and compared their complexity, compression ratio, and image quality using QCIF images.

In fractal image coding, the QCIF image to be encoded is typically divided into 4-by-4 or 8-by-8 picture elements (pel) based on nonoverlapping range blocks (RB), which perfectly tile the original image [420]. Other block sizes can also be used, but perfect tiling of the frames must be ensured. Every RB is then represented by the contractive affine transformation [423] of a larger, typically quadruple-sized domain block (DB) taken from the same frame of the original image. In general, the larger the pool of domain blocks, the better the image quality, but the higher the computational complexity and bit rate, requiring a compromise. For gray-scale coding of two-dimensional images, a third dimension, representing the brightness of the picture, must be added before affine transformation takes place. Jacquin suggested that for the sake of reduced complexity it is attractive to restrict the legitimate affine transforms to the following manipulations [422]:

1. Linear translation of the block.

2. Rotation of the block by 0, 90, 180, and 270 degrees.

3. Reflection about the diagonals, vertical, and horizontal axis.

4. Luminance shift of the block.

5. Contrast scaling of the block.

The Collage Theorem [420] and practical contractivity requirements facilitate the following (DB, RB) size combinations: $(16 \times 16, 8 \times 8), (16 \times 16, 4 \times 4), (8 \times 8, 4 \times 4)$, and our codecs attempt to match every RB with every DB of the same frame, allowing rotations by $0°, 90°, 180°$, or $270°$. Adopting the MSE expression of

$$\text{MSE} = \sqrt{\frac{1}{p^2} \sum_{i=1}^{p} \sum_{k=1}^{p} (X_{ik} - (aY_{ik} - b))^2} \qquad (11.3)$$

as a block-matching distortion measure, where p is the RB size, the optimum contrast scaling factor a and luminance shift b for the contracted DBs Y_{ik} and RBs X_{ik} can be derived by minimizing the MSE defined above leading to:

$$0 = \sum_{i=1}^{p} \sum_{k=1}^{p} (aY_{ik}^2 - bY_{ik} - X_{ik}Y_{ik}) \qquad (11.4)$$

$$0 = \sum_{i=1}^{p} \sum_{k=1}^{p} (X_{ik} - aY_{ik} + b), \qquad (11.5)$$

where the first derivative of Equation 11.3 is set to zero with respect to a and b, respectively. This yields a set of two equations with two variables, which finally leads

11.2. ONE-DIMENSIONAL FRACTAL CODING

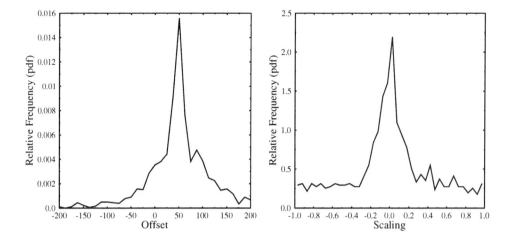

Figure 11.7: PDF of contrast scaling and luminance shift.

to:

$$b = \frac{\sum_{i=1}^{p}\sum_{k=1}^{p}X_{ik}Y_{ik}\sum_{i=1}^{p}\sum_{k=1}^{p}Y_{ik} - \sum_{i=1}^{p}\sum_{k=1}^{p}(Y_{ik}^{2}\sum_{i=1}^{p}\sum_{k=1}^{p}X_{ik})}{p^{2}\sum_{i=1}^{p}\sum_{k=1}^{p}Y_{ik}^{2} - (\sum_{i=1}^{p}\sum_{k=1}^{p}Y_{ik})^{2}} \quad (11.6)$$

$$a = \frac{\sum_{i=1}^{p}\sum_{k=1}^{p}X_{ik}}{\sum_{i=1}^{p}\sum_{k=1}^{p}Y_{ik}} + \frac{p^{2}}{\sum_{i=1}^{p}\sum_{k=1}^{p}Y_{ik}} \cdot b. \quad (11.7)$$

The optimum contrast-scale factor a and luminance-shift b have to be found for all the RBs, and their values have to be quantized for transmission to the decoder.

11.2.2 Fractal Codec Performance

In order to identify the range of design tradeoffs five different codecs, following the philosophy in [420–425] Codecs A–E, were simulated and compared in Table 11.3. The DB indices and the four different rotations of $0°, 90°, 180°$, and $270°$ were graycoded in order to ensure that a single bit error results in the neighboring codewords, while the luminance shift b and contrast scaling a were Max-Lloyd quantized using 4 bits each. The PDFs of these quantities are portrayed in Figure 11.7.

Comparison of the three basic schemes, Codecs A–C featured in Table 11.3, suggested that a RB size of 4 x 4 used in Codecs B and C was desirable in terms of image quality, having a peak signal-to-noise ratio (PSNR) of 5–6 dB higher than Codec A which had a RB size of 8 x 8. The PSNR is the most frequently used objective video-quality measure defined as

$$PSNR = 10log_{10}\frac{\sum_{i=1}^{p}\sum_{j=1}^{q}max^{2}}{\sum_{i=1}^{p}\sum_{j=1}^{q}(f_{n}(i,j) - \tilde{f}_{n}(i,j))^{2}} \quad (11.8)$$

which we will often use throughout this book. However, Codec A had an approximately four times higher compression ratio or lower coding rate expressed in bits/pels

Codec	DB Size	RB Size	Classi-fication	Split	PSNR (dB)	Rate (bpp)
A	16	8	None	No	31	0.28
B	16	4	None	No	35	1.1
C	8	4	None	No	37	1.22
D	16	8/4	Twin	Yes	36	0.84
E	16	8/4	Quad	Yes	29	1.0

Table 11.3: Comparison of Five Fractal Codecs

Block Type	Description / Edge Angle	Frequency (%)
Shade	No significant gradient	18.03
Midrange	Moderate gradient, no edge	46.34
Edge	Steep gradient, edge detected	26.39 (total)
	0 deg	2.66
	45 deg	1.85
	90 deg	5.69
	135 deg	3.18
	180 deg	3.92
	225 deg	2.96
	270 deg	3.92
	315 deg	2.22
Mixed	Edge angle ambiguous	9.24

Table 11.4: Classified Block Types and Their Relative Frequencies in Codec E

(bpp). Codec C of Table 11.3 had four times more DBs than Codec B, which yielded quadrupled block-matching complexity, but the resulting PSNR improvement was limited to about 1–2 dB and the bit rate was about 20% higher due to the increased number of DB addresses. These findings were also confirmed by informal subjective assessments.

In order to find a compromise between the four times higher compression ratio of the 8×8 RBs used in Codec A and the favorable image quality of the 4×4 RBs of Codec B and C, we followed Jacquin's [422] suggestions of splitting inhomogeneous RBs in two, three, or four subblocks. Initially, the codec attempted to encode an 8×8 RB and calculated the MSE associated with the particular mapping. If the MSE was above a certain threshold, the codec split up the block into four nonoverlapping subblocks. The MSE of these subblocks was checked against the error threshold individually, and if necessary one or two 4×4 subblocks were coded in addition.

11.2. ONE-DIMENSIONAL FRACTAL CODING

However, for three or four poorly matching subblocks, the codec stored only the transforms for the four small quarter-sized subblocks. This splitting technique was used in Codecs D and E of Table 11.3.

In addition to Jacquin's splitting technique [422], the subjectively important edge representation of Codecs D and E was improved by a block classification algorithm originally suggested by Ramamurthi and Gersho [425], which was then also advocated by Jacquin [422]. Accordingly, the image blocks were divided into four classes:

1. Shade blocks taken from smooth areas of an image with no significant gradient.

2. Midrange blocks having a moderate gradient but no significant edge.

3. Edge blocks having a steep gradient and containing only one edge.

4. Mixed blocks with a steep gradient that contains more than one edge, and hence the edge orientation is ambiguous.

Codec D used a basic twin-class algorithm, differentiating only between shade and nonshade blocks, whereas Codec E used the above quad-class categorization. The relative frequencies of all registered subclasses were evaluated following the above approach [422, 425] for Codec E using our QCIF sequences, which are shown in Table 11.4.

In Codecs D and E after the classification of all DBs and RBs normal coding ensued, but with the advantage that the codec predetermined the angle the DB had to be rotated by and it attempted to match only blocks of the same class. If, for example, a RB was classified as an edge block with a certain orientation, the codec exploited this by limiting the required search to the appropriate DB pool. Furthermore, shade blocks were not fully encoded; only their mean was transmitted to the decoder, yielding a significant reduction in complexity and bit rate. Interestingly, we found the quad-class Codec E to perform worse than the twin-class codec D. This was due to classification errors and to the smaller size of the DB pools in case of Codec E, which often were too limited to provide a good DB match.

A comparison of the five QCIF videophone codecs is presented in Table 11.3. Codecs A–C represent basic fractal codecs with no RB classification or splitting. When comparing Codecs A and B using RB sizes of 8×8 and 4×4, respectively, the compression ratio of Codec A is four times higher, but its PSNR is 5 dB lower at similar complexity. The PSNR versus frame index performance of codecs A and B is portrayed in Figure 11.8. The typical subjective image quality achieved by Codec B is portrayed in Figure 11.9 as a function of the Random Collage Theorem's iteration index. It was found that five approximations are sufficient to achieve pleasant communications image quality at a PSNR in excess of 33 dB. The 1 dB PSNR advantage of Codec C does not justify its quadruple complexity. Codecs D and E employ twin- or quad-class block classification combined with RB splitting, if the MSE associated with a particular mapping is above a certain threshold. Interestingly, the less complex Codec D has a higher compression ratio and higher image quality. The lower performance of codec E is attributed to block classification errors and to the limited size of the DB pool provided by our QCIF images.

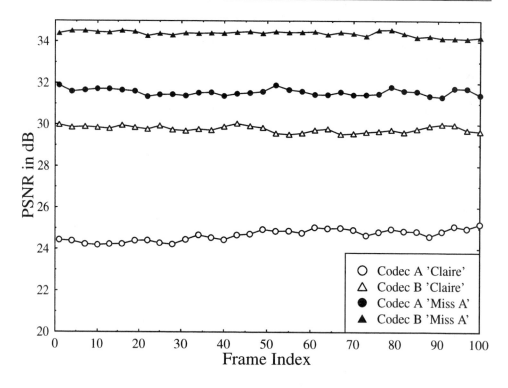

Figure 11.8: PSNR versus frame index performance of Codecs A and B for the "Claire" and "Miss America" sequences.

Bit Index	Parameter
1–2	Rotation
3–6	DB X coordinate
7–10	DB Y coordinate
11–14	Scaling
15–18	Offset

Table 11.5: Bit-Allocation per RB for Codecs A and B

The RB bit-allocation scheme of Codecs A and B is identical, but the lower-compression Codec B has four times as many RBs. The specific bit assignment of a RB in Codecs A and B is shown in Table 11.5. Bits 1–2 represent four possible rotations, bits 3–6 and 7–10 are the X and Y domain-block coordinates, respectively, while bits 11–14 are the Max-Lloyd quantized scaling gains and bits 15–18 represent offset values used in the Random Collage Algorithm. However, Codec A uses $22 \times 18 = 396$ 8×8 pels RBs associated with a rate of $R = 18/64 \approx 0.28$ bits/pel, while Codec B has $44 \times 38 = 1584$ 4×4 RBs, which is associated with a quadrupled bit rate of $R = 18/16 \approx 1.1$ bits/pel. The number of bits per frame becomes $396 \cdot 18 = 7128$ bits/frame for Codec A and 28512 for Codec B, corresponding to bit rates of

11.3. ERROR SENSITIVITY AND COMPLEXITY

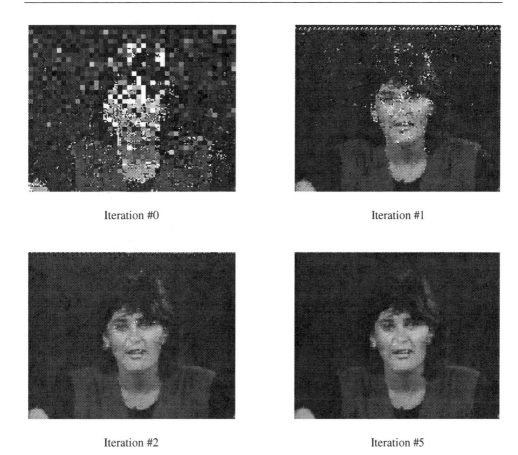

Figure 11.9: Demonstration of fractal image reconstruction from an arbitrary initial picture by the Random Collage Algorithm as a function of iteration index and the typical fractal video quality.

71.28 kbits/sec and 285.12 kbits/sec, respectively, at a scanning rate of 10 frames/sec. The associated PSNR values are about 31 and 35 dB, respectively. The compression ratio of all of the fractal codecs was more modest than that of the Discrete Cosine Transform (DCT), Vector Quantization (VQ), and Quad-tree (QT) codecs, which will be discussed in Chapters 12, 13, and 14, respectively.

11.3 Error Sensitivity and Complexity

Both Codecs A and B were subjected to bit-sensitivity analysis by consistently corrupting each bit of the 18-bit frame and evaluating the Peak Signal to Noise Ratio (PSNR) degradation inflicted, which is defined by Equation 11.8. These results are shown for both codecs in Figure 11.10. Observe from the figure that the significance of the specific coding bits can be explicitly inferred from the PSNR degradations ob-

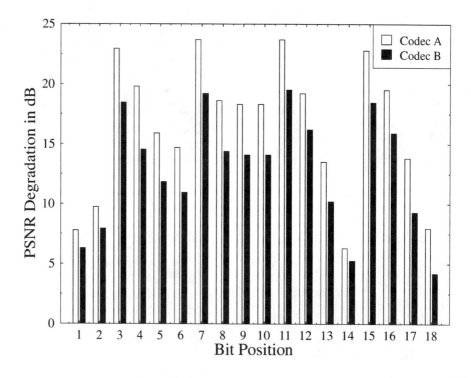

Figure 11.10: Bit sensitivities versus Bit Index for Codecs A and B using the bit-allocation of Table 11.3.

served. Generally, Codec A is more vulnerable to channel errors, since the bits for each block carry the information for $8 \times 8 = 64$ pixels rather than $4 \times 4 = 16$ pixels in the case of Codec B. Very vulnerable to channel errors are the MSBs of the DB coordinates, the scaling and offset values. These are about three times more sensitive than the corresponding LSBs. A more detailed analysis of the error sensitivities and a matching wireless transceiver design can be found in [426]. The PSNR versus BER degradation due to random bit errors, which is characteristic of the codec's behavior over wireline-based AWGN channel is depicted in Figure 11.11. On the basis of subjective video-quality degradation a 2 dB PSNR degradation, is deemed acceptable, which requires the transceiver to operate at a BER of about 10^{-3}.

Fractal codecs suffer from a high computational demand. This is due to the evaluation of Equations 11.7 and 11.6 for every block. The number of Flops per DB-RB match is proportional to $7 \cdot p^2$, leading to 99 DB $\times$ $7 \cdot 8^2$ $\times$ 396 RB $>$ 17.5 Mflop for Codecs A and B. The computational load has to be multiplied with the frame rate, which leads to a huge computational demand compared to other compression techniques.

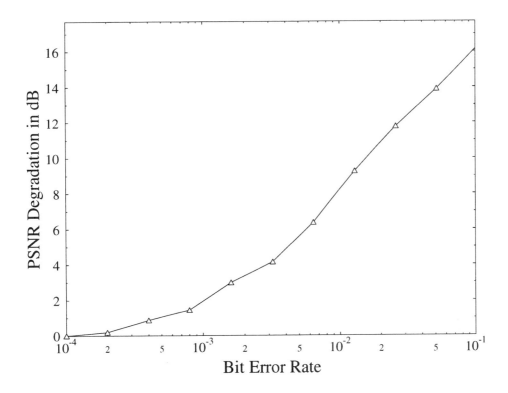

Figure 11.11: PSNR degradation versus bit error rate for Codec A.

11.4 Summary and Conclusions

A range of QCIF fractal video codecs was studied in terms of image quality, compression ratio, and complexity. Two fixed-rate codecs, Codecs A and B, were also subjected to bit-sensitivity analysis. We found that for a pleasant image quality, the rather small block size of 4 × 4 pixels is necessary. This leads to a relatively high bit rate of 71 kb/s. Furthermore, the computational demand is very high. We note furthermore that fractal coding is not amenable to motion-compensated inter-frame coding, since the motion compensation removes temporal redundancy in subsequent frames, which results in a strongly decreased intra-frame correlation. It was found that the remaining intra-frame correlation was insufficient to perform fractal coding. The following chapters will demonstrate that motion compensation in conjunction with a range of other compression techniques assists us in achieving high compression ratios.

Chapter 12

Very Low Bit-Rate DCT Codecs and Multimode Videophone Transceivers

12.1 Video Codec Outline

In video sequence coding, such as videotelephony, a combination of temporal and spatial coding techniques is used in order to remove the "predictable or redundant" image contents and encode only the unpredictable information. How this can be achieved is the subject of the forthcoming chapters. These techniques guarantee a reduced bit rate and hence lead to bit-rate economy. Let us begin our discussion with the simple encoder/decoder (codec) structure of Figure 12.1.

Assuming that the 176 × 144-pixel Quarter Common Intermediate Format (QCIF) ITU standard videophone sequence to be encoded contains a head-and-shoulder video clip, the consecutive image frames f_n and f_{n-1} typically do not exhibit dramatic scene changes. Hence, the consecutive frames are similar, a property that we refer to as being correlated. This implies that the current frame can be approximated or predicted by the previous frame, which we express as $f_n \approx \hat{f}_n = f_{n-1}$, where $\hat{f}_n$ denotes the nth predicted frame. When the previous frame f_{n-1} is subtracted from the current one, namely f_n, due to the speaker's movement, this prediction typically results in a "line-drawing-like" difference frame, as shown in Figure 12.2. Most areas of this difference frame are "flat," having values close to zero, and the variance or second moment of it is significantly lower than that of the original frame. We have removed some of the predictable components of the video frame.

This reduced variance difference signal, namely, $e_n = f_n - f_{n-1}$, is often referred to as Motion Compensated Error Residual (MCER) since the associated frame-differencing effectively attempts to compensate for the motion of the objects between consecutive video frames, yielding a reduced-variance MCER. Thus, e_n requires a

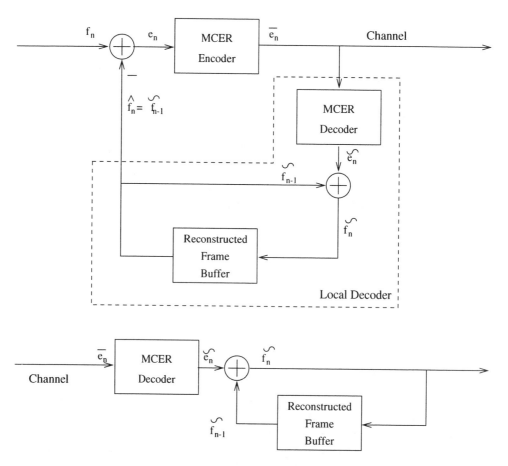

Figure 12.1: Basic video codec schematic using frame-differencing.

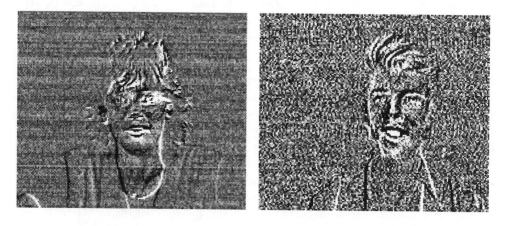

Figure 12.2: Videophone frame difference signals for two sequences.

reduced coding rate, that is, a lower number of bits in order to represent it with a certain distortion than f_n. The MCER $e_n = f_n - f_{n-1}$ can then be encoded as $\bar{e}_n$, with the required distortion using a variety of techniques, which constitute the subject of the following chapters. The quantized or encoded MCER signal $\bar{e}_n$ is then transmitted over the communications channel. In order to reproduce the original image $f_n = e_n + f_{n-1}$ exactly, knowledge of e_n would be necessary, but only its quantized version, $\bar{e}_n$ is available at the decoder, which is contaminated by the quantization distortion introduced by the MCER encoder, inflicting the reconstruction error $\Delta e_n = e_n - \bar{e}_n$. Since the previous undistorted image frame f_{n-1} is not available at the decoder, image reconstruction has to take place using their available approximate values, namely, $\bar{e}_n$ and $\tilde{f}_{n-1}$, giving the reconstructed image frame as follows:

$$\tilde{f}_n = \tilde{e}_n + \tilde{f}_{n-1}, \qquad (12.1)$$

where in Figure 12.1 we found that the locally decoded MCER residual $\tilde{e}_n$ is an equal-valued, noiseless equivalent representation of $\bar{e}_n$. This equivalence will be further elaborated during our forthcoming discussions. The above operations are portrayed in Figure 12.1, where the current video frame f_n is predicted by $\hat{f}_n = \tilde{f}_{n-1}$, an estimate based on the previous reconstructed frame $\tilde{f}_{n-1}$. Observe that (ˆ) indicates the predicted value and (˜) the reconstructed value, which is contaminated by some coding distortion. Note furthermore that the encoder contains a so-called local decoder, which is identical to the remote decoder. This measure ensures that the decoder uses the same reconstructed frame $\tilde{f}_{n-1}$ in order to reconstruct the image, as the one used by the encoder to generate the MCER e_n. Note, however, that in case of transmission errors in $\tilde{e}_n$, the local and remote reconstructed frames become different, which we often refer to as being misaligned. This phenomenon leads to transmission error propagation through the reconstructed frame buffer, unless countermeasures are employed.

To be more explicit, instead of using the previous original frame, the so-called locally decoded frame is used in the motion compensation, where the phrase "locally decoded" implies decoding it at the encoder (i.e., where it was encoded). This local decoding yields an exact replica of the video frame at the distant decoder's output. This local decoding operation is necessary, because the previous original frame is not available at the distant decoder, and so without the local decoding operation the distant decoder would have to use the reconstructed version of the previous frame in its attempt to reconstruct the current frame. The absence of the original video frame would lead to a mismatch between the operation of the encoder and decoder, a phenomenon that will become clearer during our further elaborations.

12.2 The Principle of Motion Compensation

The simple codec of Figure 12.1 used a low-complexity temporal redundancy removal or motion compensation technique, often referred to as frame-differencing. The disadvantage of frame-differencing is that it is incapable of tracking more complex motion trajectories, where different objects, for example, the two arms of a speaker, move in different directions. This section briefly summarizes a range of more efficient motion compensation (MC) techniques, discusses their computational complexity and efficiency, and introduces the notation to be used.

Image sequences typically exhibit spatial redundancy within frames and temporal redundancy in consecutive frames. Motion compensation attempts to remove the latter. If there is no drastic change of scenery, consecutive frames often depict the same video objects, but some of them may be moving in different directions. The motivation behind block-based motion compensation is to analyze the motion trajectory and to derive a set of descriptors for the changes. These descriptors characterize the location, shape, and movement of a moving object. In the ideal case, when no new information is introduced in the video scene, consecutive frames can be fully described by their appropriately motion-translated predecessors. Apart from tracking motion in the two spatial dimensions, some techniques, such as model-based coding [427, 428], also interpolate the third, temporal dimension of a given sequence so that the movement is represented by a three-dimensional vector and all three image dimensions are reproduced. Although promising results can be achieved using these techniques [429, 430], their real-time implementation has been hindered due to the associated high complexity.

In an often used practical MC implementation, the image frame is divided in a number of perfectly tiling 2×2 to 16×16-pixel blocks, which are then slid over a certain search area of Figure 12.4 surrounding the corresponding location in the previous reconstructed frame $\tilde{f}_{n-1}$, in order to identify the specific position from which each block has originated. The corresponding motion vectors (MV) describing the motion translation, also known as displacement vectors, are two dimensional and are typically restricted to integer multiples of the pixel separation, although we note that at a later stage subpixel resolution will also be considered in the context of the H.263 standard video codec. The motion vector identified this way is then usually applied to the entire group of pixels within a block, so that the predicted current block constituted by the appropriately motion translated, previously reconstructed block is subtracted from the current block about to be encoded, as portrayed in Figure 12.4. Again, the result of this operation is referred to as the Motion Compensated Error Residual (MCER). These operations are reflected in the modified codec schematic of Figure 12.3.

An MC algorithm operating with subpixel accuracy has been proposed [431]. The above motion prediction algorithm is based on two consecutive frames, that impose a minimum encoding delay of one frame duration. We refer to the previous reconstructed frame $\tilde{f}_{n-1}$, rather than the uncoded original input frame f_n during the derivation of the motion vectors, because both the encoder and the decoder must use an identical reference frame, to which they apply the motion vectors (MV) during the image reconstruction. However, the original input frame f_n is not available at the remote decoder. These issues are elaborated on in more depth later in this chapter.

Formulating the above algorithm slightly more rigorously, we find that the entire $p \times q$-sized frame f_n

$$f_n = \{(x,y) | x \epsilon [1 \ldots p], y \epsilon [1 \ldots q]\} \tag{12.2}$$

is divided into $n = (p/b)(q/b)$ smaller, $b \times b$ pixel rectangles defined as:

$$b_n = \{(x,y) | x, y \epsilon [1 \ldots b]\}. \tag{12.3}$$

12.2. THE PRINCIPLE OF MOTION COMPENSATION   479

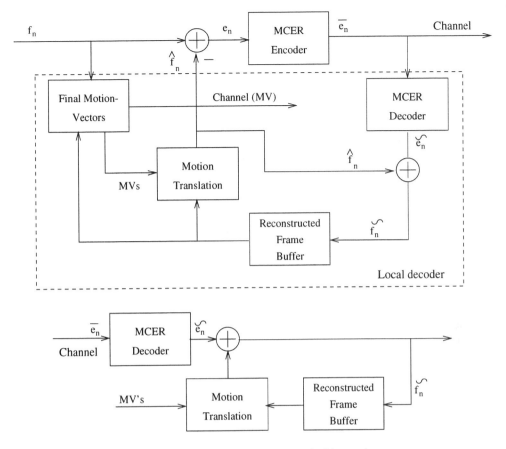

Figure 12.3: Motion-compensated video codec.

Again, this process is depicted in Figure 12.4. Then for each of the n blocks, the most similar block within the stipulated search area of the previous reconstructed or locally decoded frame, $\tilde{f}_{n-1}$ has to be found. Recall from our previous arguments that the previous locally decoded, rather than the previous original frame, has to be used in this operation, because the previous original frame is unavailable at the remote decoder. Then the vector emerging from a given pixel — for example, at the center of the block under consideration and pointing to the corresponding pixel of the most similar block within the search area of the previous reconstructed block– is defined as the motion vector $\vec{M} = M(m_x, m_y)$. The similarity criterion or distance measure used in this block-matching process has not yet been defined. A natural choice is to use the mean squared error (MSE) between the block for which the MVs are sought and the momentarily tested, perfectly overlapping block of pixels within the previous reconstructed frame $\tilde{f}_{n-1}$. This topic will be revisited in more depth in the next section.

The motion-compensated prediction error residual (MCER) is also often termed a displaced frame difference (DFD). Depending on the input sequence, the MCER

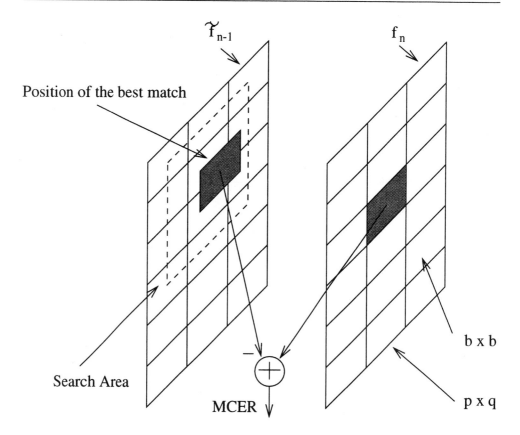

Figure 12.4: The process of motion compensation.

typically contains a zero mean signal of nonstationary nature. Figure 12.5 depicts the two-dimensional autocorrelation function (ACF) of a typical MCER and its probability density function (PDF). The ACF of the MCER suggests that there is residual spatial correlation or "predictability" over a number of adjacent pixels in both vertical and horizontal directions. Since the ACF values are above 0.5 for displacements of $X_0, Y_0 < 3$, this residual spatial correlation can be exploited using a variety of coding schemes. The PDF of the MCER signal in Figure 12.5 also suggests that most of the values of the MCER are within a range of [-10,10], assuming that the source frame luminance values are sampled with an 8-bit resolution within the range of [0..255]. However, when using the lossy coding scheme of Figure 12.3 (i.e., schemes that do not perfectly reconstruct the original image), the quantization distortion associated with the MCER encoder will accumulate in the local reconstructed frame buffer, and it will alter the statistical characteristics of the MCER, reducing the correlation and resulting in a more widely spread PDF.

12.2. THE PRINCIPLE OF MOTION COMPENSATION

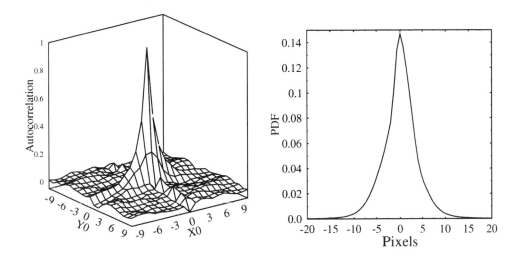

Figure 12.5: Two-dimensional autocorrelation function and the PDF of the MCER.

12.2.1 Distance Measures

At the time of writing, there is insufficient understanding of the human visual perception. Many distance criteria have been developed in order to assist in pattern matching or image processing research, but most of them fail under various circumstances. Here we introduce the three most commonly used distance measures of image processing.

The widely accepted Mean Square Error (MSE) criterion is defined as:

$$M_{mse} = MIN \sqrt{\sum_{i=1}^{b}\sum_{j=1}^{b}(f_n(x+i,y+j) - f_{n-1}(x+i-m_x, y+j-m_y))^2}, \quad (12.4)$$

where b denotes the block size, x and y represent the top left corner of the block under consideration, and m_x, m_y constitute the coordinates of the motion vector $\vec{M}$. Equation 12.4 evaluates the luminance difference of the given $b \times b$ blocks f_n and f_{n-1} on a pixel-by-pixel basis, sums the squared difference over the entire block, and after taking the square root of the sum, the MC scheme finds the position, where this expression has a minimum over the search area of Figure 12.4.

Two simplifications of this criterion are used in MC. The mean absolute difference (MAD) criterion is defined as the sum of the absolute differences rather than its second moment:

$$M_{mad} = MIN \sum_{i=1}^{n}\sum_{k=1}^{n} | f_n(x+i,y+j) - f_{n-1}(x+i-m_x, y+j-m_y) |. \quad (12.5)$$

The pel difference classification [432] (PDC) criterion of Equation 12.6 compares the matching error for every pel location of the block with respect to a preset threshold and classifies the pel as a "matching" or "mismatching" pel, yielding:

$$M_{pdc} = MIN \sum_{i=1}^{n} \sum_{k=1}^{n} T \mid f_n(x+i, y+j) - f_{n-1}(x+i-m_x, y+j-m_y) \mid, \quad (12.6)$$

where T denotes the threshold function defined as:

$$T(y) = \begin{cases} 1 & \text{if } y > 0 \\ 0 & \text{otherwise} \end{cases} . \quad (12.7)$$

This algorithm was contrived for real-time implementations, as its thresholding operation is readily implementable in hardware. However, the threshold must be set carefully and eventually adjusted for different types of image sequences. Other variants of the mentioned algorithms are also feasible. For example, for each of the above-mentioned matching criteria, a "fractional" measure may be defined that would consider only a fraction of the pixels within the two blocks under consideration. Accordingly, a 50% fractional criterion would consider only every other pixel, while halving the matching complexity.

As in Figure 12.6, the Peak Signal to Noise Ratio (PSNR) will serve as the objective quality measure, which is defined as:

$$PSNR = 10 \, log_{10} \frac{\sum_{i=1}^{p} \sum_{j=1}^{q} max^2}{\sum_{i=1}^{p} \sum_{j=1}^{q} (f_n(i,j) - \tilde{f}_n(i,j))^2}, \quad (12.8)$$

as opposed to the conventional SNR of:

$$SNR = 10 \, log_{10} \frac{\sum_{i=1}^{p} \sum_{j=1}^{q} f(i,j)^2}{\sum_{i=1}^{p} \sum_{j=1}^{q} (f_n(i,j) - \tilde{f}_n(i,j))^2}. \quad (12.9)$$

The PSNR is a derivative of the well-known Signal to Noise Ratio(SNR), which compares the signal energy to the error energy as defined in Equation 12.9. The PSNR compares the maximum possible signal energy of $max^2 = 225^2$ to the noise energy, which was shown to result in a higher correlation with the subjective quality perception of images than the conventional SNR. Figure 12.6 compares the performance for the distance criteria defined above for a codec using motion compensation only, that is, allocating zero bits for the encoding of the MCER, which is equivalent to setting MCER = 0 during the frame reconstruction. Observe in Figure 12.6 that even when no MC is used — corresponding to simple frame-differencing — the PSNR is around 25 dB for the limited-motion "Miss America" video sequence. In addition, we derived another measure from the MSE criterion, where only 50% of the pels in each block are considered. We expect the MSE criterion to lead to the best possible results, since the PSNR, is like the MSE criterion, is a mean squared distance function. Figure 12.6 also underlines the relatively poor performance of the PDC algorithm.

12.2.2 Motion Search Algorithms

Having introduced a number of pattern matching criteria, in this section we briefly highlight the so-called optimum full-search motion compensation algorithm and its

12.2. THE PRINCIPLE OF MOTION COMPENSATION

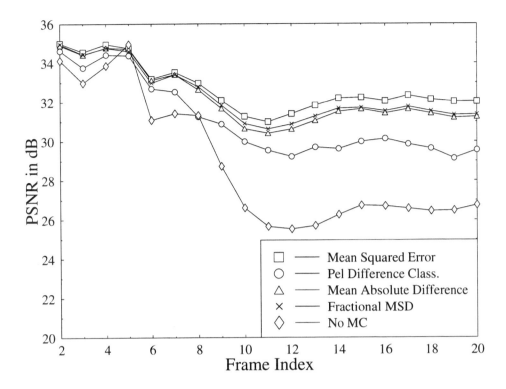

Figure 12.6: PSNR versus frame index performance comparison for various distance measures while encoding the MCER with zero bits, when using the "Miss America" sequence.

derivatives, while comparing their complexities. The gain-cost controlled motion compensation algorithm to be introduced in Section 12.2.2.6 will be used in all of our compression algorithms presented throughout Part II of the book.

12.2.2.1 Full or Exhaustive Motion Search

The full search MC algorithm determines the associated matching criterion for every possible motion vector within the given search scope and selects that particular motion vector, which results in the lowest matching error. This was portrayed in Figure 12.4. Thus, a search within a typical motion search window of 16 × 16 pixels requires 256 block comparisons. Using the MSE criterion as quality measure and a block size of 8 × 8 pels, the full search (FS) requires 13 million floating point operations (Mflop) per QCIF frame. This constitutes a computational load, which we cannot afford in typical real-time applications, such as, for example, in hand-held mobile videotelephony. Note, however, that the matching complexity is a quadratic function of the search window size. For image sequences with moderate motion activity or for sequences scanned at a high frame rate, the search window can be reduced to the size of ±2 adjacent pixel positions, without significant performance penalty, leading to a computational demand below one MFlop for the same frame size. The computational complexity associated with a single motion compensated block b_n corresponds to two additions and one multiplication for the b^2 elements of each block, which have to be executed for each of the s^2 positions within the search window $s \times s$, when using the MSE criterion of Equation 12.4, yielding:

$$Com_{fs} = s^2(3b^2 - 1). \qquad (12.10)$$

12.2.2.2 Gradient-Based Motion Estimation

The disadvantage of the exhaustive search algorithm is its complexity. The number of search steps can be reduced efficiently, if the image sequence satisfies the gradient constraint equation [433].

Let $I(x, y, t)$ be the luminance of the pixel (x, y) at the time instant t. The motion of the pixel in each spatial direction is defined as $m_x(x, y)$ and $m_y(x, y)$, respectively. At the time instant $t + \delta t$, the pixel will have moved to the location $(x + \delta x, y + \delta y)$. Using the motion vector defined as:

$$\vec{M} = M(m_x, m_y) = \frac{dx}{dt}\vec{e_x} + \frac{dy}{dt}\vec{e_y} = m_x \cdot \vec{e_x} + m_y \cdot \vec{e_y} \qquad (12.11)$$

where $\vec{e_x}$ and $\vec{e_y}$ represent the corresponding unit vectors along the x- and y-axis, we approximate the new location of the reference pixel as $(x + \delta t \cdot m_x, y + \delta t \cdot m_y)$, assuming that motion translation during the short time interval δt is linear. With these assumptions we obtain:

$$I(x, y, t) = I(x + \delta t \cdot m_x, y + \delta t \cdot m_y, t + \delta t), \qquad (12.12)$$

which physically implies that the luminance of the pixel at a new position in the current frame is the same as the luminance of the reference pixel in the previous frame.

The Taylor expansion of the right-hand side of Equation 12.12 leads to:

$$\begin{aligned} I(x, y, t) &= I(x, y, t) + \frac{\delta I(x, y, t)}{\delta x}\frac{dx}{dt} + \frac{\delta I(x, y, t)}{\delta y}\frac{dy}{dt} \\ &\quad + \frac{\delta I(x, y, t)}{\delta t} + e(x, y, t), \end{aligned} \qquad (12.13)$$

where $e(x, y, t)$ contains the higher order components of the Taylor expansion. For small δt values, that is, when $\delta t \to 0$ we have $e(x, y, t) \to 0$, and upon neglecting $e(x, y, t)$, we arrive at the gradient constraint equation:

$$\frac{\delta I(x, y, t)}{\delta x}m_x + \frac{\delta I(x, y, t)}{\delta y}m_y + \frac{\delta I(x, y, t)}{\delta t} = 0. \qquad (12.14)$$

A ramification of this equation is that an iterative pattern matching algorithm, which minimizes the MSE between segments in consecutive frames, as it is carried out in MC, will converge if Equation 12.14 is satisfied. All the iterative MC algorithms to be introduced in this section, such as the hierarchical search or the subsampling search, assume that Equation 12.14 holds for the given image sequence.

12.2. THE PRINCIPLE OF MOTION COMPENSATION

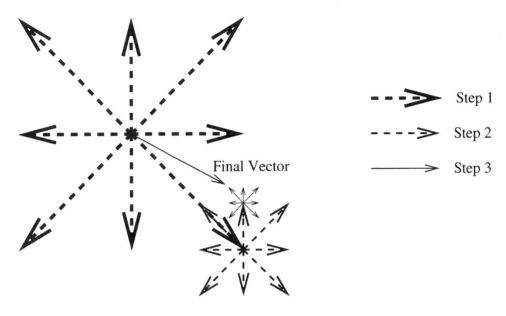

Figure 12.7: Suboptimum MV tree search.

12.2.2.3 Hierarchical or Tree Search

The so-called tree search algorithm proposed by Jain and Jain [433, 434] drastically reduces the complexity of MC, while maintaining a high-quality motion prediction. Assuming that the DFD function exhibits a dominant minimum associated with a certain MV, the search algorithm visits in the first iteration only a fraction of all possible positions in the search window.

The algorithm initially estimates the block motion vector as $(0,0)$. A set of eight correction vectors arranged on a grid, as seen in Figure 12.7, are in turn added to the estimated motion vector and a new value of the MCER is computed each time. The correction vector, which yields the lowest value of the MCER so far computed, is added to the estimated motion vector, in order to obtain a more refined estimate. The size of the grid of correction vectors is then halved. The process is repeated until the final grid size is $3 \cdot 3$, at which point the algorithm terminates with an estimate of the best motion vector. When comparing this suboptimum algorithm with the optimal full search, the motion-compensated error energy is typically only 10% higher, while the search complexity is significantly reduced, a definitely worthwhile tradeoff.

For example, the full search in a search window of 16 × 16 pels requires 256 comparisons, whereas the hierarchical search already converges after 27 comparisons. As seen in Figure 12.8, the complexity reduction is achieved without a significant loss in terms of PSNR performance. The set of curves displayed in Figure 12.8 demonstrates the performance of various search algorithms applied to the "Miss America" head-and-shoulder video sequence. A total of 50 active motion vectors was assumed out of the 396 8 × 8 QCIF blocks, while for the remaining 346 blocks frame-differencing was used. The associated MCER was not transmitted to the decoder at all, which

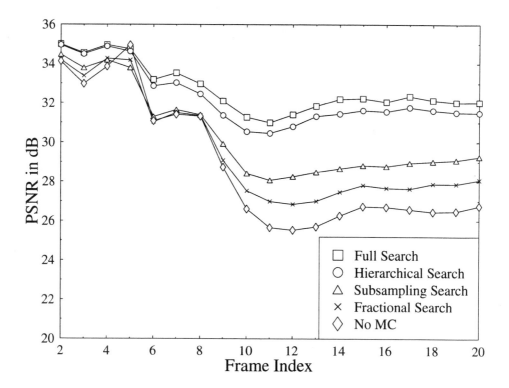

Figure 12.8: PSNR versus frame index performance comparison for various MC search algorithms using the "Miss America" sequence, where "No MC" refers to simple frame-differencing.

was equivalent to the assumption of MCER = 0. This crude assumption represented a worst-case codec performance, which was found adequate for assessing the performance of the various MC schemes.

12.2.2.4 Subsampling Search

The subsampling search algorithm [435] operates similarly to the tree search. However, instead of reducing the search window, the whole search area is subsampled by a given factor. The search in the lower resolution area is less complex since the number of possible locations and the block size are reduced according to the subsampling factor. Once the best location minimizing the MCER has been found in the subsampled domain, the resolution will be increased step-by-step, invoking the full-resolution search in the last stage of operations. The advantage is that the search window can be limited around the best position found in the previous step. The procedure continues until the original resolution is reached.

As an example, for a block size of 16×16 pels, both the incoming and previous reconstructed frame would be subsampled by a factor of 4 in each spatial direction. The subsampling factor of 4 means that a distance of 1 pel equals a distance of 4 pels

12.2. THE PRINCIPLE OF MOTION COMPENSATION

in the original frames. At the lowest resolution level, full search is applied using a search window of 3×3 pixels. In the next step the resolution is doubled in each spatial direction, and again, a full search using a search window of 3×3 pels is applied around the previously found optimum position, minimizing the MCER. The same procedure applied once again will result in the suboptimum vector in the original search window.

The complexity is even lower than that of the hierarchical search (HS), since the same number of block comparisons is performed at reduced block sizes. However, prior to any computations, the area covered by the search window must be subsampled to all required spatial resolutions. For some codecs, as, for example, quad tree codecs [435], which constitute the subject of Chapter 14, the image has to be subsampled anyway and hence no additional complexity is incurred. Although the search procedure is similar to the hierarchical search, it does not reach its performance.

12.2.2.5 Post-Processing of Motion Vectors

When using very small block sizes of 4×4 or even smaller, the motion vector field can be interpreted as a subsampled field of the "true motion" of each pixel [431]. Thus, the "true motion field" can be derived by generating an individual MV for each of the pixels on the basis of the nearest four MVs using interpolation. This is true if the correlation of adjacent motion vectors is sufficiently high and the gradient constraint of Equation 12.14 is satisfied. If Equation 12.14 is not satisfied, the MV interpolation may increase the MCER energy. Simulations revealed that for the commonly used block size of 8×8 pixels, only around 10% of the motion vectors derived by motion field interpolation resulted in an MCER gain, (i.e., a further reduction of the MCER energy). However, the reduction of the MCER energy due to this method remained generally below 3%. This marginal improvement is insufficient to justify the additional complexity of the codec.

12.2.2.6 Gain-Cost-Controlled Motion Compensation

Motion compensation reduces the MCER energy at the cost of additional complexity and channel capacity demand for the motion vectors. The contribution of each vector toward the MCER entropy reduction strongly depends on the movement associated with each vector. Comparing the MCER energy reduction with the requirement for the storage of the additional MVs leads to the concept of gain-cost controlled MC.

For most of the blocks the MCER entropy reduction, where the entropy was defined in Chapter 1, does not justify the additional transmission overhead associated with the MV. Figure 12.9 reveals this entropy reduction for various numbers of active MVs based on QCIF frames containing a total of 396 8×8 blocks. In order to evaluate the potential MC gain due to every block's motion vector, we quantified the MCER energy reduction in Equation 12.16, where b_{0n} and $b_{0(n-1)}$ denote the block under consideration in the current and previous frames, respectively, and the MC gain is measured with respect to the MCER of $[b_{0n}(i,j) - b_{0(n-1)}(i,j)]$ generated by simple frame-differencing:

$$MC_{gain} = \sum_{i=1}^{b}\sum_{j=1}^{b}(b_{0n}(i,j) - b_{0(n-1)}(i,j))^2 \qquad (12.15)$$

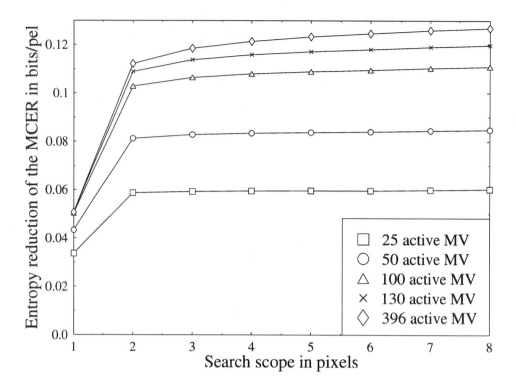

Figure 12.9: Entropy reduction versus search scope in case of limiting the number of active MVs per 8 × 8 block for QCIF head-and-shoulder video sequences.

$$-\sum_{i=1}^{b}\sum_{j=1}^{b}(b_{0n}(i,j) - b_{0(n-1)}(i+m_x, j+m_y))^2.$$

In Figure 12.9 we selected only the most efficient vectors for each experiment. All low-gain motion-passive MVs were set to [0,0], and simple frame-differencing was used at the motion-passive locations. Surprisingly, our results suggested that the most important 25 % of the MVs (i.e., using about 100 active MVs) resulted in 80–90 % of the possible entropy reduction.

The disadvantage of the motion passive/active concept is the additional requirement of indices or tables necessary to identify the active vectors. In a conceptually simple approach, we can assign a 1-bit motion activity flag to each of the 396 8×8 blocks, which constitutes a transmission overhead of 396 bits for each QCIF frame. An efficient alternative technique will be introduced in Section 12.7 in order to further compress this 396-bit binary motion activity table and hence to reduce the number of bits required. Figure 12.10 reveals the required coding rate of the MVs, when applying the above-mentioned efficient activity table compression method. From Figures 12.9 and 12.10 we concluded that the MC search window size can be limited to ±2 adjacent pixels for limited-dynamic head-and-shoulder videophone sequences,

12.2. THE PRINCIPLE OF MOTION COMPENSATION

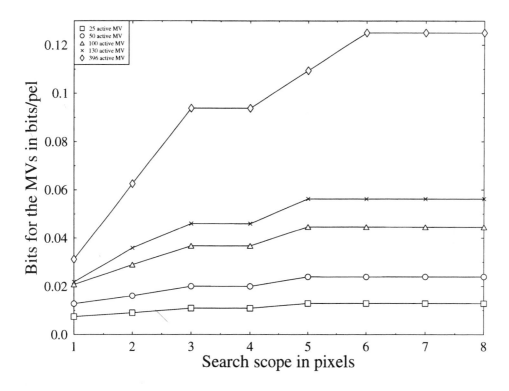

Figure 12.10: Required bit rate of active MVs versus MC search-scope for different number of active 8 × 8 blocks in QCIF head-and-shoulder videophone sequences.

while the adequate number of active motion vectors is below 100 in the case of QCIF-sized images. When using a higher proportion of active MVs, the number of bits assigned to their encoding will reduce the number of bits available for the encoding of the MCER and inevitably reduce the image quality in the case of a fixed bit-rate budget.

12.2.3 Other Motion Estimation Techniques

Various modifications of the above-mentioned algorithms have been published in the literature [436–438]. Many of them are iterative simplifications of the full search algorithm and are based on the gradient constraint criterion of Equation 12.14. Other techniques improve the accuracy of the motion prediction by interpolating the given images to a higher resolution and defining the MVs at a sub-pel accuracy [439]. Another classical approach to motion compensation is represented by the family of the pel-recursive techniques [440, 441], which will be described in Section 12.2.3.1. The grid interpolation techniques introduced quite recently are rooted in model-based techniques, and they will be highlighted in Section 12.2.3.2.

12.2.3.1 Pel-Recursive Displacement Estimation

This technique predicts the motion vectors for the current frame from previous frames. Thus, no additional information must be transmitted for the motion vectors because the prediction is based on frames, which are available at both the transmitter and the receiver at the same time. Therefore, it is affordable to carry out the prediction for every single pixel rather than grouping them into blocks. The algorithms observe the motion for each pel over the last frames and derive from these vectors the speed and orientation of the movement. These values are then used to estimate the motion vectors for the current frame. The problem associated with these algorithms is that they fail at the edges of movement, which causes a motion-compensated error residual with strongly increased nonstationary characteristics. This phenomenon limits the performance of this technique and that of the subsequent image-coding steps. Another principal problem of these so-called backward estimation techniques is that they require perfect alignment between the operations of the transmitter and receiver. A tiny misalignment between the transmitter and receiver will result in a different motion estimation at both ends and induce further, rapidly accumulating misalignment. Hence, this motion compensation technique is not suitable for mobile communications over high error rate channels.

12.2.3.2 Grid Interpolation Techniques

The so-called grid interpolation technique is based on block deformation rather than block movement. This technique divides the video frames into perfectly tiling blocks using a regular grid pattern, as seen at the left-hand side of Figure 12.11. This so-called wire frame, the image objects, and their underlying texture are appropriately deformed in order to reconstruct the next frame, which is portrayed at the right-hand side of Figure 12.11. The deformation of each patch of texture is obtained by an affine transform, which was the subject of Chapter 11 on fractal coding. This transform is defined by the motion of each grid crossing. Hence, only the motion vectors of the grid crossings have to be transmitted. This technique has been employed with some success in [442,443]. Its drawback is associated with the required interpolation of points. This is because the affine transform will not map every pixel on a valid position of the destination frame. This increases the complexity dramatically, especially during the iterative motion estimation step of the encoding process.

12.2.3.3 MC Using Higher Order Transformations

Instead of deforming the grid, we may apply MVs, which are a function of the location within the blocks. The nth order geometric transformation of:

$$mv_x(u,v) = \alpha_0 + \alpha_1 u + \ldots \alpha_n u^n + \beta_0 + \beta_1 v + \ldots \beta_n v^n$$
$$mv_y(u,v) = \gamma_0 + \gamma_1 u + \ldots \gamma_n u^n + \delta_0 + \delta_1 v + \ldots \delta_n v^n \qquad (12.16)$$

allows simultaneous translation, rotation, and change of scale [444], and it has $4(n+1)$ different parameters given by $\alpha_0 \ldots \alpha_n, \beta_0 \ldots \beta_n, \gamma_0 \ldots \gamma_n$ and $\delta_0 \ldots \delta_n$.

This is a very general approach to MC, since the motion-compensated objects may differ in their appearance, yet guarantee a low MCER for relatively large block

12.2. THE PRINCIPLE OF MOTION COMPENSATION

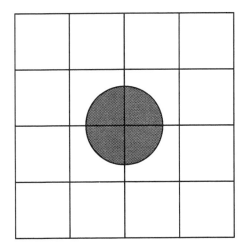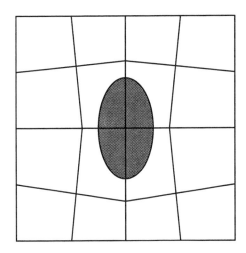

Figure 12.11: Grid interpolation example.

sizes of 16 × 16 and larger [445]. Equation 12.16 can be simplified, allowing shifting, scaling, and translation only [446], reducing it to a total of three variables. In practical terms, however, the additional complexity and bit-rate requirement of the algorithm outweigh its advantages.

12.2.3.4 MC in the Transform Domain

Block-based MC often results in *blocking artifacts* [447], which are subjectively objectionable. This problem can be mitigated by using blocks having fuzzy block borders, which can be created by interleaving the pixels of adjacent blocks according to a number of different mapping rules that are characterized by jigsaw puzzle-like block borders [448]. This kind of block shape reduces the likelihood of artifacts at the cost of a reduced MC performance. An elegant way of combatting this problem is to apply MC in a transform domain, for example, using complex wavelets [449] or the lapped orthogonal transform (LOT) [450, 451]. In the LOT, the images are transformed and then divided into overlapping blocks, hence mitigating the time-domain artifacts.

12.2.4 Conclusion

In this chapter, we have reviewed a range of block-matching motion estimation techniques and analyzed their performance in the case of various parameters. We found that upon using a rather limited proportion of active motion vectors we were able to reduce the DFD energy significantly. Another result was that small search window sizes of 4 × 4 pixels were adequate for maintaining a low complexity and near-optimum performance in case of slow-motion QCIF head-and-shoulder videophone sequences, such as the "Miss America" sequence. Such a small search window allowed us to use the highest complexity exhaustive search method, while keeping the technique's computational demand within the requirements of real-time applications. This is naturally valid only for low-activity head-and-shoulder images. Finally, we concluded that backward estimation techniques, such as pel-recursive motion compensation, are not suitable for communications over high error rate mobile channels.

12.3 Transform Coding

12.3.1 One-Dimensional Transform Coding

As is well-known from Fourier theory, signals are often synthesized by so-called orthogonal basis functions, a term, which will be augmented during our further discourse. Specifically, when using Fourier transforms, an analog time-domain signal, which can be the luminance variation along a scanline of a video frame or a voice signal, can be decomposed into its constituent frequencies.

For signals such as the above-mentioned video signal representing the luminance variation along a scanline of a video frame, orthogonal series expansions can provide a set of coefficients, which equivalently describe the signal concerned. We will make it plausible that these equivalent coefficients may become more amenable to quantization than the original signal.

For example, for a one-dimensional time-domain sample sequence $\{x(n), 0 \leq n \leq N-1\}$ a unitary transform is given in vectorial form by $\underline{X} = \underline{\underline{A}}\underline{x}$, which can also be expressed in a less compact scalar form as [26]:

$$X(k) = \sum_{n=0}^{N-1} a(k,n) \cdot x(n), \quad 0 \leq k \leq N-1 \quad (12.17)$$

where the transform is referred to as unitary, if $\underline{\underline{A}}^{-1} = \underline{\underline{A}}^{*T}$ holds. The associated inverse operation requires us to invert the matrix $\underline{\underline{A}}$ and because of the unitary property, we have $\underline{x} = \underline{\underline{A}}^{-1}\underline{X} = \underline{\underline{A}}^{*T}\underline{X}$, yielding [26]:

$$x(n) = \sum_{k=0}^{N-1} X(k) a^*(k,n), \quad 0 \leq k \leq N-1 \quad (12.18)$$

which gives a *series expansion* of the time-domain sample sequence $x(n)$ in the form of the *transform coefficients* $X(k)$. The columns of $\underline{\underline{A}}^{*T}$, that is, the vectors $\underline{a}_k^* \stackrel{\triangle}{=} \{a^*(k,n), 0 \leq n \leq N-1\}$, are the basis vectors of $\underline{\underline{A}}$, or the *basis vectors of the decomposition*. According to the above principles, the time-domain signal $x(n)$ can be equivalently described in the form of the *decomposition* in Equation 12.18, where the *basis functions* $a^*(k,n)$ are weighted by the transform coefficients $X(k)$ and then superimposed on each other, which corresponds to their summation. The transform-domain weighting coefficients $X(k)$ can be determined from Equation 12.17.

The transform-domain coefficients $X(k); k = 0 \cdots N-1$ often give a more "compact" representation of the time-domain samples $x(n)$, implying that if the original time-domain samples $x(n)$ are correlated, then in the transform domain most of the signal's energy is concentrated in a few transform-domain coefficients. To elaborate a little further — according to the Wiener-Khintchine theorem — the AutoCorrelation

Function (ACF) and the Power Spectral Density (PSD) are Fourier transform pairs. Because of the Fourier-transformed relationship of the ACF and PSD, it is readily seen that a slowly decaying autocorrelation function, which indicates a predictable signal $x(n)$ in the time domain is associated with a PSD exhibiting a rapidly decaying low-pass nature. Therefore, in case of correlated time-domain $x(n)$ sequences, the transform-domain coefficients $X(k)$ tend to be statistically small for high frequencies, that is, for high-coefficient indices and exhibit large magnitudes for low-frequency transform-domain coefficients, that is, for low tranform-domain indices. This concept will be exposed in a little more depth below, but for a deeper exposure to these issues the reader should consult Jain's excellent book [26].

12.3.2 Two-Dimensional Transform Coding

The one-dimensional signal decomposition can also be extended to two-dimensional (2D) signals, such as 2D image signals of a video frame, as follows [26]:

$$X(k,l) = \sum_{m=0}^{N-1} \sum_{n=0}^{N-1} x(m,n) \cdot a_{k,l}(m,n) \quad 0 \leq k,l \leq N-1 \quad (12.19)$$

$$x(m,n) = \sum_{k=0}^{N-1} \sum_{l=0}^{N-1} X(k,l) \cdot a_{k,l}^*(m,n) \quad 0 \leq m_1 n \leq N-1 \quad (12.20)$$

where $\{a_{k,l}^*(m,n)\}$ is a set of discrete two-dimensional basis functions, $X(k,l)$ are the 2D transform-domain coefficients, and $\underline{\underline{X}} = \{X(k,l)\}$ constitutes the transformed image.

As in the context of the one-dimensional transform, the two-dimensional (2D) time-domain signal $x(m,n)$ can be equivalently described in the form of the decomposition in Equation 12.20, where the 2D basis functions $a_{k,l}^*(m,n)$ are weighted by the coefficients $X(k,l)$ and then superimposed on each other, which again corresponds to their summation at each pixel position in the video frame. The transform-domain weighting coefficients $X(k,l)$ can be determined from Equation 12.19.

Once a spatially correlated image block $x(m,n)$ of, for example, $N \times N = 8 \times 8$ pixels is orthogonally transformed using the discrete cosine transform (DCT) matrix $\underline{A}$ defined as [26, 452]:

$$A_{mn} = \frac{2c(m)c(n)}{N} \sum_{i=0}^{N} \sum_{j=0}^{N} b(i,j) \cos\frac{(2i+1)m\pi}{2N} \cos\frac{(2j+1)n\pi}{2N}$$

$$c(m) = \begin{cases} \frac{1}{\sqrt{2}} & if(m=0) \\ 1 & otherwise \end{cases} \quad (12.21)$$

the transform-domain image described by the DCT coefficients [26, 452] can be quantized for transmission to the decoder. *The rationale behind invoking the DCT is that the frequency-domain coefficients $X(k,l)$ can typically be quantized using a lower number of bits than the original image pixel values $x(m,n)$*, which will be further augmented during our forthcoming discourse.

For illustration's sake, the associated two-dimensional 8×8 DCT [26, 452] *basis-images* are portrayed in *Figure 12.12*, where, for example, the top left-hand corner

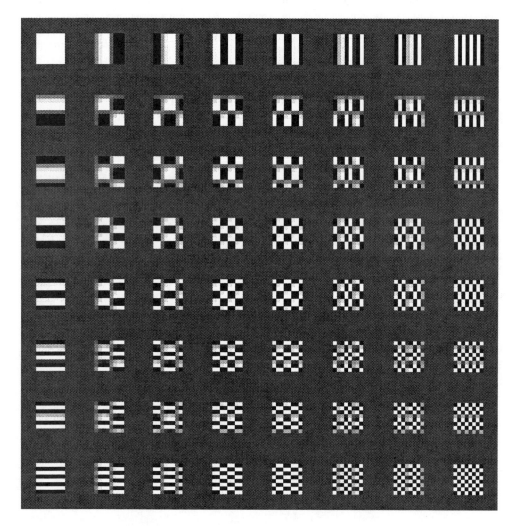

Figure 12.12: 8×8 DCT basis images. ©A. Sharaf [453].

represents the zero horizontal and vertical spatial frequency, since there is no intensity or luminance change in any direction across this basis image. Following similar arguments, the bottom right corner corresponds to the highest vertical and horizontal frequency, which can be represented using 8×8 basis images, since the luminance changes between black and white between adjacent pixels in both the vertical and horizontal directions. Similarly, the basis image in the top right-hand corner corresponds to the highest horizontal frequency but zero vertical frequency component, and by contrast, the bottom left image represents the highest vertical frequency but zero horizontal frequency. *In simple terms, the decomposition of Equations 12.19 and 12.20 can be viewed as finding the required weighting coefficients $X(k, n)$ in order to superimpose the weighted versions of all 64 different "patterns" in Figure 12.12 for the sake of reconstituting the original 8×8 video block.* In other words, each original 8×8 video block is represented as the sum of the 64 appropriately weighted 8×8 basis image.

It is plausible that for blocks over which the video luminance or gray shade does not change dramatically (i.e., at a low spatial frequency), most of the video frame's conveyed energy is associated with these low spatial frequencies. Hence, the associated low-frequency transform-domain coefficients $X(k, n)$ are high, and the high-frequency coefficients $X(k, n)$ are of low magnitude. By contrast, if there is a lot of fine detail in a video frame, such as in a finely striped pattern or in a checkerboard pattern, most of the video frame's conveyed energy is associated with high spatial frequencies. Most practical images contain more low spatial frequency energy than high-frequency energy. This is also true for those motion-compensated video blocks, where the motion compensation was efficient and hence resulted in a flat block associated with a low spatial frequency. For these blocks therefore most of the high-frequency DCT coefficients can be set to zero at the cost of neglecting only a small fraction of the video block's energy, residing in the high spatial frequency DCT coefficients. In simple terms this corresponds to gentle low-pass filtering, which in perceptual terms results in a slight blurring of the high spatial-frequency image fine details.

In other words, upon exploiting that the human eye is rather insensitive to high spatial frequencies, in particular when these appear in moving pictures, the spatial frequency-domain block is amenable to data compression. Again, this can be achieved by more accurately quantizing and transmitting the high-energy, low-frequency coefficients, while typically coarsely representing or masking out the low-energy, high-frequency coefficients. We note, however that in motion-compensated codecs there may be blocks along the edges of moving objects where the MCER does not retain the above-mentioned spatial correlation and hence the DCT does not result in significant energy compaction. Again, for a deeper exposure to the DCT, and for an example the reader is referred to [26, 452].

In critical applications, such as medical image compression, these high-frequency coefficients carry the perceptually important edge-related video information and so must be retained. On the same note, our discussion of high-quality DCT-based coding will show that if the original image contains a significant amount of high-frequency energy and hence there is less adjacent pixel correlation in the image, then this manifests itself in terms of a less concentrated set of DCT coefficients. For such pictures, the DCT typically achieves less energy compaction.

A range of orthogonal transforms have been suggested in the literature for various applications [454]. The DCT is widely used for image compression, since its compression ratio approaches that of the optimum Karhunen-Loeve (KL) transform [26], while maintaining a significantly lower implementational complexity. This is due to the fact that the KL transform's basis vectors used in the transformation are data dependent, and so they have to be computed for each data segment before transformation can take place. The simple reason for the energy compaction property of orthogonal transforms, in particular that of the DCT, is that for correlated time-domain sequences, which have a flat-centered autocorrelation function, the spatial frequency-domain representative tends to be of a low-pass nature.

The efficiency of the DCT transform has constantly attracted the interest of researchers and has led to a wide variety of DCT and DCT-based codecs. Recently, many attractive results have been published using adaptive methods [455–457] and hybrid coding [458–460]. These approaches are typically associated with a fluctuating, time-variant compression ratio, requiring buffering methods in order to allow a constant bit rate. The following approach presents a fixed-rate DCT scheme, which can dispense with adaptive feedback buffering while still guaranteeing a fixed frame rate at a constant user-selectable bit rate.

As argued earlier, most of the energy of the 8×8 block is concentrated in the low horizontal and low vertical frequency regions corresponding to the top left-hand corner region of its two-dimensional spectrum. We capitalize on this characteristic by assigning higher resolution quantizers in the frequency regions, where we locate the highest energy, whereas lower energy frequency regions are quantized with coarser quantizers or are completely neglected by setting them to zero. The issue of choosing the appropriate frequency regions for quantization and the appropriate quantizer resolution are the most crucial design steps for a DCT codec.

Next, we describe in Sections 12.3.3 and 12.3.4, how we derived a set of quantizers for the DCT transformed blocks of the MCER, which are specially tailored for very low bit rates. Then Sections 12.4–12.11 reveal the codec structure and its performance under ideal channel conditions. Finally, in Section 12.12, we examine the sensitivity to channel errors and propose two configurations for fixed and mobile video communicators.

12.3.3 Quantizer Training for Single-Class DCT

Initially, we sought to determine which of the DCT coefficients had to be retained and quantized and which could be ignored or set to zero, while not inflicting severe image-quality degradation. We commenced by setting some of the DCT coefficients to zero, whereas the remaining coefficients were left unquantized. On the basis of the energy compaction property of the DCT [26], the unquantized coefficients were typically the ones closest to the top left corner of the matrix. This location represents the direct current (DC) and low-frequency region. We found that in order to maintain the minimum required videophone quality, about six DCT coefficients had to be retained.

In a second step, we assigned the appropriate number of reconstruction levels and the reconstruction levels themselves, which were required in order to maintain the target bit rate, which in our case was around 10 kbps. A particular design

12.3. TRANSFORM CODING

3	2	1	0	0	0	0	0
2	1	0	0	0	0	0	0
1	0	0	0	0	0	0	0
0	0	0	0	0	0	0	0
0	0	0	0	0	0	0	0
0	0	0	0	0	0	0	0
0	0	0	0	0	0	0	0
0	0	0	0	0	0	0	0

Table 12.1: Bit-Allocation Tables for the Fixed-Rate DCT Coefficient Quantizers

constraint was that the number of reconstruction levels had to be an integer power of two. That is, we required $n_{rec} = 2^k \mid k = \{1, 2, 3, \ldots\}$, so that each coefficient was quantized to k bits. Initially, we allocated 8 bits for the DC coefficient, 4 bits for each of the positions [0,1] and [1,0] on the [8,8]-dimensional frequency plane, and 2 bits for each of the positions [2,1],[1,1], and [1,2]. Appropriately trained Max-Lloyd quantizers [461] were used to achieve minimum distortion quantization. We obtained the appropriate training data for the Max-Lloyd quantizers by recording the DCT coefficients of the MCER for the 30 highest-energy blocks of each QCIF frame for a 100 frame long motion-compensated sequence. This selection of training blocks considers the fact that we generally incorporate an active/passive classification scheme into the final codec. Therefore the training is also based only on blocks, which are likely to be selected by an active/passive classification scheme for quantization. The samples for each coefficient were collected, and the corresponding Max-Lloyd quantizers were derived from the training data. Perceptually pleasant videophone quality was achieved, when quantizing every 8 × 8 pixel block to a total of 22 bits. In order to achieve the maximum possible compression ratio attainable using this technique, while maintaining an adequate image quality, we iteratively reduced the number of reconstruction levels until we reached the required image quality versus bit-rate tradeoff deemed appropriate for QCIF videophony. At this stage, the entire block was quantized to a total of 10 bits, which leads to the bit-allocation depicted in Table 12.1.

12.3.4 Quantizer Training for Multiclass DCT

In the previous section, we demonstrated how we derived a set of quantizers, which achieved the maximum compression while providing an adequate quality for hand-held video communications. Noticeable degradation was caused at the edges of motion trajectories or at areas of high contrast. This was because for uncorrelated MCER sequences the DCT actually results in energy expansion rather than energy compaction [435]. Furthermore, if there is more correlation in one particular spatial direction than the other, the center of the spectral-domain energy peak moves toward the direction of higher correlation in the time domain. This happens if the block contains, for example, a striped texture. As a consequence, we added two sets of quantizers which shall cater for blocks with higher horizontal or vertical correlation. A fourth quantizer was added, which caters for the case of decreased energy

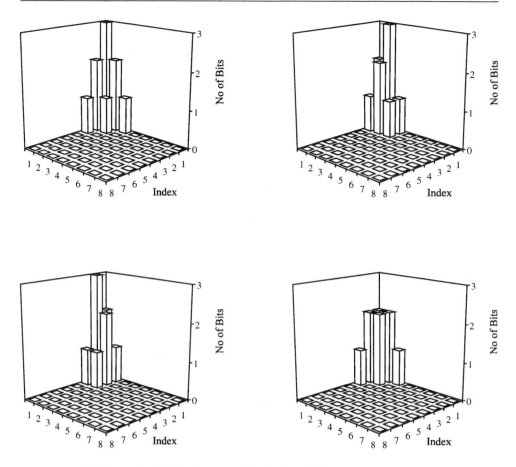

Figure 12.13: Bit-allocation for the Quad-Class DCT quantizers.

compaction. Thus, its bit allocation is more widely spread than that of the other quantizers. The bit-allocation for all four quantizers is depicted in Figure 12.13 across the 8 × 8 frequency-domain positions.

After fixing the quantizer resolutions, we had to find an appropriate training sequence for each set of quantizers. The quantizer training required a training sequence exhibiting similar statistical properties to the actual discrete cosine transformed MCER, which has to be quantized by the quantizers to be designed. This required an initial tentative quantizer for the design of the actual quantizer. This is because the design of the MCER encoder in Figure 12.1 influences the future distribution of the MCER e_n through the reconstruction frame buffer. These issues constitute the subject of our forthcoming deliberations. Hence, again, a classification algorithm was needed in order to assign each of the training blocks to one of those categories. To simulate the effect of the actual quantizer about to be designed on the MCER, which is then used to generate the quantizer training set, we decided to use a simple initial tentative quantization function, which we will refer to as a pseudo-quantizer, producing a randomly quantized quantity $\tilde{q}$ from the unquantized variable q by contaminating q using an additive, quantizer-resolution dependent fraction of the unquantized variable itself. Specifically, we attempted to simulate each quantizer by applying Equation 12.22,

$$\tilde{q} = q \times (1.0 + 1.5 \frac{(-1)^{RAND}}{n}), \qquad (12.22)$$

where q is the quantity to be quantized, $\tilde{q}$ represents its pseudo-quantized value, n equals the number of reconstruction levels, and $RAND$ is a random integer. Clearly, this pseudo-quantizer simulates the quantization error by randomly adding or subtracting a fraction of itself from the unquantized quantity q. The absolute factor of 1.5 was found empirically and should be in the range of $[1.3 \ldots 1.7]$. The magnitude of the added distortion is inversely proportional to the number of quantization levels n.

Each quantizer training block was DCT transformed, pseudo-quantized, and then inverse-transformed back to the time domain, while tentatively using all four sets of quantizer seen in Figure 12.13. The unquantized coefficients corresponding to the specific quantizer of Figure 12.13 leading to the best reconstruction in the MSE sense were collected separately for each class. Finally, the Max-Lloyd quantizers for each energy distribution class were derived from each of these classified training sets. We found that the initial set of quantizers performed well, since a retraining of those quantizers upon using the newly generated quantizers instead of the pseudo-quantizer resulted only in marginal further performance improvement.

Figure 12.14 reveals the codec's PSNR performance for a single-, dual-, and quad-class DCT codec at a constant bit rate of 1000 bits per frame, or 10 kbps at a video frame scanning rate of 10 frames per second. The dual-class codec was using the quantizer sets derived for high and normal energy compaction, which are depicted in the top left and bottom right corner of Figure 12.13. Although the dual-class scheme requires one and the quad-class scheme two more bits per DCT transformed block for the encoding of the energy distribution classifier and therefore the number of active blocks to be encoded — given the 1000 bits per frame bit-rate constraint — must be reduced, if the overall bit-rate is kept constant, the perceived image quality increased significantly and the objective PSNR also improved by up to 0.7 dB due to the classification technique. The relative frequency of selecting any one of the four quantizers varies from frame to frame and depends on the specific encoded sequences. The average relative frequency of the four quantizers was evaluated using the "Miss America" sequence for both the twin- and the quad-class DCT schemes, which is shown in Table 12.2. Observe that the highest relative frequency bottom entry of the quad-class DCT codec in Table 12.2 is associated with an extended horizontal correlation, which is also confirmed by the autocorrelation function of the typical MCER, as seen in Figure 12.5.

12.4 The Codec Outline

In order to achieve a time-invariant compression ratio associated with a constant encoded video rate, the codec is designed to switch between two modes of operation

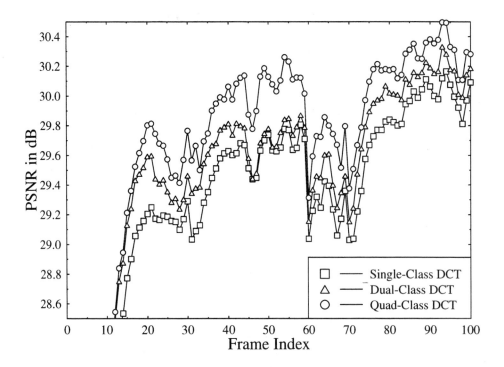

Figure 12.14: Performance comparison of multiclass DCT schemes.

Quantizer Set Type	Av. Rel. Freq. Quad-Class DCT	Av. Rel. Freq. Dual-Class DCT
Very High Correlation	14.7 %	39 %
High Correlation	29.3 %	61 %
Increased Vertical Correlation	24.0 %	-
Increased Horizontal Correlation	32.0 %	-

Table 12.2: Relative Frequency of the Various Quantizers in the Multiclass DCT Codecs in the "Miss America" Sequence

12.4. THE CODEC OUTLINE

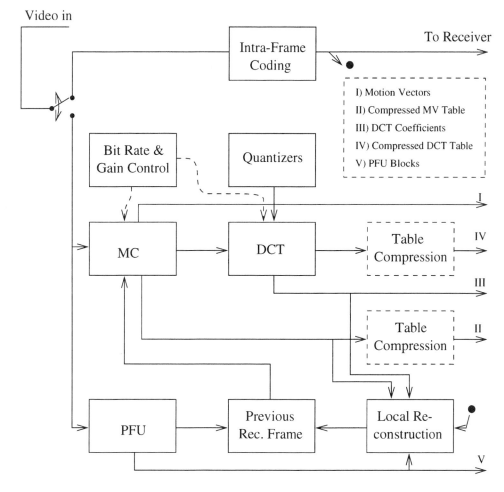

Figure 12.15: Schematic of the multiclass DCT codec.

as depicted in Figure 12.15. This is necessary because the DCT-based inter-frame codec assumes that the previous reconstructed frame is known to both the encoder and decoder. At the commencement of communications, the reconstructed frame buffers of the encoder and decoder are empty. Thus, the MCER energy for the first frame would be very high, and it would be the first frame to be encoded. The quantizers we derived using the philosophy of Section 12.3 were not trained for this particular case, and so the DC-coefficient quantizer would be driven to saturation. Therefore, we transmit a single independent intra-coded frame during the initial phase of communication in order to replenish the reconstructed frame buffers. Once switched to inter-frame mode, any further mode switches are optional and are required only if a drastic change of video scene occurs.

In the inter-frame mode of Figure 12.15 the incoming video sequence is first motion compensated, and then the resulting MCER is encoded. The active/passive decisions

Bits per Frame	Block Size
500	14×14
800	12×12
1000	10×10
1200	10×10
1500	9×9

Table 12.3: Block Sizes Used in the Startup Phase of the Intra-frame DCT Codec. The Block Size Along the Fringes of the QCIF Frames was Enlarged Slightly in Order to Perfectly Tile the Frame and to Generate the Exact Required Number of Bits per Frame.

carried out during both the MC and DCT coding stages are cost/gain controlled within the allocated constant, but programmable bit-rate budget as it was highlighted in Section 12.2.2.6. An optional table compression algorithm of Figure 12.15, which will be described at a later stage, is also incorporated in order to reduce the amount of redundancy associated with the transmission of the 396-bit motion- and DCT activity tables. Finally, a partial forced update (PFU) algorithm is included in Figure 12.15 to ensure that a realignment of the encoder's and decoder's reconstructed buffers becomes possible, while operating under high error-rate channel conditions. The following sections elaborate further on the individual blocks of Figure 12.15.

12.5 Initial Intra-Frame Coding

The objective of the intra-frame mode of Figure 12.15 is to provide an initial frame for both the encoder's and decoder's reconstructed frame buffer in order to ensure that the initial MCER frame does not saturate the DC-coefficient quantizer. Constrained by the fixed bit-rate budget of about 1000 bits/frame, the 176×144-pixel QCIF frame can only be represented very coarsely, at a resolution of about $1000/(176 \times 144) \simeq 0.04$ *bits/pixel*. This can be achieved by dividing the frame into perfectly tiling blocks and coarsely quantizing the average of each block. It was found appropriate to quantize each block average to 16 levels, which are equidistantly distributed between the absolute pixel values of 52 and 216. Since the entire intra-frame coded image must be encoded within the limits of the per-frame bit-rate budget, the block size had to be adjusted accordingly. The resulting block sizes for various target bit rates are depicted in Table 12.3. When the intra-frame coded block size did not tile the QCIF format perfectly, the codec increased the block size for those particular blocks, which were situated along the fringes of the frame.

12.6 Gain-Controlled Motion Compensation

The motion compensation process of Figure 12.15 is divided into two steps. In the first step the codec determines a motion vector for each of the 8×8 sized blocks. The search window is set to 4×4 pixels, which proved to be efficient in removing redundancies

12.7. THE MCER ACTIVE/PASSIVE CONCEPT

between consecutive frames at a cost of 4 bits per MV in the context of our QCIF head-and-shoulder videophone sequences. Since we concluded in Section 12.2.4 that MC achieves a significant MCER gain only for a fraction of the blocks, the codec stores the potential MCER energy gain for every block. The gain of a particular block B at location (x_0, y_0) in the video frame f is defined as:

$$G_{mc} = \sum_{i=1}^{b}\sum_{j=1}^{b}(f_n(x_0+i, y_0+j) - f_{n-1}(x_0+i, y_0+j))^2 \quad (12.23)$$
$$- \sum_{i=1}^{b}\sum_{j=1}^{b}(f_n(x_0+i, y_0+j) - f_{n-1}(x_0+i+m_x, y_0+j+m_y))^2.$$

The bit-rate control algorithm selects the specific number of most efficient MVs according to the available bit-rate budget. The remaining MVs are set to zero, before subtracting the current motion translated reconstructed block from the incoming one, which corresponds to simple frame-differencing, as far as the "motion-passive" blocks are concerned. At this stage we have to further clarify our active/passive block terminology in order to avoid future confusion. Our later discussion will refer to those blocks where MVs were allocated as motion-active, while to those blocks, where MV = 0 and hence frame-differencing is used, as motion-passive. On a similar note, blocks where the MCER is encoded are termed MCER-active and those where the MCER is low, and hence it is set to zero, are referred to as MCER-passive.

12.7 The MCER Active/Passive Concept

Section 12.3 introduced the philosophy of energy compaction and demonstrated how this property can be exploited in order to reduce the number of encoded bits by using the DCT. In most cases, it is sufficient to transmit a total of 10 to 12 bits for an 8×8 pixel block in order to maintain adequate videophone quality, which corresponds to a compression ratio of (25 384 pixels × 8 bits/pixel) / (396 blocks × 12 bits) ≈ 42, when neglecting the motion vectors. Further compression can be achieved when the codec intelligently allocates bits to those image regions, where it is most beneficial. From the analysis of the MCER, we know that large sections of the MCER frame are flat and there is no need to allocate bits to these MCER-passive areas. Therefore, a protocol is required in order to identify the location of MCER-active blocks, which we encode and that of the MCER-passive ones, which we neglect.

The most conceptually obvious solution would be to transmit the index of every active block. In case of QCIF images and a block size of 8×8, there are 396 blocks, which requires 9 bits to store one MCER-active block index. Assuming a total of perhaps 100 active blocks, 900 bits/frame would be necessary only to identify their locations. A more promising approach is to establish an active/passive table, which for each of the 396 blocks contains a 1-bit flag marking the corresponding block either as MCER-active or MCER-passive. This becomes advantageous compared to the previous method, if there are more than 44 active blocks. Bearing in mind that the active probability for very low bit-rate applications is usually below 15%, we note that most of the entries of the active-passive table will be marked as passive. This implies predictability or redundancy and opens further compression possibilities, such as run-length or entropy-encoding.

We decided to scan the active-passive flags on a line-by-line basis and packetize a certain number of them into one symbol and Huffman code those symbols, as highlighted in Chapter 1. It was expected that most symbols would contain a low number of 1 and a high number of 0 flags. This *a priori* knowledge could then be exploited to assign short Huffman code words to the high-probability symbols and long ones to the low-probability symbols. We found that the optimum number of bits per symbol was around five. This bit-to-symbol conversion exploits most of the latent redundancy inherent in the bit-by-bit representation, while preventing us from optimizing the codewords of the symbol-based Huffman codec to a certain activity rate. Figure 12.16 reveals the advantage of this active/passive table compression technique in case of QCIF frames and 8×8 sized blocks. However, this technique does not exploit the fact that increased motion activity often covers a group of several blocks rather than a single block implying that the uncompressed 396 activity flags are likely to be clustered in active and passive islands.

Therefore, we developed another approach in order to further compress the active/passive table. The motion activity flags were grouped into blocks of two by two, covering four original 8×8 pixel blocks. For those groups that did contain active motion flags, we assigned a 4-bit activity symbol, reflecting its activity-flag contents. These symbols were then Huffman coded. This concept of only transmitting information regarding those groups of vectors, which were active, required a second active/passive table. This second table reflected which of the grouped motion activity flags were set to active. It contained $396/4 = 99$ entries, which we again packetized to three entries per symbol, a value that was found to guarantee best coding efficiency, and these entries were then Huffman encoded. This method was found superior to the previous technique, especially, when the active rate was below 10% or 40 blocks per frame, as demonstrated by Figure 12.16. In this range of block activity ratio, on the average we needed about 5 bits per active entry in order to convey the activity information to the decoder. This corresponds to a bit-rate reduction of nearly 50%, when compared to the 9-bit "indexing" and "uncompressed table" techniques of Figure 12.16. Further potential bit-rate economies can be achieved upon invoking the error-resilient positional coding (ERPC) principle [462].

12.8 Partial Forced Update of the Reconstructed Frame Buffers

Let us now consider the partial forced update block of Figure 12.15. Communications over nonideal channels is prone to errors, which is particularly true in hostile mobile environments characterized by Rayleigh-fading channels. As a consequence of interframe coding, the video quality of the reconstructed frames is impaired not only in the affected frames, but also in consecutive frames. The errors result in a misalignment between the encoder's local reconstructed buffer and the equivalent reconstructed frame buffer at the receiver side. This misalignment will persist if no measures are

12.8. PARTIAL FORCED UPDATE

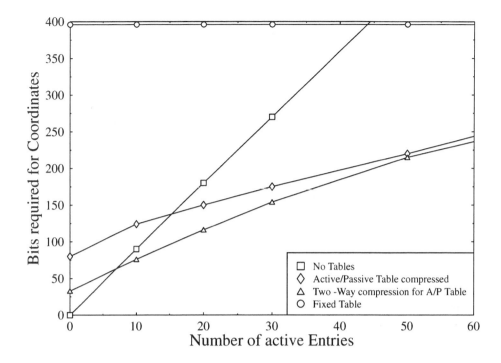

Figure 12.16: Comparison for various active/passive classifier encoding strategies in QCIF frames for 8×8 blocks.

implemented to mitigate the effects of errors. A potentially suitable remedy is to apply a so-called leakage technique, which de-weights the reconstructed frame decoded from the received information using a leakage factor [461]. Hence, this procedure inflicts some image degradation under error-free channel conditions in order to allow for the effects of channel errors to decay.

In a specific implementation of the leakage technique, the video frame or segments of the frame are periodically luminance scaled. A typical leakage factor of, for example, 0.9 would mean that the brightness of every pixel in the reconstructed frame buffer would be multiplied by 0.9, resulting in a slowly fading luminance of consecutive frames. This then would result in an increased MCER and a slightly more moderate coding performance, while improving the codec's robustness. The gradually fading segments would therefore remove the effects of channel errors over a period of time. This simple technique renders the PDF of the DC coefficient to become less highly peaked in the center, which inevitably leads to a wider dynamic range and hence typically requires more reconstruction levels for the corresponding quantizer, unless a higher coding distortion is acceptable. The extent to which the quantizers have to be adapted to the modified input data depends on the actual value of the leakage factor, which in turn depends on prevailing BER. Higher BERs require a faster luminance scaling or fading, that is, lower valued leakage factors.

Another often used procedure is to invoke a so-called forced updating technique, which forcibly realigns the reconstructed frame buffers at regular instants using typically a compromise value, which again allows the decoder to gradually taper the effects of errors. In order to incorporate the highest possible degree of flexibility, we opted for a combination of the above-mentioned leakage and replenishment techniques. Before inter-frame encoding of the MCER ensues, we select a predetermined, bit-rate dependent number of 8×8 pixel blocks in the incoming frame buffer, for which we determine the coarsely quantized average, as we detailed in Section 12.5. Then every pixel of the selected blocks in the reconstructed buffers is first scaled down by a certain factor $(1 - l)$. Then the 4-bit quantized average is superimposed, which is scaled by the leakage factor l. This partial forced updating (PFU) scheme guarantees that the average MCER energy does not change and the DC-quantizers of the subsequent DCT codec are not overloaded. The leakage factor values used in our approach are typically between 0.3 and 0.5 in order to ensure that the PFU blocks do not become blurred in the next frame. Since this Partial Forced Update (PFU) constitutes a completely separate encoding step, the subsequent encoding of the same block in the following MC and DCT encoding steps may prevent blurring completely.

The PSNR degradation inflicted by this PFU technique under error-free conditions is portrayed in Figure 12.17, which appears particularly low as long as the proportion of PFU blocks is low. When a PFU rate of about 10%, corresponding to 40 blocks is applied, the PSNR degrades by about 1 dB at a concomitantly improved codec robustness. Observe furthermore in the figure that in case of a modest proportion of replenished blocks, perhaps up to 5%, the codec is able to reduce the MCER energy and hence a slightly improved PSNR performance is achieved. We made use of the same quantizer, which was invoked for the intra-frame coding, but the size of the updated blocks is fixed to 8×8 pixels. The number of updated blocks per frame and the leakage factor determine the inflicted intentional video impairment due to PFU and the codec's ability to recover from erroneous conditions, that is, the codec's robustness. In our experiments we found that the number of updated blocks per frame had to be below 25, as also evidenced by Figure 12.17.

12.9 The Gain/Cost-Controlled Inter-Frame Codec

In the previous sections, we have described the components of the coding scheme in Figure 12.15. This section reports on how we combined the introduced components in order to create a coding scheme for QCIF images at bit rates around 10 kb/s.

The motion detector attempts to minimize the MCER between the incoming and the local reconstructed frame buffers. As a result of the conclusions of Section 12.2.4, the search window is limited to a size of 4×4 pixels, which limits the search scope to -2 in the negative and $+1$ in the positive horizontal/vertical spatial directions. The motion detector stores the best motion vector for each of the 396 blocks, as well as the corresponding gain, as defined in Section 12.6. Now the bit-rate control unit in Figure 12.15 marks inefficient motion vectors as passive and determines the corresponding active/passive tables as described in Section 12.7, computing also the resulting overall bit-rate requirement for the MVs. The codec then estimates

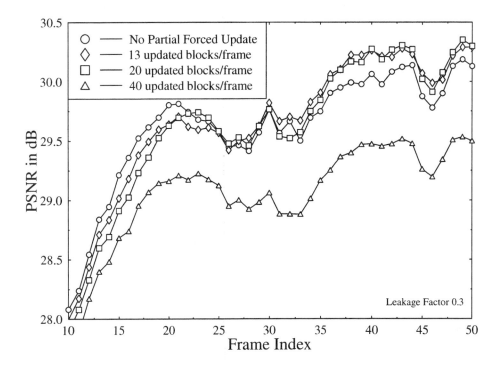

Figure 12.17: PSNR degradation as a consequence of invoking partial forced updates.

how many encoded MVs can be accommodated by the available bit-rate budget and relegates some of the active motion vectors to the motion-passive class. Then the bit-rate requirement is updated and compared to the bit-rate limit. This deactivation of motion vectors continues until we reach the predetermined bit rate for the MC.

Finally, motion compensation takes place for the motion-active blocks, while for the passive blocks simple frame-differencing is applied. The MCER is passed on to the DCT codec, which has a similar active/passive classification regime to that of the motion compensation scheme. Explicitly, each block is transformed to the frequency domain, quantized and then transformed back to the time domain in order to assess the potential benefit of the DCT-based coding in terms of overall PSNR or MSE contribution. The best set of quantizers is found by a full search — that is, by invoking all four available quantizers and evaluating their MSE performance. Now, the bit-rate control unit in Figure 12.15 determines the number of available bits for the DCT codec remaining from the total budget for the frame after reserving the required capacity for the partial forced update, the motion compensation, and the frame alignment word (FAW). The FAW is a 22-bit unique word, which allows the decoder to resynchronize at the commencement of the next video-frame using correlation techniques after losing synchronization. Again, the active/passive DCT tables are determined, and blocks attaining low MSE gains are not encoded in order to meet the overall bit-rate requirement. In order to emphasize the subjectively more important central eye and lip region of the screen, the codec incorporates the option of scaling the DCT gains and allows for the codec to gradually improve its image representation in its central section. This is particularly important when operating at 5–8 kb/s or during the first transmitted frames, when the codec builds up fine details, to reach its steady-state video quality, commencing from the coarsely quantized intraframe coded initial state.

12.9.1 Complexity Considerations and Reduction Techniques

The number of multiplications required for the direct discrete cosine transformation of the $b \times b$ matrix B is proportional to b^4 [26]. The dimensionality of the problem can be reduced to a more realistic complexity order, which is proportional to b^3 if we invoke the separable transform: $B_{dct} = TBT^t$, where T is the $b \times b$ unitary transformation matrix we defined by Equation 12.21 and T^t is the transpose of T. A single matrix multiplication of a $b \times b$ matrix requires $b^2(2b-1)$ floating point operations (Flops). In case of our 8×8 blocks ($b = 8$), 1920 Flops are necessary to evaluate one transformation. If we apply gain-controlled quad-class DCT, it is necessary to transform each block back to the time domain after choosing one of the four quantizers. If we apply the MSE criterion to select the best of the possible four quantizers, we require a total of $4b^2(2b-1) + 4(3b^2-1) \simeq 8b^3 + 12b^2$ Flops per block, which is the equivalent of 1.9 Mflops per QCIF frame. This can be reduced by about 50% upon exploiting some regularities in the matrix operations, by reducing the complexity to the order of $b^2 \log_2 b^2 = 384$ Flops per transformation, at the cost of increased data handling steps [463], or by using the so-called fast DCT [452]. This may be important for a direct implementation in silicon.

A much more efficient way of reducing the computational complexity becomes possible using a block classification algorithm prior to any encoding steps. Similar to our experience in Section 12.2.2.6 as regards MC, we found that a substantial proportion of the MCER contains flat blocks, which are unlikely to be selected by a gain-cost controlled codec. Hence, prior to the actual encoding, the encoder determines the energy contents of all blocks. Only the blocks exhibiting high energy content, that is, blocks carrying important information, are considered for the MC and DCT encoding operations. In case of the MC, the encoder first determines the frame difference, and then the motion prediction focuses on the locations where significant movement took place. After carrying out the MC step, a similar preclassification phase is invoked for the DCT. The degradation caused by the above suboptimum preselection phase is depicted in Figure 12.18 for a range of active block proportions between 20 and 50%. These results reveal that the complexity can be reduced by about 50% without noticeable PSNR impairment, when compared to the original scheme. When the proportion of active blocks is as low as 20%, the preclassification is more likely to be deceived than in the case of retaining a higher proportion of blocks. When assuming, for example, a 38% block activity rate, the codec exhibits a complexity of 2 Mflop per frame or 20 Mflops, which is only marginally higher than that of the standard so-called half-rate GSM speech codec, constituting a realistic real-time target at the time of writing.

12.10. THE BIT-ALLOCATION STRATEGY

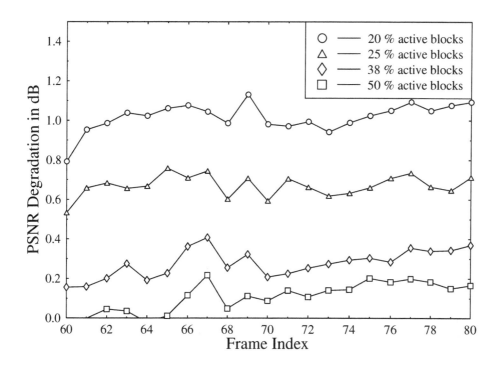

Figure 12.18: PSNR degradation caused by the preselection of active blocks in an attempt to reduce the codec's complexity.

12.10 The Bit-Allocation Strategy

Let us now consider the bit-allocation scheme of our codec portrayed in Figure 12.15. The bit-stream transmitted to the decoder consists of bits for:

- The block averages for the partial forced update.
- The table for the active MVs.
- The active MVs.
- The table for the active DCT blocks.
- The active DCT coefficients.

In our experiments we found that the best subjective and objective quality was achieved when the number of active blocks for the MC and DCT was roughly the same, although not necessarily the same blocks were processed by the two independent active/passive classification schemes. Since the 4-bit motion vectors require a lower channel capacity than the 12-bit quantized DCT coefficients, we decided to earmark between $\frac{1}{2}$ and $\frac{2}{3}$ of the available bit-rate budget to the DCT activity table and DCT coefficients, while the remaining bits were used for the MC and PFU. The PFU is usually configured to refresh 22 out of the 396 blocks in each frame. Therefore, $4 \times 22 = 88$ bits were reserved for the PFU. The actual number of encoded DCT blocks and MVs depends on the selected bit rate. It typically varied between 30 and 50 for bit rates between 8 and 12 kbps at a scanning rate of 10 frames/s.

The output of the codec contains two classes of bits — namely, the entropy-encoded MC and DCT activity tables on one hand, and the motion vectors, quantized DCT coefficients, and the partial forced updated blocks on the other. The first class of information, due to the encoding procedure, is extremely vulnerable against any corruption. A corrupted bit is likely to create a code associated with a different length, and as a result the entire frame may have to be dropped. Because this high vulnerability to channel errors is unacceptable in wireless applications, we also contrived another, more robust codec, which sacrifices coding efficiency and abandons the run-length coding concept for the sake of improved error resilience. A typical bit-allocation example for the variable-length compressed Codec I and for the more error-resilient Codec II is given in Table 12.4.

Codec	FAW	PFU	MV Ind.	MV	DCT Ind.	DCT	Total
Codec I	22	22×4	-	< 350	-	< 350	800
Codec II	22	22×4	30×9	30×4	30×9	30×12	1130

Table 12.4: Bit-Allocation of the Variable-Length Compressed Codec I at 8 kb/s and for the More Error Resilient Codec II at 11.3 kb/s

Later in this chapter, we will refer to the previously introduced codec discussed above as Codec I. In contrast, Codec II does not take advantage of the compression capabilities of run-length coding. In Codec II, we decided to transmit the index of each active DCT block and MV requiring 9 bits to identify one of the 396 indices using the enumerative method. The increased robustness of the codec is associated with an approximately 35% increased bit rate. As Figure 12.19 reveals, Codec I at 8 kb/s achieves a similar quality to that of Codec II at 11.3 kb/s.

12.11 Results

The performance of the variable-length compressed Codec I was tested at various bit rates in the range of 5 to 15 kbps. All PSNR versus frame index results presented in Figures 12.21 and 12.22 were obtained at a constant frame rate of 10 frames per second and without partial forced update. The results for the more error-resilient Codec II are similar at a 35% higher bit rate. A typical frame of the "Lab" sequence, which was one of our locally recorded high activity sequences, is depicted in Figure 12.20.

A frame rate of 10 frames/s is sufficiently high for head-and-shoulder images, as we do not expect very high motion activity. The codec builds up its steady-state video quality following the intra-frame initialization phase after about 15 frames or 1.5 seconds, which does not constitute an annoyance. The drop of the PSNR performance between frames 80 and 90 is due to the fact that "Miss America" moves her whole body, and, especially when operating at 5 kb/s, the codec cannot instantaneously

12.11. RESULTS

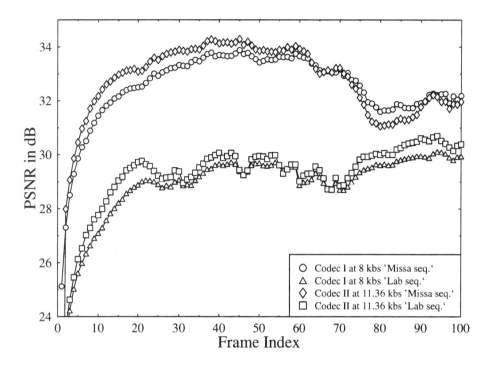

Figure 12.19: PSNR versus frame index performance for Codec I at 8 kb/s and Codec II at 11.3 kb/s.

Figure 12.20: Frame 87 of the "Lab" sequence, original (left) and 8kb/s DCT encoded (right). Both Sequences encoded at various bit rates can be viewed under the WWW address http://www-mobile.ecs.soton.ac.uk

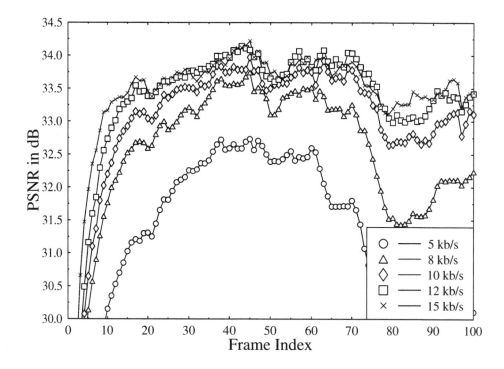

Figure 12.21: PSNR versus frame index performance of the "table-compressed" DCT Codec I at various bit rates for the "Miss America" sequence.

cope with the drastically increased motion activity. However, as soon as the motion activity surge decays, the PSNR curve recovers from its previous loss of quality. The fact that the image quality does not significantly increase, when using the higher bit rates of 15 kb/s or 12 kb/s compared to the 10 kb/s scenario, demonstrates how well the DCT scheme performs at low bit rates around 10 kb/s. The codec achieves a similar PSNR to the ITU H261 scheme presented in references [464] and [465], while operating at twice the frame rate and maintaining a similar bit rate, which implies a factor 2 higher compression ratio. The improved performance is attributable mainly to the intelligent cost-gain quantization invoked throughout the coding operation while maintaining a moderate complexity of about 20 Mflops.

12.12 DCT Codec Performance under Erroneous Conditions

The codecs presented here are suitable for a wide range of applications, including videotelephony over both fixed and wireless networks, surveillance and remote sensing etc. For example, mobile videophony over the widely spread Pan-European GSM system [41] at 13 kbps is an attractive application for the proposed codecs. Over

12.12. DCT CODEC PERFORMANCE UNDER ERRONEOUS CONDITIONS 513

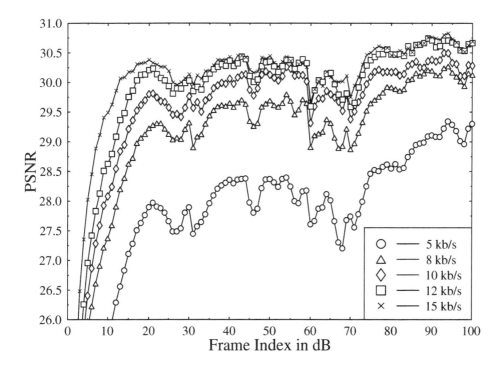

Figure 12.22: PSNR versus frame index performance DCT Codec I at various bit rates for the "Lab" sequence.

error-free transmission media, when, for example Automatic Repeat Request (ARQ) techniques can be invoked, the proposed codecs could be employed without PFU. Such a channel would provide the best performance and compression, since no bits are dedicated to PFU. However, for most practical channels the PFU must be invoked. In practical scenarios, it is essential to analyze the sensitivities of the various encoded bits to channel errors in order to be able to protect more vulnerable bits with stronger error correction codes and to design appropriate source-sensitivity matched error protection schemes.

12.12.1 Bit Sensitivity

Shannon's ideal source encoder is constituted by an ideal entropy codec, as argued in Chapter 1 on communications theory, where corrupting a single bit results in an undecodable codeword. The output bits of Shannon's ideal source codec are hence equally sensitive to bit errors. The proposed codecs still retain some residual redundancy and exhibit unequal bit sensitivities. In particular, the run-length (RL) encoded tables of Codec I are very sensitive to bit errors and need more protection than others. A common solution to this problem is to divide the bit-stream in two or more sensitivity classes and assign a different forward error correction (FEC) code to each of the classes. In case of Codec I, the sensitivity of the RL-coded bits is difficult

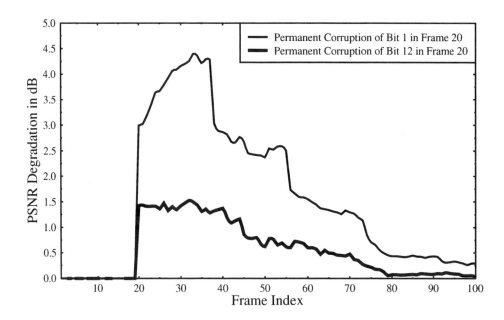

Figure 12.23: PSNR degradation profile for Bits 1 and 12 of the DCT Codecs I and II in Frame 2, where Bit 1 is the MSB of the PFU segment, while Bit 12 is a MV bit, as seen in Figure 12.24.

to quantify, for their corruption affects all other bits in the same transmission burst. Various procedures are available in order to quantify the bit sensitivity of the remaining non-RL encoded bits of Codec I and all bits of Codec II. In reference [466] a single bit of a video-coded frame was consistently corrupted, and the inflicted image peak signal-to-noise ratio (PSNR) degradation was observed. Repeating this method for all bits of a frame provided the required sensitivity figures, and on this basis bits having different sensitivities were assigned matching FEC codes. This technique does not take adequate account of the phenomenon of error propagation across image frame boundaries. Here we used the method suggested in reference [33], where each bit of the same type is corrupted in the current frame and the PSNR degradation for the consecutive frames due to the error event in the current frame is observed. As an example, Figure 12.23 depicts the PSNR degradation profile in case of corrupting Bit No. 1 of a transmission frame, namely the Most Significant Bit (MSB) of the PFU segment in frame 20 of the "Miss America" sequence, and Bit No. 12, one of the MV bits. The observable PSNR degradation over its whole duration is accumulated, averaged over all occurrences of that particular type of bit, and then weighted by the corresponding relative frequency of each bit under consideration in order to obtain the PSNR degradation sensitivity measures portrayed in Figure 12.24.

12.12. DCT CODEC PERFORMANCE UNDER ERRONEOUS CONDITIONS   515

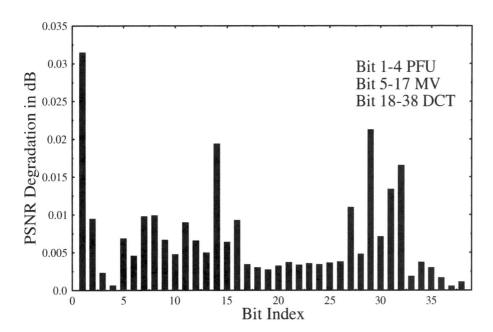

Figure 12.24: Integrated bit sensitivities for the error-resilient DCT Codec II.

12.12.2   Bit Sensitivity of Codec I and II

For Additive White Gaussian Noise (AWGN) channels we propose forward error correction (FEC). Both convolutional and block codes are feasible. Once the FEC scheme fails to remove the transmission errors, we have to differentiate between two possible error events. If the run-length encoded tables are corrupted, it is likely that a codeword of a different length is generated and the decoding process becomes impossible. This error is detectable because the decoded frame length and the preset frame length differ. Hence, a single bit error can force the decoder to drop an entire frame. If, however, one of the PFU, DCT, or MV bits is corrupted, the decoder is unable to determine that a corruption took place, but a maximum of two 8×8 blocks is affected by a single bit error. The difference in vulnerability for the run-length and non-run-length encoded bits is highlighted in Figure 12.25. If the whole bit-stream of Codec I is subjected to bit corruption, a BER of $2 \cdot 10^{-4}$ is sufficient to inflict unacceptable video degradation. If, however, the bit corruption only affects the non-run-length encoded bits, while the RL-coded bits remain intact, the codec can tolerate BERs up to $2 \cdot 10^{-3}$. In reference [467], we proposed an appropriate transmission scheme, which takes advantage of this characteristic. These schemes will be highlighted in Section 12.13. Another way of imposing the error resilience of Codec I would be the application of error resilient positional coding [462].

As evidenced by Figure 12.25, the absence of run-length encoded bits increases the error resilience of Codec II by an order of magnitude. Therefore, Codec II suits, for example, mobile applications over Rayleigh-fading channels. Further issues of

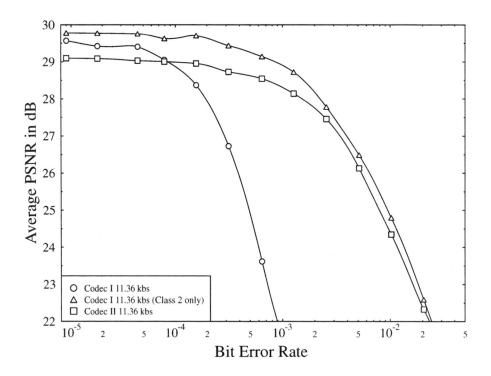

Figure 12.25: PSNR degradation versus BER for Codec I and II using random errors, demonstrating that the more robust class 2 bits of Codec I can tolerate a similar BER to Codec II.

unequal protection FEC and ARQ schemes were discussed in reference [467] and will be highlighted in Section 12.13. Having designed and analyzed our proposed DCT-based video codec, let us now consider some of the associated systems aspects in the context of a wireless videophone system.

12.13 DCT-Based Low-Rate Video Transceivers

12.13.1 Choice of Modem

The factors affecting the choice of modulation for a particular system were discussed in Chapter 5, and so we will not consider modulation aspects. Suffice to say here that the differentially coded noncoherent Star-QAM modems described in Chapter 5 exhibit typically lower complexity than their pilot symbol-assisted (PSA) coherent counterparts, but as we have shown, they inflict a characteristic 3 dB differential coding SNR penalty over AWGN channels, which also persists over Rayleigh channels. A further SNR penalty is imposed by the reduced minimum distance property of the rotationally symmetric differentially encoded StarQAM constellation. Hence, Star-QAM requires higher SNR and SIR values than the slightly more complex coherent PSA schemes.

12.13. DCT-BASED LOW-RATE VIDEO TRANSCEIVERS

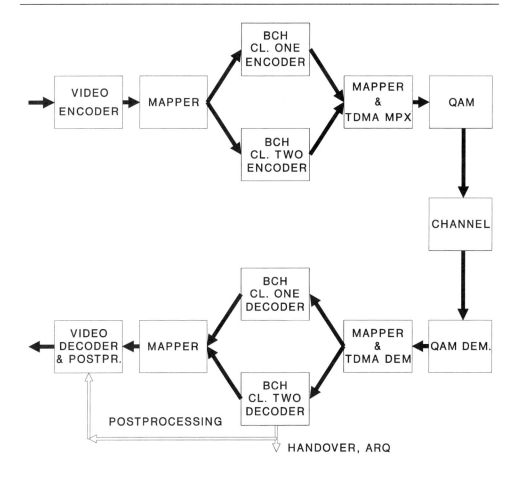

Figure 12.26: DCT-based videophone system schematic.

Therefore in our proposed video transceivers second-order switched-diversity aided coherent Pilot Symbol Assisted Modulation (PSAM) using the maximum-minimum distance square QAM constellation was employed.

12.13.2 Source-Matched Transceiver

12.13.2.1 System 1

12.13.2.1.1 System Concept The system's schematic is portrayed in Figure 12.26, where the source-encoded video bits generated by the video encoder are split in two bit-sensitivity classes, and sensitivity matched channel coding/modulation is invoked. This schematic will be detailed throughout this section. A variety of video codecs were employed in these investigations, which were loosely based on the candidate codecs presented in Table 12.4, but their bit-rates were slightly adjusted in order to better accommodate the bit-packing constraints of the binary BCH FEC codecs employed. The run length-compressed Codec I of Table 12.4 had a bit-rate of 8 kbps, and Codec

II was programmed to operate at 11.36 kbps, while a derivative of Codec I, Codec I-bis generated a rate of 8.52 kbps, as seen in Table 12.5, along with a range of other system features. The proposed system was designed for mobile videotelephony, and it had two different modes of operation, namely 4-level and 16-level quadrature amplitude modulation (QAM), which were the subject of Chapter 5. Our intention was to contrive a system, in which the more benign propagation environment of indoor cells would benefit from the prevailing higher signal-to-noise ratio (SNR) by using bandwidth-efficient 16QAM, thereby requiring only half the number of transmission packets compared to 4QAM. When the portable station (PS) is handed over to an outdoor microcell or roams in a lower SNR region toward the edge of a cell, the base station (BS) instructs the PS to lower its number of modulation levels to 4, in order to maintain an adequate robustness under lower SNR conditions. Let us now focus our attention on specific details of System 1, which was designed to accommodate the bit-stream of the error-resilient Codec II.

12.13.2.1.2 Sensitivity-Matched Modulation As shown in Figure 5.10, 16-level pilot symbol-assisted quadrature amplitude modulation (16-PSAQAM) provides two independent 2-bit subchannels having different bit error rates (BER). Specifically, the BER of the *higher integrity C1 subchannel* is a factor two to three times lower than that of *the lower quality C2 subchannel*. Both subchannels support the transmission of 2 bits per symbol. This implies that the 16-PSAQAM scheme inherently caters for sensitivity-matched protection, which can be fine-tuned using appropriate FEC codes to match the source codec's sensitivity requirements. This property is not retained by the 4QAM scheme, since it is a single-subchannel modem, but the required different protection for the source-coded bits can be ensured using appropriately matched channel codecs.

12.13.2.1.3 Source Sensitivity In order to find the appropriate FEC code for our video codec, its output stream was split in two equal sensitivity classes, namely, Class One and Class Two, according to our findings in Figure 12.24. Note that *the notation Class One and Two introduced here for the more and less sensitive video bits is different from the higher and lower integrity C1 and C2 modulation channels.* Then the PSNR degradation inflicted by both the Class One and Two video bits as well as the average PSNR degradation was evaluated for a range of BER values in Figure 12.27 using randomly distributed bit errors. These results showed that a lower BER was required by the Class One bits than by the Class Two bits. More explicitly, about a factor two lower BER was required by the Class One bits than by the Class Two bits in order for the video system to limit the PSNR degradations to 1–2 dB. These integrity requirements conveniently coincided with the integrity ratio of the C1 and C2 subchannels of our 16-PSAQAM modem, as seen in Figure 5.10. We therefore can apply the same FEC protection to both the Class One and Two video source bits and direct the Class One bits to the C1 16-PSAQAM subchannel, and the Class Two bits to the C2 subchannel.

12.13.2.1.4 Forward Error Correction Both convolutional and block codes can be successfully used over mobile radio links, but in our proposed scheme we

12.13. DCT-BASED LOW-RATE VIDEO TRANSCEIVERS

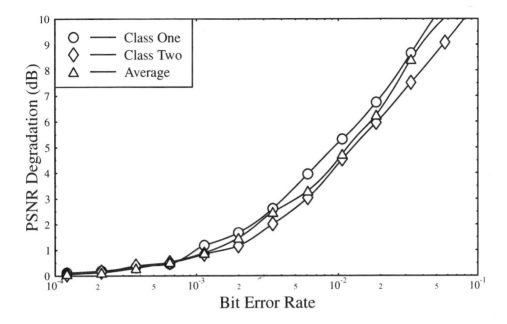

Figure 12.27: PSNR versus BER degradation of Codec II for Class One and Two bits.

have favored the binary Bose-Chaudhuri-Hocquenghem (BCH) codes introduced in Chapter 4. BCH codes combine a good burst-error correction capability with reliable error detection, a facility useful for invoking image post-enhancement in monitoring the channel's quality and in controlling handovers between traffic cells. The preferred R=71/127 ≈ 0.56-rate BCH(127,71,9) code can correct nine errors in a block of 127 bits, an error correction capability of about 7.1%. The number of channel-coded bits per image frame becomes $1136 \times 127/71 = 2032$, while the bit rate becomes 20.32 kbps at an image frame rate of 10 frames/s.

12.13.2.1.5 Transmission Format The transmission packets are constructed using one Class One BCH(127,71,9) codeword, and one Class Two BCH(127,71,9) codeword. A stronger BCH(127,50,13) codeword is allocated to the packet header, which conveys user-specific identification (ID) and dedicated control information, yielding a total of 381 bits per packet. In the case of 16QAM, these codewords are represented by 96 symbols, and after adding 11 pilot symbols using a pilot spacing of $P = 10$ as well as four ramp symbols to ensure smooth power amplifier ramping, the resulting 111-symbol packets are transmitted over the radio channel. Eight such packets represent a whole QCIF image frame; thus the signaling rate becomes 111 symb/12.5 ms ≈ 9 kBd. When using a time division multiple access (TDMA) channel bandwidth of 200 kHz, such as in the Pan-European second-generation mobile radio system known as GSM [468] and employing a Nyquist filtering (see Chapter 5) modulation excess bandwidth of 38.8%, the signaling rate becomes 144 kBd. This allows us to accommodate $144/9 = 16$ video users, or eight speech and eight video slots for videotelephony. The number of voice/videophone users accommodated in a bandwidth of 200 kHz then coincides with the number of full-rate speech users supported by the GSM system [468].

When the prevailing channel SNR does not allow 16QAM communications, 4QAM must be invoked. In this case, the 381-bit packets are represented by 191 two-bit symbols, and after adding 30 pilot symbols and four ramp symbols, the packet-length becomes 225 symb/12.5 ms, yielding a signaling rate of 18 kBd. In this case, the number of videophone users supported by System 1 becomes 8, as in the full-rate GSM speech channel which would be reduced to 4 for voice/video transmission. The system also facilitates mixed-mode operation, where 4QAM users must reserve two slots in each 12.5 ms TDMA frame toward the fringes of the cell, while in the central section of the cell, 16QAM users will require only one slot per frame in order to maximize the number of users supported. Assuming an equal proportion of 4QAM and 16QAM users, the average number of users per carrier becomes 12. The equivalent user bandwidth of the 4QAM PSs is $200 \text{ kHz}/8 = 25$ kHz, while that of the 16QAM users is $200 \text{ kHz}/16 = 12.5$ kHz.

For very high quality mobile channels or for conventional telephone lines, 64-QAM can be invoked, which further reduces the required bandwidth at the cost of a higher channel SNR demand. However, the packet format of this mode of operation is different from that of the 16QAM and 4QAM modes and so requires a different slot length. The 381-bit payload of the packet is represented by 64 six-bit symbols, and four ramp symbols are added along with 14 pilot symbols, which corresponds to a pilot spacing of $P = 5$. The resulting 82-symbol/12.5 ms packets are transmitted at a signaling rate of about 6.6 kBd, which allows us to host 22 videophone users. The equivalent user bandwidth becomes $200 \text{ kHz}/22 \approx 9.1$ kHz.

These features of the 16QAM/4QAM System 1, along with the characteristics of a range of other systems introduced in the next section, are summarized in Table 12.5.

Clearly, the required video signaling rate and bandwidth are comparable to those of most state-of-the-art mobile radio speech links, which renders our scheme attractive for mobile videotelephony in the framework of existing second- and forthcoming third-generation mobile radio systems, where an additional physical channel can be provided for the video stream. This rate can also be readily accommodated by conventional telephone subscriber loops or cordless telephone systems, such as the European DECT and CT2 or the Japanese PHS and the American PACS systems of Table 9.1 on page 343.

12.13.2.2 System 2

In this section we contrive various transceivers, which are summarized in Table 12.5, in order to expose the underlying system design tradeoffs. We again emphasize that the effects of transmission errors are particularly objectionable, if the run-length coded activity table bits of Table 12.4 are corrupted. Therefore, in System 2 of Table 12.5, which was designed to incorporate Codec I, the more sensitive run-length (RL) coded activity table bits of Table 12.4 are protected by the powerful binary Bose-Chaudhuri-Hocquenghem BCH(127,50,13) codec, while the less vulnerable remaining bits are

12.13. DCT-BASED LOW-RATE VIDEO TRANSCEIVERS

Feature	System 1	System 2	System 3	System 4	System 5
Video Codec	Codec II	Codec I-bis	Codec II	Codec I	Codec I
Video rate (kbps)	11.36	8.52	11.36	8	8
Frame Rate (fr/s)	10	10	10	10	10
C1 FEC	BCH(127,71,9)	BCH(127,50,13)	BCH(127,71,9)	BCH(127,50,13)	BCH(127,50,13)
C2 FEC	BCH(127,71,9)	BCH(127,92,5)	BCH(127,71,9)	BCH(127,50,13)	BCH(127,50,13)
Header FEC	BCH(127,50,13)	BCH(127,50,13)	BCH(127,50,13)	BCH(127,50,13)	BCH(127,50,13)
FEC-coded Rate (kbps)	20.32	15.24	20.32	20.32	20.32
Modem	4/16-PSAQAM	4/16-PSAQAM	4/16-PSAQAM	4/16-PSAQAM	4/16-PSAQAM
Re-transmitted	None	Cl. One	Cl. One & Two	Cl. One & Two	None
User Signal. Rate (kBd)	18 or 9	6.66	18 or 9	18 or 9	18 or 9
System Signal. Rate (kBd)	144	144	144	144	144
System Bandwidth (kHz)	200	200	200	200	200
No. of Video Users	8 or 16	(21-2)=19	6 or 14	6 or 14	8 or 16
Eff. User Bandwidth (kHz)	25 or 12.5	10.5	33.3 or 14.3	33.3 or 14.3	25 or 12.5
Min. AWGN SNR (dB)	7 or 15	15	6 or 13	5 or 11	8 or 12
Min. Rayleigh SNR (dB)	15 or 20	25	7 or 20	7 or 14	15 or 16

Table 12.5: Summary of DCT-based Videophone System Features. ∗ ARQs in Systems 2, 3, and 4 are Activated by Class One Bit Errors, but not by Class Two [467]. ©IEEE 1995, Hanzo, Streit

protected by the weaker BCH(127,92,5) code. Note that the overall coding rate of $R = (50 + 92)/(127 + 127) \approx 0.56$ is identical to that of System 1, but the RL-coded Class One video bits are more strongly protected. At a fixed coding rate this inevitably assumes a weaker code for the protection of the less vulnerable Class Two video bits. The 852 bits/100ms video frame is encoded using six pairs of such BCH code words, yielding a total of $6 \cdot 254 = 1524$ bits, which is equivalent to a bit rate of 15.24 kbps.

As in System 1, the more vulnerable run-length and BCH(127,50,13) coded Class One video bits are then transmitted over the higher integrity C1 16QAM subchannel. The less sensitive BCH(127,92,5) coded Class Two DCT coefficient bits are conveyed using the lower-integrity C2 16QAM subchannel. This arrangement is favored in order to further emphasize the integrity differences of the BCH codecs used, which is necessitated by the integrity requirements of the video bits.

The transmission burst is constructed by adding an additional BCH(127,50,13) codeword for the packet header conveying the user identifier (ID) as well as control information. The resulting 381 bits are again converted to 96 16QAM symbols, and 11 pilots as well as 4 ramp-symbols are added. In System 2 six such 96+11+4=111-symbol packets represent a video frame; hence, the single-user signaling rate becomes 666 symb/100 ms, which corresponds to 6.66 kBd. This allows us to accommodate now Integer[144 kBd/6.66]=21 such video users, if no time slots are reserved for packet re-transmissions. This number will have to be reduced in order to accommodate Automatic Repeat Requests (ARQs).

12.13.2.2.1 Automatic Repeat Request ARQ techniques have been successfully used in data communications [469–472] in order to render the bit- and frame-error rate arbitrarily low. However, because of their inherent delay and the additional requirement for a feedback channel required by message acknowledgments, they have not been employed in interactive speech or video communications. In state-of-the-art wireless systems, however, such as the forthcoming third-generation systems of Table 10.2 on page 394 there exists a full duplex control link between the BS and PS, which can be used for acknowledgments. The short TDMA frame length ensures a low packet delay and acknowledgment latency. Thus ARQ can be invoked.

In System 2, when the more powerful BCH codec conveying the more sensitive run-length coded Class One bits over the C1 16QAM subchannel is overloaded by channel errors, we re-transmit these bits only, using robust 4QAM. For the first transmission attempt (TX1), we use contention-free Time Division Multiple Access (TDMA). If an ARQ request occurs, the re-transmitted packets will have to vie for a number of earmarked time slots similarly to Packet Reservation Multiple Access (PRMA), which was introduced in Chapters 4 and 5. The intelligent base station (BS) detects these events of packet corruption and instructs the portable stations (PS) to re-transmit their packets during the slots dedicated to ARQ packets. Reserving slots for ARQ packets reduces the number of video users supported, depending on the prevailing channel conditions, as we will show in Section 12.14.

Although the probability of erroneous packets can be reduced by allowing repeated re-transmissions, there is a clear tradeoff between the number of maximum transmission attempts and the BCH-coded frame error rate (FER). In order to limit the num-

12.13. DCT-BASED LOW-RATE VIDEO TRANSCEIVERS

ber of slots required for ARQ attempts, which potentially reduce the number of video users supported, in System 2 we invoke ARQ only, if the more sensitive run-length coded Class One bits transmitted via the C1 16-PSAQAM channel and protected by the BCH(127,50,13) codec are corrupted. Furthermore, we re-transmit only the Class One bits, but in order to ensure a high success rate, we use 4-PSAQAM, which is more robust than 16-PSAQAM. Since only half of the information bits are re-transmitted, they can be accommodated within the same slot interval and same bandwidth, as the full packet. If there are only C2 bit errors in the packet, the corrupted received packet is not re-transmitted, which implies that typically there will be residual Class Two errors. In order to limit the number of slots dedicated to re-transmissions, we limited the number of transmission attempts to three, which implies that a minimum of two slots per frame must be reserved for ARQ. In order to maintain a low system complexity, we dispense with any contention mechanism and allocate two time slots to that particular user, whose packet was first corrupted within the TDMA frame. Further users cannot therefore invoke ARQ since there are no more unallocated slots. A further advantage is that in possession of three copies of the transmitted packet, majority decisions can be invoked as regards all video bits, if all three packets become corrupted. The basic features of System 2 designed to accommodate the run-length compressed Codec I-bis are also summarized in Table 12.5.

12.13.2.3 Systems 3–5

In order to explore the whole range of available system-design tradeoffs, we have contrived three further systems, — Systems 3–5 of Table 12.5.

System 3 of Table 12.5 uses the same video and FEC codecs as well as modems as System 1, but it allows a maximum of three transmission attempts in the case of C1 BCH(127,71,9) decoding errors. However, in contrast to System 2, which invokes the more robust 4QAM for ARQs, both the Class One and Two video bits are re-transmitted using the orignal modem mode. If there are only C2 errors, no ARQ is invoked. Employing ARQ in System 1 constitutes a further tradeoff in terms of reducing the number of subscribers supported by two, while potentially improving the communications quality at a certain BER or allowing an expansion of the range of operating channel SNRs toward lower values as seen in the last two rows of Table 12.5 for Systems 1 and 3.

System 4 employs the run-length coded source compression scheme referred to as Codec I, having a bit rate of 8 kbps or 800 bits per 100 ms video frame. This system follows the philosophy of System 3, but the Class One and Two video bits were protected by the more powerful BCH(127,50,13) code instead of the BCH(127,71,9) scheme as portrayed in row 5 of Table 12.5. The slightly reduced video rate of 8 kbps was imposed in order to accommodate the BCH(127,50,13) code in both 16QAM subchannels, while maintaining the same 20.32 kbps overall FEC-coded video rate, as Systems 1 and 3. System 4 will allow us to assess whether it is a worthwhile complexity investment to introduce run-length coding in the slightly higher-rate, but more error-resilient, Codec II in order to reduce the source bit rate and whether the increased error sensitivity of the lower-rate Codec I can be compensated for by accommodating the more complex and more powerful BCH(127,50,13) codec.

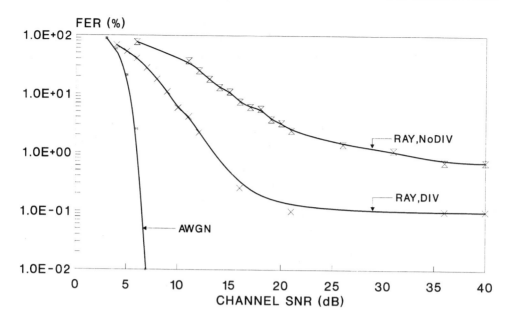

Figure 12.28: 4QAM BCH(127,71,9) FER versus channel SNR performance of the 18 kBd mode of operation of the DCT codec based System 1 of Table 12.5 over AWGN and Rayleigh (RAY) channels with diversity (DIV) and without diversity (NoDIV) [467]. ©IEEE 1995, Hanzo, Streit.

In order to maximize the number of video subscribers supported, the performance of System 4 can also be studied without ARQ techniques. We will refer to this scheme as **System 5**. Again, these system features are summarized in Table 12.5. Having designed the video transceivers, we present their performance results in the next section.

12.14 System Performance

12.14.1 Performance of System 1

In our experiments the signaling rate was 144 kBd, while the propagation frequency and the vehicular speed were 1.8 GHz and 30 mph, respectively. For pedestrian speeds, the fading envelope fluctuates less dramatically, and so our experimental conditions constitute a GSM-like urban worst-case scenario.

Here we characterize the performance of the transceiver in terms of the BCH(127,71,9) coded Frame Error Rate (FER) versus channel Signal-to-Noise Ratio (SNR), as portrayed in Figure 12.28 and 12.29 in the case of the 4QAM and 16QAM modes of operation of System 1. In these figures we displayed the FER over both AWGN and Rayleigh channels, in case of the Rayleigh channel both with and without diversity. Note that for near-unimpaired video quality, the FER must be below 1% but preferably below 0.1%. This requirement is satisfied over AWGN channels for SNRs in

12.14. SYSTEM PERFORMANCE

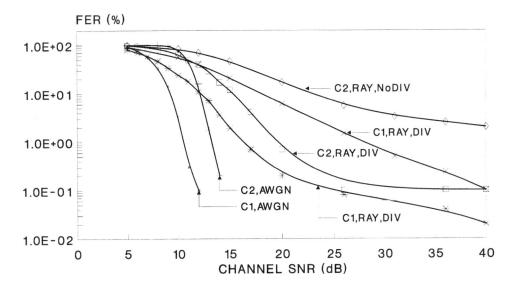

Figure 12.29: 16QAM C1 and C2 BCH(127,71,9) FER versus channel SNR performance of the 9 kBd mode of operation of the DCT codec based System 1 over AWGN and Rayleigh (RAY) channels with diversity (DIV) and without diversity (NoDIV) [467]. ©IEEE 1995, Hanzo, Streit.

excess of about 7 dB for 4QAM. In the case of 16QAM and AWGN channels, the C1 and C2 FERs are reduced to about 0.1% for SNRs above 13 and 15 dB, respectively. Observe in the figures that over Rayleigh-fading (RAY) channels with diversity (DIV) the corresponding FER values are increased to about 15 dB for 4QAM and 20 dB for 16QAM, while without diversity (NoDIV) further increased SNR values are necessitated.

The overall video PSNR versus channel SNR (ChSNR) performance of System 1 is shown in Figures 12.30 and 12.31 for the 4QAM and 16QAM modes of operation, respectively. The PSNR versus ChSNR characteristics of the 6.6 kBd 64QAM arrangement are also given for the sake of completeness in Figure 12.32. Observe in the above PSNR versus channel SNR figures that the AWGN performance was also evaluated without forward error correction (FEC) coding in order to indicate the expected performance in a conventional AWGN environment, such as telephone or satellite channels without FEC coding.

Because of its limited bandwidth efficiency gain, high SNR requirement, and incompatible slot structure, we recommend the 64QAM system only for those applications, where the bandwidth is at absolute premium, and in our further discourse we favor the 16QAM/4QAM modes of System 1. The corresponding figures suggest that best performance was achieved over AWGN channels with FEC, requiring a channel SNR (ChSNR) of about 15 and 7 dB in the case of the 16QAM and 4QAM modes of operation, respectively, in order to achieve an unimpaired image quality associated with a PSNR value of about 34 dB. Without FEC coding over AWGN channels, these SNR values had to be increased to about 20 and 12 dB, respectively. Over Rayleigh

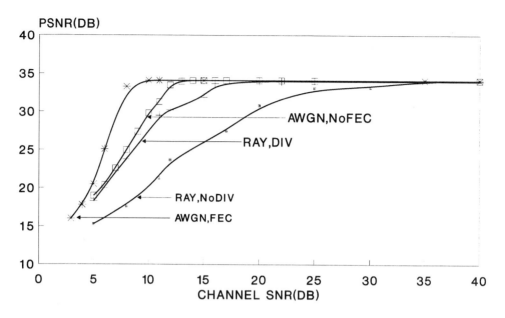

Figure 12.30: 4QAM PSNR versus channel SNR performance of the DCT codec based System 1 in its 18 kBd mode of operation over various channels [467]. ©IEEE 1995, Hanzo, Streit.

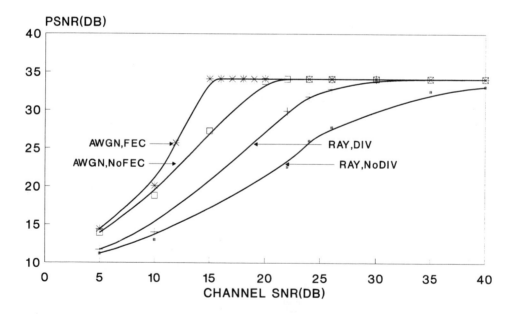

Figure 12.31: 16QAM PSNR versus channel SNR performance of the DCT codec-based System 1 of Table 12.5 in its 9 kBd mode of operation over various channels [467]. ©IEEE 1995, Hanzo, Streit.

12.14. SYSTEM PERFORMANCE

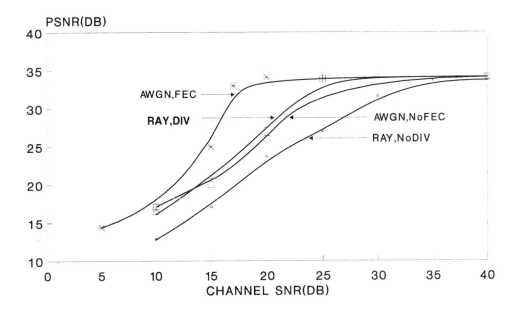

Figure 12.32: 64QAM PSNR versus channel SNR performance of the DCT codec-based System 1 of Table 12.5 in its 6.6 kBd mode of operation over various channels [467]. ©IEEE 1995, Hanzo, Streit.

channels with second-order diversity, the system required ChSNR values of about 15 and 25 dB in the 4QAM and 16QAM modes in order to reach an image PSNR within 1 dB of its unimpaired value of 34 dB. This 1 dB PSNR degradation threshold will be used in all scenarios for characterizing the near-unimpaired image quality. Finally, without diversity over Rayleigh channels, SNRs of about 25 and 33 dB were needed for near-unimpaired PSNR performance in the 4QAM and 16QAM modes.

12.14.2 Performance of System 2

12.14.2.1 FER Performance

In order to evaluate the overall video performance of System 2, 100 frames of the "Miss America" (MA) sequence were encoded and transmitted over both the best-case additive white Gaussian noise (AWGN) channel and the worst-case narrowband Rayleigh-fading channel. The BCH(127,50,13) and BCH(127,92,5) decoded frame error rate (FER) was evaluated for both the C1 and C2 bits after the first transmission attempt (TX1) over AWGN and Rayleigh channels with and without second-order diversity, as seen in Figures 12.33–12.35. These figures also portray the C1 FER after the second (TX2) and third (TX3) transmission attempts, which were carried out using 4QAM, in order to maximize the success rate of the C1 bits, representing the vulnerable run-length coded activity table encoding bits of Table 12.4.

Over AWGN channels, a C1 FER of less than 1% can be maintained for channel SNRs in excess of about 5 dB, if three transmission attempts are allowed, although at

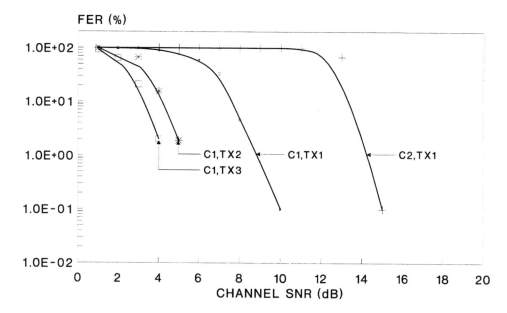

Figure 12.33: BCH(127,50,13) and BCH(127,92,5) FER versus channel SNR performance of the DCT codec-based System 2 of Table 12.5 over AWGN channels [467]. ©IEEE 1995, Hanzo, Streit.

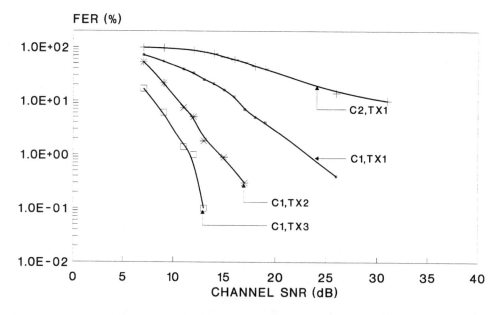

Figure 12.34: BCH(127,50,13) and BCH(127,92,5) FER versus channel SNR performance of the DCT codec-based System 2 of Table 12.5 over Rayleigh channels without diversity [467]. ©IEEE 1995, Hanzo, Streit.

12.14. SYSTEM PERFORMANCE

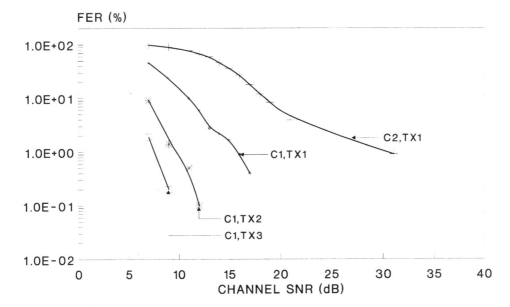

Figure 12.35: BCH(127,50,13) and BCH(127,92,5) FER versus channel SNR performance of the DCT codec-based System 2 of Table 12.5 over Rayleigh channels with diversity [467]. ©IEEE 1995, Hanzo, Streit.

such low SNRs the C2 bit errors are inflicting an unacceptably high video degradation. The corresponding C2 FER over AWGN channels becomes sufficiently low for channel SNRs above about 14–15 dB in order to guarantee unimpaired video communications, which is significantly higher than that required by the C1 subchannel. This ensures that under normal operating conditions the C1 bits are never erroneous. Over the Rayleigh channel, but without diversity, a channel SNR of about 12 dB was required with a maximum of three transmissions in order to reduce the C1 FER below 1% or FER = 10^{-2}, as shown in Figure 12.34. However, the C2 FER curved flattened out for high SNR values, which resulted in a severe "leakage" of erroneous C2 bits, and this yielded a somewhat impaired video performance. When diversity reception was used, the minimum required SNR value necessary to maintain a similar C1 FER was reduced to around 10 dB, while the C2 FER became adequately low for SNRs in excess of about 20–25 dB, as demonstrated by Figure 12.35.

12.14.2.2 Slot Occupancy Performance

As mentioned earlier, the ARQ attempts require a number of reserved time slots for which the re-transmitting MSs have to contend. When the channel SNR is too low, there is a high number of re-transmitted packets contending for too low a number of slots. The slot occupancy increase versus channel SNR performance, which was defined as the percentage of excess transmitted packets, is portrayed for a range of scenarios in Figure 12.36. For SNR values in excess of about 10, 15, and 25 dB, when using 16QAM over AWGN as well as Rayleigh channels with and without diversity,

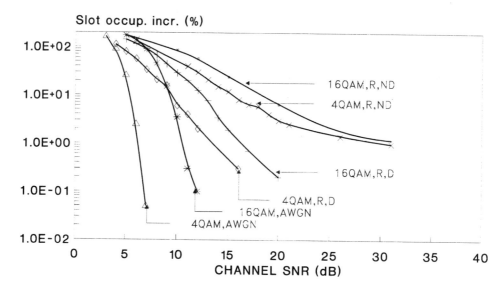

Figure 12.36: Slot occupancy increase versus channel SNR performance of the proposed DCT-based video transceivers of Table 12.5 over AWGN and Rayleigh channels with (D) and without (ND) diversity [467]. ©IEEE 1995, Hanzo, Streit.

respectively, the slot occupancy was increased due to re-transmissions only marginally. Therefore, reserving two time slots per frame for a maximum of two re-transmission attempts ensures a very low probability of packet collision during ARQ operations, while reducing the number of subscribers supported by two. In a simplistic approach, this would imply that for a channel SNR value, where the FER is below 1% and assuming 20 users, the reserved ARQ slots will be occupied only in about every fifth frame. However, we cannot earmark fewer than two slots for two additional transmission attempts. The 4QAM slot occupancy is even lower at a given channel SNR than that of the 16QAM schemes, as suggested by Figure 12.36.

12.14.2.3 PSNR Performance

The PSNR versus ChSNR performance of System 2 is characterized by Figure 12.37. Observe that over AWGN channels ChSNR values in excess of 15 dB are required for unimpaired video performance. Over Rayleigh channels with diversity, about 20 dB ChSNR is necessitated for an unimpaired PSNR performance, while without diversity the PSNR performance seriously suffers from the leaking Class Two bit errors. Overall, the 6.6 kBd System 2 has a lower robustness than the 9 kBd System 1, since its behavior is predetermined by the initially transmitted and somewhat impaired Class Two video bits, which were protected by the weaker BCH(127,92,5) code. Recall that System 1 used the BCH(127,71,9) code in both the C1 and C2 subchannels. In fact, the performance of the 6.6 kBd System 2 is more similar to that of the 6.6 kBd 64QAM system characterized in Figure 12.32, which does not use ARQ. We round that re-transmission attempts invoked in order to improve the

12.14. SYSTEM PERFORMANCE

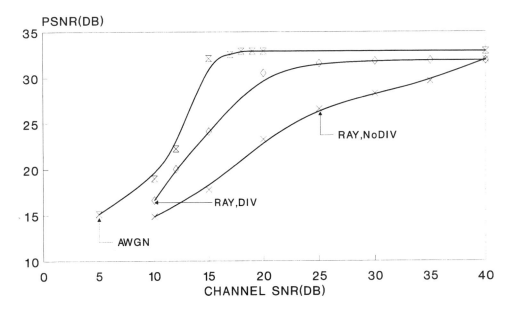

Figure 12.37: PSNR versus channel SNR performance of the 6.6 kBd DCT-based video System 2 over various channels [467]. ©IEEE 1995, Hanzo, Streit.

integrity of the initially received Class One bits only and to ensure an adequate integrity for these vulnerable bits, but without enhancing the quality of the initial 16QAM C2 subchannel failed. In other words, System 2 cannot outperform System 1. Furthermore, System 2 is inherently more complex than System 1 and only marginally more bandwidth efficient. Therefore, in contriving the remaining systems we set out to improve the noted deficiencies of System 2.

12.14.3 Performance of Systems 3-5

Our experience with System 2 suggested that it was necessary to re-transmit both the Class One and Two bits if the overloading of the C1 FEC codec indicated poor channel conditions. This plausible hypothesis was verified using System 3, which is effectively the ARQ-assisted System 1. This allowed us to assess the potential benefit of ARQs in terms of the minimum required channel SNR, while its advantages in terms of FER reduction were portrayed in Figures 12.33–12.35. The corresponding PSNR curves of System 3 are plotted in Figures 12.38 and 12.39 for its 18 kBd 4QAM and 9 kBd 16QAM modes, respectively. Re-transmission was invoked only if the C1 FEC decoder was overloaded by a preponderance of channel errors, but in these cases both the C1 and C2 subchannels were re-transmitted. Comparison with Figures 12.30 and 12.31 revealed substantial ChSNR reduction over Rayleigh channels, in particular without diversity. This was because, in case of a BCH frame error, by the time of the second or third transmission attempt, the channel typically emerged from a deep fade. Over AWGN channels, the channel conditions during any further ARQ attempts were

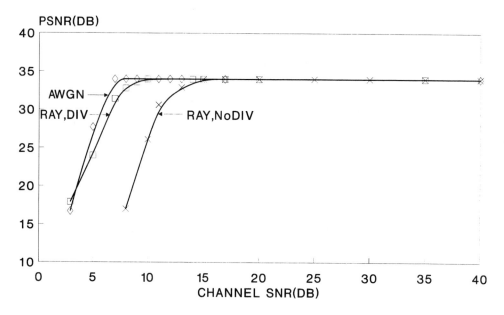

Figure 12.38: PSNR versus channel SNR performance of the 18 kBd 4QAM mode of the DCT codec-based System 3 over various channels using three transmission attempts [467]. ©IEEE 1995, Hanzo, Streit.

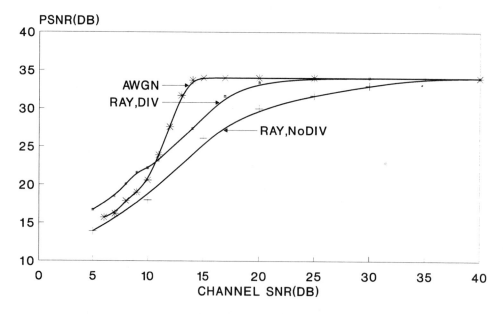

Figure 12.39: PSNR versus channel SNR performance of the 9 kBd 16QAM mode of the DCT codec-based System 3 using three transmission attempts over various channels [467]. ©IEEE 1995, Hanzo, Streit.

12.14. SYSTEM PERFORMANCE

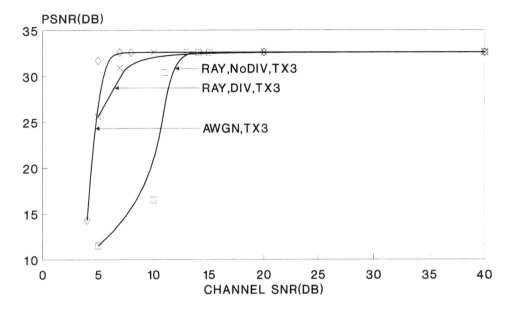

Figure 12.40: PSNR versus channel SNR performance of the 18 kBd 4QAM mode of the DCT codec-based System 4 using three transmission attempts over various channels [467]. ©IEEE 1995, Hanzo, Streit.

similar to those during the previous ones. Hence, ARQ offered more limited ChSNR reduction. The minimum required ChSNR values for the 4QAM mode over AWGN and Rayleigh channels with and without diversity are 7, 8, and 13 dB, while value for the 16QAM mode are 13, 18, and 27 dB, respectively, as also shown in Table 12.5.

Employment of ARQ was more crucial in System 4 than in System 3, because the corrupted run-length coded activity tables of the video codecs of Table 12.4 would inflict severe quality degradations for the whole video frame. The corresponding PSNR curves are portrayed for the 18 and 9 kBd 4QAM and 16QAM operating modes in Figures 12.40 and 12.41 over various channels, which can be contrasted with the results shown for System 5 without ARQ in Figures 12.42 and 12.43. Again, over Rayleigh channels the ChSNR requirement reductions due to ARQ are substantial, in particular without diversity, where the received signal typically emerges from a fade by the time ARQ takes place. Over AWGN channels the benefits of ARQ are less dramatic but still significant. This is because during re-transmission each packet faces similar propagation conditions, as during its first transmission. The required ChSNR thresholds for near-perfect image reconstruction in the 4QAM mode of System 4 are about 6, 8, and 12 dB over AWGN and Rayleigh channels with and without diversity, which are increased to 11, 14, and 27 dB in the 16QAM mode. In contrast, System 5 necessitates ChSNRs of 8, 13, and 25 dB, as well as 12, 16 and 27 dB under the previously stated conditions over 4QAM and 16QAM, respectively. The required ChSNR values are summarized in Table 12.5.

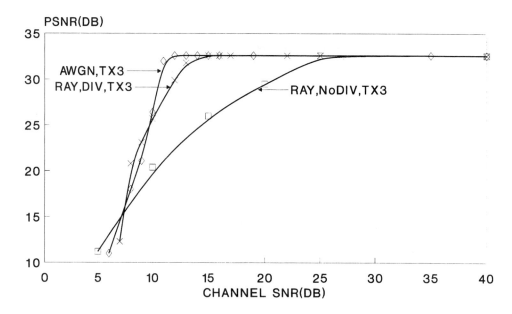

Figure 12.41: PSNR versus channel SNR performance of the 9 kBd 16QAM mode of the DCT codec-based System 4 using three transmission attempts over various channels [467]. ©IEEE 1995, Hanzo, Streit.

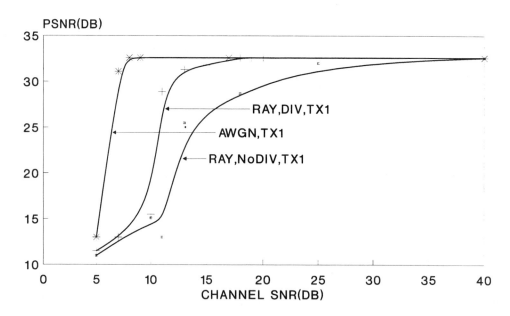

Figure 12.42: PSNR versus channel SNR performance of the 18 kBd 4QAM mode of the DCT codec-based System 5 over various channels [467]. ©IEEE 1995, Hanzo, Streit.

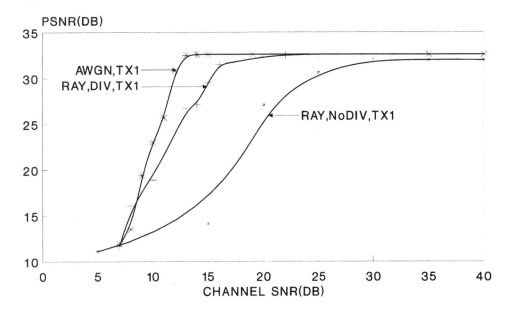

Figure 12.43: PSNR versus channel SNR performance of the 9 kBd 16QAM mode of the DCT codec-based System 5 over various channels [467]. ©IEEE 1995, Hanzo, Streit.

12.15 Summary and Conclusions

In this chapter, initially DCT-based QCIF video codecs were designed, which were then incorporated into various videophone transceivers. A range of bandwidth-efficient, fixed-rate mobile videophone transceivers have been presented, which retain the features summarized in Table 12.5. The video source rate can be fixed to any arbitrary value in order to be able to accommodate the videophone signal by conventional second generation (2G) mobile radio speech channels, such as, for example, that of the Pan-European GSM system [468], the Japanese PDC [473], and the American IS-54 [295] as well as IS-95 systems [296] of Table 9.1 on page 343 at bit rates between 6.7 and 13 kbps. The emerging third-generation (3G) systems of Table 10.2 on page 394 are more flexible than their 2G counterparts, and so they are more amenable to videotelephony.

In System 1 the 11.36 kbps Codec II of Table 12.4 was used, which has a lower rate than the 13 kbps speech rate of the GSM system. After BCH(127,71,8) coding, the channel rate becomes 20.32 kbps. When using an adaptive transceiver, which can invoke 16QAM and 4QAM depending on the channel conditions experienced, the signaling rate becomes 9 and 18 kBd, respectively. Accordingly, 16 or 8 videophone users can be accommodated in the GSM bandwidth of 200 kHz, which implies video user bandwidths of 12.5 and 25 kHz, respectively. Over line-of-sight AWGN channels, SNR values of about 15 and 7 dB are required when using 16QAM and 4QAM, respectively, in order to maintain unimpaired PSNR values of about 34 dB. An increased channel SNR of about 20 and 15 dB is needed over the diversity-assisted Rayleigh scenario.

In System 2 we have opted for the 8.52 kbps Codec I-bis videophone scheme, maintaining a PSNR of about 33 dB for the "Miss America" sequence. The source-coded bit-stream was sensitivity-matched binary BCH(127,50,13) and BCH(127,92,5) coded and transmitted using pilot assisted 16QAM. Because of the lower source-coded rate of 8.52 kbps of Codec I-bis, the single-user signaling rate of System 2 was reduced to 6.66 kBd, allowing us to accommodate $21 - 2 = 19$ videotelephone users in the 200 kHz GSM bandwidth. If the signal-to-interference ratio (SIR) and signal-to-noise ratio (SNR) values are in excess of about 15 dB, and 25 dB over the AWGN and diversity-assisted nondispersive Rayleigh-fading channels, respectively, pleasant videophone quality is maintained. The implementation complexity of System 1 is lower than that of System 2, while System 2 can accommodate more users, although it is less robust, as also demonstrated by Table 12.5. This is because only the more vulnerable C1 bits are re-transmitted.

On the basis of our experience with System 2, the fully ARQ-assisted System 3 was contrived, which provided a better image quality and a higher robustness, but was slightly less bandwidth-efficient than Systems 1 and 2 due to reserving two time slots for ARQ. Furthermore, the question arose as to whether it was better to use the more vulnerable run-length coded but reduced-rate Codec I of Table 12.4 with stronger and more complex FEC protection, as in Systems 4 and 5, or to invoke the slightly higher rate Codec II with its weaker and less complex FEC. In terms of robustness, System 5 proved somewhat more attractive than System 4, although the performances of the non-ARQ-based System 1 and System 5 are rather similar.

Overall, using schemes similar to the proposed ones, mobile videotelephony is realistic over existing mobile speech links, such as the Pan-European GSM system [468], the Japanese PDC [473], and the American IS-54 [295], as well as IS-95 systems [296] of Table 9.1 on page 343 at bit rates between 6.7 and 13 kbps. Similar schemes can be contrived for the 3G systems of Table 10.2 on page 394. Our future work in this field will be targeted at improving the complexity/quality balance of the proposed schemes using a variety of other video codecs, such as parametrically assisted quadtree and vector quantized codecs. A further important research area to be addressed is devising reliable transceiver reconfiguration algorithms. In the next chapter, we examine a range of vector-quantized video codecs and videophone systems.

Chapter 13

Very Low Bit-Rate VQ Codecs and Multimode Video Transceivers

13.1 Introduction

Vector quantization (VQ) is a generalization of scalar quantization [474]. Extensive studies of vector quantizers have been performed by many researchers using VQ for analysis purposes [475,476] or for image compression. In the latter field of applications various advances, such as adaptive [477–479], multistage [480–483], or hybrid coding [484–486], have been suggested. In contrast to these contributions, we will show that VQ may be rendered attractive even when using large vectors (Section 13.2) and small codebooks (< 512) in the codec's inter-frame mode. Furthermore, we will compare the computationally demanding full search VQ with the mean-shape gain VQ to be highlighted in Section 13.3.1, with adaptive VQ in Section 13.3.2, and with classified VQ in Section 13.3.3. Finally, we propose complexity reduction techniques and focus on bit-sensitivity issues. The chapter concludes with the system-design study of a VQ-codec based videophone system similar to the DCT-based system of the previous chapter. In this chapter we adopt a practically motivated approach to vector quantization; for a deeper exposure to VQ principles, the monograph by Gersho and Gray [474] is recommended.

13.2 The Codebook Design

The codebook design is a crucial issue for every VQ design. Unfortunately, there is no practical algorithm that leads to the optimal codebook C for a given training sequence S. Practical algorithms are usually based on a two-stage method. First, an initial codebook is derived, which is then improved by a second, so-called generalized

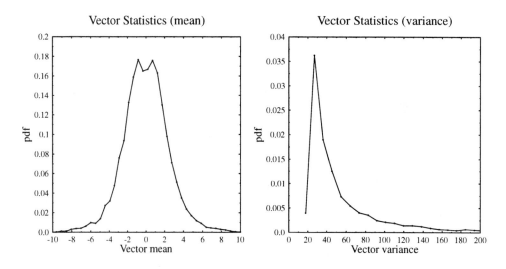

Figure 13.1: Statistical analysis of the training sequence.

Max-Lloyd algorithm [474].

The choice of the VQ training sequence is also critical, since it should contain video data, which are characteristic of a wide range of inputs. In our experiments we calculated the Displaced Frame Difference (DFD) signal for various input sequences, while only allowing for 35 active motion vectors out of the 396 8 × 8 blocks of a Quarter Common Intermediate Format (QCIF) frame to be generated. This implies that the VQ will be used in very low bit-rate image coding, where an active/passive block classification is necessary. The DFD energy was analyzed, and the 35 blocks containing the highest energy were copied into the training sequence. The size of the training sequence was chosen to be sufficiently extensive in order to guarantee a sufficient statistical independence between the codebook and the training sequence. The statistical properties of the training sequence are portrayed in terms of the Probability Density Function (PDF) of its mean and variance in figure 13.1. The mean and variance are highly peaked between [-2...2] and [20...40], respectively. Note in the figure that the narrow, but highly peaked, PDF of the DFD mean underlines the efficiency of MC, since in the vast majority of cases its mean is close to zero and its variance implies a low DFD energy.

Initially the codebook was designed using the so-called 'pruning' method [474], where a small codebook was derived from a large codebook or from a training sequence. The algorithm was initialized with a single-vector codebook. Then the vectors of the training sequence were compared to those in the codebook using the component-wise squared and summed distance as a similarity criterion. If this distance term exceeds a given threshold, the vector is classified as new and accordingly copied into the codebook. Hence, the threshold controls the size of the codebook, which contains all "new" vectors. In our case this method did not result in an adequate codebook, since the performance of a large codebook of 512 8 × 8-pixel entries, which had an unacceptably high complexity, was found to be poor.

13.2. THE CODEBOOK DESIGN

Algorithm 3 *The pairwise nearest neighbor algorithm summarized here reduces the initial training set to the required codebook size step by step, until a desired size is achieved.*

1. Assign each training vector to a so-called cluster of vectors.

2. Evaluate the potential distortion penalty associated with tentatively merging each possible pairs of clusters.

3. Carry out the actual merging of that specific pair of clusters, which inflicts the lowest excess distortion due to merging.

4. Repeat Steps 2 and 3, until the required codeboook size is arrived at.

In a second attempt, we invoked the pairwise nearest neighbor (PNN) algorithm [474]. This approach shrinks a given codebook step-by-step until a desired size is achieved. Initially, each training vector is assigned to a separate cluster, which results in as many clusters as the number of training vectors. Then for two candidate clusters the distortion penalty incurred by merging these two clusters is determined. This is carried out for all possible pairs of clusters, and finally the pair with the minimum distortion penalty is merged with a single cluster. This process continues until the codebook is shrunk to the desired size. The algorithm's complexity increases drastically with the codebook size. This technique became impractical for our large training set. We therefore simplified this algorithm by limiting the number of tentative cluster combinations as follows. Instead of attempting to merge each possible pair of clusters, where the number of combinations exhibits a quadratic expansion with increasing codebook size, in our approach a single cluster is preselected and tentatively combined with all the others in order to find the merging pair associated with the lowest distortion. The PNN procedure is summarized in Algorithm 3. Following this technique, the codebook size was initially reduced, and then the full PNN algorithm was invoked in order to create a range of codebooks having sizes between 4 and 1024. In the second step of the codebook generation, we used the generalized Max-Lloyd algorithm [474] instead of the PNN algorithm in order to enhance our initial codebooks, which resulted in a slight reduction of the average codebook distortion. This underlines the fact that our initial codebooks used by the PNN algorithm were adequate and the suboptimum two-stage approach caused only a negligible performance loss. As an example, our sixteen-entry codebook constituted by 8×8 vectors is depicted in Figure 13.2, where the vector components were shifted by 127 and multiplied by 30 in order to visually emphasize their differences. The corresponding 128-entry codebook is portrayed in Figure 13.3.

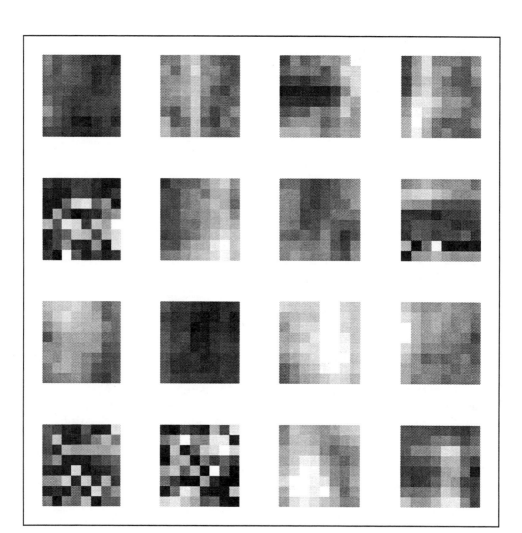

Figure 13.2: Enhanced sample codebook with sixteen 8×8 vectors.

13.3. THE VECTOR QUANTIZER DESIGN

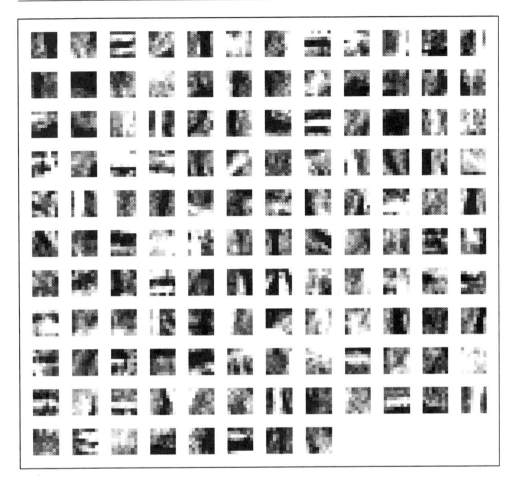

Figure 13.3: Enhanced sample codebook with 127 8 × 8 vectors.

13.3 The Vector Quantizer Design

After creating codebooks having a variety of sizes, we had to find the best design for the vector quantizer itself. The basic concept of the image codec is the same as that described in Section 12.4, which is depicted in Figure 13.4.

Since the first frame used in inter-frame coding is unknown to the decoder, we initialize the buffers of the local and remote decoder with the coarsely quantized intracoded frame we defined in Section 12.5. Once switched to the inter-frame mode, the optional PFU of Section 12.8 may periodically update the local and remote buffers, before the gain-controlled motion compensation of Section 12.6 takes place. The resulting MCER is then passed to the VQ codec. The VQ determines the closest codebook entry for each of the 396 8 × 8 blocks in the MCER frame and calculates

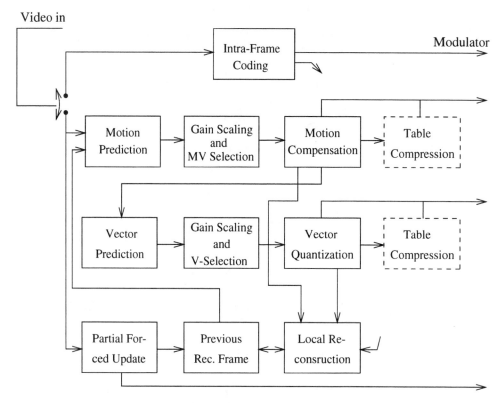

Figure 13.4: Basic schematic of the VQ codec.

their potential encoding gain, which is defined in Equation 13.1:

$$VQ_{gain} = \sum_{i=1}^{8}\sum_{j=1}^{8}(b_n(i,j))^2 - \sum_{i=1}^{8}\sum_{j=1}^{8}(b_n(i,j) - C_m(i,j))^2, \quad (13.1)$$

where b_n denotes the block to be encoded and C_m denotes the best codebook match for the block b_n. A second gain- and bit-rate-controlled algorithm decides which of the VQ blocks are active. The VQ indices are stored and transmitted using the activity table concept of Section 12.7. Since we experienced similar VQ block activities to those of the DCT-based codec in Section 12.7, we were able to use the same Huffman codes for the active/passive tables. The VQ frame was also locally decoded and fed back to the motion compensation scheme for future predictions.

As mentioned earlier, a coarsely quantized intra-frame is transmitted at the beginning of the encoding procedure and at drastic changes of scene as highlighted in Section 12.5. Recall that the DCT codec of Section 12.4 used a 12 bit per encoded 8×8 block. Furthermore, we found that in the context of the DCT-based codec of Section 12.4 the reconstructed 8×8 block means could range up to +/- 40. By contrast, the block means represented in the VQ 256-codebook, for example, covers only the interval of $[-4.55...6.34]$. This explains why we sometimes experience a better

13.3. THE VECTOR QUANTIZER DESIGN

Figure 13.5: Subjective performance of the VQ at 8 kb/s, 256-codebook, 10 frames/s, frame 87 of the "Lab" sequence, PSNR 30.33 dB; left: original; right: frame coded using 800 bits. A range of video sequences encoded at various bit rates can be viewed under the WWW address http://www-mobile.ecs.soton.ac.uk

PSNR performance when the PFU is employed, for it helps to restore the original block means.

Figure 13.5 demonstrates the impressive performance at 8 kb/s and 10 frames/s, but it also reveals granular noise around the mouth and some blockiness in the smooth regions of the background. The codec's failure in the mouth region can be explained by the fact that at this particular location new information is introduced and the MC fails. The following VQ fails to reconstruct the lip movement and results in a delayed motion of the mouth. Surprisingly good is the reproduction of the sharp edges along the waistcoat and the collar of the speaker. Subjectively, the VQ coded frame appears to be slightly sharper than the frame of the 8 kb/s DCT codec in Figure 12.20, and the background seems to be more adequately reproduced, although some disturbing artifacts are visible. This is partially due to the fact that a quantized DCT causes a bandwidth limitation, whereas the VQ codec maintains a better spatial resolution.

The VQ codec outperformed the DCT scheme in terms of subjective and objective quality at the cost of a higher complexity, an issue to be elaborated on explicitly during our further discourse. For each of the 396 blocks in a QCIF frame, the VQ searches through the entire codebook and compares every possible match by summing the squared differences for each pixel within the codebook entry block and the block to be quantized. Since the block size is fixed to 8×8, it is readily seen that 191 Flops are necessary to perform one block match. In the case of the 128-sized codebook, a total of 396 blocks $\times$ 128 codebook entries $\times$ 191 Flops = 13 Mflops have to be executed. The 256-sized codebook requires 26 Mflops. The worst-case DCT complexity of 3.8 Mflops seems to be more modest compared to the higher complexity of the VQ.

In order to explore the range of design tradeoffs, we tested several VQ schemes, in which we incorporated adaptive codebooks, classified VQ [425], and mean-shape gain VQ. Before commenting on these endeavors, let us first report our findings as

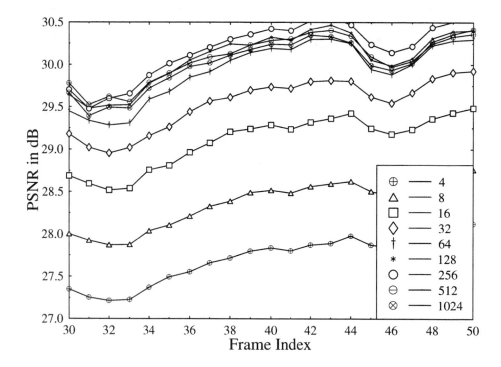

Figure 13.6: PSNR performance for various VQ codebook sizes in the range of 4-1024 entries at a constant bit rate of 1000 bits/frame or 10 kbps.

regards the practical range of codebook sizes. As mentioned, we opted for a fixed, but programmable, bit-rate scheme and varied the codebook size between 4 and 1024. As Figure 13.6 reveals, the best peak signal-to-noise ratio (PSNR) performance was achieved using codebook sizes in the range of 128 and 512. For these investigations, we used a locally recorded high-activity head-and-shoulders videophone sequence, which we refer to as the "Lab" sequence, since the well-known low-activity "Miss America" (MA) sequence was inadequate for evaluating the VQ performance. Observe that the 256-entry codebook results in the best PSNR performance. This corresponds to a VQ data rate of 0.125 bit per pixel or 8 bit per 8 × 8 block. This is because increasing the codebook (CB) size beyond 256 requires more bits for the CB address, which inevitably reduces the number of active blocks under the constraint of a fixed bit-rate budget. However, a codebook size of 128 is preferable, it halves the codec's complexity without significantly reducing the video quality. The corresponding visually enhanced codebook is depicted in Figure 13.3.

13.3.1 Mean and Shape Gain Vector Quantization

In an attempt to increase the VQ codec's video quality without increasing the codebook size, we also experimented with a so-called mean- and variance-normalized VQ

13.3. THE VECTOR QUANTIZER DESIGN

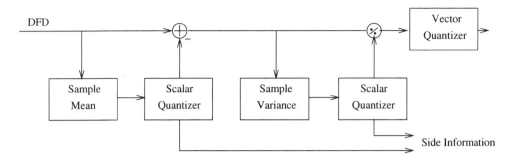

Figure 13.7: Block diagram for a Mean Shape Gain Vector Quantizer.

scheme. The approach is based on normalizing the incoming blocks to zero mean and unity variance. As seen in Figure 13.7, the Mean and Shape Gain Vector Quantizer (MSVQ) codec first removes the quantized mean of a given block, and then it normalizes the block variance to near-unity variance by its quantized variance. These two operations require $3n^2 - 2$ Flops per block. Given a certain CB size for the mean and variance, multiplying the CB entry with the quantized gain and adding the quantized mean values can be viewed as increasing the variety of the CB entries, which is equivalent to virtually expanding the codebook size by the product of the number of reconstruction levels of the mean and variance quantizers. Hence, a 2 bit quantizer for both the mean and variance would virtually expand the codebook to sixteen times its original size. The appropriate reconstruction levels were obtained by employing a Max-Lloyd quantizer [461].

We assessed the performance of the MSVQ using codebook sizes of 64 and 256 and quantizers, which ranged from 1 to 32 reconstructions levels, while maintaining a constant bit rate. Using a 1-level MSVQ scheme corresponds to the baseline VQ with no scaling at all. As demonstrated by Table 13.1, the MSVQ codec's performance was slightly degraded in comparison to the baseline codec, except for the 64-entry codebook with a 2-level mean and gain quantizer, where a minor improvement was achieved. This implies that the additional reconstruction precision of the MSVQ increased the bit-rate requirement for each block, allowing fewer blocks to be encoded and finally resulting in a worse overall performance at a constant bit rate. As a comparison, it should be noted that the 64-entry/2-level mean- and gain-scaled scheme has an almost identical performance to that of the 256-entry, 1-level arrangement, while retaining a lower complexity. Following our previous arguments, the 64-entry/2-level mean- and gain-scaled arrangement virtually expands the size of the 64-entry CB by a factor of 4 to 256 — justifying a similar performance to that of the 256-entry VQ scheme, which dispensed with invoking the MSVQ principle. From these endeavors we concluded that our codebooks obtained by the modified PNN algorithm are close to the optimum codebooks; therefore, the virtual codebook expansion cannot improve the performance significantly. The advantage of the 64-entry MSVQ scheme was that we maintained the performance of the 256-entry VQ arrangement with a complexity reduction of around 70%. Having reviewed the MSVQ principles in this section, let us now consider adaptive vector quantization.

No. of Quant. Levels	1	2	4	8	16	32
Av. PSNR for 64-entry CB	28.22	28.35	27.09	26.61	26.64	26.62
Av. PSNR for 256-entry CB	28.60	28.16	27.68	27.46	27.56	27.49

Table 13.1: Average PSNR versus Number of Quantization Levels for the MSVQ at a Constant Bit Rate of 8kb/s Using a 64- and 256-Entry Codebook (CB) and an Identical Number of Quantization Levels for the Mean- and Gain-scaling Scheme

13.3.2 Adaptive Vector Quantization

In adaptive vector quantization (AVQ) the codebook is typically updated with vectors, which occur frequently but are not represented in the codebook. Unfortunately, the increased flexibility of the method is often associated with an increased vulnerability against channel errors. Our adaptive approach is depicted in Figure 13.8. The codec is based on two codebooks implemented as First-in-First-out (FIFO) pipelines, one of which is referred to as the active codebook and the other one as passive. The VQ itself has access only to the active codebook and can choose any of the vectors contained in that codebook. Vectors selected by the VQ codec are then written back to the beginning of the pipeline. After each encoding step, a certain number of vectors are taken from the end of the pipeline and fed back to the beginning of the passive codebook. The same number of vectors are taken from the end of the pipeline in the passive codebook and entered at the beginning of the active codebook. This method forces a continuous fluctuation between the two codebooks but allows frequently used vectors to remain active. The principle is vulnerable to channel errors, but a forced realignment of the codebooks at both the encoder and at the remote decoder at regular intervals is possible, increasing the codec's robustness.

Our performance evaluation experiments were again based on the "Lab" sequence and the 256-sized codebook, which was split into an active codebook of 64 vectors and a remaining passive codebook of 192 vectors. The variable in this scenario was the number of changed vectors after each encoding step, which was fixed to 4, 8, 16, and 32 vectors per step. The corresponding PSNR versus frame index results are depicted in Figure 13.9 along with the performance of the standard 256-entry VQ. As in all other experiments, the comparison was based on a constant bit rate, specifically at 1000 bits/frame. Using this concept, we were able to increase the performance of the 64-entry VQ to a level similar to that of the standard 256-entry VQ, when applying a high vector exchange rate of 32 vectors per step. The lower vector exchange rates still improved the performance, when compared to a standard 64-entry VQ, but the PSNR gain was more modest, amounting to about 0.2 dB. Hence, adaptive VQ with a high vector replenishment rate of 32 vectors is also an attractive candidate to reduce the complexity of the VQ at a concomitant increased vulnerability to channel errors. Following the above brief notes on adaptive vector quantization, let us now consider classified vector quantization as another means of reducing the codec's complexity.

13.3. THE VECTOR QUANTIZER DESIGN 547

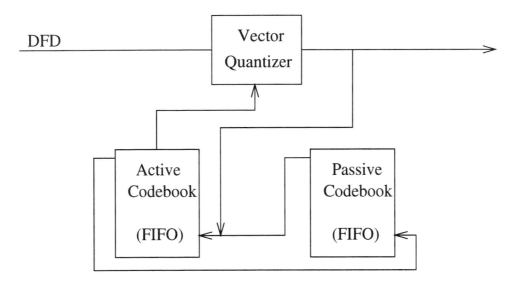

Figure 13.8: Block diagram of the adaptive VQ.

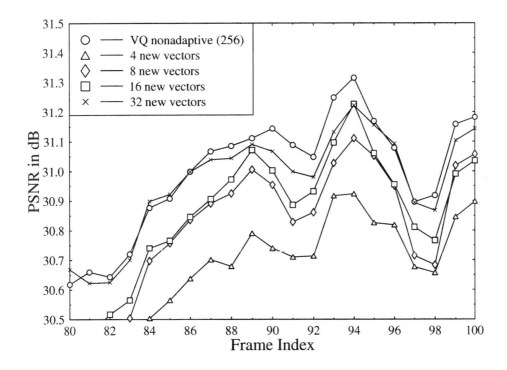

Figure 13.9: PSNR performance of VQ with and without adaptive codebooks at a constant bit rate of 10 kb/s.

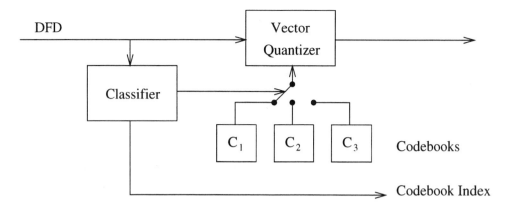

Figure 13.10: Block diagram of a classified VQ.

13.3.3 Classified Vector Quantization

Classified vector quantization (CVQ) [425] has been introduced as a means of reducing the VQ complexity and preserving the edge integrity of intra-frame coded images. As the block diagram of Figure 13.10 outlines, a CVQ codec is a combination of a classifier and an ordinary VQ using a series of codebooks. The incoming block is classified into one of n classes, and thus only the corresponding reduced-size codebook c_n is searched, thereby reducing the search complexity. In reference [425] the image blocks to be vector quantized were classified as so-called shade, midrange, mixed, and edge blocks. The edge blocks were further sorted according to their gradient across the block. Specifically, blocks that exhibited no significant luminance gradient across the 8×8-pixel area were classed as shade-blocks. By contrast, those that had a strong gradient were deemed to be edge-blocks. Midrange blocks had moderate gradients but no definite edge, while mixed blocks were typically those that did not fit the above categories. The same principle was invoked in our fractal codec in Chapter 11, and the relative frequencies of the associated blocks were summarized in Table 11.4.

In the present case the VQ's input was the DFD signal, for which the above classification algorithm is not applicable. In our approach a smaller codebook, derived from the same training sequence as the unclassified codebook C, can be viewed as a set of "centroids" for the codebook C. Hence, codebook C can be split into n codebooks c_n by assigning each vector in C to one of the centroids. The encoding procedure then consists of a two-stage VQ process and may be seen as a tree-structured VQ (TSVQ) [425]. The classifier is a VQ in its own right, using a codebook filled with the n centroids. Once the closest centroid has been found, the subcodebook containing the associated vectors is selected for the second VQ step. We carried out various experiments, based on the optimum 256-entry codebook, with classifiers containing n = 16, 64, and 128 centroids, as revealed in Figure 13.11. The PSNR performance loss due to using this suboptimum two-stage approach is less than 0.3 dB. This is surprising since the number of block comparisons was reduced from 256 for the standard VQ to about 64 for the CVQ. The complexity reduction cannot be exactly quantified, because the codebooks c_n do not necessarily contain N/n vectors.

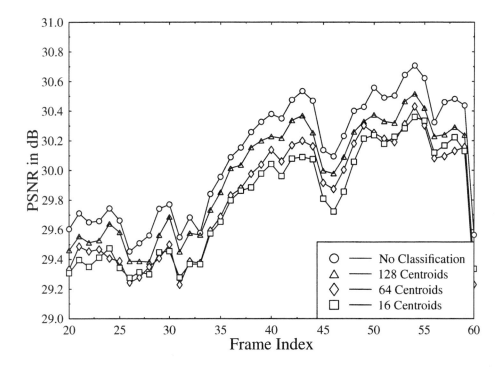

Figure 13.11: PSNR versus frame index performance of classified VQ codecs when using the "Lab" sequence at 10 kb/s.

13.3.4 Algorithmic Complexity

Although VQ has been shown [474] to match or outperform any other schemes that map an incoming vector onto a finite number of states, often other techniques such as transform coding (MPEG/H261) are preferred. This is because of the complexity of VQ, which increases drastically with increasing block and codebook sizes. The previous sections have shown that VQ is feasible for even such a large block size as 8×8 pixels using relatively small codebooks. This section will show that the amalgamation of various simplifications allows us to reduce the complexity of an entire codec down to 10 Mflops at a frame rate of 10 frames per second. We commence with a focus on the nearest neighbor rule, which has to be invoked for every possible MCER-block/codebook vector combination.

If we define a block as in Equation 12.3, then the MSE is defined as

$$d_{MSE} = \sum_{i=1}^{b}\sum_{j=1}^{b}(x(i,j) - y(i,j))^2, \qquad (13.2)$$

where $\vec{x}$ is the input block and $\vec{y}$ is the codebook vector. The search for the codebook vector, which results in the lowest MSE is known as the nearest neighbor rule.

Equation 13.2 can be expanded as:

$$d_{MSE} = \sum_{i=1}^{b}\sum_{j=1}^{b} x(i,j)^2 + \sum_{i=1}^{b}\sum_{j=1}^{b} y(i,j)^2 - 2\sum_{i=1}^{b}\sum_{j=1}^{b} x(i,j)y(i,j). \tag{13.3}$$

The first term of Equation 13.3 depends only on the input vector; therefore it is constant for every comparison, not affecting the nearest neighbour decision [487]. The second term depends only on the codebook entries, and so it can be determined prior to the encoding process. The last term of Equation 13.3 depends on both the incoming vector and the codebook entries. The evaluation of Equation 13.3 implies a complexity of $2b^2 + 1$ Flops, whereas that of Equation 13.2 is associated with $3b^2 - 1$ Flops, corresponding to a complexity reduction of about 30%.

When using a CVQ, the average number of codebook comparisons for each MCER can be reduced from 256 to 32. In addition the block preclassification technique presented in Section 12.9.1 may be employed, which limits the search to those blocks that contain high MCER energy. The VQ appeared to be less sensitive to such active block limitations (as evidenced by comparing Figures 13.12 and 12.18) than the previously studied DCT codec. When the preclassification is applied to both the MC and VQ at an activity rate of 25%, the total preclassification complexity increases the codec complexity by 0.81 MFlop per video frame. The overall VQ process complexity becomes about 1 Mflop/video frame or 10 Mflops at 10 frames/s. The overall PSNR penalty compared to the optimum full search scheme is typically less than 1 dB.

13.4 Performance under Erroneous Conditions

As observed in Section 12.12, an increased data compression ratio typically results in an increased vulnerability to channel errors. Since the codec's vulnerability increases drastically upon the introduction of run-length coding, we again design two codecs as it will become explicit during our forthcoming elaborations.

13.4.1 Bit-Allocation Strategy

Both codecs are completely reconfigurable and allow bit-rate adjustments after every transmitted frame. This is a very useful characteristic, since mobile channels typically exhibit time-variant channel capacities. We now focus our attention on two fixed configurations at 1131 bits per frame, although other bit rates and configurations are possible. The schemes are based on the low-complexity CVQ with a preselection of blocks.

The transmission burst of Codec I was composed of the bits of the FAW, PFU, the MC, and VQ bits. The 22-bit FAW is inserted to support the video decoder's operation in order to regain synchronous operation after loss of frame synchronization, as we saw in Chapter 12 in the context of the DCT-based codec. The PFU was set to periodically refresh 22 blocks per video frame. Therefore, every 18 frames or 1.8 seconds, the update refreshes the same blocks of the 396-block QCIF frame. This periodicity is signaled to the decoder by transmitting the inverted FAW. The gain-cost-controlled MC algorithm was configured so that the resulting bit-rate contribution of the MVs

13.4. PERFORMANCE UNDER ERRONEOUS CONDITIONS

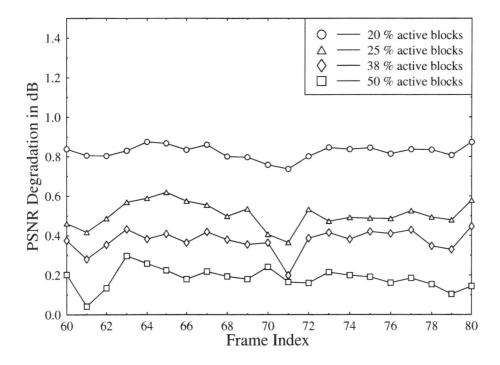

Figure 13.12: PSNR degradation performance of active/passive classified VQ codecs at 10 kbps upon using different fractions of active blocks.

in the QCIF frame did not exceed 500 bits. The remaining bits of the 1131-bit long transmission burst was devoted to the VQ. These parameters are reflected in Table 13.2. Because it is impossible to generate exactly 1131 bits per QCIF frame, we allow up to 10 bits for padding in order to achieve a constant bit rate of 11.31 kbps.

Recall that in the previous chapter Codecs I and II had an identical video quality, but their bit rates differed by about 30% and hence we had an 8 kbps and an 11.36 kbps DCT codec. Here our philosophy is slightly different, for we exploit the VLC principle in order to compress an approximately 30% higher original bit rate to 11.31 kbps and to support a higher video quality by the more error-sensitive Codec I. Logically, the identical-rate 11.31 kps Codec II then exhibits a lower video quality but a higher error resilience. Thus, in Codec I the indices of the active MV and VQ blocks have to be transmitted. The index of MV and VQ blocks ranges from 1 to 396, requiring 9 bits each, so that a total of $9 + 4 = 13$ bits for each MV and $9 + 8 = 17$ bits for each VQ block are required, as seen in Table 13.2. The use of indices instead of the run-length encoded activity tables is associated with redundancy, which manifested itself in a decreased compression ratio of about 30% to 40% or a performance loss of 1 to 2 dB, depending on the actual bit rate.

Codec	FAW	PFU	MV Ind.	MV	VQ Ind.	VQ	Padding	Total
I	22	22×4	-	< 500 VLC	-	VLC	< 10	1131
II	22	22×4	38×9	38×4	31×9	31×8	0	1131

Table 13.2: The 11.31 kbps VQ Codec Bit-Allocation Tables Both With and Without VLC

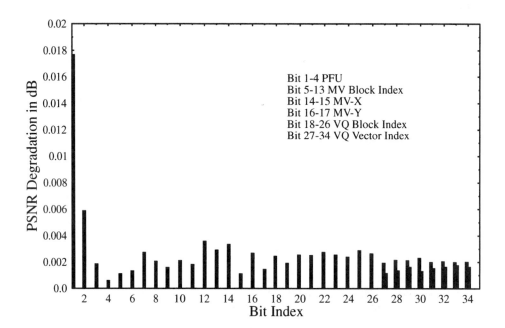

Figure 13.13: Integrated bit sensitivities for the non-VLC Codec II.

13.4.2 Bit Sensitivity

In order to prepare the codec for source-sensitivity matched protection, the codec bits were subjected to bit-sensitivity analysis. Here we have to separate those parts of the transmission burst that are VLC and those that are not. If a VLC bit is corrupted, the decoding process is likely to lose synchronization and the decoding of the frame becomes impossible. The decoder detects these bits and reacts by dropping the corrupted transmission burst and keeping the current contents of the reconstructed frame buffers. Therefore, the VLC bits of Codec I are classified as very sensitive.

In the case of non-VLC bits, we distinguish between various sensitivities. Using the same approach as in Section 12.12.1, we determined the integrated and averaged PSNR degradation for each non-VLC bit type, which we portrayed in Figure 13.13.

The VQ-index sensitivity depends on the codebook contents, as well as on the order of the codebook entries. We decreased the sensitivity of the VQ indices by reordering the codebook entries using the simulated annealing algorithm [148]. This technique was invoked several times in order to prevent the algorithm from being trapped in a suboptimum local minimum. The improved robustness is revealed by

13.4. PERFORMANCE UNDER ERRONEOUS CONDITIONS 553

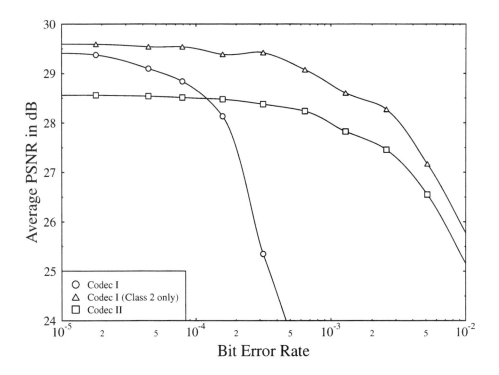

Figure 13.14: PSNR versus BER for the VLC-based Codec I and for the non-VLC Codec II.

comparing the two sets of PSNR degradation bars associated with the corresponding bit positions 27–34 in Figure 13.13.

The behavior of both codecs is portrayed in Figure 13.14. Under erroneous conditions here, we inflicted random bit errors and measured the PSNR performance at various BERs. As we expect, Codec I outperforms codec II at low BERs, for it has a higher error-free performance. However, since Codec I comprises vulnerable VLC bits, it breaks down for BERs exceeding $2 \cdot 10^{-4}$. In order to emphasize that this high vulnerability is due to the VLC bits, Figure 13.14 also depicts the performance of Codec I when corruption only affects the non-VLC parts of the transmission burst. In case of Codec II, the breakdown BER is at around $2 \cdot 10^{-3}$.

As a result of Figure 13.14, we propose Codec I for channels at high-channel SNRs and for channels with AWGN characteristics. The more robust Codec II suffers about one dB PSNR degradation under error-free conditions, while exhibiting an order of magnitude higher error resilience than Codec I.

The full examination of these VQ codecs over mobile channels in a GSM-like PRMA environment was discussed in [488], the results of which are provided in the next section.

13.5 VQ-Based Low-Rate Video Transceivers

13.5.1 Choice of Modulation

In recent years the increasing teletraffic demands of mobile systems have led to compact frequency reuse patterns, such as micro- and picocellular structures, which exhibit more benign propagation properties than conventional macrocells. Specifically, because of the low-power constraint and smaller coverage area, the transmitted power is typically channeled into street canyons. This mitigates both the dispersion due to far-field, long-delay multipath reflections and co-channel interferences. This is particularly true for indoor channels, where interferences are further mitigated by the room partitions. Under these favorable propagation conditions, multilevel modulation schemes have been invoked in order to further increase the teletraffic delivered, as in the American IS-54 [295] and the Japanese Digital Cellular (JDC) systems [473] of Table 9.1 on page 343. Hence, similarly to the DCT-based videophone systems of Section 12.13, we incorporated the VQ-based codecs designed earlier in this chapter in coherent pilot symbol assisted (PSA) 4- and 16QAM modems, which were the subjects of Chapter 5.

Again, similarly to the DCT-based videophones of Section 12.13, we created a range of reconfigurable source-sensitivity matched videophone transceivers, which have two modes of operation — namely, a more robust, but less bandwidth efficient four-level Quadrature Amplitude Modulation (4QAM) mode of operation for outdoors applications, accommodating eight users in a bandwidth of 200 kHz, as in the GSM system [468] and a less robust but more bandwidth efficient 16QAM mode supporting 16 video users in indoor cells or near the BS. In other words, indoor cells exploit the prevailing higher signal-to-noise ratio (SNR) and signal-to-interference ratio (SIR), thereby allowing these cells to invoke 16QAM and thereby requiring only half as many TDMA slots as 4QAM. When the portable station (PS) is roaming in a lower-SNR outdoor cell, the intelligent base station (BS) reconfigures the system in order to operate at 4QAM. The philosophy of these systems is quite similar to that of Section 12.13. Let us now consider the choice of error correction codecs.

13.5.2 Forward Error Correction

For convolutional codes to operate at their highest achievable coding gain, it is necessary to invoke soft-decision information. Furthermore, in order to be able to reliably detect decoding errors, they are typically combined with an external, concatenated error detecting block code, as in the GSM system [468]. As mentioned in Chapter 4, high minimum-distance block codes possess an inherently reliable error detection capability, which is a useful means of monitoring the channel's quality in order to activate Automatic Repeat Requests (ARQ), handovers, or error concealment. In the proposed systems we have opted for binary Bose-Chaudhuri-Hocquenghem (BCH) codes, which were characterized in Chapter 4. Let us now consider the specific video transceivers investigated.

In contrast to the 11.31 kbps codecs of Table 13.2, we followed the philosophy of our candidate DCT-based videophone schemes of Section 12.13 and designed a more error-resilient, 11.36 kbps codec, namely, Codec II, and a lower-rate variable-length

13.5. VQ-BASED LOW-RATE VIDEO TRANSCEIVERS

coded scheme, the 8 kbps Codec I. Although in Table 13.2 both the Type I and II codecs had a bit rate of 11.31 kbps, because of their programmable nature in these videophone systems we configured Codec I to operate at 8 kbps in order to derive comparable results with our similar DCT-based schemes. The system's schematic was the same as that for the DCT-based systems given in Figure 12.26.

In Systems 1, 3, and 6, which are charaterized in Table 13.3, we used the $R=71/127 \approx 0.56$-rate BCH(127,71,9) code in both 16QAM subchannels, which, in conjunction with the 11.36 kbps video Codec II, resulted in $1136 \times 127/71 = 2032$ bits/frame and a bit rate of 20.32 kbps at an image frame rate of 10 frames/s. These system features are summarized in Table 13.3. In contrast, in Systems 2, 4, and 5 we employed the 8 kbps video Codec I, which achieved this lower rate essentially due to invoking RL coding, in order to compress the motion- and VQ-activity tables. Since erroneous RL-coded bits corrupt the entire video frame, their protection is crucial. Hence, in Systems 2, 4, and 5 we decided to use the stronger BCH(127,50,13) code, which after FEC coding yielded the same 20.32 kbps bit rate as the remaining systems. This will allow us to assess in Section 13.6, whether it is a worthwhile complexity investment in terms of increased system robustness to use vulnerable RL coding in order to reduce the bit rate and then accommodate a stronger FEC code at the previous 20.32 kbps overall bit rate.

Source Sensitivity: Before FEC coding, the output of Codec II was split into two equal sensitivity classes, Class One and Two. The bit stream of Codec II was split into two sensitivity classes according to our findings in Figure 13.13. Note, again, that the notation Class One and Two introduced here for the more and less sensitive video bits is different from the higher and lower integrity C1 and C2 modulation channels. In the case of Codec I, all RL encoded bits were assigned to C1 and then the remaining bits were assigned according to their sensitivities to fill the capacity of the C1 subchannel, while relegating the more robust bits to C2. Let us now consider the architecture of System 1.

13.5.3 Architecture of System 1

Transmission Format: Similarly to the equivalent DCT-codec based systems of Table 12.5, the transmission packets of System 1 are constituted by a Class One BCH(127,71,9) codeword transmitted over the C1 16QAM subchannel, a Class Two BCH(127,71,9) codeword conveyed by the C2 subchannel, and a stronger BCH(127,50,-13) codeword for the packet header conveying the MS ID and user-specific control information. The generated 381-bit packets are represented by 96 16QAM symbols. Then 11 pilot symbols are inserted according to a pilot spacing of $P = 10$, and 4 ramp symbols are concatenated, over which smooth power amplifier ramping is carried out in order to mitigate spectral spillage into adjacent frequency bands. Eight 111-symbol packets represent a whole image frame, and so the signaling rate becomes 111 symb/12.5 ms $\approx$ 9 kBd. When using a time division multiple access (TDMA) channel bandwidth of 200 kHz, such as in the Pan-European second-generation mobile radio system known as GSM and a Nyquist modulation excess bandwidth of 38.8%, the signaling rate becomes 144 kBd. This allows us to accommodate $144/9 = 16$ video users, which coincides with the number of so-called half-rate speech users supported

Feature	System 1	System 2	System 3	System 4	System 5	System 6
Video Codec:	Codec II	Codec I	Codec II	Codec I	Codec I	Codec II
Video rate (kbps)	11.36	8	11.36	8	8	11.36
Frame Rate (fr/s)	10	10	10	10	10	10
C1 FEC	BCH(127,71,9)	BCH(127,50,13)	BCH(127,71,9)	BCH(127,50,13)	BCH(127,50,13)	BCH(127,71,9)
C2 FEC	BCH(127,71,9)	BCH(127,92,5)	BCH(127,71,9)	BCH(127,50,13)	BCH(127,50,13)	BCH(127,71,9)
Header FEC	BCH(127,50,13)	BCH(127,50,13)	BCH(127,50,13)	BCH(127,50,13)	BCH(127,50,13)	BCH(127,50,13)
FEC-coded Rate (kbps)	20.32	20.32	20.32	20.32	20.32	20.32
Modem	4/16-PSAQAM	4/16-PSAQAM	4/16-PSAQAM	4/16-PSAQAM	4/16-PSAQAM	4/16-PSAQAM
Retransmitted	None	Cl. One	Cl. One & Two	Cl. One & Two	None	Cl. One
User Signal. Rate (kBd)	18 or 9	9	18 or 9	18 or 9	18 or 9	9
System Signal. Rate (kBd)	144	144	144	144	144	144
System Bandwidth (kHz)	200	200	200	200	200	200
No. of Users	8-16	(16·2)=14	6-14	6-14	8-16	(16·2)=14
Eff. User Bandwidth (kHz)	25 or 12.5	14.3	33.3 or 14.3	33.3 or 14.3	33.3 or 14.3	14.3
Min. AWGN SNR (dB) 4/16QAM	5/11	11	4.5/10.5	6/11	8/12	12
Min. Rayleigh SNR (dB) 4/16QAM	10/22	15	9/18	9/17	13/19	17

Table 13.3: Summary of Vector-quantized Video System Features ∗ ARQs in Systems 2, 3, and 4 Are Activated by Class One Bit Errors But Not by Class Two [488] ©IEEE 1997, Streit, Hanzo

by the GSM system [468] of Table 9.1 on page 343.

If the channel conditions degrade and 16QAM communications cannot be supported, the BS instructs the system to switch to 4QAM, which requires twice as many time slots. The 381-bit packets are now conveyed by 191 4QAM symbols, and after adding 30 pilot symbols and four ramp symbols the packet length becomes 225 symb/12.5 ms, resulting in a signaling rate of 18 kBd. Hence, the number of videophone users supported by System 1 is reduced to 8, as in the full-rate GSM speech channel. The system also facilitates mixed-mode operation, where 4QAM users must reserve two slots in each 12.5 ms TDMA frame toward the fringes of the cell, while in the central section of the cell 16QAM users will only require one slot per frame in order to maximize the number of users supported. Assuming an equal proportion of 4QAM and 16QAM users, the average number of users per carrier becomes 12. The equivalent user bandwidth of the 4QAM PSs is 200 kHz/8 = 25 kHz, while that of the 16QAM users, is 200 kHz/16 = 12.5 kHz. The characteristics of the whole range of our candidate systems are highlighted in Table 13.3.

13.5.4 Architecture of System 2

As already noted in Section 13.5.2, the philosophy behind System 2 was to contrive a scheme that allowed an assessment of the value of Codec I in a system's context. Specifically, we sought to determine whether it pays dividends in robustness terms to invest complexity and hence reduce the 11.36 kbps rate of Codec II to 8 kbps and then accommodate a more powerful but more complex BCH(127,50,13) code within the same 20.32 kbps bit-rate budget. The sensitive RL and BCH(127,50,13) coded Class One bits are then conveyed over the lower BER C1 16QAM subchannel. Furthermore, in order to ensure the error-free reception of the RL-coded Class One bits, Automatic Repeat Request (ARQ) is invoked, if the C1 BCH(127,50,13) code indicates their corruption. Similarly to System 1, System 2 can also accommodate 16 time slots for the 9 kBd 4QAM users, but if we want to improve the integrity of the RL-coded video bits by ARQ, some slots will have to be reserved for the ARQ packets. This proportionately reduces the number of users supported, as the channel conditions degrade and requires more ARQ attempts. The underlying tradeoffs will be analyzed in Section 13.6.

Automatic Repeat Request: ARQ techniques have been successfully used in data communications [469–472] in order to render the bit and frame error rate arbitrarily low. However, because of their inherent delay and the additional requirement for a feedback channel for message acknowledgment, they have not been employed in interactive speech or video communications. In modern wireless systems, however, such as the 3G systems of Table 10.2 on page 394, there exists a full duplex control link between the BS and PS, which can be used for acknowledgments, and the short TDMA frame length ensures a low packet delay. Hence ARQ can be invoked.

System 2 was contrived so that for the first transmission attempt (TX1) we use contention-free Time Division Multiple Access (TDMA) and 16QAM to deliver the video packet. If the Class One bits transmitted over the C1 16QAM subchannel are corrupted and an ARQ request occurs, only the Class One bits are re-transmitted using the "halved-capacity" but more robust 4QAM mode of operation. These packets have to contend for a number of earmarked time slots, similarly to Packet Reservation Multiple Access (PRMA) [29], as highlighted in Chapter 6. If, however, there are only C2 bit errors in the packet, it is not re-transmitted. Since the Class One bits are delivered by the lower BER C1 16QAM subchannel, often they may be unimpaired when there are Class Two errors. To strike a compromise between the minimum required channel SNR and the maximum number of users supported, we limited the number of transmission attempts to three. Hence, two slots per frame must be reserved for ARQ. For the sake of low system complexity, we dispense with any contention mechanism and allocate two time slots to that particular user, whose packet was first corrupted within the TDMA frame. Further users cannot therefore invoke ARQ, because there are no more unallocated slots. A further advantage of this scheme is that since there are three copies of the transmitted packet, majority decisions can be invoked on a bit-by-bit basis, if all three packets became corrupted. The basic features of System 2 designed to accommodate Codec I are also summarized in Table 13.3.

13.5.5 Architecture of Systems 3–6

With the aim of exploring the whole range of design options, we have created four other systems, namely, Systems 3–6.

System 3 is similar to System 1, for it is constructed of the 11.36 kbps Codec II, BCH(127,71,9) FEC codecs, and reconfigurable 16QAM/4QAM modems, but it relies on ARQ assistance. In contrast to System 2, where only the Class One bits were re-transmitted using 4QAM, here both the Class One and Two bits are re-transmitted using the same modulation scheme, as during the first attempt. System 3 also limits the maximum number of transmission attempts to three in case of C1 BCH(127,71,9) decoding errors and therefore reduces the number of users by two but improves the transmission integrity. In the 16QAM and 4QAM modes, 14 and 6 users can be supported, respectively.

System 4 is similar to System 2, for it employs the 8 kbps Codec I in conjunction with the robust BCH(127,50,13) FEC codecs and the 16QAM/4QAM modems, but instead of Class One only re-transmission, it invokes full packet re-transmissions, if the Class One bits are corrupted. The number of video users is again reduced to 14 and 6 in the 16QAM and 4QAM modes, respectively. **System 5** is a derivative of System 4, which dispenses with ARQ and hence can serve 16/8 users in the 16QAM/4QAM modes of operation, respectively. Lastly, **System 6** is related to System 1, for it is constructed of the same components but allows ARQs in the same fashion as System 2, where in the case of Class One errors only these bits are re-transmitted using the previously introduced 16QAM/4QAM/4QAM TX1/TX2/TX3 regime. Therefore, System 6 can support 14 users.

13.6 System Performance

13.6.1 Simulation Environment

Before commenting on the system's performance, let us consider the simulation environment and the system parameters. The 11.36 and 8 kbps VQ video codecs, the pre-

13.6. SYSTEM PERFORMANCE

viously mentioned BCH codecs, and pilot-assisted diversity, and ARQ-aided 4QAM and 16QAM modems were simulated, including the AWGN and Rayleigh channel. The receiver carried out the inverse functions of the transmitter. Simple switched diversity was implemented. The criterion for deciding on which diversity channel to decode was the minimum fading-induced phase shift of the pilot symbols with respect to their transmitted phase rather than the maximum received signal power, since this has resulted in slightly better BER performance. As already mentioned in Section 12.13.2.2, the ARQ scheme, which was invoked in various situations in the different systems of Table 13.3, was controlled by the error detection capability of the BCH codes used. This was possible, since the error detection capability was extremely reliable in case of the powerful codes used and hence allowed us to dispense with Cyclic Redundancy Coding (CRC).

In interactive communications ARQ techniques have not been popular because of their delay and the required feedback channel for message acknowledgment. In intelligent low-delay picocells, however, it is realistic to assume a near-instantaneous full duplex control link between the BS and PS, for example, for ARQ acknowledgments, which in our case were always received error freely. In order to compromise between traffic capacity and robustness, only two slots per frame were reserved for ARQ, limiting the number of transmission attempts to three, where we allocated these two time slots to that particular user, whose packet was first corrupted within the TDMA frame. Since in the ARQ-assisted arrangements re-transmission took place within the last two slots of the same TDMA frame, no extra delay was inflicted. As regards the video packet of the last user in the frame, whose information was transmitted during the third slot from the end of the frame, it is not realistic, however, to expect a near-instantaneous acknowledgment even in low-delay picocells, implying the lack of ARQ assistance for this user. The specific criterion for invoking ARQ depended on which system of Table 13.3 was used.

Our results refer to the previously mentioned signaling rate of 144 kBd, while the propagation frequency and the vehicular speed were 1.8 GHz and 30 mph, respectively. These results are also comparable to those of the similar DCT-codec-based performance figures of Table 12.5. In the case of lower walking speeds, the signal's envelope fades more slowly. Therefore the pilots provide a better fading estimate, and the PSAM modem has a lower BER. On the other hand, in slow-fading conditions the fixed-length BCH codecs have to face a somewhat more pessimistic scenario, since in some codewords — despite interleaving — there may be a high number of errors, and the error correction capability may be overloaded. In contrast, in some codewords there may be only a low number of errors, which does not fully exploit the BCH codec's correction power. Throughout these experiments noise-limited, rather than interference-limited, operation was assumed, which is a realistic assumption in partitioned indoor picocells or in the six-cell microcellular cluster used, for example, in the Manhattan model of Chapter 17 in reference [29]. If, however, the decoded BCH frame error rate due to either interference or noise, which is monitored by the BS, becomes excessively high for 16QAM communications, the BS may instruct the system to switch to 4QAM, thereby reducing the number of users supported. The consideration of these reconfiguration algorithms is beyond the scope of this chapter. Having summarized the simulation conditions, let us now concentrate on the

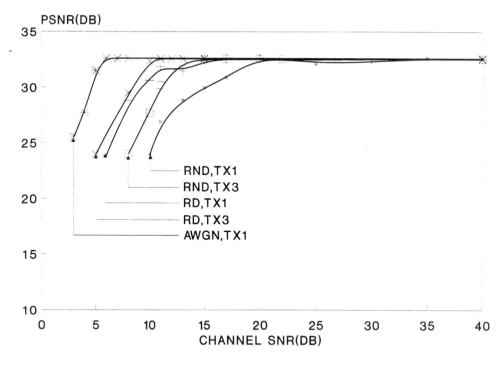

Figure 13.15: PSNR versus channel SNR performance of the VQ codec-based video Systems 1 and 3 of Table 13.3 in their 4QAM 18 kBd mode of operation over various channels [488]. ©IEEE 1997, Streit, Hanzo.

performance aspects of the candidate systems.

13.6.2 Performance of Systems 1 and 3

Recall from Table 13.3 that a common feature of Systems 1, 3, and 6 was that these all used the 11.36 kbps video codec associated with the BCH(127,71,9) FEC codec, although Systems 3 and 6 used ARQ in addition in order to enhance their robustness at the price of reserving two more time slots and hence reducing the number of users supported. The PSNR versus channel SNR (ChSNR) performance of System 1 is portrayed in Figures 13.15 and 13.16 for its 4QAM/16QAM 17.2/9 kBd modes of operation, respectively. Observe in these figures that the system's performance was evaluated for the best-case additive white Gaussian noise (AWGN) channel and the worst-case Rayleigh (R) channel, both with diversity (D) and with no diversity (ND), using one transmission attempt (TX1). However, in order to be able to assess the benefits of ARQ in terms of required ChSNR reduction, Figures 13.15 and 13.16 also display the corresponding results for System 3, using three transmission attempts (TX3).

Each system was deemed to deliver nearly unimpaired video if its PSNR was reduced by less than 1 dB owing to channel impairments. These corner-SNR values are tabulated for ease of reference in Table 13.3 for each system studied. Considering

13.6. SYSTEM PERFORMANCE

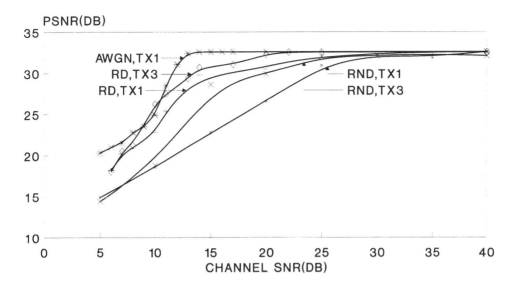

Figure 13.16: PSNR versus channel SNR performance of the VQ codec-based video Systems 1 and 3 of Table 13.3 in their 9 kBd mode of operation over various channels [488]. ©IEEE 1997, Streit, Hanzo.

System 1 first, in its 4QAM/18kBd mode of operation over AWGN channels a channel SNR of about 5 dB was required, which had to be increased to around 10 dB and 18 dB for the RD, TX1, and RND, TX1 scenarios. When opting for the doubled bandwidth efficiency but less robust 16QAM/9kBd mode of System 1, the corresponding ChSNR values were 12, 23 and 27 dB, respectively. These corner-SNR values are also summarized in Table 13.3. This tradeoff becomes explicit in Figure 13.17.

Comparing the above values in Figure 13.15 to the corresponding ChSNRs of System 3 reveals an approximately 6 dB reduced corner-SNR, when using 4QAM with no diversity, but this gain significantly eroded in the inherently more robust diversity-assisted scenario. In the 16QAM/9kBd mode, Figure 13.16 shows approximately 3 dB ARQ gain with no diversity, which is increased to about 4–5 dB, when using the diversity-aided arrangement.

13.6.3 Performance of Systems 4 and 5

Systems 4 and 5 combine the lower rate, but more error-sensitive, RL-coded 8 kbps Codec I with the more robust BCH(127,50,13) codec. These systems are more complex to implement, and hence we can assess whether the system benefited in robustness terms. The corresponding PSNR versus ChSNR curves are plotted in Figures 13.18 and 13.19 for the 4QAM/18kBd and the 16QAM/9kBd modes of operation, respectively. Considering the 4QAM/17.2 kBd mode of System 4 first, where a maximum of three transmission attempts are used, if any of the C1 or C2 bits are corrupted, an AWGN performance curve similar to that of System 3 is observed. Over Rayleigh channels with diversity, about 9 dB ChSNR is necessitated for near-unimpaired per-

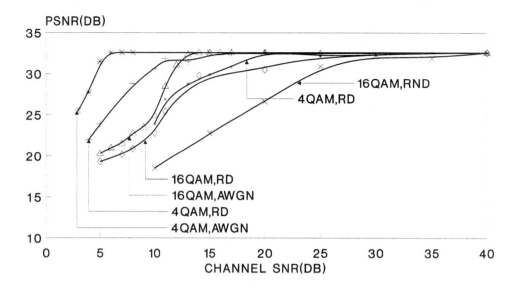

Figure 13.17: PSNR versus channel SNR performance of the VQ codec-based video System 1 of Table 13.3 in its 4QAM/16QAM 18/9 kBd modes of operation over various channels [488]. ©IEEE 1997, Streit, Hanzo.

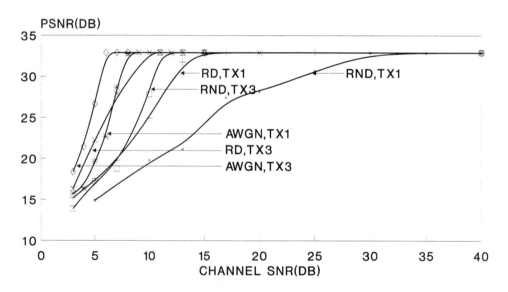

Figure 13.18: PSNR versus channel SNR performance of the 4QAM/18kBd modes of the VQ codec-based video Systems 4 and 5 of Table 13.3 over various channels [488]. ©IEEE 1997, Streit, Hanzo.

13.6. SYSTEM PERFORMANCE

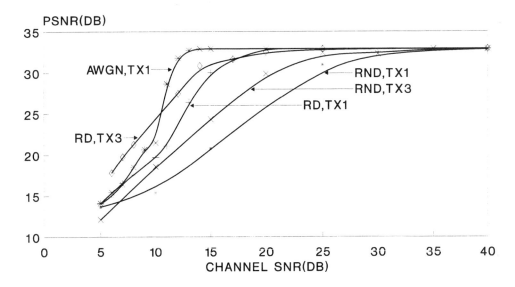

Figure 13.19: PSNR versus channel SNR performance of the 16QAM/9kBd mode of the VQ codec-based video Systems 4 and 5 of Table 13.3 over various channels [488]. ©IEEE 1997, Streit, Hanzo.

formance, increasing to around 11 dB with no diversity. In the 16QAM mode, the AWGN performance is similar to that of System 3. The ChSNR over Rayleigh channels with diversity must be in excess of about 17 dB, increasing to around 24 dB without diversity. As expected, System 5 is typically less robust than the ARQ-assisted System 4. It is interesting to observe that the RL-coded Systems 4 and 5, despite more robust FEC coding, tend to be less robust than the corresponding schemes without RL-coding, such as System 1 and 3, even though the latter systems also exhibit a reduced complexity. Another powerful solution for the RL-coded codecs is simply to drop corrupted packets and not update the corresponding video frame segment. This, however, requires a feedback acknowledgment flag in order to inform the encoder's local reconstruction frame buffer as to this packet dropping event. This issue will be discussed in great detail in Section 19.3.6 in the context of the ITU's H.263 codec.

13.6.4 Performance of Systems 2 and 6

The situation is slightly different when only the Class One bits are re-transmitted in case of erroneous decoding. These systems were contrived to constitute a compromise between robustness and high user capacity. For the RL-coded System 2, it is vital to invoke ARQ, whereas for System 6 it is less critical. However, System 6 uses the weaker BCH(127,71,9) code. The PSNR results of Figure 13.20 reveal that due to the stronger FEC code, System 2 is more robust than System 6. However, for all studied scenarios the degradation of System 2 was more rapid with decreasing ChSNR values than that of System 6. The minimum required ChSNR values are given in Table 13.3.

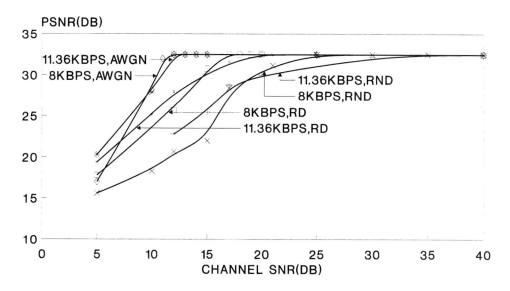

Figure 13.20: PSNR versus channel SNR performance of the VQ codec-based video Systems 2 and 6 of Table 13.3 [488]. ©IEEE 1997, Streit, Hanzo.

As expected, in all systems, regardless of the mode of operation invoked, best performance was always achieved over AWGN channels, followed by the diversity-assisted Rayleigh (RD) curves, while the least robust curves represented the system's performance with no diversity (ND). The ARQ attempts did improve the performance of all systems, but the lowest ARQ gains were offered over AWGN channels, since due to the random error distribution in every new attempt, the video packet was faced with similar channel conditions, as during its first transmission. The highest ARQ gains were typically maintained over Rayleigh channels with no diversity (ND), since during re-transmission, the signal envelope had a fair chance of emerging from a deep fade, increasing the decoding success probability.

13.7 Summary and Conclusions

Following the comparative study of a variety of VQ-based constant-rate video codecs, an 8 and an 11.36 kbps scheme were earmarked for further study in the context of various systems. It was demonstrated that the proposed VQ codecs slightly outperformed the identical-rate DCT codecs. The various studied reconfigurable VQ-codec-based mobile videophone arrangements are characterized by Table 13.3. The video codecs proposed are programmable to any arbitrary transmission rate in order to host the videophone signal by conventional mobile radio speech channels, such as the Pan-European GSM system, the IS-54, or IS-95 systems as well as the Japanese digital cellular system of Table 9.1 on page 343.

In Systems 1, 3, and 6 of Table 13.3, the 11.36 kbps Codec II was invoked. The BCH(127,71,9) coded channel rate was 20.32 kbps. Upon reconfiguring the 16QAM

13.7. SUMMARY AND CONCLUSIONS

and 4QAM modems depending on the BCH decoded frame error rate due to the prevailing noise and interference conditions experienced, signaling rates of 9 and 17.2 kBd, respectively, can be maintained. Hence, depending on the noise and interference conditions, between 8 and 16 video users can be accommodated in the GSM bandwidth of 200 kHz, which implies minimum and maximum user bandwidths of 12.5 and 25 kHz, respectively.

In Systems 2, 4, and 5 of Table 13.3, the RL-coded 8 kbps Codec I performed similarly to Codec II in terms of PSNR, but it was more error sensitive. The lower rate allowed us, and the lower robustness required us, to use the stronger BCH(127,50,13) FEC codecs. Further system features are summarized in Table 13.3.

As expected, the 4QAM/17.2kBd mode of operation of all systems is more robust than the 16QAM/9kBd mode, but it can only support half the number of users. Explicitly, without ARQ, eight or sixteen users can be supported in the 4QAM/17.2kBd or 16QAM/9kBd modes. When ARQ is used, these numbers are reduced by two in order to reserve slots for re-transmitted packets. A compromise scheme is constituted by Systems 2 and 6, which re-transmit only the corrupted Class One bits in case of decoding errors and halve the number of bits per modulation symbol, facilitating higher integrity communications for fourteen users. Both Systems 2 and 6 slightly outperform in terms of robustness the 16QAM modes of Systems 1 and 5, but not their 4QAM mode, since the initially erroneously received 16QAM Class Two bits are not re-transmitted and hence remain impaired. Systems 2, 4, and 5, which are based on the more sensitive Codec I and the more robust BCH(127,50,13) scheme, tend to have a slightly more robust performance with decreasing channel SNRs than their corresponding counterparts using the Codec II and BCH(127,71,9) combination, namely Systems 6, 3 and 1. However, once the error correction capability of the BCH(127,50,13) codec is overloaded, the corresponding PSNR curves decay more steeply, which is explained by the violently precipitated RL-coded bit errors. This becomes quite explicit in Figure 13.20. Clearly, if best SNR performance is at a premium, the Codec I and BCH(127,50,13) combination is preferable, while in terms of low complexity, the Codec I and BCH(127,71,9) arrangement is more attractive.

Careful inspection of Table 13.3 provides the system designer with a vast plethora of design options, while documenting their expected performance. Many of the listed system configuration modes can be invoked on a time-variant basis and so do not have to be preselected. In fact, in a true "software radio" [308], these modes will become adaptively selectable in order to comply with the momentary system optimization criteria.

The most interesting aspects are highlighted by contrasting the minimum required channel SNR values with the systems' user capacity, in particular in the purely 4QAM- and entirely 16QAM-based scenarios. System 2 is an interesting scheme, using 4QAM only during the Class One ARQ attempts, thereby ensuring a high user capacity, although the 11 and 15 dB minimum channel SNR requirements are valid only for one of the fourteen users, whose operation is 4QAM/ARQ assisted. The remaining thirteen users experience a slightly extended error-free operating SNR range in comparison to the 16QAM System 1 mode due to the employment of the more robust BCH(127,50,13) Class One codec. The comparison of System 1 to the otherwise identical, but ARQ-aided, System 3 suggests that over AWGN channels only very limited

SNR gains can be attained, but over Rayleigh channels up to 4 dB SNR reduction can be maintained in the 16QAM mode. In general, the less robust the modem scheme, the higher the diversity and AGC gains, in particular over Rayleigh channels. Hence we concluded that both the diversity and ARQ gains are substantial and that the slightly reduced user capacity is a price worth paying for the increased system robustness. This is slightly surprising perhaps, but it is often beneficial to opt for the lower-rate, higher-sensitivity Codec I, if the additional system complexity is acceptable, since due to the stronger accommodated BCH(127,50,13) codec, an extended operating range is maintained. All in all, in terms of performance, System 4 constitutes the best arrangement, with a capability of supporting between 6 and 14 users at minimum channel SNRs between 9 and 17 dB. Here we refrain from describing all potentially feasible systems; this is left for the interested reader to explore.

In summary, the proposed systems are directly suitable for mobile videophony by replacing a speech channel of conventional second-generation wireless systems, such as the GSM, IS-54, IS-95, and the Japanese systems. Alternatively, these video codecs can be used in the context of the emerging 3G systems of Table 10.2 on page 394. Our future work is aimed at improving the complexity/quality characteristics of these systems using parametrically enhanced coding. In Chapter 14, we will concentrate on quad-tree-based coding techniques and on matching multimode videophone transceivers.

Low Bit-Rate Parametric Quad-Tree-Based Codecs and Multimode Videophone Transceivers

14.1 Introduction

Quad-tree decomposition is a technique known from the field of image analysis [26]. A practical quad-tree structured intra-frame codec was presented by Strohbach [435], after which Zhang [489] proposed a codec suitable for ISDN applications. Other efforts concentrated on intra-frame coding [490–493], achieving bit rates as low as 0.25 bpp. These codecs share the characteristic that the images are initially segmented into equal-sized blocks, which are then subjected to quad-tree structured coding, generating variable-sized segments that can be represented, for example, by a constant luminance level at the cost of a low loss of sharpness or resolution. In this chapter, we design a low-rate quad-tree decomposed videophone codec, which in its cost-gain quantized approach is similar to the VQ- and DCT-based codecs of the previous two chapters. Furthermore, the codec incorporates a model-based parametric coding enhancement option for the subjectively critical eye-mouth region of head-and-shoulder videophone sequences. Lastly, the proposed codec will be incorporated in a multimode video transceiver, similar to those of the VQ- and DCT-based transceivers of the previous two chapters.

14.2 Quad-Tree Decomposition

Quad-trees (QT) represent a subclass of region growing techniques [26] in which the image (in our case the MCER), is described by the help of variable-size sectors characterized by similar features, for example, by similar gray levels. The stylized MCER frame portrayed in Figure 14.1 is described in terms of two sets of parameters: the structure or spatial distribution of similar regions and their gray levels. This figure is described in more depth later in this chapter. Note that the information describing the QT structure is potentially much more sensitive to bit errors than the gray level coding bits.

Before QT decomposition takes place, the 176 × 144 pixel QCIF MCER frame can be optionally divided into smaller, for example, 16 × 16 pixel rectangular blocks, perfectly tiling the frame. This initial decomposition may prove useful if the original image frame is not rectangular, although it may hamper coding efficiency by limiting the size of the largest homogeneous blocks to a maximum of 16 × 16 pixels. Creating the QT regions is a recursive operation that can follow either a top-down or a bottom-up approach. For example, according to the latter approach and considering each individual pixel, two or more neighbors are merged together, providing a certain similarity criterion is satisfied. This criterion may, for instance be, a similar gray level. This merging procedure is repeated until no more regions satisfy the merging criterion, and hence no more merging is possible.

Similarly, the QT regions can be obtained in a top-down approach, as portrayed in Figure 14.1, in which the MCER is initially divided into four sections, since the MCER frame is too inhomogeneous to satisfy the similarity criterion. Then two of the quadrants become homogeneous, whereas two of them need a number of further splitting or decomposition operations in order to satisfy the similarity criterion. This splitting operation can be continued, until the pixel level is reached and no further splits are possible. The variable-length QT-code at the bottom of the figure will be derived after elaborating on the similarity criterion.

Denoting the gray levels of the quadrants of a square by $m1 \ldots m4$, we compute their mean according to $m = (m_1 + m_2 + m_3 + m_4)/4$. If the absolute difference of all four pixels and the mean gray level is less than the system parameter σ, then these pixels satisfy the merging criterion. A simple merging criterion can be formulated as follows [435]:

$$(|m - m_1| < \sigma) \cap (|m - m_2| < \sigma) \cap (|m - m_3| < \sigma)$$
$$\cap (|m - m_4| < \sigma) = \text{True}, \qquad (14.1)$$

where $\cap$ represents the logical AND operation.

If the system parameter σ is reduced, the matching criterion is expected to become more stringent and hence less merging will take place, which is likely to increase the required encoding rate at a concomitant improvement of the MCER's representation quality. In contrast, an increased σ value is expected to allow more merging to take place and therefore reduce the bit rate. Note, however, that the number of bits generated, and accordingly the bit rate, is inherently variable. In order to circumvent this problem and to introduce a gain-cost-controlled approach, we will later derive Algorithm 4.

14.2. QUAD-TREE DECOMPOSITION

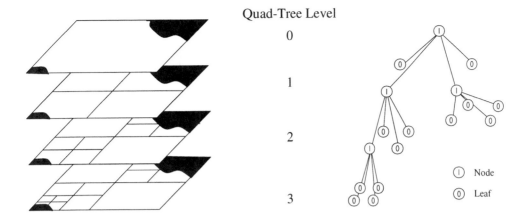

Quad-Tree Code : 1 | 0 1 1 0 | 0 1 0 0 | 0 0 0 0 | 0 0 0 0

Figure 14.1: Regular QT decomposition example, the corresponding quad tree, and its quad-tree code.

If the merging criterion is satisfied, the mean gray level m becomes the gray level of the merged quadrant in the next generation, and so on. At this stage it is important to note that the quality control threshold σ does not need to be known to the QT decoder. Therefore, the image representation quality can be rendered position-dependent within the video frame being processed, which allows perceptual weighting to be applied to the subjectively important image sections, such as the eyes and lips, without increasing the complexity of the decoder or the transmission rate.

Pursuing the top-down QT decomposition approach of Figure 14.1, we now derive the variable-length QT-code given at the bottom of the Figure. The QCIF MCER frame constitutes a node in the QT which is associated with a binary 1 in this example. After QT splitting, this *node* gives rise to four further nodes, which are classified on the basis of the similarity criterion. Specifically, if all the pixels at this level of the QT differ from the mean m by less than the threshold σ, then they are considered to be a *leaf node* in the QT which is denoted by a binary 0. Hence they do not have to be subjected to further similarity tests; they can simply be represented by the mean value m. In our example, a four-level QT decomposition was used (0–3), but the number of levels and/or the similarity threshold σ can be arbitrarily adjusted in order to achieve the required image quality and/or bit-rate target.

If, however, the pixels constituting the current node to be classified differ by more than the threshold σ, the pixels forming the node cannot be adequately represented by their mean m, and thus they must be further split until the threshold condition is met. This repetitive splitting process can be continued until there are no more nodes to split, since all the *leaf nodes* satisfy the threshold criterion. Consequently, the QT structure describes the contours of similar gray levels in the frame difference signal. The derivation of the QT-code now becomes explicit from Figure 14.1, where each

Figure 14.2: Quad-tree segmentation example with and without overlaid MCER and the original video frame. This decomposition example can be viewed in motion under http://www-mobile.ecs.soton.ac.uk/

parent-node is flagged with a binary one classifier, while the *leaf nodes* are denoted by a binary zero classifier and the flags are read from top to bottom and left to right. Observe in the example of Figure 14.1 that the location and size of 13 different blocks can be encoded using a total of 17 bits.

For the sake of completeness, we note that an alternative decomposition technique is constituted by the Binary Tree (BT) decomposition. The segmentation for a typical frame of the "Lab" sequence is exemplified in Figure 14.2, where the QT structure is portrayed with and without the overlaid MCER frame and the original video frame. In the eye and lip regions a more stringent similarity match was required than in the background, which led to a finer tree decomposition. In our experience, typically similar performance is achieved using BT-based arrangements and QT codecs. Hence only a brief exposure to BT is offered here. Logically, in BT decomposition every BT splitting step leads only to two subblocks. Therefore, typically twice as many splitting steps are necessary, to reach the same small block sizes, as in the case of QT codecs. Furthermore, the decomposition process is not as regular as for QT codecs. A parent-block may be divided both horizontally and vertically. However, in our experiments we did not allow several consecutive splits to take place in the same dimension, since this would require an extra 1-bit signaling bit. Instead, we invoked consecutive vertical/horizontal or horizontal/vertical decompositions, which allowed us to avoid the transmission of additional tree information in order to describe whether horizontal or vertical splitting was used.

In the case of high image-quality requirements, the simple block average representation of the QT decomposed blocks can be substituted by more sophisticated techniques, such as vector quantization (VQ), discrete cosine transformed (DCT) [459], or subband (SB) decomposed representations. Naturally, increasing the number of hierarchical levels in the QT decomposition leads to blocks of different sizes and applying VQ or DCT to blocks of different sizes increases the codec's complexity. Hence, the employment of these schemes for more than two to three hierarchical levels becomes impractical. Having reviewed the principles of QT decomposition, let us now consider the design of our candidate codecs.

14.3 Quad-Tree Intensity Match

With low implementational complexity in mind, we contrived two candidate codecs. In the first one we used the above mentioned zero-order mean- or average-based decomposition, whareas in the second one we attempted to model the luminance intensity profile over the block by a first-order linear approximation corresponding to a luminance plane sloping in both x and y directions.

14.3.1 Zero-Order Intensity Match

In order to achieve a fixed bit rate, we limited the number of bits per QCIF frame to 1000, which corresponds to a transmission rate of 10 kbps at a frame rate of 10 frames/sec. The associated compression ratio is 176×144 pixels $\times$ 8 bits/pixel/1000 bits $\approx$ 203, corresponding to a relative coding rate of about 0.04 bit/pixel. The statistical evaluation of the Probability Density Functions (PDF) of the average values m of the variable-sized blocks portrayed in Figure 14.3 revealed that for various block sizes quantizers having different mean values and variances were required. As the mean-value PDFs revealed, the mean of the blocks toward the top of the QT, namely, at QT Levels 2–4, which cover a large picture area, is more likely to be close to zero than the mean of smaller blocks. This fact exhibited itself in a more highly peaked and hence less spread PDF for the smaller picture areas, such as, for example, those at QT-levels 2–4 in Figure 14.3. Observe that for clarity of visualization the horizontal PDF scales of Levels 2–4 and 5–7 are about an order of magnitude different. The mean of the smallest blocks at QT Levels 5–7 tends to fluctuate over a wide range, yielding a near-uniform PDF for Level 7. Table 14.1 summarizes the intervals over which the block means fluctuate, as the block size is varied. Therefore, it was necessary to design separate quantizers for each QT hierarchy level.

With this objective in mind, a two-stage quantizer training was devised. First the unquantized mean values for each QT-decomposed block size were recorded using a training sequence, based on MCER frames, in order to derive an initial set of quantizers. During the second, true training stage, this initial set of quantizers was used tentatively in the QT codec's operation, in order to record future unquantized block averages generated by the codec operated at the chosen limited, constant bit rate, namely, at 1000 bits/frame. This two-stage approach was necessary for obtaining realistic training data for the Max-Lloyd training [461] of the final quantizers.

In order to achieve the best codec performance under the constraint of generating 1000 bits per frame, we derived codebooks for a range of different numbers of reconstruction levels. Figure 14.4 characterizes the codec's performance for various quantizers, ranging from 2- to 64-level schemes. Observe in the figure that the two-level and four-level quantizers were found to have the best performance. This was because the 10 kbps bit-rate limit severely restricted the number of hierarchical levels in the QT decomposition process, when more bits were allocated for quantizing the block averages. Hence, the codec failed to adequately decompose the inhomogeneous regions due to allocating an excessive number of bits to the quantization of their mean values.

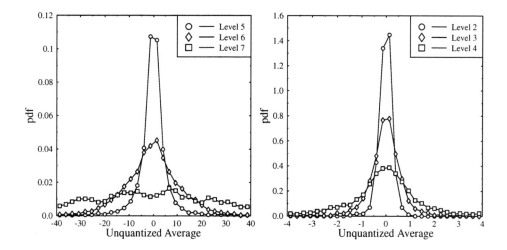

Figure 14.3: Probability density function of the block averages at Levels 2 to 7.

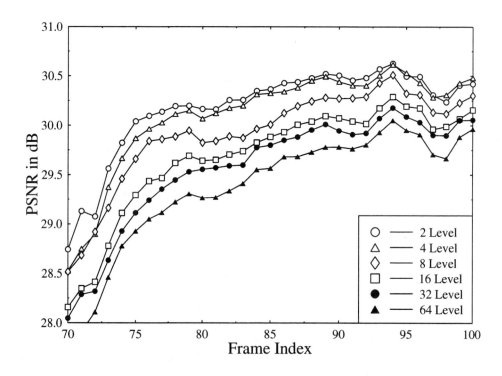

Figure 14.4: Performance for various QT codec quantizers at a constant bit rate of 1000 bits/frame.

14.3. QUAD-TREE INTENSITY MATCH

Level	0	1	2	3	4	5	6
Block Size	176×144	88×72	44×36	22×18	11×9	5/6×4/5	2/3×2/3
Range +/-	0.22/0.36	0.57/0.79	2.9/2.5	8.5/7.7	20.6/20.7	57.6/50.4	97.3/86.6

Table 14.1: Maximum Quantizer Ranges at Various Hierarchy Levels of the QT Codec

Approximation	8 kb/s	10 kb/s	12 kb/s
Zero-order PSNR (dB)	27.10	28.14	28.46
First-order PSNR (dB)	25.92	27.65	27.82

Table 14.2: PSNR Performance Comparison for Zero- and First-order QT Models at Various Constant Bit Rates

14.3.2 First-Order Intensity Match

Following our endeavors using the zero-order model of constant-luminance MCER representation within the QT decomposed subblocks, we embarked on studying the performance of the first-order luminance-intensity approximation corresponding to a sloping luminance plane across the MCER QT-block, which was defined by:

$$b(x, y) = a_0 + a_{x1}x + a_{y1}y. \tag{14.2}$$

Specifically, the block's luminance is approximated by a plane sloping in both x and y directions where the coefficients a_{x1} and a_{y1} are characteristic of the luminance slope. The squared error of this linear approximation is given by:

$$e = \sum_{x=1}^{b}\sum_{y=1}^{b}(b(x,y) - (a_0 + a_{x1}x + a_{y1}y))^2 \tag{14.3}$$

where the constants a_0, a_{x1}, and a_{y1} are determined by setting the partial derivative of Equation 14.3 with respect to all three variables to zero and solving the resulting three-dimensional problem for all decomposition steps. In order to assess the performance of this scheme, Max-Lloyd quantizers [461] were designed for each constant, and the PSNR performance of a range of quantizers using different numbers of quantization levels was tested. We found that the increased number of quantization bits required for the three different coefficient quantizers was too high to facilitate a sufficiently fine QT-based MCER decomposition given the fixed bit-rate budget imposed. Thus, the PSNR performance of the first order scheme became inferior to that of the zero-order model under the constraint of a fixed bit rate. Specifically, Table 14.2 reveals that the PSNR performance of the first-order model is at least 0.5 dB worse than that of the simple zero-order model. Consequently, for our further experiments the latter scheme was preferred.

14.3.3 Decomposition Algorithmic Issues

The crucial element of any QT or BT codec is the segmentation algorithm controlled by the similarity threshold σ. Most codecs presented in the recent literature [435, 489, 492] invoke the previously outlined thresholding process in order to decide whether a tree should be further decomposed. This principle results in an approximately time-invariant video quality and a fluctuating time-variant bit rate. This bit-rate fluctuation can be smoothed by applying adaptive buffer feedback techniques and controlling the similarity threshold σ [435]. Since our proposed videophone system was designed for constant-rate mobile radio channels, we prefer a constant selectable bit-rate scheme. A further requirement in very low bit-rate video communications is following the principle of gain-cost control, which is difficult to achieve in the framework of QT schemes, since the gain of a potential decomposition becomes known only after the decomposition took place.

In our preferred approach, the codec develops the QT or BT structure down to a given maximum number of decomposition levels and then determines the gain of each decomposition step, as summarized in Algorithm 4. Exposition of the algorithm is further aided by referring to Figure 14.1. The decomposition gain referred to in Step 2 of the algorithm is defined as the difference between the mean squared video reconstruction error of the parent block and the total MSE contribution associated with the sum of its four child blocks. The question that arose was whether the potential gain due to the current additional decomposition step should be weighted by the actual area of the blocks under consideration. Toward the bottom of the QT the subblocks represent small image sections, while toward the top the QT blocks correspond to larger areas. Our simulations showed that for most of the cases such a weighting of the gains was disadvantageous. This somewhat surprising result was an indirect consequence of the constant bit-rate requirement, since the adaptive codec often found it more advantageous to resolve some of the fine image details at an early stage — despite the restricted size of the QT-segment, if the bit-rate budget allowed it — and then save bits during future frames over the area in question.

Returning to Algorithm 4, we find that the purpose of Steps 3 and 4 is to introduce a bit-rate-adaptive, computationally efficient way of pruning the QT to the required resolution. This allows us to incorporate an element of cost-gain quantized coding, while arriving at the required target bit rate without many times tentatively decomposing the image in various ways in an attempt to find the optimum fixed bit-allocation scheme. The algorithm typically encountered four to five such fast QT pruning recursions before branching out to Step 5, which facilitated a slower converging fine-tuning phase during the bit-allocation optimization.

Algorithm 4 allowed us to eliminate the specific child blocks or leaves from the tree that resulted in the lowest decomposition gains. During this QT pruning process, the elimination of leaves converted some of the nodes to leaves, which were then considered for potential elimination during future coding steps. Therefore, the list of leaves associated with the lowest decomposition gains had to be updated before each QT pruning step. During the fast pruning phase — constituted by Steps 3 and 4 — in a computationally efficient but suboptimum approach, we deleted leaves on the basis of comparing them to the average gain of the entire set of leaves, rather than deleting the leaves associated with the lowest decomposition gain one-by-one. All leaves with a gain lower than the average gain were deleted in a single step if the potentially required number of coding bits was more than twice the target number. The slower one-by-one pruning constituted by Step 5 was then invoked before concluding the bit-allocation in order to fine-tune the number of bits allocated. This suboptimum

14.3. QUAD-TREE INTENSITY MATCH 575

Algorithm 4 *This algorithm adaptively adjusts the required QT resolution, the number of QT description bits, and the total number of encoding bits required in order to arrive at the target bit rate.*

1. Develop the full tree from minimum to maximum number of QT levels (e.g., 2--7).

2. Determine the decomposition gains associated with all decomposition steps for the full QT.

3. Determine the average decomposition gain over the full set of leaves.

4. If the potentially required number of coding bits is more than twice the target number of bits for the frame, then delete all leaves having less than average gains and repeat Step 3.

5. Otherwise delete leaves on an individual basis, starting with the lowest gain leaf, until the required number of bits is attained.

deletion process was repeated until the tree was pruned to the required size and the targeted number of coding bits was allocated.

In an attempt to explore any eventual residual redundancy or predictability within the QT, which could lead to further potential bit-rate reduction, we examined the tree structure. The philosophy of the following procedure was to fix the minimum and maximum tree depth. The advantage of such a restriction is twofold, since both the number of bits required for the QT and the complexity can be reduced. Under this QT structure limitation, bits can be saved at both the top and bottom of the tree. If a certain minimum depth of the tree is inherently assumed, the tree information of the top levels becomes *a priori* knowledge. The same happens as regards the bottom level if the tree depth is limited. In that case all decomposed subblocks at the last level are leaf nodes or, synonymously, child nodes, implying that no further splitting is allowed to take place. There is no need to transmit this *a priori* information. Hence, the decomposition complexity is limited since the tree needs to be developed only for the tree branches, which do not constitute *a priori* knowledge.

In Table 14.3 we list the average number of nodes associated with each tree level. Observe from the table that $15.3/16 = 95.6\%$ of the time the QT depth exceeds three levels. It is therefore, advantageous to restrict the minimum tree depth to three without sacrificing coding efficiency. Note from Table 14.3 that only about $52.9/16\,384 = 0.32\%$ of the potentially possible Level 7 leaves are ever decomposed. Thus, if we restrict the maximum tree depth to six for our QCIF codec, the codec's complexity is reduced, which is associated with a concomitant average PSNR reduction from 33.78 dB to 32.79 dB for the first 100 "Miss America" frames.[1]

As seen from Table 14.3, limiting the QT depth to six levels increases the number

---

[1] The "Miss America" sequence encoded at 11.36 kbps can be viewed under the following WWW address: http://www-mobile.ecs.soton.ac.uk

QT Level	0	1	2	3	4	5	6	7
Av. No. of nodes/leaves	1	4	15.3	28.4	47.6	66.9	72.6	52.9
Max. No. of nodes/leaves	1	4	16	64	256	1024	4096	16384

Table 14.3: Average Number of Quad-tree Nodes and Leaves per Hierarchical Level within the Tree at 1200 Bits per Frame

of Level 6 leaves by the number of relegated Level 7 leaves, which now have to be allocated additional reconstruction level coding bits. On the other hand, because of this restriction the (72.6 + 52.9) = 125.5 Level 6 leaves do not have to be specifically flagged as leaves in the variable-length QT code, which results in bit-rate savings. When using a four-level quantizer for the QT luminance and a bit-rate budget of 1000 bits/frame, on the basis of simple logic one would expect a similar performance with and without Level 7 decomposition, since we could save 125.5 QT description bits, due to our *a priori* knowledge that these are all leaf nodes. These bits then could be invested to quantize INT[125.5/2]=62 newly created leaf nodes using 2 bits per luminance value. Nonetheless, the 1 dB PSNR reduction encountered due to removing Level 7 from the QT suggested that the adaptive codec reacted differently, although a complexity reduction was achieved. Hence, if video quality was at premium, the seventh QT level was necessary to develop important fine details in the frame.

The achievable PSNR performance of the proposed QT codec at various bit rates is characterized by Figure 14.5 in the case of the "Lab" and "Miss America" sequences. Furthermore, PSNR versus frame-index performance of the model-based parametric coding enhancement (to be introduced in Section 14.4) is also portrayed in the figure, although its objective PSNR effects appear to be mitigated by the relatively small area of the eyes and lips, where it is applied. (A clearer indication of its efficiency will be demonstrated in Figure 14.12, where its PSNR improvement will be related to its actual area.) Having described the proposed QT codec, let us now discuss the proposed parametric enhancement scheme.

14.4 Model-Based Parametric Enhancement

Although in videophony the high-quality representation of the subjectively most important eye and lip regions is paramount, owing to the paucity of bits imposed by the system's bandlimitation this cannot always be automatically ensured. This is because the QT cannot be fully developed down to the pixel level, that is, to Level 7, even if the eye and lip regions were easily and unambiguously detectable and the total bit-rate budget of 1000 bits would be invested in these regions. When using 1000 bits/frame, only too small a fraction of the 25 384-pixel frame could be adequately modeled. Hence, in these highly active spots, we propose to employ model-based parametric coding enhancement. The penalty associated with this technique is that the decoder needs to store the eye and lip codebooks. Alternatively, these codebooks must be transmitted to the decoder during the call-setup phase. Note, however, that the proposed technique is general and that it can be invoked in conjunction with any other coding technique, such as the DCT- and VQ-based codecs of the previous two chapters.

14.4. MODEL-BASED PARAMETRIC ENHANCEMENT

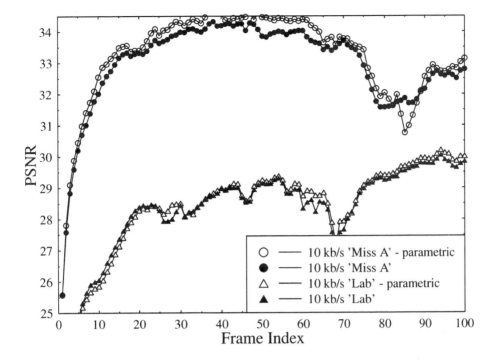

Figure 14.5: PSNR performance of the QT codec with and without parametrical enhancement at 10 kb/s.

In the next two sections we describe our approach to the eye and mouth detection and the parametric codebook or database training, while Section 14.4.3 highlights the encoding process.

14.4.1 Eye and Mouth Detection

Eye and mouth detection has been the aim of various research interests such as model-based coding, vision-assisted speech recognition [494], and lip-reading [495]. A reliable detection of the eye and mouth location is the most crucial step for all of the above techniques, and a simple procedure is proposed in the following detection steps, which are also summarized in Algorithm 5.

Step 1: It is critical for the reliable operation of the proposed eye-mouth detection algorithm that the face is the largest symmetrical object in the image frame and that its axis of symmetry is vertical. Initially, we generate a black and white two-tone image from the incoming video frame in order to detect this symmetry and to simplify the detection process. This two-tone image is free from gradual brightness changes but retains all edges and object borders. This was achieved by smoothing the picture using a simple two-dimensional averaging Finite Impulse Response (FIR) filter and then subjecting the smoothed image to frame-differencing and thresholding. Our

Algorithm 5 *This algorithm summarizes the parametric coding enhancement steps.*

1. Generate the binary image using smoothing, frame-differencing, and thresholding (see Figure 14.6).

2. Identify the axis of symmetry yielding the maximum number of symmetric pixels (see Equation 14.4).

3. Identify the position of eyes, nostrils, lips, and nose using the scaled template (see Figure 14.7).

4. Find the best matching eye and lip codebook entries and send their position, luminance shift, and codebook index.

experiments were based on 3 × 3- and 5 × 5-order simple two-dimensional averaging filters, where all 9 or 25 filter coefficients were set to $\frac{1}{9}$ and $\frac{1}{25}$, respectively, in order to preserve energy. The filtered image was then subtracted from the incoming frame and finally thresholded, which led to a binary image $f_{bin}(x,y)$, similar to the one shown at the left-hand side (LHS) of Figure 14.6. A threshold of 8.0 has been found suitable for this operation. This concludes the first step of the parametric coding algorithm summarized in Algorithm 5.

Step 2: The next step is to find the axis of symmetry for the face. Assuming tentatively that the axis of symmetry is x_0, symmetry to this axis is tested by counting the number of symmetric pixels in the two-tone image. The specific x_0 value yielding the highest number of symmetric pixels is then deemed to be the axis of symmetry. Once this axis is known, the pixel-symmetric two-tone image $f_{sym}(x,y)$ at the right-hand side (RHS) of Figure 14.6 is generated from the binary image $f_{bin}(x,y)$ as follows:

$$f_{sym}(x,y) = f_{bin}(2x_0 - x, y)f_{bin}(x,y). \quad (14.4)$$

Step 3: The pixel-symmetric image is then used to localize the eyes and the mouth. Initially, we attempted to locate the eye and the mouth as the two most symmetrical objects, assuming a given axis of symmetry x_0 in the symmetric frame $f_{sym}(x,y)$ at the RHS of Figure 14.6. This technique resulted in a detection probability of around 80% for the "Miss America" and "Lab" sequences. An object was deemed to be correctly detected if its true location in the original frame was detected with a precision of +/- 4 pixels. In most of the cases, the location of the eyes was correctly identified, but often the algorithm erroneously detected the chin as the mouth.

In a refined approach, we contrived a more sophisticated template, which consisted of separate areas for the eyes and the nose and a combined area for the nostrils and the mouth, as seen at the LHS of Figure 14.7. We expect many "contrast pixels" in the binary symmetric image $f_{sym}(x,y)$ at the location of the eyes, mouth, and nostrils, while the nose is not conspicuously represented in Figure 14.6. The template was scalable in the range of 0.8 to 1.1 in order to cater for a range of face sizes and distances measured from the camera. This template was then vertically slid along the previously determined axis of symmetry, and at each position the number of symmetric pairs of

14.4. MODEL-BASED PARAMETRIC ENHANCEMENT

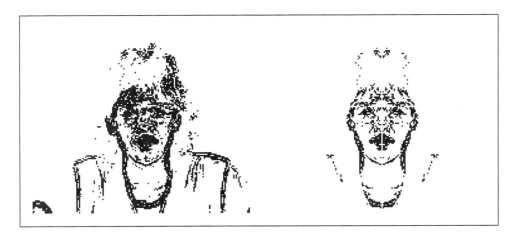

Figure 14.6: Binary (LHS) and binary-symmetric (RHS) frame of the "Miss America" sequence generated by Steps 1 and 2 of Algorithm 5.

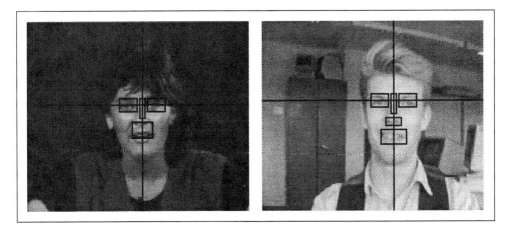

Figure 14.7: Image frames with overlaid initial (left) and improved (right) templates used in Algorithm 5.

contrast pixels appearing at both sides of the axis of symmetry within the template overlaid on the binary symmetric image $f_{sym}(x, y)$ of Figure 14.6 was determined. An exception was the nose rectangle, where the original luminance change was expected to be gradual, leading to no contrast pixels at all. This premise was amalgamated with our identification procedure by reducing the total number of symmetric contrast pixels detected within the confines of the template by the number of such contrast pixels found within the nose rectangle. Clearly, a high number of symmetric contrast pixels in the currently assumed nose region weakened the confidence that the current template position was associated with the true position. Finally, the template location resulting in the highest number of matching symmetric pixels was deemed to be the true position of the template.

Although through these measures the eye and lip detection probability was increased by another 10%, in some cases the nostrils were mistaken for the mouth. We further improved the template by adding a separate rectangular area for the nostrils, as depicted at the RHS of Figure 14.7. This method reached a detection probability of 97%.

In our previous endeavors, the eye and lip detection technique outlined was tested using image sequences, where the speaker keeps a constant distance from the camera and so the size of the face remains unchanged. However, the algorithm can adapt to the more realistic situation of encountering a time-variant face size by scaling the template. The magnification of the template was allowed to vary in the range of 0.8 to 1.1 due to limiting the associated detection complexity. The algorithm was tested using various head-and-shoulder sequences, and we found that it performed well, as long as the speaker was sufficiently close to the camera. Even the fact that a speaker wore glasses did not degrade the detection performance, although difficulties appeared in the case of individuals with a beard. In these situations, the parametric QT codec enhancement had to be disabled on grounds of reduced MSE performance, which was signaled to the decoder using a one-bit flag. Again, the above parametric coding steps are summarized in Algorithm 5; Step 4 will be discussed in more detail later in this chapter.

14.4.2 Parametric Codebook Training

A critical issue as regards the parametric codec's subjective performance is the training of the eye and lip codebooks. Large codebooks have better performance and higher complexity than small ones. The previously described eye and lip identification algorithm was also invoked for training the codebooks, which can then be manually edited in order to remove redundant entries and thus reduce the codebook search complexity. Initially, we generated a single codebook for the eyes and derived the second eye codebook by mirroring the captured codebook entries. As this could lead to pattern-matching problems, when the head was inclined, we proceeded with separate codebooks for each eye. A sample codebook of 16 eye and lip entries for the "Lab" sequence is portrayed in Figure 14.8, where the contrast of the entries was enhanced for viewing convenience. Let us now consider the parametric encoding algorithm itself.

14.4. MODEL-BASED PARAMETRIC ENHANCEMENT

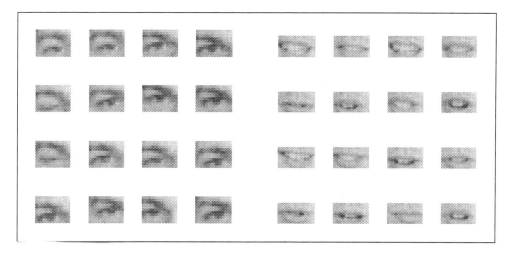

Figure 14.8: Eye and lip codebooks for the parametric coding of the "Lab" sequence.

14.4.3 Parametric Encoding

Step 4: The parametric encoder detects the eyes and mouth using the template-matching approach described in Algorithm 4. Once the template position is detected, the codec attempts to fit every codebook entry into the appropriate locations by sliding it over a window of +/- 3 pixels in each spatial direction. The MSE for each matching attempt corresponding to the full search of the codebook and to all the legitimate positions is compared, and the best-matching entry is chosen. A luminance shifting operation allows the codec to appropriately adjust the brightness of the codebook entries as well. This is necessary, because the brightness during codebook generation and encoding may differ. An example of the subjective effects of parametric coding and enhancement (PC) is portrayed in Figure 14.9, where we attempted to match the eye and mouth entries of the "Miss America" sequence into the "Lab" sequence in order to assess the codec's performance outside the training sequence. This scenario would represent the worst-case situation, when a prestored parametric codebook is used, which does not contain entries from the current user. A better alternative is to train the codebook during the call setup phase. Observe that there are no annoying artifacts, although the character of the face appears somewhat different. Hence, if the system protocols allow, it is preferable to opt for codebook training before the commencement of the communication phase, which would incur a slight call setup delay.

The optional parametric coding (PC) steps were included here for completeness also in the QT-based video codec's schematic in Figure 14.10, if PC affected the video quality advantageously in MSE terms. If the parametric enhancement was deemed to be successful, then for each modeled object the exact position, luminance shift, and codebook index were transmitted. This required, including the 1-bit PC enable flag, a total of 70 bits, as detailed in Table 14.4. Assuming arbitrary, independent eye and lip locations, for each object a 15-bit position identifier was required, yielding a total of

Figure 14.9: Parametric coding example: The uncoded "Lab" sequence with the superimposed entries from the "Miss America" eye/lip codebook (LHS) and the original frame (RHS). Both sequences encoded at various bit rates can be viewed under the WWW address http://www-mobile.ecs.soton.ac.uk

Type	Location of the Object	Lumin. Shift	Codeb. Entry	PC Flag	Total
No. of bits	3×15 = 45	3×4 = 12	3×4 = 12	1	69

Table 14.4: Bit-Allocation for the Parametric QT Codec Enhancement

45 bits. When using sixteen independent luminance shifts for the three objects, a total of 12 bits was necessary, similarly to the object codebook indices. This 70-bit segment contained some residual redundancy, since the object positions and luminance shifts are correlated both to each other and to their counterparts in consecutive frames, which would allow us to reduce the above 45-bit location identifier to around 32 bits and the total number of bits to about 57, but this further compression potential was not exploited here. Let us now consider the features of our parametrically enhanced QT codec.

14.5 The Enhanced QT Codec

The parametrically enhanced QT codec of Figure 14.10 is based on the same MC compensated approach, as described in the respective sections of Chapter 12. Here we focus on the parametric coding and QT-specific issues. After initializing the reconstructed frame buffers with the intra-coded frame, the PFU of Figure 14.10 step may be invoked, after which the optimal parametric eye and mouth detection takes place. The PC is optional for two reasons. First, PC requires *a priori* knowledge or training of the PC codebooks before the call setup; and second, the eye and mouth detection might be unsuccessful. In those cases, the encoder sends a single-bit flag in order to signal to the decoder that the PC is disabled. The success of PC is determined

14.6. PERFORMANCE UNDER ERRONEOUS CONDITIONS

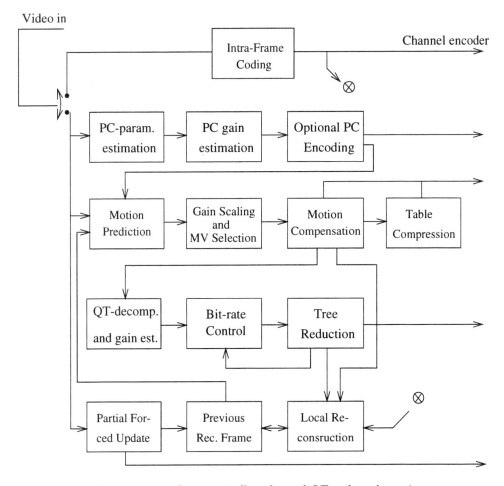

Figure 14.10: Parametrically enhanced QT codec schematic.

by the overall MSE reduction. An increase of the PSNR in the eye–mouth region enabled the PC, whereas a PSNR decrease was interpreted as a false identification of the eye–mouth region and hence the PC-flag sent to the decoder indicated that no PC was in operation. The motion compensation (see Section 12.6) was then invoked, and the MCER was passed to the previously described QT codec.

14.6 Performance and Considerations under Erroneous Conditions

The objective Peak Signal to Noise Ratio (PSNR) at various bit rates of the codec is portrayed in Figure 14.11 both with and without parametric coding enhancement. Observe that, due to the relatively small, 470-pixel area of the eye and lip regions and because of the optional employment of the parametric enhancement, the overall

584   CHAPTER 14. QT CODECS AND MULTIMODE VIDEO TRANSCEIVERS

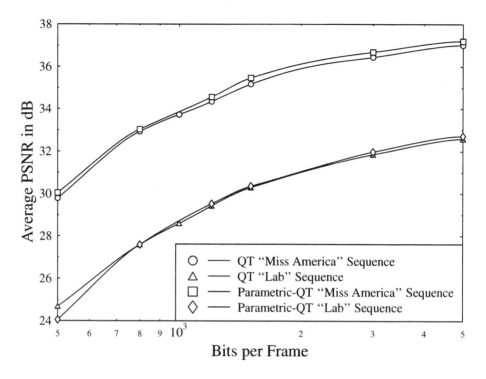

Figure 14.11: PSNR versus the number of bits per frame performance of the proposed QT codec with and without parametric enhancement.

objective PSNR video quality improvement appears more limited than its subjective improvements exemplified in Figure14.12. [2] The objective PSNR improvement due to the parametric enhancement is de-weighted by the factor of the template-to-frame area ratio, namely, by $470/(176 \times 144) = 0.0185$. The PSNR improvement of the template area becomes more explicit in Figure 14.13, which is the equivalent of Figure 14.5 related to this smaller 470-pixel image frame section where the PC enhancement is active. Observe in Figure 14.5 that for the "Miss America" sequence PSNR values in excess of 34 dB are possible.

Compared to the codecs based on the VQ and the DCT of the previous two chapters, we found that the QT-based codec required a higher bit rate. The previously described codecs supported bit rates as low as 8 kb/s. Figure 14.11 reveals that the minimum required bit rate for the QT codec lies around 10 kbs.

14.6.1   Bit Allocation

We presented two versions of the DCT and VQ codecs introduced in the previous two chapters. Codec I employed run-length encoded activity tables for the active blocks,

---

[2]Both sequences encoded at various bit rates can be viewed under the WWW address http://www-mobile.ecs.soton.ac.uk

14.6. PERFORMANCE UNDER ERRONEOUS CONDITIONS

Figure 14.12: Comparison of the QT codec without (left) and with (right) PC at 10kb/s. Both sequences encoded at various bit rates can be viewed under the WWW address http://www-mobile.ecs.soton.ac.uk

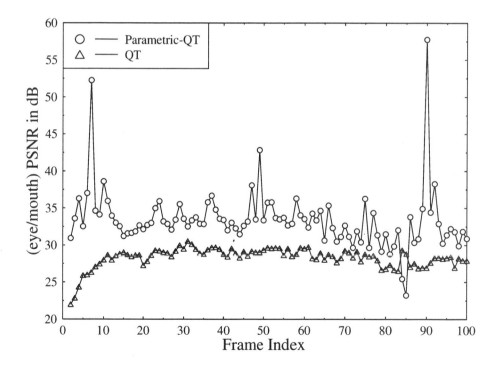

Figure 14.13: PSNR versus frame index performance for the parametrically enhanced Codec 1 in the 470-pixel PC template region using the "Lab" sequence.

Parameter	FAW	MC	PC	PFU	QT	Total
No. of bits	16	< 500	1 or 70	80	>500	≤1136

Table 14.5: Bit-Allocation Table for the 11.36 kbps QT-codec Using Optional PC Enhancement

while Codec II used simpler, but error-resilient, active block indexing. Explicitly, Codec II exhibited a higher error resilience at the cost of a 30 to 40% higher bit rate. Since the tree code for the QT representation is a variable-length code, its vulnerability is as high as that of the run-length encoded bit-streams of the Type I codecs in Chapters 12 and 13. Here we propose a single QT-based codec, where the MVs are encoded using the active/passive concept from Section 12.7.

The typical bit-allocation of the codec is summarized in Table 14.5. A total of less than 500 bits is usually allocated to the motion compensation activity table and the motion vectors, 1 or 70 bits are earmarked for parametric coding, 80 bits for partial forced updates, 16 bits for the FAW, while a minimum of 500 bits should be assigned to variable-length QT coding using the previously highlighted QT pruning, as well as to the four-level luminance mean quantization. Let us now briefly consider the bit error sensitivity of the QT codec.

14.6.2 Bit Sensitivity

We employed the technique described in Section 12.12.1 to resolve the integrated bit sensitives for each bit, which are depicted in Figure 14.14. Note that the sensitivity of the four PFU bits and that of the $2 + 2 = 4$ MV bits follow the natural trend from the Most Significant Bits (MSB) toward the Least Significant Bits (LSB). As Figure 14.14 demonstrates, the bit sensitivity of the QT reconstruction level bits increases toward smaller block sizes or deeper QT levels. This is because the average luminance of smaller blocks has a more spread distribution, as we have seen in Figure 14.3, and hence their quantization levels are more sparsely allocated, which results in an increased vulnerability. The variable-length quad-tree code itself can be interpreted as run-length-encoded information, since a corrupted bit affects the entire tree-decoding process.

In [496] we reported our results on a video transceiver designed for the previously described QT codecs, which is discussed next.

14.7 QT-Codec-Based Video Transceivers

14.7.1 Channel Coding and Modulation

Trellis-coded modulation (TCM) or block-coded modulation (BCM) [29, 497] uses modulation constellation expansion, such as from 2 bits/symbol 4-QAM to 4 bits/symbol 16-QAM, in order to accommodate the parity bits of a half-rate FEC codec without expanding the bandwidth of the transmitted signal. This constellation expansion is invoked in order to "absorb" the channel-coding bits while maintaining the same signaling rate. TCM schemes usually achieve higher coding gain over AWGN channels

14.7. QT-CODEC-BASED VIDEO TRANSCEIVERS

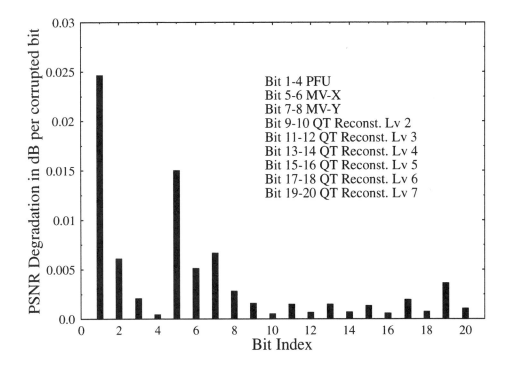

Figure 14.14: Integrated bit sensitivities of the QT codec.

than isolated coding and modulation, but this is not easily achieved over Rayleigh-fading channels, where coding gains are usually achieved only at quite high channel SNR values. This is a consequence of the hostile channel's tendency to overwhelm the error correction capability of the incorporated FEC scheme, which then often precipitates more errors than the number of errors without FEC. This is particularly true for the inherently reduced-resilience multilevel QAM schemes of Chapter 5, which are capable of accommodating the increased number of bits due to TCM. Furthermore, the bits generated by our DCT, VQ, and QT video codecs have unequal bit sensitivities, and therefore unequal protection channel coding [466] must be invoked. In the spirit of references [29, 466, 498], here we employ source-sensitivity matched unequal protection joint-source/channel-coding and modulation schemes. The QT-coded video bits are mapped in two different protection classes, namely, Class One and Two. Both streams are binary Bose-Chaudhuri-Hocquenghem (BCH) coded, as detailed in the channel-coding chapter, namely, Chapter 4 and transmitted using 4- or 16QAM, as described in our modulation chapter, Chapter 5. We employed simple Time Division Multiple Access (TDMA).

Similarly to our videophone systems designed for the DCT- and VQ-based video codecs, in order for the transmitted signal to fit into a GSM-like bandwidth of 200 kHz, while using a Nyquist excess bandwidth of 38.8% corresponding to a Nyquist rolloff factor of 0.388, the signaling rate was limited to 144 kBd. Again, the philosophy of the transceivers employed complied with that of the DCT- and VQ-based transceivers of the previous two chapters. Coherent detection was achieved using pilot symbol-assisted modulation (PSAM), where channel sounding symbols with known phase and magnitude were inserted in the transmitted TDMA burst in order to inform the receiver about the channel's estimated momentary phase rotation and attenuation. The inverse complex-domain pilot-assisted de-rotation and de-attenuation can then be applied to the data symbols in order to remove the effects of the fading channel.

After extracting the pilot symbols from the received TDMA burst, a complex channel attenuation and phase rotation estimate must be derived for each received symbol, using the simple linear or polynomial interpolation techniques highlighted in Chapter 5. As in the context of the previous two chapters on DCT- and VQ-based video transceivers, we found that the maximum-minimum distance rectangular 16QAM constellation exhibits two independent 2-bit subchannels, having different BERs. The lower integrity C2 subchannel showed a factor two to three times higher BER than the higher quality C1 subchannel. Although these integrity differences can be fine-tuned using different BCH codes in the subchannels, for our system it was found appropriate to maintain this integrity ratio for the Class One and Class Two QT-coded bits.

14.7.2 QT-Based Transceiver Architectures

The QT-based video codecs employed in our videophone transceiver were based on the coding concepts previously outlined earlier in this chapter, and their bit-allocation scheme was similar to that of Table 14.5. In order to comply with the bit-packing constraints of the BCH codecs employed in the videophone transceivers, QT Codec 1 was configured to operate at 11.36 kbps, while its slightly modified version, QT Codec 1-bis, generated a bit rate of 11 kbps. Again we note that in contrast to the previous Codec I and Codec II schemes of the DCT- and VQ-based systems of the previous two chapters, which represented the more bandwidth-efficient and the more error-resilient codecs, respectively, the QT codec is inherently more vulnerable. This is indicated by our different notation, where we refrained from using the convention Codec I, opting instead for Codec 1 and Codec 1-bis. The video transceiver's schematic is similar to that shown in Figure 12.26.

For the QT-based video codecs Codec 1 and Codec 1-bis, four different transceivers were contrived in order to explore the range of system design tradeoffs. The features of these schemes, which are similar to those of the DCT- and VQ-based systems of Tables 12.5 and 13.3 of the previous two chapters, are summarized in Table 14.6. In System 1 we employed the $R = 71/127 \approx 0.56$-rate BCH(127,71,9) code in both 16QAM subchannels. After BCH coding the 11.36 kbps QT-coded video sequence generated by Codec 1, the bit rate became $11.36 \times 127/71 = 20.32$ kbps. These system features are tabulated in Table 14.6. In System 2, on the other hand, we opted for the stronger BCH(127,50,13) code, which after FEC coding the 11 kbps stream produced by Codec 1-bis yielded a bit rate of 27.94 kbps. Since System 2 has a higher transmission rate, it will inevitably support a lower number of users than System 1. This comparison will allow us to decide whether it is worthwhile in terms of increased system robustness using a stronger BCH code at the cost of reducing the

14.7. QT-CODEC-BASED VIDEO TRANSCEIVERS

Feature	System 1	System 2	System 3	System 4
Video Codec	Codec 1	Codec 1-bis	Codec 1	Codec 1-bis
Video rate (kbps)	11.36	11	11.36	11
Frame Rate (fr/s)	10	10	10	10
C1 FEC	BCH(127,71,9)	BCH(127,50,13)	BCH(127,71,9)	BCH(127,50,13)
C2 FEC	BCH(127,71,9)	BCH(127,50,13)	BCH(127,71,9)	BCH(127,50,13)
Header FEC	BCH(127,50,13)	BCH(127,50,13)	BCH(127,50,13)	BCH(127,50,13)
FEC-coded Rate (kbps)	20.32	27.94	20.32	27.94
Modem	4/16-PSAQAM	4/16-PSAQAM	4/16-PSAQAM	4/16-PSAQAM
ARQ	No	No	Yes	Yes
User Signal. Rate (kBd)	18 or 9	12.21 or 24.75	18 or 9	12.21 or 24.75
System Signal. Rate (kBd)	144	144	144	144
System Bandwidth (kHz)	200	200	200	200
No. of Users	8 or 16	5 or 11	6 or 14	3 or 9
Eff. User Bandwidth (kHz)	25 or 12.5	40 or 18.2	33.3 or 14.3	66.7 or 22.2
Min. AWGN SNR (dB) 4/16QAM	7.5/13	7.5/12	6/12	6/11
Min. Rayleigh SNR (dB) 4/16QAM	20/20	15/18	8/14	8/14

Table 14.6: Summary of QT-based Video System Features [496] ©IEEE, 1996, Streit, Hanzo

number of users supported.

The transmission packets used in System 1 are comprised by a Class One BCH(127,71,9) codeword conveyed over the C1 16QAM subchannel, plus a Class Two BCH(127,71,9) codeword transmitted over the C2 subchannel. A BCH(127,50,13) codeword having a stronger error correction capability is assigned to the packet header. The 381-bit packets are converted to 96 16QAM symbols, and 11 pilot symbols are inserted with a pilot spacing of $P = 10$. Lastly, four ramp symbols are concatenated and were invoked in order to allow smooth power amplifier on/off ramping, which mitigates spectral spillage into adjacent frequency bands. Eight 111-symbol packets are required to transmit an entire 1136-bit image frame; hence, the signaling rate becomes 111 symb/12.5 ms $\approx$ 9 kBd. This allows us to support 144 kBd/9 kBd = 16 video users, which is identical to the number of half-rate speech users accommodated by the GSM system [41]. Hence eight combined video/speech users can be supported. The packet format of System 2 is identical to that of System 1, but eleven packets are required to transmit an entire 1100-bit/100ms frame. Thus, the signaling rate becomes 11 × 111symb/100 ms = 12.21 kBd. Then the number of users supported by System 2 at the 144 kBd TDMA rate is INT[144 kBd/12.21] = 11, where INT[•] represents the integer part of [•].

When the channel signal-to-noise ratio (SNR) becomes insufficient for 16QAM communications, since, for example, the portable station (PS) moves away from the base station (BS), the BS reconfigures the transceivers as 4QAM schemes, which require twice as many time-slots but a lower SNR value. The 381-bit packets are now converted to 191 4QAM symbols and after inserting 30 pilot symbols and four ramp symbols, the packet length becomes 225 symb/12.5 ms, yielding a signaling rate of 18 kBd. Hence, the number of videophone users supported by System 1 is reduced to eight, as in the full-rate GSM speech channel which is equivalent to four combined video/speech users. The signaling rate of the 4QAM mode of System 2 becomes 11 × 225symb/100ms = 24.75 kBd, and the number of users is INT[144/24.75] = 5.

The system also supports mixed-mode operation, where the more error-resilient 4QAM users must reserve two slots in each 12.5 ms TDMA frame while roaming near the fringes of the cell. By contrast, in the central section of the cell, 16QAM users will only require one slot per frame in order to maximize the number of users supported. Assuming an equal proportion of System 1 users in their 4QAM and 16QAM modes of operation, we see that the average number of users per carrier becomes 12. The equivalent user bandwidth of the 4QAM PSs is 200 kHz/8 = 25 kHz, while that of the 16QAM users is 200 kHz/16 = 12.5 kHz. The characteristics of the whole range of our candidate systems are highlighted in Table 14.6.

Systems 3 and 4 are identical to Systems 1 and 2, respectively, except for the fact that the former schemes can invoke Automatic Repeat Requests (ARQ), when the received video bits are erroneous. In the past the employment of ARQ schemes was limited to data communications [469–472], where longer delays are tolerable than in interactive videotelephony. In modern wireless systems, such as, for example, the 3G systems of Table 10.2 on page 394 however, there is a full duplex control link between the BS and PS, which can be used for message acknowledgments, while the short TDMA frame length ensures a low packet delay. Hence, ARQ can be realistically invoked. Therefore, Systems 3 and 4 are expected to have a higher robustness than

Systems 1 and 2, which dispense with ARQ assistance. The penalty must be paid in terms of a reduced number of users supported since the ARQ attempts occupy some of the time slots.

It is unrealistic to expect the system to operate at such low channel SNR values, where the probability of packet corruption and hence the relative frequency of ARQ attempts is high, since then the number of users supported and the teletraffic carried would become very low. In order to compromise, the number of transmission attempts was limited to three in our system, which required two earmarked slots for ARQ. Furthermore, in any frame only one user was allowed to invoke ARQ, namely, the specific user whose packet was first corrupted within the frame. For the remaining users, no ARQ was allowed in the current TDMA frame. An attractive feature of this ARQ scheme is that if three copies of the transmitted packet are received, majority logic decisions can be invoked on a bit-by-bit basis, should all three received packets be corrupted. The basic features of Systems 3 and 4 are summarized in Table 14.6.

Similarly to our DCT- and VQ-based video transceivers, here we also opted for a twin-class scheme using the BCH(127,71,9) or the BCH(127,50,13) codecs in both the higher integrity C1 and lower integrity C2 16QAM subchannels for the transmission of the more sensitive and less sensitive video bits, respectively. In the case of Codec 1, there are $1136/2 = 568$ Class One and 568 Class Two bits, while for Codec 1-bis there are 550 Class One and Class Two bits and 550, respectively. Although the variable-length QT code bits nearly fill the capacity of the C1 subchannel, up to 68 of the high-sensitivity PFU and MV MSBs of Figure 14.14 can be directed to the higher integrity C1 subchannel.

14.8 QT-Based Video-Transceiver Performance

Let us now turn to the overall video system performance evaluated in terms of the PSNR versus channel SNR curves depicted in Figures 14.15–14.18 for Systems 1–4. In all the figures, six performance curves are displayed — three curves with ARQ using three transmissions (TX3) and three curves without ARQ, that is, using a single transmission attempt (TX1). Both the 4QAM and 16QAM mode of operation of Systems 1–4 are featured over AWGN channels, Rayleigh-fading channels with diversity (RD) and with no diversity (RND). Perceptually unimpaired video performance was typically achieved by these systems, when the PSNR degradation was less than about 1 dB. Therefore, in Table 14.6 we summarized the minimum required channel SNR values to ensure a PSNR degradation of less than 1 dB, as the systems' operating channel SNR.

The full-scale exploration of the various system design tradeoffs using Figures 14.15–14.18 and Table 14.6 is left to the reader for the sake of compactness. We will restrict our comments to a description of some prevalent trends, which in general follow our expectations, rather than comparing each system to its counterparts. As anticipated, the most robust performance was achieved by all systems over AWGN channels, typically requiring a channel SNR of about 11–12 dB, when using the 16QAM mode of operation. In the 4QAM mode, the minimum required channel SNR was around 6–7 dB. The benefits of using ARQ over AWGN channels were limited to a chan-

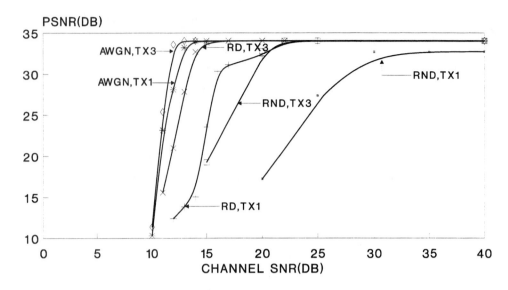

Figure 14.15: PSNR versus channel SNR performance of the 16QAM mode of operation of the QT codec-based video Systems 1 and 3 in Table 14.6 [496]. ©IEEE, 1996, Streit, Hanzo.

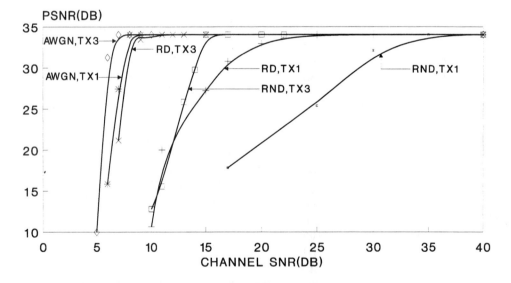

Figure 14.16: PSNR versus channel SNR performance of the 4QAM mode of operation of the QT codec-based video Systems 1 and 3 in Table 14.6 [496]. ©IEEE, 1996, Streit, Hanzo.

14.8. QT-BASED VIDEO-TRANSCEIVER PERFORMANCE

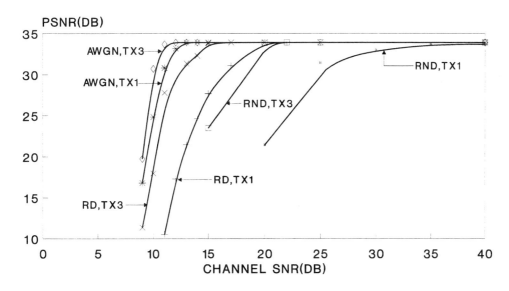

Figure 14.17: PSNR versus channel SNR performance of the 16QAM mode of operation of the QT codec-based video Systems 2 and 4 in Table 14.6 [496]. ©IEEE, 1996, Streit, Hanzo.

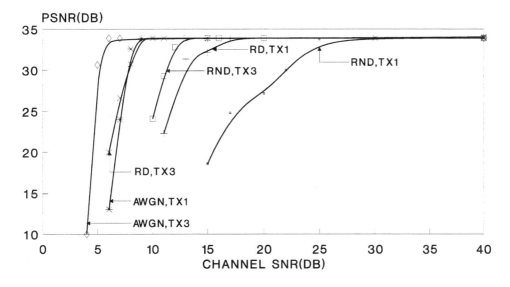

Figure 14.18: PSNR versus channel SNR performance of the 4QAM mode of operation of the QT codec-based video Systems 2 and 4 in Table 14.6 [496]. ©IEEE, 1996, Streit, Hanzo.

nel SNR reduction of about 1 dB in almost all modes of operation while reducing the number of users supported by two. This was a consequence of the stationary, time-invariant channel statistics, which inflicted similar channel impairments during re-transmission attempts. In contrast, for Rayleigh channels, typically the error-free reception probability of a packet was substantially improved during the ARQ attempts, effectively experiencing time diversity. Hence during re-transmissions the PS often emerged from the fade by the time ARQ was invoked. The employment of the stronger BCH(127,50,13) code failed to substantially reduce the required channel SNRs over Gaussian channels.

Over Rayleigh channels, the effect of diversity was extremely powerful in the case of all systems, typically reducing the minimum channel SNR requirements by about 10 to 12 dB in the 4QAM mode and even more in the 16QAM modes of operation for both the weaker and stronger BCH codes. This was a ramification of removing the residual BER of the modems. In the 4QAM mode without diversity, the stronger BCH(127,50,13) code reduced the channel SNR requirement by about 8 dB and with diversity by 5 dB, when compared to the systems using the weaker BCH code. On the same note, in the 16QAM mode without diversity the BCH(127,50,13) code managed to eradicate the modem's residual BER. With diversity, its supremacy over the higher user-capacity, lower complexity BCH(127,71,9) coded system eroded substantially to about 2 dB. This indicated that in System 1 the weaker BCH code struggled to remove the modem's residual BER, but with diversity assistance the employment of the stronger code became less important.

Similarly dramatic improvements were observed when using ARQ, especially when no diversity was employed or the weaker BCH code was used. Again, this was due to removing the modem's residual BER. Specifically, in the 4QAM mode without diversity and with the BCH(127,71,9) code, an SNR reduction of about 16 dB was possible due to ARQ, which reduced to around 12 dB in case of the stronger code. Remaining in the 4QAM mode but invoking diversity reduced the ARQ gain to about 7 dB for the inherently more robust BCH(127,50,13) code, while the weaker code's ARQ gain in System 3 was only slightly less with diversity than without, amounting still to about 12 dB. Similar findings pertain to the 16QAM ARQ gains when comparing Systems 1 and 3 as well as 2 and 4, which were of the order of 10 dB without diversity and about 5 dB with diversity for both BCH codes.

Both diversity and ARQs mitigate the effects of fading, and in broad terms they achieve similar robustness. However, the teletraffic reduction due to the reserved time-slots is a severe disadvantage of the ARQ assistance. Diversity, on the other hand, can readily be employed in small handsets operating at frequencies in excess of 1 GHz, as was demonstrated by the JDC second-generation Japanese digital voice handset of Table 9.1 on page 343, which is an attractive shirt-pocket-sized consumer product. When using a combination of diversity and ARQ, near-Gaussian performance was maintained over fading wireless channels effectively by all four systems but, as evidenced by Table 14.6, at a concomitant low carried teletraffic.

Consequently, the number of users supported by the different systems and operating modes varies over a wide range — between 3 and 16 — and the corresponding effective user bandwidth becomes 66.7 kHz and 12.5 kHz, respectively. The improved robustness of some of the ARQ-assisted schemes becomes less attractive in the light of their significantly reduced teletraffic capacity. It is interesting to compare in robustness terms the less complex 4QAM mode of System 1 supporting eight users with the more complex 16QAM mode of System 4, which can accommodate a similar number of users, namely nine. The 4QAM mode necessitates a channel SNR of about 7 dB over AWGN channels, while the more complex 16QAM requires about 11 dB, indicating a clear preference in favor of the former. Interestingly, the situation is reversed over the diversity-assisted Rayleigh channel, requiring SNR values of 20 and 14 dB, respectively. A range of further interesting system design aspects can be inferred from Figures 14.15–14.18 and Table 14.6 in terms of system complexity, robustness, and carried teletraffic.

14.9 Summary of QT-Based Video Transceivers

This chapter has presented a range of reconfigurable QT-based, parametrically enhanced wireless videophone schemes. The various systems contrived are portrayed in Table 14.6, and their PSNR versus channel SNR performances are characterized by Figures 14.15–14.18. The QT codecs contrived have a programmable bit rate, which can vary over a wide range, as demonstrated by Figure 14.11, and hence they lend themselves to a variety of wireless multimedia applications. Because of this video quality/bit-rate flexibility, these video codecs are ideal for employment in intelligent wireless adaptive multimode terminals. Before these multimode terminals become a practical market reality, the proposed video codec can be used for wireless videophony over fixed-rate mobile radio speech channels, such as the Pan-European GSM system [468] at a rate of 13 kbps, although preferably higher video rates should be maintained in order to ensure a sufficiently high video quality for high-dynamic video sequences. In this respect, cordless telephone systems, such as the Digital European Cordless Telecommunications (DECT) of Table 9.1 on page 343 scheme are preferable, since their speech rate is 32 kbps and so bandwidth is not at such a premium. Even higher system flexibility is offered by the emerging 3G systems of Table 10.2 on page 394. A further potential application can be found in surveillance systems. Our future work is targeted at improving the complexity, video quality, and robustness of the video codec and the overall system performance by contriving adaptive system reconfiguration algorithms. Let us now briefly summarize our findings presented in the context of proprietary fixed-rate QCIF videophone codecs in this part of the book and compare them to some of the existing standard-based video codecs.

14.10 Summary of Low Rate Video Codecs and Transceivers

In Chapter 11, we studied several fractal codecs in terms of image quality, computational complexity, and compression ratio tradeoffs. Two of the candidate codecs, a 0.28 bpp and a 1.1 bpp fractal codec, were also subjected to bit-sensitivity analysis. We noted a drawback of the fractal technique in terms of its incompatibility with motion compensation, which prohibits fractal codecs from achieving high compression ratios.

In Section 12.1, we reviewed a range of popular MC techniques and analyzed their efficiency in light of the bit-rate investment required for the MVs. We found that the best performance was achieved when using cost/gain-controlled MV encoding. In Chapter 12, we used these results in order to contrive a fully cost–gain controlled DCT-based coding scheme, which exhibits a constant but programmable bit rate. Two different schemes — the more bandwidth-efficient but more error-sensitive Codec I and the higher bit-rate but more robust Codec II — were proposed for various applications. In Chapter 13, we derived a set of VQ codecs, which guaranteed similar or better performances. Lastly, in Chapter 14 we presented a range of quad-tree-based schemes, which incorporated a parametric encoding enhancement option for the critical eye–mouth regions.

All codecs have been subjected to bit-sensitivity analysis in order to assess their performance under erroneous conditions and to assist in the design of source-matched FEC schemes for various channels. The associated transmission issues have been discussed in depth in [426, 467, 488, 496], respectively, for the various DCT-, VQ-, and QT-based video codecs. The performances of the various transceivers contrived were compared in Sections 12.13, 13.5, and 14.7 and in Tables 12.5, 13.3, and 12.5, respectively.

Let us finally compare the DCT-, VQ-, and QT-based codecs proposed to widely used standard codecs, namely, the MPEG-2 [499], H261 [500], and H263 [192] codecs. The latter standard schemes are variable-rate codecs, which make extensive use of variable-length compression techniques, such as RL-coding and entropy coding [26]. The H.261 and H.263 coding family will be detailed in the next part of this book. These standard codecs also require the transmission of intra-frame coded frames, which is carried out at selectable regular intervals. In contrast to our fixed-rate, distributed approach, however, these standard arrangements include a full intra-coded (I) frame, yielding a regular surge in the bit rate. This is unacceptable in fixed-rate mobile radio systems. In addition to the I frames, the MPEG-2 codec uses two more modes of operation, namely, predicted frames (P) and bidirectional (B) coding modes. Motion estimation is used in both frame types, while the P frames rely on additional differential coding strategies, invoked with reference to the surrounding I and B frames.

In the experiments we portrayed in Figure 14.19, we stipulated a fixed bit rate of 10 kbps for our three prototype codecs and adjusted the parameters of the H261, H263, and MPEG-2[3] codecs, in order to provide a similar video quality associated with a similar average PSNR performance. The corresponding PSNR curves are displayed in Figure 14.20. Observe in Figure 14.19 that the number of bits/frame for our proposed codecs is always 1000 and averages about twice as high at 22 kbps for the H261 codec, exhibiting a random fluctuation for the H261 codec. The MPEG codec exhibits three different characteristic bit rates, corresponding to the I, B, and P frames in decreasing order from around 8000 bits/frame to about 1800 and 1300, respectively. The H263 standard superseded the H261 and shows drastic improvements in terms of bit rate at a similar PSNR performance.

---

[3]The sequences may be viewed under the World Wide Web address http://www-mobile.ecs.soton.ac.uk/jurgen/vq/vq.html

14.10. SUMMARY OF LOW-RATE CODECS/TRANSCEIVERS 597

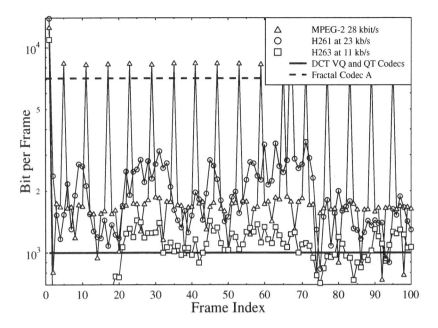

Figure 14.19: Bit-rate comparison of the previously presented proprietary 10 kbps codecs and three standard codecs, namely, H.261, H.263, and MPEG-2.

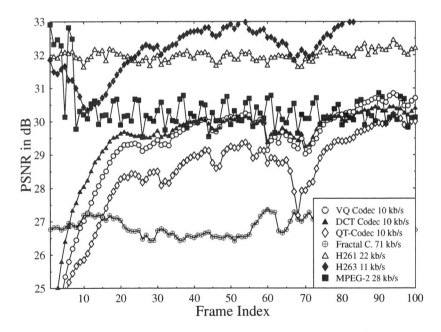

Figure 14.20: Performance of the previously presented proprietary 10 kbps codecs and three standard codecs, namely, H.261, H.263, and MPEG-2.

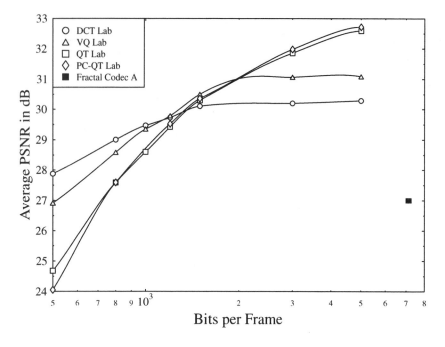

Figure 14.21: Performance versus bit-rate comparison of the proposed programmable-rate codecs.

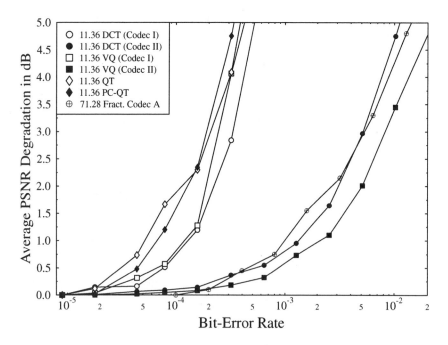

Figure 14.22: PSNR degradation versus BER for various fixed-rate codecs.

14.10. SUMMARY OF LOW-RATE CODECS/TRANSCEIVERS

A direct comparison of five codecs in Figures 14.19–14.22 reveals the following findings:

- Our programmable-rate proprietary codecs achieve a similar performance to the MPEG-2 codec at less than half the bit rate. The H-261 codec at 22 kb/s outperforms our codecs by about 2 dB in terms of PSNR. Note furthermore that our fixed-rate DCT and VQ codecs require about 20 frames to reach their steady-state video quality owing to the fixed bit-rate limitation, which is slightly prolonged for the QT codec. At the same average bit rate, the H263 standard shows a better PSNR performance than our codecs. The subjective difference seems to be less drastic than the objective measure indicates.

- The delay of our codecs and that of the H-261 codecs is in principle limited to one frame only. The delay of the MPEG-2 and H263 codecs may stretch to several frames owing to the P frames. In order to smooth the teletraffic demand fluctuation of the MPEG-2 and H263 codecs, typically adaptive feedback-controlled output buffering is used, which further increases the delay by storing the bits that cannot be instantaneously transmitted in a buffer.

- The error resilience of those specific codecs, which use the active/passive table concept, is very limited, as is that of the standard codecs. These arrangements have to invoke ARQ assistance over error-prone channels or the packet dropping regime of Section 19.3.6. Hence, in these codecs single-bit errors can corrupt an entire frame or several frames in the case of the MPEG-2 codec. Error-resilient techniques such as strong Forward Error Correction (FEC) or Error-Resilient Positional Codes (ERPC) [462,501] are required to protect such schemes. These problems are avoided by the less bandwidth efficient non-run-length encoded DCT, VQ, and QT schemes, which therefore exhibit a strongly improved error resilience.

In conclusion, compression ratios as high as 200 were achieved for ten frames/s QCIF sequences by the proposed programmable-rate proprietary codecs, while maintaining peak signal-to-noise ratios (PSNR) around 30 dB, at implementational complexities around 20 MFLOP. The fractal and quad-tree codecs were inferior to the DCT and VQ codecs, with the VQ exhibiting a somewhat improved high-frequency representation quality in comparison to the DCT codec. As stated before, the various system design tradeoffs become explicit from Sections 12.13, 13.5, and 14.7 and Tables 12.5, 13.3, and 12.5, respectively. In the following part of the book, we will concentrate on the H261/H263 coding family.

Part III

High-Resolution Image Coding

Chapter 15

Low-Complexity Techniques

15.1 Introduction and Video Formats

Having compared a wide range of QCIF-sized monochrome codecs designed for very low bit rates, we now embark on high-quality color coding of higher resolution pictures spanning the range from CIF to High Definition Television (HDTV) resolution. We will also focus on the error resilience of the codecs and attempt to evaluate their suitability for mobile video transmission.

Later in this chapter, we will evaluate two low-complexity methods before we exploit our previous results concerning motion compensation and DCT in the high-quality domain in the context of Chapter 16. The results in this and the following chapter are based on experiments using the CIF "Miss America" sequence, the "Football" and "Susie" sequences and the "Mall" sequence at HDTV resolution. Their spatial resolutions are listed in Table 15.1 among other image formats.

Each sequence has been chosen to test the codecs' performance in particular scenarios. The "Miss America" sequence is of low motion and provides an estimate of the

Resolution	Dimension	Pixel/sec at 30 f/s	Applications
Sub-QCIF	128 × 96	0.37 M	Handheld mobile video and video conferencing via public phone networks
QCIF	176 × 144	0.76 M	
CIF	352 × 288	3.04 M	Videotape recorder quality
CCIR 601	720 × 480	10.40 M	TV
4CIF	704 × 576	12.17 M	
HDTV 1440	1440 × 960	47.00 M	Consumer HDTV
16CIF	1408 × 1152	48.66 M	
HDTV	1920 × 1080	62.70 M	Studio HDTV

Table 15.1: Image Formats, Their Dimensions, and Typical Applications

maximum achievable compression ratio of a codec. The "Football" sequence contains pictures of high motion activity and high contrast. All sequences were recorded using interlacing equipment. *Interlacing* is a technique that is often used in image processing in order to reduce the required bandwidth of video signals, such as, for example, in conventional analog television signals, while maintaining a high frame-refresh rate, hence avoiding flickering and video jerkiness. This is achieved by scanning the video scene at half the required viewing-rate — which potentially halves the required video bandwidth and the associated bit rate — and then displaying the video sequence at twice the input scanning rate, such that in even-indexed video frames only the even-indexed lines are updated before they are presented to the viewer. By contrast, in odd-indexed video frames only the odd-indexed lines are updated before they are displayed, relying on the human eye and brain to reconstruct the video scene from these halved scanning rate even and odd video fields. Therefore, every other line of the interlaced frames remains un-updated.

For example, for Frame 1 of the interlaced "Football" sequence in Figure 15.1 we observe that a considerable amount of motion took place between the two re-coding instants of each frame, which correspond to the even and odd video fields. Furthermore, the "Susie" sequence was used in our experiments in order to verify the color reconstruction performance of the proposed codecs, while the "Mall" sequence was employed in order to simulate HDTV sequences with camera panning. As an example, a range of frames for each QCIF video sequence used is shown in Figure 15.2. QCIF resolution images are comprised of 176 × 144 pixels and are suitable for wireless handheld videotelephony. The 4×CIF resolution images are suitable for digital television, which are sixteen times larger than QCIF images. A range of frames for the 4×CIF video sequences is shown in Figure 15.1. Finally, in Figure 15.3 we show a range of frames from the 1280 by 640-pixel "Mall" sequence. However, since the 16CIF resolution is constituted by 1408 × 1152 pixels, a black border was added to the sequences before they were coded.

We processed all sequences in the YUV — color space [26] where the incoming picture information consists of the luminance (Y) plus two color difference signals referred to as Chrominance U (Cr_u) and Chrominance V (Cr_v). The conversion of the standard Red-Blue-Green (RGB) representation to the YUV format is defined in Equation 15.1:

$$\begin{bmatrix} Y \\ U \\ V \end{bmatrix} = \begin{pmatrix} 0.299 & 0.587 & 0.114 \\ -0.146 & -0.288 & 0.434 \\ 0.617 & -0.517 & -0.100 \end{pmatrix} \begin{pmatrix} R \\ G \\ B \end{pmatrix}. \qquad (15.1)$$

It is common practice to reduce the resolution of the two color difference signals by a factor of 2 in each spatial direction, which inflicts virtually no perceptual impairment and reduces the associated source data rate by 50%. More explicitly, this implies that instead of having to store and process the luminance signal and the two color difference signals at the same resolution, which would potentially increase the associated bit rate for color sequences by a factor of three, the total amount of color data to be processed is only 50% more than that of the associated gray-scale images. This implies that there are only one Cr_u and one Cr_v pixel for every four luminance pixels allocated.

The coding of images larger than the QCIF size multiplies the demand in terms of

15.1. INTRODUCTION AND VIDEO FORMATS

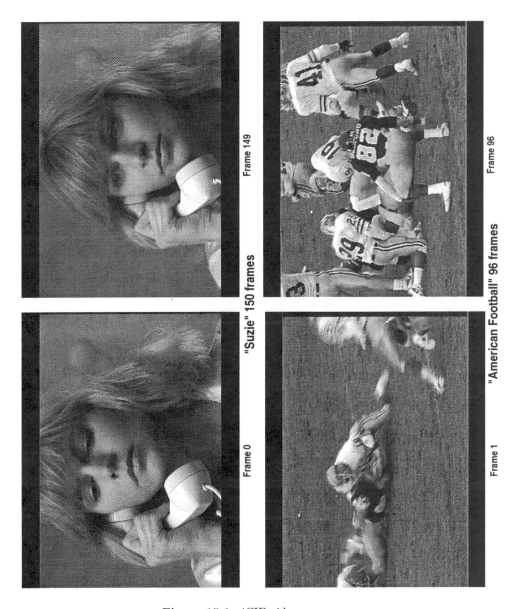

Figure 15.1: 4CIF video sequences.

Figure 15.2: QCIF video sequences.

15.1. INTRODUCTION AND VIDEO FORMATS

Figure 15.3: 16CIF "Mall" video sequence.

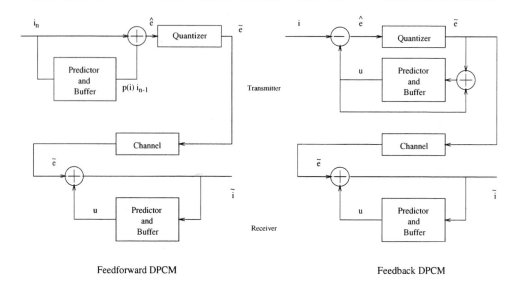

Figure 15.4: Feedforward (left) and feedback (right) DPCM schematic.

computational complexity, bit rate, and required buffer size. This might cause problems, considering that a color HDTV frame requires a storage of 6 MB per frame. At a frame rate of 30 frames/s, the uncompressed data rate exceeds 1.4 gigabits per second. Hence, for real-time applications the extremely high bandwidth requirement is associated with an excessive computational complexity. Constrained by this complexity limitation, we now examine two inherently low-complexity techniques and evaluate their performance.

15.2 Differential Pulse Code Modulation

Differential Pulse Code Modulation (DPCM) is one of the simplest techniques in image coding, and though known for a very long time, it still attracts some interest as recent publications [502, 503] demonstrate. It requires virtually no buffering of the incoming signal and only a single multiplication and subtraction step per pixel.

15.2.1 Basic Differential Pulse Code Modulation

DPCM is a predictive technique whose aim is to remove mutual redundancy between adjacent samples. The idea is to predict the current sample i_n from the previous sample i_{n-1}. Instead of transmitting the sample i_n, we transmit the error of the prediction $e_n = i_n - p(i) \cdot i_{n-1}$, where $p(i)$ is the prediction function. If the prediction is successful, the energy $E(e_n)$ of the signal e_n is lower than that of the incoming signal. Hence we can reduce the number of reconstruction levels of the quantizer. The simple schematic of forward and backward predictive coding is depicted in Figure 15.4, and their differences will become explicit during our further discussions. The estimates of the feedforward scheme are based on unquantized input samples, which — following

15.2. DIFFERENTIAL PULSE CODE MODULATION

a somewhat simplistic argument — would result in better predictions at the encoder than the estimates of the backward predictive scheme that are based on quantized pixels. However, since unquantized pixels are not available at the decoder, the forward predictive scheme is typically outperformed by the backward predictive arrangement.

In video coding the prediction function $p(i)$ is often constituted by a constant a, which is referred to as a predictor coefficient. Hence the prediction error signal is defined as $e_n = i_n - a \cdot i_{n-1}$. If we aim to minimize the energy of e_n given by:

$$E(e_n) = E(i_n - a \cdot i_{n-1}) = \frac{1}{m} \sum_{n=1}^{m} (i_n - a \cdot i_{n-1})^2 \qquad (15.2)$$

we set the first derivative of Equation 15.2 with respect to a to zero, leading to:

$$\begin{aligned} \frac{\delta E(e_n)}{\delta a} &= -2 \cdot \frac{1}{m} \sum_{n=1}^{m} (i_n - a \cdot i_{n-1}) i_{n-1} \\ &= -2 \cdot \left[\frac{1}{m} \sum_{n=1}^{m} i_n i_{n-1} - a \cdot \sum_{n=1}^{m} i_{n-1}^2 \right] = 0. \qquad (15.3) \end{aligned}$$

If we define the autocorrelation of i_n as $\rho_x(i_n)$, this finally leads to:

$$2 \cdot [\rho_1(i_n) - a\rho_0(i_n)] = 0 \implies a = \frac{\rho_1(i_n)}{\rho_0(i_n)}, \qquad (15.4)$$

which is the one-step correlation of the video signal. The prediction gain of the operation is obtained by evaluating Equation 15.2 for the optimum predictor coefficient a, which is given by:

$$gain[dB] = 10 \cdot \log_{10} \frac{E(i_n)}{E(e_n)} = 10 \cdot \log_{10} \frac{1}{2(1-a^2)} \qquad (15.5)$$

Typical values for the adjacent-pixel correlation a in the case of images are in the range of $[0.9..0.99]$, resulting in a prediction gain ranging from 4.2 dB to 14 dB. The benefits of DPCM are exemplified in Figure 15.5, where we demonstrate how well this simple predictive technique compresses the widely spread PDF of the incoming video signal i_n shown at the left of the figure to values between -5 and 5 seen at the right of Figure 15.5. Note that the value of the parameter a affects the error resilience of the codec due to the predictive nature of the scheme. In the extreme case of $a = 1.0$, which is equivalent to using the previous pixel value to predict the current one, the reconstructed signal is fed back in the predictive loop without amplitude attenuation. Thus, eventual channel errors will be accumulated, which continuously affects the result. Practical values of a in video coding are in the range of $a \in [0.85..0.95]$, which assists in reducing the error propagation.

The disadvantage of feedforward prediction is that the quantization error results in a misalignment of the encoder's and decoder's reconstruction buffers because the decoder predicts the signal on the basis of the quantized signal. In order to overcome this problem, feedback prediction combined with local decoding is preferred in most

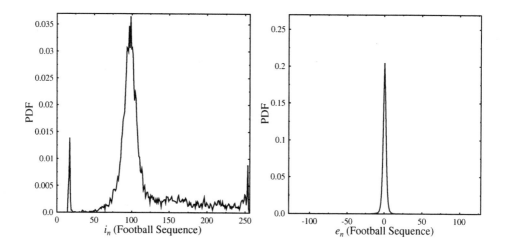

Figure 15.5: PDF of a typical frame (left) and the PDF of the prediction error signal (right).

applications, as seen at the right-hand side of Figure 15.4, unless the prediction residual can be noiselessly encoded. A simple possible implementation of such a forward predictive codec is based on the fact that the prediction residual signal is represented by a finite set of integer values, which can be noiselessly entropy coded.

15.2.2 Intra/Inter-Frame Differential Pulse Code Modulation

In order to achieve a reasonable compressed bit rate below 4.5 bpp, we require coarse quantizers with eight or fewer reconstruction levels. Initially, we collected training data from our test sequences and trained a Max-Lloyd quantizer [26] for various quantizer resolutions. This process was carried out using both intra-frame and inter-frame training sequences. However, we found that extreme positive and negative samples of the training prediction error data resulted in inadequate quantizers, since the quantizer steps were too large. For a limited number of quantizers, it constituted no problem to find better quantizers by manually adjusting the reconstruction and decision levels, which led to the results depicted in Figure 15.6, where we optimized the quantizers for the "Miss America" sequence in case of both intra- and inter-frame coding. Figure 15.6 reveals that for intra-frame coding, quantizers with three to eight levels resulting in a bit rate of $1.5 \cdot \log_2(3) = 2.38$ to $1.5 \cdot \log_2(8) = 4.5$ bits per pixel are necessary to achieve a good quality, where the 1.5 multiplier is due to the 50% increase required by the subsampled color information. Furthermore, the video performance depends on the training sequence and the sequence to be coded. Observe, for example, that the performance results for the "Mall" sequence were significantly reduced in comparison to the "Miss America" based results.

We obtained similar results for inter-frame coded images. We opted to use simple frame-differencing as motion compensation and retrained the quantizers, since the

15.2. DIFFERENTIAL PULSE CODE MODULATION

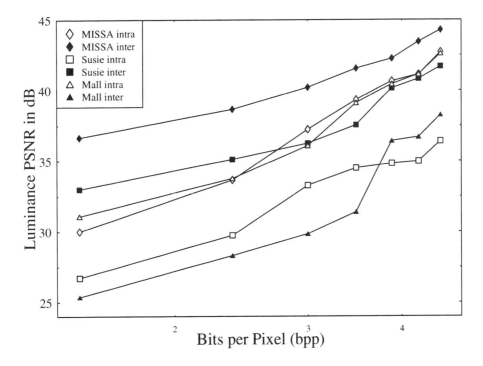

Figure 15.6: Intra-frame and inter-frame DPCM coding performance for various bit rates using symmetrical and asymmetrical quantizers for color images.

statistics of the frame difference signal are different from those of the video signal itself. As demonstrated by Figure 15.6, the PSNR increases in excess of 5 dB are possible in case of the "Susie" and "Miss America" sequence, which are similar. Again we found that the quantizers derived were suboptimum for other sequences, such as the "Mall" sequence.

15.2.3 Adaptive Differential Pulse Code Modulation

The performance of DPCM can be improved if we use a more sophisticated predictor. Jayant [504] was the first to design quantizers, which are adaptive to the incoming samples, leading to the concept of adaptive DPCM or ADPCM. The idea is that the input data is observed, and if, for example, the quantizer is continuously overloaded, since the encoded samples are always assigned the highest reconstruction level, the quantizer step size will be scaled up by a certain factor. Conversely, the quantizer will be scaled down in case of the complementary event. The predictor adaptation is based on the reconstructed samples; hence, no additional side information has to be transmitted.

The prediction might be based on the previously transmitted sample [504] or, more often, on several surrounding pixels, as, for example, suggested by [505–507]. Below we illustrate the performance of the simplest case of predicting the quantizer step-size

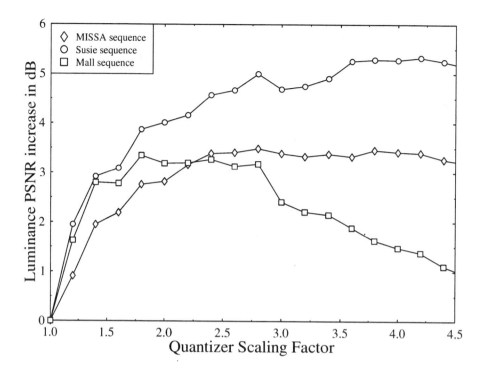

Figure 15.7: Intra-frame coded ADPCM PSNR performance improvement for various quantizer scaling factors.

scaling on the basis of the previous sample. Remembering that a three-level quantizer results in a color-coded bit rate of more than 2.4 bpp when using bit packing or 3 bpp without, we attempted to investigate whether it was possible to improve the coding performance at this bit rate according to our required quality. Optimum step-size gain factors for Gaussian sources have been derived by Jayant [504], but because of the dissimilarity of standard distributions and that of images, an empirical solution is preferred. We forced the quantizer step size to be scaled up by a given scaling factor or to be scaled down by the reciprocal value. The results of the corresponding experiments show that a PSNR improvement in excess of 4 dB is possible, as evidenced by Figure 15.7. However, remembering the poor performance of the simple DPCM codec in Figure 15.6, we concluded that despite its implementational simplicity, DPCM and ADPCM are not suitable for high-quality, high-compression video coding. The possible PSNR improvements due to using more sophisticated two-dimensional predictors are limited to the range of about one dB [505] and so will not result in sufficient improvements necessary for this technique to become attractive for our purposes. Let us now concentrate on another low-complexity technique referred to as *block truncation coding*.

15.3 Block Truncation Coding

Block Truncation Coding (BTC) attracted continuous interest from researchers after it was proposed by Mitchell and Delp [508]. The BTC algorithm is very simple and requires limited memory, as only a few image scanlines have to be stored before encoding takes place. These are very useful features for real-time encoding of high-resolution image frames, such as HDTV sequences. BTC is also known for its error resilience, since no RL-encoding or similar variable-length coding methods are used.

15.3.1 The Block Truncation Algorithm

BTC is a block-based coding method based on the typical block sizes of 4×4 or 8×8, where every single pixel in the block is quantized with the same two-level quantizer. Hence, the attributes of BTC codecs are strongly dependent on the properties of this two-level quantizer. There are several approaches to designing the BTC quantizers.

According to a simple and plausible approach, the BTC arrangement's reconstruction levels are determined on the basis of assuming an identical mean and variance or first and second moment for the original and encoded blocks. This allows us to derive a set of two simultaneous equations for finding the higher and lower reconstruction levels, respectively. Following this approach, we define the ith moment of the block as:

$$\overline{X^i} = \frac{1}{b^2} \sum_{k=1}^{b} \sum_{l=1}^{b} b^i_{(k,l)}, \tag{15.6}$$

where the first moment is the mean $\overline{X}$ and the second moment is the energy $\overline{X^2}$ of the block to be encoded. Assuming that X_{th} is a given threshold and assigning each pixel to the low or high reconstruction levels q_L or q_H, we define each BTC encoded pixel $\hat{b}_{(k,l)}$ of the block as follows:

$$\hat{b}_{(k,l)} = \begin{cases} q_L & \text{if } b_{(k,l)} < X_{th} \\ q_H & \text{otherwise.} \end{cases} \tag{15.7}$$

If we aim to preserve the first and second moment $\overline{X^1}$ and $\overline{X^2}$ of the original uncoded block, we obtain a set of two equations [509]:

$$\overline{X} = \frac{1}{b^2} \sum_{k=1}^{b} \sum_{l=1}^{b} \hat{b}_{(k,l)} = \frac{1}{b^2}((b^2 - n)q_L + nq_H) \tag{15.8}$$

$$\overline{X^2} = \frac{1}{b^2} \sum_{k=1}^{b} \sum_{l=1}^{b} \hat{b}^2_{(k,l)} = \frac{1}{b^2}((b^2 - n)q_L^2 + nq_H^2), \tag{15.9}$$

where n denotes the number of pixels exceeding the threshold X_{th}. Solving Equations 15.8 and 15.9 for q_L and q_H leads to:

$$q_L = \overline{X} - \sigma\sqrt{\tfrac{n}{b^2-n}} \tag{15.10}$$

$$q_H = \overline{X} + \sigma\sqrt{\tfrac{b^2-n}{n}} \tag{15.11}$$

Sequence	"Susie"	"Football"	"Mall"
PSNR (dB) 4×4 BTC moment method	37.83	31.80	37.68
PSNR (dB) 4×4 BTC MAD method	38.17	32.14	38.17
PSNR (dB) 8×8 BTC moment method	34.40	28.92	33.70
PSNR (dB) 8×8 BTC MAD method	34.77	29.33	34.04

Table 15.2: PSNR Upper Bounds for BTC Using Unquantized q_L and q_H Values for the "Susie," "Football," and "Mall" Sequences

$$\sigma = \sqrt{\overline{X^2} - \overline{X}^2},$$

where σ represents the standard deviation of pixel values in the block. The only unknown variable remains the threshold X_{th}. Delp argued [509] that setting X_{th} to $\overline{X}$ was a plausible choice. In fact, it is not too arduous a task to attempt setting the threshold to all possible levels and evaluate the associated coding performance. When we carried out this experiment, we found that there was only a negligible potential performance advantage due to optimizing the decision threshold.

Another possible technique of designing the BTC reconstruction levels would be to optimize x_L and x_H in the MSE or MAD sense. Accordingly, q_L is set to the median of all pixels below the threshold, and correspondingly, q_H is set to the median of all pixels exceeding the threshold X_{th} as seen in:

$$q_L = \frac{1}{b^2 - n} \sum_{k=1}^{b} \sum_{l=1}^{b} b_{(k,l)} \quad \forall \ b_{(k,l)} < X_{th} \quad (15.12)$$

$$q_H = \frac{1}{n} \sum_{k=1}^{b} \sum_{l=1}^{b} b_{(k,l)} \quad \forall \ b_{(k,l)} \geq X_{th}. \quad (15.13)$$

We examined the upper bound performance of BTC using both of these methods by leaving the q_L and q_H values unquantized. We obtained the results shown in Table 15.2. The upper bound values for the PSNR are around 38 dB in case of a block size of 4×4 and 34 dB for a block size of 8×8, which shows the limitations for this technique. This is not an impressive PSNR performance, since the binary quantized values q_L and q_H require a bit rate of one bit per pixel. Hence, the minimum bit rate is in excess of 1.5 bits per pixel for color blocks, where the chrominance information is stored at half the luminance resolution. Other optimization methods such as third-moment optimization [510, 511] exhibit similar performances.

15.3.2 Block Truncation Codec Implementations

The bit-stream generated by the BTC codec contains two different types of data:

(1) the 1 bit per pixel bit map conveying the q_L and q_H information as to whether a given pixel is above or below the threshold; and

(2) the quantized values of either σ and $\overline{X}$ or q_L and q_H.

A large body of research has been devoted to determining the optimum selection of quantizers [509], and Delp found that allocating more than 8 bits for each quantized value results in no significant performance improvement. In a second step, Delp jointly

15.3. BLOCK TRUNCATION CODING

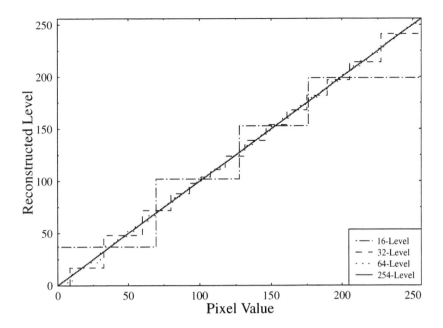

Figure 15.8: Trained Max-Lloyd quantizers characteristics for q_H at a block size of 4×4 for various quantizer resolutions.

quantized σ and $\overline{X}$ and achieved satisfying BTC performance results at around 10 bits per σ and $\overline{X}$ pair. This bit assignment results in a bit rate of

$$\frac{4 \times 4 \text{ bit} + 10 \text{ bit}}{4 \times 4 \text{ pixel}} = 1.625 \text{ bpp}$$

for monochrome 4×4 blocks and

$$\frac{8 \times 8 \text{ bit} + 10 \text{ bit}}{8 \times 8 \text{ pixel}} = 1.16 \text{ bpp}$$

for 8×8 monochrome blocks, respectively. In case of color images and subsampled Cr_u and Cr_v planes, these values increase by a factor of 1.5 to 2.44 bpp and 1.73 bpp, respectively.

15.3.3 Intra-Frame Block Truncation Coding

In our further investigations, we initially focused on intra-frame BTC schemes and opted for a direct quantization of q_L and q_H. We collected sample data from three image sequences separately for our two different block sizes of 8×8 and 4×4 and used appropriately trained Max-Lloyd quantizers [512] for various resolutions ranging from 2 to 8 bits. The optimized Max-Lloyd quantizers for q_H and a block size of 4×4 are portrayed in Figure 15.8. As an example, the set of q_L and q_H reconstruction levels

Recon. Lv.	1	2	3	4	5	6	7	8
4×4 Intra q_H	25	71	95	106	125	157	186	222
4×4 Intra q_L	18	46	72	87	98	127	160	203
4×4 Inter q_H	-76	-39	-13	5	13	37	66	103
4×4 Inter q_L	-105	-68	105	-12	-3	14	37	70

Table 15.3: An eight-level (3-bit) Quantizer Reconstruction Table for 4×4 Sized Blocks, Trained for Inter- and Intra-coded Frames.

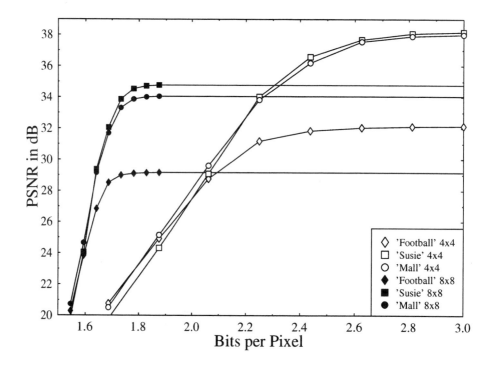

Figure 15.9: PSNR versus bit-rate intra-frame coded BTC codec performance for block sizes of 4 and 8×8 using the 4CIF and 16CIF sequences of Figures 15.1 and 15.3.

obtained for the 3-bit, eight-level quantizer is summarized in Table 15.3. Observe in the figure that the quantizers are very similar to simple linear quantizers in the range of [0..255]. Therefore, if the training sequence is large and statistically independent from the source sequences, the quantizers for BTC become almost uniform quantizers. The quantizers for q_L share the same characteristics with a negative x-axis offset of about one step size.

The results in terms of PSNR versus bit-rate performance are shown in Figure 15.9. This figure suggests that the reconstructed quality saturates for quantizer resolutions in excess of 5 bits for q_L and q_H each. This coincides with the estimated quantizer resolution found in [509], where σ and $\overline{X}$ were jointly quantized using 10 bits. In

15.3. BLOCK TRUNCATION CODING

Sequence	"Susie"	"Football"	"Mall"
PSNR 4 × 4 BTC MAD in dB	38.18	31.93	38.00
PSNR 8 × 8 BTC MAD in dB	34.78	29.31	34.05

Table 15.4: PSNR Upper Bounds for Inter-frame BTC Derived by Using the Unquantized q_L and q_H Values for the "Susie," the "Football," and the "Mall" Sequences

contrast to the above jointly quantized scheme, in our implementation of the BTC codec the q_L and q_H values are obtained from a simple lookup table, which does not require the evaluation of Equations 15.10 and 15.11 and therefore exhibits a reduced complexity. Furthermore, the error resilience is improved because a bit error for jointly quantized values will affect the entire block, whereas in the latter case only n or $b^2 - n$ pixels are affected, depending on whether q_L or q_H was corrupted. The optimum tradeoff in terms of quality and bit rate lies around 2.6 bpp for a block size of 4 × 4, resulting in a PSNR of up to 38 dB. For 8 × 8 blocks the optimum coding rate is around 1.8 bpp, and the associated maximum PSNR was around 34 dB. These values are very close to the upper bounds reflected in Table 15.2 and hence underline that our quantization method is quite efficient. On the other hand, we will see in Chapter 16 that the codec is readily outperformed by a DCT-based codec. Let us now consider the performance of inter-frame BTC schemes.

15.3.4 Inter-Frame Block Truncation Coding

BTC coding is one of the few coding techniques that does not inflict any spatial frequency limitations. Other methods, such as DCT or Subband Coding (SBC), capitalize on the fact that most of the energy of a given source tends to reside in the lower spatial frequency domain. Therefore higher resolution quantizers are allocated in the lower frequency band in subband coding or for the low-frequency DCT coefficients in DCT coding. BTC may be considered an amplitude-variant, frequency-invariant filter. This is explained by the fact that any two-level black-and-white pattern will be reconstructed as a two-level pattern, which differs from the original one only by an offset and a scaling factor. On the other hand, blocks that contain more than two different pixel values will contain an irreversible coding error, even if σ and $\overline{X}$ are not quantized. This is the reason for the relatively low PSNR upper bounds seen in Table 15.4.

In inter-frame coding, we are encoding a highly nonstationary MCER signal, and hence we have to retrain the quantizers on this basis. A typical 3-bit BTC reconstruction level quantizer is exemplified in Table 15.3 for both intra- and inter-frame coding. Surprisingly, the ranges of the quantizers for inter- and intra-frame coding are rather similar. One might expect that the quantizer's range is more limited owing to the lower variance and to the more limited-range PDF of the MCER. On the other hand, the PDF of a typical MCER sequence potentially may cover a dynamic range twice as wide as the PDF of a normal image. These effects seem to cancel each other, when applying a Max-Lloyd quantizer. The PSNR versus bit-rate performance of the inter-frame BTC codec is depicted in Figure 15.10, which is slightly worse than that of the intra-frame codec, clearly indicating that the simple BTC algorithm is unsuitable

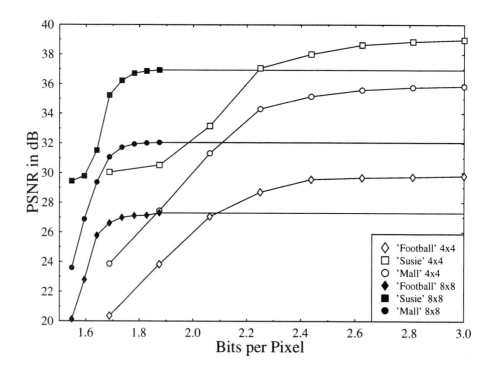

Figure 15.10: PSNR versus bit-rate performance of the inter-frame coded BTC codec for block sizes of 4 and 8 × 8.

for encoding the nonstationary MCER.

15.4 Subband Coding

Split-band or subband coding (SBC) techniques were first applied in the field of speech coding by Crochière [513, 514]. They became well established in this field before their potential for image coding was discovered [515–518]. Recently, the technique has attracted the interest of many researchers in the field of HDTV video coding [519–521], since its complexity is lower than that of other techniques, such as VQ- and DCT-based arrangements.

The concept of subband coding, as depicted in Figure 15.11, is similar to that of transform coding [26, 454] (TC), which was the topic of Chapter 12 in the context of DCT-based compression. As we have seen, in TC-based compression the temporal-domain video signal is transformed to the spatial frequency domain using an orthogonal transformation such as the DCT, where the highest-energy coefficients are quantized and transmitted to the decoder. As seen in Figure 15.11, the subband codec invokes a simpler frequency-domain analysis in that the signal is split by a bank of filters in a number of different frequency bands [522]. These bands can then be quantized independently, according to their perceptual importance. Note that in

15.4. SUBBAND CODING

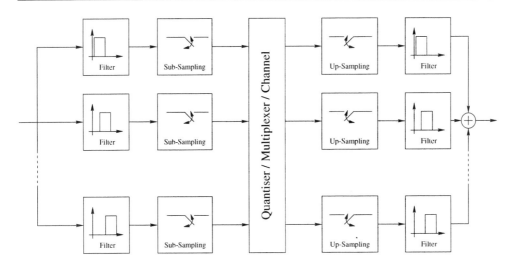

Figure 15.11: Schematic of a split-band codec.

its simplest implementation this filtering can be carried out after scanning the two-dimensional video signal on a line-by-line basis to a one-dimensional signal, where scanning is seamlessly continued at the end or beginning of each scanline by reading the pixel immediately below the last pixel scanned rather than returning always to the beginning of the next scanline in zig-zag fashion. This scanning pattern ensures that the adjacent pixels are always correlated.

Since the resulting filtered bandpass signals are of lower bandwidth than the original signal, they can be represented using a lower sampling rate than the full-band signal and so we can subsample them. This operation is also shown in Figure 15.11, which is followed by the quantization of the subband signals. This simple frequency-domain analysis allows the codec to identify which frequency band most of the video signal's energy resides in, facilitating the appropriate allocation of the encoding bits to the most important bands. If the full-band image contains low energy in a certain frequency band, the output of the corresponding filter also has low energy and vice versa. The subband signals can be losslessly encoded using entropy-coding schemes [523], which leads to relatively small compression ratios of up to 2.0. Higher compression ratios are possible if we apply quantizers optimized for the individual bands [524]. The original signal is recovered by superimposing the appropriately upsampled and filtered subband signals. The design of the band-splitting filters is crucial, since the overlapping spectral lobes result in aliasing distortion, which cannot be retrospectively removed.

In this chapter, we briefly review the appropriate filtering techniques used to prevent spectral domain aliasing, which are referred to as quadrature mirror filters (QMF). This discussion is followed by the design of the required subband quantizers. Instead of attempting to provide an in-depth treatment of subband coding, here we endeavor to design a practical codec in order to be able to relate its characteristics to those of other similar benchmarkers studied. Hence readers who are more interested in the practical codec design aspects rather than in the mathematical background of

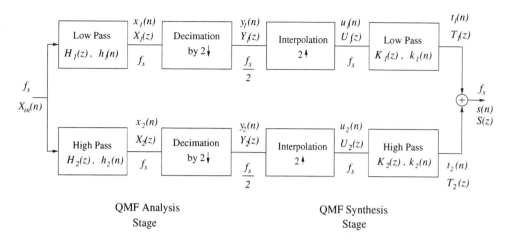

Figure 15.12: QMF analysis/synthesis arrangement.

QMFs may skip the next subsection.

15.4.1 Perfect Reconstruction Quadrature Mirror Filtering

15.4.1.1 Analysis Filtering

The success of QMF-based band-splitting hinges on the design of appropriate analysis and synthesis filters, which do not interfere with each other in their transition bands, that is, avoid introduction of the **aliasing distortion** induced by subband overlapping due to an insufficiently high sampling frequency, that is, due to unacceptable subsampling. If, on the other hand, the sampling frequency is too high, or for some other reason the filter bank employed generates a spectral gap, again the signal reporduction quality suffers. In simplistic approach this would imply employing filters having a zero-width transition band, associated with an infinite-steepness cutoff slope. Clearly, this would require an infinite filter order, which is impractical. As a practical alternative, Esteban and Galand [525] introduced an ingenious band-splitting structure referred to as a Quadrature Mirror Filter (QMF), and their construction will be detailed later. QMFs have a finite filter order and remove the spectral aliasing effects by cancellation in the overlapping frequency-domain transition bands.

As mentioned before, Quadrature Mirror Filters (QMF) were introduced by Esteban and Galand [525], while Johnston [526] designed a range of QMFs for a variety of applications. The principle of QMF analysis/synthesis filtering can be highlighted following their deliberations and considering the twin-channel scheme portrayed in Figure 15.12, where the subband signals are initially unquantized for the sake of simplicity. The corresponding spectral-domain operations can be viewed in Figure 15.13. Furthermore, for the sake of simplicity here we consider a one-dimensional band-splitting arrangement, assuming that the two-dimensional video signal was scanned in a one-dimensional sequence, as mentioned above.

If most of the energy of the signal is confined to the frequency $f_s/2$, it can be

15.4. SUBBAND CODING

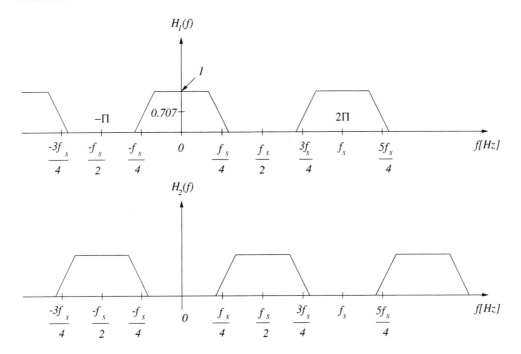

Figure 15.13: Stylized spectral domain transfer function of the lower- and higher-band QMFs.

bandlimited to this range and sampled at $f_s = 1/T = \omega_s/2\pi$, to produce the QMF's input signal $x_{in}(n)$, which is input to the QMF analysis filter of Figure 15.12. As seen in the figure, this signal is filtered by the low-pass filter $H_1(z)$ and the high-pass filter $H_2(z)$ in order to yield the low-band signal $x_1(n)$ and the high-band signal $x_2(n)$, respectively. Since the energy of $x_1(n)$ and $x_2(n)$ is now confined to half of the original bandwidths bandwidth of $x(n)$, the sampling rate of the subbands can be halved by discarding every second sample to produce the *decimated signals* $y_1(n)$ and $y_2(n)$.

In the subband synthesis stage of Figure 15.12 the decimated signals $y_1(n)$ and $y_2(n)$ are *interpolated* by inserting a zero-valued sample between adjacent samples, in order to generate the up-sampled sequences $u_1(n)$ and $u_2(n)$. These are then filtered using the z-domain transfer functions $K_1(z)$ and $K_2(z)$ in order to produce the discrete-time sequences $t_1(n)$ and $t_2(n)$, which now again have a sampling frequency of f_s and the filtering operation reintroduced nonzero samples in the positions of the previously injected zero in the process of interpolation. Finally, the $t_1(n)$ and $t_2(n)$ are superimposed on each other, delivering the recovered video signal $s(n)$.

Esteban and Galand [525] have shown that if the low-pass (LP) filter transfer functions $H_1(z), K_1(z)$ and their high-pass (HP) counterparts $H_2(z), K_2(z)$ satisfy certain conditions, perfect signal reconstruction is possible, provided the subband signals are unquantized. Let us assume that the transfer functions obey the following

constraint:
$$\left|H_1\left(e^{j\omega T}\right)\right| = \left|H_2\left(e^{j(\frac{\omega_s}{2}-\omega)T}\right)\right| \qquad (15.14)$$

where ω is the angular frequency, $2\pi = \omega_s$ and the imposed constraint implies a mirror-symmetric magnitude response around $f_s/4$, where the 3dB-down frequency responses, corresponding to $|H(\omega)| = 0.5$ cross at $f_s/4$. This can be readily verified by the following argument referring to Figure 15.12. Observe that $H_1(\omega)$ is equal to the $\omega_s/2 = \pi$-shifted version of the mirror image $H_2(-\omega)$, which becomes explicit by shifting $H_2(-\omega)$ to the right by $\omega_s/2 = \pi$ at the bottom of the figure. Furthermore, it can also be verified in the figure that by shifting $H_1(\omega)$ to the left by $\omega_s/2 = \pi$ the following relationship holds:

$$\left|H_2\left(e^{j\omega T}\right)\right| = \left|H_1\left(e^{-j(\frac{\omega_s}{2}-\omega)T}\right)\right|. \qquad (15.15)$$

Upon exploiting that:
$$\begin{aligned}
e^{-j(\frac{\omega_s}{2}-\omega)T} &= e^{-j(\pi-\omega T)} \\
&= \cos(\pi - \omega T) - j\sin(\pi - \omega T) \\
&= -\cos(\omega T) - j\sin(\omega T) \\
&= -e^{j\omega T}
\end{aligned} \qquad (15.16)$$

Equation 15.15 can also be written as:
$$\left|H_2\left(e^{j\omega T}\right)\right| = \left|H_1\left(-e^{j\omega T}\right)\right|, \qquad (15.17)$$

and upon taking into account that $z = e^{j\omega T}$, in z-domain we have $H_1(z) = H_2(-z)$. Following a similar argument, it can also be easily shown that the corresponding HP filters $K_1(z)$ and $K_2(z)$ also satisfy Equation 15.17.

Let us now show how the original full-band signal can be reproduced using the required filters. The z-transform of the LP-filtered signal $x_1(n)$ can be expressed as:

$$X_1(z) = H_1(z)X(z) \qquad (15.18)$$

or alternatively as:
$$X_1(z) = a_0 + a_1 z^{-1} + a_2 z^{-2} + a_3 z^{-3} + a_4 z^{-4} + \ldots \qquad (15.19)$$

where $a_i, i = 1, 2, \ldots$, are the z-transform coefficients. Upon decimating $x_1(n)$, we arrive at $y_1(n)$, which can be written in z-domain as:

$$Y_1(z) = a_0 + a_2 z^{-1} + a_4 z^{-2} + \ldots, \qquad (15.20)$$

where every other sample has been discarded and the previous even samples now become adjacent samples, which corresponds to halving the sampling rate. Equation 15.20 can also be decomposed to the following expression:

$$\begin{aligned}
Y_1(z) &= \frac{1}{2}\left[a_0 + a_1 z^{-\frac{1}{2}} + a_2(z^{-\frac{1}{2}})^2 + a_3(z^{-\frac{1}{2}})^3 + a_4(z^{-\frac{1}{2}})^4 + \ldots\right] \\
&+ \frac{1}{2}\left[a_0 + a_1(-z^{-\frac{1}{2}}) + a_2(-z^{-\frac{1}{2}})^2 + a_3(-z^{-\frac{1}{2}})^3 + \ldots\right] \\
&= \frac{1}{2}\left[X_1(z^{\frac{1}{2}}) + X_1(-z^{\frac{1}{2}})\right],
\end{aligned}$$

$$(15.21)$$

15.4. SUBBAND CODING

which represents the decimation operation in z-domain.

15.4.1.2 Synthesis Filtering

The original full-band signal is reconstructed by interpolating both the low-band and high-band signals, filtering them and adding them, as shown in Figure 15.12. Considering the low-band signal again, $y_1(n)$ is interpolated to give $u_1(n)$, whereby the injected new samples are assigned zero magnitude, yielding:

$$\begin{aligned} U_1(z) &= a_0 + 0 \cdot z^{-1} + a_2 z^{-2} + 0 \cdot z^{-3} + a_4 z^{-4} + \ldots \\ &= Y_1(z^2). \end{aligned} \quad (15.22)$$

From Figure 15.12, the reconstructed low-band signal is given by:

$$T_1(z) = K_1(z) U_1(z) \quad (15.23)$$

When using Equations 15.18 to 15.23, we arrive at:

$$\begin{aligned} T_1(z) &= K_1(z) U_1(z) \\ &= K_1(z) Y_1(z^2) \\ &= K_1(z) \frac{1}{2} [X_1(z) + X_1(-z)] \\ &= \frac{1}{2} K_1(z) [H_1(z) X(z) + H_1(-z) X(-z)]. \end{aligned} \quad (15.24)$$

Following similar arguments in the lower branch of Figure 15.12 as regards the high-band signal, we arrive at:

$$T_2(z) = \frac{1}{2} K_2(z) [H_2(z) X(z) + H_2(-z) X(-z)]. \quad (15.25)$$

Upon adding the low-band and high-band signals, we arrive at the reconstructed signal:

$$\begin{aligned} S(z) &= T_1(z) + T_2(z) \\ &= \frac{1}{2} K_1(z) [H_1(z) X(z) + H_1(-z) X(-z)] \\ &\quad + \frac{1}{2} K_2(z) [H_2(z) X(z) + H_2(-z) X(-z)]. \end{aligned}$$

This formula can be rearranged in order to reflect the partial system responses due to $X(z)$ and $X(-z)$:

$$\begin{aligned} S(z) &= \frac{1}{2} [H_1(z) K_1(z) + H_2(z) K_2(z)] X(z) \\ &\quad + \frac{1}{2} [H_1(-z) K_1(z) + H_2(-z) K_2(z)] X(-z), \end{aligned} \quad (15.26)$$

where the second term reflects the aliasing effects due to decimation-induced spectral overlap around $f_s/4$, which can be eliminated following Esteban and Galand [525], if we satisfy for the following constraints:

$$K_1(z) = H_1(z) \tag{15.27}$$

$$K_2(z) = -H_1(-z) \tag{15.28}$$

and invoke Equation 15.17, satisfying the following relationship:

$$H_2(z) = H_1(-z). \tag{15.29}$$

Upon satisfying these conditions, Equation 15.26 can be written as:

$$\begin{aligned} S(z) &= \frac{1}{2}\left[H_1(z)H_1(z) - H_1(-z)H_1(-z)\right]X(z) \\ &+ \frac{1}{2}\left[H_1(-z)H_1(z) - H_1(z)H_1(-z)\right]X(-z), \end{aligned}$$

simplifying the aliasing-free reconstructed signal's expression to:

$$S(z) = \frac{1}{2}\left[H_1^2(z) - H_1^2(-z)\right]X(z). \tag{15.30}$$

If we exploit that $z = e^{j\omega T}$, we arrive at:

$$S\left(e^{j\omega T}\right) = \frac{1}{2}\left[H_1^2\left(e^{j\omega T}\right) - H_1^2\left(-e^{j\omega T}\right)\right]X\left(e^{j\omega T}\right),$$

and from Equation 15.16 by symmetry we have:

$$-e^{-j\omega T} = e^{j\left(\frac{\omega_s}{2}+\omega\right)T}, \tag{15.31}$$

leading to:

$$S\left(e^{j\omega T}\right) = \frac{1}{2}\left[H_1^2\left(e^{j\omega T}\right) - H_1^2\left(e^{j\left(\frac{\omega_s}{2}+\omega\right)T}\right)\right]X\left(e^{j\omega T}\right). \tag{15.32}$$

15.4.1.3 Practical QMF Design Constraints

Having considered the analysis/synthesis filtering, we see that the elimination of aliasing becomes more explicit in this subsection. Let us now examine how the imposed filter design constraints can be satisfied. Esteban and Galand [525] have proposed an elegant solution in case of finite impulse response (FIR) filters, having a z-domain transfer function given by:

$$H_1(z) = \sum_{n=0}^{N-1} h_1(n)z^{-n}, \tag{15.33}$$

15.4. SUBBAND CODING

where N is the FIR filter order. Since $H_2(z)$ is the mirror-symmetric replica of $H_1(z)$, below we show that its impulse response can be derived by inverting every other tap of the filter impulse response $h_1(n)$. Explicitly, from Equation 15.29 we have

$$\begin{aligned} H_2(z) &= H_1(-z) \\ &= \sum_{n=0}^{N-1} h_1(n)(-z)^{-n} \\ &= \sum_{n=0}^{N-1} h_1(n)(-1)^{-n} z^{-n} \\ &= \sum_{n=0}^{N-1} h_1(n)(-1)^n z^{-n}, \end{aligned} \qquad (15.34)$$

which obeys the above-stated symmetry relationship between the low-band and high-band impulse responses.

According to Esteban and Galand, the low-band transfer function $H_1(z)$, which is a symmetric FIR filter, can be expressed by its magnitude response $H_1(\omega)$ and a linear phase term, corresponding to the filter-delay $(N-1)$, as follows:

$$H_1\left(e^{j\omega T}\right) = H_1(\omega) e^{-j(N-1)\pi(\omega/\omega_s)}. \qquad (15.35)$$

Upon substituting this linear-phase expression in the reconstructed signal's expression in Equation 15.32 and taking into account that $2\pi/\omega_s = 2\pi/(2\pi f_s) = T$, we arrive at:

$$\begin{aligned} S\left(e^{j\omega T}\right) &= \frac{1}{2} \Big[H_1^2(\omega) e^{-j2(N-1)\pi(\omega/\omega_s)} \\ &\quad - H_1^2\left(\omega + \frac{\omega_s}{2}\right) e^{-j2(N-1)\pi\left(\frac{\omega}{\omega_s}+\frac{1}{2}\right)} \Big] X\left(e^{j\omega T}\right) \\ S\left(e^{j\omega T}\right) &= \frac{1}{2} \Big[H_1^2(\omega) - H_1^2\left(\omega + \frac{\omega_s}{2}\right) e^{-j(N-1)\pi} \Big] \\ &\quad e^{-j(N-1)2\pi(\omega/\omega_s)} X\left(e^{j\omega T}\right). \end{aligned} \qquad (15.36)$$

As to whether the aliasing can be perfectly removed, we have to consider two different cases, depending on whether the filter order N is even or odd.

1. **The filter order N is even.**

    In this case we have:
    $$e^{-j(N-1)\pi} = -1, \qquad (15.37)$$

    since the expression is evaluated at odd multiples of π on the unit circle. Hence, the reconstructed signal's expression in Equation 15.36 can be formulated as:

    $$S\left(e^{j\omega T}\right) = \frac{1}{2}\left[H_1^2(\omega) + H_1^2\left(\omega + \frac{\omega_s}{2}\right)\right] e^{-j(N-1)\omega T} X\left(e^{j\omega T}\right). \qquad (15.38)$$

$H_1(z)$ is a symmetric FIR filter of even order: $h_1(n) = h_1(N-1-n)$, $n = 0\ldots(N-1)$
$H_2(z)$ is an antisymmetric FIR filter of even order: $h_2(n) = -h_2(N-1-n)$, $n = 0\ldots(N/2)-1$
Mirror symmetry: $H_2(z) = H_1(-z)$ $h_2(n) = (-1)^n h_1(n)$ $n = 0\ldots(N-1)$
$K_1(z) = H_1(z)$
$K_2(z) = -H_2(z)$
All-pass criterion: $(H_1^2(\omega) + H_1^2(\omega + \frac{\omega_s}{2})) = 1$

Table 15.5: Conditions for Perfect Reconstruction QMF

In order to satisfy the condition of a perfect all-pass system, we have maintained:

$$H_1^2(\omega) + H_1^2(\omega + \frac{\omega_s}{2}) = 1 \tag{15.39}$$

yielding:

$$S\left(e^{j\omega T}\right) = \frac{1}{2} e^{-j(N-1)\omega T} X\left(e^{j\omega T}\right), \tag{15.40}$$

which can be written in the time domain as:

$$s(n) = \frac{1}{2} x(n - N + 1). \tag{15.41}$$

In conclusion, if the FIR QMF filter order N is even, the reconstructed signal is an $N-1$-sample delayed and $\frac{1}{2}$-scaled replica of the input video signal, implying that all aliasing components have been removed.

2. **The filter order N is odd.**

For an odd filter-order N, we have:

$$e^{-j(N-1)\pi} = 1, \tag{15.42}$$

since the exponential term is evaluated now at even multiples of π; hence, the reconstructed signal's expression is formulated as:

$$S\left(e^{j\omega T}\right) = \frac{1}{2} \left[H_1^2(\omega) - H_1^2(\omega + \frac{\omega_s}{2})\right] e^{-j(N-1)\omega T} X\left(e^{j\omega T}\right). \tag{15.43}$$

Observe that due to the symmetry of $H_1(\omega)$ we have $H_1(\omega) = H_1(-\omega)$. Thus, the square-bracketed term becomes zero at $\omega = -\omega_s/4$, and the reconstructed signal $S\left(e^{j\omega T}\right)$ is different from the transmitted signal. As a consequence, perfect-reconstruction QMFs have to use even filter orders.

In conclusion, the conditions for perfect reconstruction QMFs are summarized in Table 15.5. Johnston [526] has proposed a set of perceptually optimized real QMF filter designs, which process real-time signals. A range of complex quadrature mirror filters (CQMFs) potentially halving the associated computational complexity have been suggested by Nussbaumer [527] and Galand [528].

15.4. SUBBAND CODING

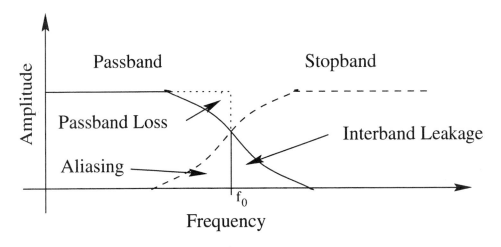

Figure 15.14: Example of two-band split using QMF filters.

15.4.2 Practical Quadrature Mirror Filters

When designing and implementing practical QMFs having a finite filter order and hence a nonzero width frequency-domain transition region between the frequency-domain passband and stopband, a certain amount of energy will always spill into the neighboring frequency regions, as illustrated in Figure 15.14. Unless special attention is devoted to this problem, it causes aliasing, manifesting itself in terms of video distortion. *Recall from the previous subsection that in case of even filter orders, while assuming infinite filter coefficient precision and unquantized subband signals, the QMF analysis/synthesis scheme of Figure 15.11 facilitates perfect signal reconstruction. However, in reality only the first condition is practically realizable, and hence practical subband video codecs often suffer from aliasing effects.*

Unfortunately, the perfect reconstruction conditions can only be satisfied for $N = 2$ and $h_{low}(1) = \frac{1}{2}$ [461]. However, this two-tap filter has a very poor stopband rejection and so it often fails to adequately reject the signal energy leakage from the adjacent subband, when the subband signals are quantized, as suggested by the slowly decaying slope of the two-tap QMF in Figure 15.15. (These effects will be studied in the context of Figure 15.17, quantitatively.) The choice of near-perfect QMFs is usually based on a tradeoff between a low interband leakage and a low ripple in the passband, a topic extensively studied in references [526, 529–531].

QMFs split the band into a higher and lower band of equal bandwidth. They may be applied recursively in order to obtain further subbands. Therefore, they are directly suitable for one-dimensional data, such as speech signals. If only the lower subband is split further in consecutive steps, the resultant scheme is referred to as a tree-structured QMF [532]. In image coding, the source signal is two-dimensional, which has to be split into four components, namely, the vertical and horizontal low and high bands, respectively. In order to solve the problem using a one-dimensional filter set, we may scan the two-dimensional video signal on the basis of scanlines to a one-dimensional signal and then first apply the QMFs horizontally and obtain the

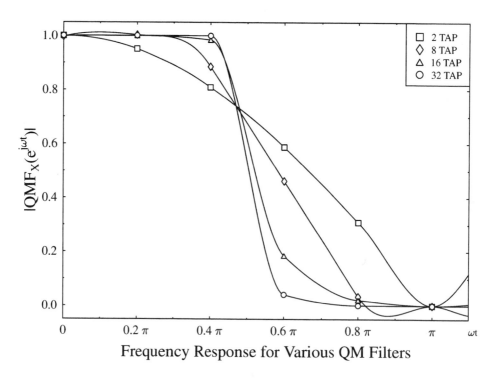

Figure 15.15: Frequency responses for various QM filters.

low-horizontal and high-horizontal frequency bands. Each of these two bands is then vertically scanned and again QMF filtered, using the result in all four combinations of horizontal and vertical band combinations. The lowest frequency subband contains a vertically and horizontally low-pass filtered version of the original video frame and therefore contains most of the energy. The efficiency of the spectral analysis and hence that of subband coding may be increased with the aid of further consecutive band-split steps. These are commonly applied to the lowest band only [466], resulting in multiband schemes having 4, 7, 10, or more bands. Other design options are suggested in references [533,534].

A typical example of a 10-band split frame of the "Football" sequence is depicted in Figure 15.16. The benefit of more highly resolved band-splitting is that the codec can take advantage of the individual frequency-domain characteristics of each band. If the interband leakage due to using low-order QMFs is significant, each band will contain aliasing from other bands, which may defeat the benefits of frequency-selective encoding. The two-tap QMF filter is characterized by a weak stopband rejection and hence may not be adequate in certain applications. Higher-order, near-perfect QMFs have a far better stopband rejection, but since they do not satisfy the perfect-reconstruction QMF criteria specified in Table 15.5, they inflict impairments, even when the subband signal components are unquantized. This problem can be overcome, when using relatively long filters, as demonstrated by Figure 15.17 in case of

15.4. SUBBAND CODING

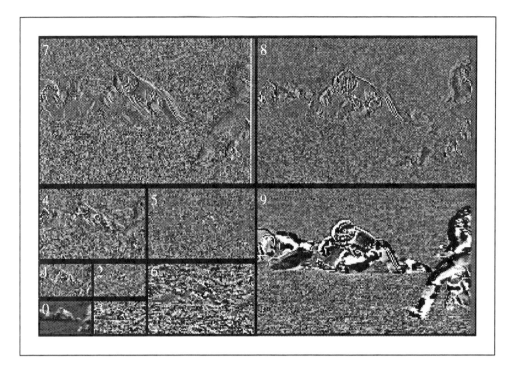

Figure 15.16: Ten-Band split example of the 16CIF "American Football" sequence of Figure 15.3. Band 0 represents the lowest horizontal (H) and vertical (V) frequency bands, while band 8 indicates the highest H and V frequency bands. Similarly, band 9 portrays the highest horizontal frequency band associated with low vertical energy contents.

unquantized subband signals for a range of different filter orders.

Specifically, for two-tap QMFs the reconstructed video signal's PSNR exceeds 20 × log (256/1) ≈ 48 dB, a PSNR value, which corresponds to the lowest possible error in the reconstructed signal, when using an 8-bit resolution or 256 possible integer values to represent the luminance component. In other words, the PSNR degradation due to this low-order filter in case of unquantized subband signals is very low. For longer near-perfect QMF filters and unquantized subband signals, the PSNR performance depends on the number of band splits, the actual filter length, and the accuracy of the arithmetic operations. Figure 15.17 is based on high-precision 64-bit arithmetics and reveals that a minimum QMF length of 16 taps is necessary for maintaining a relatively low PSNR penalty due to spectral spillage in case of nonperfect QMFs. The actual PSNR achieved is dependent on the number of subbands, but values in excess of about 35 dB are possible at a filter order of 8. For very high-quality coding we propose employing a 32-band filter, which almost reaches the reconstruction qualities of the ideal two-tap filter, achieving PSNRs in excess of about 55 dB. The cost of the improved stopband rejection is a 16-times more complex filter than the perfect-reconstruction two-tap scheme. The employment of long filter impulse responses has

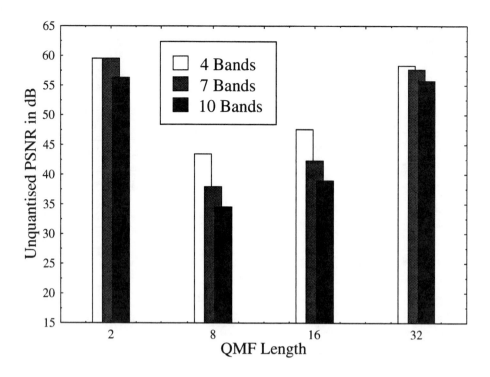

Figure 15.17: Reconstructed video quality versus QMF order without quantization for four-, seven- and ten-band subband intra-frame coding at QMF lengths of 2, 8, 16, and 32 coefficients for the "Susie" sequence using 64-bit number representation accuracy.

been proven to be important in the context of speech coding [526, 535, 536], but we found that in image coding the difference between a two-tap filter and higher-order filters having 16 taps or more may not be very significant. This difference may be attributed to the different sensitivity of the human eye and ear to errors in the frequency domain.

15.5 Run-Length-Based Intra-Frame Subband Coding

Traditional subband coding as proposed by Gharavi [518] applies a combination of a linear subband quantizer with a dead zone and variable-length coding (VLC). The motivation behind this technique will be detailed later in this chapter. Figure 15.18 shows the PDF of subbands 1 to 9, which exhibit a highly peaked distribution in the range of $[-5..5]$, suggesting the presence of low-energy, noise-like components. Band 0 is not included in the figure, since it contains the lowest frequencies and the DC component. Thus, its PDF is spread across the whole dynamic range of

15.5. RUN-LENGTH-BASED INTRA-FRAME SUBBAND CODING

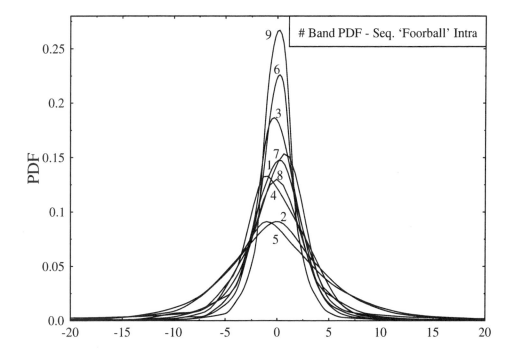

Figure 15.18: PDF of Band 1 to Band 9 of an intra-frame 10-band SBC for the 16CIF "Football" sequence of Figure 15.3.

[0..255]. Indeed, these noise-like components are typically due to camera-noise, and hence their exact encoding is actually detrimental with regard to the reconstructed video quality. Hence, the potential for compression lies in the fact that quantizers exhibiting a dead zone in this low-amplitude, high-probability region may be applied without significantly impairing the reconstructed video quality. The corresponding quantizer characteristic is shown in Figure 15.19.

Figure 15.20 characterizes the codec's PSNR performance, when experimenting with a variety of dead zone intervals. In this experiment, all values of the higher frequency bands 1 to 9 in the range [−dead zone ... dead zone] were set to zero, whereas all remaining values outside the dead zone were left unquantized. This figure confirms Gharavi's results [518] showing that a dead zone of ±4 guarantees a good reconstructed video quality having PSNR values in excess of about 35 dB. For very high-quality reproduction, the dead zone has to be of a narrower width, around ±2. Because of the high-probability subband PDF peaks near zero, a quantizer having such a dead-zone will set the majority of the subband video samples to zero, introducing a grade of predictability concerning the zero sample, which is synonymous with redundancy. This redundancy may be exploited using variable-length codes or run-length codes, as we will show below. Suitable run-length codes are, for example, the B codes [489], which are described in Section 16.4.1.

Following a range of investigations using various numbers of quantization levels

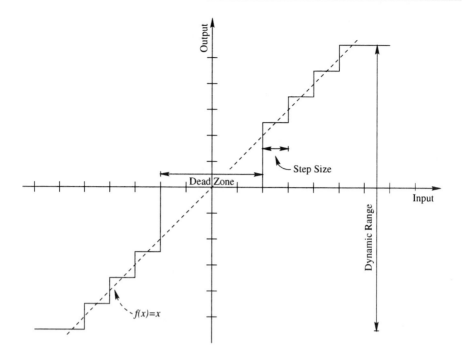

Figure 15.19: Dead-zone based quantizer characteristic.

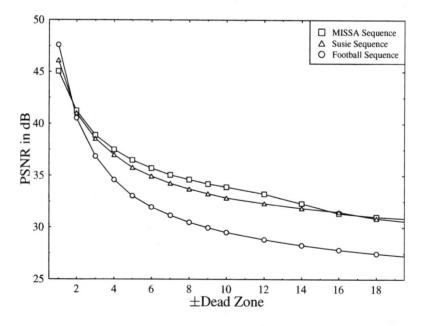

Figure 15.20: Dead zone based intra-frame subband codec PSNR versus dead zone width, setting all values of the higher frequency bands within the dead zone to zero for the 4CIF and 16CIF sequences of Figures 15.1 and 15.3.

15.5. RUN-LENGTH-BASED INTRA-FRAME SUBBAND CODING

Reconstr. Levels	Entropy in bpp	Lumin. PSNR(dB)	Cr_u PSNR(dB)	Cr_v PSNR (dB)
3	0.59	33.00	42.79	42.85
5	0.60	33.85	42.75	42.82
9	0.63	37.29	43.83	43.95
17	0.66	38.90	43.99	44.09
33	0.70	39.46	44.02	44.11

Table 15.6: Intra-frame Subband Codec Performance for Various Quantizers with a Dead Zone of $[-4\ldots4]$ and a Dynamic Range of ± 40, for the "Susie" Sequence

Sequence	Format	Luminance PSNR (dB)	Cr_u PSNR (dB)	Cr_v PSNR (dB)	% of '0' values	Entr. in bpp
"Missa"	CIF	39.82	38.40	38.33	86.3	0.79
"Susie"	4CIF	39.46	44.02	44.11	88.4	0.70
"Football"	4CIF	36.23	40.89	40.16	74.5	1.29

Table 15.7: Subband Codec Performance with a 33-Level Quantizer

as shown in Table 15.6, we concluded that good results are achieved with a 17- or 33-level linear quantizer, characterized by a ± 4 dead zone and reconstruction range of ± 40. This quantizer may be used for all higher frequency bands [537]. The lowest frequency band containing the DC component is characterized by a widely spread PDF, covering the range of $[0..255]$ and hence has been processed by invoking different compression methods, such as vector quantization [537, 538] or DPCM [518]. Because of its reduced size after subsampling, in a 10-band scheme, where there are three consecutive subsampling steps, the bit-rate contribution of the lowest subband corresponds to an unquantized data rate of $\frac{1}{4} \cdot \frac{1}{4} \cdot \frac{1}{4} \cdot 8$ bit $= 0.125$ bpp, which is negligible compared to the data rate of the other, less subsampled and hence larger-area bands. Furthermore, the high-quality reconstruction of the lowest band is vital, and hence the potential compression gains in this band seem to be limited. We opted for quantizing the lowest band with an 8-bit linear quantizer in the range of $[0..255]$. When applying such a scheme, we achieved good PSNR results. The 4CIF "Susie" and the CIF "Miss America" sequences of Figures 15.2 and 15.1 result in PSNR values just below 40 dB. The average entropy (see Section 1.4) for all bands remains with 0.70 and 0.79 for the "Susie" and "Miss America" sequences below 1 bpp, similar to those reported earlier by other researchers, for example, [518, 537]. The corresponding PSNR values are around 40 dB for both the luminance and chrominance values for the sequences tested. These results are summarized along with the associated PSNR values for various CIF to CCIR 601 resolution video sequences in Table 15.7.

15.5.1 Max-Lloyd-Based Subband Coding

The drawback of the variable-length coding scheme is its error-sensitive nature, which we attempted to avoid throughout this chapter. Therefore, we endeavored to explore

Band #	Number of quantizer levels per band										PSNR (dB)	Luminance BPP
	0	1	2	3	4	5	6	7	8	9		
1	2	0	0	0	0	0	0	0	0	0	17.067	0.0156
2	4	0	0	0	0	0	0	0	0	0	19.941	0.0313
3	8	0	0	0	0	0	0	0	0	0	22.642	0.0469
4	16	0	0	0	0	0	0	0	0	0	25.619	0.0625
5	32	0	0	0	0	0	0	0	0	0	26.540	0.0781
6	64	0	0	0	0	0	0	0	0	0	26.972	0.0938
7	64	2	0	0	0	0	0	0	0	0	27.196	0.1094
8	64	4	0	0	0	0	0	0	0	0	27.899	0.1250
9	64	8	0	0	0	0	0	0	0	0	28.323	0.1406
10	128	8	0	0	0	0	0	0	0	0	28.502	0.1563
11	128	8	2	0	0	0	0	0	0	0	28.737	0.1719
12	128	8	4	0	0	0	0	0	0	0	29.204	0.1875
13	128	8	8	0	0	0	0	0	0	0	29.511	0.2031
14	128	8	8	0	2	0	0	0	0	0	29.735	0.2656
15	128	8	8	0	4	0	0	0	0	0	30.177	0.3281
16	128	8	8	0	8	0	0	0	0	0	30.550	0.3906
17	128	8	8	0	8	2	0	0	0	0	30.768	0.4531
18	128	8	8	0	8	4	0	0	0	0	31.195	0.5156
19	128	8	8	0	8	8	0	0	0	0	31.492	0.5781
20	128	8	16	0	8	8	0	0	0	0	31.711	0.5938
21	128	8	16	0	8	8	2	0	0	0	31.930	0.8438
22	128	8	16	0	8	8	0	4	0	0	32.928	1.0938
23	128	8	16	0	8	8	0	8	0	0	33.444	1.3438
24	256	8	16	0	8	8	0	8	0	0	33.615	1.3594
25	256	8	16	0	8	16	0	8	0	0	33.886	1.4219
26	256	16	16	0	8	16	0	8	0	0	34.147	1.4375
27	256	16	16	0	8	16	0	8	2	0	34.433	1.6875
28	256	16	16	0	8	16	0	8	4	0	35.150	1.9375
29	256	16	16	0	8	16	0	8	8	0	35.733	2.1875
30	256	16	16	0	8	16	0	8	16	0	36.117	2.4375
...	...	...	...	...	...	...	...	...	...	...	...	...
40	256	16	16	8	16	32	4	32	16	2	40.015	3.4844
60	256	128	128	16	64	64	16	64	256	16	45.687	5.9063
70	256	128	128	64	128	256	32	256	256	64	47.003	7.1875
80	256	256	256	256	256	256	256	256	256	256	47.050	8.0000

Table 15.8: Number of Quantizer Levels per Subband for the 10-Band Scheme and PSNR as well as as Well as Bit Rate in BPP (monochrome) for the 4CIF "Susie" Sequence

the potential of Max-Lloyd quantizers, individually trained, according to the significance and PDF of each subband. In other words, we attempted to exploit the fact that some bands are perceptually more important than others. We chose the resolution of the quantizer for each band based on a compromise in order to satisfy our bit-rate and quality requirements. In order to achieve such a quantizer allocation scheme, we first derived Max-Lloyd quantizers for each band at all typical resolutions, that is, for $2, 4, 8 \ldots 256$ reconstruction levels, one set for each of the luminance and chrominance components.

Initially, we allocated a 1-bit quantizer for the most important band #0, whereas all other bands were quantized to 0. The overall bit rate generated by such a scheme is due to its small relative size of a $\frac{1}{64}$th of the original frame, yielding a bit rate of $\frac{1}{64} \cdot 1$ bit $= 0.0156$ bpp for the luminance information. As a consequence, the reconstructed PSNR of our example using the "Susie" sequence was fairly low at 17 dB, as seen in Table 15.8.

In order to increase the video quality at the lowest possible bit-rate investment, we tentatively increased the quantizer resolution of the band 0 quantizer by 1 bit and assigned an additional two-level quantizer for Band 1, while noting the PSNR improvement and bit-rate increase in both cases. The same relative figure of merit was then evaluated for each of the remaining nine bands, and the additional bit was finally allocated for the band, which resulted in the best figure of merit, increasing

15.5. RUN-LENGTH-BASED INTRA-FRAME SUBBAND CODING

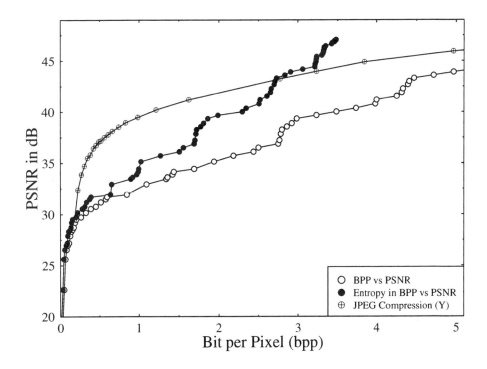

Figure 15.21: PSNR versus bits/pixel performance of the 10-band intra-frame SBC codec for the 4CIF "Susie" sequence of Figure 15.1 in comparison to the Joint Photographic Expert Group (JPEG) codec [539].

the bit rate and PSNR, as depicted in Figure 15.21 and Table 15.8. Continuing in the same fashion, we continuously increased the quantizer resolutions and derived a bit-allocation table up to 8 bpp for monochrome images or 12 bpp in case of color images, when assigning a 256-level quantizer to each of the luminance and chrominance bands. As an example, we refer to line 19 of Table 15.8, where assigning a 7-bit/128-level quantizer for band 0 and a 3-bit/8-level quantizer for each of the bands 1, 2, 4, and 5 results in a PSNR of 31.492 dB and a bit rate of 0.5781 bpp (0.87 bpp) for a monochrome (colour) frame of the "Susie" sequence.

Although we used Max-Lloyd quantizers, the quantized signals still contained redundancy. We therefore determined the entropy of each of the quantized bands and determined the information contents of all bands together in terms of bpp, which is also shown as entropy versus bit-rate curve in Figure 15.21. This remaining redundancy is not sufficiently high to justify the introduction of further compression schemes, which would strongly increase the error sensitivity.

The full quantizer allocation table is summarized in Table 15.8. Good subjective video qualities are achievable for bit rates in excess of 2.1 bpp for color images, which is associated with PSNRs in excess of 34dB. However, this PSNR performance is below that of the DCT-based standard JPEG codec seen in the figure, as well as

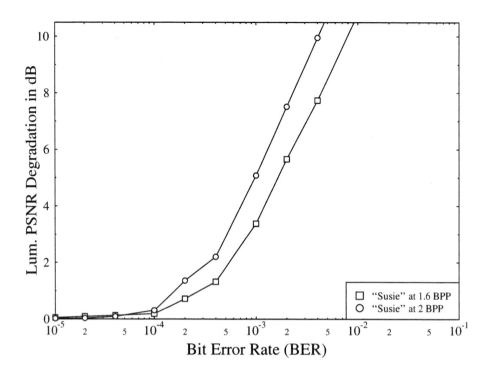

Figure 15.22: PSNR degradation versus BER for the 10-band subband codec, using the 4CIF "Susie" sequence of Figure 15.1, at 1.6 bpp and 2bpp.

below that of the DCT codec, which will be presented in the next chapter. The advantage of subband coding without RL compression is its error resilience. In case of RL-compression, the transmission errors inflicted propagate and are spread across a certain frequency range rather than being confined to a certain location of a subband.

In order to quantify the effect of transmission errors, we introduced random bit errors and evaluated the objective PSNR degradation measure as a function of the BER for two different bit rates, as seen in Figure 15.22. The BER at which the video quality becomes rather corrupted is around 10^{-3}, but the difference in comparison to block-based coding methods, such as the DCT, for example, is the variable size of the distorted areas, corresponding to the different subbands. For block-based methods using a constant block size, such as the schemes presented in Chapters 11–13, a single bit error inflicts video degradation in that particular fixed-size block. By contrast, in the case of subband coding, a single-frequency band at a certain location is affected. For our typical 10-band codec, bands 0 to 3 are down-sampled by a factor of 8, corresponding to a 64-fold reduction in size. Hence, in SBC a 1-bit error may impair a subband, corresponding to an area of 64 pels, which is equivalent to an 8×8 block. In the case of bands 4 to 6, the affected area shrinks to a size of 4×4 and a smaller area of 2×2 is affected, if the highest bands are corrupted. An example of a frame subjected to a BER of 10^{-3} is given in Figure 15.23.

Figure 15.23: Subband encoded frame of the "Susie" sequence at 2 bpp and subjected to a BER of 10^{-3}, yielding a reduced PSNR of 28.3 dB.

The above scheme is not amenable to further compression due to motion compensation. Motion compensation would decrease the entropy of the original video signal, but it would not allow the use of more coarse quantizers, because of the unstationarity of the MCER. This problem is discussed in more detail in Chapter 16.

15.6 Summary and Conclusions

In this chapter we studied a number of low-complexity video compression methods. The simplest techniques, such as DPCM and ADPCM, do not achieve a satisfactory video quality at bit rates below 3 bpp for either intra- or inter-frame coded scenarios. DPCM and BTC were shown to achieve reasonable results at a bit rate of 2.5 bpp, reaching PSNRs in excess of 38 dB, which proved adequate for our purposes. These results will be compared with the performance of various other techniques of the following chapters, which have higher compression ratios at the cost of a higher computational demand and lower error resilience.

The simplicity of both methods facilitates their employment as part of a hybrid or hierarchical coding scheme. DPCM is often used in conjunction with run-length coding or subband coding [530]. In the latter case, it is used to efficiently encode each subband. BTC coding might be combined with quad-tree splitting [540] or vector quantization [541].

Subband coding is a convenient multiresolution technique that allowed us to invest the bit-rate budget in those bands, where it was most beneficial in terms of PSNR improvements, as was suggested by Table 15.8. The associated PSNR improvements were characterized by Figure 15.21. Let us now focus our attention on high-quality DCT-based coding in the forthcoming chapter.

Chapter 16

High-Resolution DCT Coding

16.1 Introduction

In this chapter, based on our findings from Chapter 12 we apply some of the principles in the context of high-quality image coding. Specifically, the next section explores the corresponding requirements as regards the MC algorithm, before designing a switched intra/inter-frame coded DCT codec.

At the time of writing HDTV coding is often implemented using the MPEG1 [542] and MPEG2 [543] standards, as well as the H263 [192] standard. All of these codecs employ similar compression methods for the quantization of the DCT coefficients, which is based on zig-zag scanning and run-length coding. This method guarantees a near-constant video quality at the cost of a widely fluctuating bit rate and a bit-stream, which is vulnerable to bit corruption. This problem may be overcome only using strong error correction and/or error resilient entropy coding (EREC) [501].

Our aim in this chapter is to avoid run-length coding methods, where possible, which reduce the achievable compression ratio but improve the robustness against channel errors. Therefore we need bit allocation tables and appropriate quantizers, which can be derived by adopting an approach similar to that detailed in Section 12.3. For very low bit rates we considered bit-allocation schemes with less than 15 bits per 8×8 block and were able to determine the best allocation scheme by simple trial and error investigations. For the range of video qualities and video sequences considered in this section, we needed a more general approach leading to bit-allocation schemes requiring up to 3 bpp, in order to provide the required HDTV quality.

16.2 Intra-Frame Quantizer Training

Determining the appropriate quantizers requires a range of different bit-allocation schemes and corresponding quantizers for various bit rates, up to 2–3 bpp. The aim is similar to that in Section 12.3, where we exploited the energy compaction characteristics of the DCT.

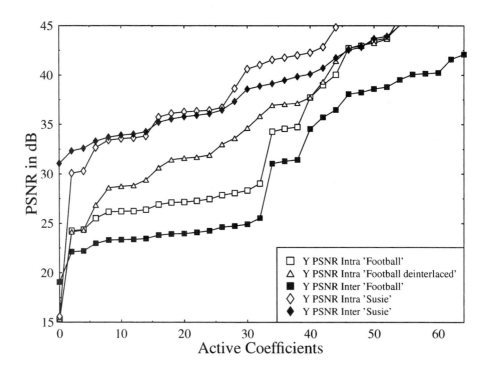

Figure 16.1: PSNR performance of the DCT versus the number of active DCT coefficients for various high-resolution video clips without motion compensation, when applying a zonal mask around the low spatial frequency top-left corner of the DCT block without quantization of the DCT coefficients.

In order to evaluate the potential of such schemes, in Figure 16.1 we examined the PSNR performance of the DCT versus the number of active coefficients per 8×8 block without coefficient quantization, when applying a simple technique called zonal masking. The zonal masking was realized by zig-zag scanning the coefficients according to Figure 16.2, followed by a cancellation or obliteration of all masked coefficients, which were set to zero.

Figure 16.1 underlines the fact that simple masking of the top-left corner [26] is not advantageous, since typically more than 30 out of 64 coefficients must be retained in order to be able to exceed a PSNR of 30 dB before quantization for the considered high-resolution video clips. We also endeavored to evaluate the effects of camera interlacing, since most state-of-the-art cameras record images in a two-phase, interlaced fashion. First, all even lines are recorded, and then in a second scan, the remaining odd lines are stored. This method was invented in order to artificially double the frame rate and to increase the viewing comfort. Since both half-frames are taken with a delay of half the frame scanning interval duration, significant motion may take place between these instants. This leads to a horizontal striped pattern at the edges of moving objects, which will increase the energy of the stored coefficients for horizontal

16.2. INTRA-FRAME QUANTIZER TRAINING

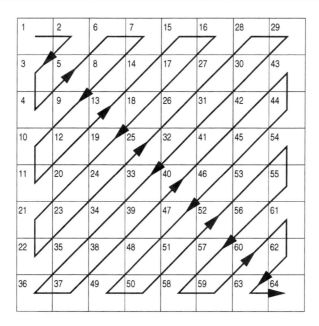

Figure 16.2: Zig-zag scanning of DCT coefficients.

frequencies, causing the distinct PSNR step-around coefficient 34 in Figure 16.1.

The question that arises is whether there are suitable fixed masking patterns, that will allow satisfactory quality. In order to evaluate the upper bound performance of a masking-type quantization scheme, we determined individual optimum masks for each block, which *retained the required number of highest energy coefficients*. Figure 16.3 reveals the PSNR versus number of retained coefficients performance, if an individual optimum mask is determined for each block before the masking takes place. The results indicate that for intra-frame-oriented DCT coding, about 10 to 15 coefficients are necessary to guarantee a good video reproduction quality. This explains why the scanning method used in the various standards, such as JPEG [539], MPEG-X [542, 543], and H.26X [192, 500], is so efficient.

An automated algorithm was necessary, which allowed us to determine all required bit-allocation tables. Starting from a 0 bpp quantizer, we used the pseudo-quantizer from Equation 12.22 on page 499 in order to emulate the distortion of the quantization process. In order to identify the DCT coefficient position, where the bit for a 1-bit per 8×8 block quantizer should be allocated for achieving the lowest distortion, the algorithm first allocated the bit at position (0,0) within the block. An entire picture was DCT transformed and quantized. The coefficient (0,0) was quantized with a two-level pseudo-quantizer, and all other coefficients were forced to zero. Finally, the inverse transformation took place, and the resulting PSNR was noted. The same DCT quantization, inverse DCT process was started again, with the bit allocated at the next possible position, for example (0,1), and so on. The bit was eventually allocated to the DCT coefficient position, which resulted in the highest PSNR, and at this stage we arrived at the one bit per block bit-allocation table. Because of the importance of

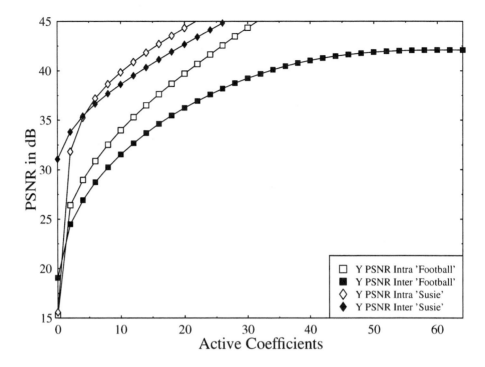

Figure 16.3: PSNR versus the number of active DCT coefficients per 8 × 8 block performance of the DCT, when applying a mask retaining the required number of highest energy coefficients in each block without quantization for the 4CIF sequences of Figure 15.1.

the DC coefficient, the first bit was naturally allocated for this particular coefficient, namely, for the coefficient at position (0,0).

In the next step the process was repeated, and the PSNR benefit of a 1-bit increase at each of the 64 possible positions was tested. This implied that the number of levels of the pseudo-quantizer at a certain bit position was doubled, or, if no bit was allocated at that position before, the number of levels was set to two, yielding a 1-bit quantizer. Following this principle, we generated the bit allocation tables from 1 bit per block up to 256 bits per block. Example bit-allocations for four reconstruction qualities are depicted in Figure 16.4, which was trained on the "Football" sequence, but we found that the bit-allocation varied very little with different sequences.

It was expected that the resolution of quantizers would shrink with the distance of each coefficient from the top-left corner of the block. For Figure 16.4, we therefore zig-zag scanned the coefficients according to Figure 16.2 in order to determine how well the allocation table correlated with these expectations. As the figure confirms by comparing the active DCT coefficient positions with Figure 16.2, the allocation table follows the expected tendency except that further quantizers are allocated around positions 20–23 and 35–38. As seen from Figure 16.2, these coefficients are located

16.2. INTRA-FRAME QUANTIZER TRAINING

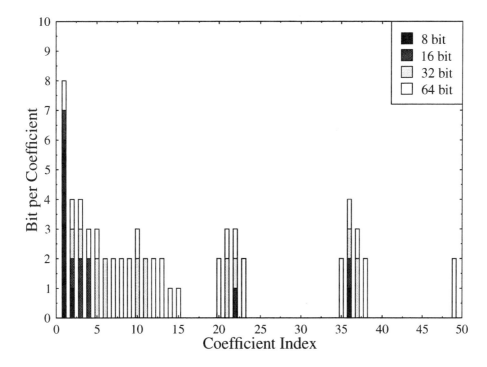

Figure 16.4: DCT coefficient quantizer bit-allocation for the zig-zag scanned DCT intra-frame coefficients for 8, 16, 32, and 64 bits per block schemes.

at the very left of the block and are caused by high vertical frequencies, that is, horizontal stripes due to interlacing, as mentioned earlier.

Based on the above tentative quantizer design, appropriate final quantizers had to be trained. Hence, we gathered training data using the above initial quantizers and trained the required Max-Lloyd quantizers, as proposed by Noah [512]. The quantizers should be trained using coefficients, which contain a range of distinct patterns, since, for example, blocks from a flat background will detrimentally narrow the dynamic range of quantizers. This results in increased quantizer overload distortion and should be avoided. Therefore, during the training process we chose from every training sequence those blocks for quantizer training which had the highest variance. A total of more than 50,000 training entries guaranteed the statistical independence of the quantizers from the training sequences. The conventional Max-Lloyd algorithm was slightly modified so that the generated quantizers were almost symmetrical to the y-axis and were of the mid-tread type [461].

Figure 16.5 reveals the performance of the quantizers. Good still-image quality was guaranteed for bit rates in excess of 0.75 bpp, where the luminance PSNR was around 27 dB, while the Cr_u and Cr_v components exhibited a PSNR close to 40 dB. Since the spatial frequency-domain spectrum of the Cr_u and Cr_v components is more compact, their reproduction quality expressed in terms of PSNR is superior

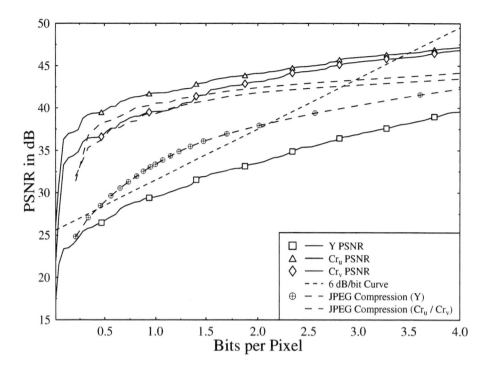

Figure 16.5: PSNR versus bit-rate performance of the DCT quantizers for rates up to 4 bits per pixel in comparison to the JPEG codec for the 4CIF "Football" sequence of Figure 15.1.

to that of the Y component at the same bit rate associated with a certain quantizer resolution. Because of this property most codecs allocate a lower fraction of the bits to the chrominance components. The performance of the JPEG standard was also characterized in Figure 16.5 in order to provide a benchmarker. The JPEG codec achieves a better performance due to using different quantizers for different blocks and exploiting the benefits of run-length encoding. Further performance examples are given in Figures 16.6 and 16.7 using different video clips, which follow similar trends.

16.3 Motion Compensation for High-Quality Images

In Chapter 12, we found that we were able to restrict the MC search to a window of 4 x 4 pixels. This was possible for relatively low-activity head-and-shoulders videophone sequences, where the image size was fixed to the comparatively small QCIF. Furthermore, we allowed the image quality to drop for short intervals due to the paucity of bits imposed by the constant bit-rate requirement.

In high-resolution image coding, however, we have to provide an adequate image quality at all times. In addition, we have to cater for image sequences of varying

16.3. MOTION COMPENSATION FOR HIGH-QUALITY IMAGES

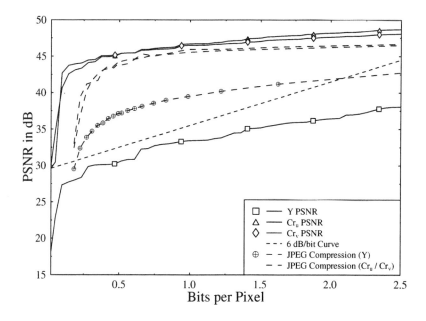

Figure 16.6: PSNR performance of the DCT quantizers for bit rates of up to 4 bit per pixel in comparison to the JPEG codec for the 4CIF "Susie" sequence of Figure 15.1.

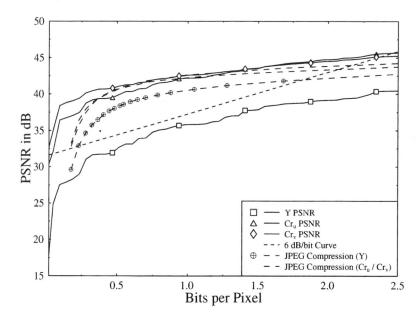

Figure 16.7: PSNR performance of the DCT quantizers for bit rates of up to 4 bits per pixel in comparison to the JPEG codec for the 16CIF "Mall" sequence of Figure 15.3.

	0	1	2	4	8	16	32
Susie Frame #1/#2	6.05	5.32	2.02	0.17	0.13	0.09	0.03
Susie Frame #71/#72	10.55	10.09	9.81	7.27	2.68	0.15	0.08
Football	9.81	8.98	8.14	6.74	4.58	2.03	0.39

Table 16.1: PSNR Degradation for the Full-Search Algorithm, When Using Various MC Search Scopes in the Range of 0–32 Pixels Measured Relative to a Search Scope of 64 for the 4CIF "Susie" and "Football" Sequences of Figure 15.1.

natures, as, for example, a news-presenter or a football game clip. The motion becomes more complex, and hence, unlike in Part II of this book, we cannot assume that the camera is stationary. The camera might pan or zoom and cause global motion of the image, and so we have to accept that the movement of objects in terms of pixel-separation between consecutive video frames depends on the type of sequence as well as on the resolution. Therefore, it is important to know the statistical properties of the sequences processed so that the search window and the search algorithm are chosen accordingly.

Table 16.1 demonstrates the PSNR video degradation due to limiting the MC search scope to values between 0–32 pixels for two specific sequences. In case of the 4CIF "Susie" sequence and frames No. 1 and No. 2, a relatively small search scope of 4 pixels in any direction exploits all the temporal redundancy, leading to only marginal PSNR improvements for further MC scope extensions. During later frames in the sequence, when "Susie" suddenly moves her head, the search scope needs to be extended to more than 8 pixels. For the higher activity "Football" sequence, however, a search scope of 64 is necessitated. This reflects the high-motion activity of the sequence, causing two problems. First, there are $(2 \times 64)^2 = 16384$ possible MVs, requiring a 14-bit MV identifier, and the computational demand for a CCIR 601-sized image exceeds 14 billion integer operations! At the time of writing, the implementation of such an algorithm in a real-time codec appears unrealistic. Employment of the full search (FS) algorithm at this search scope might be reserved for applications, such as video on demand, where the encoding process takes place independently from the decoding, as in a non-real-time scenario.

For real-time or quasi-real-time applications, the computational demand must be reduced. In Figures 16.8 and 16.9, we characterized the performance of various search strategies for the "Football" sequence. Figure 16.8 characterizes the PSNR video performance of a hypothetical codec for the first 20 frames of the "Football" sequence, when no MCER was transmitted. This uses three different MC search strategies, namely, the optimum full search (FS) and two suboptimum algorithms, the hierarchical search (HS) and the fractional search (FC). These methods were introduced, among others, in Section 12.2.2; for the sake of simplicity, we have concentrated here on these three common methods.

Figure 16.8 reveals that the hierarchical search of Section 12.2.2 is unsuitable for CCIR 601-sized images exhibiting high-motion content. Its performance is up to 4.5 dB worse than that of the FS algorithm. This is because during the initial search step, the whole search window is evaluated for only 5 (HS 5) or 9 (HS 9) block matches. A search window of 32 offers $(2 \cdot 32)^2 = 4096$ search positions, and motion

16.3. MOTION COMPENSATION FOR HIGH-QUALITY IMAGES 647

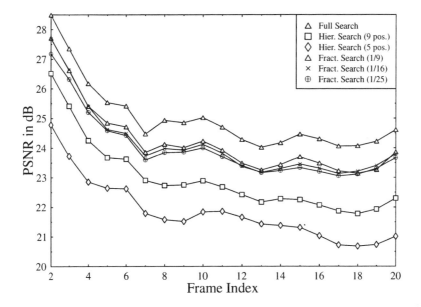

Figure 16.8: PSNR versus frame index performance comparison for motion compensation using various search techniques, a search scope of 32, 16CIF "Football" sequence, no MCER coding, assuming that the first frame was available at the decoder.

over this large area cannot be adequately estimated by just five or nine tentative motion compensation attempts. We enhanced the performance of the algorithm by increasing the number of block matches during each step to 16 (HS 16), 25 (HS 25), 36 (HS 36), 49 (HS 49), and even 64 (HS 46), as seen in Figure 16.9. This causes the PSNR penalty inflicted by suboptimum MC to drop from 4.5 dB for HS 5 to about 1.5 dB for HS 64. This is associated with an increase of the computational demand from 30 Mega Integer Operations (MIOP) to 180 MIOP. We also evaluated the potential of the fractional search technique at a quarter of the complexity (FC 1/4), a ninth (FC 1/9), a 25th (FC 1/25), and a 49th (FC 1/49) of the complexity of the exhaustive search. This leads to the range of search complexities and PSNR degradations depicted in Figure 16.9.

Figures 16.10–16.12 show the inter- and intra-frame correlation of motion vectors for the 4CIF "Football" and "Susie" sequences of Figure 15.1. The inter-frame MV correlation in this context is defined as the correlation of MVs for the corresponding blocks in consecutive frames. A high inter-frame correlation of MVs could be exploited by a telescopic MV scheme, where one transmits the difference of the MV of a certain block with respect to the previous MV at the block position under consideration. Such a scheme is an optional feature of the MPEG codecs [542, 543]. It is shown that the inter-frame correlation of MVs is below 20% for consecutive frames, even in case of the high-motion "Football" sequence. The telescopic MV scheme has been abandoned for both the H.261 [500] and H.263 codecs [192], they both rely on exploiting the intra-

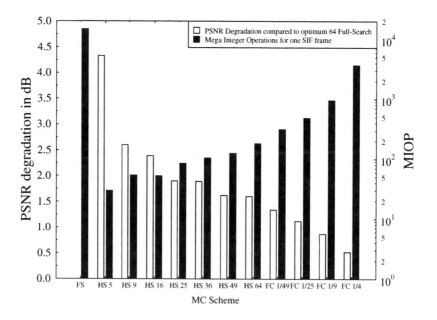

Figure 16.9: PSNR degradation and complexity for the first motion compensated frame, when applying various search methods. No MCER coding was used, and the search scope was 64 for the "Football" sequence.

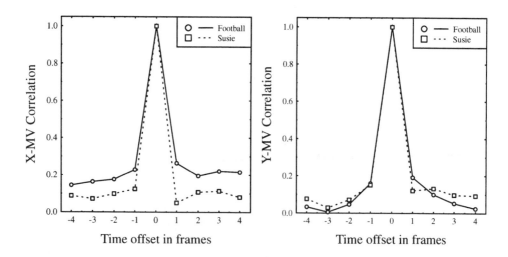

Figure 16.10: Inter-frame correlation of the MVs versus time offset for consecutive frames of the 4CIF "Football" and "Susie" sequences of Figure 15.1. The inter-frame MV correlation in this context is defined as the correlation of MVs for the corresponding blocks in consecutive frames.

16.3. MOTION COMPENSATION FOR HIGH-QUALITY IMAGES

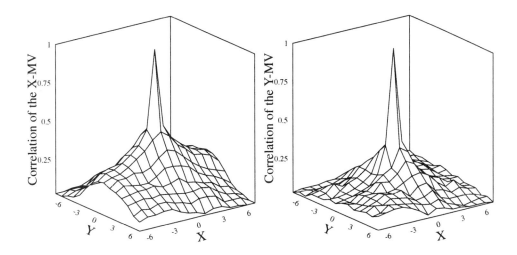

Figure 16.11: Intra-frame correlation of the MVs versus block offset for the 4CIF "Football" sequence of Figure 15.1.

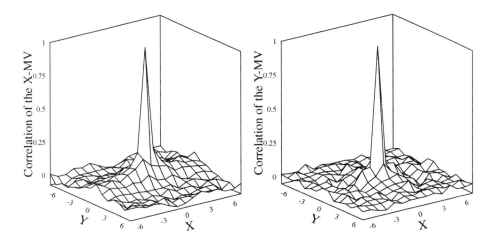

Figure 16.12: Intra-frame correlation of the MVs versus block offset for the 4CIF "Susie" sequence of Figure 15.1.

frame MV correlation, as depicted in Figures 16.11 and 16.12.

The H.261 scheme transmits the difference vector with respect to the MV at the left of the current block, while the H.263 arrangement determines a prediction vector from three surrounding MVs before encoding the difference of the predicted MV and the MV under consideration, $MV_{(x,y)}$. These issues are discussed in more depth in Section 18.2.3. Briefly, the MV of the block to the left, $MV_{(x-1,y)}$, the MV of the corresponding block in the previous row $MV_{(x,y-1)}$, and that of the next block in

SEQUENCE	Y-Cor.	U-Cor.	V-Cor.
Missa #1	0.98	0.99	0.95
Missa #1/#2 MCER	0.02	0.70	0.75
Susie #1	0.98	0.98	0.92
Susie #1/#2 MCER	0.56	0.14	-0.03
Football #1	0.96	0.93	0.96
Fb #1/#2 MCER	0.93	0.83	0.89

Table 16.2: The First Autocorrelation Coefficients for Three Sample Images and for Typical Frame Differences of the Respective Sequences

the previous row $MV_{(x+1,y-1)}$ are used to derive the MV estimate. Although the exploitable correlation is below 40%, typically around 20%, since the H.261 codec uses error-sensitive variable-length coding (VLC) anyway, it is useful to exploit this potential redundancy. It does not dramatically impair the overall codec resilience. This is not true in the case of our proposed codecs refraining from using run-length and variable-length coding, since the increased error sensitivity would overshadow the potential bit-rate savings. Having reviewed the potentially suitable MC techniques for high-resolution video compression, let us now concentrate on the high-resolution encoding of the MCER.

16.4 Inter-Frame DCT Coding

16.4.1 Properties of the DCT transformed MCER

The DCT is commonly used due to its so-called energy-compaction property, if the block exhibits spatial correlation in the time domain. The process of motion compensation removes most of the temporal redundancy inherent in the corresponding positions of consecutive frames and a substantial part of the spatial redundancy, which manifests itself in the form of the similarity of the adjacent luminance and chrominance pixels. In order to quantify how much of the spatial redundancy is removed, in Table 16.2 we summarized the first autocorrelation coefficient for certain frames of three sequences and their respective frame differences. We observed that in all three cases, the adjacent-pixel correlation of the sequences is higher than 0.95, and this is the reason why DCT is such a successful method in still-image compression. This is a consequence of the Wiener-Khintchine theorem, stating that the autocorrelation function (ACF) and the power spectral density (PSD) are Fourier Transform (FT) pairs. Since the DCT is closely related to the FT, a flat ACF is associated with a compact DCT spectrum located in the vicinity of the zero frequency. However, the autocorrelation properties of the MCER — which in this case is constituted simply by the frame difference of two consecutive frames — are rather different, since the correlation has been mainly removed, returning reduced or in some cases negative correlation values, as evidenced by the table. Clearly, MC may remove almost all adjacent-pixel autocorrelation.

Strobach [435] found that transform coding is not always efficient for encoding

16.4. INTER-FRAME DCT CODING

the MCER, since it sometimes results in energy expansion rather than energy compaction due to the temporal decorrelation process. In reference [435] he introduced an efficiency-of-transform test, which is based on the statistical moments of the MCER signal. His approach was as follows.

Equation 16.1 represents the energy preservation property of the DCT in time and frequency domains:

$$\sum_{i=0}^{b}\sum_{j=0}^{b} b_{time}^2(i,j) = \sum_{i=0}^{b}\sum_{j=0}^{b} b_{freq}^2(i,j), \qquad (16.1)$$

where $b_{time}(i,j)$ and $b_{freq}(i,j)$ correspond to the time- and frequency-domain representations of a given block. However, the higher moments of these time- and frequency-domain coefficients are different in general, which is formulated as:

$$\sum_{i=0}^{b}\sum_{j=0}^{b} b_{time}^k(i,j) \neq \sum_{i=0}^{b}\sum_{j=0}^{b} b_{freq}^k(i,j) \text{ for } k > 2. \qquad (16.2)$$

If the transformation applied results in significant energy compaction, concentrating most of the energy in a few low-frequency coefficients, then the frequency-domain higher-order statistical moments will become even more dominant after raising the high coefficient values to a high power. Therefore, Strobach argues that it is useful to define the higher-order energy compaction coefficients ρ_k as follows:

$$\rho_k = \frac{\sum_{i=0}^{b}\sum_{j=0}^{b} b_{freq}^k(i,j)}{\sum_{i=0}^{b}\sum_{j=0}^{b} b_{time}^k(i,j)} \text{ for k} > 2 \text{ (k: even)}, \qquad (16.3)$$

where again $b_{time}(i,j)$ and $b_{freq}(i,j)$ represent a given block in the time and transform domain, repectively. The transformation is deemed efficient and compacts the bulk of the energy in a low number of high-energy coefficients, if $\rho_k > 1$; otherwise it results in energy expansion. In order to circumvent this problem, Strobach proposes time-domain MCER compression methods such as vector quantization or quad-tree decomposition.

In order to better understand the statistical properties of the MCER signal, we analyzed the blocks in both the time and transform domain and classified each pixel or coefficient into a high-energy and a low-energy class by a threshold decision given by:

$$thres(x) = \begin{cases} 1 & \text{if } x > x_0 \\ 0 & \text{otherwise} \end{cases} \quad x_0 \in [4..15], \qquad (16.4)$$

where the threshold x_0 was in the range of $[4...15]$ for both the time and frequency domain. For each block we counted the high-energy coefficients or pixels in each class and noted them as sum S_{high}:

$$S_{high} = \sum_{i=0}^{b}\sum_{j=0}^{b} thres|b_{time}(i,j)|. \qquad (16.5)$$

Seq.	MC-Scope	% of Energy Expanding Blks.	% of Energy in Energy Exp. Blks.
Susie	± 0	4.7	4.37
	± 2	7.2	14.12
	± 4	5.2	7.28
Football	± 0	6.8	1.28
	± 4	8.6	1.48
	± 16	8.0	7.17
	± 64	9.63	4.62
Missa	± 0	23.1	12.3
	± 2	30.6	23.80
	± 4	28.59	22.61

Table 16.3: Percentage of energy expanding blocks and their relative energy contributions for the MCER of the 4CIF "Susie" and "Football" sequences and the CIF "Missa" sequence

This sum S_{high} is evaluated in both the time and frequency domain, and the difference ρ_{thres}:

$$\rho_{thres} = S_{high}(freq.) - S_{high}(time) , \qquad (16.6)$$

gives us an indication as to whether the DCT transform was efficient ($\rho_{thres} > 0$) or inefficient ($\rho_{thres} < 0$). For the evaluations below, we skipped those flat-blocks, which resulted in a time-domain count of four or less.

In Table 16.3 we listed the proportion of blocks exhibiting energy expansion on the basis of the previously defined ρ_{thres} efficiency measure and also summarized their total contribution to the energy of the MCER, when using a range of different MC search scopes. Observe in the table that a MC search scope of ± zero corresponds to no MC at all, implying that simple frame-differencing was applied. Surprisingly, about 5% to 9% of the blocks were classified as energy expanding, and their contribution to the overall energy of the MCER is up to 30.6%, which is quite significant. These results varied only insignificantly, when the energy classification threshold x_0 was kept within the range of [4..15]. The results shown here are based on an x_0 value of 5. The DCT's inability to compact the MCER energy increases with the increasing efficiency of the MC, since the more spatial and temporal redundancy is removed, the more unstationary and uncorrelated the MCER becomes.

In order to further explore the efficiency of the DCT, we analyzed the PDF of our test-of-efficiency criterion ρ_{thres} in Figures 16.13 and 16.14 for both the first frame and the first frame difference of the 4CIF "Susie" sequence. The marked areas to the left and right of the efficiency threshold in Figure 16.13 represent the blocks which exhibit energy expansion, or energy compaction in the latter case. The excellent compaction properties of intra-frame DCT coding are clearly demonstrated by Figure 16.13. In this case most of the energy was compacted to less than ten coefficients for the majority of the blocks and typically positive ρ_{thres} values were recorded.

However, the energy compaction performance is drastically reduced, when the

16.4. INTER-FRAME DCT CODING

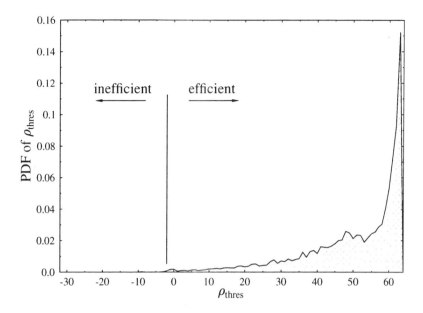

Figure 16.13: Energy compaction efficiency of the intra-frame DCT transform for frame 1 of the 4CIF "Susie" sequence of Figure 15.1, quantified in terms of the distribution of ρ_{thres}.

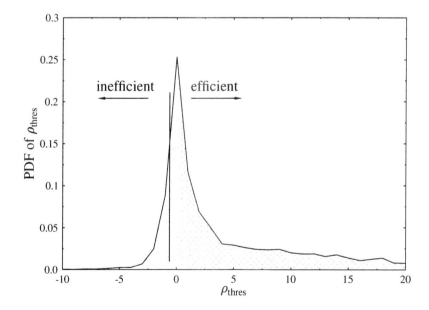

Figure 16.14: Energy compaction efficiency of the DCT transform for the frame difference of frame 1 and 2 of the 4CIF "Susie" sequence of Figure 15.1, quantified in terms of the distribution of ρ_{thres}.

DCT is applied to a MCER frame, as evidenced by Figure 16.14, which shows that for about 20% of the blocks, the DCT fails to compact the energy. This expansion however is typically limited to an increase of five additional positions in total. In other words, the DCT of a MCER frame may result in energy expansion, but it does not result in a catastrophic spread of the energy across the block. This explains why the DCT coding is still favored for MCER coding in the MPEG and H.26x standards. On the other hand, for about 30% of the blocks the DCT compacts the energy by more than five coefficients, and hence the overall DCT compression of MCER images is still deemed advantageous.

In order to further elaborate on this energy expansion property of the DCT, when applied to the reduced-correlation MCER, we refer below to the performance of the H.261 standard. Using a H.261 codec, we compressed 20 frames of the "Miss America" sequence and analyzed the encoded bit stream. The first frame was intra-frame coded, and it was found that every block required an average coding rate of 0.5 bit per pixel. All active blocks of the remaining 19 frames were inter-frame coded, and they yielded an average bit rate of 1.1 bits per pixel, implying the surprising fact that the inter-frame coding in this particular scenario turned out to be less efficient than intra-frame coding and hence increased the required bit rate.

In order to further explore this fundamental problem, we tested the performance of a technique that we referred to as the scanning method. Accordingly, all blocks of a given frame were DCT transformed and quantized using the H.263 quantizers [192]. The H.263 standard offers 32 quantizers, each of linear type and a resolution of 8 bits. For each block, we chose that particular quantizer from the possible 32 characteristics, which gave a mean absolute error just below 4, resulting in a mean PSNR of $20 \cdot \log (256/4) = 36$ dB. The quantized coefficients were then scanned according to the standard's zig-zag scanning scheme in Figure 16.2, subsequently run-length encoded, and the resulting run-length was noted for each block. For the sake of easy implementation we chose the B codes as run-length codes [532].

As a brief excursion from our original topic, namely, the energy compaction-expansion problem of the DCT, Table 16.4 shows how each run-length and color maps to a unique B1 code, where C represents the run color, black or white. As a simple example, the bit-stream ...0,0,0,0,0,1,1,1,0,0,1... would be coded as [black run = 5] [white run = 3] [black run = 2] [white run = 1], yielding the codes ...[0100][1010][01][10]..., where the run color 0 is coded as 0 and color 1 as 1, respectively. Often the modified B codes are used where a certain run-length is used to signal a special event, such as end-of-line (EOL) and escape (ESC).

The run-length decoding algorithm is straightforward. Black and white run-lengths are read from the input bit-stream. The run-lengths are used to reconstruct a scanline of, for example, the coded picture. After each line is coded, an EOL code is inserted into the bit-stream; thus, once a full line has been decoded, an EOL is read. The redundancy introduced by the EOL code, coupled with the fact that the sum of run-lengths on a given scanline is constant, gives the coding technique some error detection capability.

Returning to the topic of the DCT's energy compaction-expansion property, we then evaluated the run-length probability density function (PDF), and its integral, the cumulative distribution function (CDF), was derived and depicted in Figure 16.15

16.4. INTER-FRAME DCT CODING

Run-length	B1 code
1	C0
2	C1
3	C0C0
4	C0C1
5	C1C0
6	C1C1
7	C0C0C0
8	C0C0C1
9	C0C1C0
⋮	⋮

Run-length	Modified B1 code
1	C0
2	C1
3	C0C0
4	C0C1
5	C1C0
EOL,ESC	C1C1
6	C0C0C0
7	C0C0C1
8	C0C1C0
⋮	⋮

Table 16.4: Mapping Run-lengths and Colors to B1 Codewords (left) and Modified B1 Codewords (right) (C refers to the run color)

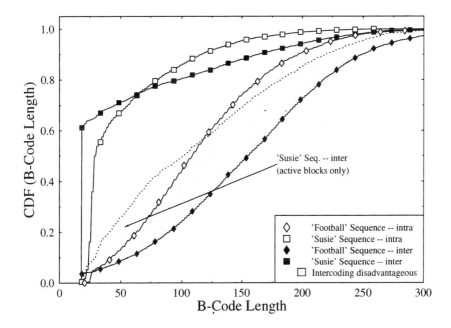

Figure 16.15: CDF of the B-code length of the scanned and quantized coefficients for the 4CIF "Football" and "Susie" sequences in both inter- and intra-frame coding.

— both with intra-frame and inter-frame coding for the "Football" and the "Susie" sequences.

Naturally, for high coding efficiency, short B-coded codewords were expected. Our first observation concerning Figure 16.15 is that the "Susie" sequence results in a substantially higher proportion of short B-coded words and higher coding efficiency than the "Football" clip both in intra- and inter-coded modes. This is due to the "flat" background of the image and its lower motion activity. For any given run-length that particular technique of the intra- and inter-coding schemes is more successful, which results in a higher proportion of short B-coded words, that is, exhibiting a higher CDF value.

The important issue in Figure 16.15 is the B-code length ratio for inter- and intra-mode of operation. For the 4CIF "Susie" sequence, for example, 80% of the inter-frame blocks results in a B-code length of 100 or less, whereas 85% of the intra-coded blocks results in a B-code length of 100 or less. In other words, intra-coding would lead to a better compression ratio.

Note in Figure 16.15 that for the "Susie" sequence 60% of the inter-frame coded blocks resulted in the shortest possible run-length of 18. This happens, if all coefficients are set to zero (i.e., it was not necessary to encode the corresponding block to fulfill the quality criterion). Omitting those "passive" blocks, we generated the new CDF for the "Susie" sequence marked as a dotted line in Figure 16.15. Since this CDF is always below the CDF for the intra-frame mode, this experimental fact corroborates Strobach's claims. Similar results were also valid for the "Football" sequence.

The ITU Study Group 15, which contrived the H.263 standard [192], also produced a document referred to as the Test Near Model version 5, which is commonly termed TMN5 [544]. This document provides a skeleton outline of a codec, which meets the H.263 standard. In the TMN5 document, the study group proposed a solution to overcome the intra/inter coded decision problem using an inter/intra-coded decision for each macroblock b_{MB}, which is composed of four luminance blocks b_n. The chrominance information is not relevant as to the inter/intra-coded decision. This decision is based on:

$$SAD_n(x,y) = \sum_{i=1,j=1}^{b,b} |b_{\text{frame n}}(i,j) - b_{\text{frame n-1}}(i,j)|$$
$$\forall\ x,y \in [-15,15], b = 8, 16 \tag{16.7}$$

$$SAD_{inter} = min\left[SAD_{b_{MB}}, \sum_{n=1}^{4} SAD_{b_n}\right] \tag{16.8}$$

$$SAD_{intra} = \sum_{i=1,j=1}^{16,16} |b_{MB} - \overline{b_{MB}}| \tag{16.9}$$

$$\text{where}\quad \overline{b_{MB}} = \sum_{i=1,j=1}^{16,16} |b_{MB}|$$

$$\text{Mode} = \begin{cases} \text{Inter} & \text{if } SAD_{intra} < SAD_{inter} - 500 \\ \text{Intra} & \text{otherwise} \end{cases} \tag{16.10}$$

which are defined in the TMN5 document [544]. Specifically, Equation 16.7 defines the Sum of Absolute Difference (SAD) criterion for blocks in consecutive frames. The SAD is then evaluated for a whole macroblock and for the four luminance blocks of a macroblock according to Equation 16.8, where the $\text{SAD}_{\text{inter}}$ is the minimum of the SAD for the macroblock and the sum of the SAD for the four luminance blocks.

16.4. INTER-FRAME DCT CODING

The absolute difference of the intra-coded macroblock relative to its mean ($\overline{b_{MB}}$) is derived in Equation 16.9. Finally, the results of both the intra- and inter-coded SAD in Equations 16.8 and 16.9 are then compared in Equation 16.10. Basically, the smoothness of the potential inter- and intra-coded macro-block candidates decides upon the mode.

However, this intra- versus inter-frame classification constitutes a suboptimum approach, since the optimum technique would be to compare the generated bit rate and the resulting reconstructed quality in case of both inter- and intra-frame coding a block. In order to find an attractive solution to this problem, we explored the possibility of joint MC and DCT-based encoding, which is the topic of the following section.

16.4.2 Joint Motion Compensation and Residual Encoding

Inter-frame coding is based on a two-stage approach, exploiting temporal redundancy in the MC step and capitalizing on the spatial frequency-domain redundancy in the subsequent MCER encoding step. The first encoding step minimizes the energy of the MCER, and the second stage attempts to reconstruct the best possible replica of the MCER, using the DCT. Previous sections have shown that an increase in the efficiency of the MC results in a decreased efficiency with regard to the DCT encoding of the MCER, inspiring us to attempt finding the joint optimum. Rather than finding the optimum MVs by decreasing the MCER energy as much as possible, we evaluated the reconstruction error after DCT transformation for each legitimate MV. This process is associated with a huge complexity increase, as the number of required DCT transformations is multiplied by the number of pixels covered by the search window. Nonetheless, at this stage we were motivated by identifying only the order of magnitude of the potential video quality and/or bit-rate benefits.

In a simple approach, we used DCT-coefficient masking without quantization of the unmasked coefficients, and Figure 16.16 portrays the results for both the jointly optimized and the traditional approaches. The potential PSNR gain is about 2 dB, if more than 30 DCT coefficients are used for a block-size of 8×8, although better results are achieved for the highest qualities, when aiming for PSNRs in excess of 40 dB. Here, we found that the jointly optimized approach was capable of identifying the MV, for which the best overall quality was achieved.

In order to obtain results, when applying quantization to the DCT coefficients, we generated bit-allocation tables, as described in Section 16.2. As was expected, we found that because of the uncorrelated nature of the MCER, we were unable to obtain satisfactory results when using pure inter-frame DCT coding. Even a multiclass quantization scheme with 16 different quantizer allocations for different frequency-domain energy distributions was inferior in PSNR terms to the intra-coding approach. A quantization scheme, in which the quantizers are allocated beforehand is in most cases simply not sufficiently flexible to encode the DCT coefficients of a MCER block. Here the algorithm used by the various standard codecs, such as the JPEG, MPEG, and H.26x schemes, which identifies and scans the nonzero DCT coefficients for run-length coding is more suitable, although it exhibits a high error sensitivity.

In conclusion, the results of this and the previous section prompted us to employ

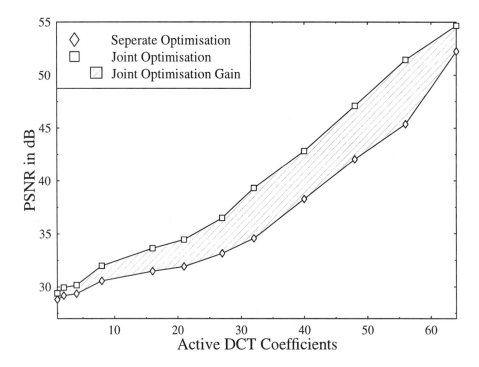

Figure 16.16: PSNR versus number of active DCT coefficients performance comparison of the joint MC-DCT coding and of the traditional two-stage approach using the 4CIF "Football" sequence of Figure 15.1 for frame 1, search scope 32, 1-64 active DCT coefficients, assuming that all blocks are active.

a combined inter/intra-coded approach for our proposed high-quality codec, being able to switch between inter- and intra-coded modes of operation on a block-by-block basis.

16.5 The Proposed Codec

An attractive codec is composed of the same blocks as our low-rate videophone codecs proposed in Part II of the book. The initial coarse intra-frame codec of Section 12.5, which relied on the block averages is replaced here by the intra-frame DCT codec from this chapter, which is seen at the top of Figure 16.17. Since no previous image information is available, every luminance and chrominance block is transformed and the quantized coefficients are transmitted to the receiver and also passed to the local decoder. The quantizer bit-allocation is selected from the range of quantizer configurations we derived earlier according to the bit-rate and quality requirements. Hence, the codec generates a high initial bit-rate according to the relationship of *Quantizer Index* $\times$ 1.5/64, which typically lies around 0.75 bpp. This information is transmitted and loaded into the frame buffer of the local decoder.

16.5. THE PROPOSED CODEC

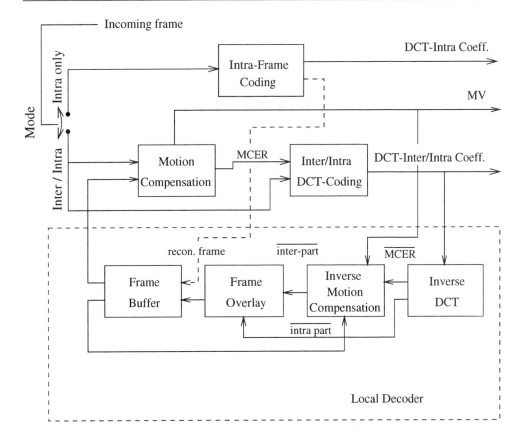

Figure 16.17: Schematic of the intra/intercoded DCT codec.

In contrast to our previous low-rate, low-quality codecs, we used combined inter-/intra-frame coding, as depicted in Figure 16.17. Applying this mode, the incoming frame is passed to the MC codec and the resulting MCER is passed to the inter/intra-frame coded DCT codec, along with the unmodified incoming frame. These operations are highlighted in simple terms in Sections 16.5.1 and 16.5.2 using Figures 16.17–16.18. Here we will only briefly allude to them. After the inter/intra-frame coding decision has been carried out, the actual DCT encoding and quantization takes place, and all relevant parameters are transmitted to the decoder. The parameters are also decoded locally and passed to the local decoder. Then the inverse DCT step is carried out, which results in the reconstructed MCER and the reconstructed intra-coded frame. Recall that two modes are possible, and the reconstructed intra-coded frame contains only the actually intra-coded blocks, while it contains blank blocks in the remaining positions. The same is valid for the reconstructed MCER, where only those blocks selected for inter-frame DCT coding are present. The MCER is passed through the inverse MC step. This reconstructed inter-coded frame is overlaid on the reconstructed intra-coded frame, giving the final reconstructed frame.

Further details of the encoding strategy will be described in the forthcoming Sections.

16.5.1 Motion Compensation

Let us first discuss the MC step, where we have to incorporate the results from Section 16.4.1. Previously, we attempted to combine MC and DCT using our transform efficiency measure, which may be disadvantageous for two reasons. First, the intraframe coding block might lead to a better PSNR than the jointly optimized MC and DCT steps at the same bit rate, since a substantial bit-rate contribution must be allocated to the encoding of the MVs, thereby disadvantaging the MCER coding due to the paucity of bits. Second, we unnecessarily increase the vulnerability of the encoded bit-stream, since the MVs are sensitive to channel errors and result in error propagation through the reconstructed frame buffer. Specifically, a single MV, which can take 32 horizontal and vertical positions, for example, requires $5 + 5 = 10$ bits. Similarly, a high number of bits is required for the encoding of the block index, signaling the position of each block for which MVs are transmitted. For example, when using 1408 x 1152 pixel 16CIF resolution images, there is a total of 25344 8 x 8 blocks, requiring a 15-bit block index, unless specific coding techniques are employed to reduce this number. Hence a total of up to 25 bits/block may be necessary for the MC information of an 8 x 8 = 64 pixel block, which would severely limit the achievable compression ratio. The associated bit-rate contribution would be $25/64 \approx 0.39$ bits/pixel. Therefore, we have to carefully consider the active MV selection criterion.

Thus, during the MC process, we first check whether our PSNR or MSE quality requirements are satisfied without MC or DCT coding, as seen in the flowchart of Figure 16.18. Hence, we defined a block quality criterion, which is a more sophisticated version of the criterion that is used by the H.263 codec. More specifically, this quality criterion is stated in Equation 16.13 using two thresholds, T_{mean}, for testing the maximum mean distortion, D_{mean}, in Equation 16.11 and T_{peak}, for checking the value of the maximum peak distortion D_{peak} in Equation 16.12. Then in the context of the blocks b_1 and b_2 we have:

$$D_{mean} = \frac{1}{b^2} \sum_{i=1,j=1}^{b,b} |b_1(i,j) - b_2(i,j)| \tag{16.11}$$

$$D_{peak} = MAX(|b_1(i,j) - b_2(i,j)|) \quad \forall\, x,y \in [1,b] \tag{16.12}$$

$$\text{Quality} = \begin{cases} \text{adequate} & \text{if } (D_{mean} < T_{mean}) \wedge (D_{peak} < T_{peak}) \\ \text{inadequate} & \text{otherwise} \end{cases} \tag{16.13}$$

The video quality is deemed adequate, if both distortion terms are below their respective thresholds and inadequate otherwise.

Initially, the corresponding blocks in consecutive frames are subjected to our quality test expressed in terms of Equation 16.13. If the criterion is satisfied, since these blocks are sufficiently similar, it is not necessary to encode the corresponding block. This block is marked as passive, which is shown in Figure 16.18. If a block does not

16.5. THE PROPOSED CODEC

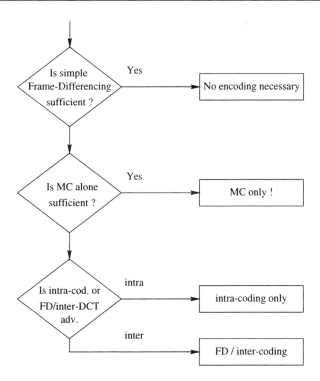

Figure 16.18: Flowchart of the block-encoding decisions.

satisfy the criterion, a motion vector is determined and the criterion is applied again to the corresponding blocks in consecutive frames but after motion compensation. If after MC the quality criterion is met, it is signaled to the subsequent inter/intra-frame coded codec that no further encoding is necessary for the block concerned. If the MC step above fails to provide the required quality, either intra-frame DCT coding is invoked or frame-differencing (FD) combined with inter-frame-oriented DCT coding is employed, as seen at the bottom of Figure 16.18. This approach is adopted, when the additional bit-rate requirement of the MVs along with MCER coding would be excessive.

16.5.2 The Inter/Intra-DCT Codec

The DCT codec of Figure 16.17 processes both the MCER and the incoming frame. Previous evaluations of the MC have shown which of the blocks are deemed to be motion passive or have been successfully subjected to MC according to the flowchart of Figure 16.18. Those blocks are ignored at this stage, reducing the computational demand. The remaining blocks failed our preset quality criterion of Equation 16.13 and must be encoded by the DCT codec. In the previous section we stated that joint MC and inter-frame coding would unnecessarily inflate our bit rate. At this stage, we determined whether frame-differencing and inter-frame DCT coding, or simple intra-frame coding leads to a better reconstructed video quality. The outcome of this decision is signaled to the decoder using a one-bit flag.

16.5.3 Frame Alignment

Because of the high video quality requirements and in contrast to the codecs proposed in Part II of the book, we cannot maintain a constant bit rate per transmitted frame. Hence, we have to signal the length of the MV and DCT data segments to the decoder. This could be easily accomplished by a small header segment at the beginning of every frame, storing the number of MVs and DCT encoded blocks. This side-information would be very vulnerable to channel errors, since even a single bit error causes misalignment of the encoder and decoder, and forces the whole decoding process to collapse. Another, more error-resilient method is to insert frame alignment words [545] (FAW) constituted by unique bit patterns of a given length N, which are detected by correlation of the bit-stream and the FAW at the decoder side. If the detection process is successful, the bit-stream is separated in different substreams for the MV and the DCT data, and the correct number of MVs and DCT blocks is recovered for each video frame after identifying the length of each stream. It is important to ensure that the bit-stream generated by the codec is not strongly correlated with the FAW, which could easily lead to frame synchronization problems in case of corrupting a few bits of the FAW. Therefore, the codec checks the encoded output and eventually rearranges the transmitted data, so that the Hamming distance of the generated data and the FAW remains sufficiently high. This can be achieved, for example, by inserting additional bits into the stream, which are then recognized and removed at the decoder. This is possible, because the proposed codec allows us to swap entire blocks of bits in the bit stream. In contrast to reference [546], we do not use bit stuffing, as for this method is vulnerable to channel errors.

In this subsection, we attempted to determine the minimum necessary length of the FAW, so that it becomes highly unlikely that the FAW is emulated by the video data. The longer the FAW, the less likely it will be emulated by the encoded data, reducing the false FAW detection probability and hence the probability of false frame synchronization. However, the longer the FAW, the more complex the correlation process required for its detection. This complexity problem is particularly grave if the decoder has no prior knowledge as to the likely position of the frame commencement, since then it is required to continuously search for it. A further problem is that the received FAW may contain a number of transmission errors, and it must not be erroneously rejected due to this phenomenon, forcing us to accept it, if a few bits are corrupted. However, this increases the probability of its emulation by video-coded data.

The bit-stream under consideration contains two different types of data, namely, the FAW and the video-encoded data. Assuming an n-tuple FAW and a given number of arbitrary random error events, below we quantify the probability of emulating the FAW by the video data.

In the simplest transmission scenario, the FAW is not corrupted and the probability $p_0(n)$ that this error-free FAW of length n is emulated perfectly by random data is given by:

$$p_0(n) = \frac{1}{2^n}. \qquad (16.14)$$

16.5. THE PROPOSED CODEC

Since data corruption is unavoidable over practical channels, we have to allow for a certain number of errors in the FAW. The number of possible bit patterns, which emulate the FAW with exactly one bit error, is simply $C_1(n) = n$. Hence, the probability $p_1(n)$ corresponding to this event is given by:

$$p_1(n) = \frac{n}{2^n}. \tag{16.15}$$

The problem is slightly less obvious, when evaluating the probability $p_2(n)$ for two simultaneous bit errors, which can be solved with reference to the probability $p_1(n)$ given above. For the given FAW having n bits $(1, 2, 3, 4, \ldots, n)$, we assume that bit 1 is corrupted. The remaining $n - 1$ bits then contain exactly one error, leading to $C_1(n-1) = n-1$ possibilities as regards the error position. We can now assume that bit 2 is corrupted and bit 1 is assumed to be uncorrupted, since it was considered to be corrupted above. Hence, the second bit error is in one of the bit positions from 3 to n, which leads to $C_1(n-2) = n-2$ possible combinations. Following this principle, we see that the number of combinations for exactly two bit errors $C_2(n)$ is defined as:

$$C_2(n) = \sum_{i=1}^{n-1} C_1(i) = (i-1) + (i-2) + \ldots + 1 = \frac{n(n-1)}{2} \tag{16.16}$$

This approach may be generalizd for the case of $C_x(n)$ as follows:

$$C_x(n) = \sum_{i=1}^{n-1} C_{(x-1)}(i), \tag{16.17}$$

which is a recursive formula and makes it easy to evaluate all required probabilities $p_i(n)$. The probability $P_x(n)$, that a FAW of length n is emulated by data, when allowing up to x bit errors, is then defined by summing all possible error probabilities $p_i(n)$ for $i = 1, 2, \ldots x$:

$$P_x(n) = \sum_{i=1}^{n} p_i(n). \tag{16.18}$$

The probability of the FAW being emulated by random video data, while tolerating a certain number of errors, is depicted in Figure 16.19.

It is now possible for us to choose the required length of the FAW. At data rates around 1 bpp, an encoded HDTV frame is up to 2 Mbits long. Allowing a misalignment once every 10^7 fames ($\approx$ once every four days at 30 frames/s), we require a misalignment probability of less than $2 \cdot 10^{-13}$, which is satisfied by a FAW length of 48 bits and longer, as seen in Figure 16.19. When opting for an even longer, 64-bit word and allowing up to five errors in the detection process, the equivalent BER is about $7.8 \cdot 10^{-2}$. Since high-quality video communications is impossible at this BER, we are certain to have obtained a safe FAW length estimate. In our scheme two different FAWs are used in order to signal the beginning of the MV and DCT information segments. Therefore, we transmit 128 extra bits per frame, which results in a marginal increment of the data rate.

So far we have determined the probability of the FAW being emulated by random data. This is an upper bound estimate for the FAW being emulated, since the codec

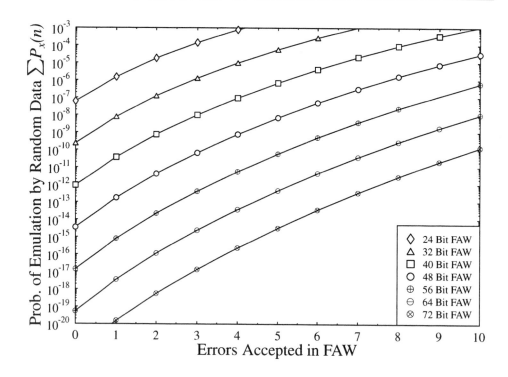

Figure 16.19: Probability of corrupted FAWs being emulated by random video data.

actually arranges the bit-stream so that the correlation of the uncorrupted data with the FAW is low. Now we have to choose the appropriate FAW so that FAW detection leads to the right frame alignment.

The choice of the appropriate FAW is based on its autocorrelation properties. Explicitly, the FAW must exhibit a high main peak to secondary peak ratio in order for the correlation process not to mistake a high secondary peak for a corrupted, and hence reduced, main peak, which would inevitably lead to a frame synchronization error. The set of best sequences in this sense for a certain length is then usually found by computer search, where all possible sequences of the required length are generated and the one exhibiting the highest main-to-side lobe ratio is deemed to be the best. We chose our FAW following Al-Subbagh's work [547], who listed the best patterns up to a word length of 32, and extended the optimum 32-bit FAW to 64 bits by simply repeating it. Their autocorrelation displayed in Figure 16.20 shows that the secondary peaks are substantially lower than the main peak. Explicitly, their difference amounts to 28. Therefore, up to 13 one-bit errors may be tolerated within the FAW without mistaking the secondary ACF peak for the main one.

Having considered the algorithmic details of the proposed HDTV codec, we can now proceed with the characterization of the bit-allocation strategy.

16.5. THE PROPOSED CODEC

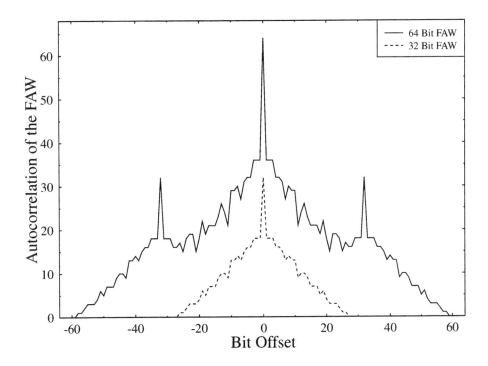

Figure 16.20: Autocorrelation function of the 32- and 64-bit FAWs used for the simulations.

Mode	FAW	MV Index	MV	DCT Index	DCT Coeff.
Intra	64	-	-	-	32 - 64
Inter	2 × 64	13-15	8...12	14-16 + 1 (inter/intra flag)	32 - 64

Table 16.5: Bit-Allocation Table for CCIR 601-sized Frames to HDTV-sized Frames

16.5.4 Bit-Allocation

The resulting bit-stream is composed of intra-frame coded DCT coefficient data, MVs, and combined inter/intra-DCT data as shown in Table 16.5. In the intra-frame coded mode of operation, we transmit the DCT coefficients of every single block at a rate of 32 − 64 bits per 8 × 8 block, so that the initial frame's contribution is in the range of 0.75 to 1.5 bpp or 259 to 518 kbits for CCIR 601 images. In the inter-frame coded mode, the MVs, the DCT coefficients, the inter/intra-coded flag and their respective indices are to be sent. For 720 × 480 pixel CCIR 601 images there are 5400 8 × 8 blocks and MVs, requiring a 13-bit block index. For a HDTV resolution of 1920 × 1080 pixels there are 32400 blocks, requiring a 15-bit index. The number of DCT blocks is increased by 50% for the chrominance blocks; hence one extra bit is necessary for their indexing. The MV search scope must be restricted to less than 32 for computational complexity reasons, which leads to a MV coordinate of 6 + 6 = 12 bits at most.

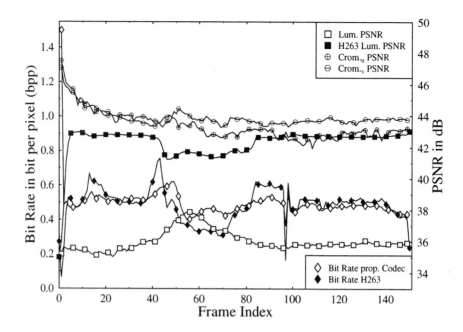

Figure 16.21: Performance of the proposed DCT codec for the 4CIF "Susie" sequence of Figure 15.1, using a MV search scope of 16.

These information segments, composed of the MV and DCT data, are concatenated to form a single bit-stream, separated by the FAW discussed above. At the decoder we detect the FAWs, which indicate the beginning, the end, and the type of the transmitted data segments.

16.5.5 The Codec Performance

The PSNR versus frame index performance figures for the 4CIF "Football" and "Susie" sequences of Figure 15.1 are given in Figures 16.21 and 16.22. For both simulations we selected quality thresholds of $T_{mean} = 4$ and $T_{peak} = 8$ for the luminance component, resulting in a PSNR of about 36 dB. Initially, we applied the same criterion to the chrominance blocks, and consequently the chrominance PSNR dropped from the initial 47 dB PSNR of Figure 16.21 to about the same value as the luminance PSNR. We found that the subjective color reproduction suffered tremendously as a result. In our experiments we needed an enhanced chrominance quality of about 1.5, according to our quality criterion of Equation 16.13 compared to the luminance in order to achieve a subjectively similar luminance and chrominance reproduction. This resulted in a PSNR improvement of about 8 dB, which was rated as a subjectively pleasant quality. Apparently, the human eye seems to be more sensitive to luminance shifts than brightness shifts. We found that especially for the 4CIF "Susie" sequence a good color reproduction was crucial. Note furthermore, that the quality of the "Susie" sequence is about 6–8 dB better at a significantly lower bit rate

16.5. THE PROPOSED CODEC

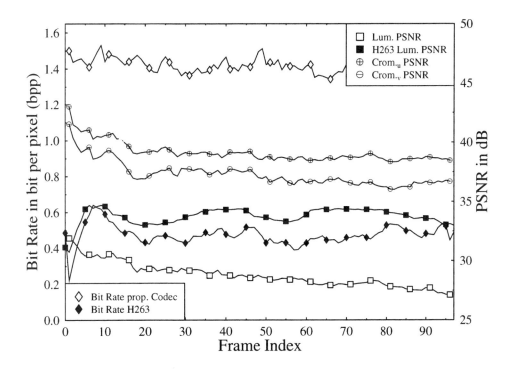

Figure 16.22: Performance of the proposed DCT codec for the 4CIF "Football" sequence of Figure 15.1, for a MV search scope of 32.

than that of the high-activity 4CIF "Football" sequence. Comparing the proposed codec with the performance of the H.263 standard codec, we found that in the case of the 4CIF "Susie" sequence at almost the same bit rate the H.263 codec outperformed our more error resilient non-RL coded codec by about 6–7 dB in terms of the PSNR.

Results for the "Football" sequence demonstrate a worst-case scenario. Our codec detects high activity in about 80% of the blocks, forcing almost every block to be fully encoded. Accordingly, the bit rate increases to about 1.5 bpp. The H.263 codec performs better in terms of bit rate and PSNR values. The subjective image-quality difference was found to be less significant as the objective PSNR difference.

16.5.6 Error Sensitivity and Complexity

The previous section characterized the performance of the non-RL coded DCT codec, which was outperformed by the H.263 standard codec. The advantage of our codec is its error resilience. Figure 16.23 characterizes the error sensitivity of the generated bit-stream, which shows error resilience up to a BER of 10^{-4}. A more quantitative characterization of the bit sensitivities ensues from Figure 16.24 for the luminance and chrominance DCT coefficient encoding of bits of 1–64 of the 8×8 blocks. Bit positions 70–83 are the 14 MV bits of the blocks, where the MSB at position 70 inflicts the highest Mean Absolute Difference (MAD) degradation, while the 13 block location

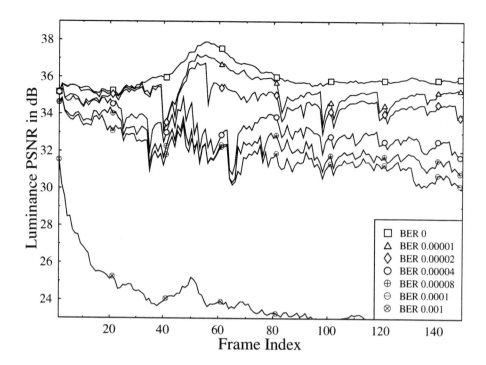

Figure 16.23: PSNR versus frame-index performance of the non-RL coded DCT codec under erroneous conditions, for a MV search scope of 16, and an average bit rate of 0.55 bpp at various BERs.

indices exhibit a more uniform MAD degradation, which was computed by consistently corrupting the bits concerned and evaluating the associated MAD. Examples of images after exposure to a BER of 10^{-3} and 10^{-4} are given in Figures 16.25 and 16.26.

Figure 16.25 shows an even distribution of the errors, whereas at the ten times higher BER of Figure 16.26 many of the errors are concentrated around the fringes of the frame. This is due to the varying encoding probability of the blocks in each frame and is explained as follows. Blocks of the background are less frequently coded than those of active locations of the image. The latter blocks are more frequently subjected to intra-frame coding and are therefore more often updated. By constrast, if the index of a MV or DCT block is corrupted, the corrupted index points to a new, false location which might belong to those locations, which are less frequently updated. Apparently, this effect becomes more dominant at higher BERs, and the problem might be solved by forcing the codec into intra-coded mode in order to encode every block after a preset number of frames, if it is used at such a high BER.

The complexity in terms of computational demand depends strongly on the MC parameters, search scope, and search algorithm, as well as on the activity within the frame and, of course, on the frame size. As shown in Section 16.3, the computational demand of the MVs easily reaches the region of 10^9 Flops, when applying Equa-

16.6. SUMMARY AND CONCLUSIONS

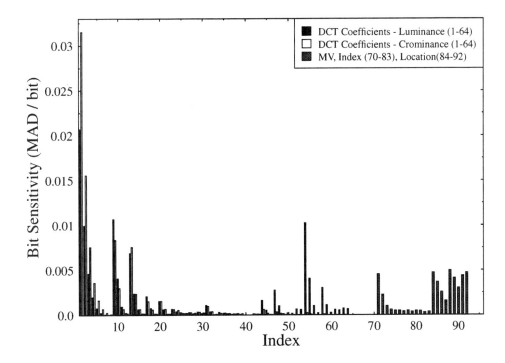

Figure 16.24: Bit sensitivities of the MVs, intra- and inter-frame-coded DCT bits, for the 4CIF "Susie" sequence, MV window size of 16 × 16, and for an average bit rate of 0.55 bpp.

tion 12.10 at 5400 blocks per CCIR 601 frame. For real-time applications we propose small search windows at the size of 4 × 4 or the use of simple frame-differencing. This will slightly increase the bit rate, but the computational demand is reduced to that of the DCT transform. The optimized DCT transformation requires 768 Flop, which results in a computational demand of 768 × 8100 blocks = 6.2 MFlop per color CCIR 601 frame. In the worst case, this is carried out for both the inter- and intra-coded frame during each inter-frame coding step, thus doubling the computational demand. As the signals have to be locally decoded for the inter- and intra-coded decisions, the inverse DCT has to be executed, doubling the computational demand again. Hence, an upper bound complexity estimate for the inter/intra-frame codec for a MV search scope of zero is 25 Mflop. Since some of the operations may be executed in parallel, a real-time implementation in silicon is possible.

16.6 Summary and Conclusions

In this chapter the exploitation of the temporal redundancy was combined with DCT-based coding. We attempted to capitalize on our results from Part II of the book in the field of high-resolution image coding, where a persistent high quality is required.

Figure 16.25: Picture quality at a constant BER of 10^{-4} for frame 100 of the "Susie" sequence at 0.65 bpp.

Figure 16.26: Picture quality at a constant BER of 10^{-3} for frame 100 of the "Susie" sequence at 0.65 bpp.

16.6. SUMMARY AND CONCLUSIONS

In the context of these codecs, we had to accept a variable bit rate, since the initial, fully intra-coded frames often impose a higher bit rate than the inter-coded frames. The proposed intra-frame subband codec achieved a PSNR of about 33.4 dB at a bit rate of about 2 bits/pixel. In Section 16.4.1 we explored the characteristics of the MCER and then contrived an intra/inter-coded DCT codec, achieving a PSNR of 36 dB at about 0.6 bit/pixel for color sequences.

Both Parts II and III of this book proposed methods for error-resilient image coding with a breakdown BER of the presented codecs in the range of 10^{-4} to 10^{-2}, which are more error resilient than existing standard video codecs, such as the H.26x and MPEG schemes. As more and more public services are provided over mobile channels, such codecs will become more important and stimulate further research. The codecs presented in Part II of this book were designed to allow direct replacement of mobile radio voice codecs in second-generation wireless systems, such as the Pan-European GSM, the American IS-54 and IS-95, as well as the Japanese systems, operating at 13, 8, 9.6, and 6.7 kbps, respectively. They are also amenable to transmission over third-generation wireless systems. The results in Part III of the book are applicable to HDTV and WLAN-type systems. As demand for these multimedia mobile services emerges, error-resilient schemes at all rates, qualities, and resolutions will be required in order to satisfy the demands of tomorrow's customers. In the next part of the book, commencing with the next chapter, we consider the standard H.261 and H.263 video codecs and design interactive wireless transmission systems.

Part IV

Video Systems Based on Standard Video Codecs

Chapter 17

An ARQ-Assisted H.261-Based Reconfigurable Multilevel Videophone System

17.1 Introduction

This chapter investigates the properties of the H.261 video codec and designs a wireless system facilitating its employment in a mobile radio environment. Initially, a detailed investigation of the properties of the H.261 standard codec [500] was carried out.

Before designing a wireless transceiver facilitating the employment of the H.261 codec in a mobile environment, several investigations of particular coding aspects had to be made. A thorough investigation of the error effects on the H.261-encoded data stream was also undertaken, and different schemes for improving the error resilience of the H.261-based system were studied. Finally, an error-resilient H.261 system was designed and investigated in a wireless mobile scenario.

17.2 The H.261 Video Coding Standard

17.2.1 Overview

The International Telecommunications Union's (ITU) H.261 standard [500] specifies a video codec for audiovisual communications services. It was designed for data rates at multiples of 64 Kbps, thus it is well suited to Integrated Services Digital Network (ISDN) lines, which also have the low Bit Error Rate (BER) required for its operation. A so-called hybrid coding algorithm is employed and inter-frame prediction is used to reduce the inter-frame temporal redundancy, while transform coding is used

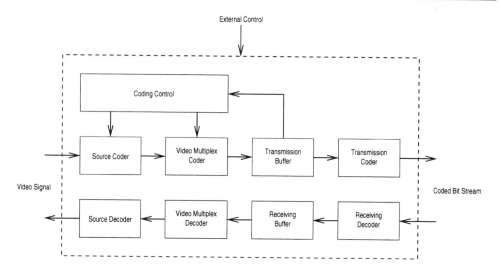

Figure 17.1: Block diagram of the H.261 scheme.

to reduce the spatial redundancy exhibiting itself in the spatial frequency domain. The motion translation information extracted by motion compensation is represented by motion vectors (MV). The system was designed for bit rates in the range of approximately 40 Kbps to 2 Mbps. The block diagram of the H.261 codec is shown in Figure 17.1. Other sources of information about the H.261 video codec are, for example, T.Turletti's report on an H.261 video codec for use on the Internet [548] and British Telecom's audiovisual telecommunications book [549].

17.2.2 Source Encoder

The source encoder operates on noninterlaced pictures at approximately 30 frames per second (fps). These pictures are encoded as luminance and two color difference components (Y, C_B, C_R). Two picture resolution formats are specified, the 352×288-pixel Common Intermediate Format (CIF) and the 176×144 Quarter Common Intermediate Format (QCIF). The color differences are sampled at half the rate of the luminance, as shown in Figure 17.2. In order to be able to adjust the frame rate for a variety of applications having different quality and/or bit-rate requirements, the H.261 scheme can disable the transmission of 0–3 frames between transmitted ones.

The video encoding algorithm consists of three main elements: motion prediction, block transformation, and quantization. A video frame can be encoded in one of two modes, either in the intra-frame coded mode, where the original video frame is encoded or in the inter-frame coded mode, where the motion-compensated frame difference related to the previous decoded frame is encoded. In each case, the frame is divided into blocks of 8×8 pixels. Four adjacent luminance blocks and two color difference blocks are then combined to form a so-called *macroblock*. Therefore a macroblock consists of 16×16 luminance pixels and 8×8 chrominance pixels, as shown in Figure 17.2. The block diagram of the H.261 source encoder is shown in

17.2. THE H.261 VIDEO CODING STANDARD

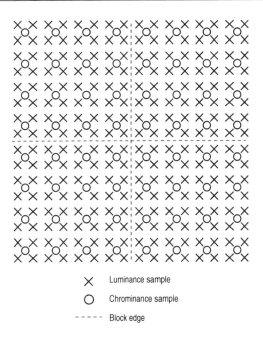

Figure 17.2: Positioning of luminance and chrominance samples in a macroblock containing four luminance blocks.

Figure 17.3.

In the simplest implementation of the inter-frame coded mode, the previous frame can be used as the prediction for the next frame, a technique often referred to as frame-differencing. However, the inter-frame prediction can be improved by the optional use of full motion compensation relying on motion vectors and/or on spatial filtering. The motion vector-based motion compensation is optional in an H.261 encoder but required in all decoders. A distinct motion vector can be defined for each macroblock. The motion vector defines a translation in two dimensions, and therefore it has two coordinates for the x and y directions, respectively. Each motion vector can assume integer values in the range of ±15, but they obey the restriction that the pixels referenced must be within the coded picture area. Accordingly, they can be encoded using 4 bits for each dimension, and therefore $(4 + 4) = 8$ bits are required for each motion vector. The prediction can also be modified by the use of a two-dimensional spatial filter [500] that operates on blocks of 8 × 8 pixels, mitigating the blockiness of the reconstructed video at low bit rates. The filter can only be used for all or none of the blocks in a macroblock.

The prediction error of the 8 × 8 pixel blocks is then transform coded. A separable two-dimensional discrete cosine transform (DCT) is used. Its output is clipped to the range of -255 − +256. The final element of the source coder is quantization of the DCT coefficients. There is a specific quantizer for the intra-frame coded zero-frequency or direct current (DC) component and 31 different quantizers for all other DCT coefficients. Selecting one of the 31 quantizers constitutes a convenient way of

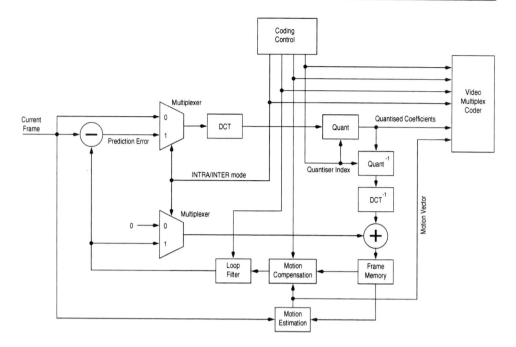

Figure 17.3: H.261 source encoder.

controlling the output bit rate. The quantizers are numbered 1 to 31 from the finest to the coarsest. All quantizers are linear, and apart from the DC-quantizer they all exhibit a central dead-zone around zero. The DC-quantizer has a step size of 8, while the higher-frequency or non-DC quantizers have even-valued step sizes in the range of 2–62. Generally, the coarser the quantizer the lower the bit rate and the transmitted image quality. The quantizer dead-zone is wider than the step size, so that camera noise does not contaminate the signal to be quantized.

The DCT coefficients of the motion prediction error are quantized. The encoder incorporates a local decoder in order to ensure that the motion compensation scheme operates using the reconstructed picture at both ends in its future motion estimation steps rather than the original frame at the encoder and the reconstructed frame at the remote decoder, which would result in misalignment between their operation. As seen in Figure 17.3, the quantized DCT coefficients are passed through an inverse quantizer in the local decoder. The combined transfer function of the quantizer and inverse quantizer is shown in Figure 17.4. The global shape of the transfer function is approximately linear. However, as the figure shows, there is a saturation limit, depending on the required bit rate and video quality, which translates into a quantizer index. The full dynamic range cannot be quantized, if a low quantizer index is required ($index < 7$), as it becomes clear from Figure 17.4, since the output is clipped to a maximum value for quantizers with an index less than 7. The figure also shows that the stepsize of the quantizer is $2n$ where n is the quantizer index. The dead-zone of the quantizer is defined as $6(n-1) + 3$, which is again introduced in order to remove low-level camera noise as well as to reduce the bit rate. Therefore the size of

17.2. THE H.261 VIDEO CODING STANDARD

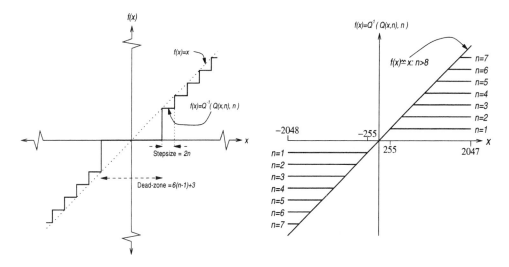

$Q(x,n)$ is the quantizing function.
$Q^{-1}(x,n)$ is the inverse quantizing function.
x denotes the DCT coefficients, n is the index of the quantizer used in the range 1--31.

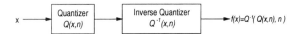

Figure 17.4: H.261 Quantizer transfer function.

the dead-zone increases, as the step size and quantizer index increase. The coarsest quantizers have the largest dead-zones and step sizes. The finest quantizers have the smallest dead-zone and step size, but they have a restricted dynamic range.

The reconstructed video frames are clipped to the range of 0–255 in order to avoid arithmetic overflow in other parts of the encoder or decoder.

17.2.3 Coding Control

The coding control block adjusts the required performance of the codec by modifying the coding parameters in order to attain the required image quality versus bit-rate compromise. Several parameters can be changed in order to modify the behavior of the codec. These can include pre-processing before source encoding, changing the quantizer, changing the block selection criterion, and temporal subsampling.

Specifically, not every block of a frame needs to be updated in each frame, because some have little or no differences in consecutive video frames. The block selection criterion evaluates for each block the difference between the current input block and the previous locally decoded frame. If it is above a certain threshold, the relevant parameters of the block are transmitted. By controlling the threshold, more or less blocks are transmitted, and hence the bit rate of the codec is modified. Temporal subsampling is performed by discarding whole frames, when there is little or no dif-

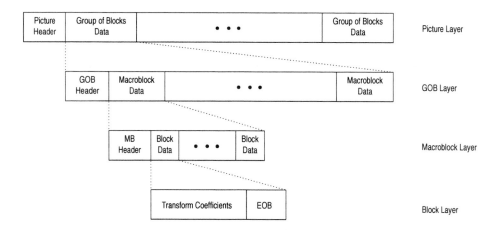

Figure 17.5: Hierarchical structure of a H.261 frame.

ference since the previous frame. Furthermore, the coding control block ensures that each macroblock is forcibly updated with an intra-frame coded macroblock at least once every 132 times it is transmitted.

17.2.4 Video Multiplex Coder

The video multiplex coder groups the source-coded data into groups and codes them into symbols to be transmitted. The video multiplex coder also attempts to further reduce the residual redundancy in the data stream and organizes the source-coded data into a hierarchical structure. This structure is shown in Figure 17.5. The data stream is then converted into a bit-stream to be transmitted. Most of the transmitted bit-stream comprises variable-length codewords, which are used to remove any remaining redundancy. Although these variable-length codes are vulnerable against channel errors, the transmission syntax was constructed to include unique words at various hierarchal levels, which facilitate resynchronization at various stages. Let us now consider how the video stream to be transmitted is constructed.

17.2.4.1 Picture Layer

As seen in Figure 17.5, this layer consists of a picture layer header followed by the Group of Block (GOB) data. The picture layer header is detailed in Figure 17.6, where the self-descriptive transmission packetization format becomes explicit. As seen in Figure 17.5, similarly to the picture layer, all other layers of the hierarchy commence with a self-contained header, describing the forthcoming information, followed by the encoded information of the next lower layer.

The picture-layer header information commences with the Picture Start Code (PSC), which is a specific 20-bit codeword facilitating the recognition of the start of a new frame. This unique word allows the codec to resynchronize, using a simple pattern recognition technique by identifying this code in the received bit-stream after the loss of synchronization due to channel errors.

17.2. THE H.261 VIDEO CODING STANDARD

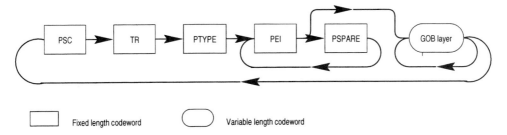

Figure 17.6: Picture layer structure for the H.261 codec.

PSC	Picture Start code (20 bits) – fixed
TR	Temporal reference (5 bits)
PTYPE	Information concerning the complete picture, (6bits) such as split screen, freeze, and CIF/QCIF.
PEI	Extra insertion information (1 bit) – set if PSPARE to follow.
PSPARE	Extra information (0/8/16... bits) – not used, always followed by PEI.

Table 17.1: Summary of the Picture Layer Information in the H.261 Codec

The Temporal Reference (TR) is a 5-bit counter labeling the video frames. If there are a number of nontransmitted frames between transmitted ones, it will be incremented by their number in order to reflect the index of the input frames rather than that of the transmitted ones. The temporal reference word is followed by the 6-bit Picture Type (PTYPE) code, which reflects a number of picture type options, such as whether the frame it refers to is a CIF or QCIF frame, whether the picture freeze option needs to be set or reset, and so on. As demonstrated by Figure 17.6, the next information field is the Extra Insertion (PEI) bit, which signals to the decoder whether further bits are to follow. This option allows the codec to incorporate future services by appending an arbitrary multiple number of 8 bits, as long as PEI is set to logical one. Following this field, the GOB layer information is appended, which will be considered in the next section. The role and the associated number of bits of the various segments are summarized in Table 17.1.

17.2.4.2 Group of Blocks Layer

At the next hierarchical level of Figure 17.5, the picture is divided into Groups of Blocks (GOB), each constituted by 176 x 48 luminance pels. A CIF and QCIF frame divided into GOBs is shown in Figure 17.7. The GOB layer consists of the GOB header seen in Figure 17.5, followed by the GOB information constituted by the macroblock information. The GOB header commences with the GOB Start Code (GBSC), followed by a number of other control words and macroblock information concerning all macroblocks. The GOB header is characterized in detail in Figure 17.8.

The GOB layer information commences with a 16-bit unique word, the so-called GOB Start Code (GBSC), again, similarly to the PSC word, in order to allow the

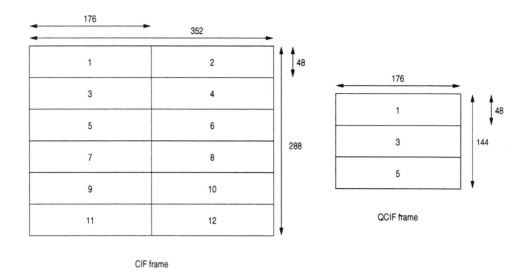

Figure 17.7: CIF and QCIF frames divided into H.261 GOBs.

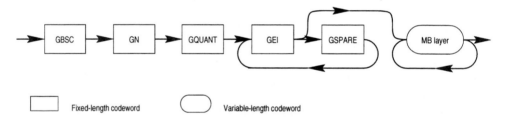

Figure 17.8: Group of blocks layer structure for the H.261 codec.

GBSC	Group of blocks start code (16 bits) – fixed.
GN	Group number (4 bits) – binary representation of 1–12, 13–15 reserved. Indicates GOB position in Figure 17.7.
GQUANT	Group quantizer information (5 bits) – codeword to indicate quantizer used for group. Active until overridden by MQUANT.
GEI	Same function and size as PEI in Table 17.1.
GSPARE	Same function and size as PSPARE in Table 17.1.

Table 17.2: Summary of GOB Layer Information in the H.261 Codec

17.2. THE H.261 VIDEO CODING STANDARD

decoder to resynchronize at the start of the next GOB word in case of loss of synchronization induced by channel errors. The position of any specific GOB in the CIF and QCIF frames is reflected by the Group Number (GN), which is a 4-bit word, allowing the encoding of 16 GOB positions, although only 12 indices are required in Figure 17.7. This segment is followed by a 5-bit field referred to as Group Quantizer (GQUANT), specifying the quantizer to be used in the current GOB. This 5-bit identifier allows the codec to encode up to 32 different quantizers. This quantizer is used until it is superseded by any other quantizer selection information, such as the Macroblock Quantizer (MQUANT) code of the macroblock layer. Similarly to the picture layer, the GOB layer also exhibits a spare capacity for future services, which is signaled to the decoder using the Extra Insertion (GEI) flag followed by the $n \times 8$-bit GOB-layer spare bits, (GSPARE). These fields are summarized in Table 17.2, indicating also the number of bits assigned. From this transmission format, it is clear that the information stream has a self-contained descriptive structure, which is attractive in terms of compactness and flexibility but is also rather vulnerable to transmission errors.

17.2.4.3 Macroblock Layer

Each GOB is divided into 33 so-called Macroblocks (MB) in an 11×3 arrangement, each constituted by 16 x 16 luminance pels or four 8 x 8 blocks, as portrayed in Figure 17.2. Recall, however, that this will correspond to an 8 x 8 block in terms of the subsampled chrominance values. The MB layer header is described in Figure 17.9.

The MB address (MBA) of Figure 17.9 is a variable-length codeword specifying the position of each macroblock in a GOB. For the first MB of each GOB, the MB address is the absolute address given in Figure 17.7, while for the remaining ones it is a variable-length coded relative address. The lower the index, the shorter the codeword assigned to it. The macroblock type (MTYPE) segment conveys information concerning the nature of a specific MB, whether it is an intra- or inter-frame coded MB and whether full motion compensation was invoked or simple frame-differencing. The 5-bit optional macroblock quantizer (MQUANT) field of Figure 17.9 specifies a new quantizer to be invoked, which therefore supersedes the previous GQUANT/MQUANT code. If the previously included MTYPE code specified the presence of motion compensation (MC), the MQUANT segment is followed by the variable-length motion vector data (MVD). The MVD is encoded relative to the Motion Vector (MV) of the previous MB, where possible, which is again prone to errors, but reduces the bit rate, since MBs close to each other are likely to have similar MVs, yielding a difference close to zero. Therefore, the highest probability MVs' coordinates are assigned a short codeword, while rare vectors are allocated long codewords.

Observe furthermore in Figure 17.9 that the variable-length coded block pattern (CBP) is optional. It conveys a pattern to the decoder, which indicates the index of those blocks in the MB, for which at least one transform coefficient (TC) was encoded. Since for some blocks no TCs will be transmitted and some actively encoded block patterns are more frequent than others, this measure again results in some coding economy. The MB-layer codes and their roles are summarized in Table 17.3, along with the number of bits assigned. Let us now consider the lowest hierarchical layer

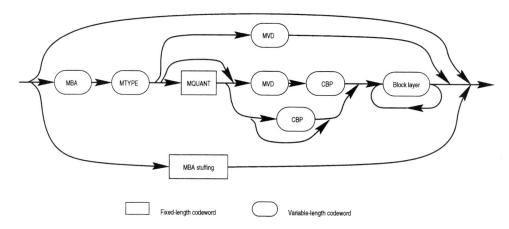

Figure 17.9: Macroblock layer structure for the H.261 codec.

MBA	Macroblock address (variable length) — Indicates position in GOB. For first transmitted MB, MBA is the absolute address, subsequent MBAs are the difference in address.
MTYPE	Type Information (variable length) — Intra/Inter, Motion Compensation, Filter, MQUANT flag.
MQUANT	Macroblock quantizer information (5 bits). Overrides GQUANT for this and future macroblocks.
MVD	Motion vector data (variable length) — included if flag set in MTYPE. MVD = vector for MB — vector for previous MB. The previous motion vector is zero if not available. Coded with variable-length code for horizontal, then vertical vector component.
CBP	Coded block pattern (variable length) — present if indicated by flag in MTYPE. Codeword specifies those blocks in macroblock, for which at least one transform coefficient is transmitted.

Table 17.3: Summary of MB Layer Information in the H.261 Codec

constituted by the blocks, where block activity control and transform coding are carried out.

17.2.4.4 Block Layer

A macroblock is comprised of four luminance blocks and one each of the two subsampled color difference blocks. The transmitted data of a block consists of codewords for the transform coefficients (TCOEFF), followed by an end of block (EOB) marker. This structure is portrayed in Figure 17.10. The coding parameters are summarized in Table 17.4.

In the intra-frame coded mode, the TCOEFF field is always present for the four luminance and two color difference blocks. In the inter-frame coded modes the macro-

17.2. THE H.261 VIDEO CODING STANDARD

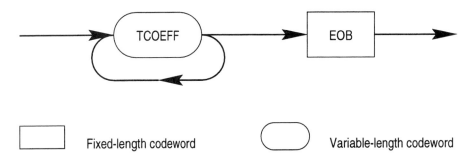

Figure 17.10: Block layer structure for the H.261 codec.

TCOEFF	Transform coefficients – Data is present for all six blocks in a macroblock, if intra-mode. In other cases, MTYPE and CBP indicate whether data is transmitted. Coefficients in the 8 by 8 block are reordered in a zig-zag pattern from top-left to bottom-right. The most common combinations are transmitted with variable-length codes, the other combinations are encoded as a 20-bit word $(6 + 6 + 8)$.
EOB	A fixed-length code that represents the end of the coefficients in the block.

Table 17.4: Summary of Block-layer Codes in the H.261 Codec

block type, namely, MTYPE, and the coded block pattern, namely CBP, indicate for the decoder these blocks, that have nonzero coefficients. The transform coefficients of the active encoded blocks are scanned in a diagonal zig-zag pattern, commencing at the top left corner of the 8 x 8 transform coefficient matrix, as shown in Figure 17.11. This scanning procedure orders them in a more-or-less descending order of their magnitudes, compounding the highest values at the beginning of the scanned sequence. Hence at the beginning of this stream, the length of consecutive zero coefficient runs is likely to be zero, but toward the end of the stream there will be many nonzero coefficients, separated by long runs of zero coefficients. This property can be advantageously exploited to further reduce the required bit rate as follows. The most frequently encountered combination of successive zero run-lengths denoted by RUN and their associated magnitude denoted by LEVEL are assigned a variable-length code from a standardized coding table. Those combinations, which are infrequent and hence were not included in the standard table, are then subsequently encoded using a 20-bit code. This 20-bit code contains a 6-bit escape code to inform the variable-length decoder that a fixed-length code is being used. The 6-bit escape code is followed by 6 bits that indicate the run-length and 8 bits for the coefficient value following the specified length zero run.

In simulations, 93% of the run-lengths of the quantized DCT coefficients are in the set of common run-lengths which are coded with variable-length codewords. On average, these variable-length codes require approximately 6 bits per run-length. This is much less than the 20 bits required for the other run-lengths. This is why the other

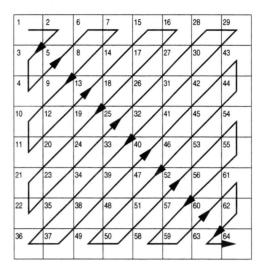

Figure 17.11: Run-length-coding block parameters along a zig-zag path.

run-lengths, which occur in only 7% of the cases, require 20% of the bits needed to code all the quantized DCT run-lengths. This can be seen in the bar graph of Figure 17.12.

The transmission coder of Figure 17.1 provides a modest error correcting capability using a simple binary BCH codec, which is only suitable for low-BER ISDN lines. The transmission coder also carries out framing before transmission.

17.2.5 Simulated Coding Statistics

In order to find the relative importance of the different coding parameters or codewords in the H.261 bit-stream, video-coding simulations were conducted. These results also assisted us in the design of the proposed error-resilient H.261 videophone system. In order to show the relative importance of all the coding parameters, two simulations were conducted:

- Coding with a fixed quantizer.

- Coding with a variable quantizer.

The default mode of the video codec was designed to use a fixed quantizer and motion vectors. The error-resilient videophone system to be described in Section 17.4 was used for the second type of simulation. This system used a variable quantizer to control the bit rate and did not use motion vectors. Both simulations used the same gray-scale "Miss America" video sequence, with just one initial intra-frame. The following sections describe the results for these simulations.

17.2. THE H.261 VIDEO CODING STANDARD

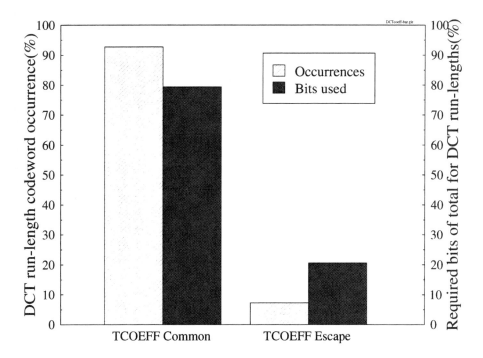

Figure 17.12: A comparison of the occurrence and bits required for coding parameters (codewords) containing run-length coded DCT coefficients.

17.2.5.1 Fixed-Quantizer Coding

The results of this section show the statistics of the coding parameters or codewords when a fixed quantizer is used. The bar graph in Figure 17.13 shows the probability that each type of coding parameter occurs in a representative coded video sequence. More than half of the coding parameters in a video sequence are the commonly occurring transform coefficient run-lengths (TCOEFF-COMMON) which use variable-length codes. Only 2.7% of the coding parameters used are the DCT coefficient run-lengths that need to be fixed-length coded (TCOEFF-ESCAPE).

The design of the error-resilient videophone system to be highlighted in Section 17.4 used the results of Figure 17.14 to estimate the probability that a random bit error would corrupt a particular coding parameter. As can be seen from the bar graph, over 60% of random bit errors would corrupt a TCOEFF-COMMON coding parameter. The effect of an error on such a parameter is severe due to the double effect of both the variable-length coding itself and the run-length coding used to represent the runs of zero-valued transform coefficients.

Also note that while the fixed-length coded DCT coefficient run-lengths constitute only 2.7% of the total coding parameters, they require over 10% of the bits used to code the video sequence. The bar graph in Figure 17.15 shows the average number of

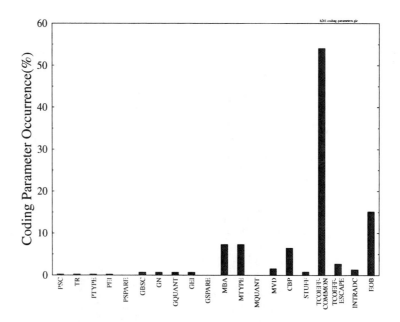

Figure 17.13: Probability of each type of coding parameter (codeword) in the bit-stream, for a fixed-quantizer simulation.

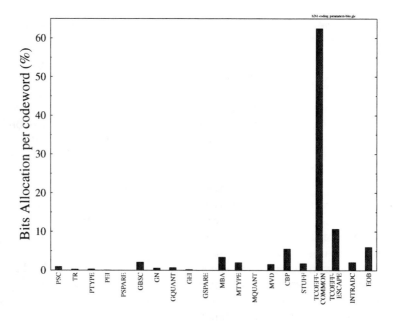

Figure 17.14: Probability of a bit being allocated to a particular coding parameter (codeword), for a fixed-quantizer simulation upon weighting the probabilities of Figure 17.13 by the number of bits allocated to the coding parameters.

17.2. THE H.261 VIDEO CODING STANDARD

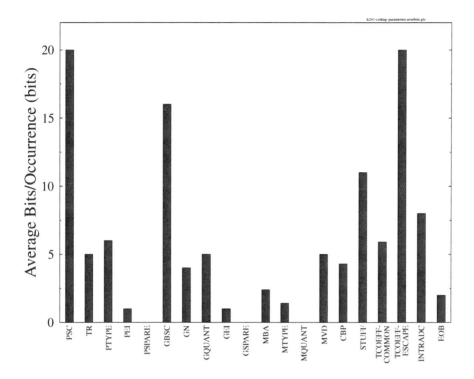

Figure 17.15: Average number of bits used by each coding parameter (codeword), for a fixed-quantizer simulation.

bits required for each coding parameter, where some of these coding parameters are fixed length, but the majority are variable-length codewords.

17.2.5.2 Variable Quantizer Coding

The results of this section show the statistics of the coding parameters or codewords when a variable quantizer is used. The results were generated using the error-resilient videophone system to be described in Section 17.4. These simulations did not use motion vectors — only frame-differencing — but the MQUANT coding parameter was used to vary the quantizer and thereby to control the output bit rate of the video codec.

The bar graph in Figure 17.16 shows the relative quantities of various coding parameters that can be contrasted with the fixed-quantizer results of Figure 17.13. The coding parameter MQUANT was not used by the fixed-quantizer simulations. However, the bar graph shows that it is a relatively infrequently occurring coding parameter. The MQUANT coding parameter is used to change the quantizer employed for the DCT coefficients.

The only major difference between these variable-quantizer simulations and the fixed-quantizer simulations is that more TCOEFF-ESCAPE coding parameters are

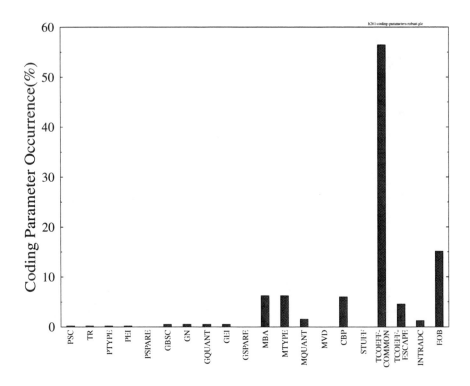

Figure 17.16: Probability of each type of coding parameter (codeword) in the bit-stream, for a variable-quantizer simulation.

used. The TCOEFF-ESCAPE coding parameter is used to transmit run-length coded DCT coefficients, which are not very common. This implies that controlling the bit rate and packetizing the bit-stream may cause more uncommon DCT run-lengths to occur.

The bar graph in Figure 17.17 shows the allocation of bits for the various coding parameters in the coded bit-stream. Again, as in the fixed-quantizer simulation of Figure 17.14, a random bit error would have a 60% chance of corrupting a TCOEFF-COMMON coding parameter. The increase in the generation of TCOEFF-ESCAPE coding parameters in the variable-quantizer simulation caused the probability of a random bit error corrupting a TCOEFF-ESCAPE codeword to increase to 17% from 10% in the fixed-quantizer simulations.

Figure 17.18 is a bar graph of the average bits per coding parameter for the variable-quantizer simulation, which can be contrasted with the fixed-quantizer results of Figure 17.15. In the next section, we consider the problems associated with employment of the H.261 codec in wireless environments.

17.2. THE H.261 VIDEO CODING STANDARD

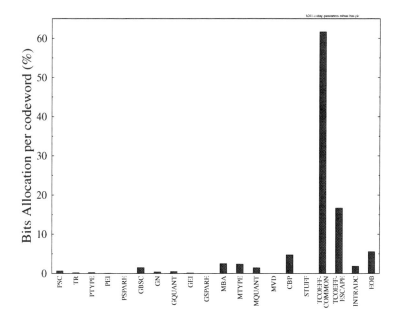

Figure 17.17: Probability of a bit being allocated to a particular coding parameter (codeword), for a variable-quantizer simulation.

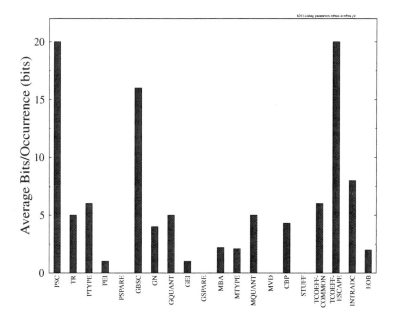

Figure 17.18: Average number of bits used by each coding parameter (codeword), for a variable-quantizer simulation.

17.3 Effect of Transmission Errors on the H.261 Codec

17.3.1 Error Mechanisms

The H.261 standard was designed mainly for ISDN lines, with a channel capacity of at least 64 kbps and a low BER. For a mobile environment, the error rate is typically significantly higher, and according to the Shannon-Hartley law, the available channel capacity is dependent on the channel conditions. Furthermore, a 64 kbps radio channel would be very inefficient and costly in terms of bandwidth. Therefore, in order to enable the employment of the H.261 codec in a mobile environment, the bit rate has to be reduced and the error resilience has to be improved. The video quality versus bit-rate balance of the codec can be controlled by the appropriate choice of the quantizers invoked. Improving the error resilience is more of a problem due to the extensive employment of bandwidth-efficient but vulnerable variable-rate coding techniques.

The transmission errors that occur in a H.261 bit-stream can be classified into three classes, namely, detected, detected late, and undetected. A detected error is one that is recognized immediately, since the received bit pattern was not expected for the current codeword. Errors that are detected late are caused by the error corrupting a codeword to another valid codeword, with the result that within a few codewords an error is detected. The final type of error is an undetected error, which also corrupts a legitimate codeword to another valid codeword, but no error is detected later.

Detected errors cause little noticeable performance degradation, but errors that are not detected or detected late cause the bit-stream to be decoded incorrectly. An incorrectly decoded bit-stream usually corrupts one or more macroblocks, and the error spreads to larger areas within a few frames due to motion compensation. These precipitated errors are only corrected by the affected macroblocks being updated in intra-mode.

Generally, when an error is detected, the H.261 decoder will search through the incoming data stream for a picture start codeword (PSC) or group of blocks start code (GBSC) and when this is found, the decoding process is restarted. This implies that all blocks following an error in a frame are lost. If the encoder is in its inter-frame coded mode, the local decoder of the encoder and the remote decoder will become misaligned, precipitating further degradation until the affected macroblocks are updated in intra-frame coded mode, or in other words, until an intra-coded frame arrives, facilitating resynchronization. In order to recover from transmission errors, the update of macroblocks in intra-mode is of paramount importance. However, if the intra-frame coded macroblocks are corrupted, the resulting degradation can be worse than that which the intra-frame coded macroblock was attempting to correct.

A further problem associated with employing the H.261 codec in a mobile environment is packet dropping, which can be inflicted, for example, by contention-based multiple access schemes. Dropping a H.261 transmission packet would be equivalent to a large number of bit errors. One solution to mitigate this problem is employing Automatic Repeat Request (ARQ) techniques. However, invoking ARQs regardless of the number of transmission attempts until a packet is received correctly is impractical.

17.3.2 Error Control Mechanisms

17.3.2.1 Background

In order to solve the problems associated with employment of the H.261 video codec in a mobile environment, the following issues have to be addressed:

- Reliable detection of errors in the bit-stream in order to prevent incorrect decoding and error propagation effects.

- Increased resynchronization frequency in order to assist the decoder's recovery from error events, which would reduce the number of blocks that are lost, following an error in the bit-stream.

- Identifying mechanisms for dropping segments of the bit-stream without propogating errors.

- Contriving synchronization techniques for maintaining the synchronous relationship of the local and remote decoders even in the presence of errors.

A variety of solutions for addressing one or more of these problems are described below. Most radio systems make use of forward error correction (FEC) codecs and/or Cyclic Redundancy Check-sums (CRC) applied to the transmitted radio packets. These techniques can assist in ensuring reliable error detection and hence curtailing error propagation effects.

17.3.2.2 Intra-Frame Coding

The H.261 standard relies on the intra-frame coded macroblocks in order to mitigate and ultimately to remove the effects of transmission errors that have occurred. Therefore, by increasing the number of intra-frame coded macroblocks in the coded bit-stream, the error recovery is accelerated. Increasing the number to the ultimate limit of having all frames coded as intra-frame coded macroblocks curtails the duration of any error that occurred in the current frame to this particular frame. The disadvantage of this method is that the bit rate increases dramatically. Simulations have shown that for the same image quality, this bit-rate increase is of the order of a factor of 10. Since bandwidth in a mobile scenario is always at premium, the image quality that could be maintained with this method under the constraint of a fixed bit-rate budget is drastically reduced.

17.3.2.3 Automatic Repeat Request

Automatic Repeat Request (ARQ) techniques can be invoked in order to improve the BER of the channel, at the expense of increased delay. However, if the channel becomes extremely hostile, then either the data will never get through or the delay will increase to a level that is impractical for an interactive videophone system. If ARQ was combined with a regime of allowing packets in the data-stream to be dropped after a few transmission attempts, then the maximum delay could be bounded, while limiting the channel capacity reduction inflicted by ARQs. However, ARQs require a low-rate acknowledgment channel, and hence specific system-level measures are required in a mobile scenario in order to support their employment.

ARQ has been used for several video-coding systems, designed for mobile scenarios. Both [550] and [467] used ARQ and interleaving to improve the error resilience of video transmitted over mobile channels.

17.3.2.4 Reconfigurable Modulations Schemes

Employment of a reconfigurable modulation scheme lends the system a higher grade of flexibility for varying channel conditions. For example, a scheme using 4-level Quadrature Amplitude Modulation (4QAM, 16QAM, and 64QAM) can operate at the same signaling rate or symbol rate, while transmitting 2, 4, and 6 bits/symbol. This allows a 1:2:3 ratio of available channel capacities under improving channel conditions, which can benefit the video codec by allowing a concomitantly higher transmission rate and better video quality in a fixed bandwidth.

Explicitly, for a high-quality channel, the more vulnerable but higher capacity 64QAM mode of operation can be invoked, providing higher quality video due to the extra capacity. When the channel is poor, the more robust 4QAM scheme can be employed; due to the lower channel capacity, the image quality is lower. However, the 4QAM scheme is more robust to channel errors and hence ensures operation at a lower BER. Therefore, a system could switch between modulation schemes depending on channel conditions. The image quality that is dependent on the bit rate therefore increases under more benign channel conditions. The combination of ARQ with reconfigurable modulation schemes would allow for rapid response to time-variant channel conditions, invoking the more robust modulation schemes during ARQ attempts that would maintain a higher chance of getting through uncorrupted.

17.3.2.5 Combined Source/Channel Coding

When invoking ARQ or reconfigurable modulation schemes, the range of channel SNRs for which the H.261 bit-stream can be transmitted without errors increases. However, this does not improve the ability of the decoder to cope with an erroneous bit-stream.

In order to reduce the image-quality degradations due to transmission errors, it is beneficial to detect the transmission errors externally without relying on the variable-length codeword decoder's ability to detect them. This prevents any potential degradation caused by incorrect decoding of the data stream. If, however, there are uncorrectable transmission errors, the H.261 codec will attempt to resynchronize with the local decoder of the encoder by searching for a PSC or GBSC in the data stream. The decoding process can then be resumed from the PSC or GBSC. However, the data between the error and resynchronization point is lost. This loss of synchronization inflicts prolonged video quality degradation because the encoder's local decoder and the remote decoder become misaligned. This degradation lasts until the local decoder and remote decoder can be resynchronized by updating the affected macroblocks in intra-frame coded mode.

If the encoder's local decoder could identify the lost data stream segment, it could compensate for it by keeping the local decoder and remote decoder aligned. This

17.3. EFFECT OF TRANSMISSION ERRORS ON THE H.261 CODEC 695

would ensure that the image segment obliterated by the error could be kept constant in the forthcoming frame. There is a small image-quality degradation due to this obsolete, but error-free, image segment, but this is readily replenished within a few frames, and it is small compared with loss of decoder synchronization or incorrect decoding.

By using a low-rate feedback channel, which can be simply superimposed on the reverse-direction packet of a duplex system, the encoder can be notified of packet losses inflicted by the mobile network. By including this packet-loss channel feedback message in the encoder's coding control feedback loop, both decoders can take adequate account of it and maintain their synchronous relationship. The source-sensitivity matched joint optimization of the source encoder and channel encoder is extremely beneficial in a mobile scenarios. For the channel feedback information to be exploited efficiently, the round-trip delay must be short.

When the source encoder is informed that a packet has been lost, it can adjust the contents of the local decoder in order to reflect this condition. As a result, the decoder will use the corresponding segment of the previous frame for the specific part of the frame that was lost, which is nearly imperceptible in most cases. The lost segment of the frame will then be updated in future frames, and hence within a few frames the effects of the error will decay.

17.3.3 Error Recovery

In order to identify the most efficient error control mechanisms, it was necessary to investigate the effects of transmission errors on the H.261 data stream for its employment in a mobile environment. Therefore, the methods of recovering from these errors were investigated, and the effects of bit errors on different coding parameters in the H.261 data stream were studied.

Many of the codec parameters in the H.261 data stream are encoded using variable-length binary codes. This implies that when an error occurs, the decoder is oblivious of where the bit-stream representing the next codec parameter starts. The decoder then has to search through the data stream for a symbol or coding parameter that can be easily and uniquely recognized. Two symbols were designed to be uniquely identifiable: the picture start code (PSC) and the group of blocks start code (GBSC). Both symbols begin with the same initial 16 bits so that the decoder can search for both simultaneously. The other variable-length codewords were designed so that a series of symbols could not be misinterpreted as a PSC or GBSC. Naturally, because of transmission errors, this situation can still be induced, but in case of long unique words the probability of this occurrence is sufficiently low.

The decoder can resume decoding at either of these symbols. However, if an error was not detected previously in the data stream, causing incorrect decoding to occur, it may not be sufficiently safe to resynchronize at the GBSC. For example, when bits in the PTYPE symbol are corrupted, the error cannot be detected by the decoder. Specifically, if, for example, the QCIF/CIF bit in the PTYPE code is corrupted, then the expected frame size is different, hosting three or twelve GOBs. Thus, resuming decoding at a GBSC may cause misplacement of the forthcoming GOBs, inflicting more severe video degradation than would have occurred upon waiting for the next PSC.

In a QCIF frame, for example, there are only four possible resynchronization points per frame, of which three are GBSC points and the fourth is the PSC. When an error occurs, a large segment of a frame can be lost before resynchronization can take place. Upon finding a mechanism of increasing the number of resynchronization points, the error effects are expected to decay more rapidly. If the decoder is restarted at a GBSC, the maximum loss is a GOB or a third of a QCIF frame. If the decoder waits for a PSC before restarting the decoding operations, then the maximum loss is a whole frame.

17.3.4 Effects of Errors

Since a single bit error can cause the loss of the remainder of the current GOB or frame, it could be argued that the effect of bit errors on different codec parameters in the H.261 data stream was more or less equal and in general very detrimental. The effect of a single error could be quantified on the basis of how much of the GOB or frame is lost.

However, if the H.261 data stream was rendered more resilient, the effect of errors on different symbols or codec parameters would vary more widely. It is relatively easy to prognose the worst-case effect of a single bit error on a particular codec parameter, but it is harder to quantify the Peak Signal to Noise Ratio (PSNR) degradation in terms of dB, which it would inflict. The only way to identify the quantitative effect of a single bit error on a symbol type or codec parameter is to simulate an error and evaluate the degradation caused. However, it is not always feasible to simulate a single transmission error and ensure that the resulting degradation is close to the worst case for that symbol or codec parameter. This is because the amount of degradation is dependent on the specific effects and mechanism of incorrectly decoding the bits of the variable-length codeword following or including the single transmission error. Let us now analyze the qualitative effects of the specific error events.

17.3.4.1 Qualitative Effect of Errors on H.261 Parameters

This section lists all the H.261 codec parameters or symbols and analyzes the typical and worst-case degradations if such a symbol is corrupted.

PSC The loss of a picture start code may in theory inflict the loss of a whole frame. However, the codec may be able to resynchronize at the beginning of the next GOB, upon receiving the next GBSC. Specifically, resynchronization becomes possible if the other parameters of the corrupted picture layer header, such as the picture type code PTYPE, are the same as in the previous frame. This allows the codec to reuse the corresponding parameters from the previous one.

TR The temporal reference is an indication of when the frame was coded with respect to the last one, and the effect of its corruption is not explicit in our codec implementation.

17.3. EFFECT OF TRANSMISSION ERRORS ON THE H.261 CODEC 697

PTYPE The picture-type parameter contains specific bits that can cause significant video degradation, while other bits hosted by it may have little or no effect. The most vulnerable bit is the QCIF/CIF flag bit. If this bit is corrupted, the decoder could switch from QCIF mode to CIF and thus cause a loss of at most one frame before receiving the next PTYPE bit in the forthcoming frame.

PEI The extra insertion bit is a very vulnerable one. It was designed for future expansion of the standard, but corruption of this bit leads to the loss of a minimum of 8 bits, since it indicates whether any PSPARE bits will follow. If it is corrupted, the next 8 bits may be misinterpreted as PSPARE bits and hence are not decoded as the next GBSC sequence and vice versa. The maximum degradation yielded by a PEI error is a whole frame, since the PEI/PSPARE syntax can loop for many iterations. The maximum loss possible is a whole frame, but the typical corruption of this bit would cause the loss of the group of blocks (GOB) following the corrupted PEI bit.

PSPARE Corruption of the PSPARE bits has no effect in the current version of the H.261 standard, since these bits are not actively utilized but may be used in the future.

GBSC Loss of a group of blocks startcode will yield the loss of a whole GOB, but resynchronization is possible at the next GBSC.

GN The group number specifies the position of the GOB within the frame about to be decoded. If this value is corrupted to another valid GOB position, the effect of the error can be large because the whole of the GOB will be misplaced. The ramifications of this in an intra-coded frame would be very grave, leading to a substantial degradation of image quality. However, if the group number is corrupted to an invalid value, then the following GOB will be lost. Loss of a GOB is preferable to misplacement in terms of image degradation that could be caused.

GQUANT Corruption of the group quantizer parameter will result in employing the wrong quantizer for decoding at the remote decoder. However, the effect on image degradation will depend on how significant the change in quantizer value is due to the corruption. The amount of degradation is also dependent on the number of macroblocks to follow in the current GOB.

GEI Another very vulnerable bit, that can be used for future expansion of the standard, for applications unknown at the time of writing is the group of blocks layer extra insertion bit. For the bits of the GEI the error effects are the same as for those of the PEI. Corruption of this bit can also result in the corruption of the first macroblock address MBA in the given GOB, which inflicts further image degradation.

GSPARE Corruption of the group of block layer spare bits GSPARE has no effect at present, but it may affect future versions of the H.261 standard, as we have stated in the context of the PSPARE symbol.

MBA The macroblock address (MBA) is the first of the variable-length coded parameters. The first MBA in a GOB indicates the absolute position of that macroblock within the GOB. Further MBAs in a GOB represent positions relative to the last MBA. Therefore, corruption of a MBA causes all further macroblocks in the GOB to be misplaced. Corruption of this symbol or codec parameter often leads to incorrect decoding owing to its variable-length coding, which can inflict further precipitated image-quality degradation. Furthermore, incorrect decoding of the MBA often leads to the decoder losing synchronization and having to resynchronize, thereby losing the rest of the GOB. Hence, the corruption of this symbol leads to one or more of the following phenomena: misplacements, incorrect decoding, or loss of the rest of the GOB.

MTYPE Corruption of the variable-length coded macroblock-type parameter, reflecting the intra/inter-frame coded, motion-compensated, filtered, nature of the macroblock can also lead to incorrect decoding and often to the loss of the rest of the GOB. However, it can also change the decoder from inter- or intra-frame coded modes, turn on or off motion vector compensation, and loop filtering. Incorrect decoding in one of these incorrect modes can lead to significant degradation of the image quality.

MQUANT The macroblock quantizer parameter is similar to the GQUANT parameter in terms of its error sensitivity and video quality degradation effects. Corruption of the macroblock quantizer identifier will result in employing a different quantizer and inverse quantizer at the encoder and decoder, respectively. The effect of its corruption on image degradation will depend on how different the choice of the quantizer and inverse quantizer is due to the corruption. The amount of degradation is also proportional to the number of macroblocks to follow, before another MQUANT or GQUANT parameter is encountered.

MVD The motion vector displacement parameter is a variable-length coded quantity. Therefore, the corruption of this symbol will typically result in incorrect decoding of the rest of the GOB, forcing the current macroblock to use the wrong macroblock-sized segment for prediction, which does not necessarily map to particular legitimate macroblock position. This type of error will then spread through the reconstruction buffer of the decoder to nearby macroblocks within a few consecutive frames. This can only be corrected by the affected macroblocks being updated in intra-frame coded mode.

CBP The coded block pattern parameter explicitly specifies for which of the blocks we can expect transform coefficient values in the block layer. Corruption of this variable-length coded parameter often leads to erroneous decoding and loss of the rest of the GOB.

INTRADC The INTRADC parameters are used to encode the DC component of the intra-coded frames. This is a fixed-length coded symbol, and hence the amount of image-quality degradation is dependent on the difference between the corrupted value and uncorrupted value. Because this is concerned with encoding the intra-frame coded macroblocks, any video degradation due to an

17.3. EFFECT OF TRANSMISSION ERRORS ON THE H.261 CODEC

error in this parameter will typically be prolonged until the affected macroblock is updated with another intra-frame coded macroblock.

TCOEFF The DCT transform coefficient parameters are split into two classes: a variable-length coded class for frequently encountered coefficient sequences and a fixed-length 20-bit class for those that are not included in the run-length/value coding tables. The variable-length class symbols will generally cause incorrect decoding and loss of the rest of the GOB when they are corrupted. The fixed-length 20-bit coded transform coefficients will inflict an error only within the specific block concerned, when they are corrupted.

EOB The end-of-block symbol signifies the end of the block layer and the start of another macroblock, GOB, or frame. When an EOB symbol is corrupted, it can at worst cause the loss of a frame but most often the loss of part or all of a GOB.

Having considered the qualitative effects of transmission errors as regards the typical and worst-case video degradations, let us now consider the quantitative effects of error events.

17.3.4.2 Quantitative Effect of Errors on a H.261 Data Stream

The only practical way of identifying the quantitative effect of errors is to corrupt the specific symbol concerned in the H.261 data stream and quantify the inflicted video degradation. The image degradation can be quantified by finding the difference between the decoded PSNR of the corrupted and uncorrupted H.261 data streams. However, it cannot be ensured that the inflicted one is the worst possible degradation. Nonetheless, this technique does give an estimate of the severity of such corruption compared to that of other parameters. These results can be invoked in order to classify symbols into sensitivity classes dependent on their vulnerability to errors.

In order to investigate the effect of errors on particular H.261 codec parameters, wide-ranging simulations were carried out using a 50-frame "Miss America" (MA) sequence. Simulations were conducted for the following four situations:

- Error in intra-coded frame (frame 1/50).

- Error in inter-coded frame (frame 2/50).

- Error in inter-coded frame using motion vectors (frame 2/50).

- Error in inter-coded frame at low quality (frame 2/50).

The following H.261 parameters were corrupted during these investigations: PEI, GN, GQUANT, MBA, MTYPE, MVD. These symbols were selected because they were the ones for which the effects of the corruption were difficult to estimate. The following H.261 symbols were not corrupted in these investigations because the effects of their corruption could be estimated reasonably well without simulations: PSC, TR, PTYPE, PSPARE. Furthermore, the MQUANT, GBSC, GEI, and GSPARE parameters were not corrupted because their effects would be similar to those of

GQUANT, PSC, PEI, and PSPARE. The remaining H.261 parameters, namely, CBP, INTRADC, TCOEFF, and EOB, were not corrupted, since their position was not readily identified in the data stream. Let us now highlight the quantitative error effects in intra-coded frames.

17.3.4.2.1 Errors in an Intra-Coded Frame These simulations were carried out in order to demonstrate the difference between the effect of errors in inter- and intra-coded frames. Let us begin our in-depth investigations in the intra-coded frame mode, concentrating on the following error events:

- Corrupted group number (GN) causing a GOB misplacement.
- Corrupted most significant bit of group quantizer (GQUANT).
- Corrupted picture extra information bit (PEI) causing loss of first GOB.
- Corrupted macroblock addresses (MBA) leading to the loss of the rest of the GOB, considering the 1st, 18th, and 32nd macroblock in the GOB.

The codec was configured to transmit an intra-coded frame every 50 frames. The graph of peak signal-to-noise ratio (PSNR) degradation versus frame index for the "Miss America" sequence is shown in Figure 17.19. The top curve represents the case when the GN parameter of the intra-coded frame was corrupted, misplacing the fifth group of blocks to the first position. These positions can be located with the aid of Figure 17.7 for our QCIF frame. The corresponding subjective effects can be inspected at the top right corner of Figure 17.20, which shows the misplaced GOB scenario. As Figure 17.19 shows, the PSNR degradation is very severe and persists until the arrival of the next intra-frame coded macroblocks.

When the most significant bit of the group quantizer (GQUANT) for the third group of blocks GOB3 is corrupted, the objective PSNR degradation in Figure 17.19 is also rather high — in excess of 15 dB and nondecaying. However, its subjective effects are less annoying, manifesting themselves mainly in terms of a bright segment in the center of MA's face at the bottom left corner of Figure 17.20.

The corrupted PEI parameter of the picture layer header results in a misinterpreted GOB1 segment, which was deemed to be a PSPARE segment due to the corruption of the PEI bit. This is portrayed in the center of Figure 17.20. The associated PSNR degradation is around 20 dB and nondecaying.

Lastly, when the macroblock address MBA of the third GOB is corrupted, the extent of the video quality degradation is dependent on the position of the corrupted macro-block. If, for example, the MBA of the last but one MB, namely, that of the 32nd, is corrupted, the damage inflicted is limited to the (1/33)rd of the third GOB, which has a similar effect to that shown for the 18th macroblock in the center of the bottom right subfigure of Figure 17.20. In contrast, if the first MBA is corrupted, the rest of the GOB is lost and becomes black. When the 18th MBA is perturbed by a transmission error, about half of GOB3 is lost, as demonstrated at the bottom right corner of Figure 17.20. The corresponding PSNR degradations are shown in Figure 17.19.

17.3. EFFECT OF TRANSMISSION ERRORS ON THE H.261 CODEC 701

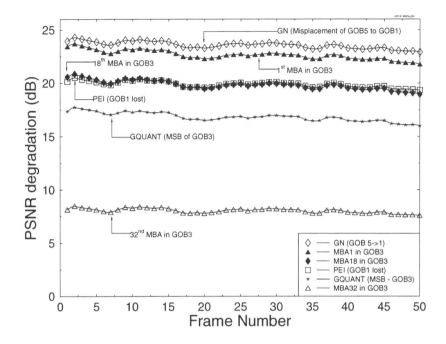

Figure 17.19: Image degradation due to corruption of H.261 coding parameters in an intra-coded frame.

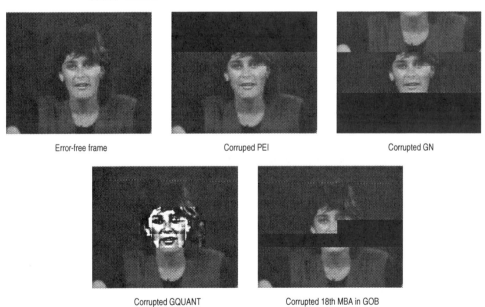

Figure 17.20: Image degradation due to corruption of H.261 coding parameters in an intra-coded frame.

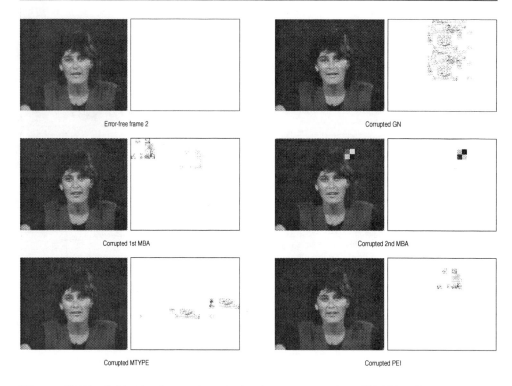

Figure 17.21: Subjective Image degradation due to corruption of H.261 coding parameters in an inter-coded frame.

All in all, as these graphs reveal, intra-frame errors result in image degradations that diminish very little with time. The effects of these errors will last until the affected macroblocks are retransmitted in intra-frame coded mode. The GOB misplacement caused by the GN error causes the worst degradation. As expected, the PSNR degradations due to MBA errors demonstrated that the position of the error in the GOB governs the amount of degradation when the remainder of the GOB is lost. Following this discourse on the effects on errors in intra-frame coded mode, let us now consider the inter-frame coded scenario.

17.3.4.2.2 Errors in an Inter-Coded Frame Most types of errors were investigated in more depth in inter-coded frames than in the case of intra-coded frames because inter-coded frames are more frequent. These investigations were conducted using the second frame of the "Miss America" sequence. The corrupted frames in Figure 17.21 are shown as pairs of pictures: at the left, the corrupted decoded frame, and at the right, the difference between the corrupted frame and the uncorrupted frame. In order to augment the visibility of errors and artifacts in the error frame, the largest luminance value in the error frame was always scaled up to 255, which was represented as a black pixel. Thus, the nature of real image degradation is best observed in the image frame itself, whereas the error images assist in observing the

17.3. EFFECT OF TRANSMISSION ERRORS ON THE H.261 CODEC

extent of error propagation and the size of the affected areas, but the shade of different error images cannot be compared with each other.

The difference image sometimes shows two areas of difference for a single transmission error. This is typically due to blocks being misplaced, where one of the corrupted areas is, where the blocks have moved from, and the second is where they moved to. The error events investigated for inter-coded frames were:

- Corrupted picture extra information bit (PEI) causing the loss of the first GOB.

- Corrupted group number (GN) causing a GOB misplacement.

- Corrupted first macroblock address (MBA) in a GOB causing all macroblocks in the GOB to be misplaced due to erroneous decoding.

- Corrupted second macroblock address (MBA) in a GOB. Here the decoder does not lose synchronization.

- Corrupted macroblock type (MTYPE) causes an error that affects all remaining macroblocks in the current GOB.

The corrupted frames are displayed in Figure 17.21 while the corresponding peak signal-to-noise ratio degradation versus video frame index is shown in Figure 17.22. The effect of the corrupted GQUANT parameter shown in Figure 17.22 was omitted from the corrupted frames in Figure 17.21, since the quantizers are investigated in greater depth at a later stage. At this stage, we note that these results analyzing the ramifications of corrupted parameters in inter-coded frames are not worst-case results. In fact, since in these studies the second MA frame was used, which follows immediately the intra-coded frame, only a very limited amount of new information was introduced during this inter-coded frame. Hence, if a GOB was corrupted, it did not result in subjectively annoying artifacts over the whole area of the GOB, since the MBs in the background region of the frame contained virtually no new information and therefore no transform coefficients. In general, the higher the number of macroblocks in the GOB where the error occurs, the worse the video degradation. During these investigations most of the errors occurred in the first GOB, which generally has fewer macroblocks per GOB than the other two GOBs in a typical head-and-shoulders QCIF frame.

Beginning with the corrupted PEI bit, it is expected that similarly to the intra-frame coded scenario, due to misinterpreting the forthcoming GOB data as PSPARE bits, the information of GOB1 is lost. However, both the subjective and objective effects of this event are relatively limited, as supported by Figures 17.21 and 17.22. Again, this was due to the virtually unchanged background area of the frame. Had the frame been more motion-active, there would have been more active MBs in the background, which would have resulted in more annoying artifacts, such as those that we have seen for the PEI bit in the case of the intra-coded frame in Figure 17.19.

When the group number GN of GOB3 is erroneously decoded as GOB1, the error effects are more severe, but not as annoying, as in the case of corrupting the same parameter of an intra-coded frame, which is demonstrated at the top right corner of Figure 17.21. Again, this leads to corruption in two GOBs, but its subjective effects

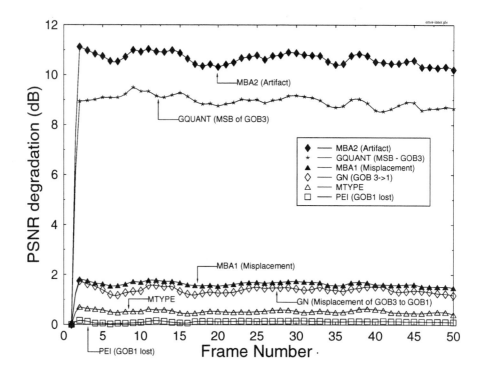

Figure 17.22: Image degradation in terms of PSNR due to corruption of H.261 coding parameters in an inter-coded frame.

are more mitigated than in Figure 17.19 owing to the previously mentioned sparsity of active MBs near the edges of the frame.

Similarly, the previously mentioned double-error pattern is observed in the left-center subfigure of Figure 17.21 due to the corrupted MBA1 parameter of GOB1, which is associated with misplacing all the MBs in GOB1. The corresponding PSNR degradation is around 2 dB in Figure 17.22. The objective PSNR reduction due to incorrectly decoding MBA2 of GOB1 is seen to be quite dramatic in Figure 17.22 — around 11 dB. This type of error also tends to spread to adjacent macroblocks within a few frames. The perceptual error can be clearly seen in Figure 17.21. The macroblock classifier MTYPE conveys the intra/inter-coded classifier, motion vector information, and so on, and its corruption affects all remaining MBs in GOB3. The subjective effects of the GQUANT MSB error in GOB3 are not demonstrated here, although the associated PSNR degradation of Figure 17.22 is rather severe.

In summary, the errors in an inter-coded frame do not affect the image quality as adversely as intra-coded frame errors, which is due mainly to the lower number of motion-active MBs. However, as with intra-coded frame errors, the effects diminish very little over time, persisting until the reception of the re-transmission of the affected macroblocks in intra-frame coded mode.

17.3. EFFECT OF TRANSMISSION ERRORS ON THE H.261 CODEC

17.3.4.2.3 Errors in Quantizer Indices Since the quantizer indices are the most commonly used fixed-length coded symbols in a H.261 data stream, a deeper investigation of the effect of errors on these specific bits was carried out. These simulations were conducted using inter-coded frames.

In general, a quantizer selection error will have a graver effect when there are more macroblocks in a GOB. For the "Miss America" QCIF sequence, generally there are more macroblocks per GOB in the central GOB (GOB3) than in the bottom GOB (GOB5). It was also found that there were usually more macroblocks per GOB in the bottom GOB (GOB5) than in the top GOB (GOB1). Further simulations were necessary to evaluate whether the number of MBs in a group of blocks made any difference to the vulnerability of a specific bit in the quantizer selection parameter GQUANT. If it did, then the GQUANT parameters in GOB3 could be classified as more vulnerable than those in GOB1 in head-and-shoulders videophony, since we argued above that the most MBs are likely to be in GOB3.

The 5-bit GQUANT parameter has 32 values associated with 32 different quantizers with bit 0 assigned to the least significant bit (LSB) and bit 4 to the most significant GQUANT bit (MSB). The following GQUANT bit errors were injected:

- Corrupted MSB (Bit 4) of GOB1 (top GOB in QCIF frame).
- Corrupted MSB of GOB3 (middle GOB in QCIF frame).
- Corrupted MSB of GOB5 (bottom GOB in QCIF frame).
- Corrupted Bit 3 of GOB1.
- Corrupted Bit 2 of GOB1.
- Corrupted LSB (Bit 0) of GOB1.

The peak signal-to-noise ratio degradation versus frame index plot for these bits is shown in Figure 17.23. The corresponding corrupted frames are portrayed in Figure 17.24. The graph confirms that the amount of PSNR degradation increases with the significance of the bit corrupted. Furthermore, as we hypothesized before, the GQUANT parameter of GOBs hosting more MBs may be more significant. This can be confirmed by inspecting Figure 17.23, where the sensitivity order GOB3, GOB5, GOB1 is confirmed. This tendency is also explicit in Figure 17.24. The three LSBs of GQUANT were found to be relatively robust, as shown by Figure 17.23.

17.3.4.2.4 Errors in an Inter-Coded Frame with Motion Vectors The H.261 codec optionally employs motion vectors in order to improve the efficiency of frame-differencing by invoking full correlative motion compensation. The effect of corrupting the motion vector displacement (MVD) parameter was investigated throughout the following simulation studies. Specifically, two sets of simulations were carried out in order to compare the sensitivity of a motion vector compensated H.261 data stream to an ordinary frame-differencing based H.261 data stream. The simulations were conducted by coding the video sequence in the same way as in the inter-coded frame simulations, except for the fact that the motion vectors were turned on. The following error events were simulated:

706 CHAPTER 17. H.261 RECONFIGURABLE VIDEOPHONE

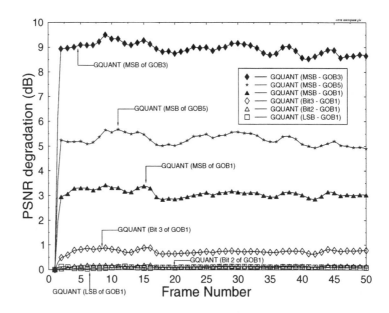

Figure 17.23: Image degradation in terms of PSNR due to corruption of the H.261 GQUANT parameter in an inter-coded frame.

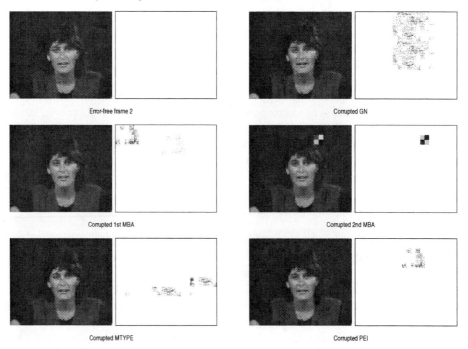

Figure 17.24: Subjective image degradation due to corruption of the H.261 GQUANT parameter in an inter-coded frame.

17.3. EFFECT OF TRANSMISSION ERRORS ON THE H.261 CODEC

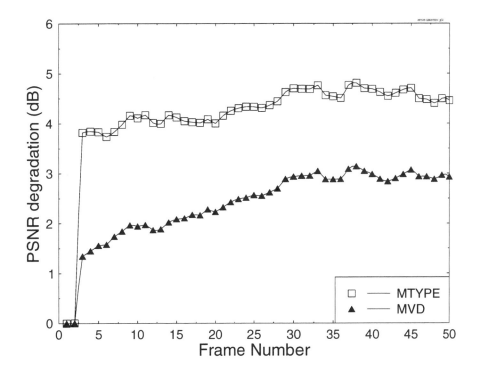

Figure 17.25: Image degradation in terms of PSNR due to corruption of H.261 motion vectors in an inter-coded frame.

- The macroblock type (MTYPE) parameter was corrupted, so that a macroblock dispensing with using motion vectors was incorrectly decoded as a data stream that contains motion vectors. This prompted a macroblock to be predicted using a specific assumed MV, although this should not have happened.

- A motion vector displacement (MVD) was corrupted, displacing a block differently from what was used by the local decoder of the encoder, which may result in prolonged error propagation.

The resultant peak signal-to-noise ratio degradation versus frame index performance of the codec is shown in Figure 17.25 as regards both the MTYPE and MVD parameters. The associated corrupted frames in which the error occurred are shown in Figure 17.26.

When MVD errors occur, the use of motion vectors allows the effects of the errors to spread over the picture. This error-spreading phenomenon implies that after an error the degradation spreads further rather than being confined to the MB it was originally related to. This becomes explicit in Figure 17.25, where the PSNR degradation increases rather than decreases. Observe that the effect of the MTYPE error is in this case more detrimental than that of the MVD error in both Figures 17.25 and

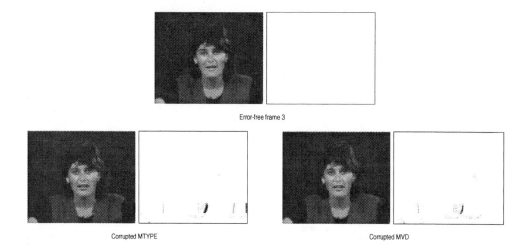

Figure 17.26: Subjective image degradation due to corruption of H.261 motion vectors in an inter-coded frame.

17.26. In conclusion of our deliberations on the impairments introduced by motion vector corruption, we note, however, that the effect of a motion vector error can be much more aggravated than these illustrations suggest. In the next section, we discuss error effects in very low-rate coding.

17.3.4.2.5 Errors in an Inter-Coded Frame at Low Rate In our next investigations the codec was constrained to operate under very low-rate conditions, using coarse quantizers. These simulations were conducted in order to ascertain whether the effect of errors was better or worse at low rates associated with lower image qualities. The error events simulated included the following:

- Corrupted MSB of GQUANT in GOB3.

- Corrupted Group number (GN) causing GOB misplacement.

- Corrupted picture extra information bit (PEI) resulting in the loss of GOB1.

The peak signal-to-noise ratio degradation versus frame index performance of the codec under these conditions is shown in Figure 17.27. The corresponding corrupted frames in which each error occurred are displayed in Figure 17.28. The effect of errors on the image quality appears to be more limited at this low quality than in the previous scenarios, which is plausible on the grounds of being more masked by the course quantization used. As we have seen before, these error effects also persist until the reception of the next intra-coded frame. The image degradation inflicted by the GQUANT error increases with future frames, which is likely to be due to the fact that in the case of low-quality coding the macroblocks are updated less often. This implies that typically it would take more frames than usual for the degradation to peak and then start to decay slightly. In conclusion, our low-rate simulations suggest

17.3. EFFECT OF TRANSMISSION ERRORS ON THE H.261 CODEC 709

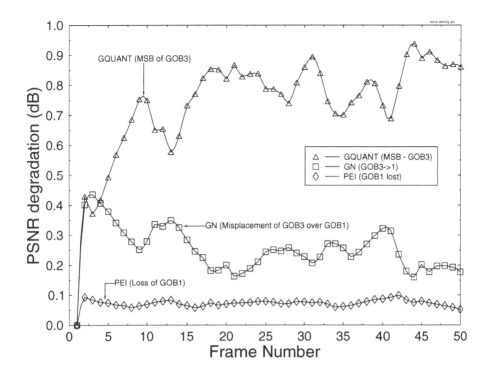

Figure 17.27: Image degradation in terms of PSNR due to corruption of H.261 codec parameters in an inter-coded frame at low bit rate.

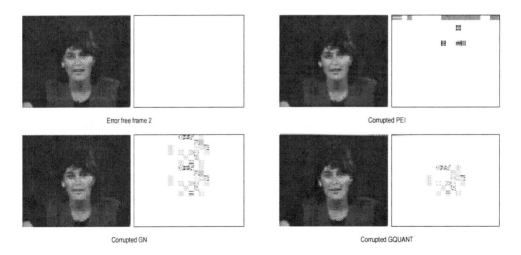

Figure 17.28: Subjective image degradation due to corruption of H.261 codec parameters in an inter-coded frame at low bit rate.

that the effect of errors is less dramatic in terms of image degradation, since many error effects are masked by coarse quantization.

In this section we have studied the objective and subjective image-quality degradations caused by transmission errors corrupting various H.261 codec parameters. Errors contaminating the intra-coded frames are particularly detrimental, and both the intra- and inter-coded frames have a persistent degradation due to errors, which typically does not decay until the retransmission of the affected macroblocks in intra-frame coded mode. With this experience, we then embarked on designing a wireless video system, which is the topic of our next section.

17.4 A Wireless Reconfigurable Videophone System

17.4.1 Introduction

In recent years, there has been increased research activity in the field of mobile videophony [194, 464, 467, 550–554]. Some authors have investigated the employment of the H.261 video codec over mobile channels [194, 550–554]. The other systems [464, 467] used proprietary video codecs. The H.261-based system developed by Redmill [553] used a technique called Error-Resilient Entropy Code (EREC) which was developed by Cheng [462, 552]. Matsumara [554] used an error-resilient syntax similar to EREC.

17.4.2 Objectives

The purpose of the error sensitivity investigations concerning the H.261 video codec was to use the knowledge gained in order to design an H.261-based videophone system for a mobile environment with the following aims:

- Better error resilience than that of the standard H.261 codec.

- A data stream that conforms to the H.261 standard or is amenable to conversion with a simple transcoder.

- A reconfigurable system so that it can adjust to different channel conditions.

In order to meet the above objectives, a flexible wireless system architecture was designed that exhibited the following features:

- A low-rate feedback channel was used for informing the encoder about the events of packet loss or corruption, for ensuring that the encoder can remain synchronized with the remote decoder. This is a requirement for any ARQ system.

- Each transmitted radio packet contained an integer number of macroblocks, which implies that the decoding process can resynchronize at the start of the next packet after an error.

17.4. A WIRELESS RECONFIGURABLE VIDEOPHONE SYSTEM

- However, not allowing macroblock splitting between packets reduces the bit packing efficiency. This problem was mitigated by using a specific macroblock packing algorithm.

- FEC codecs were employed in order to decrease the channel BER and to balance the effects of the QAM subchannels' differing BERs.

- The system was rendered reconfigurable by being able to encode the video sequence for different bit rates. When the bit rate was reduced, the packet size was also reduced so that it could be transmitted using a more robust modulation scheme at the same signaling rate.

- Any corrupted packets were dropped in order to prevent erroneous decoding and error propagation, which would badly corrupt the received video sequence and result in error propagation effects.

17.4.3 Bit-Rate Reduction of the H.261 Codec

In order to facilitate the programmable-rate operation of the H.261 codec, its performance was studied using various quantizers. The bit rate of the H.261 codec may be reduced in a number of ways.

- Opting for coarse quantization. Quantizers are numbered from fine (1) to coarse (31).

- Increasing the threshold associated with the active macroblock selection criterion, so that fewer macroblocks are transmitted. This implies that only the macroblocks with the most significant changes are transmitted.

- Reducing the frame rate, although a fixed frame rate is preferable.

Of these methods, changing the quantizer ensured the widest range of operation. Therefore, this technique was invoked in order to control the bit rate in the multiple bit-rate coding algorithm. Figure 17.29 shows the image quality in terms of PSNR and bits per pixel versus different quantizers for 100 frames of the commonly used "Miss America" (MA) video sequence. The larger the quantizer index, the coarser the quantization of the DCT coefficients: For this low-activity sequence, a PSNR in excess of 30 dB can be maintained using an average of about 0.02 bits/pixel. However, as is true of many block-based video codes, the image quality is degraded at low bit rates, due mainly to "blocking" effects, where the macroblock boundaries become visible. Much work has been done to improve this by using pre- and post-filtering, particularly by Ngan and Chai [555, 556].

17.4.4 Investigation of Macroblock Size

At this stage of the system design, it is beneficial to evaluate the histogram of the number of bits generated by each macroblock. The packetization algorithm mapping the video information onto transmission packets aims to pack an integer number of

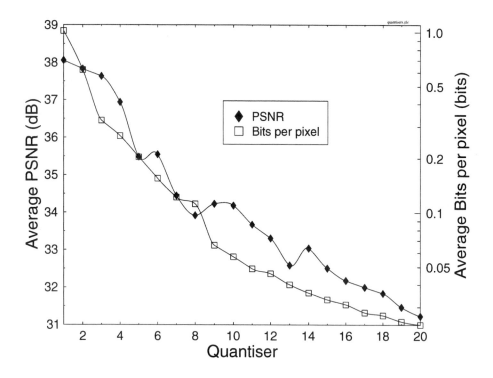

Figure 17.29: The effect of different quantizers on the H.261 coded image quality and bit rate.

macroblocks into a packet with the least amount of unused space in the packet. The graphs in Figure 17.30 show the histograms of the macroblock size in bits for various quantizers, intra-coded frame rates, both with and without motion vectors. In the latter case, simple frame-differencing is used. No macroblock length limitations or restrictions were applied. Therefore, if a packet contains one macroblock, the packet size is the macroblock size. At a later stage, we will evaluate the same histograms with packet-length restrictions, implying that some large macroblocks' bit-streams are shortened to fit them into the packets. All histograms were generated by coding a 100-frame "Miss America" video sequence.

All the histograms have a similar shape, where the leading edge rapidly rises to a peak and then tails off slowly for increasing macroblock sizes. Generally the coarser quantizers having a higher quantizer index generate a higher number of smaller macroblocks than the lower-index quantizers, moving the histogram peak toward smaller macroblock sizes. These intra-coded frames seem to generate several macroblocks with a size of 65 bits, which is associated with a distinct little spike in the histograms. When many intra-coded frames are encoded, this peak becomes more obvious. A further tendency is that when motion vectors are used, more small macroblocks are generated than in the case of a sequence-coded otherwise identically without motion

17.4. A WIRELESS RECONFIGURABLE VIDEOPHONE SYSTEM

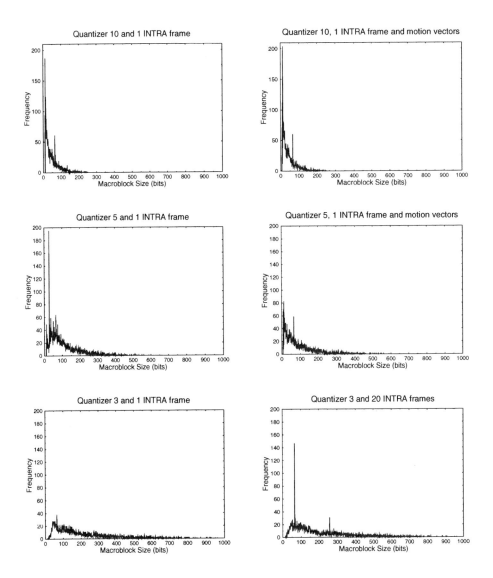

Figure 17.30: Histogram of packet/macroblock sizes for an "unrestricted" H.261 codec using various quantizers for encoding the 100-frame "Miss America" sequence using one or twenty intra-coded frames for generating the statistics, both with and without motion vectors.

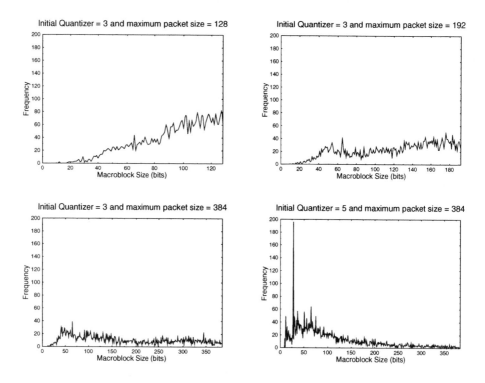

Figure 17.31: Histogram of macroblock sizes when imposing a maximum packet length, using different initial quantizers, without motion vectors.

vectors. This renders the motion vector-based histogram similar to one with a coarser quantizer but without MVs.

In order to design the packetization algorithm, it is necessary to study the effects of shortening the macroblock bit-streams in order to fit them into a transmission packet. This shortening can be achieved by increasing the quantizer index for the macroblock using the MQUANT or GQUANT coding parameter. All macroblocks are initially coded with the specified quantizer. If, however, the number of bits produced by the currrent macroblock is higher than the maximum packet size, then the quantizer index is continuously increased, until the number of bits produced for the current macroblock becomes smaller than the size limit imposed by the packet length. The effect of macroblock shortening is shown in the histograms of Figure 17.31, where the packet length was set to 128, 192, 384, and 384, while the initial quantizers were 3 and 5.

As seen in these figures, this technique changes the shape of the histograms. This is because many of the originally longer macroblocks are being shortened, causing the shortened macroblocks to have sizes just below the required packet-length limit. This procedure removes the low-probability histogram tails corresponding to the long macroblock bit-streams and increases the relative frequency of the macroblock sizes toward the required packet length. When a coarse quantizer is used, the size limit has little or no effect, since typically the generated macroblock bit-streams are relatively short. In contrast, when a fine quantizer is used in conjunction with a small packet-size limit, the histogram tends to a ramp-type shape, with the relative frequency of a given macroblock size expressed in terms of the number of bits increasing for larger macroblocks sizes, as seen in the top two graphs of Figure 17.31. In this context, the macroblock size is referring to the number of bits generated rather than the number of pixels convered by the macroblock, which is always 16 x 16.

17.4.5 Error Correction Coding

Forward error correction codes can reduce the number of transmission errors at the expense of an increased bit rate. Therefore the use of FEC can increase the range of operating channel SNR, for which transmissions are error-free. When the error rate becomes too high, a FEC is said to become overloaded, a condition that precipitates a very sharp increase in terms of BER.

When using Quadrature Amplitude Modulation (QAM) schemes, as argued in Chapter 5, certain bits in the nonbinary QAM transmission symbols have a higher probability of errors than others [557]. For example, 6 bits can be transmitted in a 64QAM symbol, and careful mapping of the bit patterns to the various 64QAM symbols allows some bits to be better protected than others. The symbols are grouped into classes according to the probability of an error. These different integrity bit classes are also often referred to as subchannels. By using different strength FEC codes on each QAM subchannel, it is possible to equalize the probability of errors on the subchannels. This means that all subchannels' FEC codes should break down at approximately the same channel SNR. This is desirable if all bits to be transmitted are equally important. Since the H.261 data stream is mainly variable-length coded, one error can cause a loss of synchronization. Therefore, in this respect most bits are equally important, and so equalization of the QAM subchannels' BER is desirable for the H.261 data stream. The specific FEC schemes used are listed in Table 17.7.

17.4.6 Packetization Algorithm

Our proposed packetization algorithm had the following objectives:

- It is important to pack an integer number of macroblocks into a packet with the least amount of wasted space.

- If the receiver asks for a packet to be re-transmitted with a smaller packet size, the previously transmitted macroblocks have to be re-encoded at a lower bit rate in order to fit into the shorter transmission packet. This enables the use of ARQ in the transmission system. ARQ issues are addressed later, especially in Section 19.5.1.

- If the transmitter is informed by the receiver acknowledgment that a packet was dropped, the encoder has to modify its parameters in order to ensure that the local decoder and remote decoder remain in synchronization.

The default action of the packetization algorithm was to encode each macroblock using a certain fixed quantizer, storing the encoded bit-stream in a buffer. When the buffer was filled with video-encoded bits above the transmission packet size, the buffer contents up to the end of the previous completed macroblock would be transmitted. The remainder of the buffer contents would then be moved to the start of the buffer, before encoding the next macroblock. The fixed quantizer was set depending on the required packet size, and the employment of smaller packets inevitably required a coarser quantizer. The quantizers for different packet sizes were set so that a more robust, but lower capacity, modulation scheme in conjunction with a smaller packet size would require approximately the same number of packets per video frame as a less robust but a higher capacity modem scheme. This would allow the system to exploit the prevailing benign channel conditions by transmitting at a higher video quality, while dropping the number of modulation levels when the channel quality degrades. This would be accompanied by a concomitantly lower video quality.

In some ARQ transmission systems, the re-transmission attempts are carried out at the original bit rate. However, in our intelligent reconfigurable transceivers, the ARQ attempts are invoked at a lower bit rate but identical signaling rate, using a more robust modulation scheme. Hence, the re-transmission attempts have a higher success rate than the initial attempts. In order to enable the employment of ARQ and reconfigurable modulation schemes, the packetization algorithm had to be able to re-encode a packet at a lower target bit rate, in order to fit the video bits into a smaller packet size. For the sake of this re-encoding procedure, the previous history of various coding parameters had to be stored before each macroblock was encoded. This history had to be remembered by the transceiver for each macroblock until it had been successfully transmitted, or it was dropped.

17.4.6.1 Encoding History List

The history of coding parameters was implemented as a bidirectionally linked list, having a structure as shown in Figure 17.32. Each element of the list represents a macroblock and contains the coding parameters of the codec before the macroblock was coded.

Each new macroblock is added to the tail of the list, encoded, and the coded bit-stream is appended to a buffer. When the buffer fullness exceeds the defined limit constituting a transmission packet, the codec is reset to its state before encoding the current macroblock that resulted in exceeding the packet length. The buffer contents containing all but the current macroblock are transmitted. Then the linked list elements for the transmitted macroblocks are removed from the head of the list. The current, untransmitted macroblock then becomes the head of the linked list.

The algorithm has two pointers to the linked list: *head* points to the head of the list and *current* points to the next element/macroblock to be encoded. When a new macroblock is added to the list, it is placed in the blank list element at the tail of the list, and a new blank list element is created.

17.4. A WIRELESS RECONFIGURABLE VIDEOPHONE SYSTEM

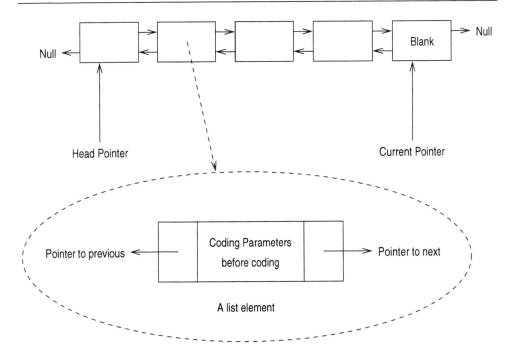

Figure 17.32: Structure of bidirectionally linked list for storing history of coding parameters.

17.4.6.2 Macroblock Compounding

The initial packetization algorithm was not very efficient, since if one macroblock took the buffer occupancy just 1 bit over the packet size limit, then no attempt was made to fit it in by encoding the macroblock in such a way as to generate fewer bits. Thus, packets had much more wasted space than they needed to, which rendered the whole system less efficient and caused more packets to be transmitted.

In order to render the packetization more efficient, a method for shortening macroblocks was designed. Macroblocks were shortened by re-encoding them with a coarser quantizer. Nonetheless, this procedure sometimes forced the macroblock bit-stream to become longer, since the encoded macroblock had to incorporate an MQUANT parameter if the previously valid quantizer had to be superceded. However, in the majority of cases, the macroblocks were encoded with fewer bits, when a coarser quantizer was used. When a macroblock was encoded which overfilled the buffer, there were two options: (1) to shorten the macroblocks in order to fit the new one in the packet buffer, or (2) to transmit the current packet buffer and leave the macroblock for the next packet. The threshold for deciding on which option to opt for was defined as follows. If the buffer overflow due to the current packet was more than half the size of the current macroblock, then the macroblock was transmitted in the next packet. Examples of the compounding decision threshold are shown in Figure 17.33.

When a macroblock is shortened by using a coarser quantizer, all the following

718 CHAPTER 17. H.261 RECONFIGURABLE VIDEOPHONE

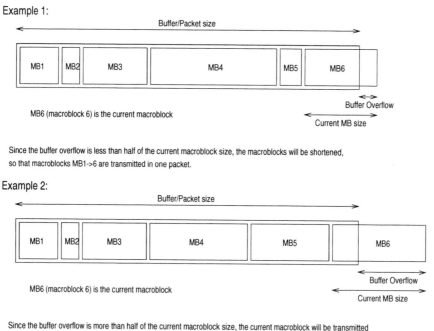

Figure 17.33: Examples of the decision threshold for compounding macroblocks.

macroblocks have to be re-encoded, since earlier those were encoded using the previously employed quantizer. Hence, without re-encoding them, the decoder would employ the most recently received MQUANT code in order to identify the inverse quantizer to be used for their decoding, which would be the correct one for the shortened last MB bit-stream, but not for the others in the packet. An alternative would be to inject an MQUANT code after the shortened and re-encoded macroblock, which would reflect the index of the previously tentatively used quantizer for the remaining MB strings in the packet. This would, however, increase the length of the packet, which may result in packet-length overflow again. After some investigations, we found that re-encoding all the MB strings in the packet was the most attractive option to use. However, this extra computational burden makes an optimal macroblock compounding algorithm more computationally demanding.

In order to reduce the associated implementational complexity, the following suboptimal algorithm was contrived for compounding the macroblocks, which proved to be fast in most cases. Initially, the fixed quantizer's index is increased by one step in order to render it coarser and generate a lower number of bits. The last macroblock is then re-encoded. If the buffer is still overflowing, the last two macroblocks are re-encoded, followed by the last three macroblocks, and so on. This procedure continues until all the macroblocks in the buffer have been re-encoded with the new coarser quantizer, or until the buffer is no longer overflowing. If all macroblocks in the buffer have been re-encoded, and the buffer is still overflowing, then the quan-

17.4. A WIRELESS RECONFIGURABLE VIDEOPHONE SYSTEM

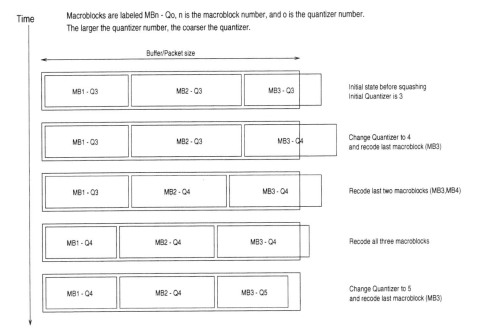

Figure 17.34: Example of macroblock compounding.

tizer index is increased by another step and the re-encoding procedure resumes with the last macroblock in the buffer. Once the compounding algorithm has shortened the macroblocks sufficiently, such that the buffer is no longer overflowing, then the procedure is concluded. An example of macroblock compounding is shown in Figure 17.34. The figure is self-explanatory; hence we will refrain from detailing the individual re-encoding steps seen in the figure.

In a second phase, the algorithm will attempt to fill the small unused space in the buffer created by the compounding algorithm with a new macroblock. If this macroblock does not fit in the packet, further compounding may be attempted. This "squashing" algorithm improved the efficiency of the packetization algorithm.

17.4.6.3 End of Frame Effect

For implementational reasons, it is not possible to re-encode macroblocks from a previous video frame. Therefore, all buffers and the history list need to be reset at the end of a video frame. This means that the last packet of a frame may be very inefficiently filled with video bits. In order to overcome this problem and increase the packing efficiency at the end of a frame, the following solution was advocated. If the buffer of the last packet within the frame was not sufficiently full, its coded data could be left in the buffer and its entries removed from the history list. This meant that these bits could not be re-encoded or dropped. However, this was an acceptable compromise, since the end of this packet was then filled with the picture start code (PSC) of the next frame, and the picture start code was the only symbol that could not be dropped without catastrophic video quality degradation and hence had to be retransmitted.

17.4.6.4 Packet Transmission Feedback

When a packet is transmitted, the packetization algorithm receives feedback from the receiver concerning the success/failure of the packet. The algorithm is also aware of the required size of the next packet. Therefore, the transmitter can change packet size for different modulation schemes. The initial quantizer used by the packetization algorithm is set according to the requested packet size expressed in terms of bits. This is basically a "Stop and Wait ARQ" protocol [472, 558]. A more complex ARQ protocol, such as, for example, the "Selective Repeat ARQ" scheme [5, 472], would be much more complex to implement. Hence, in order to gauge the achievable performance gains, here the simpler "Stop and Wait ARQ" protocol is used. The transceiver uses three types of feedback messages.

- Packet received without error.
- Packet received with error, retransmission request.
- Packet received with error, dropping request.

If the packet is received without errors, the corresponding history list elements are removed from the list and the buffer is cleared. If the packet was received with errors and a retransmission is requested, the following actions are taken. First, the codec is reset to the state which it was in at the start of the history list. Then all the macroblocks are re-encoded. If the re-transmission is requested with a smaller packet size, then the initial quantizer will be coarser. When the recently cleared buffer becomes full, the new packet is transmitted.

If a packet is received with error and a request is made to drop it, then the following actions are taken. In case the packet contained a picture start code (PSC), then a re-transmission is invoked. However, if the packet did not contain a PSC, then the codec is reset to the state stored in the head of the history list. All the macroblocks in the packet are removed from the history list.

17.4.6.5 Packet Truncation and Compounding Algorithms

In order to show the benefits of the proposed packetization algorithm, it is necessary to carry out comparisons with and without the algorithm. Recall that the graphs shown in Figure 17.31 portray the probability density function (PDF) of the various macroblock sizes occurring, when applying a simple macroblock size truncation algorithm. These graphs show that the packing efficiency is very low due to the large proportion of packets, which are well below the packet size limit. This is in contrast to an optimum packetization scheme, which would be characterized by a histogram exhibiting a peak at the required packet length and zero relative frequencies otherwise. The corresponding relative frequency graphs, when using the proposed packetization

17.5. H.261-BASED WIRELESS VIDEOPHONE SYSTEM PERFORMANCE

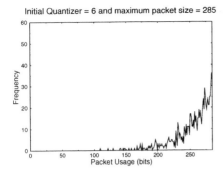
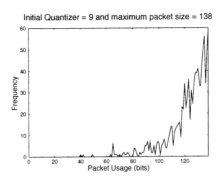

Figure 17.35: Histograms of packet usage using the compounding packetizing algorithm of Section 17.4.6.2.

algorithm, for example, for packet lengths of 285 and 138 bits, are displayed in Figure 17.35. Here we introduce the informal term *packet usage*, which we define as the number or percentage of useful video bits in each packet, where a full packet would have a packet usage of 100% or the same number of bits as the packet size. Upon returning to the figure, the associated curves exhibit a higher concentration of longer packets near the packet size limit, demonstrating the increased packing efficiency of the compounding algorithm over the simple truncation technique of Figure 17.31.

For the sake of better illustration, Figures 17.31 and 17.35 are combined and shown as cumulative density functions (CDF) in Figure 17.36. The CDF graphs show that the packet compounding algorithm guarantees a much more efficient packing than the simple truncation algorithm. This manifests itself explicitly in terms of the rapid early rise of the CDFs in case of the simple truncation algorithm, indicating a high probability of short packets associated with low packing efficiency. In contrast, the compounding packetization exhibits a very low probability tail for low packet usage, while the majority of the registered occurrences were in the vicinity of the maximum allowed packet length. Again, this is in harmony with what would be expected from an ideal packing algorithm, where all packets would be completely filled by useful data bits.

17.5 H.261-Based Wireless Videophone System Performance

17.5.1 System Architecture

The performance of the proposed error-resilient H.261-based reconfigurable videophone system was evaluated in a wireless environment over the best-case stationary Gaussian and the worst-case Rayleigh-fading channels using the proposed reconfigurable modulation and ARQ schemes. The specific system parameters employed are summarised during our later discussions in Table 17.10 on page 732.

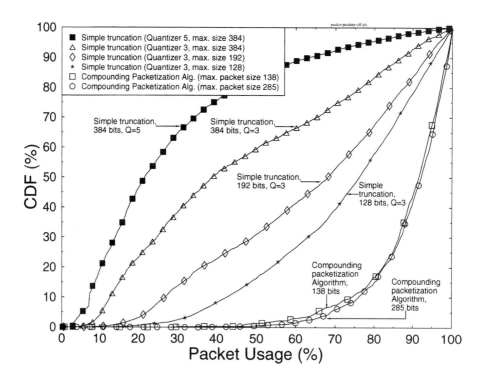

Figure 17.36: CDFs of the packet usage with simple truncation according to Figure 17.31 and with the packet compounding algorithm of Section 17.4.6.2.

This section briefly describes the specific system parameters employed during the performance evaluation of the H.261-based videophone system proposed. The reconfigurable modem invoked one of three modulation schemes, namely, 64-, 16-, and 4-QAM combined with a range of binary Bose-Chaudhuri-Hocquenghem (BCH) forward error correcting codes in order to increase the system's robustness. The system could reconfigure itself depending on the prevailing channel conditions, similar to a scheme proposed by Sampei [559] *et al.* An intelligent system could switch modulation scheme either under network control or, for example, by exploiting the inherent channel quality information in Time Division Duplex (TDD) systems upon evaluating the received signal level (RSSI) and assuming reciprocity of the channel [110, 134, 560]. A more reliable alternative to the use of RSSI as a measure of channel quality is the received BER. In interference-limited, dispersive environments, where reciprocity does not apply, the receiver attaches a channel-quality side-information message to the reverse-direction message transmitted. The reception of this channel quality information will however suffer from some latency. This reconfiguration could take place arbitrarily frequently, even on a packet-by-packet basis, depending on the channel conditions. The image quality is dependent on the bit rate and therefore increases with better channel conditions.

17.5. H.261-BASED WIRELESS VIDEOPHONE SYSTEM PERFORMANCE

Modulation Scheme	Bits/Symbol	Number of Subchannels
64QAM	6	3
16QAM	4	2
4QAM	2	1

Table 17.5: Basic Characterization of Modulation Schemes

Modulation Scheme	Bit Rate (Kbit/s)	Bits/Pixel@10Frames/s
4QAM	11.76	0.0464
16QAM	23.52	0.0928
64QAM	35.60	0.1405

Table 17.6: System Capacity in Terms of Bit-rates and Bits/Pixel at 10 Frames/s for Each Modulation Scheme

For the sake of mitigating the effect of fading, linearly interpolated pilot symbol assisted modulation (PSAM) was used. The number of bits per symbol and the previously mentioned number of associated modem subchannels are summarized for each modulation scheme in Table 17.5.

All results were generated assuming a single user and no co-channel base stations, which would be realistic in benign indoor picocells, where the partitioning walls and ceilings contribute to the required co-channel attenuation, resulting in a noise-, rather than interference-limited scenario. Simulations were done using a propagation frequency of 1.9GHz, with mobile handsets moving at a vehicular speed of 13 Km/h. Second-order diversity was also used to mitigate the effects of the multipath fading. The pilot symbol spacing was $P = 10$. In order for the system to be able to use the proposed reconfigurable modulation schemes in conjunction with ARQ, it was necessary for all transmission packets to host the same number of symbols for each modulation scheme. This was achieved by using a higher number of bits per packet for 64QAM and a lower number of bits for 16- and 4QAM, where the ratio of the number of bits per transmission packet was determined by the number of bits per symbol for each modulation scheme. In our simulations, we used a conventional TDMA frame structure [9], having 8 slots per frame and a frame duration of 12.5 ms. The user baud rate was 11.84 Kbaud, and therefore the system baud-rate was 94.72 Kbaud. The system capacity in terms of bit rates for each modulation scheme is summarized in Table 17.6, and these values are further justified at a later stage in Table 17.10 on page 732. The table also shows the average bits/pixel required by the codec to achieve an average frame rate of 10 Hz.

An appropriate target quantizer was then chosen for each modulation scheme that produced approximately the required bit rate. The graph in Figure 17.37 shows how the bit rate varies for the quantizer used with each modulation scheme. Ideally, the target bit rate for each modulation scheme should approximate the average bit rate of the codec. This is not possible for all the modulation schemes because of the limited number of quantizers, and the fixed symbol rate. The ideal solution is to implement

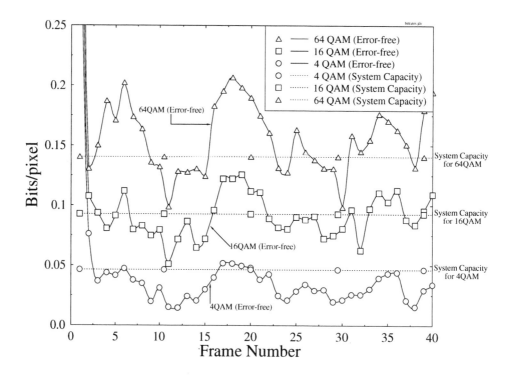

Figure 17.37: Bit rate in terms of bits/pixel versus frame index for each modulation scheme used in the H.261 videophone system.

a rate control for the video codec. This is used in our H.263 videophone system described in Section 19.3. An alternative is to use the histogram of the video bit rate to predict the future bit rate, a method suggested by Schwartz [202] et al.

The packet sizes expressed in terms of the number of bits for each modulation scheme were found by initially setting the packet size for 4QAM. The packet sizes for 16- and 64QAM were then set accordingly in order to keep the same number of transmission symbols per packet. The unrestricted macroblock length histograms of Figure 17.30 were invoked to determine the quantizer index chosen for 4QAM in order to set the appropriate packet size. Specifically, the transmission packet length in terms of bits was chosen so that the majority of macroblocks would fit into a single packet without shortening.

As mentioned earlier, the modem subchannels in 16- and 64QAM exhibit different BERs. This property can be exploited in order to provide more sensitive bits with increased protection from errors, if the source bits are mapped onto them using identical FEC codes. However, we have argued that owing to their inherent sensitivity to transmission errors, all the variable-length coded H.261 bits must be free from errors. Hence, the different subchannel error rates had to be equalized using different FEC codes. This was ensured by using stronger FEC codes over the higher BER

17.5. H.261-BASED WIRELESS VIDEOPHONE SYSTEM PERFORMANCE 725

Modulation Scheme	FEC Codes Used
4QAM	BCH(255,147,14)
16QAM	Class 1: BCH(255,179,10)
	Class 2: BCH(255,115,21)
64QAM	Class 1: BCH(255,199,7)
	Class 2: BCH(255,155,13)
	Class 3: BCH(255,91,25)

Table 17.7: FEC Codes Used for 4-, 16- and 64QAM in the H.261 Videophone System

Modulation Scheme	H.261 Packet Length (bits)	Initial Quantizer Index
4QAM	147	9
16QAM	294	6
64QAM	445	4

Table 17.8: H.261 Packet Size and Initial Quantizer Used for 4-, 16-, and 64QAM

modem subchannels. Taking also into account the practical issues of bit-packing requirements, the specific FEC codecs were selected on the basis of satisfying the BER equalization criterion as closely as possible. The different binary BCH FEC codes chosen for the various modulation modes are shown in Table 17.7, while Table 17.8 summarizes the packet sizes derived from the FEC codes and the initial quantizer indices used for each modulation scheme. Having summarized the system parameters of our H.261-based videophone, let us now characterize the system's performance.

17.5.2 System Performance

The ultimate aim of these performance investigations was to produce graphs of video PSNR versus channel SNR in a variety of system scenarios, both with and without ARQ over Gaussian and Rayleigh channels. The video sequences corresponding to the results in this section are available on the World Wide Web.[1] The "Miss America" video sequence was used, and the ARQ attempts are described as A-B-C, where A is the initial transmission, B is the first, and C is the second re-transmission attempt. The range of transmission scenarios used is summarized in Table 17.9.

The PSNR versus frame index curves of Figure 17.38 show the decoded video quality for the three different modes of operation over a perfect error-free channel. These curves demonstrate that the image quality improves when using the higher bit rates facilitated by the higher-order modulation schemes. Observe also that the first frame, which is intra-frame coded, has a lower quality than the following frames due to the relative paucity of bits, since the packetization algorithm imposes a more stringent bit-rate constraint on the comparatively long intra-frame coded macroblock

---
[1] http://www-mobile.ecs.soton.ac.uk/peter/robust-h261/robust.html

Transmission Scenarios	Channel SNR for 1dB PSNR Degradation	
	AWGN	Rayleigh
4QAM	5.78	7.8
16QAM	11.6	13.5
64QAM	17.46	20.09
ARQ: 64-16-4QAM	16.18	15.46
ARQ: 16-4-4QAM	9.06	8.81
ARQ: 4-4-4QAM	5.18	5.35

Table 17.9: Summary of H.261-based Videophone System Performance

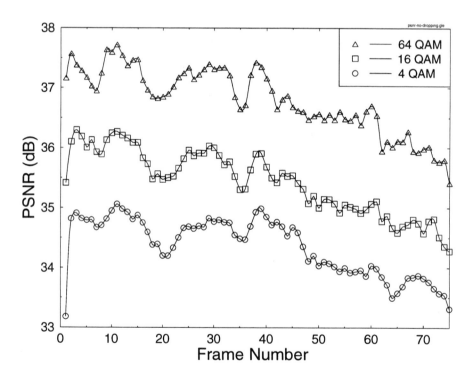

Figure 17.38: PSNR versus frame-index comparison between modulation schemes over error-free channels.

17.5. H.261-BASED WIRELESS VIDEOPHONE SYSTEM PERFORMANCE

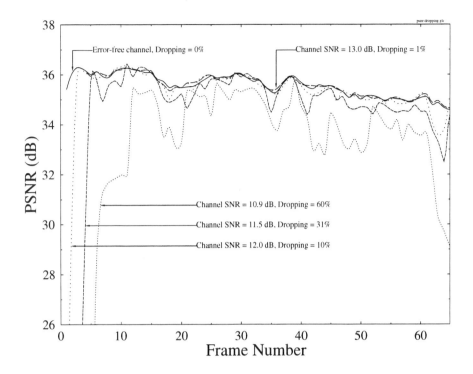

Figure 17.39: PSNR versus frame-index performance of the H.261-based videophone in its 16QAM modem mode, for various packet dropping rates over Gaussian channels.

bit-streams. An advantage of this, however, is that the inevitable bit-rate increase for intra-coded frames is mitigated. Again, the video sequences of Figure 17.38 can be viewed on the WWW.[2] Packet loss concealment techniques were also investigated by Ghanbari and Seferidis [561] using a H.261 codec.

Let us now turn to the system's performance over the best-case Gaussian channel encountered in stationary wireless scenarios. Figure 17.39 demonstrates the effect of video packet dropping over Gaussian channels exhibiting different channel SNRs, when using the 16QAM mode of operation and channel SNRs of 20, 13, 12, 11.5, and 10.9 dB, which were associated with packet dropping rates of 0, 1, 10, 31, and 60%, respectively. The decoded video sequences of these results for different dropping rates can be viewed on the WWW.[3] When a packet is dropped, the macroblocks contained in the dropped packet will not be updated for at least one video frame. The dropped macroblocks will be transmitted in future video frames until the dropped macroblocks are received without errors. In case of adverse channel conditions and excessive packet dropping, some macroblocks may not be updated for several video frames. Observe

---

[2]http://www-mobile.ecs.soton.ac.uk/peter/robust-h261/robust.html#MODULATION
[3]http://www-mobile.ecs.soton.ac.uk/peter/robust-h261/robust.html#DROPPING

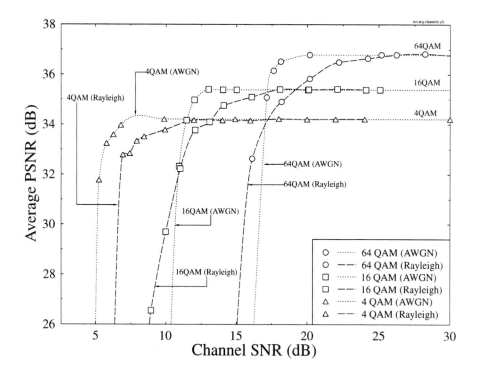

Figure 17.40: PSNR versus channel SNR performance of the H.261-based videophone for 4-, 16-, and 64QAM over both Gaussian and Rayleigh-fading channels.

in the figure that when a packet containing the initial intra-coded frame is lost, its effect on the objective PSNR and subjective video quality is more detrimental than in the case of inter-coded frame errors. This is because the decoder cannot use the corresponding video contents of the previous inter-coded frame to replenish the intra-frame coded dropped macroblocks. As seen in the figure, after the initial few frames, every macroblock has been transmitted at least once; hence, the PSNR gradually improves to around 30–40 dB. Figure 17.39 also demonstrates that for a 1% packet dropping probability, where 1 in every 100 packets is lost, the PSNR is almost identical to the error-free case, but the higher the dropping rate, the longer it takes to build the PSNR up to the normal operating range of around 35 dBs. However, even with 60% dropping, where only two in five packets are received, the PSNR varies within 4–5 dB of the error-free performance. At these high dropping rates, the delayed updating of the macroblocks can be observed, and its subjective effect is becoming objectionable.

In order to be able to observe the effect of various channel conditions on the different modes of operation of the proposed videophone system, the average PSNR versus channel SNR performance was evaluated. The performance of the 4QAM, 16QAM, and 64QAM modes dispensing with ARQ assistance is shown in Figure 17.40. These graphs demonstrate the sharp reduction of PSNR, as the channel quality de-

17.5. H.261-BASED WIRELESS VIDEOPHONE SYSTEM PERFORMANCE

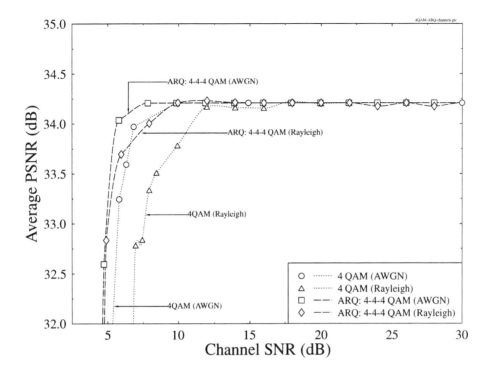

Figure 17.41: PSNR versus channel SNR performance of the H.261-based videophone for 4QAM with and without ARQ over both Gaussian and Rayleigh-fading channels.

grades. As expected, the curves corresponding to Gaussian channel conditions have a much sharper reduction of PSNR than those associated with Rayleigh-fading channels, which is a consequence of the typically "average" rather than time-variant behavior of Gaussian channels, associated with the same average number of errors per transmission packet. In contrast, Rayleigh channels exhibit a more "bursty" error distribution, associated with occasional bad and good bimodal behavior. This is the reason for the more slowly decaying PSNR curves for Rayleigh channels. This graph may also be used to estimate the channel conditions, for which a decision to switch to a more robust modulation scheme becomes effective. The Gaussian performance curves also characterize the expected video performance over conventional telephone channels.

Let us now continue our system performance study by analyzing the benefits of ARQ assistance. Figure 17.41 portrays the system's PSNR versus channel SNR performance for 4QAM with and without ARQ, while the minimum channel SNR values for maintaining less than 1 dB PSNR degradation associated with near-unimpaired video quality are summarized for all investigated systems in Table 17.9. The operating channel SNR range over Rayleigh channels is extended to lower values more substantially than in case of Gaussian channels, when using ARQ. This is a conse-

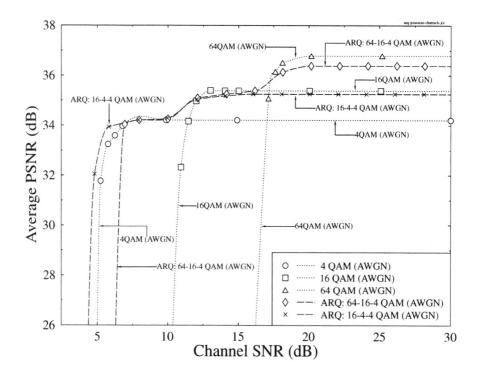

Figure 17.42: PSNR versus channel SNR performance of the H.261-based videophone for 4-, 16-, and 64QAM with and without ARQ over Gaussian channels.

quence of the fact that over Gaussian channels each video packet experiences a similar dropping probability during each re-transmission attempt, while over fading channels the re-transmitted packet may have a higher probability of error-free reception due to the portable station emerging from a fade during the re-transmission attempt.

The performance of the more elaborate ARQ schemes, such as, for example, those using 64-, 16-, and 4QAM in successive ARQ attempts is analyzed in Figures 17.42 and 17.43 in contrast to the schemes without ARQ. Again, the corresponding minimum required channel SNRs for all systems studied are presented in Table 17.9. The graphs demonstrate that the PSNR performance of the ARQ schemes gracefully degrades in the medium SNR range, tending toward the corresponding performance curves of the simple nonaided arrangements, as the channel SNR decreases. This is because while during the first transmission attempt the highest order modulation is used, due to their failed attempts, many of the successful transmissions were actually carried out at the lower-rate, lower-quality mode. Nonetheless, in subjective video quality terms it is less objectionable to view an uncorrupted, but initially lower quality, video sequence than an originally more pleasant but corrupted sequence. In other words, during the second and third transmission attempts, the high-level modulation schemes become unusable as the channel degrades, and when using the lower-order modem modes,

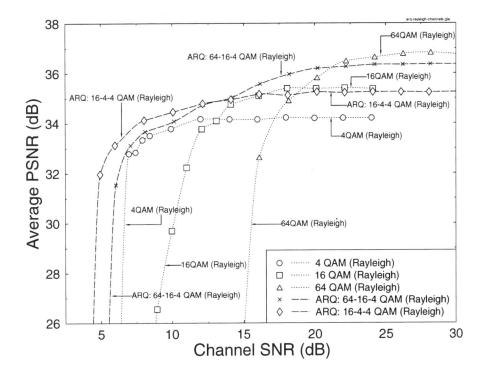

Figure 17.43: PSNR versus channel SNR performance of the H.261-based videophone for 4-, 16-, and 64QAM with and without ARQ over Rayleigh channels.

the video quality inevitably degrades owing to the more stringent bit-rate constraint imposed by the constant symbol-rate requirement. Furthermore, the ARQ-assisted PSNR performance is slightly below that of the corresponding schemes dispensing with ARQ assistance, but using the same modulation scheme, as during the first attempt of the ARQ-aided arrangements. This is because the packetization algorithm attempts to reduce the delay of the codec by minimizing the effect caused by the end of a frame, as discussed in Section 17.4.6.3. Specifically, the algorithm minimizes the end-of-frame effect by only allowing the bit-stream to be left in the transmission buffer if it is shorter than the minimum packet size. The minimum packet size is governed by the lowest order modulation scheme used in the ARQ system. This effect occurs only when the ARQ scheme contains two or more different modulation modes. The above phenomenon can be observed, for example, in case of the 64-16-4 and 16-4-4 ARQ schemes in Figures 17.42 and 17.43.

17.6 Summary and Conclusions

In this chapter the H.261 [500] standard and its sensitivity to transmission errors were analyzed. An appropriate wireless videophone system was designed for the

Features	H.261 Multirate System		
Modem	4QAM	16QAM	64QAM
Bits/Symbol	2	4	6
Number of subchannels	1	2	3
Pilot assisted modulation	yes	yes	yes
Video codec	H.261	H.261	H.261
Frame rate (Fr/s)	10	10	10
Video resolution	QCIF	QCIF	QCIF
Color	No	No	No
C1 FEC	BCH (255,147,14)	BCH (255,179,10)	BCH (255,199,7)
C2 FEC	BCH N/A	BCH (255,115,21)	BCH (255,155,13)
C3 FEC	BCH N/A	BCH N/A	BCH (255,91,25)
Data symbols/TDMA frame	128	128	128
Pilot symbols/TDMA frame	14	14	14
Ramp symbols/TDMA frame	4	4	4
Padding symbols/TDMA frame	2	2	2
Symbols/TDMA frame	148	148	148
TDMA frame length (ms)	12.5	12.5	12.5
User symbol rate (kBd)	11.84	11.84	11.84
Slots/frame	8	8	8
No. of users	8	8	8
System symbol rate (kBd)	94.72	94.72	94.72
System bandwidth (kHz)	200	200	200
Effective user bandwidth (kHz)	25	25	25
Coded bits/TDMA frame	147	294	445
Coded bit rate (kbit/s)	11.76	23.52	35.6
Vehicular speed (m/s)	13.4	13.4	13.4
Propagation frequency (GHz)	1.8	1.8	1.8
Normalized Doppler frequency	6.2696×10^{-4}	6.2696×10^{-4}	6.2696×10^{-4}
Pathloss model (Power Law)	3.5	3.5	3.5
Max. PSNR degradation (dB)	-1	-1	-1
Min. AWGN SNR (dB)	5.78	11.6	17.46
Min. AWGN BER (%)	28	25	24
Min. Rayleigh SNR (dB)	7.8	13.5	20.09
Min. Rayleigh BER (%)	29	26	17

Table 17.10: Summary of System Features for the H.261 Reconfigurable Wireless Videophone System

transmission of H.261-encoded video signals over mobile radio channels. Using error concealment and recovery techniques, the system was rendered amenable to operation under hostile channel conditions associated with high packet dropping rates. The proposed reconfigurable system allows the system to adapt to a range of channel conditions. The minimum required channel SNR values for near-unimpaired video communications are listed for the range of system scenarios studied in Table 17.9. A summary of features of the H.261 wireless reconfigurable videophone system is given in Table 17.10.

Chapter 18

Comparative Study of the H.261 and H.263 Codecs

18.1 Introduction

In 1990 the International Telecommunications Union (ITU) published the H.261 video coding standard [500] designed to support audio-visual services over Integrated Services Digital Networks (ISDNs) at rates of multiples of 64 Kbit/s. Hence, it is sometimes referred to as a $p \times 64$ Kbit/s standard, where p is in the range of 1 to 30. More recent advances in video coding have been incorporated in the ITU-T H.263 Recommendation [192]. Following our previous chapter on the H.261 codec, in this chapter we describe the H.263 standard and highlight the differences between the H.261 and H.263 standards, while endeavoring to present a comparative study of their performance differences. Throughout this chapter familiarity with the H.261 standard is assumed.

As we have seen, the H.261 coding algorithm is a hybrid of inter-picture prediction, transform coding, and motion compensation. The coding algorithm was designed to support bit rates between 40 Kbits/s and 2 Mbits/s. The inter-frame prediction removes the temporal redundancy associated with consecutive video frames, while the consecutive transform coding of the MCER exploits the spatial frequency domain redundancy of the video stream in order to reduce the required bit rate [26, 461]. Motion vectors are used to help the codec compensate for motion translation between adjacent video frames. In order to remove further redundancy from the transmitted bit-stream, variable-length (VL) coding is used [26, 461]. Recall that the H.261 codec supports two different resolutions, namely, QCIF and CIF. Let us now consider the H.263 standard.

The H.263 ITU-T standard [192] is similar to the H.261 recommendation [500], but due to a number of measures introduced, it has a higher coding efficiency and it is capable of encoding a wider variety of input video formats, which will be discussed in Table 18.1. The main differences between the H.261 and H.263 coding algorithms are

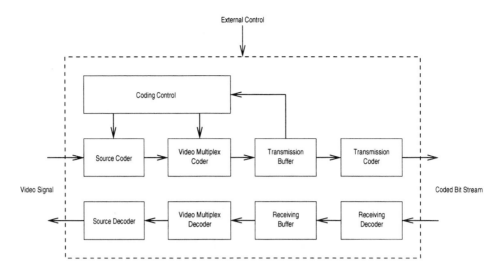

Figure 18.1: Block diagram of the video codecs.

Picture Format	Lumin. Cols.	Lumin. Lines	H.261 Support	H.263 Support	Uncompressed Bit Rate (Mbit/s)			
					10 Frame/s		30 Frame/s	
					Gray	Color	Gray	Color
SQCIF	128	96	No	Yes	1.0	1.5	3.0	4.4
QCIF	176	144	Yes	Yes	2.0	3.0	6.1	9.1
CIF	352	288	Optional	Optional	8.1	12.2	24.3	36.5
4CIF	704	576	No	Optional	32.4	48.7	97.3	146.0
16CIF	1408	1152	No	Optional	129.8	194.6	389.3	583.9

Table 18.1: Picture Formats Supported by the H.261 and H.263 Codecs

as follows. In the H.263 scheme, half-pixel precision [439] is used for motion compensation, whereas H.261 used full-pixel precision [439] and a loop filter. Some parts of the hierarchical structure of the data stream are now optional; hence, the codec can be configured for a lower data rate or better error recovery. Furthermore, in the H.263 arrangement four negotiable coding options are now included in order to improve the coding performance: Unrestricted Motion Vectors, Syntax-based Arithmetic Coding, Advanced Prediction, and Predicted and Bidirectional (P-B) frames. These options are described in further detail later in this chapter. The H.263 codec's schematic is identical to that of the H.261 scheme, which was shown in Figure 17.1 and is repeated in Figure 18.1 for convenience. As noted earlier, the H.263 scheme supports five different video resolutions, which are discussed in Table 18.1. In addition to QCIF and CIF which were supported by H.261, the H.263 scheme can also encode Sub-QCIF (SQCIF), 4CIF, and 16CIF video frames, which are summarized along with their associated number of pixels or resolutions in Table 18.1. As can be seen from the table, the SQCIF exhibits approximately half the resolution of QCIF, while 4CIF and 16CIF have four and sixteen times the resolution of CIF, respectively. The ability to support 4CIF and 16CIF resolutions implies that the H.263 codec could "compete" with other higher bit-rate video coding schemes, such as the Motion Picture Expert Group (MPEG) standard codec, MPEG2, for example, in High-Definition Television (HDTV) appplications. By contrast, for low-rate wireless videotelephony, the QCIF and SQCIF modes are realistic in bit-rate terms.

18.2 The H.263 Coding Algorithms

18.2.1 Source Encoder

The H.263 codec outline is identical to that of the H.261 scheme, and is repeated for convenience in Figure 18.1. The source encoders of both the H.261 and H.263 codecs operate on noninterlaced pictures at approximately 30 frames/s. The pictures are encoded in terms of three components, a luminance component, and two color difference components, which are denoted by (Y, C_b, C_r). The color components are subsampled at a quarter of the spatial resolution of the luminance component.

Again, the picture formats supported by the codecs are summarized in Table 18.1 along with their uncompressed bit rate at frame scanning rates of both at 10 and 30 frames/sec for both gray and color video. As mentioned, the H.261 standard defines two picture resolutions, QCIF and CIF, while the H.263 codec has the ability to support five different resolutions. All H.263 decoders must be able to operate in SQCIF and QCIF modes and optionally support CIF, 4CIF, and 16CIF.

Let us now concentrate on the main elements of the H.263 codecs, namely, interframe prediction, block transformation, and quantization.

18.2.1.1 Prediction

Both the H.261 and H.263 codecs employ inter-frame prediction based on framedifferencing, which may be improved with the aid of full motion compensation using motion vectors. When motion prediction is applied, the associated mode of operation is referred to as inter-frame coding. The coding mode is called intra-frame coding, when temporal motion prediction is not used. The H.263 coding standard has a negotiable coding option, referred to as P-B mode. This adds a new type of coding mode, where the B pictures can be partly bidirectionally predicted, which requires picture storage at the decoder, introducing latency, as will be discussed in Section 18.2.4.4. Let us now continue our discussions by considering the H.263 motion compensation scheme.

18.2.1.2 Motion Compensation and Transform Coding

Motion compensation is optional in the encoder, but it is required in all decoders in order to be able to decode the received sequence of any encoder. Both the H.261 and H.263 schemes generate one motion vector (MV) per macroblock. Some of the most important differences between the H.261 and H.263 source coding algorithms are concerned with motion compensation. The H.261 standard allows the optional use of a loop filter in order to improve the performance of the motion compensation.

This can be turned on or off for each macroblock. In contrast, the H.263 scheme does not use a loop filter since it has more advanced motion compensation.

A further difference is that the H.263 arrangement also has an advanced prediction mode (which will be discussed in Section 18.2.4.3), allowing for one or four motion vectors per macroblock. When the H.263 P-B prediction mode (described at a later stage in Section 18.2.4.4) is used, an additional delta motion vector can be transmitted for each macroblock. Although the H.261 motion compensation scheme opts for using full pixel resolution, the H.263 codec employs a more accurate half-pixel precision for the motion vectors. This typically results in a reduced motion compensated prediction error residual. The motion vectors are usually restricted, so that all pixels referenced by them are within the coded picture area. This restriction can be removed when using the unrestricted motion vector mode to be detailed in Section 18.2.4.1 in case of the H.263 codec.

In order to remove the spatial frequency-domain redundancy, transform coding is used. Both the H.261 and H.263 schemes employ an 8 × 8 pixel discrete cosine transform (DCT). The DCT operates on either the intra-frame coded picture or in the inter-coded mode on the motion-compensated prediction error residual.

18.2.1.3 Quantization

After the picture or the MCER has been transform coded, the resulting coefficients are quantized. As we have seen for the H.261 codec, the bit rate can be reduced by using a coarser quantizer, which typically results in longer runs of zero-valued quantized coefficients. Longer runs of zero-valued quantized transform coefficients can be more efficiently coded with run-length coding.

The quantizers used by the H.261 and H.263 codecs are the same, which were briefly characterized in Figure 17.4. There is a specific quantizer for the DC coefficient of intra-frame coded blocks, and there are 31 different quantizers for all other transform coefficients. Accordingly, the GQUANT and MQUANT parameters are encoded by 5 bits, as we have seen for the H.261 scheme. The DC quantizer is a linear quantizer with no dead-zone. The other quantizers are nominally linear with a dead-zone near zero and with varying step sizes, as can be seen in Figure 17.4. The thirty one quantizers are numbered, and the smaller the quantizer index, the finer the quantizer. The fine quantizers cannot represent the full dynamic range of the transform coefficients, but they provide a high resolution in a more limited range.

In both codecs, the coding control block of Figure 18.1 regulates the performance of the codec by modifying the various codec parameters in order to attain goals such as a certain image quality or bit rate. The coding control algorithms are not defined in the standards, since they are dependent on the application in which the codec is used.

18.2.2 Video Multiplex Coder

Concentrating now on the video multiplex coder, we see that its main function is to convert the quantized transform coefficients and motion vectors into symbols to be transmitted. The multiplex coder also imposes a structure onto the data generated

18.2. THE H.263 CODING ALGORITHMS

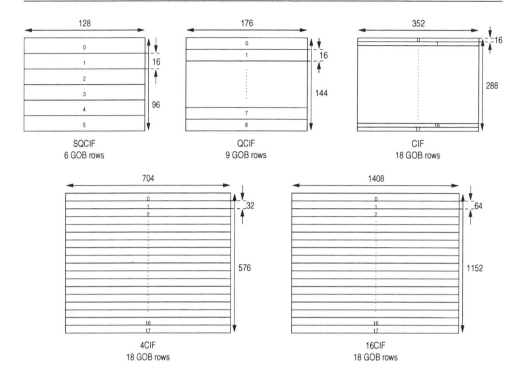

(All measurements are in Luminance pixels)

Figure 18.2: GOB structure for H.263 video codec, for frames of different resolutions.

by the source encoder in order to aid its error recovery.

Each video frame is divided into blocks of 8 × 8 pixels. Similarly to the H.261 codec, in the H.263 scheme four adjacent luminance and two color difference blocks are grouped together, which are referred to as macroblocks, as we portrayed it in the context of the H.261 codec in Figure 17.2. Each macroblock represents 16 × 16 pixels. A range of macroblocks are then grouped together in order to form a 176 × 48 pixel group of blocks (GOB), which were displayed in Figure 17.7 for the H.261 scheme in case of CIF and QCIF frames, defining a GOB as a group of 33 macroblocks in an 11 × 3 arrangement. Therefore, there are three GOBs in a QCIF frame. By contrast, the H.263 codec defines a GOB as a row of macroblocks for SQCIF, QCIF- and CIF, two rows of macroblocks for 4CIF, and four rows of macroblocks for 16CIF. The layout of the GOB in H.263 frames is shown in Figure 18.2, with a macroblock being 16 × 16 constituted by luminance pixels. These different resolutions are summarized in Table 18.1. The aim of grouping blocks of the frame into a hierarchical structure is to aid error-recovery.

The video multiplex schemes of the H.261 and H.263 codecs are arranged in an identical hierarchical structure based on four layers. From top to bottom of the hierarchy, these layers are: the Picture-, Group of Blocks- (GOB), Macroblock-, and

Block-layers, which were shown in Figure 17.5.

We again emphasize that the remainder of this chapter is written assuming familiarity with the H.261 codec. Hence, we will concentrate on describing the differences between the H.263 and H.261 multiplex codecs.

18.2.2.1 Picture Layer

Some elements of the H.261 and H.263 picture layers are common in both standards. The information conveyed by these layers was summarized for the H.261 codec in Section 17.2.4.1, Figure 17.6, and Table 17.1, and their inspection is necessary in order to follow our deliberations. Briefly, the common H.261 and H.263 features are as follows:

The Picture Start Code(PSC) : a unique start code designed to aid resynchronization after a transmission error.

The Temporal Reference(TR) : a coding parameter expressed in terms of time as regards to when the current frame was coded relative to previous frames.

The Picture Type Information(PTYPE) : a parameter concerned with the format of the complete picture, specifying features, such as split screen, freeze, and CIF/QCIF operation.

In the H.263 standard, the picture start code (PSC) must be byte aligned, which is not required for H.261. The Temporal Reference is a 5-bit encoded value in H.261, which was increased to 8 bits in the H.263 scheme. The PTYPE symbol is constituted by 6 bits in the H.261 recommendation, which was increased to 13 bits in the H.263 scheme. This is required because of the range of extra options available in the H.263 standard. The only significant change to PTYPE in the H.263 codec is a Picture Coding Type flag, which is set to either intra- or inter-frame coded. A variety of other important changes to the picture layer in the H.263 scheme were concerned with the following parameters:

Quantizer (PQUANT) — The quantizer can now be changed in the picture layer, while in the H.261 scheme this could only be carried out in the GOB and macroblock layers.

Stuffing (STUF) — Bit Stuffing can now be invoked after a frame has been coded. This is useful for fitting the H.263 data into transmission packets. In the H.261 codec, bit stuffing was available only at the macroblock layer.

The remaining coding parameters of the H.263 picture layer are concerned with a variety of new options that the H.261 scheme did not support, such as the P-B prediction mode (see Section 18.2.4.4).

18.2.2.2 Group of Blocks Layer

This section relies on our previous discussions concerned with the H.261 GOB layer, which were summarized in Table 17.2 and Figure 17.8. As we have seen in the case

18.2. THE H.263 CODING ALGORITHMS

of the picture layers, some of the coding parameters in the Group of Blocks layer are also common to the H.261 and H.263 codecs, which are as follows.

The Group of Block Start Code (GBSC) was incorporated in order to aid resynchronization after a transmission error.

The Group Number (GN) indicates which group of blocks is to follow.

The GOB Quantizer (GQUANT) conveys a request to change the quantizer commencing from the current GOB.

The changes concerning these symbols in the H.263 codec are minor. In the H.263 scheme, the group of blocks start code is byte aligned, like the picture start code, while as we noted before the H.261 codec's start codes need not be byte aligned. Byte-aligning the start code makes it easier to search for a start code after a transmission error, thereby potentially reducing hardware costs. The group number coding symbol has increased from 4 bits in the H.261 scheme to 5 bits in the H.263 arrangement. This was necessary because the GOBs were made smaller, and therefore they are more numerous, as shown in Figure 18.2. The GQUANT symbol is the same in both the H.261 and H.263 codecs.

The only other significant coding parameter in the H.263 group of blocks layer is the GOB frame ID (GFID). This symbol aids resynchronization after a transmission error in order to enable the decoder to identify whether the current GOB found is from the current frame or from a forthcoming frame. This reduces the likelihood of a whole GOB being decoded incorrectly, since this may cause catastrophic image degradation.

In the H.263 standard, the Group of Block layer header does not need to be coded for the first Group of Blocks, since a Picture layer header implies it. All other GOB headers in the frame can also be empty, depending on the encoding strategy. This implies that a video sequence can be coded with few GOB headers in order to reduce the bit rate or with more to increase the error-recovery capabilities. This renders the H.263 codec much more flexible than the H.261 scheme.

18.2.2.3 H.261 Macroblock Layer

In Section 17.2.4.3, Table 17.3, and Figure 17.9, we summarized the features of the macroblock layer in the H.261 codec. The H.263 schemes MB layer is very different. Although the picture and GOB layers in the H.263 scheme can be thought of as extensions of the corresponding H.261 layers, this is not true for the macroblock layer. The macroblock layers of both the H.261 and the H.263 codecs were designed to remove some redundancy from the data received from the source encoder by transmitting symbols containing the differential changes of parameters. However, the H.263 macroblock layer makes more intensive use of this differential coding. Recall from Table 17.3 that the H.261 macroblock consists of five different coding parameters, which are summarized now for convenience:

Macroblock Address (MBA) — used to indicate the position of the macroblock in the current group of blocks.

Macroblock Type (MTYPE) — signaling the coding mode, in which the macroblock was coded (INTER/INTRA), and whether motion compensation or the loop filter was used for the current macroblock.

Quantizer (GQUANT/MQUANT) — employed to convey a change of the current quantizer for this and forthcoming macroblocks.

Motion Vector (MVD) — containing the values of the horizontal and vertical components of the motion vector for the current macroblock.

Coded Block Pattern (CBP) — transmitted in order to inform the decoder, blocks that are active in the current macroblock.

For the first macroblock in a group of blocks, the MBA symbol contains the absolute address of the macroblock. Subsequent macroblocks in the group of blocks contain the differential macroblock address in the MBA parameter. Specifically, the differential macroblock address is the difference between the current macroblock address and the address of the previously transmitted macroblock. The MBA parameter is coded with variable-length codewords, where the shorter codewords are assigned to the smallest values. This implies that consecutive adjacent macroblocks, which are most common, use fewer bits for the MBA than those that are far apart.

The MTYPE codec parameter in the H.261 standard is variable length coded like the MBA parameter, and the most common modes are assigned short codewords. The fixed-length encoded MQUANT parameter changes the absolute value of the quantizer. The motion vector displacement (MVD) transmits the difference in the motion vectors between the current and previous macroblock. If the previous macroblock did not contain a motion vector, then the previous motion vector is assumed to be zero. The MVD symbol is variable length coded, with the shorter codes given to smaller differences.

Lastly, the coded block pattern parameter (CBP) indicates which blocks are active within a macroblock. There are 64 different patterns, since each macroblock can contain up to four luminance blocks and up to two chrominance blocks. The active blocks are transmitted in sequential order after the macroblock header. Use of a CBP symbol removes the need for each active block to have a block address, which would be rather wasteful. The CBP is coded with variable-length codewords, where the common active block patterns are given the shorter codes.

18.2.2.4 H.263 Macroblock Layer

As mentioned before, the H.263 macroblock layer is different from the H.261 macroblock layer. As seen in Figure 18.3, eight different types of coding parameters can be transmitted in the H.263 macroblock header, as follows:

The coded macroblock indicator (COD) is a one-bit flag, which is set to 0 for all active macroblocks, while it is set to logical 1 in order to indicate that no further information is to be transmitted in the current macroblock. The COD bit is present in only those video frames for which the PTYPE codec parameter implies inter-frame coding.

18.2. THE H.263 CODING ALGORITHMS

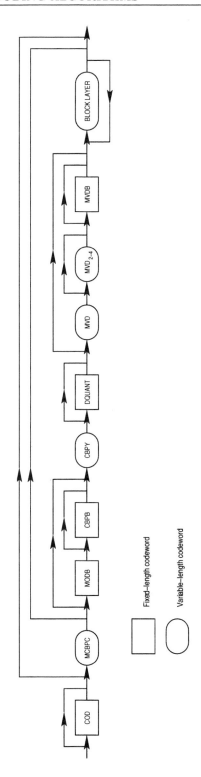

Figure 18.3: H.263 macroblock layer structure.

The MB-type and chrominance block pattern (MCBPC) is a variable-length coded parameter, and it is always present in coded macroblocks, conveying information about the type of the current macroblock and as to which chrominance blocks are to be transmitted. Depending on whether the coding mode is INTRA (I) or motion predicted INTER (P), two different variable-length coding tables can be used, which improves the coding efficiency by taking advantage of the different relative frequencies of the various coded block patterns. The assigned codewords also depend on one of five different types of MB, depending on the presence or absence of some of the MB layer's parameters. For details of the specific coding tables, the interested reader is referred to the H.263 Recommendation [192]. Without elaborating on the specific codeword allocations for the Type 0-4 MBs, we note that there is a stuffing code for both the I and P frames, which indicates that the forthcoming section of the MB layer is skipped. This code is then naturally discarded by the decoder.

Macroblock mode for B blocks (MODB) represents the macroblock type for the bidirectionally predicted macroblocks or B blocks, which can occur in conjunction with any of the MB types 0–4 mentioned in the previous paragraph, but only when PTYPE implies that the P-B prediction mode is used. This variable-length coded parameter indicates the presence or absence of the forthcoming Coded Block Pattern for B blocks (CBPB) parameter and that of the Motion Vector Data for B macroblocks (MVDB), described later in this chapter.

Coded block pattern for B blocks (CBPB) indicates the specific pattern of the coded blocks for bidirectionally predicted macroblocks, which again occur only in P-B mode. As mentioned above, its presence is indicated by the preceding variable-length coded MODB parameter. Since CBPB is a 6-bit parameter, it can differentiate among 64 various coded patterns.

The coded block pattern for luminance (CBPY) parameter is variable-length encoded and conveys the coded block pattern of the active luminance blocks in the current macroblock, for which a minimum of one transform coefficient other than the DC component is transmitted. In both the intra- and inter-frame coded modes, there are four luminance blocks per MB, yielding a total of sixteen possible coded luminance patterns. Again, the most likely patterns are allocated shorter codewords, while the less likely ones are assigned longer ones. For details of the specific coding table, the reader is referred to the Recommendation [192].

The differential quantizer (DQUANT) codec parameter is used to differentially change the quantizer used for this and future macroblocks with respect to the previously specified PQUANT or GQUANT values of the picture layer or GOB layer. Since it is encoded by two bits, it can modify the previously stipulated quantizer index by ±2. Furthermore, as the quantizer index is limited to the range [1-31], DQUANT values requiring quantizer indices outside this range are clipped.

Motion vector data (MVD) contains the variable-length coded horizontal and vertical components of the motion vector of the current inter-frame coded macroblock. Furthermore, for P-B frames, the MVD is also included in the intraframe coded mode. The horizontal and vertical motion vector components are encoded independently, and they are represented using a half-pixel resolution. Again, the variable-length coding table assigns between 1 and 13 bits to the various vectors on the basis of their relative frequencies in order to maintain high coding efficiency. Specifically, instead of coding the motion vectors directly, a motion vector prediction process is invoked, and the difference between the actual and predicted vector is variable-length encoded. There are 128 possible motion translation vector differences in the coding table. However, there are only sixty-four codewords in the table. Each codeword can be interpreted as one of two differences. Since the new motion vector components are constituted by the sum of the predicted vector component plus the transmitted difference, where the sum must be in the range $-16 \leq n \leq 15.5$, only one of the two possible differences for each codeword can be valid.

Motion vector data for B macroblock (MVDB) is a codec parameter describing the motion vector for bidirectionally predicted (B) macroblocks, which occur only when the P-B prediction mode is used. The MVDB parameter is included in the P-B frames only if its presence was previously indicated by the MODB parameter.

Here we curtail the description of the various MB layer parameters and note that the parameters MODB, CBPB, and MVDB will be discussed further in Section 18.2.4.4 concerning the P-B prediction mode. Recall that the codec's philosophy was to introduce the COD flag in order to mark the active MBs, for which then a coded block pattern was incorporated. This implies that each nonactive macroblock requires only a 1 bit macroblock header per macroblock.

In the H.263 codec, the macroblock type and the coded block pattern for chrominance have been combined into one parameter, namely, the MCBPC. The macroblock-type MCBPC informs the decoder as to what other macroblock layer parameters to expect for the current macroblock. The coded block pattern for luminance, namely, CBPY, is a separate codec parameter. The function of these symbols is similar to that of the corresponding coded block pattern parameters in the H.261 codec, where again, both are variable-length coded. The chrominance and luminance coded block patterns are seperated because the luminance changes more frequently than the chrominance. Therefore, a combined coded block pattern like CBP in H.261 may be needlessly transmitting chrominance information every time the luminance changes.

An interesting feature of the H.263 codec is that the quantizer information at the macroblock layer is differentially coded. In the H.261 scheme, the quantizer could be set to an absolute value at the macroblock layer. The H.263 quantizer information at the macroblock level is conveyed using the DQUANT symbol. The quantizer index can only be changed in steps of $-2, -1, +1, +2$ with respect to its previous value in the H.263 arrangement, or it can be left unaltered by simply not transmitting the DQUANT parameter.

Recall that the motion vector coordinates in the H.263 codec are transmitted in the MVD codec parameter. We also remind the reader that in the H.261 codec the difference between the motion vector and the previous vector is transmitted in the MVD symbol. This can be thought of as using the previous motion vector as a prediction of the current vector and the MVD symbol transmitting the prediction error. In the H.263 scheme, a more complex motion vector predictor is used, which is based on up to three previous motion vectors. This new motion vector predictor is described in more detail in Section 18.2.3.1. The prediction error at the output of this predictor is transmitted in the MVD codec parameter. Having detailed the MB layer, let us now consider the lowest hierarchical level, the block layer.

18.2.2.5 Block Layer

In both the H.261 and H.263 coding schemes, a block contains the quantized transform coefficients of an 8 × 8 group of luminance or chrominance pixels. A macroblock can contain up to four luminance blocks and up to two chrominance blocks. In case of the H.261 scheme, the active blocks of each macroblock are transmitted in sequential order, separated by an end-of-block symbol. The encoded block pattern in the macroblock layer informs the decoder as to which blocks were deemed to be active. In the H.263 codec, the active blocks are transmitted in sequential order, as in the H.261 arrangement. However, there are no end-of-block symbols between the blocks. Instead, a flag bit referred to as LAST was introduced, which is set to logical 1 only in the last block of a macroblock; otherwise it is 0. Although this LAST flag bit is rather vulnerable, the bit-rate contribution of the 2-bit end-of-block codec parameter of the H.261 scheme is higher than that of the 1-bit LAST flag of the H.263 codec.

The transform coefficients of the active blocks are encoded using the same method in both the H.261 and H.263 codecs. The first component in an intra-frame coded block is the DC component, which is encoded differently from the other coefficients, transmitting a fixed-length binary code representing its value.

For all other transform coefficients, run-length coding with variable-length codewords is used, employing a top-left to bottom-right oriented zig-zag scanning pattern in order to increase the length of the runs of zero-valued DCT coefficients. Following this scanning operation, the most common runs of zero coefficients followed by their specific nonzero values are assigned variable-length codewords on the basis of their relative frequency. Again, the length of the variable-length codewords is inversely proportional to their relative frequency. The remaining, less typical run-length and coefficient value combinations are encoded with an escape code followed by the run-length encountered and the transform coefficient value, which is encoded using fixed-length binary codewords.

In case of the H.261 codec, there are 64 common run-length codewords, and other run-lengths are encoded using 20 bits, consisting of 6 bits escape, 6 bits run-length, and an 8-bit transform coefficient value. As regards the H.263 scheme, there are 44 common run-length codewords for the last block in a macroblock, where the LAST flag was set to logical 1 and a total of 58 common run-length codewords for the other blocks. In contrast to the 20-bit H.261 representation of uncommon run/value combinations, in the H.263 standard these combinations are encoded employing 22 bits. Specifically, the 7-bit escape code is followed by the 1-bit LAST flag in order to indicate whether it is the last block in a macroblock, while 6 bits are assigned to the run-length value and an 8-bit codeword to the transform coefficient value.

18.2.3 Motion Compensation

A common feature of both codecs is that motion compensation is used to produce the motion-compensated prediction error residual and hence to reduce the bit rate of the codec. Both codecs rely on motion vectors in order to achieve improved motion compensation in comparison to simple frame-differencing.

Both codecs specify in their respective recommendations how the motion vectors are conveyed. However, the motion vector search algorithm is not defined. Hence, significant research efforts have been devoted to finding motion vector search algorithms, which are fast, reduce the bit rate of the coders, and can be easily implemented in hardware, particularly, for example, by Chung *et al.* [562].

In the video multiplex layer, both the H.261 and H.263 schemes transmit a differential rather than absolute motion vector, encoded by the MVD codec parameter. The differential motion vector is the difference between the predicted and actual vector for a macroblock. The H.261 arrangement uses a simple predictor, where the predicted motion vector is given by the motion vector of the macroblock to the left of the current one. If the macroblock to the left of the current one does not exist or does not have a motion vector, then the predicted motion vector is 0.

18.2.3.1 H.263 Motion Vector Predictor

The H.263 motion compensation uses a more complex motion vector predictor. The prediction of the motion vector is a function of up to three previous motion vectors. The three candidate predictive motion vectors are the motion vectors of the macroblocks to the left, above, and above-right relative to the current MB. The arrangement of the motion vector and its candidate predictors is shown in Figure 18.4. The candidate predictors are modified when they are adjacent to a group of blocks boundary, a situation that is portrayed in Figure 18.4.

Step 1 The candidate predictive motion vector, MV1, is set to zero if it is outside the GOB or the picture frame containing the motion vector MV to be predicted.

Step 2 Then the candidate predictive motion vectors MV2 and MV3 are set to MV1, which may now be 0 if the macroblock being predicted is at the top edge of a GOB or picture.

Step 3 Then the candidate predictor MV3 is set to 0 if it is outside the GOB or picture frame containing the motion vector MV to be predicted.

Step 4 If any of the candidate predictive motion vectors' macroblocks were coded in intra-mode or were inactive and therefore are not encoded, then the candidate predictor would be 0.

The motion vectors have two components: horizontal and vertical. Each component of the predicted vector is found separately. The predicted motion vector is the median value of the three candidate predictors, as formally stated by Equation 18.1:

746 CHAPTER 18. COMPARISON OF THE H.261 AND H.263 CODECS

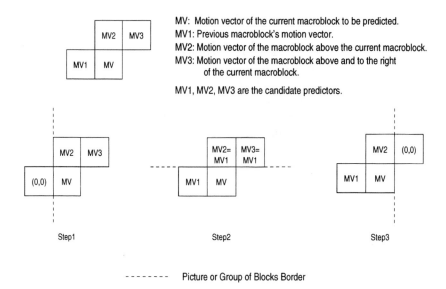

Figure 18.4: H.263 motion vector prediction scenarios.

$$P_{Horiz} = Median(MV1_{Horiz}, MV2_{Horiz}, MV3_{Horiz})$$
$$P_{Vert} = Median(MV1_{Vert}, MV2_{Vert}, MV3_{Vert}). \quad (18.1)$$

The difference between the predicted and actual motion vector is encoded by the MVD codec parameter, as suggested by Equation 18.2:

$$MVD_{Horiz} = MV_{Horiz} - P_{Horiz}$$
$$MVD_{Vert} = MV_{Vert} - P_{Vert}. \quad (18.2)$$

This motion vector predictor is much more effective than the one used by the H.261 codec. It therefore reduces the bit rate because the differential motion vectors are typically smaller, requiring a lower number of bits for their variable-length encoding.

18.2.3.2 H.263 Subpixel Interpolation

As mentioned earlier, the H.263 codec uses half-pixel precision for its motion vectors, while the H.261 standard used full-pixel precision. Half-pixel precision allows the motion vector to more closely compensate for motion, thereby reducing the error residual and bit rate after inter-frame prediction. In order to predict the half-pixel values, the H.263 scheme invokes bilinear interpolation, which is portrayed in Figure 18.5.

Assuming that the pixel-spaced luminance values A, B, C, D are known, the half-pixel-spaced luminances are computed by averaging the pixel-spaced values surrounding the half-pixel-spaced positions. This bilinear interpolation effectively quadruples

18.2. THE H.263 CODING ALGORITHMS

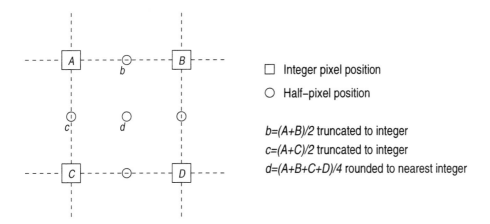

Figure 18.5: H.263 bilinear interpolation for subpixel prediction.

the original resolution and facilitates the "oversampled" high-resolution motion compensation.

Having described the various hierarchical layers of the H.263 codec and their relationship with the corresponding layers of the H.261 scheme, let us now concentrate on the H.263 codec's negotiable options.

18.2.4 H.263 Negotiable Options

The H.263 standard specifies four negotiable coding options, which can be selected between the encoder and decoder when a connection is set up. The H.263 encoder and decoder can optionally support these configurations, and the negotiations can ascertain which options are supported by both parties. The negotiable options are additional coding modes, which can improve the system's performance at the expense of extra computation; therefore, they are more demanding and expensive in hardware terms. These options are defined below.

18.2.4.1 Unrestricted Motion Vector Mode

In the H.261 scheme and in the baseline H.263 coding modes, motion vectors are limited to reference pixels within the currently coded picture area, where the term "baseline" refers to an H.263 codec refraining from using the negotiable coding modes. In the unrestricted motion vector mode, this limitation is removed. This allows motion vectors to point outside the corresponding coded picture area in the previous decoded frame. When a block position referenced by a motion vector is outside the current picture area, a pixel on the edge of the picture area is used instead of the nonexisting pixel. This edge pixel is the last full pixel inside the picture area along the path of the motion vector. Edge pixels are found on a pixel by pixel basis and are calculated separately for each component of the motion vector. The mapping of pixels outside the coded picture area onto edge pixels is shown in Figure 18.6. In the figure, the shaded block in the currently coded frame was found to be best matched with the

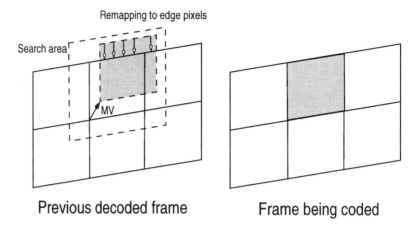

Figure 18.6: An example of the unrestricted motion vector pointing outside the coded picture area. Pixels outside the picture area are remapped to pixels on the edge of the picture area.

shaded block in the previously decoded frame; hence, the offset of this block is given by the motion vector MV.

For example, if the coordinates of the referenced pixel (which may not refer to a valid pixel position within the confines of the frame to be coded) are (x, y), then the coordinates are modified using Equation 18.3, into coordinates of the edge pixel (x', y'):

$$x' = \begin{cases} 0 & \text{if } x < 0 \\ 175 & \text{if } x > 175 \\ x & \text{otherwise,} \end{cases} \quad y' = \begin{cases} 0 & \text{if } y < 0 \\ 143 & \text{if } y > 143 \\ y & \text{otherwise.} \end{cases} \quad (18.3)$$

If pixel (x, y) exists, then $x = x'$ and $y = y'$. Clearly, Equation 18.3 is valid only for QCIF pictures of 176 by 144 pels resolution. However, the numbers 175 and 143 in the equation need to be changed for other picture sizes.

Experiments have shown that in this mode a significant gain is achieved if there is movement around the edges of the picture, especially for smaller picture resolutions. The results of Whybray et al. [193] show approximately an additional 9% bit-rate reduction, when using the unrestricted motion vector mode in the context of the "Foreman" video sequence. The "Foreman" sequence contains a large amount of camera shake. The camera shake introduces movement at the edge of the picture, which the unrestricted motion vector mode handles well.

In addition to allowing motion vectors to point outside the coded picture area, the unrestricted motion vector mode allows the range of motion vectors to be expanded. In the default H.263 mode, motion vectors are restricted to the range $[-16, 15.5]$. However, in the unrestricted motion vector mode, the range of motion vectors is extended to $[-31.5, 31.5]$ with a few restrictions. If the predictor (P) of a motion vector (MV) is in the range $[-15.5, 16]$, the range of motion vectors is limited to

18.2. THE H.263 CODING ALGORITHMS

[−16, 15.5] around the predictor (P). If, however, the motion vector predictor (P) is outside the range [−15.5, 16], then motion vectors in the range [0, 31.5] or [−31.5, 0] can be reached, depending on the sign of the predictor (P). These relationships are summarized in Equation 18.4:

$$\begin{array}{rlll} -31.5 \leq & MV & \leq 0 & \text{if } -31.5 \leq P \leq -16 \\ -16 + P \leq & MV & \leq 15.5 + P & \text{if } -15.5 \leq P \leq 16 \\ 0 \leq & MV & \leq 31.5 & \text{if } 16.5 \leq P \leq 31.5. \end{array} \quad (18.4)$$

The full range of motion vectors is split into these three ranges, so that the same coding table can be used, as in the baseline mode; only the equations to generate the motion vector from the predictor and motion vector difference are different. The extension of the motion vector range in this mode allows large motion vectors to be used, which is especially useful in the case of camera movement.

18.2.4.2 Syntax-Based Arithmetic Coding Mode

When the syntax-based arithmetic coding (SAC) is used, the coding parameters are no longer coded with variable-length codewords but with syntax-based arithmetic coding [12]. The reconstructed frames will remain the same; however, the bit rate is reduced. The performance of the syntax-based arithmetic coding (SAC) mode was evaluated by Whybray et al. [193], who found bit-rate savings between 1.6% and 6.3% for different video sequences.

SAC is a variant of arithmetic coding. It can be used to losslessly encode the video stream as an alternative to the widely used variable-length coding. The variable-length codewords are calculated from the entropy of the data. Since "conventional" codewords have to contain an integer number of bits, the rounding of the entropy values to fit a particular integer codeword length introduces inefficiency. Arithmetic coding almost completely eliminates this inefficiency by effectively allowing fractional bits per codeword. This is achieved by estimating the probability of a particular codeword in the data stream, where these estimates are referred to as the model.

The disadvantage of SAC is that it is difficult to implement. Furthermore, since each codeword is effectively constituted by a fractional number of bits, it is not possible to recognize particular symbols in the data stream. Another problem is that errors are not detected at all, unlike in the case of variable-length codewords, where errors are detected within a few codewords after an error. Lastly, if the above-mentioned model and the statistics of the actual data being coded are mismatched, the coding efficiency is reduced, which can lead to a high bit rate.

18.2.4.2.1 Arithmetic coding [12]
is based on an abstract generic concept, that subdivides its number space incrementally into nonoverlapping intervals. This somewhat abstract concept will be "demystified" at the end of this section using a tangible practical example. Each interval corresponds to a particular event, and the size of the interval is proportional to the probability of the particular event occurring. An "event" in the context of syntax-based arithmetic coding is the encountering of a specific coding parameter, which has to be coded. The probability of encountering that particular coding parameter is proportional to the size of the subinterval for

that event or coding parameter. Below we first introduce the abstract concept, which again will be made more explicit at a later stage using an example.

The arithmetic coder uses the sequence of events to incrementally subdivide the intervals. The coder then outputs as many bits as required to distinguish the subinterval or sequence of events that occurred. In general the arithmetic coders output these bits as soon as they become known. A comprehensive overview of arithmetic coding was given by Howard and Vitter [12]. The basic algorithm is as follows:

1. Each encoding stage starts with the "Current Interval," which is initially $[0, 1)$.

2. The current interval is subdivided into subintervals, one for each possible event that could occur next. The size of each event's subinterval is proportional to the model's estimated probability for that event at this particular instant.

3. The subinterval corresponding to the event that actually occurs next becomes the new "Current Interval."

4. The coder outputs a sufficient number of bits to distinguish the final current interval from all other possible final intervals.

An example of arithmetic coding is shown in Figure 18.7. In this example, the first event is either "a" or "b" with equal probability of $\frac{1}{2}$. For example, event "a" could mean that a COD coding parameter is being transmitted, and event "b" could mean that a MCBPC coding parameter is to be transmitted. In our example, event "a" was chosen since a COD coding parameter was to be transmitted. The subinterval "a" now becomes the current interval. The new current interval now has two possible next events, "c" (with probability $\frac{1}{3}$) and "d" (with probability $\frac{2}{3}$). At this stage event "d" was assumed to be encountered and became the new current interval. This new interval is $[\frac{1}{6}, \frac{1}{2}]$, which is further subdivided for events "e" (probability $\frac{1}{4}$) and "f" (probability $\frac{3}{4}$). If the final event encoded is event "f," then the interval for this event is $[\frac{1}{4}, \frac{1}{2}]$, as can be seen in Figure 18.7.

The arithmetic encoder has to transmit binary data to inform the decoder that the sequence of events was "a", "d", "f." This is accomplished by transmitting bits to define what the final interval was, which in this case was $[\frac{1}{4}, \frac{1}{2}]$. The interval is encoded by sending the value of the largest binary fraction that is fully contained inside the interval, which uniquely identifies this interval using the lowest possible number of bits. The binary fractions seen at the bottom of Figure 18.7 are allocated as follows: .1 is $\frac{1}{2}$, .01 is $\frac{1}{4}$, .101 is $\frac{1}{2} + \frac{1}{8} = \frac{5}{8}$, and so on. The interval can therefore be encoded by recursively halving the number space, until a binary fraction interval is fully contained inside the interval to be coded. Hence, for our example in Figure 18.7 the sequence of events "a", "d", "f" has the interval $[\frac{1}{4}, \frac{1}{2}]$, which corresponds to the binary fraction .01. Therefore, the example sequence is encoded with the bits "01," implying that the final interval contains the binary fraction .01. If the sequence of events was "a", "d", "e," the final interval would be $[\frac{1}{6}, \frac{1}{4}]$. The largest binary fraction contained in that interval is .0011; hence, the encoded sequence of bits is "0011." For further details on the associated coding steps, the interested reader is referred to [12].

18.2. THE H.263 CODING ALGORITHMS

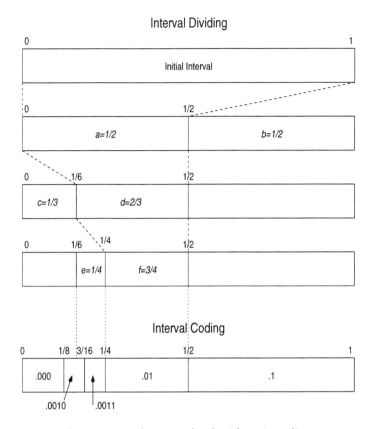

Figure 18.7: An example of arithmetic coding.

18.2.4.3 Advanced Prediction Mode

The advanced prediction mode is another negotiable option of the H.263 codec. This mode supports overlapped block motion compensation and the possibility of using four motion vectors per macroblock. In this mode, the motion vectors are allowed to cross picture boundaries, as in the unrestricted motion vector mode described in Section 18.2.4.1. The extended motion vector range feature is not part of the advanced prediction mode. It can, however, be used with the advanced prediction mode if the unrestricted motion vector mode is also active. The overlapping motion compensation of the advanced prediction mode can only be used for the prediction of P-pictures, but not for B-pictures when the P-B mode (described in Section 18.2.4.4) is used.

The advanced prediction mode generally gives a considerable improvement by improving the subjective quality of the video. The results obtained by Whybray *et al.* [193] demonstrated that the bit rate increased for some video sequences when advanced prediction was used. However, it was noticed that the subjective quality of the video had improved, mainly because of the reduced blocking artifacts. This video quality increase is not clearly seen in the PSNR results, however. The bit rate generally was reduced only for inherently higher bit-rate simulations. For example,

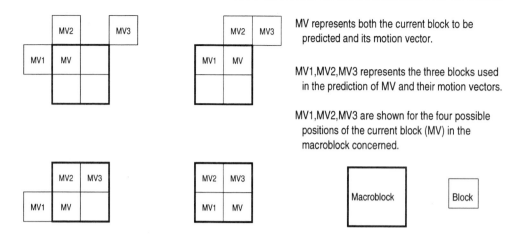

Figure 18.8: Redefinition of candidate predictive motion vectors MV1, MV2, and MV3 for advanced prediction mode using four motion vectors per macroblock.

the most dramatic bit-rate reduction of 26% was obtained for the "Suzie" sequence. However, this technique may also increase the bit rate. For example, a 16% bit-rate increase was experienced for the "Foreman" sequence.

Other techniques for reducing the blocking artifacts, which particularly occur at low bit rates, have been suggested by Ngan *et al.* [555, 556].

18.2.4.3.1 Four Motion Vectors per Macroblock In the case of the H.261 scheme and the baseline H.263 coding modes, only one motion vector can be specified for each macroblock. In the advanced prediction mode, either one or four motion vectors can be transmitted for each macroblock. When one vector is transmitted, this is equivalent to four vectors of the same value. When four vectors per macroblock are transmitted, each motion vector references one luminance block and a quarter of the color blocks in the macroblock.

The use of four motion vectors per macroblock instead of one improves the motion prediction at the expense of an initially higher bit rate caused by the transmission of the extra motion vectors. This higher bit rate may be offset by a reduced-motion compensated error residual, which can be coded more efficiently by DCT coding.

The motion vectors are calculated and encoded in a similar fashion to the baseline coding mode. The only difference is that a different set of candidate predictive motion vectors are used. The new candidate predictive motion vectors are shown in Figure 18.8.

The candidate predictor MV3 for the top left block of the macroblock drawn in bold appears to be out of place. However, it cannot assume the obvious position adjacent to MV2 because if there was only one motion vector for the macroblock containing MV2, then MV2 and MV3 would be the same, and this would affect the quality of the motion vector prediction.

18.2.4.3.2 Overlapped Motion Compensation for Luminance

When the advanced prediction mode is used, each pixel in an 8 × 8 predicted luminance block is a weighted sum of three pixel values. These three pixels are found using three motion vectors, namely, the motion vector of the current luminance block and the motion vectors of two of the four adjacent blocks, termed *"remote vectors."* The corresponding four remote vectors are the motion vectors for the luminance blocks above, below, left, and right of the current luminance block. The two selected motion vectors are derived on a pixel by pixel basis for each pixel in the predicted block, using the motion vectors of the two nearest-neighbor blocks with respect to the position of the current pixel within the predicted block. For example, the 16 pixels in a 4 × 4 pixel group in the top-left of each predicted block use the motion vectors from the block above and to the left of the current block. As a further example, the pixel at the bottom-right of the predicted block is predicted from the motion vector of the current block, the motion vector of the block below, and the motion vector of the block to the right of the current block. If any of the chosen adjacent blocks do not exist, because the current block is at the border of the picture, then the motion vector of the current block is used instead. If the chosen adjacent block was not encoded or was encoded in intra-mode, then the remote vector is set to 0.

Having highlighted how the three motion vectors used for overlapped luminance motion compensation are derived, let us now explain how they are used in order to produce the predicted pixel values. Each pixel in the predicted block is derived from the weighted sum of three pixels. For example, the three pixels associated with the top-left pixel of the predicted block are the current pixel offset by the motion vector of the current block; the current pixel offset by the motion vector of the block above the current block; and the current pixel offset by the motion vector of the block to the left of the current one. If all three motion vectors used in the prediction are zero, then the three pixels used for predicting the current pixel are constituted by the pixel itself. Generally, however, the motion vectors are not zero, and therefore the predicted pixel can be generated by the weighted sum of up to three different pixels.

Specifically, the three pixels used for prediction are weighted using the associated weighting masks of the current and the four adjacent blocks shown in Figure 18.9. The predicted pixel value is then generated as the sum of the weighted pixels divided by 8. The five weighting masks portrayed in the figure actually overlap, and the sum of the weights for any pixel is 8. Therefore, if the three motion vectors used for prediction are 0, then the current pixel is weighted by three masks, summed and divided by 8, in order to produce the predicted pixel value, which in this case would be the same as the value of the current pixel.

Let us clarify the whole process with an example for the pixel to the left of the bottom-right pixel of the current block, where the corresponding weighting mask values are circled in Figure 18.9. The three motion vectors used for the prediction of this pixel are the motion vector associated with the current block, $[\Delta x_c, \Delta y_c]$, the motion vector of the block below the current block, $[\Delta x_b, \Delta y_b]$, and the motion vector of the block to the right of the current block, $[\Delta x_r, \Delta y_r]$. Using this notation, the

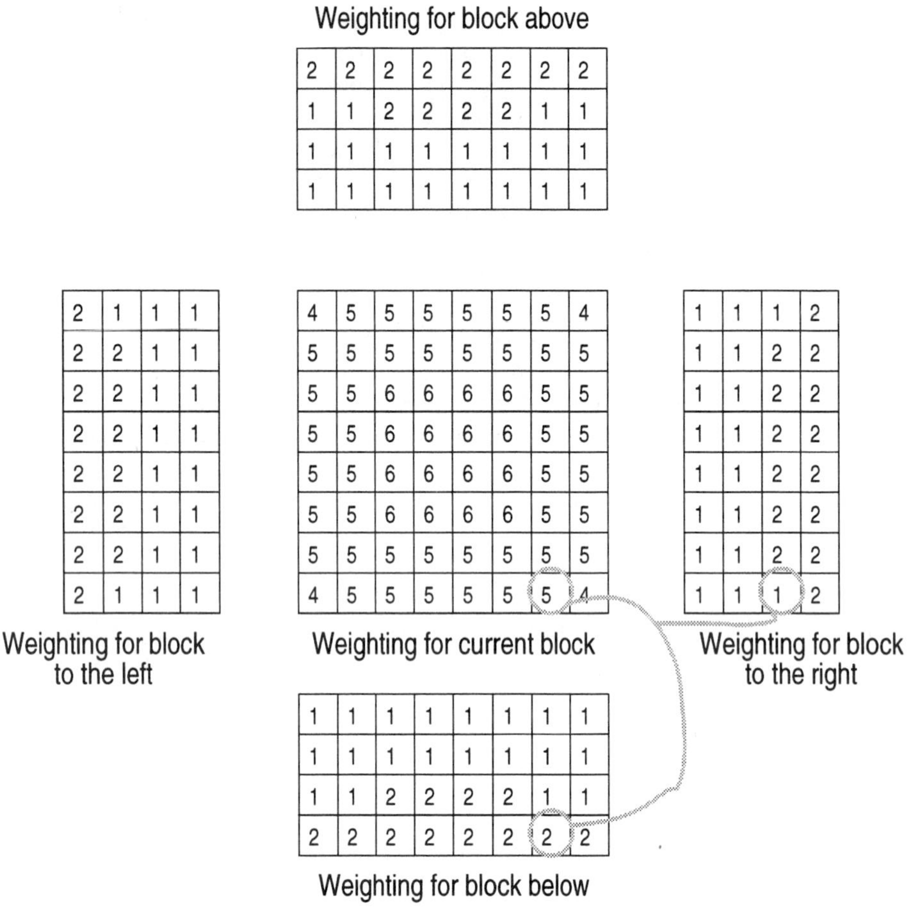

CHAPTER 18. COMPARISON OF THE H.261 AND H.263 CODECS

Weighting for block above

2	2	2	2	2	2	2	2
1	1	2	2	2	2	1	1
1	1	1	1	1	1	1	1
1	1	1	1	1	1	1	1

2	1	1	1
2	2	1	1
2	2	1	1
2	2	1	1
2	2	1	1
2	2	1	1
2	2	1	1
2	1	1	1

4	5	5	5	5	5	5	4
5	5	5	5	5	5	5	5
5	5	6	6	6	6	5	5
5	5	6	6	6	6	5	5
5	5	6	6	6	6	5	5
5	5	6	6	6	6	5	5
5	5	5	5	5	5	5	5
4	5	5	5	5	5	5	4

1	1	1	2
1	1	2	2
1	1	2	2
1	1	2	2
1	1	2	2
1	1	2	2
1	1	2	2
1	1	1	2

Weighting for block to the left   **Weighting for current block**   **Weighting for block to the right**

1	1	1	1	1	1	1	1
1	1	1	1	1	1	1	1
1	1	2	2	2	2	1	1
2	2	2	2	2	2	2	2

Weighting for block below

Figure 18.9: Overlapping motion compensation for luminance blocks using weighting masks.

associated operations are summarized as follows:

$$L_{pred} = \frac{1}{8} \times \begin{bmatrix} [5 \times P(x + \Delta x_c, y + \Delta x_c)] + \\ [2 \times P(x + \Delta x_b, y + \Delta x_b)] + \\ [1 \times P(x + \Delta x_r, y + \Delta x_r)] \end{bmatrix} \quad (18.5)$$

where function $P(x,y)$ indicates the pixel luminance at the current coordinates, namely $[x,y]$.

18.2.4.4 P-B Frames Mode

The definition of the previously mentioned Predicted (P) and Bidirectional (B) frames borrowed from the Motion Pictures Expert Group (MPEG) standard [499] becomes explicit in Figure 18.10. A P-B frame is constituted by two pictures encoded as one

18.2. THE H.263 CODING ALGORITHMS

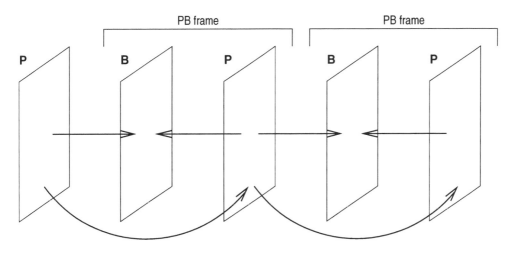

Figure 18.10: Prediction in P-B frames mode.

unit, — namely by a P frame, which is predicted from the last decoded P frame, and a B picture, which is predicted both from the last decoded P picture and the P picture currently being decoded. The terminology B picture is justified because parts of it may be bidirectionally predicted from past and future P pictures, again as shown in Figure 18.10.

In the MPEG codec there can be several B pictures between P pictures, but in the H.263 standard only one B picture is allowed in order to reduce the decoding delay that P-B frames cause. In the P-B frame mode, a macroblock contains 12 blocks rather than just $4 + 2 = 6$ blocks. First, the six P blocks are transmitted as normal, followed by the six B blocks.

Motion vectors can be used to compensate for motion in B frames in a similar way to the usual P frames. The motion vectors for a P-B frame are shown in Figure 18.11.

The motion vector MV in the figure is the normal motion vector between two P frames, as in other H.263 modes. The forward (MVF) and backward (MVB) motion vectors are derived from MV but can be adjusted by the use of an optional delta motion vector. When a delta motion vector is not used, it is assumed that the motion vector changes linearly, between P frames. Therefore the forward and backward vectors can be found by interpolating the motion vector MV. If the motion vectors change linearly then $MVF - MVB = MV$. The forward and backward motion would be equal and opposite if the time between the three frames (P,B,P) were the same. If the time between the three frames is different, then the forward and backward vector has to be adjusted, depending on the temporal references of the frame. The coding parameter TRB contains the number of untransmitted frames at 29.97 Hz between the B frame and the previous P frame ($T2 - T1$ in the figure). The equations for calculating the forward and backward motion vectors are shown in Equation 18.6:

$$MVF = \frac{TR_B \times MV}{TR_D} + MV_D$$

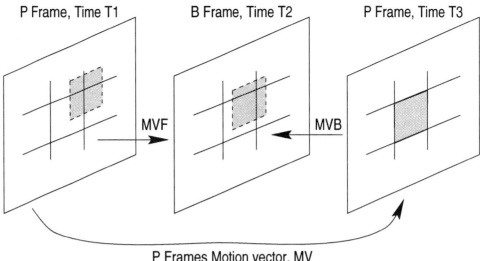

Figure 18.11: Motion vectors used in P-B prediction mode.

$$MVB = MVF - MV_D - MV, \qquad (18.6)$$

where, TR_B is the time between frames $(T1-T2)$, which is transmitted in the TRB coding parameter. TR_D is the time between the current and previous P frames $(T3-T1)$, while MV_D is the delta motion vector, which is optional. If it is not used, it is set to 0. Note that a positive delta motion vector increases the forward motion vector and reduces the backward vector.

The prediction for B blocks can be partly bidirectional. The part of each B block, which is bidirectionally encoded, is dependent on the backwards-oriented motion vector. The pixels in the B block referenced by the backward motion vector, which are inside the corresponding P macroblock, are bidirectionally predicted, while all other pixels use forward prediction only. If the whole B block were bidirectionally predicted, then the macroblock could not be decoded until all the macroblocks containing the pixels referenced by the backward motion vector were received, contributing additional latency to the decoding process. The bidirectionally predicted pixels use the average of the pixels referenced by the forward and backward motion vector, as the prediction value for the B-coded pixel. This prediction value is added to the decoded MCER in order to find the value of the B-coded pixel. The remaining pixels that use forward prediction employ the pixel values from the previous P frame as the prediction values for the pixels. An example showing the area of the B macroblocks that can be bidirectionally predicted is shown in Figure 18.12.

The P-B mode of the H.263 video codec can be utilized in two ways. First, it can be used to double the frame rate for a modest increase in bit rate. Increasing the frame rate of a coded video sequence gives a significant subjective improvement

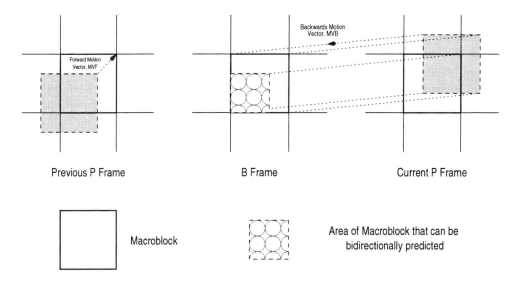

Figure 18.12: Motion vectors used in P-B prediction mode.

of quality. The second way to use P-B mode is to maintain the same frame rate but achieve greater compression by the use of the more efficient B frames. However, P-B mode cannot be used at low frame rates combined with a large amount of motion. This is because the process of interpolating the motion vector becomes inaccurate.

The investigations by Whybray et al. [193] demonstrated that the bit rate increased when P-B mode was used to double the frame rate. For video sequences with low to medium amounts of motions and a reasonable frame rate, it was found that the increase in bit rate became more moderate at higher bit rates. The increase in bit rate for these sequences ranged from 36% to 4.5%. For high-motion sequences at low frame rates, the bit-rate increase was as much as 61%. It was noted that the PSNR of the P- and B- frames was very similar, but this may be due to the fixed quantizer that was used for these simulations. In general, the PSNR of the B-frame is typically less than that for the P-frames.

Following our comparative description of the H.261 and H.263 codecs, in the next section we provide a range of performance results for these schemes.

18.3 Performance Results

18.3.1 Introduction

The H.261 codec used in our investigations was a software implementation derived from the INRIA Videoconferencing System (IVS).[1] The H.263 codec was a modified version of a software implementation developed by Telenor Research and Development [563]. An enchanced version of the source code is now available from the

---

[1] http://www.inria.fr/rodeo/personnel/Thierry.Turletti/ivs.html

University of British Columbia's World Wide Web site.[2]

Simulations were carried out with both the H.261 and H.263 codecs in order to comparatively study their performance. All simulations used well-known video sequences, such as the "Miss America," "Suzie," and "Carphone" clips and were performed with transmission frame rates of 10 and 30 frames/s.

The performance of both codecs depends very much on the quantizers used. The upper and lower limits of performance are found when the finest or coarsest quantizers are employed.

The H.261 simulations used gray scale video sequences that were generated by discarding the color information, retaining just the luminance information. The H.263 simulations invoked in comparison with the H.261 codec used the same gray scale video sequences. In this section, results are shown in graphical and tabular forms. To make the results comparable in tabular form, results were interpolated from the graphical curves in order to obtain results at the same bit rate or PSNR. Let us now turn to their performance comparison.

18.3.2 H.261 Performance

The H.261 simulations completed used the QCIF gray scale "Miss America" video sequences at 10 and 30 frames/s (fps). All simulations used inter-frame prediction with an intra-coded frame once every 132 frames. The intra-coded frame update rate of once every 132 frames was selected in order to meet the minimum macroblock forced update requirement of the codec. Simulations were carried out using the H.261 codec in one of the following three modes:

- No motion compensation.

- Motion compensation using motion vectors (MV).

- Motion compensation using motion vectors and loop filtering (MV+LF).

The simulations were performed at a wide range of bit rates. The results of these simulations are shown in Figures 18.13 and 18.14, where Figure 18.13 characterizes the average PSNR versus bit-rate performance. The coded video sequences can also be seen on the WWW,[3] where readers can judge the associated perceptual video quality for themselves.

As can be seen in Figure 18.13 for a given bit rate, the simulations without motion compensation have typically the lowest image quality. This is a consequence of constraining the quantizers to maintain a given bit rate, while having to quantize the motion-compensated prediction residual (MCPR) acquired by frame-differencing rather than by full motion compensation. Clearly, frame-differencing has a lower complexity than full motion compensation, and it can also dispense with transmitting the motion vectors, thereby potentially reducing the bit rate. However, the MCPR has a higher variance, which is thus less amenable to quantization, even when investing the extra bits, which would have been dedicated to the encoding of the MVs. Pairwise

---

[2] http://www.ee.ubc.ca/spmg/research/motion/h263plus
[3] http://www-mobile.ecs.soton.ac.uk/peter/h261/h261.html#Examples

18.3. PERFORMANCE RESULTS

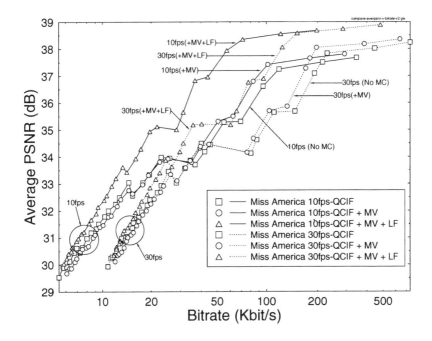

Figure 18.13: H.261 Image quality (PSNR) versus coded bit rate.

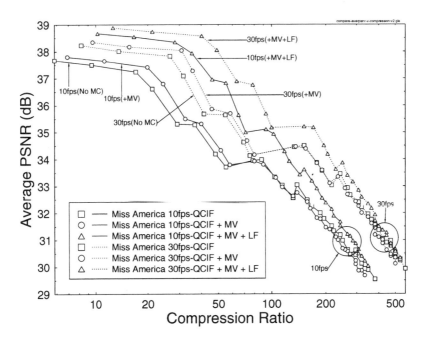

Figure 18.14: H.261 Image quality (PSNR) versus compression ratio.

Fixed PSNR (dB)	Percentage Bit-Rate Reduction (%)				
	10 fps		30 fps		
	MV	MV+LF	NoMC	MV	MV+LF
31	-8.28	10.96	-76.81	-78.81	-63.25
33	-23.55	19.53	-61.08	-67.05	-41.31
35	8.98	60.57	-85.16	-73.10	33.34
37	18.65	59.46	-72.33	-48.48	10.19

Table 18.2: Relative Reduction (%) in Bit Rate Required to Achieve Given PSNRs, Compared with Simulations at 10 Frames/s Using no Motion Compensation

comparisons in Figure 18.13 reveal that the overall effect of MV-assisted full MC was found advantageous in terms of improved motion tracking, and it increased the image PSNR slightly. However, when motion vectors and loop filtering were used jointly, the image quality was substantially improved.

The image quality expressed in PSNR saturated at high bit rates, around 100 Kbit/s. However, in the bit-rate range between 10 and 100 Kbit/s, the PSNR improved quasilinearly, as the bit rate was increased. The corresponding gradient expressed in terms of PSNR improvement versus bit-rate increment was about 10 dB/decade. When maintaining a constant bit-rate, the 30 fps curves typically exhibit a PSNR penalty due to the imposed bit-rate constraint, since on the basis of simple logic at 30 fps a threefold bit-rate increase is expected in comparison to 10 fps. This is a very coarse estimate, since due to the higher inter-frame correlation the MCPR has a typically reduced variance in comparison to 10 fps coding, which mitigates the above threefold bit-rate increment estimate, but the bit-rate penalty of 30 fps transmissions is certainly substantial. In many applications, it may be advantageous to vary the frame scanning rate under network and user control in order to arrive at the best scheme for a specific application.

Table 18.2 summarizes the decrease in bit rate achieved for the same PSNR at both 10 and 30 fps, with and without motion vectors (MV) and loop filtering (LF). The reductions are relative to the 10 frames/s simulations without motion compensation (MC). The table shows the performance improvement provided by motion-vector-assisted full motion compensation combined with loop filtering. At high PSNRs, the simulations at 30 frame/s using motion vectors and loop filtering outperform the baseline simulation at 10 fps without motion compensation. For example, at a PSNR of 37 dB and a scanning rate of 30 frames/s with motion vectors and loop filtering, the required bit rate is 10.19% less than at 10 frames/s without motion compensation. In contrast, at 30 frames/s without motion compensation a 72% increase in bit rate is required. A range of further interesting tradeoffs becomes explicit by scrutinizing Figure 18.13 and Table 18.2.

Viewing the codec's performance from a different angle, Table 18.3 shows the increase in image quality expressed in PSNR achieved at a range of fixed bit rates. The gains are relative to the 10 frames/s simulations without motion compensation.

18.3. PERFORMANCE RESULTS

Fixed Bit Rate (Kbit/s)	PSNR Increase (dB)				
	10 fps		30 fps		
	MV	MV+LF	NoMC	MV	MV+LF
20	0.00	1.54	-0.98	1.15	-0.51
30	0.00	1.51	-0.57	-0.55	0.39
50	0.40	2.55	-0.44	-0.44	0.35
100	0.62	1.73	-1.56	-1.23	0.32
190	0.22	1.22	-0.41	0.40	1.18

Table 18.3: Increase in PSNR (dB) Achieved at Given Bit Rates, Compared with Simulations at 10 Frames/s Without Motion Compensation

Again, the table characterizes the performance improvement provided by motion vector assisted full MC combined with loop filtering. For example, observe in Figure 18.13 and Table 18.3 that at a fixed bit rate of 50 Kbit/s, the 30 frames/s mode of operation without motion compensation has a 0.44 dB loss of PSNR compared with the 10 frames/s scenario without motion compensation. However, simulations conducted at 10 frames/s with motion-vector-assisted full motion compensation and loop filtering has a 2.55 dB increase in PSNR compared with the scenario at the same frame rate without motion compensation.

Figure 18.14 plots the results of the above investigations from a different perspective, as average PSNR versus the compression ratio. Again, the curves suggest a fairly linear relationship over a rather wide range, resulting in an approximately 10 dB/decade decay in PSNR terms as a result of increasing the compression ratio from around 50 to 500. As before, the improvement in performance achieved when using motion vectors and loop filtering is clearly shown in Figure 18.14. The graphs suggest that an approximately quadrupled compression ratio can be achieved for the same image quality, when using the more sophisticated MV-assisted loop-filtered full motion compensation. A rule of thumb was that at high bit rates, turning on motion vectors and loop filtering allowed the frame rate to be tripled, as well as reducing the bit rate. These results are compared with similar performance figures of the H.263 codec in Section 18.3.3.

18.3.3 H.261/H.263 Performance Comparison

In this section, we compare the H.261 performance figures derived in Section 18.3.2 with the corresponding characteristics of the H.263 codec using the same QCIF gray scale "Miss America" video sequences. The H.263 experiments used none of the negotiable coding options. In this mode, the H.263 codec uses motion vectors and inter-frame prediction for motion compensation. The appropriate macroblock-forced update algorithm was invoked in order to ensure that each macroblock was coded in intra-mode, after a maximum of 132 inter-frame coded updates.

Similarly to our H.261 investigations, the H.263 experiments were conducted at 10 and 30 frames/s and at a wide range of bit rates. Figure 18.15 shows a comparison

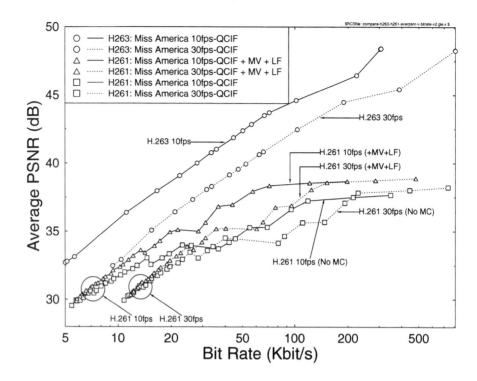

Figure 18.15: Image quality (PSNR) versus coded bit rate, for H.261 and H.263 simulations using gray scale QCIF "Miss America" video sequences at 10 and 30 frames/s.

of the H.261 and H.263 video codecs in terms of image quality expressed in PSNR versus bit rate (Kbit/s). As can be seen from the graphs, the performance of the H.263 codec is significantly better than that of the H.261 scheme. Furthermore, the useful operating bit-rate range of the H.261 codec was also extended by the H.263 scheme, ensuring an approximately 9 dB/decade PSNR improvement across the bit-rate range of 5–500 Kbit/s.

Table 18.4 shows the decrease in terms of bit rate achieved for a range of specific fixed PSNR values when using the H.263 codec rather than the H.261 scheme. Specifically, the bit-rate reductions are relative to the H.261 mode of operation at 10 frames/s without motion compensation. The table includes a column for the best H.261 performance, which uses motion-vector-assisted full MC and loop filtering. This table shows that the H.263 codec outperforms the H.261 scheme, even when the frame rate is three times that of the H.261 arrangement.

Viewing the higher performance of the H.263 codec from a different angle, Table 18.5 summarizes the increase in image quality measured in terms of PSNR achieved at a range of different fixed bit-rate values. The gains are relative to the H.261 scenario at 10 frames/s without motion compensation. Again, it is clear from the table

18.3. PERFORMANCE RESULTS

Fixed PSNR (dB)	Percentage Bit-rate Reduction (%)				
	10fps		30fps		
	H.261 (MV+LF)	H263	H.261 (NoMC)	H.261 (MV+LF)	H263
33	19.53	62.28	-61.08	-41.31	26.90
35	60.57	82.99	-85.16	33.34	70.30
37	59.46	87.86	-72.33	10.19	77.78

Table 18.4: Relative Reduction (%) in Bit Rate Required to Achieve Given PSNR Values, for the H.261 and H.263 Codecs Compared with the H.261 Performance at 10 Frames/s Using no Motion Compensation

Fixed Bit Rate (Kbit/s)	PSNR Increase (dB)				
	10fps		30fps		
	H.261 (MV+LF)	H263	H.261 (NoMC)	H.261 (MV+LF)	H263
20	1.54	5.20	-0.98	-0.51	2.76
30	1.51	6.43	-0.57	0.39	3.98
50	2.55	7.50	-0.44	0.35	4.91
100	1.73	7.81	-1.56	0.32	5.59
190	1.22	8.49	-0.41	1.18	7.04

Table 18.5: Increase in PSNR (dB) Achieved at the Same Bit Rate, for H.261 and H.263 Simulations Compared with the H.261 Simulation at 10 Frames/s Without Motion Compensation

that the H.263 scheme outperforms the H.261 codec.

Figure 18.16 provides a comparison of the H.261 and H.263 video codecs in a different context, expressed as image PSNR versus compression ratio. Again, as can be seen from the graph, the H.263 codec exhibits a higher compression performance than the H.261 codec. As expected from the corresponding PSNR versus bit-rate curves, the H.263 codec's compression ratio performance curves are near-linear on this log-log scaled graph, resulting in a predictable PSNR versus compression ratio relationship.

Recall that these investigations were conducted without the H.263 codec using the negotiable options that can be invoked in order to increase the image quality or to reduce the bit rate. Observe, furthermore, that the H.263 performance graphs in Figures 18.15 and 18.16 never intersect the H.261 curves, implying that the H.263 PSNR performance at 30 frames/s is better than the H.261 at 10 frames/s in the investigated bit-rate range.

In addition to the results presented above, example video sequences can be viewed on the WWW for both H.261[4] and H.263[5] video codecs, where the associated sub-

---
[4] http://www-mobile.ecs.soton.ac.uk/peter/h261/h261.html#Examples
[5] http://www-mobile.ecs.soton.ac.uk/peter/h263/h263.html#Examples

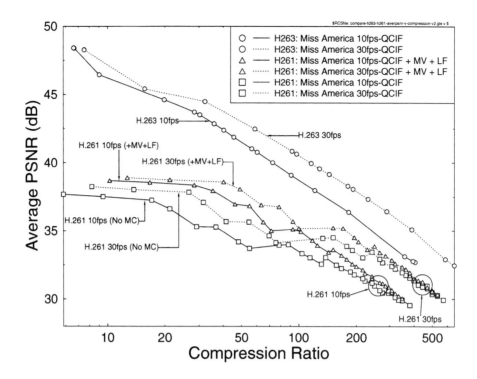

Figure 18.16: Image quality (PSNR) versus compression ratio performance for H.261 and H.263 simulations using gray scale QCIF "Miss America" video sequences at 10 and 30 frames/s.

jective quality can be judged. Having contrasted the H.261 and H.263 PSNR versus bit-rate performances, let us now study further aspects of the H.263 arrangement.

18.3.4 H.263 Codec Performance

In this section we report our findings on the tradeoffs between image quality and bit rate for different image sizes, frame rates, and video sequences, when using the H.263 codec. All simulations were conducted using well-known video sequences at both 10 and 30 frames/s. The video sequences used are summarized in Table 18.6. The image sizes used were described in Table 18.1 on page 734. Video sequences at 10 frames/s were generated from the 30 frame/s versions. The SQCIF images were generated by re-sampling the QCIF sequences. The original "Mall" sequence was a High-Definition Television (HDTV) video sequence at a resolution of 2048 x 1048 pixels. In order to convert this to 16CIF format, a black border was added to the top and bottom of the 16CIF frame, and the left and right edges were cropped to fit the 16CIF frame format.

18.3. PERFORMANCE RESULTS

Video Sequence	Size/s	Frame rates (frame/s)	Gray or Color
"Miss America"	QCIF	10, 30	Gray
"Miss America"	SQCIF, QCIF, CIF	10, 30	Color
"Carphone"	QCIF	10, 30	Color
"Suzie"	SQCIF, QCIF, 4CIF	10, 30	Color
"American Football"	4CIF	10, 30	Color
"Mall"	16CIF	10, 30	Color

Table 18.6: Video Sequences Used for H.263 Simulations

Fixed PSNR(dB)	Percentage Bit-Rate Reduction (%)	
	10 fps-Gray	30 fps-Gray
34	6.54	6.90
38	15.35	17.33
42	25.06	31.73
46	45.38	41.50
48	51.17	42.38

Table 18.7: Relative Reduction (%) in Bit Rate Required to Achieve a Range of Fixed PSNR Values, Compared with the Simulations at the Same Frame Rate in Color.

18.3.4.1 Gray-Scale versus Color Comparison

The previous H.261/H.263 comparison simulations were conducted using gray scale video sequences. In this section we investigate the extra bit rate required to support color video coding. The gray scale video sequence was generated by extracting the luminance information from the color video sequence.

The results of our gray versus color investigations are presented in Figures 18.17 and 18.18. The corresponding graphs characterize the image quality expressed in terms of PSNR versus bit rate and compression ratio, respectively. These graphs suggest that the gray scale and color PSNR performances are rather similar, but their difference becomes slightly more dominant with increasing bit rates. These tendencies imply that assuming a constant bit-rate, only a small fraction of the overall bit rate has to be allocated to convey the color information. Therefore, the PSNR penalty due to reducing the bit rate and resolution of the luminance component is marginal. The previously observed near-linear PSNR versus bit rate and compression ratio relationship is also maintained for color communications.

Table 18.7 shows the relative reduction in bit rate for the same image quality when using gray scale video rather than color. Observe in the table that the required bit-rate difference is comparatively small for lower PSNR values, but this discrepancy increases for higher bit rates, where higher video quality is maintained. This confirms our previous observation as regards earmarking only a small fraction of the available bit rate to encoding the inherently lower resolution chrominance information.

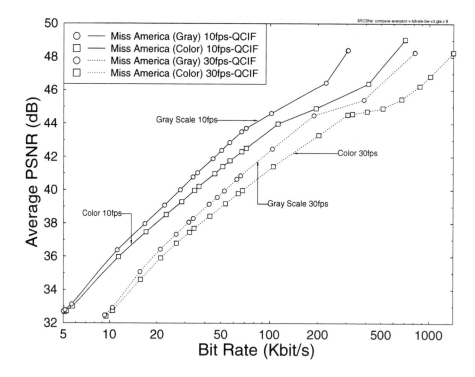

Figure 18.17: Image quality (PSNR) versus coded bit rate performance of the H.263 codec for the "Miss America" sequence in color and gray scale at 10 and 30 frames/s.

Fixed	PSNR Increase (dB)	
Bit Rate(Kbit/s)	10 fps-Gray	30 fps-Gray
10	0.44	0.16
30	0.83	0.63
100	1.01	1.10
200	1.14	1.29

Table 18.8: Increase in PSNR(dB) Achieved at a Range of Fixed Bit Rates, Compared with Simulations at the Same Frame Rate in Color

Table 18.8 portrays the same tendency from a different perspective, expressed in terms of image quality gain (PSNR) for gray scale video for the same bit rate as color video at the same framerate. Again, this table suggests that the gray scale PSNR performance is only marginally better for low bit rates, but the difference increases at higher bit rates. Overall, the perceived subjective video quality improvements due to using color, rather than gray-scale representation, are definitely attractive, unequivocally favoring the color mode of operation.

18.3. PERFORMANCE RESULTS

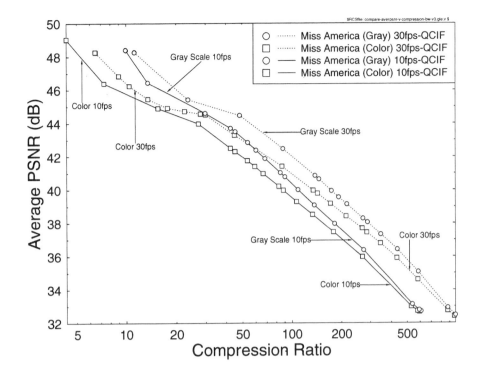

Figure 18.18: Image quality (PSNR) versus compression ratio performance of the H.263 codec for the "Miss America" sequence in color and gray scale at 10 and 30 frames/s.

18.3.4.2 Comparison of QCIF Resolution Color Video

The investigations in this section were conducted in order to characterize the performance of the H.263 codec for a range of different video sequences. All simulations were carried out using color video sequences scanned at 10 and 30 frames/s. The video sequences used were "Miss America," "Suzie," and "Carphone." Some example QCIF video sequences can also be viewed on the World Wide Web.[6]

The results of these experiments are plotted in Figures 18.19 and 18.20. As expected on the basis of its low-motion activity, the graphs suggest that the "Miss America" sequence is the most amenable to compression, followed by "Suzie" and then the "Carphone" sequence. The achievable compression ratios span approximately one order of magnitude range, when comparing the lowest and highest activity sequences. The achievable relative bit-rate reduction expressed in percentages is tabulated in Table 18.9 for the above video clips. Viewing these motion activity differences from a different angle, the PSNR reductions produced by the fixed bit-rate constraints are summarized in Table 18.10. The negative PSNR gains reflect the above-stated

---
[6]http://www-mobile.ecs.soton.ac.uk/peter/h263/h263.html#Examples

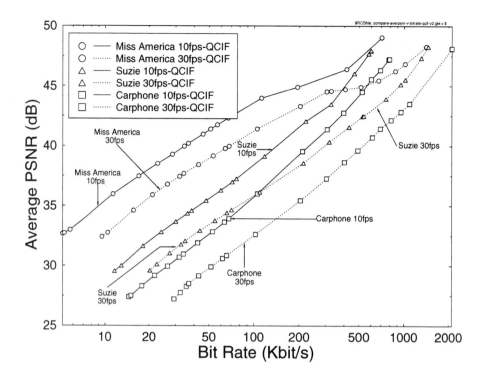

Figure 18.19: Image quality (PSNR) versus coded bit-rate performance of the H.263 codec for QCIF resolution at 10 and 30 frames/s.

tendencies and underline the importance of specifying the test sequence used in experimental studies. Let us now concentrate on analyzing the codec's performance in case of different video frame resolutions.

18.3.4.3 Coding Performance at Various Resolutions

The objective of this section is to evaluate the performance tradeoffs for different video resolutions. Our investigations were conducted at both 10 and 30 frames/s, at resolutions of SQCIF, QCIF, and CIF. The SQCIF video sequences were generated by subsampling the QCIF video sequences. The QCIF and CIF video sequences originated from different sources and therefore are not identical.

The corresponding PSNR results are portrayed in Figures 18.21 and 18.22 as a function of the bit rate and compression ratio, respectively, when using the "Miss America" sequence. The near-linear nature of the curves was maintained for the SQCIF and QCIF resolutions, but for the CIF resolution the coding performance curves became more nonlinear. The quadrupled number of pixels present in the CIF format resulted in an approximately fourfold increase of the bit rate. Furthermore, as in our previous investigations, the 30 fps scenarios typically required a factor of 2 higher bit rates in order to maintain a certain fixed PSNR. The same tendencies were observed

18.3. PERFORMANCE RESULTS

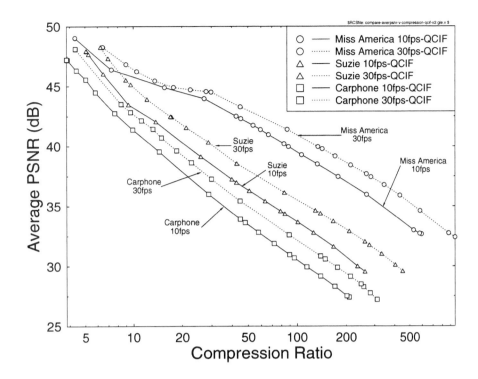

Figure 18.20: Image quality (PSNR) versus compression ratio performance of the H.263 codec for QCIF resolution at 10 and 30 frames/s.

Fixed	Percentage Bit-Rate Reduction (%)				
PSNR	Miss America	Suzie		Carphone	
(dB)	30 fps	10 fps	30 fps	10 fps	30 fps
33	-93.8	-343.3	-678.2	-857.0	-1922.6
34	-82.9	-331.9	-670.7	-805.8	-1893.8
36	-89.6	-387.8	-823.8	-821.6	-1949.2
38	-88.9	-363.6	-846.7	-724.4	-1683.7
40	-98.3	-328.2	-760.3	-577.1	-1439.1
42	-121.1	-257.3	-679.3	-466.5	-1187.5
44	-129.0	-210.6	-570.2	-325.8	-945.7
46	-127.1	-30.0	-206.4	-83.0	-358.8

Table 18.9: Relative Reduction (%) in Bit Rate Required to Achieve Fixed PSNRs for QCIF Resolution Video for the H.263 Codec, Compared with the Corresponding PSNR for the "Miss America" Sequence at 10 Frames/s

Fixed Bit Rate (Kbit/s)	PSNR Increase (dB)				
	Miss America	Suzie		Carphone	
	30 fps	10 fps	30 fps	10 fps	30 fps
10	-2.67				
20	-2.37	-6.02		-9.23	
30	-2.25	-5.80	-8.08	-9.02	-12.13
70	-2.39	-5.54	-7.82	-8.40	-11.38
100	-2.32	-5.20	-7.76	-7.84	-11.04
200	-1.69	-3.47	-6.68	-5.74	-9.60
400	-1.59	-1.47	-5.06	-3.49	-7.68
500	-2.29	-0.63	-4.97	-2.99	-7.47

Table 18.10: Increase in PSNR(dB) Achieved at a Range of Fixed Bit Rates for QCIF Resolution Video Coded with the H.263 Codec, Compared with the Simulation of "Miss America" at 10 Frames/s

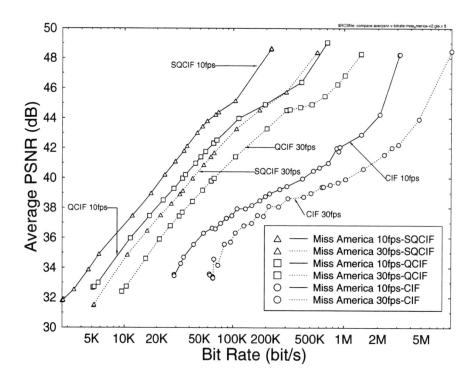

Figure 18.21: Image quality (PSNR) versus coded bit rate for H.263 "Miss America" simulations at 10 and 30 frames/s.

18.3. PERFORMANCE RESULTS

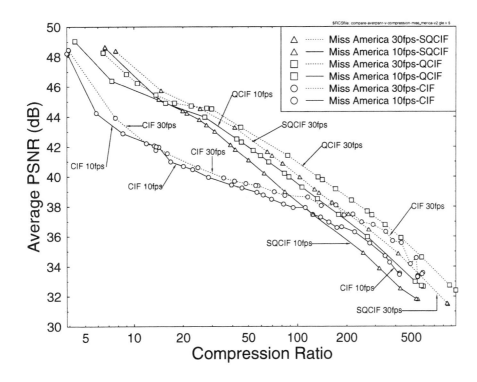

Figure 18.22: Image quality (PSNR) versus compression ratio, for H.263 "Miss America" simulations at 10 and 30 frames/s.

in terms of the compression ratio in Figure 18.22. As before, we also summarized the relative bit-rate reduction expressed in percentages with respect to the previously employed QCIF, 10 fps scenario, which is portrayed for the various PSNRs, resolutions, and frame rates in Table 18.11. The corresponding PSNR increment/reduction due to employing lower/higher resolutions, while maintaining certain fixed bit-rates, was presented in Table 18.12 in order to help gauge the associated tradeoffs.

Again, the CIF performance curves are more nonlinear than those of the lower resolution scenarios. A range of further interesting conclusions emerges from these graphs and tables. As an example, it is interesting to note in Figure 18.21 that by changing the resolution from QCIF to SQCIF, the frame rate can be tripled for just a 25% increase in bit rate, or a 0.5 dB loss of PSNR around a PSNR of 40 dB.

Let us now verify the above findings, which we arrived at on the basis of the "Miss America" sequence, also using the "Suzie" sequence, while replacing the CIF resolution with 4CIF representations. The remaining experimental conditions were unchanged. The corresponding results are plotted in Figures 18.23 and 18.24 in terms of PSNR versus the bit rate and compression ratio, respectively. Note, for example, in Figure 18.23 that the 30 fps SQCIF performance curve virtually overlaps with the 10 fps QCIF curve, indicating that the doubled number of pixels of the QCIF format

Fixed PSNR (dB)	Percentage Bit-Rate Reduction (%)				
	SQCIF		QCIF	CIF	
	10 fps	30 fps	30 fps	10 fps	30 fps
33	32.98	-32.54	-93.83		
34	37.29	-21.22	-82.94	-315.07	-780.63
36	26.76	-27.32	-89.57	-339.18	-772.36
38	30.60	-25.84	-88.94	-577.74	-945.79
40	34.88	-25.03	-98.29	-1159	-3042
42	39.86	-23.82	-121.12	-1389	-4259
44	44.12	-27.28	-129.01	-1614	-4110
46	62.74	7.71	-127.09	-613.04	-1808
48	66.12	10.01	-125.56	-419.78	-1388

Table 18.11: Percentage Reduction in Bit Rate Required to Achieved the Same PSNR, for "Miss America" Video Sequences Coded with H.263 Codec Compared with the QCIF Resolution Simulation at 10 Frames/s

Fixed Bit Rate (Kbit/s)	PSNR Increase (dB)				
	SQCIF		QCIF	CIF	
	10 fps	30 fps	30 fps	10 fps	30 fps
7	1.74	-1.05			
10	1.43	-0.80	-2.67		
20	1.47	-0.81	-2.37		
50	1.90	-0.79	-2.37	-5.33	
70	1.79	-0.69	-2.39	-5.85	-8.03
100	1.50	-0.55	-2.32	-5.97	-7.53
200	3.06	-0.16	-1.69	-6.14	-7.20
400		0.40	-1.59	-6.46	-7.59
600			-2.88	-7.41	-8.81

Table 18.12: Increase in PSNR(dB) Achieved at the Same Bit Rate, for "Miss America" Video Sequences Coded with H.263 Codec Compared with the QCIF Resolution Simulation at 10 Frames/s

18.3. PERFORMANCE RESULTS

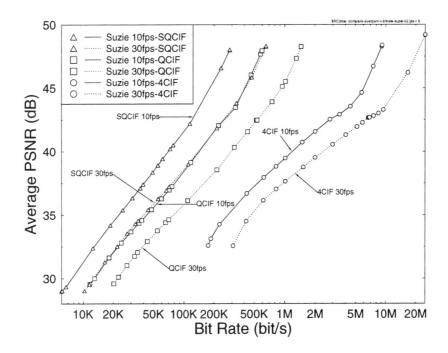

Figure 18.23: Image quality (PSNR) versus coded bit rate for H.263 "Suzie" simulations at 10 and 30 frames/s.

requires a three times lower frame rate under the constraint of identical PSNR.

Similarly to the Miss America experiments, we also presented the relative bit-rate reduction with respect to our reference scenario of QCIF resolution, 10 fps scanning rate, which is portrayed for the various fixed PSNRs, resolutions, and frame rates in Table 18.13. In contrast, Table 18.14 portrays the potential PSNR improvement or degradation at various fixed bit rates, resolutions, and frame rates. Again, comparing the SQCIF performance at 30 frames/s with the QCIF results at 10 frames/s produce interesting revelations. The results show that changing from QCIF to SQCIF resolution allows the frame rate to increase from 10 to 30 frames/s without any significant change in the bit-rate requirements or loss of image quality.

In the final set of experiments of this chapter, compared the high-resolution H.263 modes of 4CIF and 16CIF representations. As before, our experiments were carried out at 10 and 30 frames/s. The previously employed 4CIF "Suzie" video sequence, a 4CIF "American Football" video sequence, and our cropped HDTV "Mall" video sequence were used for the 16CIF experiments. The associated performance curves are shown in Figures 18.25 and 18.26, where the 4CIF curves were repeated from Figures 18.23 and 18.24. Since the amount of image fine detail is quite different in the "Suzie," "American Football," and "Mall" sequences, the corresponding performance curves are not strictly comparable, but they give a coarse estimate of the expected bit rates in relative terms. As can be seen for the graphs in Figures 18.25 and 18.26,

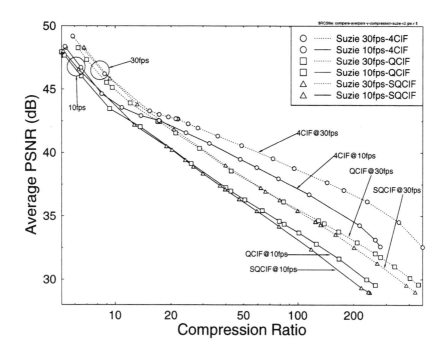

Figure 18.24: Image quality (PSNR) versus compression ratio for H.263 "Suzie" simulations at 10 and 30 frames/s.

Fixed PSNR (dB)	Percentage Bit-Rate Reduction (%)				
	SQCIF		QCIF	4CIF	
	10 fps	30 fps	30 fps	10 fps	30 fps
30	38.56	1.15	-70.55		
32	41.43	2.03	-71.86		
34	45.08	5.80	-78.44	-557	-1075
36	48.33	6.76	-89.38	-556	-969
38	50.74	4.36	-104	-585	-1170
40	52.94	0.86	-101	-720	-1558
42	49.46	-1.56	-118	-981	-2315
44	53.82	2.55	-116	-1310	-3055
46	52.87	-4.49	-136	-1385	-3381
47	53.14	-4.95	-135	-1361	-3448

Table 18.13: Percentage Reduction in Bit Rate Required to Achieve the Same PSNR, for the Various Resolution "Suzie" Video Sequences Coded with H.263 Codec Compared with the QCIF Resolution Simulation at 10 Frames/s

18.3. PERFORMANCE RESULTS

Fixed Bit Rate (Kbit/s)	PSNR Increase (dB)				
	SQCIF		QCIF	4CIF	
	10 fps	30 fps	30 fps	10 fps	30 fps
20	2.44	0.09			
40	2.72	0.26	-2.17		
70	3.12	0.25	-2.28		
100	3.17	0.17	-2.56		
200	3.86	0.02	-3.21	-7.93	
300		0.16	-3.07	-7.93	
500		-0.33	-4.34	-9.36	-11.33

Table 18.14: Increase in PSNR(dB) Achieved at the Same Bit Rate for the Various Resolution "Suzie" Video Sequences Coded with H.263 Codec Compared with the QCIF Resolution Simulation at 10 Frames/s

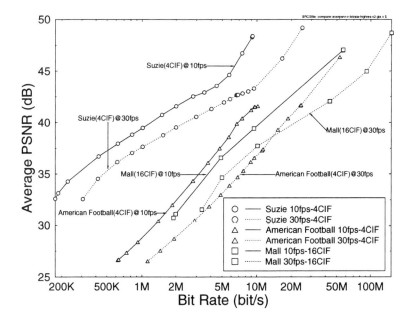

Figure 18.25: Image quality (PSNR) versus coded bit rate for H.263 high-resolution (4CIF and 16CIF) simulations at 10 and 30 frames/s.

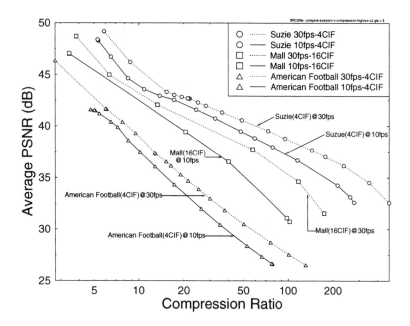

Figure 18.26: Image quality (PSNR) versus compression ratio for H.263 high-resolution (4CIF and 16CIF) simulations at 10 and 30 frames/s.

the 4CIF "American Football" sequence results in a similar performance to the 16CIF "Mall" sequence, primarily because of the large amount of motion in the "American Football" sequence. Before closing, let us finally summarize our experiences from this chapter.

18.4 Summary and Conclusions

In this chapter, we described the differences between the H.261 and H.263 video codecs, and a range of comparative performance curves were plotted. The differences between the two codecs in terms of source coding, multiplex coding, and motion compensation were described. From the range of algorithmic differences, motion compensation probably has the greatest effect. Multiplex coding also contributes to the substantial performance gain of the H.263 codec over the H.261 scheme. The H.263 codec's negotiable options were also described briefly, and more details can be found in [192]. A variety of comparative performance curves were plotted, demonstrating how the performance of the H.263 codec varied after some of its parameters were altered.

Chapter 19

A H.263 Videophone System for Use over Mobile Channels

19.1 Introduction

Having studied the differences and similarities between the H.261 and H.263 codecs, we designed a transceiver incorporating the ITU H.263 [192] video codec for visual communications in a mobile environment. We capitalized on our experience gained from our similar experiments using the H.261 [500] video codec. The software implementation of an H.263 codec was a modified version of the scheme developed by Telenor Research and Development [563]. The experimental conditions are summarized in Table 19.1. In this chapter, we review the H.263 codec, which was originally designed for benign Gaussian channels, and we design a system in order to facilitate its employment over hostile wireless channels.

19.2 H.263 in a Mobile Environment

19.2.1 Problems of Using H.263 in a Mobile Environment

The errors that occur in the H.263 bitstream can be divided into three classes: detected, detected-late, and undetected errors. A detected error is one that is recognized immediately, since the received bit was not expected for the current codeword. Errors that are detected late are caused by the error corrupting one codeword to another valid codeword. However, such errors are detected within the life span of a few codewords. Lastly, undetected errors also corrupt a valid codeword to another valid codeword, but no error is detected.

Detected errors result in picture quality degradation, but errors that are not detected or detected late cause the bit-stream to be decoded incorrectly. An incorrectly decoded bit-stream usually corrupts one or more macroblocks, and the error spreads to larger areas within a few frames. These types of errors can be corrected only by the retransmission of the affected macroblock in intra-frame coded mode.

Generally, when an error is detected, the H.263 decoder searches the incoming data stream for a picture start codeword. If this is found, the decoding process is resumed. This means that all blocks following an error in a video frame are lost. If the encoder is in inter-frame coded mode, the local decoder of the encoder and the remote decoder will become different, which we often refer to as being misaligned, causing further degradation until all the affected blocks are updated with intra-frame coded blocks. In order to recover from transmission errors, the updates using intra-frame coded blocks are very important. However, if intra-frame coded blocks are corrupted, the resulting degradation can become more perceptually objectionable than that which the intra-frame coded block was supposed to correct.

Another problem in using the H.263 codec in a mobile environment involves packet dropping. Packet dropping is common in hostile mobile scenarios, especially in conjunction with contention-based statistical multiple access schemes (PRMA). Dropping an H.263 packet would be equivalent to making a large number of bit errors. A feasible solution to this problem — especially in distributive video applications, where the latency is not a serious impediment — is using Automatic Repeat Request (ARQ) attempts. But to keep re-transmitting until a packet is received correctly is impractical, for it would require an excessive number of slots, inevitably disadvantaging other contending users.

19.2.2 Possible Solutions for Using H.263 in a Mobile Environment

The following problems have to be solved in order to use H.263 video codecs in a mobile environment:

- Reliably detecting errors in the bit-stream in order to prevent incorrect decoding.

- Increasing the relative frequency of instances, where the decoding can re-start after an error. This reduces the number of blocks that are lost following an error event in the bit-stream.

- Identifying ways of dropping segments of the bit-stream without causing video decoding errors.

- Ensuring that the encoder's local decoder and remote decoder remain synchronized, even when there are transmission errors.

Various solutions to one or more of these problems are described here. Most radio systems make use of forward error correction (FEC) codecs and/or error detecting cyclic redundancy check (CRC) sums on the transmitted packets.

Significant research efforts have been devoted to the area of robust video coding [194, 195, 464, 467, 550–554]. Some authors use proprietary video codecs [464, 467], while others employ the H.261 or other standard video codecs. A technique referred to as Error-Resilient Entropy Coding (EREC), developed by Cheng, Kingsbury, *et*

19.2. H.263 IN A MOBILE ENVIRONMENT

al. [462,552], was used by Redmill et al. [553] to design an H.261-based system. Matsumara et al. [554] used a similar technique to design a H.263-based system. EREC reduces the effects of errors but still requires low bit error rates in order to maintain perceptually unimpaired video quality. An EREC based technique is due to be used in a mobile version of the H.263 video coding standard, currently known as draft recommendation AV.26M.

19.2.2.1 Coding Video Sequences Using Exclusively Intra-Coded Frames

As mentioned, the H.263 standard relies on the intra-frame coded macroblocks in order to remove the effects of any errors. Therefore, increasing the number of intra-frame coded macroblocks in the coded bit-stream, enhances error recovery. Increasing the number of intra-frame coded macroblocks to the ultimate limit of having all frames coded as intra-frames has the advantage of resynchronizing the local decoder and the remote decoder at the start of every frame. Hence, any transmission errors can only persist for one frame duration.

The disadvantage of this method, however, is that the bit rate increases dramatically. Simulations have shown that in order to maintain a given image quality, the bit rate has to be increased by at least an order of magnitude in comparison to an inter-frame coded scenario. Since the bandwidth in mobile systems is at premium, the low bit-rate requirement imposed drastically reduces the image quality maintained.

19.2.2.2 Automatic Repeat Requests

As a more attractive means of improving robustness, Automatic Repeat Request (ARQ) [472], can be used to improve the BER of the channel at the expense of increased delay and reduced useful teletraffic capacity. However, if the channel becomes very hostile, then either the video data will never get through or the delay and teletraffic penalty will increase to a level where communications become impractical for an interactive videophone system. If ARQ was combined with a regime of allowing packets in the data stream to be dropped after a few retransmission attempts, then the maximum delay could be bounded. ARQ has been used in several video coding systems designed for mobile environments [467, 550].

19.2.2.3 Multimode Modulation Schemes

The employment of reconfigurable modulation schemes renders the system more flexible in terms of coping with time-variant channel conditions. For example, a scheme using 4QAM, 16QAM, and 64QAM [29] can operate at the same Baud rate, for 2, 4, and 6 bits/symbol, respectively. This allows a 1:2:3 ratio of available channel bit rates for the various schemes.

For a high-quality, high-capacity channel 64QAM can be used, supporting higher quality video communications due to the increased channel capacity. When the channel is poor, the 4QAM scheme can be invoked, constraining the image quality to be lower due to the lower channel capacity. However, the 4QAM scheme is more robust to channel errors and so guarantees a lower BER. Therefore, an intelligent system could switch the modulation scheme either under network control, or, for example, by exploiting the inherent channel quality information in Time Division Duplex (TDD) systems upon evaluating the received signal level (RSSI) and assuming reciprocity of the channel [110,134,560]. A more reliable alternative to the use of RSSI as a measure of channel quality is the received BER. This reconfiguration could often take place arbitrarily, even on a packet-by-packet basis, depending on the channel conditions. The image quality is dependent on the bit rate and therefore increases with better channel conditions.

The combination of ARQ with multimode modulation schemes can facilitate a rapid system response to changing channel conditions. Re-transmission attempts can be carried out using the more robust, lower capacity modulation schemes that have a higher chance of delivering an inherently lower quality but uncorrupted packet.

19.2.2.4 Combined Source/Channel Coding

It follows from our previous arguments that using ARQs or multiple modulation schemes on their own increases the range of operating channel SNRs, for which the H.263 bit-stream can be transmitted error-free. However, these measures do not improve the decoder ability is to cope with a corrupted bit-stream.

In order to reduce the possible video degradations due to channel errors, it is more beneficial and reliable to detect channel errors externally to the video codec rather than relying on the variable-length codeword decoder's ability in the video codec. This can prevent any potential video degradations inflicted by incorrect decoding of the data stream. After a transmission error has been detected, either the codec will attempt to resynchronize with the data stream by searching for a picture start code (PSC) or group of blocks start code (GBSC) in the data stream. The decoding process can then restart from the correctly identified PSC or GBSC. However, the data between the transmission error and resynchronization point is lost. This data loss causes video degradation because the encoder's local decoder and the remote decoder become misaligned. Hence, they operate on the basis of different previous reconstructed frame buffers. This degradation lasts until the local decoder and remote decoder can be resynchronized by a re-transmission of the affected macroblocks in intra-frame coded mode.

If the encoder's local decoder could be informed when a segment of the data stream is lost, it could compensate for this loss by keeping the local decoder and remote decoder in alignment, simply by freezing their contents. This implies that the image segment obliterated by the transmission error can be updated in future frames. A small short-term image quality degradation occurs because part of the image is not being updated. This is mitigated within a few frames, and it is small compared with the loss of decoder synchronization or incorrect decoding.

When using transmission packet acknowledgment feedback information, the encoder can be notified of loss of packets in the mobile network. By including this channel feedback in the encoder's coding control feedback loop, the contents of transmitted packets do not need to be stored. The close coupling of the source decoder and channel decoder is possible in a mobile videophone application. For the channel feedback to operate effectively, the round-trip delay must be short. For microcellular mobile systems, the round-trip delay is negligible, and the algorithmic and packetization delay is also typically low. This packet acknowledgment feedback information can therefore be provided by the receiver.

When the source encoder is informed that a packet has been lost, it can infer the effect of this loss on the decoder's reconstructed frame buffer and adjust the contents of its local decoder in order to match this and maintain their synchronous relationship. Consequently, the decoder will use the corresponding segment of the previous frame for the areas of the frame that are lost. The effect of this is imperceptible in most cases. The lost section of the frame will then be updated in future frames. Thus, provided that no further errors occur, within a few video frames, the effects of the transmission error would have receded.

19.3 Design of an Error-Resilient Reconfigurable Videophone System

19.3.1 Introduction

Having considered the behavior of the H.263 codec under erroneous conditions, let us now familiarize ourselves with the design principles of an error-resilient reconfigurable videophone transceiver. Our objective is to ensure better error resilience than that of the standard H.263 codec, which is achieved by adhering to the above suggested principles. Furthermore, the bit-stream generated is expected to conform to the H.263 standard or to lend itself to low-complexity transcoding. The system's multimode reconfigurable operation guarantees near-optimum performance under time-variant propagation and teletraffic conditions. The system designed exhibits the following features:

- Transmission packet acknowledgment feedback is used in order to inform the encoder, when packets have been lost, so that it can adjust the contents of its local reconstructed frame buffer and keep synchronized with that of the decoder. This feature allows the packets to be dropped without ARQ attempts, thereby reducing any potential ARQ delay.

- Each transmitted packet contains a resynchronization pointer that points to the end of the last macroblock within the packet. This "address" can be used to resynchronize the encoder and decoder if the next packet is corrupted and the remainder of the partial macroblock is lost.

- FECs are used to decrease the channel BER and to balance the effects of the QAM subchannels' differing BERs [29].

- The system is rendered reconfigurable in order to be able to encode the video sequence for achieving different bit rates. When the affordable bit rate is reduced, the packet size is also reduced so that it can be transmitted using a more robust modulation scheme at the same Baud rate.

- Any received packets containing transmission errors are dropped in order to prevent erroneous decoding, which can badly corrupt the received video sequence.

- The transmitted bit-stream conforms to the H.263 standard, and so there is no need for a transcoder.

19.3.2 Controling the Bit Rate

In this section, we highlight the concept of controling the bit rate. As with most video codecs using variable-length coding techniques, the bit rate of the H.263 codec is inherently time variant. However, most existing mobile radio systems transmit at a fixed bit rate. Our proposed multimode system maintains a constant signaling rate, leading to a different constant bit rate for each modulation scheme invoked.

A straightforward bit-rate control algorithm was used in order to maintain a fixed bit rate and fixed video frame rate. This algorithm modifies the quantizer in the video codec in order to maintain the target frame rate and bit rate. In these investigations, a fixed frame rate of 10 frames/s was employed. The target bit rate was set according to the past history of the modulation schemes used and to the packet dropping frequency experienced. The target bit rate was updated after every dropped packet or successful transmission according to Equation 19.1:

$$Target\ Bit\ Rate = \frac{S_4 B_4 + S_{16} B_{16} + S_{64} B_{64}}{S_4 + S_{16} + S_{64} + D_4 + D_{16} + D_{64}} \qquad (19.1)$$

where S_n is the number of successful packet transmissions, when using the n-QAM modulation scheme, D_n is the number of dropped packets in the n-QAM mode of operation, while B_n represents the various fixed bit rates for error-free transmission, again using the n-QAM modulation scheme.

The bit-rate control algorithm constitutes a simple way of controlling the target bit rate. The rate-control algorithm in the video codec modifies the video codecs' parameters to meet the target bit-rate requirement. This rate-control algorithm, combined with the equation to set the target bit rate, provided an adequate bit-rate control algorithm, which adapted to varying channel conditions.

If the videophone system is operating in only a 4QAM mode, the target bit rate is simplified, as shown in Equation 19.2:

$$Target\ Bit\ Rate = \frac{S_4 B_4}{S_4 + D_4}. \qquad (19.2)$$

If the channel is nearly error-free, then L_4 will be virtually zero and the target bit rate will tend to the fixed bit rate for the 4QAM transmission mode (B_4). If, however, the channel causes a 50% frame error rate, then the number of dropped transmissions (D_4) will be similar to the number of successful transmissions (S_4). This will cause the target bit rate to reduce to approximately 50% of the fixed bit rate for the 4QAM scheme (B_4). Therefore, the target bit rate reduces in proportion to the channel frame error rate. If the video codec can meet the target bit-rate requirements, the frame rate of the transmitted video should be nearly constant.

The videophone system can use modulation scheme switching or ARQ to improve the performance in the deep fades that occur in Rayleigh fading. When switching or ARQ is used, the target bit rate is proportional not only to the frame error rate, but also to the switching or ARQ changes. The fixed bit rate for each modulation

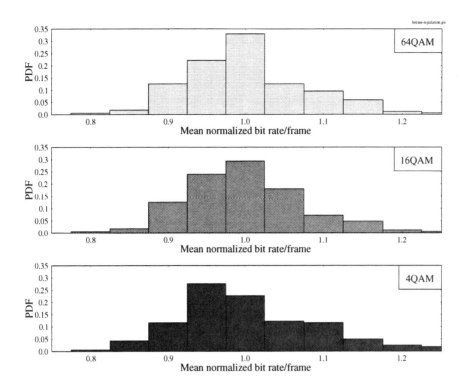

Figure 19.1: Bit-rate control histogram for transmissions at 4-, 16-, and 64QAM using the QCIF resolution "Miss America" Sequence, in error-free channel conditions.

scheme is weighted by the number of successful transmissions using that particular modulation mode. This system ensures that the target bit rate adjusts, so that the frame rate of the transmitted video remains as constant as possible.

The operation of this bit-rate control algorithm is characterized by the normalized bit-rate histograms of Figure 19.1 in the transceiver's different modes of operation using 64QAM, 16QAM, and 4QAM, respectively. As seen in this figure, in virtually all scenarios the instantaneous bit rate was within ±20% of the target mean bit rate, maintaining the target frame rate within this range. The corresponding frame rate versus time behavior of the system is characterised in Figure 19.2. Observe that after a slight initial delay, which is essentially due to the intra-frame coded bit-rate surge of the first frame, the frame rate reaches the target of 10 frames/s, typically within less than 1s.

Much research has been conducted in the area of rate control for video codecs. Ding and Liu [564] gave a good introduction to rate-control algorithms and their application to the MPEG video codec. Schuster and Katsaggelos [565] introduced an optimal algorithm and showed its application to H.263. Martins, Ding, and Feig [566] introduced a variable bit rate, variable-frame-rate-control algorithm, and applied it to an H.263 video codec. The variable bit rate/frame rate is particularly suitable for very

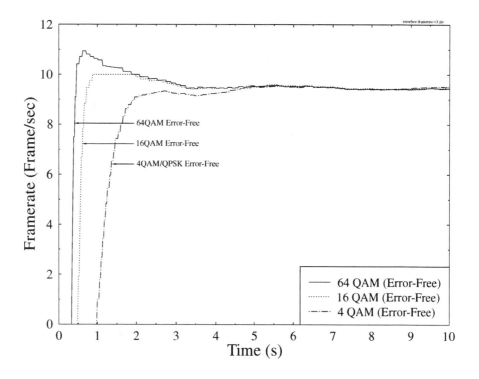

Figure 19.2: Framerate versus time behavior of the frame-rate control algorithm, for transmissions at 4, 16, and 64QAM using QCIF resolution Miss America Sequence, in error-free channel conditions.

low bit rates. A rate-control algorithm designed specifically for H.263 was suggested by Wiegand et al. [567]. It optimized the codec parameters for each group of blocks (GOB) and found PSNR improvements of about 1 dB over more basic rate-control algorithms.

19.3.3 Employing FEC Codes in the Videophone System

Since forward error correction codes can reduce the number of transmission errors at the expense of an increased Baud rate, they also expand the useful channel SNR range, for which transmissions are error-free. As we have stated before, in Quadrature Amplitude Modulation (QAM) schemes, certain bits of a symbol have a higher probability of errors than others [557]. The bits of a symbol can be grouped into different integrity classes according to their probability of error, which are also often referred to as QAM subchannels. By using different-strength FEC codes on each QAM subchannel, it is possible to equalize their probability of errors, implying that all subchannels' FEC codes become overloaded by transmission errors at approximately the same channel SNR. This is desirable if all bits to be transmitted are equally important. Since the H.263 data stream is mainly variable length coded, a single error

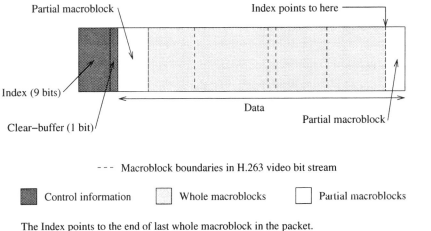

Figure 19.3: Structure of a transmission packet generated by the modified H.263 video codec.

can cause a loss of synchronization. Therefore, in this case most bits are equally important, and so the equalization of the QAM subchannels' BER is desirable for the H.263 data stream. Pelz [464] used a proprietary video codec and exploited the QAM subchannels to help classify the bit-stream into classes.

19.3.4 Transmission Packet Structure

Our error-resilient H.263 codec generates video packets and conveys them to the error correction encoders. The structure of the packets generated by the H.263 video codec is shown in Figure 19.3.

As seen in the figure, the packets have two main constituent segments: the data and control information. The control information consists of 10 bits. As seen in Figure 19.3, 9 bits of the 10-bit control information are allocated to an index, which points to the end of the last whole macroblock in the packet. This packet may contain the bits representing a number of MBs plus a fraction of the most recently encoded MB, which was allocated to the current transmission packet in order not to waste channel capacity. The remainder of the partially transmitted MB is then transmitted at the beginning of the next transmission packet, as indicated in Figure 19.3. This pointer is used to ensure that the decoder only decodes whole macroblocks. The partial macroblock after the indexed point is buffered at the decoder, until the remainder of the partial macroblock is received. The control information segment also contains a 1 bit flag, which is used to inform the decoder if the decoder's received signal buffer containing an already error-freely received partial macroblock has to be cleared before appending the current packet. The encoder sets this flag in order to inform the decoder to drop the already error-freely received partial macroblock from its received signal buffer, when the remainder of the corresponding MB's information was lost due

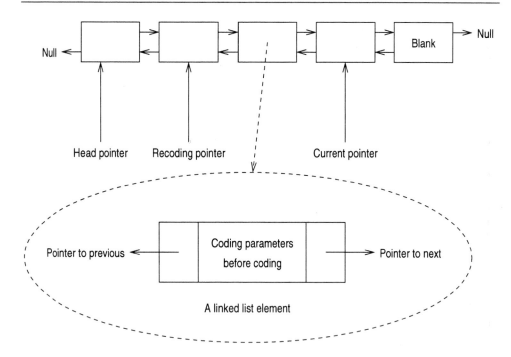

Figure 19.4: Structure of bidirectionally linked list for storing history of coding parameters.

to packet dropping. Explicitly, by clearing the received signal buffer, this mechanism ensures that the first part of the error-freely received partial macroblock stored in the receiver's decoding buffer is removed, if the rest of the packet was corrupted.

19.3.5 Coding Parameter History List

In order to enable the re-encoding of the macroblocks for re-transmission or dropping, the history of various coding parameters has to be stored for every macroblock in the encoder's buffer. This is necessary, because the encoder has to remember, for example the unquantized DCT coefficients, the MB position, the current MB coding mode, and so on, until the error-free arrival of this specific MB's encoded information at the decoder. If the information is corrupted, for example, during its first transmission attempt, it may have to be re-encoded using more coarse quantizers, which would not be possible without this coding parameter history list. The history of coding parameters was implemented as a bidirectionally linked list with a structure as shown in Figure 19.4. Each element of the list represents a macroblock and contains the coding parameters of the codec before the macroblock was coded. Initially, then, a brief general description is given, which will depend substantially on the specific application scenarios to be considered in Section 19.3.6.1.

Each new macroblock is added to the tail of the list, then encoded, and the coded bit-stream is appended to the transmission buffer. The packetization algorithm has three pointers to the linked list. *Head* points to the head of the list, *Recode* points to the first macroblock to be re-encoded for re-transmission, and *Current* points to the next element/macroblock to encode. Again, the role of these pointers will become clearer with reference to our example considered in Section 19.3.6.1. Here, only a few general comments concerning their employment are offered. For example, the *Recode* pointer points to the same macroblock as the *Head* pointer, when the decoder has no partial macroblocks in its buffer. When the decoder has a partial macroblock in its buffer, the *Recode* pointer points to the second macroblock in the list. This is because the partial transmitted macroblock, which is at the head of the list, can only be dropped or the remainder of the macroblock transmitted. When a new macroblock is added to the list, it is placed in the blank list element at the tail of the list and a new blank list element is created.

When the transmission buffer becomes filled above the defined limit corresponding to a transmission packet, its contents up to the limit is transmitted. If the transmission is successful, then the linked list elements for the fully transmitted macroblocks are removed from the head of the list. The partially transmitted macroblock then becomes the head of the linked list. The *Recode* and *Current* pointers point to the blank list element at the end of the list.

If the transmission fails, but a re-transmission is requested, then the codec is instructed to reset its state to what it was, before the macroblock pointed to by the *Recode* pointer was encoded. The remaining macroblocks are then re-encoded until the buffer becomes filled up to the new transmission packet size. This packet is then transmitted in the usual way.

If the transmission failed and re-transmission was not requested, then the codec is reset to the state before the macroblock pointed to by the *Head* pointer was encoded. The macroblocks in the history list are then re-encoded, as if they were empty. When this operation is complete, the encoding of new macroblocks continues as usual. However, when the next packet is transmitted, the clear buffer flag for the packet is set in order to clear the receiver's decoding buffer from the partially transmitted macroblock that may be in it.

The coding parameters saved in the history list are as follows:

- The unquantized DCT coefficients.

- Macroblock index.

- Buffer size before encoding the macroblock.

- Various quantizer identifiers used by the bit-rate control algorithm.

- Current macroblock coding mode.

Now that the coding parameters' history list and the transmission packet structure have been discussed, the following section describes the packetization algorithm.

19.3.6 The Packetization Algorithm

The objectives of the packetization algorithm are to:

- Pack the H.263 bit-stream into the data portion of the transmission packet, while setting the control information part of the packet so that the decoder can recover from packet losses.

- Exploit the feedback information in order to adjust the codec's bit rate to adapt to time-variant channel conditions.

- Generate variable-size packets, so that ARQ relying on different modulation schemes can be easily implemented.

- Ensure that the encoder's local decoder and the decoder are kept in synchronization after a packet is dropped.

The H.263 codec produces a bit-stream, maintaining a fixed frame rate for the target bit rate, which is adjusted depending on the channel conditions. The packetizing algorithm operates at the macroblock layer. An example of operating scenarios for the packetization algorithm is given in the next section.

19.3.6.1 Operational Scenarios of the Packetizing Algorithm

In order to augment our discussions above, in this section we consider a few operational scenarios of the packetization algorithm, which are shown in Figure 19.5. At the commencement of communications, both the transmission buffer and the history list are empty, as suggested by the figure. When the video encoder starts generating macroblock bits, the transmission buffer is filled by MB1-MB3, where MB3 is seen to "overfill" the buffer. The status of the pointers is also reflected in the second line of the figure. Assuming that this packet was transmitted successfully, the "overspilt" segment P of MB3 is transmitted at the beginning of the next packet, as displayed in the third line of Figure 19.5. The error-freely received and complete MB1 and MB2 can then be decoded, and their parameters can be removed from the history list. MB3 is now pointed to by the *Head* pointer, which implies that it constitutes the *Head* of the coding parameter history list. As mentioned earlier, the pointer *Recode* holds the position of the first MB to be re-encoded for a potential re-transmission. Lastly, the pointer *Current* identifies the MB to be encoded, which is in this case MB5. At this stage the second packet is filled up and ready for transmission, but MB5 "overfills" the packet, and hence the remaining segment P is assigned to the forthcoming packet, as seen in the fourth line of Figure 19.5.

Let us now consider an erroneous transmission scenario in which ARQ is invoked. Assume furthermore that the packet has to be re-transmitted using a more robust but lower transmission capacity modulation scheme, imposing a smaller packet size as shown in Figure 19.5. Since a substantial part of MB3 has already been received, error-free, MB3 is not re-encoded, but the length of MB4 and MB5 must be reduced. The position of the pointers is unaltered, as suggested by Figure 19.5. Hence the MBs starting at the position indicated by the pointer *Recode* are re-encoded using more coarse quantizers, which generate the required target bit rate. This operation is concerned with MB4 and MB5. Accordingly, the coding parameter history list is reset to its state, before MB4 was originally encoded, since the same unquantized DCT coefficients, MB index, MB coding mode, and so on, must be used, as during the

19.3. ERROR-RESILIENT VIDEOPHONE DESIGN

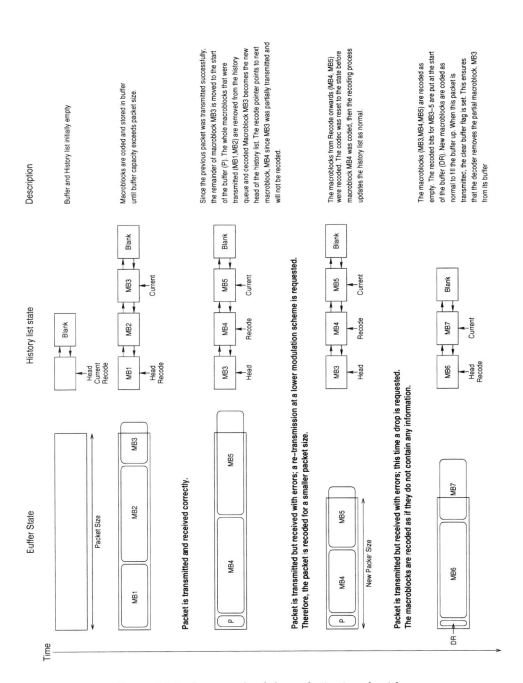

Figure 19.5: An example of the packetization algorithm.

first encoding operation. The re-encoded MB4 and MB5 now "overfill" the shortened transmission packet constrained by the lower-order modulation scheme.

Let us assume that this shortened packet transmitted using the most robust modulation scheme available is also corrupted by the channel, and the maximum number of re-transmission attempts has expired. In this case, the packet is dropped in order to prevent error propagation in the local reconstructed frame buffer. In the H.261 scheme, macroblocks could simply be dropped, since each macroblock contained an address. However, in the H.263 scheme, for every macroblock some information has to be transmitted, where inactive macroblocks are coded using a 1 bit codeword. Due to the packet drop request, the MBs are re-encoded as empty MBs conveying no information, requiring a 1 bit flag for each dropped macroblock. These bits are represented by the segment DR at the bottom of Figure 19.5. Then the forthcoming MBs, MB6 and MB7, are encoded as usual and assigned to the next transmission packet. When this packet is transmitted, the clear buffer flag is set in order to inform the decoder that the partially received MB3 has to be removed from the decoder's buffer.

This example demonstrates all possible transmission scenarios. It is worth noting that dropping macroblocks as described above, has the effect of some parts of the picture not being updated. However, in most cases the dropped macroblocks will be updated in the next frame. The effect of dropping is not noticeable, unless the frame contains a large amount of motion and the packet dropping probability is very high. These effects can be more easily appreciated by viewing the decoded video, which is available on the WWW.[1] Packet-loss concealment techniques were investigated by Ghanbari and Seferidis [561] using a H.261 codec.

Having described the video codec and the packetization algorithm, in the forthcoming section we concentrate our attention on the overall system's performance.

19.4 H.263-Based Video System Performance

19.4.1 System Environment

The performance of the error-resilient H.263-based reconfigurable videophone system was evaluated in a mobile radio environment. These investigations identified the performance limits of the videophone system. Our experiments were carried out over Gaussian and Rayleigh-fading channels, using reconfigurable modulation schemes and ARQ. The simulation conditions were as follows: a propagation frequency of 1.8 GHz, vehicular speed of 13.4 m/s (or 30 mph), and a signaling rate of 94.72 kBaud. The corresponding normalized Doppler frequency was 6×10^{-4}.

Our reconfigurable modulation scheme was able to switch between 64-, 16-, and 4-level coherent, pilot-assisted QAM modulation [29], supported by a range of binary Bose-Chaudhuri-Hocquenghem (BCH) forward error correcting (FEC) codes [35], in order to increase the system's robustness. The number of transmitted bits/symbol and the associated number of subchannels are shown in Table 19.1 for each modulation scheme.

---

[1]http://www-mobile.ecs.soton.ac.uk/peter/robust-h263/robust.html#DROPPING

19.4. H.263-BASED VIDEO SYSTEM PERFORMANCE

Modem-mode Specific System Parameters			
Features	H.263 Multirate System		
Modem	4-QAM	16-QAM	64-QAM
Bits/symbol	2	4	6
Number of subchannels	1	2	3
C1 FEC	BCH(255,147,14)	BCH(255,179,10)	BCH(255,199,7)
C2 FEC	N/A	BCH(255,115,21)	BCH(255,155,13)
C3 FEC	N/A	N/A	BCH(255,91,25)
Bits / TDMA frame	147	294	445
Transmission bit rate (kbit/s)	11.76	23.52	35.6
Packetization header (bits)	10	10	10
Video bits / TDMA frame	137	284	435
Useful video bit rate (kbit/s)	10.96	22.72	34.8
Min, PSNR threshold (dB)	-1	-1	-1
Min. AWGN SNR (dB)	7.0	12.7	18.6
Min. AWGN BER (%)	4.0	2.5	3.2
Min. Rayleigh SNR (dB)	11.5	17.3	23.2
Min. Rayleigh BER (%)	4.4	3.3	3.4

Modem-mode Independent System Parameters	
Features	General System
Pilot-Assisted Modulation	yes
Video Codec	H.263
Frame rate (Fr/s)	10
Video resolution	QCIF
Color	Yes
User data symbols / TDMA frame	128
User pilot symbols / TDMA frame	14
User ramping symbols / TDMA frame	4
User padding symbols / TDMA frame	2
User symbols / TDMA frame	148
TDMA frame length (ms)	12.5
User symbol rate (kBd)	11.84
Slots/frame	8
No. of users	8
System symbol rate (kBd)	94.72
System bandwidth (kHz)	200
Effective user bandwidth (kHz)	25
Vehicular speed (m/s)	13.4
Propagation frequency (GHz)	1.8
Normalized doppler frequency	6.2696×10^{-4}
Path-loss model	Power law exponent 3.5

Table 19.1: Summary of System Features for H.263 Wireless Reconfigurable Videophone System. Minimum required SNR and BER derived from Figure 19.9.

All our performance investigations were carried out for a single portable station (PS) and base station (BS) and for a pilot symbol separation of ten symbols. In order to be able to use our reconfigurable modulation schemes with ARQ assistance, all packets must have the same number of symbols for each modulation scheme. This is achieved by using a higher number of bits per packet for 64QAM and a lower number for 4QAM, where the ratio of the number of bits is determined by the number of bits/symbol for each modulation scheme.

Suitable values for the bit rates of the three modulation schemes were set. Specifically, the packet size, expressed in terms of the number of bits for each modulation scheme, was found by initially setting the packet size for 4QAM. Then the packet sizes for 16- and 64QAM expressed in terms of bits were set to keep the same number of symbols per packet, resulting in a bit rate ratio of 1:2:3 for the 4-, 16-, and 64-QAM arrangements.

The previously mentioned modulation subchannels in 16- and 64QAM exhibit different BER performances. This is usually exploited to provide more sensitive bits with increased protection against channel errors. Since all the video bits in the proposed H.263-based videophone system are sensitive to transmission errors, no source sensitivity matched FEC coding is employed. Hence, the subchannel error rates have to be equalized, which can be achieved using stronger FEC codes in the weaker subchannels. The FECs were chosen to equalize the different BERs of the specific modulation subchannels as closely as possible. The specific FEC codes chosen for the various modulation schemes are summarized in Table 19.1. The table also portrays the corresponding packet sizes derived on the basis of the FEC codes for each modulation scheme in the H.263-based system. All the simulation parameters and results are also summarized in Table 19.1, which we will frequently refer to throughout this chapter.

19.4.2 Performance Results

19.4.2.1 Error-Free Transmission Results

Having described the system environment, let us now focus our attention on the achievable system performance. Our ultimate aim was to produce graphs of video PSNR versus channel SNR. Our investigations were carried out both with and without ARQ in order to assess to what extent ARQ can improve system performance. The "Miss America" video sequence was H.263 encoded, packetized and transmitted over Gaussian and Rayleigh channels with and without ARQ assistance. Our ARQ-assisted investigations are described by the ARQ-regime A-B-C, where A characterizes the initial transmission, B the first retransmission attempt, and C the second. Our experiments entailed the following transmissions:

- 4QAM
- 16QAM
- 64QAM
- ARQ: 4-4-4QAM

19.4. H.263-BASED VIDEO SYSTEM PERFORMANCE

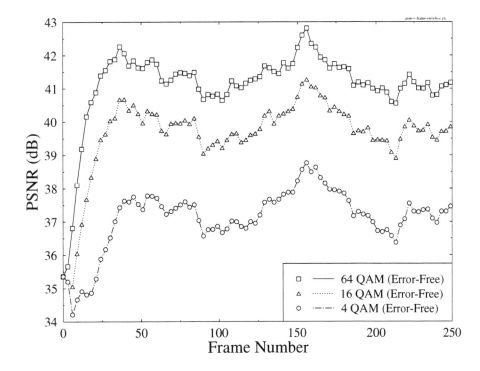

Figure 19.6: PSNR versus frame index comparison between modulation schemes over error-free Gaussian channels, using the system of Table 19.1.

In order to establish the potential benefits of reconfigurable transmissions, Figure 19.6 characterizes the decoded video quality in terms of PSNR versus frame index for the three modulation schemes invoked under error-free channel conditions. Observe that the image quality improves with the higher bit rates achieved by the help of the higher channel capacity modulation schemes. This graph also shows how the image quality improves following the initial intra-coded frame, which requires a substantially higher number of bits than inter-coded frames. However, at the commencement of transmission, the intra-coded frame is encoded using a rather coarse quantizer at a comparatively low quality in order to reduce the associated bit-rate peak. This bit-rate peak is mitigated by the bit-rate control and packetization algorithm at the cost of a concomitantly reduced image quality. An approximately 3 dB PSNR advantage is observed, when using 16QAM instead of 4QAM, which is further increased by another 1.5 dB, when opting for 64QAM facilitated by the prevailing friendly channel conditions.

19.4.2.2 Effect of Packet Dropping on Image Quality

Let us now consider the more realistic scenario of imperfect channel conditions. Figure 19.7 characterizes the effect of video packet dropping on image quality as a func-

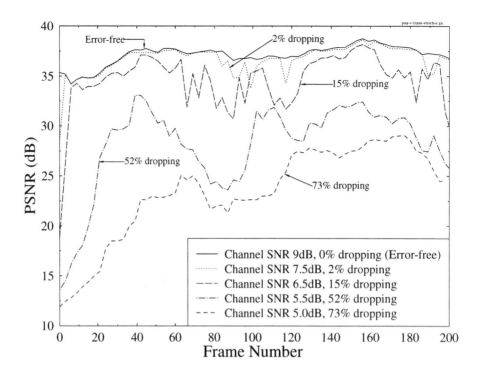

Figure 19.7: PSNR versus frame index comparison for various packet dropping rates for 4QAM over Gaussian channels, using the system of Table 19.1.

tion of the video frame index for a range of packet dropping probabilities between 0 and 73%. When a packet is dropped, it implies that the macroblocks contained in that packet will not be updated in the encoder's and the decoder's reconstruction buffer. These MBs will be updated in the next frame as long as the corresponding packet is not dropped. In the event of excessive packet dropping, some macroblocks may not be updated for several frames. When a packet containing the initial intra-coded frame is lost, its effect on the PSNR is more obvious. This is because the decoder cannot use the previous frame for dropped macroblocks, since that is typically an inter-frame coded one. After the initial few frames, every macroblock has been transmitted at least once. The PSNR then improves to around 35 dB. For 2% packet dropping, where 2 in every 100 packets are lost, the PSNR is almost identical to the error-free case. The higher the dropping rate, the longer it takes to build the PSNR up to the normal operating range of around 35 dB. However, even with 52% packet dropping, where only about half the packets are received, the PSNR averages around 28 dB. When the video contains a large amount of motion, the packet losses make the macroblocks that were not updated more perceptually obvious, which ultimately results in a lower PSNR. Toward high dropping rates, the delayed updating of the macroblocks became more objectionable, but the effect was not catastrophic. The

19.4. H.263-BASED VIDEO SYSTEM PERFORMANCE

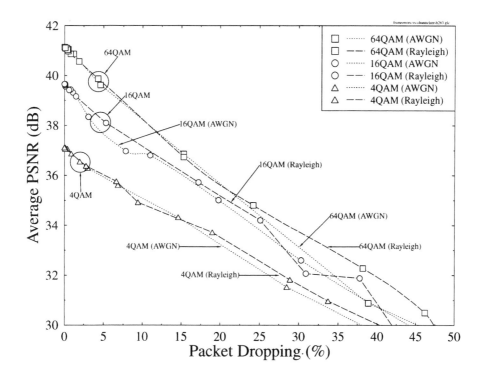

Figure 19.8: Average PSNR versus packet dropping rate for 4-, 16-, and 64QAM over Gaussian and Rayleigh channels, using the system of Table 19.1.

decoded video sequences with packet dropping can be viewed on the WWW,[2] to help judge the subjective quality loss caused by packet dropping.

Figure 19.8 shows the average PSNR of the decoded video versus the packet dropping rate for 4-, 16-, and 64QAM over Gaussian and Rayleigh-fading channels. The PSNR degradation is approximately linear as a function of the increase in packet dropping probability.

19.4.2.3 Image Quality versus Channel Quality without ARQ

In order to show the effect of varying channel quality on the videophone system, the average PSNR versus channel SNR performance of the system was evaluated in Figure 19.9 for all three modulation modes over both Gaussian and Rayleigh channels.

This graph shows the sharp reduction of PSNR as the FEC schemes become overloaded by channel errors in the low SNR region. This effect is especially pronounced for Gaussian channels, regardless of the modulation mode used, since the channel errors are more uniformly distributed over time than in the case of Rayleigh-channels. Accordingly, this breakdown is less sharp for Rayleigh fading channels because of the

---
[2]http://www-mobile.ecs.soton.ac.uk/peter/robust-h263/robust.html#DROPPING

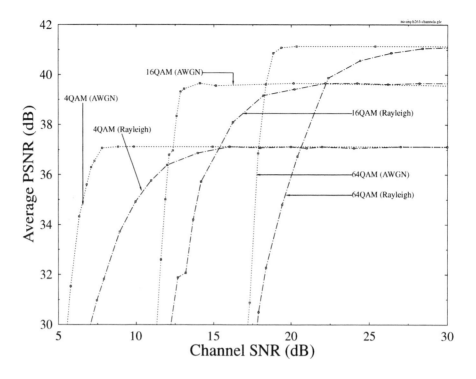

Figure 19.9: PSNR versus channel SNR for 4, 16, and 64QAM over both Gaussian and Rayleigh-fading channels, using the system of Table 19.1.

bursty nature of the errors, overwhelming the FEC scheme only occasionally. This graph can be used to estimate the conditions for which a switch to a more robust modulation scheme becomes effective. These effects are shown from a different perspective, in terms of frame error rate versus channel SNR, in Figure 19.10.

19.4.2.4 Image Quality versus Channel Quality with ARQ

When using ARQ-assisted scenarios, the number of frame errors can be reduced, thereby improving the decoded image quality. The improvement of the frame error rate versus channel SNR performance due to ARQ is shown in Figure 19.11. It can be seen that the use of ARQ extends the range of channel SNRs, for which the system can be used, while maintaining a reasonable image quality. Observe in the figure that ARQ gives a greater improvement in Rayleigh channels than over Gaussian channels. This is because over Gaussian channels the packet typically faces similar channel conditions during every transmission attempt, whereas over Rayleigh channels the receiver may have emerged from a fade, by the time re-transmission takes place. This provides better reception chances for re-transmitted packets over fading channels.

The same tendencies are reinforced in Figure 19.12 in terms of average PSNR versus channel SNR with and without ARQ.

19.4. H.263-BASED VIDEO SYSTEM PERFORMANCE

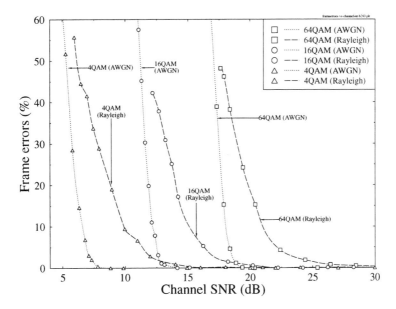

Figure 19.10: Frame errors versus channel SNR for 4-, 16-, and 64QAM over both Gaussian and Rayleigh fading channels, using the system of Table 19.1.

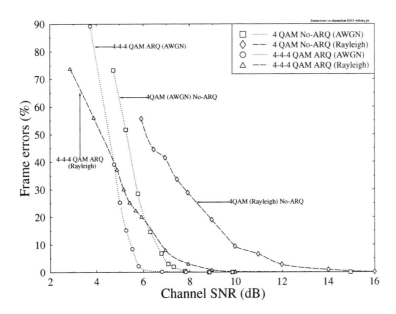

Figure 19.11: Frame errors versus channel SNR for 4QAM over both Gaussian and Rayleigh fading channels, with and without ARQ. The 4-4-4ARQ regime attempts up to three transmissions at 4QAM, using the system of Table 19.1.

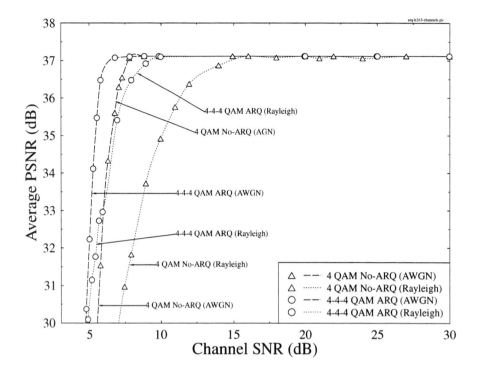

Figure 19.12: PSNR versus channel SNR for 4QAM over both Gaussian and Rayleigh fading channels, with and without ARQ. The 4-4-4ARQ regime invokes up to three transmissions at 4QAM, using the system of Table 19.1.

19.4.3 Comparison of H.263 and H.261-Based Systems

In this section, we compare the proposed H.263-based videophone transceiver's performance to that of a similar H.261-based scheme. The previously described H.261 system of Section 17.4 was slightly less advanced and used the less efficient H.261 video coding standard. Figure 19.13 portrays the PSNR versus channel SNR performance of these schemes over Gaussian channels, and Figure 19.14 shows the corresponding results over Rayleigh-fading channels.

Observe in the figures that the H.263-based video system substantially outperforms the H.261 scheme in terms of decoded image quality. This is because of the improvements in video compression achieved by H.263 codec and the more efficient transmission of packets in the H.263 system. The PSNR advantage of the H.263 scheme is between 3 and 5 dB for the 4-, 16-, and 64QAM modulation modes.

Note, however, that at very low channel SNRs, the H.261 system has a slightly higher image quality. This is because H.263 makes more intensive use of motion vectors in order to increase compression. However, the employment of motion vectors is a problem at very high packet dropping rates because the image quality degrades more rapidly. Similar trends are also observed for Rayleigh-fading channels in Figure 19.14,

19.4. H.263-BASED VIDEO SYSTEM PERFORMANCE

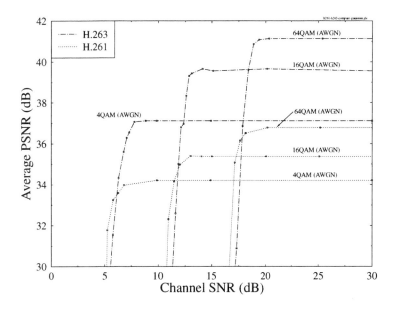

Figure 19.13: A comparison of mobile videophone systems based on the H.261 and H.263 video coding standards, over Gaussian radio channels, using the system of Table 19.1.

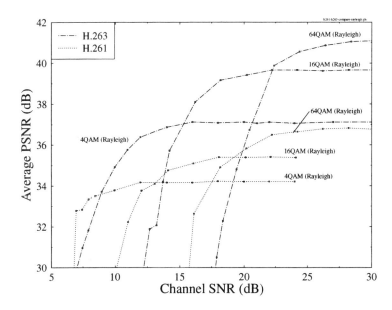

Figure 19.14: A comparison of mobile videophone systems based on the H.261 and H.263 video coding standards over Rayleigh-fading radio channels, using the system of Table 19.1.

although the video degradations become more apparent at higher channel SNRs than in the case of Gaussian channels. These degradations are more gracefully decaying, however.

19.4.3.1 Performance with Antenna Diversity

A common technique used to improve reception in mobile radio systems is to employ antenna diversity. The technique is conceptually simple. The radio receiver has two or more aerials, which are separated by some spatial distance. Provided that the distance between the aerials is sufficiently high in order to ensure that their received signals are decorrelated, if the signal received by one aerial is deeply faded, the signal from the other aerials is likely not to be The disadvantage of this technique is the expense and inconvenience of two aerials, in addition to the cost of extra circuitry. The receiver demodulates the signal from that antenna, which is deemed to result in the best reception. Several criteria [568] can be used to decide which antenna signal to demodulate; a few are as follows:

- Maximum data symbol power
- Maximum pilot symbol power
- Minimum phase shift between pilots

Alternatively, the received signals of all the antennas can be coherently combined for best performance. The best but most complex technique to optimally combine the received signals of the antennas is to employ maximum ratio combining [569].

The most common diversity technique is switched diversity, in which the antenna whose signal has the highest received signal power (maximum symbol power) is chosen by the receiver in order to demodulate the received signal. If pilot symbol-assisted modulation (PSAM) is used, then a better choice is the maximum pilot power, since the received pilot's signal strength can be used to estimate the current fading of the channel. Therefore, by demodulating the signal from the antenna with the highest received pilot symbol power, the least faded channel is chosen. An alternative antenna selection criterion in conjunction with PSAM is to choose the antenna that has the least channel-induced phase shift between consecutive pilots.

When using switched antenna diversity, the bit and frame error rates are reduced for the same SNR compared to the no-diversity scenario. Alternatively, the SNR required to maintain the target bit or frame error rate is lower, when using antenna diversity. The SNR gain becomes higher for marginally impaired channels, but the improvement is negligible for poor channels, where the frame error rate becomes very high. Figure 19.15 shows the frame error rate for all the modulation schemes of the multirate system, with and without diversity. The graph demonstrates that at any given SNR the frame error rate is lower with diversity. Furthermore, it can be seen that to maintain a desired frame error rate performance, the SNR can be consistently reduced when diversity is used. However, it can be observed that this SNR improvement reduces at high frame error rates. The SNR improvement margin for a 5% frame error rate is shown in Table 19.2 for various modulation modes.

19.4. H.263-BASED VIDEO SYSTEM PERFORMANCE

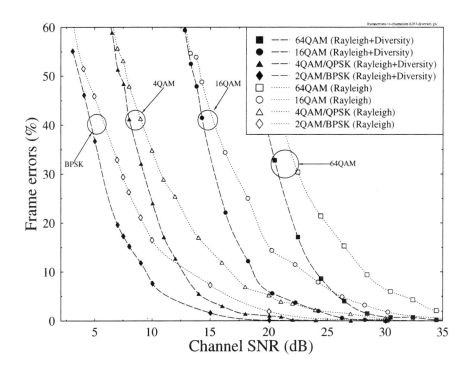

Figure 19.15: Frame errors versus channel SNR for 64QAM, 16QAM, 4QAM, and BPSK over Rayleigh-fading channels with and without dual antenna diversity, using the system of Table 19.1.

Mode	Required SNR without Diversity (dB)	Required SNR with Diversity (dB)	SNR Margin Improvement (dB)
BPSK	17.1	12.2	4.9
4QAM/QPSK	20.2	14.4	5.8
16QAM	26.2	20.9	5.3
64QAM	31.6	26.0	5.6

Table 19.2: Channel SNR Required for 5% Frame Error Rate with and without Dual Antenna Diversity, and SNR Margin Improvement with Diversity, for BPSK, 4QAM, 16QAM and 64QAM over Rayleigh Channels.

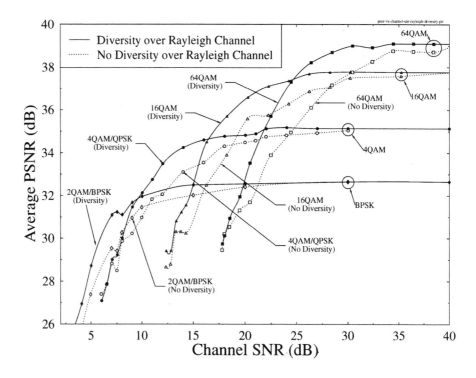

Figure 19.16: PSNR versus channel SNR for 64QAM, 16QAM, 4QAM, and BPSK over Rayleigh-fading channels, with and without dual antenna diversity, when using the system of Table 19.1.

The corresponding graph of the decoded video PSNR versus channel SNR is shown in Figure 19.16. The graphs demonstrate the improvement in video quality in terms of PSNR, when using dual antenna diversity.

We have shown results using the mobile radio systems whose parameters were summarized in Table 19.1 on page 791. In order to demonstrate the wide applicability of our error-resilient video system, we also investigated the performance of our scheme in the context of the Pan-European DECT cordless telephone system, an issue discussed in the next section.

19.4.3.2 Performance over DECT Channels

The Digital European Cordless Telecommunications (DECT) system [262,570] was defined as a digital replacement for analog cordless telephones, which are commonly used in indoor residential and office scenarios, although its employment today is beyond these fields. The DECT system is TDMA/TDD-based with 12 channels per carrier, operating in the 1880–1900 MHz band. Systems based on Time Division Multiple Access (TDMA) multiplex users onto the same carrier frequency by dividing time into slots and allocating a slot every TDMA frame to each user. In the case of DECT,

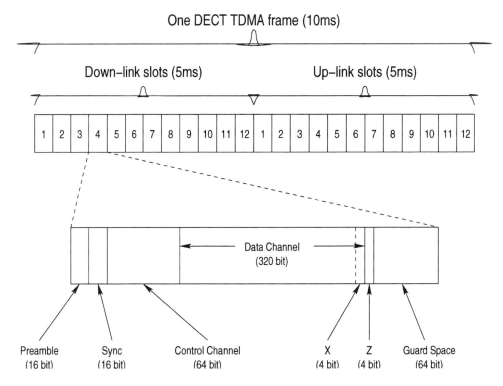

Figure 19.17: The Digital European Cordless Telecommunications (DECT) system's TDMA/TDD frame structure [262].

there are 12 time slots for 12 users on each carrier frequency. Time Division Duplex (TDD) systems use the same carrier frequency for up-link and down-link, by dividing frames into up-link and down-link portions. These up-link and down-link sections of the TDD frame are usually further divided by using TDMA. The DECT standard employs Adaptive Differential Pulse Coded Modulation (ADPCM) at 32 Kbps for encoding the voice signal. Since the system was defined for indoor microcellular environments, it does not use channel coding or equalization. The TDMA/TDD frame structure is shown in Figure 19.17. The role of the control fields in Figure 19.17 becomes clear with reference to [45,262]. As can be seen in the figure, the data channel's "payload" is 320 bits. Therefore the up-link/down-link data capacity is 320 bits every 10 ms, corresponding to a bit rate of 32 Kbps.

Using simulated DECT channel error profiles related to the scenarios described in [570], we investigated our H.263-based videophone system over these DECT channels. The corresponding DECT channel parameters are summarized in Table 19.3. The DECT system uses Gaussian Minimum Shift Keying (GMSK) modulation, as in the GSM system [41]. We simulated the videophone system with and without channel coding over these DECT channels, and the simulation parameters are shown in Table 19.4. For the uncoded simulations, the video bit rate is 32 Kbps. However, only a single-bit error is required to cause the loss of a video packet. For the FEC-coded

Feature	DECT Channel 1	DECT Channel 2
Modulation	GMSK	GMSK
Carrier frequency	1.9GHz	1.9GHz
Wavelength	15.8cm	15.8cm
Doppler frequency	2.5Hz	2.5Hz
Vehicular speed	0.39m/s	0.39m/s
SNR	20dB	30dB
BER	2.11×10^{-2}	2.11×10^{-3}

Table 19.3: DECT Channel Parameters, Channels Derived from [570]

Feature	DECT-based System Parameters	
	Uncoded	Coded
Channel data rate	32 Kbps	32 Kbps
C1 FEC	n/a	BCH(255,171,11)
C2 FEC	n/a	BCH(63,45,3)
Net video bits/TDMA-frame	320	216
Net video rate	32 Kbps	21.6 Kbps

Table 19.4: DECT-Like System Parameters

simulations, the BCH codes were chosen to match the amount of error correction used in previous simulations. This resulted in a reduced video bit rate of 21.6 Kbps for the FEC-coded simulations in order to arrive at a FEC-coded rate of 32 Kbps.

Using the DECT Channels 1 and 2 of Table 19.3 and the artificial error-free channel, the 32 Kbps videophone system was simulated with and without channel coding, assuming that video users are assigned an additional time slot per frame. An attractive feature of the DECT system is that up to 23 time slots can be allocated to a single user. This could support video rates up to 22×32 Kbps = 704 Kbps, while reserving one 32 Kbps time slot for the speech information. In practice because of the system's internal overhead the maximum achievable single-user data rate is about 500 Kbps. The resultant video quality in terms of the average PSNR is plotted against the frame error rate in Figure 19.18. When there is no FEC coding, any bit error in a TDMA frame causes a frame error. When FEC coding is used, a transmission frame error occurs only if the number of bit errors overloads the error correcting capacity of the BCH codes used. The figure shows how the transmission frame error rate increases, as the channel changes from error-free to the 0.2% BER Channel(2), and then to the 2% BER Channel(1), where the three measurement points in Figure 19.18 correspond to these BERs. When FEC coding is used, the video quality is inherently reduced due to the reduced video bit rate because of accommodating the FEC parity bits. However, the proportion of erroneous transmission frames is also reduced at both BER = 0.2% and 2%. It was found that the effect of transmission frame errors was not objectionable up to an FER of about 10%, which is supported by the PSNR versus FER curves of Figure 19.18. This implies that dropping one in ten frames does

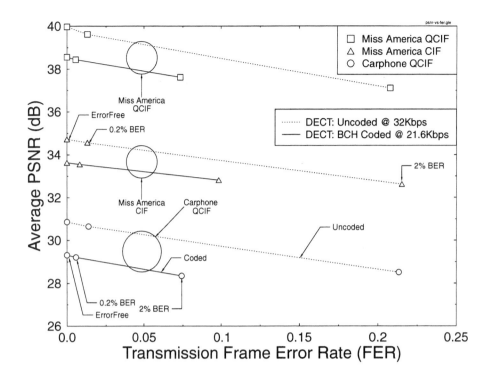

Figure 19.18: Simulation of the H.263 videophone system over DECT-like channels, with and without channel coding, and in error-free conditions. Using the "Miss America" video sequence at QCIF and CIF resolutions, and the "Carphone" sequence at QCIF resolution.

not significantly impair the video quality as long as the acknowledge flag feedback keeps the encoder's and decoder's reconstruction frame buffer in synchronization. The implementation of this will be the subject of Section 19.5.

In subjective video-quality terms, it is often more advantageous to change to a more robust modulation or coding scheme, once the PSNR degradation exceeds about 1 dB. This is demonstrated by Figure 19.19. This figure shows the average decoded video quality in terms of PSNR (dB) versus channel SNR, for 16QAM, 4QAM, and BPSK modes of operation. The 5% FER switching levels are also shown in the figure, which shows that the switching occurs when the higher PSNR but corrupted video becomes subjectively more objectionable than the inherently lower PSNR but unimpaired video.

The DECT simulations show an improvement in frame error robustness over previous simulations, increasing the maximum tolerable FER from 5% to 10%. This is because the DECT TDMA frame length of 10 ms is half that of the previous simulations, which used a frame length of 20 ms. Therefore, each TDMA packet contains on average a smaller proportion of the video frame. Hence, the loss of a packet has a

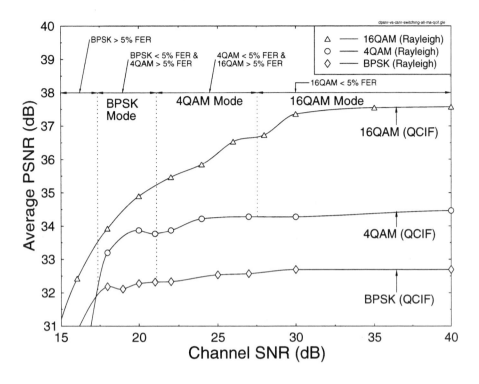

Figure 19.19: Simulation of the H.263 videophone system over Rayleigh-faded channels at QCIF resolution.

reduced effect on the PSNR and on the viewer's perception of the video quality.

Using FEC coding accordingly improves the range of channel SNRs, where the videophone system can operate. Figure 19.18 also shows that the videophone quality in the presence of transmission errors is fairly independent of the video sequence and resolution used, as evidenced by contrasting the more motion-active "Carphone" sequence and the CIF resolution "Miss America" sequence with the QCIF resolution Miss America clip. In the next section, we discuss the issues related to the transmission feedback mechanism that is used in our error-resilient mobile video system.

19.5 Transmission Feedback

The proposed wireless videophone system requires a feedback channel from the receiver to the transmitter in order to acknowledge the loss of packets and to keep the local and remote decoders in synchronization. This feedback channel can also be used to convey the ARQ requests, when ARQ is used. This feedback mechanism can be implemented in a number of ways, depending on the application. For two-way videotelephony the most efficient mechanism is to superimpose the up-link feedback information on the down-link data stream and vice versa. Normally, the up-link and

19.5. TRANSMISSION FEEDBACK

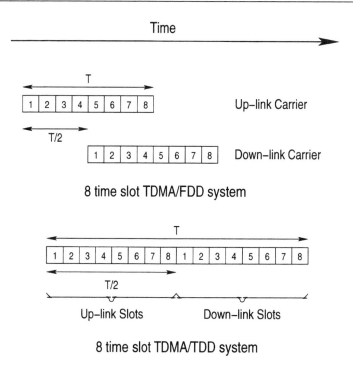

Figure 19.20: Up-link and down-link time slots for TDMA/FDD and TDMA/TDD systems, where the TDMA frame length is T seconds, and the offset between the corresponding up-link and down-link time slots is T/2 seconds.

down-link slots are offset by half a TDMA frame in time in both FDD and TDD systems. In Frequency Division Duplex (FDD), the up-link and down-link information is transmitted on different carriers. However, to reduce the complexity of the handset, the corresponding up-link and down-link slots are offset in time by half a TDMA frame in order to allow the handset time to process the information it receives and to reduce high-power transmitter leakage into the receiver, which is processing low-level received signals. In Time Division Duplex (TDD), the up-link and down-link information is transmitted on the same carrier; however, the carrier is divided in up link and down-link time slots. Again, these slots are normally offset by half a TDMA frame in time as in the DECT system shown in Figure 19.17. The up-link and down-link slot arrangements for a TDMA/FDD and a TDMA/TDD system are shown in Figure 19.20.

For videotelephony, where the feedback information is superimposed on the reverse channel, the delay before the feedback information is received is half the TDMA frame length. If the wireless video system is used for applications such as remote surveillance where video is only transmitted back to a central control room, the feedback information is conveniently implemented using a broadcast message from the base station. The proposed wireless video system is not suitable for broadcast video applications, such as television, because of the need of a reverse channel for feedback

Modem-mode Independent System Parameters	
Features	General System
User data symbols/TDMA frame	128
User pilot symbols/TDMA frame	14
User ramping symbols/TDMA frame	2
User padding symbols/TDMA frame	2
User symbols/TDMA frame	146
TDMA frame length (ms)	20
User symbol rate (kBd)	7.3
Slots/frame	18
No. of users	9
System symbol rate (kBd)	131.4
System bandwidth (kHz)	200
Eff. user bandwidth (kHz)	11.1
Vehicular speed	13.4m/s or 30mph
Propagation frequency (GHz)	1.8
Fast-fading normalized doppler frequency	6.2696×10^{-4}
Log-normal shadowing standard deviation (dB)	6
Path-loss model	Power law 3.5
Base station separation (km)	1

Table 19.5: Summary of System Features for the Quadruple-mode Reconfigurable Mobile Radio System

in this system. For broadcast video, the video stream has to be rendered more robust and able to recover from errors.

From now on, we will concentrate on interactive videotelephony applications, where there are both up-link and down-link video streams. In this case, the feedback signaling is superimposed on the reverse channel's video stream. We note that the system parameters used for investigating the transmission feedback effects are different from those used in the early part of this chapter, namely, in Table 19.1 on page 791.

Specifically, the transmission feedback investigations were conducted in the context of a similar quadruple-mode scheme, which also provided a BPSK mode. The channel parameters shown in Table 19.5 are the same as those in Chapters 7 and 20. Similarly, the error correction codes are also identical to those in later chapters; however, the useful video bit rate was reduced due to accommodating the ML-code protected transmission feedback flag. The error correction codes and the useful video bit rates of each of the four modes are shown in Table 19.6.

An example of this transmission feedback signaling and timing is shown in Figure 19.21 in the context of our system described in Tables 19.5 and 19.6. Upon receiving a transmission packet, the transceiver has two tasks to perform before the next transmission. First, the video data stream has to be demodulated and the FEC decoded in order to check whether any error occurred since this event has to be signaled in the next transmission burst to the transmitter.

19.5. TRANSMISSION FEEDBACK

Features		Multirate System			
	PSA-BPSK	4-PSAQAM	16-PSAQAM	64-PSAQAM
Modem	1	2	4	6
Bits/symbol				
No. of modem sub-channels	1	1	2	3
C1 FEC	BCH(127,85,6)	BCH(255,171,11)	BCH(255,191,8)	BCH(255,199,7)
C2 FEC	N/A	N/A	BCH(255,147,14)	BCH(255,163,12)
C3 FEC	N/A	N/A	N/A	BCH(255,131,18)
Padding bits	5	9	18	27
Total bits	132	264	528	792
Payload bits	90	180	356	520
Feedback bits	27	27	27	27
Video data bits	63	153	329	493
Video bit rate (Kbit/s)	3.15	7.65	16.45	24.65

Table 19.6: Summary of System Features for the Reconfigurable Mobile Radio System, Including Transmission Feedback Implemented Using the Majority Logic Code ML(27,1,13)

810 CHAPTER 19. H.263 MOBILE VIDEOPHONE SYSTEM

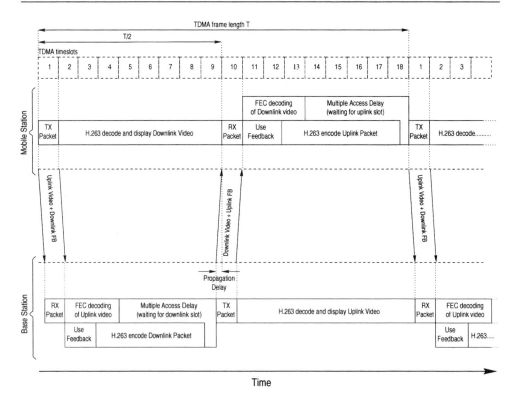

Figure 19.21: Transmission feedback timing, showing the feedback signaling superimposed on the reverse channel video data stream. The tasks that have to be performed in each time interval are shown for both the mobile station and the base station.

The second task is to produce the next video packet for the forthcoming transmission burst, as indicated in Figure 19.21. The next packet of video data cannot be encoded until the feedback signaling for the previous transmission is received. This is because the motion vectors used in the next video data packet may depend on blocks transmitted in the previous packet. Once the feedback acknowledgment flag of the previous packet's transmission is received, which was superimposed on the reverse channel video data, the effect of the loss or success of the previous packet can be taken into account in the local decoder of the H.263 codec. If the previous packet was lost, the effective changes made to the local decoder for this packet are discarded. If the previous packet was received without errors, then the local decoder changes pertinent to the previous packet are made permanent. If ARQ is used, then the transmission feedback signaling may have requested a re-encoding of the packet at the same or a lower bit rate. In this case the reconstruction buffer changes of the previous packet are discarded, and the codec parameters are modified to reduce the video bit rate temporarily. Once the feedback signaling was invoked in order to synchronize the reconstruction frame buffers of the local and remote H.263 codecs, the H.263 codec starts to encode the next video packet, as indicated in Figure 19.21.

19.5. TRANSMISSION FEEDBACK

The time between receiving a packet and transmitting a new packet is approximately half the TDMA frame length, T. Since we are considering microcells, the propagation delays are very low; for example, for a distance of 1 Km the delay is 3.3 μs. Within the time period $T/2$ between receiving and transmitting packets, the transceiver has to calculate the feedback signaling for the received packet and to generate a new video packet for transmission. These two tasks can be done in parallel or in series, but they must be completed within the $T/2$ time period. After a packet is transmitted, there is another $T/2$ time period before the next received packet arrives. This time period, as shown in Figure 19.21, is used to decode the H.263 video data received in the previous packet and to update the video display.

19.5.1 ARQ Issues

The employment use of ARQ results in several difficulties inconjunction with the use of transmission feedback signaling. If ARQ is used, a transmission feedback message is required for each transmission. In systems where a user is allocated one slot per TDMA frame, ARQ introduces latency and slot allocation problems, causing the video frame rate to fluctuate. Furthermore, use of additional slots for ARQ constrains the throughput bit rate of the system, reducing in video quality. Therefore, for systems with only one slot per TDMA frame, ARQ is not recommended because it reduces the video quality and causes a fluctuating frame rate. For these systems, it is better not to attempt to re-transmit lost packets with an ARQ request, but instead to invoke re-transmission at a higher level by transmitting the corresponding macroblocks in later packets. This policy is followed in our system.

The type of systems, in what ARQ can be used effectively are high-rate, short-frame duration wireless LAN-type systems. In these systems, a video user typically transmits less than one packet per TDMA frame. In other words, on average a video user is allocated a time slot once every "n" TDMA frames. In such systems, the ARQ re-transmission can be implemented by requesting additional time slots during the "n" TDMA frame period. Except in highly loaded systems, this will reduce the ARQ latency, yielding a more stable video frame rate. Hence upon using additional time slots for the re-transmissions, the video bit rate for the user will remain fairly constant, resulting in reduced video-quality degradation.

For our simulations we have used "Stop and Wait"-type feedback systems, similar to "Stop and Wait" ARQ [472,558]. It would be possible to invoke more complex types of feedback/ARQ systems, such as "selective repeat" [5,472]. But since this would make the packet "dropping" more complex, particularly for the motion compensation, this was not implemented for this initial version of our system.

19.5.2 Implementation of Transmission Feedback

The transmission feedback messages need to be very robust to channel errors. If a transmission feedback message is received incorrectly, the local and remote decoders lose synchronization. A feedback message received incorrectly will cause the video picture to become corrupted, and this type of corruption can be corrected only by the affected macroblocks being updated in intra-frame coded mode. However, due to motion compensation, additional macroblocks surrounding the affected blocks usually need to be intra-updated as well.

Therefore, initially the best way to implement the feedback message was deemed to use a very strong block code to encode the feedback message. The following three short block codes from the BCH family were considered:

- BCH(7,4,1)
- BCH(15,5,3)
- BCH(31,6,7)

When there is a deep fade, as experienced in mobile radio channels, however, the bit error-rate within the TDMA transmission burst can be very high. (Henceforth we will refer to the bit-error ratio (BER) within a TDMA transmission burst as the In-Burst BER.) The BCH(15,5,3) code needs an In-Burst BER of higher than $\frac{3}{15}$ or 20% to become overloaded and cause the feedback message to be received in error. In our simulations we found that this BCH code would become overloaded quite often, when deep fades occurred in the mobile radio channel. A more robust coding method for the transmission of feedback messages is therefore needed, and we eventually decided on a majority logic code.

19.5.2.1 Majority Logic Coding

Majority logic codes use a fairly simple principle. The information message is transmitted several times, and at the receiver all the messages are demodulated. The correct message is decided by majority voting. For example, in order to transmit a 1-bit message, which happens to be a logical 1, this 1-bit message is transmitted, say, nine times. Therefore, the coding has caused the 1-bit message to be turned into a 9-bit encoded message. The receiver decides the intended 1-bit message based on whether it receives more logical 0s or more 1s. Therefore, the message is received correctly as long as no more than four logical 1s are corrupted by the channel and changed to logical 0s. Therefore, the code can be described as ML(9,1,4) in the same notation as the BCH codes. This means that a 1-bit message is encoded into 9 bits, and the code can cope with up to 4 bit errors. If five or more errors are caused by the channel, then majority voting will cause the decoded message to be incorrect. This code can cope with an In-Burst BER of less than $\frac{4}{9}$ or 44%, which is much better than the above BCH codes. However, these codes are unaware of when they become overloaded, unlike the BCH codes. By increasing the number of encoded bits, the maximum In-Burst BER the code can cope with can be brought closer to the 50% limit. For example, the ML(31,1,15) code can cope with an In-Burst BER of up to 48.4%. This is much better than the correction capability of the BCH(31,6,7) code. However, the majority logic code is unaware of the event, when it becomes overloaded.

Majority Logic Codes can be characterized mathematically using the binomial distribution. The probability of n bit errors in k bits is given by:

$$P_n^k = \binom{k}{n} p^n (1-p)^{(k-n)}, \tag{19.3}$$

19.5. TRANSMISSION FEEDBACK

where p is the probability of a bit error or the bit-error rate and the binomial coefficient $\binom{k}{r}$ is defined as:

$$\binom{k}{r} = \left(\frac{k!}{r!(k-r)!}\right) \qquad (19.4)$$

which gives the number of combinations of r picks from k, not reusing picked values therefore, the probability of less than or equal to n errors in k is given by:

$$P^k_{(i \leq n)} = \sum_{i=0}^{i=n} \left(P_i^k\right) = \sum_{i=0}^{i=n} \left(\binom{k}{i} p^i (1-p)^{k-i}\right) \qquad (19.5)$$

where p is again the bit-error rate. Hence, for a majority logic code of 9 bits, ML(9,1,4), the probability of four or fewer bit errors in 9 bits is $P^9_{(i \leq 4)}$. If the number of encoded bits is an even number, say 10, then the total probability of correct decoding is given by the probability of four or fewer bit errors in 10 bits ($P^{10}_{(i \leq 4)}$), plus half the probability of 5 bit errors in 10 (P^{10}_5). This additional term for even-length codes occurs, because the majority voting can result in a draw, and the outcome is then decided by chance. Therefore, for a 10 bit code there is a 50% chance of correct decoding, even when there are 5 bit errors. We describe this code as ML(10,1,4+). This means that the code can correct 4 bit errors and sometimes one more. The general formula for the correct decoding probability (P_{CD}) of majority logic codes is given by:

$$P_{CD} \text{of ML(k,1,?)} = \begin{cases} P^k_{(i \leq (\frac{k-1}{2}))} & k \text{ is odd} \\ P^k_{(i \leq (\frac{k}{2}-1))} + \frac{P^k_{(k/2)}}{2} & k \text{ is even.} \end{cases} \qquad (19.6)$$

The probability of correct decoding of the 5-, 9-, 18-, and 27-bit majority logic codes was evaluated for the range of bit-error rates 0% to 50% in Figure 19.22.

In order to find the length of majority logic code required for our system, the histogram of In-Burst bit error rates was calculated as described in the context of Tables 19.5 and 19.6. This system experienced both Rayleigh fast fading and shadow fading. The combination of Rayleigh and shadow fading can cause very deep fades, generating very high In-Burst BER. The probability of a majority logic code in error versus the In-Burst BER was normalized by the probability of the occurrence of such an In-Burst BER, which is shown in Figure 19.23.

Figure 19.23 shows the chance of an error in a majority code for the system described in Tables 19.5 and 19.6. Even for the weakest 5-bit Majority Logic codes the probability of error is on average once in 400 packets for the worst-case situation. Our proposed system will change to a more robust modulation scheme, request a handover, or drop the call when the channel becomes very hostile. Hence, such high bit error rates will be avoided in all but a few cases.

The transmission feedback can then be implemented using a majority logic code on the reverse channel. The video data for the reverse link and the transmission feedback information majority logic code can be transmitted together in the same packet, as suggested above.

In order to evaluate the performance of this transmission feedback scheme in our mobile video system, we used the multirate scheme defined in Tables 19.5 and

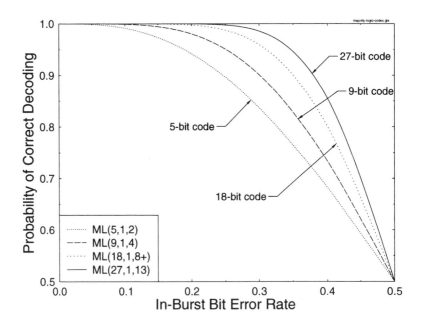

Figure 19.22: Numerical evaluation of the probability of correct decoding (P_{CD}) of majority logic codes.

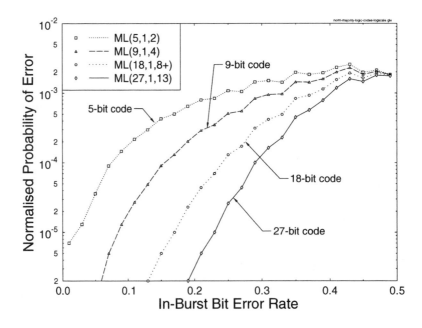

Figure 19.23: Numerical evaluation of probability of error in majority logic codes, normalized by the chance of occurrence of the given In-Burst BER.

19.5. TRANSMISSION FEEDBACK

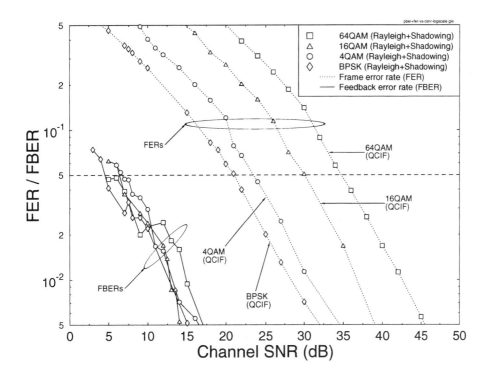

Figure 19.24: Frame error rate (FER) and feedback error rate (FBER) versus channel SNR, for the parameters defined in Table 19.6 and channel parameters of Tables 19.5 and 19.6.

19.6. We used a 27-bit majority logic code, ML(27,1,13), to encode the transmission feedback information bit. These 27 encoded bits were transmitted and exploited previously unused padding bits, where they were available. These padding bits were uncoded, while the remaining bits were transmitted within the FEC-encoded portion of the data packet. Because we are using systematic BCH coding, even if the error correction becomes overloaded, the majority logic bits can typically be recovered. The employment of the protected transmission feedback flag reduced the effective video source bit rate given in Table 7.1 on page 257 to the values shown in Table 19.6, which particularly affected the BPSK mode of operation. In this mode the video bit rate is limited to about 3 Kbit/s, which means that the video will have to be of SQCIF resolution in order to maintain a reasonable video frame rate.

Figure 19.24 shows the Frame Error Rate (FER) versus channel SNR performance for the four modes of operation listed in Table 19.6. The figure also displays the feedback error rate versus channel SNR. In order to distinguish between them, the feedback error rate is referred to as FBER. Previously, we found that with the aid of our acknowledgment technique the video quality does not degrade significantly at FERs below 5%. For these simulations, the feedback error rate (FBER) was zero, since for the channel conditions where the FER is below 5%, we have FBER = 0.

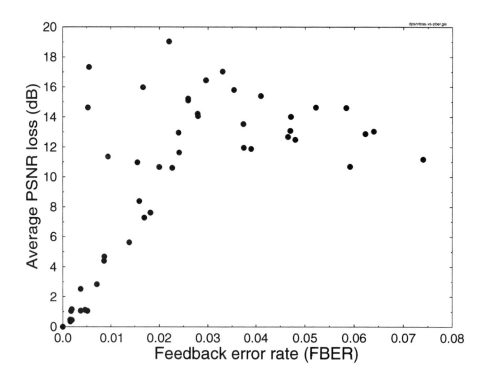

Figure 19.25: Average video-quality degradation in terms of PSNR (dB) versus feedback error rate (FBER).

Therefore, if the system switches to a more robust modulation scheme or requests a handover before the 5% FER threshold is reached, then there are no feedback errors.

When feedback errors do occur, which is fairly infrequently, this effect can be removed during the next intra-frame transmission or during the intra-update of the macroblocks affected. The rate at which these intra-updates occur is dependent on the additional robustness required. However, the more frequent the intra-updates, the lower the video quality, since the intra-blocks consume a larger fraction of the available bit rate.

Figure 19.25 shows the PSNR degradation associated with different rates of feedback errors. The effect of the feedback errors is not removed by intra-frame coded replenishments in these simulations. In the majority of cases, the average PSNR was reduced to about 15 dB.

19.6 Summary and Conclusions

This chapter described the design and performance of an H.263-based videophone system. The system performance was also compared to that of the previously characterised H.261 arrangement of section 17.4. The main system features were summarized

19.6. SUMMARY AND CONCLUSIONS

in Table 19.1. In Section 19.2, we described the problems associated with using the H.263 video codec in a mobile environment. We discussed potential solutions to these problems, together with the advantages and disadvantages of each solution.

In Section 19.3, we described the design of our error-resilient reconfigurable videophone system, using transmission feedback in order to keep the local and remote decoders in synchronization. A packetization scheme is used to aid the resynchronization of the video data stream and to help conceal the loss of packets. Forward error correction (FEC) is used to decrease the channel BER and thereby to improve the error resilience. The system is designed to be reconfigurable in order to be able to adapt to time-variant channel conditions. This improves the video quality in periods of good channel conditions. A bit-rate control algorithm is used to allow the video codec to vary its target bit rate based on the current channel conditions and to control the choice of the modulation scheme used by the reconfigurable system. The system can invoke ARQ in order to re-transmit lost packets. However, because of the system's reconfigurability, the re-transmissions can use a more robust modulation scheme, increasing the probability of success for the re-transmission. The system does not allow the standard H.263 decoder to decode erroneous data, and so errors cannot appear in the decoded video. The transmitted bit-stream conforms to the H.263 standard, and no transcoder is necessitated. The receiver employs a standard H.263 decoder, which is fed by the de-packetization system.

Section 19.4.2 characterized the performance achieved by our error-resilient reconfigurable videophone system, which could cope with very high packet-loss rates. Specifically, it was found that an average packet-loss rate of 5% was the highest rate the system could comfortably cope with, before the effects of the packet loss became perceptually obvious. However, the system would recover from packet-loss conditions as soon as the channel quality improved, or the system was reconfigured to a more robust modulation scheme. We investigated how system performance improved with use of ARQ and antenna diversity. We also studied the error-resilient reconfigurable videophone system using the DECT cordless telephone system and found that the system performed in a similar manner. However, due to the higher TDMA frame rate in the DECT scheme, it was found that the maximum average packet-loss rate could be increased to 10%.

Finally, in Section 19.5 we discussed the issues concerned with the transmission feedback that was used in the error-resilient video system. We described how we protected this transmission feedback using majority logic-coding. We then presented performance results using the majority logic-based transmission feedback system. These results showed that the transmission feedback was error-free within the operational SNR range of the video system that is governed by the reconfigurable modulation scheme switching. In the next chapter, we investigate the effects of co-channel interference, which is the main limiting factor in radio spectrum reuse efficiency. The results from this chapter show the potential benefits of our reconfigurable modulation system.

Chapter 20

Error Rate Based Power Control

20.1 Background

In recent years a great deal of research has been conducted in the field of transmission power control algorithms, in particular, Code Division Multiple Access (CDMA) based mobile radio systems. However, a substantial amount of research has also been carried out on using power control to reduce co-channel interference in TDMA systems. Articles of particular note include those by Zander [288, 289], Zorzi [571], Leung [572], Ariyavisitakul et al. [573], Pichna et al. [574] and Lee et al. [575]. A power control algorithm needs an accurate and recent estimation of the radio channel's quality. The systems that rely on the Received Signal Strength Indicator (RSSI) to measure channel quality must ensure the accuracy of the channel estimation in an interference-limited environment. Various techniques to reduce inaccuracy have been suggested, in particular an averaging technique by Austin and Stüber [576]. Power control schemes based on error rate give a reliable channel quality estimation but have some limitations.

An attractive power control algorithm based on a combination of BER and RSSI estimates was proposed by Chuang and Sollenberger [577]. In addition, Kumar et al. [578] proposed a power control algorithm based on bit error rate measurements. In this chapter we quantify the benefits of using a bit error rate (BER) and frame error rate (FER)-based power control algorithm in an interference-limited environment.

20.2 Power Control Algorithm

In our proposed BER-based power control algorithm, the main channel quality indicator was the BCH-coded frame error flag (FEF). A frame error is declared when one of the BCH codes in each TDMA frame becomes overloaded due to the preponderance of channel errors. This implies that some bits transmitted in the corresponding TDMA frame have been corrupted by the channel. In order for the transmitter to determine whether a transmission burst was received correctly, an acknowledgment has to be sent from the receiver, which is associated with a delayed indication of the channel quality. If this delay is too high, the frame error flag may be of little use. This delay is one of the disadvantages of BER-based techniques in comparison to systems that use a RSSI-reading carried out by the receiver in order to set the transmission power. However, the RSSI-based systems assume reciprocity of the up-link and down-link, an assumption that is also affected by interference.

In this chapter, we highlight the rationale behind the proposed power control algorithm and formalize its description through a flowchart. The true number of bit errors in a transmission burst is known only to the receiver, if the channel coding block code of our proposed system has not become overloaded by the excessive number of transmission errors. This is true for both convolutional and block codes, although convolutional codes are oblivious to being overloaded, while block codes are capable of detecting these events with a high probability. Hence, block codes are more attractive in this application. When the channel coding is overloaded, a frame error (FE) results. Hence, in the proposed scheme we opted for binary Bose-Chaudhuri-Hocquenghem forward error correction (FEC) coding. In the 4QAM mode of operation, the FEC used is BCH (255,171,11), correcting up to $t = 11$ errors, and if the number of errors is higher than $t = 11$, then a frame error occurs. Note, however, that the algorithm is generic, regardless of the FEC code used or the transmission integrity required.

In addition to the frame error flag, our proposed power control algorithm uses the actual number of errors corrected by the channel coding as an additional indicator of channel quality. If the number of channel errors in the received BCH-coded frame is zero before channel decoding, the correcting power of the BCH code is not exploited. Thus, the instantaneous channel quality is considered to be higher than the target quality, which is due to the excessive transmission power level. Therefore, reducing the transmission power should be considered. By contrast, if the number of errors in the frame is higher than the correcting capability of the FEC code, then a transmission frame error has occurred, and increasing the transmission power will become urgent. However, if the number of errors contained in the BCH-coded frame can be corrected by the FEC, three possible situations should be considered. First, if the number of errors is near to the error correction capability of the code, where a frame error would occur, or the number of errors in successive frames has been increasing, then the transmission power should be increased. Second, if the number of errors in the frame is low and has been decreasing in previous frames, then the power should be reduced. Finally, when the frame is not error-free, but the errors are correctable by the FEC, it is logical to keep the transmission power constant.

The amount of time needed to delay an action, before the power control algorithm increases or decreases the power, depends on many factors, such as the modulation scheme employed, the channel conditions, and the target Frame Error Rate (FER). The power control algorithm proposed in this contribution exhibits a variable step size and has been tested with a power control delay of one TDMA frame. A discussion of the optimization of the algorithmic parameters for different modulation schemes and scenarios is provided in Section 20.6. Based on the above rationale, the power control

20.2. POWER CONTROL ALGORITHM

Feature	Value
Delay (TDMA frames)	1
Minimum step size	1 dB
Maximum step size	16 dB
Max. TxPower	30 dBm (1 Watt)
TxPower Dynamic Range	64 dB

Table 20.1: Power Control Algorithm's Features

Parameter Name	Type	Comment
NearFrameError (NFE)	%	If the number of bit errors is greater than a given percentage of the number of correctable errors, then the event is classified as a near frame-error.
NearErrorFree (NEF)	%	If the number of bit errors is less than a given percentage of the number of correctable errors, then the event is classified as a near error-free frame.
IncPowCount (IPC)	Number	Number of successive frame errors (FECNT) to initiate transmission power increments.
DecPowCount (DPC)	Number	Number of successive error-free frames (NECNT) to trigger transmission power decrements.
IncPowStepSize (IPSS)	Binary Function	Function of the successive frame error count (FECNT), IPSS decides when to increase step size and by how much.
DecPowStepSize (DPSS)	Binary Function	Function of the successive error-free frame count (NECNT), DPSS determines when to increase step size and by how much.

Table 20.2: Power Control Algorithm Parameters

algorithm's main features are shown in Table 20.1. We used a typical maximum transmission power of 1 Watt and a dynamic range of 64 dB, as in the Global System of Mobile communications, known as GSM [41].

The simplified flowchart of the power control algorithm is shown in Figure 20.1. The algorithm has a set of variable parameters that can be modified in accordance with varying channel conditions, modulation schemes, or other factors. These parameters are summarized in Table 20.2, and the role of each parameter will become explicit later in the chapter.

Before the commencement of transmission, the Mobile Station (MS) receives a control signal from the Base Station (BS), informing the MS of the number of errors corrected or whether a frame error occurred. If there was a frame error, or the number

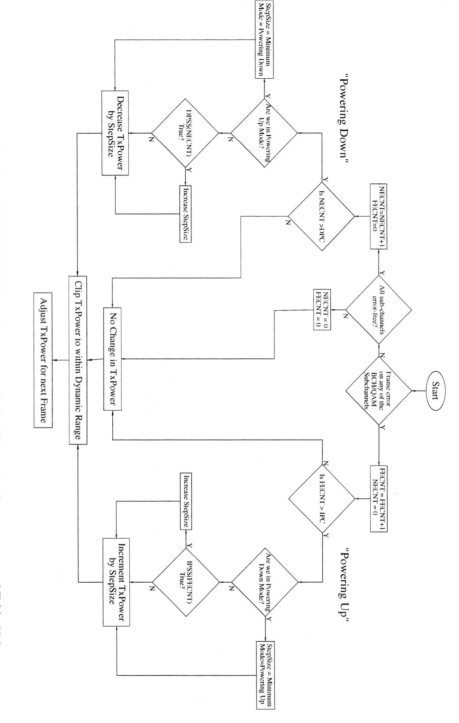

Figure 20.1: Flowchart of the BER/FER-based power control algorithm using the parameters of Table 20.2.

20.2. POWER CONTROL ALGORITHM

of errors almost caused a frame error, the frame error counter is incremented. If the number of bit errors in a BCH-coded frame was close to the FEC overload condition, which in our current example was $t = 11$, this event was considered a NearFrameError (NFE) condition, as seen in Table 20.2. The parameter frame error count (FECNT) registers the number of frame errors that have occurred in successive frames. When the frame error count is incremented, the no error counter (NECNT) is reset, where the no error count was defined as the number of consecutive error-free frames received in a row.

When a received video packet did not contain any errors even before FEC was invoked, then the no error counter (NECNT) was incremented, while the frame error counter (FECNT) was reset. The transmission bursts received with a low number of errors, constituting a low proportion of the FEC code's error correction capability, were classified as error free, where the corresponding NearErrorFree (NEF) threshold of Table 20.2 was a further optimized algorithmic parameter.

If the previous transmission burst contained channel errors, which were corrected by the channel decoder, but the number of errors was lower than what would be considered as a NearFrameError, and higher than what would be classed as a NearErrorFree frame, then the frame error count (FECNT) and no error count (NECNT) are reset and the transmission power is left unchanged. These conditions correspond to the central three branches of the flowchart in Figure 20.1.

Following a certain number of successive frame errors, the MS decides to increase the transmission power. The number of erroneous frames, IncPowerCount (IPC) in Table 20.2, which is required to initiate power boosting, is another optimized parameter of the algorithm. Upon powering up, the MS initially starts increasing the power by the smallest possible step size. However, if frame errors continue to occur, the step size is increased. The step size increase is a function of the frame error counter value, and this binary function, which we refer to as IncPowerStepSize (IPSS) in Table 20.2 is a further optimized parameter of the algorithm. The right-hand section of the flowchart in Figure 20.1 corresponds to the powering-up mode of the algorithm.

When the MS is informed by the error rate feedback channel that the last N frames have been received error free, the handset decreases the transmission power, which in this case is deemed to be unnecessarily high. The number of error-free frames, 'DecPowCount' (DPC) in Table 20.2, encountered before the handset powers down is yet another optimized parameter of the algorithm. Similarly to powering up, the initial reduction of power is carried out using the smallest step size. If, however, after powering down, the next few frames are still error-free, then the power reduction step size is increased. The step size increment is governed by the function DecPowerStepSize (DPSS) of Table 20.2, which is dependent on the number of successive error-free frames received. This binary function is also a fundamental optimized parameter of the algorithm. The left-hand section of the flowchart shown in Figure 20.1 corresponds to the powering down mode of the algorithm.

We also note that the absolute dynamic range of the algorithm is limited by the maximum transmission power of 30 dBm and by the 64 dB dynamic range, which were specified in Table 20.1. In summary, the parameters that govern the behavior of the power control algorithm are shown in Table 20.2, while its operation is summarized in the flowchart of Figure 20.1. Further details of its inner workings can be inferred by referring to the flowchart. In the next section, let us now briefly characterize the performance of the power control algorithm.

20.3 Performance of the Power Control

The power control algorithm was simulated using the 4QAM mode of our multirate system over a slow and fast fading channel as described in Table 7.1 on page 257. The worst-case scenario of a single interferer was employed to cause co-channel interference. Examples of the best- and worst-case situations are shown in Figures 20.2 and 20.3. The best case is when both the interferer and the user are close to their corresponding BSs, and the worst case is when they are at the edge of their cells, respectively. The worst-case scenario was characterized by Figure 7.4(b). Specifically, Figure 20.2 shows the transmission power variation versus time, demonstrating that the transceiver is operating close to the minimum power setting of -34 dBm. Both figures display the slow fading, the signal-to-interference+noise-ratio (SINR) averaged over each time-slot, the Frame Error Flag (FEF), and the transmission power.

The worst-case situation seen in Figure 20.3 shows more clearly how the power control reacts to the fluctuating SINR. After a one-frame latency, the power control attempts to compensate for degrading SINR values. The frame error flag (FEF) indicates when, for example, a rapid power boost is required to compensate for a deep fade, but due to the inherent latency, the scheme cannot respond sufficiently promptly. Observe in the figures that the FEF suggests a moderate FEC overloading event frequency, regardless of the path-loss and slow fading experienced. It can also be seen that the transmission power is limited to the maximum power of 30 dBm at one point.

To demonstrate the performance of the power control algorithm for different interference conditions, simulations were carried out for different combinations of user and single co-channel interferer positions. As previously, the co-channel base station distance was 1 km. The user distance was simulated in the range of 10–200 m, with increasing distances getting closer to the co-channel base station. The interfering user offset from its base station was varied in the range of -200m to +200m. The distance of the interferer from the interfered base station is given by the interferer offset plus the co-channel base station distance. A negative interferer distance therefore indicates that the interferer is less than 1 km from the interfered base station. For these simulations, the maximum interference occurs when the interferer offset is -200m. When power control is used, both the user and interferer employ the same power control algorithm.

20.3.1 Frame Error Rate Performance

As expected, the power control algorithm of Figure 20.1 maintains a near-constant BCH-coded frame error rate across the whole cell area, which is demonstrated by Figures 20.4(a) and 20.4(b). Specifically, Figure 20.4(a) portrays the global three-dimensional (3D) view, while Figure 20.4(b) represents the contour plot of constant FER trajectories. According to Figure 7.31, the target FER was adjusted to around

20.3. PERFORMANCE OF THE POWER CONTROL

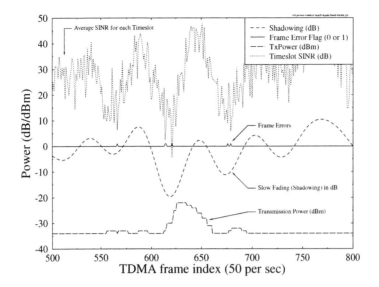

Figure 20.2: Various waveforms associated with the error rate based power control algorithm of Figure 20.1 for the *best case* situation when both interferer and user are close to their base stations, using 4QAM and the channel conditions of Table 7.1.

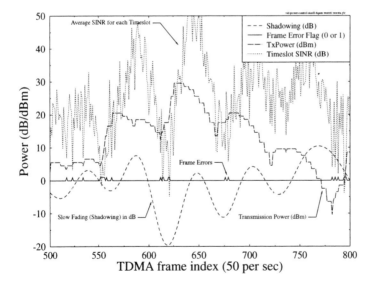

Figure 20.3: Various waveforms associated with the error rate-based power control algorithm of Figure 20.1 for the *worst-case* situation when both interferer and user are at the edge of their cells, using 4QAM and the channel conditions of Table 7.1.

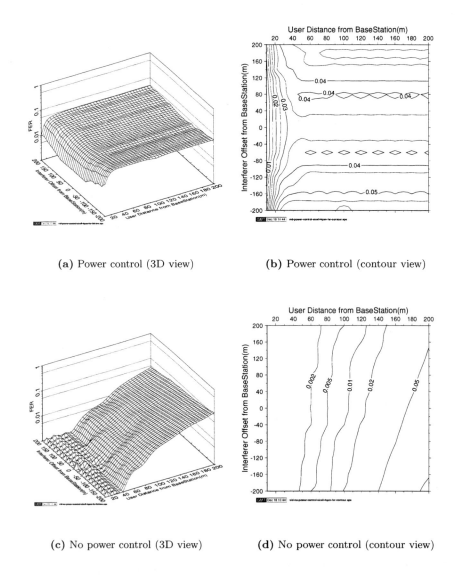

(a) Power control (3D view) **(b)** Power control (contour view)

(c) No power control (3D view) **(d)** No power control (contour view)

Figure 20.4: Simulated frame error rate (FER) with and without power control, versus user and interferer distance from the base station using 4QAM and the channel conditions of Table 7.1.

5%, as required by the video transceiver of Section 19.3, to maintain the target PSNR. However, in the extreme vicinity of the BS the power could not be reduced below the $30 - 64 = -34$ dBm level, which resulted in the reduced FER observed in the figure.

When power control is not used, the (FER) is dependent mainly on the user's distance from its base station, as shown in Figures 20.4(c) and 20.4(d). Toward the fringes of the cell, the FER becomes unacceptably high without power control. As can

20.3. PERFORMANCE OF THE POWER CONTROL

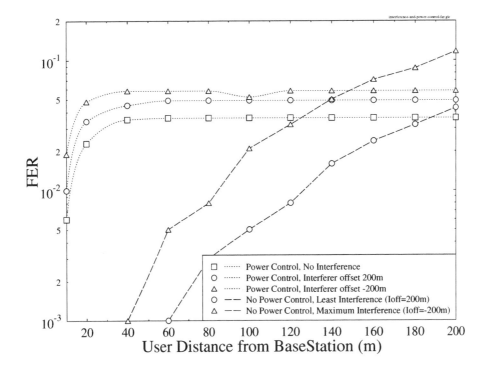

Figure 20.5: Simulated frame error rate (FER) versus user distance results with and without power control, and for power control without interference using 4QAM and the channel conditions of Table 7.1.

be seen from the graphs with and without power control, the low frame error rates near the center of the cell are traded off against a more constant FER throughout the cell, when power control is employed. The benefits of power control in terms of interference reduction are further demonstrated later in this chapter.

In our next experiment, we tested the performance of the power control scheme with various amounts of interference by varying the interferer's distance offset from its serving BS. The FER results for all these simulations are shown in Figure 20.5. The maximum interference is generated at an interferer offset of 200m, both with and without power control, yielding the highest FER. The minimum interference without power control corresponds to an interferer offset of +200m. We note, however, that the position of the interferer inducing the least possible interference is different with and without power control. Specifically, when power control is used, it can be shown that the minimum interference case is for an interferer offset of near zero, that is, when the interfer is close to its serving BS and hence transmits at a low power. The higher frame error rate of the interfered scenarios is likely to have been caused by the less predictable nature of interference in comparison to the noninterfered but noise-contaminated scenario. When the power control algorithm adjusts the power on the basis of bit error rate (BER), there is an inherent prediction latency, which renders the interfered scenario less predictable. Hence the bit errors are less "evenly" distributed over time. This bursty error distribution reduces the efficiency of the BCH decoder, which is often overloaded by error bursts, while it "wastes" channel bandwidth and decoding complexity, when error-free codewords are received.

20.3.2 Signal-to-Interference Ratio Performance

In this section, we characterize the effect that our power control algorithm has on the signal-to-interference ratio (SIR).

Figures 20.6(a) and 20.6(b) portray the simulated SIR performance without power control. Note that both axes in most of the graphs in this section have been reversed to visually enhance the results. As expected, the SIR reduces with increased user distance from the serving BS, as well as with reduced interferer separation. The simulated SIR performance is quite close to the theoretical SIR performance without power control, which is plotted in Figure 20.7. The theoretical SIR graph was based on the single interferer SIR model of Equation 7.3 on page 252. The shape of the theoretical and simulated performance profiles is also very similar. However, the simulated SIR is less than the theoretical value, mainly because of the fast and slow fading used in the simulation, which is not accounted for in the theoretical model.

The effect of the power control algorithm on the SIR is shown in Figures 20.6(c) and 20.6(d). With the power control in use, the user's position now has little or no effect on the SIR. This is because the user's transmission power is increased, as they move further away from their base stations. As can be seen from the figure, the SIR is now mainly dependent on the position of the interferer. Note, however, that the minimum SIR with power control is higher than without power control.

Figure 20.8 shows the improvement in SIR with power control, when related to the no power control scenario. There is a small area where the SIR is better without power control, where the user was close to the base station. When using power control, these users experience a reduced SIR due to a general reduction in transmitted power. However, their mobile phone will have a longer battery recharge perioid as a result of this power transmitted. This figure has not had the scaling of its axes reversed, unlike the rest of the graphs in this section.

20.3.3 Signal-to-Interference-Plus-Noise Ratio Performance

In this section, we discuss the effect of power control on the SINR defined in Equation 7.10. From our simulations we found that the SINR gives a good measure of the expected system performance in both noise- and interference-limited scenarios. The instantaneous SINR provides a good indication as to whether a particular symbol will be received in error.

Figures 20.9(a) and 20.9(b) show the simulated average SINR without power control. Note that the axis reversal is again used in order to give clarity to the 3D surface graphs. Since our simulations have concentrated on an interference-limited environment, this graph is very similar to the SIR profile without power control, which was shown in Figures 20.6(a) and 20.6(b). When power control is used, the SINR profile becomes near-constant, as shown in Figures 20.9(c) and 20.9(d). This is expected,

20.3. PERFORMANCE OF THE POWER CONTROL

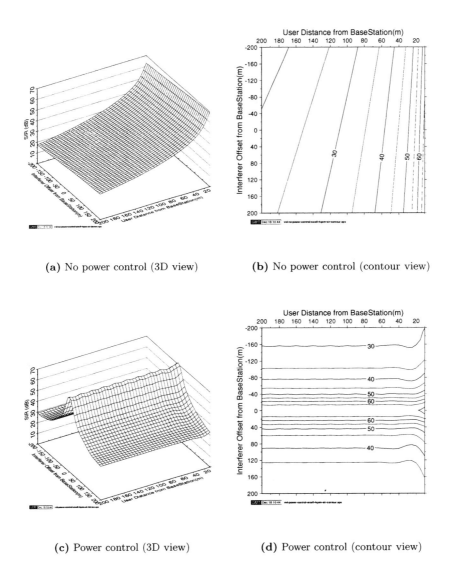

(a) No power control (3D view) (b) No power control (contour view)

(c) Power control (3D view) (d) Power control (contour view)

Figure 20.6: Simulated average signal-to-interference ratio (SIR) with and without power control, versus user and interferer distance from the base station using 4QAM and the channel conditions of Table 7.1.

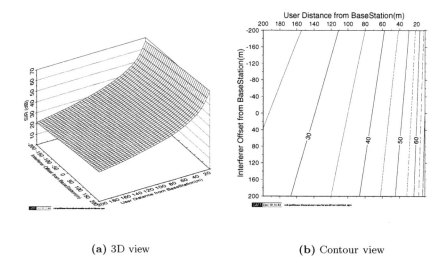

(a) 3D view (b) Contour view

Figure 20.7: Theoretical signal-to-interference ratio (SIR) without power control, versus user and interferer distance from the base station using 4QAM and the channel conditions of Table 7.1.

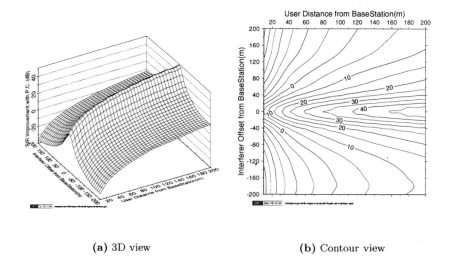

(a) 3D view (b) Contour view

Figure 20.8: The improvement in average signal-to-interference ratio (SIR) with power control, versus user and interferer distance from the base station, when related to the no power control scenario using 4QAM and the channel conditions of Table 7.1.

20.3. PERFORMANCE OF THE POWER CONTROL

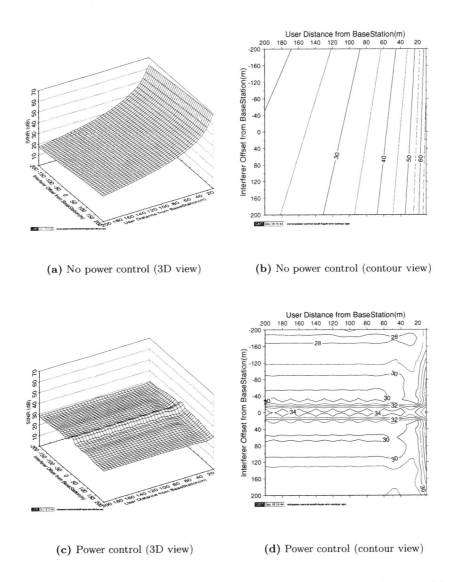

(a) No power control (3D view)

(b) No power control (contour view)

(c) Power control (3D view)

(d) Power control (contour view)

Figure 20.9: Simulated average signal to interference-plus-noise ratio (SINR) with and without power control, versus user and interferer distance from the base station using 4QAM and the channel conditions of Table 7.1.

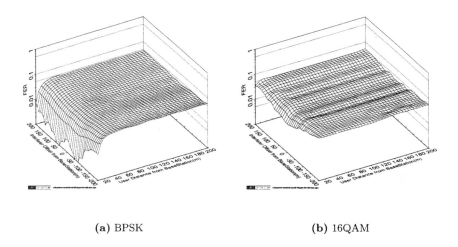

(a) BPSK (b) 16QAM

Figure 20.10: Frame error rate (FER) with power control, versus user and interferer distance from base station using BPSK and 16QAM for the channel conditions of Table 7.1.

since the power control algorithm is attempting to keep the frame error rate constant. In Section 7.6.2, we showed that the SINR can be used to approximate the error rate. Hence, by keeping the frame error rate constant, the power control algorithm keeps the SINR fairly constant as well. In the next section, we consider the application of the power control algorithm of Figure 20.1 to a multimode system.

20.4 Multimode Performance

The initial simulations of the power control algorithm were carried out using the 4QAM mode of our multimode scheme described in Table 7.1. To confirm that the algorithm would work in a multimode scheme, simulations were performed for the BPSK and 16QAM modes of operation. The frame error rate (FER) performance of the power-controlled BPSK and 16QAM modes is shown in Figure 20.10. This should be compared to the 4QAM performance shown in Figure 20.4(a). Note that there is very little difference between the FER performance in all three modes of operation. The only differences occur in the extreme vicinity of the BS, where the transmission power could not be reduced below the $30-64 = -34$ dBm minimum level, resulting in a lower FER than the targeted 5%. This phenomenon is less pronounced for the least robust modulation scheme, 16QAM. These results were obtained using the same set of parameters for the power control algorithm, which is summarized in Tables 20.1 and Table 20.4, and are elaborated on at a later stage. Therefore, the algorithm appears to perform fairly independently of the modulation scheme employed.

Figure 20.11 shows the system's FER performance for the BPSK, 4QAM, and 16QAM modes of operation. These curves were obtained by extracting the results

20.4. MULTIMODE PERFORMANCE

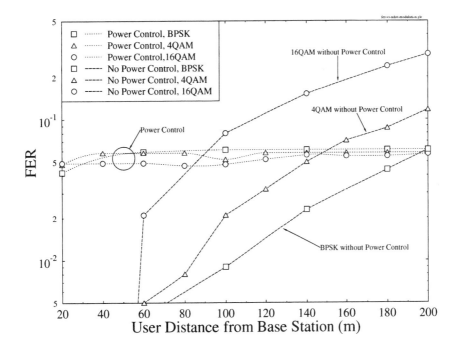

Figure 20.11: Frame error rate (FER) with and without power control, versus user distance from base station, for the scenario subjected to maximum interference that can be caused by a single interferer, both using BPSK, 4QAM, and 16QAM for the channel conditions of Table 7.1.

from Figures 20.10 and 20.4(a) for the worst-case situation of an interferer offset of -200m. The figure also shows the FER performance with and without power control. This allows the fraction of the cell area, where power control gives better performance to be calculated. For the 16QAM mode, power control gives better BER performance for user distances in excess of 90 m. Assuming a circular cell area, the power control achieves an improved BER performance in 80% of the cell area. In the BPSK mode, power control does not give an FER improvement; however, the average transmission power of the mobile is lower, thereby increasing the battery life and reducing interference. The area of the cell, where the use of power control achieves an improved FER performance is dependent on the modulation scheme and the target FER chosen.

From the above findings we conclude that the proposed BER-based power-control algorithm was capable of assisting the operation of different modulation schemes. However, when the vehicular speed or fading frequency changes, parameter changes are needed in order to maintain the target FER performance. Although for BPSK the power control seems to degrade the FER performance, we will show that maintaining the required target FER allows the portables to operate at significantly reduced transmitted power and battery consumption. The subject matter of the average transmission power, when using power control, is considered in the next section.

20.5 Average Transmission Power

In the absence of power control, the average transmitted power is identical to the individual power of each portable. When using power control, the average transmitted power for a transceiver varies within the cell and depends on the modulation mode used. The corresponding simulation results are shown in Figure 20.12 for BPSK, 4QAM, and 16QAM. Explicitly, these figures demonstrate that the average transmitted power varies with user and interferer distances from their respective base stations. As expected, the average required transmitted power increases when users move away from their base stations, but the position of the interferer also has an effect. It should be noted that the maximum average power is less than 30 dBm, the power that we used for the fixed power simulations. We also note in the 16QAM subfigures at the bottom that the position of the interferer (measured on the vertical axis) has more of an observable effect in conjunction with the least robust scheme, namely, 16QAM. The lowest transmitted power is required when the interferers are close to their own BS, hence transmitting at low powers, as can be seen in Figure 20.12(d). This manifests itself in Figure 20.12(c), for example, by observing the "0dB" trajectory, which provides coverage up to about 60 m at an interferer offset of 200 m, but only about up to 40 m at -200 m.

The average transmission power versus user distance from the BS for interferer offsets of +200 and -200 meters are plotted for the BPSK, 4QAM, and 16QAM modes of operation in Figure 20.13. This figure showed that the average power P_{av} in terms of user distance could be modeled using Equation 20.1:

$$P_{av}(dB) = 35\log_{10}(d) + \beta, \qquad (20.1)$$

where d is the user distance from the base station and β is an offset to be derived later in this discussion, which is dependent on the modulation scheme and interferer position. As expected, the $35\log_{10}(d)$ term in the equation is due to the inverse 3.5 power law, which we used as our path-loss model in Table 7.1 on page 257.

The average power in the cell, assuming a uniform geographical portable distribution and a circular cell area, can be calculated using the mean value theorem. This theorem states that the average value for a two-dimensional function can be found by dividing the volume of the corresponding enclosed space by the base area. In this case, we assume that the base area is that of a circle, that is, πR^2. The volume is the integral of Equation 20.1 for the range of possible user radii and angles from the base station. Therefore, the average power in the cell is given by:

$$\begin{aligned}
\text{Average Power in Cell (dB)} &= \frac{1}{\pi R^2} \int_{\phi=0}^{2\pi} \int_{r=0}^{R} (35\log_{10} r + \beta)\, r\, dr\, d\phi \\
&= 35\log_{10} R + \beta - \frac{35}{2 \times \ln 10} \\
&= 35\log_{10} R + \beta - 7.6 \text{ dB} \qquad (20.2)
\end{aligned}$$

This equation implies that the average transmission power in the cell is 7.6 dB less than the average power used by a user at the edge of the cell. In the case of the 3.5 path-loss power-law model we used, this is equivalent to the average power of a user

20.5. AVERAGE TRANSMISSION POWER

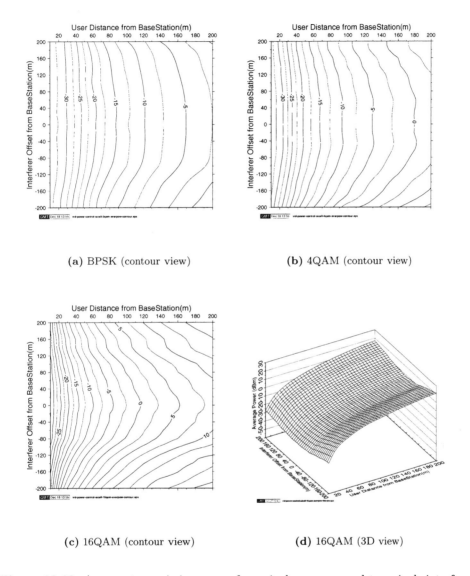

(a) BPSK (contour view)

(b) 4QAM (contour view)

(c) 16QAM (contour view)

(d) 16QAM (3D view)

Figure 20.12: Average transmission power for a single user exposed to a single interferer, both using the same modulation scheme and the power control algorithm of Figure 20.1, versus user and interferer distance from base station, under the channel conditions of Table 7.1.

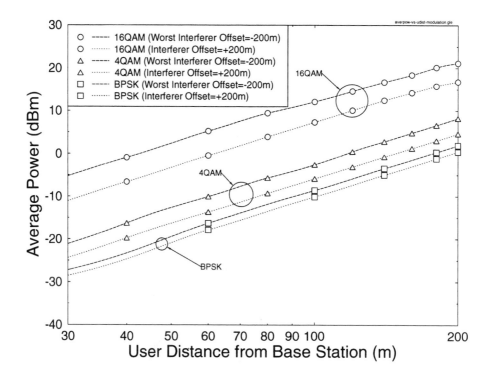

Figure 20.13: Average transmission power versus user distance from base station, for the interferer offsets of 200 m and -200 m (worst-case) and for BPSK, 4QAM, and 16QAM modes of operation, under the channel conditions of Table 7.1.

at a distance of $0.61 \times R$ from the BS, where R is the cell radius. Using the offset β found from Figure 20.13, the average transmission power in the cell is tabulated in Table 20.3 for the BPSK, 4QAM, and 16QAM modes of our transceiver. Since β is dependent on both the modulation scheme and the interferer position, as shown in Figure 20.13, the table includes results for the maximum and minimum average power that can be expected in a cell. The minimum average power is required when the interferer is close to its base station. Therefore, it transmits at a low power, reducing the amount of co-channel interference. The maximum average power is necessitated when the interferer is at the edge of its cell and at the closest point to the interfered BS.

In conclusion, Table 20.3 shows that even for the most demanding 16QAM mode, the maximum required average power in the cell is about 15 dB less than 30 dBm. In comparison to the minimum required average power of 0.68dB, the transmitted power reduction is about 30 dB. The simulations without power control used a fixed transmission power of 30 dBm and achieved worst performance at the edge of the cells, as shown in Figure 20.11.

Figure 20.14 shows the average required transmission power versus FER for the

20.5. AVERAGE TRANSMISSION POWER

Modulation Scheme	Minimum Average Power in Cell (dBm)	Maximum Average Power in Cell (dBm)
BPSK	-9.82	-5.67
4QAM	-5.98	0.64
16QAM	0.68	15.02

Table 20.3: Average Transmission Power in the Cell in dBm, for Various Modulation Schemes and Interferer Positions

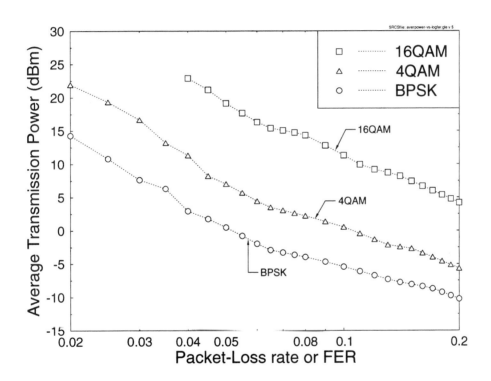

Figure 20.14: Average transmission power (dBm) versus transmission FER for all three modes of operation for users at 200 m from their BS moving at 28 mph, with a single interferer at the closest point in a co-channel cell, corresponding to the worst-case situation. The maximum transmission power is limited to 30dBm.

Parameter Name	Value
IncPowCount (IPC)	1
DecPowCount (DPC)	10
NearErrorFree (NEF)	0%
NearFrameError (NFE)	18%
IncPowStepSize (IPSS)	2
DecPowStepSize (DPSS)	5

Table 20.4: Parameter Values Used for the Previous Simulations Under the Conditions Described in Table 7.1

three modes of operation in the worst case, that is, for a user at the edge of the serving cell and for a single interferer in a position of the co-channel cell, where it produces maximum interference. The figure allows the trade-offs in terms of required transmission power, FER, and modulation mode to be interpreted. A user experiencing too high a FER can increase the transmission power, in order to reduce the FER. Alternatively, the user can switch to a more robust modulation scheme, resulting in a considerable reduction in the FER. Changing to a more robust modulation scheme and reducing power can maintain the same frame error rate, at the consequential loss of video quality. The reader can infer a range of further interesting system design aspects from this figure.

In the next section, we investigate the effect of varying the parameters of the power control algorithm. This assists in finding the set of near-optimal parameters for any propagation scenario.

20.6 Optimization of Power Control Parameters

This section finds the optimal set of power control algorithm parameters (see Table 20.2 on page 821), for all possible scenarios. Because of the computational complexity and the amount of time required to invoke standard methods of optimization, such as Powell's algorithm or simulated annealing [579], no optimization techniques were applied. In order to find a near-optimal set of parameters for a particular scenario, a heuristic method was used. This method varied one or two parameters, keeping the remainder of them constant. The knowledge inferred from these experiments may lead to an algorithm with self-adjusting parameters.

Our experiments showed that the power control algorithm's performance was independent of the modulation scheme used, but parameter changes were required if the vehicular speed or fading frequency changed. The parameter values used for our previous simulations under the channel conditions described in Table 7.1 are shown in Table 20.4.

Past experiments had shown that the crucial parameters were IPC and DPC in Table 20.2, since these determined the delay before the system could respond to potential error events and power up or down. In addition, the parameters NEF and NFE of Table 20.2 determined to what extent the bit error rate was used to predict

BCH-coded frame errors. The final two parameters, IPSS and DPSS of Table 20.2, govern how quickly the power control algorithm increases and decreases the step size. In the next section, we study the effect of jointly optimizing the IPC and DPC parameters for a variety of vehicular speeds and modulation schemes. This was accomplished by simulating carefully chosen combinations of the pair of parameters being optimized. When these results were plotted on a graph, each pair was evenly spaced on the axis. The order of the parameter pairs on the axis was chosen to give a monotonous increase or decrease in the value of the quantity monitored, which was either the FER or the transmission power.

20.6.1 Joint Optimization of IPC and DPC Parameters

In this section, we present the results related to the joint optimization of the IPC and DPC parameters. The IPC parameter indicates the required number of consecutive frame errors (or near frame errors) before the algorithm increases the transmission power. The DPC parameter of Table 20.2 stores the number of error-free (or near error-free) consecutive frames to wait before decreasing the transmission power. These parameters were varied, and their effect on the frame error rate and average transmission power was evaluated. Simulations were performed at different vehicular speeds using various modulation schemes in order to quantify the performance effects for different parameter settings.

Figure 20.15 shows the trade-off between the FER and the average transmission power for various combinations of IPC and DPC. Specifically, Figure 20.15(a) displays the performance of the BPSK mode at three different vehicular speeds. The figure demonstrates that typically, parameter changes are required in order to maintain the target FER, as the vehicular or fading speed changes. Observe in the figure that the required average transmission power increases as the vehicular speed increases. This is because, at faster vehicular speeds, the power control algorithm is less able to track the shadow fading of the channel. As the tracking accuracy of the algorithm decreases, the transmission power changes tend to "overshoot" the required power, leading to increased average transmission power. The corresponding performance trade-off graphs using 4QAM and 16QAM modulation are very similar; the only difference is the additional transmission power required for the less robust modulation schemes.

The performance trade-offs at 14 mph for the three different modulation schemes are shown in Figure 20.15(b). The figure shows that the FER is fairly independent of the modulation mode used. Furthermore, this figure can also be used to estimate the additional transmission power required to maintain the same FER, when using a higher level modulation scheme. Again, this figure shows the performance trade-offs at 14 mph, while the corresponding graphs at 28 and 56 mph are very similar, apart from the additional transmission power required at higher vehicular speeds.

The graphs shown in Figure 20.15 are reproduced in a different form in Figure 20.16. These graphs allow the FER versus average transmission power trade-off achieved by varying IPC and DPC to be more conveniently evaluated. Both graphs in Figure 20.16 show the average transmission power versus the FER. The corresponding graphs for 4QAM, 16QAM, 28 mph, and 56 mph follow similar trends, and hence

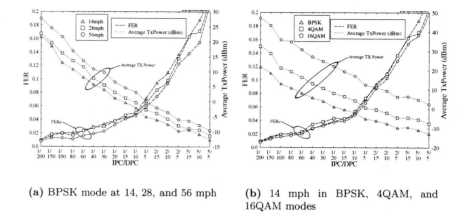

(a) BPSK mode at 14, 28, and 56 mph

(b) 14 mph in BPSK, 4QAM, and 16QAM modes

Figure 20.15: Frame error rate (FER) and average transmission power trade-offs, versus different combinations of IPC and DPC parameters, for a single-interferer at -200 m from its base station using the channel conditions of Table 7.1.

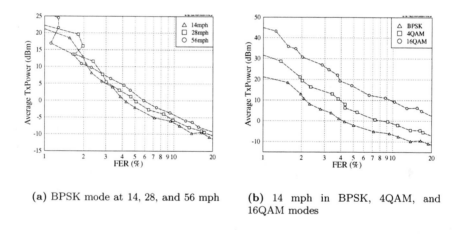

(a) BPSK mode at 14, 28, and 56 mph

(b) 14 mph in BPSK, 4QAM, and 16QAM modes

Figure 20.16: Average transmission power versus frame error rate (FER) for different combinations of IPC and DPC parameters, for a single-interferer at -200 m from its base station using the channel conditions of Table 7.1.

they are not shown for reasons of space.

Figure 20.16(a) shows the performance in BPSK mode at 14, 28, and 56 mph. This graph demonstrates that the required average transmission power increases at higher vehicular speeds due to increased inaccuracies in tracking the slow fading. It also shows that by small power or other parameter adjustments, the target FER can be maintained at different vehicular speeds. Figure 20.16(b) shows the performance at 14 mph in BPSK, 4QAM, and 16QAM modes. This figure allows the excess power required for higher level modulations schemes to be quantified for different target frame error rates.

To summarize this section, we have shown that the target FER can be varied from 1 to 20% using different combinations of values for the IPC and DPC parameters. We have also demonstrated that the performance of the power control algorithm is affected by the vehicular speed and by use of different modulation schemes. Furthermore, we have provided graphs that allow the maximum average transmission power to be estimated for each scenario. In the next section, we investigate the effect of jointly optimizing the NEF and NFE parameters for various modulation schemes and vehicular speeds. The NEF and NFE parameters of Table 20.2 determine to what extent the pre-decoding bit error rate within the transmission bursts is used to predict frame errors.

20.6.2 Joint Optimization of NEF and NFE

The power control algorithm operates by using the BCH frame error rate to predict when to change the transmission power. The power control algorithm's estimation of the required transmission power can be augmented by using the bit error rate to improve its accuracy. When the channel coding is not overloaded, an estimate of the bit error rate within the transmission bursts can be made. This bit error rate measurement can be used to help predict frame errors and when to power down. In our simulations, we employ BCH(n, k, t) block codes, which can correct t errors in a transmission burst. If no errors are corrected in a transmission burst, the bit error rate for this burst is zero. If the number of errors corrected is close to t, then the bit error rate is nearly high enough to cause a frame error. Therefore, the number of corrected errors gives a measure of the channel error rate and an estimate of how close to a frame error the system currently is. In order to make the system's operation independent of a particular BCH code, the number of corrected errors was re-mapped to a percentage, where 0% is error-free and 100% implies that the number of corrected errors was the maximum number of errors that the block code used can correct. The BCH codes used for our simulations are shown in Table 7.1 on page 257.

The NEF parameter is a percentage value of the maximum number of corrected errors. If the number of corrected errors in a particular transmission burst is less than or equal to NEF, the burst is classified as "near error-free." The rest of the power control algorithm then proceeds on the assumption that the burst was error-free. If the number of corrected errors in a transmission burst is higher than or equal to the NFE parameter, the burst is classified as a "near frame error" and the rest of the algorithm acts as if a frame error did occur.

In our simulations, we employed different combinations of the NEF and NFE parameters for a variety of modulation schemes and vehicular speeds. The NEF/NFE combination of 100%/100% implies that the bit error rate for a transmission burst is discarded and not used by the power control algorithm. An NEF/NFE combination of 0%/100% implies that the bit error rate is not used to predict frame errors, but a sequence of purely error-free frames is required to initiate a reduction in transmission power. The NEF/NFE combination of 25%/75% implies that if the number of corrected errors is consistently less than or equal to 25% of the correcting capacity of the

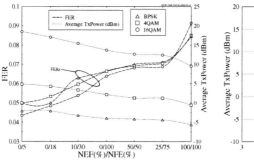 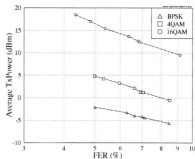

(a) Frame error rate (FER) and average transmission power tradeoffs versus different combinations of NEF and NFE parameters

(b) Average transmission power versus frame error rate (FER) for different combinations of NEF and NFE parameters

Figure 20.17: Frame error rate (FER) and average transmission power at 14 mph in BPSK, 4QAM, and 16QAM modes for a single interferer at -200 m from its base station using the channel conditions of Table 7.1.

code, then a reduction in power is required. Furthermore, if the number of corrected errors is higher than or equal to 75% of the error correction capacity of the code, then a frame error nearly occurred. If this behavior continues, then the transmission power should be increased to reduce the chance of a future frame error.

Figure 20.17 shows the effect of varying the NEF and NFE parameters on the frame error rate (FER) and average transmission power for various modulation schemes. These graphs characterize the performance at a vehicular speed of 14 mph; however, the performance is similar at 28 mph and 56 mph. At higher vehicular speeds, there is a small increase in the average transmission power, as was found when optimizing the IPC and DPC parameters.

Figure 20.17(a) shows the FER versus transmission power trade-off for different combinations of NEF and NFE. Observe in the figure that when the bit error rate information is discarded and not used by the power control algorithm, corresponding to the NEF/NFE combination of 100%/100%, the FER is significantly increased. When the bit error rate information is valid, the FER performance can be improved at the expense of a small increase in the average transmission power.

By biasing the algorithm's behavior toward increasing power, rather than decreasing when there are some bit errors, a reduced FER can be maintained. As an example, consider the NEF/NFE range of 0%/5% – 0%/100%, which implies stipulating NEF = 0%, while varying NFE from 5% to 100%. The more conservative we are in terms of reporting an NFE condition, say at 5% or 18% of the maximum number of correctable errors, the higher the required power, but the lower the FER and vice versa. However, the gradient of the transmission power curves is fairly low.

Figure 20.17(b) shows the average transmission power versus frame error rate

(FER). These graphs characterize the extra power required for higher order modulation schemes and lower target FERs. When this graph was plotted on top of the corresponding graph for IPC/DPC shown in Figure 20.16(b), it was found that the NEF/NFE graph overlapped with a section of the IPC/DPC graph, although this graph is not shown here for reasons of space economy. Since the NEF/NFE parameter changes have a smaller range of operation in terms of FER, the NEF/NFE parameters could be used to fine tune the performance. In an algorithm with self-adapting parameters, the IPC/DPC pair could set the FER to a value close to the target FER, and then the NEF and NFE parameters could be varied to fine-tune the performance to the exact target FER.

Earlier we showed that using the short-term bit error rate to assist in predicting transmission power changes is worthwhile, since the FER in Figure 20.17(b) is decreased significantly for a small increase in average transmission power. We also showed that the NEF/NFE parameters could be used to fine-tune the performance of the power control algorithm. Furthermore, in an adaptive scheme it may be possible to maintain the target FER, when the modulation scheme or vehicular speed changes just by changing the NEF/NFE parameters. Therefore, the NEF/NFE parameters are well suited to tuning the FER performance in a power control algorithm that adapts to changing propagation conditions.

20.6.3 Joint Optimization of IPSS and DPSS

The IPSS and DPSS parameters of Table 20.2 and Figure 20.1 determine how promptly the step size of the power control algorithm is increased after a succession of frame errors or error-free frames. For example, after a succession of frame errors, the power is increased. However, if the frame errors still persist after several increases in transmission power, the step size of the power control algorithm is increased. Increasing the step size allows the power control algorithm to increase the power more rapidly. The value of the IPSS or DPSS parameter is given by the number of successive power changes, before the step size is increased by 1dB. The step size of the algorithm is limited to the range of 1–16dB. The transmission power is only increased after the number of successive frame errors specified by the IPC parameter. Therefore, the minimum number of frame errors between a change in the step size is $IPC \times IPSS$. The effects of changing the IPSS and DPSS parameters are very dependent on the values to which the IPC and DPC parameters are set. This renders optimizing the IPSS and DPSS parameters an arduous task.

For example, if the IPSS parameter was set to 2 and the IPC parameter was set to 1, the sequence of transmission power changes in response to a series of successive frame errors would be +1, +1, +2, +2, +3, +3, ..., +15, +16, +16, +16. These power changes are clipped if they take the transmission power outside its dynamic range of 64 dB. Our experiments revealed that a suitable combination of the IPSS and DPSS parameters was an IPSS of 2 and a DPSS of 5. This leads to fast powering up and a slower, more cautious powering down. The weighting of the algorithm toward powering up rather than down is quite normal, since the desired behavior is to keep the transmission power slightly higher than the required power that would cause a frame error.

In our simulations we used an IPC value of 1 and a DPC value of 10. Therefore, for an IPSS/DPSS combination of 2/5, two consecutive frame errors (or near frame errors) would be required to increase the step size when powering up, and 50 consecutive error-free (or near error-free) frames would be required to increase the step size when powering down.

Since the IPSS and DPSS parameters are difficult to optimize owing to their interdependence with the IPC and DPC parameters, only a small set of simulations were conducted to find the effect of varying the IPSS and DPSS parameters. It was found that varying the IPSS and DPSS parameters could influence the frame error rate with about the same range of control as the NEF/NFE parameters. However, the performance was closely related to the IPC and DPC parameters used.

In a future version of the power control algorithm, the IPSS and DPSS parameters may become fixed or dependent on IPC and DPC in order to reduce the complexity of the algorithm. The proposed regime could be used in an adaptive modulation scheme, where the power control is used to smooth out the transient error effects of changing the modulation scheme.

20.6.4 Conclusions from Optimizing the Power Control Algorithm Parameters

We conducted three experiments in order to quantify the effect of changing the various power control parameters and augment our understanding of how the algorithm could be optimized for different scenarios. We found that the IPC and DPC parameters had the most dominant effect on the algorithm's frame error rate (FER) performance. However, to obtain a lower FER, an increase in transmission power is required. Changing the NEF and NFE parameters influenced the performance to a lesser degree, nonetheless sufficiently to be able to compensate for different vehicular speeds or modulation schemes. The final two parameters, IPSS and DPSS, were not investigated to the same degree as the others, since they are dependent on the values set for the IPC and DPC parameters, and hence they can be merged.

These experiments give us a deeper understanding of how the algorithm operates. With this knowledge, an enhanced version of the algorithm may be contrived that could adapt to various channel conditions. The target FER could be set approximately by setting the IPC and DPC parameters. The algorithm could then maintain the target FER by adapting the NEF and NFE parameters. In a future version of the algorithm, the IPSS and DPSS parameters could be fixed in order to reduce the algorithm's complexity. However, a better solution would be to set IPSS and DPSS to fixed values, which are dependent on the values set for IPC and DPC. In the next section, we investigate the performance of the power control algorithm, when the vehicular speed is reduced to that of a pedestrian.

20.7 Power Control Performance at Various Vehicular Speeds

In this section, we investigate the effect of vehicular speed on the performance of the power control algorithm. As the vehicular speed increased, the average transmission power increased in order to maintain the target FER, as shown in Figure 20.15(a). This is because at higher vehicular speeds the algorithm is less able to track the slow fading. The algorithm is usually configured to be biased toward powering up rather than powering down. This causes the increase in transmission power as the speed increases. The simulations performed in the previous parameter optimization section used vehicular speeds of 14, 28, and 56 mph. Therefore, it was decided to perform simulations at a much lower pedestrian speed of 3 mph, where the power control algorithm would be better able to track the slow fading.

20.7.1 Power Control Results for Pedestrians

At pedestrian velocities the shadow fading is fairly slow, typically fluctuating at 0.1 Hz. Therefore, long simulations were required in order to have a sufficient number of slow fades during each simulation. In addition, simulations were performed under otherwise identical conditions for a vehicular speed of 28 mph. These 28 mph simulations were performed to act as benchmarkers; hence, they were approximately 10 times shorter in terms of real time in order to provide results over the same number of slow-fading cycles.

Figure 20.18 shows the BCH-coded frame error rate (FER) performance at both 3 and 28 mph. The results are presented in two different forms to aid presentation. The 3D form allows the FER performance across the cell area to be conveniently viewed. The contour form of the results facilitates comparisons. It can be seen that the FER performance is very similar at 3 mph and 28 mph.

The signal-to-interference ratio (SIR) is also very similar at 3 mph and 28 mph, as shown in Figure 20.19. The power control algorithm, however, uses less transmission power at 3 mph than at 28 mph. This is because the algorithm is able to track the slow fading better. The comparison of the average transmission powers shown in Figure 20.20 reveals that, at the edge of the cell, the average transmission power is 3 to 4 dB less for the lower speed scenario.

20.7.2 Channel Fading

In our simulated mobile radio system, we used time division multiple access (TDMA). This implies that any mobile handset, which has access to a TDMA slot, can only evaluate the bit error rate (BER) performance of the channel during its own timeslot. Hence, the power control algorithm can only estimate the channel quality on a discontinuous basis. Simulations were performed in order to quantify how these discontinuous channel estimates affect the operation of the power control algorithm. In order to portray this discontinuous effect, a method for presenting the results was contrived, which we will refer to as sparsed-time methodology. Figure 20.21 shows graphs of the Rayleigh channel fading magnitude versus time and sparsed-time. The

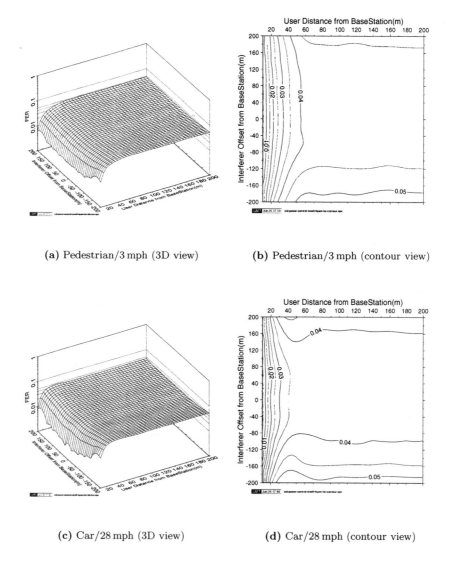

(a) Pedestrian/3 mph (3D view)   (b) Pedestrian/3 mph (contour view)

(c) Car/28 mph (3D view)   (d) Car/28 mph (contour view)

Figure 20.18: Simulated BCH frame error rate (FER) at 3 and 28 mph, using 4QAM and the channel conditions described in Table 7.1.

20.7. POWER CONTROL PERFORMANCE AT VARIOUS SPEEDS 847

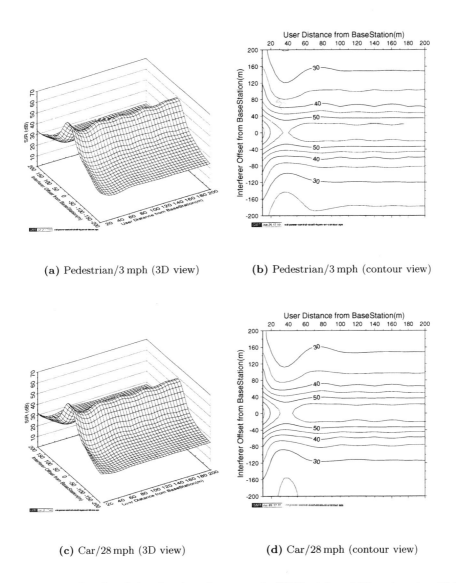

(a) Pedestrian/3 mph (3D view) (b) Pedestrian/3 mph (contour view)

(c) Car/28 mph (3D view) (d) Car/28 mph (contour view)

Figure 20.19: Simulated signal-to-interference ratio (SIR) at 3 and 28 mph, using 4QAM and the channel conditions described in Table 7.1.

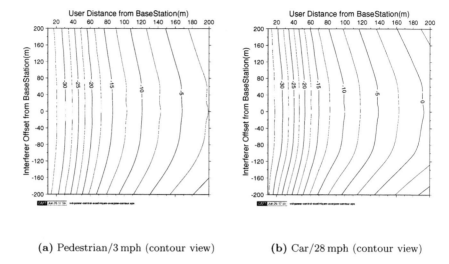

(a) Pedestrian/3 mph (contour view) (b) Car/28 mph (contour view)

Figure 20.20: Average transmission power (dB) at 3 and 28 mph, using 4QAM and the channel conditions of Table 7.1.

channel fading during a particular time-slot is highlighted. This highlighted fading is then plotted versus the "sparsed-time" in the lower graph. The channel fading during a time-slot is expanded in order to fit the whole transmission frame duration in the sparsed-time graph.

Figure 20.22 shows the channel fading versus "sparsed-time" at a variety of different vehicular speeds. The graphs also show the slow shadow fading on the same scale to help emphasize the effect of the vehicular speed. The figure shows the discontinuous nature of the fading experienced by a TDMA user. The discontinuity between a user's TDMA bursts is shown with a dotted line. At lower speeds, the fading within a TDMA burst is fairly constant. However, as expected, the fading in a TDMA burst becomes more variable at higher vehicular speeds.

In the next section, we show how the power control algorithm attempts to track the slow shadow fading at different vehicular speeds.

20.7.3 Tracking of Slow Fading

Earlier we showed that the power control algorithm can maintain the target frame error rate (FER) at a variety of different vehicular speeds. We also demonstrated that at lower vehicular speeds, the average transmission power is reduced. This is because of the power control's increased ability to track the slow-fading envelope of the channel response at lower vehicular speeds more closely, since the fading frequency reduces in proportion to the vehicular speed.

The power control algorithm varies the transmission power in response to channel errors. In this section, we present graphs showing how the transmission power changes with respect to the slow-fading envelope of the channel response. Figure 20.23(a)

20.7. POWER CONTROL PERFORMANCE AT VARIOUS SPEEDS 849

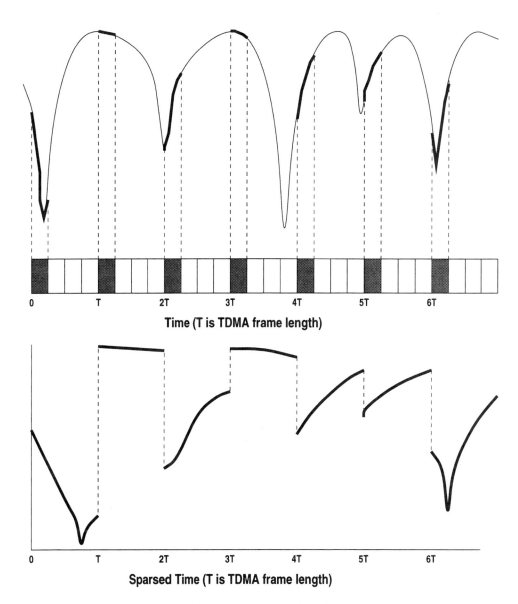

Figure 20.21: Rayleigh channel fading magnitude plotted versus time and "sparsed-time" for a TDMA system with four time slots per frame, and a frame duration of T

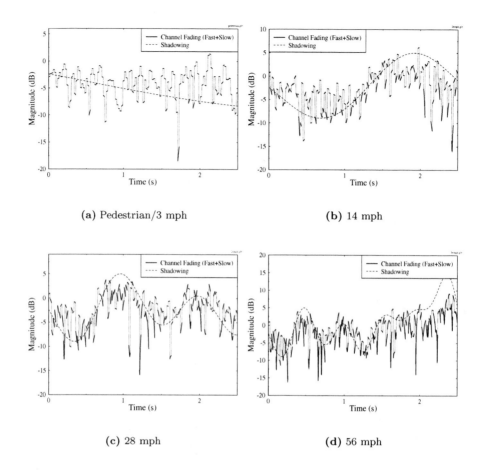

Figure 20.22: Simulated fast fading and slow shadow fading magnitude versus "sparsed-time" at a variety of different vehicular speeds.

portrays how the transmission power varies with respect to the slow shadow fading at a vehicular speed of 3 mph. The figure shows that the power control algorithm is attempting to render the transmission power the inverse of the shadow fading. In Figure 20.23(b), the transmission power is plotted with the inverse of the shadow fading. It can be seen that the transmission power is tracking the slow fading. The tracking of the slow shadow fading is not perfect, however, and there is some power overshooting, especially at higher vehicular speeds. Furthermore, since there is a fixed TDMA-frame delay in the system, the transmission power is adjusted after the channel attenuation has changed. This latency can be seen, for example, between TDMA frames 7600 and 7800 in the enlarged area of the graph. The detrimental effect of the power control latency becomes more apparent at higher vehicular speeds, as shown in Figure 20.24.

As the vehicular speed increases, the power control algorithm is less able to track

20.7. POWER CONTROL PERFORMANCE AT VARIOUS SPEEDS 851

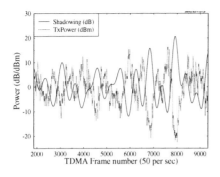
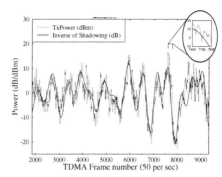

(a) Transmission power and slow fading versus time

(b) Transmission power and inverse of slow fading versus time

Figure 20.23: Simulated performance of the power control algorithm for a vehicular speed of 3 mph, showing how the algorithm attempts to track the slow fading.

the slow fading. At higher vehicular speeds the inherent system delay becomes more crucial. Figure 20.24 shows how the power control algorithm reacts to slow fading at 14, 28, and 56 mph. The graphs show the transmission power and the inverse of the slow shadow fading versus time. The system latency in changing the transmission power can be seen for a vehicular speed of 14 mph shown in Figure 20.24(a). At 56 mph shown in Figure 20.24(c), the power control algorithm is unable to track the slow fading well, resulting in a more constant transmission power. In the next section, we calculate the power control error histograms for the simulations that we used to demonstrate the tracking of the slow fading by the power control algorithm.

20.7.4 Power Control Error

In this section, we estimate the power control error of our algorithm. The power control error is defined as the difference between the actual transmitted power and the minimum required transmission power. We will use the assumption that the minimum required power is equal to the inverse of the slow fading of the channel. In reality, it will be slightly higher in order to cope with the effects of fast fading and interference.

From our simulations performed in the previous section, the histograms of the difference between the transmission power and the inverse of the slow fading of the channel were calculated. These power control error histograms are shown in Figure 20.25 for vehicular speeds of 3, 14, 28, and 56 mph. A positive power control error indicates an excess of transmission power, while a negative power control error indicates that the transmission power was less than the minimum required power. Again, the power control algorithm is biased toward increasing the transmission power in order to maintain a low frame error rate (FER). Therefore, the mean or average power control error observed in the figure is always greater than zero.

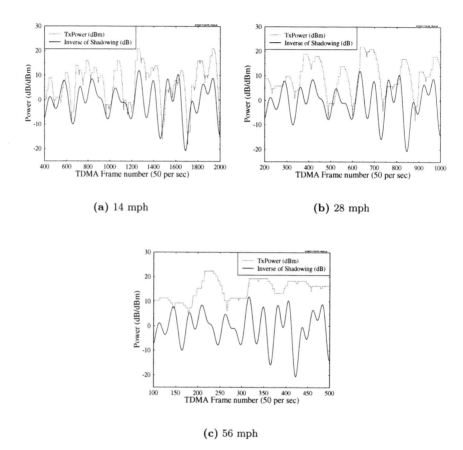

Figure 20.24: Simulated performance of the power control algorithm for vehicular speeds of 14, 28, and 56 mph, showing how the algorithm attempts to track the slow-fading portraying both the transmission power and the inverse of the slow "shadow" fading versus time.

20.7. POWER CONTROL PERFORMANCE AT VARIOUS SPEEDS

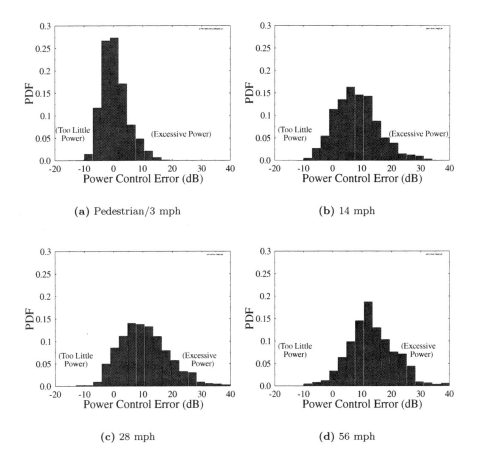

Figure 20.25: Histogram of the power control error for the proposed power control algorithm at vehicular speeds of 3, 14, 28, and 56 mph, using 4QAM and the channel conditions described in Table 7.1.

In order to enable the power control error at different vehicular speeds to be compared, the cumulative distribution function (CDF) of the power control error for the previous four vehicular speeds is plotted in Figure 20.26. The mean and variance of the power control error at the four vehicular speeds are shown in Table 20.5. The histograms and the table show that the mean power control error increases as the speed increases. This increase is due to the generic trend that the power control algorithm is less able to track the slow fading at higher vehicular speeds.

In the next section, we present results obtained for a particular cellular structure, with more interferers than the single-interferer situation that we have considered so far.

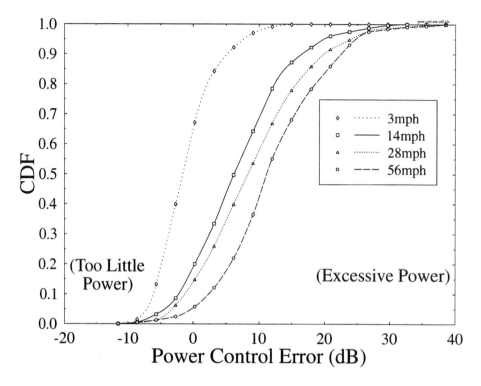

Figure 20.26: Cumulative distribution function (CDF) of the power control error for the proposed power control algorithm at vehicular speeds of 3, 14, 28, and 56 mph, using 4QAM and the channel conditions described in Table 7.1.

Vehicular Speed	Power Control Error (dB)		
	Mean	Variance	Standard Deviation
3mph	0.44	19.83	4.45
14mph	8.22	54.62	7.39
28mph	10.41	67.44	8.21
56mph	13.33	60.72	7.79

Table 20.5: Power Control Error Statistics of the Proposed Power Control Algorithm at Vehicular Speeds of 3, 14, 28 and 56 mph, Using 4QAM and the Channel Conditions Described in Table 7.1

20.8 Multiple Interferers

All our simulations using the power control algorithm have used a single interferer. Because we wanted to provide a worst-case performance study and because of the extra computation time required to simulate multiple interferers. In this section, we present results using multiple interferers. The multiple-interferer results used a simulated seven-cell reuse cluster employing an omnidirectional antenna, with all the interferers at the edge of their serving cell and at the closest point to the interfered cell. These conditions were chosen in order to simulate the worst-case condition for multiple co-channel interferers. The same conditions were used in Section 7.5, where we produced results for multiple interferers without power control.

The power control algorithm needs parameter changes in order to maintain our target frame error rate (FER) of 5% with multiple interferers. When the power control algorithm was used with the same set of parameters as for single interferers, the FER across the cell became 6.2%. In order to reduce the frame error rate to the target of 5%, the DPC parameter was increased from 10 to 12. The DPC parameter change increased the delay before the algorithm reduced the transmission power. When the power control algorithm was used with the modified parameter, the FER across the cell decreased to 4.9%.

When the power control algorithm was used in the single-interferer simulations, the FER in the close vicinity of the base station was less than the target FER because the transmission power could not be lowered below the minimum power limit. This resulted in a lower FER, when the transmission power was clipped at the minimum of −34 dBm and was therefore higher than the power required to maintain the target FER. This effect is also seen with multiple-interferer simulations. In the multiple-interferer simulations, the transmission power was in some instances limited to the maximum transmission power, when a user was close to the edge of a cell. This caused the FER near the edge of the cell to rise above the target FER. Therefore, to maintain the target FER across the whole cell area, the maximum transmission power had to be increased. It was found that a 10 dBm increase in the maximum transmission power to 40 dBm allowed the target frame error rate to be maintained across the cell area.

Therefore, the results shown below are for multiple interferers and were produced using the same parameters as we have used in previous power control simulations, except for an increase of the DPC parameter from 10 to 12 and an increase in the maximum transmission power to 40 dBm.

20.8.1 Frame Error Rate Performance

The frame error rate performance of the power control algorithm with multiple interferers is shown in Figures 20.27(a) and 20.27(b). This simulation was performed using the slightly modified set of the power control algorithm parameters described earlier. Across the majority of the cell area the FER is 4.9%. The only exception is observed in the close vicinity of the base station, where the FER is less, due to clipping of the transmission power. By allowing a larger dynamic range for the transmission power, this effect can be rendered smaller.

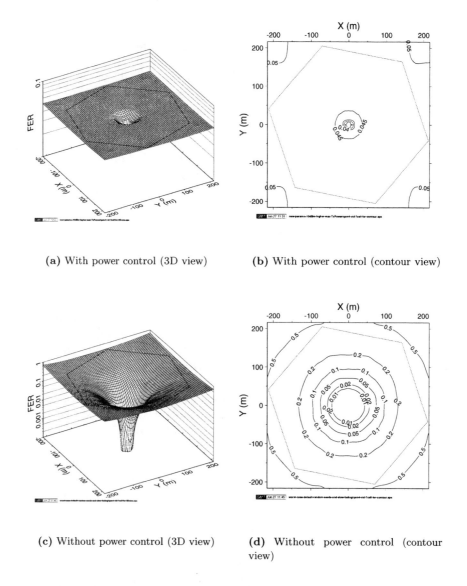

(a) With power control (3D view)   (b) With power control (contour view)

(c) Without power control (3D view)   (d) Without power control (contour view)

Figure 20.27: Simulated frame error rate (FER) performance with and without the power control algorithm over a hexagonal cell, employing a reuse factor of 7, path-loss exponent of 3.5, slow-fading frequency of 1 Hz, and standard deviation of 6 dB as well as 4QAM. The interference is generated by six video users in the co-channel cells. The interferers are at the edge of their cell at the closest point to the interfered cell, representing the worst case in terms of interference.

20.8. MULTIPLE INTERFERERS

Figure 20.27 shows the frame error rate performance with and without power control. When no power control is used, the frame error rate is dependent on the distance from the base station. When power control is employed, the FER within the cell is less than the target FER of 5%. Using Figure 20.27(d), we can see that when no power control is employed, the FER is only below the 5% limit for users up to 75 m from the base station. Therefore, the area of the cell for which the FER is less than 5% is about 14% of the hexagonal cell area.

In this section, we have shown how the FER performance improved with the use of power control in an environment having multiple interferers. In the next section, we show the performance improvement of the SIR, SINR, and the normalized channel capacity (NCC) with power control.

20.8.2 Further Effects of Power Control on System Performance

When the power control algorithm is used, the average signal-to-interference ratio (SIR) and signal-to-interference-plus-noise ratio (SINR) remain fairly constant across the whole cell area. The only variation in the SIR and SINR occurs close to the base station, where the transmission power cannot be reduced any further, resulting in a lower frame error rate (FER) and a higher SIR and SINR. These simulations represent an interference-limited scenario; hence, the SIR is approximately equal to the SINR. The SIR across the cell area is shown in Figure 20.28(a).

For our simulations, the SIR and SINR across the majority of the cell was 30 dB, when using power control. When no power control was used, the SIR and SINR decreased as users moved away from their base station. In the corresponding simulation without power control, this resulted in a SIR of 8dB at the edge of the cell, as seen in Figure 7.12(b).

Using Equations 20.1 and 20.2 of Section 20.5, we found that the average transmission power in the cell, when using power control, was 14.6 dB. The average transmission power in the cell for the simulations not using power control was 30 dB. Therefore, the power control gives better SIR and FER results while requiring a reduced transmission power.

In Section 7.5.3, we suggested a method for estimating the maximum possible normalized channel capacity (NCC) based on the SINR, as defined in Equation 7.15. Using this method, we found that the average normalized channel capacity across the cell was 9 bits/s/Hz, as shown in Figure 20.28(b). When no power control is used, the normalized channel capacity (NCC) decreases with distance from the base station. For simulations without power control under a similar set of conditions, the NCC was found to be about 2 bit/s/Hz at the edge of the cell; see Figure 7.17(b).

In this section, we have described the improvement with power control with respect to the SIR, SINR, NCC, and average transmission power. These results are tabulated in Table 20.6.

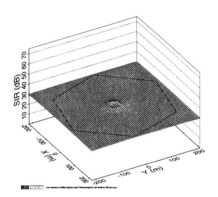 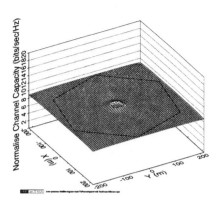

(a) Signal-to-interference ratio (SIR) across the cell area using power control

(b) Normalized channel capacity (NCC) across the cell area using power control

Figure 20.28: SIR and normalized channel capacity across a hexagonal cell, employing a reuse factor of 7, path-loss exponent of 3.5, slow-fading frequency of 1 Hz, and standard deviation of 6 dB, as well as 4QAM. The interference is generated by six video users in the co-channel cells. The interferers are at the edge of their cell at the closest point to the interfered cell, constituting the worst case in terms of interference.

	Without Power Control	With Power Control
SIR	8 dB	30.25 dB
SINR	8 dB	30 dB
NCC (bits/s/Hz)	1.7	9.1
Average Power	30 dBm	14.6 dBm

Table 20.6: Simulation results at edge of the cell with and without power control for a hexagon cell, employing a reuse factor of 7, path-loss exponent of 3.5, slow-fading frequency of 1Hz, and standard deviation of 6 dB, as well as 4QAM. The interference is generated by six video users in the co-channel cells. The interferers are at the edge of their cell at the closest point to the interfered cell, constituting the worst case in terms of interference.

20.9 Summary and Conclusions

In this chapter, we have proposed a power control algorithm based on an error rate in order to reduce co-channel interference and to maintain the required transmission integrity. To operate effectively, power control algorithms require an accurate and recent measure of channel quality. Since the algorithm uses the bit- and frame error rate for channel quality measurements, it is more reliable than the Received Signal Strength Indicator (RSSI)-based algorithms, especially in interference-limited scenarios.

The proposed algorithm was described using the flowchart in Figure 20.1. The algorithm has adjustable parameters, in order to allow the performance to be tuned for different conditions and performance requirements. These parameters were listed in Table 20.2. There are several benefits of using the power control algorithm, in addition to reducing the required transmission power and co-channel interference. Normally, in interference-limited environments without power control, the frame error rate (FER) performance is good close to the base station and poor at the edges of the cell. The power control algorithm allows these two extreme cases to be traded for a more constant FER performance across the whole cell. The power control algorithm has a similar effect on the signal-to-interference ratio (SIR), where the low SIR at the edge of the cell is increased while reducing the SIR at the center of the cell.

The power control algorithm has also been tested in conjunction with different modulation schemes, and the performance has been very similar in each case. This is because the FER-based algorithm is independent of the modulation scheme used. We simulated the power control algorithm and were able to estimate the average power within a cell, when power control was used. As expected, the power control algorithm gave an improved performance at a reduced average power requirement in comparison to the fixed-power scenarios, which operated without power control. This was demonstrated by Figure 20.12 and Table 20.3. We were also able to show in Figures 20.18–20.20 how the vehicular speed or channel fading frequency affected the power control algorithm. When the vehicular speed was low or the channel was fading slowly, the power control algorithm operated more effectively, as demonstrated by Figures 20.23(b) and 20.24. As the vehicular speed or fading frequency increased, the average power of the power-controlled mobiles increased, since the power control algorithm was less able to track the channel fading. However, as the vehicular speed or fading frequency increased, the FER and SIR performance remained fairly constant.

Even though the power control algorithm is of relatively low complexity, it is not straightforward to tune the algorithmic parameters in order to maintain the target performance requirements. In Section 20.6, we demonstrated the effect of modifying some of the algorithmic parameters described in Table 20.2. These results can be used to set the parameters for maintaining a certain desired performance. It was found that the IPC and DPC parameters have the most pronounced influence on the performance of the algorithm. The other parameters have less effect.

Finally, in this chapter we evaluated the performance of the power control algorithm with multiple interferers. This required a few parameter changes, owing to the more variable channel conditions caused by the interferers. We found that even in worst-case conditions, the algorithm was capable of maintaining a constant FER across the whole cell area. For the same channel conditions without power control, the FER was worse in 86% of the cell area. As can be seen in Table 20.6, the SIR and SINR are much better at the edge of the cell when power control is used. Furthermore, the table shows that the estimated maximum possible normalized channel capacity is 4.5 times better at the edge of the cell with power control. The power control performance improvements were achieved at half the transmission power of the scenarios without power control.

Lastly, we showed that the power control algorithm is suitable for multimode terminals, where the modulation schemes are varied according to the prevailing channel conditions.

Chapter 21

Adaptive Single-Carrier, Multicarrier, and CDMA-based Video Systems

P.J. Cherriman, L. Hanzo, T. Keller,
E-L. Kuan, C.S. Lee, S. Vlahoyiannatos,
C.H. Wong, B.L. Yeap

21.1 Turbo-Equalized H.263-Based Videophony for GSM/GPRS[1] [2]

21.1.1 Motivation and Background

The existing second-generation (2G) wireless systems [87] — which were also highlighted in Chapter 9 — constitute a mature technology. In the context of 2G systems, and, with the advent of videotelephony, attractive value-added services can be offered to a plethora of existing users. Although the 2G systems have not been designed with video communications in mind, with the aid of the specially designed error-resilient, fixed-rate video codecs proposed in Chapters 12–14 it is nonetheless realistic to provide videophone services over these low-rate schemes. Specifically, in this chapter we designed a suite of fixed-rate, proprietary video codecs capable of operating at a video scanning or refreshment rate of 10 frames/s over an additional speech channel of the 2G systems. These video codecs were capable of maintaining sufficiently

---

[1]This section is based on **P. Cherriman, B. L. Yeap, and L. Hanzo**: Turbo-Equalised H.263-based video telephony for GSM/GPRS; submitted to IEEE Tr. on CSVT, 2000.

[2]©2000 IEEE. Personal use of this material is permitted. However, permission to reprint/republish this material for advertising or promotional purposes or for creating new collective works for resale or redistribution to servers or lists, or to reuse any copyrighted component of this work in other works, must be obtained from IEEE.

Simulation Parameters	
Channel model	COST-207 hilly terrain
Carrier frequency	900 MHz
Vehicular speed	30 mph
Doppler frequency	40.3 Hz
Modulation	GMSK, $B_n = 0.3$
Channel coding	Convol.(n,k,K) = (2,1,5)
Octal generator polynomials	23, 33
Channel interleavers	Random (232, 928)
Turbo-coding interleavers	Random (116, 464)
Max turbo-equalizer iterations	10
No. of TDMA frame per packet	2
No. of slots per TDMA frame	1, 4
Convolutional decoder algorithm	LOG-MAP
Equalizer algorithm	LOG-MAP

Table 21.1: System parameters

low bit rates for the provision of videophony over an additional speech channel in the context of the operational 2G wireless systems [87] provided that low-dynamic head-and-shoulders video sequences of the 176 × 144-pixel Quarter Common Intermediate Format (QCIF) or 128 × 96-pixel sub-QCIF video resolution are employed. We note, however, that for high-dynamic sequences the 32 kbps typical speech bit rate of the cordless telephone systems [87], such as the Japanese PHS, the Digital European Cordless Telephone (DECT), or the British CT2 system, is more adequate in terms of video quality. Furthermore, the proposed programmable video codecs are capable of multirate operation in the third generation (3G) Universal Mobile Telecommunications System (UMTS) or in the IMT2000 and cdma2000 systems, which were summarized in Chapter 10.

Chapters 12–14 used constant video rate proprietary video codecs and reconfigureable Quadrature Amplitude Modulation (QAM)-based [580] transceivers. In this chapter we advocate constant-envelope Gaussian Minimum Shift Keying (GMSK) [87]. Specifically, we investigated the feasibility of H.263-based videotelephony in the context of an enhanced turbo-equalized GSM-like system, which can rely on power-efficient class-C amplification. The associated speech compression and transmission aspects are beyond the scope of this book [94].

The outline of this section is as follows. Section 21.1.2 summarizes the associated system parameters and system's schematic, while Section 21.1.3 provides a brief overview of turbo equalization. Section 21.1.4 characterizes the system both in terms of turbo equalization performance and video performance. Lastly, Section 21.1.5 provides our conclusions. Let us now consider the outline of the system.

21.1.2 System Parameters

The associated video system parameters for the GSM system are summarized in Table 21.1, and the system's schematic is portrayed in Figure 21.1. An advanced feature of the system is its employment of joint channel decoding and channel equalization,

21.1. TURBO-EQUALISED H.263-BASED VIDEOPHONY FOR GSM/GPRS

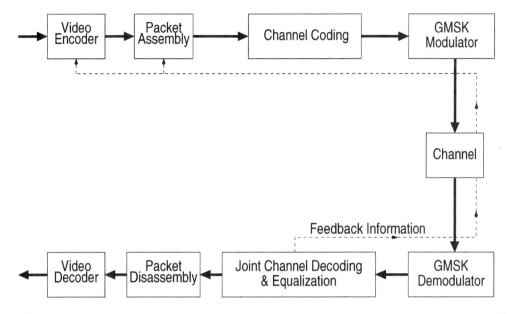

Figure 21.1: System schematic for turbo-equalized video system.

which is referred to as turbo equalization. The fundamental principles and motivation for using turbo equalization are described in Section 21.1.3.

The system uses the GSM frame structure of Chapter 9 and the COST-207 Hilly Terrain (HT) channel model, whose impulse response is shown in Figure 21.2. Each transmitted packet is interleaved over two GSM TDMA frames in order to disperse bursty errors.

The GPRS system allows the employment of multiple time-slots per user. We studied both a GSM-like system using 1 slot per TDMA frame and a GPRS-like arrangement with four slots per TDMA frame. In this scenario, the user is assigned half the maximum capacity of an eight-slot GPRS/GSM carrier. The bit rates associated with one and four slots per TDMA frame are shown in Table 21.2.

The effective video bit rates that can be obtained in conjunction with half-rate convolutional coding are 10 and 47.5 Kbit/s for the one and four slots per TDMA frame scenario, respectively. Again, the system's schematic is shown in Figure 21.1. The basic prerequisite for the duplex video system's operation (as seen in Figure 21.1) is use of the channel decoder's output is used for assessing whether the received packet contains any transmission errors. If it does, the remote transmitter is instructed (by superimposing a strongly protected packet acknowledgment flag on the reverse-direction message) to drop the corresponding video packet following the philosophy of [199]. This prevents the local and remote video reconstructed frame buffers from being contaminated by channel errors.

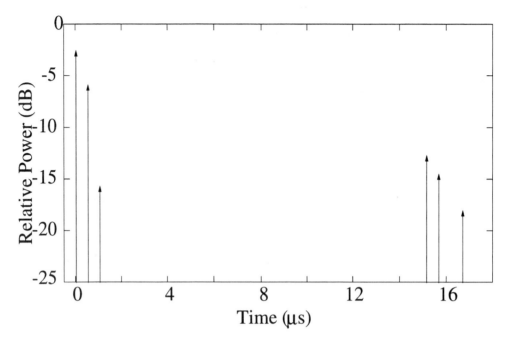

Figure 21.2: COST207-Hilly terrain channel model.

Bit rates, etc.		
Slots/TDMA frame	1	4
Coded bits/TDMA slot	116	116
Data bits/TDMA slot	58	58
Data bits/TDMA frame	58	232
TDMA frame/packet	2	2
Data bits/packet	116	464
Packet header (bits)	8	10
CRC (bits)	16	16
Video bits/packet	92	438
TDMA frame length	4.615 ms	4.615 ms
TDMA frames/s	216.68	216.68
Video packets per sec	108.34	108.34
Video bit rate (kbps)	10.0	47.5
Video frame rate (fps)	10	10

Table 21.2: Summary of System-Specific Bit Rates

21.1.3 Turbo Equalization

Turbo equalization [581] was proposed by Douillard, Picart, Jézéquel, Didier, Berrou, and Glavieux in 1995 for a serially concatenated rate $R = \frac{1}{2}$, convolutional-coded Binary Phase Shift Keying (BPSK) system. Specifically, Douillard *et al.* demonstrated that the turbo equalizer could mitigate the effects of intersymbol interference (ISI), provided that the channel impulse response (CIR) is known. Instead of performing the equalization and error correction decoding independently, better performance can be achieved by considering the channel's memory, when performing joint equalization and decoding iteratively. Gertsman and Lodge [582] then showed that the iterative process of turbo equalizers can compensate for the degradations caused by imperfect channel estimation. In the context of noncoherent detection, Marsland *et al.* [583] demonstrated that turbo equalization offered better performance than Dai's and Shwedyk's noncoherent, hard-decision-based receiver using a bank of Kalman filters [584]. Different iteration termination criteria [585], such as cross-entropy [586], were also investigated in order to minimize the number of iteration steps for the turbo equalizer. A turbo-equalization scheme for the Global System of Mobile Communications (GSM) was also proposed by Bauch and Franz [587], who investigated different approaches for overcoming the dispersion of the *a priori* information due to the interburst interleaving scheme used in GSM. Further research into combined turbo coding using convolutional constituent codes and turbo equalization has been conducted by Raphaeli and Zarai [588].

The basic philosophy of the original turbo-equalization technique derives from the iterative turbo-decoding algorithm consisting of two Soft-In/Soft-Out (SISO) decoders. This structure was proposed by Berrou *et al.* [28, 381]. Before proceeding with our in-depth discussion, let us briefly define the terms *a priori*, *a posteriori*, and extrinsic information, which we employ throughout this section.

A *priori* The *a priori* information associated with a bit v_m is the information known before equalization or decoding commences, from a source other than the received sequence or the code constraints. *A priori* information is also often referred to as intrinsic information to contrast it with extrinsic information.

Extrinsic The extrinsic information associated with a bit v_m is the information provided by the equalizer or decoder based on the received sequence and on the *a priori* information of all bits with the exception of the received and *a priori* information explicitly related to that particular bit v_m.

A *posteriori* The *a posteriori* information associated with a bit is the information that the SISO algorithm provides taking into account all available sources of information about the bit u_k.

The turbo equalizer of Figure 21.3 consists of a SISO equalizer and a SISO decoder. The SISO equalizer in the figure generates the *a posteriori* probability upon receiving the corrupted transmitted signal sequence and the *a priori* probability provided by the SISO decoder. However, at the initial iteration stages (i.e., at the first turbo-equalization iteration) no *a priori* information is supplied by the channel decoder. Therefore, the *a priori* probability is set to $\frac{1}{2}$, since the transmitted bits are assumed

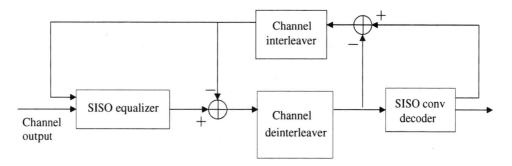

Figure 21.3: Structure of original turbo equalizer introduced by Douillard *et al.* [581].

to be equiprobable. Before passing the *a posteriori* information generated by the SISO equalizer to the SISO decoder of Figure 21.3, the contribution of the decoder in the form of — *a priori* information — accruing from the previous iteration must be removed in order to yield the combined channel and extrinsic information. This also minimizes the correlation between the *a priori* information supplied by the decoder and the *a posteriori* information generated by the equalizer. The term "combined channel and extrinsic information" indicates that they are inherently linked. In fact, they are typically induced by mechanisms, which exhibit memory. Hence, they cannot be separated. The removal of the *a priori* information is necessary, to prevent the decoder from "reprocessing" its own information, which would result in the positive feedback phenomenon, overwhelming the decoder's current reliability-estimation of the coded bits, that is, the extrinsic information.

The combined channel and extrinsic information is channel-deinterleaved and directed to the SISO decoder, as depicted in Figure 21.3. Subsequently, the SISO decoder computes the *a posteriori* probability of the coded bits. Note that the latter steps are different from those in turbo decoding, which only produces the *a posteriori* probability of the source bits rather than those of all channel-coded bits. The combined deinterleaved channel and extrinsic information are then removed from the *a posteriori* information provided by the decoder in Figure 21.3 before channel interleaving in order to yield the extrinsic information. This approach prevents the channel equalizer from receiving information based on its own decisions, which was generated in the previous turbo-equalization iteration. The extrinsic information computed is then employed as the *a priori* input information of the equalizer in the next channel equalization process. This constitutes the first turbo-equalization iteration. The iterative process is repeated until the required termination criteria are met [585]. At this stage, the *a posteriori* information of the source bits, which has been generated by the decoder, is utilized to estimate the transmitted bits.

Recent work by Narayanan and Stüber [589] demonstrates the advantage of employing turbo equalization in the context of coded systems invoking recursive modulators, such as Differential Phase Shift Keying (DPSK). Narayanan and Stüber emphasized the importance of a recursive modulator and show that high-iteration gains can be achieved, even when there is no ISI in the channel (i.e., for transmission over the nondispersive Gaussian channel). The advantages of turbo equalization as well as

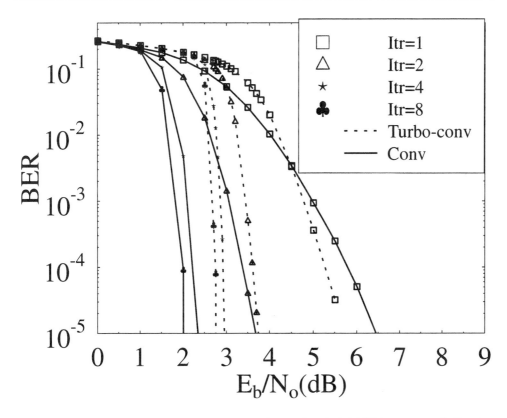

Figure 21.4: Comparison of the $R = 0.5$ convolutional-coded GMSK system with $R = 0.5$ convolutional-coding-based turbo-coded GMSK scheme, transmitting over the nondispersive Gaussian channel employing convolutional and turbo-coding-based turbo equalization, which performs eight turbo equalization iterations at the receiver.

the importance of a recursive modulator motivated our research on turbo equalization of coded partial response GMSK systems [87], since GMSK is also recursive in its nature. In our investigations, we have employed convolutional coding for the proposed turbo-equalized GSM-like video system as it has been shown in reference [590] that convolutional-coded GMSK systems are capable of providing large iteration gains — that is, gains in SNR performance with respect to the first iteration — with successive turbo-equalization iterations. We also observed that the convolutional-coded GMSK system employing turbo equalization outperformed the convolutional-coding based turbo-coded GMSK scheme, as demonstrated by Figure 21.4. Specifically, over a nondispersive Gaussian channel the convolutional-coded GMSK system had an approximately 0.8 dB better E_b/N_o performance than the corresponding convolutional-coding based turbo-coded GMSK scheme at BER = 10^{-4}. Although not explicitly shown here, similar findings were valid for dispersive Rayleigh-fading channels, where the advantage of the convolutional-coded system over the convolutional-coding-based turbo-coded scheme was approximately 1.0 dB.

These results were surprising since the more complex turbo-coded system was expected to form a more powerful encoded system compared to the convolutional-coded scheme. Below, we offer an interpretation of this phenomenon by considering the performance of both codes after the first turbo-equalization iteration. Specifically, over the nondispersive Gaussian channel in Figure 21.4, the convolutional-coded scheme yielded a lower BER than that of the turbo-coded system at E_b/N_o values below 4.5 dB, indicating that the *a posteriori* LLRs of the bits produced by the convolutional decoder had a higher reliability than that of the corresponding turbo-coded scheme. Consequently, upon receiving the higher-confidence LLR values from the decoder, the equalizer in the convolutional-coded scheme was capable of producing significantly more reliable LLR values in the subsequent turbo-equalization iteration, when compared to the turbo-coded system. After receiving these more reliable LLR values, the decoder of the convolutional-coded system will generate even more reliable LLR values. Hence, the convolutional-coded system outperformed the turbo-coded scheme after performing eight turbo-equalization iterations. Motivated by these trends — which were also confirmed in the context of dispersive Rayleigh-fading channels [590] — we opted for a convolutional-coded rather than turbo-coded GSM-like videophone system, which employs turbo equalization in order to enhance the video performance of the system.

21.1.4 Turbo-equalization Performance

Let us now characterize the performance of our video system. Figure 21.5 shows the bit error ratio (BER) versus channel SNR for the one- and four-slot scenarios, after one, two and ten iterations of the turbo equalizer. The figure shows the BER performance improvement upon each iteration of the turbo equalizer, although there is only a limited extra performance improvement after five iterations. The figure also shows that the four-slot scenario has a lower bit error ratio than the one-slot scenario. This is because the four-slot scenario has a longer interleaver, which renders the turbo-equalization process more effective due to its increased time diversity.

Let us now consider the associated packet-loss ratio (PLR) versus channel SNR performance in Figure 21.6. The PLR is a more pertinent measure of the expected video performance, since — based on the philosophy of [199] and on the previous two chapters — our video scheme discards all video packets, which are not error-free. Hence, our goal is to maintain as low a PLR as possible. Observe in Figure 21.6 that the associated iteration gains are more pronounced in terms of the packet-loss ratio than in bit error ratio.

It should also be noted that for low SNRs the packet-loss performance of the four-slot system is inferior to that of the one-slot system, while the bit error ratio is similar or better at the same SNRs. This is because the probability of having a single-bit error in the four-slot video packet is higher due to its quadruple length. This phenomenon is further detailed in Section 21.1.4.2.

21.1. TURBO-EQUALISED H.263-BASED VIDEOPHONY FOR GSM/GPRS

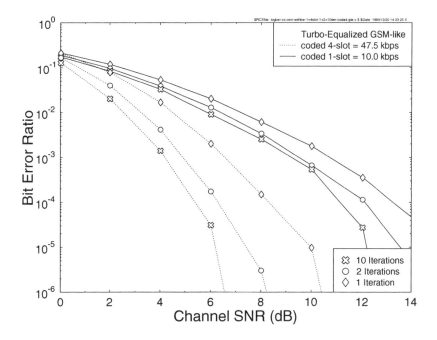

Figure 21.5: BER versus channel SNR for one and four slots per TDMA frame, and for 1, 2, and 10 turbo-equalizer iterations, over the channel of Figure 21.2.

21.1.4.1 Video Performance

The PLR performance is directly related to the video quality of our video system. Figure 21.7 shows the associated average PSNR versus channel SNR performance, demonstrating that an improved video quality can be maintained at lower SNRs as the number of iterations increases. In addition, the higher bit rate of the four-slot system corresponds to a higher overall video quality. Up to 6 dB SNR-gain can be achieved after 10 turbo-equalization iterations, as seen in Figure 21.7.

Figure 21.7 characterizes the performance of the highly motion-active "Carphone" sequence. However, the performance improvements are similar for the low-activity "Miss America" video sequence, as seen in Figure 21.8. Observe that the lower-activity "Miss America" video sequence is represented at a higher video quality at the same video bit rate. A deeper insight into the achievable video-quality improvement in conjunction with turbo equalization can be provided by plotting the video quality measured in PSNR (dB) versus time, as seen in Figure 21.9 for the "Miss America" video sequence using the four-slot system at a channel SNR of 6 dB for one, two, and ten iterations of the turbo equalizer.

Specifically, the bottom-trace of the figure shows how the video quality varies in the one-iteration scenario, which is equivalent to conventional equalization. The sudden reductions in video quality are caused by packet-loss events, which result in parts of the picture being "frozen" for one or possibly several consecutive video frames. The

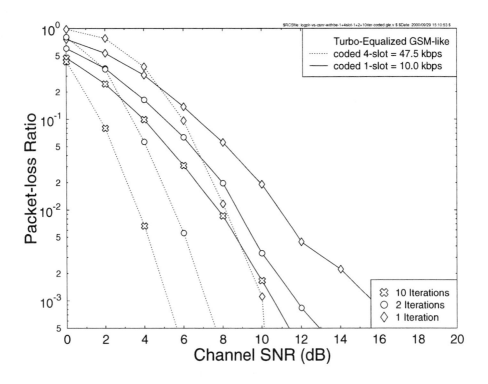

Figure 21.6: Video packet-loss ratio versus channel SNR for one and four slots per TDMA frame, and for 1, 2, and 10 turbo-equalizer iterations, over the channel of Figure 21.2 using convolutional coding.

sudden increases in video quality are achieved when the system updates the "frozen" part of the video picture in subsequent video frames. The packet-loss ratio for this scenario was 10% at the stipulated SNR of 6dB.

The video quality improved significantly with the aid of two turbo-equalizer iterations, while the packet-loss ratio was reduced from 10% to 0.7%. In the time interval shown in Figure 21.9, there are eight lost video packets, six of which can be seen as sudden reductions in video quality. However, in each case the video quality recovered with the update of the "frozen" picture areas in the next video frame.

We have found that the maximum acceptable PSNR video-quality degradation with respect to the perfect-channel scenario was about 1 dB, which was associated with nearly unimpaired video quality. In Table 21.3 we tabulated the corresponding minimum required channel SNRs that the system can operate at for a variety of scenarios, extracted from Figure 21.7. As the table shows, the minimum operating channel SNR for the one- and four-slot system using one iteration is 9 dB and 7.5 dB, respectively. This corresponds to a system using conventional equalization. A system using two turbo-equalization iterations can reduce these operating SNRs to 7.2 dB and 5.2 dB, respectively. The minimum operating SNRs can be reduced to as low as

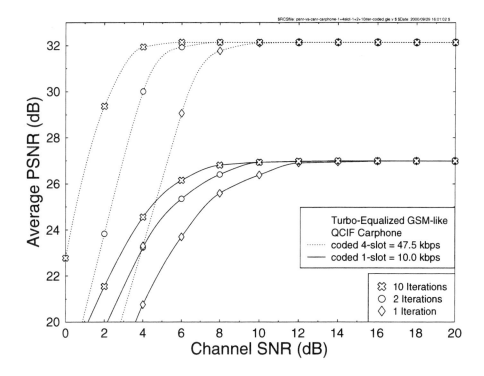

Figure 21.7: Video quality in PSNR (dB) versus channel SNR for one and four slots per TDMA frame, and for 1, 2, and 10 iterations of the turbo-equalizer upon using the highly motion active "Carphone" video sequence over the channel of Figure 21.2 using convolutional coding.

Channel SNR for 1 dB Loss of PSNR		
Slots/TDMA Frame	1	4
1 Iteration	9.0 dB	7.5 dB
2 Iterations	7.2 dB	5.2 dB
3 Iterations	6.44 dB	3.9 dB
4 Iterations	6.37 dB	3.7 dB
10 Iterations	5.8 dB	3.4 dB

Table 21.3: Minimum Required Operating Channel SNR for the QCIF "Carphone" Sequence over the Channel of Figure 21.2

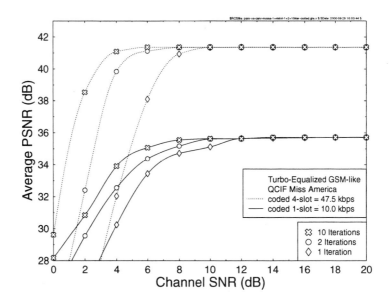

Figure 21.8: Video quality in PSNR (dB) versus channel SNR for one and four slots per TDMA frame, and for 1, 2, and 10 turbo-equalizer iterations, using the low-activity "Miss America" video sequence over the channel of Figure 21.2 using convolutional coding.

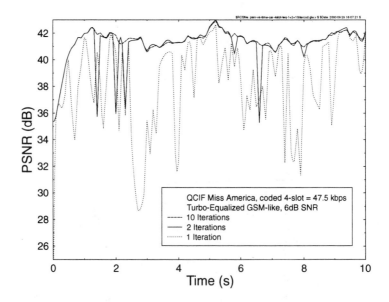

Figure 21.9: Video quality in PSNR (dB) versus time using four slots per TDMA frame, and for 1, 2, and 10 iterations of the turbo-equalizer for the low-activity "Miss America" video sequence using convolutional coding.

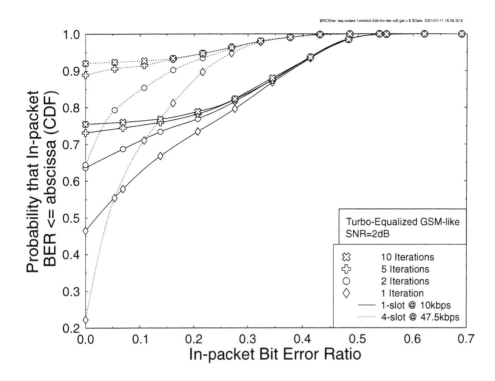

Figure 21.10: CDF of the "in-packet" BER at a channel SNR of 2 dB over the channel of Figure 21.2, and for various numbers of iterations for the turbo-equalized one- and four-slot systems using convolutional coding.

5.8 dB and 3.4 dB for the one- and four-slot systems, respectively, when invoking ten iterations.

21.1.4.2 Bit Error Statistics

In order to demonstrate the benefits of turbo equalization more explicitly, we investigated the mechanism of how turbo equalization reduces the bit error and packet-loss ratios. We found that the distribution of the bit errors in video packets after each iteration provided interesting insights. Hence, the CDF of the number of bit errors per video packet was evaluated. In order to allow a fair comparison between the one- and four-slot system, we normalized the number of bit errors per packet to the video packet size, thereby producing the CDF of "in-packet" BER.

Figure 21.10 shows the CDF of the "in-packet" BER for a channel SNR of 2 dB and for one, two, five, and ten iterations for both the one- and four-slot systems. The value of the CDF for an "in-packet" BER of zero is the probability that a packet is error-free and so can be interpreted as the packet success ratio (PSR). The packet-loss ratio is equal to 1 minus the PSR. For example, the four-slot system in Figure 21.10 at one iteration has a packet success ratio of 0.22, which corresponds to a packet-loss

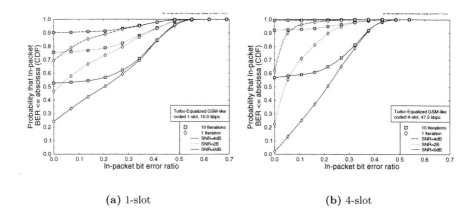

(a) 1-slot **(b)** 4-slot

Figure 21.11: CDF of "in-packet" BER performance over the channel of Figure 21.2 for the turbo-equalized one- and four-slot system for channel SNRs of 0 dB, 2 dB, and 4 dB and 1 and 10 iterations using convolutional coding.

ratio of 78%.

Both the one- and four-slot systems increase the PSR as the number of iterations increases. For example, the four-slot system increases the PSR from 22% to 92% as the number of iterations is increased from one to ten. This corresponds to a reduction in the packet-loss ratio from 78% to 8%. However, the CDF of "in-packet" BER can provide further insight into the system's operation. It can be seen in Figure 21.10 that the turbo-equalizer iterations reduce the number of packets having "in-packet" BERs of less than 30%. However, the probability of a packet having an "in-packet" BER higher than 35% is hardly affected by the number of iterations, since the number of bit errors is excessive, hence overwhelming even the powerful turbo equalization.

Figures 21.5 and 21.6, show that the four-slot system always has a lower BER than the one-slot system, although at low SNRs the PLR is higher for the four-slot system. The CDF in Figure 21.10 can assist in interpreting this further. The CDF shows that the PSR improves more significantly for the four-slot system than for the one-slot system as the number of iterations increases. This is because the four-slot system allows the employment of a longer interleaver, thereby improving the efficiency of the turbo equalizer. However, the CDF also underlines a reason for the lower BER of the four-slot system across the whole range of SNRs, demonstrating that the probability of packets having a high "in-packet" bit error rate is lower for the four-slot system. Since packets having a high "in-packet" BER have a more grave effect on the overall BER than those packets having a low "in-packet" BER, this explains the inferior overall BER performance of the one-slot system.

Figure 21.11 shows the CDF of "in-packet" BER for conventional equalization and with the aid of ten turbo-equalizer iterations for 0 dB, 2 dB, and 4 dB channel SNRs. Figure 21.11(a) represents a one-slot system, and Figure 21.11(b) a four-slot system. The figures also show the packet-loss ratio performance improvement with the aid of turbo equalization.

21.1.5 Summary and Conclusions

In this section the performance of turbo-equalized GSM/GPRS-like videophone transceivers was studied over dispersive fading channels as a function of the number of turbo-equalization iterations. Iteration gains in excess of 4 dB were attained, although the highest per iteration gain was achieved for iteration indices below 5. As expected, the longer the associated interleaver, the better the BER and PLR performance. In contrast to our expectations, the turbo-coded, turbo-equalized system was outperformed by the less complex convolutional coded turbo-equalized system. In conclusion, GPRS/GSM are amenable to videotelephony, and turbo equalization is a powerful means of improving the system's performance. Our future work will improve the system's performance invoking the forthcoming MPEG4 video codec, using space-time coding and burst-by-burst adaptive turbo equalization.

21.2 Wideband Burst-by-Burst Adaptive QAM-Based Wireless Videophony[3][4]

21.2.1 Introduction

In contrast to our previous chapters, where we used statically reconfigurable narrowband multimode modems, in this chapter we will consider a range of dynamically or near-instantaneously reconfigurable modems. These near-instantaneously reconfigurable transceivers are also often referred to as burst-by-burst adaptive modems, which were discussed in some depth in Chapter 5. Their philosophy is that a higher-order modulation mode is invoked, when the channel quality is favorable, in order to increase the system's bits per symbol capacity. Conversely, a more robust lower order modulation scheme is employed when the channel exhibits inferior channel quality, in order to improve the mean bit error ratio (BER) performance. A practical scenario, in which adaptive modulation can be applied is as follows: A reliable, low-delay feedback path is created between the transmitter and receiver, for example, by superimposing the estimated channel quality perceived by the receiver on the reverse-direction messages of a duplex interactive channel. The transmitter then adjusts its modem mode according to this perceived channel quality.

Recent developments in adaptive modulation over a narrowband channel environment have been pioneered by Webb and Steele [89], who utilized the modulation adaptation in a Digital European Cordless Telephone like (DECT) system. The concept of variable-rate adaptive modulation was further developed by Sampei *et al.* [559], showing promising advantages compared to fixed modulation in terms of spectral efficiency, BER performance, and robustness against channel delay spread. In another

---

[3]This section is based on **P. Cherriman, C. H. Wong, and L. Hanzo**, Wideband Burst-by-burst Adaptive H.263-assisted Wireless Video Telephony, submitted to IEEE Transactions on Circuits and Systems for Video Technology, 1999.

[4]©1999 IEEE. Personal use of this material is permitted. However, permission to reprint/republish this material for advertising or promotional purposes or for creating new collective works for resale or redistribution to servers or lists, or to reuse any copyrighted component of this work in other works, must be obtained from IEEE.

paper, the numerical upper bound performance of adaptive modulation in a slow Rayleigh flat-fading channel was evaluated by Torrance *et al.* [110] Subsequently, the optimization of the switching threshold levels using Powell minimization was proposed in order to achieve a targeted performance [111]. In addition, Matsuoka *et al.* [591] as well as Goldsmith and Chua [138, 592] studied modulation in conjunction with channel coding and power control techniques.

In the narrowband channel environment, the quality of the channel was determined by the short-term signal-to-noise ratio (SNR) of the received burst, which was then used as a criterion in order to choose the appropriate modulation mode for the transmitter, based on a list of switching threshold levels, l_n [89, 110, 119, 559]. However, in a wideband environment, this criterion is not an accurate measure of the quality of the channel, where the existence of multipath components produces not only power attenuation of the transmission burst but also intersymbol interference. Subsequently, a new criterion has to be defined to estimate the wideband channel quality in order to choose the appropriate modulation scheme.

In addition, the wideband channel-induced degradation is mitigated not only by the employment of adaptive modulation but also by equalization. In following this line of thought, we can formulate a two-step methodology in reducing the effects of the dispersive wideband channel. In the first step, the equalization process eliminates most of the intersymbol interference based on a Channel Impulse Response (CIR) estimate derived using the channel sounding midamble. Consequently, the signal-to-noise and residual interference ratio at the output of the equalizer is calculated.

We found that the residual channel-induced intersymbol interference (ISI) at the output of the decision feedback equalizer (DFE) was near-Gaussian distributed and that if there were no decision feedback errors, the pseudo-SNR at the output of the DFE, γ_{dfe} could be calculated as [593]:

$$\begin{aligned} \gamma_{dfe} &= \frac{\text{Wanted Signal Power}}{\text{Residual ISI Power} + \text{Effective Noise Power}} \\ &= \frac{E\left[|S_k \sum_{m=0}^{N_f-1} C_m h_m|^2\right]}{\sum_{q=-(N_f-1)}^{-1} E\left[|f_q S_{k-q}|^2\right] + N_o \sum_{m=0}^{N_f-1} |C_m|^2} \end{aligned} \quad (21.1)$$

where C_m and h_m denote the DFE's feedforward coefficients and the channel impulse response, respectively. The transmitted signal and the noise spectral density are represented by S_k and N_o, respectively. Lastly, the number of DFE feedforward coefficients is denoted by N_f. By utilizing the pseudo-SNR at the output of the equalizer, we are ensuring that the system performance is optimized by employing equalization and adaptive quadrature amplitude modulation [29] (AQAM)in a wideband environment, according to the following switching regime:

$$\text{Modulation Mode} = \begin{cases} BPSK & \text{if } \gamma_{DFE} < f_1 \\ 4QAM & \text{if } f_1 < \gamma_{DFE} < f_2 \\ 16QAM & \text{if } f_2 < \gamma_{DFE} < f_3 \\ 64QAM & \text{if } \gamma_{DFE} > f_3, \end{cases} \quad (21.2)$$

where $f_n, n = 1...3$ are the pseudo-SNR thresholds levels, which are set according to the system's integrity requirements.

Parameter	Value
Carrier frequency	1.9 GHz
Vehicular speed	30 mph
Doppler frequency	85 Hz
Norm. Doppler freq.	3.27×10^{-5}
Channel type	COST 207 Typical Urban (see Figure 21.12)
No. of impulse response taps	4
Data modulation	Adaptive QAM (BPSK, 4QAM, 16QAM, 64QAM)
Receiver type	Decision feedback equalizer Number of forward filter taps = 35 Number of backward filter taps = 7

Table 21.4: Modulation and Channel Parameters

Having reviewed the background of burst-by-burst adaptive modems, we now focus our attention on video issues. Färber, Steinbach, and Girod at Erlangen University contrived various error-resilient H.263-based schemes [594]; and Sadka, Eryurtlu, and Kondoz [595] from Surrey University proposed various improvements to the H.263 scheme. The philosophy of our proposed schemes follows that of the narrowband, statically configured multimode system introduced in [199], employing an adaptive rate control and packetization algorithm, supporting constant Baud-rate operation. **In this section, we employed wideband burst-by-burst adaptive modulation in order to quantify the video performance benefits of such systems.** It is an important element of the system that when the BCH codes protecting the video stream are overwhelmed by the plethora of transmission errors, we refrain from decoding the video packet in order to prevent error propagation through the reconstructed frame buffer [199]. Instead, these corrupted packets are dropped and the reconstructed frame buffer will not be updated until the next packet replenishing the specific video frame area arrives. The associated video performance degradation is fairly minor for packet dropping or frame error rates (FER) below about 5%. These packet-dropping events are signaled to the remote decoder by superimposing a strongly protected 1-bit packet acknowledgment flag on the reverse-direction packet, as outlined in [199].

This section is structured as follows. Subsection 21.2.2 introduces the video transceiver parameters, which is followed by Subsection 21.2.3, focusing on the video performance analysis of the proposed burst-by-burst adaptive system. Lastly, Subsection ch-switch characterizes the effects of the adaptive modem mode switching thresholds on the system's video performance.

21.2.2 Adaptive Video Transceiver

In this section, we used 176 × 144 pixel QCIF-resolution, 30 frames per sec scanned video sequences encoded at bit rates resulting in high perceptual video quality. Table 21.4 shows the modulation and channel parameters employed. The COST207 [337]

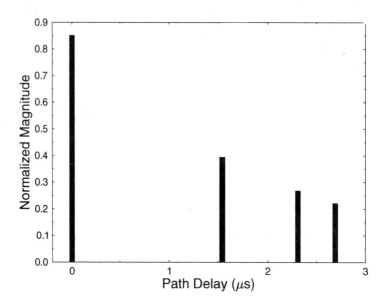

Figure 21.12: Normalized channel impulse response for the COST 207 [337] four-path Typical Urban channel.

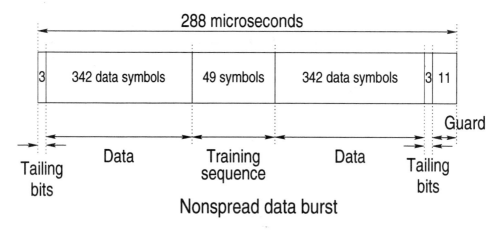

Figure 21.13: Transmission burst structure of the FMA1 nonspread data burst mode of the FRAMES proposal [596].

four-path typical urban (TU) channel model was used and its impulse response is portrayed in Figure 21.12. We used the Pan-European FRAMES proposal [596] as the basis for our wideband transmission system, the frame structure of which is shown in Figure 21.13. Employing the FRAMES Mode A1 (FMA1) so-called nonspread data burst mode required a system bandwidth of 3.9 MHz, when assuming a modulation excess bandwidth of 50%. A range of other system parameters are shown in

21.2. ADAPTIVE QAM-BASED WIRELESS VIDEOPHONY

Features	Value
Multiple access	TDMA
No. of slots/frame	16
TDMA frame length	4.615ms
TDMA slot length	$288\mu s$
Data symbols/TDMA slot	684
User data symbol rate (KBd)	148.2
System data symbol rate (MBd)	2.37
Symbols/TDMA slot	750
User symbol rate (KBd)	162.5
System symbol rate (MBd)	2.6
System bandwidth (MHz)	3.9
Eff. user bandwidth (kHz)	244

Table 21.5: Generic System Features of the Reconfigurable Multimode Video Transceiver, Using the Nonspread Data Burst Mode of the FRAMES Proposal [596] Shown in Figure 21.13

Table 21.5.

The proposed video transceiver is based on the H.263 video codec [192], which was the subject of Chapters 18 and 19. The video coded bit-stream was protected by binary Bose-Chaudhuri-Hocquenghem (BCH) coding, as discussed in Chapter 4 [9], combined with an intelligent burst-by-burst adaptive wideband multimode Quadrature Amplitude Modulation (QAM) modem, which was considered in depth in Chapter 5 [29]. The adaptive QAM scheme was configurable either under network control or under transceiver control, in order to operate as a 1, 2, 4, and 6 bits/symbol scheme, while maintaining a constant signaling rate. This allowed us to support an increased throughput expressed in terms of the average number of bits per symbol, when the instantaneous channel quality was high, leading ultimately to an increased video quality in a constant bandwidth.

The transmitted bit rate for all four modes of operation is shown in Table 21.6. The unprotected bit rate before approximately half-rate BCH coding is also shown in the table. The actual useful bit rate available for video is slightly less than the unprotected bit rate due to the required strongly protected packet acknowledgment information and packetization information. The effective video bit rate is also shown in the table.

21.2.3 Burst-by-Burst Adaptive Videophone Performance

The proposed burst-by-burst adaptive modem maximizes the system capacity available by using the most appropriate modulation mode for the current instantaneous channel conditions. We found that the pseudo-SNR at the output of the channel equalizer was an adequate channel quality measure in our burst-by-burst adaptive wideband modem. Figure 21.14 demonstrates how the burst-by-burst adaptive modem changes its modulation modes every transmission burst, that is, every 4.615 ms,

Features	Multirate System			
Mode	BPSK	4QAM	16QAM	64QAM
Bits/symbol	1	2	4	6
FEC	Near Half-rate BCH			
Transmission Bit rate (kbit/s)	148.2	296.4	592.8	889.3
Unprotected Bit rate (kbit/s)	75.8	151.7	303.4	456.1
Effective Video rate (kbit/s)	67.0	141.7	292.1	446.4
Video fr. rate (Hz)	30			

Table 21.6: Operational-mode Specific Transceiver Parameters

based on the fluctuating pseudo-SNR. The right-hand-side vertical axis indicates the associated number of bits per symbol.

By changing to more robust modulation schemes automatically, reduction in the channel quality allows the packet-loss ratio, or synonymously, the FER, to be reduced, which results in increased perceived video quality. In order to judge the benefits of burst-by-burst adaptive modulation, we considered two scenarios. In the first, the adaptive modem always chose the perfectly estimated AQAM modulation mode in order to provide a maximum upper bound performance. In the second scenario, the modulation mode was based on the perfectly estimated AQAM modulation mode for the previous burst, which corresponded to a delay of one Time Division Multiple Access (TDMA) frame duration of 4.615 ms. This second scenario represents a practical burst-by-burst adaptive modem, where the one-frame channel quality estimation latency is due to superimposing the receiver's perceived channel quality on a reverse-direction packet, for informing the transmitter concerning the best mode to be used

The probability that the adaptive modem will use each modulation mode for a particular average channel SNR is portrayed in Figure 21.15 in terms of the associated modem mode probability density functions (PDFs). It can be seen at high-channel SNRs that the modem mainly uses the 64QAM modulation mode, while at low-channel SNRs the BPSK mode is the most prevalent one.

Figure 21.16 shows the transmission FER (or packet-loss ratio) versus channel SNR for the 1-, 2-, 4-, and 6-bit/symbol fixed modulation schemes, as well as for the ideal and for the one-frame delayed realistic scenarios using the burst-by-burst adaptive QAM (AQAM) modem. A somewhat surprising fact is [597] — which is not explicitly shown here due to lack of space — that at low SNRs AQAM can maintain a lower BER than BPSK, since under inferior instantaneous channel conditions it exhibits the corresponding BPSK BER But the average number of transmitted AQAM bits is higher than that of BPSK, resulting in a reduced average AQAM BER. At high SNRs the AQAM BER curve asymptotically joins the 64QAM BER curve, since 64QAM is the predominant mode used. The associated FER curve obeys similar tendencies in terms of having the BPSK and 64QAM FER curves as asymptotes at low and

21.2. ADAPTIVE QAM-BASED WIRELESS VIDEOPHONY

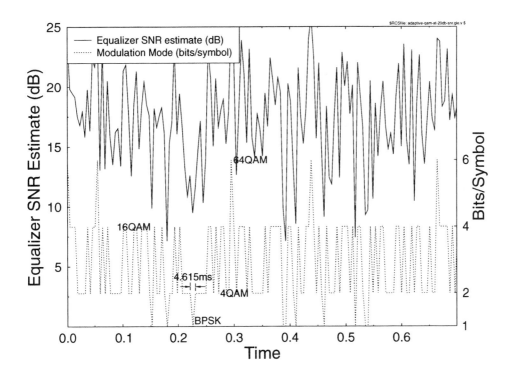

Figure 21.14: Adaptive burst-by-burst modem in operation for an average channel SNR of 20 dB, where the modulation mode switching is based on the SNR estimate at the output of the equalizer, using the channel parameters defined in Table 21.4.

high SNRs, respectively. However, in contrast to the BER, the AQAM FER cannot be lower than that of BPSK, since the same number of frames is transmitted in both cases. The substantial advantage of AQAM is that due to its higher number of bits/symbol the number of bits transmitted per frame is higher, resulting in an increased video quality.[5] The FER curves are portrayed on a logarithmic scale in Figure 21.17, where, for the sake of comparison, we also showed the associated FER curve for statically reconfigured modem modes switching at 5% transmission FER, as will be detailed below.

The statically reconfigured modem was invoked in Figure 21.17 as a benchmarker in order to indicate how a system would perform, which is unable to act on the basis of the near-instantaneously varying channel quality. As can be inferred from Figure 21.17, such a statically reconfigured transceiver switches its mode of operation from a lower-order modem mode, such as, for example, BPSK, to a higher-order mode, such as 4QAM, when the channel quality has improved sufficiently for the

---
[5]The associated performance results typically degrade upon increasing the Doppler frequency and improve upon reducing it, since the effects of channel estimation latency become less significant. This phenomenon was quantified in [597].

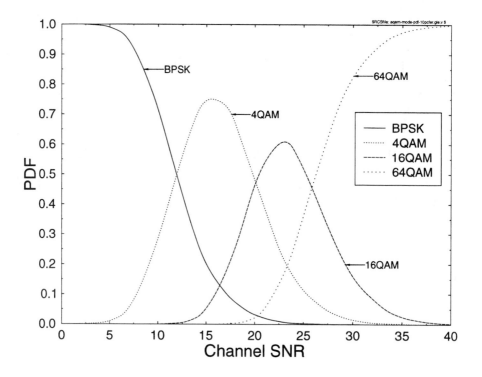

Figure 21.15: PDF of the adaptive modem being in a particular modulation mode versus channel SNR.

4QAM mode's FER to become lower than 5% upon reconfiguring the transceiver in this 4QAM mode. Again, as also seen in Figure 21.16 on a nonlogarithmic scale, Figure 21.17 shows on a logarithmic scale that the "one-frame channel estimation delay" AQAM modem manages to maintain a similar FER performance to the fixed-rate BPSK modem at low SNRs. However, we will see in Figure 21.18 that AQAM provides increasingly higher bit rates, reaching four times higher values than BPSK for high-channel SNRs, where the employment of 64QAM is predominant. In this high SNR region, the FER curve asymptotically approaches the 64QAM FER curve for both the realistic and the ideal AQAM scheme. But this is not visible in the figure for the ideal scheme, since this occurs at SNRs outside the range of Figure 21.17. Again, the reason for this performance discrepancy is the occasionally misjudged channel quality estimates of the realistic AQAM scheme. In addition, Figure 21.17 indicates that the realistic AQAM modem exhibits a near-constant 3% FER at medium SNRs. The issue of adjusting the switching thresholds in order to achieve the target FER will be addressed in detail in Section 21.2.4, and the thresholds invoked will be detailed with reference to Table 21.7. Suffice to say at this stage that the average number of bits per symbol — and potentially also the associated video quality — can be increased upon using more aggressive switching thresholds. However, this results in

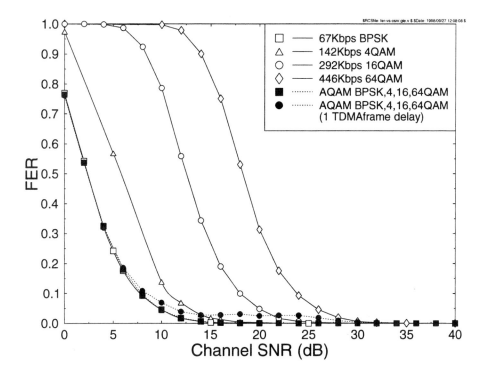

Figure 21.16: Transmission FER (or packet-loss ratio) versus channel SNR comparison of the four fixed modulation modes (BPSK, 4QAM, 16QAM, 64QAM) and that of the adaptive burst-by-burst modem (AQAM). AQAM is shown with a realistic one-TDMA frame delay between channel estimation and mode switching, and also with a zero delay version for indicating the upper bound performance. The channel parameters were defined in Table 21.4.

an increased FER, which tends to decrease the video quality, as will be discussed in Section 21.2.4.

A comparison of the effective video bit rate versus channel SNR is shown in Figure 21.18 for the four fixed modulation schemes and the ideal and realistic AQAM modems. The effective video bit rate is the average bit rate provided by all the successfully transmitted video packets. It should be noted that the realistic AQAM modem has a slightly lower throughput, since sometimes the incorrect modulation mode is chosen, which may result in packet loss. At very low-channel SNRs, the throughput bit rate converges to that of the fixed BPSK mode, since the AQAM modem is almost always in the BPSK mode at these SNRs, as demonstrated in Figure 21.15.

Having shown the effect of the burst-by-burst adaptive modem on the transmission FER and effective bit rate, let us now demonstrate these effects on the decoded video quality, measured in terms of the Peak Signal-to-Noise Ratio (PSNR). Figure 21.19 shows the decoded video quality in terms of PSNR versus channel SNR for both the ideal and realistic adaptive modem, as well as for the four modes of the statically

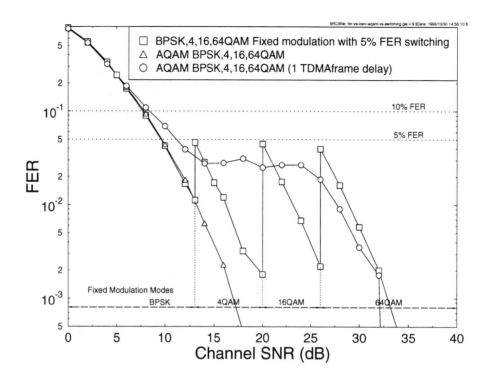

Figure 21.17: Transmission FER (or packet-loss ratio) versus channel SNR comparison of the four fixed modulation modes (BPSK, 4QAM, 16QAM, 64QAM) with 5% FER switching and adaptive burst-by-burst modem (AQAM). AQAM is shown with a realistic one-TDMA frame delay between channel estimation and mode switching, and a zero delay version is included as an upper bound. The channel parameters were defined in Table 21.4.

configured multimode modem. It can be seen that the ideal adaptive modem, which always selects the perfect modulation modes, has a better or similar video quality for the whole range of channel SNRs. For the statically configured multimode scheme the video quality degrades when the system switches from a higher-order to a lower-order modulation mode. The ideal adaptive modem, however smooths out the sudden loss of video quality, as the channel degrades. The nonideal adaptive modem has a slightly lower video-quality performance than the ideal adaptive modem, especially at medium SNRs, since it sometimes selects the incorrect modulation mode due to the estimation delay. This can inflict video packet loss and/or a reduction of the effective video bit rate, which in turn reduces the video quality.

The difference between the ideal burst-by-adaptive modem, using ideal channel estimation and the nonideal, realistic burst-by-burst adaptive modem, employing a nonideal delayed channel estimation can be seen more clearly in Figure 21.20 for a range of video sequences. Observe that at high-and low-channel SNRs, the video-quality performance is similar for the ideal and nonideal adaptive modems. This is

21.2. ADAPTIVE QAM-BASED WIRELESS VIDEOPHONY

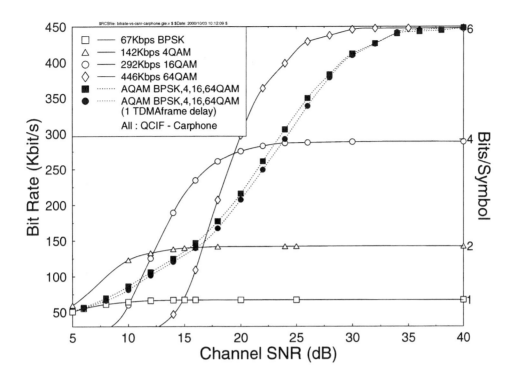

Figure 21.18: Video bit rate versus channel SNR comparison of the four fixed modulation modes (BPSK, 4QAM, 16QAM, 64QAM) and adaptive burst-by-burst modem (AQAM). AQAM is shown with a realistic one-TDMA frame delay between channel estimation and mode switching, and also as a zero delay version for indicating the upper bound. The channel parameters were defined in Table 21.4.

because the channel estimation delay has little effect, since at low- or high-channel SNRs the adaptive modem is in either BPSK or 64QAM mode for the majority of the time. More explicitly, the channel quality of a transmission frame is almost always the same as that of the next; hence, the delay has little effect at low and high SNRs.

The video-quality versus channel-quality trade-offs can be more explicitly observed in Figure 21.21. This figure portrays the decoded video quality in PSNR versus the packet-loss ratio or transmission FER. The ideal and practical adaptive modem performance is plotted against that of the four fixed modulation schemes in the figure. It can been seen that the video quality of the adaptive modems degrades from that achieved by the error-free 64QAM modem toward the BPSK modem performance as the packet-loss ratio increases. The practical adaptive modems' near-constant FER performance of 3% at medium SNRs can be clearly seen in the figure which is associated with the reduced PSNRs of the various modem modes, while having only minor channel error-induced impairments. In order to augment the interpretation of this figure, we note that, although the objective video quality of the various fixed QAM

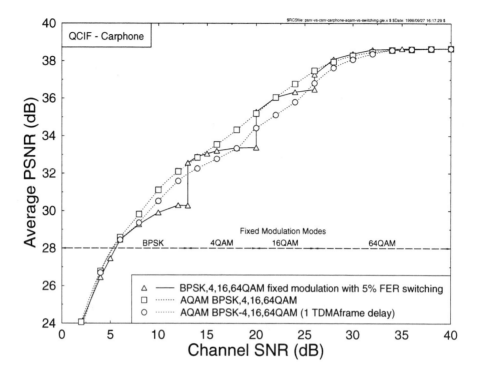

Figure 21.19: Decoded video quality (PSNR) versus channel SNR comparison of the four fixed modulation modes (BPSK, 4QAM, 16QAM, 64QAM) with 5% transmission FER switching and that of the adaptive burst-by-burst modem (AQAM). AQAM is shown with a realistic one-TDMA frame delay between channel estimation and mode switching, and a zero delay version for indicating the upper bound. The channel parameters were defined in Table 21.4.

modes expressed in PSNR appears to be higher than that of the AQAM schemes, the associated perceived video quality of AQAM is significantly more pleasing. This is because the channel-induced PSNR degradation is significantly more objectionable than the PSNR reduction imposed by invoking the inherently lower bit-rate, lower-quality, but more robust AQAM modes. Again, the philosophy here was that the AQAM scheme maintained the required near-constant target FER, which was necessary for high perceived video quality at the cost of invoking reduced rate, but more robust, modem modes under hostile channel conditions.

21.2.4 Switching Thresholds

The burst-by-burst adaptive modem changes its modulation modes based on the fluctuating channel conditions expressed in terms of the SNR at the equalizer's output. The set of switching thresholds used in all the previous graphs is the "Standard" set shown in Table 21.7, which was determined on the basis of the required channel SINR

21.2. ADAPTIVE QAM-BASED WIRELESS VIDEOPHONY

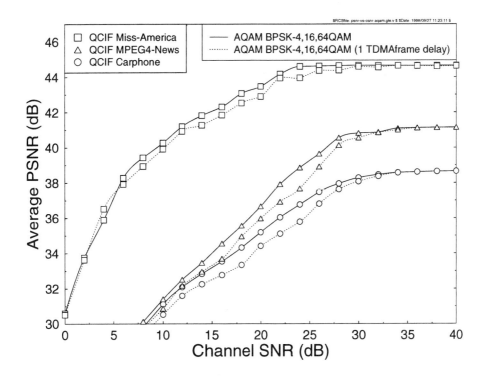

Figure 21.20: Decoded video quality (PSNR) versus channel SNR for the adaptive burst-by-burst modem (AQAM). AQAM is shown with a realistic one-TDMA frame delay between channel estimation and mode switching, and a zero delay version indicating the upper bound. Results are shown for three video sequences using the channel parameters that were defined in Table 21.4.

	BPSK	4QAM	16QAM	64QAM
Standard	<10 dB	≥10 dB	≥18 dB	≥24 dB
Conservative	<13 dB	≥13 dB	≥20 dB	≥26 dB
Aggressive	<9 dB	≥9 dB	≥17 dB	≥23 dB

Table 21.7: SINR Estimate at Output of the Equalizer Required for Each Modulation Mode in the Proposed Burst-by-Burst Adaptive Modem, That Is, Switching Thresholds

for maintaining the specific target video FER.

In order to investigate the effect of different sets of switching thresholds, — we defined two new sets of thresholds a more conservative set and a more aggressive set — employing less robust but more bandwidth-efficient modem modes at lower SNRs. The more conservative switching thresholds reduced the transmission FER at the expense of a lower effective video bit rate. The more aggressive set of thresholds increased the effective video bit rate at the expense of a higher transmission FER.

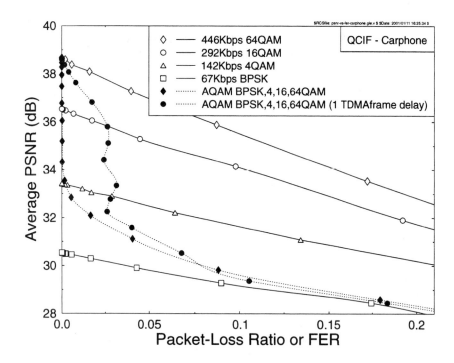

Figure 21.21: Decoded video quality (PSNR) versus transmission FER (or packet-loss ratio) comparison of the four fixed modulation modes (BPSK, 4QAM, 16QAM, 64QAM) and that of the adaptive burst-by-burst modem (AQAM). AQAM is shown with a realistic one-TDMA frame delay between channel estimation and mode switching, and a zero delay version indicating the upper bound. The channel parameters were defined in Table 21.4.

The transmission FER performance of the realistic burst-by-burst adaptive modem, which has a one-TDMA frame delay between channel quality estimation and mode switching, is shown in Figure 21.22 for the three sets of switching thresholds of Table 21.7. The more conservative switching thresholds reduce the transmission FER from about 3% to about 1% for medium-channel SNRs. The more aggressive switching thresholds increase the transmission FER from about 3% to 4–5%. However, since FERs below 5% are not objectionable in video-quality terms, this FER increase is an acceptable compromise for a higher effective video bit rate. The effective video bit rate for the realistic adaptive modem with the three sets of switching thresholds is shown in Figure 21.23. The more conservative set of switching thresholds reduces the effective video bit rate but also reduces the transmission FER. The aggressive switching thresholds increase the effective video bit rate but also increase the transmission FER. Therefore, the optimal switching thresholds should be set so that the transmission FER is deemed acceptable in the range of channel SNRs considered.

The switching thresholds can be adjusted, for example, using Powell's optimization for uncoded transmissions [111] in order to achieve the required target BER. However,

21.2. ADAPTIVE QAM-BASED WIRELESS VIDEOPHONY

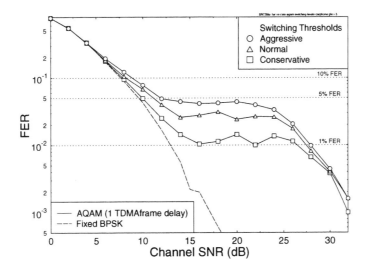

Figure 21.22: Transmission FER (or packet-loss ratio) versus channel SNR comparison of the fixed BPSK modulation mode and the adaptive burst-by-burst modem (AQAM) for the three sets of switching thresholds described in Table 21.7. AQAM is shown with a realistic one-TDMA frame delay between channel estimation and mode switching. The channel parameters were defined in Table 21.4.

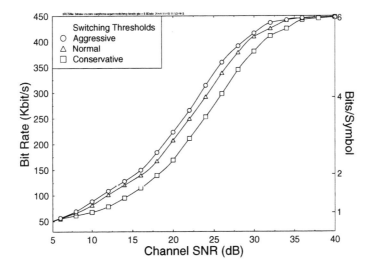

Figure 21.23: Video bit rate versus channel SNR comparison for the adaptive burst-by-burst modem (AQAM) with a realistic one-TDMA frame delay between channel estimation and mode switching for the three sets of switching thresholds as described in Table 21.7. The channel parameters were defined in Table 21.4.

Features	Multirate System			
Mode	BPSK	4QAM	16QAM	64QAM
Bits/symbol	1	2	4	6
FEC	Half-rate turbo coding with CRC			
Transmission Bit rate (kbit/s)	140.4	280.8	561.6	842.5
Unprotected Bit rate (kbit/s)	66.3	136.1	275.6	415.2
Effective Video rate (kbit/s)	60.9	130.4	270.0	409.3
Video fr. rate (Hz)	30			

Table 21.8: Operational-Mode Specific Turbo-coded Transceiver Parameters

the coded AQAM system's BER/FER performance is not analytically tractable, and so no closed-loop optimization was invoked. The thresholds were adjusted experimentally. We also note that instead of the equalizer's pseudo-SNR, we also investigated using the channel codec's BER estimates for controlling the switching process, which resulted in a similar system performance. Let us now consider the performance improvements achievable when employing powerful turbo codecs.

21.2.5 Turbo-coded Video Performance

In this subsection, we demonstrate the additional performance gains that are achievable when turbo coding is used in preference to BCH coding. The generic system parameters of the turbo-coded reconfigurable multimode video transceiver are the same as those used in the BCH-coded version summarized in Table 21.5. Turbo-coding schemes are known to perform best in conjunction with square-shaped turbo-interleaver arrays, and their performance is improved upon extending the associated interleaving depth, since then the two constituent encoders are fed with more independent data. This ensures that the turbo decoder can rely on two quasi-independent data streams in its efforts to make as reliable decisions as possible. A turbo-interleaver size of 18 × 18 was chosen, requiring 324 bits for filling the interleaver. The required recursive systematic convolutional component codes had a coding rate of $\frac{1}{2}$ and a constraint length of $K = 3$. After channel coding, the transmission burst length became 648 bits, which allowed the decoding of all AQAM transmission bursts independently. The operational-mode specific transceiver parameters are shown in Table 21.8 and should be compared to the corresponding BCH-coded parameters of Table 21.6.

The turbo-coded parameters result in a 10% lower effective throughput bit rate compared to BCH-coding under error-free conditions. However, we will show that under error-impaired conditions the turbo-coding performance becomes better, resulting in a reduced video packet-loss ratio. This reduced video packet-loss ratio results in an increased effective throughput for a wide range of channel SNRs. Figure 21.24 shows the effective throughput video bit rate versus channel SNR for the proposed AQAM modem using either BCH or turbo coding, when the delay between the channel-quality

21.2. ADAPTIVE QAM-BASED WIRELESS VIDEOPHONY

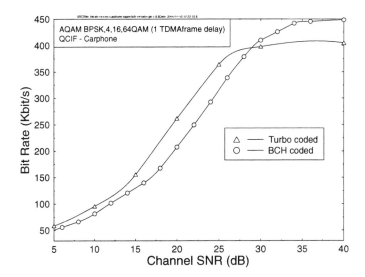

Figure 21.24: Effective throughput video bit rate versus channel SNR comparison of the adaptive burst-by-burst modems (AQAM) using either BCH or turbo coding. The AQAM modems have a one-TDMA frame delay between channel estimation and mode switching. The channel parameters were defined in Table 21.4.

estimate and mode switching of the AQAM modem was one-TDMA frame duration.

At high-channel SNRs and virtually error-free conditions, the BCH-coded modem slightly outperforms the turbo-coded modem in throughput bit rate terms, which were 446 and 409 Kbps, respectively. However, as the channel quality degrades, the lower packet-loss ratio performance of the turbo-coded AQAM modem results in a higher effective throughput bit rate below channel SNRs of about 28 dB.

The associated FER or video packet loss ratio (PLR) performance versus channel SNR is shown in Figure 21.25. Observe in this figure that the BCH-coded and turbo-coded AQAM modems exhibit similarly shaped FER performance curves, both of which obey a similar FER performance curve to that of the corresponding 64QAM modem mode at high SNRs, while at low SNRs the FER curve follows that of the BPSK modem mode. For medium-channel SNRs, the FER performance curve becomes near-constant, since the modem modes are adaptively adjusted, in order to maintain the required target FER. Observe, however that the turbo-coded AQAM modem requires approximately a 3 dB lower channel SNR for achieving the required FER target. Hence, if the minimum acceptable video packet-loss ratio is shown by the 5% limit drawn in the figure, the turbo-coded modem can maintain the required PLR for channel SNRs in excess of 8 dB, while the BCH-coded modem requires channel SNRs in excess of 11 dB.

Having shown that the turbo-coded AQAM modem outperforms the BCH-coded modem in terms of video packet-loss performance, which results in a higher effective throughput video bit rate for all except very high-channel SNRs, we demonstrate the

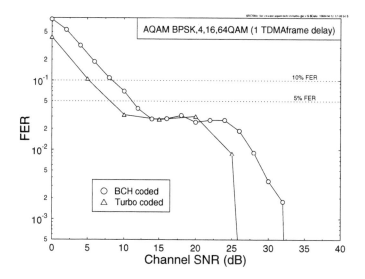

Figure 21.25: Transmission FER (or packet-loss ratio) versus channel SNR comparison of the adaptive burst-by-burst modem (AQAM) using the standard switching thresholds of Table 21.7, and either BCH or turbo coding. The channel parameters were defined in Table 21.4, and again, there was a one-TDMA frame delay between channel estimation and mode switching.

associated effects on the video quality expressed in terms of the PSNR. Figure 21.26 shows the PSNR video quality versus channel SNR for the AQAM modems using either BCH or turbo coding. The higher throughput of the BCH-coded modem at high-channel SNRs results in a slight advantage over the turbo-coded modem in terms of PSNR. However, the reduced video packet-loss ratio of the turbo-coded modem below channel SNRs of 30 dB resulted in a higher effective throughput, increasing the video quality of the turbo-coded modem. In conclusion, we have shown that the more complex turbo-coded AQAM modem outperformed the BCH-coded modem, since it reduced the video packet-loss ratio and hence increased both the effective throughput and the video quality for all but high-channel SNRs.

21.2.6 Summary and Conclusions

In this section, we have proposed a wideband burst-by-burst adaptive modem, which employs the mean squared error at the output of the channel equalizer as the quality measure for controlling the modem modes. Furthermore, we have quantified the achievable video performance gains due to employing the proposed wideband burst-by-burst adaptive modem. When our adaptive packetizer is used in conjunction with the adaptive modem, it continually adjusts the video codec's target bit rate in order to exploit the instantaneous bit-rate capacity provided by the adaptive modem.

We have also shown that a burst-by-burst adaptive modem exhibits a better performance than a statically configured multimode scheme. The delay between the

21.2. ADAPTIVE QAM-BASED WIRELESS VIDEOPHONY

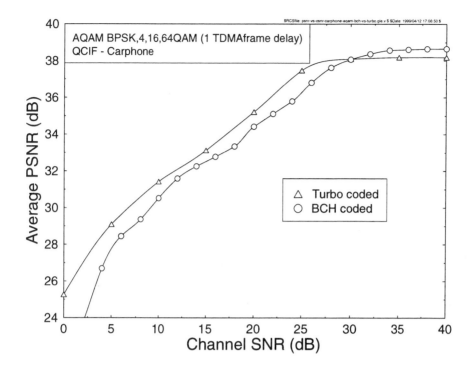

Figure 21.26: Decoded video quality (PSNR) versus transmission FER (or packet-loss ratio) comparison of the realistic adaptive burst-by-burst modems (AQAM) using either BCH or turbo coding. The channel parameters were defined in Table 21.4.

channel estimation and modulation mode switching was shown to have a considerable effect on the performance achieved by the adaptive modem. This performance penalty can be mitigated by reducing the modem mode switching latency. However, at low vehicular speeds (i.e., at low Doppler frequencies) the switching latency is less crucial, and the practical adaptive modem can achieve a performance that is close to that of the ideal adaptive modem benefiting from instantaneous modem mode switching, which we used to quantify the expected upper bound performance. We have also demonstrated, how the transmission FER performance is affected by changing the switching thresholds. Therefore, the system can be tuned to the required FER performance using appropriate switching thresholds. Our future work will concentrate on improving the system performance by invoking more complex turbo-coding schemes.

Motivated by the recent dominance of CDMA-based wireless systems such as the UMTS, IMT-2000, and cdma2000 systems, in the next section we study the video performance of various joint-detection-based multi-user wireless videophone systems.

21.3 A UMTS-Like Videophone System: Turbo-Coded Burst-by-Burst Adaptive Joint Detection CDMA and H.263-Based Videophony[6][7]

21.3.1 Motivation and Video Transceiver Overview

While the third-generation wireless communications standards are still evolving, they have become sufficiently mature for the equipment designers and manufacturers to complete the design of prototype equipment. One of the most important services tested in the field trials of virtually all dominant players in the field is interactive videotelephony at various bit rates and video qualities. Motivated by these events, the goal of this section is to quantify the expected video performance of a UMTS-like videophone scheme, while also providing an outlook on the more powerful burst-by-burst adaptive transceivers of the near future.

In this study, we transmitted 176 × 144 pixel Quarter Common Intermediate Format (QCIF) and 128 × 96 pixel Sub-QCIF (SQCIF) video sequences at 10 frames/s using a reconfigurable Time Division Multiple Access/Code Division Multiple Access (TDMA/CDMA) transceiver, which can be configured as a 1-, 2-, or 4-bit/symbol scheme. The H.263 video codec [192] exhibits an impressive compression ratio, although this is achieved at the cost of a high vulnerability to transmission errors, since a run-length coded stream is rendered undecodable by a single-bit error. In order to mitigate this problem, when the channel codec protecting the video stream is overwhelmed by the transmission errors, we refrain from decoding the corrupted video packet in order to prevent error propagation through the reconstructed video frame buffer [199]. We found that it was more beneficial in video-quality terms, if these corrupted video packets were dropped and the reconstructed frame buffer was not updated, until the next video packet replenishing the specific video frame area was received. The associated video performance degradation was found to be perceptually unobjectionable for packet-dropping or transmission frame error rates (FER) below about 5%. These packet dropping events were signaled to the remote decoder by superimposing a strongly protected 1-bit packet acknowledgment flag on the reverse-direction packet, as outlined in [199]. Bose-Chaudhuri-Hocquenghem (BCH) [9] and turbo error correction codes [28] were used, and again, the CDMA transceiver was capable of transmitting 1, 2, and 4 bits per symbol, where each symbol was spread using a low spreading factor (SF) of 16, as seen in Table 21.9. The associated parameters will be addressed in more depth during our later discourse. Employing a low spreading factor of 16 allowed us to improve the system's multi-user performance with the aid of joint-detection techniques [353]. We also note that the implementation of

---

[6]This section is based on P. Cherriman, E.L. Kuan, and L. Hanzo: Burst-by-burst Adaptive Joint-detection CDMA/H.263 Based Video Telephony, submitted to IEEE Transactions on Cicuits and Systems for Video Technology, 1999.

[7]©1999 IEEE. Personal use of this material is permitted. However, permission to reprint/republish this material for advertising or promotional purposes or for creating new collective works for resale or redistribution to servers or lists, or to reuse any copyrighted component of this work in other works, must be obtained from IEEE.

21.3. UMTS-LIKE BURST-BY-BURST ADAPTIVE CDMA VIDEOPHONY

Parameter	
Multiple access	TDMA/CDMA
Channel type	COST 207 Bad Urban
Number of paths in channel	7
Normalized Doppler frequency	3.7×10^{-5}
CDMA spreading factor	16
Spreading sequence	Random
Frame duration	4.615 ms
Burst duration	577 μs
Joint-detection CDMA receiver	Whitening matched filter (WMF) or minimum mean square error block decision feedback equalizer (MMSE-BDFE)
No. of slots/frame	8
TDMA frame length	4.615ms
TDMA slot length	577μs
TDMA slots/video packet	3
Chip periods/TDMA slot	1250
Data symbols/TDMA slot	68
User data symbol rate (kBd)	14.7
System data symbol rate (kBd)	117.9

Table 21.9: Generic System Parameters Using the Frames Spread Speech/Data Mode 2 Proposal [596]

the joint-detection receivers is independent of the number of bits per symbol associated with the modulation mode used, since the receiver simply inverts the associated system matrix and invokes a decision concerning the received symbol, regardless of how many bits per symbol were used. **Therefore, joint-detection receivers are amenable to amalgamation with the above 1-, 2-, and 4-bit/symbol modem, since they do not have to be reconfigured each time the modulation mode is switched.**

In this performance study, we used the Pan-European FRAMES proposal [596] as the basis for our CDMA system. The associated transmission frame structure is shown in Figure 21.27, while a range of generic system parameters is summarized in Table 21.9. In our performance studies, we used the COST207 [337] seven-path bad urban (BU) channel model, whose impulse response is portrayed in Figure 21.28.

Our initial experiments compared the performance of a whitening matched filter (WMF) for single-user detection and the minimum mean square error block decision feedback equalizer (MMSE-BDFE) for joint multi-user detection. These simulations were performed using four-level Quadrature Amplitude Modulation (4QAM), invoking both binary BCH [9] and turbo-coded [28] video packets. The associated bit rates are summarized in Table 21.10. The transmission bit rate of the 4QAM modem mode was 29.5 Kbps, which was reduced due to the approximately half-rate BCH or turbo coding, plus the associated video packet acknowledgment feedback flag error control [598] and video packetization overhead to produce effective video bit rates

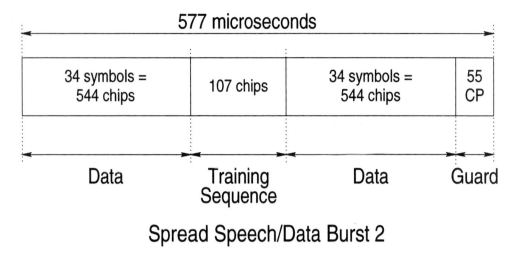

Figure 21.27: Transmission burst structure of the FMA1 spread speech/data mode 2 of the FRAMES proposal [596].

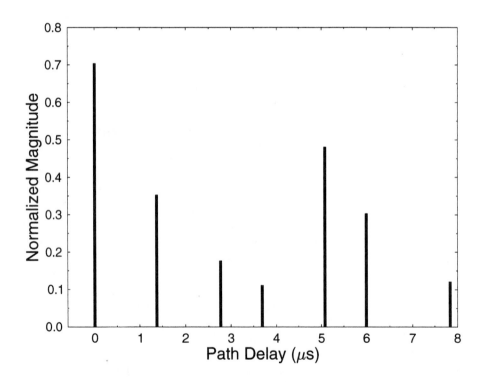

Figure 21.28: Normalized channel impulse response for the COST 207 [337] seven-path Bad Urban channel.

21.3. UMTS-LIKE BURST-BY-BURST ADAPTIVE CDMA VIDEOPHONY

Features	BCH coding	Turbo Coding
Modulation	4QAM	
Transmission bit rate (kbit/s)	29.5	
Video rate (kbit/s)	13.7	11.1
Video frame rate (Hz)	10	

Table 21.10: FEC-protected and Unprotected BCH and Turbo-Coded Bit Rates for the 4QAM Transceiver Mode

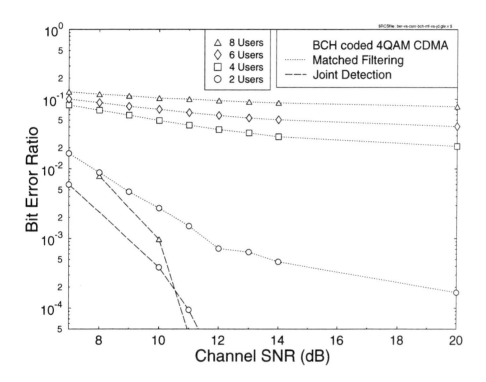

Figure 21.29: BER versus channel SNR 4QAM performance using BCH-coded, 13.7 Kbps video, comparing the performance of matched filtering and joint detection for two to eight users.

of 13.7 Kbps and 11.1 Kbps, respectively. A more detailed discussion on the video packet acknowledgment feedback error control and video packetization overhead will be provided in Section 21.3.2 with reference to the convolutionally coded multimode investigations.

Figure 21.29 portrays the bit error ratio (BER) performance of the BCH coded video transceiver using both matched filtering and joint detection for two to eight users. The bit error ratio is shown to increase as the number of users increases, even upon employing the MMSE-BDFE multi-user detector (MUD). However, while the

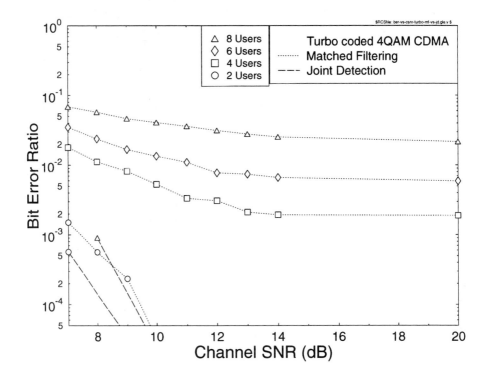

Figure 21.30: BER versus channel SNR 4QAM performance using turbo-coded 11.1 Kbps video, comparing the performance of matched filtering and joint detection for two to eight users.

matched filtering receiver exhibits an unacceptably high BER for supporting perceptually unimpaired video communications, the MUD exhibits a far superior BER performance.

When the BCH codec was replaced by the turbo-codec, the bit error ratio performance of both matched filtering and the MUD receiver improved, as shown in Figure 21.30. However, as expected, matched filtering was still outperformed by the joint-detection scheme for the same number of users. Furthermore, the matched filtering performance degraded rapidly for more than two users.

Figure 21.31 shows the video packet-loss ratio (PLR) for the turbo-coded video stream using matched filtering and joint detection for two to eight users. The figure clearly shows that the matched filter was only capable of meeting the target packet-loss ratio of 5% for up to four users, when the channel SNR was in excess of 11 dB. However, the joint detection algorithm guaranteed the required video packet loss ratio performance for two to eight users in the entire range of channel SNRs shown. Furthermore, the two-user matched-filtered PLR performance was close to the eight-user MUD PLR.

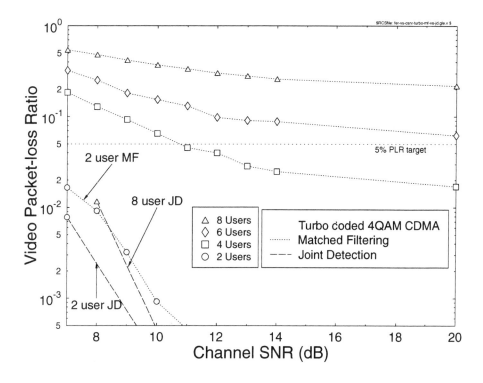

Figure 21.31: Video packet-loss ratio versus channel SNR for the turbo-coded 11.1 Kbps video stream, comparing the performance of matched filtering and joint detection for two to eight users.

21.3.2 Multimode Video System Performance

Having shown that joint detection can substantially improve our system's performance, we investigated the performance of a multimode convolutionally coded video system employing joint detection, while supporting two users. The associated convolutional codec parameters are summarized in Table 21.11.

We now detail the video packetization method employed. The reader is reminded that the number of symbols per TDMA frame was 68 according to Table 21.9. In the 4QAM mode this would give 136 bits per TDMA frame. However, if we transmitted one video packet per TDMA frame, then the packetization overhead would absorb a large percentage of the available bit rate. Hence we assembled larger video packets, thereby reducing the packetization overhead and arranged for transmitting the contents of a video packet over three consecutive TDMA frames, as indicated in Table 21.9. Therefore, each protected video packet consists of $68 \times 3 = 204$ modulation symbols, yielding a transmission bit rate of between 14.7 and 38.9 Kbps for BPSK and 16QAM, respectively. However, in order to protect the video data, we employed half-rate, constraint-length nine convolutional coding, using octal generator polynomials of 561 and 753. The useful video bit rate was further reduced due to

Features	Multirate System		
Mode	BPSK	4QAM	16QAM
Bits/symbol	1	2	4
FEC	Convolutional Coding		
Transmitted bits/packet	204	408	816
Total bit rate (kbit/s)	14.7	29.5	58.9
FEC-coded bits/packet	102	204	408
Assigned to FEC-coding (kbit/s)	7.4	14.7	29.5
Error detection per packet	16 bit CRC		
Feedback bits/packet	9		
Video packet size	77	179	383
Packet header bits	8	9	10
Video bits/packet	69	170	373
Unprotected video rate (kbit/s)	5.0	12.3	26.9
Video frame rate (Hz)	10		

Table 21.11: Operational-Mode Specific Transceiver Parameters for the Proposed Multimode System

the 16-bit Cyclic Redundancy Checking (CRC) used for error detection and the 9-bit repetition-coded feedback error flag for the reverse link. This results in video packet sizes of 77, 179, and 383 bits for each of the three modulation modes. The useful video capacity was reduced further by the video packet header of between 8 and 10 bits, resulting in useful or effective video bit rates ranging from 5 to 26.9 Kbps in the BPSK and 16QAM modes, respectively.

The proposed multimode system can switch among the 1-, 2-, and 4-bit/symbol modulation schemes under network control, based on the prevailing channel conditions. As seen in Table 21.11, when the channel is benign, the unprotected video bit rate will be approximately 26.9 Kbps in the 16QAM mode. However, as the channel quality degrades, the modem will switch to the BPSK mode of operation, where the video bit rate drops to 5 Kbps, and for maintaining a reasonable video quality, the video resolution has to be reduced to SQCIF (128 x 96 pels).

Figure 21.32 portrays the packet loss ratio for the multimode system in each of its modulation modes for a range of channel SNRs. The figure shows that above a channel SNR of 14 dB the 16QAM mode offers an acceptable packet-loss ratio of less than 5%, while providing an unprotected video rate of about 26.9 Kbps. If the channel SNR drops below 14 dB, the multimode system is switched to 4QAM and eventually to BPSK, when the channel SNR is below 9 dB, in order to maintain the required quality of service, which is dictated by the packet-loss ratio. The figure also shows the acknowledgment feedback error ratio (FBER) for a range of channel SNRs, which has to be substantially lower than the video PLR itself. This requirement is satisfied in the figure, since the feedback errors only occur at extremely low-channel SNRs, where the packet-loss ratio is approximately 50%. It is therefore assumed that the multimode system would have switched to a more robust modulation mode, before the feedback acknowledgment flag can become corrupted.

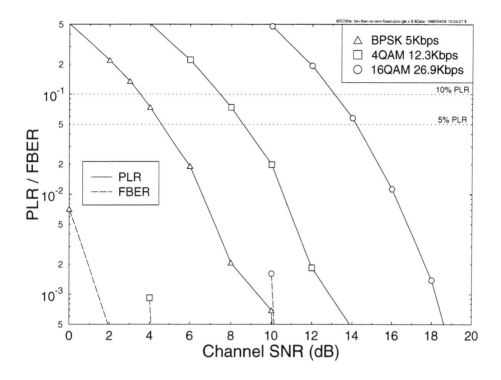

Figure 21.32: Video packet-loss ratio (PLR) and feedback error ratio (FBER) versus channel SNR for the three modulation schemes of the two-user multimode system using joint detection.

The video quality is commonly measured in terms of the peak signal-to-noise-ratio (PSNR). Figure 21.33 shows the video quality in terms of the PSNR versus the channel SNRs for each of the modulation modes. As expected, the higher throughput bit rate of the 16QAM mode provides a better video quality. However, as the channel quality degrades, the video quality of the 16QAM mode is reduced. Hence, it becomes beneficial to switch from the 16QAM mode to 4QAM at an SNR of about 14 dB, as suggested by the packet-loss ratio performance of Figure 21.32. Although the video quality expressed in terms of PSNR is superior for the 16QAM mode in comparison to the 4QAM mode at channel SNRs in excess of 12 dB, because of the excessive PLR the perceived video quality appears inferior to that of the 4QAM mode, even though the 16QAM PSNR is higher for channel SNRs in the range of 12–14 dB. More specifically, we found that it was beneficial to switch to a more robust modulation scheme when the PSNR was reduced by about 1 dB with respect to its unimpaired PSNR value. This ensured that the packet losses did not become subjectively obvious, resulting in a higher perceived video quality and smoother degradation, as the channel quality deteriorated.

The effect of packet losses on the video quality quantified in terms of PSNR is

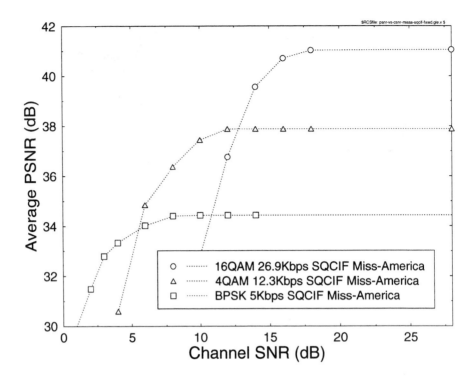

Figure 21.33: Decoded video quality (PSNR) versus channel SNR for the modulation modes of BPSK, 4QAM, and 16QAM supporting two users with the aid of joint detection. These results were recorded for the "Miss America" video sequence at SQCIF resolution (128 × 96 pels).

portrayed in Figure 21.34. The figure shows that the video quality degrades as the PLR increases. In order to ensure a seamless degradation of video quality as the channel SNR is reduced, it is best to switch to a more robust modulation scheme when the PLR exceeded 5%. The figure shows that a 5% packet-loss ratio results in a loss of PSNR when switching to a more robust modulation scheme. However, if the system did not switch until the PSNR of the more robust modulation mode was similar, the perceived video quality associated with the originally higher rate, but channel-impaired, stream became inferior.

21.3.3 Burst-by-Burst Adaptive Videophone System

A burst-by-burst adaptive modem maximizes the system's throughput by using the most appropriate modulation mode for the current instantaneous channel conditions. Figure 21.35 exemplifies how a burst-by-burst adaptive modem changes its modulation modes based on the fluctuating channel conditions. The adaptive modem uses the SINR estimate at the output of the joint detector to estimate the instantaneous channel quality and therefore to set the modulation mode.

21.3. UMTS-LIKE BURST-BY-BURST ADAPTIVE CDMA VIDEOPHONY

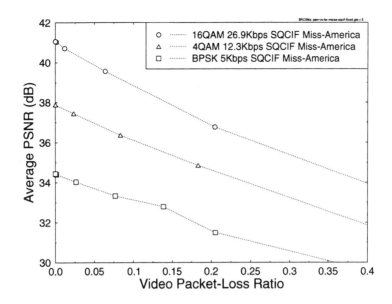

Figure 21.34: Decoded video quality (PSNR) versus video packet-loss ratio for the modulation modes of BPSK, 4QAM, and 16QAM, supporting two users with the aid of joint detection. The results were recorded for the "Miss America" video sequence at SQCIF resolution (128 × 96 pels).

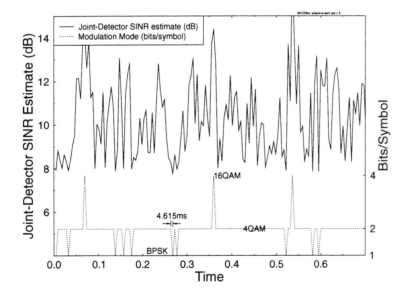

Figure 21.35: Example of modem mode switching in a dynamically reconfigured burst-by-burst modem in operation, where the modulation mode switching is based on the SINR estimate at the output of the joint detector.

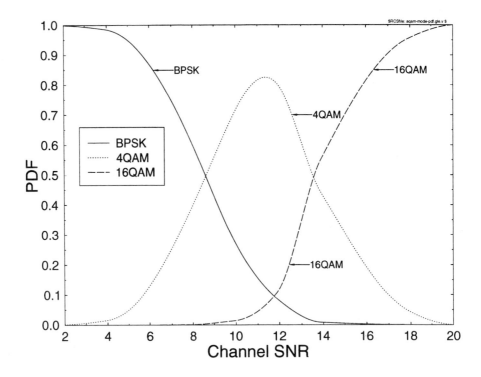

Figure 21.36: PDF of the various adaptive modem modes versus channel SNR.

The probability of the adaptive modem using each modulation mode for a particular channel SNRs is portrayed in Figure 21.36. At high-channel SNRs, the modem mainly uses the 16QAM modulation mode, while at low-channel SNRs the BPSK mode is most prevalent.

The advantage of dynamically reconfigured burst-by-adaptive modem over the statically switched multimode system previously described is that the video quality is smoothly degraded as the channel conditions deteriorate. The switched multimode system results in more sudden reductions in video quality, when the modem switches to a more robust modulation mode. Figure 21.37 shows the throughput bit rate of the dynamically reconfigured burst-by-burst adaptive modem, compared to the three modes of the statically switched multimode system. The reduction of the fixed modem modes' effective throughput at low SNRs is due to the fact that under such channel conditions an increased fraction of the transmitted packets have to be dropped, reducing the effective throughput. The figure shows the smooth reduction of the throughput bit rate, as the channel quality deteriorates. The burst-by-burst modem matches the BPSK mode's bit rate at low-channel SNRs and the 16QAM mode's bit rate at high SNRs. The dynamically reconfigured burst-by-burst adaptive modem characterized in the figure perfectly estimates the prevalent channel conditions, although in practice the estimate of channel quality is not perfect and it is inherently delayed. However,

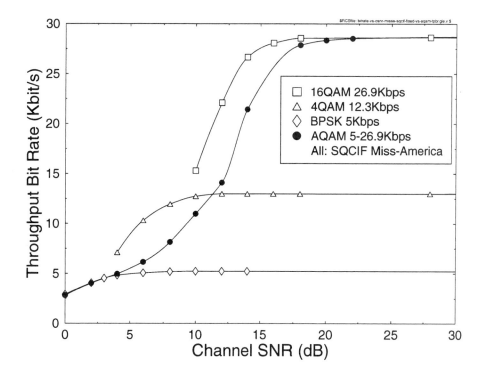

Figure 21.37: Throughput bit rate versus channel SNR comparison of the three fixed modulation modes (BPSK, 4QAM, 16QAM) and the adaptive burst-by-burst modem (AQAM), both supporting two users with the aid of joint detection.

as we have shown in Section 21.2 nonperfect channel estimates result in only slightly reduced performance when compared to perfect channel estimation.

The smoothly varying throughput bit rate of the burst-by-burst adaptive modem translates into a smoothly varying video quality as the channel conditions change. The video quality measured in terms of the average peak signal-to-noise ratio (PSNR) is shown versus the channel SNR in Figure 21.38 in contrast to that of the individual modem modes. The figure demonstrates that the burst-by-burst adaptive modem provides equal or better video quality over a large proportion of the SNR range shown than the individual modes. However, even at channel SNRs, where the adaptive modem has a slightly reduced PSNR, the perceived video quality of the adaptive modem is better since the video packet-loss rate is far lower than that of the fixed modem modes.

Figure 21.39 shows the video packet-loss ratio versus channel SNR for the three fixed modulation modes and the burst-by-burst adaptive modem with perfect channel estimation. Again, the figure demonstrates that the video packet-loss ratio of the adaptive modem is similar to that of the fixed BPSK modem mode. However, the adaptive modem has a far higher bit-rate throughput, as the channel SNR increases.

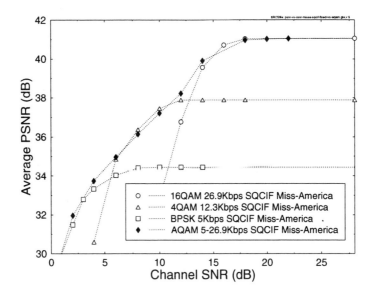

Figure 21.38: Average decoded video quality (PSNR) versus channel SNR comparision of the fixed modulation modes of BPSK, 4QAM and 16QAM, and the burst-by-burst adaptive modem — both supporting two users with the aid of joint detection. These results were recorded for the "Miss America" video sequence at SQCIF resolution (128 × 96 pels).

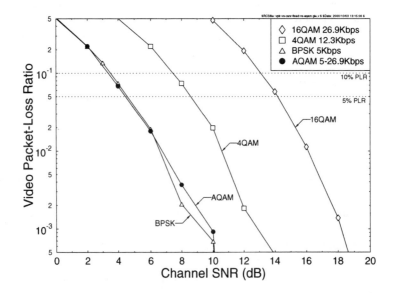

Figure 21.39: Video packet-loss ratio (PLR) versus channel SNR for the three modulation schemes of the multimode system, compared to the burst-by-burst adaptive modem. Both systems substain two users using joint detection.

21.3. UMTS-LIKE BURST-BY-BURST ADAPTIVE CDMA VIDEOPHONY

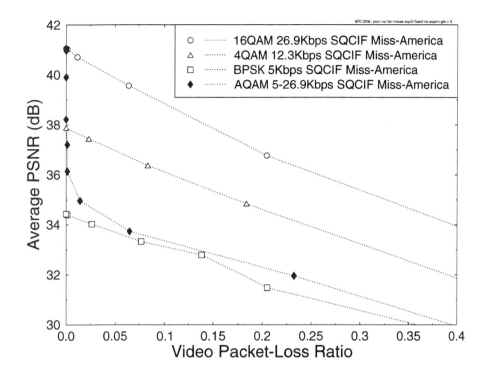

Figure 21.40: Decoded video quality (PSNR) versus video packet-loss ratio comparison of the fixed modulation modes of BPSK, 4QAM, and 16QAM, and the burst-by-burst adaptive modem. Both supporting two users with the aid of joint detection. These results were recorded for the "Miss America" video sequence at SQCIF resolution (128 × 96 pels).

The burst-by-burst adaptive modem gives an error performance similar to that of the BPSK mode, but with the flexibity to increase the bit-rate throughput of the modem, when the channel conditions improve. If imperfect channel estimation is used, the bit rate throughput of the adaptive modem is reduced slightly. Furthermore, the video packet-loss ratio seen in Figure 21.39 is slightly higher for the AQAM scheme due to invoking higher-order modem modes, as the channel quality increases. However, we have shown in our previous report on wideband video transmission [599] that it is possible to maintain the video packet-loss ratio within tolerable limits for the range of channel SNRs considered.

The interaction between the video quality measured in terms of PSNR and the video packet loss ratio can be seen more clearly in Figure 21.40. The figure shows that the adaptive modem slowly degrades the decoded video quality from that of the error-free 16QAM fixed modulation mode, as the channel conditions deteriorate. The video quality degrades from the error-free 41 dB PSNR, while maintaining a near-zero video packet-loss ratio, until the PSNR drops below about 36 dB PSNR. At this point, the further reduced channel quality inflicts an increased video packet-loss rate, and the video quality degrades more slowly. The PSNR versus packet-loss ratio performance then tends toward that achieved by the fixed BPSK modulation mode. However, the adaptive modem achieved better video quality than the fixed BPSK modem even at high packet-loss rates.

21.3.4 Summary and Conclusions

In conclusion, the proposed joint-detection assisted burst-by-burst adaptive CDMA-based video transceiver substantially outperformed the matched-filtering based transceiver. The transceiver guaranteed a near-unimpaired video quality for channel SNRs in excess of about 5 dB over the COST207 dispersive Rayleigh-faded channel. The benefits of the multimode video transceiver clearly manifest themselves in terms of supporting unimpaired video quality under time-variant channel conditions, where a single-mode transceiver's quality would become severely degraded by channel effects. The dynamically reconfigured burst-by-burst adaptive modem gave better perceived video quality due to its more graceful reduction in video quality, as the channel conditions degraded, than a statically switched multimode system.

Following our discussions on joint-detection assisted CDMA-based burst-by-burst adaptive interactive videotelephony, in the next two sections we concentrate on a range of multicarrier modems. The last section of the chapter considers distributive broadcast video transmission, based also on multicarrier modems.

21.4 H.263/OFDM-Based Video Systems for Frequency-Selective Wireless Networks[8] [9]

21.4.1 Background

In our previous video systems, single-carrier or serial modems were used. In recent years the multicarrier orthogonal frequency division multiplex (OFDM), or parallel modems, have also received considerable attention, especially in distributive, or synonymously, in broadcasting systems. In this section, we show that OFDM systems are also attractive in interactive videotelephony.

A rudimentary introduction to OFDM was provided in Chapter 5. OFDM was originally proposed by Chang in 1966 [157], where instead of estimating the wideband dispersive channel's impulse response, as in conventional equalized serial modems [29], the channel is rendered nondispersive by splitting the information to be transmitted into a high number of parallel, low-rate, nondispersive channels [29]. In this case, there

---

[8]This section is based on: P. Cherriman, T. Keller, and L. Hanzo, "Orthogonal frequency division multiplex transmission of H.263 encoded video over highly frequency-selective wireless networks." *IEEE Transactions on Circuits and Systems for Video Technology*, August 1999 [600].

[9]©1999 IEEE. Personal use of this material is permitted. However, permission to reprint/republish this material for advertising or promotional purposes or for creating new collective works for resale or redistribution to servers or lists, or to reuse any copyrighted component of this work in other works must be obtained from IEEE.

21.4. H.263/OFDM-BASED VIDEO SYSTEMS

Video Clip	Size	Frame/s	Color
Miss America	QCIF	10, 30	Gray
Miss America	SQCIF, QCIF, CIF	10, 30	Color
Carphone	QCIF	10, 30	Color
Suzie	SQCIF, QCIF, 4CIF	10, 30	Color
Football	4CIF	10, 30	Color
Mall	16CIF	10, 30	Color
CIF: 288 × 352 pixel Common Intermediate Format			
QCIF: 144 × 176 pixel Quarter CIF			
SQCIF: 96 × 128 pixel Sub-QCIF			
4CIF: 576 × 704 pixel 4×CIF			
16CIF: 1152 × 1408 pixel 16×CIF			

Table 21.12: Video Sequences Used for H.263 Simulations

is no need to estimate the channel's impulse response, since for the low-rate subchannels it can be considered nondispersive. Over the past three decades, this technique has been regularly revisited by a number of researchers [157, 164, 165, 167, 171, 601]. Despite its conceptual elegance, until recently its employment has been mostly limited to military applications due to implementational difficulties. However, it has recently been adopted as the new pan-European digital audio broadcasting (DAB) standard [176] and digital terrestrial television broadcast DTTB standard [177] — now known as (DVB-T) — as well as for a range of other high-rate applications, such as 155 Mbps Wireless Asynchronous Transfer Mode (WATM) local area networks. These wide-ranging applications underline the significance of OFDM modems as an alternative technique to conventional serial modems with channel equalization in order to combat signal dispersion [157, 164, 165, 167, 171, 601].

The outline of this section is as follows. In Section 21.4.2, the ability of the H.263 codec to support a wide range of video services is analyzed. Sections 21.4.2.1 and 21.4.2.2 provide an overview of the two proposed systems, while Section 21.4.3 characterizes the propagation environment of both our UMTS-like and WATM-oriented systems. Section 21.4.4 is concerned with the system's video aspects, and the overall system performance is evaluated in Section 21.4.5.

The image quality versus bit-rate performance of the H.263 codec for various image resolutions, frame rates, and video sequences was analyzed in the corresponding earlier sections. The video sequences used in this section are summarized in Table 21.12. We note, however, that for cellular and cordless systems only SQCIF and QCIF resolutions are realistic in terms of their minimum required bit rates, while for higher-rate local area networks CIF, 4CIF, and 16CIF resolutions cater for substantially increased video quality. Let us begin our discussion with a brief system overview.

21.4.2 System Overview

In order to explore the range of potential applications for our H.263 / OFDM system, we will investigate two different schemes. Specifically, a higher-rate wireless asynchronous transfer mode (WATM) system will be studied in Section 21.4.2.1, and a

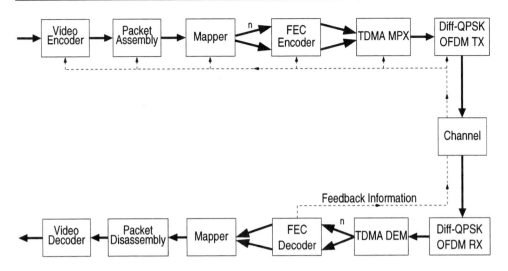

Figure 21.41: H.263/OFDM video transceiver schematic.

lower-rate cellular type system will be investigated in Section 21.4.2.2.

The schematic of both proposed wireless systems follows the structure of Figure 21.41, and both support interactive videophone calls. The video signal is compressed using the H.263 video compression standard [192]. As argued before, the H.263 standard achieves high compression ratios, but the resulting bit-stream is extremely sensitive to channel errors. This sensitivity is not a serious problem over benign wireline-based channels, such as conventional ATM links, but it is an impediment when used over wireless networks. Several solutions have been suggested in the literature for overcoming this, using Automatic Repeat Request (ARQ) [550], dual-level coding [194], and the use of a feedback channel [195]. A range of further robust video schemes was proposed in [198, 199, 467, 496, 602].

As seen in Figure 21.41, our system uses a feedback channel in order to inform the encoder of the loss of previous packets, as in the case of our single-carrier video systems. As before, we do not retransmit the corrupted packets, because this would reduce the system's teletraffic capacity by occupying additional transmission slots, while increasing the video delay. Earlier we showed that simply dropping the corrupted packets at both the local and remote decoder results in an extremely high-error resilience, in particular in high frame-rate systems, where 30 frames/s high-rate transmissions are facilitated. The rationale behind this is that un-updated video frame segments can only persist at 30 fps for 33 ms. This allows for the reconstruction frame buffer contents of the local and remote decoders to remain identical, which is essential for preventing error propagation through the reconstructed frame buffer. Then, when the instantaneous channel quality improves, the corrupted picture segments of the reconstructed frame buffers are replenished with more up-to-date video information. Similarly to our single-carrier videophone systems, the feedback channel is implemented by superimposing the packet dropping request on the reverse link, as shown in Figure 21.42. This figure shows how the feedback acknowledgment is

21.4. H.263/OFDM-BASED VIDEO SYSTEMS 911

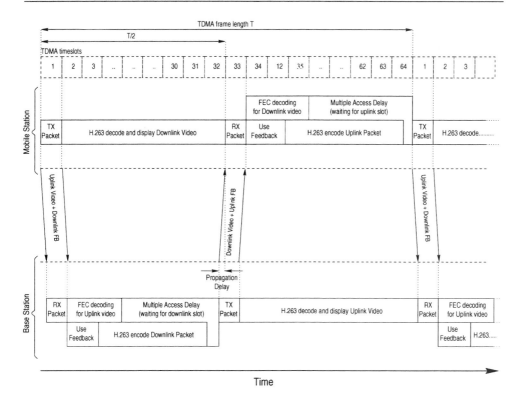

Figure 21.42: Transmission feedback timing diagram showing the feedback signaling superimposed on the reverse channel video data stream. The tasks that need to be performed in each time interval are shown for both the mobile station and the base station.

implemented in the context of the proposed Time Division Multiple Access (TDMA), Time Division Duplex (TDD) system using 32 time-slots, where one video packet was transmitted in each TDD frame. A number of further interesting details concerning the system's operation become explicit by studying the figure in more depth, which we have to refrain from here due to lack of space.

As demonstrated by Figure 21.41, the H.263 encoded bit-stream is passed to the packet assembly block, which was detailed in Section 19.3.6. The packetizer's function is to assemble the video packets for transmission, taking into account the packet acknowledgment feedback information. The packet disassembly block of Figure 21.41 ensures that always an error-free H.263 bitstream is output to the video decoder, discarding any erronously received packet and using only error-free packets to update the reconstructed frame buffer. Since the transmission packets contain typically fractions of video macroblocks at the beginning and end of the packets, a corrupted packet implies that the previously received partial macroblocks have to be discarded. The loss of the packet is then signaled via the feedback channel to the video encoder and packet assembly blocks.

The lost macroblocks are not re-transmitted, but strongly error-protected acknowl-

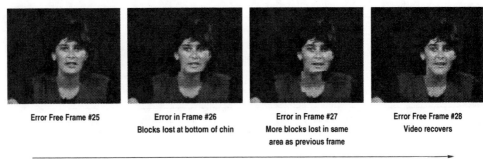

Figure 21.43: Typical expected video quality of the proposed video transceiver for 16 Kbit/s at 10 fps, for a frame error rate (or packet dropping rate) of 5%. This figure also shows the error recovery and concealment used in frames 26 and 27.

edgment flags are inserted into the video bitstream to signify the macroblocks that have not been updated. This requires one bit per lost macroblock in the next reverse-direction packet of the given user. The decoded video stream is error-free, although certain parts of some video frames may be "frozen" for a frame duration due to lost packets. These areas will usually be updated in the next video frame, and the effect of the lost packet will no longer be visible. Again, this packet loss has a prolonged effect for 100 ms at 10 fps, which is more preceivable than the losses at 30 fps. These operations were also indicated earlier in the context of Figure 21.42, while aspects of the acknowledgment flag protection and the associated probability of correct acknowledgment flag reception are quantified in Figure 19.22 and in Figure 21.52, which will be discussed at a later stage. An example of a typical scenario, portraying the error-free frame 25, a lost block in frame 26, which is not updated during frame 27, but finally replenished in frame 28, is shown in Figure 21.43.

The packetized video stream is then Forward Error Correction (FEC) coded, mapped to the allocated TDMA time-slot, and transmitted using Differential Quarternary Phase Shift Keying (D-QPSK) between adjacent subcarriers of the Orthogonal Frequency Division Multiplex (OFDM) scheme employed [29]. Again, it is important to strongly protect the binary acknowledgment flag from transmission errors, which prevents the remote decoder from updating the local reconstruction buffer, if the received packet was corrupted. Following a range of considerations, we opted for using a repetition code, which was superimposed on the forthcoming reverse-direction packet in the proposed Time Division Duplex (TDD) scheme of Table 21.13. The repetition-coded flag is then Majority Logic Decision (MLD) decoded at the receiver. The probability of correctly decoding 5-, 9-, 18-, and 27-bit majority logic codes was numerically evaluated for the range of bit error rates 0% to 50% using a random error distribution in Figure 19.22. On the basis of the results, we opted to use the strongest MLD code of MLD(27,1,13), repeating the flag 27 times, which was hence able to correct up to 13 transmission errors or a channel BER of about 50%.

In the next section, we briefly consider the specific system parameters used, which

21.4. H.263/OFDM-BASED VIDEO SYSTEMS

Feature	Value
TDMA/TDD frame length	171 μs
Slots/frame	64 (62 useable)
Slot length	2.667 μs
OFDM carriers	512 (511 used)
Modulation	Differential-QPSK
Coded bits/slot	1022 bits
FEC (1/2 rate)	BCH(255,131,18)
Pre-FEC bits/slot	524 bits
System bandwidth	225 MHz
System symbol rate (symbols/sec)	186×10^6
Normalised Doppler frequency	1.235×10^{-5}

Table 21.13: Summary of the Median-like WATM System Parameters ©IEEE, Cherriman, Keller, Hanzo, 1998 [600]

closely resemble those proposed by the Pan-European Wireless Asynchronous Transfer Mode (WATM) consortium referred to as Median.

21.4.2.1 The WATM System

The system employed in our experiments resembles the ACTS MEDIAN WATM proposal, which operates in the 60GHz band utilizing OFDM as modulation technique [29]. The MEDIAN-like WATM system parameters are listed in Table 21.13. Channel access in the MEDIAN WATM system is based on Time Division Multiple Access/Time Division Duplex (TDMA/TDD) frames having a duration of 170.7μs. This frame is split into 64 time-slots of 2.667μs duration. Two of these time-slots are reserved for networking functions, leaving 62 for useful information transfer.

In order to avoid implementationally complex equalization at the FEC-coded sampling rate, OFDM is employed as the proposed baseband modulation technique [29]. Again, a rudimentary introduction to OFDM was provided in Chapter 5. Differential Quarternary Phase Shift Keying (D-QPSK) between adjacent subcarriers is used as the frequency domain modulation scheme. If all 512 subcarriers are used, with one subcarrier being used as phase reference, then 1022 bits can be transmitted using a single OFDM symbol. Figure 21.44 portrays the Median WATM frame- and slot structure. Each time-slot contains one OFDM symbol, preceded by a cyclic extension of 64 samples in order to combat interference in wideband channels [29]. The role of the quasi-periodic cyclic extension of the time-domain OFDM symbol is to generate a waveform, which appears to be periodic for the duration of the bandlimited channel's memory. Although this time-domain signal-segment may become impaired due to channel dispersion and transients, the useful information-related segment remains intact. A cyclic post-amble is also appended to the OFDM data samples in order to simplify symbol timing synchronization [603]. For our simulations, all 512 subcarriers per OFDM symbol were actively used, resulting in the maximum throughput of 1022 bits per OFDM symbol.

Variable bit-rate users can be accommodated by allocating groups of time-slots

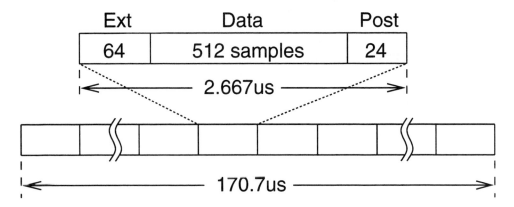

Figure 21.44: Schematic plot of the ACTS MEDIAN WATM frame structure. A time frame contains 64 time-slots of 2.667μs duration. Each time-slot holds the data samples of a 512-point IFFT OFDM symbol, 64 samples of the cyclic extension, and a cyclic post-amble of 24 samples. ©IEEE, Cherriman, Keller, Hanzo, 1998 [600].

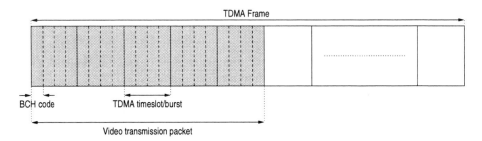

- Rate of BCH code overloading is the BCH error rate.
- Rate of at least one BCH code in each TDMA timeslot/burst becomes overloaded is TDMA burst error rate.
- Rate of at least one BCH code in each video transmission packet becomes overloaded is Frame error rate (FER) or packet error rate.

Figure 21.45: Stylized TDMA frame structure for the WATM system transmitting 4CIF resolution video, where each video transmission packet is formed using five time-slots and 20 BCH codewords per active TDMA frame and each time-slot contains 4 BCH codewords. ©IEEE, Cherriman, Keller, Hanzo, 1998 [600].

per frame, as seen in Table 21.14 in case of high-rate users, or by skipping time frames in case of low-rate users. The shaded area of Figure 21.45 defines the payload of a TDMA frame in the 4CIF scenario of the table, hence, this payload must be received errorfree. If it is corrupted, this event is defined as a TDMA frame error, and the relative frequency of these events defines the transmission frame error rate (FER) of the system. As seen in Figure 21.45 and Table 21.14, in this case we need five time-slots for supporting the associated 10.2 Mbps video rate, but other video resolutions require a different number of time-slots. Their FER is defined on the basis of the success or failure of all the slots of a specific video user in a TDMA frame, since in this regime we cannot selectively re-transmit the payload of each time-slot due

21.4. H.263/OFDM-BASED VIDEO SYSTEMS

Feature	Video Resolution			
	QCIF	CIF	4CIF	16CIF
Luminance resolution (pixels)	176x144	352x288	704x576	1408x1152
Chrominance resolution (pixels)	88x72	176x144	352x288	704x576
Packet separation (in No. of TDMA frames)	30	6	1.5	1
Packet rate (packets/s)	195	975	3900	5448
Bits/time-slot	1022	1022	1022	1022
Time-slots per active TDMA frame	5	2	5	7
Bits per active TDMA frame (packet size)	5110	2044	5110	7154
Channel bit rate	1Mbps	2Mbps	20Mbps	41.8Mbps
FEC	20×BCH(255,131,18)	8×BCH(...)	20×BCH(...)	28×BCH(...)
Pre-FEC bits per active TDMA frame	2620	1048	2620	3668
Pre-FEC bit rate	511Kbps	1Mbps	10.2Mbps	21.5Mbps
Feedback control bits	26	24	26	29
H.263 packetization header bits	13	12	13	14
Video bits per active TDMA frame	2581	1012	2581	3625
Useful video bit rate	503Kbps	1Mbps	10Mbps	21.2Mbps

Table 21.14: Summary of Video Parameters for the WATM System ©IEEE, Cherriman, Keller, Hanzo, 1998, [600]

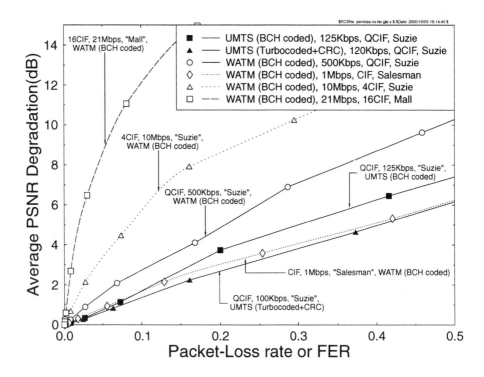

Figure 21.46: PSNR degradation versus video packet-loss rate or transmission frame error rate for the scenarios described in Tables 21.14 and 21.15. ©IEEE, Cherriman, Keller, Hanzo, 1998, [600].

to having only one acknowledgment flag per TDMA frame per user. This results in larger video frame sections remaining un-updated due to the increased payload per TDMA frame at high video rates. Again, the required number of slots per TDMA frame for each video mode was summarized for the various modes in Table 21.14, while reference [186] proposed an efficient statistical multiplexing scheme for variable rate scenarios, where the number of slots can be varied on a frame-by-frame basis. The associated PSNR degradation of the various user scenarios of Table 21.14 was quantified in Figure 21.46, where the more dramatic PSNR degradation of the larger video frame sizes becomes explicit for any given transmission frame error rate.

21.4.2.2 The UMTS-type Framework

The alternative transmission scheme used in our investigations was partially inspired by the ACTS FRAMES [604, 605] Mode 1 Universal Mobile Telecommunications System (UMTS) proposal. This proposal entails a time-frame structure of 4.615 ms, split into eight time-slots of 577 μs duration each. However, instead of using the originally proposed Direct Sequence Code Division Multiple Access (DS-CDMA) scheme with a chip rate of 2.17 Mchips/s, we have employed 1024-subcarrier OFDM, as it is shown

21.4. H.263/OFDM-BASED VIDEO SYSTEMS

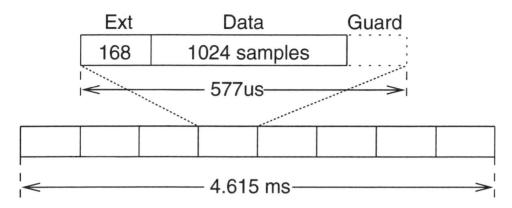

Figure 21.47: The ACTS FRAMES Mode 1-like frame structure, where a time-slot of 4.615 ms duration is split into eight time-slots of 577μs. Each time-slot holds a 1024-point IFFT OFDM symbol with a cyclic extension of 168 samples. ©IEEE, Cherriman, Keller, Hanzo, 1998, [600].

in Figure 21.47. Hence, we refer to this system as a Frames-like scheme. The modified time-slot contains a 1024-sample OFDM symbol, which is preceded by a cyclic extension of 168 samples length and followed by a guard interval of 60 samples. In order to maintain the Frames UMTS bandwidth of 1.6 MHz, the 1024 subcarrier OFDM symbol contains 410 unused, virtual subcarriers and 614 information bearer subcarriers, therefore reducing the bandwidth to 1.3 MHz and allowing for a modulation excess bandwidth within the band of 1.6 MHz. Let us now consider the corresponding channel models in the next section.

21.4.3 The Channel Model

The channel model employed for the WATM system experiments was a five-path, Rayleigh-fading indoors channel. The impulse response shown in Figure 21.48 was obtained by ray-tracing for a $100 \times 100\text{m}^2$ hall or warehouse environment, and every path in the impulse response was faded independently according to a Rayleigh-fading narrowband channel with a normalized Doppler frequency of $f'_d = 1.235 \cdot 10^{-5}$, corresponding to the 60 GHz propagation frequency and a worst-case indoor speed of 30 mph. Normalization of f'_d was carried out with respect to the OFDM symbol duration, which was constituted by 1024 transmitted samples rather than relative to the 1024 times shorter original sample duration.

A transmission rate of 155 Mbps was used, which is applicable to Wireless Asynchronous Transfer Mode (WATM) systems. A seven-path channel, corresponding to the four walls, ceiling and floor plus the line-of-sight (LOS) path was employed. The LOS path and the two reflections from the floor and ceiling were combined into one single path in the impulse response. The worst-case impulse response associated with the highest path length and delay spread was experienced in the farthest corners of the hall, which was determined using inverse second power law attenuation and the speed of light for the computation of the path delays. The corresponding frequency response

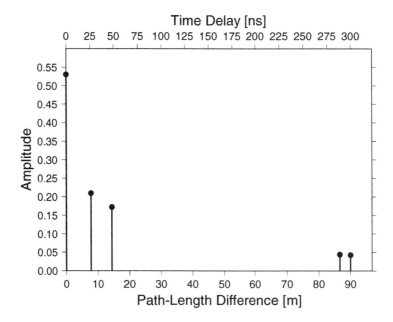

Figure 21.48: WATM five-path impulse response. ©IEEE, Cherriman, Keller, Hanzo, 1998, [600].

was plotted in Figure 21.49 for our 512-channel system, as a function of both the time-domain OFDM symbol index and frequency-domain subchannel index. Observe the very hostile frequency selective fading in the figure, which is efficiently combated by the OFDM modem, since for each of the 512 narrow subchannels the channel can be considered more or less flat fading. The residual frequency-domain subchannel transfer function "tilt" or distortion can be equalized using a simple frequency-domain pilot-assisted equalizer, where known pilot symbols are inserted periodically in the OFDM spectrum and the channel's frequency-domain transfer function is linearly interpolated and equalized between them.

The channel model used for the UMTS-type system experiments was based on a COST 207 [337] Bad Urban conformant seven-path impulse response shown in Figure 21.50. Again, each of the impulses was faded independently according to a Rayleigh narrowband fading channel with a normalized Doppler frequency of $f'_d = 89.39\text{Hz}/2.17\text{MHz} = 4.12 \cdot 10^{-5}$, where the carrier frequency and vehicular velocity were set to 2 GHz and 30 mph, respectively.

21.4.4 Video-Related System Aspects

21.4.4.1 Video Parameters of the WATM System

Our high-rate WATM system constitutes an ideal medium for high-resolution video transmission. In order to assess the system's ability to support various application scenarios, we investigated four different resolution video systems ranging from QCIF to 16CIF frame formats, as seen in Table 21.14. The video packetizer operated most

21.4. H.263/OFDM-BASED VIDEO SYSTEMS

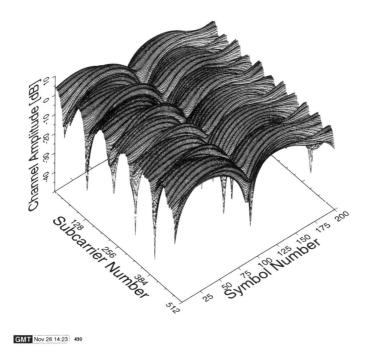

Figure 21.49: Frequency response of the 512-subcarrier WATM OFDM system at 155 Mbps. ©IEEE, Cherriman, Keller, Hanzo, 1998, [600].

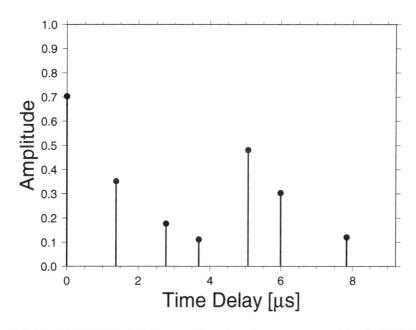

Figure 21.50: COST 207 Bad Urban compliant impulse response, used for the UMTS [337].

efficiently when the transmission packet generation rate per TDMA frame was neither too high nor too low. If the packet generation rate per TDMA frame is too high, each packet may contain less than a whole macroblock, leading to an increased buffering in the de-packetizer [199]. If the packet generation rate is too low, then each packet contains a high number of macroblocks, and therefore when a transmission packet is corrupted, a large proportion of the video frame is lost. Therefore, the packet generation rate for each video resolution was adjusted experimentally, taking into account that as the video resolution was increased fourfold, corresponding to increasing the video resolution, for example, from QCIF to CIF, the number of macroblocks per TDMA frame or per time unit was increased by the same factor, resulting in a corresponding increase in terms of the packet generation rate. These aspects were discussed in the context of defining the FER at the end of Section 21.4.2.1. The multiplier factor associated with the BCH codewords in Table 21.14 indicates the number of the BCH codewords per TDMA frame, which are jointly acknowledged by one protected feedback flag bit.

Again, the packet generation rate for each of the four video resolutions used is shown in Table 21.14 in terms of both the number of TDMA frames between video packets or, synonymously, the TDMA packet separation, as well as in terms of packets per second, which are matched to the bit-rate requirement of each mode. For example, for QCIF resolution video, the packet generation rate is 195 per second, which corresponds to one packet every 30 TDMA frames. After setting the packet generation rates for each video mode, the video target bit rates were set to give high-quality video for the majority of video sequences. As mentioned previously, the OFDM system can transmit 1022 bits in every time-slot. Hence for the QCIF mode, with one time-slot every 30 TDMA frames, the channel bit rate would be $1022 \times 195 = 200$ Kbit/s. Upon using half-rate channel coding, the video bit rate is constrained to 100 Kbit/s. Since we required around 500 Kbit/s for high-quality QCIF video for a wide range of video sequences, we decided to use five time-slots once every 30 TDMA frames, which corresponded to 20 BCH(255,131,18) codewords, as seen in Table 21.14. The corresponding 4CIF frame structure is illustrated in Figure 21.45. Therefore, the channel bit rate became $5 \times 1022 \times 195 = 1$ Mbit/s, providing a bit rate of 500 Kbit/s for video source coding. The interested reader is referred to our Web page[10] for some examples of coded sequences, which can be viewed using an MPEG player.

For the WATM investigations we decided to use binary Bose-Chaudhuri-Hocquenghem (BCH) block coding [606], since it is capable of both error correction and error detection. For all the modes we used the near half-rate code of BCH(255,131,18). The corresponding pre-FEC bit rates for the various modes are shown in Table 21.14. In conjunction with this channel coding scheme, the pre-FEC bit rate for the QCIF mode is 511Kbit/s.

The videophone system requires some additional overhead for its operation, since the feedback information for the reverse link is concatenated with the information packet, requiring a maximum of 29 additional bits/packet. In addition, the H.263 packetization adds a header to each packet [199], which is dependent on the number of bits in each packet. For our system, this header was between 12 and 14 bits per packet. Therefore, the number of useful video source bits in each packet used

---

[10]http://www-mobile.ecs.soton.ac.uk/peter/robust-h263/robust.html

21.4. H.263/OFDM-BASED VIDEO SYSTEMS

Feature	Value	
	BCH coding	Turbo Coding
Modulation	Differential-QPSK	
TDMA frame length	4.615 ms	
Slots/frame	8	
Slot length	577 μs	
OFDM carriers	1024 (612 used + 2 pilots)	
System bandwidth	1.6 MHz	
System symbol rate (symbols/sec)	2.17×10^6	
Normalized Doppler frequency	4.267×10^{-5}	
Coded bits/slot	1224 bits	
Feedback control bits	27	
H.263 packetization header bits	11	
Channel coding ($\approx$ 1/2 rate)	4 × BCH(255,131,18) + 2 × BCH(127,64,10) + BCH(63,30,6)	Turbo Coding using 612 bit random interleaver + 16bit CRC
Pre-FEC bits per time-slot	618	594
Pre-FEC bit rate	134 Kbit/s	129 Kbit/s
Video bits per time-slot (FEC)	580	556
Useful video bit rate (FEC)	126 Kbit/s	120 Kbit/s

Table 21.15: Summary of UMTS-like Parameters ©IEEE, Cherriman, Keller, Hanzo, 1998, [600]

for video transmission was about 40 bits less than the actual number of bits/packet. The corresponding useful video source bit rate for each of the modes is shown in Table 21.14, which was 503 Kbit/s for QCIF resolution video, when taking into account the associated transmission overheads.

21.4.4.2 Video parameters of the UMTS scheme

Our UMTS-type scheme was also designed for a range of bit rates and services. We opted to use the high bit-rate slot type, of which there can be a maximum of eight in each TDMA frame, since 8 × 577μs = 4.615 ms. The OFDM system designed for this scenario contained 614 active information subcarriers and 410 passive virtual subcarriers. After allocating one subcarrier as the reference for the differential decoder, it conveyed 1224 bits per time-slot. This yields a channel bit rate of 265 Kbit/s or approximately 130 Kbit/s before half-rate FEC coding. This bit rate is suitable for high-quality QCIF video or lower quality CIF video. Hence, we limited our investigations to QCIF resolution. In addition, we invoked two different types of channel coding — BCH blocks codes [606], and turbo coding [28] — which were also specified in Table 21.15. Since our system also required an error detection facility, the use of block codes is convenient because of their inherent error correction and detection capabilities. The parameters for the UMTS-type scheme are summarized in Table 21.15. Since turbo coding cannot provide error detection, to this effect a 16-bit Cyclic Redundancy Checking (CRC) code was used. Given the 1224 bits/slot payload per TDMA frame, before channel coding the number of bits per TDMA frame

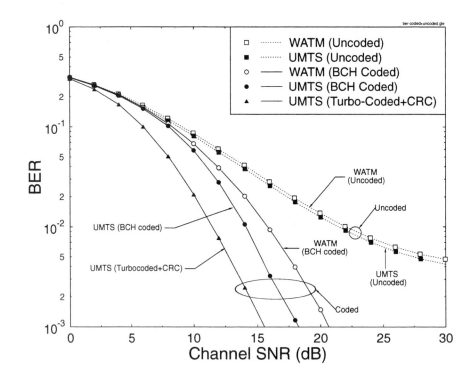

Figure 21.51: Uncoded channel BER and channel-decoded BER of the WATM and UMTS systems over the wideband channels characterized by the impulse responses shown in Figures 21.48 and 21.50, respectively. ©IEEE, Cherriman, Keller, Hanzo, 1998, [600].

was constrained to 618 for BCH coding and 594 for turbo coding. Half-rate turbo coding was used, however; two termination bits per slot were required for the convolutional encoders. Therefore, the number of pre-FEC bits per transmission packet was $1224/2 - 2(termination) - 16(CRC) = 594$. This led to a pre-FEC bit rate of 134 Kbit/s for BCH coding and 129 Kbit/s for turbo coding. The additional system overhead required 27 bits per packet for the reverse link's acknowledgment flag and 11 bits for the H.263 packetization header [199]. This led to a video bit rate of 126 Kbit/s for BCH coding and 120 Kbit/s for turbo coding. Having highlighted the salient video-specific system features, let us now consider the achievable system performance in the next section.

21.4.5 System Performance

Figure 21.51 portrays the bit error rate (BER) performance of both candidate systems over the wideband channels characterized by the impulse responses shown in Figures 21.48 and 21.50, respectively. Since the number of subcarriers was sufficiently high in both systems for narrowband subchannel conditions to prevail, the

21.4. H.263/OFDM-BASED VIDEO SYSTEMS

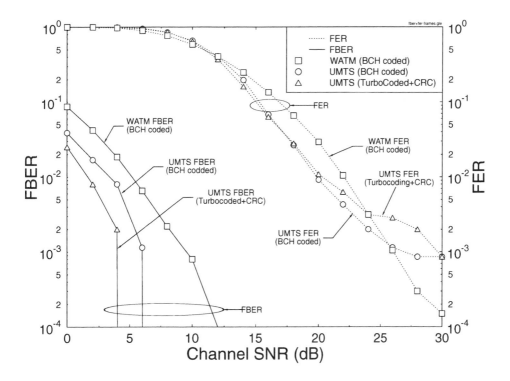

Figure 21.52: FER and FBER after channel decoding versus channel SNR for the WATM system using BCH codes, and for the UMTS-type scheme using BCH codes or turbo coding plus CRC over the wideband channels characterized by the impulse responses shown in Figures 21.48 and 21.50, respectively. The WATM results are typical and are shown for the CIF mode of operation. ©IEEE, Cherriman, Keller, Hanzo, 1998, [600].

modem BER curves are fairly similar, regardless of the different Doppler frequencies. Nevertheless, the slightly lower uncoded BER of the UMTS-type scheme manifested itself in a further improved FEC-decoded BER. Finally, as expected, the similar-rate turbo codec outperformed the BCH codec in terms of BER.

Figure 21.52 portrays the FER and the acknowledgment flag Feedback Error Rate (FBER) performance of both systems. Despite the BER differences of the systems, their FER performances are fairly similar. This indicates that the UMTS-like scheme's lower average BER actually results in a similar FER, despite its lower in-burst BER, when a BCH codeword was overwhelmed by an excessive number of channel errors. However, the lowest in-burst BER of the turbo codec translated in a substantially reduced acknowledgment flag feedback error rate after MLD decoding.

Figure 21.53 shows the video quality in terms of the average peak-signal-to-noise ratio (PSNR) versus the channel SNR for QCIF resolution video transmitted over the WATM scheme. The figure shows the video quality of a range of video sequences from the highly motion active "Foreman" and "Carphone" sequences to the lower-activity

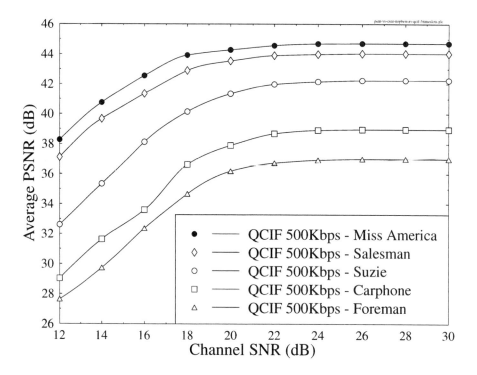

Figure 21.53: QCIF video quality in PSNR versus channel SNR at 500 Kbit/s and 30 fps over the WATM System, for various video sequences, using the impulse response of Figure 21.48. ©IEEE, Cherriman, Keller, Hanzo, 1998, [600].

"Miss America" sequence, which is more amenable to compression. For all the video sequences the PSNR starts to drop, when the channel SNR falls below about 20dB. Due to lack of space, the frame error rate (FER) versus channel signal-to-noise ratio (CSNR) performance of the system is not explicitly characterized in this treatise, but our records show that the frame error rate around this CSNR value is about 3%. The corresponding visual quality appears unimpaired, and the effects of transmission packet dropping do not become evident for CSNRs in excess of 16 dB. At a channel SNR of 16 dB, the frame error rate is 17%. However, the effects of this packet loss are only becoming just noticeable at a CSNR of 16 dB. The effect of the packet loss is that parts of the picture are "frozen," but usually for only one video frame duration of about 30 ms at 30 frames/s, which is not sufficiently long for these artifacts to become objectionable. However, if the part of the picture that was lost contains a moving object, the effect of the loss of the packet becomes more obvious. Therefore, for more motion active sequences, the effect of packet loss is more pronounced.

In order to portray the expected system performance in other application scenarios, where higher video quality is expected, in Figure 21.54 we portrayed the average PSNR versus channel SNR performance for a range of video resolutions from CIF to

21.4. H.263/OFDM-BASED VIDEO SYSTEMS

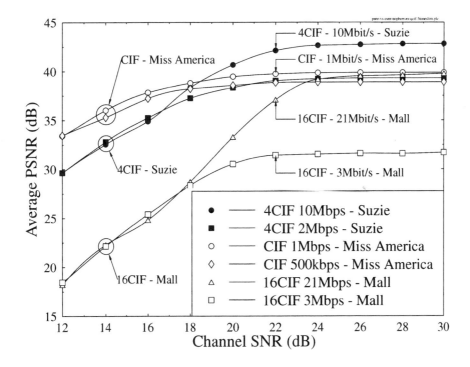

Figure 21.54: Video quality in PSNR versus channel SNR for CIF, 4CIF, and 16CIF resolution video at 30 fps over the WATM system for various video sequences using the channel impulse response shown in Figure 21.48. ©IEEE, Cherriman, Keller, Hanzo, 1998, [600].

16 x CIF HDTV quality. At CIF resolution the "Miss America" sequence was encoded at both 500 kbps and 1 Mbps. For 4CIF resolution the "Suzie" sequence was encoded at 2 and 10 Mbps, while for 16CIF resolution the "Mall" video clip was transmitted at 3 and 21 Mbps. This figure shows results for using multiple time-slots per active TDMA frame, as suggested by Table 21.14, down to just a single time-slot per active TDMA frame. Notice that the high and low bit-rate modes for each resolution seem to converge to a similar PSNR, when the channel SNR is low, which is a consequence of the higher PSNR degradations caused by a given FER in high-resolution modes due to their larger packet size, as seen in Table 21.14 and Figure 21.46.

The 16CIF scenarios seem to be more vulnerable to packet loss. This is because the packet generation rate is not four times that of the 4CIF simulations and therefore each 16CIF video packet contains approximately 2.5 times more macroblocks per transmission packet than the CIF and 4CIF resolution video packets. The effect of packet loss is therefore more noticeable, and this is manifested in the faster reduction of the PSNR, as the channel SNR degrades.

Focusing our attention on the UMTS-like scheme, since the FER of the turbo-coded scheme in Figure 21.52 was not substantially lower than that of the BCH-coded

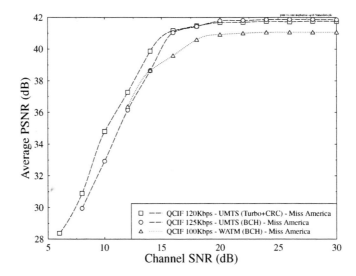

Figure 21.55: Video quality in PSNR versus channel SNR for the QCIF "Miss America" video sequence at 30 fps, using both the WATM system and the UMTS-type system with turbo coding and CRC or BCH coding, using the impulse responses shown in Figures 21.48 and 21.50, respectively. ©IEEE, Cherriman, Keller, Hanzo, 1998, [600].

arrangement, the corresponding video PSNR performances are also quite similar, as evidenced by Figures 21.55 and 21.56. This is a consequence of the system's high error resilience ensured by dropping the corrupted video packets. Hence, in practical systems operating on the basis of packet dropping, where the FER is a more important system parameter than the BER, the added complexity of turbo coding may not be justified. A similar-rate WATM performance curve is also shown in the figure.

21.4.6 Summary and Conclusions

In this section, the expected video performance of a WATM and that of a UMTS-type system were quantified in a variety of applications scenarios, using a range of video resolutions and bit rates. The high-efficiency H.263 video codec was employed to compress the video signal. The video formats used were summarized in Table 21.14 along with their associated target bit-rate figures. The proposed system ensures robust video communications using the WATM and the UMTS-type framework in a highly dispersive Rayleigh-fading environment. Even at a vehicular speed of 30 mph, the system requires channel SNRs in excess of only about 16 dB for near-unimpaired video transmission. Despite the different propagation conditions, the BER and FER modem performance of both systems was quite similar. Furthermore, due to the high error resilience of the video system, the increased complexity of the turbo codec was not justified in video performance terms, although the acknowledgment flag FBER was significantly reduced.

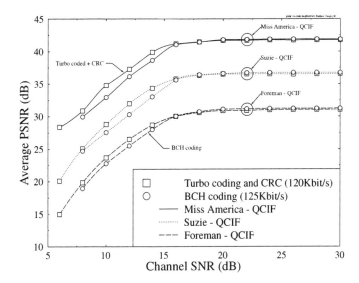

Figure 21.56: Video quality in PSNR versus channel SNR for a variety of QCIF resolution video sequences at 30fps over the UMTS-type system with turbo coding and CRC or BCH coding using the COST 207 Bad Urban impulse response of Figure 21.50. ©IEEE, Cherriman, Keller, Hanzo, 1998, [600].

21.5 Subband-Adaptive Turbo-Coded OFDM-Based Interactive Videotelephony[11] [12] [13]

21.5.1 Motivation and Background

In Sections 21.2 and 21.3 a wideband burst-by-burst adaptive QAM and a CDMA-based video system was proposed, while the previous section considered the transmission of interactive video using OFDM transceivers in various propagation environments. In this section, burst-by-burst adaptive OFDM is proposed and investigated in the context of interactive videotelephony.

As mentioned earlier, burst-by-burst adaptive quadrature amplitude modulation [29] (AQAM) was devised by Steele and Webb [29, 134] in order to enable the transceiver to cope with the time-variant channel quality of narrowband fading channels. Further related research was conducted at the University of Osaka by Sampei and his

---

[11] This section is based on P.J. Cherriman, T. Keller, and L. Hanzo: Subband-adaptive Turbo-coded OFDM-based Interactive Video Telephony, submitted to IEEE Transactions on Circuits and Systems for Video Technology, July 1999.

[12] Acknowledgment: The financial support of the Mobile VCE, EPSRC, UK and that of the European Commission is gratefully acknowledged.

[13] ©1999 IEEE. Personal use of this material is permitted. However, permission to reprint/republish this material for advertising or promotional purposes or for creating new collective works for resale or redistribution to servers or lists, or to reuse any copyrighted component of this work in other works must be obtained from IEEE.

colleagues, investigating variable coding rate concatenated coded schemes [591]; at the University of Stanford by Goldsmith and her team, studying the effects of variable-rate, variable-power arrangements [137]; and at the University of Southampton in the United Kingdom, investigating a variety of practical aspects of AQAM [597,607]. The channel's quality is estimated on a burst-by-burst basis, and the most appropriate modulation mode is selected in order to maintain the required target bit error rate (BER) performance, while maximizing the system's Bit Per Symbol (BPS) throughput. Though use of this reconfiguration regime, the distribution of channel errors becomes typically less bursty, than in conjunction with nonadaptive modems, which potentially increases the channel coding gains [608]. Furthermore, the soft-decision channel codec metrics can also be invoked in estimating the instantaneous channel quality [608], regardless of the type of channel impairments.

A range of coded AQAM schemes was analyzed by Matsuoka et al. [591], Lau et al. [609] and Goldsmith et al. [138]. For data transmission systems, which do not necessarily require a low transmission delay, variable-throughput adaptive schemes can be devised, which operate efficiently in conjunction with powerful error correction codecs, such as long block length turbo codes [28]. However, the acceptable turbo interleaving delay is rather low in the context of low-delay interactive speech. Video communications systems typically require a higher bit rate than speech systems, and hence they can afford a higher interleaving delay.

The above principles — which were typically investigated in the context of narrow-band modems — were further advanced in conjunction with wideband modems, employing powerful block turbo-coded, wideband Decision Feedback Equalizer (DFE) assisted AQAM transceivers [151,608]. A neural-network Radial Basis Function (RBF) DFE-based AQAM modem design was proposed in [610], where the RBF DFE provided the channel-quality estimates for the modem mode switching regime. This modem was capable of removing the residual BER of conventional DFEs, when linearly nonseparable received phasor constellations were encountered.

These burst-by-burst adaptive principles can also be extended to Adaptive Orthogonal Frequency Division Multiplexing (AOFDM) schemes [611] and to adaptive joint-detection-based Code Division Multiple Access (JD-ACDMA) arrangements [612]. The associated AQAM principles were invoked in the context of parallel AOFDM modems by Czylwik et al. [613], Fischer [614], and Chow et al. [615]. Adaptive subcarrier selection has also been advocated by Rohling et al. [616] in order to achieve BER performance improvements. Due to lack of space without completeness, further significant advances over benign, slowly varying dispersive Gaussian fixed links — rather than over hostile wireless links — are due to Chow, Cioffi, and Bingham [615] from the United States, rendering OFDM the dominant solution for asymmetric digital subscriber loop (ADSL) applications, potentially up to bit rates of 54 Mbps. In Europe OFDM has been favored for both Digital Audio Broadcasting (DAB) and Digital Video Broadcasting [617,618] (DVB) as well as for high-rate Wireless Asynchronous Transfer Mode (WATM) systems due to its ability to combat the effects of highly dispersive channels [619]. The idea of "water-filling" — as allocating different modem modes to different subcarriers was referred to — was proposed for OFDM by Kalet [171] and later further advanced by Chow et al. [615]. This approach was adapted later in the context of time-variant mobile channels for duplex wireless links, for example, in [611]. Finally, various OFDM-based speech and video systems were proposed in [620, 621], while the co-channel interference sensitivity of OFDM can be mitigated with the aid of adaptive beam-forming [622, 623] in multi-user scenarios.

The remainder of this section is structured as follows. Section 21.5.1 outlines the architecture of the proposed video transceiver, while Section 21.5.5 quantifies the performance benefits of AOFDM transceivers in comparison to conventional fixed transceivers. Section 21.5.6 endeavors to highlight the effects of more "aggressive" loading of the subcarriers in both BER and video quality terms, while Section 21.5.7 proposed time-variant rather than constant rate AOFDM as a means of more accurately matching the transceiver to the time-variant channel quality fluctuations, before concluding in Section 21.5.8.

21.5.2 AOFDM Modem Mode Adaptation and Signaling

The proposed duplex AOFDM scheme operates on the following basis:

- *Channel quality estimation* is invoked upon receiving an AOFDM symbol in order to select the modem mode allocation of the next AOFDM symbol.

- *The decision concerning the modem modes for the next AOFDM symbol* is based on the prediction of the expected channel conditions. Then the transmitter has to select the appropriate modem modes for the groups or subbands of OFDM subcarriers, where the subcarriers were grouped into subbands of identical modem modes in order to reduce the required number of signaling bits.

- *Explicit signaling or blind detection of the modem modes* is used to inform the receiver as to what type of demodulation to invoke.

If the channel quality of the up-link and down-link can be considered similar, then the channel-quality estimate for the up-link can be extracted from the down-link and vice versa. We refer to this regime as open-loop adaptation. In this case, the transmitter has to convey the modem modes to the receiver, or the receiver can attempt blind detection of the transmission parameters employed. By contrast, if the channel cannot be considered reciprocal, then the channel-quality estimation has to be performed at the receiver, and the receiver has to instruct the transmitter as to what modem modes have to be used at the transmitter, in order to satisfy the target integrity requirements of the receiver. We refer to this mode as closed-loop adaptation. Blind modem mode recognition was invoked, for example, in [611] — a technique that results in bit rate savings due to refraining from dedicating bits to explicit modem mode signaling at the cost of increased complexity. Let us address the issues of channel quality estimation on a subband-by-subband basis in the next subsection.

21.5.3 AOFDM Subband BER Estimation

A reliable channel-quality metric can be devised by calculating the expected overall bit error probability for all available modulation schemes M_n in each subband, which is denoted by $\bar{p}_e(n) = 1/N_s \sum_j p_e(\gamma_j, M_n)$. For each AOFDM subband, the modem mode having the highest throughput, while exhibiting an estimated BER below the target value is then chosen. Although the adaptation granularity is limited to the subband width, the channel-quality estimation is quite reliable, even in interference-impaired environments.

Against this background in our forthcoming discussions, the design tradeoffs of turbo-coded Adaptive Orthogonal Frequency Division Multiplex (AOFDM) wideband video transceivers are presented. We will demonstrate that AOFDM provides a convenient framework for adjusting the required target integrity and throughput both with and without turbo channel coding and lends itself to attractive video system construction, provided that a near-instantaneously programmable rate video codec — such as the H.263 scheme highlighted in the next section — can be invoked.

21.5.4 Video Compression and Transmission Aspects

In this study we investigate the transmission of 704 x 576 pixel Four-times Common Intermediate Format (4CIF) high-resolution video sequences at 30 frames/s using subband-adaptive turbo-coded Orthogonal Frequency Division Multiplex (AOFDM) transceivers. The transceiver can modulate 1, 2, or 4 bits onto each AOFDM subcarrier, or simply disable transmissions for subcarriers that exhibit a high attenuation or phase distortion due to channel effects.

The H.263 video codec [598] exhibits an impressive compression ratio, although this is achieved at the cost of a high vulnerability to transmission errors, since a run-length coded bit stream is rendered undecodable by a single bit error. In order to mitigate this problem, when the channel codec protecting the video stream is overwhelmed by the transmission errors, we refrain from decoding the corrupted video packet in order to prevent error propagation through the reconstructed video frame buffer [199]. We found that it was more beneficial in video-quality terms if these corrupted video packets were dropped and the reconstructed frame buffer was not updated, until the next video packet replenishing the specific video frame area was received. The associated video performance degradation was found perceptually unobjectionable for packet dropping- or transmission frame error rates (FER) below about 5%. These packet dropping events were signaled to the remote video decoder by superimposing a strongly protected one-bit packet acknowledgment flag on the reverse-direction packet, as outlined in [199]. Turbo error correction codes [28] were used.

21.5.5 Comparison of Subband-Adaptive OFDM and Fixed Mode OFDM Transceivers

In order to show the benefits of the proposed subband-adaptive OFDM transceiver, we compare its performance to that of a fixed modulation mode transceiver under identical propagation conditions, while having the same transmission bit rate. The subband-adaptive modem is capable of achieving a low bit error ratio (BER), since it can disable transmissions over low-quality subcarriers and compensate for the lost throughput by invoking a higher modulation mode than that of the fixed-mode transceiver over the high-quality subcarriers.

21.5. ADAPTIVE TURBO-CODED OFDM-BASED VIDEOTELEPHONY

	BPSK mode	QPSK mode
Packet rate	4687.5 packets/s	
FFT length	512	
OFDM symbols/packet	3	
OFDM symbol duration	$2.6667 \mu s$	
OFDM time frame	80 timeslots = 213 μs	
Normalized Doppler frequency, f'_d	1.235×10^{-4}	
OFDM symbol normalised Doppler frequency, F_D	7.41×10^{-2}	
FEC coded bits/packet	1536	3072
FEC-coded video bit rate	7.2 Mbps	14.4 Mbps
Unprotected bits/packet	766	1534
Unprotected bit rate	3.6 Mbps	7.2 Mbps
Error detection CRC (bits)	16	16
Feedback error flag bits	9	9
Packet header bits/packet	11	12
Effective video bits/packet	730	1497
Effective video bit rate	3.4 Mbps	7.0 Mbps

Table 21.16: System Parameters for the Fixed QPSK and BPSK Transceivers, as well as for the Corresponding Subband-adaptive OFDM (AOFDM) Transceivers for Wireless Local Area Networks (WLANs)

Table 21.16 shows the system parameters for the fixed BPSK and QPSK transceivers, as well as for the corresponding subband-adaptive OFDM (AOFDM) transceivers. The system employs constraint length three, half-rate turbo coding, using octal generator polynomials of 5 and 7 as well as random turbo interleavers. Therefore, the unprotected bit rate is approximately half the channel-coded bit rate. The protected to unprotected video bit rate ratio is not exactly half, since two tailing bits are required to reset the convolutional encoders' memory to their default state in each transmission burst. In both modes, a 16-bit Cyclic Redundancy Checking (CRC) is used for error detection, and 9 bits are used to encode the reverse link feedback acknowledgment information by simple repetition coding. The feedback flag decoding ensues using majority logic decisions. The packetization requires a small amount of header information added to each transmitted packet, which is 11 and 12 bits per packet for BPSK and QPSK, respectively. The effective or useful video bit rates for the BPSK and QPSK modes are then 3.4 and 7.0 Mbps.

The fixed-mode BPSK and QPSK transceivers are limited to 1 and 2 bits per symbol, respectively. By contrast, the proposed AOFDM transceivers operate at the same bit rate, as their corresponding fixed modem mode counterparts, although they can vary their modulation mode on a subcarrier- by-subcarrier basis between 0, 1, 2, and 4 bits per symbol. Zero bits per symbol implies that transmissions are disabled for the subcarrier concerned.

The "micro-adaptive" nature of the subband-adaptive modem is characterized by Figure 21.57, portraying at the top a contour plot of the channel signal-to-noise ratio (SNR) for each subcarrier versus time. At the center and bottom of the figure, the

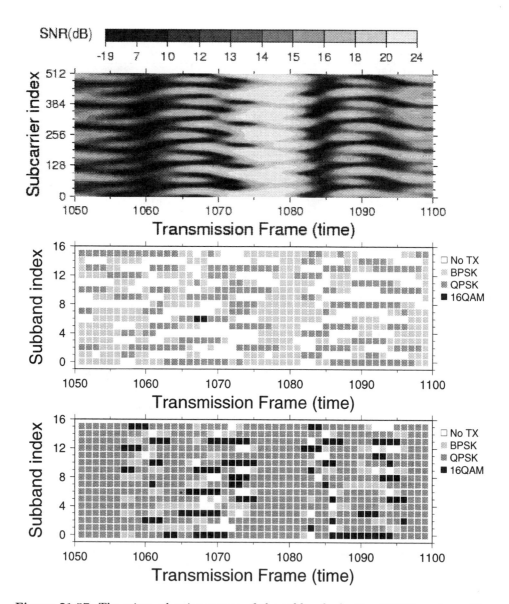

Figure 21.57: The micro-adaptive nature of the subband-adaptive OFDM modem. The top graph is a contour plot of the channel SNR for all 512 subcarriers versus time. The bottom two graphs show the modulation modes chosen for all 16 32-subcarrier subbands for the same period of time. The middle graph shows the performance of the 3.4 Mbps subband-adaptive modem, which operates at the same bit rate as a fixed BPSK modem. The bottom graph represents the 7.0 Mbps subband-adaptive modem, which operated at the same bit rate as a fixed QPSK modem. The average channel SNR was 16 dB.

21.5. ADAPTIVE TURBO-CODED OFDM-BASED VIDEOTELEPHONY

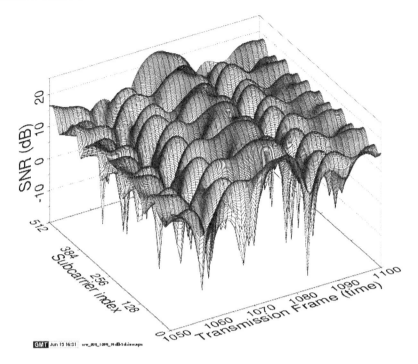

Figure 21.58: Instantaneous channel SNR for all 512 subcarriers versus time, for an average channel SNR of 16 dB over the channel characterized by the channel impulse response (CIR) of Figure 21.59.

modulation mode chosen for each 32-subcarrier subband is shown versus time for the 3.4 and 7.0 Mbps target-rate subband-adaptive modems, respectively. The channel SNR variation versus both time and frequency is also shown in three-dimensional form in Figure 21.58, which may be more convenient to visualize. This was recorded for the channel impulse response of Figure 21.59. It can be seen that when the channel is of high quality — as for example, at about frame 1080 — the subband-adaptive modem used the same modulation mode, as the equivalent fixed-rate modem in all subcarriers. When the channel is hostile — for example, around frame 1060 — the subband-adaptive modem used a lower-order modulation mode in some subbands than the equivalent fixed-mode scheme, or in extreme cases disabled transmission for that subband. In order to compensate for the loss of throughput in this subband, a higher-order modulation mode was used in the higher quality subbands.

One video packet is transmitted per OFDM symbol; therefore, the video packet-loss ratio is the same as the OFDM symbol error ratio. The video packet-loss ratio is plotted against the channel SNR in Figure 21.60. It is shown in the graph that the subband-adaptive transceivers — or synonymously termed as microscopic-adaptive (μAOFDM), in contrast to OFDM symbol-by-symbol adaptive transceivers — have a lower packet-loss ratio (PLR) at the same SNR compared to the fixed modulation mode transceiver. Note in Figure 21.60 that the subband-adaptive transceivers can operate at lower channel SNRs than the fixed modem mode transceivers, while

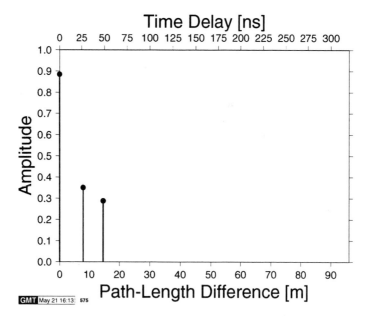

Figure 21.59: Indoor three-path WATM channel impulse response.

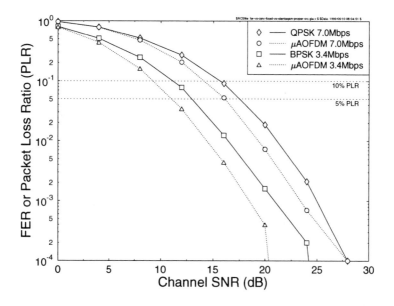

Figure 21.60: Frame Error Rate (FER) or video packet-loss ratio (PLR) versus channel SNR for the BPSK and QPSK fixed modulation mode OFDM transceivers and for the corresponding subband-adaptive μAOFDM transceiver, operating at identical effective video bit rates, namely, at 3.4 and 7.0 Mbps, over the channel model of Figure 21.59 at a normalized Doppler frequency of $F_D = 7.41 \times 10^{-2}$.

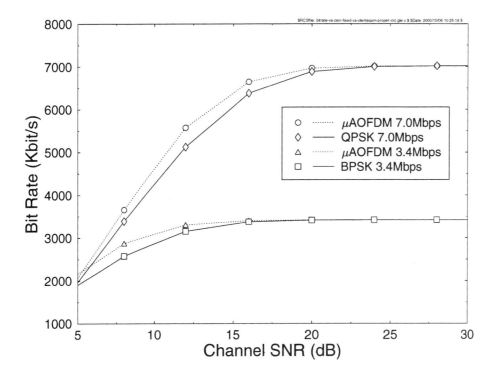

Figure 21.61: Effective throughput bit rate versus channel SNR for the BPSK and QPSK fixed modulation mode OFDM transceivers and that of the corresponding subband-adaptive or μAOFDM transceiver operating at identical effective video bit rates of 3.4 and 7.0 Mbps, over the channel of Figure 21.59 at a normalized Doppler frequency of $F_D = 7.41 \times 10^{-2}$.

maintaining the same required video packet-loss ratio. Again, the figure labels the subband-adaptive OFDM transceivers as μAOFDM, implying that the adaption is not noticeable from the upper layers of the system. A macro-adaption could be applied in addition to the microscopic adaption by switching between different target bit rates, as the longer-term channel quality improves and degrades. This issue is the subject of Section 21.5.7.

Having shown that the subband-adaptive OFDM transceiver achieved a reduced video packet loss in comparison to fixed modulation mode transceivers under identical channel conditions, we now compare the effective throughput bit rate of the fixed and adaptive OFDM transceivers in Figure 21.61. The figure shows that when the channel quality is high, the throughput bit rates of the fixed and adaptive transceivers are identical. However, as the channel degrades, the loss of packets results in a lower throughput bit rate. The lower packet-loss ratio of the subband-adaptive transceiver results in a higher throughput bit rate than that of the fixed modulation mode transceiver.

The throughput bit rate performance results translate to the decoded video-quality

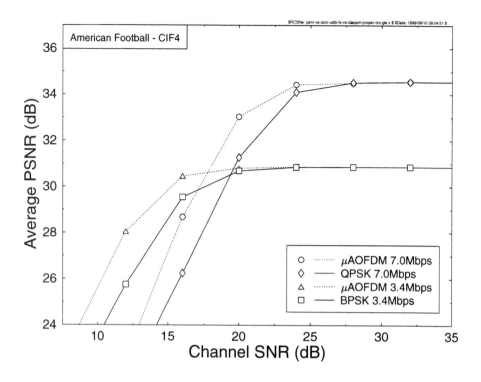

Figure 21.62: Average video quality expressed in PSNR versus channel SNR for the BPSK and QPSK fixed modulation mode OFDM transceivers and for the corresponding μAOFDM transceiver operating at identical channel SNRs over the channel model of Figure 21.59 at a normalized Doppler frequency of $F_D = 7.41 \times 10^{-2}$.

performance results evaluated in terms of PSNR in Figure 21.62. Again, for high-channel SNRs the performance of the fixed and adaptive OFDM transceivers is identical. However, as the channel quality degrades, the video quality of the subband-adaptive transceiver degrades less dramatically than that of the corresponding fixed modulation mode transceiver.

21.5.6 Subband-Adaptive OFDM Transceivers Having Different Target Bit Rates

As mentioned earlier, the subband-adaptive modems employ different modulation modes for different subcarriers in order to meet the target bit rate requirement at the lowest possible channel SNR. This is achieved by using a more robust modulation mode or eventually by disabling transmissions over subcarriers having a low channel quality. By contrast, the adaptive system can invoke less robust, but higher throughput, modulation modes over subcarriers exhibiting a high-channel quality. In the examples we have previously considered, we chose the AOFDM target bit rate

21.5. ADAPTIVE TURBO-CODED OFDM-BASED VIDEOTELEPHONY

Packet rate	4687.5 Packets/s			
FFT length	512			
OFDM symbols/packet	3			
OFDM symbol duration	2.6667μs			
OFDM time frame	80 time-slots = 213μs			
Normalized Doppler frequency, f'_d	1.235×10^{-4}			
OFDM symbol normalized Doppler frequency, F_D	7.41×10^{-2}			
FEC-coded bits/packet	858	1536	3072	4272
FEC-coded video bit rate	4.0 Mbps	7.2 Mbps	14.4 Mbps	20.0 Mbps
No. of unprotected bits/packet	427	766	1534	2134
Unprotected bit rate	2.0 Mbps	3.6 Mbps	7.2 Mbps	10.0 Mbps
No. of CRC bits	16	16	16	16
No. of feedback error flag bits	9	9	9	9
No. of packet header bits/packet	10	11	12	13
Effective video bits/packet	392	730	1497	2096
Effective video bit rate	1.8 Mbps	3.4 Mbps	7.0 Mbps	9.8 Mbps
Equivalent modulation mode		BPSK	QPSK	
Minimum channel SNR for 5% PLR (dB)	8.8	11.0	16.1	19.2
Minimum channel SNR for 10% PLR (dB)	7.1	9.2	14.1	17.3

Table 21.17: System Parameters for the Four Different Target Bit Rates of the Various Subband-adaptive OFDM (μAOFDM) Transceivers

to be identical to that of a fixed modulation mode transceiver. In this section, we comparatively study the performance of various μAOFDM systems having different target bit rates.

The previously described μAOFDM transceiver of Table 21.16 exhibited a FEC-coded bit rate of 7.2 Mbps, which provided an effective video bit rate of 3.4 Mbps. If the video target bit rate is lower than 3.4 Mbps, then the system can disable transmission in more of the subcarriers, where the channel quality is low. Such a transceiver would have a lower bit error rate than the previous BPSK-equivalent μAOFDM transceiver and therefore could be used at lower average channel SNRs, while maintaining the same bit error ratio target. By contrast, as the target bit rate is increased, the system has to employ higher-order modulation modes in more subcarriers at the cost of an increased bit error ratio. Therefore, high target bit-rate μAOFDM transceivers can only perform within the required bit error ratio constraints at high-channel SNRs, while low target bit-rate μAOFDM systems can operate at low-channel SNRs without causing excessive BERs. Therefore, a system that can adjust its target bit rate as the channel SNR changes would operate over a wide range of channel SNRs, providing the maximum possible average throughput bit rate, while maintaining the required bit error ratio.

Hence, below we provide a performance comparison of various μAOFDM transceivers that have four different target bit rates, of which two are equivalent to that of the BPSK and QPSK fixed modulation mode transceivers of Table 21.16. The system parameters for all four different bit-rate modes are summarized in Table 21.17. The modes having effective video bit rates of 3.4 and 7.0 Mbps are equivalent to the bit rates of a fixed BPSK and QPSK mode transceiver, respectively.

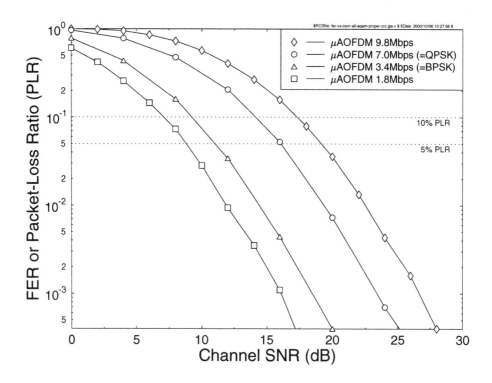

Figure 21.63: FER or video packet-loss ratio (PLR) versus channel SNR for the subband adaptive OFDM transceivers of Table 21.17 operating at four different target bit rates, over the channel model of Figure 21.59 at a normalized Doppler frequency of $F_D = 7.41 \times 10^{-2}$.

Figure 21.63 shows the Frame Error Rate (FER) or video packet-loss ratio (PLR) performance versus channel SNR for the four different target bit rates of Table 21.17, demonstrating, as expected, that the higher target bit-rate modes require higher channel SNRs in order to operate within given PLR constraints. For example, the mode having an effective video bit rate of 9.8 Mbps can only operate for channel SNRs in excess of 19 dB under the constraint of a maximum PLR of 5%. However, the mode that has an effective video bit rate of 3.4 Mbps can operate at channel SNRs of 11 dB and above, while maintaining the same 5% PLR constraint, albeit at about half the throughput bit rate and so at a lower video quality.

The trade-offs between video quality and channel SNR for the various target bit rates can be judged from Figure 21.64, suggesting, as expected, that the higher target bit rates result in a higher video quality, provided that channel conditions are favorable. However, as the channel quality degrades, the video packet-loss ratio increases, thereby reducing the throughput bit rate and hence the associated video quality. The lower target bit-rate transceivers operate at an inherently lower video quality, but they are more robust to the prevailing channel conditions and so can operate at lower

21.5. ADAPTIVE TURBO-CODED OFDM-BASED VIDEOTELEPHONY

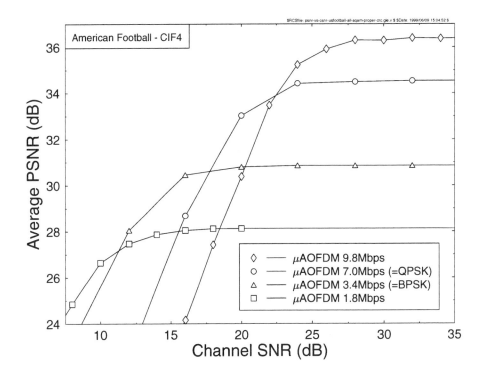

Figure 21.64: Average video quality expressed in PSNR versus channel SNR for the subband-adaptive OFDM transceivers of Table 21.17, operating at four different target bit rates, over the channel model of Figure 21.59 at a normalized Doppler frequency of $F_D = 7.41 \times 10^{-2}$.

channel SNRs, while guaranteeing a video quality, which is essentially unaffected by channel errors. It was found that the perceived video quality became impaired for packet-loss ratios in excess of about 5%.

The trade-offs between video quality, packet-loss ratio, and target bit rate are further augmented with reference to Figure 21.65. The figure shows the video quality measured in PSNR versus video frame index at a channel SNR of 16 dB as well as for an error-free situation. At the bottom of each graph, the packet-loss ratio per video frame is shown. The three figures indicate the trade-offs to be made in choosing the target bit rate for the specific channel conditions experienced — in this specific example for a channel SNR of 16dB. Note that under error-free conditions the video quality improved upon increasing the bit rate.

Specifically, video PSNRs of about 40, 41.5, and 43 dB were observed for the effective video bit rates of 1.8, 3.4, and 7.0 Mbps. The figure shows that for the target bit rate of 1.8 Mbps, the system has a high grade of freedom in choosing which subcarriers to invoke. Therefore, it is capable of reducing the number of packets that are lost. The packet-loss ratio remains low, and the video quality remains similar to

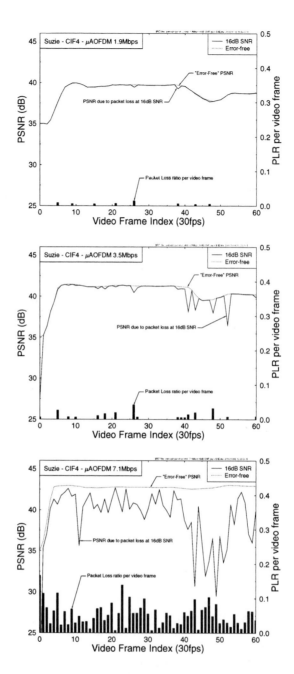

Figure 21.65: Video-quality and packet-loss ratio (PLR) performance versus video-frame index (time) comparison of subband-adaptive OFDM transceivers having target bit rates of 1.8, 3.4, and 7.0 Mbps, under the same channel conditions, at 16 dB SNR over the channel of Figure 21.59 at a normalized Doppler frequency of $F_D = 7.41 \times 10^{-2}$.

21.5. ADAPTIVE TURBO-CODED OFDM-BASED VIDEOTELEPHONY

that of the error-free situation. The two instances where the PSNR is significantly different from the error-free performance correspond to video frames in which video packets were lost. However, in both instances the system recovers in the following video frame.

As the target bit rate of the subband-adaptive OFDM transceiver is increased to 3.4 Mbps, the subband modulation mode selection process has to be more "aggressive," resulting in increased video packet loss. Observe in the figure that the transceiver having an effective video bit rate of 3.4 Mbps exhibits increased packet loss. In one frame as much as 5% of the packets transmitted for that video frame were lost, although the average PLR was only 0.4%. Because of the increased packet loss, the video PSNR curve diverges from the error-free performance curve more often. However, in almost all cases the effects of the packet losses are masked in the next video frame, indicated by the re-merging PSNR curves in the figure, maintaining a close to error-free PSNR. The subjective effect of this level of packet loss is almost imperceivable.

When the target bit rate is further increased to 7.0 Mbps, the average PLR is about 5% under the same channel conditions, and the effects of this packet-loss ratio are becoming objectionable in perceived video-quality terms. At this target bit rate, there are several video frames where at least 10% of the video packets have been lost. The video quality measured in PSNR terms rarely reaches its error-free level, because every video frame contains at least one lost packet. The perceived video quality remains virtually unimpaired until the head movement in the "Suzie" video sequence around frames 40–50, where the effect of lost packets becomes obvious, and the PSNR drops to about 30 dB.

21.5.7 Time-Variant Target Bit Rate OFDM Transceivers

By using a high target bit rate, when the channel quality is high, and employing a reduced target bit rate, when the channel quality is poor, an adaptive system is capable of maximizing the average throughput bit rate over a wide range of channel SNRs, while satisfying a given quality constraint. This quality constraint for our video system could be a maximum packet-loss ratio.

Because a substantial processing delay is associated with evaluating the packet-loss information, modem mode switching based on this metric is less efficient due to this latency. Therefore, we decided to invoke an estimate of the bit error ratio (BER) for mode switching, as follows. Since the noise energy in each subcarrier is independent of the channel's frequency domain transfer function H_n, the local signal-to-noise ratio (SNR) in subcarrier n can be expressed as

$$\gamma_n = |H_n|^2 \cdot \gamma, \qquad (21.3)$$

where γ is the overall SNR. If no signal degradation due to Inter–Subcarrier Interference (ISI) or interference from other sources appears, then the value of γ_n determines the bit error probability for the transmission of data symbols over the subcarrier n. Given γ_j across the N_s subcarriers in the jth subband, the expected overall BER for all available modulation schemes M_n in each subband can be estimated, which is denoted by $\bar{p}_e(n) = 1/N_s \sum_j p_e(\gamma_j, M_n)$. For each subband, the scheme with the

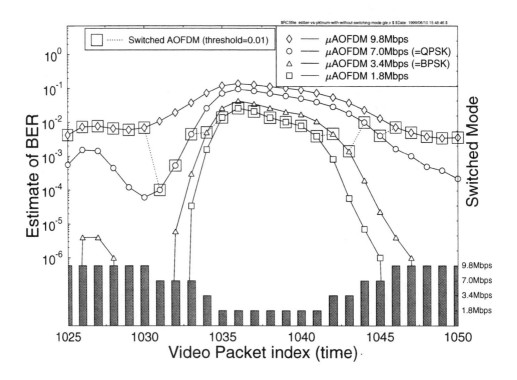

Figure 21.66: Illustration of mode switching for the switched subband adaptive modem. The figure shows the estimate of the bit error ratio for the four possible modes. The large square and the dotted line indicate the modem mode chosen for each time interval by the mode switching algorithm. At the bottom of the graph, the bar chart specifies the bit rate of the switched subband adaptive modem on the right-hand axis versus time when using the channel model of Figure 21.59 at a normalized Doppler frequency of $F_D = 7.41 \times 10^{-2}$.

highest throughput, whose estimated BER is lower than a given threshold, is then chosen.

We decided to use a quadruple-mode switched subband-adaptive modem using the four target bit rates of Table 21.17. The channel estimator can then estimate the expected bit error ratio of the four possible modem modes. Our switching scheme opted for the modem mode, whose estimated BER was below the required threshold. This threshold could be varied in order to tune the behavior of the switched subband-adaptive modem for a high or a low throughput. The advantage of a higher throughput was a higher error-free video quality at the expense of increased video packet losses, which could reduce the perceived video quality.

Figure 21.66 demonstrates how the switching algorithm operates for a 1% estimated BER threshold. Specifically, the figure portrays the estimate of the bit error ratio for the four possible modem modes versus time. The large square and the dotted

21.5. ADAPTIVE TURBO-CODED OFDM-BASED VIDEOTELEPHONY

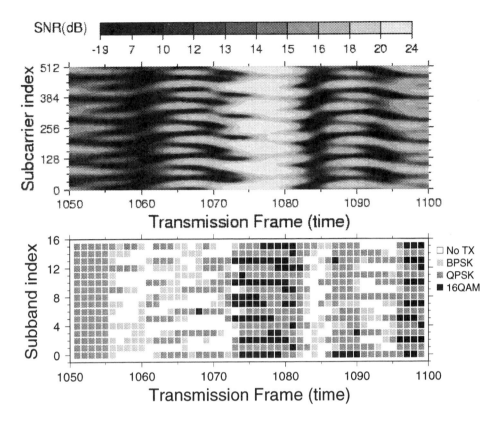

Figure 21.67: The micro-adaptive nature of the time-variant target bit-rate subband adaptive (TVTBR-AOFDM) modem. The top graph is a contour plot of the channel SNR for all 512 subcarriers versus time. The bottom graph shows the modulation mode chosen for all 16 subbands for the same period of time. Each subband is composed of 32 subcarriers. The TVTBR AOFDM modem switches between target bit rates of 2, 3.4, 7, and 9.8 Mbps, while attempting to maintain an estimated BER of 0.1% before channel coding. Average Channel SNR is 16 dB over the channel of Figure 21.59 at a normalized Doppler frequency of $F_D = 7.41 \times 10^{-2}$.

line indicate the mode chosen for each time interval by the mode switching algorithm. The algorithm attempts to use the highest bit-rate mode, whose BER estimate is less than the target threshold, namely, 1% in this case. However, if all the four modes' estimate of the BER is above the 1% threshold, then the lowest bit-rate mode is chosen, since this will be the most robust to channel errors. An example of this is shown around frames 1035–1040. At the bottom of the graph a bar chart specifies the bit rate of the switched subband adaptive modem versus time in order to emphasize when the switching occurs.

An example of the algorithm, when switching among the target bit rates of 1.8, 3.4, 7, and 9.8 Mbps, is shown in Figure 21.67. The upper part of the figure portrays the contour plot of the channel SNR for each subcarrier versus time. The lower part of the figure displays the modulation mode chosen for each 32-subcarrier subband versus time for the time-variant target bit-rate (TVTBR) subband adaptive modem. It can be seen at frames 1051–1055 that all the subbands employ QPSK modulation. Therefore, the TVTBR-AOFDM modem has an instantaneous target bit rate of 7 Mbps. As the channel degrades around frame 1060, the modem has switched to the more robust 1.8 Mbps mode. When the channel quality is high around frames 1074–1081, the highest bit-rate 9.8 Mbps mode is used. This demonstrates that the TVTBR-AOFDM modem can reduce the number of lost video packets by using reduced bit-rate but more robust modulation modes, when the channel quality is poor. However, this is at the expense of a slightly reduced average throughput bit rate. Usually, a higher throughput bit rate results in a higher video quality. However, a high bit rate is also associated with a high packet-loss ratio, which is usually less attractive in terms of perceived video quality than a lower bit-rate, lower packet-loss ratio mode.

Having highlighted how the time-domain mode switching algorithm operates, we will now characterize its performance for a range of different BER switching thresholds. A low BER switching threshold implies that the switching algorithm is cautious about switching to the higher bit-rate modes. Therefore the system performance is characterized by a low video packet-loss ratio and a low throughput bit rate. A high BER switching threshold results in the switching algorithm attempting to use the highest bit-rate modes in all but the worst channel conditions. This results in a higher video packet-loss ratio. However, if the packet-loss ratio is not excessively high, a higher video throughput is achieved.

Figure 21.68 portrays the video packet-loss ratio or FER performance of the TVTBR-AOFDM modem for a variety of BER thresholds, compared to the minimum and maximum rate unswitched modes. For a conservative BER switching threshold of 0.1%, the time-variant target bit-rate subband adaptive (TVTBR-AOFDM) modem has a similar packet-loss ratio performance to that of the 1.8 Mbps nonswitched or constant target bit-rate (CTBR) subband adaptive modem. However, as we will show, the throughput of the switched modem is always better than or equal to that of the unswitched modem and becomes far superior, as the channel quality improves. Observe in the figure that the "aggressive" switching threshold of 10% has a similar packet-loss ratio performance to that of the 9.8 Mbps CTBR-AOFDM modem. We found that in order to maintain a packet-loss ratio of below 5%, the BER switching thresholds of 2 and 3% offered the best overall performance, since the packet-loss ratio was fairly low, while the throughput bit rate was higher than that of an unswitched CTBR-AOFDM modem.

A high BER switching threshold results in the switched subband adaptive modem transmitting at a high average bit rate. However, we have shown in Figure 21.68 how the packet-loss ratio increases as the BER switching threshold increases. Therefore, the overall useful or effective throughput bit rate — that is, the bit rate excluding lost packets — may in fact be reduced in conjunction with high BER switching thresholds. Figure 21.69 demonstrates how the transmitted bit rate of the switched TVTBR-AOFDM modem increases with higher BER switching thresholds. However, when this is compared to the effective throughput bit rate, where the effects of packet loss

21.5. ADAPTIVE TURBO-CODED OFDM-BASED VIDEOTELEPHONY

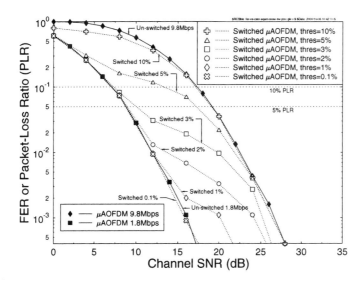

Figure 21.68: FER or video packet-loss ratio versus channel SNR for the TVTBR-AOFDM modem for a variety of BER switching thresholds. The switched modem uses four modes, with target bit rates of 1.8, 3.4, 7, and 9.8 Mbps. The unswitched 1.8 and 9.8 Mbps results are also shown in the graph as solid markers using the channel model of Figure 21.59 at a normalized Doppler frequency of $F_D = 7.41 \times 10^{-2}$

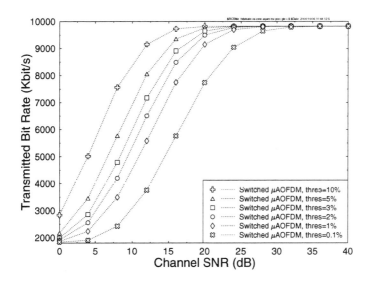

Figure 21.69: Transmitted bit rate of the switched TVTBR-AOFDM modem for a variety of BER switching thresholds. The switched modem uses four modes, having target bit rates of 1.8, 3.4, 7, and 9.8 Mbps, over the channel model of Figure 21.59 at a normalized Doppler frequency of $F_D = 7.41 \times 10^{-2}$

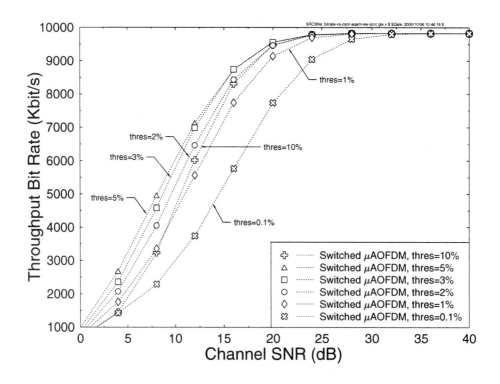

Figure 21.70: Effective throughput bit rate of the switched TVTBR-AOFDM modem for a variety of BER switching thresholds. The switched modem uses four modes, with target bit rates of 1.8, 3.4, 7, and 9.8 Mbps. The channel model of Figure 21.59 is used at a normalized Doppler frequency of $F_D = 7.41 \times 10^{-2}$

are taken into account, the tradeoff between the BER switching threshold and the effective bit rate is less obvious. Figure 21.70 portrays the corresponding effective throughput bit rate versus channel SNR for a range of BER switching thresholds. The figure demonstrates that for a BER switching threshold of 10% the effective throughput bit-rate performance was reduced in comparison to some of the lower BER switching threshold scenarios. Therefore, the BER = 10% switching threshold is obviously too aggressive, resulting in a high packet-loss ratio and a reduced effective throughput bit rate. For the switching thresholds considered, the BER = 5% threshold achieved the highest effective throughput bit rate. However, even though the BER = 5% switching threshold produces the highest effective throughput bit rate, this is at the expense of a relatively high video packet-loss ratio, which, as we will show, has a detrimental effect on the perceived video quality.

We will now demonstrate the effects associated with different BER switching thresholds on the video quality represented by the peak-signal-to-noise ratio (PSNR). Figure 21.71 portrays the PSNR and packet-loss performance versus time for a range of BER switching thresholds. The top graph in the figure indicates that for a BER

21.5. ADAPTIVE TURBO-CODED OFDM-BASED VIDEOTELEPHONY 947

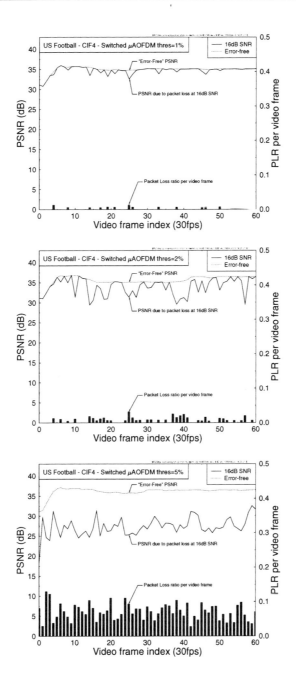

Figure 21.71: Video-quality and packet-loss ratio performance versus video-frame index (time) comparison between switched TVTBR-AOFDM transceivers with different BER switching thresholds, at an average of 16dB SNR, using the channel model of Figure 21.59 at a normalized Doppler frequency of $F_D = 7.41 \times 10^{-2}$

switching threshold of 1% the PSNR performance is very similar to the corresponding error-free video quality. However, the PSNR performance diverges from the error-free curve when video packets are lost, although the highest PSNR degradation is limited to 2 dB. Furthermore, the PSNR curve typically reverts to the error-free PSNR performance curve in the next frame. In this example, about 80% of the video frames have no video packet loss. When the BER switching threshold is increased to 2%, as shown in the center graph of Figure 21.71, the video-packet loss ratio has increased, such that now only 41% of video frames have no packet loss. The result of the increased packet loss is a PSNR curve, which diverges from the error-free PSNR performance curve more regularly, with PSNR degradations of up to 7 dB. When there are video frames with no packet losses, the PSNR typically recovers, achieving a similar PSNR performance to the error-free case. When the BER switching threshold was further increased to 3% — which is not shown in the figure — the maximum PSNR degradation increased to 10.5 dB, and the number of video frames without packet losses was reduced to 6%.

The bottom graph of Figure 21.71 depicts the PSNR and packet loss performance for a BER switching threshold of 5%. The PSNR degradation in this case ranges from 1.8 to 13 dB and all video frames contain at least one lost video packet. Even though the BER = 5% switching threshold provides the highest effective throughput bit rate, the associated video quality is poor. The PSNR degradation in most video frames is about 10 dB. Clearly, the highest effective throughput bit rate does not guarantee the best video quality. We will now demonstrate that the switching threshold of BER = 1% provides the best video quality, when using the average PSNR as our performance metric.

Figure 21.72(a) compares the average PSNR versus channel SNR performance for a range of switched (TVTBR) and unswitched (CTBR) AOFDM modems. The figure compares the four unswitched (i.e., CTBR subband adaptive modems) with switching (i.e., TVTBR subband adaptive modems), which switch between the four fixed-rate modes, depending on the BER switching threshold. The figure indicates that the switched TVTBR subband adaptive modem having a switching threshold of BER = 10% results in similar PSNR performance to the unswitched CTBR 9.8 Mbps subband adaptive modem. When the switching threshold is reduced to BER = 3%, the switched TVTBR AOFDM modem outperforms all of the unswitched CTBR AOFDM modems. A switching threshold of BER = 5% achieves a PSNR performance, which is better than the unswitched 9.8 Mbps CTBR AOFDM modem, but worse than that of the unswitched 7.0 Mbps modem, at low- and medium-channel SNRs.

A comparison of the switched TVTBR AOFDM modem employing all six switching thresholds that we have used previously is shown in Figure 21.72(b). This figure suggests that switching thresholds of BER = 0.1, 1, and 2% perform better than the BER = 3% threshold, which outperformed all of the unswitched CTBR subband adaptive modems. The best average PSNR performance was achieved by a switching threshold of BER = 1%. The more conservative BER = 0.1% switching threshold results in a lower PSNR performance, since its throughput bit rate was significantly reduced. Therefore, the best trade-off in terms of PSNR, throughput bit rate, and video packet-loss ratio was achieved with a switching threshold of about BER = 1%.

21.5. ADAPTIVE TURBO-CODED OFDM-BASED VIDEOTELEPHONY

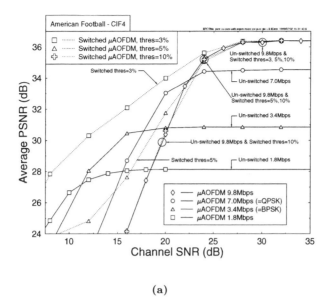

(a)

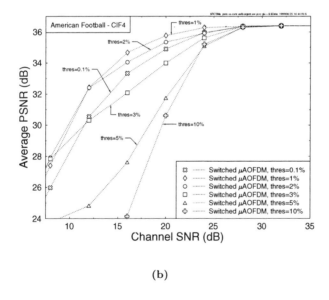

(b)

Figure 21.72: Average PSNR versus channel SNR performance for switched and unswitched subband adaptive modems. Figure (a) compares the four unswitched CTBR subband adaptive modems with switched TVTBR subband adaptive modems (using the same four modem modes) for switching thresholds of BER = 3, 5, and 10%. Figure (b) compares the switched TVTBR subband adaptive modems for switching thresholds of BER = 0.1, 1, 2, 3, 5, and 10%.

21.5.8 Summary and Conclusions

A range of AOFDM video transceivers has been proposed for robust, flexible, and low-delay interactive videotelephony. In order to minimize the amount of signaling required, we divided the OFDM subcarriers into subbands and controlled the modulation modes on a subband-by-subband basis. The proposed constant target bit-rate AOFDM modems provided a lower BER than the corresponding conventional OFDM modems. The slightly more complex switched TVTBR-AOFDM modems can provide a balanced video-quality performance, across a wider range of channel SNRs than the other schemes investigated.

21.6 Digital Terrestrial Video Broadcasting for Mobile Receivers[14]

21.6.1 Background and Motivation

Following the standardization of the pan-European digital video broadcasting (DVB) systems, we have begun to witness the arrival of digital television services to the home. However, for a high proportion of business and leisure travelers, it is desirable to have access to DVB services while on the move. Although it is feasible to access these services with the aid of dedicated DVB receivers, these receivers may also find their way into the laptop computers of the near future. These intelligent laptops may also become the portable DVB receivers of wireless in-home networks.

In recent years three DVB standards have emerged in Europe for terrestrial [624], cable-based [625], and satellite-oriented [626] delivery of DVB signals. The more hostile propagation environment of the terrestrial system requires concatenated Reed-Solomon (RS) [87,627] and rate-compatible punctured convolutional coding (RCPCC) [87, 627] combined with orthogonal frequency division multiplexing (OFDM)-based modulation [29]. By contrast, the more benign cable and satellite-based media facilitate the employment of multilevel modems using up to 256-level quadrature amplitude modulation (QAM) [29]. These schemes are capable of delivering high-definition video at bit rates of up to 20 Mbit/s in stationary broadcast-mode distributive wireless scenarios.

Recently, a range of DVB system performance studies has been presented in the literature [628–631]. Against this background, in this contribution we have proposed turbo-coding improvements to the terrestrial DVB system [624] and investigated its performance under hostile mobile channel conditions. We have also studied various video bit-stream partitioning and channel coding schemes both in the hierarchical and nonhierarchical transceiver modes and compared their performance.

---

[14]This section is based on C. S. Lee, T. Keller, and L. Hanzo, OFDM-based turbo-coded hierarchical and non-hierarchical terrestrial mobile digital video broadcasting, *IEEE Transactions on Broadcasting*, March 2000, pp. 1-22, ©2000 IEEE. Personal use of this material is permitted. However, permission to reprint/republish this material for advertising or promotional purposes or for creating new collective works for resale or redistribution to servers or lists, or to reuse any copyrighted component of this work in other works must, be obtained from the IEEE.

21.6. DVB-T FOR MOBILE RECEIVERS

The rest of this section is divided as follows. In Section 21.6.2 the bit error sensitivity of the MPEG-2 coding parameters [632] is characterized. A brief overview of the enhanced turbo-coded and standard DVB terrestrial scheme is presented in Section 21.6.3, while the channel model is described in Section 21.6.4. Following this, in Section 21.6.5 the reader is introduced to the MPEG-2 data partitioning scheme [633] used to split the input MPEG-2 video bit-stream into two error protection classes, which can then be protected either equally or unequally. These two video bit protection classes can then be broadcast to the receivers using the DVB terrestrial hierarchical transmission format [624]. The performance of the data partitioning scheme is investigated by corrupting either the high- or low-sensitivity video bits using randomly distributed errors for a range of system configurations in Section 21.6.6, and their effects on the overall reconstructed video quality are evaluated. Following this, the performance of the improved DVB terrestrial system employing the nonhierarchical and hierarchical format [624] is examined in a mobile environment in Sections 21.6.7 and 21.6.8, before our conclusions and future work areas are presented in Section 21.6.9. We note furthermore that readers interested mainly in the overall system performance may opt to proceed directly to Section 21.6.3. Let us commence our discourse in the next section by describing an objective method of quantifying the sensitivity of the MPEG-2 video parameters.

21.6.2 MPEG-2 Bit Error Sensitivity

At this stage we again note that a number of different techniques can be used to quantify the bit error sensitivity of the MPEG-2 bits. The outcome of these investigations will depend to a degree on the video material used, the output bit rate of the video codec, the objective video-quality measures used, and the averaging algorithm employed. Perceptually motivated, subjective quality-based sensitivity testing becomes infeasible due to the large number of associated test scenarios. Hence, in this section we propose a simplified objective video-quality measure based bit-sensitivity evaluation procedure, which attempts to examine all the major factors influencing the sensitivity of MPEG-2 bits. Specifically, the proposed procedure takes into account the position and the relative frequency of the MPEG-2 parameters in the bit-stream, the number of the associated coding bits for each MPEG-2 parameter, the video bit rate, and the effect of loss of synchronization or error propagation due to corrupted bits. Nonetheless, we note that a range of similar bit-sensitivity estimation techniques exhibiting different strengths and weaknesses can be devised. No doubt future research will produce a variety of similarly motivated techniques.

In this section, we assume familiarity with the MPEG-2 standard [632, 633]. The aim of our MPEG-2 error-resilience study was to quantify the average PSNR degradation caused by each erroneously decoded video codec parameter in the bit-stream, so that appropriate protection can be assigned to each parameter. First, we define three measures, namely, the peak signal-to-noise ratio (PSNR), the PSNR degradation, and the average PSNR degradation, which are to be used in our subsequent discussions. The PSNR is defined as follows:

$$\mathrm{PSNR} = 10 \, log_{10} \frac{\sum_{n=0}^{N} \sum_{m=0}^{M} 255^2}{\sum_{n=0}^{N} \sum_{m=0}^{M} \Delta^2}, \qquad (21.4)$$

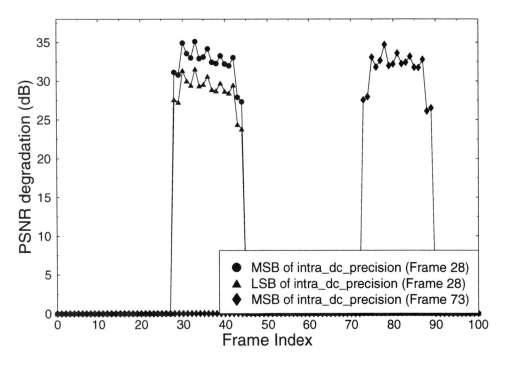

Figure 21.73: PSNR degradation profile for the different bits used to encode the intra_dc_precision parameter [632] in different corrupted video frames for the "Miss America" QCIF video sequence encoded at 30 frame/s and 1.15 Mbit/s.

where Δ is the difference between the uncoded pixel value and the reconstructed pixel value, while the variables M and N refer to the dimension of the image. The maximum possible 8-bit pixel luminance value of 255 was used in Equation 21.4 in order to mitigate the PSNR's dependence on the video material used. The PSNR degradation is the difference between the PSNR of the decoder's reconstructed image in the event of erroneous decoding and successful decoding. The average PSNR degradation is then the mean of the PSNR degradation values computed for all the image frames of the video test sequence.

Most MPEG-2 parameters are encoded by several bits, and they may occur in different positions in the video sequence. In these different positions, they typically affect the video quality differently, since corrupting a specific parameter of a frame close to the commencement of a new picture start code results in a lesser degradation than corrupting an equivalent parameter further from the resynchronization point. Hence the sensitivity of the MPEG-2 parameters is position-dependent. Furthermore, different encoded bits of the same specific MPEG-2 parameter may exhibit different sensitivity to channel errors. Figure 21.73 shows such an example for the parameter known as intra_dc_precision [632], which is coded under the picture coding extension [633]. In this example, the PSNR degradation profiles due to bit errors being inflicted on the parameter intra_dc_precision of frame 28 showed that the degradation

21.6. DVB-T FOR MOBILE RECEIVERS

is dependent on the significance of the bit considered. Specifically, errors in the most significant bit (MSB) caused an approximately 3 dB higher PSNR degradation than the least significant bit (LSB) errors. Furthermore, the PSNR degradation due to an MSB error of the intra_dc_precision parameter in frame 73 is similar to the PSNR degradation profile for the MSB of the intra_dc_precision parameter of frame 28. Due to the variation of the PSNR degradation profile for the bits of different significance of a particular parameter, as well as for the same parameter at its different occurrences in the bit-stream, it is necessary to determine the *average* PSNR degradation for each parameter in the MPEG-2 bit-stream.

Our approach in obtaining the average PSNR degradation was similar to that suggested in the literature [467, 634]. Specifically, the average measure used here takes into account the significance of the bits corresponding to the MPEG-2 parameter concerned, as well as the occurrence of the same parameter at different locations in the encoded video bit-stream. In order to find the average PSNR degradation for each MPEG-2 bit-stream parameter, the different bits encoding a specific parameter, as well as the bits of the same parameter but occurring at different locations in the MPEG-2 bit-stream, were corrupted and the associated PSNR degradation profile versus frame index was registered. The observed PSNR degradation profile generated for different locations of a specific parameter was then used to compute the average PSNR degradation. As an example, we will use the PSNR degradation profile shown in Figure 21.73. This figure presents three degradation profiles. The average PSNR degradation for each profile is first computed in order to produce three average PSNR degradation values corresponding to the three respective profiles. The mean of these three PSNR averages will then form the final average PSNR degradation for the intra_dc_precision parameter. The same process is repeated for all MPEG-2 parameters from the picture layer up to the block layer. The difference with respect to the approach adopted in [467, 634] was that while in [467, 634] the average PSNR degradation was acquired for each bit of the output bit-stream, we have adopted a simpler approach in this contribution due to the large number of different parameters within the MPEG-2 bit-stream. Figures 21.74 to 21.76 show the typical average PSNR degradations of the various MPEG-2 parameters of the picture header information, picture coding extension, slice layer macroblock layer and block layer [633], which were obtained using the 176×144 quarter common intermediate format (QCIF) "Miss America" (MA) video sequence at 30 frames/s and a high average bit rate of 1.15 Mbit/s.

The different MPEG-2 parameters or code words occur with different probabilities, and they are allocated different numbers of bits. Therefore, the average PSNR degradation registered in Figures 21.74 to 21.76 for each MPEG-2 parameter was multiplied, with the long-term probability of this MPEG-2 parameter occurring in the MPEG-2 bit-stream and with the relative probability of bits being allocated to that MPEG-2 parameter. Figures 21.77 and 21.78 show the occurrence probability of the various MPEG-2 parameters characterized in Figures 21.74 to 21.76 and the probability of bits allocated to the parameters in the picture header information, picture coding extension, as well as in the slice, macroblock, and block layers [633], respectively, for the QCIF MA video sequence encoded at 1.15 Mbit/s.

We will concentrate first on Figure 21.77(a). It is observed that all parameters

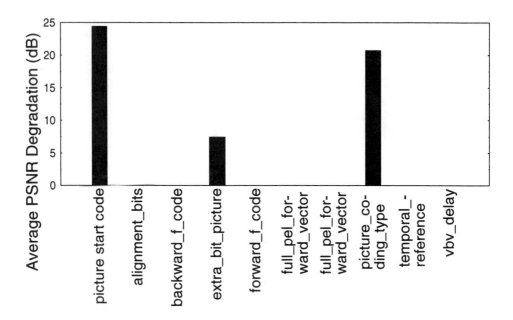

Figure 21.74: Average PSNR degradation for the various MPEG-2 parameters in picture header information for the "Miss America" QCIF video sequence encoded at 30 frame/s and 1.15 Mbit/s.

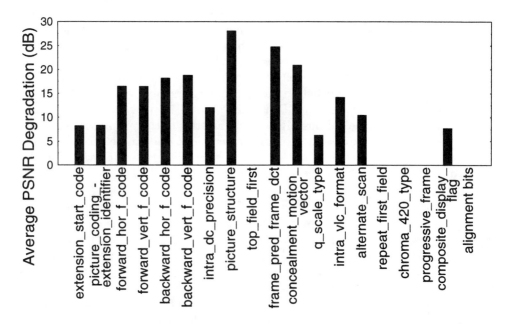

Figure 21.75: Average PSNR degradation for the various MPEG-2 parameters in picture coding extension for the "Miss America" QCIF video sequence encoded at 30 frame/s and 1.15 Mbit/s.

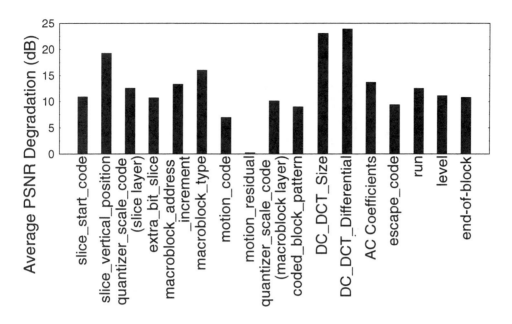

Figure 21.76: Average PSNR degradation for the various MPEG-2 parameters in slice, macroblock, and block layers for the "Miss America" QCIF video sequence encoded at 30 frame/s and 1.15 Mbit/s.

— except for the full_pel_forward_vector, forward_f_code, full_pel_backward_vector, and backward_f_code — have the same probability of occurrence, since they appear once for every coded video frame. The parameters full_pel_forward_vector and forward_f_code have a higher probability of occurrence than full_pel_backward_vector and backward_f_code, since the former two appear in both P frames and B frames, while the latter two only occur in B frames. For our experiments, the MPEG-2 encoder was configured so that for every encoded P frame, there were two encoded B frames. However, when compared with the parameters from the slice layer, macroblock layer, and block layer, which are characterized by the bar chart of Figure 21.77(b), the parameters of the picture header information and picture coding extension appeared significantly less frequently.

If we compare the occurrence frequency of the parameters in the slice layer with those in the macroblock and block layers, the former appeared less often, since there were 11 macroblocks and 44 blocks per video frame slice for the QCIF "Miss America" video sequence were considered in our experiments. The AC discrete cosine transform (DCT) [452] coefficient parameter had the highest probability of occurrence, exceeding 80%.

Figure 21.78 shows the probability of bits being allocated to the various MPEG-2 parameters in the picture header information, picture coding extension, slice, macroblock and block layers [633]. Figure 21.79 was included to more explicitly illustrate the probability of bit allocation seen in Figure 21.78(b), with the probability of allocation of bits to the AC DCT coefficients being omitted from the bar chart. Considering

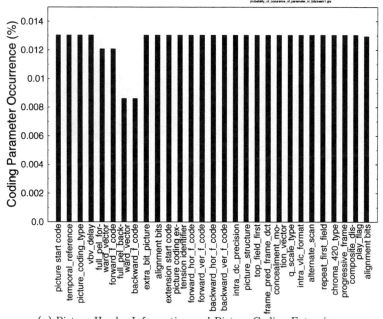

(a) Picture Header Information and Picture Coding Extension

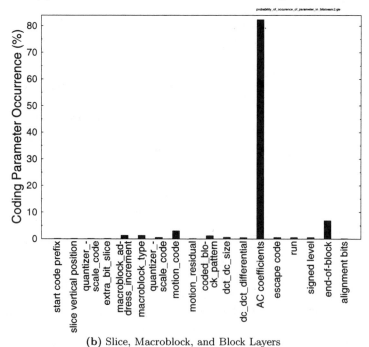

(b) Slice, Macroblock, and Block Layers

Figure 21.77: Occurrence probability for the various MPEG-2 parameters characterized in Figure 21.74 to Figure 21.76. (a) Picture header information and picture coding extension. (b) Slice, macroblock, and block layers for the "Miss America" QCIF video sequence encoded at 30 frame/s and 1.15 Mbit/s.

21.6. DVB-T FOR MOBILE RECEIVERS

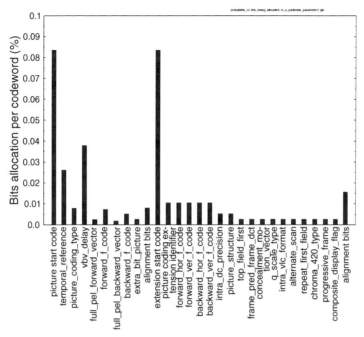

(a) Picture Header Information and Picture Coding Extension

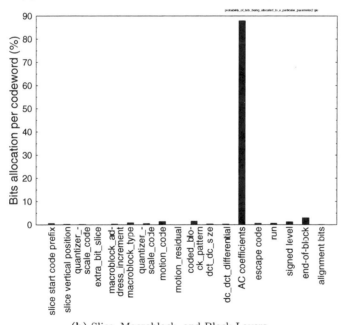

(b) Slice, Macroblock, and Block Layers

Figure 21.78: Probability of bits being allocated to parameters in (a) picture header information and picture coding extension; and (b) Slice, macroblock, and block layers for the "Miss America" QCIF video sequence encoded at 30 frame/s and 1.15 Mbit/s.

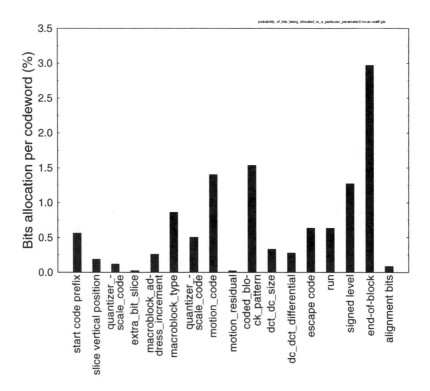

Figure 21.79: Probability of bits being allocated to the various MPEG-2 slice, macroblock, and block layer parameters, as seen in Figure 21.78(b), where the probability of bits allocated to the AC DCT coefficients was omitted in order to show the allocation of bits to the other parameters more clearly. This probability of bits allocation to the various MPEG-2 parameters is associated with the "Miss America" QCIF video sequence encoded at 30 frame/s and 1.15 Mbit/s.

Figure 21.78(a), the two dominant parameters, with the highest number of encoding bits requirement are the picture start code (PSC) and the picture coding extension start code (PCESC). However, comparing these probabilities with the probability of bits being allocated to the various parameters in the slice, macroblock, and block layers, we see that the percentage of bits allocated can still be considered minimal due to their infrequent occurrence. In the block layer, the AC DCT coefficients require in excess of 85% of the bits available for the whole video sequence. However, at bit rates lower than 1.15 Mbit/s the proportion of AC coefficient encoding bits was significantly reduced, as illustrated by Figure 21.80. Specifically, at 30 frame/s and 1.15 Mbit/s, the average number of bits per video frame is about 38,000 and a given proportion of these bits is allocated to the MPEG-2 control header information, motion information, and the DCT coefficients. Upon reducing the total bit rate budget — since the number of control header bits is more or less independent of the target bit rate — the proportion of bits allocated to the DCT coefficients is substantially reduced. This is

21.6. DVB-T FOR MOBILE RECEIVERS

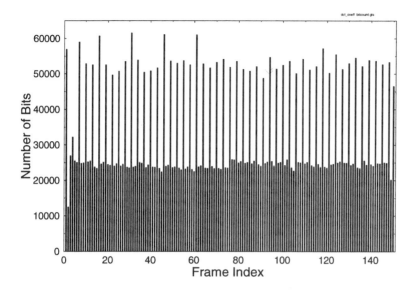

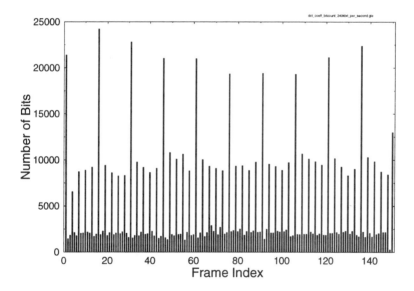

Figure 21.80: Profile of bits allocated to the DCT coefficients, when the 30 frame/s QCIF "Miss America" video sequence is coded at (a) 1.15 Mbit/s (top) and (b) 240 kbit/s (bottom). The sequence of frames is in the order I B B, P B B, P B B, P B B, and so on.

explicitly demonstrated in Figure 21.80 for bit rates of 1.15 Mbit/s and 240 kbit/s for the "Miss America" QCIF video sequence.

The next process, as discussed earlier, was to normalize the measured average PSNR degradation according to the occurrence probability of the respective MPEG-2 parameters in the bit-stream and the probability of bits being allocated to this parameter. The normalized average PSNR degradation caused by corrupting the parameters of the picture header information and picture coding extension [633] is portrayed in Figure 21.81(a). Similarly, the normalized average PSNR degradation for the parameters of the slice, macroblock, and block layers is shown in Figure 21.81(b). In order to visually enhance Figure 21.81(b), the normalized average PSNR degradation for the AC DCT coefficients was omitted in the bar chart shown in Figure 21.82.

The highest PSNR degradation was inflicted by the AC DCT coefficients, since these parameters occur most frequently and hence are allocated the highest number of bits. When a bit error occurs in the bit-stream, the AC DCT coefficients have a high probability of being corrupted. The other parameters, such as DC_DCT_size and DC_DCT_differential, exhibited high average PSNR degradations when corrupted, but registered low normalized average PSNR degradations since their occurrence in the bit-stream is confined to the infrequent intra-coded frames.

The end-of-block MPEG-2 parameter exhibited the second highest normalized average PSNR degradation in this study. Although the average number of bits used for the end-of-block is only approximately 2.17 bits, the probability of occurrence and the probability of bits being allocated to it are higher than for other parameters, with the exception of the AC DCT coefficients. Furthermore, in general, the parameters of the slice, macroblock, and block layers exhibit higher average normalized PSNR degradations due to their more frequent occurrence in the bit-stream compared to the parameters that belong to the picture header information and to the picture coding extension. This also implies that the percentage of bits allocated to these parameters is higher.

Comparing the normalized average PSNR degradations of the parameters in the picture header information and picture coding extension, we observe that the picture start code (PSC) exhibits the highest normalized average PSNR degradation. Although most of the parameters here occur with equal probability as seen in Figure 21.77(a), the picture start code requires a higher portion of the bits compared to the other parameters, with the exception of the extension start code. Despite having the same probability of occurrence and the same allocation of bits, the extension start code exhibits a lower normalized PSNR degradation than the picture start code, since its average unnormalized degradation is lower, as shown in Figure 21.74 to Figure 21.76.

In Figures 21.81 and 21.82, we observed that the video PSNR degradation was dominated by the erroneous decoding of the AC DCT coefficients, which appeared in the MPEG-2 video bit-stream in the form of variable-length codewords. This suggests that unequal error protection techniques be used to protect the MPEG-2 parameters during transmission. In a low-complexity implementation, two protection classes may be envisaged. The higher priority class would contain all the important header information and some of the more important low-frequency variable-length coded DCT coefficients. The lower priority class would then contain the remaining less

21.6. DVB-T FOR MOBILE RECEIVERS

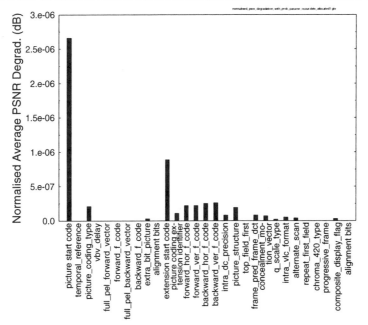

(a) Picture Header Information and Picture Coding Extension

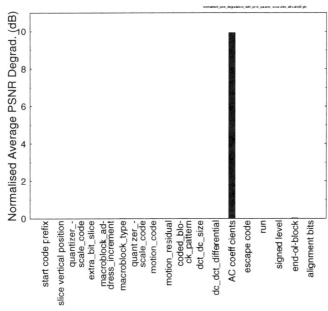

(b) Slice, Macroblock, and Block Layers

Figure 21.81: Normalized average PSNR degradation for the various parameters in (a) picture header information and picture coding extension (b) slice, macroblock, and block layers, normalized to the occurrence probability of the respective parameters in the bit-stream and the probability of bits being allocated to the parameter for the "Miss America" QCIF video sequence encoded at 30 frame/s and 1.15 Mbit/s.

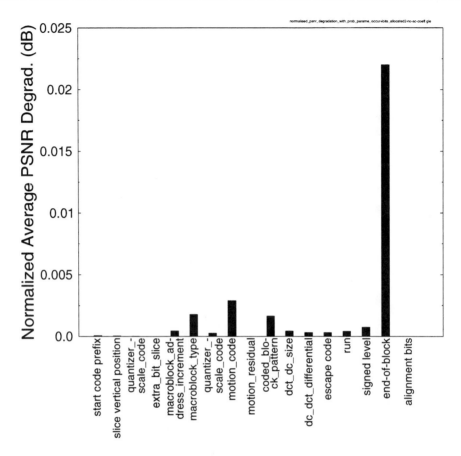

Figure 21.82: This bar chart is the same as Figure 21.81(b), although the normalized average PSNR degradation for the AC DCT coefficients was omitted in order to show the average PSNR degradation of the other parameters. This bar chart is presented for the "Miss America" QCIF video sequence encoded at 30 frame/s and 1.15 Mbit/s case.

important, higher frequency variable-length coded DCT coefficients. This partitioning process will be detailed in Section 21.6.5 together with its associated performance in the context of the hierarchical DVB [624] transmission scheme in Section 21.6.8. Let us, however, first consider the architecture of the investigated DVB system in the next section.

21.6.3 DVB Terrestrial Scheme

The block diagram of the DVB terrestrial (DVB-T) transmitter [624] is shown in Figure 21.83, which consists of an MPEG-2 video encoder, channel-coding modules, and an orthogonal frequency division multiplex (OFDM) modem [29, 635]. The bitstream generated by the MPEG-2 encoder is packetized into frames 188 bytes long. The

21.6. DVB-T FOR MOBILE RECEIVERS

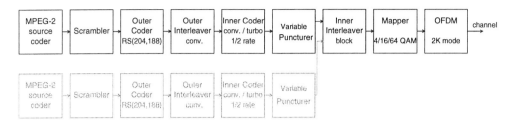

Figure 21.83: Schematic of the DVB terrestrial transmitter functions.

Convolutional Coder Parameters	
Code rate	1/2
Constraint length	7
n	2
k	1
Generator polynomials (octal format)	171, 133

Table 21.18: Parameters of the $CC(n, k, K)$ Convolutional Inner Encoder of the DVB-T Modem

video data in each packet is then randomized by the scrambler of Figure 21.83. The specific details concerning the scrambler have not been included in this chapter since they may be obtained from the DVB-T standard [624].

Because of the poor error resilience of the MPEG-2 video codec, powerful concatenated channel coding is employed. The concatenated channel codec of Figure 21.83 comprises a shortened Reed–Solomon (RS) outer code and an inner convolutional encoder. The 188-byte MPEG-2 video packet is extended by the Reed–Solomon encoder [87, 627], with parity information to facilitate error recovery in order to form a 204-byte packet. The Reed–Solomon decoder can then correct up to 8 erroneous bytes for each 204-byte packet. Following this, the RS-coded packet is interleaved by a convolutional interleaver and further protected by a half-rate inner convolutional encoder using a constraint length of 7 [87, 627].

Furthermore, the overall code rate of the concatenated coding scheme can be adapted by variable puncturing that supports code rates of 1/2 (no puncturing) as well as 2/3, 3/4, 5/6, and 7/8. The parameters of the convolutional encoder are summarized in Table 21.18.

If only one of the two branches of the transmitter in Figure 21.83 is utilized, the DVB-T modem is said to be operating in its nonhierarchical mode. In this mode, the modem can have a choice of QPSK, 16QAM, or 64QAM modulation constellations [29].

A second video bit-stream can also be multiplexed with the first one by the inner interleaver, when the DVB modem is in its hierarchical mode [624]. The choice of modulation constellations in this mode is between 16QAM and 64QAM. We employ this transmission mode when the data partitioning scheme of Section 21.6.5 is used to split the incoming MPEG-2 video bit-stream into two video bit-protection classes, with one class having a higher grade of protection or priority than the other one. The

OFDM Parameters	
Total number of subcarriers	2048 (2 K mode)
Number of effective subcarriers	1705
OFDM symbol duration T_s	224 µs
Guard interval	$T_s/4 = 56$µs
Total symbol duration (inc. guard interval)	280 µs
Consecutive subcarrier spacing $1/T_s$	4464 Hz
DVB channel spacing	7.61 MHz
QPSK and QAM symbol period	7/64 µs

Table 21.19: Parameters of the OFDM Module Used in the DVB-T Modem [624]

Turbo Coder Parameters	
Turbo code rate	1/2
Input block length	17, 952 bits
Interleaver type	Random
Number of turbo-decoder iterations	8
Turbo Encoder Component Code Parameters	
Component code encoder type	Recursive Systematic Convolutional (RSC)
Component code decoder type	log-MAP [636]
Constraint length	3
n	2
k	1
Generator polynomials (octal format)	7, 5

Table 21.20: Parameters of the Inner Turbo Encoder Used to Replace the DVB-T System's $K = 7$ Convolutional Encoder of Table 21.18 (RSC: recursive systematic code)

higher priority video bits will be mapped to the MSBs of the modulation constellation points and the lower priority video bits to the LSBs of the QAM-constellation [29]. For 16QAM and 64QAM, the two MSBs of each 4-bit or 6-bit QAM symbol will contain the more important video data. The lower priority video bits will then be mapped to the lower significance 2 bits and 4 bits of 16QAM and 64QAM, respectively [29].

These QPSK, 16QAM, or 64QAM symbols are then distributed over the OFDM carriers [29]. The parameters of the OFDM system are presented in Table 21.19.

Besides implementing the standard DVB-T system as a benchmark, we have improved the system by replacing the convolutional coder with a turbo codec [28, 381]. The turbo codec's parameters used in our investigations are displayed in Table 21.20. The block diagram of the turbo encoder is shown in Figure 21.84. The turbo encoder is constructed of two component encoders. Each component encoder is a half-rate convolutional encoder whose parameters are listed in Table 21.20. The two-component encoders are used to encode the same input bits, although the input bits of the

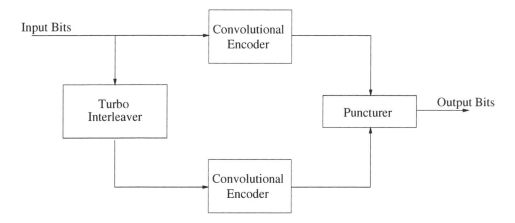

Figure 21.84: Block diagram of turbo encoder

second component encoder are interleaved before encoding. The output bits of the two-component codes are punctured and multiplexed in order to form a single-output bit-stream. The component encoder used here is known as a half-rate recursive systematic convolutional encoder (RSC) [637]. It generates one parity bit and one systematic output bit for every input bit. In order to provide an overall coding rate of $R = 1/2$, half the output bits from the two encoders must be punctured. The puncturing arrangement used in our work is to transmit all the systematic bits from the first encoder and every other parity bit from both encoders [638]. We note here that one iteration of the turbo decoder involves two Logarithmic Maximum A-Posteriori (log-MAP) [636] decoding operations, which we repeated for the eight iterations. Hence, the total turbo decoding complexity is about 16 times higher than a constraint length $K = 3$ constituent convolutional decoding. Therefore, the turbo decoder exhibits a similar complexity to the $K = 7$ convolutional decoder.

In this section, we have given an overview of the standard and enhanced DVB-T system, which we have used in our experiments. Readers interested in further details of the DVB-T system are referred to the DVB-T standard [624]. The performance of the standard DVB-T system and the turbo-coded system is characterized in Sections 21.6.7 and 21.6.8 for nonhierarchical and hierarchical transmissions, respectively. Let us now briefly consider the multipath channel model used in our investigations.

21.6.4 Terrestrial Broadcast Channel Model

The channel model employed in this study was the 12-path COST 207 [639] hilly terrain (HT) type impulse response, with a maximal relative path delay of 19.9 μs. This channel was selected in order to provide a worst-case propagation scenario for the DVB-T system employed in our study.

In the system described here, we have used a carrier frequency of 500 MHz and a sampling rate of 7/64 μs. Each of the channel paths was faded independently obeying a Rayleigh-fading distribution, according to a normalized Doppler frequency of 10^{-5} [87]. This corresponds to a worst-case vehicular velocity of about 200 km/h.

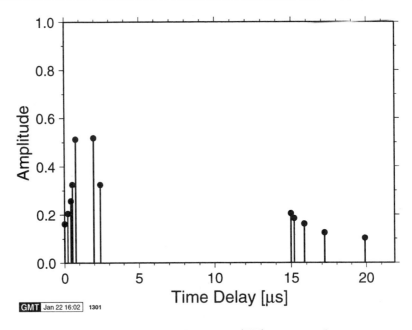

Figure 21.85: COST 207 hilly terrain (HT) type impulse response.

The unfaded impulse response is depicted in Figure 21.85. For the sake of completeness we note that the standard COST 207 channel model was defined in order to facilitate the comparison of different GSM implementations [87] under identical conditions. The associated bit rate was 271 kbit/s, while in our investigations the bit rate of DVB-quality transmissions can be as high as 20 Mbit/s, where a higher number of resolvable multipath components within the dispersion-range is considered. However, the performance of various wireless tranceivers is well understood by the research community over this standard COST 207 channel. Hence, its employment is beneficial in benchmarking terms. Furthermore, since the OFDM modem has 2048 subcarriers, the subcarrier signaling rate is effectively 2000-times lower than our maximum DVB-rate of 20 Mbit/s, corresponding to 10 kbit/s. At this subchannel rate, the individual subchannel can be considered nearly frequency-flat. In summary, in conjunction with the 200 km/h vehicular speed the investigated channel conditions constitute a pessimistic scenario.

In order to facilitate unequal error protection, the data partitioning procedure of the MPEG-2 video bit-stream is considered next.

21.6.5 Data Partitioning Scheme

Efficient bit-stream partitioning schemes for H.263-coded video were proposed, for example, by Gharavi and Alamouti [640], and were evaluated in the context of the third-generation mobile radio standard proposal known as IMT-2000 [87]. As portrayed in Figures 21.81 and 21.82, the corrupted variable-length coded DCT coefficients produce a high video PSNR degradation. Assuming that all MPEG-2 header

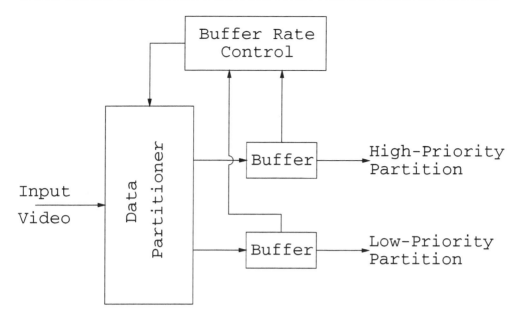

Figure 21.86: Block diagram of the data partitioner and rate controller.

information is received correctly, the fidelity of the reconstructed images at the receiver is dependent on the number of correctly decoded DCT coefficients. However, the subjective effects of losing higher spatial frequency DCT coefficients are less dramatic compared to those of the lower spatial frequency DCT coefficients. The splitting of the MPEG-2 video bit-stream into two different integrity bit-streams is termed data partitioning [633]. Recall from Section 21.6.3 that the hierarchical 16-QAM and 64-QAM DVB-T transmission scheme enables us to multiplex two unequally protected MPEG-2 video bit-streams for transmission. This section describes the details of the MPEG-2 data partitioning scheme [633].

Figure 21.86 shows the block diagram of the data partitioning scheme, which splits an MPEG-2 video bit-stream into two resultant bit-streams. The position at which the MPEG-2 bit-stream is split is based on a variable referred to as the priority breakpoint (PBP) [633]. The PBP can be adjusted at the beginning of the encoding of every MPEG-2 image slice, based on the buffer occupancy or fullness of the two output buffers. For example, if the high-priority buffer is 80% full and the low-priority buffer is only 40% full, the rate-control module will have to adjust the PBP so that more data is directed to the low-priority partition. This measure is taken to avoid high-priority buffer overflow and low-priority buffer underflow events. The values for the MPEG-2 PBP are summarized in Table 21.21 [633].

There are two main stages in updating the PBP. The first stage involves the rate-control module of Figure 21.86 in order to decide on the preferred new PBP value for each partition based on its individual buffer occupancy and on the current value of the PBP. The second stage then combines the two desired PBPs based on the buffer occupancy of both buffers in order to produce a new PBP.

PBP	Syntax Elements in High-Priority Partition
0	Low-priority partition always has its PBP set to 0.
1	Sequence, GOP, picture and slice layer information up to extra bit slice.
2	Same as above and up to macroblock address increment.
3	Same as above plus including macroblock syntax elements but excluding coded block pattern.
4 – 63	Reserved for future use.
64	Same as above plus including DC DCT coefficient and the first run-length coded DCT coefficient.
65	Same as above and up to the second run-length coded DCT coefficient.
$64 + x$	Same as above and up to x run-length coded DCT coefficient.
127	Same as above and up to 64 run-length coded DCT coefficient.

Table 21.21: Priority breakpoint values and the associated MPEG-2 parameters that will be directed to the high-priority partition [633]. A higher PBP directs more parameters to the high-priority partition. By contrast, for the low-priority partition a higher PBP implies obtaining less data

The updating of the PBP in the first stage of the rate control module is based on a heuristic approach, similar to that suggested by Aravind *et al.* [641]. The update procedure is detailed in Algorithm 1, which is discussed below and augmented by a numerical example at the end of this section.

The variable "sign" is used in Algorithm 1, in order to indicate how the PBP has to be adjusted in the high- and low-priority MPEG-2 partitions, so as to arrive at the required target buffer occupancy. More explicitly, the variable "sign" in Algorithm 1 is necessary because the MPEG-2 PBP values [633] shown in Table 21.21 indicate the amount of information that should be directed to the high-priority partition. Therefore, if the low-priority partition requires more data, then the new PBP must be lower than the current PBP. By contrast, for the high-priority partition a higher PBP implies obtaining more data.

Once the desired PBPs for both partitions have been acquired with the aid of Algorithm 1, Algorithm 2 is used to compute the final PBP for the current MPEG-2 image slice. The inner workings of these algorithms are augmented by a numerical example at the end of this section. There are two main cases to consider in Algorithm 2. The first one occurs when both partitions have a buffer occupancy of less than 50%. By using the reciprocal of the buffer occupancy in Algorithm 2 as a weighting factor during the computation of the PBP adjustment value "delta," the algorithm will favor the new PBP decision of the less occupied buffer in order to fill the buffer with more data in the current image slice. This is simply because the buffer is closer to underflow; hence, increasing the PBP according to its instructions will assist in preventing the particular buffer from underflowing. On the other hand, when both buffers experience a buffer occupancy of more than 50%, the buffer occupancy itself is used as a weighting factor instead. Now the algorithm will instruct the buffer having a higher occupancy to adjust its desired PBP so that less data is inserted into it in the current MPEG-2 image slice. Hence, buffer overflow problems are alleviated with the aid of Algorithm 1 and Algorithm 2.

Algorithm 1 Computes the desired PBP update for the high- and low-priority partitions which is then passed to Algorithm 2 in order to determine the PBP to be set for the current image slice.

Step 1: Initialize parameters
  if High Priority Partition **then**
    sign := +1
  else
    sign := −1
  end if

Step 2:
  if buffer occupancy $\geq 80\%$ **then**
    diff := 64 − PBP
  end if

  if buffer occupancy $\geq 70\%$ **and** buffer occupancy $< 80\%$ **then**
    if PBP ≥ 100 **then**
      diff := −9
    end if
    if PBP ≥ 80 **and** PBP < 100 **then**
      diff := −5
    end if
    if PBP ≥ 64 **and** PBP < 80 **then**
      diff := −2
    end if
  end if

  if buffer occupancy $\geq 50\%$ **and** buffer occupancy $< 70\%$ **then**
    diff := +1
  end if

  if buffer occupancy $< 50\%$ **then**
    if PBP ≥ 80 **then**
      diff := +1
    end if
    if PBP ≥ 70 **and** PBP < 80 **then**
      diff := +2
    end if
    if PBP ≥ 2 **and** PBP < 70 **then**
      diff := +3
    end if
  end if

Step 3:
  diff := sign × diff
  Return diff

Algorithm 2 Computes the new PBP for the current image slice based on the current buffer occupancy of both partitions

Step 1:

if   Occupancy$_{HighPriority}$ < 50%   and   Occupancy$_{LowPriority}$ < 50%
or   Occupancy$_{HighPriority}$ = 50%   and   Occupancy$_{LowPriority}$ < 50%
or   Occupancy$_{HighPriority}$ < 50%   and   Occupancy$_{LowPriority}$ = 50%
or   Occupancy$_{HighPriority}$ < 25%   and   50% < Occupancy$_{LowPriority}$ < 70%
or   50% < Occupancy$_{HighPriority}$ < 70%   and   Occupancy$_{LowPriority}$ < 25%

then

$$\text{delta} := \frac{\text{Occupancy}^{-1}_{HighPriority} \times \text{diff}_{HighPriority} + \text{Occupancy}^{-1}_{LowPriority} \times \text{diff}_{LowPriority}}{\text{Occupancy}^{-1}_{HighPriority} + \text{Occupancy}^{-1}_{LowPriority}}$$

else

$$\text{delta} := \frac{\text{Occupancy}_{HighPriority} \times \text{diff}_{HighPriority} + \text{Occupancy}_{LowPriority} \times \text{diff}_{LowPriority}}{\text{Occupancy}_{HighPriority} + \text{Occupancy}_{LowPriority}}$$

end if

Step 2:

New_PBP := Previous_PBP + ⌈delta⌉ where ⌈⌉ means rounding up to the nearest integer
Return New_PBP

The new PBP value is then compared to its legitimate range tabulated in Table 21.21. Furthermore, we restricted the minimum PBP value so that I-, P-, and B-pictures have minimum PBP values of 64, 3, and 2, respectively. Since B-pictures are not used for future predictions, it was decided that their data need not be protected as strongly as the data for I- and P-pictures. As for P-pictures, Ghanbari and Seferidis [561] showed that correctly decoded motion vectors alone can still provide a subjectively pleasing reconstruction of the image, even if the DCT coefficients were discarded. Hence, the minimum MPEG-2 bit-stream splitting point or PBP for P-pictures has been set to be just before the coded block pattern parameter, which would then ensure that the motion vectors would be mapped to the high-priority partition. Upon receiving corrupted DCT coefficients, they would be set to zero, which corresponds to setting the motion-compensated error residual of the macroblock concerned to zero. For I-pictures, the fidelity of the reconstructed image is dependent on the number of DCT coefficients that can be decoded successfully. Therefore, the minimum MPEG-2 bit-stream splitting point or PBP was set to include at least the first run-length-coded DCT coefficient.

Below we demonstrate the operation of Algorithm 1 and Algorithm 2 with the aid of a simple numerical example. We will assume that the PBP prior to the update is 75 and that the buffer occupancy for the high- and low-priority partition buffers is 40% and 10%, respectively. Considering the high-priority partition, according to the buffer occupancy of 40% Algorithm 1 will set the desired PBP update difference denoted by "diff" for the PBP to +2. This desired update is referred to as diff$_{HighPriority}$

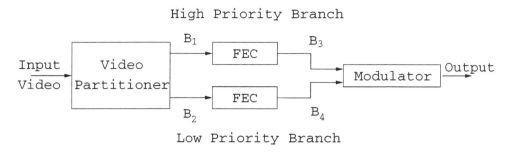

Figure 21.87: Video partitioning for the DVB-T system operating in hierarchical mode

in Algorithm 2. For the low-priority partition, according to the buffer occupancy of 10%, Algorithm 1 will set the desired update for the PBP to −2, since the sign of diff is changed by Algorithm 1. The desired PBP update for the low-priority partition is referred to as diff$_{LowPriority}$ in Algorithm 2. Since the occupancy of both partition buffers' is less than 50%, Algorithm 2 will use the reciprocal of the buffer occupancy as the weighting factor, which will then favor the desired update of the low-priority partition due to its 10% occupancy. The final update value, which is denoted by delta in Algorithm 2, is equal to −2 (after being rounded up). Hence, according to Step 2 of Algorithm 2, the new PBP is 73. This means that for the current MPEG-2 image slice more data will be directed into the low-priority partition in order to prevent buffer underflow since PBP was reduced from 75 to 73 according to Table 21.21.

Apart from adjusting the PBP values from one MPEG-2 image slice to another to avoid buffer underflow or overflow, the output bit rate of each partition buffer must be adjusted so that the input bit rate of the inner interleaver and modulator in Figure 21.83 is properly matched between the two partitions. Specifically, in the 16QAM mode the two modem subchannels have an identical throughput of 2 bits per 4-bit symbol. By contrast, in the 64QAM mode there are three 2-bit subchannels per 6-bit 64QAM symbol, although the standard [624] recommends using a higher-priority 2-bit and a lower-priority 4-bit subchannels. Hence, it is imperative to take into account the redundancy added by forward error correction (FEC), especially when the two partitions' FECs operate at different code rates. Figure 21.87 shows a block diagram of the DVB-T system operating in the hierarchical mode and receiving its input from the video partitioner. The FEC module represents the concatenated coding system, consisting of a Reed–Solomon codec [87] and a convolutional codec [87]. The modulator can invoke both 16QAM and 64QAM [29]. We shall now use an example to illustrate the choice of the various partitioning ratios summarized in Table 21.22.

We shall assume that 64QAM is selected and the high-and low-priority video partitions employ $\frac{1}{2}$ and $\frac{3}{4}$ convolutional codes, respectively. This scenario is portrayed in the third line of the 64QAM section of Table 21.22. We do not have to take the Reed–Solomon code rate into account, since both partitions invoke the same Reed–Solomon codec. Based on these facts and on reference to Figure 21.87, the input bit rates B_3 and B_4 of the modulator must be in the ratio 1:2, since the two MSBs of the 64-QAM constellation are assigned to the high-priority video partition and the remaining four bits to the low-priority video partition.

Modulation	Conv. Code Rate (High Priority)	Conv Code Rate (Low Priority)	Ratio (High - B1 : Low - B2)
16QAM	1/2	1/2	1 : 1
	1/2	2/3	3 : 4
	1/2	3/4	2 : 3
	1/2	5/6	3 : 5
	1/2	7/8	4 : 7
	2/3	1/2	4 : 3
64QAM	1/2	1/2	1 : 2
	1/2	2/3	3 : 8
	1/2	**3/4**	**1 : 3**
	1/2	5/6	3 : 10
	1/2	7/8	2 : 7
	2/3	1/2	2 : 3

Table 21.22: The Bit Rate Partitioning Ratios Based on the Modulation Mode and Code Rates Selected for the DVB-T Hierarchical Mode. The Line in Bold Corresponds to our Worked Example

At the same time, the ratio of B_3 to B_4 is related to the ratio of B_1 to B_2 with the FEC redundancy taken into account, requiring

$$\begin{aligned}
\frac{B_3}{B_4} &= \frac{2 \times B_1}{\frac{4}{3} \times B_2} \stackrel{64-QAM}{=} \frac{1}{2} \\
&= \frac{3}{2} \cdot \frac{B_1}{B_2} \stackrel{64-QAM}{=} \frac{1}{2} \\
\frac{B_1}{B_2} &= \frac{1}{2} \times \frac{2}{3} \\
&= \frac{1}{3}.
\end{aligned} \quad (21.5)$$

If, for example, the input video bit rate of the data partitioner module is 1 Mbit/s, the output bit rate of the high- and low-priority partition would be $B_1 = 250$ kbit/s and $B_2 = 750$ kbit/s, respectively, according to the ratio indicated by Equation 21.5.

In this section, we have outlined the operation of the data partitioning scheme which we used in the DVB-T hierarchical transmission scheme. Its performance in the context of the overall system will be characterized in Section 21.6.8. Let us, however, first evaluate the BER sensitivity of the partitioned MPEG-2 bit-stream to randomly distributed bit errors using various partitioning ratios.

21.6.6 Performance of the Data Partitioning Scheme

Let us consider the 16QAM modem and refer to the equally split rate $\frac{1}{2}$ convolutional coded high- and low-priority scenario as Scheme 1. Furthermore, the 16QAM rate $\frac{1}{3}$ convolutional coded high priority data and rate $\frac{2}{3}$ convolutional coded low-priority data-based scenario is referred to here as Scheme 2. Lastly, the 16QAM rate $\frac{2}{3}$ convolutional coded high-priority data and rate $\frac{1}{3}$ coded low-priority databased partitioning scheme is termed Scheme 3. We then programmed the partitioning scheme of Figure 21.87 for maintaining the required splitting ratio B_1/B_2, as seen in Table 21.23.

21.6. DVB-T FOR MOBILE RECEIVERS

	Modulation	Conv. Code Rate (High Prior. – B1)	Conv. Code Rate (Low Prior. – B2)	Ratio (High : Low) (B1 : B2)
Scheme 1	16QAM	1/2	1/2	1 : 1
Scheme 2	16QAM	1/3	2/3	1 : 2
Scheme 3	16QAM	2/3	1/3	2 : 1

Table 21.23: Summary of the Three Schemes Employed in our Investigations into the Performance of the Data Partitioning Scheme. The FEC-coded High-priority Video Bit-Stream B3, as Shown in Figure 21.87, was Mapped to the High-priority 16QAM Subchannel, while the Low-priority B4-stream to the Low Priority 16QAM Subchannel.

This was achieved by continuously adjusting the PBP using Algorithms 1 and 2. The 704×576-pixel "Football" high-definition television (HDTV) video sequence was used in these investigations.

Figures 21.88 to 21.90 show the relative frequency at which a particular PBP value occurs for each image of the "Football" video sequence for the three different schemes of Table 21.23 mentioned earlier. The reader may recall from Table 21.21 that the PBP values indicate the proportion of encoded video parameters, which are to be directed into the high-priority partition. As the PBP value increases, the proportion of video data mapped to the high-priority partition increases and vice versa. Comparing Figures 21.88 to 21.90, we observe that Scheme 3 has the most data in the high-priority partition associated with the high PBPs of Table 21.21, followed by Scheme 1 and Scheme 2. This observation can be explained as follows. We shall consider Scheme 3 first. In this scheme, the high-priority video bits are protected by a rate $\frac{2}{3}$ convolutional code and mapped to the higher integrity 16QAM subchannel. By contrast, the low-priority video bits are encoded by a rate $\frac{1}{3}$ convolutional code and mapped to the lower integrity 16QAM subchannel. Again, assuming that 16QAM is used in our experiment according to line 3 of Table 21.23, $\frac{2}{3}$ of the video bits will be placed in the high-priority 16QAM partition and the remaining video bits in the low-priority 16QAM partition, following the approach of Equation 21.5. The BER difference of the 16QAM subchannels depend on the channel error statistics, but the associated BERs are about a factor of 2–3 different [29]. In contrast to Scheme 3, Scheme 2 will have $\frac{1}{3}$ of the video bits placed in the high-priority 16QAM partition, and the remaining $\frac{2}{3}$ of the video bits mapped to the low-priority 16QAM partition, according to line 2 of Table 21.23. Lastly, Scheme 1 will have half of the video bits in the high- and low-priority 16QAM partitions, according to line 1 of Table 21.23. This explains our observation in the context of Scheme 3 in Figure 21.90, where a PBP value as high as 80 is achieved in some image frames. However, each PBP value encountered has a lower probability of being selected, since the total number of 3600 occurrences associated with investigated 3600 MPEG-2 video slices per 100 image frames is spread over a higher variety of PBPs. Hence, Scheme 3 directs about $\frac{2}{3}$ of the original video bits after $\frac{2}{3}$ rate coding to the high-priority 16QAM subchannel. This observation is in contrast to Scheme 2 of Figure 21.89, where the majority of the PBPs selected are only up to the value of 65. This indicates that about $\frac{2}{3}$ of the

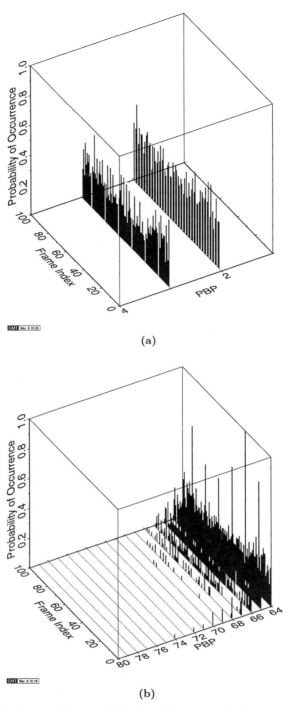

Figure 21.88: Evolution of the probability of occurrence of PBP values from one picture to another of the 704 × 576-pixel "Football" video sequence for Scheme 1 of Table 21.23.

21.6. DVB-T FOR MOBILE RECEIVERS

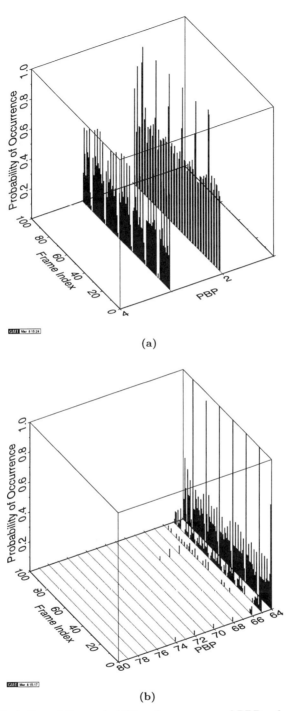

Figure 21.89: Evolution of the probability of occurrence of PBP values from one picture to another of the 704 × 576-pixel "Football" video sequence for Scheme 2 of Table 21.23.

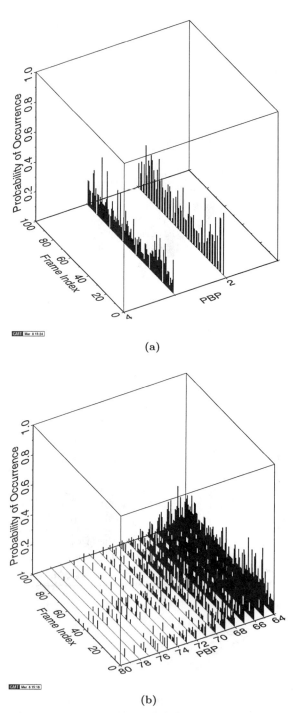

Figure 21.90: Evolution of the probability of occurrence of PBP values from one picture to another of the 704 × 576-pixel "Football" video sequence for Scheme 3 of Table 21.23.

21.6. DVB-T FOR MOBILE RECEIVERS

video bits are concentrated in the lower priority partition, as indicated in line 2 of Table 21.23.

Figures 21.91(a) to 21.93(a) show the average probability at which a particular PBP value is selected by the rate control scheme, as discussed in Section 21.6.5, during the encoding of the video sequence. Again, we observe that Scheme 3 encounters the widest range of PBP values, followed by Scheme 1 and Scheme 2, respectively. According to Table 21.23, these schemes map a decreasing number of bits to the high-priority partition in this order.

We then quantified the error sensitivity of the partitioning Schemes 1 to 3 characterized in Table 21.23, when each partition was subjected to randomly distributed bit errors, although in practice the error distribution will depend on the fading channel's characteristics. Specifically, the previously defined average PSNR degradation was evaluated for given error probabilities inflicting random errors imposed on one of the partitions, while keeping the other partition error-free. These results are portrayed in Figures 21.91(b), 21.92(b) and 21.93(b) for Schemes 1 to 3, respectively.

Comparing Figures 21.91(b) to 21.93(b), we observe that the average PSNR degradation exhibited by the three schemes of Table 21.23, when only their high-priority partitions are corrupted, is similar. The variations in the average PSNR degradation in these cases are caused by the different quantity of sensitive video bits, which resides in the high priority partition. If we compare the performance of the schemes summarized in Table 21.23 at a BER of 2×10^{-3}, Scheme 3 experienced approximately 8.8 dB average video PSNR degradation, while Schemes 1 and 2 exhibited approximately 5 dB degradation. This trend was expected, since Scheme 3 had the highest portion of the video bits, namely, $\frac{2}{3}$ residing in the high-priority partition, followed by Scheme 1 hosting $\frac{1}{2}$ and Scheme 2 having $\frac{1}{3}$ of the bits in this partition.

On the other hand, we can observe a significant difference in the average PSNR degradation measured for Schemes 1 to 3 of Table 21.23, when only the low priority partitions are corrupted by comparing the curves shown as broken lines in Figures 21.91(b) to 21.93(b). Under this condition, Scheme 2 experienced approximately 16 dB average video PSNR degradation at a BER of 2×10^{-3}. By contrast, Scheme 1 exhibited an approximately 4 dB average video PSNR degradation, while Scheme 3 experienced about 7.5 dB degradation at this BER. The scheme with the highest portion of video bits in the lower priority partition (i.e., Scheme 2) experienced the highest average video PSNR degradation. This observation correlates well with our earlier findings in the context of the high-priority partition scenario, where the partition holding the highest portion of the video bits in the error-impaired partition exhibited the highest average PSNR degradation.

Having discussed our observations for the three schemes of Table 21.23 from the perspective of the relative amount of video bits in one partition compared to the other, we now examine the data partitioning process further in order to relate them to our observations. Figure 21.94 shows a typical example of an MPEG-2 video bit-stream both prior to and after data partitioning. There are two scenarios to be considered here, namely, intra-frame and inter-frame coded macroblock partitioning. We have selected the PBP value of 64 from Table 21.21 for the intra-frame coded macroblock scenario and the PBP value of 3 for the inter-frame coded macroblock scenario, since these values have been selected frequently by the rate-control arrangement for

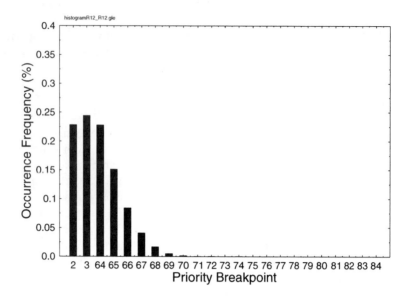

(a)

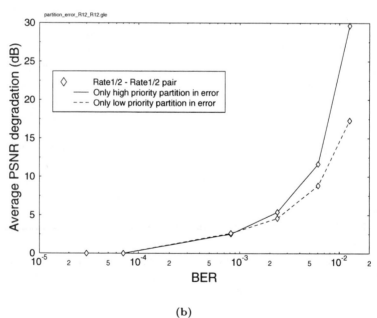

(b)

Figure 21.91: (a) Histogram of the probability of occurrence for various priority breakpoints and (b) average PSNR degradation versus BER for rate 1/2 convolutional coded high- and low-priority data in Scheme 1 of Table 21.23.

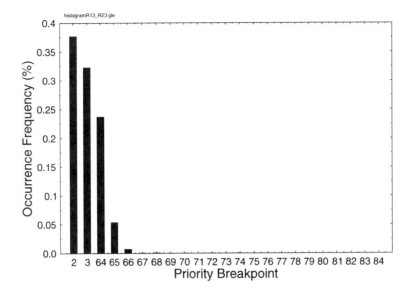

(a)

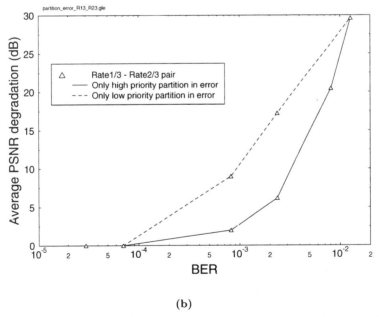

(b)

Figure 21.92: (a) Histogram of the probability of occurrence for various priority breakpoints and (b) average PSNR degradation versus BER for the rate 1/3 convolutional coded high-priority data and rate 2/3 convolutional coded low-priority data in Scheme 2 of Table 21.23.

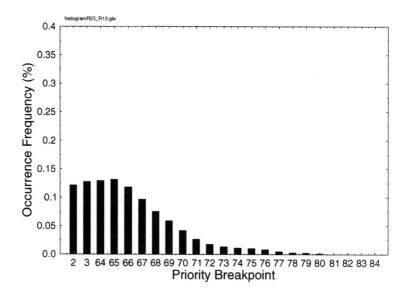

(a)

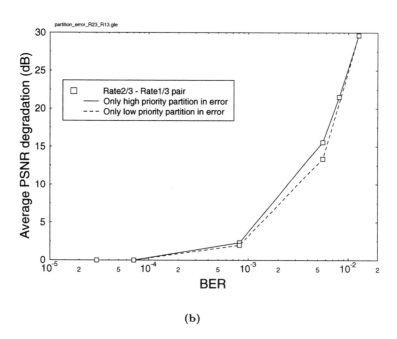

(b)

Figure 21.93: (a) Histogram of the probability of occurrence for various priority breakpoints and (b) average PSNR degradation versus BER for the rate 2/3 convolutional coded high-priority data and rate 1/3 convolutional coded low-priority data in Scheme 3 of Table 21.23.

21.6. DVB-T FOR MOBILE RECEIVERS

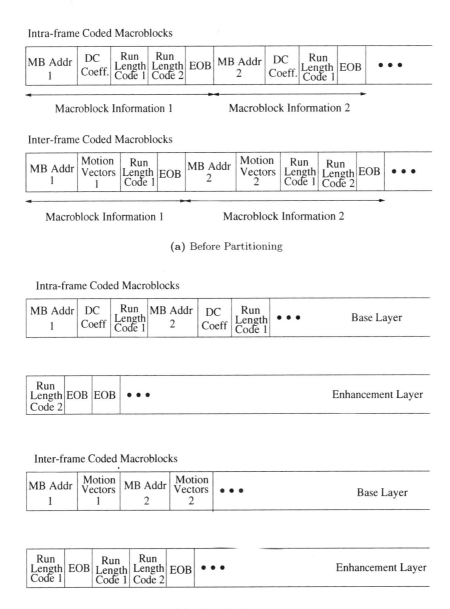

Figure 21.94: Example of video bit stream (a) before data partitioning and (b) after data partitioning for intra-frame coded macroblocks (MB) assuming a PBP of 64 and for inter-frame coded macroblocks assuming a PBP of 3.

Schemes 1 and 2. This is evident from Figures 21.88 and 21.89 as well as from Figures 21.91(a) and 21.92(a). With the aid of Table 21.21 and Figure 21.94, this implies that only the macroblock (MB) header information and a few low-frequency DCT coefficients will reside in the high-priority partition, while the rest of the DCT coefficients will be stored in the low-priority partition. These can be termed as base layer and enhancement layer, as seen in Figure 21.94. In the worst-case scenario, where the entire enhancement layer or low-priority partition data are lost due to a transmission error near the beginning of the associated low-priority bit stream, the MPEG-2 video decoder will only have the bits of the high-priority partition in order to reconstruct the encoded video sequence. Hence, the MPEG-2 decoder cannot reconstruct good-quality images. Although the results reported by Ghanbari and Seferidis [561] suggest that adequate video reconstruction is possible, provided that the motion vectors are correctly decoded, this observation is only true if the previous intra-coded frame is correctly reconstructed. If the previous intra-coded frame contains artifacts, these artifacts will be further propagated to forthcoming video frames by the motion vectors. By attempting to provide higher protection for the high-priority partition or base layer, we have indirectly forced the rate-control scheme of Section 21.6.5 to reduce the proportion of video bits directed into the high-priority partition under the constraint of a given fixed bit rate, which is imposed by the 16QAM subchannels.

In order to elaborate a little further, at a BER of 2×10^{-3}, Scheme 1 in Figure 21.91(a) exhibited a near-identical PSNR degradation for the high- and low-priority video bits. When assigning more bits to the low-priority partition, in order to accommodate a stronger FEC code in the high-priority partition, an increased proportion of error-impaired bits is inflicted in the low-priority partition. This is the reason for the associated higher error sensitivity seen in Figure 21.92(b). As such, there is a trade-off between the amount of video data protected and the code rate of the channel codec. As a comparison to the above scenarios in the context of Schemes 1 and 2, we shall now examine Scheme 3. In this scheme, more video data — namely, half the bits — can be directed into the high-priority partition, as demonstrated by Figure 21.90 due to encountering higher PBPs. This can also be confirmed with reference to Figures 21.92(b) and 21.93(b) by observing the PSNR degradations associated with the curves plotted in broken lines. If the low-priority partition is lost in Scheme 3, its effect on the quality of the reconstructed images is less detrimental than that of Scheme 2, since Scheme 3 loses only half the bits rather than 2/3. Hence, it is interesting to note that Scheme 3 experiences slightly higher average PSNR degradation than Scheme 1 at a BER of 2×10^{-3}, when only the low-priority partition is lost in both cases, despite directing only $\frac{1}{3}$ rather than $\frac{1}{2}$ of the bits to the low-priority partition. This observation can be explained as follows.

Apart from partitioning the macroblock header information and the variable-length coded DCT coefficients into the high- and low-priority partitions, synchronization information such as the picture header information [633] is replicated in the enhancement layer, as suggested by Gharavi et al. [640] as well as the MPEG-2 standard [633]. The purpose is to enable the MPEG-2 decoder to keep the base and enhancement layers synchronized during decoding. An example of this arrangement is shown in Figure 21.95. This resynchronization measure is only effective when the picture start code of both the high- and low-priority partitions is received correctly.

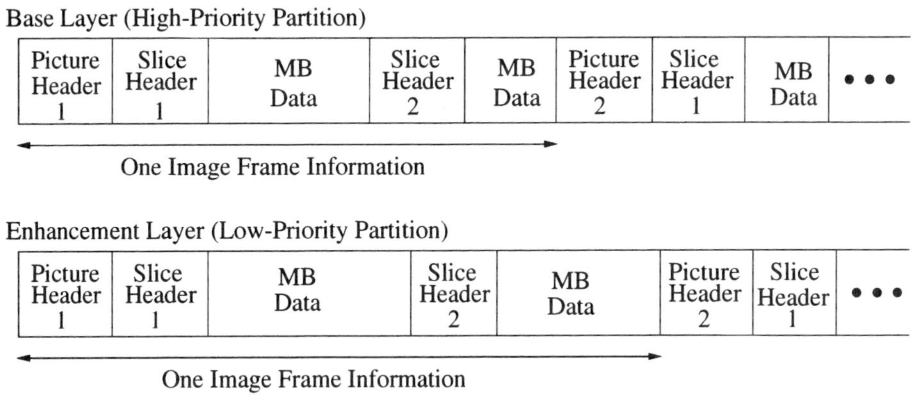

Base Layer (High-Priority Partition)

| Picture Header 1 | Slice Header 1 | MB Data | Slice Header 2 | MB Data | Picture Header 2 | Slice Header 1 | MB Data | ••• |

←——— One Image Frame Information ———→

Enhancement Layer (Low-Priority Partition)

| Picture Header 1 | Slice Header 1 | MB Data | Slice Header 2 | MB Data | Picture Header 2 | Slice Header 1 | ••• |

←——— One Image Frame Information ———→

Figure 21.95: Example of high-level bit-stream syntax structure of a data partitioned MPEG-2 video bit stream. The "MB data" shown in the diagram refers to the macroblock header information and to the variable-length coded DCT coefficients, which have been partitioned as shown in Figure 21.94.

If the picture start code in the low-priority partition is corrupted, for example, the MPEG-2 decoder may not detect this PSC, and all the data corresponding to the current image frame in the low-priority partition will be lost. The MPEG-2 decoder will then interpret the bits received for the low-priority partition of the next frame as the low-priority data expected for the current frame. As expected, because of this synchronization problem, the decoded video would have a higher average PSNR degradation than for the case where picture start codes are unimpaired. This explains our observation of a higher average PSNR degradation for Scheme 3 when only its lower priority partition was corrupted by the transmission channel. On the other hand, in this specific experiment, Scheme 1 did not experience the loss of synchronization due to corruption of its picture start code. Viewing events from another perspective, by opting for allocating less useful video bits to the low-priority partition, the probability of transmission errors affecting the fixed-length PSC within the reduced-sized low priority partition becomes higher.

These findings will assist us in explaining our observations in the context of the hierarchical transmission scheme of Section 21.6.8, suggesting that the data partitioning scheme did not provide overall gain in terms of error resilience over the nonpartitioned case. Let us, however, consider first the performance of the nonhierarchical DVB-T scheme in the next section.

21.6.7 Nonhierarchical OFDM DVBP Performance

In this section, we elaborate on our findings when the convolutional code used in the standard nonhierarchical DVB scheme [624] is replaced by a turbo code. We will invoke a range of standard-compliant schemes as benchmarks. The 704 × 576-pixel HDTV-resolution "Football" video sequence was used in our experiments. The MPEG-2 decoder employs a simple error concealment algorithm to fill in missing portions of the reconstructed image in the event of decoding errors. The concealment algorithm will select the specific portion of the previous reconstructed image, which corresponds to the missing portion of the current image in order to conceal the errors.

In Figure 21.96(a) and (b), the bit error rate (BER) performance of the various modem modes in conjunction with our diverse channel-coding schemes are portrayed over stationary, narrowband additive white Gaussian noise channels (AWGN), where the turbo codec exhibits a significantly steeper BER reduction in comparison to the convolutionally coded arrangements.

Specifically, comparing the performance of the various turbo and convolutional codes for QPSK and 64QAM at a BER of 10^{-4}, we see that the turbo code exhibited an additional coding gain of about 2.24 dB and 3.7 dB, respectively, when using half-rate codes in Figure 21.96(a) and (b). Hence, the peak signal-to-noise ratio (PSNR) versus channel signal-to-noise ratio (SNR) graphs in Figure 21.97 demonstrate that approximately 2 dB and 3.5 dB lower channel SNRs are required in conjunction with the rate $\frac{1}{2}$ turbo codec for QPSK and 64QAM, respectively, than for convolutional coding, in order to maintain high reconstructed video quality. The term *unimpaired* as used in Figure 21.97 and Figure 21.98 refers to the condition where the PSNR of the MPEG-2 decoder's reconstructed image at the receiver is the same as the PSNR of the same image generated by the local decoder of the MPEG-2 video encoder, corresponding to the absence of channel — but not MPEG-2 coding — impairments.

Comparing the BER performance of the $\frac{1}{2}$ rate convolutional decoder in Figure 21.99(a) and the log-MAP [636] turbo decoder using eight iterations in Figure 21.99(b) for QPSK modulation over the worst-case fading mobile channel of Figure 21.85, we observe that at a BER of about 10^{-4} the turbo code provided an additional coding gain of 6 dB in comparison to the convolutional code. By contrast, for 64QAM using similar codes, a 5 dB coding gain was observed at this BER.

Similar observations were also made with respect to the average peak signal-to-noise ratio (PSNR) versus channel signal-to-noise ratio (SNR) plots of Figure 21.98. For the QPSK modulation mode and a $\frac{1}{2}$ coding rate, the turbo code required an approximately 5.5 dB lower channel SNR for maintaining near-unimpaired video quality than the convolutional code.

Comparing Figure 21.99(a) and Figure 21.100(a), we note that the Reed–Solomon decoder becomes effective in lowering the bit error probability of the transmitted data further below the BER threshold of 10^{-4}. From these figures we also observe that the rate $\frac{3}{4}$ convolutional code is unsuitable for transmission over the highly dispersive hilly terrain channel used in this experiment, when 64QAM is employed. When the rate $\frac{7}{8}$ convolutional code is used, both the 16QAM and 64QAM schemes perform poorly. As for the QPSK modulation scheme, a convolutional code rate as high as $\frac{7}{8}$ can still provide a satisfactory performance after Reed–Solomon decoding.

In conclusion, Tables 21.24 and 21.25 summarize the system performance in terms of the channel SNR (CSNR) required for maintaining less than 2 dB PSNR video degradation. At this PSNR degradation, decoding errors were still perceptually unnoticeable to the viewer due to the 30 frame/s refresh rate, although the typical still frame shown in Figure 21.101 in this scenario exhibits some degradation. It is important to underline once again that the $K = 3$ turbo code and the $K = 7$ convolutional code exhibited comparable complexities. The higher performance of the turbo codec facilitates, for example, the employment of turbo-coded 16-QAM at a similar channel

21.6. DVB-T FOR MOBILE RECEIVERS

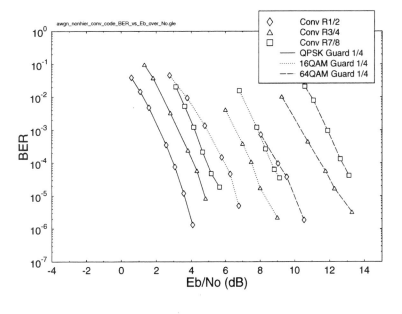

(a) Convolutional Code

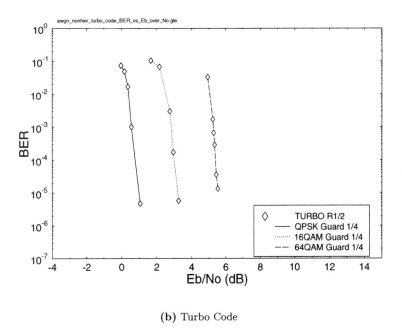

(b) Turbo Code

Figure 21.96: BER after (a) convolutional decoding and (b) turbo decoding for the DVB-T scheme over stationary nondispersive **AWGN** channels for **nonhierarchical transmission**.

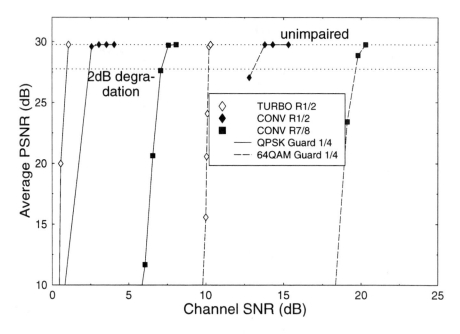

Figure 21.97: Average PSNR versus channel SNR of the DVB scheme [624] over nondispersive **AWGN** channels for **nonhierarchical transmission**.

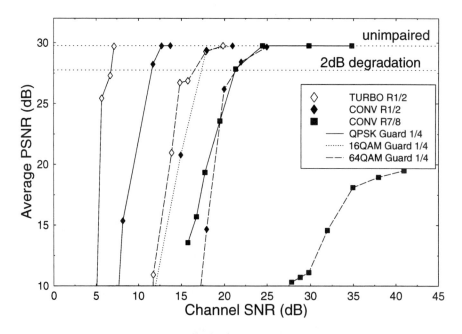

Figure 21.98: Average PSNR versus channel SNR of the DVB scheme [624] over the **wideband fading channel** of Figure 21.85 for **nonhierarchical transmission**.

21.6. DVB-T FOR MOBILE RECEIVERS

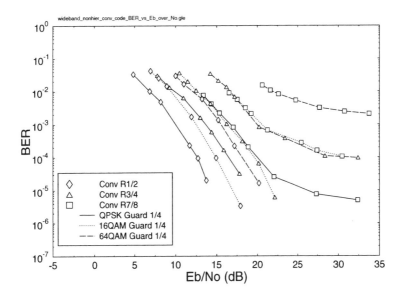

(a) Convolutional Code

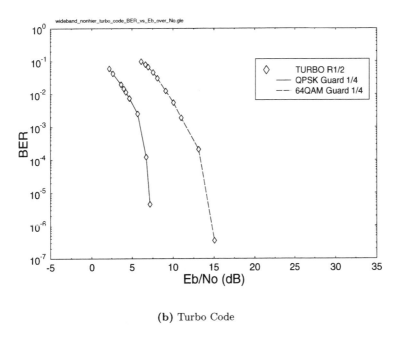

(b) Turbo Code

Figure 21.99: BER after (a) convolutional decoding and (b) turbo decoding for the DVB-T scheme over the **wideband fading channel** of Figure 21.85 for **nonhierarchical transmission**.

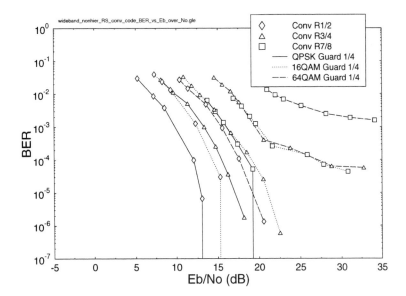

(a) RS and Convolutional Code

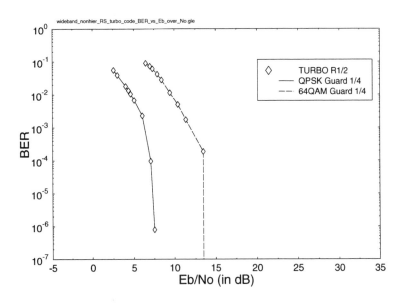

(b) RS and Turbo Code

Figure 21.100: BER after (a) RS and convolutional decoding and (b) RS and turbo decoding for the DVB-T scheme over the **wideband fading channel** of Figure 21.85 for **nonhierarchical transmission**.

21.6. DVB-T FOR MOBILE RECEIVERS

Mod.	Code	CSNR (dB)	E_b/N_0	BER
QPSK	Turbo (1/2)	1.02	1.02	6×10^{-6}
64QAM	Turbo (1/2)	9.94	5.17	2×10^{-3}
QPSK	Conv (1/2)	2.16	2.16	1.1×10^{-3}
64QAM	Conv (1/2)	12.84	8.07	6×10^{-4}
QPSK	Conv (7/8)	6.99	4.56	2×10^{-4}
64QAM	Conv (7/8)	19.43	12.23	3×10^{-4}

Table 21.24: Summary of the *Nonhierarchical* Performance Results over Nondispersive *AWGN* Channels Tolerating a PSNR Degradation of 2 dB. The BER Measure Refers to BER after Viterbi or Turbo Decoding

Mod.	Code	CSNR (dB)	E_b/N_0	BER
QPSK	Turbo (1/2)	6.63	6.63	2.5×10^{-4}
64QAM	Turbo (1/2)	15.82	11.05	2×10^{-3}
QPSK	Conv (1/2)	10.82	10.82	6×10^{-4}
64QAM	Conv (1/2)	20.92	16.15	7×10^{-4}
QPSK	Conv (7/8)	20.92	18.49	3×10^{-4}

Table 21.25: Summary of the *Nonhierarchical* Performance Results over *Wideband Fading Channels* Tolerating a PSNR Degradation of 2 dB. The BER Measure Refers to BER after Viterbi or Turbo Decoding

Figure 21.101: Frame 79 of the "Football" sequence, which illustrates the visual effects of minor decoding errors at a BER of 2.10^{-4} after convolutional decoding. The PSNR degradation observed is approximately 2 dB. The sequence was coded using a rate 7/8 convolutional code and transmitted employing QPSK modulation.

SNR, where convolutional-coded QPSK can be invoked. This in turn allows us to double the bit rate within the same bandwidth and thereby to improve the video quality. In the next section, we present the results of our investigations employing the DVB-T system [624] in a hierarchical transmission scenario.

21.6.8 Hierarchical OFDM DVB Performance

The philosophy of the hierarchical transmission mode is that the natural BER difference of a factor 2 to 3 of the 16QAM modem is exploited for providing unequal error protection for the FEC-coded video streams B3 and B4 of Figure 21.87 [29]. If the sensitivity of the video bits requires a different BER ratio between the B3 and B4 streams, the choice of the FEC codes protecting the video streams B1 and B2 of Figure 21.87 can be appropriately adjusted to equal out or to augment these differences.

Below we invoke the DVB-T hierarchical scheme in a mobile broadcasting scenario. We also demonstrate the improvements that turbo codes offer when replacing the convolutional code in the standard scheme. Hence, the convolutional codec in both the high- and low-priority partitions was replaced by the turbo codec. We have also investigated replacing only the high-priority convolutional codec with the turbo codec, pairing the $\frac{1}{2}$ rate turbo codec in the high-priority partition with the convolutional codec in the low-priority partition. Again, the "Football" sequence was used in these experiments. Partitioning was carried out using the schematic of Figure 21.87 as well as Algorithms 1 and 2. The FEC-coded high-priority video partition B3 of Figure 21.87 was mapped to the higher integrity 16QAM or 64QAM subchannel. By contrast, the low-priority partition B4 of Figure 21.87 was directed to the lower integrity 16QAM or 64QAM subchannel. Lastly, no specific mapping was required for QPSK, since it exhibits no subchannels. We note, however, that further design trade-offs become feasible when reversing the above mapping rules. This is necessary, for example, in conjunction with Scheme 2 of Table 21.23, since the high number of bits in the low-priority portion render it more sensitive than the high-priority partition. Again, the 16QAM subchannels exhibit a factor of 2 to 3 BER difference under various channel conditions, which improves the robustness of the reverse-mapped Scheme 2 of Table 21.23.

Referring to Figure 21.102 and comparing the performance of the 1/2 rate convolutional code and turbo code at a BER of 10^{-4} for the low-priority partition, we find that the turbo code, employing eight iterations, exhibited a coding gain of about 6.6 dB and 5.97 dB for 16QAM and 64QAM, respectively. When the number of turbo-decoding iterations was reduced to 4, the coding gains offered by the turbo code over that of the convolutional code were 6.23 dB and 5.7 dB for 16QAM and 64QAM, respectively. We observed that by reducing the number of iterations to four halved the associated complexity, but the turbo code exhibited a coding loss of only about 0.37 dB and 0.27 dB in comparison to the eight-iteration scenario for 16QAM and 64QAM, respectively. Hence, the computational complexity of the turbo codec can be halved by sacrificing only a small amount of coding gain. The substantial coding gain provided by turbo coding is also reflected in the PSNR versus channel SNR graphs of Figure 21.103. In order to achieve transmission with very low probability of error,

21.6. DVB-T FOR MOBILE RECEIVERS

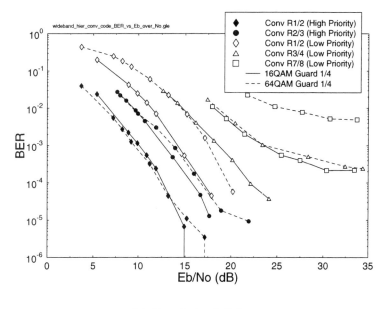

(a) Convolutional Code

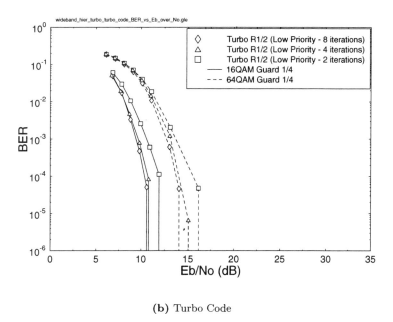

(b) Turbo Code

Figure 21.102: BER after (a) convolutional decoding and (b) turbo decoding for the **DVB-T hierarchical scheme** over the **wideband fading channel** of Figure 21.85 using the schematic of Figure 21.87 as well as Algorithms 1 and 2. In (b), the BER of the turbo- or convolutional-coded high-priority partition is not shown.

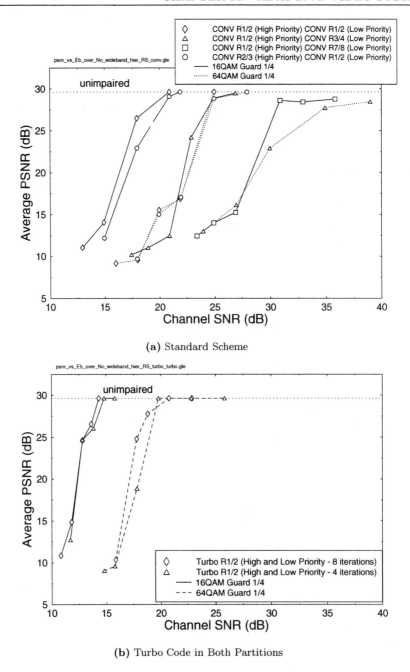

Figure 21.103: Average PSNR versus channel SNR for (a) standard DVB scheme [624] and (b) system with turbo coding employed in both partitions, for transmission over the **wideband fading channel** of Figure 21.85 for **hierarchical transmission** using the schematic of Figure 21.87 as well as Algorithms 1 and 2.

21.6. DVB-T FOR MOBILE RECEIVERS

Figure 21.103 demonstrated that approximately 5.72 dB and 4.56 dB higher channel SNRs are required by the standard scheme compared to the scheme employing turbo coding, when using four iterations in both partitions. We have only shown the performance of turbo coding for the low-priority partition in Figures 21.102(b) and 21.104(b), since the turbo or convolutional-coded high-priority partition was received with very low probability of error after Reed–Solomon decoding for the range of SNRs used.

We also observed that the rate 3/4 and rate 7/8 convolutional codes in the low-priority partition were unable to provide sufficient protection to the transmitted video bits, as becomes evident from Figures 21.102(a) and 21.104(a). In these high coding rate scenarios, due to the presence of residual errors even after the Reed–Solomon decoder, the decoded video exhibited some decoding errors, which is evidenced by the flattening of the PSNR versus channel SNR curves in Figure 21.103(a), before reaching the error-free PSNR.

A specific problem when using the data partitioning scheme in conjunction with the high-priority partition being protected by the rate $\frac{1}{2}$ code and the low-priority partition protected by the rate $\frac{3}{4}$ and rate $\frac{7}{8}$ codes was that when the low-priority partition data was corrupted, the error-free high priority data available was insufficient for concealing the errors, as discussed in Section 21.6.6. We have also experimented with the combination of rate $\frac{2}{3}$ convolutional coding and rate $\frac{1}{2}$ convolutional coding, in order to protect the high- and low-priority data, respectively. From Figure 21.103(a) we observed that the performance of this $\frac{2}{3}$ rate and $\frac{1}{2}$ rate combination approached that of the rate $\frac{1}{2}$ convolutional code in both partitions. This was expected, since now more data can be inserted into the high-priority partition. Hence, in the event of decoding errors in the low-priority data, we had more error-free high-priority data that could be used to reconstruct the received image.

Our last combination investigated involved using rate $\frac{1}{2}$ turbo coding and convolutional coding for the high- and low-priority partitions, respectively. Comparing Figures 21.105 and 21.103(a), the channel SNRs required for achieving unimpaired video transmission were similar in both cases. This was expected, since the turbo-convolutional combination's video performance is dependent on the convolutional code's performance in the low-priority partition.

Lastly, comparing Figures 21.103 and 21.98, we found that the unimpaired PSNR condition was achieved at similar channel SNRs for the hierarchical and nonhierarchical schemes, suggesting that the data partitioning scheme had not provided sufficient performance improvements in the context of the mobile DVB scheme to justify its added complexity. Again, this was a consequence of relegating a high proportion of video bits to the low integrity partition.

21.6.9 Summary and Conclusions

In this chapter, we have investigated the performance of a turbo-coded DVB system in a mobile environment. A range of system performance results was presented based on the standard DVB-T scheme as well as on an improved turbo-coded scheme. The convolutional code specified in the standard system was replaced by turbo coding, which resulted in a substantial coding gain of around 5 dB. It is important to un-

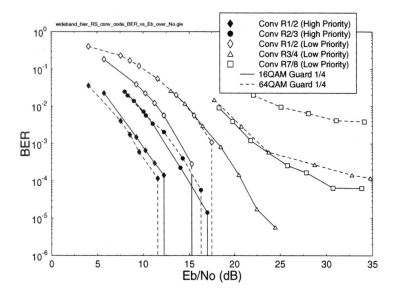

(a) RS and Convolutional Code

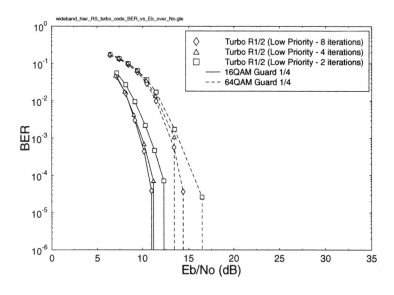

(b) RS and Turbo Code

Figure 21.104: BER after (a) RS and convolutional decoding and (b) RS and turbo decoding for the **DVB-T hierarchical scheme** over the **wideband fading channel** of Figure 21.85 using the schematic of Figure 21.87 as well as Algorithms 1 and 2. In (b), the BER of the turbo- or convolutional-coded high-priority partition is not shown.

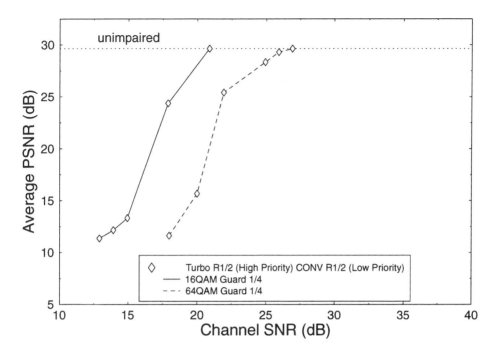

Figure 21.105: Average PSNR versus channel SNR of the DVB scheme, employing turbo coding in the high-priority partition and convolutional coding in the low-priority partition, over the **wideband fading channel** of Figure 21.85 for **hierarchical transmission** using the schematic of Figure 21.87 as well as Algorithms 1 and 2.

derline once again that the $K = 3$ turbo code and the $K = 7$ convolutional code exhibited comparable complexities. The higher performance of the turbo codec facilitates, for example, the employment of turbo-coded 16QAM at a similar SNR, where convolutional-coded QPSK can be invoked. This in turn allows us to double the video bit rate within the same bandwidth and hence to improve the video quality. We have also applied data partitioning to the MPEG-2 video stream to gauge its efficiency in increasing the error resilience of the video codec. However, from these experiments we found that the data partitioning scheme did not provide substantial improvements compared to the nonpartitioned video transmitted over the nonhierarchical DVB-T system. Our future work will focus on extending this DVB-T system study to incorporate various types of channel models, as well as on investigating the effects of different Doppler frequencies on the system. Further work will also be dedicated to trellis-coded modulation (TCM) and turbo trellis-coded modulation (TTCM) based OFDM. The impact of employing various types of turbo interleavers on the system performance is also of interest. A range of further wireless video communications issues are addressed in [598, 642]. Let us now consider a variety of satellite-based turbo-coded blind-equalized multilevel modulation-assisted video broadcasting schemes.

21.7 Satellite-Based Video Broadcasting [15]

21.7.1 Background and Motivation

In recent years, three harmonized digital video broadcasting (DVB) standards have emerged in Europe for terrestrial [624], cable-based [625], and satellite-oriented [626] delivery of DVB signals. The dispersive wireless propagation environment of the terrestrial system requires concatenated Reed–Solomon (RS) [87,627] and rate-compatible punctured convolutional coding (RCPCC) [87, 627] combined with orthogonal frequency division multiplexing (OFDM)-based modulation [29]. The satellite-based system employs the same concatenated channel coding arrangement as the terrestrial scheme, while the cable-based system refrains from using concatenated channel coding, opting for RS coding only. The performance of both of the latter schemes can be improved upon invoking blind-equalized multilevel modems [29], although the associated mild dispersion or linear distortion does not necessarily require channel equalization. However, since we propose invoking turbo-coded 4-bit/symbol 16-level quadrature amplitude modulation (16QAM) in order to improve the system's performance at the cost of increased complexity, in this section we also invoked blind channel equalizers. This is further justified by the associated high video transmission rates, where the dispersion may become a more dominant performance limitation.

Lastly, the video codec used in all three systems is the Motion Pictures Expert Group's MPEG-2 codec. These standardization activities were followed by a variety of system performance studies in the open literature [643–646]. Against this background, we suggest turbo-coding improvements to the satellite-based DVB system [626] and present performance studies of the proposed system under dispersive channel conditions in conjunction with a variety of blind channel equalization algorithms. The transmitted power requirements of the standard system employing convolutional codecs can be reduced upon invoking more complex, but more powerful, turbo codecs. Alternatively, the standard quaternary or 2-bit/symbol system's bit error rate (BER) versus signal-to-noise ratio (SNR) performance can almost be matched by a turbo-coded 4-bit/symbol 16QAM scheme, while doubling the achievable bit rate within the same bandwidth and hence improving the associated video quality. This is achieved at the cost of an increased system complexity.

The remainder of this section is organized as follows. A succinct overview of the turbo-coded and standard DVB satellite scheme is presented in Section 21.7.2, while our channel model is described in Section 21.7.3. A brief summary of the blind equalizer algorithms employed is presented in Section 21.7.4. Following this, the performance of the improved DVB satellite system is examined for transmission over a dispersive two-path channel in Section 21.7.5, before our conclusions and future work areas are presented in Section 21.7.6.

---

[15]This section is based on C. S. Lee, S. Vlahoyiannatos, and L. Hanzo, "Satellite based turbo-coded, blind-equalized 4QAM and 16QAM digital video broadcasting," *IEEE Transactions on Broadcasting*, March 2000, pp. 22–34, ©2000 IEEE. Personal use of this material is permitted. However, permission to reprint/republish this material for advertising or promotional purposes or for creating new collective works for resale or redistribution to servers or lists, or to reuse any copyrighted component of this work in other works must be obtained from the IEEE.

21.7. SATELLITE-BASED VIDEO BROADCASTING

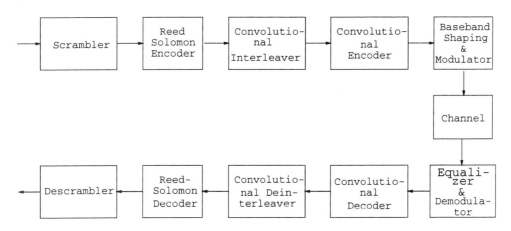

Figure 21.106: Schematic of the DVB satellite system.

21.7.2 DVB Satellite Scheme

The block diagram of the DVB satellite (DVB-S) system [626] is shown in Figure 21.106, which is composed of a MPEG-2 video encoder (not shown in the diagram), channel-coding modules, and a quadrature phase shift keying (QPSK) modem [29]. The bit-stream generated by the MPEG-2 encoder is packetized into frames 188 bytes long. The video data in each packet is then randomized by the scrambler. The details concerning the scrambler have not been included in this chapter, since these may be obtained from the DVB-S standard [626].

Because of the poor error resilience of the MPEG-2 video codec, powerful concatenated channel coding is employed. The concatenated channel codec comprises a shortened Reed–Solomon (RS) outer code and an inner convolutional encoder. The 188-byte MPEG-2 video packet is extended by the Reed–Solomon encoder [87, 627], with parity information to facilitate error recovery to form a 204-byte packet. The Reed–Solomon decoder can then correct up to 8 erroneous bytes for each 204-byte packet. Following this, the RS-coded packet is interleaved by a convolutional interleaver and is further protected by a half-rate inner convolutional encoder with a constraint length of 7 [87, 627].

Furthermore, the overall code rate of the concatenated coding scheme can be adapted by variable puncturing, not shown in the figure, which supports code rates of $\frac{1}{2}$ (no puncturing) as well as $\frac{2}{3}$, $\frac{3}{4}$, $\frac{5}{6}$, and $\frac{7}{8}$. The parameters of the convolutional encoder are summarized in Table 21.26.

In addition to implementing the standard DVB-S system as a benchmark, we have improved the system's performance with the aid of a turbo codec [28, 381]. The block diagram of the turbo encoder is shown in Figure 21.107. The turbo encoder is constructed of two component encoders. Each component encoder is a half-rate convolutional encoder whose parameters are listed in Table 21.27. The two component encoders are used to encode the same input bits, although the input bits of the second component encoder are interleaved before encoding. The output bits of the two component codes are punctured and multiplexed in order to form a single-output bit

Convolutional Coder Parameters	
Code rate	1/2
Constraint length	7
n	2
k	1
Generator polynomials (octal format)	171, 133

Table 21.26: Parameters of the $CC(n, k, K)$ Convolutional Inner Encoder of the DVB-S Modem

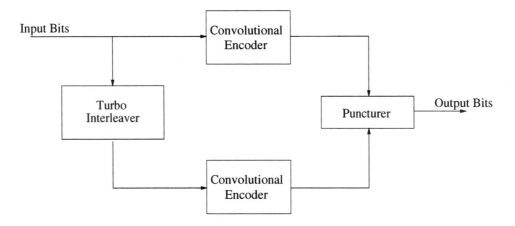

Figure 21.107: Block diagram of turbo encoder.

Turbo Coder Parameters	
Turbo code rate	1/2
Input block length	17 952 bits
Interleaver type	Random
Number of turbo decoder iterations	8
Turbo Encoder Component Code Parameters	
Component code encoder type	Convolutional Encoder (RSC)
Component code decoder type	log-MAP [636]
Constraint length	3
n	2
k	1
Generator polynomials (octal format)	7, 5

Table 21.27: Parameters of the Inner Turbo Encoder Used to Replace the DVB-S System's Convolutional Coder (RSC: Recursive Systematic Code)

stream. The component encoder used here is known as a half-rate recursive systematic convolutional encoder (RSC) [637]. It generates one parity bit and one systematic output bit for every input bit. In order to provide an overall coding rate of one-half, half the output bits from the two encoders must be punctured. The puncturing arrangement used in our work is to transmit all the systematic bits from the first encoder and every other parity bit from both encoders.

Readers interested in further details of the DVB-S system are referred to the DVB-S standard [626]. The performance of the standard DVB-S system and the performance of the turbo-coded system are characterized in Section 21.7.5. Let us now briefly consider the multipath channel model used in our investigations.

21.7.3 Satellite Channel Model

The DVB-S system was designed to operate in the 12 GHz frequency band (K-band). Within this frequency band, tropospheric effects such as the transformation of electromagnetic energy into thermal energy due to induction of currents in rain and ice crystals lead to signal attenuations [647, 648]. In the past 20 years, various researchers have concentrated on attempting to model the satellite channel, typically within a land mobile satellite channel scenario. However, the majority of the work conducted, for example, by Vogel and his colleagues [649–652] concentrated on modelling the statistical properties of a narrowband satellite channel in lower frequency bands, such as the 870 MHz UHF band and the 1.5 GHz L-band.

Our high bit-rate DVB satellite system requires a high bandwidth, however. Hence, the video bit-stream is exposed to dispersive wideband propagation conditions. Recently, Saunders *et al.* [653, 654] have proposed the employment of multipath channel models to study the satellite channel, although their study concentrated on the L-band and S-band only.

Due to the dearth of reported work on wideband satellite channel modeling in the K-band, we have adopted a simpler approach. The channel model employed in this study was the two-path (nT)-symbol spaced impulse response, where T is the symbol duration. In our studies we used $n = 1$ and $n = 2$ (Figure 21.108). This corresponds to a stationary dispersive transmission channel. Our channel model assumed that the receiver had a direct line-of-sight with the satellite as well as a second path caused by a single reflector probably from a nearby building or due to ground reflection. The ground reflection may be strong if the satellite receiver dish is only tilted at a low angle.

Based on these channel models, we studied the ability of a range of blind equalizer algorithms to converge under various path-delay conditions. In the next section, we provide a brief overview of the various blind equalizers employed in our experiments. Readers interested mainly in the system's performance may proceed directly to our performance analysis section, Section 21.7.5.

21.7.4 The Blind Equalizers

This section presents the blind equalizers used in the system. The following blind equalizers have been studied:

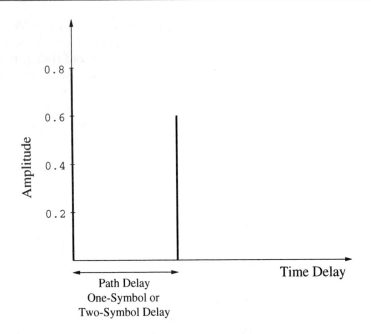

Figure 21.108: Two-path satellite channel model with either a one-symbol or two-symbol delay.

1. The modified constant modulus algorithm (MCMA) [655].

2. The Benveniste-Goursat algorithm (B-G) [656].

3. The stop-and-go algorithm (S-a-G) [657].

4. The per-survivor processing (PSP) algorithm [658].

We will now briefly introduce these algorithms.

First, we define the variables that we will use:

$$\mathbf{y}(n) = [y(n+N_1), \ldots, y(0), \ldots, y(n-N_2)]^T \tag{21.6}$$

$$\mathbf{c}^{(n)} = [c_{-N_1}, \ldots, c_o, \ldots, c_{N_2}]^T \tag{21.7}$$

$$z(n) = \left(\mathbf{c}^{(n)}\right)^T \mathbf{y}(n) = \mathbf{y}^T(n)\mathbf{c}^{(n)} \tag{21.8}$$

where $\mathbf{y}(n)$ is the received symbol vector at time n, containing the $N_1 + N_2 + 1$ most recent received symbols, while N_1, N_2 are the number of equalizer feedback and feedforward taps, respectively. Furthermore, $\mathbf{c}^{(n)}$ is the equalizer tap vector, consisting of the equalizer tap values, and $z(n)$ is the equalized symbol at time n, given by the convolution of the received signal with the equalizer's impulse response, while $()^T$ stands for matrix transpose. Note that the variables of Equations 21.6–21.8 assume complex values, when multilevel modulation is employed.

21.7. SATELLITE-BASED VIDEO BROADCASTING

The **modified CMA** (MCMA) is an improved version of Godard's well-known *constant modulus algorithm (CMA)* [659]. The philosophy of the CMA is based on forcing the magnitude of the equalized signal to a constant value. In mathematical terms, the CMA is based on minimising the cost function:

$$J^{(CMA)} = E\left[\left(|z(n)|^2 - R_2\right)^2\right], \qquad (21.9)$$

where R_2 is a suitably chosen constant and $E[]$ stands for the expectation. Similarly to the CMA, the MCMA, which was proposed by Wesolowsky [655], forces the real and imaginary parts of the complex signal to the constant values of $R_{2,R}$ and $R_{2,I}$, respectively, according to the equalizer tap update equation of [655]:

$$\mathbf{c}^{(n+1)} = \mathbf{c}^{(n)} - \lambda \cdot \mathbf{y}^*(n) \cdot \{Re[z(n)] \cdot \left((Re[z(n)])^2 - R_{2,R}\right) \\ + jIm[z(n)] \cdot \left((Im[z(n)])^2 - R_{2,I}\right)\}, \qquad (21.10)$$

where λ is the step-size parameter and the $R_{2,R}$, $R_{2,I}$ constant parameters of the algorithm are defined as:

$$R_{2,R} = \frac{E\left[(Re[a(n)])^4\right]}{E\left[(Re[a(n)])^2\right]} \qquad (21.11)$$

$$R_{2,I} = \frac{E\left[(Im[a(n)])^4\right]}{E\left[(Im[a(n)])^2\right]}, \qquad (21.12)$$

where $a(n)$ is the transmitted signal at time instant n.

The **BenvenisteGoursat** (B-G) algorithm [656] is an amalgam of the Sato's algorithm [660] and the decision-directed (DD) algorithm [29]. Strictly speaking, the decision-directed algorithm is not a blind equalization technique, since its convergence is highly dependent on the channel.

This algorithm estimates the error between the equalized signal and the detected signal as:

$$\epsilon^{DD}(n) = z(n) - \hat{z}(n), \qquad (21.13)$$

where $\hat{z}(n)$ is the receiver's estimate of the transmitted signal at time instant n. Similarly to the DD algorithm's error term, the Sato error [660] is defined as:

$$\epsilon^{Sato}(n) = z(n) - \gamma \cdot csgn(z(n)), \qquad (21.14)$$

where γ is a constant parameter of the Sato algorithm, defined as:

$$\gamma = \frac{E\left[(Re\,[a(n)])^2\right]}{E\left[|Re\,[a(n)]|\right]} = \frac{E\left[(Im\,[a(n)])^2\right]}{E\left[|Im\,[a(n)]|\right]} \qquad (21.15)$$

and $csgn(x) = sign(Re\{x\}) + jsign(Im\{x\})$ is the complex sign function. The B-G algorithm combines the above two error terms into one:

$$\epsilon^G(n) = k_1 \cdot \epsilon^{DD}(n) + k_2 \cdot |\epsilon^{DD}(n)| \cdot \epsilon^{Sato}(n), \qquad (21.16)$$

where the two error terms are suitably weighted by the constant parameters k_1 and k_2 in Equation 21.16. Using this error term, the B–G equalizer updates the equalizer coefficients according to the following equalizer tap update equations [656]:

$$\mathbf{c}^{(n+1)} = \mathbf{c}^{(n)} - \lambda \cdot \mathbf{y}^*(n) \cdot \epsilon^G(n). \tag{21.17}$$

In our investigations, the weights were chosen as $k_1 = 1$, $k_2 = 5$, so that the Sato error was weighted more heavily than the DD error.

The **stop-and-go** (S-a-G) algorithm [657] is a variant of the decision-directed algorithm [29], where at each equalizer coefficient adjustment iteration, the update is enabled or disabled depending on whether or not the update is likely to be correct. The update equations of this algorithm are given by [657]

$$\mathbf{c}^{(n+1)} = \mathbf{c}^{(n)} - \lambda \cdot \mathbf{y}^*(n) \cdot [f_{n,R} \cdot Re\{\epsilon^{DD}(n)\} + jf_{n,I} \cdot Im\{\epsilon^{DD}(n)\}], \tag{21.18}$$

where * stands for the complex conjugate, $\epsilon^{DD}(n)$ is the decision directed error as in Equation 21.13 and the binary functions $f_{n,R}$, $f_{n,I}$ enable or disable the update of the equalizer according to the following rule. If the sign of the Sato error (the real or the imaginary part independently) is the same as the sign of the decision-directed error, then the update takes place; otherwise it does not.

In mathematical terms, this is equivalent to [657]:

$$f_{n,R} = \begin{cases} 1 & \text{if } sgn(Re[\epsilon^{DD}(n)]) = sgn(Re[\epsilon^{Sato}(n)]) \\ 0 & \text{if } sgn(Re[\epsilon^{DD}(n)]) \neq sgn(Re[\epsilon^{Sato}(n)]) \end{cases} \tag{21.19}$$

$$f_{n,I} = \begin{cases} 1 & \text{if } sgn(Im\{\epsilon^{DD}(n)\}) = sgn(Im\{\epsilon^{Sato}(n)\}) \\ 0 & \text{if } sgn(Im\{\epsilon^{DD}(n)\}) \neq sgn(Im\{\epsilon^{Sato}(n)\}). \end{cases} \tag{21.20}$$

For a blind equalizer, this condition provides us with a measure of the probability of the coefficient update being correct.

The **PSP algorithm** [658] is based on employing convolutional coding. Hence, it is a trellis-based sequence estimation technique in which the channel is not known *a priori*. An iterative channel estimation technique is employed in order to estimate the channel jointly with the modulation symbol. In this sense, an initial channel is used, and the estimate is updated at each new symbol's arrival.

In our case, the update was based on the *least means squares (LMS)* estimates, according to the following channel-tap update equations [658]:

$$\hat{\mathbf{h}}^{(n+1)} = \hat{\mathbf{h}}^{(n)} + \lambda \cdot \hat{\mathbf{a}}^*(n) \cdot \left(y(n) - \hat{\mathbf{a}}^T(n)\hat{\mathbf{h}}^{(n)} \right), \tag{21.21}$$

where $\hat{\mathbf{h}}^{(n)} = (\hat{h}^{(n)}_{-L_1}, \ldots, \hat{h}^{(n)}_o, \ldots, \hat{h}^{(n)}_{L_2})^T$ is the estimated (for one surviving path) channel tap vector at time instant n, $\hat{\mathbf{a}}(n) = (\hat{a}(n+L_1), \ldots, \hat{a}(0), \ldots, \hat{a}(n-L_2))^T$ is the associated estimated transmitted symbol vector, and $y(n)$ is the actually received symbol at time instant n.

Each of the surviving paths in the trellis carries not only its own signal estimation, but also its own channel estimation. Moreover, convolutional decoding can take place jointly with this channel and data estimation procedure, leading to improved bit error

21.7. SATELLITE-BASED VIDEO BROADCASTING

	Step Size λ	No. of Equalizer Taps	Initial Tap Vector
Benveniste–Goursat	5×10^{-4}	10	$(1.2, 0, \cdots, 0)$
Modified CMA	5×10^{-4}	10	$(1.2, 0, \cdots, 0)$
Stop-and-go	5×10^{-4}	10	$(1.2, 0, \cdots, 0)$
PSP (1 sym delay)	10^{-2}	2	$(1.2, 0)$
PSP (2 sym delay)	10^{-2}	3	$(1.2, 0, 0)$

Table 21.28: Summary of the Equalizer Parameters Used in the Simulations. The Tap Vector $(1.2, 0, \cdots, 0)$ Indicates that the First Equalizer Coefficient is Initialized to the Value 1.2, While the Others are Initialized to 0.

rate (BER) performance. The various equalizers' parameters are summarized in Table 21.28.

Having described the components of our enhanced DVB-S system, let us now consider the overall system's performance.

21.7.5 Performance of the DVB Satellite Scheme

In this section, the performance of the DVB-S system was evaluated by means of simulations. Two modulation types were used, namely, the standard QPSK and the enhanced 16QAM schemes [29]. The channel model of Figure 21.108 was employed. The first channel model had a one-symbol second-path delay, while in the second one the path-delay corresponded to the period of two symbols. The average BER versus SNR per bit performance was evaluated after the equalization and demodulation process, as well as after Viterbi [87] or turbo decoding [381]. The SNR per bit or E_b/N_o is defined as follows:

$$\text{SNR per bit} = 10 \, log_{10} \frac{\bar{S}}{\bar{N}} + \delta, \qquad (21.22)$$

where $\bar{S}$ is the average received signal power, $\bar{N}$ is the average received noise power, and δ, which is dependent on the type of modulation scheme used and channel code rate (R), is defined as:

$$\delta = 10 \, log_{10} \frac{1}{R \times \text{bits per modulation symbol}}. \qquad (21.23)$$

Our results are further divided into two subsections for ease of discussion. First, we present the system performance over the one-symbol delay two-path channel in Section 21.7.5.1. Next, the system performance over the two-symbol delay two-path channel is presented in Section 21.7.5.2. Lastly, a summary of the system performance is provided in Section 21.7.5.3.

21.7.5.1 Transmission over the Symbol-Spaced Two-Path Channel

The linear equalizers' performance was quantified and compared using QPSK modulation over the one-symbol delay two-path channel model of Figure 21.109. Since

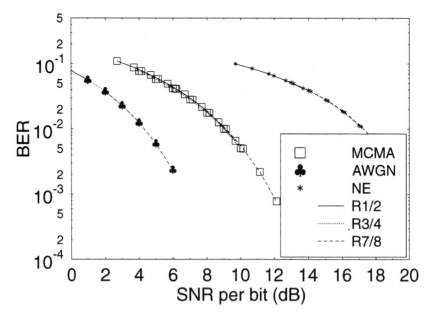

(a) After equalization and demodulation

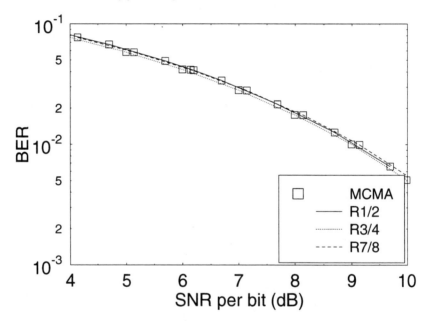

(b) Same as (a) but enlarged in order to show performance difference of the blind equalizer, when different convolutional code rates are used

Figure 21.109: Average BER versus SNR per bit performance after equalization and demodulation employing **QPSK** modulation and **one-symbol delay channel** (**NE** = nonequalized; **MCMA** = modified constant modulus algorithm).

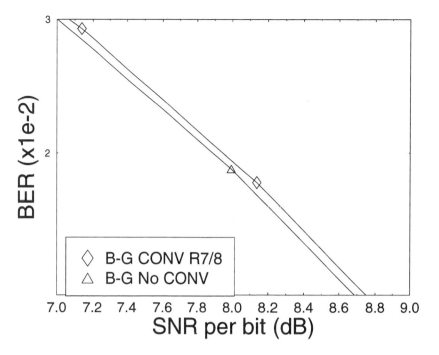

Figure 21.110: Average BER versus SNR per bit performance after equalization and demodulation employing **QPSK** modulation and the **one-symbol delay two-path channel** of Figure 21.108, for the Benveniste–Goursat algorithm, where the input bits are random (No CONV) or correlated (CONV 7/8) as a result of convolutional coding having a coding rate of 7/8.

all the equalizers have similar BER performance, only the modified CMA results are shown in the figure.

The equalized performance over the one symbol-spaced channel was inferior to that over the nondispersive AWGN channel. However, as expected, it was better than without any equalization. Another observation for Figure 21.109 was that the different punctured channel-coding rates appeared to give slightly different bit error rates after equalization. This was because the linear blind equalizers required uncorrelated input bits in order to converge. However, the input bits were not entirely random when convolutional coding was used. The consequences of violating the zero-correlation constraint are not generally known. Nevertheless, two potential problems were apparent. First, the equalizer may diverge from the desired equalizer equilibrium [661]. Second, the performance of the equalizer is expected to degrade, owing to the violation of the randomness requirement, which is imposed on the input bits in order to ensure that the blind equalizers will converge.

Since the channel used in our investigations was static, the first problem was not encountered. Instead, the second problem was what we actually observed. Figure 21.110 quantifies the equalizers' performance degradation due to convolutional coding. We can observe a 0.1 dB SNR degradation when the convolutional codec creates correlation among the bits for this specific case.

The average BER versus SNR curves after Viterbi or turbo decoding are shown in Figure 21.111(a). In this figure, the average BER over the nondispersive AWGN channel after turbo decoding constitutes the best-case performance, while the average BER of the one-symbol delay two-path MCMA-equalized rate 7/8 convolutionally coded scenario exhibits the worst-case performance. Again, in this figure only the modified CMA was featured for simplicity. The performance of the remaining equalizers was characterized in Figure 21.111(b). Clearly, the performance of all the linear equalizers investigated was similar.

As seen in Figure 21.111(a), the combination of the modified CMA blind equalizer with turbo decoding exhibited the best SNR performance over the one-symbol delay two-path channel. The only comparable alternative was the PSP algorithm. Although the performance of the PSP algorithm was better at low SNRs, the associated curves cross over and the PSP algorithm's performance became inferior below the average BER of 10^{-3}. Although not shown in Figure 21.111, the Reed–Solomon decoder, which was concatenated to either the convolutional or the turbo decoder, became effective, when the average BER of its input was below approximately 10^{-4}. In this case, the PSP algorithm performed by at least 1 dB worse in the area of interest, which is at an average BER of 10^{-4}.

A final observation in the context of Figure 21.111(a) is that when convolutional decoding was used, the associated E_b/N_o performance of the rate 1/2 convolutional coded scheme appeared slightly inferior to that of the rate 3/4 and the rate 7/8 scenarios beyond certain E_b/N_o values. This was deemed to be a consequence of the fact that the 1/2 rate encoder introduced more correlation into the bitstream than its higher rate counterparts, and this degraded the performance of the blind channel equalizers which performed best when fed with random bits.

Having considered the QPSK case, we shall now concentrate on the enhanced system which employed 16QAM under the same channel and equalizer conditions. Figures 21.112 and 21.113 present the performance of the DVB system employing 16QAM. Again, for simplicity, only the modified CMA results are given. In this case, the ranking order of the different coding rates followed our expectations more closely in the sense that the lowest coding rate of 1/2 was the best performer, followed by rate 3/4 codec, in turn followed by the least powerful rate 7/8 codec.

The stop-and-go algorithm has been excluded from these results, because it does not converge for high SNR values. This happens because the equalization procedure is activated only when there is a high probability of correct decision-directed equalizer update. In our case, the equalizer is initialized far from its convergence point, and hence the decision directed updates are unlikely to be correct. In the absence of noise, this leads to the update algorithm being permanently deactivated. If noise is present, however, then some random perturbations from the point of the equalizer's initialization activate the stop-and-go algorithm and can lead to convergence. We made this observation at medium SNR values in our simulation study. For high SNR values, the algorithm did not converge.

It is also interesting to compare the performance of the system for the QPSK and 16QAM schemes. When the one-symbol delay two-path channel model of Figure 21.108 was considered, the system was capable of supporting the use of 16QAM

21.7. SATELLITE-BASED VIDEO BROADCASTING

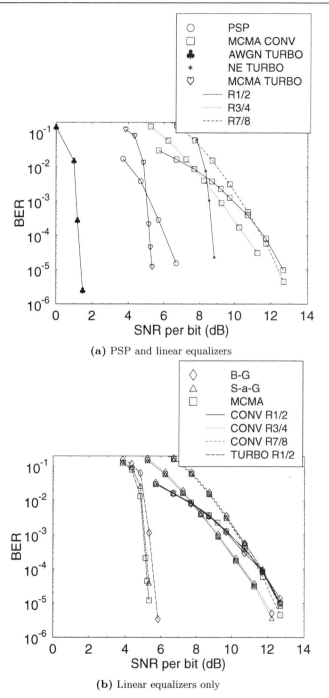

(a) PSP and linear equalizers

(b) Linear equalizers only

Figure 21.111: Average BER versus SNR per bit performance after convolutional or turbo decoding for **QPSK** modulation and **one-symbol delay channel** (**NE** = nonequalized; **B-G** = Benveniste–Goursat; **S-a-G** = stop-and-go; **MCMA** = modified constant modulus algorithm; **PSP** = per-survivor processing).

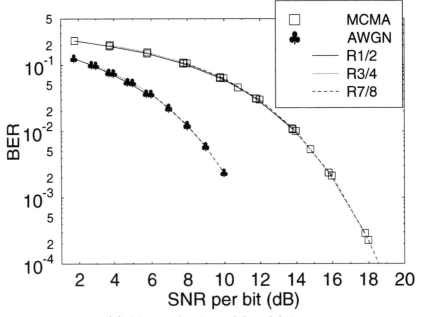

(a) After equalization and demodulation

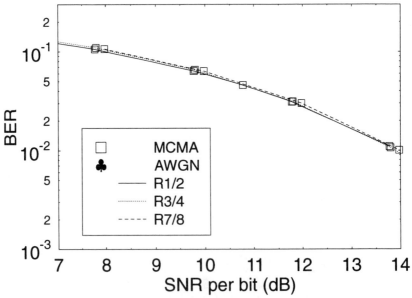

(b) Same as (a) but enlarged in order to show performance difference of the blind equalizer, when different convolutional code rates are used

Figure 21.112: Average BER versus SNR per bit after equalization and demodulation for **16QAM** over the **one-symbol delay two-path channel** of Figure 21.108 (**MCMA** = modified constant modulus algorithm).

21.7. SATELLITE-BASED VIDEO BROADCASTING

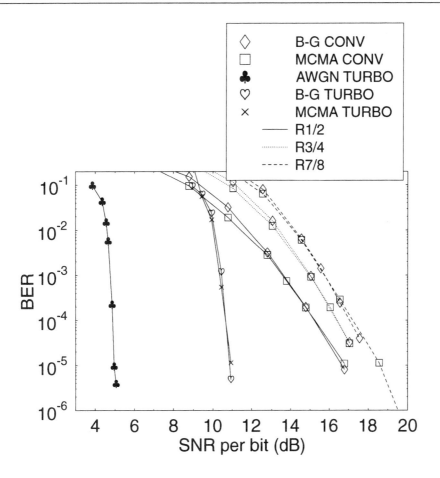

Figure 21.113: Average BER versus SNR per bit after Viterbi or turbo decoding for **16QAM** over the **one-symbol delay two-path channel** of Figure 21.108 (**B-G** = Benveniste–Goursat; **S-a-G** = stop-and-go; **MCMA** = modified constant modulus algorithm; **PSP** = per-survivor processing).

with the provision of an additional SNR per bit of approximately 4–5 dB. This observation was made by comparing the performance of the DVB system when employing the modified CMA and the half-rate convolutional or turbo code in Figures 21.111 and 21.113 at a BER of 10^{-4}. Although the original DVB satellite system only employs QPSK modulation, our simulations had shown that 16QAM can be employed equally well for the range of blind equalizers that we have used in our work. This allowed us to double the video bit rate and hence to substantially improve the video quality. The comparison of Figures 21.111 and 21.113 also reveals that the extra SNR requirement of approximately 4–5 dB of 16QAM over QPSK can be eliminated by employing turbo coding at the cost of a higher implementational complexity. This allowed us to accommodate a doubled bit rate within a given bandwidth, which improved the video quality.

21.7.5.2 Transmission over the Two-Symbol Delay Two-Path Channel

In Figure 21.114 (only for the Benveniste-Goursat algorithm for simplicity) and Figure 21.115, the corresponding BER results for the two-symbol delay two-path channel of Figure 21.108 are given for QPSK. The associated trends are similar to those in Figures 21.109 and 21.111, although some differences can be observed, as listed below:

- The "crossover point" is where the performance of the PSP algorithm becomes inferior to the performance of the modified CMA in conjunction with turbo decoding. This point is now at 10^{-4}, which is in the range where the RS decoder guarantees an extremely low probability of error.

- The rate 1/2 convolutional decoding is now the best performer, when convolutional decoding is concerned, while the rate 3/4 scheme exhibits the worst performance.

Finally, in Figure 21.116, the associated 16QAM results are presented. Notice that the stop-and-go algorithm was again excluded from the results. Furthermore, we observe a high-performance difference between the Benveniste-Goursat algorithm and the modified CMA. In the previous cases we did not observe such a significant difference. The difference in this case is that the channel exhibits an increased delay spread. This illustrated the capability of the equalizers to cope with more widespread multipaths, while keeping the equalizer order constant at 10. The Benveniste-Goursat equalizer was more efficient than the modified CMA in this case.

It is interesting to note that in this case the performance of the different coding rates was again in the expected order: the rate $\frac{1}{2}$ scheme is the best, followed by the rate $\frac{3}{4}$ scheme and then the rate $\frac{7}{8}$ scheme.

If we compare the performance of the system employing QPSK and 16QAM over the two-symbol delay two-path channel of Figure 21.108, we again observe that 16QAM can be incorporated into the DVB system if an extra 5 dB of SNR per bit is affordable in power budget terms. Here, only the B-G algorithm is worth considering out of the three linear equalizers of Table 21.28. This observation was made by comparing the performance of the DVB system when employing the Benveniste-Goursat equalizer and the half-rate convolutional coder in Figures 21.115 and 21.116.

21.7.5.3 Performance Summary of the DVB-S System

Table 21.29 provides an approximation of the convergence speed of each blind equalization algorithm of Table 21.28. PSP exhibited the fastest convergence, followed by the Benveniste-Goursat algorithm. In our simulations the convergence was quantified by observing the slope of the BER curve and finding when this curve was reaching the associated residual BER, implying that the BER has reached its steady-state value. Figure 21.117 gives an illustrative example of the equalizer's convergence for 16QAM. The stop-and-go algorithm converges significantly slower than the other algorithms, which can also be seen from Table 21.29. This happens because during start-up the algorithm is deactivated most of the time. The effect becomes more severe with increasing QAM order.

21.7. SATELLITE-BASED VIDEO BROADCASTING

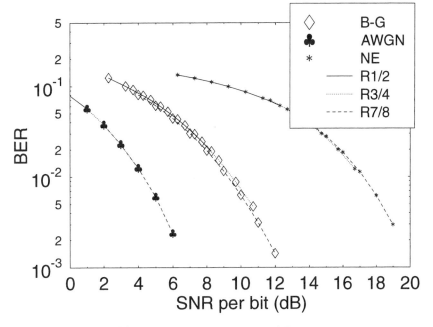

(a) After equalization and demodulation

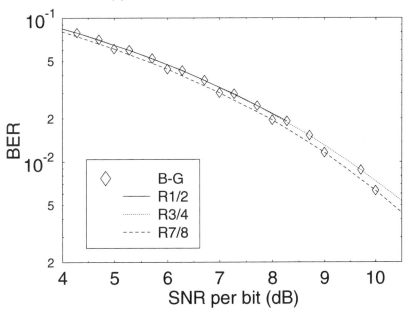

(b) Same as (a) but enlarged in order to show performance difference of the blind equalizer, when different convolutional code rates are used

Figure 21.114: Average BER versus SNR per-bit performance after equalization and demodulation for **QPSK** modulation over the **two-symbol delay two-path channel** of Figure 21.108 (**B-G** = Benveniste-Goursat).

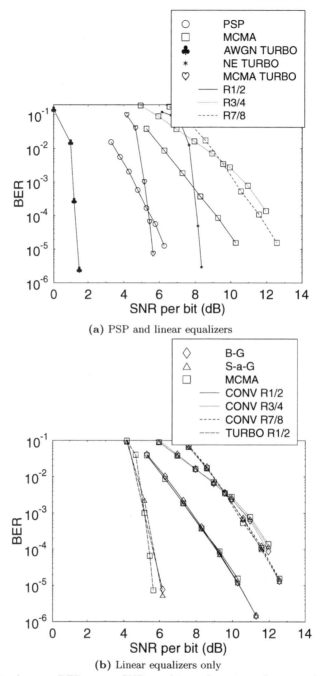

Figure 21.115: Average BER versus SNR per-bit performance after convolutional or turbo decoding for **QPSK** modulation over the **two-symbol delay two-path channel** of Figure 21.108 (**B-G** = Benveniste-Goursat; **S-a-G** = stop-and-go; **MCMA** = modified constant modulus algorithm; **PSP** = per-survivor processing).

21.7. SATELLITE-BASED VIDEO BROADCASTING

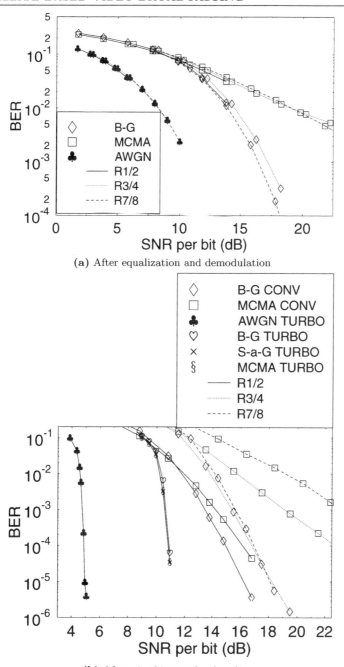

(a) After equalization and demodulation

(b) After viterbi or turbo decoding

Figure 21.116: Average BER versus SNR per-bit performance (a) after equalization and demodulation and (b) after Viterbi or turbo decoding for **16QAM** over the **two-symbol delay two-path channel** of Figure 21.108 (**B–G** = Benveniste-Goursat; **S-a-G** = stop-and-go; **MCMA** = modified constant modulus algorithm; **PSP** = per-survivor processing).

	B-G	MCMA	S-a-G	PSP
QPSK 1 sym	73	161	143	0.139
QPSK 2 sym	73	143	77	0.139
16QAM 1 sym	411	645	1393	
16QAM 2 sym	359	411	1320	

Table 21.29: Equalizer Convergence Speed (in miliseconds) Measured in the Simulations, Given as an Estimate of Time Required for Convergence When 1/2 Rate Puncturing Is Used (x sym = x-Symbol Delay Two-path Channel and x Can Take Either the Value 1 or 2)

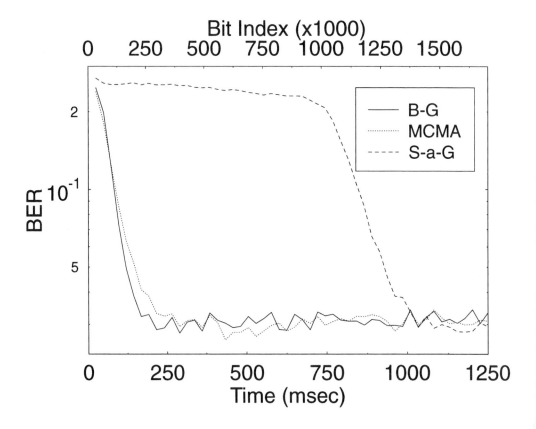

Figure 21.117: Learning curves for 16QAM, one-symbol delay two-path channel at SNR = 18 dB.

21.7. SATELLITE-BASED VIDEO BROADCASTING

Mod.	Equalizer	Code	CSNR (dB)	E_b/N_0
QPSK	PSP	R=1/2	5.3	5.3
QPSK	MCMA	Turbo (1/2)	5.2	5.2
16QAM	MCMA	Turbo (1/2)	13.6	10.6
QPSK	MCMA	Conv (1/2)	9.1	9.1
16QAM	MCMA	Conv (1/2)	17.2	14.2
QPSK	MCMA	Conv (3/4)	11.5	9.7
16QAM	MCMA	Conv (3/4)	20.2	15.4
QPSK	B-G	Conv (7/8)	13.2	10.8
16QAM	B-G	Conv (7/8)	21.6	16.2

Table 21.30: Summary of Performance Results over the Dispersive One-symbol Delay Two-Path AWGN Channel of Figure 21.108 Tolerating a PSNR Degradation of 2 dB

Figure 21.118 portrays the corresponding reconstructed video quality in terms of the average peak signal-to-noise ratio (PSNR) versus channel SNR (CSNR) for the one-symbol delay and two-symbol delay two-path channel models of Figure 21.108. The PSNR is defined as follows:

$$\text{PSNR} = 10 \, log_{10} \frac{\sum_{n=0}^{N} \sum_{m=0}^{M} 255^2}{\sum_{n=0}^{N} \sum_{m=0}^{M} \Delta^2}, \quad (21.24)$$

where Δ is the difference between the uncoded pixel value and the reconstructed pixel value. The variables M and N refer to the dimension of the image. The maximum possible 8-bit represented pixel luminance value of 255 was used in Equation 21.24 in order to mitigate the PSNR's dependence on the video material used. The average PSNR is then the mean of the PSNR values computed for all the images constituting the video sequence.

Tables 21.30 and 21.31 provide a summary of the DVB satellite system's performance tolerating a PSNR degradation of 2 dB, which was deemed to be nearly imperceptible in terms of subjective video degradations. The average BER values quoted in the tables refer to the average BER achieved after Viterbi or turbo decoding. The channel SNR is quoted in association with the 2 dB average video PSNR degradation, since the viewer will begin to perceive video degradations due to erroneous decoding of the received video around this threshold.

Tables 21.32 and 21.33 provide a summary of the SNR per bit required for the various system configurations. The BER threshold of 10^{-4} was selected here, because of this average BER after Viterbi or turbo decoding, the RS decoder becomes effective, guaranteeing near error-free performance. This also translates into near unimpaired reconstructed video quality.

Finally, in Table 21.34 the QAM symbol rate or baud rate is given for different puncturing rates and for different modulation schemes, based on the requirement of supporting a video bit rate of 2.5 Mbit/sec. We observe that the baud rate is between 0.779 and 2.73 MBd, depending on the coding rate and the number of bits per modulation symbol.

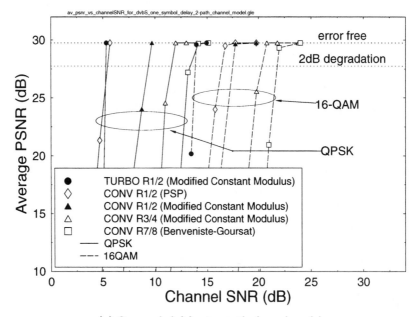

(a) One-symbol delay two-path channel model

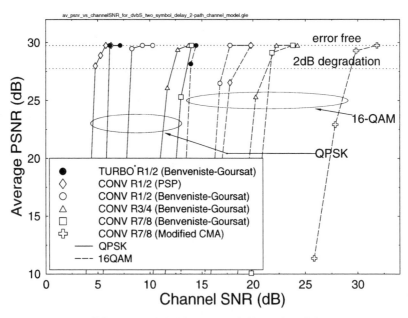

(b) Two-symbol delay two-path channel model

Figure 21.118: Average PSNR versus channel SNR for (a) the one-symbol delay two-path channel model and (b) the two-symbol delay two-path channel model of Figure 21.108 at a video bit rate of 2.5 Mbit/s using the "Football" video sequence.

21.7. SATELLITE-BASED VIDEO BROADCASTING

Mod.	Equalizer	Code	CSNR (dB)	E_b/N_0
QPSK	PSP	R=1/2	4.7	4.7
QPSK	B-G	Turbo (1/2)	5.9	5.9
16QAM	B-G	Turbo (1/2)	13.7	10.7
QPSK	B-G	Conv (1/2)	8.0	8.0
16QAM	B-G	Conv (1/2)	17.0	14.0
QPSK	B-G	Conv (3/4)	12.1	10.3
16QAM	B-G	Conv (3/4)	21.1	16.3
QPSK	B-G	Conv (7/8)	13.4	11.0
16QAM	MCMA	Conv (7/8)	29.2	23.8

Table 21.31: Summary of Performance Results over the Dispersive Two-symbol Delay Two-path AWGN Channel of Figure 21.108 Tolerating a PSNR Degradation of 2 dB

Mod.	Equalizer	Code	E_b/N_0
QPSK	PSP	R=1/2	6.1
QPSK	MCMA	Turbo (1/2)	5.2
16QAM	MCMA	Turbo (1/2)	10.7
QPSK	MCMA	Conv (1/2)	11.6
16QAM	MCMA	Conv (1/2)	15.3
QPSK	MCMA	Conv (3/4)	10.5
16QAM	MCMA	Conv (3/4)	16.4
QPSK	B-G	Conv (7/8)	11.8
16QAM	B-G	Conv (7/8)	17.2

Table 21.32: Summary of System Performance Results over the Dispersive One-symbol Delay Two-path AWGN Channel of Figure 21.108 Tolerating an Average BER of 10^{-4}, which was Evaluated after Viterbi or Turbo Decoding but before RS Decoding

21.7.6 Summary and Conclusions

In this chapter, we have investigated the performance of a turbo-coded DVB system in a satellite broadcast environment. A range of system performance results was presented based on the standard DVB-S scheme, as well as on a turbo-coded scheme in conjunction with blind-equalized QPSK/16QAM. The convolutional code specified in the standard system was substituted with turbo coding, which resulted in a substantial coding gain of approximately 4–5 dB. We have also shown that 16QAM can be utilized instead of QPSK if an extra 5 dB SNR per bit gain is added to the link budget. This extra transmitted power requirement can be eliminated upon invoking the more complex turbo codec, which requires lower transmitted power for attaining the same performance as the standard convolutional codecs.

Our future work will focus on extending the DVB satellite system to support

Mod.	Equalizer	Code	E_b/N_0
QPSK	PSP	R=1/2	5.6
QPSK	B-G	Turbo (1/2)	5.7
16QAM	B-G	Turbo (1/2)	10.7
QPSK	B-G	Conv (1/2)	9.2
16QAM	B-G	Conv (1/2)	15.0
QPSK	B-G	Conv (3/4)	12.0
16QAM	B-G	Conv (3/4)	16.8
QPSK	B-G	Conv (7/8)	11.7
16QAM	MCMA	Conv (7/8)	26.0

Table 21.33: Summary of System Performance Results over the Dispersive Two-symbol Delay Two-path AWGN Channel of Figure 21.108 Tolerating an Average BER of 10^{-4}, which Was Evaluated after Viterbi or Turbo Decoding but before RS Decoding

Punctured Rate	4QAM Baud Rate (MBd)	16QAM Baud Rate (MBd)
1/2	2.73	1.37
3/4	1.82	0.909
7/8	1.56	0.779

Table 21.34: The Channel Bit Rate for the Three Different Punctured Coding Rates and for the Two Modulations

mobile users for the reception of satellite broadcast signals. The use of blind turbo equalizers will also be investigated in comparison to conventional blind equalizers. Further work will also be dedicated to trellis-coded modulation (TCM) and turbo trellis-coded modulation (TTCM)-based orthogonal frequency division multiplexed (OFDM) and single-carrier equalized modems. The impact on the system performance by employing various types of turbo interleavers and turbo codes is also of interest. A range of further wireless video communications issues are addressed in [598].

21.8 Summary and Conclusions

In this concluding chapter, we attempted to provide the reader with a system-oriented overview in the context of various adaptive and/or iterative transceivers. In all the systems investigated we have employed the H.263 video codec, but the channel coding and transmission schemes were different. The channel codecs spanned from conventional low-complexity binary BCH and nonbinary RS codes to complex iteratively decoded turbo codecs.

In Section 21.1, we investigated the feasibility of real-time wireless videotelephony over GSM/GPRS links and quantified the achievable performance advantages with

21.8. SUMMARY AND CONCLUSIONS

the advent of turbo equalization invoking joint channel decoding and channel equalization. Interactive videotelephony is feasible over GSM/GPRS even without turbo equalization and — as expected — the attainable video performance improves when the number of GPRS time-slots is increased. Apart from the higher associated bit rate the length of the interleaver can be increased, which is also beneficial in overall error resilience terms.

In Section 21.2 our investigations were based on a burst-by-burst adaptive QAM-based wideband videophone system, which was capable of maintaining the required target transmission frame error ratio associated with unimpaired video quality perception. The channel-quality dependent bit-rate range spanned a factor of six, invoking always the highest possible throughput modulation mode, while maintaining the required target frame error ratio. This regime was capable of providing a superior subjective video quality in comparison to conventional fixed-mode transceivers. The more complex and more powerful turbo codec guaranteed a typically higher throughput than the lower-complexity binary BCH channel codec employed. These burst-by-burst adaptive AQAM systems can be also employed in conjunction with direct-sequence spreading, as it was demonstrated in Section 21.3. Explicitly, in Section 21.3 multi-user detection assisted burst-by-burst adaptive AQAM/CDMA systems were studied, which attained a near-single-user performance.

In recent years OFDM has received a tremendous attention and has been standardised for employment in a whole host of applications. Specifically, the IEEE 802.11 and the High-Performance Local Area Network standard known as HIPERLAN II employ OFDM in predominantly indoor LAN environments. Similarly, both the Pan-European terrestrial Digital Video Broadcast (DVB) system and the Digital Audio Broadcast (DAB) systems opted for advocating OFDM. Motivated by these trends, in Section 21.4 we quantified the expected video performance of a LAN-type and a personal communications type high-rate, high-quality interactive OFDM video system over strongly dispersive channels.

The concept of symbol-by-symbol adaptive OFDM systems was introduced in Section 21.5, which converted the lessons of the single-carrier burst-by-burst adaptive modems to OFDM modems. It was found that in indoor scenarios — where low vehicular/pedestrian speeds prevail and hence the channel quality does not change dramatically between the instant of channel quality estimation and the instant of invoking this knowledge at the remote transmitter — symbol-by-symbol adaptive OFDM outperforms fixed-mode OFDM. More specifically, a high-quality group of subcarriers is capable of conveying a higher number of bits per subcarrier at a given target transmission frame error ratio than the low-quality subcarrier groups. Furthermore, it was found beneficial to control the number of video bits per OFDM symbol as a function of time, since in drastically faded instances the video bit rate could be reduced and hence a less heavily loaded OFDM symbol yielded fewer transmission errors. An interesting trade-off was, however, that upon loading the OFDM symbols with more bits allowed us to use a higher turbo coding interleaver and channel interleaver, which improved the system's robustness up to a limit, resulting in an optimum OFDM symbol loading.

The last two sections of this chapter considered turbo-coded terrestrial and satellite based DVB broadcast systems, respectively. Specifically, Section 21.6 employed an

OFDM-based system, while Section 21.7 used blind-equalized single-carrier modems. An interesting conclusion of our investigations was that upon invoking convolutional turbo coding — rather than the standard convolutional codecs — the system's robustness improves quite dramatically. Hence the required SNR is reduced sufficiently, in order to be able to support, for example, 4 bit/symbol transmissions, instead of 2 bit/symbol signalling. This then potentially allows us to double the associated video bit-rate within a given bandwidth at the cost of an acceptable implementational complexity increase when replacing the standard convolutional codecs by turbo channel codecs.

21.9 Wireless Video System Design Principles

We conclude that 2G and 3G interactive cordless as well as cellular wireless videophony using the H.263 ITU codec is realistic in a packet acknowledgment assisted TDMA/TDD, CDMA/TDD, or OFDM/TDD system both with and without power control following the system design guidelines summarised below:

- In time-variant wireless environments adaptive multimode transceivers are beneficial, in particular, if the geographic distribution of the wireless channel capacity — i.e., the instantaneous channel quality — varies across the cell due to path loss, slow-fading, fast-fading, dispersion, co-channel interference, etc.

- In order to support the operation of intelligent multimode transceivers a video source codec is required that is capable of conveniently exploiting the variable-rate, time-variant capacity of the wireless channel by generating always the required number of bits that can be conveyed in the current modem mode by the transmission burst. This issue was discussed throughout this chapter.

- The H.263 video codec's local and remote video reconstruction frame buffers have to be always identically updated, which requires a reliable, strongly protected acknowledgment feedback flag. The delay of this flag has to be kept as low as possible, and a convenient mechanism for its transmission is constituted by its superposition on the reverse-direction packets in TDD systems. Alternatively, the system's control channel can be used for conveying both the video packet acknowledgment flag and the transmission mode requested by the receiver. The emerging 3G CDMA systems provide a sufficiently high bitrate to convey the necessary side information. The associated timing issues were summarized in Figure 19.21.

- The acknowledgment feedback flag can be conveniently error protected by simple majority logic decoded repetition codes, which typically provide more reliable protection than more complex similar-length error correction codes.

- For multimode operation an appropriate adaptive packetization algorithm is required that controls the transmission, storage, discarding, and updating of video packets.

- For multimode operation in cellular systems an appropriate fixed FER power-control algorithm is required, that is capable of maintaining a constant transmission burst error rate, irrespective of the modulation mode used, since it is the FER, rather than the bit error rate, which determines the associated video quality.

- We emphasize again that wireless videotelephony over 2G systems is realistic using both proprietary and standard video codecs, and the same design principles can also be applied in the context of the emerging 3G CDMA systems, as it was demonstrated in the context of our JD-ACDMA design example.

- The feasibility of using similar design principles has been shown also for high-rate WATM OFDM modems in this closing chapter.

Future video transceiver performance improvements are possible, while retaining the basic system design principles itemized above. The image compression community completed the MPEG4 video standard [662–664], while the wireless communications community has commenced research toward the fourth generation of mobile radio standards. All in all, this is an exciting era for wireless video communications researchers, bringing about new standards to serve a forthcoming wireless multimedia age.

Glossary

16CIF	Sixteen Common Intermediate Format Frames are sixteen times as big as CIF frames and contain 1408 pixels vertically and 1152 pixels horizontally
2G	Second generation
3G	Third generation
3GPP	Third Generation Partnership Project
4CIF	Four Common Intermediate Format Frames are four times as big as CIF frames and contain 704 pixels vertically and 576 pixels horizontally
AB	Access burst
ACCH	Associated control channel
ACELP	Algebraic Code Excited Linear Predictive (ACELP) Speech Codec
ACF	autocorrelation function
ACL	Autocorrelation
ACO	Augmented Channel Occupancy matrix, which contains the channel occupancy for the local and surrounding base stations. Often used by locally distributed DCA algorithms to aid allocation decisions.
ACTS	Advanced Communications Technologies and Services. The fourth framework for European research (1994–98). A series of consortia consisting of universities and industrialists considering future communications systems.
ADC	Analog–to–Digital Converter
ADPCM	Adaptive Differential Pulse Coded Modulation
AGCH	Access grant control channel
AI	Acquisition Indicator
AICH	Acquisition Indicator CHannel
ANSI	American National Standards Institute
ARIB	Association of Radio Industries and Businesses
ARQ	Automatic Repeat Request, Automatic request for retransmission of corrupted data

ATDMA	Advanced Time Division Multiple Access
ATM	Asynchronous Transfer Mode
AUC	Authentication center
AV.26M	A draft recommendation for transmitting compressed video over error-prone channels, based on the H.263 [192] video codec
AWGN	Additive White Gaussian Noise
B-ISDN	Broadband ISDN
BbB	Burst-by-Burst
BCCH	Broadcast control channel
BCH	Bose-Chaudhuri-Hocquenghem, a class of forward error correcting codes (FEC)
BCH Codes	Bose-Chaudhuri-Hocquenghem (BCH) Codes
BER	Bit error rate, the fraction of the bits received incorrectly
BN	Bit number
BPSK	Binary Phase Shift Keying
BS	A common abbreviation for base station
BSIC	Base station identifier code
BTC	Block Truncation Coding
CBER	Channel bit error rate, the bit error rate before FEC correction
CBP	Coded block pattern, a H.261 video codec symbol that indicates which of the blocks in the macroblock are active
CBPB	A fixed-length codeword used by the H.263 video codec to convey the coded block pattern for bidirectionally predicted (B) blocks
CBPY	A variable-length codeword used by the H.263 video codec to indicate the coded block pattern for luminance blocks
CC	Convolutional Code
CCCH	Commom control channel
CCITT	Now ITU, standardization group
CCL	Cross-correlation
CD	Code Division, a multiplexing technique whereby signals are coded and then combined in such a way that they can be separated using the assigned user signature codes at a later stage
CDF	Cumulative density function, the integral of the probability density function (PDF)
CDMA	Code Division Multiple Access
CELL_BAR_ACCESS	Boolean flag to indicate whether the MS is permitted
CIF	Common Intermediate Format Frames containing 352 pixels vertically and 288 pixels horizontally
CIR	Carrier to Interference Ratio, same as SIR
COD	A one-bit codeword used by the H.263 video codec that indicates whether the current macroblock is empty or nonempty
CPICH	Common PIlot CHannel

GLOSSARY

CT2	British Second Generation Cordless Phone Standard
CWTS	China Wireless Telecommunication Standard
DAB	Digital Audio Broadcasting
DAC	Digital-to-Analog Convertor
DAMPS	Pan-American Digital Advanced Phone System, IS-54
DB	Dummy burst
DC	Direct Current, normally used in electronic circuits to describe a power source that has a constant voltage, as opposed to AC power in which the voltage is a sine wave. It is also used to describe things that are constant, and hence have no frequency component.
DCA	Dynamic Channel Allocation
DCH	Dedicated transport CHannel
DCS1800	A digital mobile radio system standard, based on GSM but operating at 1.8 GHz at a lower power
DCT	A discrete cosine transform that transforms data into the frequency domain. Commonly used for video compression by removing high-frequency components of the video frames
DECT	A Pan-European digital cordless telephone standard
DL	Down-link
DPCCH	Dedicated Physical Control CHannel
DPCH	Dedicated Physical CHannel
DPCM	Differential Pulse Coded Modulation
DPDCH	Dedicated Physical Data CHannel
DQUANT	A fixed-length coding parameter used to differential change the current quantizer used by the H.263 video codec
DS–CDMA	Direct Sequence Code Division Multiple Access
DSMA-CD	Digital Sense Multiple Access-Collision Detection
DTTB	Digital Terrestrial Television Broadcast
DVB-T	Terrestrial Pan-European Digital Video Broadcast Standard
EIR	Equipment identity register
EMC	Electromagnetic Compatibility
EOB	An end-of-block variable-length symbol used to indicate the end of the current block in the H.261 video codec
EREC	Error-Resilient Entropy Coding. A coding technique that improves the robustness of variable-length coding by allowing easier resynchronization after errors
ERPC	Error-Resilient Position Coding, a relative of the coding scheme known as Error-Resilient Entropy Coding (EREC)
ETSI	European Telecommunications Standards Institute
EU	European Union
FA	First Available, a simple centralized DCA scheme that allocates the first channel found that is not reused within a given preset reuse distance

FACCH	Fast associated control channel
FACH	Forward Access CHannel
FAW	Frame Alignment Word
FBER	Feedback error ratio, the ratio of feedback acknowledgment messages that are received in error
FCA	Fixed Channel Allocation
FCB	Frequency correction burst
FCCH	Frequency Correction Channel
FD	Frequency Division, a multiplexing technique whereby different frequencies are used for each communications link
FDD	Frequency-Division Duplex, a multiplexing technique whereby the forward and reverse links use a different carrier frequency
FDM	Frequency Division Multiplexing
FDMA	Frequency Division multiple access, a multiple access technique whereby frequency division (FD) is used to provide a set of access channels
FEC	Forward Error Correction
FEF	Frame Error Flag
FER	Frame error rate
FH	Frequency hopping
FIFO	First-In First-Out, a queuing strategy in which elements that have been in the queue longest are served first
FN	TDMA frame number
FPLMTS	Future Public Land Mobile Telecommunication System
fps	Frames per second
FRAMES	Future Radio Wideband Multiple Access System
GBSC	Group of blocks (GOB) start code, used by the H.261 and H.263 video codecs to regain synchronization, playing a similar role to PSC
GEI	Functions similar to PEI but in the GOB layer of the H.261 video codec
GFID	A fixed-length codeword used by H.263 video codec to aid correct resynchronization after an error
GMSK	Gaussian Mean Shift Keying, a modulation scheme used by the Pan-European GSM standard by virtue of its spectral compactness
GN	Group of blocks number, an index number for a GOB used by the H.261 and H.263 video codecs
GOB	Group of blocks, a term used by the H.261 and H.263 video codecs, consisting of a number of macroblocks
GOS	Grade of Service, a performance metric to describe the quality of a mobile radio network
GP	Guard Period
GPS	Global Positioning System
GQUANT	Group of blocks quantizer, a symbol used by the H.261 and H.263 video codecs to modify the quantizer used for the GOB

GSM	A Pan-European digital mobile radio standard, operating at 900MHz
GSPARE	Functions similar to PSPARE but in the GOB layer of the H.261 video codec
H.261	A video coding standard [500], published by the ITU in 1990
H.263	A video coding standard [192], published by the ITU in 1996
HC	Huffman Coding
HCA	Hybrid Channel Allocation, a hybrid of FCA and DCA
HCS	Hierarchical Cell Structure
HDTV	High-Definition Television
HLR	Home location register
HO	Handover
HTA	Highest interference below Threshold Algorithm, a distributed DCA algorithm also known as MTA. The algorithm allocates the most interfered channel, whose interference is below the maximum tolerable interference threshold.
IF	Intermediate Frequency
IMSI	International mobile subscriber identity
IMT-2000	International Mobile Telecommunications-2000
IMT2000	Intelligent Mobile Telecommunications in the Year 2000, Japanese Initiative for 3rd Generation Cellular Systems
IS-54	Pan-American Digital Advanced Phone System, IS-54
IS-95	North American mobile radio standard, that uses CDMA technology
ISDN	Integrated Services Digital Network, digital replacement of the analogue telephone network
ISI	Intersymbol Interference, Inter-Subcarrier Interference
ITU	International Telecommunications Union, formerly the CCITT, standardization group
ITU-R	International Mobile Telecommunication Union – Radiocommunication Sector
JDC	Japanese Digital Cellular Standard
JPEG	"Lossy" DCT-based Still Picture Coding Standard
LFA	Lowest Frequency below threshold Algorithm, a distributed DCA algorithm that is a derivative of the LTA algorithm, the difference being that the algorithm attempts to reduce the number of carrier frequencies being used concurrently
LIA	Least Interference Algorithm, a distributed DCA algorithm that assigns the channel with the lowest measured interference that is available.
LODA	Locally Optimized Dynamic Assignment, a centralized DCA scheme, which bases its allocation decisions on the future blocking probability in the vicinity of the cell
LOLIA	Locally Optimized Least Interference Algorithm, a locally distributed DCA algorithm that allocates channels using a hybrid of the LIA and an ACO matrix

LOMIA	Locally Optimized Most Interference Algorithm, a locally distributed DCA algorithm that allocates channels using a hybrid of the MTA and an ACO matrix
LP filtering	Low-pass filtering
LP-DDCA	Local Packing Dynamic Distributed Channel Assignment, a locally distributed DCA algorithm that assigns the first channel available that is not used by the surrounding base stations, whose information is contained in an ACO matrix
LPF	Low-pass filter
LSB	Least significant bit
LSR	Linear Shift Register
LTA	Least interference below Threshold Algorithm, a distributed DCA algorithm that allocates the least interfered channel, whose interference is below a preset maximum tolerable interference level
LTI	Linear Time-invariant
MA	Abbreviation for Miss America, a commonly used head and shoulders video sequence referred to as Miss America
Macroblock	A grouping of 8 by 8 pixel blocks used by the H.261 and H.263 video codecs. Consists of four luminance blocks and two chrominance blocks
MAI	Multiple Access Interference
MAP	Maximum–A–Posteriori
MB	Macroblock
MBA	Macroblock address symbol used by the H.261 video codec, indicating the position of the macroblock in the current GOB
MBS	Mobile Broadband System
MC	Motion Compensation
MCBPC	A variable-length codeword used by the H.263 video codec to convey the macroblock type and the coded block pattern for the chrominance blocks
MCER	Motion Compensated Error Residual
MDM	Modulation Division Multiplexing
MF-PRMA	Multi-Frame Packet Reservation Multiple Access
MFlop	Mega Flop, 1 million floating point operations per second
MODB	A variable-length coding parameter used by the H.263 video codec to indicate the macroblock mode for bidirectionally predicted (B) blocks
MPEG	Motion Picture Expert Group, also a video coding standard designed by this group that is widely used
MPG	Multiple Processing Gain
MQUANT	A H.261 video codec symbol that changes the quantizer used by current and future macroblocks in the current GOB
MS	A common abbreviation for Mobile Station
MSC	Mobile switching center
MSE	Mean Square Error

GLOSSARY

MSQ	Mean Square centralized DCA algorithm that attempts to minimize the mean square distance between cells using the same channel
MTA	Most interference below Threshold Algorithm, a distributed DCA algorithm also known as HTA. The algorithm allocates the most interfered channel, whose interference is below the maximum tolerable interference level.
MTYPE	H.261 video codec symbol that contains information about the macroblock, such as coding mode, and flags to indicate whether optional modes are used, like motion vectors, and loop filtering
MV	Motion vector, a vector to estimate the motion in a frame
MVD	Motion vector data symbol used by H.261 and H.263 video codecs
MVDB	A variable-length codeword used by the H.263 video codec to convey the motion vector data for bidirectionally predicted (B) blocks
NB	Normal burst
NCC	Normalized Channel Capacity
NLF	Nonlinear filtering
NMC	Network management center
NN	Nearest-Neighbor centralized DCA algorithm; allocates a channel used by the nearest cell, which is at least the reuse distance away
NN+1	Nearest-Neighbor-plus-one centralized DCA algorithm; allocates a channel used by the nearest cell, which is at least the reuse distance plus one cell radius away
OFDM	Orthogonal Frequency Division Multiplexing
OMC	Operation and maintenance center
OVSF	Orthogonal Variable Spreading Factor
P-CCPCH	Primary Common Control Physical CHannel
PCH	Paging CHannel
PCM	Pulse code modulation
PCN	Personal Communications Network
PCPCH	Physical Common Packet CHannel
PCS	Personal Communications System, a term used to describe third-generation mobile radio systems in North America
PDF	Probability Density Function
PDSCH	Physical Down-link Shared CHannel
PEI	Picture layer extra insertion bit, used by the H.261 video codec, indicating that extra information is to be expected
PFU	Partial Forced Update
PGZ	Peterson-Gorenstein-Zierler (PGZ) Decoder
PHP	Japanese Personal Handyphone Phone System
PI	Page Indicator
PICH	Page Indicator CHannel
PLMN	Public land mobile network

PLMN_PERMITTED	Boolean flag to indicate, whether the MS is permitted
PLMR	Public Land Mobile Radio
PLR	Packet-Loss Ratio
PP	Partnership Project
PQUANT	A fixed-length codeword used by the H.263 video codec to indicate the quantizer to use for the next frame
PRACH	Physical Random Access CHannel
PRMA	Packet Reservation Multiple Access, a statistical multiplexing arrangement contrived to improve the efficiency of conventional TDMA systems, by detecting inactive speech segments using a voice activity detector, surrendering them and allocating them to subscribers contending to transmit an active speech packet
PRMA++	PRMA System allowing contention only in the so-called contention slots, which protect the information slots from contention and collisions
PSAM	Pilot symbol-assisted modulation, a technique whereby known symbols (pilots) are transmitted regularly. The effect of channel fading on all symbols can then be estimated by interpolating between the pilots.
PSC	Picture start code, a preset sequence used by the H.261 and H.263 video codecs, that can be searched for to regain synchronization after an error
PSD	Power Spectral Density
PSNR	Peak Signal to Noise Ratio, noise energy compared to the maximum possible signal energy. Commonly used to measure video image quality
PSPARE	Picture layer extra information bits, indicated by a PEI symbol in H.261 video codec
PSTN	Public switched telephone network
PTYPE	Picture layer information, used by H.261 and H.263 video codecs to transmit information about the picture, e.g. Resolution, etc.
QAM	Quadrature Amplitude Modulation
QCIF	Quarter Common Intermediate Format Frames containing 176 pixels vertically and 144 pixels horizontally
QMF	Quadrature Mirror Filtering
QN	Quater bit number
QoS	Quality of Service
QT	Quad-Tree
RACE	Research in Advanced Communications Equipment Programme in Europe, from June 1987 to December 1995
RACH	Random Access CHannel
RC filtering	Raised-cosine filtering
RF	Radio frequency
RFCH	Radio frequency channel
RFN	Reduced TDMA frame number in GSM

RING	A centralized DCA algorithm that attempts to allocate channels in one of the cells, which is at least the reuse distance away that forms a "ring" of cells
RLC	Run-Length Coding
RPE	Regular pulse excited
RS Codes	Reed-Solomon (RS) codes
RSSI	Received Signal Strength Indicator, commonly used as an indicator of channel quality in a mobile radio network
RTT	Radio Transmission Technology
RXLEV	Received signal level: parameter used in hangovers
RXQUAL	Received signal quality: parameter used in hangovers
S-CCPCH	Secondary Common Control Physical CHannel
SAC	Syntax-based arithmetic coding, an alternative to variable-length coding, and a variant of arithmetic coding
SACCH	Slow associated control channel
SB	Synchronization burst
SCH	Synchronization CHannel
SCS	Sequential Channel Search distributed DCA algorithm that searches the available channels in a predetermined order, picking the first channel found, which meets the interference constraints
SDCCH	Stand-alone dedicated control channel
SF	Spreading Factor
SINR	Signal-to-Interference plus Noise ratio, same as signal-to-noise ratio (SNR) when there is no interference.
SIR	Signal-to-Interference ratio
SNR	Signal-to-Noise Ratio, noise energy compared to the signal energy
SPAMA	Statistical Packet Assignment Multiple Access
SQCIF	Sub-Quarter Common Intermediate Format Frames containing 128 pixels vertically and 96 pixels horizontally
SSC	Secondary Synchronization Codes
TA	Timing advance
TB	Tailing bits
TC	Trellis Coded
TCH	Traffic channel
TCH/F	Full-rate traffic channel
TCH/F2.4	Full-rate 2.4 kbps data traffic channel
TCH/F4.8	Full-rate 4.8 kbps data traffic channel
TCH/F9.6	Full-rate 9.6 kbps data traffic channel
TCH/FS	Full-rate speech traffic channel
TCH/H	Half-rate traffic channel
TCH/H2.4	Half-rate 2.4 kbps data traffic channel

TCH/H4.8	Half-rate 4.8 kbps data traffic channel
TCM	Trellis code modulation
TCOEFF	An H.261 and H.263 video codec symbol that contains the transform coefficients for the current block
TD	Time Division, a multiplexing technique whereby several communications links are multiplexed onto a single carrier by dividing the channel into time periods, and assigning a time period to each communications link
TDD	Time-Division Duplex, a technique whereby the forward and reverse links are multiplexed in time.
TDMA	Time Division Multiple Access
TFCI	Transport-Format Combination Indicator
TIA	Telecommunications Industry Association
TN	Time slot number
TPC	Transmit Power Control
TR	Temporal reference, a symbol used by H.261 and H.263 video codecs to indicate the real-time difference between transmitted frames
TS	Technical Specifications
TTA	Telecommunications Technology Association
TTC	Telecommunication Technology Committee
TTIB	Transparent tone in band
UHF	Ultra high frequency
UL	Up-link
UMTS	Universal Mobile Telecommunications System, a future Pan-European third-generation mobile radio standard
UTRA	Universal Mobile Telecommunications System Terrestrial Radio Access
VA	Viterbi Algorithm
VAD	Voice activity detection
VAF	Voice activity factor, the fraction of time the voice activity detector of a speech codec is active
VE	Viterbi equalizer
VL	Variable length
VLC	Variable-length coding/codes
VLR	Visiting location register
VQ	Vector Quantization
W-CDMA	Wideband Code Division Multiple Access
WARC	World Administrative Radio Conference
WATM	Wireless Asynchronous Transfer Mode (ATM)
WLAN	Wireless Local Area Network
WN	White noise
WWW	World Wide Web, the name given to computers that can be accessed via the Internet using the HTTP protocol. These computers can provide information in a easy-to-digest multimedia format using hyperlinks.

Bibliography

[1] E. Berlekamp, *Algebraic Coding Theory*. New York: McGraw-Hill, 1968.

[2] J. Massey, "Shift-register synthesis and BCH decoding," *IEEE Transactions on Information Theory*, vol. IT-15, pp. 122–127, January 1969.

[3] R. Blahut, *Theory and Practice of Error Control Codes*. Reading, MA: Addison-Wesley, 1983.

[4] A. Michelson and A. Levesque, *Error Control Techniques for Digital Communication*. New York: John Wiley and Sons, 1985.

[5] S. Lin and D. Constello Jr., *Error Control Coding: Fundamentals and Applications*. Englewood Cliffs, NJ: Prentice-Hall, October 1982.

[6] W. Peterson and E. Weldon Jr., *Error Correcting Codes*. Cambridge, MA: MIT. Press, 2nd ed., August 1972.

[7] G. Clark Jr. and J. Cain, *Error Correction Coding for Digital Communications*. New York: Plenum Press, May 1981.

[8] K. Wong, *Transmission of Channel Coded Speech and Data over Mobile Channels*. PhD thesis, University of Southampton, 1989.

[9] R. Steele and L. Hanzo, eds., *Mobile Radio Communications*. Piscataway, NJ: IEEE Press, 1999.

[10] L. Hanzo, W. Webb, and T. Keller, *Single- and Multi-carrier Quadrature Amplitude Modulation*. New York: IEEE Press-John Wiley, April 2000.

[11] J. Torrance and L. Hanzo, "Comparative study of pilot symbol assisted modem schemes," in *Proceedings of IEE Conference on Radio Receivers and Associated Systems (RRAS'95)*, (Bath), pp. 36–41, IEE, 26–28 September 1995.

[12] P. Howard and J. Vitter, "Arithmetic coding for data compression," *Proceedings of the IEEE*, vol. 82, pp. 857–865, June 1994.

[13] C. Shannon, *Mathematical Theory of Communication*. University of Illinois Press, 1963.

[14] C. Shannon, "A mathematical theory of communication — part I," *Bell Systems Technical Journal*, pp. 379–405, 1948.

[15] C. Shannon, "A mathematical theory of communication — part II," *Bell Systems Technical Journal*, pp. 405–423, 1948.

[16] C. Shannon, "A mathematical theory of communication — part III," *Bell Systems Technical Journal*, vol. 27, pp. 623–656, 1948.

[17] H. Nyquist, "Certain factors affecting telegraph speed," *Bell System Technical Journal*, p. 617, April 1928.

[18] R. Hartley, "Transmission of information," *Bell Systems Technical Journal*, p. 535, 1928.

[19] N. Abramson, *Information Theory and Coding*. New York: McGraw-Hill, 1975.

[20] A. Carlson, *Communication Systems*. New York: McGraw-Hill, 1975.

[21] H. Raemer, *Statistical Communication Theory and Applications*. Englewood Cliffs, NJ: Prentice-Hall, 1969.

[22] P. Ferenczy, *Telecommunications Theory, in Hungarian: Hirkozleselmelet*. Budapest, Hungary: Tankonyvkiado, 1972.

[23] K. Shanmugam, *Digital and Analog Communications Systems*. New York: John Wiley and Sons, 1979.

[24] C. Shannon, "Communication in the presence of noise," *Proceedings of the I.R.E.*, vol. 37, pp. 10–22, January 1949.

[25] C. Shannon, "Probability of error for optimal codes in a Gaussian channel," *Bell Systems Technical Journal*, vol. 38, pp. 611–656, 1959.

[26] A. Jain, *Fundamentals of Digital Image Processing*. Englewood Cliffs, NJ: Prentice-Hall, 1989.

[27] A. Hey and R. Allen, eds., *R.P. Feynman: Feynman Lectures on Computation*. Reading, MA: Addison-Wesley, 1996.

[28] C. Berrou, A. Glavieux, and P. Thitimajshima, "Near shannon limit error-correcting coding and decoding: Turbo codes," in *Proceedings of the International Conference on Communications*, (Geneva, Switzerland), pp. 1064–1070, May 1993.

[29] L. Hanzo, W. Webb, and T. Keller, *Single- and Multi-Carrier Quadrature Amplitude Modulation: Principles and Applications for Personal Communications, WLANs and Broadcasting*. IEEE Press, 2000.

[30] J. Hagenauer, "Quellengesteuerte kanalcodierung fuer sprach- und tonuebertragung im mobilfunk," *Aachener Kolloquium Signaltheorie*, pp. 67–76, 23–25 March 1994.

[31] A. Viterbi, "Wireless digital communications: A view based on three lessons learned," *IEEE Communications Magazine*, pp. 33–36, September 1991.

[32] D. Greenwood and L. Hanzo, "Characterisation of mobile radio channels," in Steele and Hanzo [9], ch. 2, pp. 92–185.

[33] L. Hanzo and J. Woodard, "An intelligent multimode voice communications system for indoor communications," *IEEE Transactions on Vehicular Technology*, vol. 44, pp. 735–748, November 1995.

[34] L. Hanzo, R. Salami, R. Steele, and P. Fortune, "Transmission of digitally encoded speech at 1.2 Kbaud for PCN," *IEE Proceedings, Part I*, vol. 139, pp. 437–447, August 1992.

[35] K. Wong and L. Hanzo, "Channel coding," in Steele and Hanzo [9], ch. 4, pp. 347–488.

[36] W. Jakes, ed., *Microwave Mobile Communications*. New York: John Wiley and Sons, 1974.

[37] W. Lee, *Mobile Cellular Communications*. New York: McGraw-Hill, 1989.

[38] R. Steele, "Towards a high capacity digital cellular mobile radio system," *Proceedings of the IEE*, vol. 132, Part F, pp. 405–415, August 1985.

[39] R. Steele and V. Prabhu, "High-user density digital cellular mobile radio system," *IEE Proceedings*, vol. 132, Part F, pp. 396–404, August 1985.

[40] R. Steele, "The cellular environment of lightweight hand-held portables," *IEEE Communications Magazine*, pp. 20–29, July 1989.

[41] L. Hanzo and J. Stefanov, "The Pan-European Digital Cellular Mobile Radio System — known as GSM," in Steele and Hanzo [9], ch. 8, pp. 677–765.

[42] P. Mermelstein, "The IS-54 digital cellular standard," in Gibson [292], ch. 26, pp. 419–429.

[43] K. Kinoshita and M. Nakagawa, "Japanese cellular standard," in Gibson [292], ch. 28, pp. 449–461.

[44] L. Hanzo, "The British cordless telephone system: CT2," in Gibson [292], ch. 29, pp. 462–477.

[45] S. Asghar, "Digital European Cordless Telephone," in Gibson [292], ch. 30, pp. 478–499.

[46] J. Parsons and J. Gardiner, *Mobile Communication Systems*. London: Blackie, 1989.

[47] D. Parsons, *The Mobile Radio Propagation Channel*. London: Pentech Press, 1992.

[48] I. Wassell and R. Steele, "Wideband systems," in Steele and Hanzo [9], ch. 6, pp. 523–600.

[49] R. Edwards and J. Durkin, "Computer prediction of service area for VHF mobile radio networks," *Proc of IRE*, vol. 116, no. 9, pp. 1493–1500, 1969.

[50] M. Hata, "Empirical formula for propagation loss in land mobile radio," *IEEE Transactions on Vehicular Technology*, vol. 29, pp. 317–325, August 1980.

[51] Y. Okumura, E. Ohmori, T. Kawano, and K. Fukuda, "Field strength and its variability in VHF and UHF land mobile service," *Review of the Electrical Communication Laboratory*, vol. 16, pp. 825–873, September–October 1968.

[52] J. Proakis, *Digital Communications*. New York: McGraw-Hill, 3rd ed., 1995.

[53] M. Gans, "A power-spectral theory of propogation in the mobile-radio environment," *IEEE Transactions on Vehicular Technology*, vol. 21, pp. 27–38, February 1972.

[54] W. Press, S. Teukolsky, W. Vetterling, and B. Flannery, "Random numbers," in *Numerical Recipes in C* [148], ch. 7, pp. 289–290.

[55] R. Hamming, "Error detecting and error correcting codes," *Bell System Technical Journal*, vol. 29, pp. 147–160, 1950.

[56] P. Elias, "Coding for noisy channels," *IRE Conv. Rec. pt.4*, pp. 37–47, 1955.

[57] J. Wozencraft, "Sequential decoding for reliable communication," *IRE Natl. Conv. Rec.*, vol. 5, pt.2, pp. 11–25, 1957.

[58] J. Wozencraft and B. Reiffen, *Sequential Decoding*. MIT Press, 1961.

[59] R. Fano, "A heuristic discussion of probabilistic coding," *IEEE Transactions on Information Theory*, vol. IT-9, pp. 64–74, April 1963.

[60] J. Massey, *Threshold Decoding*. Cambridge, MA: MIT Press, 1963.

[61] A. Viterbi, "Error bounds for convolutional codes and an asymphotically optimum decoding algorithm," *IEEE Transactions on Information Theory*, vol. IT-13, pp. 260–269, April 1967.

[62] G. Forney, "The Viterbi algorithm," *Proceedings of the IEEE*, vol. 61, pp. 268–278, March 1973.

[63] J. Heller and I. Jacobs, "Viterbi decoding for satellite and space communication," *IEEE Transactions on Communication Technology*, vol. COM-19, pp. 835–848, October 1971.

[64] L. Bahl, J. Cocke, F. Jelinek, and J. Raviv, "Optimal decoding of linear codes for minimising symbol error rate," *IEEE Transactions on Information Theory*, vol. 20, pp. 284–287, March 1974.

[65] A. Hocquenghem, "Codes correcteurs d'erreurs," *Chiffres (Paris)*, vol. 2, pp. 147–156, September 1959.

[66] R. Bose and D. Ray-Chaudhuri, "On a class of error correcting binary group codes," *Information and Control*, vol. 3, pp. 68–79, March 1960.

[67] R. Bose and D. Ray-Chaudhuri, "Further results on error correcting binary group codes," *Information and Control*, vol. 3, pp. 279–290, September 1960.

[68] W. Peterson, "Encoding and error correction procedures for the Bose-Chaudhuri codes," *IRE Trans. Inform. Theory*, vol. IT-6, pp. 459–470, September 1960.

[69] D. Gorenstein and N. Zierler, "A class of cyclic linear error-correcting codes in p^m symbols," *J. Soc. Ind. Appl. Math.*, vol. 9, pp. 107–214, June 1961.

[70] I. Reed and G. Solomon, "Polynomial codes over certain finite fields," *J. Soc. Ind. Appl. Math.*, vol. 8, pp. 300–304, June 1960.

[71] E. Berlekamp, "On decoding binary Bose-Chaudhuri-Hocquenghem codes," *IEEE Transactions on Information Theory*, vol. 11, pp. 577–579, 1965.

[72] J. Massey, "Step-by-step decoding of the Bose-Chaudhuri-Hocquenghem codes," *IEEE Transactions on Information Theory*, vol. 11, pp. 580–585, 1965.

[73] Consultative Committee for Space Data Systems, *Blue Book: Recommendations for Space Data System Standards: Telemetry Channel Coding*, May 1984.

[74] W. Peterson, *Error Correcting Codes*. Cambridge, MA: MIT Press, 1st ed., 1961.

[75] V. Pless, *Introduction to the Theory of Error-correcting Codes*. New York: John Wiley and Sons, 1982.

[76] I. Blake, ed., *Algebraic Coding Theory: History and Development*. Dowden, Hutchinson and Ross, 1973.

[77] R. Lidl and H. Niederreiter, *Finite Fields*. Cambridge: Cambridge University Press, October 1996.

[78] D. Gorenstein and N. Zierler, "A class of error-correcting codes in p^m symbols," *J. Soc. Ind. Appl. Math.*, no. 9, pp. 207–214, 1961.

[79] J. Makhoul, "Linear prediction: A tutorial review," *Proceedings of the IEEE*, vol. 63, pp. 561–580, April 1975.

[80] R. Blahut, *Fast Algorithms for Digital Signal Processing*. Reading MA: Addison-Wesley, 1985.

[81] J. Schur, "Ueber potenzreihen, die im innern des einheits- kreises beschraenkt sind," *Journal fuer Mathematik*, pp. 205–232. Bd. 147, Heft 4.

[82] R. Chien, "Cyclic decoding procedure for the Bose-Chaudhuri-Hocquenghem codes," *IEEE Transactions on Info. Theory*, vol. 10, pp. 357–363, October 1964.

[83] A. Jennings, *Matrix Computation for Engineers and Scientists*. New York: John Wiley and Sons Ltd., 1977.

[84] G. Forney Jr, "On decoding BCH codes," *IEEE Transactions on Information Theory*, vol. IT-11, pp. 549–557, 1965.

[85] Y. Sugiyama, M. Kasahara, S. Hirasawa, and T. Namekawa, "A method for solving key equation for decoding goppa codes," *Information Control*, no. 27, pp. 87–99, 1975.

[86] S. Golomb, *Shift Register Sequences*. Laugana Hills, CA: Aegean Park Press, 1982.

[87] R. Steele and L. Hanzo, eds., *Mobile Radio Communications*. New York: IEEE Press-John Wiley, 2nd ed., 1999.

[88] J. Cavers, "Variable rate transmission for rayleigh fading channels," *IEEE Transactions on Communications Technology*, vol. COM-20, pp. 15–22, February 1972.

[89] W. Webb and R. Steele, "Variable rate QAM for mobile radio," *IEEE Transactions on Communications*, vol. 43, pp. 2223–2230, July 1995.

[90] L. Hanzo, "Bandwidth-efficient wireless multimedia communications," *Proceedings of the IEEE*, vol. 86, pp. 1342–1382, July 1998.

[91] S. Nanda, K. Balachandran, and S. Kumar, "Adaptation techniques in wireless packet data services," *IEEE Communications Magazine*, vol. 38, pp. 54–64, January 2000.

[92] T. Keller and L. Hanzo, "Adaptive multicarrier modulation: A convenient framework for time-frequency processing in wireless communications," *Proceedings of the IEEE*, vol. 88, pp. 611–642, May 2000.

[93] L. Hanzo, C. Wong, and P. Cherriman, "Channel-adaptive wideband video telephony," *IEEE Signal Processing Magazine*, vol. 17, pp. 10–30, July 2000.

[94] L. Hanzo, F. Somerville, and J. Woodard, "Voice compression and communications: Principles and applications for fixed and wireless channels." 2001 (For detailed contents, please refer to http://www-mobile.ecs.soton.ac.uk/).

[95] P. Cherriman, E. Kuan, and L. Hanzo, *Patent: Transceiver*. Mobile VCE, Basingstoke, 20th May 1999. Patent Application Number 9911777.2.

[96] P. Cherriman, E. Kuan, and L. Hanzo, "Turbo coded burst-by-burst adaptive joint-detection CDMA based video telephony." Submitted to IEEE Transactions on Circuits Systems and Video Technology, June 1999.

[97] L. Hanzo, P. Cherriman, and E. Kuan, "Interactive cellular and cordless video telephony: State-of-the-art, system design principles and expected performance," *Proceedings of IEEE*, September 2000.

[98] P. Cherriman, E. Kuan, and L. Hanzo, "Multi-mode joint-detection CDMA/H.263 based video telephony," in *Proceeding of ACTS Mobile Communication Summit '99*, (Sorrento, Italy), pp. 715–720, ACTS, 8–11 June 1999.

[99] E. Kuan, C. Wong, and L. Hanzo, "Burst-by-burst adaptive joint detection CDMA," in *Proceeding of VTC'99 (Spring)*, (Houston, TX), IEEE, 16–20 May 1999.

[100] G. Forney Jr, R. Gallager, G. Lang, F. Longstaff, and S. Qureshi, "Efficient modulation for band-limited channels," *IEEE Journal on Selected Areas in Communications*, vol. 2, pp. 632–647, September 1984.

[101] K. Feher, "Modems for emerging digital cellular mobile systems," *IEEE Transactions on Vehicular Technology*, vol. 40, pp. 355–365, May 1991.

[102] W. Webb, L. Hanzo, and R. Steele, "Bandwidth-efficient QAM schemes for rayleigh-fading channels," *IEE Proceedings*, vol. 138, pp. 169–175, June 1991.

[103] A. Wright and W. Durtler, "Experimental performance of an adaptive digital linearized power amplifier," *IEEETransactions on Vehicular Technology*, vol. 41, pp. 395–400, November 1992.

[104] M. Faulkner and T. Mattson, "Spectral sensitivity of power amplifiers to quadrature modulator misalignment," *IEEE Transactions on Vehicular Technology*, vol. 41, pp. 516–525, November 1992.

[105] P. Kenington, R. Wilkinson, and J. Marvill, "Broadband linear amplifier design for a PCN base-station," in *Proceedings of IEEE Vehicular Technology Conference (VTC'91)*, (St. Louis, MO), pp. 155–160, IEEE, 19–22 May 1991.

[106] R. Wilkinson et al., "Linear transmitter design for MSAT terminals," in *Proceedings of 2nd International Mobile Satellite Conference*, June 1990.

[107] S. Stapleton and F. Costescu, "An adaptive predistorter for a power amplifier based on adjacent channel emissions," *IEEE Transactions on Vehicular Technology*, vol. 41, pp. 49–57, February 1992.

[108] S. Stapleton, G. Kandola, and J. Cavers, "Simulation and analysis of an adaptive predistorter utilizing a complex spectral convolution," *IEEE Transactions on Vehicular Technology*, vol. 41, pp. 387–394, November 1992.

[109] Y. Kamio, S. Sampei, H. Sasaoka, and N. Morinaga, "Performance of modulation-level-control adaptive-modulation under limited transmission delay time for land mobile communications," in *Proceedings of IEEE Vehicular Technology Conference (VTC'95)*, (Chicago), pp. 221–225, IEEE, 15–28 July 1995.

[110] J. Torrance and L. Hanzo, "Upper bound performance of adaptive modulation in a slow Rayleigh fading channel," *Electronics Letters*, vol. 32, pp. 718–719, 11 April 1996.

[111] J. Torrance and L. Hanzo, "Optimisation of switching levels for adaptive modulation in a slow Rayleigh fading channel," *Electronics Letters*, vol. 32, pp. 1167–1169, 20 June 1996.

[112] J. Torrance and L. Hanzo, "Demodulation level selection in adaptive modulation," *Electronics Letters*, vol. 32, pp. 1751–1752, 12 September 1996.

[113] J. Torrance and L. Hanzo, "Performance upper bound of adaptive QAM in slow Rayleigh-fading environments," in *Proceedings of IEEE ICCS'96/ISPACS'96*, (Singapore), pp. 1653–1657, IEEE, 25–29 November 1996.

[114] J. Torrance and L. Hanzo, "Adaptive modulation in a slow Rayleigh fading channel," in *Proceedings of IEEE International Symposium on Personal, Indoor, and Mobile Radio Communications (PIMRC'96)*, vol. 2, (Taipei, Taiwan), pp. 497–501, IEEE, 15–18 October 1996.

[115] J. Torrance and L. Hanzo, "Latency considerations for adaptive modulation in a slow Rayleigh fading channel," in *Proceedings of IEEE VTC'97*, vol. 2, (Phoenix, AZ), pp. 1204–1209, IEEE, 4–7 May 1997.

[116] K. Feher, ed., *Digital Communications—Satellite/Earth Station Engineering*. Englewood Cliffs, NJ: Prentice-Hall, 1983.

[117] Y. Chow, A. Nix, and J. McGeehan, "Analysis of 16-APSK modulation in AWGN and rayleigh fading channel," *Electronic Letters*, vol. 28, pp. 1608–1610, November 1992.

[118] B. Sklar, *Digital Communications—Fundamentals and Applications*. Englewood Cliffs, NJ: Prentice-Hall, 1988.

[119] J. Torrance, *Adaptive Full Response Digital Modulation for Wireless Communications Systems*. PhD thesis, Department of Electronics and Computer Science, University of Southampton, 1997.

[120] J. Cavers, "An analysis of pilot symbol assisted modulation for rayleigh fading channels," *IEEE Transactions on Vehicular Technology*, vol. 40, pp. 686–693, November 1991.

[121] F. Adachi, "Error rate analysis of differentially encoded and detected 16APSK under rician fading," *IEEE Transactions on Vehicular Technology*, vol. 45, pp. 1–12, February 1996.

[122] J. McGeehan and A. Bateman, "Phase-locked transparent tone in band (TTIB): A new spectrum configuration particularly suited to the transmission of data over SSB mobile radio networks," *IEEE Transactions on Communications*, vol. COM-32, no. 1, pp. 81–87, 1984.

[123] A. Bateman and J. McGeehan, "Feedforward transparent tone in band for rapid fading protection in multipath fading," in *IEE International Conference on Communications*, vol. 68, pp. 9–13, 1986.

[124] A. Bateman and J. McGeehan, "The use of transparent tone in band for coherent data schemes," in *IEEE International Conference on Communications*, (Boston, MA), 1983.

[125] A. Bateman, G. Lightfoot, A. Lymer, and J. McGeehan, "Speech and data transmissions over a 942MHz TAB and TTIB single sideband mobile radio system," *IEEE Transactions on Vehicular Technology*, vol. VT-34, pp. 13–21, February 1985.

[126] A. Bateman and J. McGeehan, "Data transmissions over UHF fading mobile radio channels," *Proceedings of IEE*, vol. 131, no. Pt.F, pp. 364–374, 1984.

[127] J. McGeehan and A. Bateman, "A simple simultaneous carrier and bit synchronisation system for narrowband data transmissions," *Proceedings of IEE*, vol. 132, no. Pt.F, pp. 69–72, 1985.

[128] J. McGeehan and A. Bateman, "Theoretical and experimental investigation of feedforward signal regeneration," *IEEE Transactions on Vehicular Technology*, vol. VT-32, pp. 106–120, 1983.

[129] A. Bateman, "Feedforward transparent tone in band: Its implementation and applications," *IEEE Transactions on Vehicular Technology*, vol. 39, pp. 235–243, August 1990.

[130] M. Moher and J. Lodge, "TCMP—a modulation and coding strategy for rician fading channels," *IEEE Journal on Selected Areas in Communications*, vol. 7, pp. 1347–1355, December 1989.

[131] S. Sampei and T. Sunaga, "Rayleigh fading compensation method for 16-QAM in digital land mobile radio channels," in *Proceedings of IEEE Vehicular Technology Conference (VTC'89)*, (San Francisco, CA), pp. 640–646, IEEE, 1–3 May 1989.

[132] S. Haykin, *Adaptive Filter Theory*. Englewood Cliffs, NJ: Prentice-Hall, 1996.

[133] J. Cavers, "The performance of phase locked transparent tone in band with symmetric phase detection," *IEEE Transactions on Communications*, vol. 39, pp. 1389–1399, September 1991.

[134] R. Steele and W. Webb, "Variable rate QAM for data transmission over Rayleigh fading channels," in *Proceeedings of Wireless '91*, (Calgary, Alberta), pp. 1–14, IEEE, 1991.

[135] K. Arimochi, S. Sampei, and N. Morinaga, "Adaptive modulation system with discrete power control and predistortion-type non-linear compensation for high spectral efficient and high power efficient wireless communication systems," in *Proceedings of IEEE International Symposium on Personal, Indoor and Mobile Radio Communications, PIMRC'97*, (Marina Congress Centre, Helsinki, Finland), pp. 472–477, IEEE, 1–4 September 1997.

[136] M. Naijoh, S. Sampei, N. Morinaga, and Y. Kamio, "ARQ schemes with adaptive modulation/TDMA/TDD systems for wireless multimedia communication systems," in *Proceedings of IEEE International Symposium on Personal, Indoor and Mobile Radio Communications, PIMRC'97*, (Marina Congress Centre, Helsinki, Finland), pp. 709–713, IEEE, 1–4 September 1997.

[137] S.-G. Chua and A. Goldsmith, "Variable-rate variable-power mQAM for fading channels," in *Proceedings of IEEE VTC'96*, (Atlanta, GA), pp. 815–819, IEEE, 28 April–1 May 1996.

[138] A. Goldsmith and S. Chua, "Variable-rate variable-power MQAM for fading channels," *IEEE Transactions on Communications*, vol. 45, pp. 1218–1230, October 1997.

[139] A. Goldsmith, "The capacity of downlink fading channels with variable rate and power," *IEEE Transactions on Vehicular Technolgy*, vol. 46, pp. 569–580, August. 1997.

[140] A. Goldsmith and P. Varaiya, "Capacity of fading channels with channel side information," *IEEE Transactions on Information Theory*, vol. 43, pp. 1986–1992, November 1997.

[141] M.-S. Alouini and A. Goldsmith, "Capacity of rayleigh fading channels under different adaptive transmission and diversity-combining techniques," *to appear in IEEE Transactions on Vehicular Technology*, 1999. http://www.systems.caltech.edu.

[142] M.-S. Alouini and A. Goldsmith, "Area spectral efficiency of cellular mobile radio systems," *to appear IEEE Transactions on Vehicular Technology*, 1999. http://www.systems.caltech.edu.

[143] A. Goldsmith and S. Chua, "Adaptive coded modulation for fading channels," *IEEE Transactions on Communications*, vol. 46, pp. 595–602, May 1998.

[144] D. Pearce, A. Burr, and T. Tozer, "Comparison of counter-measures against slow Rayleigh fading for TDMA systems," in *IEE Colloquium on Advanced TDMA Techniques and Applications*, (London), pp. 9/1–9/6, IEE, 28 October 1996. digest 1996/234.

[145] W. Lee, "Estimate of channel capacity in Rayleigh fading environment," *IEEE Transactions on Vehicular Technology*, vol. 39, pp. 187–189, August 1990.

[146] N. Morinaga, "Advanced wireless communication technologies for achieving high-speed mobile radios," *IEICE Transactions on Communications*, vol. 78, no. 8, pp. 1089–1094, 1995.

[147] J. Woodard and L. Hanzo, "A low delay multimode speech terminal," in *Proceedings of IEEE VTC'96*, vol. 1, (Atlanta, GA), pp. 213–217, IEEE, 28 April–1 May 1996.

[148] W. Press, S. Teukolsky, W. Vetterling, and B. Flannery, *Numerical Recipes in C*. Cambridge: Cambridge University Press, 1992.

[149] J. Torrance and L. Hanzo, "Statistical multiplexing for mitigating latency in adaptive modems," in *Proceedings of IEEE International Symposium on Personal, Indoor and Mobile Radio Communications, PIMRC'97*, (Marina Congress Centre, Helsinki, Finland), pp. 938–942, IEEE, 1–4 September 1997.

[150] J. Torrance, L. Hanzo, and T. Keller, "Interference resilience of burst-by-burst adaptive modems," in *Proceeding of ACTS Mobile Communication Summit '97*, (Aalborg, Denmark), pp. 489–494, ACTS, 7–10 October 1997.

[151] C. Wong, T. Liew, and L. Hanzo, "Turbo coded burst by burst adaptive wideband modulation with blind modem mode detection," in *Proceeding of ACTS Mobile Communication Summit '99*, (Sorrento, Italy), pp. 303–308, ACTS, 8–11 June 1999.

[152] C. Wong and L. Hanzo, "Upper-bound of a wideband burst-by-burst adaptive modem," in *Proceeding of VTC'99 (Spring)*, (Houston, TX), pp. 1851–1855, IEEE, 16–20 May 1999.

[153] K. Narayanan and L. Cimini, "Equalizer adaptation algorithms for high speed wireless communications," in *Proceedings of IEEE VTC'96*, (Atlanta, GA), pp. 681–685, IEEE, 28 April–1 May 1996.

[154] J. Wu and A. Aghvami, "A new adaptive equalizer with channel estimator for mobile radio communications," *IEEE Transactions on Vehicular Technology*, vol. 45, pp. 467–474, August 1996.

[155] Y. Gu and T. Le-Ngoc, "Adaptive combined DFE/MLSE techniques for ISI channels," *IEEE Transactions on Communications*, vol. 44, pp. 847–857, July 1996.

[156] A. Clark and R. Harun, "Assessment of kalman-filter channel estimators for an HF radio link," *IEE Proceedings*, vol. 133, pp. 513–521, October 1986.

[157] R. Chang, "Synthesis of band-limited orthogonal signals for multichannel data transmission," *Bell Systems Technical Journal*, vol. 46, pp. 1775–1796, December 1966.

[158] M. Zimmermann and A. Kirsch, "The AN/GSC-10/KATHRYN/variable rate data modem for HF radio," *IEEE Transactions on Communication Technology*, vol. CCM–15, pp. 197–205, April 1967.

[159] E. Powers and M. Zimmermann, "A digital implementation of a multichannel data modem," in *Proceedings of the IEEE International Conference on Communications*, (Philadelphia), 1968.

[160] B. Saltzberg, "Performance of an efficient parallel data transmission system," *IEEE Transactions on Communication Technology*, pp. 805–813, December 1967.

[161] R. Chang and R. Gibby, "A theoretical study of performance of an orthogonal multiplexing data transmission scheme," *IEEE Transactions on Communication Technology*, vol. COM–16, pp. 529–540, August 1968.

[162] S. Weinstein and P. Ebert, "Data transmission by frequency division multiplexing using the discrete fourier transform," *IEEE Transactions on Communication Technology*, vol. COM–19, pp. 628–634, October 1971.

[163] A. Peled and A. Ruiz, "Frequency domain data transmission using reduced computational complexity algorithms," in *Proceedings of International Conference on Acoustics, Speech, and Signal Processing, ICASSP'80*, vol. 3, (Denver, CO), pp. 964–967, IEEE, 9–11 April 1980.

[164] B. Hirosaki, "An orthogonally multiplexed QAM system using the discrete fourier transform," *IEEE Transactions on Communications*, vol. COM-29, pp. 983–989, July 1981.

[165] L. Cimini, "Analysis and simulation of a digital mobile channel using orthogonal frequency division multiplexing," *IEEE Transactions on Communications*, vol. 33, pp. 665–675, July 1985.

[166] K. Kammeyer, U. Tuisel, H. Schulze, and H. Bochmann, "Digital multicarrier transmission of audio signals over mobile radio channels," *European Transactions on Telecommunications*, vol. 3, pp. 243–253, May–June 1992.

[167] F. Mueller-Roemer, "Directions in audio broadcasting," *Journal Audio Engineering Society*, vol. 41, pp. 158–173, March 1993.

[168] G. Plenge, "DAB — a new radio broadcasting system — state of development and ways for its introduction," *Rundfunktech. Mitt.*, vol. 35, no. 2, 1991.

[169] M. Alard and R. Lassalle, "Principles of modulation and channel coding for digital broadcasting for mobile receivers," *EBU Review, Technical No. 224*, pp. 47–69, August 1987.

[170] *Proceedings of 1st International Symposium, DAB*, (Montreux, Switzerland), June 1992.

[171] I. Kalet, "The multitone channel," *IEEE Transactions on Communications*, vol. 37, pp. 119–124, February 1989.

[172] H. Kolb, "Untersuchungen über ein digitales mehrfrequenzverfahren zur datenübertragung," in *Ausgewählte Arbeiten über Nachrichtensysteme*, no. 50, Universität Erlangen-Nürnberg, 1982.

[173] H. Schüssler, "Ein digitales Mehrfrequenzverfahren zur Datenübertragung," in *Professoren-Konferenz, Stand und Entwicklungsaussichten der Daten und Telekommunikation*, (Darmstadt, Germany), pp. 179–196, 1983.

[174] K. Preuss, "Ein Parallelverfahren zur schnellen Datenübertragung Im Ortsnetz," in *Ausgewählte Arbeiten über Nachrichtensysteme*, no. 56, Universität Erlangen-Nürnberg, 1984.

[175] R. Rückriem, "Realisierung und messtechnische Untersuchung an einem digitalen Parallelverfahren zur Datenübertragung im Fernsprechkanal," in *Ausgewählte Arbeiten über Nachrichtensysteme*, no. 59, Universität Erlangen-Nürnberg, 1985.

[176] ETSI, *Digital Audio Broadcasting (DAB)*, 2nd ed., May 1997. ETS 300 401.

[177] ETSI, *Digital Video Broadcasting (DVB)*, 1.1.2 ed., August 1997. EN 300 744.

[178] J. Lindner et al, "OCDM — ein übertragungsverfahren für lokale funknetze," in *Codierung fuer Quelle, Kanal und Uebertragung*, no. 130 in ITG Fachbericht, pp. 401–409, VDE Verlag, 26–28 October 1994.

[179] K. Fazel and G. Fettweis, eds., *Multi-Carrier Spread-Spectrum*. Dordrecht: Kluwer, 1997.

[180] S. Nanda, D. Goodman, and U. Timor, "Performance of PRMA: A packet voice protocol for cellular systems," *IEEE Transactions on Vehicular Technology*, vol. 40, pp. 584–598, August 1991.

[181] W. Webb, R. Steele, J. Cheung, and L. Hanzo, "A packet reservation multiple access assisted cordless telecommunications scheme," *IEEE Transactions on Vehicular Technology*, vol. 43, pp. 234–245, May 1994.

[182] R. Salami, C. Laflamme, J.-P. Adoul, and D. Massaloux, "A toll quality 8 kb/s speech codec for the personal communications system (PCS)," *IEEE Transactions on Vehicular Technology*, pp. 808–816, August 1994.

[183] M. Frullone, G. Riva, P. Grazioso, and C. Carciofy, "Investigation on dynamic channel allocation strategies suitable for PRMA schemes," *IEEE International Symposium on Circuits and Systems*, pp. 2216–2219, May 1993.

[184] M. Frullone, G. Falciasecca, P. Grazioso, G. Riva, and A. Serra, "On the performance of packet reservation multiple access with fixed and dynamic channel allocation," *IEEE Transactions on Vehicular Technology*, vol. 42, pp. 78–86, February 1993.

[185] J. Brecht, M. del Buono, and L. Hanzo, "Multiframe packet reservation multiple access using oscillation-scaled histogram-based markov modelling of video codecs," *Signal Processing: Image Communications*, vol. 12, pp. 167–182, 1998.

[186] J. Brecht and L. Hanzo, "Statistical packet assignment multiple access for wireless asynchronous transfer mode systems," in *Proceeding of ACTS Mobile Communication Summit '97*, (Aalborg, Denmark), ACTS, 7–10 October 1997.

[187] J. Leduc and P. Delogne, "Statistic for variable bit-rate digital television sources," *Signal Processing: Image Communication*, vol. 8, pp. 443–464, July 1996.

[188] S. Lam, S. Chow, and D. Yau, "A lossless smoothing algorithm for compressed video," *IEEE/ACM Transactions on networking*, vol. 4, pp. 697–708, October 1996.

[189] J. Cosmas, G. Petit, R. Lehenert, C. Blondia, K. Kontovassilis, O. Casals, and T. Theimer, "A review of voice, data and video traffic models for ATM," *European Transactions on Telecommunications*, vol. 5, pp. 139–154, March-April 1994.

[190] O. Rose and M. Frater, "A comparison of models for VBR traffic sources in B-ISDN," in *Proceedings of the IFIP TC6 Second International Conference on Broadband Communications*, (Paris, France), pp. 275–287, Chapman and Hall Ltd, London, 2–4 March 1994.

[191] D. Heymann and T. Lakshman, "Source models for VBR broadcast-video traffic," *IEEE/ACM Transactions on networking*, vol. 4, pp. 40–48, February 1996.

[192] ITU-T, *Recommendation H.263: Video Coding for Low Bitrate communication*, March 1996.

[193] M. Whybray and W. Ellis, "H.263 - video coding recommendation for PSTN videophone and multimedia," in *IEE Colloquium (Digest)*, pp. 6/1–6/9, IEE, June 1995.

[194] M. Khansari, A. Jalali, E. Dubois, and P. Mermelstein, "Low bit-rate video transmission over fading channels for wireless microcellular systems," *IEEE Transactions on Circuits and Systems for Video Technology*, vol. 6, pp. 1–11, February 1996.

[195] N. Färber, E. Steinbach, and B. Girod, "Robust H.263 video transmission over wireless channels," in *Proceedings of International Picture Coding Symposium (PCS)*, (Melbourne, Australia), pp. 575–578, March 1996.

[196] P. Cherriman and L. Hanzo, "H.261 and H.263-based programmable video transceivers," in *Proceedings of IEEE ICCS'96/ISPACS'96*, vol. 3, (Singapore), pp. 1369–1373, IEEE, 25–29 November 1996.

[197] P. Cherriman and L. Hanzo, "Robust H.263 video transmission over mobile channels in interference-limited environments," in *Proc of First International Workshop on Wireless Image/Video Communications*, (Loughborough), pp. 1–7, 4–5 September 1996.

[198] P. Cherriman and L. Hanzo, "Power-controlled H.263-based wireless videophone performance in interference-limited scenarios," in *Proceedings of IEEE International Symposium on Personal, Indoor, and Mobile Radio Communications (PIMRC'96)*, vol. 1, (Taipei, Taiwan), pp. 158–162, IEEE, 15–18 October 1996.

[199] P. Cherriman and L. Hanzo, "Programmable H.263-based wireless video transceivers for interference-limited environments," *IEEE Transactions on Circuits and Systems for Video Technology*, vol. 8, pp. 275–286, June 1998.

[200] P. Cherriman, T. Keller, and L. Hanzo, "Orthogonal frequency division multiplex transmission of H.263 encoded video over wireless ATM networks," in *Proceeding of ACTS Mobile Communication Summit '97*, (Aalborg, Denmark), pp. 276–281, ACTS, 7–10 October 1997.

[201] A. Tanenbaum, "Introduction to queueing theory," in *Computer Networks*, pp. 631–641, Englewood Cliffs, NJ: Prentice-Hall, 2nd ed., 1989.

[202] P. Skelly, M. Schwartz, and S. Dixit, "A histogram-based model for video traffic behavior in a ATM multiplexer," *IEEE/ACM Transactions Networking*, vol. 1, pp. 446–459, August 1993.

[203] D. Habibi, S. Gabrielsson, and Z. Man, "A multiplexed four layers markov model for queueing studies of MPEG traffic," in *Proceedings of IEEE ICCS'96/ISPACS'96*, vol. 3, (Singapore), pp. 1180–1184, IEEE, 25–29 November 1996.

[204] F. Panken, "Multiple-access protocols over the years: a taxonomy and survey," in *1996 IEEE International Conference on Communication Systems (ICCS)*, pp. 2.1.1–2.1.5, November 1996.

[205] V. Li and X. Qiu, "Personal communications systems," *Proceedings of the IEEE*, vol. 83, pp. 1210–1243, September 1995.

[206] D. Goodman and S. Wei, "Efficiency of packet reservation multiple access," *IEEE Transactions. on Vehicular Technology*, vol. 40, pp. 170–176, February 1991.

[207] J. Dunlop, D. Robertson, P. Cosimi, and J. D. Vile, "Development and optimisation of a statistical multiplexing mechanism for ATDMA," in *Proceedings of IEEE VTC '94*, (Stockholm, Sweden), pp. 1040–1044, IEEE, 8–10 June 1994.

[208] F. Delli Priscoli, "Adaptive parameter computation in a PRMA, TDD based medium access control for ATM wireless networks," in *Proceeding of IEEE Global Telecommunications Conference, Globecom 96*, vol. 3, (London), pp. 1779–1783, IEEE, 18–22 November 1996.

[209] A. Acampora, "Wireless ATM: a perspective on issues and prospects," *IEEE Personal Communications*, vol. 3, pp. 8–17, August 1996.

[210] "Group speciale mobile (GSM) recommendation," April 1988.

[211] V. Claus, ed., *Duden zur Informatik*. Mannheim: Dudenverlag, 1993.

[212] M. Schwartz, *Broadband Integrated Networks*. Englewood Cliffs, NJ: Prentice-Hall, March 1996.

[213] A. Safak, "Optimal channel reuse in cellular radio systems with multiple correlated log-normal interferers," *IEEE Transactions on Vehicular Technology*, vol. 43, pp. 304–312, May 1994.

[214] L.-C. Wang and C.-T. Lea, "Incoherent estimation on co-channel interference probability for microcellular systems," *IEEE Transactions on Vehicular Technology*, vol. 45, pp. 164–173, February 1996.

[215] C. Lee and R. Steele, "Signal-to-interference calculations for modern TDMA cellular communication systems," *IEE Proceedings on Communication*, vol. 142, pp. 21–30, February 1995.

[216] P. Brady, "A technique for investigating on-off patterns of speech," *Bell Systems Technical Journal*, vol. 44, pp. 1–22, January 1965.

[217] P. Brady, "A model for generating on-off speech patterns in two-way conversation," *Bell Systems Technical Journal*, vol. 48, pp. 2445–2472, September 1969.

[218] D. Appleby and Y. Ko, "Frequency hopping," in Steele and Hanzo [9], ch. 7, pp. 601–676.

[219] C. Lee, *CDMA for Cellular Mobile Radio Systems*. PhD thesis, Department of Electronics and Computer Science, University of Southampton, November 1994.

[220] W. Lee, "Spectrum efficiency in cellular," *IEEE Transactions on Vehicular Technology*, vol. 38, pp. 69–75, May 1989.

[221] R. Gejji, "Channel efficiency in digital cellular capacity," in *Proceedings of IEEE VTC '92*, vol. 2, (Denver, CO), pp. 1005–1007, 10–13 May 1992.

[222] M. Chiani, V. Tralli, and R. Verdone, "Outage and spectrum efficiency analysis in microcellular systems," in *Proceedings of IEEE VTC '93*, (Secaucus, NJ), pp. 598–601, IEEE, 18–20 May 1993.

[223] R. Haas and J.-C. Belfiore, "Spectrum efficiency limits in mobile cellular systems," *IEEE Transmissions on Vehicular Technology*, vol. 45, pp. 33–40, February 1996.

[224] M. Zorzi and S. Pupolin, "Outage probability in multiple access packet radio networks in the presence of fading," *IEEE Transactions on Vehicular Technology*, vol. 43, pp. 604–610, August 1994.

[225] N. Senarath and D. Everitt, "Combined analysis of transmission and traffic characteristics in micro-cellular mobile communications systems," in *Proceedings of IEEE VTC '93*, (Secaucus, NJ), pp. 577–580, IEEE, 18–20 May 1993.

[226] A. Abu-Dayya and N. Beaulieu, "Outage probability in the presence of correlated lognormal interferers," *IEEE Tr. on Vehicular Technology*, vol. 43, pp. 33–39, February 1994.

[227] Q. Zhang, "Outage probability in cellular mobile radio due to nakagami signal and interferers with arbitrary parameters," *IEEE Transactions on Vehicular Technology*, vol. 45, pp. 364–372, May 1996.

[228] P. Cherriman, F. Romiti, and L. Hanzo, "Channel allocation for third-generation mobile radio systems," in *Proceeding of ACTS Mobile Communication Summit '98*, vol. 1, (Rhodes, Greece), pp. 255–261, ACTS, 8–11 June 1998.

[229] J. Blogh, P. Cherriman, and L. Hanzo, "Comparative study of dynamic channel allocation algorithms." to appear in IEEE Transactions on Vehicular Technology, 1999.

[230] S. Wales, "Technique for cochannel interference suppression in TDMA mobile radio systems," *IEE Proceedings on Communication*, vol. 142, no. 2, pp. 106–114, 1995.

[231] J. Litva and T. Lo, *Digital Beamforming in Wireless Communications*. London: Artech House, 1996.

[232] L. Godara, "Applications of antenna arrays to mobile communications, part I: Performance improvement, feasibility, and system considerations," *Proceedings of the IEEE*, vol. 85, pp. 1029–1060, July 1997.

[233] L. Godara, "Applications of antenna arrays to mobile communications, part II: Beam-forming and direction-of-arrival considerations," *Proceedings of the IEEE*, vol. 85, pp. 1193–1245, August 1997.

[234] E. Sourour, "Time slot assignment techniques for TDMA digital cellular systems," *IEEE Transactions on Vehicular Technology*, vol. 43, pp. 121–127, February 1994.

[235] D. Wong and T. Lim, "Soft handoffs in CDMA mobile systems," *IEEE Personal Communications*, vol. 4, pp. 6–17, December 1997.

[236] S. Tekinay and B. Jabbari, "A measurement-based prioritisation scheme for handovers in mobile cellular networks," *IEEE Journal on Selected Areas of Communications*, vol. 10, no. 8, pp. 1343–1350, 1992.

[237] G. Pollini, "Trends in handover design," *IEEE Communications Magazine*, vol. 34, pp. 82–90, March 1996.

[238] R. Bernhardt, "Timeslot re-assignment in a frequency reuse TDMA portable radio system," *IEEE Transactions on Vehicular Technology*, vol. 41, pp. 296–304, August 1992.

[239] A. Viterbi, *CDMA: Principles of Spread Spectrum Communication*. Reading MA: Addison-Wesley, June 1995.

[240] R. Prasad, *CDMA for Wireless Personal Communications*. London: Artech House, May 1996.

[241] S. Glisic and B. Vucetic, *Spread Spectrum CDMA Systems for Wireless Communications*. London: Artech House, April 1997.

[242] S. Glisic and P. Leppanen, eds., *Wireless Communications : TDMA versus CDMA*. Dordrecht: Kluwer Academic Publishers, June 1997.

[243] A. Ross and K. Gilhousen, "CDMA technology and the IS-95 north american standard," in Gibson [292], ch. 27, pp. 430–448.

[244] ETSI, *Universal Mobile Telecommunications Systems (UMTS); UMTS Terrestrial Radio Access (UTRA); Concept evaluation*, December 1997. TR 101 146 V3.0.0.

[245] I. Katzela and M. Naghshineh, "Channel assignment schemes for cellular mobile telecommunication systems: A comprehensive survey," *IEEE Personal Communications*, pp. 10–31, June 1996.

[246] S. Tekinay and B. Jabbari, "Handover and channel assignment in mobile cellular networks," *IEEE Communications Magazine*, pp. 42–46, November 1991.

[247] B. Jabbari, "Fixed and dynamic channel assignment," in Gibson [292], ch. 83, pp. 1175–1181.

[248] J. Zander, "Radio resource management in future wireless networks: Requirements and limitations," *IEEE Communications Magazine*, pp. 30–36, August 1997.

[249] D. Everitt, "Traffic engineering of the radio interface for cellular mobile networks," *Proceedings of the IEEE*, vol. 82, pp. 1371–1382, September 1994.

[250] J. Dahlin, "Ericsson's multiple reuse pattern for DCS1800," *Mobile Communications International*, November 1996.

[251] M. Madfors, K. Wallstedt, S. Magnusson, H. Olofsson, P.-O. Backman, and S. Engström, "High capacity with limited spectrum in cellular systems," *IEEE Communications Magazine*, vol. 35, pp. 38–45, August 1997.

[252] H. Jiang and S. Rappaport, "Prioritized channel borrowing without locking: a channel sharing strategy for cellular communications," *IEEE/ACM Transactions on Networking*, vol. 43, pp. 163–171, April 1996.

[253] L. Anderson, "A simulation study of some dynamic channel assignment algorithms in a high capacity mobile telecommunications system," *IEEE Transactions on Communications*, vol. 21, pp. 1294–1301, November 1973.

[254] J. Engel and M. Peritsky, "Statistically optimum dynamic server assignment in systems with interfering servers," *IEEE Transactions on Vehicular Technology*, vol. 22, pp. 203–209, November 1973.

[255] M. Zhang and T.-S. Yum, "'Comparisons of channel assignment strategies in cellular mobile telephone systems," *IEEE Transactions on Vehicular Technology*, vol. 38, pp. 211–215, November 1989.

[256] S. Elnoubi, R. Singh, and S. Gupta, "A new frequency channel assignment algorithm in high capacity mobile communications systems," *IEEE Transactions on Vehicular Technology*, vol. 31, pp. 125–131, August 1982.

[257] M. Zhang and T.-S. Yum, "The non-uniform compact pattern allocation algorithm for cellular mobile systems," *IEEE Transactions on Vehicular Technology*, vol. 40, pp. 387–391, May 1991.

[258] S. Kuek and W. Wong, "Ordered dynamic channel assignment scheme with reassignment in highway microcell," *IEEE Transactions on Vehicular Technology*, vol. 41, pp. 271–277, August 1992.

[259] T.-S. Yum and W.-S. Wong, "Hot spot traffic relief in cellular systems," *IEEE Journal on selected areas in Communications*, vol. 11, pp. 934–940, August 1993.

[260] J. Tajima and K. Imamura, "A strategy for flexible channel assignment in mobile communication systems," *IEEE Transactions on Vehicular Technology*, vol. 37, pp. 92–103, May 1988.

[261] ETSI, *Digital European Cordless Telecommunications (DECT)*, 1st ed., October 1992. ETS 300 175-1 – ETS 300 175-9.

[262] R. Steele, "Digital European Cordless Telecommunications (DECT) systems," in Steele and Hanzo [9], ch. 1.7.2, pp. 79–83.

[263] H. Ochsner, "The digital european cordless telecommunications specification, DECT," in Tuttlebee [264], pp. 273–285.

[264] W. Tuttlebee, ed., *Cordless telecommunications in Europe: the evolution of personal communications*. London: Springer-Verlag, 1990.

[265] A. Law and L. Lopes, "Performance comparison of DCA call assignment algorithms within DECT," in *Proceedings of IEEE VTC'96*, (Atlanta, GA), pp. 726–730, IEEE, 28 April–1 May 1996.

[266] H. Salgado-Galicia, M. Sirbu, and J. Peha, "A narrowband approach to efficient PCS spectrum sharing through decentralized DCA access policies," *IEEE Personal Communications*, pp. 24–34, February 1997.

[267] R. Steele, J. Whitehead, and W. Wong, "System aspects of cellular radio," *IEEE Communications Magazine*, vol. 33, pp. 80–86, January 1995.

[268] J.-I. Chuang, "Performance issues and algorithms for dynamic channel assignment," *IEEE Journal on Selected Areas of Communications*, vol. 11, pp. 955–963, August 1993.

[269] D. Cox and D. Reudink, "The behavior of dynamic channel-assignment mobile communications systems as a function of number of radio channels," *IEEE Transactions on Communications*, vol. 20, pp. 471–479, June 1972.

[270] D. Dimitrijević and J. Vučerić, "Design and performance analysis of the algorithms for channel allocation in cellular networks," *IEEE Transactions on Vehicular Technology*, vol. 42, pp. 526–534, November 1993.

[271] D. Cox and D. Reudink, "Increasing channel occupancy in large scale mobile radio systems: Dynamic channel reassignment," *IEEE Transactions on Vehicular Technology*, vol. 22, pp. 218–222, November 1973.

[272] D. Cox and D. Reudink, "A comparison of some channel assignment strategies in large-scale mobile communications systems," *IEEE Transactions on Communications*, vol. 20, pp. 190–195, April 1972.

[273] M. Cheng and J.-I. Chuang, "Performance evaluation of distributed measurement-based dynamic channel assignment in local wireless communications," *IEEE Journal on Selected Areas of Communications*, vol. 14, pp. 698–710, May 1996.

[274] S. Grandhi, R. Yates, and D. Goodman, "Resource allocation for cellular radio systems," *IEEE Transactions on Vehicular Technology*, vol. 46, pp. 581–587, August 1997.

[275] J.-I. Chuang and N. Sollenberger, "Performance of autonomous dynamic channel assignment and power control for TDMA/FDMA wireless access," *IEEE Journal on Selected Areas of Communications*, vol. 12, pp. 1314–1323, October 1994.

[276] M. Serizawa and D. Goodman, "Instability and deadlock of distributed dynamic channel allocation," in *Proceedings of IEEE VTC'93*, (Secaucus, NJ), pp. 528–531, May 18–20 1993.

[277] Y. Akaiwa and H. Andoh, "Channel segregation — a self-organized dynamic channel allocation method: Application to TDMA/FDMA microcellular system," *IEEE Journal on Selected Areas in Communications*, vol. 11, pp. 949–954, August 1993.

[278] A. Baiocchi, F. Delli Priscoli, F. Grilli, and F. Sestini, "The geometric dynamic channel allocation as a practical strategy in mobile networks with bursty user mobility," *IEEE Transactions on Vehicular Technology*, vol. 44, pp. 14–23, February 1995.

[279] F. Delli Priscoli, N. Pio Magnani, V. Palestini, and F. Sestini, "Application of dynamic channel allocation strategies to the GSM cellular network," *IEEE Journal on Selected Areas in Communications*, vol. 15, pp. 1558–1567, October 1997.

[280] I. ChihLin and C. PiHui, "Local packing — distributed dynamic channel allocation at cellular base station," in *Proceedings of the IEEE Global Telecommunications Conference 1993*, vol. 1, (Houston, TX), pp. 293–301, 29 November – 2 December 1993.

[281] E. Del Re, R. Fantacci, and G. Giambene, "Handover and dynamic channel allocation techniques in mobile cellular networks," *IEEE Transactions on Vehicular Technology*, vol. 44, pp. 229–237, May 1995.

[282] T. Kahwa and N. Georganas, "A hybrid channel assignment scheme in large-scale, cellular-structured mobile communication systems," *IEEE Transactions on Communications*, vol. 26, pp. 432–438, April 1978.

[283] J. Sin and N. Georganas, "A simulation study of a hybrid channel assignment scheme for cellular land-mobile radio systems with erlang-c service," *IEEE Transactions on Communications*, vol. 29, pp. 143–147, February 1981.

[284] S.-H. Oh and D.-W. Tcha, "Prioritized channel assignment in a cellular radio network," *IEEE Transactions on Communications*, vol. 40, pp. 1259–1269, July 1992.

[285] D. Hong and S. Rappaport, "Traffic model and performance analysis for cellular mobile radio telephone systems with prioritizes and nonprioritized handoff procedures," *IEEE Transactions on Vehicular Technology*, vol. 35, pp. 77–92, August 1986.

[286] R. Guérin, "Queueing-blocking system with two arrival streams and guard channels," *IEEE Transactions on Communications*, vol. 36, pp. 153–163, February 1988.

[287] S. Grandhi, R. Vijayan, D. Goodman, and J. Zander, "Centralized power control in cellular radio systems," *IEEE Transactions Vehicular Technology*, vol. 42, pp. 466–468, November 1993.

[288] J. Zander, "Performance of optimum transmitter power control in cellular radio systems," *IEEE Transactions on Vehicular Technology*, vol. 41, pp. 57–62, February 1992.

[289] J. Zander, "Distributed cochannel interference control in cellular radio systems," *IEEE Transactiosn on Vehicular Technology*, vol. 41, pp. 305–311, August 1992.

[290] D. Cox and D. Reudink, "Effects of some nonuniform spatial demand profiles on mobile radio system performance," *IEEE Transactions on Vehicular Technology*, vol. 21, pp. 62–67, May 1972.

[291] M. Sonka, V. Hlavac, and R. Boyle, "Image pre-processing," in *Image Processing, Analysis and Machine Vision*, ch. 4, pp. 59–61, London: Chapman and Hall, 1st ed., 1993.

[292] J. Gibson, ed., *The Mobile Communications Handbook*. Boca Raton FL: CRC Press and IEEE Press, 1996.

[293] "Special issue: The European Path Toward UMTS," *IEEE Personal Communications: The magazine of nomadic communications and computing*, vol. 2, February 1995.

[294] European Commission, *Advanced Communications Technologies and Services (ACTS)*, August 1994. Workplan DGXIII-B-RA946043-WP.

[295] Telcomm. Industry Association (TIA), Washington, DC, *Dual-mode subscriber equipment — Network equipment compatibility specification, Interim Standard IS-54*, 1989.

[296] Telcomm. Industry Association (TIA), Washington, DC, *Mobile station — Base station compatibility standard for dual-mode wideband spread spectrum cellular systém, EIA/TIA Interim Standard IS-95*, 1993.

[297] H. Alikhani, R. Bohnke, and M. Suzuki, "BDMA-band division multiple access; a new air-interface for 3rd generation mobile system, UMTS, in europe," in *Proceeding of ACTS Mobile Communication Summit '97*, (Aalborg, Denmark), pp. 482–488, ACTS, 7–10 October 1997.

[298] P. Baier, P. Jung, and A. Klein, "Taking the challenge of multiple access for third-generation cellular mobile radio systems — a European view," *IEEE Communications Magazine*, vol. 34, pp. 82–89, February 1996.

[299] F. Adachi, M. Sawahashi, and K. Okawa, "Tree-structured generation of orthogonal spreading codes with different lengths for forward link of DS-CDMA mobile radio," *Electronic Letters*, vol. 33, pp. 27–28, January 1997.

[300] F. Adachi, K. Ohno, A. Higashi, T. Dohi, and Y. Okumura, "Coherent multicode DS-CDMA mobile Radio Access," *IEICE Transactions on Communications*, vol. E79-B, pp. 1316–1324, September 1996.

[301] F. Adachi and M. Sawahashi, "Wideband wireless access based on DS-CDMA," *IEICE Transactions on Communications*, vol. E81-B, pp. 1305–1316, July 1998.

[302] F. Adachi, M. Sawahashi, and H. Suda, "Wideband DS-CDMA for next-generation mobile communications systems," *IEEE Communications Magazine*, vol. 36, pp. 56–69, September 1998.

[303] D. Knisely, S. Kumar, S. Laha, and S. Nanda, "Evolution of wireless data services : IS-95 to cdma2000," *IEEE Communications Magazine*, vol. 36, pp. 140–149, October 1998.

[304] Telecommunications Industry Association (TIA), *The cdma2000 ITU-R RTT Candidate Submission*, 1998.

[305] D. Knisely, Q. Li, and N. Rames, "cdma2000: A third generation radio transmission technology," *Bell Labs Technical Journal*, vol. 3, pp. 63–78, July–September 1998.

[306] "Feature topic: Wireless personal communications," *IEEE Personal Communications*, vol. 2, April 1995.

[307] "Feature topic: Universal telecommunications at the beginning of the 21st century," *IEEE Communications Magazine*, vol. 33, November 1995.

[308] "Feature topic: Software radios," *IEEE Communications Magazine*, vol. 33, pp. 24–68, May 1995.

[309] "Feature topic: Wireless personal communications," *IEEE Communications Magazine*, vol. 33, January 1995.

[310] "Feature topic: European research in mobile communications," *IEEE Communications Magazine*, vol. 34, February 1996.

[311] R. Cox and P. Kroon, "Low bit-rate speech coders for multimedia communications," *IEEE Communications Magazine*, pp. 34–41, December 1996.

[312] A. Kucar, "Mobile radio: An overview," in Gibson [292], ch. 15, pp. 242–262.

[313] C. Li, C. Zheng, and C. Tai, "Detection of ECG characteristic points using wavelet transforms," *IEEE Transactions in Biomedical Engineering*, vol. 42, pp. 21–28, January 1995.

[314] "Personal handy phone (PHP) system." RCR Standard, STD-28, Japan.

[315] ETSI, Sophia Antipolis, Cedex, France, *Digital Cellular Telecommunications System (Phase 2+); High Speed Circuit Switched Data (HSCSD) — Stage 1 (GSM 02.34 Version 5.2.1)*, July 1997.

[316] A. Noerpel, Y.-B. Lin, and H. Sherry, "PACS: Personal access communications system — a tutorial," *IEEE Personal Communications Magazine*, vol. 3, pp. 32–43, June 1996.

[317] P. Vary and R. Sluyter, "MATS-D speech codec: Regular-pulse excitation LPC," in *Proceedings of the Nordic Seminar on Digital Land Mobile Radio Communications (DMRII)*, (Stockholm, Sweden), pp. 257–261, October 1986.

[318] R. Salami, L. Hanzo, R. Steele, K. Wong, and I. Wassell, "Speech coding," in Steele and Hanzo [9], ch. 3, pp. 186–346.

[319] J. Rapeli, "UMTS:targets, system concept, and standardization in a global framework," *IEEE Personal Communications*, vol. 2, pp. 20–28, February 1995.

[320] P.-G. Andermo and L.-M. Ewerbring, "A CDMA-based radio access design for UMTS," *IEEE Personal Communications*, vol. 2, pp. 48–53, February 1995.

[321] E. Nikula, A. Toskala, E. Dahlman, L. Girard, and A. Klein, "FRAMES multiple access for UMTS and IMT-2000," *IEEE Personal Communications*, vol. 5, pp. 16–24, April 1998.

[322] T. Ojanperä and R. Prasad, "An overview of air interface multiple access for IMT-2000/UMTS," *IEEE Communications Magazine*, vol. 36, pp. 82–95, September 1998.

[323] A. Toskala, J. Castro, E. Dahlman, M. Latva-aho, and T. Ojanperä, "FRAMES FMA2 wideband-CDMA for UMTS," *European Transactions on Telecommunications*, vol. 9, pp. 325–336, July–August 1998.

[324] E. Berruto, M. Gudmundson, R. Menolascino, W. Mohr, and M. Pizarroso, "Research activities on UMTS radio interface, network architectures, and planning," *IEEE Communications Magazine*, vol. 36, pp. 82–95, February 1998.

[325] M. Callendar, "Future public land mobile telecommunication systems," *IEEE Personal Communications*, vol. 12, no. 4, pp. 18–22, 1994.

[326] W. Lee, "Overview of cellular CDMA," *IEEE Transactions on Vehicular Technology*, vol. 40, pp. 291–302, May 1991.

[327] K. Gilhousen, I. Jacobs, R. Padovani, A. Viterbi, L. Weaver Jr., and C. Wheatley III, "On the capacity of a cellular CDMA system," *IEEE Transactions on Vehicular Technology*, vol. 40, pp. 303–312, May 1991.

[328] R. Pickholtz, L. Milstein, and D. Schilling, "Spread spectrum for mobile communications," *IEEE Transactions on Vehicular Technology*, vol. 40, pp. 312–322, May 1991.

[329] R. Kohno, R. Meidan, and L. Milstein, "Spread spectrum access methods for wireless communication," *IEEE Communications Magazine*, vol. 33, pp. 58–67, January 1995.

[330] V. Garg, K. Smolik, J. Wilkes, and K. Smolik, *Applications of CDMA in Wireless/Personal Communications*. Englewood Cliffs NJ: Prentice-Hall, 1996.

[331] R. Price and E. Green Jr., "A communication technique for multipath channels," *Proceedings of the IRE*, vol. 46, pp. 555–570, March 1958.

[332] B. Sklar, "Rayleigh fading channels in mobile digital communication systems part I : Characterization," *IEEE Communications Magazine*, vol. 35, pp. 90–100, July 1997.

[333] B. Sklar, "Rayleigh fading channels in mobile digital communication systems part II : Mitigation," *IEEE Communications Magazine*, vol. 35, pp. 148–155, July 1997.

[334] F. Amoroso, "Use of DS/SS signaling to mitigate rayleigh fading in a dense scatterer environment," *IEEE Personal Communications*, vol. 3, pp. 52–61, April 1996.

[335] M. Nakagami, "The m-distribution-a general formula of intensity distribution of fading," *Statistical Methods in Radio Wave Propagation*, 1960. W.C. Hoffman, Ed. New York: Pergamon.

[336] H. Suzuki, "A statistical model for urban multipath propagation," *IEEE Transactions on Communications*, vol. COM-25, pp. 673–680, July 1977.

[337] "COST 207: Digital land mobile radio communications, final report." Office for Official Publications of the European Communities, 1989. Luxembourg.

[338] M. Whitmann, J. Marti, and T. Kürner, "Impact of the power delay profile shape on the bit error rate in mobile radio systems," *IEEE Transactions on Vehicular Technology*, vol. 46, pp. 329–339, May 1997.

[339] T. Eng, N. Kong, and L. Milstein, "Comparison of diversity combining techniques for Rayleigh-fading channels," *IEEE Transactions on Communications*, vol. 44, pp. 1117–1129, September 1996.

[340] M. Kavehrad and P. McLane, "Performance of low-complexity channel coding and diversity for spread spectrum in indoor, wireless communications," *AT&T Technical Journal*, vol. 64, pp. 1927–1965, October 1985.

[341] K.-T. Wu and S.-A. Tsaur, "Selection diversity for DS-SSMA communications on nakagami fading channels," *IEEE Transactions on Vehicular Technology*, vol. 43, pp. 428–438, August 1994.

[342] L.-L. Yang and L. Hanzo, "Serial acquisition techniques for DS-CDMA signals in frequency-selective multi-user mobile channels," in *Proceeding of VTC'99 (Spring)*, (Houston, TX), IEEE, 16–20 May 1999.

[343] L.-L. Yang and L. Hanzo, "Serial acquisition of DS-CDMA signals in multipath fading mobile channels." submitted to IEEE Transactions on Vehicular Technology, 1998.

[344] R. Ziemer and R. Peterson, *Digital Communications and Spread Spectrum System*. New York: Macmillan Publishing Company, 1985.

[345] R. Pickholtz, D. Schilling, and L. Milstein, "Theory of spread-spectrum communications — a tutorial," *IEEE Transactions on Communications*, vol. COM-30, pp. 855–884, May 1982.

[346] S. Rappaport and D. Grieco, "Spread-spectrum signal acquisition: Methods and technology," *IEEE Communications Magazine*, vol. 22, pp. 6–21, June 1984.

[347] E. Ström, S. Parkvall, S. Miller, and B. Ottersten, "Propagation delay estimation in asynchronous direct-sequence code division multiple access systems," *IEEE Transactions on Communications*, vol. 44, pp. 84–93, January 1996.

[348] R. Rick and L. Milstein, "Optimal decision strategies for acquisition of spread-spectrum signals in frequency-selective fading channels," *IEEE Transactions on Communications*, vol. 46, pp. 686–694, May 1998.

[349] J. Lee, *CDMA Systems Engineering Handbook*. London: Artech House Publishers, 1998.

[350] M. Varanasi and B. Aazhang, "Multistage detection in asynchronous code-division multiple-access communications," *IEEE Transactions on Communications*, vol. 38, pp. 509–519, April 1990.

[351] S. Moshavi, "Multi-user detection for DS-CDMA communications," *IEEE Communications Magazine*, vol. 34, pp. 124–136, October 1996.

[352] S. Verdú, "Minimum probability of error for asynchronous gaussian multiple-access channel," *IEEE Transactions on Communications*, vol. 32, pp. 85–96, January 1986.

[353] E. Kuan and L. Hanzo, "Joint detection CDMA techniques for third-generation transceivers," in *Proceeding of ACTS Mobile Communication Summit '98*, (Rhodes, Greece), pp. 727–732, ACTS, 8–11 June 1998.

[354] E. Kuan, C. Wong, and L. Hanzo, "Upper-bound performance of burst-by-burst adaptive joint detection CDMA." submitted to IEEE Communications Letters, 1998.

[355] S. Verdú, *Multiuser Detection*. Cambridge: Cambridge University Press, 1998.

[356] F. Simpson and J. Holtzman, "Direct sequence CDMA power control, interleaving, and coding," *IEEE Journal on Selected Areas in Communications*, vol. 11, pp. 1085–1095, September 1993.

[357] M. Pursley, "Performance evaluation for phase-coded spread-spectrum multiple-access communication-part I: System analysis," *IEEE Transactions on Communications*, vol. COM-25, pp. 795–799, August 1977.

[358] R. Morrow Jr., "Bit-to-bit error dependence in slotted DS/SSMA packet systems with random signature sequences," *IEEE Transactions on Communications*, vol. 37, pp. 1052–1061, October 1989.

[359] J. Holtzman, "A simple, accurate method to calculate spread-spectrum multiple-access error probabilities," *IEEE Transactions on Communications*, vol. 40, pp. 461–464, March 1992.

[360] U.-C. Fiebig and M. Schnell, "Correlation properties of extended m-sequences," *Electronic Letters*, vol. 29, pp. 1753–1755, September 1993.

[361] F. Davarian, "Mobile digital communications via tone calibration," *IEEE Transactions on Vehicular Technology*, vol. VT-36, pp. 55–62, May 1987.

[362] G. Irvine and P. McLane, "Symbol-aided plus decision-directed reception for PSK/TCM modulation on shadowed mobile satellite fading channels," *IEEE Journal on Selected Areas in Communications*, vol. 10, pp. 1289–1299, October 1992.

[363] A. Baier, U.-C. Fiebig, W. Granzow, W. Koch, P. Teder, and J. Thielecke, "Design study for a CDMA-based third-generation mobile system," *IEEE Journal on Selected Areas in Communications*, vol. 12, pp. 733–743, May 1994.

[364] P. Rapajic and B. Vucetic, "Adaptive receiver structures for asynchornous CDMA systems," *IEEE Journal on Selected Areas in Communications*, vol. 12, pp. 685–697, May 1994.

[365] M. Benthin and K.-D. Kammeyer, "Influence of channel estimation on the performance of a coherent DS-CDMA system," *IEEE Transactions on Vehicular Technology*, vol. 46, pp. 262–268, May 1997.

[366] M. Sawahashi, Y. Miki, H. Andoh, and K. Higuchi, "Pilot symbol-assisted coherent multistage interference canceller using recursive channel estimation for DS-CDMA mobile radio," *IEICE Transactions on Communications*, vol. E79-B, pp. 1262–1269, September 1996.

[367] S. Sampei and T. Sunaga, "Rayleigh fading compensation for QAM in land mobile radio communications," *IEEE Transactions on Vehicular Technology*, vol. 42, pp. 137–147, May 1993.

[368] K. Yen and L. Hanzo, "Third-generation wireless systems," in Steele and Hanzo [87], ch. 10, pp. 897–964.

[369] *The 3GPP1 website.* http://www.3gpp.org.

[370] *The 3GPP2 website.* http://www.3gpp2.org.

[371] T. Ojanperä and R. Prasad, *Wideband CDMA for Third Generation Mobile Communications*. London: Artech House, 1998.

[372] E. Dahlman, B. Gudmundson, M. Nilsson, and J. Sköld, "UMTS/IMT-2000 based on wideband CDMA," *IEEE Communications Magazine*, vol. 36, pp. 70–80, September 1998.

[373] T. Ojanpera, "Overview of research activities for third generation mobile communications," in Glisic and Leppanen [242], ch. 2 (Part 4), pp. 415–446.

[374] European Telecommunications Standards Institute, *The ETSI UMTS Terrestrial Radio Access (UTRA) ITU-R RTT Candidate Submission*, June 1998. ETSI/SMG/SMG2.

[375] Association of Radio Industries and Businesses, *Japan's Proposal for Candidate Radio Transmission Technology on IMT-2000: W-CDMA*, June 1998.

[376] A. Sasaki, "Current situation of IMT-2000 radio transmission technology study in Japan," *IEICE Transactions on Communications*, vol. E81-B, pp. 1299–1304, July 1998.

[377] J. Schwarz da Silva, B. Barani, and B. Arroyo-Fernández, "European mobile communications on the move," *IEEE Communications Magazine*, vol. 34, pp. 60–69, February 1996.

[378] F. Ovesjö, E. Dahlman, T. Ojanperä, A. Toskala, and A. Klein, "FRAMES multiple access mode 2 — wideband CDMA," in *Proceedings of IEEE International Symposium on Personal, Indoor and Mobile Radio Communications, PIMRC'97*, (Marina Congress Centre, Helsinki, Finland), IEEE, 1–4 September 1997.

[379] *The UMTS Forum website.* http://www.umts-forum.org/.

[380] M. Sunay, Z.-C. Honkasalo, A. Hottinen, H. Honkasalo, and L. Ma, "A dynamic channel allocation based TDD DS CDMA residential indoor system," in *IEEE 6th International Conference on Universal Personal Communications, ICUPC'97*, (San Diego, CA), pp. 228–234, October 1997.

[381] C. Berrou and A. Glavieux, "Near optimum error correcting coding and decoding: turbo codes," *IEEE Transactions on Communications*, vol. 44, pp. 1261–1271, October 1996.

[382] A. Fujiwara, H. Suda, and F. Adachi, "Turbo codes application to DS-CDMA mobile radio," *IEICE Transactions on Communications*, vol. E81A, pp. 2269–2273, November 1998.

[383] M. Juntti, "System concept comparison for multirate CDMA with multiuser detection," in *Proceedings of IEEE Vehicular Technology Conference (VTC'98)*, (Ottawa, Canada), pp. 18–21, IEEE, 18–21 May 1998.

[384] S. Ramakrishna and J. Holtzman, "A comparison between single code and multiple code transmission schemes in a CDMA system," in *Proceedings of IEEE Vehicular Technology Conference (VTC'98)*, (Ottawa, Canada), pp. 791–795, IEEE, 18–21 May 1998.

[385] M. Simon, J. Omura, R. Scholtz, and B. Levitt, *Spread Spectrum Communications Handbook*. New York: McGraw-Hill, 1994.

[386] T. Kasami, *Combinational Mathematics and its Applications*. University of North Carolina Press, 1969.

[387] A. Brand and A. Aghvami, "Multidimensional PRMA with prioritized bayesian broadcast — a MAC strategy for multiservice traffic over UMTS," *IEEE Transactions on Vehicular Technology*, vol. 47, pp. 1148–1161, November 1998.

[388] R. Ormondroyd and J. Maxey, "Performance of low rate orthogonal convolutional codes in DS-CDMA," *IEEE Transactions on Vehicular Technology*, vol. 46, pp. 320–328, May 1997.

[389] A. Chockalingam, P. Dietrich, L. Milstein, and R. Rao, "Performance of closed-loop power control in DS-CDMA cellular systems," *IEEE Transactions on Vehicular Technology*, vol. 47, pp. 774–789, August 1998.

[390] R. Gejji, "Forward-link-power control in CDMA cellular-systems," *IEEE Transactions on Vehicular Technology*, vol. 41, pp. 532–536, November 1992.

[391] K. Higuchi, M. Sawahashi, and F. Adachi, "Fast cell search algorithm in DS-CDMA mobile radio using long spreading codes," in *Proceedings of IEEE VTC'97*, vol. 3, (Phoenix, AZ), pp. 1430–1434, IEEE, 4–7 May 1997.

[392] M. Golay, "Complementary series,," *IRE Transactions on Information Theory*, vol. IT-7, pp. 82–87, 1961.

[393] V. Tarokh, H. Jafarkhani, and A. Calderbank, "Space-time block codes from orthogonal designs," *IEEE Transactions on Information Theory*, vol. 45, pp. 1456–1467, May 1999.

[394] W. Lee, *Mobile Communications Engineering*. New York: McGraw-Hill, 2nd ed., 1997.

[395] H. Wong and J. Chambers, "Two-stage interference immune blind equaliser which exploits cyclostationary statistics," *Electronics Letters*, vol. 32, pp. 1763–1764, September 1996.

[396] C. Lee and R. Steele, "Effects of Soft and Softer Handoffs on CDMA System Capacity," *IEEE Transactions on Vehicular Technology*, vol. 47, pp. 830–841, August 1998.

[397] M. Gustafsson, K. Jamal, and E. Dahlman, "Compressed mode techniques for inter-frequency measurements in a wide-band DS-CDMA system," in *Proceedings of IEEE International Symposium on Personal, Indoor and Mobile Radio Communications, PIMRC'97*, (Marina Congress Centre, Helsinki, Finland), pp. 231–235, IEEE, 1–4 September 1997.

[398] Y. Okumura and F. Adachi, "Variable-rate data transmission with blind rate detection for coherent DS-CDMA mobile radio," *IEICE Transactions on Communications*, vol. E81B, pp. 1365–1373, July 1998.

[399] M. Raitola, A. Hottinen, and R. Wichman, "Transmission diversity in wideband CDMA," in *Proceeding of VTC'99 (Spring)*, (Houston, TX), pp. 1545–1549, IEEE, 16–20 May 1999.

[400] J. Liberti Jr. and T. Rappaport, "Analytical results for capacity improvements in CDMA," *IEEE Transactions on Vehicular Technology*, vol. 43, pp. 680–690, August 1994.

[401] J. Winters, "Smart antennas for wireless systems," *IEEE Personal Communications*, vol. 5, pp. 23–27, February 1998.

[402] T. Lim and L. Rasmussen, "Adaptive symbol and parameter estimation in asynchronous multiuser CDMA detectors," *IEEE Transactions on Communications*, vol. 45, pp. 213–220, February 1997.

[403] T. Lim and S. Roy, "Adaptive filters in multiuser (MU) CDMA detection," *Wireless Networks*, vol. 4, pp. 307–318, June 1998.

[404] L. Wei, "Rotationally-invariant convolutional channel coding with expanded signal space, part I and II," *IEEE Transactions on Selected Areas in Comms*, vol. SAC-2, pp. 659–686, September 1984.

[405] T. Lim and M. Ho, "LMS-based simplifications to the kalman filter multiuser CDMA detector," in *Proceedings of IEEE Asia-Pacific Conference on Communications/International Conference on Communication Systems*, (Singapore), November 1998.

[406] D. You and T. Lim, "A modified blind adaptive multiuser CDMA detector," in *Proceedings of IEEE International Symposium on Spread Spectrum Techniques and Application (ISSSTA'98)*, (Sun City, South Africa), pp. 878–882, IEEE, September 1998.

[407] S. Sun, L. Rasmussen, T. Lim, and H. Sugimoto, "Impact of estimation errors on multiuser detection in CDMA," in *Proceedings of IEEE Vehicular Technology Conference (VTC'98)*, (Ottawa, Canada), pp. 1844–1848, IEEE, 18–21 May 1998.

[408] Y. Sanada and Q. Wang, "A co-channel interference cancellation technique using orthogonal convolutional codes on multipath rayleigh fading channel," *IEEE Transactions on Vehicular Technology*, vol. 46, pp. 114–128, February 1997.

[409] P. Patel and J. Holtzman, "Analysis of a simple successive interference cancellation scheme in a DS/CDMA system," *IEEE Journal on Selected Areas in Communications*, vol. 12, pp. 796–807, June 1994.

[410] P. Tan and L. Rasmussen, "Subtractive interference cancellation for DS-CDMA systems," in *Proceedings of IEEE Asia-Pacific Conference on Communications/International Conference on Communication Systems*, (Singapore), November 1998.

[411] K. Cheah, H. Sugimoto, T. Lim, L. Rasmussen, and S. Sun, "Performance of hybrid interference canceller with zero-delay channel estimation for CDMA," in *Proceeding of Globecom'98*, (Sydney, Australia), pp. 265–270, IEEE, 8–12 November 1998.

[412] S. Sun, L. Rasmussen, and T. Lim, "A matrix-algebraic approach to hybrid interference cancellation in CDMA," in *Proceedings of IEEE International Conference on Universal Personal Communications '98*, (Florence, Italy), pp. 1319–1323, October 1998.

[413] A. Johansson and L. Rasmussen, "Linear group-wise successive interference cancellation in CDMA," in *Proceedings of IEEE International Symposium on Spread Spectrum Techniques and Application (ISSSTA'98)*, (Sun City, South Africa), pp. 121–126, IEEE, September 1998.

[414] S. Sun, L. Rasmussen, H. Sugimoto, and T. Lim, "A hybrid interference canceller in CDMA," in *Proceedings of IEEE International Symposium on Spread Spectrum Techniques and Application (ISSSTA'98)*, (Sun City, South Africa), pp. 150–154, IEEE, September 1998.

[415] D. Guo, L. Rasmussen, S. Sun, T. Lim, and C. Cheah, "MMSE-based linear parallel interference cancellation in CDMA," in *Proceedings of IEEE International Symposium on Spread Spectrum Techniques and Application (ISSSTA'98)*, (Sun City, South Africa), pp. 917–921, IEEE, September 1998.

[416] L. Rasmussen, D. Guo, Y. Ma, and T. Lim, "Aspects on linear parallel interference cancellation in CDMA," in *Proceedings of IEEE International Symposium on Information Theory*, (Cambridge, MA), p. 37, August 1998.

[417] L. Rasmussen, T. Lim, H. Sugimoto, and T. Oyama, "Mapping functions for successive interference cancellation in CDMA," in *Proceedings of IEEE Vehicular Technology Conference (VTC'98)*, (Ottawa, Canada), pp. 2301–2305, IEEE, 18–21 May 1998.

[418] S. Sun, T. Lim, L. Rasmussen, T. Oyama, H. Sugimoto, and Y. Matsumoto, "Performance comparison of multi-stage SIC and limited tree-search detection in CDMA," in *Proceedings of IEEE Vehicular Technology Conference (VTC'98)*, (Ottawa, Canada), pp. 1854–1858, IEEE, 18–21 May 1998.

[419] H. Sim and D. Cruickshank, "Chip based multiuser detector for the downlink of a DS-CDMA system using a folded state-transition trellis," in *Proceeding of VTC'99 (Spring)*, vol. 1, (Houston, TX), pp. 846–850, IEEE, 16–20 May 1999.

[420] M. Barnsley, "A better way to compress images," *BYTE*, pp. 215–222, January 1988.

[421] J. Beaumont, "Image data compression using fractal techniques," *BT Technology*, vol. 9, pp. 93–109, October 1991.

[422] A. Jacquin, "Image coding based on a fractal theory of iterated contractive image transformations," *IEEE Transactions on Image Processing*, vol. 1, pp. 18–30, January 1992.

[423] D. Monro and F. Dudbridge, "Fractal block coding of images," *Electronic Letters*, vol. 28, pp. 1053–1055, May 1992.

[424] D. Monro, D. Wilson, and J. Nicholls, "High speed image coding with the bath fractal transform," in *Proceedings of IEEE International Symposium of Multimedia Technologies and Future Applications* (R. Damper, W. Hall, and J. Richards, eds.), (London), pp. 23–30, Pentech Press, April 1993.

[425] B. Ramamurthi and A. Gersho, "Classified vector quantization of images," *IEEE Transactions on communications*, vol. COM-34, pp. 1105–1115, November 1986.

[426] J. Streit and L. Hanzo, "A fractal video communicator," in *Proceedings of IEEE VTC '94*, (Stockholm, Sweden), pp. 1030–1034, IEEE, 8–10 June 1994.

[427] W. Welsh, "Model based coding of videophone images," *Electronic and Communication Engineering Journal*, pp. 29–36, February 1991.

[428] J. Ostermann, "Object-based analysis-synthesis coding based on the source model of moving rigid 3D objects," *Signal Processing: Image Communication*, vol. 6, pp. 143–161, 1994.

[429] M. Chowdhury, "A switched model-based coder for video signals," *IEEE Transactions on Circuits and Systems*, vol. 4, pp. 216–227, June 1994.

[430] G. Bozdagi, A. Tekalp, and L. Onural, "3-D motion estimation and wireframe adaptation including photometric effects for model-based coding of facial image sequences," *IEEE Transactions on circuits and Systems for Video Technology*, vol. 4, pp. 246–256, June 1994.

[431] Q. Wang and R. Clarke, "Motion estimation and compensation for image sequence coding," *Signal Processing: Image Communications*, vol. 4, pp. 161–174, 1992.

[432] H. Gharavi and M. Mills, "Blockmatching motion estimation algorithms — new results," *IEEE Transactions on Circuits and Systems*, vol. 37, pp. 649–651, May 1990.

[433] J. Jain and A. Jain, "Displacement measurement and its applications in inter frame image coding," *IEEE Transactions on Communications*, vol. 29, December 1981.

[434] B. Wang, J. Yen, and S. Chang, "Zero waiting-cycle hierarchical block matching algorithm and its array architectures," *IEEE Transactions on Circuits and Systems for Video Technology*, vol. 4, pp. 18–27, February 1994.

[435] P. Strobach, "Tree-structured scene adaptive coder," *IEEE Transactions on Communications*, vol. 38, pp. 477–486, April 1990.

[436] B. Liu and A. Zaccarin, "New fast algorithms for the estimation of block motion vectors," *IEEE Transactions on Circuits and Systems*, vol. 3, pp. 148–157, April 1993.

[437] R. Li, B. Zeng, and N. Liou, "A new three step search algorithm for motion estimation," *IEEE Transactions on Circuits and Systems*, vol. 4, pp. 439–442, August 1994.

[438] L. Lee, J. Wang, J. Lee, and J. Shie, "Dynamic search-window adjustment and interlaced search for block-matching algorithm," *IEEE Transactions on Circuits and Systems for Video Technology*, vol. 3, pp. 85–87, February 1993.

[439] B. Girod, "Motion-compensating prediction with fractional-pel accuracy," *IEEE Transactions on Communications*, vol. 41, pp. 604–611, April 1993.

[440] J. Huang et al, "A multi-frame pel-recursive algorithm for varying frame-to-frame displacement estimation," in *Proceedings of International Conference on Acoustics, Speech, and Signal Processing, ICASSP'92*, vol. 3, pp. 241–244, IEEE, March 1992.

[441] N. Efstratiadis and A. Katsaggelos, "Adaptive multiple-input pel-recursive displacement estimation," in *Proceedings of International Conference on Acoustics, Speech, and Signal Processing, ICASSP'92*, vol. 3, pp. 245–248, IEEE, March 1992.

[442] C. Huang and C. Hsu, "A new motion compensation method for image sequence coding using hierarchical grid interpolation," *IEEE Transactions on Circuits and Systems for Video Technology*, vol. 4, pp. 42–51, February 1994.

[443] J. Nieweglowski, T. Moisala, and P. Haavisto, "Motion compensated video sequence iinterpolation using digital image warping," in *Proceedings of the IEEE International Conference on Acoustics, Speech and Signal Processing (ICASSP'94)*, vol. 5, (Adelaide, Australia), pp. 205–208, IEEE, 19–22 April 1994.

[444] C. Papadopoulos and T. Clarkson, "Motion compensation using second-order geometric transformations," *IEEE Transactions on Circuits and Systems for Video Technology*, vol. 5, pp. 319–331, August 1995.

[445] C. Papadopoulos, *The use of geometric transformations for motion compensation in video data compression*. PhD thesis, University of London, 1994.

[446] M. Hoetter, "Differential estimation based on object oriented mapping parameter estimation," *Signal Processing*, vol. 16, pp. 249–265, March 1989.

[447] S. Karunaserker and N. Kingsbury, "A distortion measure for blocking artifacts in images based on human visual sensitivity," *IEEE Transactions on Image Processing*, vol. 6, pp. 713–724, June 1995.

[448] D. Pearson and M. Whybray, "Transform coding of images using interleaved blocks," *IEE Proceedings*, vol. 131, pp. 466–472, August 1984.

[449] J. Magarey and N. Kingsbury, "Motion estimation using complex wavelets," in *Proceedings of the IEEE International Conference on Acoustics, Speech and Signal Processing (ICASSP'96)*, vol. 4, (Atlanta, GA), pp. 2371–2374, IEEE, 7–10 May 1996.

[450] R. Young and N. Kingsbury, "Frequency-domain motion estimation using a complex lapped transform," *IEEE Transactions on Image Processing*, vol. 2, pp. 2–17, January 1993.

[451] R. Young and N. Kingsbury, "Video compression using lapped transforms for motion estimation/compensation and coding," in *Proceedings of the SPIE Communication and Image Processing Conference*, (Boston, MA), pp. 1451–1463, SPIE, November 1992.

[452] K. Rao and P. Yip, *Discrete Cosine Transform: Algorithms, Advantages and Applications*. New York: Academic Press Ltd., 1990.

[453] A. Sharaf, *Video coding at very low bit rates using spatial transformations*. PhD thesis, Department of Electronic and Electrical Engineering, Kings College, London, 1997.

[454] R. Clarke, *Transform Coding of Images*. New York: Academic Press, 1985.

[455] A. Palau and G. Mirchandani, "Image coding with discrete cosine transforms using efficient energy-based adaptive zonal filtering," in *Proceedings of the IEEE International Conference on Acoustics, Speech and Signal Processing (ICASSP'94)*, vol. 5, (Adelaide, Australia), pp. 337–340, IEEE, 19–22 April 1994.

[456] H. Yamaguchi, "Adaptive DCT coding of video signals," *IEEE Transactions on Communications*, vol. 41, pp. 1534–1543, October 1993.

[457] K. Ngan, "Adaptive transform coding of video signals," *IEE Proceedings*, vol. 129, pp. 28–40, February 1982.

[458] R. Clarke, "Hybrid intra-frame transform coding of image data," *IEE Proceedings*, vol. 131, pp. 2–6, February 1984.

[459] F.-M. Wang and S. Liu, "Hybrid video coding for low bit-rate applications," in *Proceedings of the IEEE International Conference on Acoustics, Speech and Signal Processing (ICASSP'94)*, (Adelaide, Australia), pp. 481–484, IEEE, 19–22 April 1994.

[460] M. Ghanbari and J. Azari, "Effect of bit rate variation of the base layer on the performance of two-layer video codecs," *IEEE Transactions on Communications for Video Technology*, vol. 4, pp. 8–17, February 1994.

[461] N. Jayant and P. Noll, *Digital Coding of Waveforms, Principles and Applications to Speech and Video*. Englewood Cliffs, NJ: Prentice-Hall, 1984.

[462] N. Cheng and N. Kingsbury, "The ERPC: an efficient error-resilient technique for encoding positional information of sparse data," *IEEE Transactions on Communications*, vol. 40, pp. 140–148, January 1992.

[463] M. Narasimha and A. Peterson, "On the computation of the discrete cosine transform," *IEEE Transactions on Communications*, vol. 26, pp. 934–936, June 1978.

[464] R. M. Pelz, "An un-equal error protected px8 kbit/s video transmission for DECT," in *Proceedings of IEEE VTC '94*, (Stockholm, Sweden), pp. 1020–1024, IEEE, 8–10 June 1994.

[465] L. Hanzo, R. Stedman, R. Steele, and J. Cheung, "A portable multimedia communicator scheme," in *Proceedings of IEEE International Symposium of Multimedia Technologies and Future Applications* (R. Damper, W. Hall, and J. Richards, eds.), (London), pp. 31–54, Pentech Press, April 1993.

[466] R. Stedman, H. Gharavi, L. Hanzo, and R. Steele, "Transmission of subband-coded images via mobile channels," *IEEE Transactions on Circuits and Systems for Video Technology*, vol. 3, pp. 15–27, February 1993.

[467] L. Hanzo and J. Streit, "Adaptive low-rate wireless videophone systems," *IEEE Transactions on Circuits and Systems for Video Technology*, vol. 5, pp. 305–319, August 1995.

[468] ETSI, *GSM Recommendation 05.05, Annex 3*, November 1988.

[469] G. Djuknic and D. Schilling, "Performance analysis of an ARQ transmission scheme for meteor burst communications," *IEEE Transactions on Communications*, vol. 42, pp. 268–271, February/March/April 1994.

[470] L. de Alfaro and A. Meo, "Codes for second and third order GH-ARQ schemes," *IEEE Transactions on Communications*, vol. 42, pp. 899–910, February–April 1994.

[471] T.-H. Lee, "Throughput performance of a class of continuous ARQ strategies for burst-error channels," *IEEE Transactions on Vehicular Technology*, vol. 41, pp. 380–386, November 1992.

[472] S. Lin, D. Constello Jr., and M. Miller, "Automatic-repeat-request error-control schemes," *IEEE Communications Magazine*, vol. 22, pp. 5–17, December 1984.

[473] Research and Development Centre for Radio Systems, Japan, *Public Digital Cellular (PDC) Standard, RCR STD-27*.

[474] A. Gersho and R. Gray, *Vector Quantization and Signal Compression*. Dordrecht: Kluwer Academic Publishers, 1992.

[475] L. Torres, J. Casas, and S. deDiego, "Segmetation based coding of textures using stochastic vector quantization," in *Proceedings of the IEEE International Conference on Acoustics, Speech and Signal Processing (ICASSP'94)*, vol. 5, (Adelaide, Australia), pp. 597–600, IEEE, 19–22 April 1994.

[476] M. Jaisimha, J. Goldschneider, A. Mohr, E. Riskin, and R. Haralick, "On vector quantization for fast facet edge detection," in *Proceedings of the IEEE International Conference on Acoustics, Speech and Signal Processing (ICASSP'94)*, vol. 5, (Adelaide, Australia), pp. 37–40, IEEE, 19–22 April 1994.

[477] P. Yu and A. Venetsanopoulos, "Hierarchical finite-state vector quantisation for image coding," *IEEE Transactions on Communications*, vol. 42, pp. 3020–3026, November 1994.

[478] C.-H. Hsieh, K.-C. Chuang, and J.-S. Shue, "Image compression using finite-state vector quantization with derailment compensation," *IEEE Transactions on Circuits and systems for Video Technology*, vol. 3, pp. 341–349, October 1993.

[479] N. Nasrabadi, C. Choo, and Y. Feng, "Dynamic finite-state vector quantisation of digital images," *IEEE Transactions on Communications*, vol. 42, pp. 2145–2154, May 1994.

[480] V. Sitaram, C. Huang, and P. Israelsen, "Efficient codebooks for vector quantisation image compression with an adaptive tree search algorithm," *IEEE Transactions on Communications*, vol. 42, pp. 3027–3033, November 1994.

[481] W. Yip, S. Gupta, and A. Gersho, "Enhanced multistage vector quantisation by joint codebook design," *IEEE Transactions on Communications*, vol. 40, pp. 1693–1697, November 1992.

[482] L. Po and C. Chan, "Adaptive dimensionality reduction techniques for tree-structured vector quantisation," *IEEE Transactions on Communications*, vol. 42, pp. 2246–2257, June 1994.

[483] L. Lu and W. Pearlman, "Multi-rate video coding using pruned tree-structured vector quantization," in *Proceedings of the IEEE International Conference on Acoustics, Speech and Signal Processing (ICASSP'93)*, vol. 5, (Minneapolis, MN), pp. 253–256, IEEE, 27–30 April 1993.

[484] F. Bellifemine and R. Picco, "Video signal coding with DCT and vector quantisation," *IEEE Transactions on Communications*, vol. 42, pp. 200–207, February 1994.

[485] K. Ngan and K. Sin, "HDTV coding using hybrid MRVQ/DCT," *IEEE Transactions on Circuits and Systems for Video Technology*, vol. 3, pp. 320–323, August 1993.

[486] D. Kim and S. Lee, "Image vector quantiser based on a classification in the DCT domain," *IEEE Transactions on Communications*, pp. 549–556, April 1991.

[487] L. Torres and J. Huguet, "An improvement on codebook search for vector quantisation," *IEEE Transactions on Communications*, vol. 42, pp. 208–210, February 1994.

[488] J. Streit and L. Hanzo, "Dual-mode vector-quantised low-rate cordless videophone systems for indoors and outdoors applications," *IEEE Transactions on Vehicular Technology*, vol. 46, pp. 340–357, May 1997.

[489] X. Zhang, M. Cavenor, and J. Arnold, "Adaptive quadtree coding of motion -compensated image sequences for use on the broadband ISDN," *IEEE Transactions on Circuits and Systems for Video Technology*, vol. 3, pp. 222–229, June 1993.

[490] J. Vaisey and A. Gersho, "Image compression with variable block size segmentation," *IEEE Transactions on Signal Processing*, vol. 40, pp. 2040–2060, August 1992.

[491] M. Lee and G. Crebbin, "Classified vector quantisation with variable block-size DCT models," *IEE Proceedings, Vision, Image and Signal Processing*, pp. 39–48, February 1994.

[492] E. Shustermann and M. Feder, "Image compression via improved quadtree decomposition algorithms," *IEEE Transactions on Image Processing*, vol. 3, pp. 207–215, March 1994.

[493] F. DeNatale, G. Desoli, and D. Giusto, "A novel tree-structured video codec," in *Proceedings of the IEEE International Conference on Acoustics, Speech and Signal Processing (ICASSP'94)*, vol. 5, (Adelaide, Australia), pp. 485–488, IEEE, 19–22 April 1994.

[494] M. Hennecke, K. Prasad, and D. Stork, "Using deformable templates to infer visual speech dynamics," in *Proceedings of the 28th Asilomar Conference on Signals, Systems and Computers*, vol. 1, (Pacific Grove, CA), pp. 578–582, 30 October – 2 November 1994.

[495] G. Wolf et al., "Lipreading by neural networks: Visual preprocessing, learning and sensory integration," *Proceedings of the neural information processing systems*, vol. 6, pp. 1027–1034, 1994.

[496] J. Streit and L. Hanzo, "Quad-tree based parametric wireless videophone systems," *IEEE Transactions Video Technology*, vol. 6, pp. 225–237, April 1996.

[497] E. Biglieri and M. Luise, "Coded modulation and bandwidth-efficient transmission," in *Proceedings of the Fifth Tirrenia Internatioanal Workshop*, (Netherlands), 8–12 September 1991.

[498] L.-F. Wei, "Trellis-coded modulation with multidimensional constellations," *IEEE Transactions on Information Theory*, vol. IT-33, pp. 483–501, July 1987.

[499] ITU-T, *ISO/IEC-CD-11172 — Coding of moving pictures and associated audio for digital storage*.

[500] ITU-T, *Recommendation H.261: Video codec for audiovisual services at px64 Kbit/s*, March 1993.

[501] D. Redmill and N. Kingsbury, "Improving the error resilience of entropy encoded video signals," in *Proceedings of the Conference on Image Processing: Theory and Applications (IPTA)*, (Netherlands), pp. 67–70, Elsevier, 1993.

[502] S. Emani and S. Miller, "DPCM picture transmission over noisy channels with the aid of a markov model," *IEEE Transactions on Image Processing*, vol. 4, pp. 1473–1481, November 1995.

[503] M. Chan, "The performance of DPCM operating on lossy channels with memory," *IEEE Transactions on Communications*, vol. 43, pp. 1686–1696, April 1995.

[504] N. Jayant, "Adaptive quantization with a one-word memory," *Bell System Technical Journal*, vol. 52, pp. 1119–1144, September 1973.

[505] L. Zetterberg, A. Ericsson, and C. Couturier, "DPCM picture coding with two-dimensional control of adaptive quantisation," *IEEE Transactions on Communications*, vol. 32, no. 4, pp. 457–642, 1984.

[506] C. Hsieh, P. Lu, and W. Liou, "Adaptive predictive image coding using local characteristics," *IEE Proceedings*, vol. 136, pp. 385–389, December 1989.

[507] P. Wellstead, G. Wagner, and J. Caldas-Pinto, "Two-dimensional adaptive prediction, smoothing and filtering," *Proceedings of the IEE*, vol. 134, pp. 253–266, June 1987.

[508] O. Mitchell, E. Delp, and S. Carlton, "Block truncation coding: A new approach to image compression," in *IEEE International Conference on Communications (ICC)*, pp. 12B.1.1–12B.1.4, 1978.

[509] E. Delp and O. Mitchell, "Image compression using block truncation coding," *IEEE Transactions on Communications*, vol. 27, pp. 1335–1342, September 1979.

[510] D. Halverson, N. Griswold, and G. Wiese, "A generalized block truncation coding algorithm for image compression," *IEEE Transactions Acoustics, Speech and Signal Processing*, vol. 32, pp. 664–668, June 1984.

[511] G. Arce and N. Gallanger, "BTC image coding using median filter roots," *IEEE Transactions on Communications*, vol. 31, pp. 784–793, June 1983.

[512] M. Noah, "Optimal Lloyd-Max quantization of LPC speech parameters," in *Proceedings of International Conference on Acoustics, Speech, and Signal Processing, ICASSP'84*, (San Diego, CA), pp. 1.8.1–1.8.4, IEEE, 19–21 March 1984.

[513] R. Crochiere, S. Webber, and J. Flangan, "Digital coding of speech in sub-bands," *Bell System Technology Journal*, vol. 52, pp. 1105–1118, 1973.

[514] R. Crochiere, "On the design of sub-band coders for low bit rate speech communication," *Bell System Technology Journal*, vol. 56, pp. 747–770, 1977.

[515] J. Woods and S. O'Neil, "Subband coding of images," *IEEE Transactions on Accoustic, Sound and Signal Processing*, vol. 34, pp. 1278–1288, October 1986.

[516] J. Woods, ed., *Subband Image Coding*. Dordrecht: Kluwer Academic Publishers, March 1991.

[517] H. Gharavi and A. Tabatabai, "Subband coding of digital images using two-dimensional quadrature mirror filtering," in *Proceedings of SPIE*, 1986.

[518] H. Gharavi and A. Tabatabai, "Subband coding of monochrome and color images," *IEEE Transactions on Circuits and Systems*, vol. 35, pp. 207–214, February 1988.

[519] H. Gharavi, "Subband coding algorithms for video applications: Videophone to HDTV-conferencing," *IEEE Transactions on Circuits and Systems for Video Technology*, vol. 1, pp. 174–183, February 1991.

[520] A. Alasmari, "An adaptive hybrid coding scheme for HDTV and digital video sequences," *IEEE Transactions on consumer electronics*, vol. 41, no. 3, pp. 926–936, 1995.

[521] K. Irie et al., "High-quality subband coded for HDTV transmission," *IEEE Transactions on Circuits and Systems for Video Technology*, vol. 4, pp. 195–199, April 1994.

[522] E. Simoncelli and E. Adelson, "Subband transforms," in Woods [516], pp. 143–192.

[523] K. Irie and R. Kishimoto, "A study on perfect reconstructive subband coding," *IEEE Transactions on Circuits and Systems for Video Technology*, vol. 1, pp. 42–48, January 1991.

[524] J. Woods and T. Naveen, "A filter based bit allocation scheme for subband compression of HDTV," *IEEE Transactions on Image Processing*, vol. 1, pp. 436–440, July 1992.

[525] D. Esteban and C. Galand, "Application of quadrature mirror filters to split band voice coding scheme," in *Proceedings of International Conference on Acoustics, Speech, and Signal Processing, ICASSP'77*, (Hartford, CT), pp. 191–195, IEEE, 9–11 May 1977.

[526] J. Johnston, "A filter family designed for use in quadrature mirror filter banks," in *Proceedings of International Conference on Acoustics, Speech, and Signal Processing, ICASSP'80*, (Denver, CO), pp. 291–294, IEEE, 9–11 April 1980.

[527] H. Nussbaumer, "Complex quadrature mirror filters," in *Proceedings of International Conference on Acoustics, Speech, and Signal Processing, ICASSP'83*, (Boston, MA), pp. 221–223, IEEE, 14–16 April 1983.

[528] C. Galand and H. Nussbaumer, "New quadrature mirror filter structures," *IEEE Transactions on Acoustic Speech Signal Processing*, vol. ASSP-32, pp. 522–531, June 1984.

[529] R. Crochiere and L. Rabiner, *Multirate Digital Processing*. Englewood Cliffs, NJ: Prentice-Hall, 1993.

[530] S. Aase and T. Ramstad, "On the optimality of nonunitary filter banks in subband coders," *IEEE Transactions on Image Processing*, vol. 4, pp. 1585–1591, December 1995.

[531] V. Nuri and R. Bamberger, "Size limited filter banks for subband image compression," *IEEE Transactions on Image Processing*, vol. 4, pp. 1317–1323, September 1995.

[532] H. Gharavi, "Subband coding of video signals," in Woods [516], pp. 229–271.

[533] O. Egger, W. Li, and M. Kunt, "High compression image coding using an adaptive morphological subband decomposition," *Proceedings of the IEEE*, vol. 83, pp. 272–287, February 1995.

[534] P. Westerink and D. Boekee, "Subband coding of color images," in Woods [516], pp. 193–228.

[535] Q. Nguyen, "Near-perfect-reconstruction pseudo-QMF banks," *IEEE Transactions on signal processing*, vol. 42, pp. 65–76, January 1994.

[536] S.-M. Phoong, C. Kim, P. Vaidyanathan, and R. Ansari, "A new class of two-channel biorthogonal filter banks and wavelet bases," *IEEE Transactions on Signal Processing*, vol. 43, pp. 649–665, March 1995.

[537] E. Jang and N. Nasrabadi, "Subband coding with multistage VQ for wireless image communication," *IEEE Transactions in Circuit and Systems for Video Technology*, vol. 5, pp. 347–253, June 1995.

[538] P. Cosman, R. Gray, and M. Vetterli, "Vector quantisation of image subbands: A survey," *IEEE Transactions on Image Processing*, vol. 5, pp. 202–225, February 1996.

[539] ITU, *Joint Photographic Experts Group ISO/IEC, JTC/SC/WG8, CCITT SGVIII. JPEG technical specifications, revision 5. Report JPEG-8-R5*, January 1990.

[540] P. Franti and O. Nevalainen, "Block truncation coding with entropy coding," *IEEE Transcations on Communications*, vol. 43, no. 4, pp. 1677–1685, 1995.

[541] V. Udpikar and J. Raina, "BTC image coding using vector quantisation," *IEEE Transactions on Communications*, vol. 35, pp. 353–359, March 1987.

[542] International Standards Organization, *ISO/IEC 11172 MPEG 1 International Standard, 'Coding of moving pictures and associated audio for digital storage media up to about 1.5 Mbit/s, Parts 1–3*.

[543] International Standards Organization, *ISO/IEC CD 13818 MPEG 2 International Standard, Information Technology, Generic Coding of Moving Video and Associated Audio Information, Parts 1–3*.

[544] Telenor Research and Development, P.O.Box 83, N-2007 Kjeller, Norway, *Video Codec Test Model 'TMN 5', ITU Study Group 15, Working Party 15/1*.

[545] D. Choi, "Frame alignment in a digital carrier system — a tutorial," *IEEE Communications Magazin*, vol. 28, pp. 46–54, February 1990.

[546] ITU (formerly CCITT), *ITU Recommendation X25*, 1993.

[547] M. Al-Subbagh and E. Jones, "Optimum patterns for frame alignment," *IEE Proceedings*, vol. 135, pp. 594–603, December 1988.

[548] T. Turletti, "A H.261 software codec for videoconferencing over the internet," Tech. Rep. 1834, INRIA, 06902 Sophia-Antipolis, France, January 1993.

[549] N. Kenyon and C. Nightingale, eds., *Audiovisual Telecommunications*. London: Chapman and Hall, 1992.

[550] N. MacDonald, "Transmission of compressed video over radio links," *BT technology Journal*, vol. 11, pp. 182–185, April 1993.

[551] M. Khansari, A. Jalali, E. Dubois, and P. Mermelstein, "Robust low bit-rate video transmission over wireless access systems," in *Proceedings of International Communications Conference (ICC)*, pp. 571–575, 1994.

[552] N. Cheng, *Error resilient video coding for Noisy Channels*. PhD thesis, Department of Engineering, University of Cambridge, 1991.

[553] D. Redmill, *Image and Video Coding for Noisy Channels*. PhD thesis, Signal Processing and Communication Laboratory, Department of Engineering, University of Cambridge, November 1994.

[554] Y. Matsumura, S. Nakagawa, and T. Nakai, "Very low bit rate video coding with error resilience," in *Proceedings of International Workshop on Coding Techniques for Very Low Bitrate Video (VLBV'95)*, (Shinagawa, Tokyo, Japan), pp. L–1, 8–10 November 1995.

[555] K. Ngan and D. Chai, "Enhancement of image quality in VLBR coding," in *Proceedings of International Workshop on Coding Techniques for Very Low Bit-rate Video (VLBV'95)*, (Shinagawa, Tokyo, Japan), pp. L–3, 8–10 November 1995.

[556] K. Ngan and D. Chai, "Very low bit rate video coding using H.263 coder," *IEEE Transactions on Circuits and Systems for Video Technology*, vol. 6, pp. 308–312, June 1996.

[557] W Webb and L. Hanzo, "Square QAM," in *Modern Quadrature Amplitude Modulation: Principles and Applications for Wireless Communications* [29], ch. 5, pp. 156–169.

[558] IBM Corp., White Plains, NY, *General Information: Binary Synchronous Communication*, IBM Publication GA27-3004, 1969.

[559] S. Sampei, S. Komaki, and N. Morinaga, "Adaptive modulation/TDMA scheme for large capacity personal multi-media communication systems," *IEICE Transactions on Communications (Japan)*, vol. E77-B, pp. 1096–1103, September 1994.

[560] W. Webb and L. Hanzo, "Variable rate QAM," in *Modern Quadrature Amplitude Modulation: Principles and Applications for Wireless Communications* [29], ch. 13, pp. 384–406.

[561] M. Ghanbari and V. Seferidis, "Cell-loss concealment in ATM video codecs," *IEEE Transactions on Circuits and Systems for Video Technology*, vol. 3, pp. 238–247, June 1993.

[562] W. Chung, F. Kossentini, and M. Smith, "An efficient motion estimation technique based on a rate-distortion criterion," in *Proceedings of the IEEE International Conference on Acoustics, Speech and Signal Processing (ICASSP'96)*, (Atlanta, GA), pp. 1977–1980, IEEE, 7–10 May 1996.

[563] Telenor Research and Development, P.O.Box 83, N-2007 Kjeller, Norway, *H.263 Software Codec*. http://www.nta.no/brukere/DVC.

[564] W. Ding and B. Liu, "Rate control of MPEG video coding and recording by rate-quantization modeling," *IEEE Transactions on Circuits and Systems for Video Technology*, vol. 6, pp. 12–20, February 1996.

[565] G. Schuster and A. Katsaggelos, "A video compression scheme with optimal bit allocation between displacement vector field and displaced frame difference," in *Proceedings of the IEEE International Conference on Acoustics, Speech and Signal Processing (ICASSP'96)*, (Atlanta, GA), pp. 1967–1970, IEEE, 7–10 May 1996.

[566] F. Martins, W. Ding, and E. Feig, "Joint control of spatial quantization and temporal sampling for very low bitrate video," in *Proceedings of the IEEE International Conference on Acoustics, Speech and Signal Processing (ICASSP'96)*, (Atlanta, GA), pp. 2074–2077, IEEE, 7–10 May 1996.

[567] T. Wiegand, M. Lightstone, and D. Mukherjee, "Rate-distortion optimized mode selection for very low bit rate video coding and the emerging H.263 standard," *IEEE Transactions on Circuits and Systems for Video Technology*, vol. 6, pp. 182–190, April 1996.

[568] A. Paulraj, "Diversity techniques," in Gibson [292], ch. 12, pp. 166–176.

[569] A. Mämmelä, *Diversity receivers in a fast fading multipath channel*. PhD thesis, Department of Electrical Engineering, University of Oulu, Finland, 1995.

[570] P. Crespo, R. M. Pelz, and J. Cosmas, "Channel error profile for DECT," *IEE Proceedings on Communications*, vol. 141, pp. 413–420, December 1994.

[571] M. Zorzi, "Power control and diversity in mobile radio cellular systems in the presence of rician fading and log-normal shadowing," *IEEE Transactions on Vehicular Technology*, vol. 45, pp. 373–382, May 1996.

[572] Y.-W. Leung, "Power control in cellular networks subject to measurement error," *IEEE Transsactions on Communications*, vol. 44, pp. 772–775, July 1996.

[573] S. Ariyavisitakul and L. Chang, "Signal and interference statistics of a CDMA system with feedback power control," *IEEE Transactions on Communications*, vol. 41, pp. 1626–1634, November 1993.

[574] R. Pichna and Q. Wang, "Power control," in Gibson [292], ch. 23, pp. 370–380.

[575] T.-H. Lee, J.-C. Lin, and Y. Su, "Downlink power control algorithms for cellular radio systems," *IEEE Transactions on Vehicular Technology*, vol. 44, pp. 89–94, February 1995.

[576] M. Austin and G. Stüber, "In-service signal quality estimation for TDMA cellular systems," *Wireless Personal Communications*, vol. 2, pp. 245–254, 1995. Kluwer Academic Publishers.

[577] J.-I. Chuang and N. Sollenberger, "Uplink power control for TDMA portable radio channels," *IEEE Transactions on Vehicular Technology*, vol. 43, pp. 33–39, February 1994.

[578] P. Kumar, R. Yates, and J. Holtzman, "Power control based on bit error rate (BER) measurements," in *Proceedings of the Military Communications Conference (MILCOM)*, (San Diego, CA), 5–8 November 1995.

[579] W. Press, S. Teukolsky, W. Vetterling, and B. Flannery, "Minimization or maximization of functions," in *Numerical Recipes in C* [148], ch. 10, pp. 394–455.

[580] L. Hanzo, W. Webb, and T. Keller, *Single- and Multi-carrier Quadrature Amplitude Modulation*. New York: John Wiley-IEEE Press, April 2000.

[581] C. Douillard, A. Picart, M. Jézéquel, P. Didier, C. Berrou, and A. Glavieux, "Iterative correction of intersymbol interference: Turbo-equalization," *European Transactions on Communications*, vol. 6, pp. 507–511, 1995.

[582] M. Gertsman and J. Lodge, "Symbol-by-symbol MAP demodulation of CPM and PSK signals on Rayleigh flat-fading channels," *IEEE Transactions on Communications*, vol. 45, pp. 788–799, July 1997.

[583] I. Marsland, P. Mathiopoulos, and S. Kallel, "Non-coherent turbo equalization for frequency selective Rayleigh fast fading channels," in *Proceedings of the International Symposium on Turbo Codes & Related Topics*, (Brest, France), pp. 196–199, 3–5 September 1997.

[584] Q. Dai and E. Shwedyk, "Detection of bandlimited signals over frequency selective Rayleigh fading channels," *IEEE Transactions on Communications*, pp. 941–950, February/March/April 1994.

[585] G. Bauch, H. Khorram, and J. Hagenauer, "Iterative equalization and decoding in mobile communications systems," in *European Personal Mobile Communications Conference*, pp. 301–312, 1997.

[586] M. Moher, "Decoding via cross-entropy minimization," in *Proceedings of the IEEE Global Telecommunications Conference 1993*, (Houston, TX), pp. 809–813, 29 November – 2 December 1993.

[587] G. Bauch and V. Franz, "Iterative equalisation and decoding for the GSM-system," in *Proceedings of IEEE Vehicular Technology Conference (VTC'98)*, (Ottawa, Canada), pp. 2262–2266, IEEE, 18–21 May 1998.

[588] D. Raphaeli and Y. Zarai, "Combined turbo equalization and turbo decoding," *IEEE Communications Letters*, vol. 2, pp. 107–109, April 1998.

[589] K. Narayanan and G. Stuber, "A serial concatenation approach to iterative demodulation and decoding," *IEEE Transactions on Communications*, vol. 47, pp. 956–961, July 1999.

[590] B. Yeap, T. Liew, J. Hamorsky, and L. Hanzo, "Comparative study of turbo equalisers using convolutional codes and block-based turbo-codes for GMSK modulation," in *Proceeding of VTC'99 (Fall)*, (Amsterdam, Netherlands), pp. 2974–2978, IEEE, 19–22 September 1999.

[591] H. Matsuoka, S. Sampei, N. Morinaga, and Y. Kamio, "Adaptive modulation system with variable coding rate concatenated code for high quality multi-media communications systems," in *Proceedings of IEEE VTC'96*, vol. 1, (Atlanta, GA), pp. 487–491, IEEE, 28 April–1 May 1996.

[592] A. Goldsmith and S. Chua, "Adaptive coded modulation for fading channels," in *Proceedings of IEEE International Conference on Communications*, vol. 3, (Montreal, Canada), pp. 1488–1492, 8–12 June 1997.

[593] J. Cheung, *Adaptive Equalisers for Wideband TDMA Mobile Radio*. PhD thesis, Department of Electronics and Computer Science, University of Southampton, 1991.

[594] N. Färber, E. Steinbach, and B. Girod, "Robust H.263 compatible transmission for mobile video server access," in *Proc of First International Workshop on Wireless Image/Video Communications*, (Loughborough), pp. 122–124, 4–5 September 1996.

[595] A. Sadka, F. Eryurtlu, and A. Kondoz, "Improved performance H.263 under erroneous transmission conditions," *Electronics Letters*, vol. 33, pp. 122–124, 16 January 1997.

[596] A. Klein, R. Pirhonen, J. Skoeld, and R. Suoranta, "FRAMES multiple access mode 1 — wideband TDMA with and without spreading," in *Proceedings of IEEE International Symposium on Personal, Indoor and Mobile Radio Communications, PIMRC'97*, vol. 1, (Marina Congress Centre, Helsinki, Finland), pp. 37–41, IEEE, 1–4 September 1997.

[597] J. Torrance and L. Hanzo, "Latency and networking aspects of adaptive modems over slow indoors rayleigh fading channels," *IEEE Transactions on Vehicular Technology*, vol. 48, no. 4, pp. 1237–1251, 1998.

[598] L. Hanzo, P. Cherriman, and J. Streit, "Video compression and communications over wireless channels: From second to third generation systems, WLANs and beyond." IEEE Press, 2001 (For detailed contents please refer to http://www-mobile.ecs.soton.ac.uk/).

[599] P. Cherriman, C. Wong, and L. Hanzo, "Multi-mode H.263-assisted video telephony using wideband adaptive burst-by-burst modems," Core programme deliverable — terminals work area, Mobile VCE, 1998.

[600] P. Cherriman, T. Keller, and L. Hanzo, "Orthogonal frequency division multiplex transmission of H.263 encoded video over highly frequency-selective wireless networks," *IEEE Transactions on Circuits and Systems for Video Technology*, vol. 9, pp. 701–712, August 1999.

[601] M. Sandell, J.-J. van de Beek, and P. Börjesson, "Timing and frequency synchronisation in OFDM systems using the cyclic prefix," in *Proceedings of International Symposium on Synchronisation*, (Essen, Germany), pp. 16–19, 14–15 December 1995.

[602] J. D. Marca and N. Jayant, "An algorithm for assigning binary indices to the codevectors of a multi-dimensional quantizer," in *Proceedings of IEEE International Conference on Communications 1987*, (Seattle, WA), pp. 1128–1132, IEEE, 7–10 June 1987.

[603] T. Keller and L. Hanzo, "Orthogonal frequency division multiplex synchronisation techniques for wireless local area networks," in *Proceedings of IEEE International Symposium on Personal, Indoor, and Mobile Radio Communications (PIMRC'96)*, vol. 3, (Taipei, Taiwan), pp. 963–967, IEEE, 15–18 October 1996.

[604] K. Pehkonen, H. Holma, I. Keskitalo, E. Nikula, and T. Westman, "Performance analysis of TDMA and CDMA based air interface solutions for UMTS high bit rate services," in *Proceedings of IEEE International Symposium on Personal, Indoor and Mobile Radio Communications, PIMRC'97*, vol. 1, (Marina Congress Centre, Helsinki, Finland), pp. 22–26, IEEE, 1–4 September 1997.

[605] K. Pajukoski and J. Savusalo, "Wideband CDMA test system," in *Proceedings of IEEE International Symposium on Personal, Indoor and Mobile Radio Communications, PIMRC'97*, vol. 2, (Marina Congress Centre, Helsinki, Finland), pp. 669–673, IEEE, 1–4 September 1997.

[606] R. Blahut, "Transform techniques for error control codes," *IBM Journal on Research and Development*, vol. 23, pp. 299–315, May 1979.

[607] J. Torrance, L. Hanzo, and T. Keller, "Interference aspects of adaptive modems over slow rayleigh fading channels," *IEEE Transactions on Vehicular Technology*, vol. 48, pp. 1527–1545, September 1999.

[608] T. Liew, C. Wong, and L. Hanzo, "Block turbo coded burst-by-burst adaptive modems," in *Proceedings of Microcoll'99, Budapest, Hungary*, pp. 59–62, 21–24 March 1999.

[609] V. Lau and M. Macleod, "Variable rate adaptive trellis coded QAM for high bandwidth efficiency applications in rayleigh fading channels," in *Proceedings of IEEE Vehicular Technology Conference (VTC'98)*, (Ottawa, Canada), pp. 348–352, IEEE, 18–21 May 1998.

[610] M. Yee and L. Hanzo, "Upper-bound performance of radial basis function decision feedback equalised burst-by-burst adaptive modulation," in *Proceedings of ECMCS'99*, (Krakow, Poland), 24–26 June 1999.

[611] T. Keller and L. Hanzo, "Adaptive orthogonal frequency division multiplexing schemes," in *Proceeding of ACTS Mobile Communication Summit '98*, (Rhodes, Greece), pp. 794–799, ACTS, 8–11 June 1998.

[612] E. Kuan, C. Wong, and L. Hanzo, "Burst-by-burst adaptive joint detection CDMA," in *Proceeding of VTC'99 (Spring)*, (Houston, TX), IEEE, 16–20 May 1999.

[613] A. Czylwik, "Adaptive OFDM for wideband radio channels," in *Proceeding of IEEE Global Telecommunications Conference, Globecom 96*, (London), pp. 713–718, IEEE, 18–22 November 1996.

[614] R. Fischer and J. Huber, "A new loading algorithm for discrete multitone transmission," in *Proceeding of IEEE Global Telecommunications Conference, Globecom 96*, (London), pp. 713–718, IEEE, 18–22 November 1996.

[615] P. Chow, J. Cioffi, and J. Bingham, "A practical discrete multitone transceiver loading algorithm for data transmission over spectrally shaped channels," *IEEE Transactions on Communications*, vol. 48, pp. 772–775, 1995.

[616] H. Rohling and R. Grünheid, "Peformance of an OFDM-TDMA mobile communication system," in *Proceeding of IEEE Global Telecommunications Conference, Globecom 96*, (London), pp. 1589–1593, IEEE, 18–22 November 1996.

[617] K. Fazel, S. Kaiser, P. Robertson, and M. Ruf, "A concept of digital terrestrial television broadcasting," *Wireless Personal Communications*, vol. 2, pp. 9–27, 1995.

[618] H. Sari, G. Karam, and I. Jeanclaude, "Transmission techniques for digital terrestrial TV broadcasting," *IEEE Communications Magazine*, pp. 100–109, February 1995.

[619] J. Borowski, S. Zeisberg, J. Hübner, K. Koora, E. Bogenfeld, and B. Kull, "Performance of OFDM and comparable single carrier system in MEDIAN demonstrator 60GHz channel," in *Proceeding of ACTS Mobile Communication Summit '97*, (Aalborg, Denmark), pp. 653–658, ACTS, 7–10 October 1997.

[620] P. Cherriman, T. Keller, and L. Hanzo, "Constant-rate turbo-coded and block-coded orthogonal frequency division multiplex videophony over UMTS," in *Proceeding of Globecom'98*, vol. 5, (Sydney, Australia), pp. 2848–2852, IEEE, 8–12 November 1998.

[621] J. Woodard, T. Keller, and L. Hanzo, "Turbo-coded orthogonal frequency division multiplex transmission of 8 kbps encoded speech," in *Proceeding of ACTS Mobile Communication Summit '97*, (Aalborg, Denmark), pp. 894–899, ACTS, 7–10 October 1997.

[622] Y. Li and N. Sollenberger, "Interference suppression in OFDM systems using adaptive antenna arrays," in *Proceeding of Globecom'98*, (Sydney, Australia), pp. 213–218, IEEE, 8–12 November 1998.

[623] F. Vook and K. Baum, "Adaptive antennas for OFDM," in *Proceedings of IEEE Vehicular Technology Conference (VTC'98)*, vol. 2, (Ottawa, Canada), pp. 608–610, IEEE, 18–21 May 1998.

[624] ETSI, *Digital Video Broadcasting (DVB); Framing structure, channel coding and modulation for digital terrestrial television*, August 1997. EN 300 744 V1.1.2.

[625] ETSI, *Digital Video Broadcasting (DVB); Framing structure, channel coding and modulation for cable systems*, December 1997. EN 300 429 V1.2.1.

[626] ETSI, *Digital Video Broadcasting (DVB); Framing structure, channel coding and modulation for 11/12 GHz Satellite Services*, August 1997. EN 300 421 V1.1.2.

[627] A. Michelson and A. Levesque, *Error Control Techniques for Digital Communication*. New York: Wiley-Interscience, 1985.

[628] S. O'Leary and D. Priestly, "Mobile broadcasting of DVB-T signals," *IEEE Transactions on Broadcasting*, vol. 44, pp. 346–352, September 1998.

[629] W.-C. Lee, H.-M. Park, K.-J. Kang, and K.-B. Kim, "Performance analysis of viterbi decoder using channel state information in COFDM system," *IEEE Transactions on Broadcasting*, vol. 44, pp. 488–496, December 1998.

[630] S. O'Leary, "Hierarchical transmission and COFDM systems," *IEEE Transactions on Broadcasting*, vol. 43, pp. 166–174, June 1997.

[631] L. Thibault and M. Le, "Performance evaluation of COFDM for digital audoo broadcasting Part I: parametric study," *IEEE Transactions on Broadcasting*, vol. 43, pp. 64–75, March 1997.

[632] B. Haskell, A. Puri, and A. Netravali, *Digital Video: An Introduction To MPEG-2*. Digital Multimedia Standards Series, London: Chapman and Hall, 1997.

[633] *ISO/IEC 13818-2: Information Technology — Generic Coding of Moving Pictures and Associated Audio Information — Part 2: Video*, March 1995.

[634] L. Hanzo and J. Woodard, "An intelligent voice communications system for indoors communications," in *Proceedings of IEEE Vehicular Technology Conference (VTC'95)*, vol. 4, (Chicago), pp. 735–749, IEEE, 15–28 July 1995.

[635] P. Shelswell, "The COFDM modulation system: the heart of digital audio broadcasting," *Electronics & Communication Engineering Journal*, vol. 7, pp. 127–136, June 1995.

[636] P. Robertson, E. Villebrun, and P. Hoeher, "A comparison of optimal and sub-optimal MAP decoding algorithms operating in the log domain," in *Proceedings of the International Conference on Communications*, pp. 1009–1013, June 1995.

[637] S. Wicker, *Error Control Systems for Digital Communication and Storage*. Englewood Cliffs, NJ: Prentice-Hall, 1994.

[638] A. Barbulescu and S. Pietrobon, "Interleaver design for turbo codes," *IEE Electronic Letters*, pp. 2107–2108, December 1994.

[639] M. Failli, "Digital land mobile radio communications COST 207," tech. rep., European Commission, 1989.

[640] H. Gharavi and S. Alamouti, "Multipriority video transmission for third-generation wireless communication system," in Gharavi and Hanzo [642], pp. 1751–1763.

[641] A. Aravind, M. Civanlar, and A. Reibman, "Packet loss resilience of MPEG-2 scalable video coding algorithms," *IEEE Transaction on Circuits And Systems For Video Technology*, vol. 6, pp. 426–435, October 1996.

[642] H. Gharavi and L. Hanzo, eds., *Proceedings of the IEEE*, vol. 87, October 1999.

[643] G. Reali, G. Baruffa, S. Cacopardi, and F. Frescura, "Enhancing satellite broadcasting services using multiresolution modulations," *IEEE Transactions on Broadcasting*, vol. 44, pp. 497–506, December 1998.

[644] Y. Hsu, Y. Chen, C. Huang, and M. Sun, "MPEG-2 spatial scalable coding and transport stream error concealment for satellite TV broadcasting using Ka-band," *IEEE Transactions on Broadcasting*, vol. 44, pp. 77–86, March 1998.

[645] L. Atzori, F. D. Natale, M. D. Gregario, and D. Giusto, "Multimedia information broadcasting using digital TV channels," *IEEE Transactions on Broadcasting*, vol. 43, pp. 383–392, December 1997.

[646] W. Sohn, O. Kwon, and J. Chae, "Digital DBS system design and implementation for TV and data broadcasting using Koreasat," *IEEE Transactions on Broadcasting*, vol. 44, pp. 316–323, September 1998.

[647] J. Griffiths, *Radio Wave Propagation and Antennas — An Introduction*. Englewood Cliffs, NJ: Prentice-Hall, 1987.

[648] M. Karaliopoulos and F.-N. Pavlidou, "Modelling the land mobile satellite channel: a review," *Electronics and Communication Engineering Journal*, vol. 11, pp. 235–248, October 1999.

[649] J. Goldhirsh and W. Vogel, "Mobile satellite system fade statistics for shadowing and multipath from roadside trees at UHF and L-band," *IEEE Transactions on Antennas and Propagation*, vol. 37, pp. 489–498, April 1989.

[650] W. Vogel and J. Goldhirsh, "Multipath fading at L band for low elevation angle, land mobile satellite scenarios," *IEEE Journal on Selected Areas in Communications*, vol. 13, pp. 197–204, February 1995.

[651] W. Vogel and G. Torrence, "Propagation measurements for satellite radio reception inside buildings," *IEEE Transactions on Antennas and Propagation*, vol. 41, pp. 954–961, July 1993.

[652] W. Vogel and U. Hong, "Measurement and modelling of land mobile satellite propagation at UHF and L-band," *IEEE Transactions on Antennas and Propagation*, vol. 36, pp. 707–719, May 1988.

[653] S. Saunders, C. Tzaras, and B. Evans, "Physical statistical propagation model for mobile satellite channel," tech. rep., European Commission, 1998.

[654] S. Saunders, *Antennas and Propagation for Wireless Communication Systems Concept and Design*. New York: John Wiley and Sons, 1999.

[655] K. Wesolowsky, "Analysis and properties of the modified constant modulus algorithm for blind equalization," *European Transactions on Telecommunication*, vol. 3, pp. 225–230, May–June 1992.

[656] M. Goursat and A. Benveniste, "Blind equalizers," *IEEE Transactions on Communications*, vol. COM–28, pp. 871–883, August 1984.

[657] G. Picchi and G. Prati, "Blind equalization and carrier recovery using a "stop-and-go" decision-directed algorithm," *IEEE Transactions on Communications*, vol. COM–35, pp. 877–887, September 1987.

[658] A. Polydoros, R. Raheli, and C. Tzou, "Per-survivor processing: a general approach to MLSE in uncertain environments," *IEEE Transactions on Communications*, vol. COM–43, pp. 354–364, February–April 1995.

[659] D. Godard, "Self-recovering equalization and carrier tracking in two-dimensional data communication systems," *IEEE Transactions on Communications*, vol. COM–28, pp. 1867–1875, November 1980.

[660] Y. Sato, "A method of self-recovering equalization for multilevel amplitude–modulation systems," *IEEE Transactions on Communications*, vol. COM–23, pp. 679–682, June 1975.

[661] Z. Ding, R. Kennedy, B. Anderson, and R. Johnson, "Ill-convergence of Godard blind equalizers in data communications systems," *IEEE Transactions on Communications*, vol. COM-39, pp. 1313–1327, September 1991.

[662] Y.-Q. Zhang, F. Pereira, T. Sikora, and C. Reader (Guest Editors), "Special issue on MPEG-4," *IEEE Transactions on Circuits and Systems for Video Technology*, vol. 7, February 1997.

[663] L. Chiariglione, "MPEG and multimedia communication," *IEEE Transaction On Circuits And Systems For Video Technology*, vol. 7, pp. 5–18, February 1997.

[664] T. Sikora, "The MPEG-4 video standard verification model," *IEEE Transaction On Circuits And Systems For Video Technology*, vol. 7, pp. 19–31, February 1997.

Subject Index

Symbols
16QAM BER versus SNR performance over AWGN channels 175–179
16QAM constellation comparison 174
16QAM demodulation 178–179
16QAM demodulation in AWGN 178
16QAM square constellation 166
3GPP1 393
3GPP2 393

A
Access control 225
ACELP 201
ACF 480
Acknowledgments . . . . xxix–xxxi
Active/passive concept 503
ACTS (Advanced Communications Technology and Services) 395
ACTS program 339
Adaptive antenna 382, 399, 402, 453
Adaptive differential pulse code modulation 611–612
Adaptive link control in GSM 360–363
Adaptive modulation 190–197
Adaptive modulation performance . . 195–197
Adaptive QAM-based wireless videophony 875–893
Adaptive turbo-coded OFDM-based videotelephony 927–950
Adaptive vector quantization 546
Adaptive video systems 861–1021
Adaptive video transceiver 877–879
Adaptive VQ 546
Additive white Gaussian noise (AWGN) . . . 7, 10, 370, 371, 374, 379
Additive white Gaussian noise channel 6
ADPCM 611
Affine transformation 459
Aging 7
Algorithmic complexity 549–550
Aliasing distortion 620
Analysis filtering 620–623

Analytical 16QAM BER 175
Antenna diversity 800–802
AOFDM modem mode adaptation and signaling 929
AOFDM subband BER estimation . . 929–930
Approval 225
AR model properties 29–30
Architecture of system 1 555–557
Architecture of system 2 557–558
Architecture of systems 3–6 558
ARIB (Association of Radio Industries and Businesses) 393, 395, 454
Arithmetic coding 749, 751
ARQ . . 247, 516, 693, 694, 725, 729–731, 779, 796–798, 806, 810, 811
Assignment probability growth function . 240, 242, 243
ATM 341
Augmented channel occupancy matrix . . . 296
Automatic repeat request . 522–523, 693, 779
Autoregressive model 28–29
AV.26M 779
Average conveyed mutual information per symbol 47
Average error entropy 48
Average information loss 48
Average joint information 46
Average loss of information per symbol . . . 49
AVQ 546
AWGN 7, 515
AWGN channel 7–11

B
B-Codes 655
B-ISDN 341
Background to adaptive modulation 190–193
Bandwidth efficiency 369
Basic CDMA system 366–392
Basic differential pulse code modulation 608–610
Bayes' rule 38–39
Bayes's theorem 175

BER 93
Berlekamp-Massey algorithm 138–144
Berlekamp-Massey decoding example 144–147
Best interferer position 258
Binary phase shift keying (BPSK) . 369, 371, 376
        Probability of bit error 371
Binary symmetric channel example 34–38
Bit sensitivity 513–514, 552–553, 586
Bit sensitivity of codec I and II 515–516
Bit-allocation 584–586, 665–666
Bit-allocation strategy 509–510, 550–551
Blind equalizers 999–1003
Block truncation algorithm 613–614
Block truncation codec implementations 614–615
Block Truncation Coding (BTC) 613–618
Block-based channel coding 107–159
BM decoding example 144
Bose-Chaudhuri-Hocquenghem (BCH) codes 114
Box-Müller algorithm of AWGN generation 87
BPSK 175
Branch metric 100
Brief channel coding history 93–94
Broadcasting 198
BTC 613
        Algorithm 613
        Implementation 614
        Inter-coding 617
        Intra-coding 615
        Performance-intra 616, 618
        Quantizers 615, 616
        Upper bounds (Intra) 614, 617
Burst-by-Burst adaptive videophone performance 879–886
Burst-by-Burst adaptive videophone system 902–908
Butterworth filtering 168

C

Call dropping probability 308
Camera shake 748
Capacity of continuous channels 55–62
Capacity of discrete channels 49–52
Carrier recovery 171
CDF of SINR
        Comparison 280
        With shadowing 279, 280
        Without shadowing 279, 280
CDF of SIR
        Comparison 278
        With shadowing 278
        Without shadowing 277, 278
CDMA 288
cdma2000 .... 393, 439–452
        Channel coding 443–444
        Characteristics 439–441

Handover 450–452
Modulation 445–447
        Down-link 446–447
        Up-link 447
Physical channel 441–443
Random access 447–450
Service multiplexing 443–444
Spreading 445–447
        Down-link 446–447
        Up-link 447
Cellular cell shapes 249
Cellular concept 67–70
Central limit theorem 383
Channel 6
Channel allocation 287–337
        Centrally controlled DCA algorithms 294–295
        Channel borrowing 291–292
        Comparison of FCA and DCA 297
        Cutoff priority scheme 298
        DCA 292–297
        Deadlock definition 295
        Distributed DCA algorithms ... 295–296
        Dynamic channel allocation ... 292–297
        Effect of handovers 298–299
        Effect of transmission power control 299
        Family tree 289
        FCA 289–292, 304
        FCA vs. DCA 297
        Fixed channel allocation .. 289–292, 304
        Flexible channel allocation 292
        Guard channel schemes 298
        Hybrid borrowing 291
        Hybrid channel allocation 297–298
        Instability 295
        Interruption definition 295
        Introduction 287
        Locally distributed DCA algorithms 296–297
        Maximum consecutive outages parameter 302
        Outage SINR threshold 302, 312
        Overview 288
        Performance metrics 306–309
        Physical layer model 302, 312
        Reallocation SINR threshold .. 302, 312
        Results
            Carried traffic 311–313
            Effect of 3dB shadow fading 324–325
            Effect of shadow fading frequency 325–327
            Effect of shadow fading standard deviation 325–328
            LOLIA vs. FCA 312–314
            LOLIA vs. LOMIA vs. Interference threshold DCA 318–320
            LOLIA vs. LOMIA vs. LIA 317–318

LOLIA/LOMIA vs. reuse constraint
        314–317
    Nonuniform traffic 321–323
    Overview 332–335
    SINR profile of cell area 329–332
    Service interruption definition...... 295
    Simple borrowing 291
    Simulation parameters......... 310–311
    Static borrowing 291
Channel allocation algorithms 304–306
Channel borrowing 291–292
Channel coding 247
Channel coding and modulation 586–588
Channel coding theorem.................. 53
Channel encoder........................... 6
Channel estimation 376, 387–392
    Decision-directed 390–392
        Decision feedback structure...... 390
        Decision feedforward structure .. 391
    Pilot channel-assisted 388–389
        Structure 388
        Tone-above-band................ 388
        Tone-in-band................... 388
    Pilot-symbol assisted 389–390
        Data stream 390
Channel gain estimation in PSAM .. 183–185
Channel impairments...................... 4
Channel model 917–918
Channel segregation..................... 296
Channel segregation algorithm 201
Channel sounding 88
Channels — wideband.................... 87
Chebichev filtering 168
Chien search 132
Choice of modem.................. 516–517
Choice of modulation 161–164, 554
Chrominance............................ 604
Circuits for cyclic encoders 122–125
Class one 178
Class two 178
Classification of multiple access techniques 225–
        227
Classified vector quantization........... 548
Classified VQ 548
Clock recovery 172
Co-channel interference 189, 247–285
    Factors controlling................ 248
    Introduction 247
    Maximum SIR..................... 250
    Minimum SIR 250
Coaxial cable............................. 8
Code acquisition............... 376, 380, 385
Code Division Multiple Access (CDMA) 365,
        366
    System model...................... 378
Codebook design 537–539
Codec outline..................... 499–502
Codec performance 666–667

Coding efficiency 16
Coding memory 94
Coding rate......................... 53, 94
Coherence bandwidth 374, 375
Coherent demodulation........ 181, 376, 387
Collision detection 225
Combined source/channel coding ... 694, 780
Communications channels 7
Communications system design considerations
        4
Comparison
    Bit-rate..................... 597, 598
    Error sensitivity 598
    Performance...................... 597
Comparison of subband-adaptive OFDM and
        fixed-mode OFDM transceivers 930–
        936
Complex quadrature mirror filters 626
Complexity considerations and reduction tech-
        niques........................ 508
Constellation design..................... 173
Constellation diagram 167
Constellations.......................... 172
Constraint length 94
Contractive affine transforms............ 459
Contributors xxxi
Convolutional channel coding 93–106
Convolutional encoding 94–96
Cost/gain.............................. 502
Cr_u 604
Cr_v 604
CT2 system 341, 343
Cumulated assignment probability .. 239, 243
Cumulative density function 4
Cutoff priority scheme................... 298
CVQ................................... 548
CWTS (China Wireless Telecommunication
        Standard) 393, 395
Cyclic codes........................... 114
Cyclic encoding 116

D

DAB................................... 909
Data partitioning scheme........... 966–972
Data recovery 171–172
DCA 289, 292–297
    Centralised algorithms............. 294
    Centrally controlled algorithms 294–295
    Centrally located algorithms 294
    Comparison with FCA............. 297
    Distributed algorithms........ 295–296,
        304–306, 317–322
    First available algorithm........... 295
    Highest interference below threshold al-
        gorithm 305
    HTA/MTA algorithm.............. 305
    Interference threshold............. 320
    Least interference algorithm........ 304

Least interference below threshold algorithm........................305
LFA algorithm....................318
LIA algorithm....................304
Locally distributed algorithms 296–297, 306, 312, 314–329
Locally optimized least interference algorithm........306, 307, 314–317, 319–329
Locally optimized most interference algorithm..................306, 307, 317–324
LODA algorithm...................295
LOLIA algorithm........296, 306, 307, 314–317, 319–329
LOMIA algorithm.......296, 306, 307, 317–324
LP-DDCA algorithm...............296
LTA algorithm................305, 318
MSQ algorithm....................295
MTA algorithm....................318
Nearest neighbor algorithms.......295
NN algorithm......................295
NN+1 algorithm...................295
Ring algorithm...................295
Simulation parameters............310
DCS-1800 system...................255, 343
DCT............................475, 639
    B-encoded........................655
    Bit-allocation.................509, 665
    Codec
        Active/passive..............503, 505
        Bit sensitivity..............513, 515
        Bit-allocation...................509
        Complexity.....................508
        Cost gain controlled............506
        Degradation profile.............514
        Erroneous conditions...........512
        Error sensitivity................516
        Gain controlled.................502
        Intra-block sizes................502
        Intra-frame.....................502
        Performance....................511
        PFU degradation................507
        Preselection....................509
        Results.........................510
        Results – "Lab" sequence...511, 513
        Results – "Miss America" sequence 512
    Codec outline......................499
    Codec schematic...................501
    Complexity.......................667
    Efficiency (Inter)..................653
    Efficiency (Intra)..................653
    Error sensitivity.............667, 670
    Inter frame......................650
    Inter/Intra codec.................661
    Inter/Intra decision..............661
    Intra bit-allocation...............643
    Intra performance.................644
    Introduction.....................639
    Multiclass.......................500
    Optimum mask...................642
    Performance.................666, 667
        Erroneous conditions.......668, 669
    PFU.............................504
    Proposed codec...................658
    Quantizer allocation..........497, 498
    Quantizer training............497, 639
    Quantizer types..................500
    Schematic........................659
    Single class.....................496
    Zonal mask......................640
DCT basis images.....................494
DCT codec performance under erroneous conditions...................512–516
DCT codecs..........................475
DCT-based low-rate video transceivers . 516–524
Deadlock............................295
Decimated signals....................621
Decision theory..................175–177
Decomposition algorithmic issues... 573–576
DECT.....................341, 802–805
DECT (Digital Enhanced Cordless Telecommunications)..................395
DECT channels..................804, 805
DECT frame structure................803
DECT system........................343
Definitions..............108–111, 114–116
Demodulator.........................171
DFD................................479
Differential pulse code modulation .. 608–612
Differentially detected QAM........186–190
Digital audio.........................198
Direct sequence....................368–371
Discarded path.......................101
Discontinuous transmission in GSM.....363
Distance measures................481–482
Distance properties of FEC codes.......114
Diversity............................373
    Frequency........................375
    Multipath....................375, 376
    Probability of bit error............375
    Space...........................375
Diversity combining...................376
    n best signals (SCn)..............376
    Equal gain (EGC)............376, 377
    Maximal ratio (MRC).........376, 377
    Selection (SC)....................376
Domain block........................460
Doppler frequency....................372
Doppler spectrum..................83–84
Down-link (see also Forward link).......389
Down-Link interference............379–380

SUBJECT INDEX

Down-Link pilot-assisted channel estimation 388–389
Down-link SIR 256
Down-link spreading and modulation 446–447
DPCM 608, 611
    Adaptive 611
    ADPCM-performance 612
    Basic 608
    Inter/Intra 610
    PDF 610
    Performance 611
    Schematic 608
DS–CDMA 916
DTTB 909
DTX (discontinuous transmission) .. 402, 415
DVB satellite scheme 997–999
DVB terrestrial scheme 962–965
DVB-T 909
DVB-T for mobile receivers 950–995
Dynamic channel allocation 292–297
    Centrally controlled algorithms 294–295
    Distributed algorithms 295–296
    Locally distributed algorithms . 296–297

E

Effect of multipath channels 371–374
Effect of parameters on NCC 268
Effect of parameters on SIR 261
Effect of SINR on CBER 276
Effect of SINR on error rate 276
Effect of SINR on FER 276
Effect of SNR and SIR on CBER ... 272–274
Effect of SNR and SIR on error rate 272–274
Effect of SNR and SIR on FER 272–274
Energy compaction coefficient 651
Enhanced QT codec 582–583
Entropy 12–15
Entropy of analogue sources 56
Entropy of sources exhibiting memory . 22–25
Equalization techniques 197
Equivocation 43, 49
EREC 778
ERPC 599
Erroneous hard-decision Viterbi decoding 101–104
Error control background 693
Error control mechanisms 693
Error correction coding 715
Error detection capability 132
Error entropy 43
Error entropy via imperfect channels .. 43–49
Error evaluator polynomial 148
Error evaluator polynomial computation 153–155
Error locator polynomial 128
Error mechanisms 692
Error polynomial 127

Error positions and magnitudes in RS and BCH coding 127
Error probability computation 176
Error sensitivity and complexity 471–472, 667–669
Error-free hard-decision Viterbi decoding 98–101
Error-free soft-decision Viterbi decoding 104–106
ETSI (European Telecommunications Standards Institute) 393, 395, 454
Euclidean distance 167
European digital audio broadcasting 198
Extended m-sequences 387
Extension field 108
Eye and mouth detection 577–580

F

Fading margin 73
Fast fading 248
Fast-fading statistics 77–83
FAW 507, 586, 662
    Autocorrelation 665
    Data-emulating probability 664
FCA 289–292, 304, 311, 312, 314, 315, 321–324, 327–329
FDD 807
FDMA (Frequency Division Multiple Access) 288, 436
FEC ... 516, 599, 711, 715, 722, 724, 725, 784
FER performance 527–529
FIFO 546
Finite field 108
Finite fields 108–114
First-order intensity match 573
First-order Markov model 30–31
Fixed channel allocation .. 289–292, 304, 312, 314, 315, 321–324, 327–329
Flexible channel allocation 292
Flexible transceiver architecture 202–204
Forced termination probability 308
Forced updating 506
Formulation of the key equations ... 126–130
Forney algorithm 148–151
Forney algorithm example 151–153
Forward Error Correction (FEC) ... 518–519, 554–555
Forward link 379, 382
Four-state Markov model for a 2-bit quantizer 27–28
FPLMTS (Future Public Land Mobile Telecommunication System) 392
Fractal codec design 465–467
Fractal codec performance 467–471
Fractal encoding 459
    Affine transformation 459, 460
    Bit sensitivity 472

Bit-allocation 470
Block types 468
Codec comparison 468
Codec design 465
Collage theorem 466
Complexity 471
Conclusion 473
DB-RB mapping example 461
Domain block (DB) 460
Error sensitivity 471, 473
IFS
    Encoding 464
    Example 465
    Example decoding 465
Iterated function system (IFS) 461
Iterative reconstruction 471
One-dimensional 462
One-dimensional example 463, 464
PDF 467
Performance 467, 470
Principles 459
Random collage theorem 461
Fractal image codecs 459–473
Fractal principles 459–461
Frame alignment 662–664
Frame-differencing 677
FRAMES (Future Radio Wideband Multiple Access System) 395
Free-space 8
Frequency Division Multiple Access (FDMA) 366, 368, 375, 382, 385
Frequency hopping 367–368
Frequency-domain fading simulation 86
Fresnel-zone 74
Full or exhaustive motion search 483–484
Future Public Land Mobile Telecommunication Systems(FPLMTS) 365

G

Gain-controlled motion compensation .. 502–503
Gain-cost-controlled motion compensation 487–489
Gain/cost-controlled inter-frame codec . 506–508
Galois field arithmetic 113–114
Galois field construction 111–112
Gauss-Jordan elimination 135
Gaussian 10
Gaussian approximation 383–385
Gaussian channels 281–283, 721, 725, 727–730, 795–798
Gaussian Minimum Shift Keying in GSM 358–359
Gaussian noise 7–11
Generating model sources 28–31
Generator polynomial 94, 116

Global System for Mobile communications (GSM) 342–365
Gold sequences 386
GOS 201, 308, 309
GPS (Global Positioning System) 399
Grade of service 201, 308, 309
Gradient constraint equation 484
Gradient-based motion estimation 484
Gray encoding 167
Gray mapping 167
Gray mapping and phasor constellation 167–168
Grid interpolation techniques 490
Group delay 7
GSM 255, 256, 258, 288, 290, 341, 508
GSM (Global System for Mobile Telecommunications) . 393, 395, 400, 401, 406, 436, 437, 454
GSM system 343
Guard channel scheme 298

H

H.261
    Bit-rate adjustment 711
    Block diagram 676, 734
    Block layer 684, 685
    Coded block pattern 683, 684, 698, 699, 740
    Coding control 679
    Comparison with H.263 733–776
    Effects of errors 696
    End of block marker 684, 685, 699
    Error degradation
        Inter 702, 704–709
        Intra 701
        Motion vectors 705, 707, 708
        Quantizer 705, 706
    Error recovery 695
    Errors in Inter blocks 702
    Errors in intra blocks 700
    Extra data bits (GSPARE) ... 682, 683, 697, 699
    Extra data bits (PSPARE) ... 681, 697, 699, 700, 703
    Extra insertion bit (GEI) 682, 683, 697, 699
    Extra insertion bit (PEI) 681, 697, 699, 700, 703, 708
    Frame divided into GOBs 682
    GOB index 682, 683, 697, 699, 700, 702, 703, 708
    GOB quantizer .682, 683, 697, 699, 700, 705, 708, 714, 740
    GOB start code. 681, 682, 695–697, 699
    Group of blocks 680, 681
    Group of blocks layer 681, 682
    Hierarchical structure 680
    Intra DC parameters 698, 699

Macroblock 676, 683
Macroblock address . 683, 684, 698–700, 703, 704, 739, 740
Macroblock layer 683, 684
Macroblock motion vector 683, 684, 698, 699, 707, 740
Macroblock quantizer 683, 684, 698, 699, 704, 714, 740
Macroblock size 711, 713, 714, 724
Macroblock type 683, 684, 698, 699, 703, 704, 707, 740
Motion compensation 676, 677, 683, 684, 698, 705, 734, 745
Overview 675
Performance 758–761
Picture layer 680, 681
Picture layer header 680
Picture start code ... 680, 681, 696, 699
Picture type word ... 681, 695, 697, 699
Pixel sampling 677
Qualitative error effects 696
Quantitative error effects 699
Quantizers 679, 711, 712
Run-length coding 686, 687
Source encoder 676, 678
Statistics 686
Temporal reference 681, 696, 699
Transform coefficients 684, 685, 699
Transmission coder 686
Video multiplex coder 680
Video-coding standard 675–690
Videophone system 710–732
Zig-zag run-length coding 686
H.261 performance 758–761
Comparison to H.263 761–764
Fixed bit-rate 761
Fixed PSNR 760
PSNR versus bit-rate 759
PSNR versus compression 759
H.261/H.263 comparison 733–776
Coding algorithms 735–757
Conclusions 776
Introduction 733
Motion compensation 745–747
Overview 733
Performance 761–764
Fixed bit-rate 763
Fixed PSNR 763
PSNR vs. bit-rate 762
PSNR vs. compression 764
Results 757–776
Source encoder 735–736
Video multiplex coder 736–744
H.263 .... 656, 667
16CIF format 734, 735, 737
4CIF format 734, 735, 737
Advanced prediction 734, 736, 751, 752, 754

Bidirectional prediction 757
Bilinear interpolation 746, 747
Bit stuffing 738
Bit-rate control 782
Block diagram 734
Block layer 744
CBPB coding parameter 742
CBPY coding parameter 742
CIF format 734, 735, 737
COD coding parameter 740, 743
Coded block pattern
    B-Blocks 742
    Chrominance 742, 743
    Luminance 742
Coded macroblock indicator ... 740, 743
Coding algorithms 735
Coding control 736
Comparison with H.261 733–776
Differential quantizer 742, 743
DQUANT coding parameter .. 742, 743
Four motion vectors per MB 752
GOB frame ID 739
GOB index 739
GOB quantizer 739
GOB start code 739
GOB structure 737
Group of Blocks layer 738
Half-pixel interpolation 746
Half-pixel motion vectors 746
In a mobile environment 777
Macroblock layer 739, 740
Macroblock layer quantizer .. 742, 743
Macroblock mode (B-Blocks) 742
Macroblock motion vector 742, 743
Macroblock type 742, 743
MCBPC coding parameter 742, 743
MODB coding parameter 742
Motion compensation 734, 735, 745
Motion vector (B-Blocks) 743
Motion vector predictor 745, 746
Motion vectors 745
MVD coding parameter 743
MVDB coding parameter 743
Negotiable options 734, 747
Overlapped motion vectors 752, 754
P-B mode 734–736, 743, 754–757
Performance 764–776
Picture layer 738
Picture layer quantizer 738
Picture start code 738
Picture type word 738
Prediction 735
QCIF format 734, 735, 737
Quantization 736
Run-length coding 744
Source encoder 735
SQCIF format 734, 735, 737
Syntax-based arithmetic coding 734, 749

Temporal reference............738
Transform coding.............736
Unrestricted motion vectors .. 734, 736, 747, 748
Video multiplex coder..........736
Video-coding standard 733–776
Videophone system............781–817
Wireless problems 777
Wireless solutions.............778
Zig-zag run-length coding........744
H.263 performance..............764–776
    4CIF, 16CIF 773–776
    Comparison to H.261..........761–764
    Different resolutions..........768–776
    Fixed Bit-rate................772
    Fixed bit-rate............766, 770, 775
    Fixed PSNR.........765, 769, 772, 774
    Gray versus color.............765–766
    PSNR vs. bit-rate .. 766, 768, 770, 773, 775
    PSNR vs. compression...767, 769, 771, 774, 776
    QCIF comparison 767–768
    SQCIF, QCIF, 4CIF 771–773
    SQCIF, QCIF, CIF...........768–771
    Video sequences..............765, 909
H.263/OFDM-based video systems..908–926
Hamming distance 167
Handover prioritization............298, 319
Handovers................298–299, 450–452
Hard-decision decoding................98
HCA.............................289
Hierarchical OFDM DVB performance . 990–993
Hierarchical or tree search 485–486
Hierarchical search 485
High-resolution DCT coding 639–671
High-resolution image coding 603–672
Holding time........................225
HTA.............................305
Huffman coding.......................18–22
Hybrid borrowing 291
Hybrid channel allocation..........297–298

I

Ideal communications system.............62
IF spectrum.............................171
Impulse response........................88
IMT-2000 (International Mobile Telecommunications - 2000) ... 339, 392, 393, 395, 396, 443
Information loss.........................42
Information loss via imperfect channels 42–43
Information of a source 11–12
Information properties...................12
Information sources......................3
Information theory....................3–65
Initial intra-frame coding................502

Instability...........................295
Inter-cell handover 298, 302
Inter-frame block truncation coding . 617–618
Inter-frame correlation 646
Inter-frame DCT coding...........650–658
Inter/Intra-DCT codec 661–662
Interference cancellation...399, 402, 453–454
Interferer positions.....................255
Interim Standard-95(IS-95).........365, 366
Interlacing 604
Interleaver.............................6
International Mobile Telecommunications - 2000 (IMT-2000)...................365
Interpolation.........................621
Interruption..........................295
Intersymbol interference.................372
Intra-cell handover 298, 302
Intra-frame block truncation coding . 615–617
Intra-frame coding.............502, 693, 779
Intra-frame correlation 650
Intra-frame quantizer training 639–644
Intra/Inter-frame differential pulse code modulation.......................610–611
Introduction to GSM 342–345
Irreducible polynomial 108
IS-54..............................341, 343
IS-95..288, 343, 399, 400, 439–443, 445, 446, 455
ISDN.............................341
Issues in information theory.............3–7
Iterated function system.................461
ITU (International Telecommunication Union) 392, 393, 439, 475

J

Jakes' method...........................303
Jakes' model 303
JDC................................342, 343
Joint information.......................45
Joint motion compensation and residual encoding 657–658
JPEG..................................635

L

Leakage factor 505
Least interference algorithm...317, 319, 321, 322
Least interference below threshold algorithm 320
LFA...............................305, 320
LIA 296, 304, 317, 319, 321, 322
Linear shift-register circuits for cyclic encoders 122
Linear shift-register division circuit......125
Local decoder 477
Locally optimized least interference algorithm 312, 314–317, 319–329

SUBJECT INDEX

Locally optimized most interference algorithm 314, 317–322, 324
Logical and physical channels in GSM 346–347
LOLIA . 296, 306, 307, 312, 314–317, 319–329
LOMIA 296, 306, 307, 314, 317–322, 324
Low-complexity techniques 603–638
Lowest frequency below threshold algorithm 320
LTA 305, 320
Luminance 604

M

m-sequences 385–386
MAD 481
MAP algorithm 93
Markov model 22
Markov modeling of video sources . . . 207–209
Matched filtering 169
Max-Lloyd-based subband coding . . . 633–637
Maximum entropy of a binary source . . 13–15
Maximum entropy of a q-ary source 15
Maximum-minimum distance code 114
MBS 341
MC 478
MC in the transform domain 491
MC using higher order transformations . 490–491
MCER 475, 637, 736, 756
MCER active/passive concept 503–504
Mean absolute difference 481
Mean and shape gain vector quantization 544–545
Mean Squared Error (MSE) 481
Media Access Control 223
MF-PRMA 227, 228, 1074
MF-PRMA++ 231
Microwave channels 7
Minimum distance 172
Minimum distance of FEC codes 114
Minimum euclidean distance 173
Model-based parametric enhancement . . 576–582
Modeling of wideband channels 87–92
Modem performance in AWGN 179
Modem schematic 164–172
Modulation and demodulation 170–171
Modulation and transmission 161–204
Modulation issues 161–199
Modulation overview 161
Modulator 6, 170
Modulo polynomial operations 110
Most interference below threshold algorithm 320
Motion compensation 475, 478, 644, 660–661
    ACF 481
    Algorithm comparison 486
    Comparison 648
    Conclusion 491

Distance measures 481, 483
Example 480
Exhaustive motion search 483
Frame difference 476
Full search 483
Gain cost controlled 487
Gradient based 484
Grid interpolation 490
Grid interpolation example 491
Higher order 490
In transform domain 491
Inter-frame correlation 646
Joint 657, 658
MCER entropy 488
MV bit-rate 489
MV correlation (Inter) 648
MV correlation (Intra) 649
Other techniques 489
Pel recursive 490
Post-processing 487
Principle 477
Schematic 479
Search algorithms 482
Search techniques 647
Simple codec 476
Subsampling search 486
Tree search 485
Motion compensation for high-quality images 644–650
Motion estimation techniques 489–491
Motion search algorithms 482–489
Motion vector 478, 487
Motivation and video transceiver overview 894–898
Motivation of the book . . . . xxvii–xxix
MPEG 783
MPEG-2 bit error sensitivity 951–962
MPEG-X 641
MPEG2 735
MSE 481
MSVQ 544
MTA 305, 320
Multiframe packet reservation multiple access 227–235
Multiframe structure 227
Multimode video system performance 899–902
Multipath channels 371–374
    Frequency nonselective 375
    Frequency selective 374
    Impairments on signal 374
    Impulse response 372, 373
        COST207 373
    Resolvable paths 374, 376, 377
Multipath fading 366, 370–372
    Long term 371, 372
        Lognormal 372
    Short term 371–372
        Nakagami 372

Rayleigh 372
Rician 372
Multiple access 223–243, 378–385
  Gaussian approximation .. 383–385, 392
    Probability of bit error 384
  Interference 382
Multiple modulations schemes 779
Multiuser detection 382, 453–454
Mutual information 39–40
Mutual information example 40–41
MV 478, 487

N

N-state Markov model for discrete sources exhibiting memory 24–25
Narrowband 73
Narrowband channels — simulation 85
Narrowband fading channels 73
Nassi-Shneiderman diagram 240
NCC profile
  Multiple interferers 267
    Effect of speech or video users ... 270
  Multiple interferers (speech)
    Worst interferer positions 270
  Multiple interferers (video)
    Best interferer positions 267
    Effect of slow fading 269
    Random/Fixed fading 268
    Worst interferer positions ... 267–270
Near-far effect 383
Nearest base stations 307, 314, 316–318
Neighborhood of cells 307, 314, 316–318
Neighboring base stations . 307, 314, 316–318
Netsim mobile radio network simulator . 299, 300
Netsim screenshot 300
New call blocking probability 308
No power control
  Performance
    FER 826, 827, 833
    SINR 831
    SIR 829
    Theoretical SIR 830
Noise 7
Noise — man-made 8
Noise — natural 8
Noncoherent demodulation 376
Noncoherently detected QAM 186
Nonhierarchical OFDM DVB performance 983–990
Nonlinear filtering 169
Nonsystematic encoding 116
Nonuniform traffic 309, 321, 323
Nonuniform traffic distribution 321, 322
Nonuniform traffic model 309, 321
Nonuniform traffic performance 321–323
Normalized channel capacity 264
  Profile across cell 267

Rayleigh channels 264
Using SINR 264, 265, 267
Nyquist filtering 168–170

O

OFDM 339, 908, 909, 913, 917, 918
One-dimensional fractal coding 462–471
One-dimensional transform coding .. 492–493
Optical fiber 7
Optimization of adaptive modems .. 193–195
Optimum decision threshold 176
Optimum detection theory 169
Optimum ring ratio 173
Orthogonal Frequency Division Multiplexing 197–199
Overview of GSM 345–346
OVSF (Orthogonal Variable Spreading Factor) code 399, 423–426

P

Packet Reservation Multiple Access . 201–202
PACS system 343
Parametric codebook training 580
Parametric encoding 581–582
Partial forced update 504–506
Path loss 383
Path metric 100
Path-loss 383
Path-loss power law 248
Path-loss-based SIR model 256
PCN 339
PDC 481
PDF 480
Peak-to-average phasor power 173
Pel difference classification 482
Pel-recursive displacement estimation 490
Perfect reconstruction quadrature mirror filtering 620–626
Performance and considerations under erroneous conditions 583–586
Performance metrics 306–309
Performance of MF-PRMA 228–235
Performance of system 1 524–527
Performance of system 2 527–531
Performance of systems 1 and 3 560–561
Performance of systems 2 and 6 563–564
Performance of systems 3–5 531–533
Performance of systems 4 and 5 561–563
Performance of the data partitioning scheme 972–983
Performance of the DVB satellite scheme 1003–1015
Performance of the SPAMA protocol 242–243
Performance summary of the DVB-S system 1010–1015
Performance under erroneous conditions 550–553

SUBJECT INDEX

Personal Digital Cellular (PDC) 365
Peterson-Gorenstein-Zierler decoder . 130–133
PFU 504
PGZ decoder for RS and BCH codes 128
PGZ decoding example 133–138
PGZ decoding of RS and BCH codes 131
PGZ RS and BCH decoder 130
Phase jitter immunity 173
Phasor constellation 167
PHP 341
PHS 343
Physical channels in cdma2000 441–443
Pilot symbol-assisted modulation ... 181, 723
Pilot-symbol assisted decision-directed channel estimation 390–392
PLMR 341
PNN 539
Polynomial multiplication 122–123
Polynomial multiplication circuit 122
Post-processing of motion vectors 487
Power Control
    Multiple interferers
        NCC performance 858
Power control ... 378, 382, 383, 392, 819, 821, 822, 824–843, 845–858
    Average transmission power ... 834–837
        vs. FER 837
        vs. User distance 836
        vs. User/Interferer distance 835
    Channel fading 845, 849, 850
    Closed loop 383
    Effect of vehicular speed 845–853
    FER and average transmission power trade-off 840, 842
    FER and average transmission power trade-off 840
    Multiple interferers 855–858
        FER performance 855, 856
        NCC performance 857, 858
        SINR performance 857, 858
        SIR performance 857, 858
    Open loop 383
    Parameter optimization 838
        IPC and DPC 839, 840
        IPSS and DPSS 843
        NEF and NFE 841, 842
    Performance 824, 825
        Average transmission power 848
        FER 824, 826, 827, 832, 833, 846
        Multimode 832
        SINR 828, 831
        SIR 828–830, 847
    Performance at pedestrian speeds . 845–848, 851
        Average transmission power 848
        FER 846
        SIR 847

Power control error 851, 853, 854
    PDF 853, 854
    Statistics 854
    Tracking of slow fading ... 848, 851, 852
Power control algorithm 819, 821, 822
Power spectral density 4, 169
Practical evaluation of the Shannon-Hartley law 58–62
Practical Gaussian channels 8
Practical QMF design constraints ... 624–626
Practical quadrature mirror filters .. 627–630
Preface .... xxiii–xxix
Primitive element 108
Principle of motion compensation ... 477–492
PRMA 201
PRMA++ 227, 231
Probability density function 4
Probability of low-quality access 308
Processing gain 366, 368
Propagation environment 67–92
Propagation path-loss law 73–76
Properties 650
Properties of the DCT Transformed MCER 650
Properties of the DCT transformed MCER 657
Proposed codec 658–671
PSAM 181
PSAM performance 185–186, 189
PSAM schematic 182
PSAM system description 181–183
Pseudo quantizer 499
PSNR 482
PSNR performance 530–531

Q

Q-function 176
QAM 247, 694
QAM constellations 172–174
QAM constellations for AWGN channels . 172
QAM modem schematic 165
QAM modulation and transmission . 177–178
QAM overview 164
QAM subchannels 715
QCIF 475
QMF 619
    Analysis-synthesis schematic 620
    Band-splitting 621
QMF design 624
QT codecs 567
QT decomposition 568
QT-based transceiver architectures .. 588–591
QT-based video-transceiver performance 591–595
QT-codec-based video transceivers .. 586–591
Quad-tree 567
    Algorithmic issues 573
    Bit sensitivity 586, 587

Bit-allocation 584, 586
Bit-rate 575
Decomposition 568
Decomposition example 569
Erroneous conditions 583
Intensity match 571
    Comparison 573
    First order 573
    Zero order 571
Introduction 567
Nodes/leaves 576
Parametric enhancement 576
    Algorithm 578
    Bit-allocation 582, 586
    Codebook 581
    Codebook training 580
    Comparison 585
    Detection 577
    Encoding 581
    Example 582
    Performance 585
    Process example 579
    Schematic 583
    Template 579
PDF 572
Performance 572, 577, 584
Quantizer ranges 573
Regular decomposition 569
Segmentation 570
Quad-tree based codecs 567–599
Quad-tree decomposition 568–570
Quad-tree intensity match 571–576
Quadrature Amplitude Modulation . 164–190, 247
Quadrature mirror filtering 620
Quadrature mirror filters 620, 627
Quality criterion 660
Quality of service (QoS) 393, 399, 410
Quantization 3
Quantizer training for multi-class DCT . 497–499
Quantizer training for single-class DCT 496–497

R

RACE (Research in Advanced Communication Equipment) 395
RACE program 339
Radio wave propagation 71–92
Raised-cosine filter characteristic 168
RAKE receiver 371, 374–378
    Structure 377
Random access 447–450
Random collage theorem 461
Random interferer positions 258
Range block 460

Rayleigh channels 248, 264, 281–284, 721, 725, 728, 729, 731, 795–798, 801, 802
Reconfigurable modulation schemes . 694, 779
Reconfigurable videophone system . 710, 777, 781
Reduced-length Poisson cycles 210–215
Reed-Solomon (RS) codes 114
Reference assisted coherent QAM for fading channels 181–186
Request phase 225
Reservation rate 227
Reuse cluster
    3 cells, directional antenna 249
    7 cells, omnidirectional antenna 249
Reuse distance 253, 254, 314, 316–318
Reuse partitioning 299
Reverse link 380, 382
RGB 604
Rician channel 248
Rician K-factor 79
RL 513
RLC efficiency 32–34
Roll-off 168
RS and BCH codec performance 156–158
RS and BCH codes 114–155
RS and BCH decoding 126
RS and BCH syndrome equations 126
RS decoding 126–155
RS encoding 116–118
RS encoding example 118–120
RS(12,8,2) PGZ decoding example 133
Run-length based intra-frame subband coding 630–637
Run-length coding 31–34
Run-length coding principle 31

S

Sampling 3
Satellite channel model 999
Satellite-based video broadcasting . 996–1018
SCS 296
Search algorithms 482
Second generation ... 399, 400, 406, 439, 454, 455
Second-generation mobile systems ... 339–364
Sensitivity-matched modulation 518
Service interruption 295
Service multiplexing and channel coding 443–444
Shadow fading 324, 326, 328, 329
Shadow fading model 302–304
Shannon's channel coding theorem 53–55
Shannon's message for wireless channels 62–65
Shannon-Fano coding 16–18
Shannon-Hartley law 58, 263
Shift-register encoding example 123–125
Signal and interferer paths 249

SUBJECT INDEX

Signal-to-interference ratio 270
Signal-to-interference+noise ratio 262, 270–273
Signal-to-noise ratio 4, 270
Simple borrowing 291
Simulation environment 558–560
Simulation of narrowband fading channels 85–87
Singleton bound in FEC coding 114
SINR 262, 264, 270–273
SINR definition 270, 271
SINR model 270, 271
SINR profile
    Multiple interferers 265
SINR simulations 272, 273
SIR 254, 255, 270
SIR definition 270
SIR profile 258
    Multiple interferers
        Effect of speech or video users ... 264
    Multiple interferers (speech)
        Worst interferer positions 264
    Multiple interferers (video) 258–262
        Best interferer positions 258
        Effect of slow-fading 263
        Random interferer positions 258
        Random/Fixed shadow fading ... 262
        Typical interferer positions 258
        Worst interferer positions 258, 262–264
Skin effect 8
Slot occupancy performance 529–530
Slow fading 76, 248
Slow-fading statistics 76–77
Smoothing filter 389
SNR 270, 482
SNR definition 270
Sounding 88
Source coding 15–22
Source encoder 5, 16
Source sensitivity 518
Source-matched transceiver 517–524
SPAMA 237
SPAMA, performance of 242
Sparsed-time 845, 848–850
Spatial filtering 677
Speech and data transmission in GSM .. 347–351
Speech systems 62
Spread spectrum 366–371
    Direct sequence 367–371
        Decoding waveforms 370
        Encoding waveforms 368
        Receiver 371
        Transmitter 369
    Frequency hopping 367–368
        Fast hopping 367
        Slow hopping 367

Power spectral density 367
Spread spectrum fundamentals 366–371
Spreading and modulation 445–447
Spreading codes 385–387
Spreading sequence .. 368–370, 377–379, 382, 385–387
    m-sequence 385
        Cross correlation 386
        Shift register 385
    Autocorrelation 385
    Cross-correlation 380
    Energy 369
    Extended m-sequence 387
    Gold sequence 386
        Cross correlation 386
    Orthogonality property 382
Square 16QAM constellation 167
Star 16QAM constellation 172
State and trellis transitions 96–97
Static borrowing 291
Statistical Packet Assignment Multiple Access 237–243
Statistical packet assignment principles . 237–242
Stylized NLF waveforms 170
Stylized Nyquist filters 169
Subband coding 618–630
    10 split band example 629
    Corrupted 637
    Dead zone 632
    Dead zone quantizer 632
    Error sensitivity 636
    Max-Lloyd based 633
    PDF 631
    Performance "Susie" 635
    Run-length based 630
    Split-band codec 619
    Split-band filters 627, 628
    Without quantization 630
Subband-Adaptive OFDM Transceivers Having Different Target Bit Rates 936–941
Subsampling search 486–487
Summary of 3G systems 454–455
Summary of GSM features 363–364
Summary of low-rate codecs/transceivers 595–599
Summary of QT-based video transceivers 595
Survivor path 101
Switching thresholds 886–890
Synchronization issues in GSM 357–358
Syndrome polynomial 148
Syndromes in RS and BCH coding 127
Synthesis filtering 623–624
System 1 517–520
System 2 520–523
System concept 517–518
System overview 909–917

System parameters 255
System performance 524–533, 558–564, 922–926
Systematic RS and BCH encoding 117
Systems 3–5 523–524

T

TACS system 343
Target rate 228, 229, 233, 235
TDD 807
TDMA (Time Division Multiple Access) 288, 395, 436, 811
Terrestrial broadcast channel model . 965–966
Test sequences 604
Theoretical SIR 252–255
    Multiple interferer 252, 253
    Single interferer 252
Thermal noise 8
Third generation 392, 393, 395, 399, 400, 406, 438, 439, 452, 454, 455
    Frequency allocation 392
Third-Generation CDMA systems ... 365–455
Third-generation systems 392–455
TIA (Telecommunications Industry Association) 393, 395, 439, 454
Time division duplex (TDD) 237
Time Division Multiple Access (TDMA) .. 4, 366, 368, 375, 382, 385
Time-domain fading simulation 86
Time-variant target bit rate OFDM transceivers 941
Time-variant target bit-rate OFDM transceivers 948
Time-varying effects of interference 275, 277–280
Time-varying nature of SINR .. 275, 279, 280
Time-varying nature of SIR 275, 277, 278
TMN5 656
Traffic modeling and multiple access 205–246
Transducer 4
Transfer function 7
Transform coding 492–499
Transform coding efficiency test 651
Transmission
    Asynchronous 380, 382
    Symbol-synchronous 379
Transmission format 519–520
Transmission issues 3–456
Transmission of control signals in GSM . 351–357
Transmission over the symbol-spaced two-path channel 1003–1009
Transmission over the two-symbol delay two-path channel 1010
Transmission power control ... 299, 819, 821, 822, 824–843, 845–858
    Average power 834–837

    vs. FER 837
    vs. User distance 836
    vs. User/Interferer distance 835
    Background 819
    Channel fading 845, 849, 850
    Effect of vehicular speed 845–853
    FER and average transmission power trade-off 840, 842
    FER and average transmission power trade-off 840
    Interference reduction 819–860
    Multiple interferers 855–858
        FER performance 855, 856
        NCC performance 857, 858
        SINR performance 857, 858
        SIR performance 857, 858
    Parameter optimization 838
        IPC and DPC 839, 840
        IPSS and DPSS 843
        NEF and NFE 841, 842
    Performance 824, 825
        Average transmission power 848
        FER 824, 826, 827, 832, 833, 846
        Multimode 832
        SINR 828, 831
        SIR 828–830, 847
    Performance at pedestrian speeds . 845–848, 851
        Average transmission power 848
        FER 846
        SIR 847
    Power control error 851, 853, 854
        PDF 853, 854
        Statistics 854
    Tracking of slow fading ... 848, 851, 852
Transmission power control algorithm ... 819, 821, 822
Transmission via discrete channels 34–49
Transmitted and received spectra 171
Transparent tone in band modulation ... 181
Tree search 485
TTA (Telecommunications Technology Association) 393, 395
TTC (Telecommunication Technology Committee) 393, 395
TTIB 181
Turbo-coded video performance 890–892
Twisted pairs 7
Two-dimensional transform coding .. 493–496
Two-state Markov model example 25–27
Two-state Markov model for discrete sources exhibiting memory 22–23
Two-state speech model 260, 262

SUBJECT INDEX

U

UMTS (Universal Mobile Telecommunications System)... 339, 393, 395, 436, 916
UMTS-like videophone system...... 894–908
UMTS-type framework.............. 916–917
Uniform traffic distribution.............. 322
Uniform traffic performance............. 322
Universal Mobile Telecommunications (UMTS) 365
Up-link (*see also* Reverse link). 382, 389, 392
Up-Link interference................ 380–383
Up-Link pilot-symbol assisted channel estimation....................... 390
Up-link pilot-symbol assisted channel estimation........................ 389
Up-link SIR............................ 256
Up-link spreading and modulation....... 447
UTRA (UMTS Terrestrial Radio Access) 393, 395–439
    Cell identification.... 403, 426, 432–435
        FDD mode................. 432–434
        TDD mode................. 434–435
    Channel coding.................... 414
    Channel-coding................... 411
    Characteristics............... 395–399
    Down-link transmit diversity .. 452–453
    Frequency spectrum 395
    Handover........... 399, 403, 436–437
        Inter frequency 436–437
        Soft........................... 436
    Inter cell time synchronization..... 399, 438–439
    Modulation................... 422–429
        Down-link................. 428–429
        Up-link................... 426–428
    Multicode transmission .. 404, 420, 425, 427
    Physical channels.............. 400–410
    Power control......., 400, 402, 430–431
        Inner loop................. 430–431
        Open loop 431
    Random access 399, 404, 429–430
    Service multiplexing...... 399, 410–420
    Spreading.................... 422–429
        Down-link................. 428–429
        Up-link................... 426–428
    Transport channels............ 399–401

V

VA.. 93
VAD (Voice activity detection).......... 400
VAF.................................... 262
Vandemonde matrix..................... 129
Variable-length coding 630
Vector quantization 537
    127-entry codebook 541
    16-entry codebook 540
    Adaptive 546
    Performance 547
    Schematic........................ 547
    Basic schematic................... 542
    Bit sensitivity................ 552, 553
    Bit-allocation 550, 552
    Classified 548
        Active/passive 551
        Complexity 549
        Performance 549
        Schematic..................... 548
    Codebook design................... 537
    Codebook sizes 544
    Design............................ 541
    Erroneous conditions.............. 550
    Introduction 537
    MSVQ............................. 544
        Performance 546
        Schematic..................... 545
    PDF.............................. 538
    Performance 543
Vector quantizer design............ 541–550
Very low bit-rate DCT codecs and multimode videophone transceivers .. 475–536
Very low bit-rate VQ codecs and multimode videophone transceivers .. 537–566
Video codec outline................ 475–477
Video coding standard, H.261....... 675–690
Video compression and transmission aspects 930
Video formats 603–604
Video model matching............. 215–223
Video over wireless systems....... xxiv–xxvii
Video parameters of the UMTS scheme 921–922
Video parameters of the WATM system 918–921
Video systems based on proprietary video codecs 459–601
Video systems based on standard video codecs 675–1022
Video traffic modeling 205–223
Video-related system aspects........ 918–922
Videophone system 710, 777, 781
Viterbi algorithm.................... 98–106
VLC.............................. 630, 650
Voice activity control................... 382
Voice Activity Factor (VAF) 262
VQ 537
VQ-based low-rate video transceivers 554–558

W

W-CDMA (Wideband CDMA) 393, 395, 400
Walsh-Hadamard code 380
WATM............................ 237, 909
WATM system 913–916
Waveguides.............................. 7
Wideband CDMA...................... 452
Wideband channel models in GSM .. 359–360

Wideband channels 87–92
Wiener-Hopf equations 184
Winning path 101
Wireless communications scene ... xxiii–xxiv, 339–342
Wireless videophone 710–732, 781–817
    ARQ 779, 811
    Bit-rate adjustment 711
    Bit-rate control 782
    Combined source/channel coding ... 780
    Conclusions 816
    Effect of co-channel interference 281
    Effect of interference 280
    Encoding history 716, 717, 786
    End of frame effect 719
    Error correction coding 715
    Error resilience 710, 711
    Error-free performance ... 726, 792, 793
    Features 732, 781, 791
    Feedback channel 710, 720
    Frame errors vs. channel SNR . 797, 801
    Gaussian channels 799
    H.261/H.263 comparison 798, 799
    Intra-frame coding 779
    Macroblock compounding 717–722
    Macroblock truncation 720–722
    Modulation schemes 723, 725
    Objectives 710
    Packet size 725
    Packet structure 785
    Packetization 711, 715, 785, 788
    Packetization algorithm 787, 789
    Packetization examples 788
    Performance 721, 725, 726, 732, 790–792, 799–802, 804–806, 809
    Performance over DECT channels . 802, 804, 805
    Performance vs. channel SINR 281–284
    Performance vs. channel SNR . 728–731, 795, 796, 798, 802
    Performance with antenna diversity 800–802
    Performance with errors . 280, 281, 727, 793–795
    Problems 777
    PSNR vs. channel SNR 799
    Rayleigh channels 799
    Reconfigurable 710, 711
    Reconfigurable modulation 779
    Resynchronization 710
    Solutions 778
    Subchannel equalization 715
    Switching levels 805
    System architecture 721
    System environment 790
    Transmission feedback ... 720, 806–811, 815, 816
        Majority logic codes 809, 814
        Transmission feedback timing 810
    Use of ARQ 694, 725, 729–731, 796–798, 811
    Use of FEC 711, 715, 724, 725, 784
WLANs 341
Worst interferer position 258

Y
YUV 604

Z
Zero-order intensity match 571

Author Index

A

Aase, S.O. [530] 627, 637
Aazhang, B. [350] 381, 453
Abramson, N. [19] 6, 53
Abu-Dayya, A.A. [226] 280
Acampora, A.S. [209] 237
Adachi, F. [300] 339, 404, 420
Adachi, F. [301] 339, 395
Adachi, F. [302] 339, 395
Adachi, F. [299] 339, 399, 423, 425
Adachi, F. [121] 179, 190
Adachi, F. [398] 442
Adachi, F. [382] 411, 444
Adachi, F. [391] 432
Adelson, E.H. [522] 618
Adoul, J-P. [182] 201
Aghvami, A.H. [154] 197
Aghvami, A.H. [387] 425
Akaiwa, Y. [277] 296
Al-Subbagh, M. [547] 664
Alamouti, S.M. [640] 966, 982
Alard, M. [169] 197, 198
Alasmari, A.K. [520] 618
Alikhani, H. [297] 339
Allen, R.W. [27] 53, 54
Alouini, M-S. [141] 190, 191
Alouini, M-S. [142] 190, 191
Amoroso, F. [334] 371
Andermo, P-G. [320] 365, 395
Anderson, B.D.O. [661] 1005
Anderson, L.G. [253] 291
Andoh, H. [277] 296
Andoh, H. [366] 389, 454
Ansari, R. [536] 630
Appleby, D.G. [218] 249
Aravind, A. [641] 968
Arce, G.R. [511] 614
Arimochi, K. [135] 190, 191
Ariyavisitakul, S. [573] 819
Arnold, J.F. [489] 567, 573, 631
Arroyo-Fernández, B. [377] 395
Asghar, S. [45] 70, 292, 341, 343, 803

Atzori, L. [645] 996
Austin, M.D. [576] 819
Azari, J. [460] 496

B

Backman, P-O. [251] 290
Bahl, L.R. [64] 93
Baier, A. [363] 388, 395
Baier, P.W. [298] 339, 395
Baiocchi, A. [278] 296
Balachandran, K. [91] 161–163
Bamberger, R.H. [531] 627
Barani, B. [377] 395
Barbulescu, A.S [638] 965
Barnsley, M.F. [420] 459, 461, 466, 467
Baruffa, G. [643] 996
Bateman, A. [123] 181, 186
Bateman, A. [124] 181
Bateman, A. [125] 181, 387, 388
Bateman, A. [126] 181
Bateman, A. [129] 181, 186
Bateman, A. [122] 181, 387, 388
Bateman, A. [127] 181
Bateman, A. [128] 181
Bauch, G. [587] 865
Bauch, G. [585] 865, 866
Baum, K.L. [623] 929
Beaulieu, N.C. [226] 280
Beaumont, J.M. [421]   459, 462, 466, 467
Belfiore, J-C. [223] 266
Bellifemine, F. [484] 537
Benthin, M. [365] 389
Benveniste, A. [656] 1000–1002
Berlekamp, E.R. [71] 93, 94
Berlekamp, E.R. [1] ...94, 107, 108, 126, 138, 148
Bernhardt, R.C. [238] 288
Berrou, C. [28] 53, 61, 93, 865, 894, 895, 921, 928, 930, 964, 997
Berrou, C. [581] 865, 866
Berrou, C. [381] 411, 865, 964, 997, 1003
Berruto, E. [324] 365, 395

Biglieri, E. [497] 586
Bingham, J.A.C. [615] 928
Blahut, R.E. [80] 130, 158
Blahut, R.E. [3] 107, 108, 126, 128, 130, 138, 141, 148, 158
Blahut, R.E. [606] 920, 921
Blake, I.F. [76] 107, 128, 138
Blogh, J.S. [229] 287
Blondia, C. [189] 205, 207
Bochmann, H. [166] 197, 198
Boekee, D.E. [534] 628
Bogenfeld, E. [619] 928
Bohnke, R. [297] 339
Börjesson, P.O. [601] 909
Borowski, J. [619] 928
Bose, R.C. [66] 94
Bose, R.C. [67] 94
Boyle, R. [291] 330
Bozdagi, G. [430] 478
Brady, P.T. [216] 248, 260
Brady, P.T. [217] 248, 260
Brand, A.E. [387] 425
Brecht, J. [185] 205, 207, 209, 211–214, 216–224, 226, 228, 230, 232, 234, 236–238, 241, 244, 245
Brecht, J. [186] 205, 916
Burr, A.G. [144] 190

C

Cacopardi, S. [643] 996
Cain, J.B. [7] 107, 126, 138, 148, 152
Caldas-Pinto, J.R. [507] 611
Calderbank, A.R. [393] 432, 452
Callendar, M.H. [325] 365, 392
Carciofy, C. [183] 201
Carlson, A.B. [20] 6, 16, 31, 32, 39, 59
Carlton, S.G. [508] 613
Casals, O. [189] 205, 207
Casas, J.R. [475] 537
Castro, J.P. [323] 365, 395, 399, 402, 426
Cavenor, M.C. [489] 567, 573, 631
Cavers, J.K. [133] 186
Cavers, J.K. [120]179, 181–183, 186, 190, 389
Cavers, J.K. [108] 164
Cavers, J.K. [88] 161
Chae, J.S. [646] 996
Chai, D. [556] 711, 752
Chai, D. [555] 711, 752
Chambers, J.A. [395] 436
Chan, C. [482] 537
Chan, M. [503] 608
Chang, L.F. [573] 819
Chang, R.W. [157] 197, 908, 909
Chang, R.W. [161] 197
Chang, S. [434] 485
Cheah, C. [415] 453
Cheah, K.L. [411] 453
Chen, Y.C. [644] 996

Cheng, M.L. [273] 296, 304–306, 309
Cheng, N.T. [462] 504, 515, 599, 710, 779
Cheng, N.T. [552] 710, 778, 779
Cheung, J. [465] 512
Cheung, J.C. [593] 876
Cheung, J.C.S. [181] 201
Chiani, M. [222] 266
Chiariglione, L. [663] 1021
Chien, R.T. [82] 132
ChihLin, I. [280] 296, 306
Chockalingam, A. [389] 430
Choi, D.W. [545] 662
Choo, C. [479] 537
Chow, P.S. [615] 928
Chow, S. [188] 205, 206
Chow, Y.C. [117] 173
Chowdhury, M.F. [429] 478
Chua, S-G. [137] 190, 191, 928
Chua, S. [138] 190, 191, 876, 928
Chua, S. [143] 190, 191
Chua, S. [592] 876
Chuang, J.C-I. [268] 293
Chuang, J.C-I. [275] 296, 299, 301, 311
Chuang, J.C-I. [273] 296, 304–306, 309
Chuang, J.C-I. [577] 819
Chuang, K-C. [478] 537
Chung, W.C. [562] 745
Cimini, L.J. [165] 197, 909
Cimini, L.J. [153] 197
Cioffi, J.M. [615] 928
Civanlar, M.R. [641] 968
Clark, A.P. [156] 197
Clark, G.C. Jr [7] 107, 126, 138, 148, 152
Clarke, R.J. [454] 496, 618
Clarke, R.J. [458] 496
Clarke, R.J. [431] 478, 487
Clarkson, T.G. [444] 490
Claus, V. [211] 240
Cocke, J. [64] 93
Constello, D.J. Jr [472] ... 522, 557, 590, 720, 779, 811
Constello, D.J. Jr [5] 107, 126, 128, 138, 148, 720, 811
Cosimi, P. [207] 227, 231
Cosman, P.C. [538] 633
Cosmas, J. [570] 802–804
Cosmas, J. [189] 205, 207
Costescu, F.C. [107] 164
Couturier, C. [505] 611, 612
Cox, D.C. [272] 295
Cox, D.C. [269] 295
Cox, D.C. [271] 295, 298
Cox, D.C. [290] 309
Cox, R.V. [311] 340
Crebbin, G. [491] 567
Crespo, P. [570] 802–804
Crochiere, R.E. [513] 618
Crochiere, R.E. [514] 618

Crochiere, R.E. [529] 627
Cruickshank, D. [419] 454
Czylwik, A. [613] 928

D
Dahlin, J. [250] 290
Dahlman, E. [372] ... 395, 396, 399, 414, 429, 454
Dahlman, E. [397] 436
Dahlman, E. [321] 365, 395
Dahlman, E. [378] 395
Dahlman, E. [323] 365, 395, 399, 402, 426
Dai, Q. [584] 865
Davarian, F. [361] 387, 388
de Alfaro, L. [470] 522, 557, 590
De Marca, J.R.B. [602] 910
De Natale, F.G.B. [645] 996
De Vile, J. [207] 227, 231
deDiego, S. [475] 537
del Buono, M. [185] .. 205, 207, 209, 211–214, 216–224, 226, 228, 230, 232, 234, 236–238, 241, 244, 245
Del Re, E. [281] 296, 306
Delli Priscoli, F. [278] 296
Delli Priscoli, F. [279] 296
Delli Priscoli, F. [208] 233, 237
Delogne, P. [187] 205
Delp, E.J. [508] 613
Delp, E.J. [509] 613, 614, 616
DeNatale, F.G.B. [493] 567
Desoli, G.S. [493] 567
Di Gregario, M. [645] 996
Didier, P. [581] 865, 866
Dietrich, P. [389] 430
Dimitrijević, D.D. [270] 295
Ding, W. [564] 783
Ding, W. [566] 783
Ding, Z. [661] 1005
Dixit, S. [202] 209, 724
Djuknic, G.M. [469] 522, 557, 590
Dohi, T. [300] 339, 404, 420
Douillard, C. [581] 865, 866
Dubois, E. [194] 206, 710, 778, 910
Dubois, E. [551] 710, 778
Dudbridge, F. [423] 459, 465–467
Dunlop, J. [207] 227, 231
Durkin, J. [49] 74
Durtler, W.G. [103] 164

E
Ebert, P.M. [162] 197
Edwards, R. [49] 74
Efstratiadis, N. [441] 489
Egger, O. [533] 628
Elias, P. [56] 93
Ellis, W. [193] 206, 748, 749, 751, 757
Elnoubi, S.M. [256] 292
Emani, S. [502] 608

Eng, T. [339] 376
Engel, J.S. [254] 291, 292
Engström, S. [251] 290
Ericsson, A. [505] 611, 612
Eryurtlu, F. [595] 877
Esteban, D. [525] 620, 621, 624
Evans, B.G. [653] 999
Everitt, D. [225] 280
Everitt, D.E. [249] 289
Ewerbring, L-M. [320] 365, 395

F
Failli, M. [639] 965
Falciasecca, G. [184] 202
Fano, R.M. [59] 93
Fantacci, R. [281] 296, 306
Färber, N. [594] 877
Färber, N. [195] 206, 778, 910
Faulkner, M. [104] 164
Fazel, K. [617] 928
Fazel, K. [179] 198
Feder, M. [492] 567, 573
Feher, K. [101] 164
Feher, K. [116] 169
Feig, E. [566] 783
Feng, Y. [479] 537
Ferenczy, P. [22] 6, 42, 44, 49, 52, 55, 56, 58–60
Fettweis, G. [179] 198
Fiebig, U-C. [363] 388, 395
Fiebig, U-C.G. [360] 387
Fischer, R.F.H. [614] 928
Flangan, J.L. [513] 618
Flannery, B.P. [148] 194, 216, 552
Flannery, B.P. [54] 87
Flannery, B.P. [579] 838
Forney, G.D. [62] 93
Forney, G.D. Jr [84] 138, 148, 149
Forney, G.D. Jr [100] 164
Fortune, P.M. [34] 64
Franti, P. [540] 637
Franz, V. [587] 865
Frater, M.R. [190] 205
Frescura, F. [643] 996
Frullone, M. [184] 202
Frullone, M. [183] 201
Fujiwara, A. [382] 411, 444
Fukuda, K. [51] 75

G
Gabrielsson, S. [203] 223
Galand, C. [525] 620, 621, 624
Galand, C.R. [528] 626
Gallager, R.G. [100] 164
Gallanger, N. [511] 614
Gans, M.J. [53] 84
Gardiner, J.G. [46] 71
Garg, V.K. [330] 366

Gejji, R.R. [390].........................430
Gejji, R.R. [221].........................266
Georganas, N.D. [282]..................298
Georganas, N.D. [283]..................298
Gersho, A. [490].........................567
Gersho, A. [474]...............537–539, 549
Gersho, A. [425]......466, 467, 469, 543, 548
Gersho, A. [481].........................537
Gertsman, M.J. [582]....................865
Ghanbari, M. [561].......727, 790, 970, 982
Ghanbari, M. [460]......................496
Gharavi, H. [466].............514, 587, 628
Gharavi, H. [640]..................966, 982
Gharavi, H. [432].......................481
Gharavi, H. [642].......................995
Gharavi, H. [517].......................618
Gharavi, H. [518].........618, 630, 631, 633
Gharavi, H. [532]..................627, 654
Gharavi, H. [519].......................618
Giambene, G. [281]................296, 306
Gibby, R.A. [161].......................197
Gibson, J.D. [292]................339, 340
Gilhousen, K.S. [327]...................366
Gilhousen, K.S. [243]...................288
Girard, L. [321]...................365, 395
Girod, B. [594].........................877
Girod, B. [439]...................489, 734
Girod, B. [195]..............206, 778, 910
Giusto, D.D. [493]......................567
Giusto, D.D. [645]......................996
Glavieux, A. [28]...53, 61, 93, 865, 894, 895,
            921, 928, 930, 964, 997
Glavieux, A. [581]................865, 866
Glavieux, A. [381]...411, 865, 964, 997, 1003
Glisic, S. [241]..............288, 366, 393
Glisic, S. [242].........................288
Godara, L.C. [232]......................287
Godara, L.C. [233]......................287
Godard, D.N. [659].....................1001
Golay, M.J.E. [392]................432, 434
Goldhirsh, J. [649].....................999
Goldhirsh, J. [650].....................999
Goldschneider, J.R. [476]...............537
Goldsmith, A.J. [139]..............190, 191
Goldsmith, A.J. [138].....190, 191, 876, 928
Goldsmith, A.J. [141]..............190, 191
Goldsmith, A.J. [143]..............190, 191
Goldsmith, A.J. [142]..............190, 191
Goldsmith, A.J. [140]..............190, 191
Goldsmith, A.J. [592]...................876
Goldsmith, A.J. [137].........190, 191, 928
Golomb, S.W. [86].......................138
Goodman, D.J. [274]....................296
Goodman, D.J. [287]....................299
Goodman, D.J. [180]...............201, 226
Goodman, D.J. [276]....................296
Goodman, D.J. [206]...............226, 237
Gorenstein, D. [78]............126, 128–130
Gorenstein, D. [69].......................94
Goursat, M. [656].................1000–1002
Grandhi, S.A. [274].....................296
Grandhi, S.A. [287].....................299
Granzow, W. [363].................388, 395
Gray, R.M. [538]........................633
Gray, R.M. [474]...............537–539, 549
Grazioso, P. [184].......................202
Grazioso, P. [183].......................201
Green, E.P. Jr [331]...............371, 376
Greenwood, D. [32]..63, 71, 77, 78, 248, 303,
            374, 375
Grieco, D.M. [346]......................376
Griffiths, J. [647]......................999
Grilli, F. [278].........................296
Griswold, N.C. [510]....................614
Grünheid, R. [616]......................928
Gu, Y. [155]............................197
Gudmundson, B. [372]....395, 396, 399, 414,
            429, 454
Gudmundson, M. [324]............365, 395
Guérin, R. [286]...................298, 300
Guo, D. [415]...........................453
Guo, D. [416]...........................453
Gupta, S. [481].........................537
Gupta, S.C. [256].......................292
Gustafsson, M. [397]....................436

H

Haas, R. [223]..........................266
Haavisto, P. [443]......................490
Habibi, D. [203]........................223
Hagenauer, J. [585]................865, 866
Hagenauer, J. [30]..................63, 64
Halverson, D.R. [510]...................614
Hamming, R.W. [71]..................93, 94
Hamorsky, J. [590].................867, 868
Haralick, R.M. [476]....................537
Hartley, R.V.L. [18].....................6, 11
Hartmann, C.R.P. [66]....................94
Harun, R. [156].........................197
Haskell, B.G. [632]................951, 952
Hata, M. [50]...........................75, 76
Haykin, S. [132]..................183, 184
Heller, J.A. [63]........................93
Hennecke, M.E. [494]...................577
Hey, A.J.G. [27]....................53, 54
Heymann, D.P. [191]....................206
Higashi, A. [300]............339, 404, 420
Higuchi, K. [391].......................432
Higuchi, K. [366].................389, 454
Hirasawa, S. [85].................138, 158
Hirosaki, B. [164].................197, 909
Hlavac, V. [291]........................330
Ho, M.H. [405]..........................453
Hocquenghem, A. [65]....................94
Hoeher, P. [636]..........964, 965, 984, 998
Hoetter, M. [446].......................491

Holma, H. [604] 916
Holtzman, J. [578] 819
Holtzman, J. [409] 453
Holtzman, J.M. [359] 383
Holtzman, J.M. [384] 420
Holtzman, J.M. [356] 383
Hong, D. [285] 298
Hong, U.S. [652] 999
Honkasalo, H. [380] 398
Honkasalo, Z-C. [380] 398
Hottinen, A. [399] 452
Hottinen, A. [380] 398
Howard, P.G. [12] 749, 750
Hsieh, C-H. [478] 537
Hsieh, C. [506] 611
Hsu, C. [442] 490
Hsu, Y.F. [644] 996
Huang, C. [442] 490
Huang, C. [480] 537
Huang, C.J. [644] 996
Huang, J. [440] 489
Huber, J.B. [614] 928
Hübner, J. [619] 928
Huguet, J. [487] 550

I
Imamura, K. [260] 292
Irie, K. [523] 619
Irie., K. [521] 618
Irvine, G.T. [362] 387, 389, 390
Israelsen, P. [480] 537

J
Jabbari, B. [246] 289, 291, 292, 298, 302
Jabbari, B. [236] 287, 293, 298, 301
Jabbari, B. [247] 289, 291
Jacobs, I.M. [63] 93
Jacobs, I.M. [327] 366
Jacquin, A.E. [422] 459, 465–469
Jafarkhani, H. [393] 432, 452
Jain, A.K. [26] ... 32, 492, 493, 495, 496, 508, 567, 568, 596, 604, 610, 618, 640, 733
Jain, A.K. [433] 484, 485
Jain, J.R. [433] 484, 485
Jaisimha, M.Y. [476] 537
Jakes, W.C. [36] 67, 71, 74, 303, 372
Jalali, A. [194] 206, 710, 778, 910
Jalali, A. [551] 710, 778
Jamal, K. [397] 436
Jang, E. [537] 633
Jayant, N.S. [602] 910
Jayant, N.S. [504] 611, 612
Jayant, N.S. [461] ... 497, 505, 545, 571, 573, 627, 643, 733
Jeanclaude, I. [618] 928
Jelinek, F. [64] 93
Jennings, A. [83] 135

Jézéquel, M. [581] 865, 866
Jiang, H. [252] 291
Johansson, A.L. [413] 453
Johnson, R.C. [661] 1005
Johnston, J.D. [526] 620, 626, 627, 630
Jones, E.V. [547] 664
Jung, P. [298] 339, 395
Juntti, M.J. [383] 420

K
Kahwa, T.J. [282] 298
Kaiser, S. [617] 928
Kalet, I. [171] 197, 909, 928
Kallel, S. [583] 865
Kamio, Y. [109] 164, 190, 191, 194
Kamio, Y. [591] 876, 928
Kamio, Y. [136] 190, 191
Kammeyer, K-D. [365] 389
Kammeyer, K.D. [166] 197, 198
Kandola, G.S. [108] 164
Kang, K-J. [629] 950
Karaliopoulos, M.S. [648] 999
Karam, G. [618] 928
Karunaserker, S.A. [447] 491
Kasahara, M. [85] 138, 158
Kasami, T. [386] 422
Katsaggelos, A. [441] 489
Katsaggelos, A.K. [565] 783
Katzela, I. [245] . 289, 291, 292, 295, 298–300, 302
Kavehrad, M. [340] 376
Kawano, T. [51] 75
Keller, T. [621] 929
Keller, T. [611] 928, 929
Keller, T. [600] 908, 913–919, 921–927
Keller, T. [29] .. 59, 80, 83, 85, 173, 198, 247, 255, 256, 339, 388–390, 558, 559, 586, 587, 779, 781, 790, 876, 879, 908, 912, 913, 927, 950, 962–964, 971, 973, 990, 996, 997, 1001–1003
Keller, T. [580] 862
Keller, T. [92] 161–163, 197, 199, 204
Keller, T. [200] 206
Keller, T. [620] 929
Keller, T. [603] 913
Keller, T. [10] ... 161–166, 168–172, 179, 181, 182, 186, 189–191, 193, 194, 197, 199, 202–204
Keller, T. [150] 197, 204
Keller, T. [607] 928
Kenington, P.B. [105] 164
Kennedy, R.A. [661] 1005
Kenyon, N.D. [549] 676
Keskitalo, I. [604] 916
Khansari, M. [194] 206, 710, 778, 910
Khansari, M. [551] 710, 778
Khorram, H. [585] 865, 866
Kim, C.W. [536] 630

Kim, D.S. [486] 537
Kim, K-B. [629] 950
Kingsbury, N.G. [462] 504, 515, 599, 710, 779
Kingsbury, N.G. [501] 599, 639
Kingsbury, N.G. [447] 491
Kingsbury, N.G. [450] 491
Kingsbury, N.G. [451] 491
Kingsbury, N.G. [449] 491
Kinoshita, K. [43] 70, 342
Kirsch, A.L. [158] 197
Kishimoto, R. [523] 619
Klein, A. [298] 339, 395
Klein, A. [596] 878, 879, 895, 896
Klein, A. [321] 365, 395
Klein, A. [378] 395
Knisely, D.N. [305] 340, 439
Knisely, D.N. [303] 340, 439
Ko, Y.F. [218] 249
Koch, W. [363] 388, 395
Kohno, R. [329] 366, 383
Kolb, H.J. [172] 197
Komaki, S. [559] 722, 875, 876
Kondoz, A.M. [595] 877
Kong, N. [339] 376
Kontovassilis, K. [189] 205, 207
Koora, K. [619] 928
Kossentini, F. [562] 745
Kroon, P. [311] 340
Kuan, E.L. [353] 382, 426, 453, 894
Kuan, E.L. [96] 163, 204
Kuan, E.L. [97] 163, 204
Kuan, E.L. [354] 382, 396, 401, 403, 408, 425, 454
Kuan, E.L. [612] 928
Kuan, E.L. [99] . 163, 204, 382, 396, 403, 408, 454
Kuan, E.L. [98] 163, 204
Kuan, E.L. [95] 163, 204
Kucar, A.D. [312] 340
Kuek, S.S. [258] 292
Kull, B. [619] 928
Kumar, P.S. [578] 819
Kumar, S. [303] 340, 439
Kumar, S. [91] 161-163
Kunt, M. [533] 628
Kürner, T. [338] 373
Kwon, O.H. [646] 996

L

Laflamme, C. [182] 201
Laha, S. [303] 340, 439
Lakshman, T.V. [191] 206
Lam, S.S. [188] 205, 206
Lang, G.R. [100] 164
Lassalle, R. [169] 197, 198
Latva-aho, M. [323] .. 365, 395, 399, 402, 426
Lau, V.K.N. [609] 928
Law, A. [265] 292

Le, M.T. [631] 950
Le-Ngoc, T. [155] 197
Lea, C-T. [214] 247
Leduc, J.P. [187] 205
Lee, C.C. [215] 247, 249
Lee, C.C. [396] 436
Lee, C.C. [219] 249
Lee, J. [438] 489
Lee, J.S. [349] 380, 399, 400
Lee, L. [438] 489
Lee, M.H. [491] 567
Lee, S.U. [486] 537
Lee, T-H. [575] 819
Lee, T-H. [471] 522, 557, 590
Lee, W-C. [629] 950
Lee, W.C.Y. [145] 190, 264, 266, 284
Lee, W.C.Y. [326] 366, 367, 371, 372, 375
Lee, W.C.Y. [220] 253, 266
Lee, W.C.Y. [394] 436
Lee, W.Y.C. [37] 67, 71, 74
Lehenert, R. [189] 205, 207
Leppanen, P.A. [242] 288
Leung, Y-W. [572] 819
Levesque, A.H. [4] ... 107, 126, 128, 130, 138, 148
Levesque, A.H. [627] 950, 963, 996, 997
Levitt, B.K. [385] 422
Li, C. [313] 340
Li, Q. [305] 340, 439
Li, R. [437] 489
Li, V.O.K. [205] 226
Li, W. [533] 628
Li, Y. [622] 929
Liberti, J.C. Jr [400] 453
Lidl, R. [77] 107
Liew, T.H. [151] 197, 204, 928
Liew, T.H. [608] 928
Liew, T.H. [590] 867, 868
Lightfoot, G. [125] 181, 387, 388
Lightstone, M. [567] 784
Lim, T.J. [235] 287
Lim, T.J. [411] 453
Lim, T.J. [415] 453
Lim, T.J. [402] 453
Lim, T.J. [403] 453
Lim, T.J. [405] 453
Lim, T.J. [416] 453
Lim, T.J. [417] 453
Lim, T.J. [412] 453
Lim, T.J. [414] 453
Lim, T.J. [407] 453
Lim, T.J. [418] 453
Lim, T.J. [406] 453
Lin, J-C. [575] 819
Lin, S. [472] 522, 557, 590, 720, 779, 811
Lin, S. [5] .. 107, 126, 128, 138, 148, 720, 811
Lin, Y-B [316] 343
Lindner, J. [178] 198

AUTHOR INDEX

Liou, N. [437] 489
Liou, W. [506] 611
Litva, J. [231] 287
Liu, B. [564] 783
Liu, B. [436] 489
Liu, S. [459] 496, 570
Lo, T. [231] 287
Lodge, J.H. [130] 183, 387, 389
Lodge, J.L. [582] 865
Longstaff, F.M. [100] 164
Lopes, L.B. [265] 292
Lu, L. [483] 537
Lu, P. [506] 611
Luise, M. [497] 586
Lymer, A. [125] 181, 387, 388

M

Ma, L. [380] 398
Ma, Y. [416] 453
MacDonald, N. [550] . 694, 710, 778, 779, 910
Macleod, M.D. [609] 928
Madfors, M. [251] 290
Magarey, J. [449] 491
Magnusson, S. [251] 290
Makhoul, J. [79] 130
Mämmelä, A. [569] 800
Man, Z. [203] 223
Mann Pelz, R. [570] 802–804
Mann Pelz, R. [464] 512, 710, 778, 785
Marsland, I.D. [583] 865
Marti, J. [338] 373
Martins, F.C.M. [566] 783
Marvill, J.D. [105] 164
Massaloux, D. [182] 201
Massey, J.L. [2] ... 94, 126, 129, 138, 141, 148
Massey, J.L. [60] 93
Massey, J.L. [72] 94
Mathiopoulos, P.T. [583] 865
Matsumoto, Y. [418] 453
Matsumura, Y. [554] 710, 778, 779
Matsuoka, H. [591] 876, 928
Mattson, T. [104] 164
Maxey, J.J. [388] 425
McGeehan, J.P. [123] 181, 186
McCoohan, J.P. [124] 181
McGeehan, J.P. [125] 181, 387, 388
McGeehan, J.P. [126] 181
McGeehan, J.P. [122] 181, 387, 388
McGeehan, J.P. [127] 181
McGeehan, J.P. [128] 181
McGeehan, J.P. [117] 173
McLane, P.J. [362] 387, 389, 390
McLane, P.J. [340] 376
Meidan, R. [329] 366, 383
Menolascino, R. [324] 365, 395
Meo, A.R. [470] 522, 557, 590
Mermelstein, P. [194] 206, 710, 778, 910
Mermelstein, P. [42] 70, 341

Mermelstein, P. [551] 710, 778
Michelson, A.M. [4] .. 107, 126, 128, 130, 138, 148
Michelson, A.M. [67] 94
Michelson, A.M. [627] 950, 963, 996, 997
Miki, Y. [366] 389, 454
Miller, M.J. [472] 522, 557, 590, 720, 779, 811
Miller, S. [502] 608
Miller, S.L. [347] 376
Mills, M. [432] 481
Milstein, L.B. [389] 430
Milstein, L.B. [339] 376
Milstein, L.B. [329] 366, 383
Milstein, L.B. [345] 376
Milstein, L.B. [328] 366–368, 376
Milstein, L.B. [348] 376
Mirchandani, G. [455] 496
Mitchell, O.R. [508] 613
Mitchell, O.R. [509] 613, 614, 616
Moher, M. [586] 865
Moher, M.L. [130] 183, 387, 389
Mohr, A.E. [476] 537
Mohr, W. [324] 365, 395
Moisala, T. [443] 490
Monro, D.M. [423] 459, 465–467
Monro, D.M. [424] 459, 466, 467
Morinaga, N. [559] 722, 875, 876
Morinaga, N. [109] 164, 190, 191, 194
Morinaga, N. [591] 876, 928
Morinaga, N. [146] 190, 191
Morinaga, N. [135] 190, 191
Morinaga, N. [136] 190, 191
Morrow, R.K. Jr [358] 383
Moshavi, S. [351] 382, 453
Mueller-Roemer, F. [167] 197, 198, 909
Mukherjee, D. [567] 784

N

Naghshineh, M. [245] 289, 291, 292, 295, 298–300, 302
Naijoh, M. [136] 190, 191
Nakagami, M. [335] 372
Nakagawa, M. [43] 70, 342
Nakagawa, S. [554] 710, 778, 779
Nakai, T. [554] 710, 778, 779
Namekawa, T. [85] 138, 158
Nanda, S. [180] 201, 226
Nanda, S. [303] 340, 439
Nanda, S. [91] 161–163
Narasimha, M. [463] 508
Narayanan, K.R. [153] 197
Narayanan, K.R. [589] 866
Nasrabadi, N. [537] 633
Nasrabadi, N. [479] 537
Naveen, T. [524] 619
Netravali, A.N. [632] 951, 952
Nevalainen, O. [540] 637
Ngan, K.N. [556] 711, 752

Ngan, K.N. [555] 711, 752
Ngan, K.N. [457] 496
Ngan, K.N. [485] 537
Nguyen, Q.T. [535] 630
Nicholls, J.A. [424] 459, 466, 467
Niederreiter, H. [77] 107
Nieweglowski, J. [443] 490
Nightingale, C. [549] 676
Nikula, E. [321] 365, 395
Nikula, E. [604] 916
Nilsson, M. [372] 395, 396, 399, 414, 429, 454
Nix, A.R. [117] 173
Noah, M.J. [512] 615, 643
Noerpel, A.R. [316] 343
Noll, P. [461] ... 497, 505, 545, 571, 573, 627, 643, 733
Nuri, V. [531] 627
Nussbaumer, H.J. [528] 626
Nussbaumer, H.J. [527] 626
Nyquist, H. [17] 3, 6, 168

O

O'Leary, S. [628] 950
O'Leary, S. [630] 950
O'Neil, S.D. [515] 618
Ochsner, H. [263] 292, 343
Oh, S-H. [284] 298
Ohmori, E. [51] 75
Ohno, K. [300] 339, 404, 420
Ojanperä, T. [371] 395, 437
Ojanperä, T. [322] ... 365, 395, 402, 426, 439, 453
Ojanperä, T. [323] ... 365, 395, 399, 402, 426
Ojanperä, T. [378] 395
Ojanpera, T. [373] 395
Okawa, K. [299] 339, 399, 423, 425
Okumura, Y. [300] 339, 404, 420
Okumura, Y. [398] 442
Okumura, Y. [51] 75
Olofsson, H. [251] 290
Omura, J.K. [385] 422
Onural, L. [430] 478
Ormondroyd, R.F. [388] 425
Ostermann, J. [428] 478
Ottersten, B.E. [347] 376
Ovesjö, F. [378] 395
Oyama, T. [417] 453
Oyama, T. [418] 453

P

Padovani, R. [327] 366
Pajukoski, K. [605] 916
Palau, A. [455] 496
Palestini, V. [279] 296
Panken, F.J. [204] 225, 226
Papadopoulos, C.A. [444] 490
Papadopoulos, C.A. [445] 491
Park, H-M. [629] 950

Parkvall, S. [347] 376
Parsons, D. [47] 71, 74, 75, 248
Parsons, J.D. [46] 71
Patel, P. [409] 453
Paulraj, A. [568] 800
Pavlidou, F-N. [648] 999
Pearce, D.A. [144] 190
Pearlman, W.A. [483] 537
Pearson, D.E. [448] 491
Peha, J.M. [266] 292
Pehkonen, K. [604] 916
Peled, A. [163] 197
Pereira, F. [662] 1021
Peritsky, M.M. [254] 291, 292
Peterson, A. [463] 508
Peterson, R.L. [344] 376
Peterson, W.W. [6] .. 107, 126, 128, 129, 138, 148
Peterson, W.W. [68] 94, 126, 128, 130
Peterson, W.W. [74] 107, 126
Petit, G. [189] 205, 207
Phoong, S-M. [536] 630
Picart, A. [581] 865, 866
Picchi, G. [657] 1000, 1002
Picco, R. [484] 537
Pichna, R. [574] 819
Pickholtz, R.L. [345] 376
Pickholtz, R.L. [328] 366–368, 376
Pietrobon, S.S. [638] 965
PiHui, C. [280] 296, 306
Pio Magnani, N. [279] 296
Pirhonen, R. [596] 878, 879, 895, 896
Pizarroso, M. [324] 365, 395
Plenge, G. [168] 197, 198
Pless, V. [75] 107, 129
Po, L. [482] 537
Pollini, G.P. [237] 287
Polydoros, A. [658] 1000, 1002
Powers, E.N. [159] 197
Prabhu, V.K. [39] 68
Prasad, K.V. [494] 577
Prasad, R. [371] 395, 437
Prasad, R. [240] 288, 339, 366, 393
Prasad, R. [322] . 365, 395, 402, 426, 439, 453
Prati, G. [657] 1000, 1002
Press, W.H. [148] 194, 216, 552
Press, W.H. [54] 87
Press, W.H. [579] 838
Preuss, K. [174] 197
Price, R. [331] 371, 376
Priestly, D. [628] 950
Proakis, J.G. [52] ... 78, 79, 82, 172, 371–374, 376, 385, 386, 399, 426, 445
Pupolin, S. [224] 280
Puri, A. [632] 951, 952
Pursley, M.B. [357] 383, 384, 422

AUTHOR INDEX

Q
Qiu, X. [205] 226
Qureshi, S.U. [100] 164

R
Rabiner, L.R. [529] 627
Raemer, H.R. [21] 6, 169
Raheli, R. [658] 1000, 1002
Raina, J. [541] 637
Raitola, M. [399] 452
Ramakrishna, S. [384] 420
Ramamurthi, B. [425] 466, 467, 469, 543, 548
Rames, N.S. [305] 340, 439
Ramstad, T.A. [530] 627, 637
Rao, K.R. [452] 493, 495, 508, 955
Rao, R.R. [389] 430
Rapajic, P.B. [364] 389
Rapeli, J. [319] 365, 392, 395
Raphaeli, D. [588] 865
Rappaport, S.S. [252] 291
Rappaport, S.S. [285] 298
Rappaport, S.S. [346] 376
Rappaport, T.S. [400] 453
Rasmussen, L.K. [411] 453
Rasmussen, L.K. [415] 453
Rasmussen, L.K. [413] 453
Rasmussen, L.K. [402] 453
Rasmussen, L.K. [416] 453
Rasmussen, L.K. [417] 453
Rasmussen, L.K. [412] 453
Rasmussen, L.K. [414] 453
Rasmussen, L.K. [407] 453
Rasmussen, L.K. [418] 453
Rasmussen, L.K. [410] 453
Raviv, J. [64] 93
Ray-Chaudhuri, D.K. [66] 94
Ray-Chaudhuri, D.K. [67] 94
Reader (Guest Editors), C. [662] 1021
Reali, G. [643] 996
Redmill, D.W. [501] 599, 639
Redmill, D.W. [553] 710, 778, 779
Reed, I.S. [70] 94
Reibman, A.R. [641] 968
Reiffen, B. [58] 93
Reudink, D.O. [272] 295
Reudink, D.O. [269] 295
Reudink, D.O. [271] 295, 298
Reudink, D.O. [290] 309
Rick, R.R. [348] 376
Riskin, E.A. [476] 537
Riva, G. [184] 202
Riva, G. [183] 201
Robertson, D. [207] 227, 231
Robertson, P. [617] 928
Robertson, P. [636] 964, 965, 984, 998
Rohling, H. [616] 928
Romiti, F. [228] 287
Rose, O. [190] 205

Ross, A.H.M. [243] 288
Roy, S. [403] 453
Rückriem, R. [175] 197
Rudolph, L.D. [66] 94
Ruf, M.J. [617] 928
Ruiz, A. [163] 197

S
Sadka, A.H. [595] 877
Safak, A. [213] 247, 290
Salami, R.A. [34] 64
Salami, R.A. [182] 201
Salami, R.A. [318] 347
Salgado-Galicia, H. [266] 292
Saltzberg, B.R. [160] 197
Sampei, S. [131] 183
Sampei, S. [559] 722, 875, 876
Sampei, S. [109] 164, 190, 191, 194
Sampei, S. [591] 876, 928
Sampei, S. [135] 190, 191
Sampei, S. [136] 190, 191
Sampei, S. [367] 389
Sanada, Y. [408] 453
Sandell, M. [601] 909
Sari, H. [618] 928
Sasaki, A. [376] 395
Sasaoka, H. [109] 164, 190, 191, 194
Sato, Y. [660] 1001
Saunders, S. [654] 999
Saunders, S.R. [653] 999
Savusalo, J. [605] 916
Sawahashi, M. [301] 339, 395
Sawahashi, M. [302] 339, 395
Sawahashi, M. [299] 339, 399, 423, 425
Sawahashi, M. [391] 432
Sawahashi, M. [366] 389, 454
Schilling, D.L. [345] 376
Schilling, D.L. [328] 366–368, 376
Schilling, D.L. [469] 522, 557, 590
Schnell, M. [360] 387
Scholtz, R.A. [385] 422
Schulze, H. [166] 197, 198
Schur, J. [81] 130
Schüssler, H.W. [173] 197
Schuster, G.M. [565] 783
Schwartz, M. [202] 209, 724
Schwartz, M. [212] 242
Schwarz da Silva, J. [377] 395
Seferidis, V. [561] 727, 790, 970, 982
Senarath, N.G. [225] 280
Serizawa, M. [276] 296
Serra, A.M. [184] 202
Sestini, F. [278] 296
Sestini, F. [279] 296
Shanmugam, K.S. [23] 6, 59
Shannon, C.E. [14] .3, 6, 7, 11–15, 22, 34, 39,
      42, 44, 52, 62

Shannon, C.E. [15] 3, 6, 7, 11, 12, 15, 22, 34, 35, 39, 42–44, 49, 50, 52, 62
Shannon, C.E. [16] 3, 6, 7, 11, 12, 15, 22, 34, 39, 42, 44, 52, 55, 56, 62
Shannon, C.E. [24] .. 7, 11, 12, 15, 22, 34, 39, 42, 44, 52, 53, 57, 58, 62
Shannon, C.E. [25] .. 7, 11, 12, 15, 22, 34, 39, 42, 44, 52, 53, 62
Shannon, C.E. [13] 3, 6, 7, 11, 12, 15, 16, 22, 34, 39, 42, 44, 52, 62, 63
Sharaf, A. [453] 494
Shelswell, P. [635] 962
Sherry, H. [316] 343
Shie, J. [438] 489
Shue, J-S. [478] 537
Shustermann, E. [492] 567, 573
Shwedyk, E. [584] 865
Sikora, T. [664] 1021
Sikora, T. [662] 1021
Sim, H.K. [419] 454
Simon, M.K. [385] 422
Simoncelli, E.P. [522] 618
Simpson, F. [356] 383
Sin, J.K.S. [283] 298
Sin, K.K. [485] 537
Singh, R. [256] 292
Sirbu, M. [266] 292
Sitaram, V. [480] 537
Skelly, P. [202] 209, 724
Sklar, B. [332] 371, 372
Sklar, B. [333] 371
Sklar, B. [118] 176
Skoeld, J. [596] 878, 879, 895, 896
Sköld, J. [372] ... 395, 396, 399, 414, 429, 454
Sluyter, R.J. [317] 347
Smith, M.J.T. [562] 745
Smolik, K. [330] 366
Smolik, K.F. [330] 366
Sohn, W. [646] 996
Sollenberger, N.R. [275] ... 296, 299, 301, 311
Sollenberger, N.R. [577] 819
Sollenberger, N.R. [622] 929
Solomon, G. [70] 94
Somerville, F.C.A. [94] 162, 163, 862
Sonka, M. [291] 330
Sourour, E. [234] 287
Stapleton, S.P. [107] 164
Stapleton, S.P. [108] 164
Stedman, R. [466] 514, 587, 628
Stedman, R. [465] 512
Steele, R. [39] 68
Steele, R. [56] 93
Steele, R. [102] 164, 186
Steele, R. [215] 247, 249
Steele, R. [466] 514, 587, 628
Steele, R. [267] 293
Steele, R. [89] 161, 162, 190, 875, 876

Steele, R. [87] .. 161, 185, 358, 359, 362, 861, 862, 867, 950, 963, 965, 966, 971, 996, 997, 1003
Steele, R. [9] .. 67, 86, 91, 107, 111, 115, 126, 138, 148, 155, 156, 158, 247, 248, 255, 351, 445, 446, 723, 879, 894, 895
Steele, R. [465] 512
Steele, R. [181] 201
Steele, R. [34] 64
Steele, R. [262] 292, 802, 803
Steele, R. [48] 72
Steele, R. [396] 436
Steele, R. [134] 190, 203, 722, 780, 927
Steele, R. [318] 347
Steele, R. [38] 68
Steele, R. [40] 68
Stefanov, J. [41] ... 70, 72, 88, 201, 255, 258, 288, 296, 341–343, 347, 363, 365, 393, 400, 401, 406, 437, 512, 590, 803, 821
Steinbach, E. [594] 877
Steinbach, E. [195] 206, 778, 910
Stork, D.G. [494] 577
Strobach, P. [435] ... 486, 487, 497, 567, 568, 573, 574, 650, 651
Ström, E.G. [347] 376
Stüber, G. [576] 819
Stuber, G.L. [589] 866
Su, Y.T. [575] 819
Suda, H. [302] 339, 395
Suda, H. [382] 411, 444
Sugimoto, H. [411] 453
Sugimoto, H. [417] 453
Sugimoto, H. [414] 453
Sugimoto, H. [407] 453
Sugimoto, H. [418] 453
Sugiyama, Y. [85] 138, 158
Sun, M.J. [644] 996
Sun, S.M. [411] 453
Sun, S.M. [415] 453
Sun, S.M. [412] 453
Sun, S.M. [414] 453
Sun, S.M. [407] 453
Sun, S.M. [418] 453
Sunaga, T. [131] 183
Sunaga, T. [367] 389
Sunay, M.O. [380] 398
Suoranta, R. [596] 878, 879, 895, 896
Suzuki, H. [336] 372
Suzuki, M. [297] 339

T

Tabatabai, A. [517] 618
Tabatabai, A. [518] 618, 630, 631, 633
Tai, C. [313] 340
Tajima, J. [260] 292
Tan, P.H. [410] 453

Tanenbaum, A.S. [201] 208, 210, 301, 311
Tarokh, V. [393] 432, 452
Tcha, D-W. [284] 298
Teder, P. [363] 388, 395
Tekalp, A.M. [430] 478
Tekinay, S. [246] 289, 291, 292, 298, 302
Tekinay, S. [236] 287, 293, 298, 301
Teukolsky, S.A. [148] 194, 216, 552
Teukolsky, S.A. [54] 87
Teukolsky, S.A. [579] 838
Theimer, T. [189] 205, 207
Thibault, L. [631] 950
Thielecke, J. [363] 388, 395
Thitimajshima, P. [28] .. 53, 61, 93, 865, 894,
    895, 921, 928, 930, 964, 997
Timor, U. [180] 201, 226
Torrance, J.M. [113] .. 164, 191, 192, 196, 204
Torrance, J.M. [597] 880, 881, 928
Torrance, J.M. [110] . 164, 191, 193, 204, 722,
    780, 876
Torrance, J.M. [111] . 164, 191, 194, 204, 876,
    888
Torrance, J.M. [119] .. 180, 186, 191, 196, 876
Torrance, J.M. [114] 164, 191, 204
Torrance, J.M. [11] ... 181, 185–189, 194, 389
Torrance, J.M. [112] 164, 191, 204
Torrance, J.M. [150] 197, 204
Torrance, J.M. [149] 197, 204
Torrance, J.M. [607] 928
Torrance, J.M. [115] .. 164, 191, 195, 197, 204
Torrence, G.W. [651] 999
Torres, L. [475] 537
Torres, L. [487] 550
Toskala, A. [321] 365, 395
Toskala, A. [378] 395
Toskala, A. [323] 365, 395, 399, 402, 426
Tozer, T.C. [144] 190
Tralli, V. [222] 266
Tsaur, S-A. [341] 376
Tuisel, U. [166] 197, 198
Turletti, T. [548] 676
Tuttlebee, W.H.W. [264] 292
Tzaras, C. [653] 999
Tzou, C. [658] 1000, 1002

U
Udpikar, V. [541] 637

V
Vaidyanathan, P.P. [536] 630
Vaisey, J. [490] 567
van de Beek, J-J. [601] 909
Varaiya, P.P. [140] 190, 191
Varanasi, M.K. [350] 381, 453
Vary, P. [317] 347
Venetsanopoulos, A. [477] 537
Verdone, R. [222] 266
Verdú, S. [355] 382

Verdú, S. [352] 382, 453
Vetterli, M. [538] 633
Vetterling, W.T. [148] 194, 216, 552
Vetterling, W.T. [54] 87
Vetterling, W.T. [579] 838
Vijayan, R. [287] 299
Villebrun, E. [636] 964, 965, 984, 998
Viterbi, A.J. [61] 93
Viterbi, A.J. [239] 288, 366, 393, 425
Viterbi, A.J. [327] 366
Viterbi, A.J. [31] 63
Vitter, J.S. [12] 749, 750
Vogel, W.J. [649] 999
Vogel, W.J. [650] 999
Vogel, W.J. [652] 999
Vogel, W.J. [651] 999
Vook, F.W. [623] 929
Vučerić, J. [270] 295
Vucetic, B. [241] 288, 366, 393
Vucetic, B.S. [364] 389

W
Wagner, G.R. [507] 611
Wales, S.W. [230] 287
Wallstedt, K. [251] 290
Wang, B.M. [434] 485
Wang, F-M. [459] 496, 570
Wang, J. [438] 489
Wang, L-C. [214] 247
Wang, Q. [574] 819
Wang, Q. [431] 478, 487
Wang, Q. [408] 453
Wassell, I. [318] 347
Wassell, I.J. [48] 72
Weaver, L.A. Jr [327] 366
Webb, W.T. [102] 164, 186
Webb, W.T. [89] 161, 162, 190, 875, 876
Webb, W.T. [29] 59, 80, 83, 85, 173, 198, 247,
    255, 256, 339, 388–390, 558, 559,
    586, 587, 779, 781, 790, 876, 879,
    908, 912, 913, 927, 950, 962–964,
    971, 973, 990, 996, 997, 1001–1003
Webb, W.T. [181] 201
Webb, W.T. [557] 715, 784
Webb, W.T. [560] 722, 780
Webb, W.T. [580] 862
Webb, W.T. [134] 190, 203, 722, 780, 927
Webb, W.T. [10] 161–166, 168–172, 179, 181,
    182, 186, 189–191, 193, 194, 197,
    199, 202–204
Webber, S.A. [513] 618
Wei, L-F. [498] 587
Wei, L.F. [404] 453
Wei, S.X. [206] 226, 237
Weinstein, S.B. [162] 197
Weldon, E.J. Jr [6] .. 107, 126, 128, 129, 138,
    148
Wellstead, P.E. [507] 611

Welsh, W.J. [427] 478
Wesolowsky, K. [655] 1000, 1001
Westerink, P.H. [534] 628
Westman, T. [604] 916
Wheatley, C.E. III [327] 366
Whitehead, J. [267] 293
Whitmann, M. [338] 373
Whybray, M.W. [193] 206, 748, 749, 751, 757
Whybray, M.W. [448] 491
Wichman, R. [399] 452
Wicker, S.B. [637] 965, 999
Wiegand, T. [567] 784
Wiese, G.L. [510] 614
Wilkes, J.E. [330] 366
Wilkinson, R.J. [105] 164
Wilkinson., R.J. [106] 164
Wilson, D.L. [424] 459, 466, 467
Winters, J.H. [401] 453
Wolf., G.J. [495] 577
Wong, C.H. [151] 197, 204, 928
Wong, C.H. [152] 197, 204
Wong, C.H. [354] 382, 396, 401, 403, 408, 425, 454
Wong, C.H. [612] 928
Wong, C.H. [99] 163, 204, 382, 396, 403, 408, 454
Wong, C.H. [608] 928
Wong, C.H. [93] . 161–163, 408, 416, 425, 454
Wong, C.H. [599] 907
Wong, D. [235] 287
Wong, H.E. [395] 436
Wong, K.H.H. [56] 93
Wong, K.H.H. [8] 107, 115, 126, 138, 148
Wong, K.H.H. [35] 65, 351, 362, 790
Wong, K.H.J. [318] 347
Wong, W-S. [259] 292, 309
Wong, W.C. [258] 292
Wong, W.C. [267] 293
Woodard, J.P [621] 929
Woodard, J.P. [94] 162, 163, 862
Woodard, J.P. [634] 953
Woodard, J.P. [33] 64, 514
Woodard, J.P. [147] 193
Woods, J.W. [516] 618
Woods, J.W. [524] 619
Woods, J.W. [515] 618
Wozencraft, J.M. [57] 93
Wozencraft, J.M. [58] 93
Wright, A.S. [103] 164
Wu, J. [154] 197
Wu, K-T. [341] 376

Y

Yamaguchi, H. [456] 496
Yang, L-L. [343] 376
Yang, L-L. [342] 376
Yates, R.D. [274] 296
Yates, R.D. [578] 819

Yau, D.K.Y. [188] 205, 206
Yeap, B.L. [590] 867, 868
Yee, M.S. [610] 928
Yen, J.C. [434] 485
Yen, K. [368] 393
Yip, P. [452] 493, 495, 508, 955
Yip, W. [481] 537
You, D. [406] 453
Young, R.W. [450] 491
Young, R.W. [451] 491
Yu, P. [477] 537
Yum, T-S.P. [259] 292, 309
Yum, T-S.P. [255] 292, 295
Yum, T-S.P. [257] 292, 295

Z

Zaccarin, A. [436] 489
Zander, J. [287] 299
Zander, J. [289] 299, 819
Zander, J. [248] 289
Zander, J. [288] 299, 819
Zarai, Y. [588] 865
Zeisberg, S. [619] 928
Zeng, B. [437] 489
Zetterberg, L. [505] 611, 612
Zhang, M. [255] 292, 295
Zhang, M. [257] 292, 295
Zhang, Q.T. [227] 280
Zhang, X. [489] 567, 573, 631
Zhang, Y.-Q. [662] 1021
Zheng, C. [313] 340
Ziemer, R.E. [344] 376
Zierler, N. [78] 126, 128–130
Zierler, N. [69] 94
Zimmermann, M.S. [159] 197
Zimmermann, M.S. [158] 197
Zorzi, M. [571] 819
Zorzi, M. [224] 280

About the Authors

Lajos Hanzo received his degree in electronics in 1976 and his doctorate in 1983. During his 25-year career in telecommunications he has held various research and academic posts in Hungary, Germany and the UK. Since 1986 he has been with the Department of Electronics and Computer Science, University of Southampton, UK, where he holds the chair in telecommunications. He has co-authored five books on mobile radio communications, published over 300 research papers, organised and chaired conference sessions, presented overview lectures and been awarded a number of distinctions. Currently he is managing an academic research team, working on a range of research projects in the field of wireless multimedia communications sponsored by industry, the Engineering and Physical Sciences Research Council (EPSRC) UK, the European IST Programme and the Mobile Virtual Centre of Excellence (VCE), UK. He is an enthusiastic supporter of industrial and academic liaison and he offers a range of industrial courses. He is also an IEEE Distinguished Lecturer. For further information on research in progress and associated publications please refer to http://www-mobile.ecs.soton.ac.uk

Peter J. Cherriman graduated in 1994 with an M.Eng. degree in Information Engineering from the University of Southampton, UK. Since 1994 he has been with the Department of Electronics and Computer Science at the University of Southampton, UK, working towards a Ph.D. in mobile video networking which was completed in 1999. Currently he is working on projects for the Mobile Virtual Centre of Excellence, UK. His current areas of research include robust video coding, microcellular radio systems, power control, dynamic channel allocation and multiple access protocols. He has published five journal papers, 13 conference papers, and holds three patents. Further information and a list of publications can be found on http://www-mobile.ecs.soton.ac.uk/peter

Jürgen Streit received his Dipl.-Ing. Degree in electronic engineering from the Aachen University of Technology in 1993 and his Ph.D. in image coding from the Department of Electronics and Computer Science, University of Southampton, UK, in 1995. From 1992 to 1996 Dr. Streit had been with the Department of Electronics and Computer Science working in the Communications Research Group. His work there led to numerous publications. Since then he has joined a management consultancy working as an information technology consultant.